Plant Metal Interaction

Emerging Remediation Techniques

Plant Metal Interaction

Emerging Remediation Techniques

Edited by

Parvaiz Ahmad
Department of Botany, S.P. College, Srinagar,
Jammu and Kashmir, India

ELSEVIER

AMSTERDAM • BOSTON • HEIDELBERG • LONDON • NEW YORK • OXFORD
PARIS • SAN DIEGO • SAN FRANCISCO • SINGAPORE • SYDNEY • TOKYO

Elsevier
Radarweg 29, PO Box 211, 1000 AE Amsterdam, Netherlands
The Boulevard, Langford Lane, Kidlington, Oxford OX5 1GB, UK
225 Wyman Street, Waltham, MA 02451, USA

Notices

Knowledge and best practice in this field are constantly changing. As new research and experience broaden our understanding, changes in research methods, professional practices, or medical treatment may become necessary.

Practitioners and researchers must always rely on their own experience and knowledge in evaluating and using any information, methods, compounds, or experiments described herein. In using such information or methods they should be mindful of their own safety and the safety of others, including parties for whom they have a professional responsibility.

To the fullest extent of the law, neither the Publisher nor the authors, contributors, or editors, assume any liability for any injury and/or damage to persons or property as a matter of products liability, negligence or otherwise, or from any use or operation of any methods, products, instructions, or ideas contained in the material herein.

ISBN: 978-0-12-803158-2

British Library Cataloguing in Publication Data
A catalogue record for this book is available from the British Library

Library of Congress Cataloging-in-Publication Data
A catalog record for this book is available from the Library of Congress

For information on all Elsevier publications visit our website at http://store.elsevier.com/

Dedicated to

Hakim Abdul Hameed
(1908–1999)
Founder of Jamia Hamdard (Hamdard University)
New Delhi, India.

Contents

List of Contributors

Elsayed F. Abd-Allah
Department of Plant Production, Faculty of Food & Agricultural Sciences, King Saud University, Riyadh, Saudi Arabia

Maria Rosa Abenavoli
Dipartimento Agraria, Università Mediterranea di Reggio Calabria, Salita Melissari, Reggio Calabria, Italy

Mohammad Abass Ahanger
Stress Physiology Laboratory, School of Studies in Botany, Jiwaji University Gwalior, MP, India

Zeshan Ali
National Institute of Bioremediation, National Agricultural Research Center (NARC), Islamabad, Pakistan

Mushtaq Ahmad
National Institute of Bioremediation, National Agricultural Research Center (NARC), Islamabad, Pakistan

Parvaiz Ahmad
Department of Botany, S.P. College, Srinagar, Jammu and Kashmir, India

Fabrizio Araniti
Dipartimento Agraria, Università Mediterranea di Reggio Calabria, Salita Melissari, Reggio Calabria, Italy

Saroj Arora
Department of Botanical and Environmental Sciences, Guru Nanak Dev University, Amritsar, Punjab, India

M.M. Azooz
Department of Botany, Faculty of Science, South Valley University, Qena, Egypt;
Department of Biological Sciences, Faculty of Science, King Faisal University, Al-Hassa, Saudi Arabia

Shagun Bali
Department of Botanical and Environmental Sciences, Guru Nanak Dev University, Amritsar, Punjab, India

Sávio Bastos de Souza
Centro de Biociências e Biotecnologia, Universidade Estadual do Norte Fluminense Darcy Ribeiro, Campos dos Goytacazes, Rio de Janeiro, Brazil

Amanda Azevedo Bertolazi
Centro de Ciências e Tecnologias Agrárias, Universidade Estadual do Norte Fluminense Darcy Ribeiro, Campos dos Goytacazes, Rio de Janeiro, Brazil

Renu Bhardwaj
Department of Botanical and Environmental Sciences, Guru Nanak Dev University, Amritsar, Punjab, India

Muhammad F. Bhatti
Atta-ur-Rahman School of Applied Biosciences, National University of Sciences and Technology (NUST), Islamabad, Pakistan

Janaina Biral dos Santos
Centro de Biociências e Biotecnologia, Universidade Estadual do Norte Fluminense Darcy Ribeiro, Campos dos Goytacazes, Rio de Janeiro, Brazil

Marian Brestic
Department of Plant Physiology, Slovak University of Agriculture in Nitra, Nitra, Slovakia

Natalia Anatoljevna Burmistrova
Timiryazev Institute of Plant Physiology, Russian Academy of Sciences, Moscow, Russia

Khushboo Chaudhary
Department of Bioscience and Biotechnology, Banasthali University, Tonk, Rajasthan, India

Devendra Kumar Chauhan
Center of Advanced Study in Botany, Banaras Hindu University, Varanasi, India

Silvana Chocobar-Ponce
Cátedra de Fisiología Vegetal, Facultad de Ciencias Naturales e IML, Tucumán, Argentina

Antônio Jesus Dorighetto Cogo
Centro de Biociências e Biotecnologia, Universidade Estadual do Norte Fluminense Darcy Ribeiro, Campos dos Goytacazes, Rio de Janeiro, Brazil

Juliana Melo da Conceição
Centro de Biociências e Biotecnologia, Universidade Estadual do Norte Fluminense Darcy Ribeiro, Campos dos Goytacazes, Rio de Janeiro, Brazil

Jaime A.T. da Silva
Ikenobe, Kagawa-ken, Japan

Muriel da Silva Folli-Pereira
Laboratório de Microbiologia Ambiental e Biotecnologia, Universidade Vila Velha, Vila Velha, Espírito Santo, Brazil

Natália de Aquino Portela
Universidae Federal do Espírito Santo, Goiabeiras, Vitória, Espírito Santo, Brazil

Mehmet Demiralay
Department of Forest Engineering, Faculty of Forestry, Artvin Çoruh University, Artvin, Turkey

Murat Dikilitas
Department of Plant Protection, Faculty of Agriculture, Harran University, Şanlıurfa, Turkey

Anupam Dikshit
Biological Product Laboratory, Department of Botany, University of Allahabad, Allahabad, Uttar Pradesh, India

Nawal Kishore Dubey
D. D. Pant Interdisciplinary Research Laboratory, Department of Botany, University of Allahabad, Allahabad, India

Dilfuza Egamberdieva
Leibniz Centre for Agricultural Landscape Research (ZALF), Institute for Landcape Biogeochemistry, Müncheberg, Germany

Frederico Jacob Eutrópio
Centro de Biociências e Biotecnologia, Universidade Estadual do Norte Fluminense Darcy Ribeiro, Campos dos Goytacazes, Rio de Janeiro, Brazil

Frederico Figueira Firme
Centro de Biociências e Biotecnologia, Universidade Estadual do Norte Fluminense Darcy Ribeiro, Campos dos Goytacazes, Rio de Janeiro, Brazil

Mercedes García-Sánchez
Department of Agro-Environmental Chemistry and Plant Nutrition, Faculty of Agrobiology, Food and Natural Resources, Kamýcká, Prague, Czech Republic

Vandana Gautam
Department of Botanical and Environmental Sciences, Guru Nanak Dev University, Amritsar, Punjab, India

Zoya Ghori
Atta-ur-Rahman School of Applied Biosciences, National University of Sciences and Technology (NUST), Islamabad, Pakistan

Asiya Hameed
Department of Botany, Faculty of Science, Jamia Hamdard, New Delhi, India

Neha Handa
Department of Botanical and Environmental Sciences, Guru Nanak Dev University, Amritsar, Punjab, India

Asim Hayat
Soil Fertility Lab, Land Resources Research Institute, National Agricultural Research Center (NARC), Islamabad, Pakistan

Mirna Hilal
Cátedra de Fisiología Vegetal, Facultad de Ciencias Naturales e IML, Tucumán, Argentina

Mohammad Anwar Hossain
Department of Genetics and Plant Breeding, Bangladesh Agricultural University, Mymensingh, Bangladesh

Hira Iftikhar
Atta-ur-Rahman School of Applied Biosciences, National University of Sciences and Technology (NUST), Islamabad, Pakistan

Sameen R. Imadi
Atta-ur-Rahman School of Applied Biosciences, National University of Sciences and Technology (NUST), Islamabad, Pakistan

Sumira Jan
Department of Biotechnology, ICAR – Central Institute of Temperate Horticulture, Srinagar, Jammu and Kashmir, India

Rohit Joshi
Plant Molecular Biology Group, International Centre for Genetic Engineering and Biotechnology, New Delhi, India

Dhriti Kapoor
Department of Botanical and Environmental Sciences, Guru Nanak Dev University, Amritsar, Punjab, India

Nitika Kapoor
Department of Botany, Hansraj Mahila Maha Vidyalaya, Jalandhar, Punjab, India

Sema Karakas
Department of Soil Science and Plant Nutrition, Faculty of Agriculture, Harran University, Şanlıurfa, Turkey

Harpreet Kaur
Department of Botanical and Environmental Sciences, Guru Nanak Dev University, Amritsar, Punjab, India

Parminder Kaur
Department of Botanical and Environmental Sciences, Guru Nanak Dev University, Amritsar, Punjab, India

Satwinderjit Kaur
Department of Botanical and Environmental Sciences, Guru Nanak Dev University, Amritsar, Punjab, India

Sukhmeen Kaur
Department of Botanical and Environmental Sciences, Guru Nanak Dev University, Amritsar, Punjab, India

Alvina G. Kazi
Atta-ur-Rahman School of Applied Biosciences, National University of Sciences and Technology (NUST), Islamabad, Pakistan

Suphiya Khan
Department of Bioscience and Biotechnology, Banasthali University, Tonk, Rajasthan, India

Sukhmeen Kaur Kohli
Department of Botanical and Environmental Sciences, Guru Nanak Dev University, Amritsar, Punjab, India

Marina Sergeevna Krasavina
Timiryazev Institute of Plant Physiology, Russian Academy of Sciences, Moscow, Russia

Rajesh Kumar
Biological Product Laboratory, Department of Botany, University of Allahabad, Allahabad, Uttar Pradesh, India

Smita Kumar
Department of Biochemistry, University of Lucknow, Lucknow, Uttar Pradesh, India

Caterina Longo
Dipartimento Agraria, Università Mediterranea di Reggio Calabria, Salita Melissari, Reggio Calabria, Italy

Nina Fjodorovna Lunkova
Timiryazev Institute of Plant Physiology, Russian Academy of Sciences, Moscow, Russia

Antonio Lupini
Dipartimento Agraria, Università Mediterranea di Reggio Calabria, Salita Melissari, Reggio Calabria, Italy

Antonio Mauceri
Dipartimento Agraria, Università Mediterranea di Reggio Calabria, Salita Melissari, Reggio Calabria, Italy

Rohit Kumar Mishra
Biological Product Laboratory, Department of Botany, University of Allahabad, Allahabad, Uttar Pradesh, India

Nasar-um-Minullah
Atta-ur-Rahman School of Applied Biosciences, National University of Sciences and Technology (NUST), Islamabad, Pakistan

Vani Mishra
Nanotechnology Application Centre, University of Allahabad, Allahabad, Uttar Pradesh, India

Puja Ohri
Department of Botanical and Environmental Sciences, Guru Nanak Dev University, Amritsar, Punjab, India

Eduardo Pagano
Departamento de Biología Aplicada y Alimentos, Facultad de Agronomía, Universidad de Buenos Aires, Buenos Aires, Argentina

Anand Pandey
Biological Product Laboratory, Department of Botany, University of Allahabad, Allahabad, Uttar Pradesh, India

Madhu Pandey
Biological Product Laboratory, Department of Botany, University of Allahabad, Allahabad, Uttar Pradesh, India

Manisha Pandey
Biological Product Laboratory, Department of Botany, University of Allahabad, Allahabad, Uttar Pradesh, India

Ashwani Pareek
Stress Physiology and Molecular Biology Laboratory, School of Life Sciences, Jawaharlal Nehru University, New Delhi, India

Ashutosh Pathak
Biological Product Laboratory, Department of Botany, University of Allahabad, Allahabad, Uttar Pradesh, India

Poonam
Department of Botanical and Environmental Sciences, Guru Nanak Dev University, Amritsar, Punjab, India

Carolina Prado
Cátedra de Fisiología Vegetal, Facultad de Ciencias Naturales e IML, Tucumán, Argentina

Fernando E. Prado
Cátedra de Fisiología Vegetal, Facultad de Ciencias Naturales e IML, Tucumán, Argentina

Maria Polsia Princi
Dipartimento Agraria, Università Mediterranea di Reggio Calabria, Salita Melissari, Reggio Calabria, Italy

Afifa Qidwai
Biological Product Laboratory, Department of Botany, University of Allahabad, Allahabad, Uttar Pradesh, India

Galina Nikolaevna Raldugina
Timiryazev Institute of Plant Physiology, Russian Academy of Sciences, Moscow, Russia

Alessandro Coutinho Ramos
Centro de Biociências e Biotecnologia, Universidade Estadual do Norte Fluminense Darcy Ribeiro, Campos dos Goytacazes, Rio de Janeiro, Brazil

Bushra Rashid
Center of Excellence in Molecular Biology, University of the Punjab, Lahore, Pakistan

Nazima Rasool
Department of Botany, University of Kashmir, Jammu and Kashmir, India

Saiema Rasool
Department of Forest Management, Faculty of Forestry, University Putra Malaysia, Serdang, Selangor, Malaysia

Amandeep Rattan
Department of Botany, Hansraj Mahila Maha Vidyalaya, Jalandhar, Punjab, India

Mariana Rosa
Cátedra de Fisiología Vegetal, Facultad de Ciencias Naturales e IML, Tucumán, Argentina

Aykut Sağlam
Department of Molecular Biology and Genetics, Faculty of Sciences, Karadeniz Technical University, Trabzon, Turkey

Syed W. Shah
Atta-ur-Rahman School of Applied Biosciences, National University of Sciences and Technology (NUST), Islamabad, Pakistan

Iti Sharma
Department of Bioscience and Biotechnology, Banasthali University, Tonk, Rajasthan, India

Resham Sharma
Department of Botanical and Environmental Sciences, Guru Nanak Dev University, Amritsar, Punjab, India

Shashi Kant Shukla
Biological Product Laboratory, Department of Botany, University of Allahabad, Allahabad, Uttar Pradesh, India

Naveen K. Singh
Institute of Environment and Sustainable Development, Banaras Hindu University, Varanasi, Uttar Pradesh, India

Rajeev P. Singh
Institute of Environment and Sustainable Development, Banaras Hindu University, Varanasi, Uttar Pradesh, India

Shweta Singh
D. D. Pant Interdisciplinary Research Laboratory, Department of Botany, University of Allahabad, Allahabad, India

Sneh L. Singla-Pareek
Plant Molecular Biology Group, International Centre for Genetic Engineering and Biotechnology, New Delhi, India

Francesco Sunseri
Dipartimento Agraria, Università Mediterranea di Reggio Calabria, Salita Melissari, Reggio Calabria, Italy

Oksana Sytar
Department of Plant Physiology, Slovak University of Agriculture in Nitra, Nitra, Slovakia;
Plant Physiology and Ecology Department, Institute of Biology, Taras Shevchenko National University of Kyiv, Kyiv, Ukraine

Jiřina Száková
Department of Agro-Environmental Chemistry and Plant Nutrition, Faculty of Agrobiology, Food and Natural Resources, Kamýcká, Prague, Czech Republic

Nataliya Taran
Plant Physiology and Ecology Department, Institute of Biology, Taras Shevchenko National University of Kyiv, Kyiv, Ukraine

Rabiye Terzi
Department of Biology, Faculty of Sciences, Karadeniz Technical University, Trabzon, Turkey

Ashwani Kumar Thukral
Department of Botanical and Environmental Sciences, Guru Nanak Dev University, Amritsar, Punjab, India

Durgesh Kumar Tripathi
D. D. Pant Interdisciplinary Research Laboratory, Department of Botany, University of Allahabad, Allahabad, India;
Center of Advanced Study in Botany, Banaras Hindu University, Varanasi, India

Prabodh K. Trivedi
CSIR-National Botanical Research Institute (CSIR-NBRI), Lucknow, Uttar Pradesh, India

Bojjam Vidya Vardhini
Department of Botany, Telangana University, Dichpally, Nizamabad, Andhra Pradesh, India

Adarsh Pal Vig
Department of Botanical and Environmental Sciences, Guru Nanak Dev University, Amritsar, Punjab, India

Hina Waheed
Department of Botany, Pir Mehr Ali Shah Arid Agriculture University, Rawalpindi, Pakistan

Sarah Waseem
Atta-ur-Rahman School of Applied Biosciences, National University of Sciences and Technology (NUST), Islamabad, Pakistan

Fuat Yetişsin
Department of Molecular Biology and Genetics, Faculty of Arts and Sciences, Muş Alparslan University, Muş, Turkey

Marek Zivcak
Department of Plant Physiology, Slovak University of Agriculture in Nitra, Nitra, Slovakia

About the Editor

DR PARVAIZ AHMAD (EDITOR)

Dr Parvaiz is a senior assistant professor in the Department of Botany at Sri Pratap College, Srinagar, Jammu and Kashmir, India. He completed his postgraduate degree in botany in 2000 at Jamia Hamdard, New Delhi, India. After receiving a doctorate from the Indian Institute of Technology, Delhi, India, he joined the International Centre for Genetic Engineering and Biotechnology, New Delhi, in 2007. His main research area is stress physiology and molecular biology. He has published more than 40 research papers in peer-reviewed journals and 35 book chapters. He is also the editor of 15 volumes (one with Studium Press Pvt. India Ltd., New Delhi, India; eight with Springer NY USA; three with Elsevier USA; and three with Wiley). He is a recipient of the Junior Research Fellowship and Senior Research Fellowship awarded by the Council of Scientific and Industrial Research, New Delhi, India. Dr Parvaiz was awarded the Young Scientist Award under the Fast Track scheme in 2007 by the Department of Science and Technology, Government of India. Dr Parvaiz is actively engaged in studying the molecular and physio-biochemical responses of different agricultural and horticultural plants under environmental stress.

Preface

PLANT METAL INTERACTION: EMERGING REMEDIATION TECHNIQUES

Plants are valuable resources for all living organisms existing on planet Earth. Plants provide food, medicine, regulate water cycle, produce oxygen, and reform climate. Any major change in environment has a negative impact on the growth and development of plants. It means change in climatic conditions is having a direct or indirect impact on human beings. Climatic change (environmental stress) has a drastic impact on crop yield. Food production for future generations is a major problem because of (1) exponential increase in human population and (2) reduction in farmable land due to environmental pollution, caused by natural and anthropogenic activities. The industrial revolution has brought immense comforts to man, but at the same time many problems have also been reported. Heavy metals (cadmium, chromium, copper (Cu), lead, mercury, nickel, arsenic, and zinc) produced from industries and factories are nondegradable and have accumulated in soil and water bodies. It is estimated that 60% of farmable land is suffering from mineral problems, such as metal toxicity and nutritional and metal deficiencies. Heavy metal toxicity has become a major threat to plant growth and crop yield globally. It is not limited to plants, soil, and water only—heavy metals have also entered the human body through the food chain and caused many diseases and disorders. The mechanisms of heavy metal detoxification, accumulation, and tolerance have become the basis for using plants for remediating heavy metal–contaminated soil and water. Phytoremediation is the use of plants to uptake pollutants from the soil and water. It is a sustainable environmental cleanup technology and is used worldwide.

This edited volume *Plant Metal Interaction: Emerging Remediation Techniques* includes 25 chapters that will shed light on different heavy metals, their impact on plants, and remedial approaches. Chapter 1 is an overview on aluminum toxicity in plants, in which the author describes the impact of aluminum on plants, nutrient uptake, gene expression, and so on. Chapter 2 is about the effect of copper stress on plant growth, photosynthesis, and signal transduction. Chapter 3 discusses the effect of lead on plant and human DNA and its damages and its impact on the environment. Chapter 4 describes resistance of plants to Cu stress (transgenesis) and includes a discussion of Cu as an essential element, the toxic effects of high concentrations of Cu, maintenance of intracellular Cu homeostasis, construction of transgenic plants resistant to heavy metals, etc. Chapters 5, 6, and 7 throw light on boron toxicity, chromium toxicity, and antioxidative properties of selenium on plants. Chapter 8 is about the mitigation of heavy metal stress by brassinosteroids in plants. Chapters 9, 10, and 11 are related to plant metallothioneins and phytochelatins; biosynthesis of phytochelatins, phytochelatins, mediated redox homeostasis, and metallothioneins are also well-documented. Chapters 12, 13, and 14 describe the role of plants in detoxification and tolerance of heavy metals such as mercury and trace elements from the soil and water. Chapter 15 is related to phytoextraction—the use of plants to remove heavy metals from soil. Chapters 16, 17, 18, and 19 explain the efficient role of different plants in accumulating heavy metals from water and soil. Genetic strategies for phytoremediation potential in plants are also thoroughly explained. Chapters 20 and 21 are related to phytoremediation of heavy metals in the presence of microbes and plant growth–promoting rhizobacteria. Chapters 22 and 23 are about the molecular mechanism and role of adenosine triphosphatase (HMA2, HMA3, and HMA4) in the hyperaccumulation of heavy metals. Chapters 24 and 25 explain the impact of heavy metals on plants and their role in signaling.

This book is a compilation of different chapters on heavy metal stress in plants, phytoremediation, molecular mechanisms of phytoremediation, and signaling in plants. Each topic has been discussed to fully to ensure that readers receive the maximum information for each topic. There may be some errors in the book even after my best efforts. I need your feedback and suggestions on this (parvaizbot@yahoo.com).

I would like to thank all the contributors of this volume for their timely submission. Their collaboration and patience during the preparation of this book is unforgettable. I would also like to thank Laura Kelleher (acquisition editor, Elsevier), Emily Joy Grace Thomson (editorial project manager, Elsevier), Rajendran, Mohanapriyan (project manager, Elsevier), and the other team members of Elsevier for their help, suggestions, and timely publication of this volume.

Parvaiz Ahmad

Acknowledgements

I acknowledge all the contributors of this volume for their valuable contributions. Parvaiz Ahmad also acknowledges the Higher Education Department, Government of Jammu and Kashmir, India, for their support.

CHAPTER

1

ALUMINUM TOXICITY IN PLANTS: AN OVERVIEW

Sameen R. Imadi[1], Sarah Waseem[1], Alvina G. Kazi[1], M.M. Azooz[2,3], Parvaiz Ahmad[4]

Atta-ur-Rahman School of Applied Biosciences, National University of Sciences and Technology (NUST), Islamabad, Pakistan[1]; Department of Botany, Faculty of Science, South Valley University, Qena, Egypt[2]; Department of Biological Sciences, Faculty of Science, King Faisal University, Al-Hassa, Saudi Arabia[3]; Department of Botany, S.P. College, Srinagar, Jammu and Kashmir, India[4]

CHAPTER OUTLINE

Plant Metal Interaction. http://dx.doi.org/10.1016/B978-0-12-803158-2.00001-1

1. INTRODUCTION

Aluminum is found in nature, mostly in the form of aluminum silicates; a trace amount of element is required to carry out the biological processes by living organisms. The metal exists in the +2 and +3 states, thus having high affinity toward negatively charged ions and ionic groups such as fluoride (F^-) and hydroxyl group (OH^-), respectively (Mossor-Pietraszewska, 2001). Other groups having affinity toward aluminum for complex bond formation are carboxyl, carbonyl, and phosphate (Vitorello et al., 2005). The ionic form of aluminum with +3 charge is considered the soluble one, and thus is toxic to plants (Gupta et al., 2013). The ionic form is harmful to plants even at micromolar concentrations (Ma et al., 2001; reviewed by Silva, 2012). Some plants have evolved mechanisms to tolerate aluminum toxicity through the formation of organic acid complexes with aluminum, either internally within leaves or outside the root (Ma et al., 2001). Approximately 50% of the world's soil used for cultivating crops has turned acidic, causing a great setback to agriculture (Kochian et al., 2004; Sujatha and Mehar, 2015). In such an acidic soil, aluminum is the major mineral responsible for the inhibition of plant cell growth and division (Samac and Tesfaye, 2003) and the foremost target of the element for growth inhibition is root of plant (Poschenrieder et al., 2008).

Being very reactive, aluminum targets the cell wall, plasma membrane, nucleus, and cytoskeleton of plant root cells (Kochian et al., 2004) and affects and disrupts cell wall, plasma membrane, signal transduction pathway, and calcium homeostasis (Ma, 2007). Moreover, it is somewhat difficult to identify the signs and symptoms of aluminum toxicity in plants, which in some cases have the appearance of calcium deficiency and transport reduction (Rout et al., 2001). Alteration of calcium homeostasis by aluminum affects the signal transduction pathways through the inhibition of enzyme phospholipase C (Kochian et al., 2004). In wheat, rice, and sorghum, aluminum abundance leads to iron deficiency (Rout et al., 2001). The roots lose their fine branching ability because of monomeric aluminum toxicity (Rout et al., 2001). Under a different acidic pH range, various complex ionic forms of aluminum with hydroxyl groups are found in soil (Mossor-Pietraszewska, 2001). Such soils are deficient in calcium and magnesium, which provide the soil with basic characteristics, rendering it favorable for plant production (Vitorello et al., 2005). Aluminum polymerization in presence of various chelating agents such as phosphate raises the pH of soil, with a subsequent increase in aluminum concentration causing the loss of monomeric aluminum (Rout et al., 2001). Aluminum toxicity in plants also affects mitochondrial functioning because of the production of reactive oxygen species (ROS), thus causing ATP depletion and respiratory stress (Yamamoto et al., 2002). Thus in many ways aluminum produces harmful effects on plants, which should be analyzed in detail.

2. ALUMINUM TOXICITY

Aluminum toxicity is one of the major factors reducing the yield and quality of crops grown on acidic soils around the globe. Toxic effects of aluminum on plants are attributed to different physiological and biochemical pathways occurring in plant systems, which are believed to be initiated by the synthesis of mucopolysaccharides (Roy et al., 1988). Aluminum toxicity in plants begins with inhibiting growth, accumulating callose, distorting the cytoskeleton, and disturbing the surface charge of plasma membranes. After this, aluminum has been shown to distort H^+- adenosine triphosphatase (ATPase) activity, lipid peroxidation of membranes, and production of ROS in cytosol and ultimately in mitochondria. The damage further leads to respiratory dysfunction, opening of mitochondria permeability transition pores, collapsing of inner mitochondrial membrane potential, activation of mitochondrial protease, and finally induction of nuclear apoptosis that results in programmed cell death (Panda and Matsumoto, 2007).

2.1 EFFECT ON LEAVES

Aluminum toxicity results in thickening of epidermal layer cells in old leaves of tea plants (Matsumoto et al., 1976). When the plants are exposed to aluminum, the proline concentration in leaves increases significantly. A notable increase in malonaldehyde concentration and stimulation of super oxide dismutase and peroxidase has also been observed in aluminum-treated leaves (Guo et al., 2004). Young leaves grown under aluminum stress are seen to be small, curled along the margins, and frequently chlorotic under the leaf margins (Pavan and Bingham, 1982). A significant reduction in the leaf area is also seen in tomato plants subjected to aluminum stress (Simon et al., 1994).

2.2 EFFECTS ON SHOOTS

Shoots are indirectly affected by increased and decreased aluminum concentrations in soil (Gupta et al., 2013; Ribeiro et al., 2013). It has been reported that nitrogen, phosphorus, and iron concentrations significantly decrease as aluminum concentrations increase in the shoots of maize plants under aluminum stress. Magnesium decreases significantly at higher aluminum concentrations (>9 mg/L) (Lidon et al., 1999). Shoot growth of coffee seedlings also progressively decreases with the increase in aluminum concentrations (Pavan and Bingham, 1982).

Soya beans grown on aluminum-enriched soils produces significant inhibition of development of lateral branches of shoot. Aluminum primarily affects soot meristem, which is the area for cytokinin production and that is essential for development of lateral branches of shoots. The decrease in cytokinin level results in inhibition of development (Pan et al., 1988).

2.3 EFFECTS ON ROOTS

Aluminum has been observed as the major limiting factor in the growth process of roots in acid soils (Sujatha and Mehar, 2015). Reduction in root growth is the most common symptom of aluminum toxicity (Ali et al., 2011; He et al., 2012). When soil pH decreases to 4–4.5, aluminum is solubilized in soil water and is absorbed by roots affecting root growth (Matsumoto, 2000). Root growth is significantly affected by monomeric aluminum, but it has not been shown to be affected by total aluminum in soil solutions. As the concentration of monomeric aluminum increases in soil solutions, the growth capacity of plant root decreases (He et al., 2012). Aluminum absorbed by plant roots specifically on the root apex significantly decreases root elongation within hours of absorption (Matsumoto, 2000; Sujatha and Mehar, 2015; Kopittke et al., 2015).

As a result of aluminum toxicity, root elongation is reduced because of a reduction in mitotic activity (Roy et al., 1988). Callose formation with aluminum exposure in plants is negatively correlated to root growth (Zhang et al., 1994; Too et al., 2014; Sujatha and Mehar, 2015). A study performed on soybeans, sunflowers, subterranean clovers, and alfalfas to determine the phosphorus/aluminum and monomeric aluminum effects on them reports an increase in root elongation under high phosphorus/aluminum and calcium concentrations. Increased phosphorus/aluminum concentrations also decrease monomeric aluminum concentration from the solution. Thus, monomeric aluminum is reported as the limiting factor for root growth (Alva et al., 1986; Ali et al., 2011; Sujatha and Mehar, 2015). Lateral roots of aluminum-stressed plants are thicker, shorter, and fewer in number (Pavan and Bingham, 1982). Extension of lateral roots is more sensitive to aluminum compared with taproots, and the tips and apex of lateral roots accumulate more aluminum, consequently having a lower tolerance (Silva et al., 2001). Aluminum-induced inhibition of root growth in aluminum-sensitive plants may be due to disruption of both cell division in the meristematic region and cell expansion in the elongation zone in roots (Nichol and Oliveira, 1995; Silva, 2012).

Aluminum causes many morphological changes in root surface of different plants. It has been shown that exposure to aluminum results in decreased turgescence of epidermal cells of root tip and elongating regions, formation of a high amount of small depressions in elongating regions, destruction

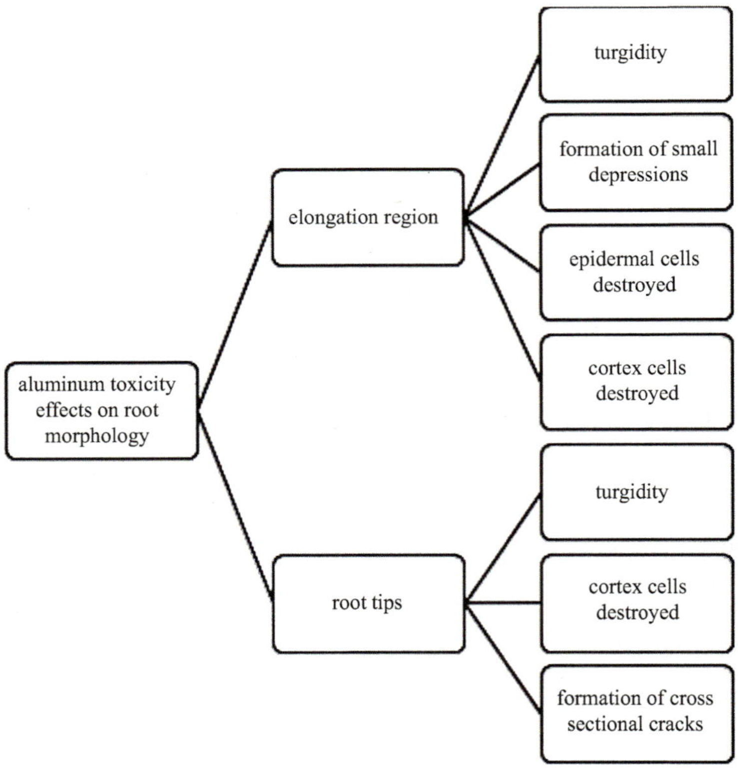

FIGURE 1.1

Effect of aluminum toxicity on root morphology.

of epidermal cells and outer cortex cells in tip and elongation region, and formation of cross-sectional cracks in the inner cortex of roots (Wagatsuma et al., 1987; Wu et al., 2014). The effect of aluminum on root morphology is shown in Figure 1.1.

2.4 EFFECTS ON PLANT PHYSIOLOGY AND MORPHOLOGY

Interaction of aluminum with different plant systems usually follows very different pathways. A potential target for aluminum is the plasma membrane as it alters the physical properties of membrane by interacting to ATPase and lipids in membranes. The primary site of infection for aluminum is the Golgi apparatus, after which it may also affect plasmalemma (Roy et al., 1988). Aluminum also affects photosynthesis because it lowers the chlorophyll content of plants (Roy et al., 1988).

Neepawa, an aluminum-tolerant cultivar of wheat exposed to 75 μM aluminum, induced callose synthesis after 6–12 h of aluminum exposure. Treatment with aluminum increases callose formation up to 86% within 30 min after an exposure of 48 h (Zhang et al., 1994). Wheat cultivars having differential aluminum tolerance when subjected to aluminum stress revealed that photosynthetic electron transport at photosystem II is greatly affected in plants where closure of the reaction center of photosystem II can be seen. This inhibition of photosynthesis is characterized by a chloroplast elemental loss and concomitant intrathylakoid acidification (Moustakas et al., 1995).

Experiments on maize plants under aluminum stress at various aluminum concentrations showed that root plasma membrane proton ATPase activity decreased after 0.33 ppm of treatment. Increased membrane permeability of plants at 1 mM of aluminum treatment has also been reported (Lidon et al., 2000). Aluminum has been also responsible for a decreased number of protoplasts in the main roots and induces abnormality, shrinking, and thickening in plasmalemma of protoplasts (Wagatsuma et al., 1987).

3. ALUMINUM UPTAKE AND TRANSPORT IN PLANTS

Plant uptake of aluminum occurs at near-neutral soil pH in soils of high organic matter content (Vondráčková et al., 2015). Aluminum–organic matter complexes can be solubilized at the pH range of 5–7, making it available to plant roots (Hargrove, 1986; Drábek et al., 2005; Kassaye et al., 2013; Vondráčková et al., 2015). *Chara corallina* is a giant alga and was checked for aluminum transport and uptake. It was observed that up to 99% of aluminum taken up by plants is accumulated in the cell wall and cytosol (Rengel and Reid, 1997). Uptake of aluminum by plants also depends on the type of group it is attached to. It has also been observed that plants exposed to aluminum chloride have a 15-fold more concentration of aluminum in their xylem sap compared with plants that are exposed to aluminum oxalate. Aluminum concentration becomes four times higher after 1 h of exposure and 10 times higher after 2 h of exposure in xylem sap. Aluminum uptake is highly inhibited by the presence of Lanthanum because Lanthanum competes with aluminum for the binding sites on the plasma membrane (Ma and Hiradate, 2000).

Studies show that aluminum is taken up by roots through endocytosis. This endocytosis may be fluid phase or receptor mediated and involve specific carriers. Internalization of aluminum takes place by involving polysaccharides and carriers (Haug and Shi, 1991). Aluminum chloride and aluminum fluoride have also been observed to be ligands for the membrane-associated G protein. Uptake of aluminum and its physiological effects are shown in Figure 1.2.

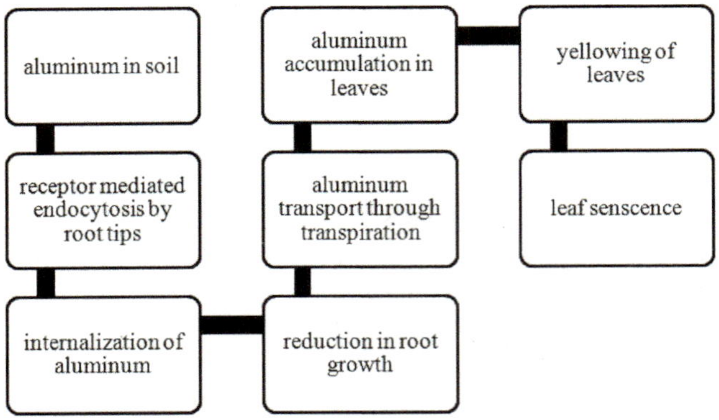

FIGURE 1.2

Aluminum uptake, transport, and physiological effects.

3.1 ALUMINUM ACCUMULATION IN TOLERANT PLANTS

It has been observed in aluminum-resistant and aluminum-sensitive soybean cultivars that aluminum accumulation and aluminum-induced callose formation is 50% in aluminum-sensitive and 25% in aluminum-resistant cultivars. Aluminum inhibits both aluminum-resistant and aluminum-sensitive soybean cultivars for the first 6 h of exposure. After that time, root growth is not inhibited in aluminum-resistant cultivars. Exudate analysis shows that both cultivars have specifically exuded citrate in exudates. Concentration of citrate in resistant cultivar exudates is reported to be high compared with that of sensitive cultivar (Yang et al., 2000).

After 7 days of seedling exposure in hydroponic culture of aluminum, it has been observed that tolerant lines accumulate less aluminum, whereas they have a high concentration of calcium, magnesium, and potassium compared with sensitive lines. In response to aluminum treatment, the proline concentration is highly increased in roots of tolerant plants; this increase can be up to three-fold. Lipid peroxidation is not observed as a result of aluminum treatment in tolerant plants (Giannakoula et al., 2008).

3.2 ALUMINUM UPTAKE AT ROOT LEVELS

The initial uptake site of aluminum is the root cap and mucilaginous layer covering epidermal cells. Aluminum ions bind to mucilage by exchange absorption on polyuronic acid and because of formation of polyhydroxy forms. This results in an increased concentration of aluminum per positive charge. Because aluminum is a polycation, it is expected to follow apoplasmic pathway to reach cortical cells, but it may also enter the stele through plasmalemma (Roy et al., 1988).

Root of aluminum-tolerant plants have much intense and vigorous growth compared with aluminum-sensitive plants. Aluminum-tolerant plants are also shown to accumulate less aluminum ion and ROS. Aluminum accumulation is highly present in the root apex, and formation of superoxides and peroxides are detected mainly in the elongation zone (Darko et al., 2004). Aluminum affects the rate of root growth more drastically compared with shoot growth. Because of this, it decreases the root/shoot ratio of plants (Pettersson and Strid, 1989).

4. PHYTOTOXICITY OF ALUMINUM AND ITS INTERACTIONS WITH NUTRIENTS

Mechanism by which aluminum interferes with ion influx in plants is not known. But aluminum has shown interaction with the uptake of many anions and cations (Nunes-Nesi et al., 2014). Aluminum has been involved in interfering with the uptake of cation minerals, whereas it enhances the uptake of anion mineral ions.

4.1 ALUMINUM INTERFERENCE WITH NITRATE IONS AND AMMONIUM IONS

Aluminum accumulation in the root apex, root tip, adjacent root cap, and mucilage has shown to be directly related to inhibition of nitrate ion uptake by root in aluminum-tolerant as well as aluminum-resistant plants. Uptake of nitrate inhibition by aluminum exposure has been shown to be of the same extent on all root regions (Lazof et al., 1994). Aluminum treatment in aluminum-sensitive barley results in enhancement of nitrate ion influx by 44% (Nichol et al., 1993). Aluminum has been found to inhibit ammonium uptake of plants up to 40% in aluminum-sensitive barley cultivars (Nichol et al., 1993). *Triticum aestivum* plants were grown in aluminum solution for 21 days at pH 4.5; the ammonium ions were depleted rapidly, followed by depletion in nitrate ions. Depletion of ammonium ions is most rapid, which results in increased pH. The rate of decline in pH is positively correlated to depletion in the concentration of ammonium ions (Taylor and Foy, 1985).

Aluminum has been shown to reduce the percentage of nitrate–nitrogen, whereas the percentage of amino acid–nitrogen is increased in the two sorghum cultivars having differential resistance to aluminum. The increase in amino acid–nitrogen percent is attributed to aluminum enhancing the degradation of proteins. The proportion of asparagine and glutamine is also changed by the presence of aluminum (Gomes et al., 1985). Plant tolerance to aluminum has been associated with differential nitrate and ammonium uptake and the pH of soil. pH decreases when the concentration of ammonium is high, but as long as the ammonium concentration decreases because of aluminum toxicity, the pH increases (Galvez and Clark, 1991). At pH 4.5, the rate of nitrate uptake significantly decreases by adding 100 μm of molar aluminum in maize plants. Removal of aluminum after 6 h of exposure resulted in increased nitrate uptake within 30 min (Durieux et al., 1993). The presence of aluminum in solution results in an increased nitrogen concentration in leaves and roots (Lee and Pritchard, 1984).

Aluminum reduced the uptake of anions, particularly nitrate, whereas the uptake of ammonium remained unaffected. This was proved in an experiment conducted on two maize cultivars that had differential tolerance to aluminum. The reduction in nitrate uptake is followed by increased hydrogen release. The reduction in nitrate uptake and increase in hydrogen release is higher in aluminum-sensitive plants compared with aluminum-tolerant plants. Reduction in nitrate uptake leads to rhizosphere stimulation by plants (Calba and Jaillard, 1997). Nitrate uptake inhibition is the primary response to aluminum treatment because in maize plants, nitrate ion uptake ceases after just 30 min of treatment (Durieux et al., 1994).

5. EFFECTS OF ALUMINUM ON GENE EXPRESSION

Some studies have shown that aluminum is able to bind with DNA and cause cytogenetic changes in plants, but it has also been shown that penetration of aluminum in mitotically active cells is very low (Roy et al., 1988). Aluminum alters the expression of genes significantly in aluminum-sensitive plants. Cell wall– and phosphate-responsive genes are highly regulated by aluminum (Maron et al., 2008).

Aluminum sensitivity in *Medicago trunculata* was shown, and it was found that the transcript accumulation of 2782 genes was significantly changed. There were 324 genes that were upregulated and 267 genes that were downregulated by at least two-fold in response to aluminum. The genes that are upregulated are involved in cell-wall modifications and responses to biotic and abiotic stresses. The genes that are significantly downregulated are involved in primary and secondary metabolism, protein synthesis, protein processing, and cell cycle. Oxidative stress and cell-wall stiffening markers for aluminum-induced genetic response are also differentially regulated (Chandran et al., 2008).

A study conducted on *Arabidopsis thaliana* showed that genes encoding peroxidase, glutathione-S transferase, blue copper binding protein, and protein homologous to reticuline, oxygen oxidoreductase enzyme, are significantly upregulated by aluminum treatment within 48 h of exposure. These genes are known to be upregulated in oxidative stress. Genes for super oxide dismutase and Bowman-Birk protease inhibitor are also highly upregulated by aluminum treatment. Therefore, it is clear that aluminum toxicity induces oxidative stress in plants (Richards et al., 1998) and aluminum-induced genes serve to protect against aluminum toxicity (Ezaki et al., 2000).

6. EFFECTS OF ALUMINUM ON PLANT METABOLISM

The efficient conversion of carbon into high-energy organic compounds through the phenomenon of photosynthesis and retrieval of this energy through respiration determine the growth rate of plant (Sulpice et al., 2010). Among the major factors of plant growth inhibition in acidic soils, aluminum toxicity limits root development by arresting cell division, extension, and transportation (Mossor-Pietraszewska, 2001; He et al., 2012). In the presence of aluminum in plant roots, the net rate of photosynthesis decreases, which seems to be a consequence of damage to thylakoid structure (Pereira et al., 2000; Ali et al., 2011; Silva, 2012; Ribeiro et al., 2013).

Pectin, present in root cells, binds to aluminum through the carboxyl group (Lidon and Barreiro, 2002). This binding ends up with callose accumulation in root cell walls, inhibition of DNA synthesis (Lidon and Barreiro, 2002) as a result of increased rigidity of the double helical configuration of DNA (Rout et al., 2001), and accumulation of up regulated resulting in programmed cell death of root cells (Lidon and Barreiro, 2002). ATP depletion and ROS production are the core players resulting from aluminum toxicity because of mitochondrial dysfunction that ends up as root growth inhibition (Yamamoto et al., 2002).

Toxic effects produced by aluminum result in ROS accumulation in the epidermal cells of roots within 10 min after exposure (Jones et al., 2006; Rigoulet et al., 2011; Xu et al., 2012). Most of the aluminum enters the plant root through the apoplast pathway; a very minute volume enters quickly via the symplast pathway, producing toxic effects on their respective targets (Kochian et al., 2004). Aluminum ions affect the metabolism of plasma membrane through strong interaction with the phospholipid component of the membrane, making it rigid and hence affecting metabolism (Yamamoto et al., 2001; Ahn and Matsumoto, 2006). It alters root respiration processes, becomes a hurdle to enzymatic activity responsible for phosphorylation of sugar, and plays an important role in depositing polysaccharides to the plant cell wall (Rout et al., 2001).

Along with other enzymes, which have roles in signaling pathway regulation, phospholipase-C is greatly regulated by the effect of aluminum on membrane lipid metabolism (Martínez-Estévez et al., 2003). One of the responses toward aluminum by plant cells regarding callose production is somehow related to plasma membrane alterations, which in turn are related to the blockage of Ca^{2+} channels (Yamamoto et al., 2001; Silva, 2012; Too et al., 2014). The β1, 3-glucan synthase, also known as the callose synthase, is found interior to the plasma membrane being activated by increased intracellular concentration of Ca^{2+} ions, resulting from plasma membrane damage (Yamamoto et al., 2001). Aluminum reduces phosphorus fixation to soil, thus decreasing its availability for roots and hence to the rest of the plant (Rout et al., 2001). Other than phosphorus, aluminum also interferes with other nutrient minerals such as calcium, magnesium, potassium, and iron (Rout et al., 2001; Nunes-Nesi et al., 2014).

7. INTERFERENCE WITH MINERAL METABOLISM

Aluminum interferes with the metabolisms of different nutrients, including copper, zinc, calcium, magnesium, potassium, phosphorus, and iron (Ribeiro et al., 2013). Generally, aluminum has negative effects on uptake of most macro- and micronutrients (Mariano and Keltjens, 2005). Most of the minerals show a significant inverse correlation with soil aluminum saturation and aluminum concentration in shoots. Aluminum also shows a positive correlation with shoot and root dry weight (Baligar et al., 1993). The correlation of aluminum with mineral ion uptake is shown in Table 1.1.

7.1 CALCIUM

Calcium ion movement in plants is unidirectional, which means its uptake is started from the roots and routed to meristematic zones and young tissue. Calcium, when deposited in the leaves, is unable to recycle it even in calcium-stressed conditions (Hanger, 1979). Calcium ions serve as secondary messengers in signal transduction and metabolic regulation. Aluminum is observed to bind all calcium-binding sites on the cell surface. At lower pH (<5.5), aluminum interferes with calcium absorption, which is a key factor affecting calcium uptake by roots. This antagonism leads to lower absorption of calcium and plants become malnourished (Roy et al., 1988). There is a strong correlation between aluminum toxicity and aluminum-induced inhibition of calcium ion uptake by plant root

Table 1.1 Correlation of aluminum with mineral ion uptake and accumulation

Mineral Ion	Nutrient	Charge	Correlation	
			Uptake	Accumulation
Calcium	Macronutrient	+2	Negative	Negative
Potassium	Macronutrient	+1	Positive	Negative
Magnesium	Macronutrient	+2	Negative	Negative
Iron	Micronutrient	+3, +2	Negative	–
Nitrogen	Macronutrient	+2	Negative	Positive
Phosphorus	Macronutrient	+3, +5	Positive	Positive

apices (Ryan and Kochian, 1993; Hossain et al., 2014). The same amount of aluminum concentration affects calcium uptake differentially in aluminum-tolerant and aluminum-sensitive plants (Huang et al., 1992).

Inhibition of calcium may be the result of aluminum exposure (Vazquez et al., 1999; Hossain et al., 2014; Ribeiro et al., 2013). A total of 100 μM aluminum has been found to inhibit the influx of calcium ion by 69% in aluminum-sensitive cultivars of barley (Nichol et al., 1993). Aluminum decreases the net uptake of calcium, and this decrease is more pronounced if bepridil is also present along with aluminum in the uptake solution. Calcium uptake is also reduced at low pH (about 4.5). Aluminum ions inhibit calcium ion uptake by binding to verapamil-specific channel sites as well as by interfering with action of guanosine 5′ triphosphate binding proteins (Rengel and Elliott, 1992).

In wheat cultivars differing in aluminum tolerance, the concentration of calcium decreases at a pH of 4.5 when they are treated with aluminum at concentrations of 37.1, 74.1, and 148 μM, respectively. Concentration of calcium in aluminum-sensitive cultivars under high aluminum exposure is within deficiency range (Moustakas et al., 1995). A combination of high concentrations of nitrogen and aluminum is shown to decrease the uptake of calcium ions in beech plants (Bengtsson et al., 1994). Passive transport of calcium ions in roots is accomplished through inhibited aluminum (Widell et al., 1994). In wheat, aluminum treatment is seen to reduce adsorption and accumulation of calcium in roots. This reduction is not overcome by supplying more calcium. Similarly, it has also been observed that an increase in aluminum concentration is not associated with the extent of inhibition if there is calcium ion uptake (Johnson and Jackson, 1964).

Aluminum reduces calcium concentration in spruce roots. Calcium concentration is unaffected in shoots under aluminum treatment (Hakan et al., 1988). It has been shown that aluminum enters the plant cells through calcium channel–like pathway (Liu and Luan, 2001). Maize plants grown under high aluminum concentrations showed a decrease in calcium uptake by up to 61% (Mariano and Keltjens, 2005). In spruce seedlings with mycorrhizal associations, the calcium concentration in the cortex cell layer of root tips was decreased after 13–17 weeks of aluminum exposure (Jentschke et al., 1991).

7.2 POTASSIUM

Aluminum treatment of aluminum-sensitive barley cultivar is shown to inhibit potassium influx by 13% (Nichol et al., 1993). High aluminum concentrations severely inhibit potassium ion uptake (Jongbloed and Borst-Pauwels, 1992; Hossain et al., 2014; Ribeiro et al., 2013). In wheat cultivars differing in aluminum tolerance, concentration of Potassium decreases at a pH of 4.5 when they are treated with aluminum at different concentrations (Moustakas et al., 1995). Active potassium uptake obtained in intact roots is also inhibited by aluminum (Widell et al., 1994).

The addition of aluminum to aluminum-sensitive wheat cultivars shows hyperpolarization of the membrane potential. Aluminum addition is also associated with a rapid pH-dependent change in cytosolic potassium ion concentration. At pH 5, the presence of potassium ion in plant growth medium can diminish the aluminum-induced decrease in cytosolic potassium ion concentration (Lindberg and Strid, 1997). Potassium concentration was reported to decrease in shoots at 10 mM aluminum concentration in spruce roots, whereas rice seedlings under aluminum stress have maximum uptake for potassium ion (Sivaguru and Paliwal, 1993).

Uptake of potassium ion by pea plants considerably decreases with aluminum chloride treatment (Matsumoto and Yamaya, 1986). Aluminum inhibits potassium ions in root cells as well as guard cells, which are caused by blocking the channels at cytoplasmic site of plasma membrane. Potassium ion inhibition in the guard cells because of aluminum action is seen only on internal exposure (Liu and Luan, 2001). Aluminum stress is seen to significantly enhance potassium ion efflux from barley roots that had been preloaded with potassium ions as a result of potassium stress (Kasai et al., 1992). Root and shoot potassium ion concentration of potassium ions is also seen to increase with increase in aluminum concentration in pine trees (Huang and Bachelard, 1993).

Aluminum exposure in *T. aestivum* species shows striking inhibition in potassium ion influx and a decrease in potassium ion concentration in roots and shoots. It has been shown that durum wheat is more tolerant to aluminum toxicity and decrease in potassium ion concentration compared with common winter wheat (Zsoldos et al., 2000). Aluminum is seen to inhibit potassium uptake also in fungi, which was confirmed in a study on three fungi species (Jongbloed et al., 1992).

7.3 MAGNESIUM

Aluminum affects magnesium uptake more than any other nutrient. Aluminum exposure induces symptoms from a lack of magnesium (Wheeler and Follett, 1991). In wheat cultivars, the level of magnesium decreases at pH 4.5 in plants treated with different concentrations of aluminum. Aluminum nutrient solution inhibits magnesium uptake by roots much stronger than calcium uptake (Van Praag et al., 1997; Bose et al., 2011, 2013). Aluminum reduces magnesium concentration in spruce roots (Hakan et al., 1988) and impairs the uptake of magnesium. Decrease in pH from 6.0 to 4.2 exerted a noncompetitive inhibition of net magnesium ion uptake, whereas competitive inhibition of net magnesium uptake was observed at pH of 4.2. Cultivar ability to retain higher affinity for magnesium might be one of the mechanisms of differential aluminum tolerance (Rengel and Robinson, 1989).

Increased aluminum concentrations are known to decrease magnesium concentrations in roots and shoots of pine trees (Huang and Bachelard, 1993). Maize plants grown under high aluminum concentrations for 14 days showed a decrease in magnesium uptake of 72% (Mariano and Keltjens, 2005). The effect of pH at 6.6 μM aluminum concentration on net magnesium uptake is much larger in 15-day-old plants compared with 35 day old plants. Competitive inhibition of potassium has been shown at an aluminum concentration with 26 μM aluminum ion activity (Rengel, 1990).

Increased aluminum concentration in soil inhibits the concentration and content of magnesium most significantly among all macro nutrients in rice plants (Fageria and Carvalho, 1982). Presence of aluminum in soil reduces the level of magnesium uptake in both mycorrhizal plants and nonmycorrhizal plants and reduction in magnesium ion uptake is more vigorous in nonmycorrhizal plants (Egerton-Warburton et al., 1993).

7.4 IRON

In wheat cultivars differing in aluminum tolerance, concentration of iron decreases at a pH of 4.5 when they are treated with aluminum at different concentrations. Among all the micronutrients, inhibition of iron uptake by exposure to high aluminum concentration is much higher. But it does not affect the root growth significantly as iron is micro nutrient (Moustakas et al., 1995).

7.5 NITROGEN

Aluminum decreases the uptake of nitrogen in plants and is reported by many workers (Purcino et al., 2003). The amount of nitrogen accumulated in roots is decreased as a result of aluminum exposure whereas the percent nitrogen increases. Aluminum also reduces the percent of nitrogen uptake by roots and its translocation to aerial parts of plants. The reduction in nitrogen translocation is from aluminum stress on plants (Gomes et al., 1985).

7.6 PHOSPHORUS

The plants that are more aluminum sensitive are less phosphorus efficient. Aluminum inhibits the uptake of phosphorus. Soils that have a high amount of aluminum in them have low amounts of phosphorus. Aluminum tolerance is directly related and positively correlated with phosphorus efficiency (Liao et al., 2006). Aluminum enhances the uptake of phosphate from soil up to 17% in aluminum-sensitive barley plants. Phosphate binding is strongly enhanced in aluminum-sensitive plants in the cell walls under the presence of aluminum (Nichol et al., 1993). The uptake of phosphate and concentration of phosphorus in roots of beech plants has been shown to increase when plants were grown in the presence of 0.1 mM aluminum. It has also been observed that an aluminum concentration of 1.0 mM is associated with decreased phosphorus concentration in both roots and shoots (Bengtsson et al., 1994).

Aluminum treatment has been shown to enhance the accumulation of phosphorus in spruce roots by more than two orders of magnitude, but aluminum also reduces the translocation of phosphorus to apical buds and mature primary needles (Cumming et al., 1986). Aluminum increases the concentration of phosphorus in roots of triticale and rye. Aluminum treatment results in smaller increases in root phosphorus in aluminum-tolerant triticale and rye compared with aluminum-sensitive cultivars (Mugwira et al., 1980).

8. INTERACTION WITH CALMODULIN

Among various metallic sensors, calmodulin is the preeminent calcium (Ca^{+2}) sensor in eukaryotes and plays its role under environmental stress conditions by activating specific targets through calcium regulation (Inostroza-Blancheteau et al., 2013). Al^{+3} has an ability to bind at Ca^{+2}-binding site of calmodulin, thus mimicking the action of calcium. In this way, aluminum blocks the enzymatic activity of phospholipase-C, which is a calcium-dependent enzyme (Jones and Kochian, 1997). Aluminum toxicity in plants affect calcium-induced metabolic processes (Sivaguru et al., 2005). Hence, for the sake of plant growth and metabolism, calcium acts as a second messenger in plant signal transduction pathways (Delhaize and Ryan, 1995).

There is an increase in intracellular calcium activity within cytosol of plant cells under aluminum stress (Sivaguru et al., 2005). There are other enzymes that are also dependent on the Ca^{+2}/calmodulin signaling pathway. Among them, the most distinct is calcium-dependent protein kinases in plants and other protozoan (Wernimont et al., 2010). A micromolar concentration of aluminum to calmodulin of approximately 3:1 plays a major role in the inhibition of calcium-magnesium ATPase activity that is responsible for the maintenance of transmembrane potential (Siegel and Hang, 1983). One of the factors accountable for affecting the calcium uptake by plants via aluminum as an antagonist is the acidic pH of approximately 5.5 (Roy et al., 1988).

Aluminum affects cellular homeostasis through the blockage of calcium ion channels leading toward inhibition of the cellular events related to cellular division and elongation by breaking down calcium signaling pathways (Jones et al., 1998). The cell-dividing capacity of plant depends upon the meristematic tissue that demands phosphate for its cell-cycle activities (Lai et al., 2007). Aluminum might also directly inhibits the H^+-ATPase activity of plasma membrane in squash roots at the time when the membrane potential is at its maximum positive (i.e., when it is depolarized) (Ahn et al., 2001).

9. EXISTENCE OF DIFFERENTIAL ALUMINUM TOLERANCE IN PLANTS

Aluminum-sensitive plants absorb more aluminum compared with aluminum-tolerant plants (Matsumoto, 2000). When the wheat seedlings are exposed to aluminum, it has been reported that the seedlings increases the expression and accumulation of malic acid, which protect the seedlings from phytotoxic aluminum toxicity. Synthesis of malic acid is continuous for the whole time of aluminum exposure and specifically stimulated by aluminum (Delhaize et al., 1993).

Malate dehydrogenase has been seen to be significantly overexpressed in transgenic alfalfa on exposure to aluminum stress, which results in enhancement of organic acid synthesis. Increased organic acid synthesis enhances the plants tolerance to aluminum stress (Tesfaye et al., 2001). Aluminum-tolerant plants form superoxide dismutase, ascorbate peroxidase, catalase, and glutathione-S transferase as a response to aluminum stress for killing ROS (Darko et al., 2004). Aluminum tolerance is also associated with citrate release in exudates. Citrate and malate excretion are shown to increase in all plants during early exposure to aluminum, but only citrate excretion is sustained in aluminum-tolerant plants (Silva et al., 2001). *T. aestivum* plants grown under aluminum stress shows that plant tolerance to aluminum is negatively correlated to decreased pH (Taylor and Foy, 1985).

Aluminum accumulation is a phenomenon in which native plants of acidic soils uptake aluminum and form an aluminum–ligand complex for translocation from roots to shoots and finally aluminum gets accumulated in leaves. Aluminum accumulator plants have been shown to prevent aluminum toxicity on leaf metabolism by isolating aluminum in sites that are insensitive to aluminum (Watanabe and Osaki, 2002). Aluminum-tolerant carrot cell lines, when provided with aluminum phosphate as the only source of phosphorus mineral ion, showed an increase in secretion of citrate in the solution compared with citrate secreted as a result of sodium phosphate or iron phosphate (Koyama et al., 1988). Buckwheat is an aluminum accumulator plant that accumulates high concentrations of aluminum in leaves without showing any toxicity. Aluminum distribution in accumulator plants highly depends on the rate and duration of transpiration. It also has been reported that aluminum loses its mobility once it gets accumulated in leaves (Shen and Ma, 2001).

10. TOLERANCE MECHANISMS OF ALUMINUM

Mechanism of aluminum tolerance involves exudation of organic acid anions, complexation of aluminum with organic acids, and subsequent detoxification of aluminum (Panda and Matsumoto, 2007). Overexpression of citrate is also positively correlated to aluminum tolerance in transgenic tobacco and papaya (De la Fuente et al., 1997). Aluminum induces dose- and time-dependent formation of ROS and subsequent protein oxidation (Boscolo et al., 2003). A better understanding of aluminum stress tolerant mechanisms could lead to development of more tolerant plants (Foy, 1988).

10.1 pH CHANGES IN ROOT ZONES AND ORGANIC ACID ANION RELEASE

The well-defined mechanism by which plants may attain tolerance to aluminum toxicity and detoxification of aluminum is pH change in the root zones. External detoxification of aluminum is performed by the release of organic acid anions from roots, resulting in alkalization of rhizosphere and nearby soil (Ma, 2007). Wheat and maize detoxify aluminum by releasing organic acid anions to change the pH through aluminum-activated anion channels in plasma membrane (Ma et al., 2001). Organic acid metabolism is also associated to plant tolerance in aluminum stress. Organic acids not only act as intermediates in carbon metabolism, but also as key components in mechanisms that are used by some plants to tolerate mineral deficiencies and metal tolerance (Lopez-Bucio et al., 2000).

10.2 UPTAKE AND DISTRIBUTION

Internal toxicity of aluminum can be reduced by formation of nontoxic aluminum complexes with organic acids and other chelators and sequestration of these complexes into the vacuoles (Ma, 2007).

10.3 SILICON TREATMENT

Silicon treatment of plants detoxifies aluminum. Silicon-treated plants exposed to aluminum showed enhanced silicon accumulation in cell-wall fractions that reduced aluminum mobility by apoplastic pathway. Silicon treatment results in formation of hydroxy aluminum silicates in apoplast of roots, which reduces aluminum toxicity (Wang et al., 2004). Studies have shown that aluminum toxicity is ameliorated by treatment to silicon. If aluminum and silicon are codisposed in plants, silicon reduces the effects of aluminum. Because of this, silicon is considered to be a beneficent mineral for plant growth (Hodson and Evans, 1995).

10.4 NITROGEN NUTRITION

Nitric oxide plays role in protecting the plants from aluminum toxicity. Nitric oxide applied to plants exposed to aluminum results in a decreased amount of ROS, lipoxygenase enzymes, and antioxidant enzymes. The decreased effects caused by aluminum toxicity have proved that nitric oxide has a role in aluminum tolerance in plants (Wang and Yang, 2005).

10.5 CALCIUM NUTRITION

One of the major solutions for aluminum excess in soil and aluminum toxicity is neutralization of soil by increased supply of calcium carbonate (Matsumoto, 1991). Calcium nutrition results in amelioration of aluminum toxicity. This was proved in an experiment on wheat genotypes. The effect of calcium is not completely deleterious to aluminum toxicity, but it eliminates the toxicity to a great extent (Edmeades et al., 1991).

11. CONCLUSIONS AND FUTURE PROSPECTS

Aluminum toxicity is a global and serious threat to plants grown in acidic soil. It causes many serious problems to plants by affecting their root, shoot, and leaf growth. Exposure to aluminum also destroys many physiological pathways in plants and results in declined uptake and accumulation of many macro- and micronutrients. Aluminum toxicity cannot be prevented because aluminum is the one of the most common minerals found in Earth's crust and is also one of the most economically important toxic metals; it also increases leaf senescence because it is a heavy metal.

This threat to the growth and development of plants can be reduced by performing a great deal more research on reducing aluminum effects by calcium and nitrogen nutrition. Ectomycorrhizal fungi can also pave a way to get rid of aluminum toxicity. The association formed by these fungi reduces the uptake of aluminum. Multidisciplinary collaborative research should be performed in the future to get rid of aluminum toxic effects. The biochemistry of aluminum and the mechanisms by which it affects the plant body need to be investigated in future. Recent techniques must be used to check aluminum tolerance in different crop plants.

REFERENCES

Ahn, S.J., Matsumoto, H., 2006. The role of the plasma membrane in the response of plant roots to aluminum toxicity. Plant Signal Behav. 1 (2), 37–45.

Ahn, S.J., Sivaguru, M., Osawa, H., Chung, G.C., Matsumoto, H., 2001. Aluminum inhibits the H^+-ATPase activity by permanently altering the plasma membrane surface potentials in squash roots. Plant Physiol. 126 (4), 1381–1390.

Ali, S., Zeng, F., Qiu, L., Zhang, G., 2011. The effect of chromium and aluminum on growth, root morphology, photosynthetic parameters and transpiration of the two barley cultivars. Biol. Plant. 55 (2), 291–296.

Alva, A.K., Edwards, D.G., Asher, C.J., Blamey, F.P.C., 1986. Effects of phosphorus/aluminum molar ratio and calcium concentration on plant response to aluminum toxicity. Am. Soc. Agron. 50 (1), 133–137.

Baligar, V.C., Shaffert, R.E., Dos Santos, H.L., Pitta, G.V.E., Bahia Filho, A.F.De C., 1993. Soil aluminium effects on uptake, influx, and transport of nutrients in sorghum genotypes. Plant Soil 150 (2), 271–277.

Bengtsson, B., Asp, H., Jensen, P., 1994. Uptake and distribution of calcium and phosphorus in beech (*Fagus sylvatica*) as influenced by aluminum and nitrogen. Tree Physiol. 14 (1), 63–73.

Boscolo, P.R.S., Menossi, M., Jorge, R.A., 2003. Aluminum-induced oxidative stress in maize. Phytochemistry 62 (2), 181–189.

Bose, J., Babourina, O., Rengel, Z., 2011. Role of magnesium in alleviation of aluminium toxicity in plants. J. Exp. Bot. 62 (7), 2251–2264.

Bose, J., Babourina, O., Shabala, S., Rengel, Z., 2013. Low-pH and aluminum resistance in *Arabidopsis* correlates with high cytosolic magnesium content and increased magnesium uptake by plant roots. Plant Cell Physiol. 54 (7), 1093–1104.

Calba, H., Jaillard, B., 1997. Effect of aluminium on ion uptake and H^+ release by maize. New Phytol. 137 (4), 607–616.

Chandran, D., Sharopova, N., Ivashuta, S., Gantt, J.S., VandenBosch, K.A., Samac, D.A., 2008. Transcriptome profiling identified novel genes associated with aluminum toxicity, resistance and tolerance in Medicago truncatula. Planta 228 (1), 151–166.

Cumming, J.R., Eckert, R.T., Evans, L.S., 1986. Effect of aluminum on ^{32}P uptake and translocation by red spruce seedlings. Can J. For. Res. 16 (4), 864–867.

Darko, E., Ambrus, H., Stefanovits-Banyai, E., Fodor, J., Bakos, F., Barnabas, B., 2004. Aluminium toxicity, Al tolerance and oxidative stress in an Al-sensitive wheat genotype and in Al-tolerant lines developed by in vitro microspore selection. Plant Sci. 166 (3), 583–591.

De la Fuente, J.M., Ramireez-Rodriguez, V., Cabrera-Ponce, J.L., Herrera-Estrella, L., 1997. Aluminum tolerance in transgenic plants by alteration of citrate synthesis. Science 276 (5318), 1566–1568.

Delhaize, E., Ryan, P.R., Randall, P.J., 1993. Aluminum tolerance in wheat (*Triticum aestivum* L.) (II. Aluminum-stimulated excretion of malic acid from root apices). Plant Physiol. 103 (3), 695–702.

Delhaize, E., Ryan, P.R., 1995. Aluminum toxicity and tolerance in plants. Plant Physiol. 107, 315–321.

Drábek, O., Mládková, L., Borůvka, L., Száková, J., Nikodem, A., Němeček, K., 2005. Comparison of water-soluble and exchangeable forms of Al in acid forest soils. J. Inorg. Biochem. 99, 1788–1795.

Durieux, R.P., Bartlett, R.J., Magdoff, F.R., 1994. Separate mechanisms of aluminium toxicity for nitrate uptake and root elongation. Plant Soil 172 (2), 229–234.

Durieux, R.P., Jackson, W.A., Kamprath, E.J., Moll, R.H., 1993. Inhibition of nitrate uptake by aluminium in maize. Plant Soil 151 (1), 97–104.

Edmeades, D.C., Wheeler, D.M., Blamey, F.P.C., Christie, R.A., 1991. Calcium and magnesium amelioration of aluminium toxicity in Al-sensitive and Al-tolerant wheat. Develop. Plant Soil Sci. 45, 755–761.

Egerton-Warburton, L.M., Kuo, J., Griffin, B.J., Lamont, B.B., 1993. The effect of aluminium on the distribution of calcium, magnesium and phosphorus in mycorrhizal and non-mycorrhizal seedlings of *Eucalyptus rudis*: a cryo-microanalytical study. Plant Soil 155–156 (1), 481–484.

Ezaki, B., Gardner, R.C., Ezaki, Y., Matsumoto, H., 2000. Expression of aluminum-induced genes in transgenic *Arabidopsis* plants can ameliorate aluminum stress and/or oxidative stress. Plant Physiol. 122 (3), 657–666.

Fageria, N.K., Carvalho, J.R.P., 1982. Influence of aluminum in nutrient solutions on chemical composition in upland rice cultivars. Plant Soil 69 (1), 31–44.

Foy, C.D., 1988. Plant adaptation to acid, aluminum-toxic soils. Comm. Soil Sci. Plant Anal. 19 (7–12), 959–987.

Galvez, L., Clark, R.B., 1991. Nitrate and ammonium uptake and solution pH changes for Al-tolerant and Al-sensitive sorghum (*Sorghum bicolor*) genotypes grown with and without aluminum. Develop. Plant Soil Sci. 45, 805–814.

Giannakoula, A., Moustakas, M., Mylona, P., Papadakis, I., Yupsanis, T., 2008. Aluminum tolerance in maize is correlated with increased levels of mineral nutrients, carbohydrates and proline, and decreased levels of lipid peroxidation and Al accumulation. J. Plant Physiol. 165 (4), 385–396.

Gomes, M.M.S., Cambraia, J., Sant'anna, R., Estevao, M.M., 1985. Aluminum effects on uptake and translocation of nitrogen in sorghum (*Sorghum bicolor*, L. Moench). J. Plant Nutr. 8 (6), 457–465.

Guo, T., Zhang, G., Zhou, M., Wu, F., Chen, J., 2004. Effects of aluminum and cadmium toxicity on growth and antioxidant enzyme activities of two barley genotypes with different Al resistance. Plant Soil. 258 (1), 241–248.

Gupta, N., Gaurav, S.S., Kumar, A., 2013. Molecular basis of aluminium toxicity in plants: a review. Amer J. Plant Sci. 4, 21–37, 21.

Hakan, A.S.P., Bengtsson, B., Jensen, P., 1988. Growth and cation uptake in spruce (*Picea abies* Karst.) grown in sand culture with various aluminium contents. Plant Soil 111 (1), 127–133.

Hanger, B.C., 1979. The movement of calcium in plants. Comm. Soil Sci. Plant Anal. 10 (1–2), 171–193.

Hargrove, W.L., 1986. The solubility of aluminum-organic matter and its implication in plant uptake of aluminum. Soil Sci. 142 (3).

Haug, A., Shi, B., 1991. Biochemical basis of aluminium tolerance in plant cells. Develop. Plant Soil Sci. 45, 839–850.

He, H.Y., He, L.F., Gu, M.H., Li, X.F., 2012. Nitric oxide improves aluminum tolerance by regulating hormonal equilibrium in the root apices of rye and wheat. Plant Sci. 2012 (183), 123–130.

Hodson, M.J., Evans, D.E., 1995. Aluminium/silicon interactions in higher plants. J. Exp. Bot. 46 (2), 161–171.

Hossain, M.A., Ashrafuzzaman, M., Hossain, A.K.M.Z., Ismail, M.R., Koyama, H., 2014. Role of accumulated calcium in alleviating aluminum injury in wheat plants. Sci. World J. 2014, 457187.

Huang, J., Bachelard, E.P., 1993. Effects of aluminium on growth and cation uptake in seedlings of *Eucalyptus mannifera* and *Pinus radiate*. Plant Soil 149 (1), 121–127.

Huang, J.W., Grunes, D.L., Kochian, L.V., 1992. Aluminum effects on the kinetics of calcium uptake into cells of the wheat root apex. Planta 188 (3), 414–421.

Inostroza-Blancheteau, C., Aquea, F., Loyola, R., Slovin, J., Josway, S., Rengel, Z., Reyes-Diaz, M., Alberdi, M., Arce-Johnson, P., 2013. Molecular characterisation of a calmodulin gene, VcCaM1, that is differentially expressed under aluminium stress in highbush blueberry. Plant Biol. 15 (6), 1013–1018.

Jentschke, G., Schlegel, H., Godbold, D.L., 1991. The effect of aluminium on uptake and distribution of magnesium and calcium in roots of mycorrhizal Norway spruce seedlings. Physiol. Plant. 82 (2), 266–270.

Johnson, R.E., Jackson, W.A., 1964. Calcium uptake and transport by wheat seedlings as affected by aluminum. SSSAJ 28 (3), 381–385.

Jones, D.L., Blancaflor, E.B., Kochian, L.V., Gilroy, S., 2006. Spatial coordination of aluminium uptake, production of reactive oxygen species, callose production and wall rigidification in maize roots. Plant Cell Environ. 29 (7), 1309–1318.

Jones, D.L., Kochian, L.V., Gilroy, S., 1998. Aluminum induces a decrease in cytosolic calcium concentration in BY-2 tobacco cell cultures. Plant Physiol. 116 (1), 81–89.

Jones, D.L., Kochian, L.V., 1997. Aluminum interaction with plasma membrane lipids and enzyme metal binding sites and its potential role in Al cytotoxicity. FEBS Lett. 400 (1), 51–57.

Jongbloed, R.H., Borst-Pauwels, G.W.F.H., 1992. Effects of aluminium and pH on growth and potassium uptake by three ectomycorrhizal fungi in liquid culture. Plant Soil 140 (2), 157–165.

Jongbloed, R.H., Tosserams, M.W.A., Borst-Pauwels, G.W.F.H., 1992. The effect of aluminium on phosphate uptake by three isolated ectomycorrhizal fungi. Plant Soil 140 (2), 167–174.

Kasai, M., Sasaki, M., Yamamoto, Y., Matsumoto, H., 1992. Aluminum stress increases K^+ efflux and activities of ATP- and PPi- dependent H^+ pumps of tonoplast-enriched membrane vesicles from barley roots. Plant Cell Physiol. 33 (7), 1035–1039.

Kassaye, Y.A., Salbu, B., Skipperud, L., Einset, J., 2013. High tolerance of aluminum in the grass species *Cynodon aethiopicus*. Acta Physiol. Plant 35, 1749–1761.

Kochian, L.V., Hoekenga, O.A., Pineros, M.A., 2004. How do crop plants tolerate acid soils? Mechanisms of aluminum tolerance and phosphorous efficiency. Ann. Rev. Plant Biol. 55, 459–493.

Kopittke, P.M., Moore, K.L., Lombi, E., Gianoncelli, A., Ferguson, B.J., et al., 2015. Identification of the primary lesion of toxic aluminum in plant roots. Plant Physiol. 167 (4), 1402–1411.

Koyama, H., Okawara, R., Ojima, K., Yamaya, T., 1988. Re-evaluation of characteristics of a carrot cell line previously selected as aluminum-tolerant cells. Physiol. Plant. 74 (4), 683–687.

Lai, F., Thacker, J., Li, Y., Doerner, P., 2007. Cell division activity determines the magnitude of phosphate starvation responses in *Arabidopsis*. Plant J. 50 (3), 545–556.

Lazof, D.B., Rincon, M., Rufty, T.W., Mackown, C.T., Carter, T.E., 1994. Aluminum accumulation and associated effects on NO_3^- influx in roots of two soybean genotypes differing in Al tolerance. Plant Soil 164 (2), 291–297.

Lee, J., Pritchard, M.W., 1984. Aluminium toxicity expression nutrient uptake, growth and root morphology of *Trifolium* repens L. cv. 'Grasslands Huia'. Plant Soil 82 (1), 101–116.

Liao, H., Wan, H., Shaff, J., Wang, X., Yan, X., Kochian, L.V., 2006. Phosphorus and aluminum interactions in soybean in relation to aluminum tolerance. Exudation of specific organic acids from different regions of the intact root system. Plant Physiol. 141 (2), 674–684.

Lidon, F.C., Azinheira, H.G., Barreiro, M.G., 2000. Aluminum toxicity in maize: biomass production and nutrient uptake and translocation. J. Plant Nutr. 23 (2), 151–160.

Lidon, F.C., Barreiro, M.D.G., 2002. An Overview into aluminum toxicity in maize. Bulg. J. Plant Physiol. 28 (3–4), 96–112.

Lidon, F.C., Barreiro, M.G., Ramalho, J.C., Lauriano, J.A., 1999. Effects of aluminum toxicity on nutrient accumulation in maize shoots: implications on photosynthesis. J. Plant Nutr. 22 (2), 397–416.

Lindberg, S., Strid, H., 1997. Aluminium induces rapid changes in cytosolic pH and free calcium and potassium concentrations in root protoplasts of wheat (*Triticum aestivum*). Physiol. Plant. 99 (3), 405–414.

Liu, K., Luan, S., 2001. Internal aluminum block of plant Inward K^+ channels. Plant Cell 13 (6), 1453–1466.

Lopez-Bucio, J., Nieto-Jacobo, M.F., Ramirez-Rodriguez, V., Herrera-Estrella, L., 2000. Organic acid metabolism in plants: from adaptive physiology to transgenic varieties for cultivation in extreme soils. Plant Sci. 160 (1), 1–13.

Ma, J.F., Hiradate, S., 2000. Form of aluminium for uptake and translocation in buckwheat (*Fagopyrum esculentum* Moench). Planta 211 (3), 355–360.

Ma, J.F., Ryan, P.R., Delhaize, E., 2001. Aluminium tolerance in plants and the complexing role of organic acids. Trends Plant Sci. 6 (6), 273–278.

Ma, J.F., 2007. Syndrome of aluminum toxicity and diversity of aluminum resistance in higher plants. Int. Rev. Cytol. 264, 225–252.

Mariano, E.D., Keltjens, W.G., 2005. Long-term effects of aluminum exposure on nutrient uptake by maize genotypes differing in aluminum resistance. J. Plant Nutr. 28 (2), 323–333.

Maron, L.G., Kirst, M., Mao, C., Milner, M.J., Menossi, M., Kochian, L.V., 2008. Transcriptional profiling of aluminum toxicity and tolerance responses in maize roots. New Phytol. 179 (1), 116–128.

Martínez-Estévez, M., Racagni-Di Palma, G., Muñoz-Sánchez, J.A., Brito-Argáez, L., Loyola-Vargas, V.M., Hernández-Sotomayor, S.M., 2003. Aluminium differentially modifies lipid metabolism from the phosphoinositide pathway in *Coffea arabica* cells. J. Plant Physiol. 160 (11), 1297–1303.

Matsumoto, H., Hirasawa, E., Morimura, S., Takahashi, E., 1976. Localization of aluminium in tea leaves. Plant Cell Physiol. 17 (3), 627–631.

Matsumoto, H., Yamaya, T., 1986. Inhibition of potassium uptake and regulation of membrane-associated Mg^{2+}-ATPase activity of pea roots by aluminium. Soil Sci. Plant Nutr. 32 (3), 179–188.

Matsumoto, H., 1991. Biochemical mechanism of the toxicity of aluminium and the sequestration of aluminium in plant cells. Develop. Plant Soil Sci. 45, 825–838.

Matsumoto, H., 2000. Cell biology of aluminum toxicity and tolerance in higher plants. Int. Rev. Cytol. 200, 1–46.

Mossor-Pietraszewska, T., 2001. Effect of aluminum on plant growth and metabolism. Acta Biochim. Pol. 48 (3), 673–686.

Moustakas, M., Ouzounidou, G., Lannoye, R., 1995. Aluminum effects on photosynthesis and elemental uptake in an aluminum-tolerant and non-tolerant wheat cultivar. J. Plant Nutr. 18 (4), 669–683.

Mugwira, L.M., Patel, S.U., Fleming, A.L., 1980. Aluminium effects on growth and Al, Ca, Mg, K and P levels in triticale, wheat and rye. Plant Soil 57 (2–3), 467–470.

Nichol, B.E., Oliveira, L.A., Glass, A.D.M., Siddiqi, M.Y., 1993. The effects of aluminum on the influx of calcium, potassium, ammonium, nitrate, and phosphate in an aluminum-sensitive cultivar of barley (*Hordeum vulgare* L.). Plant Physiol. 101 (4), 1263–1266.

Nichol, B.E., Oliveira, L.A., 1995. Effects of aluminum on the growth and distribution of calcium in roots of an aluminum-sensitive cultivar of barley (*Hordeum vulgare*). Can. J. Bot. 73 (12), 1849–1858.

Nunes-Nesi, A., Brito, D.S., Inostroza-Blancheteau, C., Fernie, A.R., Araújo, W.L., 2014. The complex role of mitochondrial metabolism in plant aluminum resistance. Trend Plant Sci. 19 (16), 399–407.

Pan, W.L., Hopkins, A.G., Jackson, W.A., 1988. Aluminum-inhibited shoot development in soybean: a possible consequence of impaired cytokinin supply. Comm. Soil Sci. Plant Anal. 19 (7–12), 1143–1153.

Panda, S.K., Matsumoto, H., 2007. Molecular physiology of aluminum toxicity and tolerance in plants. Bot. Rev. 73 (4), 326–347.

Pavan, M.A., Bingham, F.T., 1982. Toxicity of aluminum to coffee seedlings grown in nutrient solution. SSSAJ 46 (5), 993–997.

Pereira, W.E., de Siqueira, D.L., Martinez, C.A., Puiatti, M., 2000. Gas exchange and chlorophyll fluorescence in four citrus rootstocks under aluminum stress. J. Plant Physiol. 157, 513–520.

Pettersson, S., Strid, H., 1989. Effects of aluminium on growth and kinetics of K^+(^{86}Rb) uptake in two cultivars of wheat (*Triticum aestivum*) with different sensitivity to aluminium. Physiol. Plant. 76 (3), 255–261.

Poschenrieder, C., Gunse, B., Corrales, I., Barcelo, J., 2008. A glance into aluminum toxicity and resistance in plants. Sci. Total Environ. 400 (1–3), 356–368.

Purcino, A.A.C., Alves, V.M.C., Parentoni, S.N., Belele, C.L., Loguercio, L.L., 2003. Aluminum effects on nitrogen uptake and nitrogen assimilating enzymes in maize genotypes with contrasting tolerance to aluminum toxicity. J. Plant Nutr. 26 (1), 31–61.

Rengel, Z., Elliott, D.C., 1992. Mechanism of aluminum inhibition of net $^{45}Ca^{2+}$ Uptake by Amaranthus Protoplasts. Plant Physiol. 98 (2), 632–638.

Rengel, Z., Reid, R.J., 1997. Uptake of Al across the plasma membrane of plant cells. Plant Soil 192 (1), 31–35.

Rengel, Z., Robinson, D.L., 1989. Competitive Al^{3+} inhibition of net Mg^{2+} uptake by intact *Lolium multiflorum* roots. Plant Physiol. 91 (4), 1407–1413.

Rengel, Z., 1990. Competitive Al^{3+} inhibition of net Mg^{2+} uptake by intact *Lolium multiflorum* roots. Plant Physiol. 93 (3), 1261–1267.

Ribeiro, M.A.Q., de Almeida, A.A.F., Mielke, M.S., Gomes, F.P., Pires, M.V., Baligar, V.C., 2013. Aluminum effects on growth, photosynthesis, and mineral nutrition of cacao genotypes. J. Plant Nutr. 36 (8), 1161–1179.

Richards, K.D., Schott, E.J., Sharma, Y.K., Davis, K.R., Gardner, R.C., 1998. Aluminum induces oxidative stress genes in *Arabidopsis thaliana*. Plant Physiol. 116 (1), 409–418.

Rigoulet, M., Yoboue, E.D., Devin, A., 2011. Mitochondrial ROS generation and its regulation: mechanisms involved in H_2O_2 signaling. Antioxid. Redox Signal. 14, 459–468.

Rout, R.G., Samantaray, S., Das, P., 2001. Aluminum toxicity in plants: a review. Agronomie 21, 3–21.

Roy, A.K., Sharma, A., Talukder, G., 1988. Some aspects of aluminum toxicity in plants. Bot. Rev. 54 (2), 145–178.

Ryan, P.R., Kochian, L.V., 1993. Interaction between aluminum toxicity and calcium uptake at the root apex in near-isogenic lines of wheat (*Triticum aestivum* L.) differing in aluminum tolerance. Plant Physiol. 102 (3), 975–982.

Samac, D.A., Tesfaye, M., 2003. Plant improvement for tolerance to aluminum in acid soils – a review. Plant Cell Tissue Organ Cult. 75 (3), 189–207.

Shen, R., Ma, J.F., 2001. Distribution and mobility of aluminium in an Al-accumulating plant, *Fagopyrum esculentum* Moench. J. Exp. Bot. 52 (361), 1683–1687.

Siegel, N., Hang, A., 1983. Calmodulin dependent formation of membrane potential in barley root plasma membrane vesicles: a biochemical model of aluminum toxicity in plants. Physiol. Plant. 59 (2), 285–291.

Silva, I.R., Smyth, T.J., Raper, C.D., Carter, T.E., Rufty, T.W., 2001. Differential aluminum tolerance in soybean: an evaluation of the role of organic acids. Physil. Plant. 112 (2), 200–210.

Silva, S., 2012. Aluminium toxicity targets in plants. J. Bot. 2012, 8, 219462.

Simon, L., Smalley, T.J., Jones Jr., B., Lasseigne, F.T., 1994. Aluminum toxicity in tomato. Part 1. growth and mineral nutrition. J. Plant Nutr. 17 (2–3), 293–306.

Sivaguru, M., Paliwal, K., 1993. Differential aluminum tolerance in some tropical rice cultivars. II. mechanism of aluminum tolerance. J. Plant Nutr. 16 (9), 1717–1732.

Sivaguru, M., Yamamoto, Y., Rengel, Z., Ahn, S.J., Matsumoto, H., 2005. Early events responsible for aluminum toxicity symptoms in suspension-cultured tobacco cells. New Phytol. 165 (1), 99–109.

Sujatha, K., Mehar, S.K., 2015. Toxic effects of aluminium in plants. Ind. J. Plant Sci. 4 (1), 1–4.

Sulpice, R., Trenkamp, S., Steinfath, M., Usadel, B., Gibon, Y., Witucka-Wall, H., Pyl, E.-T., Tschoep, H., Steinhauser, M.C., Guenther, M., Hoehne, M., Rohwer, J.M., Altmann, T., Fernie, A.R., Stitt, M., 2010. Network analysis of enzyme activities and metabolite levels and their relationship to biomass in a large panel of *Arabidopsis* accessions. Plant Cell 22, 2872–2893.

Taylor, G.J., Foy, C.D., 1985. Mechanisms of aluminum tolerance in *Triticum aestivum* (wheat). IV. The role of ammonium and nitrate nutrition. Can J. Bot. 63 (12), 2181–2186.

Tesfaye, M., Temple, S.J., Allan, D.L., Vance, C.P., Samac, D.A., 2001. Overexpression of malate dehydrogenase in transgenic alfalfa enhances organic acid synthesis and confers tolerance to aluminum. Plant Physiol. 127 (4), 1836–1844.

Too, E.J., Carlsson, A.S., Onkware, A.O., Were, B.A., Geleta, M., Bryngelsson, T., Gudu, S., 2014. Cell membrane integrity, callose accumulation, and root growth in aluminum-stressed sorghum seedlings. Biol. Plant 58 (4), 768–772.

Van Praag, H.J., Weissen, F., Dreze, P., Cogneau, M., 1997. Effects of aluminium on calcium and magnesium uptake and translocation by root segments of whole seedlings of Norway spruce (*Picea abies* Karst.). Plant Soil 189 (2), 267–273.

Vazquez, M.D., Poschenrieder, C., Corrales, I., Barcelo, J., 1999. Change in apoplastic aluminum during the initial growth response to aluminum by roots of a tolerant maize variety. Plant Physiol. 119 (2), 435–444.

Vitorello, V.A., Capaldi, F.R., Stefanuto, V.A., 2005. Recent advances in aluminum toxicity and resistance in higher plants. Braz. J. Plant Physiol. 17 (1).

Vondráčková, S., Száková, J., Drábek, O., Tejnecký, V., Hejcman, M., Müllerová, V., Tlustoš, P., 2015. Aluminium uptake and translocation in al hyperaccumulator *Rumex obtusifolius* is affected by low-molecular-weight organic acids content and soil pH. PLoS One 10 (4), e0123351.

Wagatsuma, T., Kaneko, M., Hayasaka, Y., 1987. Destruction process of plant root cells by aluminum. Soil Sci. Plant Nutr. 33 (2), 161–175.

Wang, Y.-S., Yang, Z.-M., 2005. Nitric oxide reduces aluminum toxicity by preventing oxidative stress in the roots of *Cassia tora* L. Plant Cell Physiol. 46 (12), 1915–1923.

Wang, Y., Stass, A., Horst, W.J., 2004. Apoplastic binding of aluminum is involved in silicon-induced amelioration of aluminum toxicity in maize. Am. Soc. Plant Biol. 136 (3), 3762–3770.

Watanabe, T., Osaki, M., 2002. Mechanisms of adaptation to high aluminum condition in native plant species growing in acid soils: a review. Comm. Soil Sci. Plant Anal. 33 (7–8), 1247–1260.

Wernimont, A.K., Artz, J.D., Finerty Jr., P., Lin, Y.-H., Amani, M., Allali-Hassani, A., Saniterra, G., Vedadi, M., Tempel, W., Mackenzie, F., Chau, I., Lourido, S., Sibley, L.D., Hui, R., 2010. Structures of apicomplexan calcium-dependent protein kinases reveal mechanism of activation by calcium. Nat. Struc. Mol. Biol. 17, 596–601.

Wheeler, D.M., Follett, J.M., 1991. Effect of aluminium on onions, asparagus and squash. J. Plant Nutr. 14 (9), 897–912.

Widell, S., Asp, H., Jensen, P., 1994. Activities of plasma membrane-bound enzymes isolated from roots of spruce (*Picea abies*) grown in the presence of aluminium. Physiol. Plant. 92 (3), 459–466.

Wu, D., Shen, H., Yokawa, K., Baluška, F., 2014. Alleviation of aluminium-induced cell rigidity by overexpression of *OsPIN2* in rice roots. J. Exp. Bot. 65 (18), 5305–5315.

Xu, F.J., Li, G., Jin, C.W., Liu, W.J., Zhang, S.S., Zhang, Y.S., Lin, X.Y., 2012. Aluminum-induced changes in reactive oxygen species accumulation, lipid peroxidation and antioxidant capacity in wheat root tips. Biol. Plant. 56 (1), 89–96.

Yamamoto, Y., Kobayashi, Y., Devi, S.R., Rikiishi, S., Matsumoto, H., 2002. Aluminum toxicity is associated with mitochondrial dysfunction and the production of reactive oxygen species in plant cells. Plant Physiol. 128 (1), 63–72.

Yamamoto, Y., Kobayashi, Y., Matsumoto, H., 2001. Lipid peroxidation is an early symptom triggered by aluminum, but not the primary cause of elongation inhibition in pea roots. Plant Physiol. 125, 199–208.

Yang, Z.M., Sivaguru, M., Horst, W.J., Matsumoto, H., 2000. Aluminium tolerance is achieved by exudation of citric acid from roots of soybean (*Glycine max*). Physiol. Plant. 110 (1), 72–77.

Zhang, G., Hoddinott, J., Taylor, G.J., 1994. Characterization of 1,3-β-D-glucan (callose) synthesis in roots of *Triticum aestivum* in response to aluminum toxicity. J. Plant Physiol. 144 (2), 229–234.

Zsoldos, F., Vashegyi, A., Bona, L., Pecsvaradi, A., Szegletes, Z., 2000. Growth of and potassium transport in winter wheat and durum wheat as affected by various aluminum exposure times. J. Plant Nutr. 23 (7), 913–926.

CHAPTER

COPPER STRESS AND RESPONSES IN PLANTS

2

Aykut Sağlam[1], Fuat Yetişsin[3], Mehmet Demiralay[4], Rabiye Terzi[2]

Department of Molecular Biology and Genetics, Faculty of Sciences, Karadeniz Technical University, Trabzon, Turkey[1]; Department of Biology, Faculty of Sciences, Karadeniz Technical University, Trabzon, Turkey[2]; Department of Molecular Biology and Genetics, Faculty of Arts and Sciences, Muş Alparslan University, Muş, Turkey[3]; Department of Forest Engineering, Faculty of Forestry, Artvin Çoruh University, Artvin, Turkey[4]

CHAPTER OUTLINE

1. INTRODUCTION

Copper (Cu) is an abundant essential micronutrient element in various rocks and minerals and is required for a variety of metabolic processes in both prokaryotes and eukaryotes (Sun et al., 2014). Much of copper contains enzymes functioning as oxygen carriers (hemocyanin) or redox catalyzer (cytochrome oxidase, nitrate reductase) (Whitacre, 2011; Ferreira et al., 2015). Copper is a transition

Plant Metal Interaction. http://dx.doi.org/10.1016/B978-0-12-803158-2.00002-3

metal with three valence states: Cu°, Cu^{+1}, and Cu^{+2}. In addition, it has been classified as a heavy metal because of its the heavier density (5 g/cm^3) (Singh et al., 2011).

Cu, however, is an essential micronutrient for normal plant metabolism and plays a role in several physiological processes such as photosynthesis, respiration, carbohydrate distribution, and protein metabolism. However, excess Cu can disturb normal development by adversely affecting biochemical reactions and physiological processes in plants (Guzel and Terzi, 2013). We will discuss the effects of copper on plants and their responses to copper stress in this chapter and address how plants cope with Cu stress and what kinds of changes occur in plant morphology, physiology, and metabolism.

2. EXCESS COPPER IS TOXIC TO PLANTS

Chlorosis in leaves is one of the common initial symptoms of copper toxicity (Verma and Bhatia, 2014). The other is stunted root growth, which includes poor development, reduced branching, thickening, and dark coloration (Nair and Chung, 2015). Chlorosis usually causes cream- or white-colored spots or lesions (Lee et al., 1996). Increased copper toxicity results in forming necrotic areas in leaf tips and edges (Taylor and Foy, 1985). Leaves can wilt under excess copper before necrotic areas are formed (Yau et al., 1991).

Cu^{+} and Cu^{2+} ions block photosynthetic electron transport by inhibiting primary quinone acceptor QA, and the secondary quinone acceptor QB thereby maximal quantum yield of photosystem II (PSII) photochemistry is reduced (Zlobin et al., 2014). Excess heavy metals often change membrane permeability by causing leakage of K^{+} and other important ions. It has been reported that Cu^{2+} inhibited in vitro H^{+} transport through the plasmalemma in cucumber roots; however, the transport was induced in corn roots (Barker and Pilbeam, 2014).

3. EFFECT OF COPPER STRESS ON PLANT GROWTH

Excess Cu can affect important physiological processes in plants and cause problems in plant growth and development. Cu taken from the soil must be transported, distributed, and compartmentalized within different tissues and organelles for healthy plant growth and development (Habiba et al., 2015). On the other hand, excessive Cu is characterized by a reduced plant biomass, leaf chlorosis, inhibited root growth, bronzing, and necrosis. The effect of Cu toxicity is largely on root growth and morphology, which has particular importance for the whole plant. Because water and nutrients enter plants by the roots, any defect or malformation of the roots creates problems for plant growth and development. Copper tends to accumulate in the root tissue with little translocated to the shoots (Marschner, 2011). Retarded root growth of *Festuca arudinacea* and *Lolium perenne* plants was observed after 7 days of copper application (Zhao et al., 2010). Kolbert et al. (2012) reported that long-term excess copper leads to inhibition of stem and root development of *Arabidopsis thaliana*. Cell elongation, division, and expansion were also affected. Altered auxin homeostasis in both organs was found to be a cause of the growth alterations because of elevated nitric oxide (NO) production that inhibited PIN1-mediated auxin transport.

Liu et al. (2014) reported that a continuous decrease in maize root activities, and physiological disorders were observed during increased copper stress, which resulted in root cell membrane damage. Intracellular ion and organic substance leaked after Cu^{2+} entered root cells.

Excess copper inhibits leaf expansion, cell elongation, and cell division (Maksymiec, 1997; Panou-Filotheou and Bosabalidis, 2004). Long exposure to copper stress causes folding up of the leaf blade, increased numbers of hair on the lower surface of leaf blades, increased numbers of stomata and decreased stomata sizes, and reduced volume of mesophyll intercellular spaces. All these changes in the leaf anatomy and morphology are disorders from exposure of roots to copper stress that causes decreased transpiration (Panou-Filotheou et al., 2001).

Leaf chlorosis is a typical symptom of plants subjected to high soil Cu concentrations (Miotto et al., 2014). During leaf chlorosis, some morphological and physiological alterations take place in the chloroplasts. Mesophyll chloroplasts in Cu-treated oregano plants are reduced in volume and number (Panou-Filotheou et al., 2001). This observation is in accordance with those of other plants under Cu stress, such as rice (Lidon and Henriques, 1993), beans (Maksymiec et al., 1994), and wheat (Quartacci et al., 2000). Sanchez-Pardo et al. (2014) reported that under copper stress, the structure of the thylakoids in white lupin and soybean showed changes. Disintegration of chloroplast membranes, degradation of grana stacking, swelling of the thylakoids, and increased numbers of plastoglobuli were also observed in soybean plants.

4. EFFECTS OF COPPER STRESS ON PHOTOSYNTHESIS

Increasing levels of copper in the environment affects photosynthesis adversely. The main effects of copper on photosynthesis are related to changes in pigment compositions and ultrastructure of chloroplast, decreased net photosynthesis rate, reduced ribulose-1,5-bisphosphate carboxylase/oxygenase (RuBisCo) efficiency, and inhibition of electron transport and PSII activities (Aggarwal et al., 2012).

4.1 ULTRASTRUCTURAL CHANGES IN CHLOROPLASTS

Excess Cu adversely affects architecture, lipid, and pigment composition of thylakoid membranes and causes decreased photochemical activity of PSII (Maksymiec et al., 1995; Xu et al., 2013). Swollen thylakoids and an increase in the number of plastoglobuli have been observed in soybean plants (Sanchez-Pardo et al., 2014). Similar effects have been reported in spinach and wheat leaves after Cu application (Eleftheriou and Karataglis, 1989). The leaves of *Theobroma cacao* seedlings exposed to higher levels of Cu showed damaging effects on the thylakoid and chloroplast membranes. Absence of starch grains and swelling of chloroplast double membranes were also reported (Souza et al., 2014).

One of structural components of thylakoid membranes, acyl lipid levels, is reduced by copper stress. Significant decreases in monogalactosyl diacylglycerol (MGDG), sulfoquinovosyl diacylglycerol, phosphatidyl glycerol, and phosphatidyl choline (PC) have been observed in runner bean plants subjected to copper stress (Maksymiec et al., 1994). The content of digalactosyl diacylglycerols, MGDG, and phosphatidyl glycerol in chloroplasts of the *Hydrilla verticillata* plant increased after an hour of copper exposure (100 μM $Cu(NO_3)_2$). The content of all the lipids except phosphatidic acid and PC have decreased after 3 h of exposure (Rozentsvet et al., 2012). Decrease in acyl lipid content has been shown to result in impaired photosynthesis. Reductions in phosphatidyl glycerol and PC content under excess copper have disturbed grana stacking (Krupa, 1988). Changes in chloroplast ultrastructure are associated with decreased MGDG content. Its low level disturbs organization of PSII and thus PSII activity is reduced (Murata et al., 1990). Fatty acid unsaturation and lipid peroxidation because of copper stress reduces MGDG content, which result in inhibition of PSII (Quartacci et al.,

2000). Lipid peroxidation shows to disturbing lipid–protein–pigment complexes of chloroplasts (Panda and Panda, 2009). Declines in chlorophyll and carotenoid contents has been observed in *Elsholtzia splendens* plants subjected to excess copper for 8 weeks (Peng et al., 2013). Malec et al. (2009) reported that hydrogen peroxide, superoxide, and hydroxyl radicals, which are formed by Cu toxicity, induces degradation of chlorophyll in Cu-treated plants.

Carotenoids protect chlorophylls from peroxidation (Dall'Osto et al., 2012). When carotenoid levels are decreased by Cu treatment, this gives rise to an increase in chlorophyll destruction. The synthesis of 5-aminolevulinic acid and protochlorophyllide reductase activity are believed to be inhibited by copper (van Assche and Clijsters, 1990). Iron and magnesium deficiency induced by excess Cu inhibits synthesis of protochlorophyllide and phytoene synthesis; therefore, chlorophyll and carotenoid contents are decreased (Vijayarengan and Jose, 2014).

4.2 PHOTOSYSTEM II

The role of Cu in PSII-mediated electron transport is known (Puig, 2014). Cu has been determined in PSII-enriched fraction of spinach chloroplasts (Anderson et al., 1964). Sibbald and Green (1987) reported that about 75% of Cu in PSII preparations from barley and spinach was bound to the major antenna complex of PSII (LHCII). Cu is also involved in the water-splitting system (Orzechowska et al., 2008). On the other hand, high concentrations of copper are toxic to PSII. Extensive in vitro studies have shown that PSII is very susceptible to Cu toxicity (Baron et al., 1995). However, the precise location of the Cu binding site on PSII and the underlying mechanisms of copper inhibition are still the subject of debate. Different sites of copper action in PSII are suggested in the literature. Cu^{2+} ions inhibit both the donor and the acceptor sides of PSII, although the oxidizing side of PSII is the most sensitive site of Cu inhibition (Rouillon et al., 2006) at a reversible inhibition on PSII primary electron carrier donor Tyr_Z or on oxygen evolving complex (Burda et al., 2003). Copper ions also affect PSII electron transport on acceptor side. At higher copper concentrations, target sites of Cu^{2+} are primary quinone acceptor Q_A, pheophytin-Q_A-Fe region, and nonheme iron and secondary quinone acceptor Q_B (Mohanty et al., 1989). On the other hand, oxygen evolution is found to be induced by very low equimolar Cu^{2+}/PSII proportions (Burda et al., 2002). Apart from the oxygen-evolving complex, the level of the photoreduced cyt b_{559} is reduced and its rate of photoreduction is decreased by copper stress (Yruela et al., 1996).

4.3 RuBisCo ACTIVITY

The most abundant protein in plants is RuBisCo (Raven, 2013). RuBisCo is an enzyme that catalyzes the carbon dioxide fixation reaction in photosynthesis forming phosphoglycerate with the reaction of ribulose-1,5-bisphosphate and carbon dioxide, and also catalyzes the photorespiration forming the phosphoglycolate and phosphoglycerate with the reaction to O_2 (Parry et al., 2012). Declined RuBisCo activity has been determined in plants grown with high levels of Cu (Schafer et al., 1992). Carboxylase and oxygenase activities of RuBisCo are inhibited by Cu toxicity (Lidon and Henriques, 1991). This effect appears to be due to a metal-induced interaction with essential cysteine residues of the enzyme (Siborova, 1988). Similarly, the concentration of RuBisCo reaction sites and its carbamylation state decreased considerably in response to Cu stress in *Spartina densiflora* (Mateos-Naranjo et al., 2015). This was most likely related to the reduced concentration of the enzyme instead of its activation status.

Adversely, RuBisCo activity is increased in copper-stressed tobacco plants, whereas RuBisCo content is decreased (Son et al., 2014). Simova-Stoilova et al. (2002) have reported that RuBisCo content is reduced in barley leaves subjected to toxic copper concentrations. This reduction is probably from oxidative stress resulting from forming of reactive oxygen species (ROS) by copper toxicity.

Copper stress also causes accumulation of osmoprotectants. Proline is a well-known and studied amino acid under copper stress. Further section will handle changes in proline metabolism in plants exposed to copper.

5. PROLINE METABOLISM

Plants exposed to excess copper have been shown to accumulate proline in their tissues (Ku et al., 2012). Accumulation of proline is an adaptive response of plants against stresses. Proline is believed to be regulatory or signal molecule activating some physiological and molecular responses (Szabados and Savoure, 2010). Mechanism of proline accumulation is related to increased synthesis, decreased catabolism, or increased degradation of proteins (Kavi Kishor et al., 2005). Proline is synthesized in higher plants from glutamate via reactions, which are catalyzed by pyrroline-5-carboxylate synthetase (P5CS) and pyrroline-5-carboxylate reductase. Proline can also be formed by transamination of ornithine into proline-5-carboxylate by ornithine amino transferase. Proline catabolism in plant cells is occurred by proline dehydrogenase (PDH) and pyrroline-5-carboxylate dehydrogenase (Monteoliva et al., 2014).

Increased activity of P5CS and proline accumulation has been reported in copper-stressed tobacco plants grown in pots (Ku et al., 2012). In addition, higher pyrroline-5-carboxylate reductase activity in detached rice leaves exposed to Cu has been observed (Chen et al., 2001). The stimulation of P5CS activity in *Triticum aestivum* plants under excess copper has been reported to be from activation of the P5CS gene (Tripathi et al., 2013). Proline accumulation in *T. aestivum* plants under copper toxicity is mostly from contribution of glutamate-based proline synthesis, because the relatively slower activity of ornithine amino transferase in comparison to P5CS and pyrroline-5-carboxylate reductase activities has been determined in *T. aestivum* leaves under copper excess. The contribution of proline catabolism to proline accumulation in the copper-treated plants has also been shown by some studies assessing PDH activity. Inhibition of PDH activity in the Cu-treated *T. aestivum* plants has been reported (Tripathi et al., 2013). The decreased proline catabolism is suggested to contribute to proline accumulation in various plants (Mani et al., 2002). Similarly, Wen et al. (2013) also reported that hydrogen peroxide (H_2O_2) pretreatment enhanced proline accumulation in sweet corn seedlings under copper stress. The increase in the proline content was related to induction of both ornithine and glutamate pathways and reduced PDH pathway. Copper stress-induced increase in proline and NO levels in *Chlamydomonas reinhardtii* indicated that this intracellular NO generation was involved in Cu-induced proline accumulation and signaling based on increased activity of P5CS and upregulated expression of P5CS gene in Cu-treated algae after sodium nitroprusside (SNP) treatment (Rejeb et al., 2014). Proline is also known to be an ROS scavenger like antioxidant enzymes and nonproteinic antioxidants (ascorbic acid (AsA) and glutathione (GSH)). A well-known result of copper stress is the formation of ROS. Plants are equipped with an antioxidant defense system to cope with ROS. This part of the chapter will discuss the effects of copper on an antioxidant system.

6. ANTIOXIDANT SYSTEM

Copper in excess causes generation of ROS such as superoxide radical ($O^{\cdot -}$), H_2O_2, singlet oxygen ($^1O_2$), and hydroxyl radicals (OH) (Tie et al., 2012; Liu et al., 2015). Antioxidant enzymes such as SOD, peroxidase (POX), and catalase (CAT) are involved in the scavenging of ROS. The SOD catalyzes the dismutation of superoxide to H_2O_2 and molecular oxygen. CAT dismutases H_2O_2 into H_2O and O_2, whereas POX decomposes H_2O_2 by oxidation of cosubstrates, such as phenolic compounds and or antioxidants (Halliwell and Gutteridge, 2007). SOD and POX activities have been studied in copper-stressed turf grass cultivars *Festuca arudinacea* and *Lolium perenne* plants (Zhao et al., 2010). SOD activity is induced by increasing copper concentration in *F. arudinacea* roots. POX activity has increased at low copper levels, however, in *L. perenne,* these enzymes are not induced under copper excess. Authors suggest that tolerance of different turf grass cultivars to copper stress may depend on enhanced antioxidant capacity. Recent studies report that an increase in SOD and ascorbate peroxidase (APX) activities in rice (Thounaojam et al., 2012) and jack bean (Andrade et al., 2010) plants. Induced CAT activity is reported in *Spirodela polyrhiza* (Upadhyay and Panda 2010), *Atriplex halimus* (Brahim and Mohamed, 2011), and jack bean plants (Andrade et al., 2010) under copper stress. Similarly, enhanced levels of gene expressions of CAT (*RsCat*) and SOD (*MnSOD*) in *Withania somnifera* under excess Cu levels have been determined (Rout and Sahoo, 2013). Fidalgo et al. (2013) could not find any significant changes in level of antioxidant gene transcripts in *Solanum nigrum* plants under excess copper levels for 28 days. They concluded that this happened because Cu levels used were below the maximum limit that the plants could tolerate. Long-time exposure to copper results in inhibition of antioxidant enzyme activities. A reduction in SOD, CAT, and APX activities has been determined in *Solanum lycopersicon* plants subjected to prolonged copper treatment (Chamseddine et al., 2009).

AsA and GSH are the most important antioxidant metabolites reacting directly with different ROS and participating in the AsA–GSH cycle (Foyer and Noctor, 2011). It has been reported that the AsA–GSH cycle is involved in response to excess Cu (Yruela, 2005). AsA and GSH levels in maize leaves significantly increased under low copper concentration, then decreased as copper concentrations increased (Tie et al., 2012). Similar results have been reported by Wang et al. (2011) in maize under excess copper levels. Mostofa et al. (2014) found that exogenous SNP and GSH successfully decreased Cu-induced ROS and malondialdehyde levels, providing strong evidence that SNP and GSH substantially protects rice seedlings against oxidative stress by inducing antioxidant enzyme activities.

7. SIGNAL TRANSDUCTION IN RESPONSE TO COPPER

Plant cells possess a complex signal transduction network that is activated whenever any stress is sensed by the receptors at the plasma membrane (DalCorso et al., 2010). When the cells recognize the metal, common signaling pathway elements such as calcium fluxes and ROS, are used. Then, stress-related proteins and signaling molecules are synthetized, and specific metal-response genes that confront the stress are activated transcriptionally (Maksymiec, 2007). Here we discuss signal transduction pathways that respond to copper stress in plants.

7.1 SIGNAL PERCEPTION

A signal perception starts a signal transduction pathway after recognition of heavy metals by the cells. To date, much less knowledge is available about a primary recognition for instance by a receptor. Being a central controller in Cu homeostasis, *Chlamydomonas reinhardtii* copper response regulator Crr1 and its *Arabidopsis* homolog SQUAMOSA promoter-binding protein-like 7 (SPL7) have drawn attention in recent years (Garcia-Molina et al., 2014). As indicated previously, Cu is an essential micronutrient for plants that is necessary for photosynthesis, oxidative stress responses, cell wall lignification, and hormone signaling, etc. (Puig et al., 2007). On the other hand, cellular components such as lipids, proteins, or nucleic acids are damaged because of ROS generation because the plants are subjected to excess free Cu ions (Halliwell and Gutteridge, 1984). To keep Cu levels within a proper range, plants are equipped with a homeostatic network. During Cu deficiency in *Arabidopsis*, SPL7 directly binds to GTAC motif-containing Cu response elements located in the promoter regions of Cu-responsive genes (Yamasaki et al., 2009). Cu-metalloreductases FRO4/5 and Cu transport proteins (COPT)1/2/6 are promoted (Bernal et al., 2012), then SPL7 reconstitutes gene expression to use and redistribute Cu within the plant. Thus, translation of Cu-requiring proteins such as superoxide dismutases and laccases are repressed and Cu is delivered to plastocyanin efficiently (Yamasaki et al., 2009). Just after signal perception, plant hormone signaling pathways are activated. We will review hormone signals and their roles in copper stress responses of plants in forthcoming section.

7.2 STRESS SIGNAL TRANSDUCTION BY PLANT HORMONES

Plants synthesize variety of hormones such as abscisic acid (ABA), brassinosteroids, melatonin, or indole acetic acid (IAA) as they are subjected to copper stress (Vázquez et al., 2013). These plant hormones are related to metal-induced signaling and defense reactions. During copper exposure, an increased shoot emergence and plant biomass production has been observed in mustard plants, which are treated with 24-epi-brassinolide (EBL; 10^{-7}, 10^{-9}, and 10^{-11} M) for 8 h (Sharma and Bhardwaj, 2007). In addition, EBL has blocked Cu uptake and accumulation in the mustard seedlings.

Radish seedlings generate copper resistance after 10^{-7} M EBL treatment on seeds. In addition, root and shoot growth increases (Choudhary et al., 2010). Endogenous IAA and ABA concentrations increase in these EBL-treated radish seedlings. Antioxidant enzyme activities, proline, AsA, total phenol, and phytochelatin contents are enhanced; thus, oxidative stress is relieved. Antioxidant capacity and free radical scavenging activity of EBL-treated radish seedlings have also been studied (Choudhary et al., 2011). EBL treatment improves seedling growth, antioxidant levels, and enzyme (GPOX, CAT, SOD, and GR) activities and free radical scavenging capacities and reduces membrane damage in seedlings.

The role of the combined application of homobrassinolide and H_2O_2 to the leaves of mung bean plants subjected to copper stress has been studied (Fariduddin et al., 2014). The exogenously applied homobrassinolide and/or H_2O_2 alleviate copper toxicity in mung bean and suggest their role in maintaining copper homeostasis in these plants.

In addition, putrescine biosynthesis is enhanced under Cu-stressed wheat leaves (Groppa et al., 2007). The enhancement in putrescine concentration is due to increased activities of ornithine decarboxylase and arginine decarboxylase enzymes. Chen et al. (2013) reported that exogenous putrescine acted as a chemical messenger, induced several antioxidant enzymes (CAT, APX, and POD) and scavenged H_2O_2 in *Populus cathayana* which was exposed to copper stress.

Coapplication of EBR and Spd is very effective in detoxifying Cu in plants (Choudhary et al., 2012). The combined application of EBL and spermidine on radish seedlings affects gene expressions of phosphatidic acid synthesis enzymes, IAA and ABA metabolism; thus, Cu stress tolerance in radish seedlings is enhanced. Munzuroglu et al. (2008) have recorded significant increases in the ABA content of wheat seeds subjected to copper toxicity. In another study, Zengin and Kirbag (2007) have recorded that copper stress increases the content of the ABA in sunflower seedlings. Increased ABA decreases transpiration; thus, the transport of copper ions from roots to shoot is hindered. Similarly, Ku et al. (2012) have reported that the increase of ABA in tobacco leaves, which is induced by Cu, reduce water transport and downregulate metal absorption efficiency.

Melatonin is one of the hormones that accumulates during stress. Melatonin has roles in multiple plant developmental processes and various stress responses (Tan et al., 2012). Posmyk et al. (2008) have reported that exogenous melatonin application improved copper stress in red cabbage seedlings. Pea plants treated with melatonin survive under copper toxicity compared with nontreated plants (Tan et al., 2007).

Auxin controlling primary root elongation increases formation of lateral roots and the number of root hairs affecting plant responses to metal stresses (Yuan et al., 2013). Copper stress inhibits primary root elongation. Cu-mediated auxin redistribution is responsible for the inhibition of primary root elongation that is mediated by PIN1, but not PIN2 or AUX1 (Yuan et al., 2013). During copper stress, a reduction of the auxin responsive promoter DR5 gene is observed at *A. thaliana* primary root meristems, which are exposed to high copper concentrations. However, DR5 accumulation at just above root apex might cause the formation of lateral roots. This might also be involved in the formation of root hairs in the area adjacent to the Cu^{2+}-treated root tips of *A. thaliana* (Lequeux et al., 2010).

Plants synthesize ethylene, depending on the intensity of heavy metal stresses (Schellingen et al., 2014). Mertens et al. (1999) have reported an increasing ethylene production in 7-day-old *A. thaliana* plants exposed to 25–500 μM Cu for up to 6 h. However, Lequeux et al. (2010) have not observed any effect of 24-h copper stress (50 μM) on ethylene production in *A. thaliana* plants. They suggested that ethylene is not involved in the long-term root architecture reorganization.

In addition, increased ethylene production in 4-week-old wheat and sunflower leaves after 14 h of exposure to 1 mM of Cu has been reported by Groppa et al. (2003). Exposure to high copper concentrations (50–400 mM) for a short period (24 h) differently induces ethylene production in various plant parts of *A. thaliana* (Arteca and Arteca, 2007). Hego et al. (2014) have reported that some proteins involved in ethylene biosynthesis, for example 1-aminocyclopropane-1-carboxylate synthase, ETO-like protein, and S-adenosylmethionine synthetase (SAMS) are overexpressed in *Agrostis capillaris* roots, especially at 5 μM Cu and 1 μM Cu exposures.

Gene expressions of 1-aminocyclopropane-1-carboxylate synthase (ACS) that catalysis production of 1-aminocyclopropane-1-carboxylic acid from S-adenosyl-L-methionine and 1-aminocyclopropane-1-carboxylate oxidase that catalysis conversion of 1-aminocyclopropane-1-carboxylate synthase to ethylene have been studied in *Brassica oleracea* seedlings exposed to copper stress. Expressions of the two important enzymes of ethylene biosynthesis, ACS1 and 1-aminocyclopropane-1-carboxylate oxidase 1, are induced by 2.5 mM of $CuCl_2$ treatment. Therefore copper ions are shown to be a very effective inducer of ethylene biosynthesis (Jakubowicz et al., 2010). Ethylene, together with other plant hormones, may contribute to growth inhibition and malformations of root architecture during

Cu stress. These ethylene-regulated responses are mediated by both ROS accumulation and/or ROS signaling. Plants have antioxidant barriers against ROS attacks. An antioxidant system is known to be induced by ethylene under different heavy metal stresses such as Ni and Zn (Khan and Khan, 2014); however, there is no study about the effects of ethylene on the antioxidant enzymes of copper-stressed plants (Iqbal et al., 2013).

7.3 THE ROLES OF REACTIVE OXYGEN SPECIES

Production of ROS is one of the major consequences of copper accumulation in plants. In addition to causing oxidative stress, ROS can also serve as a signal molecule for plants to survive copper stress. H_2O_2 levels increase in response to Cu and cadmium (Cd) treatment in *A. thaliana* (Maksymiec and Krupa, 2006). This increase in H_2O_2 accumulation changes the redox status of the cell and induces the production of antioxidants and the activation of antioxidant mechanisms. H_2O_2 is known as a signal for APX induction. H_2O_2 induced messenger RNA (mRNA) for cytosolic APX in callus cultures of rice embryos (Morita et al., 1999). In tomato plants under copper excess, Wang et al. (2010) suggested that copper stress induced H_2O_2 and NO generation and that antioxidant enzyme activities were induced by NO to alleviate H_2O_2 overaccumulation. Hence, tolerance to copper stress in tomato plants was developed. Many recent studies have indicated that exogenous NO mitigated effects of toxic Cu (Zhang et al., 2009). Furthermore, exogenous NO was described as an effective substance for alleviation of Cu toxicity in many plants, such as rice, tomato and barley seedlings, and wheat seeds (Yu et al., 2005; Cui et al., 2010; Hu et al., 2015). Various mechanisms were reported to contribute to NO-induced increase of Cu tolerance. For instance, alleviation Cu toxicity was determined to be related to exogenous NO-induced antioxidant system and increased adenosine triphosphatase (ATPase) activities (Zhang et al., 2009). Tewari et al. (2008) reported that NO alleviates Cu toxicity by decreasing oxidative stress in the adventitious roots of *Panax ginseng*. Dong et al. (2014) demonstrated that in the hydroponics experiment, the low concentrations (100 μM) of NO provided the resistance toward Cu and had an ameliorating effect on Cu-stressed ryegrass. Wang et al. (2014) showed that the addition of SNP, a donor of exogenous NO, significantly alleviated photosynthetic inhibition in Cu-stressed tomatoes by recovering photosynthetic pigment content and chlorophyll a fluorescence and the improvement of net photosynthetic rate, stomatal conductance, and transpiration rate.

Mitogen-activated protein kinases (MAPKs) can perceive the changes of ROS content in signal transmission for a range of stresses and the distinct MAPK pathways involve specifically to particular metal ions.

7.4 THE MAPK CASCADE

Plants contain specific metal sensors detecting changes in status of metals (deficiency or excess) and trigger signaling cascades that activate the appropriate responses. Signal transduction pathways in plants about metal stress responses have not been identified yet. MAPKs have been shown to activate in *Medicago sativa* seedlings exposed to toxic copper concentrations (Jonak et al., 2004). MAPK gene (*OsMAPK2*) expression in rice under excess copper has been studied (Hung et al., 2005). Copper treatment induces MAPK signaling in rice (Yeh et al., 2007). In that study, Cu^{2+}-induced 42-kDa MBP kinase activities have been determined in rice roots. Accumulation of *OsMAPK2* was found to be related with ROS, extracellular calcium ions, and cantharidin-sensitive protein phosphatase(s)–mediated signaling cascades. Smeets et al. (2013) reported that oxidative signal-inducible kinase1 was required for the upregulation of antioxidant enzymes such as catalase in leaves and Fe-superoxide dismutase in *A. thaliana* roots exposed to Cu. These processes are involved in the MEKK1-MKK2-WRKY25 cascade.

In brief, MAPKs activate genes required for stress adaptation. Genes for metal transporters and synthesis of chelating compounds are activated by MAPKs.

7.4.1 Chelating Compounds

Copper in plant tissues is kept in a narrow physiological range for homeostatic equilibrium. In the case of copper toxicity, plants trap free Cu by increasing amount of chelating compounds such as chaperones, metallothioneins, and phytochelatins (PCs) (Ritter et al., 2014).

Cu chaperones are from Cu-binding domains. They have roles in detoxifying oxygen radicals and Cu trafficking and detoxification. Antioxidant protein 1 (ATX1) and ATX1-like Cu chaperone (CCH) are two Cu chaperones in *A. thaliana*. mRNA levels of *Arabidopsis CCH* were reported to decrease in response to high concentrations of copper (Mira et al., 2001). Phenotypes of *atx1* mutant and the *cchatx1* double mutant were characterized in plants under excess copper and deficiency (Shin et al., 2012). The shoot and root growth of *atx1* and *cchatx1* mutants were found to be very sensitive to excess Cu. On the other hand, *atx1* and *cchatx1* were also reported to be sensitive to Cu deficiency. It was determined that Cu tolerance and accumulation increased growth was promoted in plants overexpressing ATX1.

Arabidopsis contains the CCH for SOD (CCS), delivering Cu to Cu/zinc-containing superoxide dismutase (ZnSOD). In *Arabidopsis ccs* mutants, no measurable SOD activity was detected in chloroplast and cytosol. This suggests that CCS is necessary for SOD activity. In addition, Fe superoxide dismutase and Cu/ZnSOD activity were not observed in *ccs* mutants grown on media containing high Cu concentrations. However, no difference in phenotype and photosynthetic activities between mutant and wild type was reported. This indicated that lack of Cu/ZnSOD was not vitally important for photosynthesis (Cohu et al., 2009). In addition, *Glycine max* CCS and cuprozinc superoxide dismutase were investigated in photosynthetic cell suspensions under copper stress (Sagasti et al., 2014). Both genes were upregulated by copper stress. Furthermore, *GmCCS* was found to be present in the stroma only, but the *Glycine max* CCS and cuprozinc superoxide dismutase was present in both stroma and thylakoids despite the general idea that the SOD enzymes are typically soluble stroma proteins. The presence of the SOD enzyme on the surface of the thylakoid membranes is reasonable considering that the superoxide radical (O_2^-) is preferentially formed at the acceptor side of PSI.

Metallothioneins are Cys-rich metal chelators that play a role in many cellular processes, including the regulation of cell growth and proliferation, DNA damage repair, and scavenging of ROS (Cherian and Kang, 2006). We summarize here the importance of metallothioneins for plants subjected to copper stress. Phenotypic differences in the root of *Arabidopsis mt1a* mutants, *MT1a-RNAi*, and *mt1a-mt2b* double mutant under excess copper were reported. Decreased accumulation of copper in the roots of these mutants suggested that MT1a was an important player in copper accumulation of the roots (Grennan, 2011). Two MT genes, *BcMT1* and *BcMT2*, isolated from *Brassica campestris* were expressed in *A. thaliana*. Transgenic *Arabidopsis* plants with a reduction in ROS production developed tolerance to copper excess (Lv et al., 2013). Transgenic tobacco plant overexpressing an MT gene, *EhMT1*, from a copper accumulator plant *Elsholtzia haichowensis* developed a tolerance to copper stress. Increased Cu accumulation and POX activities and reduced H_2O_2 level were reported in

EhMT1-overexpressing plants as compared with wild type (Xia et al., 2012). *Pisum sativum* metallothionein A1 gene (*PsMT*$_{A1}$) expressed in white poplar enhanced tolerance to excess copper through sequestration of Cu ions in roots and leaves and by scavenging of ROS (Turchi et al., 2012).

Ectopic expression of MT2 gene from *Iris lactea* var. *Chinensis*, which is a copper-tolerant species, in *A. thaliana* increased tolerance to Cu stress by reducing H_2O_2 production in the transgenic plants. Root lengths of two transgenic seedlings were also induced more than that of the wild type after treatment with 50 and 100 μM Cu (Gu et al., 2015). Similarly, increased Cu tolerances and rooting efficiency were observed in transgenic *Salix matsudana* overexpressing the *Tamarix hispida ThMT3* gene (Yang et al., 2015).

Similar to metallothioneins (PCs) are polymers of g-Glu-Cys units with the terminal glycine and can bind heavy metals by thiolate coordination (Emamverdian et al., 2015). PCs are known to be induced by copper stress (Boojar and Tavakoli, 2010). Copper stress also induces a stable Cu-phytochelatin complex that has roles in detoxification of copper. However, Lee and Kang (2005) hypothesized that PCs were not a primer factor in copper detoxification because there was no difference in copper sensitivity between *Arabidopsis* mutants (*cad1*-3, which had no PCs) and wild type. It was also demonstrated that *AtPC1*-overexpressing *Arabidopsis* plants were not copper tolerant. However, *A. thaliana* developed copper tolerance when a fission yeast (*Schizosaccharomyces pombe*) phytochelatin–Cd transporter SpHMT1 was introduced into the plant (Huang et al., 2012). Beladi et al. (2011) claimed that an increased phytochelatin synthesis in Sainfoin (*Onobrychis vicifolia*) plants caused less damage in DNA by decreasing the formation of oxygen-free radicals. Fidalgo et al. (2013) showed that the production of PCs was enhanced in roots of *Solanum nigrum* when the plant was exposed to Cu. This resulted in the immobilization of excess Cu in the root and its preclusion from moving toward the shoot. All these data suggest that PCs are of importance for copper detoxification by sequestration Cu ions. Also, copper transport is as important as the sequestration for copper tolerance. There has been very rapid progress in understanding of Cu transport into and within cells in plants. Various copper transporters involved in intracellular copper homeostasis have been identified in plants (Yuan et al., 2011). Here we discuss copper transporters and their roles in responses of plants under copper stress.

7.4.2 Copper Transporters

7.4.2.1 COPT Copper Transporters

Six COPT family transporter proteins (COPT1-6) were identified in *Arabidopsis thaliana*. COPT proteins are responsible for Cu uptake and accumulation (Puig, 2014). *Arabidopsis* COPT1 is the first COPT family member to be identified and characterized (Sancenon et al., 2004). Root apex, stomata, and pollen grains are where COPT1 is mostly expressed (Sancenon et al., 2004). COPT-1 is responsible for copper acquisition in roots and pollen development (Andrés-Colás et al., 2010). Furthermore, COPT-1 was proven to have role in Cu acquisition by studies on *copt-1* lack of mutants, which exhibit decreased root growth under copper deficiency (Yamasaki et al., 2009). Reduction in plant growth and changes in the morphology of leaf, increased accumulation of copper, and sensitivity to excess copper were reported in *A. thaliana* overexpressing COPT1 (Andrés-Colás et al., 2010).

COPT2 mRNA is increased by SPL7 under Cu deficiency (Yamasaki et al., 2009). COPT2 is a cell-surface Cu transporter expressed in roots, young leaves, cotyledons, apical meristem, trichomes, and anthers (Perea-García et al., 2013). COPT2 functions in Cu acquisition and the cross-talk between iron deficiency responses and low-phosphate signaling (Perea-García et al., 2013). COPT5 is a protein that is located on vacuolar tonoplast and expressed in roots, leaves, cotyledons, and reproductive organs (Garcia-Molina et al., 2011). Under copper scarcity, impaired photosynthesis, chlorosis, and root defects are observed in *copt5* knockout mutants, even though COPT5 transcript levels are not affected by Cu concentration in media (Klaumann et al., 2011). Increased Cu concentration in roots and decreased accumulation of Cu in siliques and seeds of *copt5* knockout mutants are because of reduced vacuolar Cu export. Therefore it is postulated that COPT5 roles in strongly in Cu distribution from the root to reproductive tissues (Klaumann et al., 2011). COPT6 is described as a new member of the COPT family transporters (Peñarrubia et al., 2010). COPT6 overexpression and *copt6* T-DNA insertion mutant studies in both yeast cells and *Arabidopsis* under copper deprivation show that COPT6 responsible for Cu accumulation and distribution to leaves and seeds (Jung et al., 2012).

7.4.2.2 P_{1B}-Type Heavy Metal ATPases

P_{1B}-type heavy metal ATPases transport potentially toxic metals such as Cu^+, Cu^{2+}, Zn^{2+}, Cd^{2+}, and Pb^{2+} across cell membranes (Palmgren and Axelsen, 1998; Mattle et al., 2015). Eight members of the P_{1B}-type ATPase subfamily, also known as HMA transporters (AtHMA1–AtHMA8), are encoded by the *Arabidopsis* genome (Yruela, 2009). It has been shown that Cu regulates the expression of four members of the HMA transporters AtHMA5–AtHMA8. AtHMA5 is described as pump that is induced by excess copper and removes Cu from symplast for xylem loading (Andrés-Colás et al., 2006). The responsive to antagonist 1/HMA7 delivers Cu to ethylene receptor and important for normal vegetative development. It has no role in copper resistance of plants (Binder et al., 2010). AtHMA6/PAA1 and AtHMA8/PAA2 are responsible for delivering Cu to plastocyanin (Pilon, 2011). PAA2 works in tandem with PAA1/HMA6, which is located in the inner chloroplast envelope. To maintain adequate photosynthesis under Cu deprivation, plants prioritize the transfer of Cu to plastocyanin. The level of PAA2 increase under Cu-deficient conditions to ensure deviating Cu to plastocyanin. Its level decreases as Cu becomes available. This is proven by using *pc2* mutants, which have a low level of PAA2. This also shows that plastocyanin has a regulatory effect on the level of PAA2 (Tapken et al., 2012).

8. CONCLUSION AND FUTURE PROSPECTS

Even though a lot of work exists to understand the copper stress and detoxification in plants, our knowledge is still limited. We do not know exactly how plants perceive copper. Which signal molecules play a role in the perception of copper by cells and how the genes that function in response to copper stress are regulated protect this uncertainty. On the other hand, many chelating compounds, chaperones, and metal transporters that may have functions in copper stress responses are determined by rapidly emerging sequencing techniques, transcriptomic, metabolomic, and proteomic approaches. Signal transduction in plants under copper stress is versatile. Diversity of hormones such as ABA, IAA, ethylene, and ROS (H_2O_2, NO) affect transcript factors controlling gene expression. However, our understanding of regulatory and cross-talk mechanisms is still rudimentary. Hyperaccumulator plants are best candidates to learn the regulatory and cross-talk mechanisms. These plants accumulate many different proteins under copper stress as well as heavy metals. Dissecting the genes coding these proteins from the hyperaccumulators opens up a way to understand the copper homeostasis network, crop tolerance, and phytoremediation.

REFERENCES

Aggarwal, A., Sharma, I., Tripathi, B.N., Munjal, A.K., Baunthiyal, M., Sharma, V., 2012. Metal toxicity and photosynthesis. In: Itoh, S., Mohanty, P., Guruprasad, K.N. (Eds.), Photosynthesis: Overviews on Recent Progress and Future Perspectives. International Publishing House Limited, India.

Anderson, J.M., Boardman, N.K., David, D.J., 1964. Trace metal composition of fractions obtained by digitonin fragmentation of spinach chloroplasts. Biochem. Biophys. Res. Commun. 17, 685–690.

Andrade, S.A.L., Gratao, P.L., Azevedo, R.A., Silveira, A.P.D., Schiavinato, M.A., Mazzafera, P., 2010. Biochemical and physiological changes in jack bean under mycorrhizal symbiosis growing in soil with increasing Cu concentrations. Environ. Exp. Bot. 68, 198–207.

Andrés-Colás, N., Sancenon, V., Rodriguez-Navarro, S., Mayo, S., Thiele, D.J., Ecker, J.R., Puig, S., Penarrubia, L., 2006. The *Arabidopsis* heavy metal P-type ATPase HMA5 interacts with metallochaperones and functions in copper detoxification of roots. Plant J. 45, 225–236.

Andrés-Colás, N., Perea-Garcia, A., Puig, S., Peñarrubia, L., 2010. Deregulated copper transport affects *Arabidopsis* development especially in the absence of environmental cycles. Plant Physiol. 153, 170–184.

Arteca, R.N., Arteca, J.M., 2007. Heavy-metal-induced ethylene production in *Arabidopsis thaliana*. J. Plant Physiol. 164, 1480–1488.

Barker, A.V., Pilbeam, D.J., 2014. Handbook of Plant Nutrition. CRC Press, p. 632.

Baron, M., Arellano, J.B., Gorge, J.L., 1995. Copper and photosystem II: a controversial relationship. Physiol. Plant 94, 174–180.

Beladi, M., Kashani, A., Habibi, D., Paknejad, F., Golshan, M., 2011. Uptake and effects of lead and copper on three plan species in contaminated soils: role of phytochelatin. Afr. J. Agric. Res. 6, 3483–3492.

Bernal, M., Casero, D., Singh, V., Wilson, G.T., Grande, A., Yang, H., Dodani, S.C., Pellegrini, M., Huijser, P., Connolly, E.L., Merchant, S.S., Krämer, U., 2012. Transcriptome sequencing identifies *SPL7*-regulated copper acquisition genes *FRO4/FRO5* and the copper dependence of iron homeostasis in *Arabidopsis*. Plant Cell 24, 738–761.

Binder, B.M., Rodriguez, I.F., Bleecker, A.B., 2010. The copper transporter RAN1 is essential for biogenesis of ethylene receptors in *Arabidopsis*. J. Biol. Chem. 285, 37263–37270.

Boojar, M.M.A., Tavakoli, Z., 2010. Role of antioxidant enzyme responses and phytochelatins in tolerance strategies of *Alhagi camelorum* Fisch growing on copper mine. Acta Bot. Croat. 69, 107–121.

Brahim, L., Mohamed, M., 2011. Effects of copper stress on antioxidative enzymes, chlorophyll and protein content in *Atriplex halimus*. Afr. J. Biotechnol. 50, 10143–10148.

Burda, K., Kruk, J., Strzalka, K., Schmid, G.H., 2002. Stimulation of oxygen evolution in Photosystem II by copper(II) ions. Z. Naturforsch. 57, 853–857.

Burda, K., Kruk, J., Schmid, G.H., Strzalka, K., 2003. Inhibition of oxygen evolution in Photosystem II by Cu(II) ions is associated with oxidation of cytochrome b_{559}. Biochem. J. 371, 597–601.

Chamseddine, M., Wided, B.A., Guy, H., Marie-Edith, C., Fatma, J., 2009. Cadmium and copper induction of oxidative stress and antioxidative response in tomato (*Solanum lycopersicon*) leaves. Plant Growth Regul. 57, 89–99.

Chen, C.T., Chen, L.M., Lin, C.C., Kao, C.H., 2001. Regulation of proline accumulation in detached rice leaves exposed to excess copper. Plant Sci. 160, 283–290.

Chen, L., Wang, L., Chen, F., Korpelainen, H., Li, C., 2013. The effects of exogenous putrescine on sex-specific responses of *Populus cathayana* to copper stress. Ecotox Environ. Safe 97, 94–102.

Cherian, M.G., Kang, Y.J., 2006. Metallothionein and liver cell regeneration. Exp. Biol. Med. (Maywood) 231, 138–144.

Choudhary, S.P., Bhardwaj, R., Gupta, B.D., Dutt, P., Gupta, R.K., Biondi, S., Kanwar, M., 2010. Epibrassinolide induces changes in indole-3-acetic acid, abscisic acid and polyamine concentrations and enhances antioxidant potential of radish seedlings under copper stress. Physiol. Plant. 140, 280–296.

Choudhary, S.P., Oral, H.V., Bhardwaj, R., Yu, J.Q., Lam-Son, P., Tran, L.S.P., 2012. Interaction of brassinosteroids and polyamines enhances copper stress tolerance in *Raphanus sativus*. J. Exp. Bot. 63, 5659–5675.

Choudhary, S.P., Bhardwaj, R., Gupta, B.D., Dutt, P., Gupta, R.K., Mukesh, K., Biondi, S., 2011. Enhancing effects of 24-epibrassinolide and putrescine on the antioxidant capacity and free radical scavenging activity of *Raphanus sativus* seedlings under Cu ion stress. Acta Physiol. Plant 33, 1319–1333.

Cohu, C.M., Abdel-Ghany, S.E., Gogolin Reynolds, K.A., Onofrio, A.M., Bodecker, J.R., Kimbrel, J.A., Niyogi, K.K., Pilon, M., 2009. Copper delivery by the copper chaperone for chloroplast and cytosolic copper/zinc-superoxide dismutases: regulation and unexpected phenotypes in an *Arabidopsis* mutant. Mol. Plant 2, 1336–1350.

Cui, X.M., Zhang, Y.K., Wu, X.B., Liu, C.S., 2010. The investigation of the alleviated effect of copper toxicity by exogenous nitric oxide in tomato plants. Plant Soil Environ. 56, 274–281.

DalCorso, G., Farinati, S., Furini, A., 2010. Regulatory networks of cadmium stress in plants. Plant Signal Behav. 5, 663–667.

Dall'Osto, L., Holt, N.E., Kaligotla, S., Fuciman, M., Cazzaniga, S., Carbonera, D., Frank, H.A., Alric, J., Bassi, R., 2012. Zeaxanthin protects plant photosynthesis by modulating chlorophyll triplet yield in specific light-harvesting antenna subunits. J. Biol. Chem. 287, 41820–41834.

Dong, Y., Xu, L., Wang, Q., Fan, Z., Kong, J., Bai, X., 2014. Effects of exogenous nitric oxide on photosynthesis, antioxidative ability, and mineral element contents of perennial ryegrass under copper stress. J. Plant Interact. 9, 402–411.

Eleftheriou, E.P., Karataglis, S., 1989. Ultrastructural and morphological characteristics of cultivated wheat growing on copper-polluted fields. Botanica Acta 102, 134–140.

Emamverdian, A., Ding, Y., Mokhberdoran, F., Xie, Y., 2015. Heavy metal stress and some mechanisms of plant defense response. Sci. World J. 2015. http://dx.doi.org/10.1155/2015/756120.

Fariduddin, Q., Khan, T.A., Yusuf, M., 2014. Hydrogen peroxide mediated tolerance to copper stress in the presence of 28-homobrassinolide in *Vigna radiata*. Acta Physiol. Plant. 36, 2767–2778.

Ferreira, P.A.A., Ceretta, C.A., Soriani, H.H., Tiecher, T.L., Soares, C.R.F.S., Rossato, L.V., Nicoloso, F.T., Brunetto, G., Paranhos, J.T., Cornejo, P., 2015. *Rhizophagus clarus* and phosphate alter the physiological responses of *Crotalaria juncea* cultivated in soil with a high Cu level. Appl. Soil Ecol. 91, 37–47.

Fidalgo, F., Azenha, M., Silva, A.F., de Sousa, A., Santiago, A., Ferraz, P., Teixeira, J., 2013. Copper-induced stress in *Solanum nigrum* L. and antioxidant defense system responses. Food Energ. Sec. 2, 70–80.

Foyer, C.H., Noctor, G., 2011. Ascorbate and glutathione: the heart of the redox hub. Plant Physiol. 155, 12–18.

Garcia-Molina, A., Andrés-Colás, N., Perea-Garcia, A., Del Valle-Tascon, S., Peñarrubia, L., Puig, S., 2011. The intracellular *Arabidopsis* COPT5 transport protein is required for photosynthetic electron transport under severe copper deficiency. Plant J. 65, 848–860.

Garcia-Molina, A., Xing, S., Huijser, P., 2014. Functional characterisation of *Arabidopsis* SPL7 conserved protein domains suggests novel regulatory mechanisms in the Cu deficiency response. BMC Plant Biol. 14, 231.

Grennan, A.K., 2011. Metallothioneins, a diverse protein family. Plant Physiol. 155, 1750–1751.

Groppa, M.D., Ianuzzo, M.P., Tomaro, M.L., Benavides, M.P., 2007. Polyamine metabolism in sunflower plants under long-term cadmium or copper stress. Amino Acid 32, 265–275.

Groppa, M.D., Benavides, M.P., Tomaro, M.L., 2003. Polyamine metabolism in sunflower and wheat leaf discs under cadmium or copper stress. Plant Sci. 164, 293–299.

Gu, C.S., Liu, L.Q., Deng, Y.M., Zhu, X.D., Huang, S.Z., Lu, X.-Q., 2015. The heterologous expression of the *Iris lactea* var. *chinensis* Type2 metallothionein *IlMT2b* gene enhances copper tolerance in *Arabidopsis thaliana*. Bull. Environ. Contam. Toxicol. 94, 247–253.

Guzel, S., Terzi, R., 2013. Exogenous hydrogen peroxide increases dry matter production, mineral content and level of osmotic solutes in young maize leaves and alleviates deleterious effects of copper stress. Bot. Stud. 54, 26.

Habiba, U., Ali, S., Farid, M., Shakoor, M.B., Rizwan, M., Ibrahim, M., Abbasi, G.H., Hayat, T., Ali, B., 2015. EDTA enhanced plant growth, antioxidant defense system, and phytoextraction of copper by *Brassica napus* L. Environ. Sci. Pollut. Res. Int. 22, 1534–1544. http://dx.doi.org/10.1007/s11356-014-3431-5.

Halliwell, B., Gutteridge, J.M.C., 2007. Free Radicals in Biology and Medicine. Oxford University Press, NewYork, p. 888.

Halliwell, B., Gutteridge, J.M., 1984. Lipid peroxidation, oxygen radicals, cell damage, and antioxidant therapy. Lancet 1, 1396–1397.

Hego, E., Bes, C.M., Bedon, F., Palagi, P.M., Chaumeil, P., Barré, A., Claverol, S., Dupuy, J.W., Bonneu, M., Lalanne, C., Plomion, C., Mench, M., 2014. Differential accumulation of soluble proteins in roots of metallicolous and nonmetallicolous populations of *Agrostis capillaris* L. exposed to Cu. Proteomics 14, 1746–1758.

Hu, Y., You, J., Liang, X., 2015. Nitrate reductase-mediated nitric oxide production is involved in copper tolerance in shoots of hulless barley. Plant Cell Rep. 34, 367–379.

Huang, J., Zhang, Y., Peng, J.S., Zhong, C., Yi, H.Y., Ow, D.W., Gong, J.M., 2012. Fission yeast HMT1 lowers seed cadmium through phytochelatin-dependent vacuolar sequestration in *Arabidopsis*. Plant Physiol. 158, 1779–1788.

Hung, W.C., Huang, D.D., Yeh, C.M., Huang, H.J., 2005. Reactive oxygen species, calcium and serine/threonine phosphatase are required for copper-induced MAP kinase gene, *OsMAPK2*, expression in rice. Plant Growth Regul. 45, 233–241.

Iqbal, N., Trivellini, A., Masood, A., Ferrante, A., Khan, N.A., 2013. Current understanding on ethylene signalling in plants: the influence of nutrient availability. Plant Physiol. Biochem. 73, 128–138.

Jakubowicz, M., Galganska, H., Nowak, W., Sadowski, J., 2010. Exogenously induced expression of ethylene biosynthesis, ethylene perception, phospholipase D, and Rboh-oxidase genes in broccoli seedlings. J. Exp. Bot. 61, 3475–3491.

Jonak, C., Nakagami, H., Hirt, H., 2004. Heavy metal stress. Activation of distinct mitogen-activated protein kinase pathways by copper and cadmium. Plant Physiol. 136, 3276–3283.

Jung, H., Gayomba, S.R., Rutzke, M.A., Craft, E., Kochian, L.V., Vatamaniuk, O.K., 2012. COPT6 is a plasma membrane transporterthat functions in copper homeostasis in *Arabidopsis* and is a novel target of SQUAMOSA promoter-binding protein-like 7. J. Biol. Chem. 287, 33252–33267.

Kavi Kishor, P.B., Sangam, S., Amrutha, R.N., Sri Laxmi, P., Naidu, K.R., Rao, K.R.S.S., Rao, S., Reddy, K.J., Theriappan, P., Sreenivasulu, N., 2005. Regulation of proline biosynthesis, degradation, uptake and transport in higher plants: Its implications in plant growth and abiotic stress tolerance. Curr. Sci. 88, 424–438.

Khan, M.I., Khan, N.A., 2014. Ethylene reverses photosynthetic inhibition by nickel and zinc in mustard through changes in PS II activity, photosynthetic nitrogen use efficiency, and antioxidant metabolism. Protoplasma 251, 1007–1019.

Klaumann, S., Nickolaus, S.D., Furst, S.H., Starck, S., Schneider, S., Ekkehard Neuhaus, H., Trentmann, O., 2011. The tonoplast copper transporter COPT5 acts as an exporter and is required for interorganal location of copper in *Arabidopsis thaliana*. New Phytol. 192, 393–404.

Kolbert, Z., Peto, A., Lehotai, N., Feigl, G., Erdei, L., 2012. Long-term copper ($Cu_2{}^+$) exposure impacts on auxin, nitric oxide (NO) metabolism and morphology of *Arabidopsis thaliana* L. Plant Growth Regul. 68, 151–159.

Krupa, Z., 1988. Acyl lipids in the supramolecular chlorophyll–protein complexes of photosynthesis-isolation artifacts or integral components regulating their structure and functions? Acta Soc. Bot. Pol. 57, 401–418.

Ku, H.M., Tan, C.W., Su, Y.S., Chiu, C., Chen, C.T., Jan, F.J., 2012. The effect of water deficit and excess copper on proline metabolism in *Nicotiana benthamiana*. Biol. Plant 56, 337–343.

Lee, C.W., Jackson, M.B., Duysen, M.E., Freeman, T.P., Self, J.R., 1996. Induced micronutrient toxicity in 'Towndown' Kentucky bluegrass. Crop Sci. 36, 705–712.

Lee, S., Kang, B.S., 2005. Phytochelatin is not a primary factor in determining copper tolerance. J. Plant Biol. 48, 32–38.

Lequeux, H., Hermans, C., Lutts, S., Verbruggen, N., 2010. Response to copper excess in *Arabidopsis thaliana*: Impact on the root system architecture, hormone distribution, lignin accumulation and mineral profile. Plant Physiol. Biochem. 48, 673–682.

Lidon, F.C., Henriques, F.S., 1991. Limiting step on photosynthesis of rice plants treated with varying copper levels. J. Plant Physiol. 138, 115–118.

Lidon, F.C., Henriques, F.S., 1993. Changes in the thylakoid membrane polypeptide patterns triggered by excess Cu in rice. Photosynthetica 28, 109–117.

Liu, J.J., Wei, Z., Li, J.H., 2014. Effects of copper on leaf membrane structure and root activity of maize seedling. Bot. Stud. 55, 47.

Liu, Q., Zheng, L., He, F., Zhao, F.J., Shen, Z., Zheng, L., 2015. Transcriptional and physiological analyses identify a regulatory role for hydrogen peroxide in the lignin biosynthesis of copper-stressed rice roots. Plant Soil 387, 323–336.

Lv, Y., Deng, X., Quan, L., Xia, Y., Shen, Z., 2013. Metallothioneins BcMT1 and BcMT2 from *Brassica campestris* enhance tolerance to cadmium and copper and decrease production of reactive oxygen species in *Arabidopsis thaliana*. Plant Soil 367, 507–519.

Maksymiec, W., 2007. Signaling responses in plants to heavy metal stress. Acta Physiol. Plant 29, 177–187.

Maksymiec, W., Krupa, Z., 2006. The effects of short-term exposition to Cd, excess Cu ions and jasmonate on oxidative stress appearing in *Arabidopsis thaliana*. Environ. Exp. Bot. 57, 187–194.

Maksymiec, W., 1997. Effect of copper on cellular processes in higher plants. Photosynthetica 34, 132–342.

Maksymiec, W., Bednara, J., Baszynski, R., 1995. Responses of runner bean plants to excess copper as a function of plant growth stages: effects on morphology and structure of primary leaves and their chloroplast ultrastructure. Photosynthetica 31, 427–435.

Maksymiec, W., Russa, R., Urbanik-Sypniewska, T., Baszynski, T., 1994. Effect of excess Cu on the photosynthetic apparatus of runner bean leaves treated at two different growth stages. Physiol. Plant 91, 715–721.

Malec, P., Maleva, M., Prasad, M.N.V., Strzałka, K., 2009. Copper toxicity in leaves of *Elodea canadensis* Michx. Bull. Environ. Contam. Toxicol. 82, 627–632.

Mani, S., Van de Cotte, B., Van Montagu, M., Verbruggen, N., 2002. Altered levels of proline dehydrogenase cause hypersensitivity to proline and its analogs in *Arabidopsis*. Plant Physiol. 128, 73–83.

Marschner, H., 2011. Marschner's Mineral Nutrition of Higher Plants. Academic Press, London, p. 672.

Mateos-Naranjo, E., Gallé, A., Florez-Sarasa, I., Perdomo, J.A., Galmés, J., Ribas-Carbó, M., Flexas, J., 2015. Assessment of the role of silicon in the Cu-tolerance of the C_4 grass *Spartina densiflora*. J. Plant Physiol. 178, 74–83.

Mattle, D., Zhang, L., Sitsel, O., Pedersen, L.T., Moncelli, M.R., Tadini-Buoninsegni, F., Gourdon, P., Rees, D.C., Nissen, P., Meloni, G., 2015. A sulfur-based transport pathway in Cu^+-ATPases. EMBO Rep. 16, 728–740. http://dx.doi.org/10.15252/embr.201439927.

Mertens, J., Vangronsveld, J., Van Der Straeten, D., Poucke, M., 1999. Effects of copper and zinc on the ethylene production of *Arabidopsis thaliana*. In: Kanellis, A.K., Chang, C., Klee, H., Bleecker, A.B., Pech, J.C., Grierson, D. (Eds.), Biol Biotechnol Plant Horm Ethyl II SE - 60. Kluwer Academic Publishers, Dordrecht, pp. 333–338.

Miotto, A., Ceretta, C.A., Brunetto, G., Nicoloso, F.T., Girotto, E., Farias, J.G., Tiecher, T.L., De Conti, L., Trentin, G., 2014. Copper uptake, accumulation and physiological changes in adult grapevines in response to excess copper in soil. Plant Soil 374, 593–610.

Mira, H., Martínez-Garcia, F., Penñarrubia, L., 2001. Evidence for the plant-specific intercellular transport of the *Arabidopsis* copper chaperone CCH. Plant J. 25, 521–528.

Mohanty, N., Vass, I., Demeter, S., 1989. Copper toxicity affects Photosystem II electron transport at the secondary quinone acceptor (Q_B). Plant Physiol. 90, 175–179.

Monteoliva, M.I., Rizzi, Y.S., Cecchini, N.M., Hajirezaei, M.R., Alvarez, M.E., 2014. Context of action of proline dehydrogenase (ProDH) in the hypersensitive response of *Arabidopsis*. BMC Plant Biol. 14, 21.

Morita, S., Kaminaka, H., Masumura, T., Yanaka, K., 1999. Induction of rice cytosolic ascorbate peroxidase mRNA by oxidative stress signaling. Plant Cell Physiol. 40, 417–422.

Mostofa, M.G., Seraj, Z.I., Fujita, M., 2014. Exogenous sodium nitroprusside and glutathione alleviate copper toxicity by reducing copper uptake and oxidative damage in rice (*Oryza sativa* L.) seedlings. Protoplasma 251, 1373–1386.

Munzuroglu, O., Zengin, F.K., Yahyagil, Z., 2008. The abscisic acid levels of wheat (*Triticum aestivum* L. cv. cakmak 79) seeds that were germinated under heavy metal (Hg^{++}, Cd^{++}, Cu^{++}) stress. Gazi Univ. J. Sci. 2, 1–7.

Murata, N., Higashi, S.I., Fujimura, Y., 1990. Glycerolipids in various preparations of photosystem II from spinach chloroplasts. Biochim. Biophys. Acta 1019, 261–268.

Nair, P.M.G., Chung, I.M., 2015. Study on the correlation between copper oxide nanoparticles induced growth suppression and enhanced lignification in Indian mustard (*Brassica juncea* L.). Ecotox. Environ. Safe 113, 302–313.

Orzechowska, A., Bock, R., Piramowicz, M.O., Strzałka, K., Burda, K., 2008. Cu^{2+} binding sites in PSII. In: Allen, J.F., Gantt, E., Golbeck, J., Osmond, B. (Eds.), Photosynthesis: Energy from the Sun. Springer, Netherland, pp. 657–660.

Palmgren, M.G., Axelsen, K.B., 1998. Evolution of P-type ATPases. Biochim. Biophys. Acta 1365, 37–45.

Panda, S., Panda, S., 2009. Effect of mercury ion on the stability of the lipid-protein complex of isolated chloroplasts. Indian J. Biochem. Biophys. 46, 405–408.

Panou-Filotheou, H., Bosabalidis, A.M., 2004. Root structural aspects associated with copper toxicity in oregano (*Oreganum vulgare* sudsp. *hirtum*). Plant Sci. 166, 1497–1504.

Panou-Filotheou, H., Bosabalidis, A.M., Karataglis, S., 2001. Effects of copper toxicity on leaves of oregano (*Origanum vulgare* subsp. *hirtum*). Ann. Bot. 88, 207–214.

Parry, M.A.J., Andralojc, J., Scales, J.C., Salvucci, M.E., Cormo-Silva, A.E., Alonso, H., Whitney, S.M., 2012. Rubisco activity and regulation as targets for crop. J. Exp. Bot. 63, 695–709.

Peñarrubia, L., Andrés-Colás, N., Moreno, J., Puig, S., 2010. Regulation of copper transport in *Arabidopsis thaliana*: a biochemical oscillator? J. Biol. Inorg. Chem. 15, 29–36.

Peng, H., Kroneck, P.M.H., Küpper, H., 2013. Toxicity and deficiency of copper in *Elsholtzia splendens* affect photosynthesis biophysics, pigments and metal accumulation. Environ. Sci. Technol. 47, 6120–6128.

Perea-García, A., Garcia-Molina, A., Andrés-Colás, N., Vera-Sirera, F., Pérez-Amador, M.A., Puig, S., Peñarrubia, L., 2013. *Arabidopsis* copper transport protein COPT2 participates in the cross talk between iron deficiency responses and low-phosphate signaling. Plant Physiol. 162, 180–194.

Pilon, M., 2011. Moving copper in plants. New Phytol. 192, 305–307.

Posmyk, M.M., Kuran, H., Marciniak, K., Janas, K.M., 2008. Presowing seed treatment with melatonin protects red cabbage seedlings against toxic copper ion concentrations. J. Pineal Res. 45, 24–31.

Puig, S., 2014. Function and regulation of the plant COPT family of high-affinity copper transport proteins. Adv. Bot. 2014. http://dx.doi.org/10.1155/2014/476917.

Puig, S., Andres-Colas, N., Garcia-Molina, A., Peñarrubia, L., 2007. Copper and iron homeostasis in *Arabidopsis*: responses to metal deficiencies, interactions and biotechnological applications. Plant Cell Environ. 30, 271–290.

Quartacci, M.F., Pinzino, C., Sgherri, C.L.M., Dalla Vecchia, F., Navari-Izzo, F., 2000. Growth in excess copper induces changes in the lipid composition and fluidity of PSII-enriched membranes in wheat. Physiol. Plant 108, 87–93.

Raven, J.A., 2013. Rubisco: still the most abundant protein of Earth? New Phytol. 198, 1–3.

Rejeb, K.B., Abdelly, C., Savouré, A., 2014. How reactive oxygen species and proline face stress together. Plant Physiol. Biochem. 80, 278–284.

Ritter, A., Dittami, S.M., Goulitquer, S., Correa, J.A., Boyen, C., Potin, P., Tonon, T., 2014. Transcriptomic and metabolomic analysis of copper stress acclimation in *Ectocarpus siliculosus* highlights signaling and tolerance mechanisms in brown algae. BMC Plant Biol. 14, 116.

Rouillon, R., Piletsky, S.A., Breton, F., Piletska, E.V., Carpentier, R., 2006. Photosystem II biosensors for heavy metals monitoring. In: Teresa, M., Giardi, P.E.V. (Eds.), Biotechnological Applications of Photosynthetic Proteins: Biochips, Biosensors and Biodevices. Springer, USA, pp. 166–174.

Rout, J.R., Sahoo, S.L., 2013. Antioxidant enzyme gene expression in response to copper stress in *Withania somnifera* L. Plant Growth Regul. 71, 95–99.

Rozentsvet, O.A., Nesterov, V.N., Sinyutina, N.F., 2012. The effect of copper ions on the lipid composition of subcellular membranesin *Hydrilla verticillata*. Chemosphere 89, 108–113.

Sagasti, S., Bernal, M., Sancho, D., Castillo, M.B., Picorel, R., 2014. Regulation of the chloroplastic copper chaperone (CCS) and cuprozinc superoxide dismutase (CSD2) by alternative splicing and copper excess in *Glycine max*. Funct. Plant Biol. 41, 144–155.

Sancenon, V., Puig, S., Mateu-Andres, I., Dorcey, E., Thiele, D.J., Penarrubia, L., 2004. The *Arabidopsis* copper transporter COPT1 functions in root elongation and pollen development. J. Biol. Chem. 279, 15348–15355.

Sanchez-Pardo, B., Fernandez-Pascual, M., Zornoza, P., 2014. Copper microlocalisation and changes in leaf morphology, chloroplast ultrastructure and antioxidative response in white lupin and soybean grown in copper excess. J. Plant Res. 127, 119–129.

Schafer, C., Simper, H., Hofmann, B., 1992. Glucose feeding results in coordinated changes of chlorophyll content, ribulose-1,5-biphosphate carboxylase-oxygenase activity and photosynthetic potential in photo-autrophic suspension cultured cells of *Chenopodium rubrum*. Plant Cell Environ. 15, 343–350.

Schellingen, K., Van Der Straeten, D., Vandenbussche, F., Prinsen, E., Remans, T., Vangronsveld, J., Cuypers, A., 2014. Cadmium-induced ethylene production and responses in *Arabidopsis thaliana* rely on *ACS2* and *ACS6* gene expression. BMC Plant Biol. 14, 214.

Sharma, P., Bhardwaj, R., 2007. Effects of 24-epibrassinolide on growth and metal uptake in *Brassica juncea* L. under copper metal stress. Acta Physiol. Plant 29, 259–263.

Shin, L.J., Lo, J.C., Yeh, K.C., 2012. Copper chaperone antioxidant protein1 is essential for copper homeostasis. Plant Physiol. 159, 1099–1110.

Sibbald, P.R., Green, B.R., 1987. Copper in photosystem II: association with LHC II. Photosynth. Res. 14, 201–209.

Siborova, M., 1988. Cd^{2+} ions affect the quaternary structure of ribulose-1,5-bisphosphate carboxylase from barley leaves. Biochem. Physiol. Pflanzen 183, 371–378.

Simova-Stoilova, L., Demirevska-Kepova, K., Stoyanova, Z., Smilova, E., 2002. Effect of Cu and Mn toxicity on leaf protein pattern and rubisco quantity of young barley plants. Compt. Rend. Acad. Bulg Sci. 55, 79–84.

Singh, R., Gautam, N., Mishra, A., Gupta, R., 2011. Heavy metals and living systems: an overview. Indian J. Pharmacol. 43, 246–253.

Smeets, K., Opdenakker, K., Remans, T., Forzani, C., Hirt, H., Vangronsveld, J., Cuypers, A., 2013. The role of the kinase OXI1 in cadmium- and copper-induced molecular responses in *Arabidopsis thaliana*. Plant Cell Environ. 36, 1228–1238.

Son, J.A., Narayanankutty, D.P., Roh, K.S., 2014. Influence of exogenous application of glutathione on rubisco and rubisco activase in heavy metal-stressed tobacco plant grown in vitro. Saudi J. Biol. Sci. 21, 89–97.

Souza, V.L., Almeida, A.A.F., Souza, J.S., Mangabeira, P.A.O., Jesus, R.M., Pirovani, C.P., Ahnert, D., Baligar, V.C., Loguercio, L.L., 2014. Altered physiology, cell structure, and gene expression of *Theobroma cacao* seedlings subjected to Cu toxicity. Environ. Sci. Pollut. Res. 21, 1217–1230.

Sun, X.H., Yu, G., Li, J.T., Jia, P., Zhang, J.C., Jia, C.G., Zhang, Y.H., Pan, H.Y., 2014. A heavy metal-associated protein (AcHMA1) from the Halophyte, *Atriplex canescens* (Pursh) Nutt. confers tolerance to iron and other abiotic stresses when expressed in *Saccharomyces cerevisiae*. Int. J. Mol. Sci. 15, 14891–14906.

Szabados, L., Savoure, A., 2010. Proline: a multifunctional amino acid. Trends Plant Sci. 15, 89–97.

Tan, D.X., Hardeland, R., Manchester, L.C., Korkmaz, A., Ma, S., Rosales-Corral, S., Reiter, R.J., 2012. Functional roles of melatonin in plants, and perspectives innutritional and agricultural science. J. Exp. Bot. 63, 577–597.

Tan, D.X., Manchester, L.C., Helton, P., Reiter, R.J., 2007. Phytoremediative capacity of plants enriched with melatonin. Plant Signal Behav. 2, 514–516.

Tapken, W., Ravet, K., Pilon, M., 2012. Plastocyanin controls the stabilization of the thylakoid Cu-transporting P-type ATPase PAA2/HMA8 in response to low copper in *Arabidopsis*. J. Biol. Chem. 287, 18544–18550.

Taylor, G.J., Foy, C.D., 1985. Differential uptake and toxicity of ionic and chelated copper in Triticum aestivum. Can. J. Bot. 63, 1271–1275.

Tewari, R.K., Hahn, E.J., Paek, K.Y., 2008. Modulation of copper toxicity-induced oxidative damage by nitric oxide supply in the adventitious roots of *Panax ginseng*. Plant Cell Rep. 27, 171–181.

Thounaojam, T.C., Panda, P., Mazumdar, P., Kumar, D., Sharma, G.D., Sahoo, L., Sanjib, P., 2012. Excess copper induced oxidative stress and response of antioxidants in rice. Plant Physiol. Biochem. 53, 33–39.

Tie, S.G., Tang, Z.J., Zhao, Y.M., Li, W., 2012. Oxidative damage and antioxidant response caused by excess copper in leaves of maize. Afr. J. Biotechnol. 11, 4378–4384.

Tripathi, B.N., Singh, V., Ezaki, B., Sharma, V., Gaur, J.P., 2013. Mechanism of Cu- and Cd-induced proline hyperaccumulation in *Triticum aestivum* (Wheat). J. Plant Growth Regul. 32, 799–808.

Turchi, A., Tamantini, I., Camussi, A.M., Racchi, M.L., 2012. Expression of a *metallothionein A1* gene of *Pisum* sativum in white poplar enhances tolerance and accumulation of zinc and copper. Plant Sci. 183 (2012), 50–56.

Upadhyay, R., Panda, S.K., 2010. Zinc reduces copper toxicity induced oxidative stress by promoting antioxidant defense in freshly grown aquatic duckweed Spirodela polyrhiza L. J. Hazard. Mater 175, 1081–1084.

van Assche, F., Clijsters, H., 1990. Effects of metals on enzyme activity in plants. Plant Cell Environ. 13, 195–206.

Vázquez, M.N., Guerrero, Y.R., González, L.M., de la Noval, W.T., 2013. Brassinosteroids and plant responses to heavy metal stress. An overview. Open J. Metal 3, 34–41.

Verma, A., Bhatia, S., 2014. Analysis of some physicochemical parameters and trace metal concentration present in the soil around the area of Pariccha thermal power station in Jhansi, India. Int. J. Inno Res. Sci. Eng. Technol. 3, 10482–10488.

Vijayarengan, P., Jose, M.D., 2014. Changes in growth, pigments and phytoremediating capability of four plant species under copper stress. Int. J. Environ. Biol. 4, 119–126.

Wang, L., Yang, X., Ren, Z., Hu, X., Wang, X., 2014. Alleviation of photosynthetic inhibition in copper-stressed tomatoes through rebalance of ion content by exogenous nitric oxide. Turkish J. Bot. 38 http://dx.doi.org/10.3906/bot-1312-17.

Wang, L., Yang, L., Yang, F., Li, X., Song, Y., Wang, X., Hu, X., 2010. Involvements of H_2O_2 and metallothionein in NO-mediated tomato tolerance to copper toxicity. J. Plant Physiol. 167, 1298–1306.

Wang, Y.Z., Nie, L.H., Tie, S., Xie, D., Zhu, W., Qi, J., Yue, R., 2011. Effects of excess copper on the oxidative stress in roots of maize seedlings. Afr. J. Agric. Res. 6, 4998–5004.

Wen, J.F., Gong, M., Liu, Y., Hu, J.L., Deng, M.H., 2013. Effect of hydrogen peroxide on growth and activity of some enzymes involved in proline metabolism of sweet corn seedlings under copper stress. Sci. Hort. 164, 366–371.

Whitacre, D.M., 2011. Reviews of Environmental Contamination and Toxicology, vol. 213. Springer, USA, p. 180.
Xia, Y., Qi, Y., Yuan, Y., Wang, G., Cui, J., Chen, Y., Zhang, H., Shen, Z., 2012. Overexpression of *Elsholtzia haichowensis* metallothionein 1 (EhMT1) in tobacco plants enhances copper tolerance and accumulation in root cytoplasm and decreases hydrogen peroxide production. J. Hazard Mater. 234, 65–71.
Xu, Q., Qiu, H., Chu, W., Fu, Y., Cai, S., Min, H., Sha, S., 2013. Copper ultrastructural localization, subcellular distribution, and phytotoxicity in *Hydrilla verticillata* (L.f.). Royle. Environ. Sci. Pollut. Res. 20, 8672–8679.
Yamasaki, H., Hayashi, M., Fukazawa, M., Kobayashi, Y., Shikanai, T., 2009. SQUAMOSA promoter binding protein-like7 is a central regulator for copper homeostasis in *Arabidopsis*. Plant Cell 21, 347–361.
Yau, P.Y., Loh, C.F., Azmil, I.A.R., 1991. Copper toxicity of clove (*Syzygium aromaticum* (L.) Merr. and Perry) seedlings. MARDI Res. J. 19, 49–53.
Yang, J., Chen, Z., Wu, S., Cui, Y., Zhang, L., Dong, H., Yang, C., Li, C., 2015. Overexpression of the *Tamarix hispida ThMT3* gene increases copper tolerance and adventitious root induction in *Salix matsudana* Koidz. Plant Cell Tiss. Organ Cult. 121, 469–479.
Yeh, C.M., Chien, P.S., Huang, H.J., 2007. Distinct signalling pathways for induction of MAP kinase activities by cadmium and copper in rice roots. J. Exp. Bot. 58, 659–671.
Yruela, I., 2009. Copper in plants: acquisition, transport and interactions. Funct. Plant Biol. 36, 409–430.
Yruela, I., 2005. Copper in plants. Braz. J. Plant Physiol. 17, 145–156.
Yruela, I., Pueyo, J.J., Alonso, P.J., Picorel, R., 1996. Photoinhibition of Photosystem II from higher plants–effect of copper inhibition. J. Biol. Chem. 271, 27408–27415.
Yu, C.C., Hung, K.T., Kao, C.H., 2005. Nitric oxide reduces Cu toxicity and Cu-induced NH^{4+} accumulation in rice leaves. J. Plant Physiol. 162, 1319–1330.
Yuan, H.M., Liu, W.C., Jin, Y., Lu, Y.T., 2013. Role of ROS and auxin in plant response to metal-mediated stress. Plant Signal Behav. 8, e24671.
Yuan, M., Li, X., Xiao, J., Wang, S., 2011. Molecular and functional analyses of COPT/Ctr-type copper transporter-like gene family in rice. BMC Plant Biol. 11, 69.
Zengin, F.K., Kirbag, S., 2007. Effects of copper on chlorophyll, proline, protein and abscisic acid level of sunflower (*Helianthus annuus* L.) seedlings. J. Environ. Biol. 28, 561–566.
Zhang, Y., Han, X., Chen, X., Jin, H., Cui, X., 2009. Exogenous nitric oxide on antioxidative system and ATPase activities from tomato seedlings under copper stress. Scientia Hortic. 123, 217–223.
Zhao, S., Liu, Q., Qi, Y., Duo, L., 2010. Responses of root growth and protective enzymes to copper stress in Turfgrass. Acta Biol. Cracov Ser. Bot. 52, 7–11.
Zlobin, I.E., Kholodova, V.P., Rakhmankulova, Z.F., Kuznetsov, V.V., 2014. *Brassica napus* responses to short-term excessive copper treatment with decrease of photosynthetic pigments, differential expression of heavy metal homeostasis genes including activation of gene NRAMP4 involved in photosystem II stabilization. Photosynth. Res. 125, 141–150. http://dx.doi.org/10.1007/s11120-014-0054-0.

EFFECT OF LEAD ON PLANT AND HUMAN DNA DAMAGES AND ITS IMPACT ON THE ENVIRONMENT

3

Murat Dikilitas[1], Sema Karakas[2], Parvaiz Ahmad[3]

Department of Plant Protection, Faculty of Agriculture, Harran University, Şanlıurfa, Turkey[1]; Department of Soil Science and Plant Nutrition, Faculty of Agriculture, Harran University, Şanlıurfa, Turkey[2]; Department of Botany, S.P. College, Srinagar, Jammu and Kashmir, India[3]

CHAPTER OUTLINE

1. INTRODUCTION

Pollution of the environment by heavy metals has received considerable attention in recent years because of anthropogenic activities in soils such as metal mining, smelting, irregular agricultural and horticultural practices (i.e., excess application of fertilizers, pesticides, overirrigation), and waste disposal (Fahr et al., 2013). Heavy metals are defined as those having a specific density of more than 5 g/cm^3 (Alissa and Ferns, 2011). Heavy metal stress has become a serious environmental issue affecting human and plant lives. Industrial effluents and drainage water carry a high amount of metallic substances far from their sources into the land and water systems (Aldoobie and Beltagi, 2013). The use of intensive micronutrients and macronutrients, especially phosphorous-containing fertilizers, has dramatically increased to meet the growing food demand globally. Thus, with the increase of fertilizers, increasingly larger amounts of heavy metals have also been introduced into nonpolluted areas. One of the most toxic heavy metals involved in many fertilizers is lead, which can be absorbed

Plant Metal Interaction. http://dx.doi.org/10.1016/B978-0-12-803158-2.00003-5

and accumulated in plants. From the soil, it easily enters the food chain and poses a threat to life. The downward movement of elemental and inorganic lead compounds from soil to the groundwater is very slow because of its slow mobility, and therefore, it is easily absorbed by plants and other organisms (Rodriguez et al., 2011). Although the level of lead is low in the upper parts of plants such as seeds, leaves, and fruits because of great accumulation in roots, it can be still, however, toxic to human when edible parts are consumed. Therefore, the pollution of soil and water is directly linked to the cross-contamination of food chain and carries a considerable risk for humans especially for children and pregnant women. Sewage sludge input and pesticide treatments are also responsible for the pollution of the environment, particularly in agricultural soils. Although these practices significantly increase yields by protecting plants from the insects and pathogens and providing them with nutrients necessary for the growth and crop production, they, however, put large amounts of heavy metals into soil and water resources. Therefore, pollution from heavy metals is a matter of growing concern because of their toxicity to all forms of life. Disposal of municipal waste, rich in heavy metals, is also considered to be one of the major sources of soil pollution. For example, gas emission from power plants and vehicles, burning of fossil gases, and industrial technology discharge have played important roles for the metal contamination in our environment (Mishra and Dubey, 2005). As stated previously, heavy metals mostly accumulate in soil and water and therefore are a particular threat to crop plants and us via contaminated food source when our plants are cultivated in such environments. Lead is widely distributed in the Earth's crust; however, their occurrence is more prominent in deep layers. Erosion, earthquakes, volcanic eruptions, and the industrial need brought the heavy metals to the upper layers, even to the atmosphere via emission from vehicles. For example, natural inputs such as weathering of rocks (Gadd, 2010) and volcanic eruptions (Richardson et al., 2001) put thousands of lead into the environment. The presence of even trace amounts in the atmosphere, soil, and water can cause serious problems to all organisms, although their uptake by the organisms has been restricted.

Metal-contaminated soils not only reduce crop production, but also affect life indirectly via edible parts of crop plants. These plants may not show any symptoms of toxicity, especially at low concentrations of heavy metal accumulation. This would eventually result in transferring heavy metals into the food chain slowly and creates great potential risks for the human health as well as the entire ecosystem (Hart et al., 2006; Zhai et al., 2015; Gall et al., 2015).

Excessive heavy metals in soil can also result in decreased soil microbial activity and soil fertility (Gao et al., 2010; Yuan et al., 2015). In contrast to organic compounds, heavy metals, especially lead, cadmium, and mercury, do not decompose at all and do not disappear from soil easily, although their release to the environment is limited and restricted (Brusseau, 1997; Monteiro et al., 2012) because they have low solubility and strong binding capacity with soil colloids. Therefore, they are persistent in both water and soil and cannot be converted into harmless compounds via biological processes either (Aldoobie and Beltagi, 2013; Yuan et al., 2015). Lead's half-life is quite long compared with other elements in nature (150–5000 years) and can be retained in the environment for ever. Therefore, the effect of lead contamination in agricultural areas can be quite devastating and last for many years (El-Beltagi and Mohamed, 2010; Truta et al., 2011; Hu et al., 2012; Romero-Freire et al., 2015). Without remedial studies, high soil or water lead levels will never return to normal (Traunfeld and Clement, 2001; Saba et al., 2015). Although the toxicity of heavy metals, particularly lead, has been known for many decades, their use in household renovation, battery, automobile manufacturing, fuel combustion, building construction, solder, pewter, ceramics, and paint make expose people to heavy metal stress continuously (Kasuba et al., 2010; Garcia-Leston et al., 2012). Because of the long persistence of lead in the environment, it is ranked second of all hazardous substances by the Agency for Toxic Substances and Disease Registry (ATSDR, 2012). The International Agency for Research on Cancer classified it as a possible human carcinogen (group 2B) (IARC, 2006; Garcia-Leston et al., 2012) and its inorganic compounds as probable human carcinogens (group 2A) (IARC, 2006; Garcia-Leston et al., 2012; Kasuba et al., 2012). These classifications are based on sufficient evidence of carcinogenic activity of some lead compounds in animal studies; however, no deaths, to our knowledge, related to cancer and cancer-related diseases have been reported.

2. MECHANISMS OF HEAVY METAL STRESS AND TOLERANCE

Under natural conditions of growth and development, plants are inevitably exposed to various states of stresses such as drought, heat, chilling, high light intensity, ultraviolet (UV) radiation, heavy metals, organic chemicals, and air pollutants such as O_3 and SO_2 (Dezhban et al., 2015). As a result, levels of highly reactive oxygen species (ROS) such as superoxide radicals (O_2^-), singlet oxygen ($\cdot O_2$), hydrogen peroxide (H_2O_2), and hydroxyl radicals ($OH\cdot$) increase and result in damages to organisms (Wang et al., 2015). Although ROS are produced continuously in different cellular compartments as by-products in all aerobic organisms and normally are in balance with antioxidant molecules (Reddy et al., 2005; Kwankua et al., 2012: Girisha and Ragavendra, 2009); however, when the critical balance is disrupted because of an excess production of ROS, antioxidants in the organisms is depleted and creates disorders in organisms (Kaur et al., 2015; Qiao et al., 2015; Figure 3.1). Increased levels of ROS inhibit most cellular processes and could damage all types of biochemical molecules, including cell wall properties, membrane lipids, proteins, amino acid chains, pigments, nucleic acids, and carbohydrates (Gratao et al., 2005). They also cause membrane leakage and loss of cell metabolites out of cell protoplasm. As a result, reduction in cell growth and impaired metabolic functions are inevitable (Goncalves et al., 2007a,b; Lee et al., 2007).

Living organisms need certain metals as micronutrients for their growth and metabolic functions in their enzymatic reactions. For example, iron, zinc, copper, magnesium, manganese, and molybdenum are required by organisms in trace amounts and have been well-defined for their biological functions such as playing roles for the structure of nucleic acids, chlorophyll, and other cell components. They play significant roles for the production of secondary metabolites and enzymes as well. Their deficiency or excess amount can cause significant problems in living organisms.

However, nonessential heavy metals behave differently. Lead, mercury, aluminum, cadmium, nickel, arsenic, and chromium have no known biological functions for the physiological and biochemical processes in cells and are toxic even when present at low concentrations to living organisms (Salt et al., 1995). Elevated concentrations of these metals in the ecosystem cause considerable damages to crop plants and our health. Because they cannot be biologically or chemically degraded, they can be present in soils as free metal ions, soluble metal complexes, exchangeable metal ions, and insoluble compounds.

Pollution of the environment by heavy metals, particularly by lead, poses a serious threat to crop production and human health (Kaur et al., 2012; Evens et al., 2015). Upon release into the environment, it is accumulated in soils and easily absorbed by the plants. Roots are the first organs to face the lead. The first defense strategy is to stop the toxic ions entering the root tissues by excluding it (Mishra et al., 2006). Roots quickly show responses by forming barriers such as callose and lignification in cell-wall components. Under metal stress, the callose accumulation also inhibits cell-to-cell transport

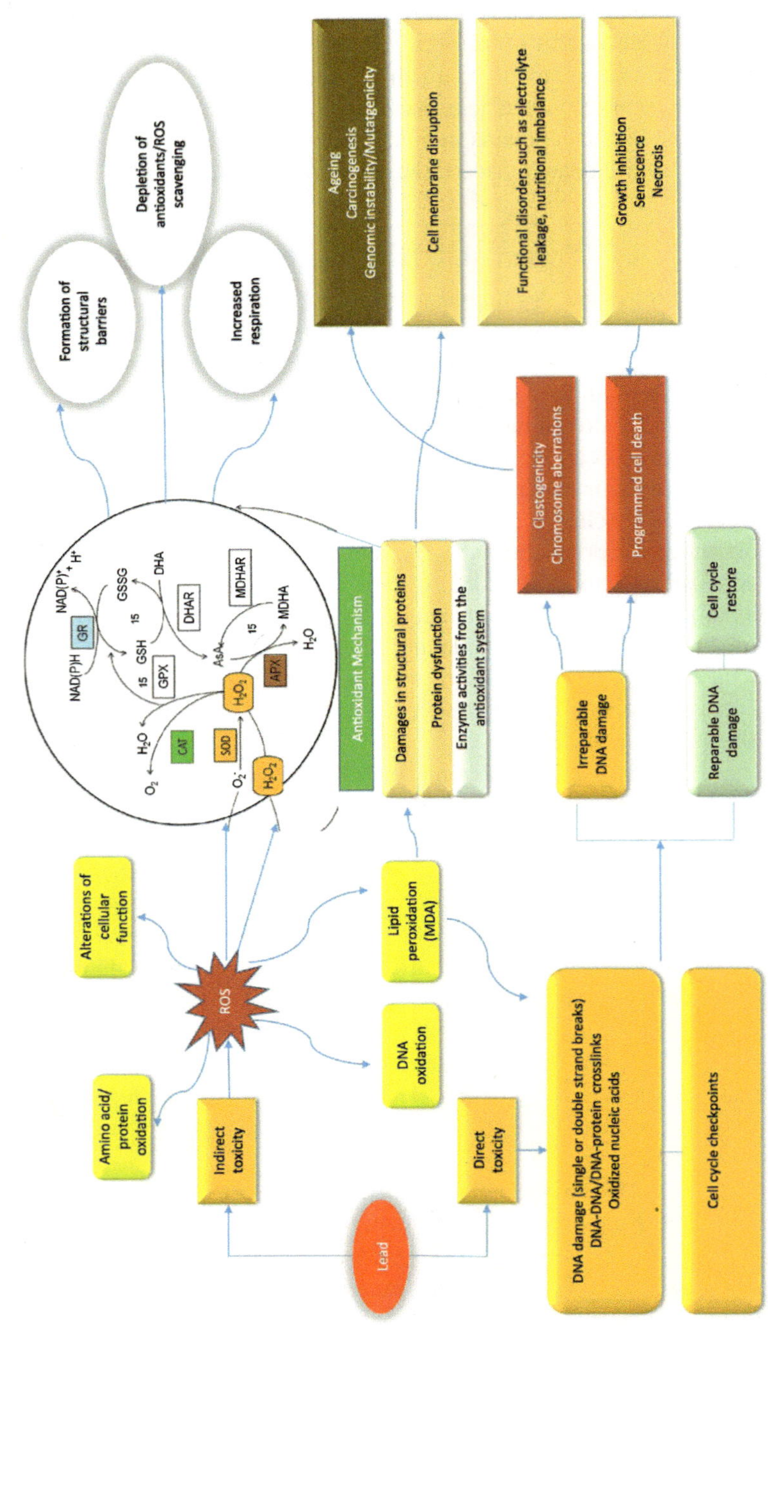

FIGURE 3.1

Effect of lead on oxidative stress and DNA damages in living organisms.

Adapted from the works of Gurer and Ercal (2000) and Monteiro et al. (2012) with permissions of Elsevier and American Chemical Society.

(Samardakiewicz et al., 2012). This could also prevent the transport of other ions. However, once the roots have been penetrated, lead may accumulate there and only a small fraction is translocated to aerial plant parts (Shahid et al., 2011). When lead gains entrance into the root cortex, it moves upward in the apoplastic space (Rodriguez et al., 2011) and also gains access to the symplastic space. It mainly moves by apoplast and follows water streams until it reaches the endodermis. Immobilization by negatively charged pectins within the cell wall (Islam et al., 2007; Arias et al., 2010), precipitation of insoluble lead salts in intercellular spaces (Islam et al., 2007), and sequestration in the vacuoles of rhizodermal and cortical cells (Kopittke et al., 2007) restrict the movement of lead in the root system. Apart from this, the endodermis also blocks the lead translocation with the Casparian strip and directs lead translocation into symplastic way. However, high concentrations of lead are able to destroy the physical barrier formed by the Casparian strip.

Although the mechanism of lead toxicity still needs to be elucidated, inhibition of seed germination, stunted growth, chlorosis, blackening of root system, decreased water potential, alterations in membrane permeability, disturbed mineral nutrition, nutritional deficiencies, impairment of enzyme activities, changes in hormonal status, damaged DNA, protein dysfunction, lipid peroxidation, increased ROS production, increased transpiration, and cell death have been reported as common symptoms (Sharma and Dubey, 2005; Pourrut et al., 2011; Sharma et al., 2014a,b; Table 3.1). Rapid exposure to a high concentration of lead or continuous exposure to low concentrations also results in morphological, physiological, and biochemical disorders as well as genetic changes in living organisms (Evens et al., 2015; Gautam et al., 2015). Lead also results in increases in the adenosine triphosphate/adenosine diphosphate ratio and decreases efficiency of photosynthesis in plants via the

Table 3.1 Effect of Lead on Human and Crop Plants

Species	Symptoms
Human	Depressed immune status, mental impairment, neuromuscular weakness, brain damage (Flora et al., 2007)
	Cancer (Kanuba et al., 2012)
	High blood pressure (Poreba et al., 2011)
	IQ loss, anemia, neurological disorders (Fewtrell et al., 2004)
	Competition for calcium, direct damage to mitochondria and mitochondrial membranes, generation of reactive oxygen species, alteration of lipid metabolism (Ahamed and Siddiqui, 2007a)
	DNA damage (Danadevi et al., 2003)
	DNA damage (Manikantan et al., 2010)
	Chromosome aberrations and birth defects (Johnson, 1998)
	Neurological, hematological, immunological, and pathological dysfunctions (Ahamed and Siddiqui, 2007b)
	Hypertension (Ahamed and Siddiqui, 2007b)
Pisum sativum	Reduction DNA synthesis (Gabara et al., 1995)
Vicia faba	Mitotic stage shortened (Hamid et al., 2010)
Nicotiana tabacum	DNA damage (Gichner et al., 2008)
Oryza sativa	Increased antioxidant enzymes (Verma and Dubey, 2003)

disruption of chloroplast ultrastructure and lipid composition in thylakoid membrane. As a result, a reduction of synthesis in chlorophyll and carotenoids is inevitable (Sharma and Dubey, 2005; Dezhban et al., 2015). Lead interferes with chlorophyll and antioxidant enzyme synthesis by either direct inhibition of enzymatic steps or by preventing the uptake of essential elements such as K^+, Mg^{2+}, Ca^{2+}, Fe^{2+}, and Zn^{2+} (Brunet et al., 2008; Bai et al., 2015; Gautam et al., 2015). Lead, like other heavy metals, could also replace the magnesium (Mg) atom in chlorophyll and impair its synthesis (Shakya et al., 2008). Decrease in uptake of essential elements by the increase of lead could also lead to nutrient imbalance, especially in micronutrient levels (Sharma and Dubey, 2005; Bai et al., 2015).

The elevated levels of malondialdehyde (MDA) and H_2O_2 could also be good markers in plants under metal stress. Although the effect of lead on protein metabolism needs to be clarified, high concentrations of lead decrease the protein pool (Piotrowska et al., 2009). This could be caused by ROS production, modification in gene expression, increased ribonuclease activity, and protein utilization by plants for the purpose of detoxification of metal stress (Gupta et al., 2009; Pourrut et al., 2011).

Based on chemical and physical properties of toxic metals, four different molecular mechanisms have been proposed.

1. Production of toxic oxygen derivatives such as ROS by autooxidation and catalysis by Fenton reaction, which accelerate generation of highly damaging of OH radicals from O_2 and H_2O_2 substrates that deplete antioxidant reserves (Matros et al., 2015).
2. Blocking of essential functional groups in biomolecules, especially in protein structures (Schutzendubel and Polle, 2002; Stobrawa and Lorenc-Plucinska, 2007). Biological function of the target molecule may also be altered because binding of lead ions to sulfhydryl groups in proteins leads to inhibition of antioxidant enzymes or disruption of protein structure (Aldoobie and Beltagi, 2013).
   For example, glutathione is a necessary substrate responsible for the metabolism of detoxification of toxins, in addition to acting as an important antioxidant molecule. The sulfhydryl complex of glutathione directly binds to toxic metals to inactivate them; however, in severe cases, heavy metals are able to inactivate the glutathione molecule that becomes unavailable as an antioxidant or as a substrate (Christie and Costa, 1984; Patrick, 2006). For example, Ahamed et al. (2005) reported that children with high blood lead levels had lower levels of δ-aminolevulinic acid dehydratase (ALAD) than those of with low blood lead levels. Depressed levels of ALAD correlated with depressed levels of glutathione. A positive correlation also existed between ALAD activity and MDA along with the increase of blood lead level (Yiin and Lin, 1995). Lead has also been shown to change the level of antioxidant enzyme levels. For example, elevations of superoxide dismutase (SOD), catalase (CAT), and glutathione peroxidase have been seen in lower levels of lead exposure, whereas the suppression of these enzymes has been characterized at higher exposure with longer periods (Li et al., 2015). Glutathione depletion and lead exposure have been shown to depress nitric oxide and cause hypertension in animal models (Vaziri et al., 2000). These enzymes might play important roles in photosynthesis, plant water status, mineral nutrition, and DNA repair pathways (Singh et al., 1997). For example, the replacement of nickel for Mg in some metalloproteins or in DNA, destroys their function and might lead to oxidative cellular or DNA damages (Rani and Goel, 2009).

3. Displacement of essential metal ions from biomolecules (Stobrawa and Lorenc-Plucinska, 2007).
4. Destabilization of nutrient and water uptake and their distribution in plants (Srivastava et al., 2014).

Although the effect of lead is toxic to all organisms, the most devastating effect of lead is in germinating seeds. It easily passes into the seed core through the seed coat and affects the process of germination (Gautam et al., 2015). Similar conclusions can be made for infants. Although the mechanisms outlined here have been made for plant responses, human response for lead toxicity shows similarities with some differences in cellular level (Gurer and Ercal, 2000).

In response to harmful ROS, aerobically growing organisms (eukaryotic and prokaryotic) have evolved multiple defense systems for protection of their cellular components (Schutzendubel and Polle, 2002). The first step is prevention of ROS formation. Protective proteins that remove ROS and secondary defense mechanism consisting of enzymes that remove and repair the products of oxidatively damaged components play important roles in this step (Matros et al., 2015). Tocopherol, a lipid-soluble nonenzymatic antioxidant, is a potential scavenger of ROS and lipid radicals (Gajewska and Sklodowska, 2007). Of four isomers of tocopherols (α-, β-, γ-, and δ-) (Hasanuzzaman et al., 2014), α-tocopherol has the highest antioxidative activity because of the presence of three methyl groups in its molecular structure. Although nonenzymatic antioxidant compounds (reduced glutathione, tocopherol, ascorbic acid, and carotenoids) have the ability to remove the ROS, the most efficient way to eliminate ROS is through the enzymatic antioxidant system (SOD, peroxidase (POD), CAT, and ascorbate peroxidase (APX)) either acting sequentially or simultaneously (Goncalves et al., 2007a,b; Bai et al., 2015). The inductions of ROS-scavenging enzymes can be used as biomarkers that reflect the affinity between antioxidative capacity and lead tolerance (Gajewska and Sklodowska, 2007; Li et al., 2015). SOD is the key enzyme located in various cell components and responsible for catalyzing the disproportionation of two O_2^- radicals to H_2O_2 and O_2 (Figure 3.1). The resulting H_2O_2 is further decomposed to water and oxygen by various antioxidant systems such as CAT, which is located in peroxisomes and glyoxysomes and mitochondria and by APX located in chloroplasts and cytosol (Huang et al., 2012; Kumar et al., 2013; Li et al., 2015). APX also decomposes peroxides by converting ascorbic acid to dehydroascorbate. Similarly, POD is involved in converting H_2O_2 to water and plays a significant role in lignification of cell walls. Similar findings were also made by Goncalves et al. (2007a,b) in cucumber seedlings under cadmium threat. Those detoxification processes are complex and highly compartmentalized in plant cells.

The antioxidant properties of plants exposed to various stress factors have been well-studied (Havaux and Kloppstech, 2001), but studies related to the effect of heavy metal-induced stress on vitamin and quality parameters in plants are still limited. It is crucially important to determine what level of lead toxicity or what stages in the crop plants affect the quality parameters such as vitamin contents and other nutritional parameters in the fruits. Because any transfer of heavy metals from leaves to fruits does not only reduce the quality parameters of fruits, it also put the lives of human and other organisms into great danger through the food chain. As much as this affects the present status of environment and human beings, it also puts our future into great danger (Bharwana et al., 2014; Hassan and Mansoor, 2014). For example, the medicinal plants providing raw materials for drug industries could endanger human life when they are collected from or cultivated in heavy metal-polluted areas. This could negatively affect the health conditions of people who use the metal-exposed herbal plants. A few plants have already been reported to accumulate heavy metals (Affholder et al., 2013; Okem et al., 2015). Therefore, it is crucially important to test the medicinal plants at first before exploiting them for medical purposes. Not only is its acute toxicity dangerous to living organisms but also its chronic activities should be considered carefully. Therefore, once they enter the ecosystem, their concentrations gradually increases and creates bigger problems for the next generation. Because plants in general do not have a sensory system to detect the properties of heavy metals, they either accumulate or exclude the metals after a certain level (Sheppard et al., 2009). Soil characteristics like low pH and abundance of organic matters can increase lead absorption (Azad et al., 2011).

Although adverse health effects caused by lead at high and moderate levels have long been recognized, its use has been restricted and limited in many countries; however, their low exposure has also great potential to create health disorders leading to cancer and cancer-related diseases in humans and other organisms. The major exposure pathways for workers are occupational inhalation and ingestion of lead-bearing dust and fumes (Kasuba et al., 2012). Thus, lead and its related compounds generate cytotoxicity and genotoxicity even at low doses in the long run. For example, indoor floor dust accounts for approximately 50% of children's total lead intake (Ahamed and Siddiqui, 2007a).

Low-level lead exposure significantly affects IQ, concentration ability, and hearing as well (Fewtrell et al., 2004; Flora et al., 2012). Again, Wang and Jia (2009) reported that DNA damages were evident in the testes of frog *Rana nigromaculata* exposed to low concentrations of $PbNO_3$. They concluded that low levels of lead exposure induced DNA damages were because of oxidative damage.

Although low concentrations also become toxic, very low concentrations might have a stimulating or inducing effect. For example, 5×10^{-8} mol/L Cd showed stimulating effects in barley seedlings (Kovacs et al., 2009). Similar findings were also made with other abiotic stress factors such as salinity (Dikilitas, 2003; Karakas, 2013) and drought (Kos et al., 2013; Son et al., 2012). Recent epidemiological studies have reported that low levels of lead exposure have a connection with several diseases such as hypertension, kidney disease, and neurodegenerative disease (Ahamed and Siddiqui, 2007b). For example, in rabbits, after inhalation exposure to lead oxide particles between 0.5 and 3.0 μg at 30 $\mu g/m^3$ for 4 days (blood lead level, 1.2–2.0 μg/dL), enhanced H_2O_2 and O_2^- production was noted (Ahamed and Siddiqui, 2007b). Ding et al. (1998) reported that low level of lead (3.2 μg/dL) induced hypertension and resulted in an increase in ROS in rats. Similarly, Wang and Jia (2009) suggested that lead had a potential to induce oxidative stress and DNA damage at very low concentrations in humans. Therefore, a low level of lead is capable of inducing more oxidative stress in the long term than that of a high level of lead in the short term; it also affects urinary δ-ALA levels, glutathione reductase, guaiacol peroxidase activities, and blood selenium levels. Although a low level of lead is toxic, its circulation in the human body is also long. For example, lead in blood has a half-life of 27 days, whereas the half-life of lead in bone is 30 years (Needleman, 2009).

3. GENOTOXICITY OF LEAD IN LIVING ORGANISMS

The capacity to withstand genotoxic damage is important to maintain genetic stability in population dynamics because maintenance of DNA integrity is essential for proper cellular and organismal function (Wurgler and Kramers, 1992). Unrepaired DNA damage can lead to mutations, cellular senescence, apoptosis, progression of cancer (Enoch and Norbury, 1995), and the process of aging (Cooke et al., 2003). Chronic effects of heavy metals not only cause oxidative stress, but they also could have the strong potential for affecting offspring of living organisms by transferring damaged

DNA. However, once stability of DNA integrity is interrupted, this could either be repaired, aggravated, or destroyed via apoptosis depending on the environmental conditions and source of stressors. If it is not repaired or destroyed, it results in mutation and eventually leads to cancer. DNA is the target of many toxic chemicals and is prone to be damaged because of constant exposure to both endogenous and exogenous sources. It could also be damaged spontaneously during metabolism. When metabolism slows down, the speed of increase in DNA damage increases. To repair damaged DNA or minimize the impact of damaged DNA, organisms are equipped with a variety of cellular defense mechanisms to prevent the structure of DNA. DNA damage results in inhibition of replication, transcription, or protein synthesis in the short term and mutation or chromosomal anomalies in the long term. High DNA damage is a contributing factor to the pathogenesis of certain diseases such as cancer and metabolic diseases. Therefore, a technique that enables the rapid determination of DNA damage and repair is of the utmost importance to determine the initial stage of genotoxicity to prevent further damages (see Section 4).

Genotoxicity in the cell occurs when there is an imbalance between DNA damage and DNA repair mechanisms. Therefore, stress factors, especially environmental stressors, affecting the genetic level of organisms are of concern because of impact on future generations and current toxicological studies. The potential repair capacity of an organism or the ability to withstand DNA damage shows differences among organisms because of variations in the organism's physiological, biochemical, and genetic backgrounds. Constantly produced ROS targets DNA molecules and affects transcriptional pathway. Increases in ROS result in further DNA damages; therefore, oxidative DNA damage is a good biomarker for monitoring the risk of development of many biotic and abiotic stress agents in advance. It is also a good biomarker for chronic effects of low concentration of toxicants. It is important to note that here not only acute effects of toxicants result in significant damages, but also chronic effects of low doses, so-called safe doses, result in similar damages on DNA, only the damaging effects is observed at a later stage, if not repaired. DNA damages are physical abnormalities in DNA such as single- and double-strand breaks. Oxidative DNA damage caused by ROS involves single- or double-stranded DNA breaks, purine, pyrimidine, and DNA cross-links. If a cell retains DNA damage, transcription of the gene can be prevented and therefore, translation into a protein would also be halted. As a result, replication would be blocked and may lead to genomic disturbances, carcinogenicity, and cell death (Figure 3.1). A cell that has accumulated a large amount of DNA damage can have three possible consequences: an irreversible state of dormancy known as senescence or aging, it could undergo apoptosis or programmed cell death, or it could undergo unregulated cell division, which can lead to the formation of tumor (Browner and Smith, 2013; Khan et al., 2015).

3.1 GENOTOXIC EFFECT OF LEAD ON CROP PLANTS

Lead induces a decrease in mitotic activity and shortens the mitotic stage and prolongs interphase, thus increasing the period of cell cycle (Patra et al., 2004) because lead is able to enter the nucleus and bind directly to the DNA or indirectly to protein (Małecka et al., 2008). After binding to DNA, lead disrupts DNA repair and replication mechanisms. Lead does not induce direct genotoxic effects until it becomes attached to naked DNA (Valverde et al., 2001). Rodriguez et al. (2011) stated that only the high level of lead (2000 mg/L) dosage induced genotoxicity. Similarly, Gichner et al. (2008) demonstrated that large concentrations of lead induced DNA strand breaks in tobacco plants. In *Vicia faba* roots, lead shortened the mitotic stage and prolonged the interphase and the cell cycle

(Hamid et al., 2010). At low concentrations (below 10 μM), lead increased the mitotic index, but decreased it at higher concentrations. The results of decline in DNA and RNA content are also evident with heavy metal stress. In plants, reduced efficiency of DNA synthesis, weaker DNA protection from damaged chromatin protein (histone), and increased DNase and RNase activities have been reported for Cd, Cu, Cr, Ni, Pb, Hg, Pt, and Zn (Schmidt, 1996; Abou Auda et al., 2002; Kumar et al., 2013).

3.2 GENOTOXIC EFFECT OF LEAD ON HUMANS

Lead has been used for many centuries and is still used for various industrial purposes because of its physical and chemical characteristics. Lead toxicity poses serious threats to human health because of occupational and environmental exposures, especially in the past 20 years. This toxic metal alters the immune response of animals and humans. People who are battery reconditioning workers, silver jewelry makers, paint, varnish, gasoline, and photocopy machine workers have been continuously exposed to lead toxicity; therefore, they are at high risk (Singh et al., 2000; IARC, 2006). Lead toxicity, in general, has cumulative deleterious effects at low levels of exposure, which can cause chronic degenerative changes in the nervous system, liver, kidneys, cardiovascular, renal, immune, and reproductive systems, as well as on bones and teeth (Hunaiti et al., 1995; Hsu and Guo, 2002). Although blood lead levels in the population show an indication of exposure, the level of lead has been declining following strict measurements; however, chronic low-level lead has been affecting a great part of the human population.

Although the mechanism of lead toxicity in humans still remains elusive, enzymatic inhibition, impaired antioxidants metabolism, and oxidative stress may play a role. As a result, adverse health effects occur in organisms through the formation of free radicals, lipid peroxidation, and depletion of protein sulfhydryls (e.g., glutathione). When an overproduction of free radicals is induced by lead exposure, these unstable molecules may not only deplete the antioxidant defenses but might also react directly with macromolecules such as lipids, proteins, and DNA, leading to the increase of peroxides, such as MDA. High blood lead levels also result in high DNA damages in lymphocytes (Figure 3.1). When the sexes were compared, women with high blood lead levels had higher DNA damages than men (Ruzica et al., 2007).

The genotoxic potential of lead has been examined in several studies. In vitro and in vivo studies indicate that even if lead compounds are not directly genotoxic, they cause genetic damage through oxidative stress (Clement et al., 2010). Lead is mobile in the blood stream and exerts adverse effects on the body. It has now been generally accepted that under the conditions of prolonged exposure, an individual's blood lead level reflects the quantity of "biologically active" form of lead in their body. The permissible limit of blood lead level is 10 μg/dL (Gurer and Ercal, 2000; Jin et al., 2006). However, adverse effects involving the nervous system can be observed even below this level in some sensitive populations. In children, this level is much lower. Hemoglobin synthesis is negatively affected at 50 μg/dL, whereas 40 μg/dL is enough to affect hemoglobin synthesis in children (Gurer and Ercal, 2000). The top limits of lead to kill the adults and children has been accepted as 150 and 120 μg/dL, respectively (Gurer and Ercal, 2000). Although lead below 10 μg/dL is accepted as a nontoxic dose in adults, this low level could result in significant permanent damages in hearing, growth, and IQ levels if chronic exposures carry on at this level (Evens et al., 2015). This has quite a remarkable effect on children younger than age 6 years because they absorb far more lead from their environments than adults. Therefore, they tend to develop permanent developmental and neurological problems while their central nervous systems are developing, whereas many of the symptoms in adults are reversed when exposure is ceased (Jin et al., 2006).

It is well known that a 90% body burden of lead is deposited in bone (Martinez et al., 2013; Evens et al., 2015). The level of lead in the bones can be reduced during growth and development of children. Distance of residence from traffic density, ethnicity, housing, poor nutrition, and ignorance about the toxicity of lead are the main factors further influencing blood lead levels in children. One of the main targets for lead toxicity is the hematological system. Lead affects this system by inhibiting heme and hemoglobin biosynthesis. In this pathway, the enzyme that is very sensitive to lead toxicity is ALAD (Martinez et al., 2013). At low blood levels, lead inhibits ALAD, one of the most important enzymes catalyzing the condensation of two molecules of δ-ALA to porphobilinogen and results in depressed heme formation. Elevated levels of ALA can be oxidized to generate ROS (Bechara, 1996; Gurer and Ercal, 2000). As a consequence, enhanced lipid peroxidation, DNA damage, and altered calcium and sulfhydryl hemostasis may occur.

Lead impregnation in children has declined significantly in many developed countries following withdrawal of leaded gasoline (Martinez et al., 2013). An appreciable downward trend in the body burden of lead in their population has been reported. However, workers with occupational lead exposure have a still higher frequency of infertility, stillbirths, miscarriages and spontaneous abortion, and reduced sperm counts and mobility (Wu et al., 2012).

4. RAPID ASSESSMENT OF LEAD DAMAGE ON DNA

Because we are surrounded by a tremendous number of toxic chemicals and exposed to their toxic effects continuously, we need to determine their toxicity on DNA structure if the offspring has the potential to be affected. Any stress factors, whether oxidative stress or genotoxicants, have the potential to result in DNA damages, depending on the severity, duration, and defense structures of the host. Therefore, we need to apply more appropriate, quick, and reliable tests to assess the damaging effects of toxicants in living organisms.

The use of commercial cytotoxicity methods with animals or tissue culture cells is associated with enormous cost and labor demands. Therefore, introduction of an easy, simple and a reliable cell-based method as a preliminary step to assess the condition of DNA would be great importance with respect to cost, labor, and time. One of the most common and reliable methods used in animal and human genotoxic studies is through measuring DNA damages via comet assay. The comet assay is widely used for specifying oxidative DNA damage at a single-cell level (Gichner et al., 2008). The assay, also known as single-cell gel electrophoresis, is a simple, sensitive, and reliable method for measuring the damages quantitatively and qualitatively at the DNA damage level. This method is frequently used to analyze and monitor the environmental health through examining human, animal, and plant cells (Figure 3.2). Recent studies showed that the effect of low concentrations of toxicants on DNA can be detected in a wide range of species (Dikilitas et al., 2015). The comet assay can detect double- or single-stranded DNA breaks at neutral and alkaline conditions (Singh et al., 1988). DNA double-strand breaks are very critical for cells and usually are lethal. For example, Danadevi et al. (2003) detected a significant increase in DNA breaks in workers exposed to lead by the comet assay. Similar tests were also undertaken on plants and microbial organisms (Gichner et al., 2008). DNA strand breaks could well correlate with cell death, aging, and cancer. The assay is based on the electrophoretic separation of relaxed strands of damaged DNA, which eventually forms a comet-like halo after the

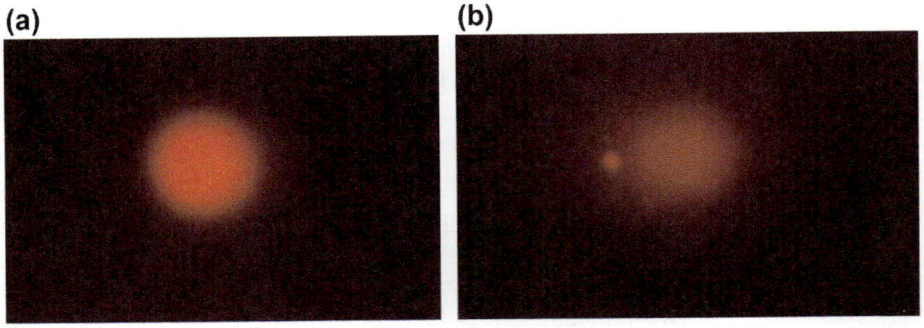

FIGURE 3.2

Heavy metal toxicity in DNA of human lymphocytes assessed by single-cell gel electrophoresis. (a) Control and (b) damaged DNA.

electrophoresis step (Kocyigit et al., 2015). The detailed protocols are given in the works of Dikilitas et al. (2015) and Kocyigit et al. (2015). The outline of the assay is as follows: it involves mixing cells with agarose, making microgels on a microscope slide, lysing cells in the microgels with salts and detergents, removing proteins from DNA by using proteinase K, unwinding/equilibrating and electrophoresing DNA, and then fixing the DNA and visualizing it with a fluorescent dye and finally analyzing the migration patterns of DNA from individual cells with an image analysis system.

Although great measures were taken to reduce the impact of lead toxicity, it is still a major environmental and health problem in many countries; therefore, any significant changes in toxicity level reflected on DNA should be measured accurately and determined carefully if the restrictions of heavy metals have reached its purpose. Because the low level of concentrations of toxic compounds could generate toxicity at cellular and DNA level potentially. Therefore, a good environmental and human health monitoring, the assay described here is one of the best methods available.

5. STIMULATING AND INTERACTIVE EFFECTS OF LEAD ON CROP PLANTS AND MICROORGANISMS AND ITS POSSIBLE TOLERANCE BY THE MICROORGANISMS

Several studies have indicated that very low concentrations of toxicants have stimulating or inducing effects on plant growth. For example, arsenic stimulated the growth of plants, whereas excessive concentrations damaged to cellular functions (Han et al., 2002). Similar findings were made with that of Cd (5×10^{-8} mol/L) in barley seedlings (Kovacs et al., 2009). Again, Li et al. (2007) showed that germination energy and germination index increased at 0–0.5 mg/kg concentrations of arsenic, whereas vitality index, length, and biomass of root and shoot were all increased at much lower concentrations of arsenic (0–1 mg/kg). Other abiotic stress factors such as short-term drought or low concentrations of salinity were also reported to have promoting effects on crop plants. For example, Dikilitas (2003) and Karakas (2013) reported that low level of NaCl salinity (25 mmol/L) resulted in slightly increased germination rates in lucerne and tomato plants compared with those of their controls germinated in nonsaline conditions. Many research findings reported that the low level of stress had encouraging or promoting effects on the organisms. For example, fungal mycelial development of

Verticillium alboatrum exhibited slightly better performance at 25 mmol/L of NaCl in liquid culture of Dox medium (Dikilitas, 2003, personal communication with Dr Mike Milton, UK).

As in the case of salinity, drought, or heat stress, low concentrations of heavy metal toxicity might also exhibit similar effects on microorganisms whether they are beneficial saprophytes or pathogens. Low concentration of heavy metals could increase the germination of conidia and mycelial development as well as affecting the biochemical behaviors of microorganisms. There is evidence that low levels of salinity result in increases in fungal enzyme and toxins (Kos et al., 2013). Drought stress, on the other hand, may have a similar positive effect on pathogen growth and proliferation: the slightly stressed fungi may grow better and become more virulent (Kos et al., 2014). Microorganisms, in general, have a high adaptation ability to harsh conditions and therefore, they exhibit higher tolerance than other organisms such as plants or human beings (Egamberdieva et al., 2014). It has been reported that fungi and yeast tolerated heavy metal stress remarkably well (Wan et al., 2015). They could easily adapt and grow under various extreme conditions of pH, temperature, heat, salinity, drought, nutrient deficiency, and high metal and pesticide toxicity (Dikilitas and Karakas, 2012). Their cell-wall material shows excellent metal-binding properties; therefore, they respond to heavy metal toxicity rapidly. They precipitate and transform them into nontoxic forms or absorb them to cell walls or entrap them to the extracellular capsules. Various studies showed that microorganisms isolated from heavy metal contaminated sites had a higher ability to adapt to the conditions of heavy metal-polluted areas (Wan et al., 2015; Zhan et al., 2015). This behavior has been turned into an advantage and used for the removal of heavy metals from polluted areas. For example, El-Morsy (2004) reported that a species isolated from polluted water, *Cunninghamela echinulata*, could be used as a biosorbent of metal ions in wastewater. In the same way, Zafar et al. (2007) reported remarkable biosorption for Cd and Cr by *Aspergillus* sp. and *Rhizopus* sp., respectively, from metal-contaminated agricultural soils. Similarly, Akar and Tunali (2006) and Sugasini et al. (2014) stated that *Aspergillus* isolates adapted to heavy metal-polluted areas were reported to be very tolerant to heavy metals in vitro and exhibited strong growth, often exceeding the control (isolates grown in agar medium without heavy metals). Their minimum inhibitory concentration (MIC) values were remarkably high; ranging from 20 to 25 mmol/L. These traits made them attractive potential candidates for their ability to remove heavy metals from contaminated environments. However, it should be noted that the microbial response to heavy metal toxicity depends on the concentration, availability, and the nature of metals and microorganisms.

It should not be ignored that the pathogens might get also stimulated and secrete more toxins or cell wall–degrading enzymes under heavy metal toxicity. In that case, it is possible that even resistant plants become predisposed to the effect of diseases. Heavy metal stress and the pathogen could act together and their additive effects could then be sufficient to cause injury as none existed before. The recovery process of the host is also impaired. Apart from the negative effects of combined stress, the possible interactions (e.g., abiotic stress promotes the biotic one) between stressors make the situation worse and more complex (Figure 3.3). The end-product released upon heavy metal toxicity such as sugar and amino acids could be a good source of substrate for the biotic stress agents that encourage the growth, sporulation, and toxin production of pathogens (Son et al., 2012). It is also known that many types of yeast and fungi produce polyols for maintaining cell integrity and function (Hoppenau et al., 2014). For example, thiamine biosynthesis has also been reported to enhance the stress tolerance of the fungus, especially, against ROS, UV radiation, and drought; therefore, it is quite possible that production of polyols under heavy metal stress could play an important role in maintaining the virulence of the pathogens.

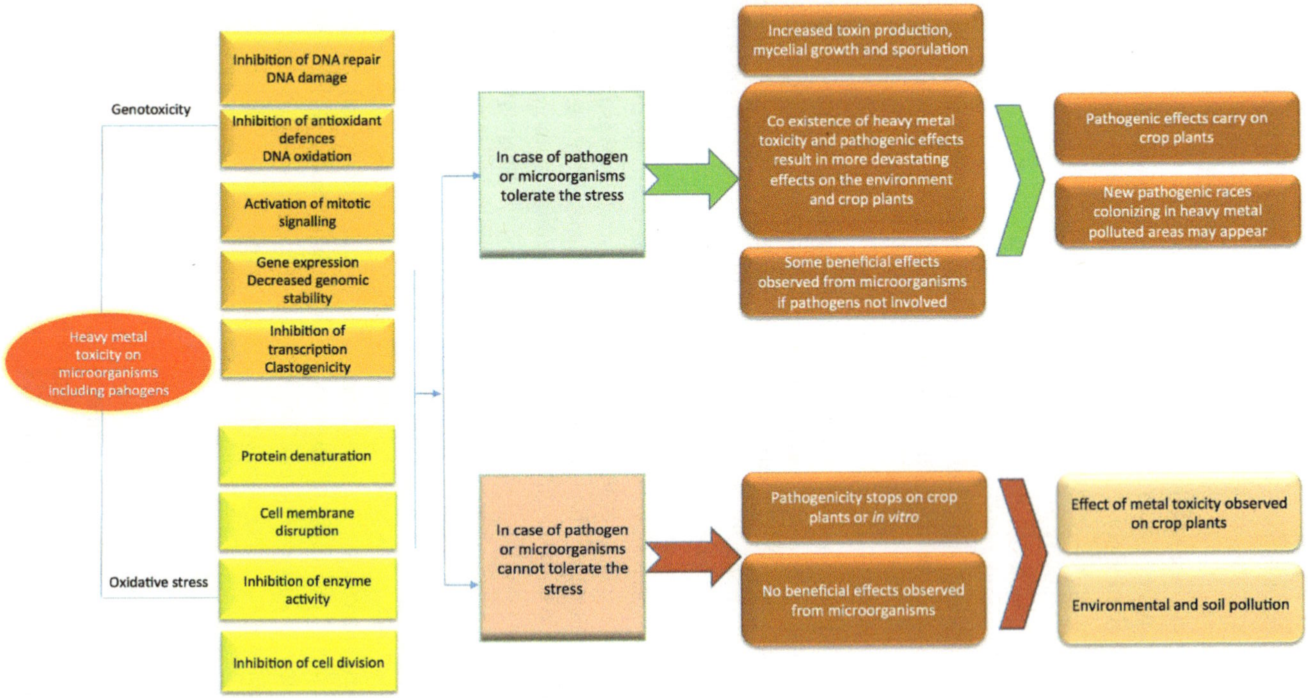

FIGURE 3.3

Major mechanisms.of lead toxicity in microorganisms and its possible interactions with pathogens.

Adapted from Beyersmann and Hartwig (2008) and Rodriguez et al. (2011) with some modifications.

Exposure to environmental chemicals and oxidative stress, particularly at low-dose levels, may produce additive or synergistic interactions not seen in single-component exposure (El-Sonbaty and El-Hadedy, 2012). For example, exposure to cadmium, lead, and UV rays occurs in many occupational settings, such as battery production, galvanization, and recycling of electric tools. This could increase the toxicity of heavy metals. For example, the interactions of ultraviolet rays of 254 nm (UVB) with cadmium or lead on *Bacillus cereus* showed that the combination of stressors resulted in DNA damage, cytotoxicity, depletion of glutathione, and formation of lipid peroxidation. UVB rays alone enhanced glutathione production, which was depleted with lead and high doses of cadmium. Lead alone did not increase DNA damage. The mechanism behind these interactions might be repair inhibition of oxidative DNA damage because a decrease in repair capacity will increase susceptibility to ROS generated by cadmium or lead.

6. IMPROVEMENT OF CROP PRODUCTION UNDER LEAD-POLLUTED AGRICULTURAL SOILS

Plants have specialized mechanisms for essential metals to transport, maintain, and sequestration within cells, and they have also a control mechanism for the excessive accumulation for nonessential metals. Therefore, they have capacity to survive in metalliferous environment and are able to accumulate increased amount of heavy metals as part of an ecophysiological adaptation mechanism. For example, *Phaseolus vulgaris* has been reported to be a good accumulator of lead and arsenic (De Temmerman et al., 2015). Similarly, *Potamogeton crispus* L. did not show oxidative stress upon response to lead toxicity (Qiao et al., 2015). Although sensitive and tolerant plants behave differently against heavy metal stress in general, three different mechanisms have been proposed, ranging from exclusion, inclusion, and accumulation of heavy metals. Inclusion and accumulation of heavy metals could be interpreted as immobilization, chelation, and compartmentalization (Jing et al., 2007; Solanki and Dhankar, 2011). However, in sensitive plants, a high concentration of these metals may inhibit enzymes involved in photosynthetic and other biochemical reactions.

During the past decade, much progress has been made in understanding molecular mechanisms related to metal contamination, transport, and defense of plants under heavy metal stress. However, for heavy metal stress, new strategies for phytoremediation and chemical approach should be developed. An effective metal phytoremediation strategy depends on the ability of plants to tolerate and accumulate metals from the environment. So far, several approaches have been applied to remediate heavy metals from contaminated soils. The remediation processes take quite a long time and energy because of the complexity of heavy metal pollution. In recent years, halophytes have been used to remediate heavy metal-polluted soils. Salt-resistant or salt-accumulator halophytes could accumulate or excrete heavy metals (Table 3.2). Phytoremediation is considered as a green alternative solution with a low-cost and environmentally friendly method for remediation of metal-contaminated soils. It offers a cheap and efficient option for cleaning sites contaminated with toxic elements. There is also a great need to find new plant species for phytoremediation and to understand the mechanisms involved in processes such as tolerance, accumulation, exclusion, and metabolism of toxic metals in plants (Jing et al., 2007). Phytoremediation of heavy metals may involve several forms: phytoextraction, rhizofiltration, phytostabilization, and phytovolatilization. Phytoextraction refers to processes of concentrating metal ions in plants in their roots and shoots; rhizofiltration refers to process of absorbing,

Table 3.2 Various Sources for Lead Remediation

Plant Species	Area of Application
Carex pendula	Wastewater (Yadav et al., 2011)
Scirpus americanus	Santos-Diaz and Barron-Cruz (2011)
Phaseolus vulgaris	Contaminated soil (De Temmerman et al., 2015)
Hirschfeldia incana	Contaminated soil (Auguy et al., 2013)
Brassica juncea	Singh et al. (2010)
Pseudomonas fluorescens	Contaminated soil (Belimov et al., 2005)
Atriplex halimus L.	Manousaki and Kalogerakis (2009)
Vitamin C and selenium intake in higher organisms	Yuan et al. (2013), Patrick (2006)
Aspergillus niger	Contaminated soil (Anahid et al., 2011)
Trichoderma harzianum	Chauhan and Rai (2009)

concentrating, or precipitating metals from effluents; and phytostabilization refers to the use of plants that reduce the mobility of heavy metals through absorption and precipitation; phytovolatilization refers to process in plants which uptake and release volatile materials such as mercury- or arsenic-containing compounds into the atmosphere.

The ideal plant for phytoextraction that should grow rapidly, produce a high amount of biomass, and be able to tolerate and accumulate high concentrations of metals in shoots. Although hyper-accumulator plants have exceptionally high metal–accumulating capacity, most of these have a slow growth rate and often produce limited amounts of biomass when the concentration of available metal in the contaminated soil is very high. Rhizobacteria has been used recently in metal-polluted areas. When they are combined with halophytes, their efficiency has been greatly increased (Hu et al., 2012). In heavily contaminated soils, where the metal content exceeds the limit of plant tolerance, this approach has a good potential to increase plant biomass and thereby stabilizing, revegetating, and remediating metal-polluted soils. Because halophytes have higher levels of activity for SOD, CAT, and POD, they have lower levels of damage to their lipid membrane from ROS under metal stress (Karakas, 2013). Generally, antioxidant systems of halophytes are more efficient than those in glycophytes at scavenging free radicals. Several poplar and willow species were also reported to accumulate or tolerate high levels of heavy metals because of their extensive root systems, fast growth rates, easy propagation, and high biomass production (Munzuroglu et al., 2008). Singh et al. (2010) showed that in vitro grown *Brassica juncea* plants could effectively phytoextract Cu from solutions. Similar findings were also made with *Salsola passerina* and *Chenopodium album* L. for removal of lead (Hu et al., 2012). They stated that *S. passerine* showed more tolerance than *C. album* and that such a tolerance (1000 mg/L lead) was associated with the stronger activity of the antioxidant enzymes and less MDA accumulation. Again, high levels of lead tolerance were also found with *Avicennia marina*, which is a lead-adapted species and could tolerate 800 mg/L of lead stress (Yan et al., 2010). For example, *Acacia farnesiana* L., one of the newly introduced heavy metal accumulators, has been reported to have a high ability to tolerate lead (EC50 = 766 mg/L lead) and accumulated more than 80% of the lead in roots (up to 51.928 mg/kg of air dried tissue) in seedlings exposed to 1000 mg/L lead (Maldonado-Magana et al., 2011).

Several hyperaccumulator plant species, such as *Brassica pekinensis* and *Pelargonium*, are capable of translocating higher concentrations of lead to aerial plant parts without incurring damage to their basic metabolic functions (Arshad et al., 2008). A specific hyperaccumulator species can accumulate more than 1000 ppm lead (Maestri et al., 2010). In fact, these plants exclude substances from root that dissolve metals in soil (Arshad et al., 2008).

To remediate the heavy metal-affected soils and improve the conditions of crop plants, plants with high tolerance to heavy metals have been used recently. For example, the Mediterranean saltbush, *Atriplex halimus* L., a C4 perennial native shrub of the Mediterranean basin, has excellent tolerance to drought and salinity and has been used to remove Pb and Cd from contaminated soils (Manousaki and Kalogerakis, 2009).

Efficacy of plants that would be used for the phytoremediation studies could be increased. For example, Yuan et al. (2013) stated that coleus plants could be useful in phytoremediation of aquatic systems contaminated with lead, especially with addition of low concentrations of selenium (Na_2SeO_3). Similarly, application of salicylic acid played a positive role in rice seedlings under lead toxicity (Jing et al., 2007). John et al. (2008) stated that *Lemna polyrrhiza* L. (duckweed) can be effectively used as a phytoremediator for wastewater polluted with more than one heavy metal at moderate concentrations (30 and 40 mg/L for cadmium and lead, respectively).

Although genotoxic studies have shown that lead is strong genotoxicant and has highly damaging effects on DNA structures, recent studies showed that supplementation of selenium with N-acetylcysteine (NAC) could improve the lead-induced biochemical oxidative stress in blood and tissue, the burden of lead on the body, and molecular alterations by recoupment in DNA damage (Sharma et al., 2014a,b). Similar findings were also made with vitamin C on tobacco plants exposed to nitrobenzene (Yuan et al., 2011). Abdul-Hamid and Salah (2013) stated that lycopene reduced the DNA damage of thyroid cells as observed from the significant decrease in the comet assay. Selenium (Se) is an important element for human and animal nutrition based on its presence in antioxidative defense systems, but it is not considered essential for plants. Nevertheless, several studies have reported the beneficial effects of Se because it increased antioxidant activity in plants and humans. In *Brassica napus* seedlings, Se was found to reverse the Cd-induced decrease in fresh mass and changes in lipid peroxidation as well as changes in the DNA methylation pattern (Filek et al., 2010). Studies in animals have also shown that Se is one of the potential antagonists to Pb, Cr, Hg, and Cd (Yuan et al., 2013).

Chemical application and agronomical crop management practices have also been used to alleviate metal toxicity with some success. Exogenous application of plant hormones provided protection to plants against abiotic stress and increased crop yield (Elobeid et al., 2012). Application of Ca^{2+} could also stabilize lipid bilayers and thus provide structural integrity to cellular membranes by binding to the phospholipids (Hepler, 2005). Ca^{2+} application was also reported to compete with lead for entry into the root cells. When Ca^{2+} was supplied in the medium, it reduced lead uptake and toxicity (Kim et al., 2002).

Phytoremediation strategies also moved on the new areas including microbial remediation. Bioremediation using microbes well adapted to diverse physiological conditions could be used for remediation of heavy metal-contaminated areas. With this approach, microorganisms found to play important roles for the bioremediation could be an alternative solution, although they cannot degrade hazardous compounds into harmless ones but they can uptake heavy metals from the polluted environment. Microorganisms have evolved several mechanisms which can immobilize, mobilize, or transform heavy metals (El-Sonbaty and El-Hadedy, 2012). For example, Anahid et al. (2011) stated that *Aspergillus niger* was found to be tolerant to heavy metal stress. Similarly, Iram et al. (2012) stated that *Aspergillus flavus*, *A. niger*, and *Fusarium* were found to be tolerant to chromium and lead. Thus, tolerant isolates could be used for the purpose of bioremediation of heavy metal-contaminated areas. Similar findings were also made by Iram et al. (2013a,b). They also stated that among the fungal strains, *A. niger* was the most tolerant against lead, with MIC of 600 mg/L, and *A. flavus* was the second best against chromium with MIC of 400 mg/L. Fungi found successful in biocontrol studies such as *Trichoderma* spp. were also used for the remediation of heavy metal-contaminated soils. For example, Chauhan and Rai (2009) stated that the ability to remove heavy metals increased with the use of *Trichoderma harzianum*. Again, Chandrakar et al. (2014) reported that fungal species *Aspergillus* sp., *Fusarium* sp., and *Cladosporium* sp. had exhibited higher tolerance in liquid media against the heavy metals lead and chromium. They suggested that they could be attractive potential candidates for removal of metals from contaminated water. Because of this property, fungi have great importance in polluted habitats and have potentials for binding heavy metals.

Although it is not the subject of this chapter, lead toxicity could also be reduced in human beings with various antioxidant supplements such as vitamin B6, zinc, vitamin C, and NAC (Gurer and Ercal, 2000).

Compared with other organ systems, the nervous system is the most sensitive target for lead toxicity (Flora et al., 2012). Both the central and peripheral nervous system become affected by lead exposure. Therefore, brain damage is a direct consequence of lead exposure. Major symptoms include headache, dullness, loss of memory, and hallucinations (Flora et al., 2012).

Lead-induced genotoxical or toxicological stress could be relieved via the use of antioxidants in nutrients intake. Vitamin C has been known to have free-radical scavenger effect and has been shown to inhibit lipid peroxidation in liver and brain tissue of lead-exposed animals. For example, a minimum of 500 mg/L intake of vitamin C via drinking water reduced the levels of ROS in lead-exposed rats (Hsu et al., 1998). Vitamin C intervention also normalized blood ALAD levels and resulted in a significant decrease in blood lead content. Again, vitamin B1 (thiamine) and vitamin C intakes, as supplements, reduced the symptoms of lead toxicity (anemia and abdominal colic) in workers with mean blood lead levels, 32.84 μg/dL (Tandon et al., 2001). However, the lead-lowering effect of vitamin C has not been proven in occupationally exposed workers with high blood lead levels. For example, daily intake of 1 g vitamin C orally once daily, 5 days a week for 20 weeks or 60 mg zinc intake daily for 8 weeks neither reduced the blood lead level nor improved the lead-affected metabolism in workers with high blood lead levels (29–76 μg/dL) exposed to long-term lead toxicity (Lauwerys et al., 1983). In addition to beneficial effects of vitamin C, it has also an inhibiting effect on lead uptake on a cellular level (Patrick, 2006). Vitamin C combined with silymarin has also been shown to reduce the hepatotoxic effect of lead poisoning (Shalan et al., 2005). Antioxidants such as vitamins C, E, B1, B6; taurine methionine; selenium; zinc; alpha-lipoic acid; and NAC have also been shown to lower ROS-generated cellular damage.

7. CONCLUSION AND FUTURE PROSPECTS

Lead continues to pose a serious threat to the environment and our health. Although the population exposure to lead has fallen down over the past decades with the introduction of low lead paint, unleaded gasoline, and safe pesticides etc. In the 1960s, a blood lead level of 60 μg/dL was considered safe (Ahamed and Siddiqui, 2007a). Because of the consciousness of lead toxicity, the acceptable blood lead level was initially reduced to 25 μg/dL and finally it was reduced to 10 μg/dL. However, an even lower lead level of 10 μg/dL is still capable of inducing oxidative stress and cause disorders in organisms. Because lead is not decomposed, it is gradually accumulated in the environment and agricultural soils. Therefore, even low exposure to lead is going to create devastating effects on population and the environment. Accumulation of lead in soil also affects our life indirectly through the food chain apart from the direct toxicity of lead. Therefore, it is important to determine its toxic potentials and effects via a reliable way on the cell and even on target molecule, DNA. Low levels of lead could also interact with other biotic and abiotic stress agents. In that case, a double effect of the stress was not seen before may have come out and low level of lead stress encourage the development of pathogenic races of microorganisms. New heavy metal–tolerant races of pathogens may result in devastating effects on crop plants as these type of interactions observed between drought and pathogens, salt and pathogens, etc. For the remediation of soil and environment from heavy metals, microorganisms, halophyte plants, and chemical application to problematic areas have been suggested. However, to increase the efficacy of microorganisms or halophytes, their combinations have also been used recently.

For future prospects, it is clearly understood that lead accumulation is a continuous process either in humans or the environment. Therefore, new biotechnological approaches such as genetically or biochemically modified plants or microorganisms, or their combinations should be taken into consideration to remove lead or other heavy metals from contaminated areas. Sensitive techniques such as comet assay should be used in every toxicological study if low levels of stress agents take part in DNA damaging.

ACKNOWLEDGEMENT

The authors are grateful to the Elsevier and American Chemical Society for giving permissions to use the figures.

REFERENCES

Abdul-Hamid, M., Salah, M., 2013. Lycopene reduces deltamethrin effects induced thyroid toxicity and DNA damage in albino rats. J. Basic Appl. Zool. 66, 155–163.

Abou Auda, M.M., Symeonidis, L., Hatzistavrou, E., Yupsanis, T., 2002. Nucleolytic activities and appearance of a new DNase in relation to nickel and manganese accumulation in *Alyssum murale*. J. Plant Physiol. 159 (10), 1087–1095.

Affholder, M.C., Prudent, P., Masotti, V., Coulomb, B., Rabier, J., Nguyen-The, B., Laffont-Schwob, I., 2013. Transfer of metals and metalloids from soil to shoots in wild rosemary (*Rosmarinus officinalis* L.) growing on a former lead smelter site: human exposure risk. Sci. Total Environ. 454, 219–229.

Ahamed, M., Siddiqui, M.K.J., 2007a. Environmental lead toxicity and nutritional factors. Clin. Nutr. 26 (4), 400–408.

Ahamed, M., Siddiqui, M.K.J., 2007b. Low level lead exposure and oxidative stress: current opinions. Clin. Chim. Acta 383, 57–64.

Ahamed, M., Verma, S., Kumar, A., Siddiqui, M.K., 2005. Environmental exposure to lead and its correlation with biochemical indices in children. Sci. Total Environ. 346, 48–55.

Akar, T., Tunali, S., 2006. Biosorption characteristics of *Aspergillus flavus* biomass for removal of Pb(II) and Cu(II) ions from an aqueous solution. Bioresour. Technol. 97, 1780–1787.

Aldoobie, N.F., Beltagi, M.S., 2013. Physiological, biochemical and molecular responses of common bean (*Phaseolus vulgaris* L.) plants to heavy metals stress. Afr. J. Biotechnol. 12 (29), 4614–4622.

Alissa, E.M., Ferns, G.A., 2011. Heavy metal poisoning and cardiovascular disease. J. Toxicol. http://dx.doi.org/10.1155/2011/870125.

Anahid, S., Yaghmaei, S., Ghobadinejad, Z., 2011. Heavy metal tolerance of fungi. Sci. Iran. C 18 (3), 502–508.

Arias, J.A., Peralta-Videa, J.R., Ellzy, J.T., Ren, M., Niveros, M.N., Gardea-Torresdey, J.L., 2010. Effects of *Glomus deserticola* inoculation on prosopis: enhancing chromium and lead uptake and translocation as confirmed by X-ray mapping, ICP-OES and TEM techniques. Environ. Exp. Bot. 68 (2), 139–148.

Arshad, M., Silvestre, J., Pinelli, E., Kallerhoff, J., Kaemmerer, M., Tarigo, A., Shahid, M., Guiresse, M., Pradere, P., Dumat, C., 2008. A field study of lead phytoextraction by various scented *Pelargonium* cultivars. Chemosphere 71 (11), 2187–2192.

ATSDR, 2012. Priority List of Hazardous Substances. Agency for Toxic Substances and Diseases Registry. Available at: http://www.atsdr.cdc.gov/.

Auguy, F., Fahr, M., Moulin, P., Brugel, A., Laplaze, L., El-Mzibri, M., 2013. Lead tolerance and accumulation in *Hirschfeldia incana*, a mediterranean *Brassicaceae* from metalliferous mine spoils. PLoS One 8 (5), e61932.

Azad, H.N., Shiva, A.H., Malekpour, R., 2011. Toxic effects of lead on growth and some biochemical and ionic parameters of sunflower (*Helianthus annuus* L.) seedlings. Curr. Res. J. Biol. Sci. 3 (4), 398–403.

Bai, X.Y., Dong, Y.J., Wang, Q.H., Xu, L.L., Kong, J., Liu, S., 2015. Effects of lead and nitric oxide on photosynthesis, antioxidative ability, and mineral element content of perennial ryegrass. Biol. Plant. 59 (1), 163–170.

Bechara, E.J., 1996. Oxidative stress in acute intermittent porphyria and lead poisoning may be triggered by 5-aminolevulinic acid. Braz. J. Med. Biol. Res. 29, 841–851.

Belimov, A., Hontzeas, N., Safronova, V.I., Demchinskaya, S.V., Piluzza, G., Bullitta, S., Glick, B.R., 2005. Cadmium-tolerant plant growth promoting bacteria associated with the roots of Indian mustard (*Brassica juncea* L. Czern.). Soil Biol. Biochem. 37, 241–250.

Beyersmann, D., Hartwig, A., 2008. Carcinogenic metal compounds: recent insight into molecular and cellular mechanisms. Arch. Toxicol. 82, 493–512.

Bharwana, S.A., Ali, S., Farooq, M.A., Iqbal, N., Hameed, A., Abbas, F., Ahmad, M.S.A., 2014. Glycine-betaine induced lead toxicity tolerance related to elevated photosynthesis, antioxidant enzymes suppressed lead uptake andoxidative stress in cotton. Turk. J. Bot. 38, 281–292.

Browner, I.S., Smith, T.J., 2013. Symptom assessment in elderly cancer patients receiving palliative care. Ann. Oncol. 24, 25–29.

Brunet, J., Repellin, A., Varrault, G., Terryn, N., Zuily-Fodil, Y., 2008. Lead accumulation in the roots of grass pea (*Lathyrussativus* L.): a novel plant for phytoremediation systems? C. R. Biol. 331, 859–864.

Brusseau, M.L., 1997. Transport and fate of toxicants in soils. In: Tarradellas, J., Bitton, G., Rossel, D. (Eds.), Soil Ecotoxicology. Lewis Publishers, New York, USA, pp. 33–53.

Chandrakar, V., Sahu, S., Khare, J., Sathpathy, S., 2014. Removal of Pb and Cr by fungi in municipal sewage water. Ind. J. Sci. Res. 4 (1), 214–217.

Chauhan, J.S., Rai, J.P.N., 2009. Phytoextraction of soil cadmium and zinc by microbes-inoculated Indian mustard (*Brassica juncea*). J. Plant Interact. 4 (4), 279–287.

Christie, N.T., Costa, M., 1984. *In vitro* assessment of the toxicity of metal compounds. IV. Disposition of metals in cells: interaction with membranes, glutathione, metallothionein, and DNA. Biol. Trace Elem. Res. 6, 139–158.

Clement, G., Yedjou, Christine, K., Tchounwou, H.S., Samuel Haile, B.S., Falicia Edwards, M.S., Paul, B., Tchounwou, ScD., 2010. N-Acetyl-Cysteine protects against DNA damage associated with lead toxicity in hepG2 cells. Ethn. Dis. 20 (1 Suppl. 1), 1–101-103.

Cooke, M.S., Evans, M.D., Dizdaroglu, M., Lunec, J., 2003. Oxidative DNA damage: mechanisms, mutation, and disease. FASEB J. 17, 1195–1214.

Danadevi, K., Rozati, R., Banu, B.S., Rau, P.H., Grower, P., 2003. DNA damage in workers exposed to lead using comet assay. Toxicology 187, 183–193.

De Temmerman, L., Waegeneers, N., Ruttens, A., Vandermeiren, K., 2015. Accumulation of atmospheric deposition of As, Cd and Pb by bush bean plants. Environ. Pollut. 199, 83–88.

Dezhban, A., Shirvany, A., Attarod, P., Delshad, M., Matinizadeh, M., Khoshnevis, M., 2015. Cadmium and lead effects on chlorophyll fluorescence, chlorophyll pigments and proline of *Robinia pseudoacacia*. J. For. Res. 26 (2), 323–329.

Dikilitas, M., Karakas, S., 2012. Behaviour of plant pathogens for crops under stress during the determination of physiological, biochemical, and molecular approaches for salt stress tolerance. In: Ashraf, M. (Ed.), Crop Production for Agricultural Improvement. Springer Publisher, pp. 417–441.

Dikilitas, M., Kocyigit, A., Bilinc, H., Taskin, A., Karakas, S., 2015. Evaluation of genotoxicity of environmentally friendly pesticides on higher cells using the single cell gel electrophoresis (SCGE) assay. Fresen. Environ. Bull. 24 (1), 39–47.

Dikilitas, M., 2003. Effect of Salinity and Its Interactions with *Verticillium albo-atrum* on the Disease Development in Tomato (*Lycopersicon esculentum* Mill.) and Lucerne (*Medicago sativa* & *M. media*) Plants (Ph.D. thesis). University of Wales, Swansea.

Ding, Y., Vaziri, N.D., Gonick, H.C., 1998. Lead-induced hypertension II: response to sequential influsions of L-arginine, superoxide dismutase, and nitroprusside. Environ. Res. 76, 107–113.

Egamberdieva, D., Hashem, D.A., Elsayed, F.A., 2014. Biological control of fungal disease by rhizobacteria under saline soil conditions. In: Ahmad, P. (Ed.), Emerging Technologies and Management of Crop Stress Tolerance, vol. 2, pp. 161–172.

El-Beltagi, H.S., Mohamed, A.A., 2010. Changes in non protein thiols, some antioxidant enzyme activity and ultrastructural alteration in radish plant (*Raphanus sativus* L.) grown under lead toxicity. Not. Bot. Horti. Agrobot. Cluj. 38 (3), 76–85.

El-Morsy, S.M., 2004. *Cunninghamella echinulata* a new biosorbent of metal ions from polluted water in Egypt. Mycologia 96 (6), 1183–1189.

Elobeid, M., Gobel, C., Feussner, I., Polle, A., 2012. Cadmium interferes with auxin physiology and lignification in poplar. J. Exp. Bot. 63 (3), 1413–1421.

El-Sonbaty, S.M., El-Hadedy, D.E., 2012. Combined effect of cadmium, lead, and UV rays on *Bacillus cereus* using comet assay and oxidative stress parameters. Environ. Sci. Pollut. Res. http://dx.doi.org/10.1007/s11356-012-1250-0.

Enoch, T., Norbury, C., 1995. Cellular responses to DNA damage: cell-cycle checkpoints, apoptosis and the roles of p53 and ATM. Trends Biochem. Sci. 20, 426–430.

Evens, A., Hryhorczuk, D., Lanphear, B.P., Rankin, K.M., Lewis, D.A., Forst, L., Rosenberg, D., 2015. The impact of low-level lead toxicity on school performance among children in the Chicago public schools: a population-based retrospective cohort study. Environ. Health 14. http://dx.doi.org/10.1186/S12940-015-0008-9.

Fahr, M., Laplaze, L., Bendaou, N., Hocher, V., El-Mzibri, M., Bogusz, D., Smoun, A., 2013. Effect of lead on root growth. Plant Sci. http://dx.doi.org/10.3389/fpls.2013.00175.

Fewtrell, L.J., Pruss-Ustan, A., Ladrigan, P., Ayuso-Mateos, J.L., 2004. Estimation of global burden disease of mild mental retardation and cardiovascular diseases from environmental lead exposure. Environ. Res. 94, 120–133.

Filek, M., Koscielniak, J., Labanowska, B.E., Bidzinska, E., 2010. Selenium-induced protection of photosynthesis activity in rape (*Brassica napus*) seedlings subjected to cadmium stress. Fluorescence EPR measurements. Photosynth. Res. 105 (1), 27–37.

Flora, S.J.S., Saxena, F.G., Mehta, A., 2007. Reversal of lead- induced neuronal apoptosis by chelation treatment in rats: Role of reactive oxygen species and intracellular Ca^{2+}. J. Pharmacol. Exp. Ther. 322, 108–116.

Flora, G., Gupta, D., Tiwari, A., 2012. Toxicity of lead: a review with recent updates. Interdiscip. Toxicol. 5 (2), 47–58.

Gabara, B., Krajewska, K., Stecka, E., 1995. Calcium effect on number, dimension and activity of nucleoli in cortex cells of pea (*Pisum sativum* L.) roots after treatment with heavy metals. Plant Sci. 111, 153–161.

Gadd, G.M., 2010. Metals, minerals and microbes: geomicrobiology and bioremediation. Microbiology 156, 609–643.

Gajewska, E., Sklodowska, M., 2007. Relations between tocopherol, chlorophyll and lipid peroxidase contents in shoots of Ni-treated wheat. J. Plant Physiol. 164, 364–366.

Gall, J.E., Boyd, R.S., Rajakaruna, N., 2015. Transfer of heavy metals through terrestrial food webs: a review. Environ. Monit. Assess. 187 (4), 201. http://dx.doi.org/10.1007/S10661-015-4436-3.

Gao, S., Ou-yanga, C., Tanga, L., Zhub, J.Q., Xua, Y., Wanga, S.H., Chen, F., 2010. Growth and antioxidant responses in *Jatropha curcas* seedling exposed to mercury toxicity. J. Hazard Mater. 182, 591–597.

Garcia-Leston, J., Roma-Torres, J., Vilares, M., Pinto, R., Prista, J., Teixeira, J.P., Mayan, O., Conde, J., Pingarilho, M., Gaspar, J.F., Pasaro, E., Mendez, J., Laffon, B., 2012. Genotoxic effects of occupational exposure to lead and influence of polymorphisms in genes involved in lead toxico kinetics and in DNA repair. Environ. Int. 43, 29–36.

Gautam, M., Singh, A.K., Johri, R.M., 2015. Effect of lead and chromium on germination and seedling growth of tomato (*Solanum lycopersicum*) and eggplant (*Solanum melongena*). Indian J. Agric. Sci. 85 (3), 126–130.

Gichner, T., Znidar, I., Szakova, J., 2008. Evaluation of DNA damage and mutagenicity induced by lead in tobacco plants. Mutat. Res. 652 (2), 186–190.

Girisha, S.T., Ragavendra, V.B., 2009. Accumulation of heavy metals in leafy vegetables grown in urban areas by using sewage water and its effect. Arc. Phytopathol. Plant Protect. 42 (10), 956–959.

Goncalves, A.C., Antas, S.E., Nunes, M.L., 2007a. Freshness and quality criteria of iced farmed senegalese sole (*Solea senegalensis*). J. Agric. Food Chem. 55, 3452–3461.

Goncalves, J.F., Becker, A.G., Cargnelutti, D., Tabaldi, L.A., Pereira, L.B., Battisti, V., Spanevello, R.M., Morsch, V.M., Nicoloso, F.T., Schetinger, M.R.C., 2007b. Cadmium toxicity causes oxidative stress and induces response of the antioxidant system in cucumber seedlings. Braz. J. Plant Physiol. 19 (3), 223–232.

Gratao, P.L., Polle, A., Lea, P.J., Azevedo, R.A., 2005. Making the life of heavy metal-stressed plants a little easier. Funct. Plant Biol. 32, 481–494.

Gupta, D., Nicoloso, F., Schetinger, M., Rossato, L., Pereira, L., Castro, G., Srivastava, S., Tripathi, R., 2009. Antioxidant defense mechanism in hydroponically grown *Zea mays* seedlings under moderate lead stress. J. Hazard Mater. 172 (1), 479–484.

Gurer, H., Ercal, N., 2000. Can antioxidants be beneficial in the treatment of lead poisoning? Free Radic. Biol. Med. 29 (10), 927–945.

Hamid, N., Bukhari, N., Jawaid, F., 2010. Physiological responses of *Phaseolus vulgaris* to different lead concentrations. Pak. J. Bot. 42 (1), 239–246.

Han, Z.X., Feng, G.Y., Lu, W.Z., 2002. Study on effects of As(III) in environment on wheat sprout and the original researcher of prevention and treatment of arsenic toxicant. Acta Bot. Boreal Occident Sin. 22 (1), 123–128.

Hart, A.J., Hester, T., Sinclair, K., Powell, J.J., Goodship, A.E., Pele, L., Fersht, N.L., Skinner, J., April 2006. The association between metal ions from hip resurfacing and reduced T-cell counts. J. Bone Joint Surg. Br. 88 (4), 449–454.

Hasanuzzaman, M., Nahar, K., Fujita, M., 2014. Role of tocopherol (vitamin E) in plants: abiotic stress tolerance and beyond. In: Ahmad, P., Rasool, S. (Eds.), Emerging Technologies and Management of Crop Stress Tolerance, A Sustainable Approach, vol. II. Elsevier, London, pp. 267–290.

Hassan, M., Mansoor, S., 2014. Oxidative stress and antioxidant defense mechanism in mung bean seedlings after lead and cadmium treatments. Turk. J. Agric. For. 38, 55–61.

Havaux, M., Kloppstech, K., 2001. The protective functions of carotenoid and flavonoid pigments against excess visible radiation at chilling temperature investigated in *Arabidopsis* npq and tt mutants. Planta 13 (6), 953–966.

Hepler, P.K., 2005. Calcium: a central regulator of plant growth and development. Plant Cell 17, 2142–2155.

Hoppenau, C.E., Tran, V.T., Kusch, H., Asshauer, K.P., Landesfeind, M., Meinicke, P., Popova, B., Braus-Stromeyer, S.A., Braus, G.H., 2014. *Verticillium dahliae* VdTHI4, involved in thiazole biosynthesis, stress response and DNA repair functions, is required for vascular disease induction in tomato. Environ. Exp. Bot. 108, 11–22.

Hsu, P.C., Guo, Y.L., 2002. Antioxidant nutrients and lead toxicity. Toxicology 180, 33–44.

Hsu, P.C., Hsu, C.C., Liu, M.Y., Chen, L.Y., Guo, Y.L., 1998. Lead-induced changes in spermatozoa function and metabolism. J. Toxicol. Environ. Health 55 (1), 45–64.

Hu, R., Sun, K., Su, X., Pan, Y.X., Zhang, Y.F., Wan, X.P., 2012. Physiological responses and tolerance mechanisms to Pb in two xerophils: *Salsola passerina* Bunge and *Chenopodium album* L. J. Hazard Mater. 205–206, 131–138.

Huang, H., Gupta, D.K., Tian, S., Yang, X.E., Li, T., 2012. Lead tolerance and physiological adaptation mechanism in roots of accumulating and non-accumulating ecotypes of *Sedum alfredii*. Environ. Sci. Pollut. Res. Int. 19 (5), 1640–1651.

Hunaiti, A., Soud, M., Khalil, A., 1995. Lead concentration and the level of glutathione, glutathione S-transferase reductase and peroxidase in the blood of some occupational workers from Irbid City, Jordan. Sci. Total Environ. 175, 95–100.

IARC, 2006. IARC monographs on the evaluation of carcinogenic risks to humans. Inorg. Org. Lead Compd. 87, 1–529.

Iram, S., Parveen, K., Usman, J., Nasir, K., Akhtar, N., Arouj, S., Ahmad, I., 2012. Heavy metal tolerance of filamentous fungal strains isolated from soil irrigated with industrial wastewater. Biologia 58 (3), 107–116.

Iram, S., Perveen, K., Shuja, N., Waqar, K., Akhtar, I., Ahmad, I., 2013a. Tolerance potential of different species of *Aspergillus* as bioremediation tool-comparative analysis. J. Bio. Environ. Sci. 3 (4), 1–10.

Iram, S., Zaman, A., Iqbal, Z., Shabbir, R., 2013b. Heavy metal tolerance of fungus isolated from soil contaminated with sewage and industrial wastewater. Pol. J. Environ. Stud. 22 (3), 691–697.

Islam, E., Yang, X., Li, T., Liu, D., Jin, X., Meng, F., 2007. Effect of Pb toxicity on root morphology, physiology and ultrastructure in the two ecotypes of *Elsholtzia argyi*. J. Hazard Mater. 147 (3), 806–816.

Jin, Y., Liao, Y., Lu, C., Li, G., Yu, F., Zhi, X., Xu, J., Liu, S., Liu, M., Yang, J., 2006. Health effects in children aged 3–6 years induced by environmental lead exposure. Ecotoxicol. Environ. Saf. 63, 313–317.

Jing, C., Cheng, Z., Li-Ping, L., Zhong-Yang, S., Xue-Bo, P., 2007. Effects of exogenous salicylic acid on growth and H_2O_2-metabolizing enzymes in rice seedlings under lead stress. J. Environ. Sci. 19, 44–49.

John, R., Ahmad, P., Gadgil, K., Sharma, S., 2008. Effect of cadmium and lead on growth, biochemical parameters and uptake in *Lemna polyrrhiza* L. Plant Soil Environ. 54 (6), 262–270.

Johnson, F.M., 1998. The genetic effects of environmental lead. Mutat. Res. 10 (2), 123–140.

Karakas, S., 2013. Development of Tomato Growing in Soils Differing in Salt Levels and Effects Companion Plants on Some Physiological Parameters and Soil Remediation (Ph.D. thesis). Harran University, Graduate School of Natural and Applied Sciences, Soil Science Plant Nutrition.

Kasuba, V., Rozgaj, R., Milic, M., Zeljezic, D., Kopjar, N., Pizent, A., Kljakovic-Gaspic, Z., 2010. Evaluation of lead exposure in battery-manufacturing workers with focus on different biomarkers. J. Appl. Toxicol. 30 (4), 321–328.

Kasuba, V., Rozgaj, R., Milic, M., Zeljezic, D., Kopjar, N., Pizent, A., Kljakovic-Gaspic, Z., Jazbec, A., 2012. Evaluation of genotoxic effects of lead in pottery-glaze workers using micronucleus assay, alkaline comet assay and DNA diffusion assay. Int. Arch. Occup. Environ. Health 85 (7), 807–818.

Kaur, G., Singh, H.P., Batish, D.R., Kohli, R.K., 2012. A time course assessment of changes in reactive oxygen species generation and antioxidant defense in hydroponically grown wheat in response to lead ions (Pb_2^+). Protoplasma 249, 1091–1100.

Kaur, G., Singh, H.P., Batish, D.R., Kohli, R.K., 2015. Adaptations to oxidative stress in *Zea mays* roots under short-term Pb_2^+ exposure. Biologia 70 (2), 190–197.

Khan, S., Shah, I.A., Muhammad, S., Malik, R.N., Shah, M.T., 2015. Arsenic and heavy metal concentrations in drinking water in Pakistan and risk assessment: a case study. Hum. Ecol. Risk Assess. 21, 1020–1031.

Kim, Y.Y., Yang, Y.Y., Lee, Y., 2002. Pb and Cd uptake in rice roots. Physiol. Plant 116, 368–372.

Kocyigit, A., Koyuncu, I., Taskin, A., Dikilitas, M., Bahadori, F., Turkkan, B., 2015. Antigenotoxic and antioxidant potentials of newly derivatized compound naringenin-oxime relative to naringenin on human mononuclear cells. Drug Chem. Toxicol. 31, 1–8.

Kopittke, P.M., Asher, C.J., Kopittke, R.A., Menzies, N.W., 2007. Toxic effects of Pb^{2+} on growth of cowpea (*Vigna unguiculata*). Environ. Pollut. 150 (2), 280–287.

Kos, J., Mastilovic, J., Hajnal, E.J., Sari, B., 2013. Natural occurrence of aflatoxins in maize harvested in Serbia during 2009–2012. Food Control 34, 31–34.

Kos, J., Hajnal, E.J., Skrinjar, M., Misan, A., Mandi, A., Jovanov, P., Milovanovic, I., 2014. Presence of *Fusarium* toxins in maize from autonomous province of Vojvodina, Serbia. Food Control 46, 98–101.

Kovacs, E., Nyitrai, P., Czovek, P., Ovari, M., Keresztes, A., 2009. Investigation into the mechanism of stimulation by low-concentration stressors in barley seedlings. J. Plant Physiol. 166 (1), 72–79.

Kumar, A., Prasad, M.N., Mohan Murali Achary, V., Panda, B.B., 2013. Elucidation of lead-induced oxidative stress in *Talinum triangulare* roots by analysis of antioxidant responses and DNA damage at cellular level. Environ. Sci. Pollut. Res. Int. 20 (7), 4551–4561.

Kwankua, W., Sengsai, S., Muangphra, P., Euawong, N., 2012. Screening for plants sensitive to heavy metals using cytotoxic and genotoxic biomarkers. Kasetsart J. Nat. Sci. 46, 10–23.

Lauwerys, R., Roels, H., Buchet, J.P., Bernard, A.A., Verhoeven, L., Konings, J., 1983. The influence of orally-administered vitamin C or zinc on the absorption of and the biological response to lead. J. Occup. Med. 25 (9), 668–678.

Lee, S., Ryu, J.Y., Kim, S.Y., Jeon, J.H., Song, J.Y., Cho, H.T., Choi, S.B., Choi, D., De Marsac, N.T., Park, Y.I., 2007. Transcriptional regulation of the respiratory genes in the Cyanobacterium *Synechocystis* sp. PCC 6803 during the early response to glucose feeding. Plant Physiol. 145, 1018–1030.

Li, C., Feng, S., Shao, Y., Jiang, L., Lu, X., Hou, X., 2007. Effects of arsenic on seed germination and physiological activities of wheat seedlings. J. Environ. Sci. 19, 725–732.

Li, N., Hou, Y.H., Ma, D.D., Jing, W.X., Dahms, H.U., Wang, L., 2015. Lead accumulation, oxidative damage and histopathological alteration in testes and accessory glands of freshwater crab, *Sinopotamon henanense*, induced by acute lead exposure. Ecotoxicol. Environ. Saf. 117, 20–27.

Maestri, E., Marmiroli, M., Visioli, G., Marmiroli, N., 2010. Metal tolerance and hyperaccumulation: costs and trade-offs between traits and environment. Environ. Exp. Bot. 68 (1), 1–13.

Maldonado-Magana, A., Favela-Torres, E., Rivera-Cabrera, F., Volke-Sepulveda, T., 2011. Lead bioaccumulation in *Acacia farnesiana* and its effect on lipid peroxidation and glutathione production. Plant Soil 339, 377–389.

Małecka, A., Piechalak, A., Morkunas, I., Tomaszewska, B., 2008. Accumulation of lead in root cells of *Pisum sativum*. Acta Physiol. Plant 30 (5), 629–637.

Manikantan, P., Balachandar, V., Sasikala, K., 2010. DNA damage in workers occupationally exposed to lead, using comet assay. Int. J. Biol. 2 (1), 103–110.

Manousaki, E., Kalogerakis, N., 2009. Phytoextraction of Pb and Cd by the mediterranean saltbush (*Atriplex halimus* L.): metal uptake in relation to salinity. Environ. Sci. Pollut. Res. Int. 16 (7), 844–854.

Martinez, S.A., Simonella, L., Hansen, C., Rivolta, S., Cancela, L.M., Virgolini, M.B., 2013. Blood lead levels and enzymatic biomarkers of environmental lead exposure in children in Cordoba, Argentina, after the ban of leaded gasoline. Hum. Exp. Toxicol. 32 (5), 449–463.

Matros, A., Peshev, D., Peukert, M., Mock, H.P., Van den Ende, W., 2015. Sugars as hydroxyl radical scavengers: proof-of-concept by studying the fate of sucralose in *Arabidopsis*. Plant J. 82 (5), 822–839.

Mishra, S., Dubey, R., 2005. Heavy metal toxicity induced alteration. In: Pessarakli, M. (Ed.), Handbook of Photosynthesis, vol. 28. CRC Press, Boca Raton, pp. 827–844.

Mishra, S., Srivastava, S., Tripathi, R.D., Kumar, R., Seth, C.S., Gupta, D.K., 2006. Lead detoxification by coon tail (*Ceratophyllum demersum* L.) involves induction of phytochelatins and antioxidant system in response to its accumulation. Chemosphere 65, 1027–1039.

Monteiro, C., Santos, C., Pinho, S., Oliveira, H., Pedrosa, T., Dias, M.C., 2012. Cadmium-induced cyto- and genotoxicity are organ-dependent in lettuce. Chem. Res. Toxicol. 25 (7), 1423–1434.

Munzuroglu, O., Zengin, F.K., Yahyagil, Z., 2008. The abscisic acid levels of wheat (*Triticum aestivum* L. cv. Cakmak 79) seeds that were germinated under heavy metal (Hg^{++}, Cd^{++}, Cu^{++}) Stress. G.U.J. Sci. 21 (1), 1–7.

Needleman, H., 2009. Low level lead exposure: history and discovery. AEP 19 (4), 235–238.

Okem, A., Southway, C., Stirk, W.A., Street, R.A., Finnie, J.F., Van Staden, J., 2015. Effect of cadmium and aluminum on growth, metabolite content and biological activity in *Drimia elata* (Jacq.) Hyacinthaceae. S. Afr. J. Bot. 98, 142–147.

Patra, M., Bhowmik, N., Bandopadhyay, B., Sharma, A., 2004. Comparison of mercury, lead and arsenic with respect to genotoxic effects on plant systems and the development of genetic tolerance. Environ. Exp. Bot. 52 (3), 199–223.

Patrick, L., 2006. Lead toxicity part II: the role of free radical damage and the use of antioxidants in the pathology and treatment of lead toxicity. Altern. Med. Rev. 11 (2), 114–127.

Piotrowska, A., Bajguz, A., Godlewska-Zylkiewicz, B., Czerpak, R., Kaminska, M., 2009. Jasmonic acid as modulator of lead toxicity in aquatic plant *Wolffia arrhiza* (Lemnaceae). Environ. Exp. Bot. 66 (3), 507–513.

Poreba, R., Gac, P., Poreba, M., Andrzejak, R., 2011. Environmental and occupational exposure to lead as a potential risk factor for cardiovascular disease. Environ. Toxicol. Phar. 31, 267–277.

Pourrut, B., Shahid, M., Dumat, C., Winterton, P., Pinelli, E., 2011. Lead uptake, toxicity, and detoxification in plants. Rev. Environ. Contamin. Toxicol. 213, 113–136.

Qiao, X.Q., Zheng, Z.Z., Zhang, L.F., Wang, J.H., Shi, G.X., Xu, X.Y., 2015. Lead tolerance mechanism in sterilized seedlings of *Potamogeton crispus* L. Subcellular distribution, polyamines and proline. Chemosphere 120, 179–187.

Rani, A., Goel, R., 2009. Strategies for crop improvement in contaminated soils using metal-tolerant bioinoculants. In: Khan, M.S., Zaidi, A., Musarrat, J. (Eds.), Microbial Strategies for Crop Improvement. Springer, Berlin, pp. 85–104.

Reddy, A.M., Kumar, S.G., Jyothsnakumari, G., Thimmanaik, S., Sudhakar, C., 2005. Lead induced changes in antioxidant metabolism of horsegram (*Macrotyloma uniflorum* (Lam.) Verdc.) and bengal gram (*Cicer arietinum* L.). Chemosphere 60 (1), 97–104.

Richardson, G.M., Garrett, R., Mitchell, I., Mah-Poulson, M., Hackbarth, T., 2001. Critical Review on Natural Global and Regional Emissions of Six Trace Metals to the Atmosphere. International Lead Zinc Research Organisation, the International Copper Association, and the Nickel Producers. Environmental Research Association.

Rodriguez, E., Azevedo, R., Fernandes, P., Santos, C., 2011. Cr (VI) induces DNA damage, cell cycle arrest and polyploidization: a flow cytometric and comet assay study in *Pisum sativum*. Chem. Res. Toxicol. 24 (7), 1040–1047.

Romero-Freire, A., Martin Peinado, F.J., van Gestel, C.A., 2015. Effect of soil properties on the toxicity of Pb: assessment of the appropriateness of guideline values. J. Hazard Mater. 289, 46–53.

Ruzica, R., Vilena, K., Mirta, M., Alica, P., Zorana, K.G., 2007. DNA damage in lymphocytes of lead workers determined by the alkaline comet assay. In: Dekant, W. (Ed.), The 44th Congress of the European Societies of Toxicology. Elsevier, Amsterdam, p. 116.

Saba, G., Parizanganeh, A.H., Zamani, A., Saba, J., 2015. Phytoremediation of heavy metals contaminated environments: screening for native accumulator plants in Zanjan-Iran. Int. J. Environ. Res. 9 (1), 309–316.

Salt, D.E., Blaylock, M., Kumar, Nanda P.B.A., Dushenkov, V., Ensley, B.D., Chet, I., Raskin, I., 1995. Phytoremediation: a novel strategy for the removal of toxic metals from the environment using plants. Biotechnology 13, 468–474.

Samardakiewicz, S., Krzesłowska, M., Bilski, H., Bartosiewicz, R., Wozny, A., 2012. Is callose a barrier for lead ions entering *Lemna minor* L. root cells? Protoplasma 249, 347–351.

Santos-Diaz, M.S., Barron-Cruz, M.C., 2011. Lead, chromium and manganese removal by *in vitro* root cultures of two aquatic macrophytesspecies: *Typhalatifolia* L. and *Scirpus americanus* Pers. Int. J. Phytorem. 13, 538–551.

Schmidt, W., 1996. Influence of chromium (III) on root associated Fe (III) reductase in *Plantago lanceolata* L. J. Exp. Bot. 47, 805–810.

Schutzendubel, A., Polle, A., 2002. Plant responses to abiotic stresses: heavy metal-induced oxidative stress and protection by mycorrhization. J. Exp. Bot. 53, 1351–1365.

Shahid, M., Pinelli, E., Pourrut, B., Silvestre, J., Dumat, C., 2011. Lead-induced genotoxicity to *Vicia faba* L. roots in relation with metal cell uptake and initial speciation. Ecotoxicol. Environ. Saf. 74 (1), 78–84.

Shakya, K., Chettri, M.K., Sawidis, T., 2008. Impact of heavy metals (copper, zinc, and lead) on the chlorophyll content of some mosses. Arch. Environ. Contam. Toxicol. 54, 412–421.

Shalan, M.G., Mostafa, M.S., Hassouna, M.M., El-Nabi, S.E., El-Refaie, A., 2005. Amelioration of lead toxicity on rat liver with vitamin C and silymarin supplements. Toxicology 206 (1), 1–15.

Sharma, P., Dubey, R.S., 2005. Lead toxicity in plants. Braz. J. Plant Physiol. 17, 35–52.

Sharma, B., Singh, S., Siddiqi, N.J., 2014a. Biomedical implications of heavy metals induced imbalances in redox systems. BioMed. Res. Int. Article ID: 640754, 26 pp. http://dx.doi.org/10.1155/2014/640754.

Sharma, S., Raghuvanshi, B.P., Shukla, S., 2014b. Toxic effects of lead exposure in rats: involvement of oxidative stress, genotoxic effect, and the beneficial role of N-acetylcysteine supplemented with selenium. J. Environ. Pathol. Toxicol. Oncol. 33 (1), 19–32.

Sheppard, C.R.C., Davy, S.K., Pilling, G.M., 2009. The Biology of Coral Reefs. Oxford University Press, Oxford, UK.

Singh, N.P., McCoy, M.T., Tice, R.R., Schneider, E.L., 1988. A simple technique for quantitation of low levels of DNA damage in individual cells. Exp. Cell Res. 175, 184–191.

Singh, R.P., Tripathi, R.D., Sinha, S.K., Maheshwari, R., Srivastava, H.S., 1997. Response of higher plants to lead contaminated environment. Chemosphere 34, 2467–2493.

Singh, T.D., Basham, M.E., Nordeen, E.J., Nordeen, K.W., 2000. Early sensory and hormonal experience modulate age-related changes in NR2B mRNA within a forebrain region controlling avian vocal learning. J. Neurobiol. 44, 82–94.

Singh, S., Singh, S., Ramachandran, V., Eapen, S., 2010. Copper tolerance and response of antioxidative enzymes in axenically grown *Brassica juncea* (L.) plants. Ecotoxicol. Environ. Saf. 73 (8).

Solanki, R., Dhankhar, R., 2011. Biochemical changes and adaptive strategies of plants under heavy metal stress. Biologia 66, 195–204.

Son, H., Lee, J., Lee, Y.W., 2012. Mannitol induces the conversion of conidia to chlamydospore-like structures that confer enhanced tolerance to heat, drought, and UV in *Gibberella zeae*. Microbiol. Res. 167, 608–615.

Srivastava, R.K., Pandey, P., Rajpoot, R., Rani, A., Dubey, R.S., 2014. Cadmium and lead interactive effects on oxidative stress and antioxidative responses in rice seedlings. Protoplasm 251 (5), 1047–1065.

Stobrawa, K., Lorenc-Plucinska, G., 2007. Changes in antioxidant enzyme activity in the fine roots of black poplar (*Populus nigra* L.) and cottonwood (*Populus deltoides* Bartr. ex Marsch) in a heavy-metal-polluted environment. Plant Soil 298, 57–68.

Sugasini, A., Rajagopal, K., Banu, N., 2014. A study on biosorption potential of *Aspergillus* sp. of tannery effluent. Adv. Biosci. Biotechnol. 5, 853–860.

Tandon, S.K., Chatterjee, M., Bhargava, A., Shukla, V., Bihari, V., 2001. Lead poisoning in Indian silver refiners. Sci. Total Environ. 281 (1–3), 177–182.

Traunfeld, J.H., Clement, D.L., 2001. Lead in Garden Soils. Homeand Garden. Maryland Cooperative Extension, University of Maryland, College Park, MD.

Truta, E., Rosu, C.M., Bara, I.C., 2011. Lead-induced genotoxicity in wheat. Analele stiintificeale Universitaji, Alexandru Ioan Cuza, Sectiunea Genetica Biologie Moleculara 12 (1).

Valverde, M., Trejo, C., Rojas, E., 2001. Is the capacity of lead acetate and cadmium chloride to induce genotoxic damage due to direct DNA-metal interaction? Mutagenesis 16 (3), 265–270.

Vaziri, N.D., Wang, X.Q., Oveisi, F., Rad, B., 2000. Induction of oxidative stress by glutathione depletion causes severe hypertension in normal rats. Hypertension 36, 142–146.

Verma, S., Dubey, R.S., 2003. Lead toxicity induces lipid peroxidation and alters the activities of antioxidant enzymes in growing rice plants. Plant Sci. 164, 645–655.

Wan, J., Zeng, G.M., Huang, D.L., Huang, C., Lai, C., Li, N.J., Wei, Z., Xu, P., He, X.X., Lai, M.Y., 2015. The oxidative stress of *Phanerochaete chrysosporium* against lead toxicity. Appl. Biochem. Biotech. 175 (4), 1981–1991.

Wang, M.Z., Jia, X.Y., 2009. Low levels of lead exposure induce oxidative damage and DNA damage in the testes of the frog *Rana nigromaculata*. Ecotoxicology 18, 94–99.

Wang, Q.Q., Wang, L.H., Han, R.B., Yang, L.Y., Zhou, Q., Huang, X.H., 2015. Effects of bisphenol A on antioxidant system in soybean seedling roots. Environ. Toxicol. Chem. 34 (5), 1127–1133.

Wu, H.M., Lin-Tan, D.T., Wang, M.L., Huang, H.Y., Lee, C.L., Wang, H.S., Soong, Y.K., Lin, J.L., 2012. Lead level in seminal plasma may affect semen quality for men without occupational exposure to lead. Reprod. Toxicol. Biol. Endochrinol. 10. http://dx.doi.org/10.1186/1477-7827-10-91.

Wurgler, F.E., Kramers, P.G.N., 1992. Environmental effects of genotoxins (ecogenotoxicology). Mutagenesis 7, 321–327.

Yadav, B.K., Siebel, M.A., VanBruggen, J.J.A., 2011. Rhizofiltration of a heavy metal (lead) containing wastewater using the wetland plant *Carexpendula*. Clean Soil Air Water 39, 467–474.

Yan, Z.Z., Ke, L., Tom, N.F.Y., 2010. Lead stress in seedlings of *Avicennia marina*, a common mangrove species in South China, with and without cotyledons. Aquat. Bot. 92, 112–118.

Yiin, S.J., Lin, T.H., 1995. Lead-catalyzed peroxidation of essential unsaturated fatty acid. Biol. Trace Elem. Res. 50, 167–172.

Yuan, Z., Cao, Y., Si, L., Wang, D., Guo, C., 2011. The effects of nitrobenzene on the genetic toxicity in tobacco seedling leaf cells by comet assay. Mol. Cell Toxicol. 7, 291–298.

Yuan, J., Hu, M., Zhou, Z., 2013. Selenium treatment mitigates the effect of lead exposure in *Coleus blumei* benth. J. Environ. Anal. Toxicol. 3, 6.

Yuan, L., Zhi, W., Liu, Y.S., Karyala, S., Vikesland, P.J., Chen, X., Zhang, H.S., 2015. Lead toxicity to the performance, viability, and community composition of activated sludge microorganisms. Environ. Sci. Technol. 49 (2), 824–830.

Zafar, S., Aqil, F., Ahmad, I., 2007. Metal tolerance and biosorption potential of filamentous fungi isolated from metal contaminated agricultural soil. Bioresour. Technol. 98, 2557–2561.

Zhai, Q.X., Narbad, A., Chen, W., 2015. Dietary strategies for the treatment of cadmium and lead toxicity. Nutrients 7 (1), 552–571.

Zhan, F.D., He, Y.M., Li, T., Yang, Y.Y., Toor, G.S., Zhao, Z.W., 2015. Tolerance and antioxidant response of a dark septate endophyte (DSE), *Exophiala pisciphila*, to cadmium stress. Bull. Environ. Contam. Tox. 94 (1), 96–102.

CHAPTER

RESISTANCE OF PLANTS TO Cu STRESS: TRANSGENESIS

4

Galina Nikolaevna Raldugina, Marina Sergeevna Krasavina, Nina Fjodorovna Lunkova, Natalia Anatoljevna Burmistrova

Timiryazev Institute of Plant Physiology, Russian Academy of Sciences, Moscow, Russia

CHAPTER OUTLINE

Plant Metal Interaction. http://dx.doi.org/10.1016/B978-0-12-803158-2.00004-7

1. INTRODUCTION

Industrial progress leads to environmental pollution with toxic compounds, specifically with heavy metals (HMs). Extraction and processing of ore, waste utilization (including batteries), production and application of fertilizers and pesticides, use of sewage for irrigation, and car exhaust contribute to pollution of soil and air. Use of copper (Cu)-bearing compounds in agriculture for combating soil pathogens results in its accumulation in the soil and plants (Vazhenin, 1985; Yruela, 2009; Fu and Wang, 2011; Saha et al., 2012; Shahid et al., 2014).

Under normal growth conditions, average concentration of Cu in plant tissues ranges from 5 to 20 mcg/g dry weight; variations within this range depend on plant species, stage of development, content of metal in the soil, and soil characteristics such as acidity (Marschner, 2002; Ravet and Pilon, 2013). When Cu concentration in the leaves is below 5 μg/g dry weight, the signs of Cu shortage become evident; at a concentration >20 μg/g, the signs of toxic effect of the metal are observed (Droppa et al., 1987; Cohu and Pilon, 2010). Physiological processes differently respond to changes in Cu concentration in the medium. For instance, root growth intensifies with an initial increase of Cu concentration; further rise leads to a suppression of growth. However, a percentage of seed germination depends on the Cu concentration much weaker (Elleuch et al., 2013).

Numerous HMs such as Cu, zinc (Zn), molybdenum, iron, and others are indispensable microelements. They are essential elements (Kholodova et al., 2011b; Ravet and Pilon, 2013; Jung et al., 2014) that at low concentrations govern normal growth and development of the organism. However, when they accumulate to high concentrations they adversely influence plant life and human health. Both necessity of HMs for plant vital functions and baneful influence of their excess depend on chemical properties of the elements.

2. COPPER AS AN ESSENTIAL ELEMENT

Cu is a trace element necessary for many different processes of plant life. Lack of this element can be determined by some features of external appearance of a plant; for example, twisting flag leaves, dried out leaf tips, reduced intensity leaf color, bent stems, and increased abortions of seeds. Many of the symptoms of Cu deficiency in plants are associated with the violations of the cell wall structure; therefore, to straw strength. By binding to the cell wall Cu support processes of lignification, increasing cell wall strength.

Plants' need for Cu is primarily from the participation of this element in enzymatic catalysis. Cu is a cofactor for some enzymes, such as nitrate reductase and certain proteases. A wide range of Cu-dependent processes largely depends on operation of this metal as a cofactor of numerous enzymes: Cu/Zn superoxide dismutase (SOD), cytochrome oxidase, polyphenol- and monophenol oxidase, ascorbate oxidase, amino oxidase, laccase, urease, etc.

Cu is also necessary for operation of cytochrome *c*-oxidase that transfers protons to terminal oxidase located on the inner mitochondrial membrane (Carr and Winge, 2003; Wilson and Vinogradov, 2014). This enzyme is a polypeptide using three atoms of Cu as cofactors. Moreover, together with Zn, Cu is a cofactor of Cu/Zn-SOD, the protein that participates in metabolism of reactive oxygen species (ROS) (Bowler et al., 1992; Opdenakker et al., 2012; Karnachuk et al., 2015). Cu/Zn-SOD catalyzes conversion of two superoxide anion radicals and two protons into hydrogen peroxide and molecular oxygen.

Furthermore, because the high-reactivity Cu binds with most amino acids and proteins and other substances. Cu is included in the composition of several Cu containing proteins (e.g., plastocyanin). It is known that Cu binds to chromosomes participating in the maintenance of their structure.

An important characteristic of Cu is the ability of Cu ions to be involved in redox reactions (López-Maury et al., 2012).

Under physiological conditions, Cu exists as Cu^{2+} and Cu^{+}. As with other metals with transient valence, Cu can accept or lose electrons, thereby affecting redox state of numerous cell components. Owing to changes in redox status, Cu ions have a high affinity for donors of electron pairs: sulfhydryl, amino groups, and carboxyl residues (Fraústo da Silva and Williams, 2001). Because Cu is included in the prosthetic group of polypeptides and pigments involved in electron transfer, this metal is necessary for activity of respiratory and photosynthetic electron transport chains. The effect of Cu on photosynthesis also occurs through the synthesis of antioxidants, removing formed during photosynthesis peroxide (H_2O_2).

More than 50% of Cu present in plant is found in chloroplasts, which underscores the urgent need for Cu in photosynthesis. A lack of Cu disturbs synthesis of chloroplasts and their functioning. Cu is a constituent of plastocyanin, which is an important component of electron transport chain of photosystem I (PSI). Plastocyanin is one of the most important Cu-bearing proteins in chloroplasts. This protein is located in the thylakoidal lumen and operates as a mobile carrier of electrons from the cytochrome-b_6/f complex to PSI (Kieselbach et al., 1998; Schubert et al., 2002; Krämer and Clemens, 2006; López-Maury et al., 2012). High levels of HMs usually suppress photosynthesis because of their direct influence on the photosynthetic apparatus including the structure of thylakoids. Some researchers report that HMs can induce liberation of proteins, lipids, and other components of thylakoidal membranes, which leads to injury of light-harvesting complexes and photosystem II (PSII) (Backor et al., 2007).

Cu can cross-link with the majority of proteins, for instance with cell wall proteins rich in hydroxyproline (Kossakowska et al., 1988). Cu^{2+} cation is often bound by nitrogen of histidine side chain of various proteins, and Cu^{+} predominantly interacts with sulfur in cysteine or methionine. As a result of such interactions, Cu becomes a component of some metalloproteins that are often involved in electron transport along photosynthetic and respiration chains (i.e., in energy conversion).

The content of Cu-dependent proteins in plants is much lower than the content of Zn-bearing proteins because much greater reactivity of Cu. In this relation, Cu ions (in contrast to Zn that is also an essential element) are less often used as structural components. At the same time, Cu-bearing proteins participate in numerous processes crucial for vital functions of the cell: photosynthesis, mitochondrial respiration, protection against oxidative stress, metabolism of cell walls, nitrogen metabolism, cell wall remodeling and lignification, flowering, seed formation and therefore in plant productivity, pathogen resistance, mobilization of iron, and biosynthesis of molybdenum cofactor (Puig et al., 2007; Yruela, 2009; Burkhead et al., 2009; Yuan et al., 2010; Lequeux et al., 2010; Cohu and Pilon, 2010; Billard et al., 2014). Cu also plays an important role in the transmission of transcriptional signals, transmission and reception of ethylene signal, oxidative phosphorylation, and mobilization of iron (Kanoun-Boule et al., 2008; Gonzalez and Giegé, 2014). In addition, Cu is necessary for the synthesis of nucleic acids, regulation of their metabolism, and oxygen superoxide scavenging (Kanoun-Boule et al., 2008; Lequeux et al., 2010; Yuan et al., 2010; Ravet and Pilon, 2013; Jung et al., 2014). It is possible that Cu indirectly participates in nitrogen assimilation, ABA biosynthesis, and many other processes (Burkhead et al., 2009; Ye et al., 2014). Another Cu-bearing protein is located in endoplasmic reticulum. It acts as an ethylene receptor and transmits ethylene signal (Rodriguez et al., 1999; Smékalová et al., 2014). This protein participates in numerous plant vital functions: growth, cell elongation, fruit ripening, leaf senescence, etc.

3. TOXIC EFFECTS OF HIGH CONCENTRATIONS OF COPPER

Metal toxicity may result from:

1. inhibition of enzymatic activity and suppression of the functions of other intracellular proteins because of binding their sulfhydryl groups to Cu ions;
2. injuries caused by oxidative stress;
3. induction of the shortage of other essential elements—Ca, Mg, K, and P (Baszynski and Krupa, 1995; Baszynski et al., 1988; Benavides et al., 2005; Yuan et al., 2011);
4. disturbance of the processes of intracellular transport and distribution of substances between different plant organs (translocation of water and nutrients) (Fodor, 2002; Poschenrieder and Barceló, 2004; Meharg, 2005; Yruela, 2005, 2009; Peñarrubia et al., 2015).

3.1 PLANT GROWTH UNDER Cu STRESS

A clear manifestation of the toxic effect of Cu stress—the inhibition of seed germination, can be caused by inhibition of the catabolism of ABA and oxidative stress (Viehweger, 2014; Ye et al., 2014). The effect of Cu excess is most distinctly manifested in suppression of plant growth (Sharma and Dubey, 2007; Thounaojam et al., 2012; Yuan et al., 2013; Muhammad et al., 2015; Wang et al., 2015), especially of root growth (Ivanova et al., 2010; Lequeux et al., 2010; Thounaojam et al., 2012; Elleuch et al., 2013). At a high content of Cu, the structure of roots changes—growth of primary roots is inhibited and the density of the distribution of the lateral roots increases. Apparently, in such conditions, plant-specific protein CMR1, required for optimal growth in vivo, in particular for cell division, suffers. Indeed, during Cu stress, a violation of the processes of cell division and expansion was detected and chromosomal aberrations appeared, the cell-cycle was disrupted, and the mitotic index decreased (Jaraniec et al., 2014). These disorders can be attributed to detected DNA under the influence of Cu excess and inhibition of functioning of the cytoskeleton—a decrease in the content of α-tubulin and the changed functioning of microtubules in meristematic cells (Qin et al., 2015). Inhibition of growth induced by excess Cu is related to the disturbance of synthesis of cellular membrane components and suppression of elongation growth. The components of cell walls become cross-linked with the molecules produced as a result of oxidative burst (Maksymiec, 1997; Maksymiec et al., 2008). Excess Cu inhibits the transport and distribution of mineral elements essential for growth (Alaoui-Sossé et al., 2004; Mateos-Naranjo et al., 2013; Cambrollé et al., 2013; Wang et al., 2015). The leaves reduce K and Mg content, apparently from transport violations from the roots (Ouzounidou et al., 1998). Sometimes there was an increase in the content of Mg, Ca, and Fe and a reduction in S content in the root (Bernal et al., 2012).

Distribution of auxin and cytokinin in the root system is impaired, and deposition of lignin is activated. As a result, apical meristem may die off (Lequeux et al., 2010). Suppression of growth is also promoted by inhibition of proteolytic and hydrolytic enzymes—amylase in particular. Because of retarded photosynthesis and disturbed mobilization of storage substances, supply of growth processes with energy and structural elements are impaired (Elleuch et al., 2013; Karmous et al., 2014).

3.2 DISTURBANCE OF PHOTOSYNTHESIS

Retardation of leaf growth by excess Cu is most distinctly pronounced in the early stages of plant growth with medium rich in Cu. Because of such inhibition, the content of pigments per area unit may rise. However, Elefterion and Karataglis (1989) have shown that as early as on the second or third day of wheat plants' exposure to Cu, the content of pigments decreased, which was related to destruction of internal structure of chloroplast grana and modification of protein-lipid complex of thylakoid membranes. Saha et al. (2012) and Dey et al. (2014) observed a decrease in the content of chlorophyll in the leaves of tea plants exposed to the excess Cu, Vijayarengan and Jose (2014) in the leaves of four plant species (mustard, maize, cotton, and watermelon), and Drzewicka-Matuszek et al. (2005) in the cells of *Spirulina laporte*. All the previously mentioned items brought about a reduction in total biomass (Arduini et al., 1995; Weck and Clijsters, 1996; Martin and Mourato, 2006; Upadhyay and Panda, 2009; Karimi et al., 2012).

Disturbance of photosynthesis may depend on direct and indirect effect on photosynthetic machinery (Mateos-Naranj et al., 2013). Excess Cu suppresses biosynthesis of all the components of biosynthesis modifying pigments and proteins of photosynthetic membranes (Lidon and Henriques, 1993; Maksymiec, 1997; Yruela et al., 2000; Maksymiec et al., 2008). A decrease in the content of lipids and changes in composition of fatty acids in thylakoid membranes was observed (Sandmann and Boger, 1980; Maksymiec, 1997). These modifications led to changes in membrane fluidity in PSII complex (Quartacci et al., 2000). These processes induced by the excess of Cu^{2+} can be associated either with destruction of polypeptide complex under the effect of oxidative stress or with interaction with ions necessary for correct operation of the complex such as Mn, Ca, and Cl (Baszynski and Krupa, 1995).

Corruption of chloroplast integrity, decrease in the content of chlorophyll, and changes in plastid membrane composition along with inhibition of iron uptake lead to chlorosis, bronzing, rolling, and necrosis of the leaves and to progressive senescence and even plant death (Foy et al., 1978; Carrier et al., 2003; Lequeux et al., 2010; Mostofa and Fujita, 2013). Another reason for such changes may be induced synthesis of ethylene that can accelerate senescence, suppress cell growth, and increase rigidity of cell wall by means of lignification (Maksymiec, 1997). Moreover, HMs affect water uptake and translocation as well as evaporation and metabolism of nutrients (Fodor, 2002; Poschenrieder and Barceló, 2004) and interfere with the uptake of Ca, Mg, K, and P (Benavides et al., 2005).

Changes in composition and structure of thylakoids contribute to disturbance of light-harvesting complexes and PSII (Barón et al., 1995; Hsu and Kao, 2004; Backor et al., 2007). Separation of charges by PSII and the water-splitting complex is inhibited (Mijovilovich et al., 2009). Mohanty et al. (1989) noted that excess Cu destabilizes the molecule of plastoquinone that is a secondary electron acceptor of PSII reaction center. On the whole, PSI is considered to be relatively resistant to the effect of HMs; however, certain disorders were detected therein (Baszynski et al., 1988; Ouzounidou, 1996). Specifically, some researchers reported suppression of PSI operation stipulated by the interaction of metal with ferredoxin (Shioli et al., 1978) and the changes in the content of plastocyanin bearing Cu atoms (Lidon and Henriques, 1993).

Some researchers investigated a relationship between Cu toxicity and photoinhibition (Yruela et al., 1996; Pätsikkä et al., 1998, 2001, 2002) (i.e., with a decrease in photosynthetic activity in the organisms exposed to a long-term high-rate illumination). It was shown that Cu heightens the effect of light. In experiments with isolated thylakoids, Cu ions heightened susceptibility to long-term illumination (Pätsikkä et al., 2001). Yruela et al. (1996) assumed that elevated toxicity of Cu ions upon the exposure to high-rate light was accounted for by the formation of hydroxyl radicals and other forms of oxygen.

3.3 CHANGE IN THE REDOX STATE OF CELLS

Toxic effects of Cu depend on their ability to easily convert from a reduced state into an oxidized state and to change valence. Oxidative stress arising upon the accumulation of free Cu ions may retard growth. It is known that Cu ions efficiently catalyze production of ROS and other free radicals (Maksymiec, 1997; Sáez et al., 2015a,b). Transition metals and especially such active metals as Cu catalyze formation of organic and inorganic peroxides, superoxide anions $O_2^{\cdot-}$), (HO^-), singlet oxygen, hydroxyl radicals, and H_2O_2. Accumulation of active radicals is also promoted by activation of lipid peroxidation (activation of lipoxygenases confirmed by accumulation of malonic dialdehyde) (Schützendübel and Polle 2002; Saha et al., 2012; Thounaojam et al., 2012; Elleuch et al., 2013; Mostofa and Fujita 2013; Shahid et al., 2014). In addition to redox activity of Cu ions per se, their excess can activate nicotinamide adenine dinucleotide phosphate (NADPH) oxidases, inhibit enzymatic activity because of affinity to SH groups on the enzyme or displacement of necessary cations from enzyme-binding centers. Increasing oxidation of tissue components can switch on a multitude of damaging effects, including injury of membranes, nucleic acids, proteins, and carbohydrates (Shahid et al., 2014). As a result, a significant amount of excess Cu can cause the death of the organism (Marschner, 2002; Prasad and Strzalka, 1999; Quartacci et al., 2000, 2001).

As with the effect of HMs, the influence of ROS depends on their concentration. Although ROS accumulation leads to a considerable disturbance of metabolism, at low concentrations they are necessary for regulation of gene expression governing such processes as cell cycle, growth, and development. ROS is involved in plant signal systems inducing activation of the mechanisms, ensuring protection against biotic and abiotic stresses and programmed cell death (Mittler et al., 2004; Shahid et al., 2014). In these cases, ROS accumulation is considered as a kind of adaptation.

For normal existence, plants must maintain ROS concentration within a physiological range regulating pro-oxidant/antioxidant balance that prevents cellular damage (Sharma et al., 2012). Protection against excessive accumulation of ROS is ensured by numerous systems destroying ROS: the systems of antioxidant protection (Foyer and Noctor, 2005; Azevedo et al., 2007; Cruz de Carvalho, 2008; Thounaojam et al., 2012; Talukdar, 2013).

Excess Cu in plants caused changes in activity and the content of certain components of antioxidant pathway, for instance:

1. specific antioxidant enzymes: SOD, catalase (CAT), ascorbate peroxidase (APX), glutathione reductase (GR), monodehydroascorbate reductase, dehydroascorbate reductase, glutathione peroxidase (GPX), guaiacol POD, glutathione-S-transferase, and ferredoxin:thioredoxin reductase (Navari-Izzo et al., 1998; Gupta et al., 1999; Wang et al., 2004; Cohu and Pilon, 2007; Thounaojam et al., 2012; Saha et al., 2012; Elleuch et al., 2013; Mostofa and Fujita, 2013; Smiri and Missaoui, 2014). Enzymes can operate separately or jointly. For instance, SOD that performs dismutation of superoxide resulting in formation of peroxide and molecular oxygen operates together with peroxidases (PODs) to destroy the produced peroxide. Therefore, most often these enzymes are activated or inhibited simultaneously.

2. low-molecular nonprotein antioxidants (nonprotein amino acids and α-tocopherols, pigments such as anthocyanins and carotenoids as well as proline, flavonoids, etc.). This group comprises both water- (glutathione, ascorbic acid, phenolic compounds) and fat-soluble (tocopherol, ubiquinone, retinol, carotenoids) antioxidants. Production of these compounds was also shown at a high concentration of Cu (Thounaojam et al., 2012; Saha et al., 2012).
3. metal-binding SH proteins (soluble and associated with membrane). Among HMs, Cu induces the greatest increase in the content of SH groups in proteins (Maleva et al., 2012).
4. nonprotein thiols (Mostofa and Fujita, 2013; Smiri and Missaoui, 2014).

All of these compounds play a key role in control over cellular level of free radicals (Apel and Hirt, 2004). Activation of antioxidant protection upon the effect of stressors directly aims at an improvement of plant resistance under extreme environmental conditions. Suppression of this system aggravates oxidative stress (Mostofa and Fujita, 2013). Therefore, in terrestrial plants from polluted ecotopes and in their seed posterity, activity of antioxidant enzymes (SOD, PODs) and proline is reliably greater than in unadapted species from conditionally pure habitats (Maleva et al., 2012).

However, the listed systems of plant protection against ROS usually do not operate simultaneously and act under different conditions. Only some antioxidants improve plants' resistance to stresses. For instance, in the leaves of rice exposed to rather high concentrations of Cu (150 μM $CuSO_4$) and a severe oxidative stress that induced a considerable accumulation of malonyl dialdehyde (MDA), SOD activity decreased by almost two times (Mostofa and Fuita, 2013). Reduced SOD activity was also observed as a result of Cu stress in the leaves of *Pistia stratiotes* (Upadhyay and Panda, 2009). There were reports that SOD and CAT are noticeably activated in susceptible plants of barnyard grass, but not in resistant plants; this means that resistance does not depend on activation of these enzymes. Another group of enzymes comprising POD, APX, and GR behaves differently: in susceptible plants, these enzymes are inhibited, which leads to oxidative stress in spite of active SOD (Abogadallah, 2010).

Activation or inhibition of antioxidant protection depends on concentration of Cu. For instance, according to Elleuch (2013), activity of APX rises almost two times at 10 μM $CuSO_4$. However, at such a concentration activities of SOD and CAT changed slightly, with appreciable changes occurring at higher Cu concentrations. Similarly, the content of total phenols and flavonoids rose at low concentrations but went down to control level upon further increase in Cu^{2+} concentration.

Exposure to high concentrations of Cu induced changes in the content of different components of antioxidant system depending on its duration: the content of low-molecular antioxidants (carotenoids and nonprotein thiols [glutathione]) rose as early as in the beginning of exposure (1–4 h). More prolonged exposure (4–8 h) on the medium supplemented with higher concentrations of metals brought about a depletion of the pool of low-molecular antioxidants with quick activation of antioxidant enzymes (SOD, CAT, and GR). Subsequently (after 12-h-long exposure), protection systems associated with the synthesis of SH compounds became activated. In some cases, low-molecular organic antioxidants more efficiently protect metabolism against ROS than antioxidant enzymes (Radyukina et al., 2008; Maleva et al., 2012). Recognizing the undeniable importance of the involvement of Cu in the redox reaction, in the data obtained for *Streptococcus pneumonia*, the toxicity of excess Cu is explained not so much by oxidative stress and metabolic disorders, but inhibition of aerobic trinucleotide synthesis (Johnson et al., 2015). Fighting against toxic effect of excess Cu involves general mechanisms of resistance: in addition to antioxidant systems, it comprises low-molecular stress-protective substances and macromolecules (Amudha and Balasubramani, 2011).

4. MAINTENANCE OF INTRACELLULAR Cu HOMEOSTASIS

In addition to activation of antioxidant protection, to avoid toxic effects induced by excess Cu in the medium, plants develop the mechanisms preventing accumulation in the cells of free Cu ions (Lequeux et al., 2010; Hossain et al., 2012; Argüello et al., 2013). Not only accumulation but also a shortage of Cu impairs plant growth and development, adversely affecting important physiological processes; therefore, operation of the systems maintaining appropriate concentration of Cu is of importance. The range of intracellular concentrations of Cu good for plant growth is very narrow. It was reported that concentration of Cu ions in the cytosol is insignificant and amounts to about 10^{-18} M. Maintenance of such a low concentration requires a coordinated operation of protective mechanisms ensuring intracellular concentration of Cu sufficient for the functioning of Cu-dependent compounds but protecting the cells against excessive accumulation of the metal. Because of the operation of these mechanisms under natural conditions, the plants rarely perish from Cu excess even on polluted soils. In intact soils, the plants more often experience a deficit of this essential element.

4.1 MAIN COMPONENTS MAINTAINING INTRACELLULAR HOMEOSTASIS OF COPPER

4.1.1 Metal Translocation

An important role in the maintenance of steady concentration of Cu ions belongs to their transport (Andres-Colas et al., 2010; Guo et al., 2013). The ions are taken up from the soil, distributed among plant organs, and compartmentalized in the cells and organelles (Xu et al., 2015). Absorption of Cu ions from the rhizosphere, where metals are always bound with various ligands (Cestone et al., 2012; Saijo and Nagasawa, 2015), probably depends on the operation of a Fe-dependent reductase (Welch et al., 1993). Initial stage of HMs' absorption consists in their uptake from the soil by root cells by means of specialized systems (transporters and chaperones) (Peñarrubia et al., 2015). Activity of these systems depends on the concentration of metals in the medium. When Cu is deficient, these systems are activated; when it is in abundance, they are suppressed. Expression of the genes encoding transporters is induced by the addition of Cu to the medium (Hall and Williams, 2003). This means that it is possible to reduce the uptake of the metal by the cells by means of inhibiting these transporters.

Membrane transporters belong to different families of proteins: cation diffusion facilitator, natural resistance–associated macrophage protein, ATP-binding cassette (ABC), Zn- and iron-regulated-like protein, and P-type ATPase are carriers. They are located in cellular and organelle membranes. Their function is to transport metals across the membranes.

ABC transporter proteins are located in the cell membrane. Cu ions are translocated to the cells at the expense of ATP hydrolysis (Williams and Mills, 2005; Grotz and Guerinot, 2006; Argüello et al., 2007; Burkhead et al., 2009; Deng et al., 2013). Hydrolysis is performed by heavy metal–transporting P-type ATPase (HMA), the P1B subfamily of the P-type ATPase superfamily (Abdel-Ghany et al., 2005).

AtHMA5–AtHMA8 and OsHMA4–OsHMA9 belong to the Cu/silver transporters. Therefore, the second opportunity to prevent the entry of Cu to the cell and its accumulation therein is to inhibit the activity of H+ ATPase (Hall and Williams, 2003; Williams and Mills, 2005; Grotz and Guerinot, 2006; Argüello et al., 2007; Burkhead et al., 2009; Deng et al., 2013). Inhibition of H+ ATPase activity in the presence of low concentrations of vanadate considerably reduced the content of Cu in the roots and shoots of *Brassica carinata* and therefore maintained growth (Vachirapatama et al., 2011; Wang et al., 2013).

Another type of ATPase associated with Cu transport (Cu-transporting P_{1B}-type ATPase [PAA2/HMA8]) is located in thylakoid membrane; it is necessary for the maturation of plastocyanin (Tapken et al., 2014). When Cu is easily available, this ATPase becomes degraded by means of a protease-Clp-protease system (Tapken et al., 2014).

Heavy metal transport P-type ATPase OsHMA5 was found in rice. In a vegetative stage, it is predominantly expressed in the roots; in the generative stage, it was also found in the nodes, peduncle, rachis, and husk. OsHMA5 promoter is located in the plasma membrane of root pericycle cells and the xylem of conducting bundles in the peduncle, rachis, and husk. It was assumed that OsHMA5 participates in Cu loading into xylem of roots and other organs both in vegetative and generative stages (Deng et al., 2013). Inhibition of OsHMA5 gene expression causes a decrease in Cu concentration in grain apparently to an extremely low level because the yield is also reduced.

According to Lee et al. (2007), OsHMA9 is expressed not only in the xylem, but also in the phloem. Transgenic plants that do not produce this transporter accumulated Cu and other HMs; this suggests that it plays a role in removing metals from the cells.

Cu is taken up in the roots by the carriers belonging to Cu transporter (COPT) family, a homolog of the yeast Cu transporter CTR1 (Sancenón et al., 2003; Yuan et al., 2011; Jung et al., 2012; Gayomba et al., 2013; Garcia-Molina et al., 2013). In *Arabidopsis*, transcription of the members of CTR/COPT family is activated when Cu is deficient; they participate either in the uptake from the soil to the root epidermal cells or in long-distance transport and distribution of Cu in photosynthesizing tissues. COPT1 protein is mainly localized in plasma membrane of root cells, especially when Cu is deficient. COPT2 protein is expressed in the shoots. These transporters govern entry of Cu to the cells. COPT3–COPT5 are located on the membranes of intracellular organelles. COPT5 participates in mobilization of Cu from the vacuoles of conducting tissues in case of profound deficiency. COPT6 is responsible for Cu mobilization from green stems and leaves and its redistribution to reproductive organs. This transporter is associated with the membranes of different cells but its main location is vascular system (Jung et al., 2012; Gayomba et al., 2013). Expression of transporters is regulated by Cu availability (Jung et al., 2012). COPT2 is activated in case of acute shortage of the metal, and COPT6 is activated when Cu is in abundance (Garcia-Molina et al., 2013; Gayomba et al., 2013; Puig, 2014).

BdCOPT3 and BdCOPT4 proteins located in plasma membrane are notable for low affinity to Cu ions. Therefore, such transporters do not promote high concentration of Cu in plants (Jung et al., 2014). COPTs carry reduced Cu ions; however, in the soil, the majority of ions occur as Cu (II). Cu (II) is reduced by ferric reductase present on plasma membrane of root cells (Welch et al., 1993).

Another family whose members carry HMs is ZIP transporters. ZIP transporters carry the cations of Cu^{2+} and other HMs to the cytoplasm (Puig et al., 2007). Expression of ZIP2 and AtIP4 genes discovered in *Arabidopsis* starts at the deficit of Cu and Zn. It was assumed that ZIP2 participates in Cu uptake by the roots of *Arabidopsis*.

Li et al. (2014) reported about the presence in common wheat of a new COPT gene family, TaCT1. TaCT1 gene is expressed in the leaves more intensely than in the roots, culm, and spikelet. Expression mostly occurs in the Golgi apparatus. Excess Cu inhibits expression, and its shortage makes expression more active; therefore, transporter participates in the maintenance of Cu homeostasis. The researchers believe that the genes expressing these proteins belong to the MFS superfamily of COPTs that so far have been identified solely in monocotyledons.

Intracellular transport of Cu ions involves one more group of proteins: Cu chaperones (Tottey et al., 2008; Cohu et al., 2009; Blaby-Haas et al., 2014). They belong to the family of soluble cytosolic metal receptors of low molecular weight. Cu chaperones play an important biological role in translocation of Cu within the cytoplasm to the place of Cu-dependent proteins localization. These proteins bind Cu ions, deliver them to intracellular compartments, and incorporate Cu into reaction centers of Cu-dependent enzymes. Thus, these proteins protect Cu from interaction with other cell components.

Three Cu chaperones (CCH, COXl7, and CCS) were identified and described in plants. In the stems of *Arabidopsis*, there was a high level of expression of CCH homologous to the yeast ATXl (Himelblau et al., 1998; Mira et al., 2001a,b). *COX17* (*AtCOX17*) gene encodes the protein that can act as an intermediary in the delivery of Cu to the mitochondria for the assembly of a functional complex of cytochrome oxidase as well as cytosolic enzymes such as Cu/Zn SODs (Balandin and Castresana, 2002). Expression of AtCOX17 genes was activated in response to the excess Cu, which suggests that this chaperone may participate in securing resistance just as metallothioneins that are also induced by high concentrations of metals (Zhou and Goldsbrough, 1995; Benatti et al., 2014; Liu et al., 2015b).

The efficiency of Cu absorption by plant roots depends on rhizosphere microorganisms. The nature of the interaction of rhizosphere microorganisms with the plant apparently depends on the properties of the soil and the plant species.

First of all, some microorganisms may change the availability of Cu for the plants in the soil by immobilization of Cu in the rhizosphere by its chelation and precipitation. But on the other hand, some microorganisms may increase the rate of absorption of Cu in the roots, accelerating transport of the metal from the root cortex into the xylem, transport of the metal with the transpiration stream, and accumulation in shoots (Tak et al., 2013; Xu et al., 2015). The rate of absorption of Cu by roots depends on mycorrhizal infection. The ectomycorrhizal fungus can serve as a filter, inhibiting the absorption of Cu. It is believed that this may be due to the presence of the fungal transporters contributing to the accumulation of Cu in the mycorrhizal sheath (Schlunk et al., 2015).

4.1.2 Binding of Cu in Plants

After the uptake from the soil, most of the Cu stays in the roots because its binding with amino- and carboxyl-ligands, which leads to restriction of translocation to the shoots (Ivanova et al., 2010; Thounaojam et al., 2012; Choudhary et al., 2012; Mostofa and Fujita, 2013; Collin et al., 2014; Qin et al., 2015). In addition to retention of Cu in the roots, this metal is weakly translocated to the above-ground organs with xylem stream. As a result, lower leaves accumulate much more Cu than the upper leaves and finally die off. Such withering of plant organs rich in Cu is another way of detoxification (Cook et al., 1997; Ivanova et al., 2010; Elleuch et al., 2013).

High binding of metal ions in the cell wall is achieved by its high cation exchange capacity (Silver and Misra, 1988). The leading role in binding HM ions, including Cu, belongs to negatively charged groups of polymer matrix (Meychik and Yermakov, 2001). The presence in cell walls of pectin substances containing nonesterified carboxyls of polygalacturonic acid, phenolic compounds, and lignin contributes to cation binding. At the same time, some papers report that long-term exposure to the excess of Cu causes changes in the composition and structure of cell walls. The excess of Cu induced elevation of the content of pectin and hemicelluloses as well as lignin, suberin, and uronic acids within the cell walls (Llugany et al., 2003; Konno et al., 2005). pH-dependent equilibrium between bound metal in the cell walls and free hydrated metal cations in transpiration flow (Clemens et al., 2002) facilitates translocation of metal ions throughout the plant.

4.1.3 Complexation with Chelating Agents

Preventing metal in the cell in a reactive form is achieved by complexation with chelating agents (Cobbett and Goldsbrough, 2002; Clemens, 2006; Chen et al., 2011; Roncarati et al., 2015). HMs can produce complexes with phytochelatins, metallothioneins, and amino acids such as asparagine, glutamine, histidine, and proline (Petit and Van de Geijn, 1978; Curvetto and Rauser, 1979; Sharma and Dietz, 2006). Organic acids (citric, fumaric, and malonic) (Senden et al., 1992; Rauser, 1999; Sarret et al., 2002; Curie et al., 2009; Mihalik et al., 2012), other compounds (glutathione, quercetin) (El Hajji et al., 2006; Pekal et al., 2011), or nicotianamine (Rauser, 1999; Curie et al., 2009; Krämer, 2010) can act as complexing agents. Complexation with nonproteogenic amino acid nicotianamine, which has a very high affinity for Cu(II) (Mijovilovich, 2009) is of special importance. In that way, the metal that arrived to the cells is detoxified (Shahid et al., 2014). The result of Cu interactions with cell components may be conversion of high-toxic Cu(II) to less toxic inorganic Cu(I)S compounds (Collin et al., 2014).

An important role in the regulation of intracellular level of Cu belongs to Cu-regulated small RNA molecules (Yamasaki et al., 2007; Abdel-Ghany and Pilon, 2008; Giele et al., 2012; Naya et al., 2014). Cu shortage switches on these regulators that inhibit expression of currently unneeded Cu proteins. This process involves Cu-responsive transcription factor SPL7 (Burkhead et al., 2009).

4.1.4 Sequestration in Cell Organelles (Vacuole) and Organs (Roots)

Liberation of ions from the cell across the plasma membrane and their translocation into the vacuole are two ways of reducing excessive toxic concentrations of metals in the cytosol. Some metal ions that were not bound with the cell wall are transferred to the cytoplasm where metals are chelated, moved to cell compartments, or released to the apoplast. For instance, in soybean cell culture, Bernal et al. (2006) showed that the chloroplast, vacuole, and cell wall are the sites where Cu accumulates. Cu(II) binds in the roots with amino and carboxyl ligands. Cu(I) forms complexes with both organic and inorganic sulfur ligands in the stems and leaves (Collin et al., 2014). Other researchers (Neumann et al., 1995; Arru et al., 2002) have shown that under the excess of Cu, the plants are notable for more voluminous vacuoles in the root tips of meristematic cells. The distribution of Cu between the leaf tissues reveals the mechanism of plant resistance: accumulation of Cu in the tissues of the vascular bundles protects photosynthetically active tissue against damage (Cestone et al., 2012). All the mentioned processes allow the plants to adapt to environmental conditions (Solanki and Dhankhar, 2011). When Cu^{2+} concentration in the medium grew to 100 μM, the plants of *Phyllostachys fastuosa* showed: (1) a considerable accumulation in the roots, (2) binding Cu(II) to amino and carboxyl ligands in the roots, (3) complexation with both organic and inorganic sulfur ligands in stems and leaves, and (4) conversion to a less toxic Cu(I), inorganic Cu(I)S compounds (Collin et al., 2014). In different plants and under different conditions, components of the system ensuring homeostasis behave differently. For instance, in *Thlaspi caerulescens* sensitive to the excess Cu reparation is predominantly associated with complexation of the majority of metal with sulfur ligands instead of reduction in the intracellular concentration of this metal (the content of Cu in the tissues even rises) (Mijovilovich et al., 2009).

Stress-regulated genes encode the following products: chaperones, osmotin-like proteins, antifreeze proteins, LEA proteins, messenger RNA-binding proteins, key enzymes governing biosynthesis and degradation of osmolytes, proteins of water channels, carriers of sugars and proline, enzymes ensuring ROS detoxification, enzymes for metabolism of fatty acids, inhibitors of proteinases, and ferritin and protein carriers of lipids as well as proteins involved in ion transport (antiporters and transporters located in the plasma membrane and the membranes of the vacuoles and organelles). Some researchers report that superexpression of transgenes encoding key enzymes of biosynthesis of osmolytes, LEA proteins and enzymes ensuring detoxification improved stress tolerance of plants (see reviews by Cushman and Bohnert, 2000; Kolodyazhna et al., 2009; Jewell et al., 2010; Yu et al., 2014).

In addition to the factors directly affecting the transport of HMs or synthesis of protective proteins and other components preserving the cells from the injuries (e.g., antistress protection), improvement of tolerance also depends on the molecules involved in signal transduction and regulation of expression of specific protective genes (Koornnef and Pieterse, 2008). Transcription factors (TFs) belong to such molecules.

5. CONSTRUCTION OF TRANSGENIC PLANTS RESISTANT TO HEAVY METALS

Affecting the processes listed previously can improve plant resistance. This goal may be achieved by means of plant treatment with appropriate substances that activate the processes elevating resistance. But the most promising instrument attracting numerous researchers is production of transgenic plants. Certain genes controlling protective mechanisms have already been identified and cloned, and the attempts to incorporate them in plants were made for example (Juraniec et al., 2014; Liu et al., 2015a).

5.1 TRANSGENIC PLANTS EXPRESSING METALLOTRANSPORTER GENES

For instance, numerous genes of metallotransporters were cloned. Among them, there are *CAX1, CAX2, NtCBP4, FRE1, FRE2, Nramp, ZIP, ABC, CDF*, and *P1B*-ATPases forming a large family of P-type ATPases. For the most part, incorporation of these genes made the plants resistant to Cd^{2+}. Resistance to Cu was associated with the incorporation of genes encoding metallothioneins. For instance, incorporation of *metallothionein 1* gene (*CcMT1*) from pigeon pea (*Cajanus cajan*, L.) into *Arabidopsis thaliana* plants considerably elevated their resistance to Cu (Sekhar et al., 2011). Resistance to elevated concentration of Cu ions was also obtained as a result of hyperexpression of *MT-II* genes isolated from human genome (human metallothionein-II) in tobacco (*N. tabacum*) and rapeseed (*Brassica napus*) (Misra and Gedamu, 1989). Transformation of poplar (*Populus alba* L.) with *PsMTA1* gene (metallothionein-like protein from *Pisum sativum*) brought about an increase in the resistance of transgenic plants to HMs (Cu and Zn) and to oxidative stress induced by paraquat. Transgenic poplars accumulated less ROS and had a lower level of DNA injury induced by joint action of 8-hydroxy-20-desoxyguanosine and Cu (Balestrazzi et al., 2009). Incorporation into plants of another metallothionein gene isolated from yeast (*CUP1*) also caused a rise in resistance to Cd^{2+} and Cu of transformed plants of tobacco (Pavlikova et al., 2004; Janouskova et al., 2005; Krystofova et al., 2012) and cauliflower (Hasegawa et al., 1997; Thomas et al., 2003; Watanabe et al., 2005).

Some amino acids, in particular cysteine and serine, act as metallothioneins. The plants of *Arabidopsis* with superexpression of gene *Atcys-3A* encoding the enzyme catalyzing cysteine biosynthesis showed a ninefold (as compared with wild type) improvement of resistance to accumulation of cadmium in plants grown on the medium with 250 μM $CdCl_2$ (Dominguez-Solis et al., 2001, 2004). Kawashima et al. (2004) produced transgenic plants of tobacco carrying a gene of O-acetyltransferase (a serine acetyltransferase *At Sat*) from *Arabidopsis*, which were much more resistant to Cd (250 and 300 mcM), Se (250 mcM), and Ni (500 mcM) than common plants and showed much greater gain in biomass and root length.

One of the possible mechanisms ensuring resistance to HMs is metal binding by organic ions. Insertion into tobacco, pawpaw, and rapeseed mutant genes of citrate synthetase from alfalfa controlled by CaMV35S promoter ensured elevation of the content of citrate and tolerance to aluminum (Zhang et al., 2000). Another work dealing with improved resistance to Al was performed by Koyama et al. (2000) who inserted a gene of mitochondrial citrate synthetase isolated from *Daucus carota* into plants of *Arabidopsis thaliana*. Subsequently, several more articles were published that described production of transgenic plants of different species carrying CS genes isolated from plants, yeast, and bacteria (Anoop et al., 2003; Barone et al., 2008; Deng et al., 2009; review Inostroza-Blancheteau et al., 2012). The level of other organic acids—malate, oxalate, succinate, and acetate—rose when a gene of malate dehydrogenase (nodule enhanced) from the nodules of alfalfa was inserted into alfalfa plants (*Medicago sativa* L.) (Tesfaye et al., 2001) and this rise also considerably improved plant resistance to Al^{3+}. Activation of transport flux of organic acids may also improve resistance of plants to the excess of metals. Delhaize et al. (2004) inserted a gene of malate transporter (*ALMT1* gene) from barley into wheat plants (*Triticum aestivum*), which also improved resistance of transgenic plants to the excess of Al^{3+}.

5.2 EXPRESSION OF PHYTOCHELATIN GENES

Plant resistance to HMs was promoted by expression of the gene encoding phytochelatin synthase responsible for the synthesis of phytochelatins. For instance, it was shown that superexpression of *AtPCS1* gene in transgenic plants of tobacco improved their resistance to Cd^{2+} (Pomponi et al., 2006) and in the plants of Indian mustard (*Brassica juncea* L.) to Cd^{2+} and As^{2+} (Gasic and Korban, 2007). Transgenic tobacco carrying the gene *OsPCS1* can grow on the medium with elevated content of Cd^{2+} because of the activated synthesis of phytochelatin synthase (Postrigan et al., 2013).

5.3 TRANSGENIC PLANTS OVEREXPRESSING ANTIOXIDATIVE GENES

As it was noted previously, HMs cause oxidative stress in plants; therefore, elevated expression of genes encoding antioxidant enzymes must improve resistance to ions of HMs. For instance, in transgenic tobacco plants, superexpression of the gene of glutathione transferase increased tolerance to Cu ions (Anoop et al., 2003). Transgenic plants of spruce pine with elevated content of antioxidant enzymes (APX, GR, and SOD) were notable for greater survival on the soils polluted with HMs (Tang et al., 2005). In some cases, greater resistance depended on incorporation of two genes within a chimeric construction. For instance, after incorporation of *cassava* CuZn-*SOD* and pea *APX* genes into tall fescue plants (*Festuca arundinacea* Schreb), some researchers (Lee et al., 2007; Luo et al., 2013) showed that transgenic plants were more resistant to oxidative stress induced by paraquat, which suggests that these plants would be also resistant to other stressors—in particular, to the excess of HMs. Identical results were obtained by other researchers (Kwon et al., 2002; Tang et al., 2006). Xia et al. (2012) created transgenic tobacco plants notable for superexpression of *EhMT1* gene from Cu-accumulator *Elsholtzia naichowensis* Sun. Dry biomass of transgenic seedlings was greater than in the wild-type plants. In the roots of transgenic plants, the level of peroxidation of lipids and H_2O_2 production were lower. This protein was assumed to bind Cu^{2+} in the cytoplasm; therefore, the formation of ROS decreased. However, activation of antioxidative enzymes not necessarily accompanied oxidative stress; sometimes, even their inhibition was observed. For instance, in the leaves of rice exposed to severe stress (150 mcM $CuSO_4$), CAT activity considerably decreased (Upadhyay and Panda, 2009; Mostofa and Fuita, 2013).

6. EXPRESSION GENES OF COMPATIBLE OSMOLYTES

Under adverse conditions, there occurs induction of genes governing the synthesis of compounds mediating plant response to abiotic stresses, including the genes encoding synthesis and degradation of compatible osmolytes—low-molecular organic compounds that even at high concentrations do not suppress cellular metabolism. These osmolytes comprise amino acids (proline, alanine), quaternary amines (betaine, glycine betaine), oligosaccharides (raffinose), sugar alcohols (mannitol, sorbitol, trehalose, and inositol), and sugars (sucrose, glucose, fructose, raffinose). Many authors have noted the accumulation of compatible osmolytes in plant tissues in almost all stress conditions—low temperatures, drought, salinity, ultraviolet (UV) irradiation, adaptation to frost, pathogenesis, and other effects (Serraj and Sinclair, 2002; Radyukina et al., 2008; Sinkevich et al., 2010; Kumar et al., 2009).

Amino acid proline is one of such osmolytes (Kuznetsov and Shevyakova 1999; Serraj and Sinclair 2002; Ashraf and Fooland 2007; Shevyakova 1983; Shevyakova et al., 2009. Misra and Saxena 2009; Yousfi et al., 2010).

6.1 PROLINE'S ROLE IN PLANT RESISTANCE

As with other osmolytes, there is a correlation between the content of proline and resistance to adverse environmental factors. Dramatic accumulation of proline in the cells protects them from numerous disastrous influences: excess insolation, UV radiation, salinization, low and high temperature, water and osmotic stress, and H_2O_2 stress (Hare and Cress, 1997; Kishor et al., 1995, 2005; Kuznetsov and Shevyakova, 1999; Ruiz and Blumwald, 2002; Hong et al., 2000; Yamada et al., 2005; Kumar et al., 2009; Kondrak et al., 2012; Hayat et al., 2012). Proline accumulates in the presence of HMs (Bassi and Sharma, 1993a,b; Mehta and Gaur, 1999). Proline accumulation is especially important for resistance to salinization, drought, and low temperatures (Kondrak et al., 2012). Treatment with exogenous proline of the leaves of the olive tree (*Olea europaea* L) removed the effect of salinization on water balance, content of chlorophyll, photosynthetic activity, and stimulated activity of antioxidant enzymes (Ben Ahmed et al., 2010). Proline treatment also had a favorable effect under drought: in this case, the content of H_2O_2 accumulated under stress decreased (Nayyar and Chander, 2004). Proline increases the resistance and under Cd stress (Xu et al., 2009a).

Correlation between the content of proline and plant resistance is distinct when stress-resistant and stress-susceptible plants are compared. Comparison of more resistant to osmotic stress (0.2 M mannitol or sorbitol) halophyte *Arabis stelleri* and less resistant *Arabidopsis* revealed higher content of proline in resistant *Arabis* (Jung et al., 2010). The same pattern was observed in cold-resistant and cold-sensitive genotypes of tomatoes: resistant plants accumulated much more proline (Liu et al., 2012).

In spite of numerous indications that a correlation exists between proline accumulation and resistance of many plant species to various stresses (Zhang et al., 1995; Igarashi et al., 1997; Bartels and Sunkar, 2005), this conclusion cannot be always considered indisputable (Legay et al., 2011). For instance, in some cases proline was more intensely expressed in labile plants that more distinctly respond to stress. This suggested that activation of proline synthesis is more likely associated with stress response and is not characteristic of plant resistance. Proline accumulation is more likely a marker of severe stress in the leaves than a component of resistance (Hanson et al., 1977).

Proline is believed to be a universal organic protector operating under various stresses (Kholodova et al., 2011b). In the cells, proline is present in a free state or as a component of structural proteins. Free proline can act as an osmolyte. Elevation of the content of free proline under stresses can occur at the expense of liberation from bound state, activation of synthesis, and retardation of degradation. Investigating the effect of proline, researchers focused their attention on proline as an osmotic agent; however, as compared with other osmotically active components (for example potassium), its concentration in the cell is usually lower. According to Grinin, in brown mustard *Brassica juncea* L. in the beginning of salinization with NaCl, peak concentration of proline was almost two-thirds of the total concentration of Na and Cl ions. The researcher admitted that in this stage of salinization, proline could considerably contribute to osmotic potential of the cell. Subsequently, intracellular concentration of proline decreased and hardly amounted to one-tenth the concentration of the ions, and proline could scarcely be considered as an important osmotic agent (Grinin, 2010). Insufficiency of concentration of proline for the maintenance of osmotic equilibrium between cytoplasm and vacuole was also shown by other researchers (Munns and Tester, 2008; Verbruggen and Hermans, 2008; Attri et al., 2010; Larher et al., 2009).

Proline plays perhaps more important role as an antioxidant (Matysik et al., 2002; Upadhyaya and Panda, 2009). Oxidative injuries are among key disturbances induced by stresses. Stresses upset normally existent equilibrium between ROS production and quenching. Correlation between proline accumulation and resistance to oxidative stress was observed (Ashraf and Fooland, 2007). Proline can quench hydroxyl radicals and singlet oxygen produced under stresses (Matysik et al., 2002). An important advantage of proline over other compatible osmolytes is its ability to promptly act under stress influences and dissipate early formed free radicals. It is interesting that, in some investigations, an inverse correlation was found between proline content and activity of other antioxidants. For instance, in the plants of facultative halophyte *Mesembryanthemum crystallinum* L. and in *Thellungiella halophila*, where much proline accumulated under saline stress, SOD activity did not change, and in the lines with low accumulation of proline activity of this enzyme rose (Radyukina et al., 2007, 2008; Shevyakova et al., 2009). This points to a possible interchangeability of proline and antioxidant enzymes in control over the level of free radicals (Radyukina et al., 2011a,b; Sorkheh et al., 2012).

Proline is an antioxidant that efficiently scavenges hydroxyl radical, singlet oxygen, and other free radicals produced under any adverse conditions (Smirnoff and Cumbes, 1989; Matysik et al., 2002; Ashraf and Foolad, 2007; Trovato et al., 2008). Proline's ability to quench $OH^{\cdot -}$ radicals (Smirnoff and Cumbes, 1989) can prevent cell death (Chen and Dickman, 2005). A correlation between proline content and GSH redox state was shown. Moreover, in the presence of proline, the ratio NADP+: NADPH is maintained (Hare and Cress, 1997). Proline also plays an important role in ensuring the activity of pentose phosphate pathway that produces reducing equivalents necessary for maintaining numerous antioxidants in reduced state.

Thus, proline is a multifunctional compound. It can act as an osmolyte, ROS scavenger, redox balancer, buffer of cytoplasmic pH, and regulator of expression of the genes of stress response (see reviews by Verbruggen and Hermans, 2008; Szabados and Savoure, 2010; Singh et al., 2010).

Rather recently, researchers noticed another important function of proline: its ability to stabilize the structure of proteins and other macromolecules and membranes disturbed under any type of stress (i.e., to act as a chaperone) (Alia and Mohanty, 1997; Shah and Dubey, 1998; Nanjo et al., 1999; Ain-Lhout et al., 2001; Diamant et al., 2001; Chattopadhyay et al., 2004; Grinin, 2010). For instance, in plants of brown mustard it was shown that under salinization proline can ensure maintenance of activity of malate dehydrogenase, which was evidence of its operation as a chemical chaperone (Grinin, 2010).

Other proteins induced by various stresses, in particular LEAs (late embryogenesis abundant) and heat shock proteins (HSPs) also act as chaperones (Bohnert and Jensen, 1996; Jouve et al., 2004; Legay et al., 2011). Genes encoding chaperones, dehydrins, and LEA proteins are more intensely and preferably expressed in resistant plants. Small chloroplastic and mitochondrial HSPs bind proteins, ensuring correct folding and preventing protein aggregation (Waters et al., 2008; Sato and Yokoya, 2008). Apparently, these proteins also support structure of actin, plays an important role in the primary response to Cu stress (Kulikova et al., 2009). The ability of osmolytes to maintain the structure membranes causes a reduction in the leakage of substances from the cells, which increases by Cu stress (Berglund et al., 2002). Cell functioning very much depends on the maintenance of intracellular concentration of potassium and therefore of membrane electrical potential (Alaoui-Sossé et al., 2004; Chen et al., 2007).

Accumulation of proline may be caused by several reasons: stimulation of its synthesis from glutamic acid, inhibition of proline oxidation, and modification of involvement of this amino acid in protein synthesis (Trotel-Aziz et al., 2000). Synthesis of proline occurs in the cytosol and chloroplasts under various stresses and is governed by Δ1-pyrroline-carboxylate reductase (Lehmann et al., 2010). Proline degradation upon recovery from stress occurs in mitochondria with the participation of proline dehydrogenase (Evers et al., 2010). Creation of plants with elevated content of proline was based on two types of modification of the enzymes of glutamate pathway: expression of pyrrolin-5-carboxylate synthetase (P5CS) transgene responsible for proline synthesis or reduction in activity of proline dehydrogenase governing its degradation (Kavi Kishore et al., 2005; Verbruggen and Hermans, 2008; Ibragimova et al., 2012). For instance, in 1995, tobacco plants were produced expressing the P5CS gene from *Vigna aconitifolia* controlled by a powerful constitutive promoter (Kavi Kishore et a1. 2005), which did not decrease accumulation of biomass and yield of seeds on saline soils. Transgenic plants were notable for a greater activity of the enzymes participating in proline synthesis and accumulated it 10–18 times more than common plants. Moreover, transformed plants better endured salinization. In addition to tobacco, identical results were obtained in wheat (Kishor et al., 1995; Hong et al., 2000; Sawahel and Hassan, 2002; Dörffling et al., 2009), *Arabidopsis* (Nanjo et al., 1999), citrus plants (Molinari et al., 2004), sugar cane (Molinari et al., 2007), rice (Su and Wu, 2004), and potato (Hamida-Sayari et al., 2005).

Overexpression of Pro biosynthetic pathway genes also considerably improved resistance to abiotic stresses (Dixon and Paiva, 1995; Kishor et al., 1995; Tateishi et al., 2005; McClinchey and Kott, 2008). Transgenic plants ensured a higher yield of potato tubers and greater resistance of rice as compared with wild-type plants (Kishor et al., 1995; Hong et al., 2000; Hmida-Sayari et al., 2005). Initially, transgenic plants had greater content of proline; they grew more considerably after 2 days of plant exposure to stresses (Xin and Browse, 2000). Proline accumulation favorably influenced water potential and stimulated production of protective pigments of xanthophyll cycle (Dobra et al., 2010). When P5CS genes from moth bean was inserted in green algae *Chlamydomonas reinhardtii,* their cells became more resistant to the effect of Cd $^{2+}$ than the cells of wild type (Siripornadulsil et al., 2002).

The plants with elevated content of proline were also obtained after insertion of a fragment of gene *pdh* in antisense orientation into the plants of tobacco or *Arabidopsis* (Kochetov et al., 2004; Szekely et al., 2008). In this case, transgenic plants were also more resistant to freezing and salinization (Nanjo et al., 1999; Xin and Browse, 1998, 2000; Tateishi et al., 2005).

The effect of compatible osmolytes is most easily understood for abiotic stresses associated with a disturbance of water-saline balance (cold, salinization, wilting). Although HMs sometimes impaired water exchange (Wang et al., 2010; Kholodova et al., 2011a,b; Bizzo et al., 2014), it was much less pronounced. However, proline improved plant resistance to HMs also. This was shown in numerous works with the exposure of various species to excess Cu ions (Bassi and Sharma, 1993a,b; Chen et al., 2001; Szafranska et al., 2011). Resistance to Cu positively correlated with the content of proline in lichen *Trebouxia erici* (Bačkor et al., 2003, 2007). The higher the proline concentration was, the more resistant the lines of transgenic lichen proved to be. Transgenic plants with elevated content of proline were tolerant to toxic concentrations of the salts of HMs: cadmium, nickel, lead, and mercury. Such plants were notable for weaker growth retardation; signs of chlorosis therein were less pronounced or essentially lacking (Kolodyazhnaya et al., 2007). Superexpression of the *P5CS* gene from kidney bean in green microalga *Chlamydomonas reinhardtii* ensured the ability to survive in the presence of toxic concentration of cadmium ions. As the content of proline correlated with the level of MDA, it was assumed that free proline produces an antioxidant effect in the presence of cadmium ions (Kolodyazhnaya et al., 2009). A group of Chinese researchers (Zhang et al., 2014a,b) produced transgenic plants of tobacco with an inserted cassette of genes *OsP5CS1* and *OsP5CS2,* which were resistant to elevated concentration of Cu^{2+} (300 μM $CuSO_4$) probably owing to accumulation of more proline in the cells. Apparently, greater content of proline improves nonspecific stress resistance.

6.2 SOLUBLE SUGARS AND PLANT RESISTANCE

As with same proline, sugars are considered components of the osmoregulatory system of the cell. Elevation of concentration of soluble sugars (glucose, fructose, raffinose, and sucrose) under adverse environmental conditions was observed by many researchers (Garg et al., 2002; Welling and Palva, 2008; Dörffling et al., 2009; Lee et al., 2012; Kulikova et al., 2015). Carbohydrates accumulated in the above-ground parts of plants with growth on solutions of HMs such as cadmium and nickel (Moya et al., 1993).

The interference in carbohydrate metabolism affects the resistance of plants to stress. For example, inhibition of invertase inhibitor leading to an increase in the content of hexoses increases the sensitivity of the plant leaves to infection *Pseudomonas syringae* (Bonfig et al., 2010). Alaoui-Sossé et al. (2004) observed noticeable accumulation of starch (155%) and sucrose in the leaves but not in roots of Cu-stressed plants. The authors suggested that excess Cu inhibits the loading of assimilates in the conducting system of plants and their outflow from the leaves. Accumulation of carbohydrate may cause inhibition of photosynthesis.

In many respects, the effect of proline is synergistic with the effect of carbohydrates. Soluble sugars also support osmotic potential of the cells and participate in redox reactions and in the maintenance of the structure of cellular macromolecules and membranes (Krasavina et al., 2014, review). It is interesting that usually these two components accumulate simultaneously. For instance, it was shown that a mutant that hyperexpresses proline also accumulates soluble sugars (mainly glucose and fructose) and soluble proteins (Dörffling et al., 2009). Elevation of the content of these osmotically active components considerably reduces osmotic potential of the mutant cells at low temperatures as compared with wild-type cells. Such changes in osmotic potential correlate with plant resistance to chilling and freezing (Vagujfalvi et al., 1999).

As with proline, sugars act as active antioxidants (Bläsing et al., 2005; Chen and Dickman, 2005; Nishizawa et al., 2008; Peshev and van den Ende, 2013; Keunen et al., 2013). Photosynthetic activity in source leaves results in temporary accumulation of sugars. These sugars can directly or indirectly combat oxidative stress, restoring homeostasis. Sugar starvation also induces oxidative stress (Bolouri-Moghaddam et al., 2010). Both a deficit and an excess of sugars may disturb respiratory metabolism and electron transport in mitochondria and chloroplasts and induce accumulation of ROS (Xiang et al., 2011).

Sugar signaling and expression of sugar-modulated genes are believed to be related to control over oxidative stress (Couée et al., 2006; Rosa et al., 2009). On the one hand, sugars are involved in metabolic pathways producing ROS; on the other hand, via the pentose phosphate cycle, they activate NADPH-generating reactions and promote ROS quenching. Interaction between signal pathways activating sugar- and ROS-dependent genes makes it possible to keep balance between the effects of carbohydrates and ROS.

When sugars are deficient, the rate of ADP regeneration considerably decreases, electron transport via cytochrome *c* oxidase is suppressed, and production of ROS in mitochondria intensifies (Dutilleul et al., 2003a,b). Active mobilization of lipids also contributes to accumulation of ROS. At a high concentration of sugars, mobilization of fatty acids is reduced and ROS accumulation is less intense. Affecting photosynthesis, mitochondrial respiration, and metabolism of fatty acids, sugars play a key role in stabilization of ROS production.

The role of osmotically active substances, including proline and sugars, in protection against dehydration of macromolecules and membrane lipid bilayers is well-known (Shao et al., 2006; Korn et al., 2008; Krasensky and Jonak, 2012). Promoting retention of intracellular water, osmolytes modify its properties; they can interfere in interaction between water and biomolecules. At low humidity and high osmolarity of the medium, water leaves the cells impairing hydration of macromolecules. Without hydration, it is impossible to preserve protein structure in active state. Direct interaction of osmolytes with polar groups of membrane phospholipids stabilizes them. For instance, molecules of sugars can replace water molecules bound to phospholipids head groups of membranes or to proteins and H-bonds to stabilize them (Bentsink et al., 2000; Hincha et al., 2006; Livingston et al., 2009). Sugars maintain membrane hydration and prevent membrane fusion preserving the space between phospholipid molecules (Valluru and van den Ende, 2008). Thereby, they maintain the structure of macromolecules and membrane functioning. In this respect, proline and sugars act likewise.

7. TRANSCRIPTION FACTORS

Besides the factors influencing directly onto HMs translocation or protective synthesis of proteins and other components, protecting the cells from damage (e.g., antistress protection) in improving the tolerance of plants involves molecules involved in signal transduction and regulation of expression of specific protective genes. The complexity of responses to stress conditions and the great number of processes participating in plant protection suggest necessity of integrated regulation of protection against various stresses. In this respect, TFs regulating expression of several groups of genes are important (Pardo, 2010; Xu et al., 2011; Atkinson and Urwin, 2012).

TFs, the most important regulators of gene expression, consist of at least four domains—a DNA-binding domain, a signal sequence for nuclear localization, a transcription activation domain, and oligomerization site—that operate in conjunction. TFs regulate many physiological and biochemical processes modulating the rate of initiation of transcription of target genes. Among regulated TF processes are response to environmental stress and participation in signaling pathways (Abe et al., 2003; Li et al., 2006; Kazuhiro et al., 2005; Prouse and Campbell, 2012, 2013). Expression of target genes is regulated by binding TFs to specific DNA sequences (*cis*-acting elements) in the promoters of target genes, thereby enhancing or repressing transcriptional activity (Prouse and Campbell, 2012, 2013). In addition, TFs can stoichiometrically react with another protein, which can form a protein–DNA complex specific to the DNA sequence.

Translocation of TFs into the cell nucleus regulates transcription. TFs participate in assembly of an elementary transcriptional complex common for a great number of genes, which is a subject of regulation. TFs can modify the process of assembly and strengthen or weaken a preinitiation complex (PIC). On the other hand, TFs participate in assembly of regulatory complexes different from PIC, which interact with the elementary complex and specifically adjust it to a gene or group of genes. Regulation occurs by means of changing catalytic activity of already assembled transcribing complex in the stages of initiation, elongation, or termination. Great diversity of TFs and *cis*-active elements accounts for a baffling combinative complexity, which ensures fine control over gene expression and accounts for a wide developmental and physiological ranges. TFs ensure a fine and precise mechanism governing myriad of process simultaneously occurring in plants.

Another important aspect of TFs operation is regulation of distribution of nutrients between different organs or cell functions, for instance, between growth processes and protective reactions (Nakashima et al., 2007; AbuQamar et al., 2009; Tsutsui et al., 2009).

Such an important role of TFs in many aspects of coordination and regulation of plant vital functions makes it promising to produce transgenic plants expressing different TFs. For instance, Xu et al. (2009b) produced transgenic plants of *Arabidopsis* with a high expression of transcription factor ACE1. At the same time, another group of researchers identified a genetic regulator, SlAIM1, which integrates plant responses to pathogens and abiotic stresses (salt and oxidative stress) by modulating responses to ABA (AbuQamar et al., 2009). Soon, it was found interaction between signaling pathways regulating resistance to Botrytis cinerea infection and cold, drought, and oxidative stresses (Sham et al., 2014). Another regulator of stress response in *Arabidopsis*, class A1 heat shock factor (HSFA1, also controls several reactions by playing a role in heat shock responses, participate in other abiotic stress responses, are involved in growth and development (Sato and Yokoya, 2008; Liu et al., 2011). Seedlings of transgenic plants grew much better on the medium with an elevated concentration of Cu than the plants of wild type; these transgenic plants also accumulated less MDA and had higher concentration of antioxidant enzymes.

TFs are classified into the family or superfamily in accordance with structure of their conservative DNA-binding domain. One such family is the family of MYB genes, whose members can be expressed in all organs of plants.

7.1 Myb FACTORS IN PLANTS

The first plant Myb-gene, C1, was detected in maize (*Zea mays* L.). It was shown that this gene encodes c-Myb–like transcription factor involved in the biosynthesis of anthocyanins in the aleurone layer of the caryopsis (Paz-Arez et al., 1987). The Myb family of TFs is the largest in composition among all the currently known in plants. In *Arabidopsis* 135 genes belonging to this family were identified, and 108 genes in grapes (Matus et al., 2008). In the past decade, this family has been the object of intensive research because its involvement of its various representatives in a variety of physiological and biochemical processes were discovered (Du et al., 2009; Xu et al., 2011; Atkinson and Urwin, 2012; An et al., 2015; Chen et al., 2015).

R2R3-MYB proteins of subgroup 12 regulate glucosinolate biosynthesis and biosynthesis of aliphatic glucosinolates. Proteins encoded by these genes determine the type of epidermal cells, control initiation of trichomes, govern distribution of root hairs and trichomes, and perform many other functions (according to Park et al., 2010). Several R2R3-MYBs participate in regulation of biosynthesis of flavonoids in all the tissues; others control biosynthesis of anthocyans in vegetative tissues or the synthesis of proanthocyanidins in seed cover and biosynthesis of tannins. *Myb* genes control biosynthesis of cell wall components: lignin in fibers and/or vessels, xylans, cellulose, and cell wall thickening (Zhong et al., 2010). AtMYB58, AtMYB63 (subgroup three) and AtMYB85 activate lignin biosynthesis, but AtMYB68 negatively influences lignin deposition in the roots. *PvMyb4* isolated from genome of millet (*Panicum virgatum*) encodes a protein also suppressing transcription of the genes participating in the synthesis of lignin (Shen et al., 2012). *AtMYB61* exerts a pleiotropic effect: this gene affects lignin deposition, production of mucilage, and stomatal aperture.

Gene *DkMyb4* from persimmon plants (*Diospyros kaki*) participated in the operation of the genes governing synthesis of proanthocyanidins (Akagi et al., 2009; Dubos et al., 2010). Gene *AtMyb4* is a key regulator of phenylpropanoid pathway suppressing expression of the gene encoding cinnamate 4-hydroxylase; as a result, plants become more resistant to various stress agents (Jin et al., 2000). *AtMyb4* regulates operation of the gene expressing SAD2, an importin-like protein, which confers resistance to UV on the plants of *Arabidopsis* (Zhao et al., 2007).

Such functions of MYB proteins are related to regulation of secondary metabolism in plants (Borevitz et al., 2000; Jin et al., 2000; Nesi et al., 2001; Baudry et al., 2004; Patzlaff et al., 2003; Gomez-Maldonado et al., 2004; Goicoechea et al., 2005), control over cell morphogenesis (Lee et al., 1999; Lee and Schiefelbein, 2001; Higginson et al., 2003), regulation of meristem formation and development of flowers and seeds (Penfield et al., 2001; Schmitz et al., 2002; Shin et al., 2002; Steiner-Lange et al., 2003), control over cell cycle (Ito et al., 2001; Araki et al., 2004), and protective reactions and responses to stresses (Abe et al., 2003; Denekamp and Smeekens, 2003; Nagaoka and Takano, 2003; Feng and Andreasson, 2004). In other words, Myb proteins are a key factor regulating the development of plants, their metabolism, and responses to biotic and abiotic stresses (Dubos et al., 2008, 2010).

Many members of the family of *myb* genes, for instance, *Atmyb4*, *Atmyb6*, *Atmyb7*, *Atmyb44*, *Atmyb73*, *Atmyb77*, and *AtmybCDC5*, are expressed constitutively (Yanhui et al., 2006; Li et al., 2006; Cominelli et al., 2005; Vannini et al., 2004, 2006). Others are expressed only under stress. Some *myb* genes participate in drought response. For instance, R2R3-MYB transcription factor in *Arabidopsis Atmyb2*, is induced under dehydration operating as an activator of transcription in the course of expression of ABA-induced genes (Abe et al., 2003). Other *myb* genes may be induced by light. For instance, the genes *Atmyb4*, *Atmyb21*, *Atmyb60,* and *Atmyb61* isolated from *Arabidopsis* were induced by light (Li et al., 2006; Cominelli et al., 2005). MYB factors govern the responses to dehydration, intense illumination, low temperature, and shortage of oxygen (Soltész et al., 2012; Vannini et al., 2004, 2006, 2007). They participate in signal transduction of salicylic acid (SA) (Raffaele and Rivas, 2006), ABA, gibberellins, and jasmonic acid (JA) (Lee and Schiefelbein, 2001; Borevitz et al., 2000; Jin and Martin, 1999).

The genes of this family may be induced in response to simultaneous action of several factors. For instance, Denekamp and Smeekens (2003) showed that the gene *Atmyb102* is induced both by osmotic and saline stress as well as by injury of plant tissues, and genes *Atmyb15*, *Atmyb41*; *Atmyb44*, *Atmyb73,* and *Atmyb77* are activated upon injury (Cheong et al., 2002), under the effect of light (Ma et al., 2005) as well as at salinization, drought, and cold (Fowler and Thomashow, 2002; Kamei et al., 2005; Agarwal et al., 2006; Vannini et al., 2004, 2006). Integrated response of *myb* TFs to several stresses may be demonstrated in *myb96* that is induced by drought and creates ABA-dependent resistance by means of biosynthesis of cuticular waxes (Seo et al., 2011). Moreover, it is necessary for ABA-dependent SA biosynthesis and elevates expression of PR genes related to pathogen resistance (Seo and Park, 2010). Expression in transgenic plants of TaPIMP1 MYB factor from wheat ensures resistance to pathogens, drought, and salt stresses (Liu et al., 2011).

Transgenic plants of *Arabidopsis* and tomatoes expressing the rice *myb*4 gene also respond to a wide range of stress agents (Vannini et al., 2006, 2007). Such a complexity of plants' responses to stress allows plants to develop a comprehensive nonspecific resistance (Schenke et al., 2011; Zhang et al., 2014). For instance, the *myb* factor BOS1 is necessary for response to both biotic and abiotic stresses (Mengiste et al., 2003).

This makes it possible to finely regulate protective reactions under environmental conditions to which the plant is exposed. To make built-in TF transgenic plants operate properly and improve resistance to specific stresses, appropriate stress-inducible promoters are sometimes used (Kasuga et al., 1999; Qiu and Yu, 2009).

Improvement of resistance to excess Cu ions may be also caused by the incorporation into plants with Tfs genes regulating expression of numerous genes responsible for plant response to stress effects. For instance, transgenic tobacco plants bearing the TF gene *LbDREB* (cloned from *Limonium bicolor*) that plays an important role in stress resistance of plants showed their greater resistance to treatment with high concentrations of Cu ions. Under stress conditions, transgenic plants accumulated more biomass, had higher concentration of proline, and much lower peroxidation of lipids recorded by accumulation of MDA (Ban et al., 2011).

7.2 PROTEIN Osmyb4

One of the most widespread transfactors in plants is an MYB family of TFs (Du et al., 2009). A common feature of MYB proteins is a functional DNA-binding domain conserved in animals, yeast, and plants and usually consisting of several (from one to three) imperfect repeats (R1, R2, and R3) (Du et al., 2009). Transfactor protein MYB4 was recently isolated from coleoptile of rice (cv. Arborio) and the gene encoding its synthesis of *Osmyb4* (accession number Y11414) was identified (Pandolfi et al., 1997). This gene encoding a protein with two R2 and R3 repeats was used for transformation of several species of plants (Vannini et al., 2004, 2006, 2007; Laura et al., 2010; Pasquali et al., 2008; Docimo et al., 2013; Gomaa et al., 2012; Soltész et al., 2012; Aydin et al., 2014). It was found that transgenic plants of *Arabidopsis* as well as the rapeseed plant expressing this gene were more resistant to low above-zero and below-zero temperatures (Vannini et al., 2004; Gomaa et al., 2012). Other types of transgenic plants with a gene *Osmyb4* were resistant to drought (Mattana et al., 2005; Vannini et al., 2006; Pasquali et al., 2008; Soltész et al., 2012). Incubation of a perennial ornamental plant *Osteospermum ecklonis* during 10 days at 4 °C or freezing overnight (at −5 °C) was endured only by transgenic plants with a high extent of gene expression (Laura et al., 2010). Plant resistance to salinization, UV light, ozone, and pathogens (viruses, bacteria, and fungi) also improved (Vannini et al., 2006). The gene was induced by cold and conferred on transgenic plants of apple, maize, tomato, sage, and tobacco resistance to chill (Mattana et al., 2005; Mattana, 2010).

Ectopic gene expression in different plants improved their resistance to other abiotic stresses (Vannini et al., 2004, 2006; Pasquali et al., 2008; Laura et al., 2010; Park et al., 2010; Aydin et al., 2014). Recently, it was reported that transgenic barley expressing *Osmyb4*-gene under the control of the *Arabidopsis* cold-inducible promoter p*Cor15a* was more resistant not only to freezing, but to hypoxia as well (Soltész et al., 2012). Resistance to freezing was not constitutive, but turned out to be induced by a short-term exposure to low above-zero temperatures (i.e., by cold acclimation) (Laura et al., 2010). *Arabidopsis* is capable of adapting to low temperatures, and ectopic expression of *Osmyb4* therein brought about physiological reactions resembling the process of adaptation (Vannini et al., 2004). In different plant species, the gene *Osmyb4* can cause resistance to different types of stress. For instance, transgenic tomato plants were more resistant to drought and viral infection but not to cold (Vannini et al., 2004, 2007). The gene *Osmyb4* in different plant species can cause resistance to various stresses.

Transgenic potato plants with both a constitutive promoter p*CaMV35S* and a cold-inducible promoter p*COR15*a were significantly more tolerant of salinity than the wild type, according to measured growth parameters. Transgenic plants grown under a high concentration of boric acid (3 mM) exhibited greater survival rates than nontransgenic control plants. On the other hand, the transgenic plants did not show an improvement in freezing tolerance (Aydin et al., 2014).

A wide range of protective processes activated in plants superexpressing the *Osmyb4*-gene is related to its upstream position in a long and branched transcriptional chain. Therefore, concurrently with *Osmyb4*, other transcriptional clusters are activated (Park et al., 2010; Aydin et al., 2014). Such a position of the gene in the sequence of responses results in pleiotropy of its action. Apparently, in cooperation with other TFs *Osmyb4* coactivates or coinhibits expression of other specific genes that are expressed downstream of the Myb4 transcription factor, including those performing protective functions (Koornnef and Pieterse, 2008). Numerous *Osmyb4*-dependent chains are regulated by ABA, ROS, ethylene, JA, or SA. Apparently, all of these determine the level of tolerance and the nature of cold-induced response of transgenic plants (Pasquali et al., 2008; Vannini et al., 2007).

We worked with transgenic plants of canola *Brassica napus* L. expressing the *Osmyb4* gene under control of stress-induced promoter p*Cor15a*. Under normal conditions, this promoter is inactive but it is activated by cold, drought, and probably by other stress agents. As in articles cited previously, transgenic plants were more resistant to low above-zero temperatures (4 °C). It is significant that in our experiments with rapeseed, the same as in the experiments conducted by Soltész et al. (2012) with barley, transgenic plants did not slow down their growth in contrast to dwarf plants in the experiments by Mattana et al. (2005) in which the *Osmyb4* gene was under constitutive promoter p*CaMV35S* (Mattana et al., 2005). It is unexpected that transgenic tomato carrying *Osmyb4* under the p*Cor15a* promoter did not improve frost resistance (Vannini et al., 2007). Apparently, different responses of individual species depend on whether or not the chain contains downstream genes necessary for operation of specific genes of resistance (Soltész et al., 2012).

Among the causes inducing improvement of resistance to cold, freezing, and drought, the main is accumulation of compatible osmotica. It was shown that the plants expressing *Osmyb4* gene exposed to the effect of low above-zero and below-zero temperatures accumulated more sugars (glucose, fructose, and sucrose), glycine betaine, and proline (Laura et al., 2010; Gomaa et al., 2012). After 5 or 10 days of cold exposure, wild-type plants accumulated glucose more intensely; in transgenic plants with high expression, the accumulation was lower. Thus, it was concluded that improved resistance was not associated with glucose accumulation. The opposite case was observed in respect to sucrose and proline: their content rose most intensely in the lines with high gene expression, and exactly these compounds could act as agents improving cold resistance (Laura et al., 2010). This conclusion agrees with the data obtained for red raspberry. In the course of cold acclimation, a considerable accumulation of sucrose was observed in the tissues of this plant, and this accumulation correlated with the induction of cold resistance (Palonen et al., 2000).

In our experiments with wild-type canola, we also observed an intense accumulation of sugars at low above-zero temperature. However, transgenic plants expressing the *Osmyb4* gene greatly improved their resistance to chilling but did not accumulate extra sugars (Gomaa et al., 2012). Such a lack of relationship between improved resistance to low temperatures induced by the *Osmyb4* gene expression and accumulation of sugars was discovered in rice (Park et al., 2010). The genes involved in osmoregulation (P5CS, trehalose phosphatase, galactinol synthase) were activated upon chilling, but they did not respond to *Osmyb4* hyperexpression induced by cold (Park et al., 2010).

Discrepancy between our results and the data obtained by the group of Mattana (Vannini et al., 2004) could depend on the fact that in our experiments the *Osmyb4*-gene was expressed under the control of p*Cor15a* promoter and in appearance the experimental plants looked like control plants. In the experiments performed by the group of Mattana, expression was controlled by constitutive promoter p*CaMV35S* and transgenic plants were dwarf. Probably, this fact influenced assimilated partitioning: a weaker efflux upon steady level of photosynthesis could lead to accumulation of assimilates in source leaves of dwarf plants.

Proline content rose in transgenic plants even without stress influences and continued to grow within several-days-long stay at low temperatures. It was assumed that Osmyb4 protein affects proline metabolism (Laura et al., 2010). For instance, proline accumulation in transgenic plants could be accounted for by the fact that in such plants the level of messenger RNAs coding for D1-pyrroline-5-carboxylate synthase was higher and the level of D1-pyrroline-5-carboxylate dehydrogenase was lower (Mattana et al., 2005).

We did not find available reports pointing to changes in sensitivity to HMs of transgenic plants carrying the *Osmyb4*-gene. In our experiments, transgenic canola plants turned out to be more resistant to the medium containing 150 μM Cu. Seven-day-long exposure of wild-type plants to such a concentration of Cu ions caused a slight delay in growth, but a 15-day-long stay of such plants on the medium containing Cu ions caused their death. However, transgenic plants remained viable (unpublished).

After 7 days on the solution that was free of Cu ions, the leaves of transgenic plants contained more sucrose and fructose than common plants (120% and 180%, respectively). Accumulation of sugars depended on Cu concentration in the solution. It was most pronounced in response to 100 μM

Cu, when the content of all the sugars rose almost twice. At a concentration of 50 and 150 μM, $CuSO_4$ essentially did not change the content of sugars. After longer incubation on 100–150 μM Cu solution (for 15 days), common plants died and transgenic plants survived. The content of sugars therein did not differ from the level in plants on 100 μM. The change in the sugar content depends on the species of plants: in the resistant-to-stress Cu aquatic plant *Salvinia auriculata* Aubl. at the same Cu concentration in the medium, sugar content decreased markedly (Bizzo et al., 2014).

The content of proline in the leaves of transgenic plants not exposed to the effect of HMs essentially did not differ from that in common plants. There were no differences in the content of proline in the leaves of transgenic plants incubated for 7 days on the medium containing 250 μM or 50 μM $CuSO_4$. However, on the medium with 100 μM Cu, the content of proline rose significantly. One can conclude that elevation of the content of sugars and especially of proline in the leaf tissues of transgenic plant improves canola resistance to the effect of Cu.

One of the reasons for improved viability of transgenic plants may be a lesser extent of oxidative stress developing therein. The genes ensuring redox homeostasis of the cell are expressed: many glutaredoxin GPXs, phospholipid hydroperoxide GPX, L-APX POD, and various components of the glutathione redox system, such as glutathione S-transferase, glutathione synthetase, and lactoglutathione lyase (Park et al., 2010). Total antioxidase activity in transgenic plants was elevated. Many genes involved in the radical scavenging system are susceptible to the effect of Osmyb4.

The article prepared for publication by Raldugina et al. reports that exposure to $CuSO_4$ considerably elevated the content of MDA in the leaves of wild-type and transgenic plants of rapeseed expressing the *Osmyb4* gene under the control of pCor15a promoter induced by low above-zero temperatures. Nevertheless, the leaves of transgenic plants accumulated less MDA (up to 12.5 as compared with 18.6 nmol g^{-1} FW in wild-type plants). This may point to a less pronounced oxidative stress. Most often, oxidative stress is accompanied by a brief or prolonged increase in the activity of SOD and PODs. However, we did not observe it in our experiments in spite of intense manifestation of oxidative stress in canola plants. Moreover, in rapeseed plants grown with excess $CuSO_4$, there was a reliable decrease in the activity of SOD and POX as compared with incubation on standard medium. This could point to a profound disturbance of metabolic processes caused by long exposure (15 days) to HMs associated with their accumulation in the leaves of canola plants at high concentrations.

Weaker oxidative stress could partially depend on the fact that in spite of a considerable inhibition of the activity of SOD and guaiacol POD in the presence of 100 μM $CuSO_4$ manifested in both groups of plants, in the leaves of transgenic plants it was less pronounced.

Another pathway leading to improvement of resistance is phenylpropanoid metabolic process. All phenylpropanoids are produced from cinnamic acid formed from phenyl alanine by phenylalanine ammonia lyase. Expression of this enzyme is activated in *Osmyb*-transgenic plants (Vannini et al., 2004). Production of some aromatic compounds rises accordingly. This pathway produces a great number of phenolic compounds playing an important role in protection against various biotic and abiotic stresses (Dubos et al., 2008; Docimo et al., 2008, 2013). For instance, in addition to participation in regulation of reproductive processes in plants, MYB is involved in regulation of biosynthesis of flavonoids that are important components of the system ensuring resistance to biotic and abiotic stresses. The genes of phenylpropanoid and isopropanoid metabolism regulated by ABA, JA, SA, ethylene, ROS, and wounding are expressed (from Park et al., 2010). The work performed with

Arabidopsis, tobacco, and sage showed an elevated expression of *PAL2* and synthesis of flavonoids with antioxidant functions (Docimo et al., 2008).

Osmyb4 is a conservative regulator of the phenylpropanoid pathway (Dubos et al., 2008, 2010; Park et al., 2010). It activates transcription of several genes of this pathway encoding phenylalanine ammonia-lyase, cinnamic acid 4-hydroxylase, 4-coumarate: Co A ligase1, and 4-coumarate: Co A ligase2 (from Park et al., 2010). Transgenic plants of canola accumulated much more low-molecular compounds with antioxidase properties including flavonoids and anthocyans.

Thus, *Osmyb4* controls a complex train of processes of metabolism, development, and protection. Essentially all the kinds of stress effects have a common component: oxidative stress. Therefore, creation of antistress protection improves overall plant resistance. Components of such protection are not the same in different plants under different stress conditions.

8. CONCLUSION AND FUTURE PROSPECTS

In spite of increasing environmental pollution and urgency of improvement of plant resistance to such contamination, how to raise the level of plant resistance is not worked out. An important aspect of combating pollution with HMs is recultivation of ruined soils. Numerous ecologists are engaged in selecting plants fit for remediation. To solve this problem, more details on how detoxification of metals in soils, mechanisms of Cu uptake by roots, and its further distribution in the plants must be understood. The possibility of carrying of the metal up to the above-ground organs of plants and stimulation of this process is an important aspect of development of methods of remediation of contaminated soils. Cu distribution in different plants should be investigated for the selection of species most suitable for practical use in remediation. Influence of microorganisms on the transport of Cu is an important, but not yet explored point. The results of such investigation could be used when changing of the microbiological composition of soils and the possibility of its participation in the phytoremediation. This aspect seems to be promising. It is still unclear the mechanisms of competition between Cu ions and other ions during uptake of them by roots. The aspect of change is also significant but still poorly developed in the ionic composition of the various plant organs in conditions of excess Cu in the environment. Meanwhile, the function of many metabolic reactions that determine not only the growth and harvest but also the tolerance of the plant are dependent of the content of the necessary elements potassium, phosphorus, and nitrogen. The influence of Cu ions on hormonal balance also requires further development because this issue is also poorly understood. Application of genetic engineering requires knowledge of the mechanisms of influence of pollutants on physiological processes in plants, which are also poorly investigated. As to other HMs, plant response to Cu stress is scantily known. Certain genes responsible for the processes most vulnerable to Cu (photosynthesis, mitochondrial respiration, and growth) were identified. However, because the phenomenon of resistance is complex and the agents interacting in nature are numerous, expression of single genes will not ensure creation of plants capable of surviving under varying environmental condition. The way out of this situation may be to search for common genes expressed in the recruitment of different stress factors. The most promising prospect is construction of plants expressing the genes of transfactor proteins that regulate operation of a great number of genes responsible for different aspects of metabolism. This chapter briefly describes transgenic plants expressing the *OsMyb* gene. The work of earlier investigators showed that this protein improves plant resistance to such factors as cold and drought; however, its importance for improving the resistance of plants to HMs has not yet been studied. We conducted the first experiments, which showed increased resistance of transgenic canola to the excess Cu in the medium. Such studies are useful to extend to other HMs.

REFERENCES

Abdel-Ghany, S.E., Müller-Moul, P., Niyogi, K.K., Pilon, M., Shikanai, T., 2005. Two P-type ATPases are required for copper delivery in *Arabidopsis thaliana* chloroplasts. Plant Cell 17, 1233–1251.

Abdel-Ghany, S.E., Pilon, M., 2008. MicroRNA-mediated systemic down-regulation of copper protein expression in response to low copper availability in Arabidopsis. J. Biol. Chem. 283, 15932–15945.

Abe, H., Urao, T., Ito, T., Seki, M., Shinozaki, K., Yamaguchi-Shinozaki, K., 2003. Arabidopsis AtMYC2 (bHLH) and AtMYB2 (MYB) function as transcriptional activators in abscisicK signaling. Plant Cell 15, 63–78.

Abogadallah, G.M., 2010. Antioxidative defense under salt stress. Plant Signal Behav. 5, 369–374.

AbuQamar, S., Luo, H., Laluk, K., Mickelbart, M.V., Mengiste, T., 2009. Crosstalk between biotic and abiotic stress responses in tomato is mediated by the AIM1 transcription factor. Plant J. 58, 347–360.

Agarwal, M., Hao, Y., Kapoor, A., Dong, C.H., Fuji, H., Zheng, X., Zhu, J.K., 2006. A R2R3 type MYB transcription factor is involved in the cold regulation of CBF genes and in acquired freezing tolerance. Biol. Chem. 281, 37636–37645.

Ain-Lhout, F., Zunzunegui, M., Diaz Barradas, M.C., Tirado, R., Clavijio, A., Garcia Novo, F., 2001. Comparison of proline accumulation in two subjected to natural and experimental water deficit. Plant Soil 230, 175–183.

Akagi, T., Ikegami, A., Tsujimoto, T., Kobayashi, S., Sato, A., Kono, A., Yonemori, K., 2009. DkMyb4 is a Myb transcription factor involved in proanthocyanidin biosynthesis in persimmon fruit. Plant Physiol. 151, 2028–2045.

Alia, S.P.P., Mohanty, P., 1997. Involvement of proline in protecting thylakoid membranes against free radical-induced photodamage. J. Photochem. Photobiol. 38, 253–257.

Alaoui-Sossé, B., Genet, P., Vinit-Dunand, F., Toussaint, M.-L., Epron, D., Badot, P.-M., 2004. Effect of copper on growth in cucumber plants (*Cucumis sativus*) and its relationships with carbohydrate accumulation and changes in ion contents. Plant Sci. 166, 1213–1218.

Amudha, J., Balasubramani, G., 2011. Recent molecular advances to combat abiotic stress tolerance in crop plants. Biotech. Mol. Biol. Rev. 6, 31–58.

An, X.H., Tian, Y., Chen, K.Q., Liu, X.J., Liu, D.D., Xie, X.B., Cheng, C.G., Cong, P.H., Hao, Y.J., 2015. MdMYB9 and MdMYB11 are involved in the regulation of the JA-induced biosynthesis of anthocyanin and proanthocyanidin in apples. Plant Cell Physiol. 56 (4), 650–662.

Andres-Colas, N., Perea-Garcıa, A., Puig, S., Penarrubia, L., 2010. Deregulated copper transport affects Arabidopsis development especially in the absence of environmental cycles. Plant Physiol. 153, 170–184.

Anoop, V.M., Basu, U., McCammon, M.T., McAlister-Henn, L., Taylor, G.J., 2003. Modulation of citrate metabolism alters aluminum tolerance in yeast and transgenic canola overexpressing a mitochondrial citrate synthase. Plant Physiol. 132, 2205–2217.

Apel, K., Hirt, H., 2004. Reactive oxygen species: metabolism, oxidative stress, and signal transduction. Annu. Rev. Plant Biol. 55, 373–399.

Araki, S., Ito, M., Soyano, T., Nishihama, R., Machida, Y., 2004. A-myb is expressed in bovine vascular smooth muscle cells during the late G1-to-S phase transition and cooperates with c-*myc* to mediate progression to S phase. Biol. Chem. J. 279, 32979–32988.

Arduini, I., Godbold, D.L., Onnis, A., 1995. Influence of copper on root growth and morphology of *Pinus pinea* L and *Pinus pinaster* Ait Seedlings. Tree Physiol. 15, 411–415.

Argüello, J.M., Eren, E., González-Guerrero, M., 2007. The structure and function of heavy metal transport P_{1B}-ATPases. Biometals 20, 233–248.

Argüello, J.M., Raimunda, D., Padilla-Benavides, T., 2013. Mechanisms of copper homeostasis in bacteria. Front. Cell Infect. Microbiol. 3, 73.

Arru, L., Regnoni, S., Baroncini, M., Bonatti, P.M., Perata, P., 2002. Copper localization in *Cannabis sativa* L grown in copper-rich solution. Euphytica 140, 33–38.

Ashraf, M., Fooland, M.R., 2007. Roles of glycine betaine and proline in improving plant abiotic stress resistance. Environ. Exp. Bot. 59, 206–216.

Atkinson, N.J., Urwin, P.E., 2012. The interaction of plant biotic and abiotic stresses: from genes to the field. J. Exp. Bot. 63, 3523–3543.

Attri, P., Venkatesu, P., Lee, M.-J., 2010. Influence of osmolytes and denaturants on the structure and enzyme activity of chymotrypsin. J. Phys. Chem. B 114, 1471–1478.

Aydin, G., Yucel, M., Chan, M.T., Oktem, H.A., 2014. Evaluation of abiotic stress tolerance and physiological characteristics of potato (*Solanum tuberosum* L. cv. Kennebec) that heterologously expresses the rice *Osmyb4* gene. Plant Biotechnol. Rep. 8, 295–304.

Azevedo, M.M., Carvalho, A., Pascoal, C., Rodrigues, F., Cássio, F., 2007. Responses of antioxidant defenses to Cu and Zn stress in two aquatic fungi. Sci. Total Environ. 15, 233–243.

Backor, M., Fahselt, D., Davidson, R.D., Wu, C.T., 2003. Effects of copper on wild and tolerant stains of the lichen photobiont *Trebouxia erici* (Chlorophyta) and possible tolerance mechanisms. Arch. Environ. Contam. Toxicol. 45, 159–167.

Backor, M., Vaczi, P., Bartak, M., Budova, J., Dzubaj, A., 2007. Uptake photosynthetic characteristics and membrane lipid peroxidation levels in the lichen photobiont *Trebouxia erici* exposed to copper and cadmium. Bryologist 110, 100–107.

Balandin, T., Castresana, C., 2002. AtCOX17 an Arabidopsis homolog of the yeast copper chaperone COX171. Plant Physiol. 129, 1852–1857.

Balestrazzi, A., Botti, S., Zelasco, S., Biondi, S., Franchin, C., Calligari, P., Racchi, M., Turchi, A., Lingua, G., Berta, G., Carbonera, D., 2009. Expression of the PsMT (A1) gene in white poplar engineered with the MAT system is associated with heavy metal tolerance and protection against 8-hydroxy-2'-deoxyguanosine mediated-DNA damage. Plant Cell Rep. 28, 1179–1192.

Ban, Q., Liu, G., Wang, Y., 2011. A DREB gene from Limonium bicolor mediates molecular and physiological responses to copper stress in transgenic tobacco. J. Plant Physiol. 168, 449–458. http://www.ncbi.nlm.nih.gov/pubmed/20951468.

Barón, M., Arellano, J.B., López Gorgé, J., 1995. Copper and photosystem II: a controversial relationship. Physiol. Plant 94, 174–180.

Barone, P., Rosellini, D., La Fayette, P., Bouton, J., Veronesi, F., Parrott, W., 2008. Bacterial citrate synthase expression and soil aluminum tolerance in transgenic alfalfa. Plant Cell Rep. 27, 893–901.

Bartels, D., Sunkar, R., 2005. Drought and salt tolerance in plants. Crit. Rev. Plant Sci. 24, 23–58.

Bassi, R., Sharma, S.S., 1993a. Changes in proline content accompanying the uptake of zinc and copper by *Lemna minor*. Ann. Bot. (Lond) 72, 151–154.

Bassi, R., Sharma, S.S., 1993b. Proline accumulation in wheat seedlings exposed to zinc and copper. Phytochem 33, 1339–1342.

Baszynski, T., Krupa, Z., 1995. Some aspects of heavy metal toxicity towards photosynthetic apparatus-direct an indirect effects on light and dark reactions. Acta Physiol. Plant 17, 177–191.

Baszynski, T., Tukendorf, A., Ruszkowska, M., Skorzynska, E., Maksymiec, W., 1988. Characteristics of the photosynthetic apparatus of cooper non-tolerant spinach exposed to excess copper. J. Plant Physiol. 132, 708–713.

Baudry, A., Heim, M.A., Dubreucq, B., Caboche, M., Weisshaar, B., Lepiniec, L., 2004. TT2 TT8 and TTG1 synergistically specify the expression of BANYULS and proanthocyanidin biosynthesis in *Arabidopsis thaliana*. Plant J. 39, 366–380.

Ben Ahmed, C., Ben Rouina, B., Sensoy, S., Boukhris, M., Ben Abdullah, F., 2010. Exogenous proline effects on photosynthetic performances and antioxidant defense system of young olive. J. Agric. Food Chem. 58, 4216–4222.

Benatti, R.M., Yookongkaew, N., Meetam, M., Guo, W.J., Punyasuk, N., AbuQamar, S., Goldsbrough, P., 2014. Metallothionein deficiency impacts copper accumulation and redistribution in leaves and seeds of Arabidopsis. New Phytol. 202 (3), 940–951.

Benavides, M.P., Gallego, S.M., Tomaro, M.L., 2005. Cadmium toxicity in plants. Brazil J. Plant Physiol. 17, 21–34.

Bentsink, L., Alonso-Blanco, C., Vreugdenhil, D., Tesnier, K., Groot, S.P.C., Koornneef, M., 2000. Genetic analysis of seed soluble oligosaccharides in relation to seed storability of Arabidopsis. Plant Physiol. 124, 1595–1604.

Berglund, A., Quartacci, M.F., Calucci, L., Navari-Izzo, F., Pinzino, C., Liljenberg, C., 2002. Alterations of wheat root plasma membrane lipid composition induced by copper stress result in changed physicochemical properties of plasma membrane lipid vesicles. Biochim. Biophys. Acta 1564, 466–472.

Bernal, M., Casero, D., Singh, V., Wilson, G.T., Grande, A., Yang, H., Dodani, S.C., Pellegrini, M., Huijser, P., Connolly, E.L., Merchant, S.S., Krämer, U., 2012. Transcriptome sequencing identifies SPL7-regulated copper acquisition genes FRO4/FRO5 and the copper dependence of iron homeostasis in Arabidopsis. Plant Cell 24 (2), 738–761.

Bernal, M., Sanches-Testillano, P., Risueno, M.D., Yruela, I., 2006. Excess copper induces structural changes in cultured photosynthetic soybean cells. Funct. Plant Biol. 33, 1001–1012.

Billard, V., Ourry, A., Maillard, A., Garnica, M., Coquet, L., Jouenne, T., Cruz, F., Garcia-Mina, J., Yvin, J.-C., Etienne, P., 2014. Copper-deficiency in *Brassica napus* induces copper remobilization, molybdenum accumulation and modification of the expression of chloroplastic proteins. PLoS ONE 9 (10), e109889. http://dx.doi.org/10.1371/journal.pone.0109889.

Bizzo, A.L.T., Intorne, A.C., Gomes, P.H., Suzuki, M.S., dos Santos Esteves, B., 2014. Short-term physiological responses to copper stress in *Salvinia auriculata*. Aubl. Acta Limnol. Brasil 26, 268–277.

Blaby-Haas, C.E., Padilla-Benavides, T., Stübe, R., Argüello, J.M., Merchant, S.S., 2014. Evolution of a plant-specific copper chaperone family for chloroplast copper homeostasis. Proc. Nat. Acad. Sci. 111, E5480–E5487.

Bläsing, O.E., Gibon, Y., Günther, M., Hehne, M., Morcuende, R., Osuna, D., et al., 2005. Sugars and circadian regulation make major contributions to the global regulation of diurnal gene expression in Arabidopsis. Plant Cell 17, 3257–3281.

Bohnert, H.J., Jensen, R.G., 1996. Metabolic engineering for increased salt tolerance—the next step: comment. Aust. J. Plant Physiol. 23, 661–666.

Bolouri-Moghaddam, M.R., Roy, K.L., Li, X.L., Rolland, F., van den Ende, W., 2010. Sugar signaling and antioxidant network connections in plant cells. FEBS J. 277, 2022–2037.

Bonfig, K.B., Gabler, A., Simon, U.K., Luschin-Ebengreuth, N., Hatz, M., Berger, S., Muhammad, N., Zeier, J., Sinha, A.K., Roitsch, T., 2010. Post-translational derepression of invertase activity in source leaves via down-regulation of invertase inhibitor expression is part of the plant defense response. Mol. Plant 3, 1037–1048.

Borevitz, J.O., Xia, Y., Blount, J., Dixon, R.A., Lamb, C., 2000. Activation tagging identifies a conserved MYB regulator of phenylpropanoid biosynthesis. Plant Cell 12, 2383–2394.

Bowler, C., Vam Montagu, M., Inze, D., 1992. Superoxide dismutase and stress tolerance. Ann. Rev. Plant Physiol. Plant Mol. Biol. 43, 83–116.

Burkhead, J.L., Reynolds, K.A.G., Abdel-Ghany, S.E., Cohu, C.M., Pilon, M., 2009. Copper homeostasis. New Phytol. 182, 799–816.

Cambrollé, J., Mancilla-Leytón, J.M., Muñoz-Vallés, S., Figueroa-Luque, E., Luque, T., Figueroa, M.E., 2013. Effects of copper sulfate on growth and physiological responses of Limoniastrum monopetalum. Environ. Sci. Pollut. Res. Int. 20, 8839–8847. http://www.ncbi.nlm.nih.gov/pubmed/23744160.

Carr, H.S., Winge, D.R., 2003. Assembly of cytochrome c oxidase within the mitochondrion. Acc. Chem. Res. 36, 309–316.

Carrier, P., Baryla, A., Havaux, M., 2003. Cadmium distribution and microlocalization in oilseed rape (*Brassica napus*) after long-term growth on cadmium-contaminated soil. Planta 216, 939–950.

Cestone, B., Vogel-Mikuš, K., Quartacci, M.F., Rascio, N., Pongrac, P., Pelicon, P., Vavpetič, P., Grlj, N., Jeromel, L., Kump, P., Nečemer, M., Regvar, M., Navari-Izzo, F., 2012. Use of micro-PIXE to determine spatial distributions of copper in *Brassica carinata* plants exposed to $CuSO_4$ or CuEDDS. Sci. Total Environ. 427–428, 339–346.

Chattopadhyay, M.K., Kern, R., Mistou, M., Dandekar, A.M., Uratsu, S., Richarme, G., 2004. The chemical chaperone proline relieves the thermosensitivity of a dnaK deletion mutant at 42 °C. J. Bacteriol. 186, 8149–8152.

Chen, C., Chen, Y., Tang, I., Liang, H., Lai, C., Chiou, J., 2011. Arabidopsis SUMO E3 ligase SIZ1 is involved in excess copper tolerance. Plant Physiol. 156, 2225–2234.

Chen, C., Dickman, M.B., 2005. Proline suppresses apoptosis in the fungal pathogen *Colletotrichum trifolii*. Proc. Natl. Acad. Sci. U.S.A. 102, 3459–3464.

Chen, C.T., Chen, L., Lin, C.C., Kao, C.H., 2001. Regulation of proline accumulation in detached rice leaves exposed to excess copper. Plant Sci. 160, 283–290.

Chen, Z., Cuin, T.A., Zhou, M., Twomey, A., Naidu, B.P., Shabala, S., 2007. Compatible solute accumulation and stress-mitigating effects in barley genotypes contrasting in their salt tolerance. J. Exp. Bot. 58, 4245–4255.

Chen, T., Li, W., Hu, X., Guo, J., Liu, A., Zhang, B., 2015. A cotton myb transcription factor, GbMYB5, is positively involved in plant adaptive response to drought stress. Plant Cell Physiol. 56 (5), 917–929.

Cheong, Y.H., Chang, H.S., Gupta, R., Wang, X., Zhu, T., Luan, S., 2002. Transcriptional profiling reveals novel interactions between wounding, pathogen, abiotic stress, and hormonal responses in Arabidopsis. Plant Physiol. 129, 661–677.

Choudhary, S.P., Oral, H.V., Bhardwaj, R., Yu, J.-Q., Tran, L.-S.P., 2012. Interaction of brassinosteroids and polyamines enhances copper stress tolerance in *Raphanus sativus*. J. Exp. Bot. 63, 5659–5675.

Clemens, S., Palmgren, M.G., Krämer, U., 2002. A long way ahead: understanding and engineering plant metal accumulation. Trends Plant Sci. 7, 309–315.

Clemens, S., 2006. Toxic metal accumulation, responses to exposure and mechanisms of tolerance in plants. Biochimie 88, 1707–1719.

Cobbett, C., Goldsbrough, P., 2002. Phytochelatins and metallothioneins: roles in heavy metal detoxification and homeostasis. Annu. Rev. Plant Biol. 53, 159–182.

Cohu, C.M., Pilon, M., 2007. Regulation of superoxide dismutase expression by copper availability. Physiol. Plant 129, 747–755.

Cohu, C.M., Abdel-Ghany, S.E., Gogolin-Reynolds, K.A., Onofrio, A.M., Bodecker, J.R., Kimbrel, J.A., Niyogi, K.K., Pilon, M., 2009. Copper delivery by the copper chaperone for chloroplast and cytosolic copper/zinc-superoxide dismutases: regulation and unexpected phenotypes in an Arabidopsis mutant. Mol. Plant 2, 1336–1350.

Cohu, C.M., Pilon, M., 2010. Cell biology of copper. In: Hell, R., Mendel, R.R. (Eds.), Plant Cell Monographs: Cell Biology of Metals and Nutrients, vol. 17. Springer, Heidelberg, pp. 55–74.

Collin, B., Doelsch, E., Keller, C., Cazevieille, P., Tella, M., Chaurand, P., Panfili, F., Hazemann, J.L., Meunier, J.D., 2014. Evidence of sulfur-bound reduced copper in bamboo exposed to high silicon and copper concentrations. Environ. Pollut. 187, 22–30.

Cominelli, E., Galbiati, M., Vavasseur, A., Conti, L., Sala, T., Vuylsteke, M., Leonhardt, N., Dellaporta, S.L., Tonell, C., 2005. A guard-ctll-specific MYB transcription factor regulates stomatal movements and plant drought tolerance. Curr. Biol. 15, 1196–1200.

Cook, C.M., Kostidou, A., Vardaka, E., Lanaras, T., 1997. Effects of copper on the growth, photosynthesis and nutrient concentration of *Phaseolus* plants. Photosynthetica 34, 179–193.

Couée, I., Sulmon, C., Gouesbet, G., El Amrani, A., 2006. Involvement of soluble sugars in reactive oxygen species balance and responses to oxidative stress in plants. J. Exp. Bot. 57, 449–459.

Cruz de Carvalho, M.H., 2008. Drought stress and reactive oxygen species: production, scavenging and signaling. Plant Signal Behav. 3, 156–165.

Curie, C., Cassin, G., Couch, D., Divol, F., Higuchi, K., Jean, M., Misson, J., Schikora, A., Czernic, P., Mari, S., 2009. Metal movement within the plant: contribution of nicotianamine and yellow stripe 1-like transporters. Ann. Bot. 103, 1–11.

Curvetto, N.R., Rauser, W.E., 1979. Isolation and characterization of copper binding proteins from roots of *Agrostis gigantea* tolerant to excess copper. Plant Physiol. 63, 55–59.

Cushman, J.C., Bohnert, H.J., 2000. Genomic approaches to plant stress tolerance. Curr. Opin. Plant Biol. 3, 117–124.

Delhaize, E., Ryan, P.R., Hebb, D.M., Yamamoto, Y., Sasaki, T., Matsumoto, H., 2004. Engineering high-level aluminum tolerance in barley with the ALMT1 gene. Proc. Nat. Acad. Sci. U.S.A. 101, 15249–15254.

Denekamp, M., Smeekens, S.C., 2003. Integration of wounding and osmotic stress signals determines the expression of the *AtMYB102* transcription factor gene. Plant Physiol. 32, 1415–1423.

Deng, F., Yamaji, N., Xia, J., Ma, J.F., 2013. A member of the heavy metal P-type ATPase OsHMA5 is involved in xylem loading of copper in rice. Plant Physiol. 163, 1353–1362.

Deng, W., Luo, K., Li, Z., Yang, Y., Hu, N., Wu, Y., 2009. Overexpression of *Citrus junos* mitochondrial citrate synthase gene in *Nicotiana benthamiana* confers aluminum tolerance. Planta 230, 355–365.

Dey, S., Mazumder, P.B., Paul, S.B., 2014. Effect of copper on growth and chlorophyll content in tea plants (*Camellia Sinensis* (L.) O. Kuntze). Int. J. Res. Appl. Nat. Soc. Sci. 2, 223–230.

Diamant, S., Eliahu, N., Rosenthal, D., Goloubinoff, P., 2001. Chemical chaperones regulate molecular chaperones in vitro and in cells under combined salt and heat stress. J. Biol. Chem. 276, 39586–39591.

Dixon, N., Paiva, L., 1995. Stress-induced phenylpropanoid metabolism. Plant Cell 7, 1085–1097.

Dobra, J., Motyka, V., Dobrev, P., Malbeck, J., Prasil, I.T., Haisel, D., Gaudinova, A., Havlova, M., Gubis, J., Vankova, R., 2010. Comparison of hormonal responses to heat, drought and combined stress in tobacco plants with elevated proline content. J. Plant Physiol. 167, 1360–1370.

Docimo, T., Mattana, M., Fasano, R., Consonni, R., de Tommasi, N., Coraggio, I., Leone, A., 2013. Ectopic expression of the *Osmyb4* rice gene enhances synthesis of hydroxycinnamic acid derivatives in tobacco and clary sage. Biol. Plant 57, 179–183.

Docimo, T., Coraggio, I., de Tommasi, N., Leone, A., 2008. Enhancing phenylpropanoid secondary metabolites in *Nicotiana tabacum* and *Salvia sclarea* by overexpression of a rice Myb4 transcription factor. Planta Med. 74, 87.

Dominguez-Solis, J.R., Gutierrez-Alcala, G., Vega, J.M., Romero, L.C., Gotor, C., 2001. The cytosolic O-acetylserine(thiol)lyase gene is regulated by heavy metals and can function in cadmium tolerance. J. Biol. Chem. 276, 9297–9302.

Dominguez-Solis, J.R., Lopez-Martin, M.C., Ager, F.J., Ynsa, M.D., Romero, L.C., Gotor, C., 2004. Increased cysteine availability is essential for cadmium tolerance and accumulation in *Arabidopsis thaliana*. Plant Biotechnol. J. 2, 469–476.

Dörffling, K., Dörffling, H., Luck, E., 2009. Improved frost tolerance and winter hars in proline overaccumulating winter wheat mutants obtained by in vitro-selection is associated with increased carbohydrate, soluble protein and abscisic acid (ABA) levels. Euphytica 165, 545–556.

Droppa, M., Masojidek, J., Rózsa, Z., Wolak, A., Horváth, L.I., Farkas, T., Horváth, G., 1987. Characteristics of Cu deficiency-induced inhibition of photosynthetic electron transport in spinach chloroplasts. Biochim. Biophys. Acta 891, 75–84.

Drzewicka-Matuszek, A., Skalna, A., Karocki, G., Stochel, L.F., 2005. Effects of heavy central metal on the ground and excited states of chlorophyll. J. Biol. Inorg. Chem. 10, 453–462.

Du, H., Zhang, L., Liu, L., Tang, X.F., Yang, W.J., Wu, Y.M., Huang, Y.B., Tang, Y.X., 2009. Biochemical and molecular characterization of plant MYB transcription factor family. Biochemistry (Mosc) 74, 1–11. http://www.ncbi.nlm.nih.gov/pubmed/19232042.

Dubos, C., Le Gourrierec, J., Baudry, A., Huep, G., Lanet, E., Debeaujon, I., Routaboul, J.-M., Alboresi, A., Weisshaar, B., Lepiniec, L., 2008. MYBL2 is a new regulator of flavonoid biosynthesis in *Arabidopsis thaliana*. Plant J. 55, 940–953.

Dubos, C., Stracke, R., Grotewold, E., Weisshaar, B., Martin, C., Lepiniec, L., 2010. MYB transcription factors in Arabidopsis. Trends Plant Sci. 15, 573–581.

Dutilleul, C., Driscoll, S., Cornic, G., De Paepe, R., Foyer, C.H., Noctor, G., 2003a. Functional mitochondrial complex I is required by tobacco leaves for optimal photosynthetic performance in photorespiratory conditions and during transients. Plant Physiol. 313, 264–275.

Dutilleul, C., Garmier, M., Nocto, G., Mathieu, C., Chétrit, P., Foyer, C.H., 2003b. Leaf mitochondria modulate redox regulation in plants whole cell redox homeostasis, set antioxidant capacity and determine stress resistance through altered signaling and diurnal regulation. Plant Cell 15, 1212–1226.

El Hajji, H., Nkhili, E., Tomao, V., Dangles, O., 2006. Interactions of quercetin with iron and copper ions: complexation and autoxidation. Free Radic. Res. 40, 303–320.

Elefterion, E.P., Karataglis, S., 1989. Ultrastructural and morphological characteristics of cultivated wheat growing on copper-polluted fields. Bot. Acta 102, 134–140.

Elleuch, A., Chaâbene, Z., Grubb, D.C., Drira, N., Mejdoub, H., Khemakhem, B., 2013. Morphological and biochemical behavior of fenugreek (*Trigonella foenum-graecum*) under copper stress. Ecotoxicol. Environ. Saf. 98, 46–53.

Evers, D., Lefèvre, I., Legay, S., Lamoureux, D., Hausman, J.F., Gutierrez-Rosales, R.O., Tincopa-Marca, L.R., Hoffmann, L., Bonierbale, M., Schafleitner, R., 2010. Identification of drought responsive compounds in potato through a combined transcriptomic and targeted metabolite approach. J. Exp. Bot. 61, 2327–2343.

Feng, C., Andreasson, E., 2004. Arabidopsis MYB68 in development and responses to environmental cues. Plant Sci. 167, 1099–1107.

Fodor, F., 2002. Physiological responses of vascular plants to heavy metals. In: Prasad, M.N.V., Strzalka, K. (Eds.), Physiology and Biochemistry of Metal Toxicity and Tolerance in Plants. Kluwer Academic, Dortrech, UK, pp. 149–177.

Fowle, S., Thomashow, M.F., 2002. Arabidopsis transcriptome profiling indicates that multiple regulatory pathways are activated during cold acclimation in addition to the CBF cold response pathway. Plant Cell 14, 1675–1690.

Foy, C.D., Chaney, R.L., White, M.C., 1978. Physiology of metal toxicity in plants. Annu. Rev. Plant Physiol. Plant Mol. Biol. 29, 511–566.

Foyer, C.H., Noctor, G., 2005. Redox homeostasis and antioxidant signaling: a metabolic interface between stress perception and physiological responses. Plant Cell 17, 1866–1875.

Fraústo da Silva, J.J.R., Williams, R.J.P., 2001. The Biological Chemistry of the Elements: The Inorganic Chemistry of Life, second ed. Oxford University Press, New York, 575 pp.

Fu, F., Wang, Q., 2011. Removal of heavy metal ions from wastewaters: a review. J. Environ. Manage. 92, 407–418.

Garcia-Molina, A., Andrés-Colás, N., Perea-García, A., Neumann, U., Dodani, S.C., Huijser, P., Peñarrubia, L., Puig, S., 2013. The Arabidopsis COPT6 transport protein functions in copper distribution under copper-deficient conditions. Plant Cell Physiol. 54, 1378–1390.

Garg, A.K., Kim, J.-K., Owens, T.G., Ranwala, A.P., Choi, Y.D., Kochian, L.V., Wu, R.J., 2002. Trehalose accumulation in rice plants confers high tolerance levels to different abiotic stresses. Proc. Nat. Acad. Sci. 99, 15898–15903.

Gasic, K., Korban, S.S., 2007. Transgenic Indian mustard (*Brassica juncea*) plants expressing an Arabidopsis phytochelatin synthase (AtPCS1) exhibit enhanced as and Cd tolerance. Plant Mol. Biol. 64, 361–369.

Gayomba, S.R., Jung, H.I., Yan, J., Danku, J., Rutzke, M.A., Bernal, M., Krämer, U., Kochian, L.V., Salt, D.E., Vatamaniuk, O.K., 2013. The CTR/COPT-dependent copper uptake and SPL7-dependent copper deficiency responses are required for basal cadmium tolerance in *A. thaliana*. Metallomics 5, 1262–1275.

Giele, H., Remans, T., Vangronsveld, J., Cuypers, A., 2012. MicroRNAs in metal stress: specific roles or secondary responses? Int. J. Mol. Sci. 13, 15826–15847.

Goicoechea, M., Lacombe, E., Legay, S., Mihaljevic, S., Rech, P., 2005. EgMYB2, a new transcriptional activator from Eucalyptus xylem, regulates secondary cell wall formation and lignin biosynthesis. Plant J. 43, 553–567.

Gomaa, A.M., Raldugina, G.N., Burmistrova, N.A., Radionov, N.V., Kuznetsov, VIV., 2012. Response of transgenic rape plants bearing the *Osmyb4* gene from rice encoding a trans-factor to low above-zero temperature. Russ. J. Plant Physiol. 59 (1), 105–114.

Gomez-Maldonado, J., Avila, C., de la Torre, F., Canas, R., Canovas, F.M., 2004. Functional interactions between a glutamine synthetase promoter and MYB proteins. Plant J. 39, 513–526.

Gonzalez, D., Giegé, P., 2014. Biogenesis of the oxidative phosphorylation machinery in plants. From gene expression to complex assembly. Front. Plant Sci. 5, 225.

Grinin, A.L., 2010. Plant Resistance to Salinity and the Possible Role of Proline. Abstract of Cand Dissert, Moscow, 25 pp.

Grotz, N., Guerinot, M.L., 2006. Molecular aspects of Cu, Fe and Zn homeostasis in plants. Biochim. Biophys. Acta 1763, 595–608.

Guo, J., Xu, L., Su, Y., Wang, H., Gao, S., Xu, J., Que, Y., 2013. ScMT2-1-3, a metallothionein gene of sugarcane, plays an important role in the regulation of heavy metal tolerance/accumulation. Biomed. Res. http://dx.doi.org/10.1155/904769.

Gupta, M., Cuypers, A., Vangronsveld, J., Clijsters, H., 1999. Copper affects the enzymes of the ascorbate-glutathione cycle and its related metabolites in the roots of *Phaseolus vulgaris*. Physiol. Plant 106, 262–267.

Hall, J., Williams, E., 2003. Transition metal transporters in plants. J. Exp. Bot. 54, 2601–2613.

Hamida-Sayari, A., Gargouri-Bouzid, R., Bidani, A., Jaoua, L., Savoure, A., Jaoua, S., 2005. Overexpreccion of Δ^1-pyrroline-5-carboxylate synthetase increases proline production and confers salt tolerance in transgenic potato plants. Plant Sci. 169, 746–752.

Hanson, A.D., Nelsen, C.E., Everson, E.H., 1977. Evaluation of free proline accumulation as an index of drought resistance using two contrasting barley cuitivars. Crop Sci. 17, 720–726.

Hare, P.D., Cress, W.A., 1997. Metabolic implications of stress-induced proline accumulations in plants. Plant Growth Regulat. 21, 79–102.

Hasegawa, I., Terada, E., Sunairi, M., Wakita, H., Shinmachi, F., Noguchi, A., Nakajima, M., Yazaki, J., 1997. Genetic improvement of heavy metal tolerance in plants by transfer of the yeast metallothionein gene (CUP1). Plant Soil 196, 277–281.

Hayat, S., Haya, Q., Alyemeni, M.N., Wani, A.S., Pichtel, J., Ahmad, A., 2012. Role of proline under changing environment: a review. Plant Signal Behav. 7, 1456–1466.

Higginson, T., Li, S.F., Parish, R.W., 2003. AtMYB103 regulates tapetum and trichome development in *Arabidopsis thaliana*. Plant J. 35, 177–192.

Himelblau, E., Mira, H., Lin, S.J., Culotta, V.C., Penarrubia, L., Amasino, R.M., 1998. Identification of a functional homolog of the yeast copper homeostasis gene *ATX1* from Arabidopsis. Plant Physiol. 117, 1227–1234.

Hincha, D.K., Zuther, E., Hundertmark, M., Heyer, A.G., 2006. The role of compatible solutes in plant freezing tolerance: a case study on raffinose. In: Chen, T.H.H., Uemura, M., Fujikawa, S. (Eds.), Cold Hardiness in Plants: Molecular Genetics, Cell Biology and Physiology. Oxford University Press, New York, pp. 203–218.

Hmida-Sayari, A., Gargoui-Bouzid, R., Bidani, A., Jaoua, L., Savouré, A., Jaoua, S., 2005. Overexpression of Δ1-pyrroline-5-carboxylate synthetase increases proline production and confers salt tolerance in transgenic potato plants. Plant Sci. 169, 746–752.

Hong, Z.L., Lakkineni, K., Zhang, Z.M., Verma, D.P.S., 2000. Removal of feedback inhibition of Δ(1)-pyrroline-5-carboxylate synthetase results in increased proline accumulation and protection of plants from osmotic stress. Plant Physiol. 22, 1129–1136.

Hossain, M.A., Piyatida, P., da Silva, J.A.T., Fujita, M., 2012. Molecular mechanism of heavy metal toxicity and tolerance in plants: central role of glutathione in detoxification of reactive oxygen species and methylglyoxal and in heavy metal chelation. J. Bot. 1, 1–37.

Hsu, Y.T., Kao, C.H., 2004. Cadmium toxicity is reduced by nitric oxide in rice leaves. Plant Growth Regulat. 42, 227–238.

Ibragimova, S.S., YaS, K., Gerasimova, S.V., Kochetov, A.V., 2012. Partial suppression of gene encoding proline dehydrogenase enhances plant tolerance to various abiotic stresses. Russ. J. Plant Physiol. 59, 88–96.

Igarashi, Y., Yoshiba, Y., Sanada, Y., Yamaguchi-Shinozaki, K., Wada, K., Shinozaki, K., 1997. Characterization of the gene for Δ1-pyrroline-5-carboxylate synthetase and correlation between the expression of the gene and the salt tolerance in *Oryza sativa* L. Plant Mol. Biol. 33, 857–865.

Inostroza-Blancheteau, C., Rengel, Z., Alberdi, M., de la Luz Mora, M., Aquea, F., Arce-Johnson, P., Reyes-Díaz, M., 2012. Molecular and physiological strategies to increase aluminum resistance in plants. Mol. Biol. Rep. 9, 2069–2079.

Ito, M., Araki, S., Matsunaga, S., Itoh, T., Nishihama, R., Machida, Y., Doonan, J.H., Watanabe, A., 2001. A novel *cis*-acting element in promoters of plant B-type cyclin genes activates M phase–specific transcription. Plant Cell 13, 1891–1905.

Ivanova, E.M., Kholodova, V.P., Kuznetsov, VlV., 2010. Biological effects of high copper and zinc concentrations and their interaction in rapeseed plants. Russ. J. Plant Physiol. 57, 806–814.

Janouskova, M., Pavlıkova, D., Macek, T., Vosatka, M., 2005. Arbuscular mycorrhiza decreases cadmium phytoextraction by transgenic tobacco with inserted metallothionein. Plant Soil 272, 29–40.

Jewell, M.C., Campbell, B.C., Godwin, I.D., 2010. Transgenic plants for abiotic stress resistance. In: Kole, C., et al. (Eds.), Transgenic Crop Plants. Springer-Verlag, Berlin Heidelberg, pp. 67–132.

Jin, H., Cominelli, E., Bailey, P., Parr, A., Mehrtens, F., Jones, J., Tonelli, C., Weisshaar, B., Martin, C., 2000. Transcriptional repression by AtMYB4 controls production of UV-protecting sunscreens in Arabidopsis. EMBO J. 19, 6150–6161.

Jin, H., Martin, C., 1999. Multifunctionality and diversity within the plant MYB-gene family. Plant Mol. Biol. 41, 577–585.

Johnson, M.D.L., Kehl-Fie, T.E., Rosch, J.W., 2015. Copper intoxication inhibits aerobic nucleotide synthesis in *Streptococcus pneumoniae*. Metallomics 7, 786–794.

Jouve, L., Hoffmann, L., Hausman, J.F., 2004. Polyamine, carbohydrate, and proline content changes during salt stress exposure of aspen (*Populus tremula* L.): involvement of oxidation and osmoregulation metabolism. Plant Biol. 6, 74–80.

Jung, H.I., Gayomba, S.R., Rutzke, M.A., Craft, E., Kochian, L.V., Vatamaniuk, O.K., 2012. COPT6 is a plasma membrane transporter that functions in copper homeostasis in *Arabidopsis* and is a novel target of SQUAMOSA promoter-binding protein-like 7. J. Biol. Chem. 287, 33252–33267.

Jung, H.I., Gayomba, S.R., Yan, J., Vatamaniuk, O.K., 2014. *Brachypodium distachyon* as a model system for studies of copper transport in cereal crops. Front. Plant Sci. 5, 236–242.

Jung, Y., Park, J., Choi, Y., Yang, J.G., Kim, D., Kim, B.G., et al., 2010. Expression analysis of proline metabolism-related genes from halophyte *Arabis stelleri* under osmotic stress conditions. J. Integr. Plant Biol. 52, 891–903.

Juraniec, M., Lequeux, H., Hermans, C., Willems, G., Nordborg, M., Schneeberger, K., Salis, P., Vromant, M., Lutts, S., Verbruggen, N., 2014. Towards the discovery of novel genetic component involved in stress resistance in *Arabidopsis thaliana*. New Phytol. 201 (3), 810–824.

Kamei, A., Seki, M., Umezawa, T., Ishida, J., Satou, M., Akiyama, K., Zhu, J.K., Shinozaki, K., 2005. Analysis of gene expression profiles in Arabidopsis salt overly sensitive mutants sos2-1 and sos3-1. Plant Cell Environ. 28, 1267–1273.

Kanoun-Boule, M., Albuquerque, M.B., Nabais, C., Freitas, H., 2008. Copper as an environmental contaminant: phytotoxicity and human health implications. In: Prasad, M.N.V. (Ed.), Trace Elements: Environmental Contamination, Nutritional Benefits and Health Implications. John Wiley Sons, USA, pp. 653–678.

Karimi, P., Khavari-Nejad, R., Niknam, V., Ghahremaninejad, F., Najafi, F., 2012. The effects of excess copper on antioxidative enzymes, lipid peroxidation, proline, chlorophyll, and concentration of Mn, Fe, and Cu in *Astragalus neo-mobayenii*. Scientific World J. 2012, 1–6.

Karmous, I., Jaouani, K., El Ferjani, E., Chaoui, A., 2014. Responses of proteolytic enzymes in embryonic axes of germinating bean seeds under copper stress. Biol. Trace Elem. Res. 160, 108–115.

Karnachuk, O., Gavrilov, S., Avakyan, M., Podosokorskaya, O., Yu, F., Bonch-Osmolovskaya, E., Kublanov, I., 2015. Diversity of copper proteins and copper homeostasis systems in *Melioribacter roseus*, a facultatively anaerobic thermophilic member of the new phylum *Ignavibacteriae*. Microbiology 84, 134–143.

Kasuga, M., Liu, Q., Miura, S., Yamaguchi-Shinozaki, K., Shinozaki, K., 1999. Improving plant drought, salt, and freezing tolerance by gene transfer of a single stress-inducible transcription factor. Nat. Biotechnol 17, 287–291.

Kavi Kishore, P.B., Sangam, S., Amrutha, R.N., Laxmi, P.S., Naidu, K.R., Rao, K.R.S.S., Rao, S., Reddy, K.J., Theriappan, P., Sreenivasulu, N., 2005. Regulation of proline biosynthesis, degradation, uptake and transport in higher plants: its implications in plant growth and abiotic stress tolerance. Curr. Sci. 88, 424–438.

Kawashima, C.G., Noji, M., Nakamura, M., Ogra, Y., Suzuki, K.T., Saito, K., 2004. Heavy metal tolerance of transgenic tobacco plants over-expressing cysteine synthase. Biotechnol. Lett. 26, 153–157.

Kazuhiro, M., Soichi, K., Taku, D., Junko, T., Yoshihiro, O., 2005. DcMYB1 acts as a transcriptional activator of the carrot phenylalanine ammonia-lyase gene (DcPAL1) in response to elicitor treatment, UV-B irradiation and the dilution effect. Plant Mol. Biol. 59, 739–752.

Keunen, E., Peshev, D., Vangronsveld, J., Van den Ende, W., Cuypers, A., 2013. Plant sugars are crucial players in the oxidative challenge during abiotic stress: extending the traditional concept. Plant Cell Environ. http://dx.doi.org/10.1111/pce.12061.

Kholodova, V., Volkov, K., Abdeyeva, A., Kuznetsov, VlV., 2011a. Water status in *Mesembryanthemum crystallinum* under heavy metal stress. Environ. Exp. Bot. 71, 382–389.

Kholodova, V.P., Ivanova, E.M., Kuznetsov, VlV., 2011b. Initial step of copper detoxification: outside and inside of the plant cell. Soil Biology. In: Sherameti, I., Varma, A. (Eds.), Detoxification of Heavy Metals. Series: Soil Biology, 30. Springer–Verlag, Berlin Heidelberg, pp. 143–167.

Kieselbach, T., Hagman, A., Andersson, B., Schröder, W.P., 1998. The thylakoid lumen of chloroplasts. Isolation and characterization. J. Biol. Chem. 273, 6710–6716.

Kishor, P.B.K., Hong, Z., Miao, G.-H., Hu, C.A., Verma, D.P.S., 1995. Overexpression of Δ1-pyrroline-5-carboxylase synthetase increases proline production and confers osmotolerance in transgenic plants. Plant Physiol. 108, 1387–1394.

Kishor, P.B.K., Sangam, S., Amrutha, R.N., Laxmi, P.S., Naidu, K., Rao, K.R.S.S., 2005. Regulation of proline biosynthesis, degradation, uptake and transport in higher plants: its implications in plant growth and abiotic stress tolerance. Curr. Sci. 88, 424–438.

Kochetov, A.V., Titov, S.E., YaS, K., Komarova, M.L., Koval, V.S., Makarova, N.N., Yuu, Il'yinskyi, Trifonova, E.A., Shumny, V.K., 2004. Tobacco transformants bearing antisense suppressor of proline dehydrogenase gene, are characterized by higher proline content and cytoplasm osmotic pressure. Russ. J. Genet. 40, 216–218.

Kolodyazhnaya, YaS., Kutsokon, N.K., Levenko, B.A., Syutikova, O.S., Rakhmetov, D.B., Kochetov, A.V., 2009. Transgenic plants tolerant to abiotic stresses. Cytol. Genet. 43, 72–81.

Kolodyazhnaya, YaS., Titov, S.E., Kochetov, A.V., Trifonova, E.A., Romanova, A.V., Komarova, M.L., Koval, V.S., Shumny, V.K., 2007. Tobacco transformants expressing antisense sequence of proline dehydrogenase gene possess tolerance to heavy metals. Russ. J. Genet. 43, 825–828.

Kondrák, M., Marincs, F., Antal, F., Juhász, Z., Bánfalvi, Z., 2012. Effects of yeast trehalose-6-phosphate synthase on gene expression and carbohydrate contents of potato leaves under drought stress conditions. BMC Plant Biol. 12, 74–86.

Konno, H., Nakat, T., Nakashim, S., Katoh, 2005. *Lygodium japonicum* fern accumulates copper in the cell wall pectin. J. Exp. Bot. 56, 1923–1931.

Korn, M., Peterek, S., Mock, H.P., Heyer, A.G., Hincha, D.K., 2008. Heterosis in the freezing tolerance, and sugar and flavonoid contents of crosses between *Arabidopsis thaliana* accessions of widely varying freezing tolerance. Plant Cell Environ. 31, 813–827.

Koornnef, A., Pieterse, C.M.J., 2008. Cross talk in defence signaling. Plant Physiol. 146, 839–844.

Kossakowska, A., Falkowski, Y., Levadowska, J., 1988. Effect of aminoacids on toxity of heavy metals to phytoplankton. Bull Environ. Contam. Toxicol. 40, 532–538.

Koyama, H., Kawamura, A., Kihara, T., Hara, T., Takita, E., Shibata, D., 2000. Overexpression of mitochondrial citrate synthase in *Arabidopsis thaliana* improved growth on a phosphorus-limited soil. Plant Cell Physiol. 41, 1030–1037.

Krämer, U., Clemens, S., 2006. Functions and homeostasis of zinc, copper, and nickel in plants. Curr. Genet. 14, 215–271.

Krämer, U., 2010. Metal hyperaccumulation in plants. Ann. Rev. Plant Biol. 61, 517–534.

Krasavina, M.S., Burmistrova, N.A., Raldugina, G.N., 2014. The role of carbohydrates in plant resistance to abiotic stresses. In: Ahmad, P. (Ed.), Emerging Technologies and Management of Crop Stress Tolerance, 1. Elsevier Inc, pp. 227–268.

Krasensky, J., Jonak, C., 2012. Drought, salt, and temperature stress-induced metabolic rearrangements and regulatory networks. J. Exp. Bot. 63, 1593–1608.

Krystofova, O., Zitka, O., Krizkova, S., Hynek, D., Shestivska, V., Adam, V., Hubale, J., Mackova, M., Macek, T., Zehnalek, J., Babula, P., Havel, L., Kizek, R., 2012. Accumulation of cadmium by transgenic tobacco plants (*Nicotiana tabacum* L.) carrying yeast metallothionein gene revealed by electrochemistry. Int. J. Electrochem Sci. 7, 886–907.

Kulikova, A.L., Kholodova, V.P., Kuznetsov, VlV., 2009. Actin is involved in early plant responses to heavy metal stress and associates with molecular chaperons in stress environments. Doklady Biol. Sci. 424, 49–52.

Kulikova, A.L., Kuznetsova, N.A., Burmistrova, N.A., 2015. Change in growth and physiological parameters in soybean seedlings in response to toxic action of cooper. Russ. J. Plant Physiol. 62 (4), 488–498.

Kumar, G., Purty, R.S., Sharma, M.P., Singla-Pareek, S.L., Pareek, A., 2009. Physiological responses among *Brassica* species under salinity stress show strong correlation with transcript abundance for SOS pathway-related genes. J. Plant Physiol. 166, 507–520.

Kuznetsov, VlV., Shevyakova, N.I., 1999. Proline under stress conditions; biological role, metabolism and regulation. Fiziologiya Rastenii 46, 321–336.

Kwon, S.Y., Jeong, Y.J., Lee, H.S., Kim, J.S., Cho, K.Y., Allen, R.D., Kwak, S.S., 2002. Enhanced tolerances of transgenic tobacco plants expressing both superoxide dismutase and ascorbate peroxidase in chloroplasts against methyl viologen-mediated oxidative stress. Plant Cell Environ. 25, 873–882.

Larher, F.R., Lugana, R., Gagneul, D., Guyotc, S., Monniera, C., Lespinassed, Y., Bouchereau, A., 2009. A reassessment of the prevalent organic solutes constitutively accumulated and potentially involved in osmotic adjustment in pear leaves. Environ. Exp. Bot. 66, 230–241.

Laura, M., Consonni, R., Locatelli, F., Fumagalli, E., Allavena, A., Coraggio, I., Mattana, M., 2010. Metabolic response to cold and freezing of *Osteospermum ecklonis* overexpressing *Osmyb4*. Plant Physiol. Biochem. 48, 764–771.

Lee, J.H., Yu, J.D., Kim, S.J., Choi, D., Lee, H.J., 2012. Intraspecies differences in cold hardiness, carbohydrate content and β-amylase gene expression of *Vaccinium corymbosum* during cold acclimation and deacclimation. Tree Physiol. 6, 1–8.

Lee, M.M., Schiefelbein, J., Werewolf, A., 1999. MYB-related protein in *Arabidopsis* is a position-dependent regulator of epidermal cell patterning. Cell 99, 473–483.

Lee, M.M., Schiefelbein, J., 2001. Developmentally distinct MYB genes encode functionally equivalent proteins in Arabidopsis. Development 128, 1539–1546.

Lee, S., Kim, Y.Y., Lee, Y., An, G., 2007. Rice P1B-type heavy-metal ATPase, OsHMA9, is a metal efflux protein. Plant Physiol. 145, 831–842.

Legay, S., Lefevre, I., Lamoureux, D., Barreda, C., Luz, R.T., Gutierrez, R., et al., 2011. Carbohydrate metabolism and cell protection mechanisms differentiate drought tolerance and sensitivity in advanced potato clones (*Solanum tuberosum* L). Funct. Integr. Genomics 11, 275–291.

Lehmann, S., Funck, D., Szabados, L., Rentsch, D., 2010. Proline metabolism and transport in plant development. Amino Acids 39, 949–962.

Lequeux, H., Hermans, C., Lutts, S., Verbruggen, N., 2010. Response to copper excess in *Arabidopsis thaliana*: impact on the root system architecture, hormone distribution, lignin accumulation and mineral profile. Plant Physiol. Biochem. 48, 673–682.

Li, H., Fan, N., Li, L., Wey, B., Li, G., Gu, L., Wang, X., Zhang, X., 2014. Identification and characterization of a novel copper transporter gene family TaCT1 in common wheat. Plant Cell Environ. 37, 1561–1573.

Li, J., Yang, X., Wang, Y., Li, X., Gao, Z., Pei, M., Chen, Z., Qu, L.J., Gu, H., 2006. Two groups of MYB transcription factors share a motif which enhances trans-activation activity. Biochem. Biophys. Res. Commun. 341, 1155–1163.

Lidon, F.C., Henriques, F.S., 1993. Oxygen metabolism in higher plant chloroplasts. Photosynthetica 29, 249–279.

Liu, D., An, Z., Mao, Z., Ma, L., Lu, Z., 2015a. Enhanced heavy metal tolerance and accumulation by transgenic sugar beets expressing Streptococcus thermophilus StGCS-GS in the presence of Cd, Zn and Cu alone or in combination. PLoS ONE 10 (6). http://dx.doi.org/10.1371/journal.pone.0128824.

Liu, H.-C., Liao, H.-T., Charng, Y.-Y., 2011. The role of class A1 heat shock factors (HSFA1s) in response to heat and other stresses in *Arabidopsis*. Plant Cell Environ. 34, 738–751.

Liu, H., Ouyang, B., Zhang, J., Wang, T., Li, H., Zhang, Y., Yu, C., Ye, Z., 2012. Differential modulation of photosynthesis, signaling, and transcriptional regulation between tolerant and sensitive tomato genotypes under cold stress. PLoS ONE 7, e50785.

Liu, J., Shi, X., Qian, M., Zheng, L., Lian, C., Xia, Y., Shen, Z., 2015b. Copper-induced hydrogen peroxide upregulation of a metallothionein gene, OsMT2c, from *Oryza sativa* L. confers copper tolerance in *Arabidopsis thaliana*. J. Hazard Mat. 294, 99–108.

Livingston 3rd, D.P., Hincha, D.K., Heyer, A.G., 2009. Fructan and its relationship to abiotic stress tolerance in plants. Cell Mol. Life Sci. 66, 2007–2023.

Llugany, M., Lombini, A., Poschenrieder, C., Dinelli, E., Barcelo, J., 2003. Different mechanisms account for enhanced copper resistance in *Silene armeria* ecotypes from mine spoil and serpentine sites. Plant Soil 251, 55–63.

López-Maury, L., Giner-Lamia, J., Florencio, F.J., 2012. Redox control of copper homeostasis in cyanobacteria. Plant Signal Behav. 7, 1712–1714. http://www.ncbi.nlm.nih.gov/pmc/articles/PMC3578916/.

Luo, X., Wu, J., Li, Y., Nan, Z., Guo, X., Wang, Y., Zhang, A., Wang, Z., Xia, G., Tian, Y., 2013. Synergistic effects of *GhSOD1* and *GhCAT1* overexpression in cotton chloroplasts on enhancing tolerance to methyl viologen and salt stresses. PLoS ONE. http://dx.doi.org/10.1371/journal.pone.0054002.

Ma, L., Sun, N., Liu, X., Jiao, Y., Zhao, H., Deng, X.W., 2005. Organ-specific expression of Arabidopsis genome during development. Plant Physiol. 138, 80–91.

Maksymiec, W., Drążkiewicz, M., Skórzyńska-Polit, E., 2008. Responses of higher plants to heavy metal stress. In: Khan, N.A., Singh, S. (Eds.), Abiotic Stress and Plant Responses. International Publishing House, New Delhi–Bangalore–Mumbai, pp. 139–163.

Maksymiec, W., 1997. Effect of copper on cellular processes in higher plants. Photosynthetica 34, 321–342.

Maleva, M.G., Nekrasova, G.F., Borisova, G.G., Chukina, N.V., Ushakova, O.S., 2012. Effect of heavy metals on photosynthetic apparatus and antioxidant status of elodea. Russ. J. Plant Physiol. 59, 190–197.

Marschner, H., 2002. Mineral Nutrition of Higher Plants. Academic Press, London, UK, 889 pp.

Martin, L.L., Mourato, M.P., 2006. Effect of excess copper on tomato plants: growth parameters, enzyme activities, chlorophyll, and mineral content. J. Plant Nutrit. 29, 2179–2198.

Mateos-Naranjo, E., Andrades-Moreno, L., Cambrollé, J., Perez-Martin, A., 2013. Assessing the effect of copper on growth, copper accumulation and physiological responses of grazing species *Atriplex halimus*: ecotoxicological implications. Ecotoxicol. Environ. Saf. 90, 136–142.

Mattana, M., Biazzi, E., Consonni, R., Locatelli, F., Vannini, C., Provera Coraggio, I., 2005. Overexpression of Osmyb4 enhances compatible solute accumulation and increases stress tolerance of *Arabidopsis thaliana*. Physiol. Plant 125, 212–223.

Mattana, M., 2010. Metabolic response to cold and freezing of *Osteospermum ecklonis* overexpressing Osmyb4. Plant Physiol. Biochem. 48, 764–771.

Matus, J.T., Aquea, F., Arce-Johnson, P., 2008. Analysis of the grape MYB R2R3 subfamily reveals expanded wine quality-related clades and conserved gene structure organization across *Vitis* and *Arabidopsis* genomes. BMC Plant Biol. 8, 83.

Matysik, J., Alai Bhalu, B., Mohanty, P., 2002. Molecular mechanisms of quenching of reactive oxygen species by proline under stress in plants. Curr. Sci. 82, 525–532.

McClinchey, S.L., Kott, L.S., 2008. Production of mutants with high cold tolerance in spring canola (*Brassica napus*). Euphytica 162, 18–27.

Meharg, A.A., 2005. Mechanisms of plant resistance to metal and metalloids ions and potential biotechnological applications. Plant Soil 274, 163–174.

Mengiste, T., Chen, X., Salmeron, J.M., Dietrich, R.A., 2003. The BOTRYTIS SUSCEPTIBLE1 gene encodes an R2R3MYB transcription factor protein that is required for biotic and abiotic stress responses in Arabidopsis. Plant Cell 15, 2551–2565.

Mehta, S.K., Gaur, J.P., 1999. Heavy-metal-induced proline accumulation and its role in ameliorating metal toxicity in *Chlorella vulgaris*. New Phytol. 143, 253–259.

Meychik, N.R., Yermakov, I.P., 2001. Ion exchange of plant root cell walls. Plant Soil 234, 181–193.

Mihalik, J., Henner, P., Frelon, S., Camilleri, V., Fevrier, L., 2012. Citrate assisted phytoextraction of uranium by sunflowers: study of fluxes in soils and plants and resulting intra-planta distribution of Fe and U. Environ. Exp. Bot. 77, 249–258.

Mijovilovich, A., Leitenmaier, B., Meyer-Klaucke, W., Kroneck, P.M., Götz, B., Küpper, H., 2009. Complexation and toxicity of copper in higher plants. II. Different mechanisms for copper versus cadmium detoxification in the copper-sensitive cadmium/zinc hyperaccumulator *Thlaspi caerulescens* (Ganges Ecotype). Plant Physiol. 151, 715–731.

Mira, H., Martinez-Garcia, F., Penarrubia, L., 2001a. Evidence for the plant specific intercellular transport of the Arabidopsis copper chaperone CCH. Plant J. 25, 521–528.

Mira, H., Vilar, M., Perez-Raya, E., Penarrubia, L., 2001b. Functional and conformational properties of the exclusive C-domain from the Arabidopsis copper chaperone CCH. Biochem. J. 15, 545–549.

Misra, N., Saxena, P., 2009. Effect of salicylic acid on proline metabolism in lentil grown under salinity stress. Plant Sci. 177, 181–189.

Misra, S., Gedamu, L., 1989. Heavy metal tolerant transgenic *Brassica napus* L. and *Nicotiana tabacum* L. plants. Theor. Appl. Genet. 78, 161–168.

Mittler, R., Vanderauwera, S., Gollery, M., Van Breusegem, F., 2004. Reactive oxygen gene network of plants. Trends Plant Sci. 9, 490–498.

Mohanty, N., Vass, I., Demeter, S., 1989. Cooper toxicity affects photosystem II electron transport at the secondary quione acceptor QB. Plant Physiol. 90, 175–179.

Molinari, H.B.C., Marur, C.J., Bespalhok, J.C., Kobayashi, A.K., Pileggi, M., Pereira, F.P.P., Vieira, L.G.E., 2004. Osmotic adjustment in transgenic citrus rootstock Carrizo citrange (Citrus sinensis Osb. _Poncirus trifoliate L. Raf.) overproducing proline. Plant Sci. 167, 1375–1381.

Molinari, H.B.C., Marur, C.J., Daros, E., Freitas de Campos, M.R., Portela de Carvalho, J.F.R., Filho, J.C.B., Pereira, L.F.P., Vieira, L.G.E., 2007. Evaluation of the stress-inducible production of proline in transgenic sugarcane (Saccharum spp.): osmotic adjustment, chlorophyll fluorescence and oxidative stress. Physiol. Plant 130, 218–229.

Mostofa, M.G., Fujita, M., 2013. Salicylic acid alleviates copper toxicity in rice (*Oryza sativa* L.) seedlings by up-regulating antioxidative and glyoxalase systems. Ecotoxicology 22, 959–973.

Moya, J.L., Ro, R., Picazo, I., 1993. Influence of cadmium and nickel on growth, net photosynthesis and carbohydrate distribution in rice plants. Photosynth. Res. 36, 75–80.

Muhammad, A., Ali, S., Rizwan, M., Ibrahim, M., Abbas, F., Farid, M., Zia-ur-Rehman, M., Irshad, K., Bharwana, S., 2015. The effect of excess copper on growth and physiology of important food crops: a review. Environ. Sci. Pollut. Res. http://dx.doi.org/10.1007/s11356-015-4496-5.

Munns, R., Tester, M., 2008. Mechanisms of salinity tolerance. Annu. Rev. Plant Biol. 59, 651–681.

Nagaoka, S., Takano, T., 2003. Salt tolerance-related protein STO binds to a Myb transcription factor homologue and confers salt tolerance in Arabidopsis. J. Exp. Bo. 54, 2231–2237.

Nakashima, K., Tran, L.-S.P., Nguyen, D.V., Fujita, M., Maruyama, K., Todaka, D., Ito, Y., Hayashi, N., Shinozaki, K., Yamaguchi-Shinozaki, K., 2007. Functional analysis of a NAC-type transcription factor OsNAC6 involved in abiotic and biotic stress-responsive gene expression in rice. Plant J. 51, 617–630.

Nanjo, T., Kobayashi, K., Yoshiba, Y., Sanada, Y., Wada, K., Tsukaya, H., Kakubari, Y., Yamaguchi-Shinozaki, K., Shinozaki, K., 1999. Biological functions of proline in morphogenesis and osmotolerance revealed in antisense transgenic *Arabidopsis thaliana*. Plant J. 18, 185–193.

Navari-Izzo, F., Quartacci, M.F., Pinzino, C., Dalla Vecchia, F., Sgherr, C.L.M., 1998. Thylakoid-bound and stromal antioxidative enzymes in wheat treated with excess copper. Phy. Plant 104, 630–638.

Naya, L., Paul S,Valdés-López, O., Mendoza-Soto, A.B., Nova-Franco, B., Sosa-Valencia, G., Reyes, J.L., Hernández, G., 2014. Regulation of copper homeostasis and biotic interactions by microRNA 398b in common bean. PLoS ONE 9. http://dx.doi.org/10.1371/journal.pone.0084416 eCollection.

Nayyar, H., Chander, S., 2004. Protective effects of polyamines against oxidative stress induced by water and cold stress in chickpea. J. Agron. Crop Sci. 190, 355–365.

Nesi, N., Jond, C., Debeaujon, I., Caboche, M., Lepiniec, L., 2001. The Arabidopsis TT2 gene encodes an R2R3 MYB domain protein that acts as a key determinant for proanthocyanidin accumulation in developing seed. Plant Cell 13, 2099–2114.

Neumann, D., Nieden, U., Lichtenberger, O., Leopold, I., 1995. How does *Armeria maritima* tolerate high heavy metal concentrations? Plant Physiol. 146, 704–717.

Nishizawa, A., Yabuta, Y., Shigeoka, S., 2008. Galactinol and raffinose constitute a novel function to protect plants from oxidative damage. Plant Physiol. 147, 1251–1263.

Opdenakker, K., Remans, T., Vangronsveld, J., Cuypers, A., 2012. Mitogen-activated protein (MAP) kinases in plant metal stress: regulation and responses in comparison to other biotic and abiotic stresses. Int. J. Mol. Sci. 13, 7828–7853.

Ouzounidou, G., 1996. The use of photoacoustic spectroscopy in assessing leaf photosynthesis under copper stress: correlation of energy storage to photosystem II fluorescence parameters and redox change of P700. Plant Sci. 113, 229–237.

Ouzounidou, G., Ilias, I., Tranopoulou, H., Karatalgis, S., 1998. Amelioration of copper toxicity by iron on spinach physiology. J. Plant Nutr. 21, 2089–2101.

Palonen, P., Buszard, D., Donnelly, D., 2000. Changes in carbohydrates and freezing tolerance during cold acclimation of red raspberry cultivars grown in vitro and in vivo. Physiol. Plant 110, 393–401.

Pandolfi, D., Solinas, G., Valle, G., Coraggio, I., 1997. Cloning of a cDNA encoding a novel myb gene (accession no. Y11414) highly expressed in cold stressed rice coleoptiles. Plant Physiol. 114, 747.

Pardo, J.M., 2010. Biotechnology of water and salinity stress tolerance. Curr. Opin. Biotechnol. 21, 185–196.

Park, M.-R., Yun, K.-Y., Mohanty, B., Herath, V., Xu, F., Wijaya, E., Bajic, V.B., Yun, S.-J., De Los Reyes, B.G., 2010. Supra-optimal expression of the cold-regulated OsMyb4 transcription factor in transgenic rice changes the complexity of transcriptional network with major effects on stress tolerance and panicle development. Plant Cell Environ. 33, 2209–2230.

Pasquali, G., Biricolti, S., Locatelli, F., Baldoni, E., Mattana, M., 2008. *OsMyb4* expression improves adaptive responses to drought and cold stress in transgenic apples. Plant Cell Rep. 27, 1677–1686.

Pätsikkä, E., Aro, E.-M., Tyystjärvi, E., 1998. Increase in the quantum yield of photoinhibition contributes to copper toxicity *in vivo*. Plant Physiol. 117, 619–627.

Pätsikkä, E., Aro, E.M., Tyystjarvi, E., 2001. Mechanism of copper-enhanced photoinhibition in thylakoid membranes. Physiol. Plant 113, 142–150.

Pätsikkä, E., Kairavuo, M., Sersen, F., Aro, E.M., Tyystjarvi, E., 2002. Excess copper predisposes photosystem II to photoinhibition in vivo by out-competing iron and causing a decrease in leaf chlorophyll. Plant Physiol. 129, 1359–1367.

Patzlaff, A., McInnis, S., Courtenay, A., Surman, C., Newman, L.J., Smith, C., Bevan, M.W., Mansfield, S., Whetten, R.W., Sederoff, R.R., Campbell, M.M., 2003. Characterisation of a pine MYB that regulates lignification. Plant J. 36, 743–754.

Pavlıkova, D., Macek, T., Mackova, M., Szakova, J., Balık, J., 2004. Cadmium tolerance and accumulation in transgenic tobacco plants with a yeast metallothionein combined with a polyhistidine tail. Int. Biodeter. Biodegrad. 54, 233–237.

Paz-Ares, J., Ghosal, D., Wienand, U., Peterson, P.A., Saedle, H., 1987. The regulatory cl locus of *Zea mays* encodes a protein with homology to myb proto-oncogene products and with structural similarities to transcriptional activators. EMBO 6, 3553–3558.

Pekal, A., Biesaga, M., Pyrzynska, K., 2011. Interaction of quercetin with copper ions: complexation, oxidation and reactivity towards radicals. Biometals 24, 41–49.

Peñarrubia, L., Romero, P., Carrió-Seguí, A., Andrés-Bordería, A., Moreno, J., Sanz, A., 2015. Temporal aspects of copper homeostasis and its crosstalk with hormones. Front. Plant Sci. 6 http://dx.doi.org/10.3389/fpls.2015.00255.

Penfield, S., Meissner, R.C., Shoue, D.A., Carpita, N.C., Bevan, M.W., 2001. *MYB61* is required for mucilage deposition and extrusion in the Arabidopsis seed coat. Plant Cell 13, 2777–2791.

Peshev, D., Van den Ende, W., 2013. Sugars as antioxidants in plants. In: Tuteja, N., Gill, S.S. (Eds.), Crop Improvement under Adverse Conditions. Springer-Verlag, Berlin, Heidelberg, Germany, pp. 285–308.

Peti, C.M., Van de Geijn, S.C., 1978. In vivo measurement of cadmium transport and accumulations in the stems of intact tomato plants. Planta 138, 137–143.

Pomponi, M., Censi, V., Di Girolamo, V., De Paolis, A., di Toppi, L.S., Aromolo, R., Costantino, P., Cardarelli, M., 2006. Overexpression of Arabidopsis phytochelatin synthase in tobacco plants enhances Cd^{2+} tolerance and accumulation but not translocation to the shoot. Planta 223, 180–190.

Poschenrieder, C., Barceló, J., 2004. Water relations in heavy metal stressed plants. In: Prasad, M.N. (Ed.), Heavy Metal Stress in Plants, third ed. Springer, Berlin, Germany, pp. 249–270.

Postrigan, B.N., Knyazev, A.V., Kuluev, B.R., 2013. Effect of cadmium on promoter activity of rice phytochelatin synthase gene in transgenic tobacco plants. Russ. J. Plant Physiol. 60, 701–705.

Prasad, M.N.V., Strzalka, K., 1999. Impact of heavy metals on photosynthesis. In: Prasad, M.N.V., Hagemeyer, J. (Eds.), Heavy Metal Stress in Plants. Springer Publishing, Berlin, pp. 117–138.

Prouse, M.B., Campbell, M.M., 2012. The interaction between MYB proteins and their target DNA binding sites. Biochim. Biophys. Acta 1819, 67–77.

Prouse, M.B., Campbell, M.M., 2013. Interactions between the R2R3-MYB transcription factor, at MYB61, and target DNA binding sites. PLoS ONE 8, e65132.

Puig, S., Andres-Colas, N., Garcia-Molina, A., Penarrubia, L., 2007. Copper and iron gomeostasis in Arabidopsis: responses to metal deficiencies, interactions and biotechological applications. Plant Cell Environ. 30, 271–290.

Puig, S., 2014. Function and regulation of the plant COPT family of high-affinity copper transport proteins. Adv. Bot. http://dx.doi.org/10.1155/476917.

Qiu, Y., Yu, D., 2009. Over-expression of the stress-induced OsWRKY45 enhances disease resistance and drought tolerance in Arabidopsis. Environ. Exp. Bot. 65, 35–47.

Qin, R., Ce, W., Chen, D., Björn, L.O., Li, S., 2015. Copper-induced root growth inhibition of *Allium cepa* var. *agrogarum* L. involves disturbances in cell division and DNA damage. Environ. Toxicol. Chem. 34, 1045–1055.

Quartacci, M.F., Cosi, E., Navari-Izzo, F., 2001. Lipids and NADPH-dependent superoxide production in plasma membrane vesicles from roots of wheat grown under cooper deficiency or excess. J. Exp. Bot. 52, 77–84.

Quartacci, M.F., Pinzino, C., Sgherri, C.L.M., Dalla Vecchia, F., Navari-Izzo, F., 2000. Growth in excess copper induces changes in the lipid composition and fluidity of PSII-enriched membranes in wheat. Physiol. Plant 108, 87–93.

Radyukina, N.L., Ivanov, Yu, V., Kartashov, A.V., Shevyakova, N.I., Rakitin, V.Y., Khryanin, V.N., Kuznetsov, VlV., 2007. Inducible and constitutive mechanisms of salt stress resistance in *Geum urbanum* L. Russ. J. Plant Physiol. 54, 692–698.

Radyukina, N.L., Ivanov, YuV., Kartashov, A.V., 2011a. Regulation of gene expression governing proline metabolism in *Thellungiella salsuginea* by NaCl and paraquat. Russ. J. Plant Physiol. 58, 643–652.

Radyukina, N.L., Shashukova, A.V., Mapelli, S., 2011b. Effect of common sage plant treatment with polyamines on the contents of proline and free and conjugated polyamines under oxidative stress. Russ. J. Plant Physiol. 58, 673–680.

Radyukina, N.L., Shashukova, A.V., Shevyakova, N.I., Kuznetsov, VlV., 2008. Proline involvement in the common sage antioxidant system in the presence of NaCl and paraquat. Russ. J. Plant Physiol. 55, 649–656.

Raffaele, S., Rivas, S., 2006. An essential role for salicylic acid in AtMYB30- mediated controls of the hypersensitive cell death program in Arabidopsis. FEBS Lett. 580, 3498–3504.

Rauser, W.E., 1999. Structure and function of metal chelators produced by plants. Cell Biochem. Biophys. 31, 19–48.

Ravet, K., Pilon, M., 2013. Copper and iron homeostasis in plants: the challenges of oxidative stress. Antioxid. Redox Signal 19, 919–932.

Rodriguez, F.I., Esch, J.J., Hall, A.E., Binde, B.M., Schaller, G.E., Bleecker, A.B., 1999. A copper cofactor for the ethylene receptor ETR1 from Arabidopsis. Science 283, 996–998.

Roncarati, F., Sáez, C.A., Greco, M., Gledhill, M., Bitonti, M.B., Brown, M.T., 2015. Response differences between *Ectocarpus siliculosus* populations to copper stress involve cellular exclusion and induction of the phytochelatin biosynthetic pathway. Aquat. Toxicol. Feb. 159, 167–175.

Rosa, M., Prado, C., Podazza, G., Interdonato, R., González, J.A., Hilal, M., et al., 2009. Soluble sugars—metabolism, sensing and abiotic stress: a complex network in the life of plants. Plant Signal Behav. 4, 388–393.

Ruiz, J.M., Blumwald, E., 2002. Salinity-induced glutathione synthesis in *Brassica napus*. Planta 214, 965–969.

Sáez, C.A., González, A., Contreras, R.A., Moody, A.J., Moenne, A., Brown, M.T., 2015a. A novel field transplantation technique reveals intra-specific metal-induced oxidative responses in strains of *Ectocarpus siliculosus* with different pollution histories. Environ. Pollut. 199, 130–138.

Sáez, C.A., Roncarati, F., Moenne, A., Moody, A.J., Brown, M.T., 2015b. Copper-induced intra-specific oxidative damage and antioxidant responses in strains of the brown alga *Ectocarpus siliculosus* with different pollution histories. Aquat. Toxicol. 159, 81–89.

Saha, D., Mandal, S., Saha, A., 2012. Copper induced oxidative stress in tea (*Camellia sinensis*) leaves. J. Environ. Biol. Sep. 33, 861–866.

Saijo, T., Nagasawa, A., 2015. A new detection tool for bioavailable copper utilizing transgenic plants carrying recombinant yeast ACE1 transcription factor and GFP reporter genes. Soil Sci. Plant Nutrit. 61. http://dx.doi.org/10.1080/00380768.2014.989803.

Sancenón, V., Puig, S., Mira, H., Thiele, D.J., Penarrubia, L., 2003. Identification of a copper transporter family in *Arabidopsis thaliana*. Plant Mol. Biol. 51, 577–587.

Sandmann, G., Boger, P., 1980. Copper deficiency and toxicity in *Scenedesmus*. Z. Pflanzenphysiol. 98, 53–59.

Sarret, G., Saumitou-Laprade, P., Bert, V., Proux, J., Hazemann, J.L., Traverse, A., Marcus, M.F., Manceau, A., 2002. Forms of Zn accumulated in the hyperaccumulator *Arabidopsis halleri*. Plant Physiol. 130, 1815–1826.

Sato, Y., Yokoya, S., 2008. Enhanced tolerance to drought stress in transgenic rice plants overexpressing a small heat-shock protein, sHSP17.7. Plant Cell Rep. 27, 329–334.

Sawahel, W.A., Hassan, A.H., 2002. Generation of transgenic wheat plants producing high levels of the osmoprotectant proline. Biotechnol. Lett. 24, 721–725.

Schenke, D., Büttcher, C., Scheel, D., 2011. Crosstalk between abiotic ultraviolet-B stress and biotic (flg22) stress signalling in Arabidopsis prevents flavonol accumulation in favor of pathogen defence compound production. Plant Cell Environ. 34, 1849–1864.

Schlunk, I., Krause, K., Wirth, S., Kothe, E., 2015. A transporter for abiotic stress and plant metabolite resistance in the ectomycorrhizal fungus *Tricholoma vaccinum*. Environ. Sci. Pollut. Res. Int. [supplied by publisher].

Schmitz, G., Tillmann, E., Carriero, F., Fiore, C., Theres, K., 2002. The tomato *Blind* gene encodes a MYB transcription factor that controls the formation of lateral meristems. Proc. Natl. Acad. Sci. U.S.A. 99, 1064–1069.

Schubert, M., Petersson, U.A., Haas, B.J., Funk, C., Schröde, W.P., Kieselbach, T., 2002. Proteome map of the chloroplast lumen of *Arabidopsis thaliana*. J. Biol. Chem. 277, 8354–8365.

Schützendübel, A., Polle, A., 2002. Plant responses to abiotic stresses: heavy metal-induced oxidative stress and protection by mycorrhization. J. Exp. Bot. 53, 1351–1365.

Sekhar, K., Priyanka, B., Reddy, V.D., Rao, K.V., 2011. Metallothionein 1 (CcMT1) of pigeon pea (Cajanus cajan, L.) confers enhanced tolerance to copper and cadmium in *Escherichia coli* and *Arabidopsis thaliana*. Environ. Exp. Bot. 72, 131–139.

Senden, M.H.M.N., Van Paassen, F.J.M., Van der Meer, A.J.G.M., Wolterbeek, H.T., 1992. Cadmium–citric acid xylem cell wall interactions in tomato plants. Plant Cell Environ. 15, 71–79.

Seo, P.J., Park, C.M., 2010. MYB96-mediated abscisic acid signals induce pathogen resistance response by promoting salicylic acid biosynthesis in Arabidopsis. New Phytol. 186, 471–483.

Seo, Y.-S., Chern, M., Bartley, L.E., Han, M., Jung, K.-H., Lee, I., Walia, H., Richter, T., Xu, X., Cao, P., Bai, W., Ramanan, R., Amonpant, F., Arul, L., Canlas, P.E., Ruan, R., Park, C.-J., Chen, X., Hwang, S., Jeon, J.-S., Ronald, P.C., 2011. Towards establishment of a rice stress response interactome. PLoS Genet. 7, 1–12.

Serraj, R., Sinclair, T.R., 2002. Osmolyte accumulation: can it really help increase crop yield under drought conditions? Plant Cell Environ. 25, 333–341.

Shah, K., Dubey, R.S., 1998. Effect of cadmium on proline accumulation and ribonuclease activity in rice seedlings: role of proline as a possible enzyme protectant. Biologia Plant 40, 121–130.

Shahid, M., Pourrut, B., Dumat, C., Nadeem, M., Aslam, M., Pinelli, E., 2014. Heavy-metal-induced reactive oxygen species: phytotoxicity and physicochemical changes in plants. Rev. Environ. Contam. Toxicol. 232, 1–44.

Sham, A., Al-Azzawi, A., Al-Ameri, S., Al-Mahmoud, B., Awwad, F., Al-Rawashdeh, A., Iratni, R., AbuQamar, S., 2014. Transcriptome analysis reveals genes commonly induced by botrytis cinerea infection, cold, drought and oxidative stresses in arabidopsis. PLoS ONE. http://dx.doi.org/10.1371/journal.pone.0113718.

Shao, H.B., Shao, M.A., Liang, Z.S., 2006. Osmotic adjustment comparison of 10 wheat (*Triticum aestivum L.*) genotypes at soil water deficits. Colloids Surf. B Biointerfaces 47, 132–139.

Sharma, P., Bhushan, J.A., RSh, D., Pessarakli, M., 2012. Reactive oxygen species, oxidative damage, and antioxidative defense mechanism in plants under stressful conditions. J. Bot. http://dx.doi.org/10.1155/2012/217037.

Sharma, P., Dubey, R.S., 2007. Involvement of oxidative stress and role of antioxidative defense system in growing rice seedlings exposed to toxic concentrations of aluminum. Plant Cell Rep. 26, 2027–2038.

Sharma, S.S., Dietz, K.-J., 2006. The significance of amino acids and amino acid-derived molecules in plant responses and adaptation to heavy metal stress. J. Exp. Bot. 57, 711–726.

Shen, H., He, X., Poovaiah, C.R., Wuddineh, W.A., Ma, J., Mann, D.G.J., Wang, H., Jackson, L., Tang, Y.C., Stewart Jr., N., Chen, F., Dixon, R.A., 2012. Functional characterization of the switchgrass (*Panicum virgatum*) R2R3-MYB transcription factor *PvMYB4* for improvement of lignocellulosic feedstocks. New Phytol. 193, 121–136.

Shevyakova, N.I., Bakulina, E.A., Kuznetsov, VlV., 2009. Proline antioxidant role in the common ice plant subjected to salinity and paraquat treatment inducing oxidative stress. Russ. J. Plant Physiol. 56, 663–669.

Shevyakova, N.I., 1983. Metabolism and physiological role praline in plants under conditions of water and stress. Fiziologiya Rastenii 30, 768–783.

Shin, B., Choi, G., Yi, H., Yang, S., Cho, I., Kim, J., Lee, S., Paek, N.C., Kim, J.H., Song, P.S., Choi, G., 2002. AtMYB21, a gene encoding a flower-specific transcription factor, is regulated by COR1. Plant J. 30, 23–32.

Shioli, Y., Tamai, H., Sasa, T., 1978. Inhibition of photosystem II in the green alga *Ankistrodesmus falcatus* by copper. Phytology 44, 434–438.

Silver, S., Misra, T.K., 1988. Plasmid-mediated heavy metal resistances. Annu. Rev. Microbiol. 42, 717–743.

Singh, V., Bhatt, I., Aggarwal, A., Tripathi, B.N., Munjal, A.K., Sharma, 2010. Proline improves copper tolerance in chickpea (*Cicer arietinum*). Protoplasma 245, 173–181.

Sinkevich, M.S., Naraykina, N.V., Trunova, T.I., 2010. Involvement of sugars in the antioxidant defense against paraquat-induced oxidative stress in potato transformed with yeast invertase gene. Doklady Biol. Sci. 434, 338–340.

Siripornadulsil, S., Traina, S., Verma, D.P., Sayre, R.T., 2002. Molecular mechanisms of proline-mediated tolerance to toxic heavy metals in transgenic microalgae. Plant Cell 14, 2837–2847.

Smékalová, V., Doskočilová, A., Komis, G., Šamaj, J., 2014. Crosstalk between secondary messengers, hormones and MAPK modules during abiotic stress signalling in plants. Biotechnol. Adv. 32, 2–11.

Smiri, M., Missaoui, T., 2014. The role of ferredoxin:thioredoxin reductase/thioredoxin m in seed germination and the connection between this system and copper ion toxicity. J. Plant Physiol. 171, 1664–1670.

Smirnoff, N., Cumbes, Q.J., 1989. Hydroxyl radical scavenging activity of compatible solutes. Phytochemistry 37, 1057–1060.

Solanki, R., Dhankhar, R., 2011. Biochemical changes and adaptive strategies of plants under heavy metal stress. Biologia 66, 195–204.

Soltész, A., Vágújfalvi, A., Rizza, F., Kerepesi, I., Galiba, G., Cattivelli, L., Coraggio, I., Crosatti, C., 2012. The rice *Osmyb4* gene enhances tolerance to frost and improves germination under unfavourable conditions in transgenic barley plants. J. Appl. Genet. 53, 133–143.

Sorkheh, K., Shiran, B., Khodambashi, M., Rouhi, V., Mosavei, S., Sofo, A., 2012. Exogenous proline alleviates the effects of H_2O_2-induced oxidative stress in wild almond species. Russ. Plant Physiol. 59, 788–798.

Steiner-Lange, S., Unte, U.S., Eckstein, L., Yang, C., Wilson, Z.A., Schmelzer, E., Dekker, K., Saedler, H., 2003. Disruption of Arabidopsis thaliana MYB26 results in male sterility due to non-dehiscent anthers. Plant J. 34, 519–528.

Su, J., Wu, R., 2004. Stress-inducible synthesis of proline in transgenic rice confers faster growth under stress conditions than that with constitutive synthesis. Plant Sci. 166, 941–948.

Szabados, L., Savoure, A., 2010. Proline: a multifunctional amino acid. Trends Plant Sci. 15, 89–97.

Szafranska, K., Cvikrova, M., Kowalska, U., Gorecka, K., Gorecki, R., Martincova, O., Janas, K.M., 2011. Influence of copper ions on growth, lipid peroxidation, and proline and polyamines content in carrot rosettes obtained from anther culture. Acta Physiol. Plant 33, 851–859.

Szekely, G., Abraham, E., Cseplo, A., Rigo, G., Zsigmond, L., Csiszá, J., 2008. Duplicated P5CS genes of *Arabidopsis* play distinct roles in stress regulation and developmental control of proline biosynthesis. Plant J. 53, 11–28.

Tak, H.I., Ahmad, F., Babalola, O.O., 2013. Advances in the application of plant growth-promoting rhizobacteria in phytoremediation of heavy metals. Rev. Environ. Contam. Toxicol. 223, 33–52.

Talukdar, D., 2013. Studies on antioxidant enzymes in *Canna indica* plant under copper stress. J. Environ. Biol. 34, 93–98.

Tang, L., Kwon, S.Y., Kim, S.H., Kim, J.S., Choi, J.S., Cho, K.Y., Sung, C.K., Kwak, S.,S., Lee, H.S., 2006. Enhanced tolerance of transgenic potato plants expressing both superoxide dismutase and ascorbate peroxidase in chloroplasts against oxidative stress and high temperature. Plant Cell Rep. 25, 1380–1386.

Tang, W., Peng, X., Newton, R.J., 2005. Enhanced tolerance to salt stress in transgenic loblolly pine simultaneously expressing two genes encoding mannitol-1-phosphate dehydrogenase and glucitol-6-phosphate dehydrogenase. Plant Physiol. Biochem. 43, 139–146.

Tapken, W., Kim, J., Nishimura, K., van Wijk, K.J., Pilon, M., 2014. The Clp protease system is required for copper ion-dependent turnover of the PAA2/HMA8 copper transporter in chloroplasts. New Phytol. http://dx.doi.org/10.1111/nph.13093.

Tateishi, Y., Nakagawa, T., Esaka, M., 2005. Osmotolerance and growth stimulation of transgenic tobacco cells accumulating free proline by silencing proline dehydrogenase expression with double-stranded RNA interference technique. Physiol. Plant 125, 224–234.

Tesfaye, M., Temple, S.J., Allan, D.L., Vance, C.P., Samac, D.A., 2001. Overexpression of malate dehydrogenase in transgenic alfalfa enhances organic acid synthesis and confers tolerance to aluminum. Plant Physiol. 127, 1836–1844.

Thomas, J.C., Davlas, E.C., Malick, F.K., Endreszl, C., Williams, C.R., Abbas, M., Petrella, S., Swisher, K., Perron, M., Edwards, R., Osenkowski, P., Urbanczyk, N., Wiesend, W.N., Murray, K.S., 2003. Yeast metallothionein in transgenic tobacco promotes copper uptake from contaminated soils. Biotechnol. Prog. 19, 273–280.

Thounaojam, T.C., Panda, P., Mazumdar, P., Kumar, D., Sharma, G.D., Sahoo, L., Panda, S.K., 2012. Excess copper induced oxidative stress and response of antioxidants in rice. Plant Physiol. Biochem. 53, 33–39.

Tottey, S., Waldron, K.J., Firbank, S.J., Reale, B., Bessant, C., Sato, K., Cheek, T.R., Gray, J., Banfield, M.J., Dennison, C., Robinson, N.J., 2008. Protein-folding location can regulate manganese-binding versus copper- or zinc-binding. Nature 455, 1138–1142.

Trotel-Aziz, P., Niogret, M.F., Larher, F., 2000. Proline level is partly under the control of abscisic acid in canola leaf discs during recovery from hyper-osmotic stress. Physiol. Plant 110, 376–383.

Trovato, M., Mattiol, R., Costantino, P., 2008. Multiple roles of proline in plant stress tolerance and development. Rend. Lincei-Sci. Fis. 19, 325–346.

Tsutsui, T., Kato, W., Asada, Y., Sako, K., Sato, T., Sonoda, Y., Kidokoro, S., Yamaguchi-Shinozaki, K., Tamaoki, M., Arakawa, K., 2009. DEAR1, a transcriptional repressor of DREB protein that mediates plant defense and freezing stress responses in Arabidopsis. J. Plant Res. 122, 633–643.

Upadhyay, R.K., Panda, S.K., 2009. Copper-induced growth inhibition, oxidative stress and ultrastructural alterations in freshly grown water lettuce (*Pistia stratiotes* L.). C. R. Biol. 332, 623–632.

Vachirapatama, N., Jirakiattikul, Y., Dicinoski, G., Townsend, A.T., Haddad, P.R., 2011. Effect of vanadium on plant growth and its accumulation in plant tissues. Songklanakarin. Sci. Technol. 33, 255–536.

Vagujfalv, A., Kerepesi, I., Galiba, G., 1999. Frost hardiness depending on carbohydrate changes during cold acclimation in wheat. Plant Sci. 144, 85–92.

Valluru, R., van den Ende, W., 2008. Plant fructans in stress environments: emerging concepts and future prospects. J. Exp. Bot. 59, 2905–2916.

Vannini, C., Campa, M., Iriti, M., Genga, A., Faoro, F., Carravieri, S., Rotino, G.L., Rossoni, M., Spinardi, A., Bracale, M., 2007. Evaluation of transgenic tomato plants ectopically expressing the rice *Osmyb4* gene. Plant Sci. 173, 231–239.

Vannini, C., Iriti, M., Bracale, M., Locatelli, F., Faoro, F., Croce, P., Pirona, R., di Maro, A., Coraggio, I., Genga, A., 2006. The ectopic expression of the rice *OsMYB4* gene in Arabidopsis increases tolerance to abiotic, environmental and biotic stresses. Physiol. Mol. Plant Pathol. 69, 26–42.

Vannini, C., Locatelli, F., Bracale, M., Magnani, E., Marsoni, M., Osnato, M., Mattana, M., Baldoni, E., Coraggio, I., 2004. Overexpression of the rice *OsMyb4* gene increases chilling and freezing tolerance of *Arabidopsis thaliana* plants. Plant J. 37, 115–127.

Vazhenin, I.G., 1985. On the normalization of soil pollution emissions from industry. Chem. Agron. 6, 42–45.

Verbruggen, N., Hermans, C., 2008. Proline accumulation in plants: a review. Amino Acids 35, 753–759.

Viehweger, K., 2014. How plants cope with heavy metals. Bot. Stud. 55, 35–47.

Vijayarengan, P., Jose, M.D., 2014. Changes in growth, pigments and phytoremediating capability of four plant species under copper stress. Int. J. Environ. Biol. 4, 119–126.

Wang, H., Wang, T., You, L., Zhong, G., Shi, G., 2013. Effects of vanadate supply on plant growth, Cu accumulation, and antioxidant capacities in *Triticum aestivum* L. Environ. Geochem. Health 35, 585–592.

Wang, L., Li, P., Brutnell, T.P., 2010. Exploring plant transcriptomes using ultra high-throughput sequencing. Brief. Funct. Genomics 9, 118–128.

Wang, R., Wang, J., Zhao, L., Yang, S., Song, Y., 2015. Impact of heavy metal stresses on the growth and auxin homeostasis of Arabidopsis seedlings. Biometals 28, 123–132. http://www.ncbi.nlm.nih.gov/pubmed/25416404.

Wang, S.H., Yang, Z.M., Yang, H., Lu, B., Li, S.Q., Lu, Y.P., 2004. Copper-induced stress and antioxidative responses in roots of *Brassica juncea* L. Bot. Bull Acad. Sin 45, 203–212.

Watanabe, M., Shimachi, F., Noguchi, A., Hasegawa, I., 2005. Introduction of yeast metallothionein gene (CUP1) into plant and evaluation of heavy metal tolerance of transgenic plant at the callus stage. Soil Sci. Plant Nutr. 51, 129–133.

Waters, E.R., Aevermann, B.D., Sanders-Reed, Z., 2008. Comparative analysis of the small heat shock proteins in three angiosperm genomes identifies new subfamilies and reveals diverse evolutionary patterns. Cell Stress Chaperones 13, 127–142.

Weck, J.E.J., Clijsters, H.M.M., 1996. Oxidative damage and defense mechanisms in primary leaves of *Phaseolus vulgaris* as a result of root assimilation of toxic amounts of copper. Physiol. Plant 96, 506–512.

Welch, R.M., Norvell, W.A., Schaefer, S.C., Shaff, J.E., Kochian, L.V., 1993. Induction of iron(II1) and copper(I1) reduction in pea (*Pisum sativum* L.) roots by Fe and Cu status: does the root-cell plasmalemma Fe(II1)-chelate reductase perform a general role in regulating cation uptake? Planta 190, 555–556.

Welling, A., Palva, E.T., 2008. Molecular control of cold acclimation in trees. Physiol. Plant 127, 167–181.

Wilson, D., Vinogradov, S., 2014. Mitochondrial cytochrome *c* oxidase: mechanism of action and role in regulating oxidative phosphorylation. J. Appl. Physiol. 117, 1431–1439. http://dx.doi.org/10.1152/japplphysiol.00737.2014.

Williams, L.E., Mills, R.F., 2005. P1B-ATPases – an ancient family of transition metal pumps with diverse functions in plants. Trends Plant Sci. 10, 491–502.

Xia, Y., Qi, Y., Yuan, Y., Wang, G., Cui, J., Chen, Y., Zhang, H., Shen, Z., 2012. Overexpression of Elsholtzia haichowensis metallothionein 1 (EhMT1) in tobacco plants enhances copper tolerance and accumulation in root cytoplasm and decreases hydrogen peroxide production. J. Hazard. Mater. 233/234, 65–71.

Xiang, L., Le Roy, K.L., Bolouri-Moghaddam, M.-R., Vanhaecke, M., Lammens, W., Rolland, F., 2011. Exploring the neutral invertase-oxidative stress defence connection in *Arabidopsis thaliana*. J. Exp. Bot. 62, 3849–3862.

Xin, Z., Browse, J., 1998. Eskimo1 mutant of Arabidopsis are constitutively freezing tolerant. Proc. Nat. Acad. Sci. U.S.A. 95, 7799–7804.

Xin, Z., Browse, J., 2000. Cold comfort farm: the acclimation of plants to freezing temperatures. Plant Cell Environ. 23, 893–902.

Xu, J., Heng Xia, Y., Li, X., 2009a. Protective effects of proline against cadmium toxicity in micropropagated hyperaccumulator, *Solanum nigrum* L. Plant Cell Rep. 28, 325–333.

Xu, J., Tian, Y.S., Peng, R.H., Xiong, A.S., Zhu, B., Jin, X.F., Gao, J.J., Hou, X.L., Yao, Q.H., 2009b. Yeast copper-dependent transcription factor ACE1 enhanced copper stress tolerance in Arabidopsis. BMB Rep. 42, 752–757.

Xu, Z.S., Chen, M., Li, L.C., Ma, Y.Z., 2011. Functions and application of the AP2/ERF transcription factor family in crop improvement. J. Integr. Plant Biol. 53, 570–585.

Xu, C., Chen, X., Duan, D., Peng, C., Le, T., Shi, J., 2015. Effect of heavy-metal-resistant bacteria on enhanced metal uptake and translocation of the Cu-tolerant plant, *Elsholtzia splendens*. Environ. Sci. Pollut. Res. Int. 22 (7), 5070–5081.

Yamada, M.H., Morishita, K., Urano, N., Shiozaki, K., Yamaguchi-Shinozaki, K., Shinozaki, K., Yoshiba, Y., 2005. Effects of free proline accumulation in petunias under drought stress. J. Exp. Bot. 56, 1975–1981.

Yamasaki, H., Abdel-Ghany, S.E., Cohu, C.M., Kobayashi, Y., Shikanai, T., Pilon, M., 2007. Regulation of copper homeostasis by micro-RNA in *Arabidopsis*. J. Biol. Chem. 282, 16369–16378.

Yanhui, C., Xiaoyuan, Y., Kun, H., Meihua, L., Jigang, L., Zhaofeng, G., Zhiqiang, L., Yunfei, Z., Xiaoxiao, W., Xiaoming, Q., Yunping, S., Xiaohui, D., Jingchu, L., Xing-Wang, D., Zhangliang, C., Hongya, G., Li-Jia, Q., 2006. The MYB transcription factor superfamily of Arabidopsis: expression analysis and phylogenetic comparison with the riceMYB family. Plant Mol. Biol. 60, 107–124.

Ye, N., Li, H., Zhu, G., Liu, Y., Liu, R., Xu, W., Jing, Y., Peng, X., Zhang, J., 2014. Copper suppresses abscisic acid catabolism and catalase activity, and inhibits seed germination of rice. Plant Cell Physiol. 55, 1–9.

Yousfi, N., Slama, I., Ghnaya, T., Savoure, A., Abdelly, C., 2010. Effects of water deficit stress on growth, water relations and osmolyte accumulaton in *Medicago truncatula* and *M. laciniata* populations. Compt Rend Biol. 333, 205–213.

Yruela, I., Alfonso, M., Barón, M., Picorel, R., 2000. Copper effect on the protein composition of photosystem II. Physiol. Plant 110, 551–557.

Yruela, I., Pueyo, J.J., Alonso, P.J., Picorel, R., 1996. Photoinhibition of photosystem II from higher plants: effect of copper inhibition. J. Biol. Chem. 271, 27408–27415.

Yruela, I., 2005. Copper in plants. Braz. J. Physiol. 17, 145–156.

Yruela, I., 2009. Copper in plants: acquisition, transport and interactions. Funct. Plant Biol. 36, 409–430.

Yu, P., Yuan, J., Deng, X., Ma, M., Zhang, H., 2014. Subcellular targeting of bacterial CusF enhances Cu accumulation and alters root to shoot Cu translocation in Arabidopsis. Plant Cell Physiol. 55, 1568–1581.

Yuan, M., Li, X., Xiao J,Wang, S., 2011. Molecular and functional analyses of COPT/Ctr-type copper transporter-like gene family in rice. BMC Plant Biol. http://dx.doi.org/10.1186/1471-2229-11-69.

Yuan, M., Wang, S., Chu, Z., Li, X., Xu, C., 2010. The bacterial pathogen *Xanthomonas oryzae* overcomes rice defenses by regulating host copper redistribution. Plant Cell 22, 3164–3176.

Yuan, H.M., Xu, H.H., Liu, W.C., Lu, Y.T., 2013. Copper regulates primary root elongation through PIN1-mediated auxin redistribution. Plant Cell Physiol. 54, 766–778.

Zhang, C.S., Lu, Q., Verma, D.P., 1995. Removal of feedback inhibition of delta 1-pyrroline-5-carboxylate synthetase, a bifunctional enzyme catalyzing the first two steps of proline biosynthesis in plants. J. Biol. Chem. 270, 20491–20496.

Zhang, J., Van Toai, T., Huynh, L., Preiszner, J., 2000. Development of flooding-tolerant *Arabidopsis thaliana* by autoregulated cytokinin production. Mol. Breed 6, 135–144.

Zhang, L., Pan, Y., Lv, W., Xiong, Z.T., 2014. Physiological responses of biomass allocation, root architecture, and invertase activity to copper stress in young seedlings from two populations of Kummerowia stipulacea (maxim.) Makino. Ecotoxicol Environ. Saf. 104, 278–284.

Zhang, X., Tang, W., Liu, J., Liu, Y., 2014a. Co-expression of rice OsP5CS1 and OsP5CS2 genes in transgenic tobacco resulted in elevated proline biosynthesis and enhanced abiotic stress tolerance. Chin. J. Appl. Environ. Biol. 20, 717–722.

Zhang, L., Wang, Y., Sun, M., Wang, J., Kawabata, S., Li, Y., 2014b. BrMYB4, a suppressor of genes for phenylpropanoid and anthocyanin biosynthesis, is down-regulated by UV-B but not by pigment-inducing sunlight in turnip cv. Tsuda. Plant Cell Physiol. 55 (12), 2092–2101.

Zhao, J., Zhang, W., Zhao, Y., Gong, X., Guo, L., Zhu, G., Wang, X., Gong, Z., Schumaker, K.S., Guo, 2007. SAD2, an importin-like protein, is required for UV-B response in Arabidopsis by mediating MYB4 nuclear trafficking. Plant Cell 19, 3805–3818.

Zhong, R., Lee, C., Ye, Z.-H., 2010. Evolutionary conservation of the transcriptional network regulating secondary cell wall biosynthesis. Trends Plant Sci. 15, 625–632.

Zhou, J., Goldsbrough, J., 1995. Structure, organization and expression of the metallothionein gene family in Arabidopsis. Mol. Gen. Genet. 248, 318–328.

CHAPTER

5

BORON TOXICITY AND TOLERANCE IN PLANTS: RECENT ADVANCES AND FUTURE PERSPECTIVES

Maria Polsia Princi, Antonio Lupini, Fabrizio Araniti, Caterina Longo, Antonio Mauceri, Francesco Sunseri, Maria Rosa Abenavoli

Dipartimento Agraria, Università Mediterranea di Reggio Calabria, Salita Melissari, Reggio Calabria, Italy

CHAPTER OUTLINE

Plant Metal Interaction. http://dx.doi.org/10.1016/B978-0-12-803158-2.00005-9

1. INTRODUCTION

1.1 BORON CHEMISTRY

Belonging to third group in the periodic table (Tariq and Mott, 2007), boron (B) is considered as a typical metalloid element like silicon (Si), arsenic (As) and germanium (Ge), having properties intermediate between the metals and the nonmetals (Nable et al., 1997). The small atom size (4.39 cm^3/mol) together with three valence electrons and high ionization energy defines its unique and complex chemical properties (Greenwood and Earnshaw, 1984). Boron is electron-deficient, possessing a vacant *p*-orbital and it is normally found in the B^{3+} state. It is usually assigned a +3 valence because it combines with more electronegative elements. Furthermore, it has tendency to form anionic rather than cationic complexes.

Boron is not present on Earth in its elemental form; it occurs in nature in combination with oxygen as borates such as the borax $Na_2B_4O_7 \bullet 10H_2O$, less frequently as boric acid [$B(OH)_3$ or H_3BO_3], and very rarely associated with fluorine, as anion BF_4^- (Power and Woods, 1997). On the contrary, in plant and animal cells, at a pH of about 7.5, in the cytoplasmic compartment, more than 99.95% of B exists in the H_3BO_3 form, whereas the remaining part is present as borate $[B(OH)_4^-]$.

Boron occurs in aqueous solution as H_3BO_3, which hydrolyzes reversibly to the borate ion $[B(HO)_4^-]$, according to the reaction by Baes and Mesmer (1976):

$$B(OH)_3 + H_2O = B(OH)_4^- + H^+ \leftrightarrow pK_a = 9.2$$

In accordance with electronic B configuration, at pH < 7, H_3BO_3 acts as a very weak Lewis acid, and its activity seems to be related to the acceptance of OH^- rather than the donation of H^+. Therefore, in neutral or slightly acid soils, B is located mainly in the form of undissociated boric acid (Raven, 1980).

Both boric acid and borate can quickly react with many different types of molecules (Brown et al., 2002). In particular, they are able to form esters and complexes with a wide variety of mono-, di-, and poly-hydroxyl compounds. Boric acid binds to many sugars containing a furanoid rather than a piranoid ring, such as ribose (the main constituent of the RNA) (Loomis and Durst, 1992; Goldbach, 1997), apiose, mannitol, mannan, and polymannuronic acid (essential constituents in plant cell walls). O'Neill et al. (2004) demonstrated that borate forms a cross-link with apiose residues of rhamnogalacturonans II (RGII), important components of a plant cell wall, which is essential for normal leaf expansion in *Arabidopsis thaliana*. Loomis and Durst (1992), however, demonstrated that B forms complexes with ribonucleotides, which was a probable cause of B toxicity. Recently, the ability of B to stabilize ribose and to form borate ester nucleotides makes it a "prebiotic element," which provides an essential contribution to the "pre-RNA world" (Grew et al., 2011; Scorei, 2012).

Furthermore, boric acid forms stable borates with organic *cis*-diols and with some *o*-diphenols, such as caffeic and hydroxiferulic acids, which are important precursors for the lignin biosynthesis in dicots (Bolaños et al., 2004).

Both of these B complexes with sugars and organic *cis*-diols could be considered essential in the living world. Indeed, the stabilization of these molecules could represent a most probable defense mechanism of genetic molecules, providing them thermal and chemical stability in hostile environments throughout the evolution of life (Grew et al., 2011; Scorei, 2012).

In addition, the ubiquitous presence of OH groups within biological molecules allows the formation/dissociation of many B complexes, which generally occur spontaneously through rapid kinetics mainly influenced by pH (Power and Woods, 1997). Furthermore, the stability depends on the nature of the constitutive molecular groups of B complexes. For example, the coenzyme NAD^+ shows a greater ability to form more stable borate complexes than its reduced form NADH (Brown et al., 2002). This could indirectly affect some enzymatic activities depending on NAD^+ coenzyme, causing significant metabolic disorders (Wimmer et al., 2003).

1.2 BORON: SOURCE AND DISTRIBUTION

Boron is widely distributed in nature and its major reservoirs in the biosphere are shown in Table 5.1 (Kot, 2008). Its turnover and the extent of flow through which B moves among the different environmental compartments seem to be less clear. The inorganic forms of B originate from both natural and anthropogenic sources and they are usually found in water, soil, and atmosphere. In the Earth's crust, boron occurs mainly in silicate minerals at the approximate concentration of 10 mg/kg. It can be found as borosilicate in igneous, metamorphic, and sedimentary rocks, and its distribution among rock classes and types was proposed by Krauskopf (1972) (Table 5.1). Natural weathering of sedimentary rocks is thought to be the primary source of B compounds in water and soil, whereas it is predominantly released into the atmosphere from oceans (65–85%), volcanoes, and geothermal steams (Jansen, 2003). B-rich deposits around the globe are located in California, Australia, China, Russia, and Argentina, although Turkey ranks the first with almost 73% of the global B reserve.

The B level in atmosphere averages $\sim$20 ng/m^3 with a range of 0.5–80 ng/m^3. Because borates exhibit low volatility, B would not be significantly present in the atmosphere as a vapor. Atmospheric borates and boric acid emissions in particulate (<1–45 μm in size) or vapor form occur as a result of boric acid volatilization from the sea, volcanic activity, mining operations, glass and ceramic manufacturing, agricultural activities, and coal-fired power plants.

In soils, B occurs at concentrations ranging from 10 to 300 mg/kg of soil, whereas in surface water B concentrations range from 0.001 to 2 mg/L (Howe, 1998). Both these issues will be discussed further in the following sections.

Table 5.1 Boron Major Reservoirs in the Biosphere

Rock Class	Rock Type	Boron (mg/kg)	Mineral Class	Mineral Type	
Igneous	Granite	15	Hydrous borates	Borax	$Na_2B_4O_7 \cdot 10H_2O$
	Basalt	5		Kernite	$Na_2B_4O_7 \cdot 4H_2O$
				Colemanite	$Ca_2B_6O_{11} \cdot 5H_2O$
Sedimentary	Limestone	20		Ulexite	$NaCaB_5O_9 \cdot 8H_2O$
	Sandstone	35			
	Shale	100	Anhydrous borate	Ludwigite	Mg_2FeBO_5
	Soils	7–80		Kotoite	$Mg_3(BO_3)_2$
			Complex borosilicates	Tourmaline	$Na(Mg, Fe, Mn, Li, Al)_3 Al_6Si_6O_{18}(BO_3)_3(OH,F)_4$
				Axinite	$(Ca, Mn, Fe)_3Al_2BO_3Si_4O_{12}(OH)$

In seawater, B concentration ranges from 0.5 to 9.6 mg/kg with an average of 4.6 mg/kg (Power and Woods, 1997). High levels of B are also present in domestic wastewater (0.5–2 mg/L) and in groundwater (0.3–100 mg/L; e.g., in Mediterranean countries, approximately 8 mg/L). These values can significantly increase near areas with geothermal activity because of its high concentration in thermal waters (0.2–72 mg/L). For example, B concentration in water supplies can reach values of 3–13 mg/L in exposed areas of Cornia Valley (Italy) (European Union research project BOREMED, http://boremed.brgm.fr/). BOREMED showed that the contamination of B in groundwater is probably due to the Pleistocene geothermal area and not to human activity.

Boron concentrations are also found in drinking-water (0.01–15 mg/L) due either to the leaching from surrounding geology and wastewater discharges or to the difficulty in removing it by conventional drinking-water treatment methods. Moreover, B is also present in irrigation water causing serious problems in agriculture (Polat et al., 2004).

Finally, B is also found in foods and vegetables consumed by humans. Diets rich in fruits, vegetables, legumes, and nuts may provide small amount of B for the normal functioning of several body parts, including the brain, skeleton, and immune system (Nielsen and Forrest, 2008). An intake of more than 1 mg per day is desirable, but probably not more than 13 mg per day (Nielsen, 1997). Neurological effects, weight loss, testicular atrophy, and skeleton malformation have been reported in animals with excessive B intake (Yazbeck et al., 2005). It is an essential element also for diatoms, cyanobacteria, other marine algal flagellate (Loomis and Durst, 1992), and animal species.

Anthropogenic B sources (power and chemical plants; manufacturing facilities) are considered to contribute a lesser amount to the environment (7–18%) than natural processes (Nable et al., 1997). On the contrary, fertilizers, herbicides, and industrial wastes are among significant soil contamination sources. The occurrence of B contamination in water can derive from industrial wastewater and municipal sewage as well as indirectly from air deposition and soil runoff (Wood and Nicholson, 1998).

1.3 BORON IN SOILS

Boron occurs in soils as H_3BO_3 and partially as $B(HO)_4^-$ distributed unevenly in the solution and in organic and mineral fractions. In particular, whereas the dissociate form $H_2BO_3^-$ becomes predominant only at pH > 9.2. In soils, B is considered the most mobile element and often it is deficient among all the trace elements. A study conducted by FAO on micronutrients in soils revealed that B deficiency was the most common problem, affecting at least 8 million hectares worldwide (Tariq and Mott, 2007). Boron deficiency is found primarily in humid regions with well-drained soils or in sand soils as reported in some regions of China, Japan, and the United States (Tanaka and Fujiwara, 2008). High rainfall together with high B solubility in soil solution may be the major reasons for B deficiency (Shorrocks, 1997). However, at slightly high concentrations, B may become toxic for plants because the range between B deficiency and toxicity is relatively narrow (Paull et al., 1991), making B management in plant–soil system difficult. Both B deficiency and toxicity are related to plant disorders and crop yield losses. Soils excessively fertilized with B, irrigated with sewage debris or salt water, may contain toxic B concentrations. As a rule, B toxicity occurs when soils contain concentrations greater than 12 mg/kg B (Hall, 2010), and it is usually confined to areas with less than 550 mm of annual rainfalls. Low rainfall in dry regions also means that limited soluble B is leached from the root zone and soil profile (McDonald et al., 2010). The amount of water required to leach B is approximately three times that required for sodium chloride leaching (Moore, 2004). B toxic soils occur in Australia (Western Australia, South Australia, and Victoria), Jordan, Malaysia, Peru, Chile (North), India, Israel, Mediterranean areas (Turkey, Morocco), and the United States (California) (Nable et al., 1997; Kot, 2008; Tanaka and Fujiwara, 2008).

In most soils, the main B sources are the primary minerals such as tourmaline [Na (Al, Fe, Li, Mg, Mn)$_3$Al$_6$ Si$_6$O$_{18}$ (BO$_3$)$_3$ (OH, F)$_4$] and volcanos' volatile emanations (Chesworth, 1991). Other common B-containing minerals are: the ulexite Na Ca[B$_5$O$_6$(OH)$_6$]*5H$_2$O, the borax Na$_2$[B$_4$O$_5$(OH)$_4$]*8H$_2$O, the less-soluble mineral colemanite Ca$_2$[B$_3$O$_4$(OH)$_3$]*2H$_2$O, and the less hydrated borax kernite Na$_2$[B$_4$O$_5$(OH)$_4$]*2H$_2$O. Because of the limited solubility of such B-containing minerals in soils and their resistance to weathering, B is not readily available to plants (Nable et al., 1997); indeed, the B adsorbed on the surfaces of the colloidal soils does not affect the amount of B responsible for toxicity in plants (Keren et al., 1985). Groundwater, for the addition of artificial residues from mining processes, fertilizers, or residues of fossil fuels (Nable et al., 1997), contributes to raise the B level in soils. However, the major B source in soils is probably irrigation water (Keles et al., 2004). The threshold B concentration in irrigation water has been established for sensitive (0.3 mg/L) and tolerant (2 mg/L) crops, taking into account the physical–chemical soil properties and B interaction in the soil (Keren, 1996).

Total, acid, and water-soluble are the three B pools present in soils. The total B content has a slight bearing on the status of available B to plants, which results in about 10% of the total B content in soils (Power and Woods, 1997). Water-soluble B content frequently ranges from 7 to 80 μg/g (Krauskopf, 1972), providing a general indication of B supply to plants. In relation to water-soluble B concentrations, Fleming (1980) defines three categories: deficient (<1 μg/mL), sufficient (1–5 μg/mL), and toxic (>5 μg/mL) B for normal plant development. Shorrocks (1997) later reviewed this classification, indicating that water-soluble B in soils >0.5 μg/g is already sufficient for plant growth of many crops. However, water-soluble B depends on soil system, crop species, lime application, irrigation management, and environmental conditions (Tariq and Mott, 2007).

Boron sorption–desorption processes regulate water-soluble B availability, acting as source sink for plant uptake in soil solution that in turn is influenced by soil's physical–chemical properties (Chen et al., 2002). Among these, the pH is retained the main factor affecting B adsorption in agricultural soils (Soares et al., 2008) as well as soil texture, soil moisture, clay content, Al and Fe (hydr) oxides, clay minerals, calcium carbonate, and organic matter (Goldberg, 1992; Goldberg et al., 2008; Arora and Chahal, 2010).

A positive correlation between B adsorption on clay minerals, hydroxyl-Al, and pH values increase in soils has been reported (Keren, 1996; Goldberg, 1992). Indeed, at pH lower than 7.0, the predominant H$_3$BO$_3$ form shows a relatively low affinity to the clay, whereas in the alkaline pH range, $B(HO)_4^-$ species increased quickly reaching the maximum of adsorption around pH 9.0 (Elrashidi and O'Connor, 1982). Different empirical models have been applied to describe adsorption reactions such as Langmuir and Freundlich isotherm equations (Goldberg, 2003). Recently, analyzing some soils of Paraná (Brazil), Steiner and Lana (2013) confirmed that B adsorption was dependent on and increasing as a function of pH, but it was also affected by soil properties such as organic matter, clay, and aluminum oxide content.

Excessive moisture in the soil can cause significant B loss related to leaching, whereas in clay soils, B is more easily retained partly because of the ability of clays to form stable complexes with the microelement (Kot, 2008). On a weight basis, illite is the most reactive among the common clay minerals, whereas kaolinite is the least reactive (Keren, 1996). The presence of organic matter in the soil provides an important reserve of B because of the presence of this element in many organic compounds. Some authors have shown a higher presence of B in soils in organic matter than in the mineral fraction, but also a positive influence of organic fertilization in B absorption (Yermiyahu et al., 2001).

In arid or semiarid areas, B toxicity is frequently associated with salt stress (Goldberg, 1992) as observed in the Lluta Valley (Northern Chile) and San Joaquin Valley (California) (Wimmer et al., 2003). Interactive effects between salt stress and B plant nutrition have been clearly established, but the results are often contrasting, indicating antagonistic or synergistic interactions even within the same plant species (Yermiyahu et al., 2008). Bingham et al. (1987) found that plant responses to B were independent from salinity levels in the soil. On the other hand, salinity seemed to alleviate B toxicity decreasing total shoot B concentrations (Diaz and Grattan, 2009). Wimmer et al. (2003) found that salinity can magnify B toxicity symptoms in several plant species because these combined stresses significantly increased soluble B concentrations at the intra- and intercellular levels. So far, conclusive considerations on mutual relationship between salt stress and B toxicity are lacking yet (Yermiyahu et al., 2008).

2. BORON IN PLANTS

The presence of B in plants was reported for the first time in 1910, but only later Katherine Warington (1923) claimed the importance of B for growth and development in broad beans and other legumes. At the same time, B requirements for six dicots and one monocot were also demonstrated (Sommer and Lipman, 1926). Currently, based on the B requirement for their growth and development, plants can be divided into four classes: (1) lactifers (latex-forming species); (2) legumes; (3) the remaining dicots and rather all monocots families; and (4) the Poaceae, which considered the least B-demanding (Bonilla et al., 2009). Excluding lactifers, the higher B requirement in all other plants was presumably from a higher content of *cis*-diol configuration compounds within the cell wall, such as pectins and polygalatturans (Loomis and Durst, 1992). Therefore, such differences could support the diversity in the B-requirement among plant species for reaching an optimal growth (Marschner, 1995).

Nowadays, B is considered an essential micronutrient for normal plant growth (Emebiri et al., 2009), unevenly distributed within plants, and especially found in tissues of reproductive organs (Saleem et al., 2011). Crop species and cultivars show varying ranges at which B is considered adequate. For example, in monocots, B concentrations range from 1 to 6 mg/kg, whereas in most dicots they are from 20 to 70 mg/kg. However, the range of B required for an optimal plant growth is very narrow (Bingham et al., 1987; Moore, 2004). Both deficient and toxic B levels caused plant disorders reducing the yield and quality of crop harvests.

Boron is involved in many important processes in higher plants such as: (1) the transport of sugars and carbohydrate metabolism; (2) the cell wall synthesis and the lignification process; (3) the maintenance of plasma membrane integrity and its function; (4) the stimulation of the nucleic acids metabolism; (5) the indoleacetic acid metabolism; (6) the ascorbate/glutathione cycle; (7) the phenolic compounds metabolism; (8) the pollen tube formation; (9) the nitrogen metabolism; (10) the photosynthesis; and (11) several enzymatic activities (Moore, 2004; Rehman et al., 2006; Reid, 2010) (Figure 5.1).

However, the widely recognized B primary role is as structural component conferring stability to cell wall and plasma membrane (Warington, 1923).

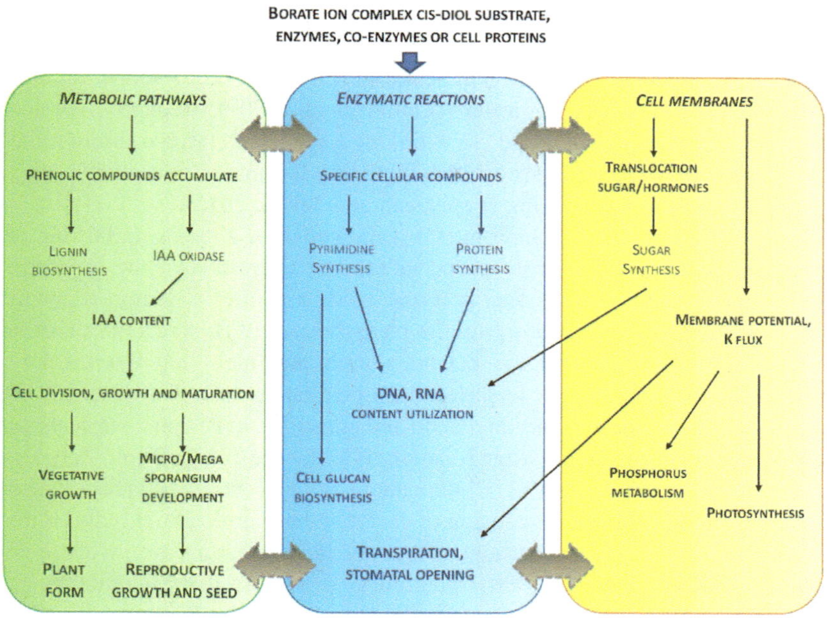

FIGURE 5.1

Physiological and molecular boron targets.

2.1 BORON IN THE CELL WALL AND MEMBRANES

Ninety percent of B content in plant is localized in the cell wall fraction (Loomis and Durst, 1992; Blewins and Lukaszewski, 1998). Such evidence highlights the contribution of B in the organization of cell wall proteins, pectins, and/or precursors during plant growth and development (Hu and Brown, 1997). Indeed, in B-deficient plants, the structural abnormalities in the composition of cell wall and middle lamella caused a growth block on the apical meristems of root and stem (Brown et al., 2002), on the pollen tubes, as well as the fragility in developing leaves and petioles (Loomis and Durst, 1992; Shorrocks, 1997; Goldbach, 1997).

The B functional role in cell wall organization was first demonstrated after the isolation of a B–polysaccharide complex from radish root cell walls on which an RGII polysaccharide was later characterized (O'Neil et al., 1996). In particular, RGII was cross-linked by 1:2 borate-diol diesters to form the dimeric RGII via *cis*-diol groups of two apiose residues forming a stable three-dimensional network (O'Neil et al., 1996). At the same time, the presence of RGII–B complex in the cell wall of another 22 plant species has been demonstrated (Matoh et al., 1996). In absence of B, Fleischer et al. (1999) observed wider formation of cell wall holes than normality inside the three-dimensional structure because of a lack of dB-RGII, suggesting a new B functional role in the cell wall. The presence of an improper pore, in B-deficient cells, may affect physiological important processes such as the incorporation and transport of polymers into the wall. In this respect, the contribution of B–RGII complexes to the porosity and strength of the cell wall has been stated (Dannel et al., 2002). Moreover, O'Neil et al. (2004), using a *mur1 Arabidopsis* mutant characterized by abnormal sugar composition of

RGII, defined the importance of B–RGII for the normal leaf expansion, at molecular level. Hence, the B requirement was strongly associated to the RGII content of the cell walls in different plant species (Reid et al., 2004) and a reduced production of pectic substances or precursors of the cell wall in plants exposed to B deficiency has been reported (Bonilla et al., 2009). Recently, the location and the mechanisms of RGII synthesis and borate cross-linking have been clarified through the identification of key enzymes for RGII synthesis and regulators including B transporters, which were required for an efficient formation of RGII cross-linking (Funakawa and Miwa, 2015).

Boron deficiency in organisms lacking in cell walls underlined the essential role of this microelement in plant growth and development beyond its role in cell wall structure (Läuchli, 2002). Many authors have speculated that B plays a structural role inside the plasma membrane, which may explain the large number of B effects on it. Several studies have shown that B affects the structure and function of the membrane and especially of the plasma membrane, including ion uptake (Blewins and Lukaszweski, 1998) and stimulation of ferricyanide-dependent H^+ release (Goldbach et al., 1990). On the contrary, B deficiency reduced rubidium (Rb^+) and phosphorus (P) uptake in *Vicia faba,* sunflower, and maize roots, which were restored after B supply (Goldbach, 1984). Boron deficiency and excess also inhibited adenosine triphosphate (ATP)-dependent H^+ pumping and vanadate-sensitive ATPase activity (Ferrol et al., 1993). In cell suspension of carrot and tomato cells, B reduced the ferricyanide-induced proton release mediated by vanadate suppression, suggesting the involvement of a plasma membrane proton (Goldbach et al., 1990). Therefore, B stabilized the membrane molecules with *cis*-diol groups (Bolaños et al., 2004), but also regulated its function. Recently, B deficiency has been observed to cause a nitrate content decrease in tobacco because of the lower net nitrate uptake rate because of root plasma membrane H^+-ATPase (*PMA2*) transcript reductions (Camacho-Cristóbal et al., 2008).

Moreover, it has been hypothesized that B may be involved in the structure of so-called "membrane rafts", particularly "lipid rafts", physiologically active membrane fractions with relevant functions in signal transduction and useful as binding sites for glucosilfosfatil-inositol proteins (Brown et al., 2002). The "lipid rafts" are characterized by high concentrations of glycolipids and glycoproteins, providing a significant number of B complexing sites. In addition, these fractions contain either sugars such as galactose and mannose or amino acids such as serine and tyrosine, which are able to link with B. For this reason, B seemed to play a specific function in stability, integrity and function of "membrane rafts." Recently, a glycosylinositol phosphorylceramides (GIPCs), the major sphingolipids in lipid rafts, able to form a GIPC–B–RGII complex (Borner et al., 2005), has been characterized in rose cell culture using thin-layer chromatography and mass spectrometry approaches (Voxeur and Fry, 2014). They concluded that: (1) B played a structural role in plasma membrane; (2) a high B level, disrupting the membrane components, was responsible for membrane phytotoxic effect; (3) GIPCs facilitated B-dependent RGII dimerization process; and finally (4) GIPC–B–RGII gave, for the first time, the molecular explanation of the wall-membrane attachment sites (Voxeur and Fry, 2014).

2.2 BORON AS SIGNAL

Although signal transduction pathways and plant sensing for mineral deficiencies are well known for macronutrients (Schachtman and Shin, 2007), the knowledge for many micronutrients, such as B, are more limited. None of the proposed hypotheses fully explains how so many decisive pathways for plant development respond in the short term to B deficiency.

The first evidence of B as signal molecule required for quorum sensing was shown in bacteria (Chen et al., 2002), suggesting a similar role for B in both animals and plants. Although many studies explained how the variation in B concentrations inside plant cells could trigger a cascade of signals, which in turn altered the membrane-bound proteins conformation (reviewed by Goldbach and Wimmer, 2007), the role of B as signal molecule has not been clearly demonstrated yet.

This hypothesis could be supported by the rapid increase of proteins such as actin and tubulin within the membranes of plant root cells of *Arabidopsis* and maize under B deprivation. This resulted in the alteration of the polymerization pattern and consequently in the cytoskeleton assembly (Yu et al., 2003). Interestingly, B deficiency could induce the expression of genes in response to stress such as *NIP5;1*, a member of the major intrinsic protein (MIP) family, that encodes an essential protein for B absorption when its availability is limited (Takano et al., 2006). In this respect, a rapid signal movement from the cell wall to the cytoplasm that triggered the induction of the *NIP5;1* or BOR family genes under B deprivation was already hypothesized (Kobayashi et al., 2004). The possible role of B as signal was suggested because of its direct or indirect interaction with transcription factors (TFs) (González-Fontes et al., 2008). According to the target gene and the TF types (activators or repressors), this complex could regulate the expression of several genes, explaining either the diverse B effects on so many physiological processes, and how a negligible amount of B into the protoplast can be decisive for the normal plant development (González-Fontes et al., 2008). For example, several TF genes belonging to MYB, WRKY, and bZIP families were up- and downregulated in response to short B deprivation (González-Fontes et al., 2013).

The possible role of B as signal molecule in combination with Ca^{2+} has been observed in plants, animals, and humans (Bolaños et al., 2004; González-Fontes et al., 2014). Under B deficiency, Ca^{2+} addition reduced the negative effect on nodulation of N_2-fixing legume–rhizobia symbiosis and the expression of some nodulation genes in *Medicago truncatula*, without reconverting the abnormal cell wall structure in nodules (Redondo-Nieto et al., 2003). It was also suggested that B deficiency caused an oxidative damage from reactive oxygen species (ROS) accumulation in the elongation zone of *Arabidopsis* root, effects similar to that observed under Ca^{2+} deficiency (Oiwa et al., 2013). Considering that B and Ca^{2+} shared a key role in stabilizing cell-wall structures and regulating genes expression involved in several plant processes (Camacho-Cristóbal et al., 2011), it was possible to suppose that B and Ca^{2+} could interplay in signaling events under B deficiency in plants (González-Fontes et al., 2008, 2014). Recently, an intermediary role for Ca^{2+} and Ca^{2+}-related proteins in the transduction pathway triggered by B deprivation has been proposed. In particular, cyclic nucleotides (adenosine-3′,5′-cyclic monophosphate or cyclic guanosine monophosphate) appeared to be involved, as observed in different stresses (Ma, 2011), playing a major role to stimulate cyclic nucleotide-gated ion channels that allow Ca^{2+} to enter the cytosol. Afterward, the increase of Ca^{2+} could trigger many physiological responses in plants (González-Fontes et al., 2014).

2.3 BORON UPTAKE BY ROOTS: PASSIVE AND ACTIVE MECHANISMS

Boron is taken up by plant roots from the soil solution as uncharged boric acid (Marschner, 1995), an exception compared with all other mineral nutrients generally absorbed by roots in ionic form (Miwa and Fujiwara, 2010). The boric acid uptake mechanism in higher plants has been controversial for more than 30 years, but now evidence supports either passive or active processes (Dannel et al., 2000, 2002; Brown et al., 2002; Tanaka and Fujiwara, 2008). Based on the high permeability of boric acid across lipid bilayers of biological membranes, its uptake has been considered a passive process for a long time (Raven, 1980). The author first postulated the boric acid passive diffusion which was based on a theoretical lipid permeability coefficient of $B(OH)_3$ (8×10^{-6} cm/s), also claiming that active transport to maintain boric acid distribution across a membrane away a thermodynamic equilibrium was likely to be energetically expensive. Recently, the higher permeability of membranes to B than other solutes were also observed in giant algal cells (Reid, 2014).

Several studies on boric acid absorption have been reported with contrasting results. Until 1990, the general opinion was that the B absorption rate by roots was influenced by its concentration in the external solution, the formation of B complex within the cell wall, and plant water flux (Hu and Brown, 1997). This idea was closely related to the patterns of B deficiency and toxicity symptoms (Marschner, 1995).

Later, through direct measurements of B membrane permeability (Pf_B), Dordas and Brown (2000) confirmed the theoretical values indicated by Raven (1980) in artificial liposomes, but not in plasma membrane. Indeed, in isolated membranes of squash roots, the Pf_B values were slower than either the theoretical prediction of Raven (1980) or artificial liposomes (Dordas et al., 2000). Similar low Pf_B values were found in charophyte algae cells (Stangoulis et al., 2001). This discrepancy between calculated and experimental Pf_B was probably due to different lipid membrane composition, which defined the properties and permeability of membranes. This hypothesis was confirmed by the results obtained with mutant lines of *A. thaliana* differing in membrane lipid composition (Dordas and Brown, 2000). Furthermore, Dordas et al. (2000), using mercuric chloride and phloretin, two channel blockers, observed a partial inhibition (30–39%) of boric acid permeation across plasma membrane vesicles of squash roots, which was restored by 2-mercaptoethanol. From these results, they stated that B entered into plant cells in part by passive diffusion through the plasma membrane lipid bilayer and in part through channel-mediated transport (Dordas et al., 2000). In addition, they first supported the hypothesis that B can be taken up through facilitated diffusion via an MIP superfamily, as discussed later (Dordas et al., 2000). Brown et al. (2002) pointed out that boric acid passive permeation would be adequate to provide the B requirement for both canola and tobacco under an adequate B supply (10 μM B), but not under a limited B supply (1 μM B). In addition, they sustained the role of membrane proteins in the facilitation of B transmembrane movement without precluding the existence, at low B levels, of an active B transport mechanism needed to satisfy B plant requirements (Brown et al., 2002).

A major shortcoming of the passive uptake hypothesis was inside the unexplained observed differences in boric acid uptake among plant species or cultivars in field experiments. Indeed, susceptible varieties to B excess acquired seven times as much B as tolerant ones, even if they grow under the same conditions (Hu and Brown, 1997). Nable et al. (1997) found that two barley cultivars grown under identical conditions dramatically differed in B content (112 vs 710 mg/kg) in the youngest expanded leaf blade. However, the different water use efficiency proposed by Passioura (1997) to justify the same behavior observed in wheat cultivars could not explain these different responses.

Dannel et al. (2000) first demonstrated the existence of active boric acid transport in sunflower plants grown under a low B supply (1 μM B), but not under a high B level (100 μM B). They suggested a saturable carrier mediated under a low B level and a nonsaturable linear diffusion under a high B level involving either the root uptake process or xylem loading (Dannel et al., 2000). Similar results were also obtained using the charophyte algae *Chara coralline* (Stangoulis et al., 2001).

In recent years, experimental evidence has clarified that there are three different physiological and molecular mechanisms for boric acid transport from soil solution into root cells and xylem loading, depending on B availability:

1. Passive diffusion across plasma membrane operates under adequate or relatively high B supply;
2. Facilitated transport by nonselective membrane channels (NIPs) belonging to the MIP family operates under limited B supply;
3. Energy-dependent high-affinity transport against concentration gradients, mediated by selective B transporters (BOR) operates in response to low B supply.

Therefore, under adequate or B excess, plants showed a passive diffusion across lipid bilayers to B absorption into root cell to satisfy plant microelement requirement. Nevertheless, under limited B availability, a facilitated membrane transport of boric acid through an MIPs channel and an energy-dependent high-affinity transport system, mediated by BOR transporters, are required for B transport into the roots and toward xylem (Takano et al., 2006; Choi et al., 2007; Tanaka and Fujiwara, 2008; Miwa and Fujiwara, 2010) (Figure 5.2).

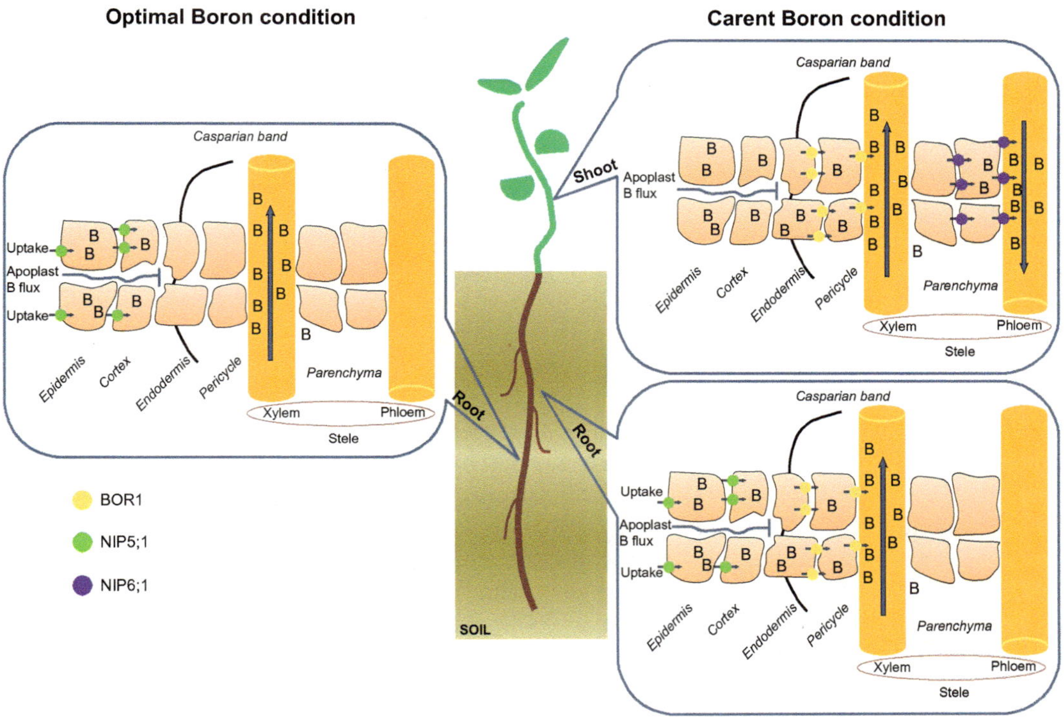

FIGURE 5.2

Scheme for different function of boron transporters under optimal and deficient B conditions in plants.

2.3.1 Facilitated B Transport System: The Role of Aquaporin-like Channels

Plant aquaporins are hydrophobic transmembrane proteins with six membrane domains also known as MIPs. Based on both sequence homology and localization, they are clustered into four subfamilies: plasma membrane (PIPs), tonoplast (TIPs), nodulin 26 like (NIPs), and small basic intrinsic proteins (SIPs) (Johanson and Gustavsson, 2002; Chaumont et al., 2005). Aquaporins are essentially channel proteins that mediate the movement of water and different low molecular solutes such as urea and glycerol (Kaldenhoff and Fischer, 2006), but they are also involved in many physiological processes (Ma et al., 2006).

Dordas et al. (2000) first suggested that aquaporin-like channels were involved in boric acid transport, demonstrating that its permeation across the plasma membrane was partially reduced by $HgCl_2$ and phloretin, two channel blockers. These results were later confirmed through in vivo experiments on squash roots (Dordas and Brown, 2001). In addition, they demonstrated that some small solutes such as urea and glycerol could competitively suppress boric acid uptake up to 54%, confirming the involvement of membrane channels to facilitate B acid movement. Similar results were also obtained by Fitzpatrick and Reid (2009) using metabolic inhibitors in barley roots. However, the possibility to express aquaporin channel proteins of different species in *Xenopus laveis* oocytes led to improved knowledge of boric acid absorption. Dordas et al. (2000) first observed that the permeability to B increased by expressing a maize aquaporin (*Zm*-PIP1) in oocytes. In particular, this gene belongs to a PIPs subfamily including subgroups PIP1 and PIP2 characterized by very low and high water channel activity, respectively (Chaumont et al., 2005). Nuttall (2000) also demonstrated the role of PIP1b, PIP2a, and PIP2b in *Xenopus* oocytes, where its expression increased the permeability to boric acid. Later, Fitzpatrick and Reid (2009), using yeast complementation assays, demonstrated that HvPIP1;3 and HvPIP1;4 expressions increased boric acid transport in barley roots. These results supported that at least some boric acid flux occurred through a channel-like protein. Recently, *OsPIP2;4* and *OsPIP2;7* from rice have been found to mediate B permeability in yeast and *Arabidopsis*, showing also a role in tolerance to B toxicity (Kumar et al., 2014).

Moreover, members of NIPs family have also been involved in B transport. In particular, NIPs were first localized in the peribacteroid membrane of soybean nodule cells, but their location in nonlegume plants is not known (Chaumont et al., 2005). NIPs are classified into two groups based on similarity of their aromatic/arginine (ar/R) region compared to the archetypal Nodulin 26: group I, which possesses the conserved ar/R region of Nodulin 26, whereas group II are divergent for ar/R tetrad. This latter includes NIP5;1, NIP6;1, and NIP7;1.

A transcriptome analysis under a B-limited supply led to identify *AtNIP5;1* gene that codes for a protein channel localized in the plasma membrane of root epidermal, especially in the root elongation zone, where it was strongly upregulated under B deficiency (Takano et al., 2006). Furthermore, NIP5;1 facilitated boric acid flux in *Xenopus* oocytes by heterologous expression of *AtNIP5;1*. The authors also showed a reduced B uptake together with a severe plant growth delay in transfer DNA insertion *AtNIP5;1* mutant under limited B supply (Takano et al., 2006). Later, *OsNIP3;1* homolog to *AtNIP5;1* has been identified as boric acid channel in rice (Hanaoka and Fujiwara, 2007).

More recently, another member of the group II *AtNIP6;1* involved in B transport under B deficiency has been isolated in *Arabidopsis* (Tanaka et al., 2008). AtNIP6;1 is responsible of xylem-phloem transfer of boric acid at the nodal regions and its transcript accumulation appears highest in both young rosette leaves and shoot apices but not in roots. They concluded that *AtNIP6;1* might play a different role from *AtNIP5;1* in B transport for its tissue specific localization (Tanaka et al., 2008).

Schnurbusch et al. (2010a) demonstrated that *HvNIP2;1* aquaporin from barley could facilitate B transport when expressed in *Xenopus* oocytes, which is able to increase the plasma membrane permeability. The control of its expression could limit B toxicity in barley.

Finally, AtTIP5;1 aquaporin is the only protein belonging to the TIPs subfamily involved in a B transport pathway via vacuolar compartmentation in *Arabidopsis*, which plays an important role in B toxicity tolerance (Pang et al., 2010).

2.3.2 Boron Transporters

The isolation and characterization of the *Arabidopsis* mutant *bor1-1* (high B-requiring), which is sensitive to B deficiency, was the first evidence that *BOR1* could be directly or indirectly involved in B metabolism in higher plants (Noguchi et al., 1997). In particular, B-uptake analysis indicated that *bor1-1* mutant was unable to tolerate a reduced B delivery to shoots, because of impaired xylem loading, showing a severe delay in plant growth (Noguchi et al., 1997). Thereafter, *AtBOR1*, the first B transporter, was identified through map-based cloning in *Arabidopsis*, where it was characterized as a membrane protein with homology to bicarbonate transporters in animals (Takano et al., 2002). Expressed in the pericycle cells of root stele, *AtBOR1* was responsible for xylem loading and essential for protecting shoots from B deficiency. Indeed, under a low B supply, mutant *bor1-1* showed a lower B concentration in xylem sap than wild-type, confirming that BOR1 acts as borate exporter to the xylem against B concentrations. Previously, the same authors had proved that BOR1 was also involved at least in the preferential B distribution to young leaves under low B supply (Takano et al., 2001).

BOR1 is a member of the solute carrier (SLC4) family of transporters, which are classified into three main classes: anion exchangers (AEs), sodium-coupled bicarbonate transporters (NCBTs), and borate/boron transporters (BOR-type) (Frommer and Wirén, 2002; Reid, 2014). Databases reported the existence of seven predicted proteins in *Arabidopsis* similar to BOR1, in which an encoding gene exhibited high similarity to expressed sequence tag clones from different plant species, including angiosperms and gymnosperms. This indicated that *AtBOR1* belongs to a group of highly conserved membrane proteins in plants (Frommer and Wirén, 2002; Miwa et al., 2013). For many years, *AtBOR1* was considered a borate/chloride anion exchanger and compared with BAND3 (AEs), a prototype anion exchanger of bicarbonate and chloride in red blood cells (Takano et al., 2002). However, the phylogenetic analysis of the SLC4 family showed that *AtBOR1* shared the same clade with a human bicarbonate transporter-related protein, *HsBTR1*, belonging to sodium-coupled bicarbonate transporters (Frommer and Wirén, 2002; Park et al., 2004). Although sequence similarity between BOR-type and *HsBTR1* was quite low, it was higher than that found between *BOR1* and AEs (Parker and Boron, 2013). Another *BOR1* homolog, *YNL275w*, which operates as an efflux B transporter in *Saccharomyces cerevisiae*, has been found to be able to maintain the soluble B concentration in wild-type yeast cells 13 times less than their mutant counterparts (Takano et al., 2002). A low similarity between *AtBOR1* and *ScBOR1p*, another yeast B transporter, and the related protein sequences was also reported (Reid, 2014). More recent phylogenetic analysis showed that BORs could be classified into two clades (I and II) in vascular plants, presumably consistent with low- and high-B tolerance evolved in each lineage by adaptation to their environments (Wakuta et al., 2015).

Although the nature of cotransporter responsible for borate anion outward movement from plant cell cytoplasm has not been reasonably identified, there is much evidence supporting the hypothesis of ion H^+ as possible driver for borate cotransport in plants (Reid et al., 2000; Reid, 2014). The most recent work reported the ability of BOR-type transporters to produce a concentration gradient in plant cells for which it was necessary the energy source to produce the electrochemical potential. Studies on barley and yeast cells showed that B efflux was unaffected by addition of Na^+, Cl^-, or bicarbonate, whereas it increased at low pH values (Reid, 2014).

To efficiently regulate B transport and maintain its homeostasis, a sophisticated regulation mechanism of BOR1 proteins in response to B availability in the soil was probably used by plants. Indeed, under low B supply, BOR1 proteins are sorted by the early endosome for recycling into the plasma membrane where are accumulated; under high B supply, BOR1 proteins are directed to the late endosome and transported to the vacuole for degradation (Takano et al., 2005, 2010). Conversely, the same authors observed that *AtBOR1* messenger RNA accumulation was not affected by B availability, suggesting that it was constitutively expressed but the expression regulated at posttranscriptional level (Takano et al., 2005).

The B exporter role of BOR1 for efficient xylem loading under B-limited conditions was also supported by the identification of its localization in the plasma membrane of endodermis cells facing the root stele, or in all root cells on which endodermis was absent (Takano et al., 2010; Miwa et al., 2013).

In the *Arabidopsis* genome, there are six *BOR1* paralogs. The most similar paralog to *BOR1* is *BOR2*, which encodes an efflux B transporter localized in plasma membrane cells facing the stele. *AtBOR2* is highly expressed in epidermal cells but not in endodermis of root elongation zones, complementing the distribution of *AtBOR1*, which appears essential for root growth and RGII–B cross-linking in cell walls under B-limited conditions (Miwa et al., 2013; Reid, 2014). Another paralog is *AtBOR4* localized into the plasma membrane of the outer side of root epidermal cells, whose overexpression determines an efficient B efflux from roots under toxic B level (Miwa et al., 2007).

In monocots and dicots, many BOR1-like genes have been identified. In rice, *OsBOR1* expression of efflux B transporter was detected, involved in B uptake and xylem loading (Nakagawa et al., 2007).

Recently, Reid (2014) clearly underlined three main functions of B transporters: (1) pumping of B into cell walls; (2) radial transport of B across roots and shoots; and (3) avoidance of toxicity, which contributes to maintain B homeostasis in plants.

2.4 BORON MOVEMENT IN PLANTS

In many plant species, once B has been absorbed by roots, it is loaded into the xylem and translocated to shoots via the transpiration stream and then accumulated in older leaves without being retranslocated (Brown et al., 2002). For this reason, B-deficiency symptoms were first evident in the growing root and shoot tissues (Stangoulis et al., 2001).

Under adequate B supply, xylem loading occurred by passive mechanisms involving B simple or facilitated diffusion through lipid bilayer and channels, respectively (Dannel et al., 2002). On the other hand, under B deficiency, an active transport system via B transporters has been postulated. BOR1 was identified as the first transporter involved in the xylem loading in *Arabidopsis* (Takano et al., 2002), a gene similar to *AtBOR1* was detected in rice (Nakagawa et al., 2007). *AtBOR1*, *AtBOR2*, and *NIP5;1* seemed localized to one side facing the vascular system as a low-resistance symplastic pathway involved in B xylem loading (Miwa et al., 2013; Reid, 2014). However, the NIP6;1 transporter was also involved in B distribution in shoots. In particular, a marked *NIP6;1* promoter activity in phloem region was observed suggesting a specific role of NIP6;1 transporter in B distribution into young growing tissues (Tanaka et al., 2008).

However, phloem also plays a role in providing B to sites that do not lose water readily such as both vegetative and reproductive tissues, depending on species (Matoh and Ochiai, 2005). These plants commonly showed B concentrations higher in young leaves compared with the oldest ones under B deficiency. The mechanism of B transport through phloem resulted in the formation of B-diol complex with sugar alcohols (sorbitol, mannitol, or dulcitol), in general used for the phloem translocation of photosynthates (Hu et al., 1997). Interestingly, Brown et al. (1999) observed that transgenic tobacco plants with elevated sorbitol production had a higher ability to transport B by phloem toward the young tissues, compared with plants without sorbitol (Bellaloui et al., 2003). Several studies observed that B retranslocation has important effect on the expression of B deficiency and toxicity symptoms. Nable et al. (1997) reported that most species in which B is phloem mobile are susceptible to B toxicity. Recently, plants such as canola and wheat, which translocate sucrose as its primary photoassimilate, can transport boric acid preferentially to young tissues within the phloem (Stangoulis et al., 2010).

3. BORON TOXICITY IN PLANTS: CAUSES AND CONSEQUENCES

Boron toxicity is a significant disorder that affects both crop yield and quality. The typical visible symptom of boron toxicity consist in leaves discoloration as well as reduced plant vigor, delayed plant development, decreased number, and size and weight of fruits (Muntean, 2009).

The mechanisms related to boron toxicity are still unclear (Nuttall, 2000; Reid et al., 2004). However, considering the B ability to bind compounds with multiple hydroxyl groups in the *cis*-configuration, ribose appeared to be the probable candidate for toxicity-related effects. In particular, *cis* hydroxyls on the ribose side of energy-carrying molecules such as ATP, nicotinamide adenine dinucleotide (reduced form, NADH), or nicotinamide adenine dinucleotide phosphate (reduced form, NADPH) could be sites of B binding. However, because of the insensitivity of both photosynthesis and respiration to very high B concentrations, it is unlikely that binding to energy-carrying molecules could be considered the actual cause of toxicity (Reid et al., 2004). Nozawa et al. (2006) identified several ribosomal proteins and TFs from *Arabidopsis*, which may prevent B from binding, conferring B tolerance in yeast. They suggested that B could interfere with transcription and/or translation by binding to *cis* hydroxyls on ribose molecules that are exposed during gene splicing and/or in the transfer RNA, thereby protecting transcription and translation (Nozawa et al., 2006; Reid, 2010).

3.1 SYMPTOMS OF B TOXICITY

Boron toxicity symptoms vary among crops species, frequently they were confused with spot-type net blotch, a common leaf disease, and especially at early stages, the symptoms are barely distinguishable from those of other toxic ions in plants. However, its typical visible toxicity symptoms are generally restricted to older leaves as a yellowing, spotting, or drying of leaf tissue at the tips and edges. In boron-immobile species, these symptoms then spread among the lateral veins toward the midrib because of B accumulation transported through the transpiration stream (Nable et al., 1997; Reid et al., 2004; Reid and Fitzpatrick, 2009). In dicots, which generally have reticulate venation, B toxicity is observed around the leaf margins, whereas in grasses, such as wheat and barley, characterized by parallel-veined leaves, the toxic effect develops black patches in leaf tips where the veins terminate (Roessner et al., 2006). For this reason, leaf burn and necrosis have been extensively used for evaluating B toxicity in different crops (Sutton et al., 2007).

Another specific visible symptom of B toxicity is the reduction of leaf area (Roessner et al., 2006), coupled with leaf cupping, and observed in some plant species, probably correlated with the inhibition of cell wall expansion, through disturbance of cross-linking (Suarez, 2012).

Moreover, Cervilla et al. (2012) examined different abiotic-stress indicators to select the parameters most indicative of B toxicity in two tomato genotypes, characterized by different sensitivity to B excess. They suggested that O_2^- and anthocyanins levels in leaves, GPX activity, chlorophyll b, and proline content are the best indicators for B stress in tomato. In B-mobile species (e.g., *Prunus, Malus, Pyrus*), B accumulation has been observed in developing sinks rather than at the end of the transpiration stream. In these plants, the symptoms of B toxicity are expressed as fruit disorders (gummy nuts, internal necrosis), bark necrosis, and death of the cambial tissues, and stem die-back (Brown and Hu, 1996). In particular, in stone-fruit trees, B toxicity caused the reduction of flower bud formation, poor fruit set, and malformed fruit that is particularly poor in flavor (Suarez, 2012). In contrast, in rice, a B-mobile species, B toxicity caused similar foliar symptoms as barley (Bellaloui et al., 2003).

A direct relationship between B content in leaves and severity of toxicity symptoms has been demonstrated. Leaf B concentrations of sensitive and tolerant species have been reported to vary extremely up to 10-fold (Furlani et al., 2003). For this reason, the diagnosis of B toxicity has been extensively evaluated by tissue B-content analysis in leaves rather than in shoots (Reid, 2013).

Moreover, critical toxicity values of tissue B concentrations have been established in many plant species, because B concentrations also greatly varied in relation to different plant tissues and/or plant developmental stages.

3.2 EFFECTS OF BORON EXCESS IN PLANTS

To explain B toxicity mechanisms, a large amount of data about the negative impacts of B excess on important biochemical and physiological processes during plant life cycle have been reported.

3.2.1 Boron and Root System

A primary phenotypic effect of B toxicity is root growth inhibition often concomitant with a decreased plant dry weight (Turan et al., 2009) and an increased B level in root tissues. The reduction of root growth has been observed in different crops such as tomato (Cervilla et al., 2009), wheat (Turan et al., 2009), and grapevine (Gunes et al., 2006). In particular, B toxicity caused an abnormal cell division in root meristem of broad bean (Liu et al., 2000) and a formation of hypodermis together with a progressive suberin deposition in cortical cell wall of soybean root (Ghanati et al., 2005). However, the lignification was not considered an essential factor for B-induced root growth inhibition in tomato (Cervilla et al., 2009). Further, Reid et al. (2004) reported a localized inhibitory response to high B concentration in the wheat root tips but not in mature root zones. Boron excess induced cytotoxic effects on root tip cells during mitosis similar to that of colchicine, forming bridges, fragments, and stickiness in chromosomes and micronuclei development (Liu et al., 2000; Konuk et al., 2007). Shimotohno et al. (2015), using a mathematical model, predicted that the highest B concentration is near the tip and lower in the more proximal region of the meristem zone. Furthermore, the model predicted that B taken up in the root tip was not efficiently translocated to shoots, suggesting that B root tip–absorbed was probably used for local root growth, whereas the more mature root regions were responsible in B transporting toward the shoot (Shimotohno et al., 2015).

Recently, Aquea et al. (2012) reported the molecular basis of root growth inhibition caused by B toxicity in *Arabidopsis*. They observed that B toxicity induced the expression of genes involved in abscisic acid signaling, abscisic acid response, and cell-wall modifications. Furthermore, B repressed the expression of genes encoding water transporters, concluding that B toxicity triggered a water-stress response associated with root growth inhibition. Considering the role of the root system in B-excess response, genotypic variation in root elongation has been well used as an indicator of B tolerance (Hayes and Reid, 2004; Choi et al., 2006). Indeed, Choi et al. (2007) showed that B tolerance in barley is associated with root morphological changes, leading to an increase in branching and finer root development, which allows a better soil exploitation as result of osmotic adjustment. More recently, Princi et al. (2013) reported that short-term B-excess treatment had an evident effect on different root morphological traits. In particular, under B excess, tolerant tomato showed a longer and thinner root system compared to susceptible one.

3.2.2 Boron and Photosynthesis Process

High B concentration induced damaging to the photosynthesis, although the mechanisms of toxicity are still unclear. Under high B stress, the edge of the leaf died (Fang, 2001), the photosynthetic area, chlorophyll content, and consequently the photosynthetic rate were reduced (Han et al., 2009; Chen et al., 2013). A contrasting result was reported only in barley leaves where photosynthesis was not particularly sensitive to B excess because it was unaffected by 50 mM and scarcely inhibited (23%) at a 100 mM B concentration (Reid et al., 2004). Landi et al. (2013) found that B excess caused a Chl a/b ratio decreasing together with a downregulation of photosystem II photochemical efficiency in cucurbits (*Cucumis sativus* L. and *Cucurbita pepo* L.). In many species, B excess significantly reduced Fv/Fm ratio (maximum quantum yield of chlorophyll fluorescence), which indicated that leaves were photoinhibited, a condition that can lead to ROS generation (Velez Ramirez et al., 2011). This event could also explain the decrease in chlorophyll content (Han et al., 2009; Chen et al., 2012). Furthermore, the inhibition in electron transport rate was also concomitant to the reduced activity of some enzymes involved in CO_2 assimilation (carboxylase/oxygenase, ribulose-1,5-bisphosphate, and fructose-1,6-bisphosphate phosphatase), determining a reduction in NADPH and ATP utilization (Han et al., 2009). Recently, Chen et al. (2013), through a proteomic approach, investigated protein profiles in *Arabidopsis* leaves in response to B excess. Interestingly, proteins involved in both light and CO_2 fixation reactions of photosynthesis process were affected by B excess, before the appearance of visible symptoms in leaves and the decrease in chlorophyll content, total cell protein, or growth.

3.2.3 Boron and Antioxidant Pathways

Abiotic stress generally promotes oxidative stress, which causes ROS accumulation, such as hydroxyl radicals (OH^-), superoxide radicals ($O_2^{\,-}$), and hydrogen peroxide (H_2O_2), responsible for proteins, nucleic acids, and lipids damage that eventually leads to the cell death. B excess induced ROS accumulation in barley (Karabal et al., 2003) and oxidative damage by lipid peroxidation and hydrogen peroxide accumulation in grapevine (*Vitis vinifera*) and *Artemisia annua* (Gunes et al., 2006). Antioxidant molecules such as ascorbate and glutathione (nonenzymatic antioxidant activity) and enzymes such as ascorbate peroxidase, catalase, and superoxide dismutase were considered an important defense mechanism against free radicals (Sharma et al., 2012; Hossain et al., 2015). Thus, they have been studied in different crops under B excess (Karabal et al., 2003; Cervilla et al., 2007; Aftab et al., 2011). This condition inhibited the formation of tocopherol in orange (*Citrus x sinensis L. Osbeck*), where it also boosted ascorbate, glucose, and fructose concentrations (Keles et al., 2004). Furthermore, in two citrus species, *C. sinensis* and *C. grandis*, Sang et al. (2015) demonstrated that B toxicity differentially increased the abundances of several protein species involved in photosynthesis, energy metabolism, antioxidation, detoxification, proteolysis, and cell transport. In apple rootstock, glutathione and ascorbate content increased with increasing B concentrations in the culture medium. The authors also sustained that a decline in leaves proline content, important for ROS detoxification, could contribute to greater lipid peroxidation under B excess (Molassiotis et al., 2006).

Finally, B excess–tolerant chickpea (*Cicer arietinum* L.) and basil (*Ocimum basilicum* L.) genotypes seemed to cope with B stress–enhancing antioxidant machinery, but the signaling and coordination of responses remain unclear yet (Landi et al., 2013).

3.2.4 Boron and Nitrogen Metabolism

Plants require large amounts of nitrogen (N) for the biosynthesis of amino acids, nucleic acids, proteins, and secondary metabolites. In agricultural soils, nitrate is the dominant form of nitrogen, essential for plant growth and productivity (Crawford, 1995). Once nitrate is acquired from roots, it is reduced to nitrite (NO_2^-) by nitrate reductase (NR), the primary step in the nitrate assimilation process, which in turn is converted in ammonium (NH_4^+) by nitrite reductase and then incorporated into carbon skeletons for amino acids biosynthesis. Boron affected N metabolism in many crops. Both NR and glutamate dehydrogenase (GDH) activities were affected in leaf and root tissues of barley and wheat under B toxicity (Mahboobi et al., 2002). They found a reduction of NR activity (16%) in leaf and root tissues of both tolerant and sensitive species, together with an increase (30% in leaf and 81% in root tissues) of GDH activity. The response of nitrogen metabolism to B toxicity in tomato has been also investigated (Cervilla et al., 2009). The authors found that glutamine synthase, glutamate synthetase, and GDH increased in tomato leaves under B toxicity, whereas a significant decrease on NR and nitrite reductase activities was observed. They concluded that B toxicity caused an inhibition of nitrate reduction increasing ammonium assimilation in tomato (Cervilla et al., 2009).

Recently, it has been suggested that B excess can also affected nitrate uptake by roots, the first key step of nitrogen metabolism. In sensitive tomato, B excess reduced net nitrate uptake affecting the PM H^+-ATPase activity (Princi et al., 2013).

Finally, the possibility to alleviate B stress through improving N fertilization has been evaluated. Tepe and Aydemir (2011) showed that NH_4^+ supply in lentil and barley had less oxidative damage and yield reduction under B stress in comparison with plants supplied with NO_3^- and urea.

3.2.5 Boron and Carbohydrate Metabolism

Alterations in sucrose levels are highly common in plant responses to various environmental stresses (Rosa et al., 2009), including B stress. Several studies showed that B had a variable effect on plant glycosides biosynthesis including sucrose. For example, a decline in glucose in both leaf and root sugar beet sap under B toxicity was observed (Bonilla et al., 1980). Furthermore, B inhibited the formation of starch from sugar. An increase of reducing sugars (RS) have been also found in the root tip under B excess in soil (Marschner, 1995; McDonald et al., 2003). Recently, an increase of invertase activity within the root tip together with a concomitant increase in RS content, glucose and fructose, were observed in tolerant barley under B toxicity. This change in carbohydrate metabolism would support root development maintaining plant growth under B toxicity (Choi et al., 2007). Recently, genome regions (quantitative trait locus (QTL)) associated with RS content have been mapped using a segregant population derived from a cross between B-susceptible and B-tolerant barley cultivars, grown under high B supply (Huynh et al., 2009). The relationship between B tolerance and a high level of RS in the root tip under B excess could confirm the role of RS in B tolerance.

4. BORON TOLERANCE

4.1 EARLY CONSIDERATIONS

Soil amendment such as leaching B with water and application of organic compounds to inactivate or immobilize B in soil has been considered the main approach to solve B toxicity issue for many years. Nowadays, it appears not feasible on a large scale in B-toxic areas because of economic and practical aspects. On the contrary, the most realistic and effective method to increase crop yields in B-rich soils could be the selection of B excess–tolerant genotypes (Nable et al., 1997). Genetic variation for B excess tolerance has been assessed in many crop species since the 1980s until now (Schnurbusch et al., 2010b; Bogacki et al., 2013). Therefore, tolerant varieties can be rather easily bred, offering a most hopeful approach to minimizing decreases in crop yield in areas with high B concentrations in the soil.

4.2 TOLERANCE MECHANISMS REVISITED

Physiological mechanisms related to B excess tolerance are not well-understood yet. The mechanisms of tolerance in vascular plants include B uptake from soil and its mobility within plant, B accumulation at the end of transpiration stream, tissue B contents, and concentration gradient within a leaf (Reid et al., 2004). Boron tolerance model assumes (1) the existence of binding B compounds once it reaches to toxic concentrations within the cell; (2) the B compartmentation; and (3) an active B efflux by transporters (Hayes and Reid, 2004). Moreover, B accumulation at lower concentrations in tolerant cultivars compared with sensitive ones underlined the predominant role of efflux-type borate transporter(s) in roots rather than internal tolerance mechanisms (B binding complexes or B compartmentation in vacuoles) (Reid, 2007). Taken together, the basis of tolerance to B excess postulates a more limited tissue B concentration involving both B uptake reduction or an active efflux of the micronutrient, at least partly, from the roots (Reid, 2014).

As previously discussed, *AtBOR1* and *AtNIP5;1* are required for an efficient B uptake when its availability in the soil is limited (Takano et al., 2002, 2006). However, when in excess B uptake is mainly regulated through the transcriptional regulation of *AtNIP5;1* (Takano et al., 2006) or by the endocytosis and degradation of *AtBOR1* (Takano et al., 2005). Besides, *AtBOR1* overexpression does not result in a better plant growth under toxic B concentrations (Miwa et al., 2007). Further, the degradation of *AtNIP5;1* messenger RNA under B excess is controlled by the 5′ untranslated region of *AtNIP5;1*, suggesting that both *AtBOR1* and *AtNIP5;1* appear not to be involved in B tolerance (Tanaka et al., 2011). On the contrary, Mishra et al., 2015 sustained that BOR1 and NIP5;1 activity could be used as markers to identify plant genotypes with increased tolerance to B stress in barley.

Many other B transporters, as well as aquaporins, have been recently identified in many plants, for some of which the involvement in B tolerance mechanisms has been proposed (Miwa and Fujiwara, 2010).

Miwa et al. (2007) found that *AtBOR4*, one of the six *BOR1* paralogs in the *Arabidopsis* genome, showed a B efflux activity in yeast cells. BOR4 protein was detected on the outer (soil-facing) membranes of root epidermal cells in *Arabidopsis*, by using green fluorescent protein construct. This localization is important for B directional export to the soil, avoiding high B concentration in growing cells and xylem. *AtBOR4* overexpression improved significantly plant growth under B excess, suggesting it avoids the posttranslational degradation mechanism (reported for *AtBOR1*), being on the contrary highly inducible in this condition (Miwa et al., 2007, 2014). Further, transgenic rice plants expressing *AtBOR4* showed a high tolerance to B toxicity (Kajikawa et al., 2011). The growth enhancement was attributed to the effective B export from the roots, as a consequence of the optimal B concentration maintained within the plant. Thus, the difference in *BOR1* and *BOR4* genes regulation suggests the presence of complex mechanisms for the perception and control of B homeostasis.

Recently, aquaporin isoforms, involved in water and ion transport, appeared to improve tolerance toward many abiotic stresses (Pang et al., 2010). Indeed, the overexpression of *AtTIP5;1*, a tonoplast aquaporin, resulted in increased tolerance to moderately high B levels in the growing medium being involved in borate compartmentation in the vacuole (Pang et al., 2010). Further, two aquaporin rice genes, *OsPIP2;4* and *OsPIP2;7*, have been found to be involved in B permeability and tolerance (Kumar et al., 2014). Both genes, responsible for exporting B from roots under B excess, were downregulated in shoots and strongly upregulated in roots, whose higher expression avoided B toxicity. Furthermore, efflux B assay in roots indicated that, after 1 h of exposure, ^{10}B was excluded from roots in *Arabidopsis* transgenic plants overexpressing *OsPIP2;4* or *OsPIP2;7* (Kumar et al., 2014).

Recently, a gene encoding an NAC-like TF with a single nucleotide polymorphism between the sensitive and tolerant rice cultivars has been identified using recombinant inbred lines (Ochiai et al., 2011). The change in a single nucleotide appeared to provide tolerance to B toxicity in rice by disruption of the gene, which was named BET1 (Boron Excess Tolerant 1). This mechanism could be independent from B efflux because differences in root and/or shoot B concentrations between sensitive and tolerant rice cultivars were not observed (Ochiai et al., 2011).

To identify novel mechanism involved in B tolerance, two *Arabidopsis* mutants, defecting in genes related to B absorption, have been also studied (Sakamoto et al., 2011). Thus, *heb1-1* and *heb2-1* (*hypersensitive to excess B*) mutants, showing growth defects only under B excess, lacked the ability to encode for two subunits of the chromosomal protein complex known as "condensin II." Although both *heb* mutants contained less B than wild-type, their sensitivity to B excess was much greater. These findings confirmed the existence of tolerance mechanisms different from the B efflux. Condensin II seemed to act in DNA double-strand break improvements and to maintain the replication process, both functions considered to be required for plant B excess tolerance (Sakamoto et al., 2011).

Throughout RNA-Seq analysis, Tombuloglu et al. (2015) profiled differential transcripts in responses to B excess in tolerant barley. They revealed that genes related to cell wall, plasma membrane and cytoskeleton construction, Ca^{2+}/calmodulin system, phospholipase activity, and signal transduction played a crucial role under B excess (Tombuloglu et al., 2015).

4.3 GENETIC VARIATION FOR B SENSITIVITY/TOLERANCE IN PLANTS

Crop species have varying ranges at which B concentration is considered adequate, and the threshold between B deficiency and toxic level is often very narrow depending on different plant tissues (leaf, root, shoot, or whole plant) and different growth stages. Therefore, inside each plant species, the evaluation of genotypes for its critical B concentration is important for crop yield and tolerant B selection. Three wide categories of tolerance have been established namely sensitive, semitolerant, and tolerant (Ayvaz, 2002). The sensitive species can tolerate up to 0.5 mg/L of B, whereas the tolerant ones tolerate up to 4 mg/L (Batar et al., 2009). Apple, beans, figs, grapes, and peaches are considered the most sensitive crops to B excess; barley, maize, peas, potato, tobacco, and tomato are classified as semitolerant, whereas alfalfa, carrot, cotton, sugar beet, and turnip appear the most tolerant (Mengel and Kirkby, 2001). Further, the B-sensitive plants exhibit a strong reactivity to high or low concentrations of the microelement, whereas the tolerant ones show adaptability to a wide range of B concentrations without evident growth decreases (Ozturk et al., 2010).

To improve a crop species for B excess tolerance, the selection within genetic variation of novel genotypes able to adapt to area with B at high concentrations has become necessary. Nevertheless, what are the reliable criteria for B excess tolerance screening? The physiological basis of B tolerance proposed by Nable et al. (1990) postulated that tolerant varieties showed lower B concentrations in their leaf tissues than sensitive ones, probably because of a reduced B uptake in both roots and shoots. Boron excess tolerance commonly implies limited or not evident B toxic symptoms, low tissue B concentrations, and high growth or yield under soil B excess. Therefore, efficient phenotypic assays for B-tolerant screening under controlled conditions include leaf symptom expression, relative root length, shoot dry weight, and B concentration in root or leaf tissues measurements (Jefferies et al., 2000; Schnurbusch et al., 2008). However, the results concerning B concentration and content in genotypes contrasting for B-excess tolerance are not always in accordance. Although tolerant barley genotypes with low tissue B concentrations under B excess has been found (Rehman et al., 2006), tolerant wheat genotypes with high tissue B concentrations have been also identified (Torun et al., 2006). These studies confirmed the wide range of intraspecific variations in response to B excess in different crop species, some of which are listed in Table 5.2.

Further, what is the relationship between B deficiency and toxicity for each genotype? Are the B deficiency tolerant genotypes also susceptible to B excess and vice versa? Furlani et al. (2003) reported two wheat cultivars with considerable B efficiency utilization, able to produce the highest shoot, and spike and grain dry matter under B deficiency among several tested varieties. In their experiments, IAC287 showed also a B excess tolerance because the typical toxic symptoms were not observed up to 32.4 mM B in growing media.

The genetic variation in response to B excess in crops, such as barley and wheat (Torun et al., 2006; Hayes et al., 2013; Pallotta et al., 2014), lentil (Yau and Erskine, 2000; Kaur et al., 2014), rice (Ochiai et al., 2008), citrus (Guo et al., 2014), and alfalfa (Bogacki et al., 2013) has been more recently used for QTL analyses. This approach allowed (1) to map genome regions including genes involved in B excess tolerance; (2) to understand physiological, genetic, and molecular mechanisms of tolerance; and (3) to breed B excess tolerant genotypes by molecular-assisted breeding.

4.4 QTL AND ISOLATING GENES INVOLVED IN B TOLERANCE

One of the first examples of QTL analysis showed that the B excess tolerance of "japonica" rice cultivars was greater than "indica" because of a major QTL that accounted for the phenotypic variation (Ochiai et al., 2008). This difference was evident even though B content in root and shoot of both tolerant and susceptible rice genotypes did not significantly vary, highlighting the potential role of molecular tools for selecting novel B-tolerant genotypes (Ochiai et al., 2008).

In any case, QTL detection has been also useful for isolating genes involved in this genetic complex trait. The identification of QTL regions and cloning genes conferring B toxicity tolerance is potentially

Table 5.2 Boron Toxicity Tolerant Lines or Cultivars in Different Crop Species

Crops	Name of Line, Cultivar (Origin)	References
Barley	Anadolu (Turkey) Baluchistan (Pakistan), ICB 104041 (Afghanistan), Tadmor (Syria), Tokak (Turkey), Walfajr (Iran) Sahara 3763 (Algeria)	Mahboobi et al. (2002) Yau (2002) Emebiri et al. (2009)
Rapeseed	WWY Sarson (Australia), Local (India) Pactol; Star (Turkey)	Kaur et al. (2006) Ozturk et al. (2010)
Durum wheat	ICDW 7674 (Afghanistan), Candeal de Grao Escuro 7746, Senatore Cappelli (Italy) AUS10105 (India), AUS10110 (India), Niloticum (Iraq)	Yau and Ryan (2008) Schnurbusch et al. (2008)
Bread wheat	IAC287 (Brasil) Benvenuto Inca (Argentina), Iraq 22 (Iraq), India 126 (India), LinCalel (Argentina), Klein Granador (Argentina), Turkey 1473 (Turkey) Shi#4414/Crow's (Syria) Greek (Greece), BT-Schomburgk (Australia) Halberd (Australia)	Furlani et al. (2003) Chantachume et al. (1995) Yau and Ryan (2008) Wimmer et al. (2005) Jefferies et al. (2000)
Lentil	ILL 0213A, ILL 2024 (Afghanistan) ILL 1765 (Afghanistan), ILL 5883 (Syria)	Hobson et al. (2006) Yau and Erskine (2000)
Alfalfa	Angel, Caliph, Harbinger, Herald, Paraggio (Australia)	Howie (2012)
Pea	SA 132, SA 310 (Afghanistan)	Bagheri et al. (1994)
Rice	IR42, IR46, IR48, IR54, IR9884-54 (Philippines) Nipponbare (Japan)	Dobermann and Fairhurst (2000) Ochiai et al. (2011)
Tomato	Kosaco (Spain) Losna (Italy)	Cervilla et al. (2012) Princi et al. (2013)

the major challenge for the development of varieties able to grow in high soil B levels. In barley, four QTLs associated with B toxicity tolerance were detected on chromosome 2, 3, 4, and 6H. Thus, *HvBot1*, an *AtBOR1*-like gene, was detected in QTL of chromosome 4H and then cloned (Sutton et al., 2007). It was the first B toxicity tolerance gene identified in plants playing a role in limiting the net B uptake into the root and in the disposal of B from leaves. B tolerance mechanism in the tolerant cultivar Sahara, related to an increase in copy number of *HvBot1* gene and abundance of messenger RNA transcript, has been demonstrated (Sutton et al., 2007).

Another QTL on barley chromosome 3H was identified to control relative root length at toxic B concentrations having a lesser effect than 4H chromosome QTL, but they cooperate additively to confer tolerance. Moreover, a gene encoding an NIP-like aquaporin—*HvNIP2;1*—has been identified in barley and mapped to B tolerance QTL on 6H (Schnurbusch et al., 2010a). Finally, Hassan et al. (2010) found that chromosome 2H QTL region encoding S-adenosylmethionine decarboxylase precursor (SAMDC), involved in antioxidative response. Yeast overexpressing was able to provide tolerance to high B.

In bread wheat, at least three unlinked Bo1, Bo2, and Bo3 genes controlled tolerance to B toxicity mapped on chromosomes 4 and 7 (Paull et al., 1991). They additively controlled yield and tissue B concentrations under excess B condition (Jefferies et al., 2000) and one of these genes mapped in the 7B chromosome was considered to play the main role in crop yield under B toxicity (Nable et al., 1997).

Recently, Pallotta et al. (2014) described the identification of near-identical, root-specific B transporter genes underlying two major effects (QTL) for B tolerance in wheat, Bo1 and Bo4. They showed that tolerance to high B concentration was associated with multiple genomic changes including dispersed gene duplication, tetraploid introgression, and variation in gene structure and transcript level. A distinct pattern of gene variant distribution correlated to B levels in soils from different geographical regions was also observed. These findings could support wheat breeders with molecular tools to select for variants of tolerance gene required to specific environments. Thus, the characterization of B tolerance in wheat well highlighted the powerful of the new genomic technologies to define key adaptive processes underpinning crop improvement (Pallotta et al., 2014).

5. CONCLUSION AND FUTURE PROSPECTS

Boron is an essential microelement for growth and development of higher plants, although when occurs in excess it becomes toxic for sensitive plants. In alkaline/saline soils in association with low rainfall conditions or irrigated with B-rich water and/or overfertilized, B often accumulates to toxic levels for plants. Thus, soils containing B toxic levels are a worldwide problem for crop yields and quality, especially in arid and semiarid areas. B requirement is highly variable among plant species and the range between scarce and toxic B levels is smaller than for any other nutrient element. B toxicity affects many physiological processes including inhibition of cell division, disruption of cell wall, rapid inhibition of root elongation, metabolic disturbance, and oxidative stress.

Because it is difficult to remediate soil with a high B level, the suitable strategy to cope this important disorder could be the utilization of tolerant cultivars. A wide genetic variation in tolerance to B toxicity among crop species and genotypes within each species is largely known and it should be exploited. Therefore, the most recent advances in understanding the physiological and molecular mechanisms how tolerant genotypes evade toxicity are here reviewed.

Tolerance to B toxicity appeared to be associated with a reduced B accumulation in both root and shoot tissues in comparison with nontolerant varieties. The basis for B tolerance has been assigned at least in part to an active B efflux from root cells by ion transporters. Several genes encoding for B transport and movement in plant have been cloned in *A. thaliana* (*AtBOR4*), rice (*OsBOR1-4*), barley (*HvBot1*, *HvNIP2;1*), maize (*ZmBot1*), and wheat (*TaBOR1*). The consensus function of these genes was to reduce the expression of multifunctional channel proteins, limiting B net entry into roots and B disposal for leaves. However, overexpression of these genes did not positively affect crop yield in field trials in B-contaminated soils or in rain-fed semiarid environments, where low water availability and high salinity usually coexist. These aspects deeply limited the importance of B efflux transporters overexpression through a transgenic approach for improving B tolerance. Furthermore, mechanisms of tolerance to high B appeared until now confusing because of some evidence of reduced expression of the same gene family in tolerant genotypes.

Recently, an increased B tolerance through overexpressing the tonoplast channel (*AtTIP5;1*) was reported in *Arabidopsis*. Thus, more emphasis should be placed on the role of vacuole and tonoplast boron transporters in preventing B accumulation in the cytoplasm reducing its toxicity.

However, the adequate management of B content is usually difficult for plants. Indeed, at low B levels, plants are not able to stand up for their weak cell wall and to reproduce resulting in poor seed set or sterility, whereas a slight B increase can determine toxicity because of plants' inability to easily excrete it.

Thus, as sustained by several authors, the comprehension of what (boric acid or metabolites), where (local or systemic), and how (binding protein) plants perceived B status at cellular level allowed them to address B toxicity. More recently, B tolerance has been demonstrated to be associated with multiple genomic changes including tetraploid introgression, dispersed gene duplication, and variation in gene structure and transcript level in wheat, confirming the high complexity of this quantitative trait.

In conclusion, future research programs for increasing B tolerance should tackle the problem in a multidisciplinary approach by the integration of the physiological dissection of B-tolerance traits and the genetic and genomics tools, such as QTL and microarrays. Furthermore, the challenge in expanding the B excess tolerance sources to breeders depends on the identification and exploitation of novel variation with varying levels of tolerance to high B concentrations in the soil.

ACKNOWLEDGEMENTS

This research was supported by MIUR-PRIN 2009 (Ministero dell'Istruzione, dell'Università e della Ricerca, Italy, Project "Physiological response of vegetables crops to boron excess").

REFERENCES

Aftab, T., Khan, M.M.A., Idrees, M., Naeem, M., Moinuddin, N.H., 2011. Methyl jasmonate counteracts boron toxicity by preventing oxidative stress and regulating antioxidant enzyme activities and artemisinin biosynthesis in *Artemisia annua* L. Protoplasma 248 (3), 601–612.

Aquea, F., Federici, F., Moscoso, C., Vega, A., Jullian, P., Haseloff, J., Arce-Johnson, P., 2012. A molecular framework for the inhibition of Arabidopsis root growth in response to boron toxicity. Plant Cell Environ. 35, 719–734.

Arora, S., Chahal, D.S., 2010. Effect of soil properties on boron adsorption and release in arid and semi-arid Benchmark soils. Commun. Soil Sci. Plant Anal. 4, 2532–2544.

Ayvaz, M., 2002. Bazi arpa çesitlerinde borun büyüme ve gelişme üzerine etkileri, yüksek lisans tezi. Biyoloji Anabilim Dali S 21–32.

Baes Jr., C.F., Mesmer, R.E., 1976. The hydrolysis of Cations. John Wiley & Sons. Inc., p. 104–111.

Bagheri, A., Paull, J.G., Rathjen, A.J., 1994. The response of Pisum sativum L. germplasm to high concentrations of soil boron. Euphytica 75 (1-2), 9–17.

Batar, T., Koksal, N.S., Yersel, S.A., 2009. Production and characterization of wall plaster with borax and paper wastes and perlite additives. Ekoloji 18 (72), 45–53.

Bellaloui, N., Yadavc, R.C., Chern, M.S., Hu, H., Gillen, A.M., Greve, C., Dandekar, A.M., Ronald, P.C., Brown, P.H., 2003. Transgenically enhanced sorbitol synthesis facilitates phloem-boron mobility in rice. Physiol. Plant 117, 79–84.

Bingham, F.T., Strong, J.E., Rhoades, J.D., Keren, R., 1987. Effects of salinity and varying boron concentrations on boron uptake and growth of wheat. Plant Soil 97, 345–351.

Blewins, D.G., Lukaszewski, K.M., 1998. Boron in plant structure and function. Ann. Rev. Plant Physiol. Plant Mol. Biol. 49, 481–500.

Bogacki, P., Peck, D.M., Nair, R.M., Klaus, J.H., Oldach, H., 2013. Genetic analysis of tolerance to Boron toxicity in the legume *Medicago truncatula*. BMC Plant Biol. 13 (54). http://www.biomedcentral.com/1471-2229/13/54.

Bolaños, L., Lukaszewski, K., Bonilla, I., Blevins, D., 2004. Why boron? Plant Physiol. Biochem. 42, 907–912.

Bonilla, I., Cadahia, C., Carpena, O., Hernando, V., 1980. Effects of boron on nitrogen metabolism and sugar levels of sugar beet. Plant Soil 57, 3–9.

Bonilla, I., Blevins, D., Bolaños, L., 2009. Boron functions in plants: looking beyond the cell wall. In: Taiz, L., Zeiger, E. (Eds.), Plant Physiology, fourth ed. IOP Publishing. physics web. http://4e.plantphys.net. (accessed December 2009.).

Borner, G.H.H., Sherrier, D.J., Weimar, T., Michaelson, L.V., Hawkins, N.D., Macaskill, A., Napier, J.A., Beale, M.H., Lilley, K.S., Dupree, P., 2005. Analysis of detergent-resistant membranes in Arabidopsis. Evidence for plasma membrane lipid rafts. Plant Physiol. 137, 104–116.

Brown, P.H., Hu, H., 1996. Phloem mobility of boron is species dependent: evidence for phloem mobility in sorbitol-rich species. Ann. Bot. 77, 497–505.

Brown, P.H., Bellaloui, N., Hu, H., Dandekar, A., 1999. Transgenically enhanced sorbitol synthesis facilitates phloem boron transport and increases tolerance of tobacco to boron deficiency. Plant Physiol. 119, 17–20.

Brown, P.H., Bellaloui, N., Wimmer, M.A., Bassil, E.S., Ruiz, J., Hu, H., Pfeffer, H., Dannel, F., Römheld, V., 2002. Boron in plant biology. Plant Biol. 4, 205–223.

Camacho-Cristóbal, J.J., Herrera-Rodríguez, M.B., Beato, V.M., Rexach, J., Navarro-Gochicoa, M.T., Maldonado, J.M., González-Fontes, A., 2008. The expression of several cell wall-related genes in Arabidopsis roots is down-regulated under boron deficiency. Environ. Exp. Bot. 63, 351–358.

Camacho-Cristóbal, J.J., Rexach, J., Herrera-Rodríguez, M.B., Navarro-Gochicoa, M.T., González-Fontes, A., 2011. Boron deficiency and transcript level changes. Plant Sci. 181, 85–89.

Cervilla, L.M., Blasco, B., Ríos, J.J., Romero, L., Ruiz, J.M., 2007. Oxidative stress and antioxidants in tomato (*Solanum lycopericum*) plants subjected to boron toxicity. Ann. Bot. 100, 747–756.

Cervilla, L.M., Blasco, B., Ríos, J.J., Rosales, M.A., Rubio-Wilhelmi, M.M., Sánchez-Rodríguez, E., Romero, L., Ruiz, J.M., 2009. Response of nitrogen metabolism to boron toxicity in tomato plants. Plant Biol. 11, 671–677.

Cervilla, L.M., Blasco, B., Rios, J.J., Rosales, M.A., Sanchez-Rodriguez, E., Rubio-Wilhelmi, M.M., Romero, L., Ruiz, J.M., 2012. Parameters symptomatic for boron toxicity in leaves of tomato plants. J. Bot. 1–17.

Chantachume, Y., 1995. Genetic studies on the tolerance of wheat to high concentrations of boron. Ph.D. thesis, Department of Plant Science, Waite Agricultural Research Institute, University of Adelaide, Australia.

Chaumont, F., Moshelion, M., Daniels, M., 2005. Regulation of plant aquaporin activity. Biol. Cell 97, 749–764.

Chen, X., Schauder, S., Potier, N., Van Dorsselaer, A., Pelczer, I., Bassler, B.L., Hughson, F.M., 2002. Structural identification of a bacterial quorum-sensing signal containing boron. Nature 415, 545–549.

Chen, L.S., Han, S., Qi, Y.P., Yang, L.T., 2012. Boron stresses and tolerance in citrus. Afr. J. Biotechnol. 11, 5961–5969.

Chen, M., Mishra, S., Heckathorn, S.A., Frantz, J.M., Krause, C., 2013. Proteomic analysis of *Arabidopsis thaliana* leaves in response to acute boron deficiency and toxicity reveals effects on photosynthesis, carbohydrate metabolism, and protein synthesis. J. Plant Physiol. 171 (3–4), 235–242.

Chesworth, W., 1991. Geochemistry of micronutrients. In: Mortvedt, J.J., Luxmoore, R.J. (Eds.), Micronutrients in Agriculture, pp. 1–30.

Choi, E.Y., McNeill, A.M., Coventry, D., Stangoulis, J.C.R., 2006. Whole plant response of crop and weed species to high subsoil boron. Aust. J. Agr Res. 57, 761–770.

Choi, E.Y., Kolesik, P., Mcneill, A., Collins, H., Zhang, Q., Huynh, B.L., Graham, R., Stangoulis, J., 2007. The mechanism of boron tolerance for maintenance of root growth in barley (*Hordeum vulgare* L.). Plant Cell Environ. 30, 984–993.

Crawford, N.M., 1995. Nitrate: nutrient and signal for plant growth. Plant Cell 7, 859–868.

Dannel, F., Pfeffer, H., Römheld, V., 2000. Characterization of root boron pools, boron uptake and boron translocation on sunflower using stable isotopes 10 B and 11 B. Aust. J. Plant Physiol. 27, 397–405.

Dannel, F., Pfeffer, H., Römheld, V., 2002. Update on boron in higher plant - uptake, primary translocation and compartmentation. Plant Biol. 4, 193–204.

Diaz, F.J., Grattan, S.R., 2009. Performance of tall wheatgrass (*Thinopyrum ponticum* cv. Jose) irrigated with saline-high boron drainage water: implications on ruminant mineral nutrition. Agric. Ecosyst Environ. 131, 128–136.

Dobermann, A., Fairhurst, T., 2000. Rice. Nutrient Disorders and Nutrient Management. Handbook Series. Potash and Phosphate Institute. Atlanta, GA, Potash and Phosphate Institute of Canada, and International Rice Research Institute, Los Banos, the Philippines, p. 191.

Dordas, C., Brown, P.H., 2000. Permeability of boric acid across lipid bilayers and factors affecting it. J. Membr. Biol. 175, 95–105.

Dordas, C., Chrispeels, M.J., Brown, P.H., 2000. Permeability and channel-mediated transport of boric acid across membrane vesicles isolated from squash roots. Plant Physiol. 124, 1349–1361.

Dordas, C., Brown, P.H., 2001. Evidence of channel mediated transport of boric acid in squash (Cucurbita pepo). Plant Soil 235, 95–101.

Elrashidi, M.A., O'Connor, G.A., 1982. Boron sorption and desorption in soils. Soil Sci. Soc. Am. J. 46, 27–31.

Emebiri, L., Michael, P., Moody, D., 2009. Enhanced tolerance to boron toxicity in two-rowed barley by marker-assisted introgression of favorable alleles derived from Sahara 3771. Plant Soil 314, 77–85.

Fang, Y.H., 2001. Study on effect of high boron stress on photosynthesis of oilseed rape. Plant Nutr. Fertil. Sci. 7 (l), 109–112.

Ferrol, N., Belver, A., Roldan, M., Rodriguez-Rosales, M.P., Donaire, J.R., 1993. Effects of boron on proton transport and membrane properties of sunflower (*Helianthus annuus* L.) cell microsomes. Plant Physiol. 103, 763–769.

Fitzpatrick, K.L., Reid, R.J., 2009. The involvement of aquaglyceroporins in transport of boron in barley roots. Plant Cell Environ. 32, 1357–1365.

Fleischer, A., O'Neill, M.A., Ehwald, R., 1999. The pore size of nongraminaceaous plant cell walls is rapidly decreased by borate ester cross-linking of the pectic polysaccharide rhamnogalacturonan II. Plant Physiol. 121, 829–838.

Fleming, G.A., 1980. Essential micronutrients. I: Boron and molybdenum. In: Davies, B.E. (Ed.), Applied Soil Trace Elements. John Wiley and Sons, New York, pp. 155–197.

Frommer, W.B., von Wirén, N., 2002. Plant biology: Ping-pong with boron. Nature 420, 282–283.

Funakawa, H., Miwa, K., 2015. Synthesis of borate cross-linked rhamnogalacturonan II. Front. Plant Sci. 6, 223.

Furlani, A.M.C., Carvalho, C.P., De Freitas, J.G., Verdial, M.F., 2003. Wheat cultivar tolerance to boron deficiency and toxicity in nutrient solution. Sci. Agric. 60, 359–370.

Ghanati, F., Morita, A., Yokota, H., 2005. Deposition of suberin in roots of soybean induced by excess boron. Plant Sci. 168 (2), 397–405.

Goldbach, H.E., 1984. Influence of boron nutrition on net uptake and efflux of ^{32}P and ^{14}C-glucose in *Helianthus annuus* root and cell cultures of *Daucus carota*. J. Plant Physiol. 118, 431–438.

Goldbach, H.E., Hartmann, D., Rötzer, T., 1990. Boron is required for the ferricyanide-induced proton release by auxins in suspension cultured cells of *Daucus carota* and *Lycopersicon esculentum*. Physiol. Plant 80, 114–118.

Goldbach, H.E., 1997. A critical review on current hypotheses concerning the role of boron in higher plants: suggestions for further research and methodological requirements. J. Trace Microprobe Tech. 15, 51–91.

Goldbach, H.E., Wimmer, M., 2007. Boron in plants and animals: Is there a role beyond cell-wall structure? J. Plant Nutr. Soil Sci. 170, 39–48.

Goldberg, S., 1992. Use of surface complexation models in soil chemical systems. Adv. Agron. 47, 233–239.

Goldberg, S., 2003. Modeling boron adsorption isotherms and envelopes using the constant capacitance model. Soil Sci. Soc. Am. J. 3, 676–680.

Goldberg, S., Suarez, D.L., Shouse, P.J., 2008. Influence of soil solution salinity on boron adsorption by soils. Soil Sci. 173, 368–374.

González-Fontes, A., Rexach, J., Navarro-Gochicoa, M.T., Herrera-Rodríguez, M.B., Beato, V.M., Maldonado, J.M., Camacho-Cristóbal, J.J., 2008. Is boron involved solely in structural roles in vascular plants? Plant Signal Behav. 3, 24–26.

González-Fontes, A., Rexach, J., Quiles-Pando, C., Herrera-Rodríguez, M.B., Camacho-Cristóbal, J.J., Navarro-Gochicoa, M.T., 2013. Transcription factors as potential participants in the signal transduction pathway of boron deficiency. Plant Signal Behav. 8 (11), e26114, 1–3.

González-Fontes, A., Navarro-Gochicoa, M.T., Camacho-Cristóbal, J.J., Herrera-Rodríguez, M.B., Quiles-Pando, C., Rexach, J., 2014. Is Ca^{2+} involved in the signal transduction pathway of boron deficiency? New hypotheses for sensing boron deprivation. Plant Sci. 217, 135–139.

Greenwood, N.N., Earnshaw, A., 1984. Chemistry of the elements. John Wiley & Sons, Inc.

Grew, E.S., Bada, J.L., Hazen, R.M., 2011. Borate minerals and origin of the RNA world. Orig. Life Evol. Biosph. 41, 307–316.

Gunes, A., Soylemezoglu, G., Inal, A., Bagci, E.G., Coban, S., Sahin, O., 2006. Antioxidant and stomatal responses of grapevine (*Vitis vinifera* L) to boron toxicity. Sci. Hortic. 110, 279–284.

Guo, P., Qi, Y.P., Yang, L.T., Ye, X., Jiang, H.X., Huang, J.H., Chen, L.S., 2014. cDNA-AFLP analysis reveals the adaptive responses of citrus to long-term boron-toxicity. BMC Plant Biol. 14, 284.

Hall, D., 2010. Boron. Department of Agriculture and Food Wester Australia 1–2.

Han, S., Tang, N., Jiang, H.X., Yang, L.T., Li, Y., Chen, L.S., 2009. CO_2 assimilation, photosystem II photochemistry, carbohydrate metabolism and antioxidant system of citrus leaves in response to boron stress. Plant Sci. 176, 143–153.

Hanaoka, H., Fujiwara, T., 2007. Channel-mediated boron transport in rice. Plant Cell Physiol. 48, 227.

Hayes, J.E., Reid, R.J., 2004. Boron tolerance in barley is mediated by efflux of boron from the roots. Plant Physiol. 136, 3376–3382.

Hayes, J.E., Pallotta, M., Baumann, U., Berger, B., Langridge, P., Sutton, T., 2013. Germanium as a tool to dissect boron toxicity effects in barley and wheat. Funct. Plant Biol. 40 (6), 618–627.

Hassan, M., Oldach, K., Baumann, U., Langridge, P., Sutton, T., 2010. Genes mapping to boron tolerance QTL in barley identified by suppression subtractive hybridization. Plant Cell Environ. 33, 188–198.

Hobson, K., Armstrong, R., Nicolas, M., Connor, D., Materne, M., 2006. Response of lentil (*Lens culinaris*) germplasm to high concentrations of soil boron. Euphytica 151, 371–382.

Hossain, M.F., Shenggang, P., Meiyang, D., Zhaowen, M., Karbo, M.B., Bano, A., Xiangru, T., 2015. Photosynthesis and antioxidant response to winter rapeseed (*Brassica napus* L.) as affected by boron. Pak J. Bot. 47 (2), 675–684.

Howe, P.D., 1998. A review of boron effects in the environment. Biol. Trace Elem. Res. 66, 153–166.

Howie, J.H., 2012. Boron tolerance in annual medics (Medicago spp.). Crop and Pasture Sci. 63, 886–892.

Hu, H., Brown, P.H., 1997. Absorption of boron by plant roots. Plant Soil 193, 49–58.

Hu, H., Penn, S.G., Lebrilla, C.B., Brown, P.H., 1997. Isolation and characterization of soluble boron complexes in higher plants (The mechanism of phloem mobility of boron). Plant Physiol. 113, 649–655.

Huynh, B.L., Pallotta, M., Choi, E.Y., Garnett, T., Graham, R., Stangoulis, J., 2009. Quantitative trait loci for reducing sugar concentration in the barley root tip under boron toxicity. In: The Proceedings of the International Plant Nutrition Colloquium XVI. UC Davis: Department of Plant Sciences, UC Davis. Retrieved from: https://escholarshiporg/uc/item/7d49s8vk.

Jansen, L.H., 2003. Boron, elemental. In: Kirk-othmer Encyclopedia of Chemical Technology. John Wiley & Sons, Inc.

Jefferies, S.P., Pallotta, M.A., Paull, J.G., Karakousis, A., Kretschmer, J.M., Manning, S., Islam, A.K.M.R., Langridge, P., Chalmers, K.J., 2000. Mapping and validation of chromosome regions conferring boron toxicity tolerance in wheat (*Triticum aestivum* L.). Theoret. Appl. Genet. 101, 767–777.

Johanson, U., Gustavson, S., 2002. A new subfamily of major intrinsic proteins in plants. Mol. Biol. Evol. 19, 456–461.

Kajikawa, M., Fujibe, T., Uraguchi, S., Miwa, K., Fujiwara, T., 2011. Expression of the *Arabidopsis* borate efflux transporter gene, AtBOR4, in rice affects the xylem loading of boron and tolerance to excess boron. Biosci. Biotechnol. Biochem. 75 (12), 2421–2423.

Kaldenhoff, R., Fischer, M., 2006. Aquaporins in plants. Acta Physiol. 187, 169–176.

Karabal, E., Yucel, M., Oktem, H.A., 2003. Antioxidant responses of tolerant and sensitive barley cultivars to boron toxicity. Plant Sci. 164, 925–933.

Kaur, S., Nicolas, M.E., Ford, R., Norton, R.M., Taylor, P.W.J., 2006. Selection of Brassica rapa genotypes for tolerance to boron toxicity. Plant Soil 285, 115–123.

Kaur, S., Cogan, N.O., Stephens, A., Noy, D., Butsch, M., Forster, J.W., Materne, M., 2014. EST-SNP discovery and dense genetic mapping in lentil (*Lens culinaris* Medik) enable candidate gene selection for boron tolerance. Theor. Appl. Genet. 127, 703–713.

Keles, Y., Öncel, I., Yenice, N., 2004. Relationship between boron content and antioxidant compounds in Citrus leaves taken from fields with different water source. Plant Soil 265, 345–353.

Keren, R., 1996. Boron. In: Sparks, D.L. (Ed.), Methods of Soil Analysis, Part 3. Chemical Methods. Soil Science Society of America. Book series no. 5.

Keren, R., Bingham, F.T., Rhoades, J.D., 1985. Effect of clay content in soil on boron uptake and yield of wheat. Soil Sci. Soc. Am. J. 49, 1466–1470.

Kobayashi, M., Mutoh, T., Matoh, T., 2004. Boron nutrition of cultured tobacco BY-2 cells. IV Genes induced under low boron supply. J. Exp. Bot. 55, 1441–1443.

Konuk, M., Liman, R., Ciğerci, İ.H., 2007. Determination of genotoxic effect of boron on *Allium cepa* root meristematic cells. Pakistan J. Bot. 39 (1), 73–79.

Kot, S.F., 2008. Boron sources, speciation and its potential impact on health. Rev. Environ. Sci. Biotechnol. 8 (1), 3–28.

Krauskopf, K.B., 1972. Geochemistry of micronutrients. In: Mortvedt, J.J., Cox, F.R., Shuman, L.M., Welch, R.M. (Eds.), Micronutrients in Agriculture. Soil Science Society. Amercan Inc., Madison, Wisconsin, USA, pp. 7–40.

Kumar, K., Mosa, K., Chhikara, S., Musante, C., White, J., Dhankher, O., 2014. Two rice plasma membrane intrinsic proteins, OsPIP2;4 and OsPIP2;7, are involved in transport and providing tolerance to boron toxicity. Planta 239, 187–198.

Landi, M., Pardossi, A., Remorini, D., Guidi, L., 2013. Antioxidant and photosynthetic response of a purple-leaved and a green-leaved cultivar of sweet basil (*Ocimum basilicum*) to boron excess. Environ. Exp. Bot. 85, 64–75.

Läuchli, A., 2002. Functions of boron in higher plants: recent advances and open questions. Plant Biol. 4, 190–192.

Liu, D., Jiang, W., Zhang, L., Li, L., 2000. Effects of boron ions on root growth and cell division of broadbean (*Vicia faba* L.). Isr. J. Plant Sci. 48 (1), 47–51.

Loomis, W.D., Durst, R.W., 1992. Chemistry and biology of boron. BioFactors 3, 229–239.

Ma, J.F., Tamai, K., Yamaji, N., Mitani, N., Konishi, S., Katsuhara, M., Ishiguro, M., Murata, Y., Yano, M., 2006. A silicon transporter in rice. Nature 440, 688–691.

Ma, W., 2011. Roles of Ca^{2+} and cyclic nucleotide gated channel in plant innate immunity. Plant Sci. 4, 342–346.

Mahboobi, H., Yucel, M., Oktem, H.A., 2002. Nitrate reductase and glutamate dehydrogenase activities of resistant and sensitive cultivars of wheat and barley under boron toxicity. J. Plant Nutr. 25, 1829–1837.

Marschner, H., 1995. Functions of mineral nutrients: micronutrients. In: Marschner, H. (Ed.), Mineral Nutrition of Higher Plants, second ed. Academic Press, London, pp. 313–404.

Matoh, T., Kawaguchi, S., Kobayashi, M., 1996. Ubiquity of a boraterhamno-galacturonan II complex in the cell walls of higher plants. Plant Cell Physiol. 37, 636–640.

Matoh, T., Ochiai, K., 2005. Distribution and partitioning of newly taken-up boron in sunflower. Plant Soil 278, 351–360.

McDonald, G.K., Stangoulis, J.C.R., Genc, Y., Lewis, J., Robin, D.G., 2003. Boron toxicity, micronutrient deficiency and salt: overcoming the trifecta of nutritional problems in alkaline soils. In: The Meeting on 'Genetic Solutions for Hostile Crops', CSIRO Plant Industry, Canberra, Australia.

McDonald, G.K., Eglinton, J.K., Barr, A.R., 2010. Assessment of the agronomic value of QTL on chromosomes 2H and 4H linked to tolerance to boron toxicity in barley (*Hordeum vulgare* L.). Plant Soil 326 (1–2), 275–290.

Mengel, K., Kirkby, E.A., 2001. Boron. In: Principles of plant nutrition, 5th ed. Kluwer Academic Publishers, Dordrecht/Boston/London, Netherlands, pp. 621–638.

Mishra, S., Heckathorn, S., Krause, C., 2015. The levels of Boron-Uptake proteins in roots are correlated with tolerance to Boron stress in barley. Crop Sci. 55 (4), 1741–1748.

Miwa, K., Takano, J., Omori, H., Seki, M., Shinozaki, K., Fujiwara, T., 2007. Plants tolerant of high boron levels. Science 318, 1417.

Miwa, K., Fujiwara, T., 2010. Boron transport in plants: co-ordinated regulation of transporters. Ann. Bot. 105, 1103–1108.

Miwa, K., Wakuta, S., Takada, S., Ide, K., Takano, J., Naito, S., Omori, H., Matsunaga, T., Fujiwara, T., 2013. Roles of BOR2, a boron exporter, in cross linking of rhamnogalacturonan II and root elongation under boron limitation in Arabidopsis. Plant Physiol. 163, 1699–1709.

Miwa, K., Aibara, I., Fujiwara, T., 2014. *Arabidopsis thaliana* BOR4 is upregulated under high boron conditions and confers tolerance to high boron. Soil Sci. Plant Nutr. 60, 349–355.

Molassiotis, A., Sotiropoulos, T., Tanou, G., Diamantidis, G., Therios, I., 2006. Boron-induced oxidative damage and antioxidant and nucleolytic responses in shoot tips culture of the apple rootstock EM9 (*Malus domestica* Borkh). Environ. Exp. Bot. 56, 54–62.

Moore, G., 2004. In: Moore, G. (Ed.), Soil Guide: A Handbook for Understanding and Managing Agricultural Soils. Department of Agriculture, Western Australia.

Muntean, D.W., 2009. Boron, the Overlooked Essential Element. Soil and Plant Laboratory Inc., Bellevue, WA 98009. P.O Box 1648. http://www.soilandplantlaboratory.com/pdf/articles/BoronOverlookedEssential.pdf.

Nable, R.O., Cartwright, B., Lance, R.C.M., 1990. Genotypic differences in boron accumulation in barley: relative susceptibilities to boron deficiency and toxicity. In: El Bassam, N. (Ed.), Genetic Aspects of Plant Mineral Nutrition. Kluwer Academic Publishers, Dordrecht, The Netherlands, pp. 243–251.

Nable, R.O., Banuelos, G.S., Paull, J.G., 1997. Boron toxicity. Plant Soil 193, 181–198.

Nakagawa, Y., Hanaoka, H., Kobayashi, M., Miyoshi, K., Miwa, K., Fujiwara, T., 2007. Cell-type specificity of the expression of OsBOR1, a rice efflux boron transporter gene, is regulated in response to boron availability for efficient boron uptake and xylem loading. Plant Cell 19, 2624–2635.

Nielsen, F.H., 1997. Boron in human and animal nutrition. Plant Soil 193, 199–208.

Nielsen, H., Forrest, 2008. Is boron nutritionally relevant? Nutr. Rev. 66 (4), 183–191.

Noguchi, K., Yasumori, M., Imai, T., Matsunaga, T., Oda, H., Hayashi, H., Chino, M., Fujiwara, T., 1997. bor1-1, an *Arabidopsis thaliana* mutant that requires a high level of boron. Plant Physiol. 115, 901–906.

Nozawa, A., Takano, J., Kobayashi, M., Von Wirén, N., Fujiwara, T., 2006. Roles of BOR1, DUR3 and FPS1 in B transport and B tolerance in *Saccharomyces cerevisiae*. FEMS Microbiol. Lett. 262, 216–222.

Nuttall, C.Y., 2000. Boron Tolerance and Uptake in Higher Plants (Ph.D. thesis). University of Cambridge, Cambridge, UK

Ochiai, K., Uemura, S., Shimizu, A., Okumoto, Y., Matoh, T., 2008. Boron toxicity in rice (*Oryza sativa* L.) I. Quantitative trait locus (QTL) analysis of tolerance to boron toxicity. Theor. Appl. Genet. 117, 125–133.

Ochiai, K., Shimizu, A., Okumoto, Y., Fujiwara, T., Matoh, T., 2011. Suppression of a NAC-like transcription factor gene improves boron-toxicity tolerance in rice (*Oryza sativa* L.). Plant Physiol. 156, 1457–1463.

Oiwa, Y., Kitayama, K., Kobayashi, M., Matoh, T., 2013. Boron deprivation immediately causes cell death in growing roots of *Arabidopsis thaliana* (L.) Heynh. Soil Sci. Plant Nutr. 59 (4), 621–627.

O'Neil, M.A., Warrenfeltz, D., Kates, K., Pellerin, P., Doco, T., Darvill, A.G., Albersheim, P., 1996. Rhamnogalatturonan-II, a pectic polysaccharide in the walls of growing plant cell, forms a dimmer that is covalently cross-linked by a borate ester. J. Biol. Chem. 271, 22923–22930.

O'Neill, M.A., Ishii, T., Albersheim, P., Darvill, A.G., 2004. Rhamnogalacturonan II: structure and function of a borate cross-linked cell wall pectic polysaccharide. Annu. Rev. Plant Biol. 55, 109–139.

Ozturk, M., Sakcali, S., Gucel, S., Tombuloglu, H., 2010. Boron and plants. In: Ashraf, M., Ozturk, M., Ahmad, M.S.A. (Eds.), Plant Adaptation and Phytoremediation. Springer, The Netherlands, pp. 275–312.

Pallotta, M., Schnurbusch, T., Hayes, J., Hay, A., Baumann, U., Paul, J., Langridge, P., Sutton, T., 2014. Molecular basis of adaptation to high soil boron in wheat landraces and elite cultivars. Nature. http://dx.doi.org/10.1038/nature13538.

Pang, Y., Li, L., Ren, F., Lu, P., Wei, P., Cai, J., Xin, L., Zhang, J., Chen, J., Wang, X., 2010. Overexpression of the tonoplast aquaporin AtTIP5;1 conferred tolerance to boron toxicity in Arabidopsis. J. Genet. Genomics 37, 389–397.

Park, M., Li, Q., Shchynikov, N., Zeng, W., Muallem, S., 2004. NaBC1 is a ubiquitous electrogenic Na^+-coupled borate transporter essential for cellular boron homeostasis and cell growth and proliferation. Mol. Cell 16, 331–341.

Parker, M.D., Boron, W.F., 2013. The divergence, actions, roles, and relatives of sodium-coupled bicarbonate transporters. Physiol. Rev. 93 (2), 803–959.

Passioura, J.B., 1997. Grain yield, harvest index and water use of wheat. J. Aust. Inst. Agr Sci. 43, 117–120.

Paull, J.G., Rathjen, A.J., Cartwright, B., 1991. Major gene control of tolerance of bread wheat (*Triticum aestivum* L.) to high concentrations of soil boron. Euphytica 55, 217–228.

Polat, H., Vengosh, A., Pankratov, I., Polat, M., 2004. A new methodology for removal of boron from water by coal and fly ash. Desalination 164 (2), 173–188.

Power, P.P., Woods, W.G., 1997. The chemistry of boron and its speciation in plants. Plant Soil 193, 1–13.

Princi, M.P., Lupini, A., Araniti, F., Sunseri, F., Abenavoli, M.R., 2013. Short-term effects of boron excess on root morphological and functional traits in tomato. In: XVII International Plant Nutrition Colloquium- Boron Satellite Meeting–Proceedings Book–17–18 August 2013–Istanbul, Turkey, pp. 1150–1151. http://www.plantnutrition.org/files/downloads/2013ipnc-b-proceedings.pdf.

Raven, J.A., 1980. Short- and long-distance transport of boric acid in plants. New Phytol. 84, 231–249.

Rehman, S., Park, T.I., Kim, Y.J., Seo, Y.W., Yung, S.J., 2006. Inverse relationship between boron toxicity tolerance and boron contents of barley seed and root. J. Plant Nutr. 29, 1779–1789.

Redondo-Nieto, M., Wilmot, A., El-Hamdaoui, A., Bonilla, I., Bolaños, L., 2003. Relationship between boron and calcium in the N_2-fixing legume–rhizobia symbiosis. Plant Cell Environ. 26, 1905–1915.

Reid, R.J., Mimura, T., Ohsumi, Y., Walker, N.A., Smith, F.A., 2000. Phosphate uptake in Chara: membrane transport via Na/Pi cotransport. Plant Cell Environ. 23, 223–228.

Reid, R.J., Hayes, J.E., Post, A., Stangoulis, J.C.R., Graham, R.D., 2004. A critical analysis of the causes of boron toxicity in plants. Plant Cell Environ. 25, 1405–1414.

Reid, R., 2007. Identification of Boron Transporter Genes Likely to be Responsible for Tolerance to Boron Toxicity in Wheat and Barley. Plant Cell Physiol. 48 (12), 1673–1678.

Reid, R., 2010. Can we really increase yields by making crop plants tolerant to boron toxicity? Plant Sci. 178, 9–11.

Reid, R., Fitzpatrick, K., 2009. Influence of leaf tolerance mechanisms and rain on boron toxicity in barley and wheat. Plant Physiol. 151, 413–420.

Reid, R., 2013. Boron toxicity and tolerance in crop plants. In: Tuteja, N., Gill, S. (Eds.), Crop Improvement under Adverse Conditions. Springer, New York, pp. 333–346.

Reid, R., 2014. Understanding the boron transport network in plants. Plant Soil 385, 1–13.

Roessner, U., Patterson, J.H., Forbes, M.G., Fincher, G.B., Langridge, P., Bacic, A., 2006. An investigation of boron toxicity in barley using metabolomics. Plant Physiol. 142, 1087–1101.

Rosa, M., Prado, C., Podazza, G., Interdonato, R., González, J.A., Hilal, M., Prado, F.E., 2009. Soluble sugars-metabolism, sensing and abiotic stress. A complex network in the life of plants. Plant Signal Behav. 4, 388–393.

Sakamoto, T., Inui, Y.T., Uraguchi, S., Yoshizumi, T., Matsunaga, S., Mastui, M., Umeda, M., Fukui, K., Fujiwara, T., 2011. Condensin II alleviates DNA damage and is essential for tolerance of boron overload stress in Arabidopsis. Plant Cell 23, 3533–3546.

Saleem, M., Khanif, Y.M., Fauziah, I., Samsuri, A.W., Hafeez, B., 2011. Importance of boron for agriculture productivity. Int. Res. J. Agric. Sci. Soil Sci. 1 (8), 293–300.

Sang, W., Huang, Z.R., Qi, Y.P., Yang, L.T., Guo, P., Chen, L.S., 2015. An investigation of B-toxicity in leaves of two citrus species differing in boron-tolerance using comparative proteomics. J. Proteomics 123, 128–146.

Schachtman, D.P., Shin, R., 2007. Nutrient sensing and signaling: NPKS. Annu. Rev. Plant Biol. 58, 47–69.

Schnurbusch, T., Langridge, P., Sutton, T., 2008. The Bo1-specific PCR marker AWW5L7 is predictive of boron tolerance status in a range of exotic durum and bread wheats. Genome 51, 963–971.

Schnurbusch, T., Hayes, J., Hrmova, M., Baumann, U., Ramesh, S.A., Tyerman, S.D., Langridge, P., Sutton, T., 2010a. Boron toxicity tolerance in barley through reduced expression of the multifunctional aquaporin HvNIP2;1. Plant Physiol. 153, 1706–1715.

Schnurbusch, T., Hayes, J., Sutton, T., 2010b. Boron toxicity tolerance in wheat and barley: Australian perspectives. Breed. Sci. 60, 297–304.

Scorei, R., 2012. Is boron a prebiotic element? A mini-review of the essentiality of boron for the appearance of life on earth. Orig Life Evol. Biosph. 42 (1), 3–17.

Sharma, P., Jha, A.B., Dubey, R.S., Pessarakli, M., 2012. Reactive Oxygen Species, Oxidative Damage, and Antioxidative Defense Mechanism in Plants under Stressful Conditions. J. Bot. 2012, 26.

Shimotohno, A., Sotta, N., Sato, T., De Ruvo, M., Marée, A.F.M., Grieneisen, V.A., Fujiwara, T., 2015. Mathematical modelling and experimental validation of spatial distribution of boron in the root of *Arabidopsis thaliana* identify high boron accumulation in the tip and predict a distinct root tip uptake function. Plant Cell Physiol. 56, 620–630.

Shorrocks, V.M., 1997. The occurrence and correction of boron deficiency. Plant Soil 193, 121–148.

Soares, M.R., Casagrande, J.C., Alleoni, L.R.F., 2008. Adsorção de boro em solos ácricos em função da variação do pH. Revista Brasileira de Ciencias do Solo 32, 111–120.

Sommer, A.L., Lipman, C.B., 1926. Evidence on the indispensable nature of zinc and boron for higher green plants. Plant Physiol. 1, 231–249.

Stangoulis, J.C.R., Reid, R.J., Brown, P.H., Graham, R.D., 2001. Kinetic analysis of boron transport in *Chara*. Planta 213, 142–146.

Stangoulis, J., Tate, M., Graham, R., Bucknall, M., Palmer, L., Boughton, B., Reid, R., 2010. The mechanism of boron mobility in wheat and canola phloem. Plant Physiol. 153, 876–881.

Steiner, F., Lana, M.C., 2013. Effect of pH on boron adsorption in some soils of Paraná, Brazil. Chilean J. Agric. Res. 73 (2), 181–186.

Suarez, D.L., 2012. Irrigation water quality assessments. In: Wallender, W.W., Tanji, K.K. (Eds.), ASCE Manual and Reports on Engineering Practice No. 71 Agricultural Salinity Assessment and Management, second ed. ASCE, Reston, VA, pp. 343–370.

Sutton, T., Baumann, U., Hayes, J., Collins, N.C., Shi, B.J., Schnurbusch, T., Hay, A., Mayo, G., Pallotta, M., Tester, M., Langridge, P., 2007. Boron-toxicity tolerance in barley arising from efflux transporter amplification. Science 318, 1446–1449.

Takano, J., Yamagami, M., Noguchi, K., Hayashi, H., Fujiwara, T., 2001. Preferential translocation of boron to young leaves in *Arabidopsis thaliana* regulated by the *BOR1* gene. Soil Sci. Plant Nutr. 47 (2), 345–357.

Takano, J., Noguchi, K., Yasumori, M., Kobayashi, M., Gajdos, Z., Miwa, K., Hayashi, H., Yoneyama, T., Fujiwara, T., 2002. Arabidopsis boron transporter for xylem loading. Nature 420, 282–283.

Takano, J., Miwa, K., Yuan, L., von Wirén, N., Fujiwara, T., 2005. Endocytosis and degradation of BOR1, a boron transporter of *Arabidopsis thaliana*, regulated by boron availability. Proc. Nat. Acad. Sci. USA 102, 12276–12281.

Takano, J., Wada, M., Ludewig, U., Schaaf, G., von Wirén, N., Fujiwara, T., 2006. The Arabidopsis major intrinsic protein NIP5;1 is essential for efficient boron uptake and plant development under boron limitation. Plant Cell 18, 1498–1509.

Takano, J., Tanaka, M., Toyoda, A., Miwa, K., Kasai, K., Fuji, K., Onouchi, H., Naito, S., Fujiwara, T., 2010. Polar localization and degradation of Arabidopsis boron transporters through distinct trafficking pathways. Proc. Natl. Acad. Sci. USA 107, 5220–5225.

Tanaka, M., Fujiwara, T., 2008. Physiological roles and transport mechanisms of boron: perspectives from plants. Eur. J. Physiol. 456, 671–677.

Tanaka, M., Wallace, I.S., Takano, J., Roberts, D.M., Fujiwara, T., 2008. NIP6;1 is a boric acid channel for preferential transport of boron to growing shoot tissues in Arabidopsis. Plant Cell 20, 2860–2875.

Tanaka, M., Takano, J., Chiba, Y., Lombardo, F., Ogasawara, Y., Onouchi, H., Naito, S., Fujiwara, T., 2011. Boron-dependent degradation of NIP5;1 mRNA for acclimation to excess boron conditions in Arabidopsis. Plant Cell 23, 3547–3559.

Tariq, M., Mott, C.J.B., 2007. The significance of boron in plant nutrition and environment-a review. J. Agron. 6 (1), 1–10.

Tepe, M., Aydemir, T., 2011. Antioxidant responses of lentil and barley plants to boron toxicity under different nitrogen sources. Afr. J Biotechnol. 53 (10), 10882–10891.

Tombuloglu, G., Tombuloglu, H., Sakcali, M.S., Unver, T., 2015. High-throughput transcriptome analysis of barley (*Hordeum vulgare*) exposed to excessive boron. Gene 557, 71–81.

Torun, A.A., Yazici, A., Erdem, H., Çakmak, I., 2006. Genotypic variation in tolerance to boron toxicity in 70 durum wheat genotypes. Turk J. Agric. For. 30, 49–58.

Turan, M., Taban, N., Taban, S., 2009. Effect of calcium on the alleviation of boron toxicity and localization of boron and calcium in cell wall of wheat. Notulae Botanicae Horti Agrobotanici Cluj-Napoca 37 (2), 99–103.

Velez-Ramirez, A.I., van Ieperen, W., Vreugdenhi, D., Millenaar, F.F., 2011. Plants under continous light. Trends Plant Sci. 16, 310–318.

Voxeur, A., Fry, S.C., 2014. Glycosylinositol phosphorylceramides from Rosa cell cultures are boron-bridged in the plasma membrane and form complexes with rhamnogalacturonan II. Plant J. 79 (1), 139–149.

Wakuta, S., Mineta, K., Amano, T., Toyoda, A., Fujiwara, T., Naito, S., Takano, J., 2015. Evolutionary Divergence of plant borate exporters and critical amino acid residues for the polar localization and Boron-Dependent vacuolar sorting of AtBOR1. Plant Cell Physiol. 56 (5), 852–862.

Warington, K., 1923. The effect of boric acid and borax on the broad bean and certain other plants. Ann. Bot. 37, 629–672.

Wimmer, M.A., Mühling, K.H., Läuchli, A., Brown, P.H., Goldbach, H.E., 2003. The interaction between salinity and boron toxicity affects the subcellular distribution of ions and proteins in wheat leaves. Plant Cell Environ. 26, 1267–1274.

Wimmer, M.A., Bassil, E.S., Brown, P.H., Läuchli, A., 2005. Boron response in wheat is genotype dependent and related to boron uptake, translocation, allocation, plant phenological development and growth rate. Funct. Plant Biol. 32, 507–515.

Wood, J., Nicholson, K., 1998. Boron determination in coal ash by fluoroborate ion-selective electrode. In: Nicholson, K. (Ed.), Energy and the Environment. Geochemistry of Fossil, Nuclear and Renewable Resources. Environ. Geochem.

Yasbeck, C., Kloppmann, W., Cottier, R., Sahuquillo, J., Debotte, G., Huel, G., 2005. Health impact evaluation of boron in drinking water: a geographical risk assessment in Northern France. Environ. Geochem. Health 27, 419–427.

Yau, S.K., Erskine, W., 2000. Diversity of boron-toxicity tolerance in lentil. Genet. Resour. Crop Evol. 47, 55–61.

Yau, S.K., 2002. Comparison of European with West Asian and North African winter barleys in tolerance to boron toxicity. Euphytica 123, 307–314.

Yau, S.K., Ryan, J., 2008. Boron toxicity tolerance in crops: a viable alternative to soil Amelioration. Crop Sci. 48 (3), 854–865.

Yermiyahu, U., Keren, R., Chen, Y., 2001. Effects of composted organic matter on boron uptake by plants. Soil Sci. Soc. Am. J. 65, 1436–1441.

Yermiyahu, U., Ben-Gal, A., Keren, R., Reid, R.J., 2008. Combined effect of salinity and excess boron on plant growth and yield. Plant and Soil 304, 73–87.

Yu, Q., Baluska, F., Jasper, F., Menzel, D., Goldbach, H.E., 2003. Short-term boron deprivation enhances levels of cytoskeletal proteins in maize, but not zucchini, root apices. Physiol. Plant 117, 270–278.

CHAPTER

CHROMIUM AND THE PLANT: A DANGEROUS AFFAIR?

6

Fernando E. Prado[1], Mirna Hilal[1], Silvana Chocobar-Ponce[1], Eduardo Pagano[2], Mariana Rosa[1], Carolina Prado[1]

Cátedra de Fisiología Vegetal, Facultad de Ciencias Naturales e IML, Tucumán, Argentina[1]; *Departamento de Biología Aplicada y Alimentos, Facultad de Agronomía, Universidad de Buenos Aires, Buenos Aires, Argentina*[2]

CHAPTER OUTLINE

1. INTRODUCTION

Chromium (Cr) is a transition metal belongs to group VI of Periodic Table of Elements. It is a steely-gray, bright, hard, and brittle metal. Physicochemical properties of chromium metal are: atomic number 24, atomic weight 51.996 atomic mass unit, density 7.19 g/cm^3, melting point 1907 °C and boiling point 2671 °C. Its electronic configuration, $1s^2 2s^2 2p^6 3s^2 3p^6 3d^5 4s^1$, makes it very resistant to atmospheric oxidants, but it will oxidize to form a thin layer that acts as a protective coating to prevent its own corrosion and also the corrosion of other metals (Emsley, 2011). Chemically, Cr is a redox-type element that is able to exist in a wide range of oxidation states (−2 to +6), which control its geochemistry and toxicity (Daulton and Little, 2006). Most stable forms of Cr occurring in the environment are Cr(0) (native form), Cr(III) (trivalent form) and Cr(VI) (hexavalent form) (Zayed and Terry, 2003). Chromium(0) rarely occurs in nature, but some sites containing uncombined native chromium have been reported (Motzer, 2005). Uncombined native Cr was first discovered in the Xizang site in the Tibetan region of China (Zhu and Liu, 1981); more reliable samples are from the Udachnaya diamonds mine located in the Siberian region of Russia. This mine is a kimberlite pipe, rich in diamonds, and the reducing environment helped produce both elemental Cr and diamonds (Gorshkov et al., 1996). Combined native Cr(0) occurs in some mineral alloys such as: ferchromide ($Cr_3Fe_{1-0.6}$), chromferide ($Fe_3Cr_{1-0.6}$), and isovite [$(Cr,Fe)_{23}C_6$] from Russia and tongbaite (Cr_3C_2)

Plant Metal Interaction. http://dx.doi.org/10.1016/B978-0-12-803158-2.00006-0

and zhanghengite (Cu,Zn,Fe,Al,Cr) from China. None of these native Cr(0) minerals has been used as a metallurgical ore source. Conversely, in nature, Cr compounds are most stable as trivalent form and occur in many minerals such as chromite ($FeCr_2O_3$), ferrochromite ($FeCr_2O_4$), magnesiochromite ($MgCr_2O_4$), zincochromite ($ZnCr_2O_4$), bentorite [$Ca_6(Cr,Al)_2(SO_4)_3(OH)_{12}\cdot 26H_2O$], tarapacaite ($K_2CrO_4$), uvarovite [$Ca_3Cr_2(SiO_4)_3$], stichtite [$Mg_6Cr_2CO_3(OH)_{16}\cdot 4H_2O$], and vauquelinite [$CuPb_2CrO_4PO_4OH$], among others (Babula et al., 2008). Minerals containing Cr(VI) are less abundant than those containing Cr(III). Of the 82 Cr-containing minerals naturally occurring, only 24 are Cr(VI)-bearing minerals (Motzer, 2005). They include anhydrous chromates, compound chromates, compound phosphates, compound borates, compound iodates, and multiple oxides, among others. Crocoite ($PbCrO_4$) is the most abundant Cr(VI)-bearing mineral. Less abundant minerals containing Cr(VI) are hemihedrite [$Pb_{10}Zn(CrO_4)_6(SiO_4)_2F_2$], deanesmithite ($Hg_2{}^{+}Hg_3{}^{2+}Cr^{6+}O_5S_2$), and edoylerite ($Hg_3{}^{2+}Cr^{6+}O_4S_2$) (Barthelmy, 2002). Chromate mineral is also found in nitrate-rich evaporite deposits called "caliche" in the Chilean Atacama Desert. Chromate is largely confined to iodine-bearing nitrate ores that consist of caliche-containing sodium nitrate ($NaNO_3$), lautarite [$Ca(IO_3)_2$], other iodates, and dietzeite [$Ca_2(IO_3)_2(CrO_4)$] (Barthelmy, 2002). Chromite is the world's principal source of Cr metal and its main producers are South Africa, Kazakhstan, India, and Turkey (Brown et al., 2013). The Cr industry consumes approximately 30 million tons of chromite per year, 90% of which is consumed as ferrochrome dedicated to stainless steel production with China and Western Europe as main consumers (Roskill, 2014). Interestingly, some precious stones (e.g., ruby, emerald) and semi-precious stones (e.g., spinel, tanzanite, green tourmaline) owe their colors to chromium, either from Cr being an important part of their chemistry or just as a trace element (Matlins, 2010).

2. CHROMIUM IN THE ENVIRONMENT

As a result of geochemical processes occurring in Cr-containing rocks (e.g., weathering, diagenetic reactions, Mn(IV/III)-dependent oxidations, Fe(III)-dependent reductions, volcanic eruptions), Cr(III) and Cr(VI) in a much lower concentration, are naturally present in freshwater and seawater, sediments, soils, and even in the atmosphere (Nagajyoti et al., 2010). Natural occurrence of Cr(VI) in soils and waters is primarily controlled by solubility and dissolution rates of Cr(III)-bearing minerals. The surrounding pH and accessibility of aqueous solutions of Cr(III) minerals to mixed Mn(IV/III)-oxides forming coat mineral surfaces and pore spaces in sediments and soils derived from ultramafic rocks such as serpentinite commonly enriched in Cr(III), also affect the occurrence of Cr(VI) species (Oze et al., 2007). The level of Cr(VI) in soils, waters, and sediments is also controlled by itself through both enzymatic and nonenzymatic reductions (Fendorf et al., 2000). Physicochemical and geochemical properties of Cr(VI)-containing environments also affect the sorption/complexation of Cr(VI) and its concentration/availability (Jardine et al., 2013). On the other hand, since the twentieth century, Cr has been considered as strategic metal (metal that is vital to modern technology and industry) with a sustained increase in consumption. Either as a chemical compound or in metallic form, Cr is used worldwide in dozens of industrial processes such as paint pigments, textile dyes, leather tanning, wood impregnation, stainless steel manufacturing, alloys, chrome plating, catalysts, refractory bricks, and magnetic tapes, producing thousands of consumer products (Papp and Lipin, 2010). Increasing demand of metallurgical and industrial Cr-containing products has produced a significant increase of both production and consumption of Cr ores (Brown et al., 2013). In this context, the increased production and demand of Cr-containing goods has produced a sustained increase of Cr-polluted sites around the world with increasing risks to human health and also to subsistence of aquatic and terrestrial ecosystems. In fact, only as a result of anthropogenic activities are substantial amounts of Cr present in many world regions despite not having either Cr mining or Cr-related industries (Jacobs and Testa, 2005). From polluted ecosystems, Cr, mainly as Cr(VI), can be easily incorporated into plants and animals, and finally entering in the human food chain (Jardine et al., 2011).

Oxidation states of Cr occurring in the natural environment i.e., Cr(III) and Cr(VI), are quite different in terms of speciation, mobility, availability, and toxicity (EPA, 1998a,b; Oliveira, 2012). Main aqueous Cr(III) species are Cr^{3+}, $Cr(OH)^{2+}$, and $Cr(OH)_4^-$, whereas polymeric species such as $Cr_2(OH)_2^{4+}$, $Cr_3O_4(OH)_4^{3-}$, $Cr_4(OH)_4^{5+}$, and $Cr_2O_2(OH)_4^{2-}$ are scarce in natural waters. In addition, Cr(III) easily forms complexes with organic matter and precipitates as $Cr(OH)_3nH_2O$ above pH 5 or coprecipitates with other elements (Kimbrough et al., 1999). In the pH range encountered in natural systems, Cr(III) mostly occurs in the most insoluble form $Cr(OH)_3$ (Faisal and Hasnain, 2006). Conversely, Cr(VI) is present in anionic forms that are soluble over a wide pH range. Hydrolysis of Cr(VI) produces different pH-dependent species such as $HCrO_4^-$, CrO_4^{2-}, and $Cr_2O_7^{2-}$ (Kotaś and Stasicka, 2000). Although under high concentrations of dissolved oxygen or in the presence of Mn(IV/III)-oxides, Cr(III) can be oxidized to Cr(VI), the most common redox reaction that takes place in many natural environments is the reduction of Cr(VI) to Cr(III) (Barrera-Díaz et al., 2012).

Unlike what occurs in soils, waters, and sediments, levels of Cr that naturally are found in the atmosphere are much lower. They range from 1 ng/m^3 in rural areas to 10 ng/m^3 in polluted urban areas (Nriagu et al., 1988). However, Cr amounts at any particular time and location depend on meteorological factors (Bielicka et al., 2005). Erosion of soils and rocks linked to wind-dependent continental fluxes of dust is the main natural source of atmospheric Cr. Another important source of atmospheric Cr is the forest fires. Gaseous and particulate emissions from volcanoes are minor natural sources of Cr in the atmosphere (Stanin and Pirnie, 2005). Chromium is present in the atmosphere primarily in the form of particles and droplet aerosols, but in gaseous form practically does not exist (Seigneur and Constantinou, 1995). In the dynamic of atmospheric particulate matter, major mechanisms of particle/droplet aerosol deposition include inertial impaction, gravitational sedimentation (settling), and diffusion. Inertial impaction occurs with fast-moving particles larger than 3 μm. Gravitational settling is a function of particle mass and time, with the rate of settling proportional to particle size and mass. Diffusion occurs with particles smaller than 1 μm. Size of Cr-containing particles and droplet aerosols ranged from 0.1 to 50 μm. Because Cr-containing atmospheric particles are transported over different distances by the wind before they fall from the air onto terrestrial and water surfaces either by inertial impaction (dry deposition) or by rainout within clouds and washout below clouds (wet deposition), the size of particle/droplet becomes the major regulator of the fate of atmospheric Cr (IPCS, 2009). In this regard, acid rain may facilitate the removal of acid-soluble Cr compounds from the atmosphere. Furthermore, the distance covered by particles or droplet aerosols depends on meteorological factors, topography, and vegetation. Therefore, the atmosphere has become a major pathway for long-range transfer of Cr to different ecosystems (Theodosi et al., 2010). The oxidation state of atmospheric Cr is not well-defined quantitatively. It has been assumed that the heat of forest fires or volcanic eruptions can oxidize an unknown proportion of the element to Cr(VI). In this oxidation state, Cr probably stays stable in the atmosphere until it settles down and comes into contact with organic matter, which will eventually reduce it to Cr(III) (IPCS, 2009). As with Cr speciation in water and soil, the chemistry of atmospheric Cr is affected by the pH value. It is more acidic (pH ~ 2) in typical atmospheric aerosols compared with pH values found in mostly aquatic environments (Stanin and Pirnie, 2005). In atmospheric droplets, the pH value is close to 5, similar to pH values occurring in natural water systems (Seigneur and Constantinou, 1995).

3. SPECIATION AND AVAILABILITY OF Cr(VI) AND Cr(III) TO PLANTS

Speciation and availability of Cr in both unpolluted and polluted environments relies heavily on its redox chemistry and interactions with dissolved organic matter (DOM), and also with particulate matter (Fendorf et al., 2000; Avudainayagam et al., 2003). Naturally occurring DOM is a complex amorphous mixture of carbohydrates, proteins, lipids, and lignin derived from microbial, algal, plant, and animal detritus and exudates, and the dissolution of sediment and soil organic matter (Sutton and Sposito, 2005). Humic substances (HS) are the most active fractions of DOM and represent the naturally most abundant pool of organic matter, also being ubiquitous in soils, waters, and sediments (Rupiasih and Vidyasagar, 2008). Currently, it is assumed that HS are of paramount importance in sustaining plant growth and controlling both the fate of environmental pollutants and the biogeochemistry of organic carbon into the global ecosystem (Piccolo, 2001). Humic substances are produced by polymerizations of low-molecular-weight (LMW) compounds derived mainly from both biotic and abiotic degradation of lignin, cellulose, polyphenols, and amino acids (Huang and Hardie, 2009). Regardless of their origin, HS are a yellow-brownish to black color in appearance; acidic, paramagnetic, polymeric, polydisperse (variable composition and molecular weight); and extremely complex in structure (Rupiasih and Vidyasagar, 2008). Most of difficulties encountered to define the chemical structure and reactivity of HS derive from their large chemical heterogeneity and geographical variability (Semenov et al., 2013). In this way, chemical structure of HS is not only highly complex, but it is also a function of the different general properties of the ecosystem in which it is formed, such as vegetation, climate, and topography (Piccolo, 2001). Polymeric assembly and supramolecular association of HS are also debated extensively in many studies (Chen et al., 2011; Mouvenchery et al., 2012). Both the number of chemical and biological reactions involved in their genesis and lack of repetitive molecular sequences constitute main controversial points. In fact, many authors consider HS as polymeric macromolecular material with high molecular mass (100–300 kDa) derived mainly from biotic decomposition of lignin and from abiotic reactions catalyzed by inorganic catalysts such as primary minerals, layer silicates, metal oxides, hydroxides, and oxyhydroxides (Trevisan et al., 2010). In this model, simple (though heterogeneous) monomeric units progressively build up into high-molecular-weight polymers by random condensation and oxidation processes (Schulten and Leinweber, 2000). Linear or branched polymeric chains adopt different conformational folds that increase their length and resistance to microbial degradation in soils and sediments (Alvarez-Puebla et al., 2005; Chang et al., 2014). To the contrary, other authors have proposed a supramolecular association of heterogeneous molecules held together by hydrophobic interactions (van der Waals, π–π, ion-dipole) and hydrogen bonds that stabilize the structure of molecular aggregates. Chemically, these aggregates consist of sugar, fatty acids, polypeptides, aliphatic chains, and aromatic rings (Schaumann and Thiele-Bruhn, 2011; Nebbioso and Piccolo, 2012). In this model, the stability of HS aggregates seems to be dynamic and influenced by ionic strength and pH of surrounding solution. At alkaline pH, aggregates occur in a disperse form because intramolecular hydrogen bonds are disrupted, whereas in strong acidic conditions ($pH < 3$), HS collapse into smaller aggregates because protonation of carboxyl groups tends to increase the number of intra- and intermolecular hydrogen bonds inside macromolecular HS aggregates (Carletti et al., 2010). Furthermore, the term HS is also used to define fractions obtained on the basis of solubility characteristics. Based on the solubility in strong acids and bases, HS are classified in humic acids (HA), fulvic acids (FA), and humins. Humic acids are dark brown to black in color and correspond to insoluble fraction in strong acidic conditions ($pH < 2$), but are soluble at higher pH values. Fulvic acids are light yellow to yellow-brown in color and constitute the fraction soluble under all pH conditions. Humins are black in color and represent the fraction insoluble at all pH values (Chen et al., 2011). Although this classification does not give indication on chemical behavior or structural features of HS, strong evidence has been gathered for supramolecular structure of HA and FA, where smaller biomolecules are associated through weak dispersive forces (hydrophobic or hydrogen bonds), responsible for their apparent large molecular sizes, whereas the humins fraction may be considered as a mixture of organic matter and clay, linked by covalent or strong hydrogen bonds (Dudal and Gérard, 2004). Similarly, differences between HA and FA can be explained by variations in molecular weight, number of oxygen-containing functional groups (e.g., carboxyl, phenolic-OH), and extent of polymerization (Güngör and Bekbölet, 2010). Molecular weight of HA ranges between 5 and 100 kDa, whereas FA molecular weight ranges between 0.3 and 5 kDa. Moreover, HA contains less oxygen-containing functional groups than FA (Zavarzina, 2011). Structurally, HA is formed by a hydrophobic aromatic core that is highly substituted with alcohol, carboxyl, phenolic-OH, hydroxyl, carbonyl, amide, amine, and quinone functional groups, and also having side aliphatic chains. This framework core is embedded in a periphery of hydrolyzable carbohydrate-protein fragments (Leita et al., 2009). Because of their polyfunctionality, HA represent a strongly pH-dependent reservoir of electron donors/acceptors acting as reducing agents. Furthermore, they take part in acid-base (protolytic), ion exchange, and complexation reactions, and also participate in hydrogen bonds and van der Waals interactions (Perminova and Hatfield, 2005). In fact, HA can interact with practically all chemicals released in the environment including, among others, heavy metals, petroleum and chlorinated hydrocarbons, pesticides, nitroaromatic explosives, azo dyes, and actinides to form complexes of different solubility and chemical and biochemical stability. Therefore, HA can affect both bioavailability and transport of organic and inorganic contaminants in soils, sediments, and waters (Borges et al., 2005; Campitelli et al., 2006).

One of the most significant properties of HS and its major components (i.e., HA and FA) is their ability to facilitate the electron transfer in biogeochemical redox reactions as well as in redox reactions involving organic and inorganic pollutants (Ratasuk and Nanny, 2007). Despite the origin of HS, under reducing conditions, it may accept or donate electrons. In temporary anoxic systems such as bogs, sediments, and capillary fringes (soil zone of variable moisture above water table), HS may accept electrons directly from microorganisms or from reduced species (e.g., Fe^{2+}, HS^-). According with Fendorf et al. (2000), under anaerobic conditions, Fe^{2+} dominates the reduction of Cr(VI) at pH values higher than 5.5, whereas HS^- (hydrogen sulfide anion) dominates at a lower pH. In contrast, bacterial reduction occurs mainly in aerobic conditions. After aeration, reduced HS may donate electrons to O_2, resulting in the regeneration of an electron-accepting pool (Bauer and Kappler, 2009). These redox buffer properties of HS may reduce the flow of electrons to other electron acceptors. Humic substances can also serve as electron shuttle from microorganisms either to solid Fe(III) oxides and hydroxides promoting their reduction to Fe(II), or to another organic and inorganic pollutants (Kappler et al., 2004). Although the electron-accepting property of HS has been ascribed primarily to different quinone-like moieties (quinoid components) (Aeschbacher et al., 2011), other nonquinoid components of HS (N- and S-containing moieties), including complexed metals such as Fe(III), seem to be also involved (Fimmen et al., 2007). Under oxic conditions, HS exhibit an electron donating property and may act as antioxidants in the environment (Aeschbacher et al., 2012). In this way, HS can transfer electrons to free radicals to inhibit oxidative radical chain reactions that lead to its own oxidative breakdown. This fact increases the recalcitrance of HS in the environment (Rimmer, 2006). Electron-donating capacity of HS also permits the oxidation of radical pollutant intermediates and their parent compounds in the environment. Moreover, in aquatic environments, the antioxidant property of HS indirectly also affects the phototransformation rates of organic and inorganic pollutants. Phenolic and polyphenolic moieties derived from lignin and tannins have been ascribed as major electron-donating components in HS (Kim et al., 2014). Another important trait of HS derived from its polyelectrolitic capacity is related to its buffering capacity for a wide pH range in soils, sediments, and waters (Campitelli et al., 2006). The buffering capacity of HS arises essentially from dissociation of acidic functional groups such as carboxyl and phenolic-OH. At pH < 7, carboxyl acid–type groups are predominant, whereas phenolic acid–type groups are dominant at higher pH values. As a result, the pH dependence of proton binding shows a smooth curve, with progressively increasing dissociation over a very wide pH range (Gustafsson, 2001). Dissociation of functional groups is strongly influenced by the surrounding pH. It determines the ionization extent of the HS particle and controls both density and availability of negative charges (Ritchie and Perdue, 2003). Under strong acidic conditions (pH < 2), almost all carboxyl and phenolic-OH groups on HS surface are protonated (H^+/HS), and the density of negative charges is very low. In this condition, HS acquires a condensed aggregate conformation by coiling and folding its supramolecular structure through weak intramolecular attractive interactions that take place in HS particles. At pH 4–6, the surface charge of HS is close to zero, which minimizes intramolecular interactions, making the macromolecular structure of HS more extended and dispersed (Prado et al., 2011). At a higher pH, deprotonation of HS becomes significant, making that particle surface acquires a negative charge. This favors the dispersion and extension of HS structure through unfolding and uncoiling that take place because of electrostatic repulsion between negative charges. In this condition, there are more deprotonated carboxylic acid and phenol groups available for specific binding sites, and then the formation of strongly bound metal-HS complexes is facilitated. Moreover, a more open configuration of HS at higher pH allows a faster diffusion of cations toward more available negative binding sites (Baker and Khalili, 2007). Under very alkaline conditions, negative charges on HS surface significantly increase, which leads to increased repulsion and dispersion of HS particles. This condition promotes the formation of micelle-like structures from the amphiphilic features of HS (Prado et al., 2011). Because in natural environments, the pH of unpolluted and polluted soils and waters generally is higher than 2, a considerable fraction of HS functional groups is dissociated, and then HS becomes negatively charged (Kabata-Pendias, 2011). In this context, the concentration of Cr^{3+} and $CrOH^{2+}$ near functional groups enhances significantly, and then the sorption (complexation) of Cr(III) on HS particle is increased (Fukushima et al., 1995). Between pH 3.5 and 5, nearly 50% of Cr(III) species is adsorbed electrostatically onto HS, whereas the remaining 50% is bound to dissociated carboxyl groups. At pH > 5, the Cr(III) begins to precipitate as $Cr(OH)_3nH_2O$, and then its solubility and mobility decreases strongly. When considering Cr(III) complexation with HA and FH individually, a different trend is observed. Cr(III) bound to FH produces a soluble complex that contributes to Cr(III) mobility, whereas the Cr(III)-HA complex is insoluble and immobile. However at a higher pH, the solubility of HA increases and this contributes to increase the solubility and mobility of Cr(III) bound to HS (Carrillo-González et al., 2006). Speciation of Cr(III) depends upon pH value with both soluble $Cr^{3+}$and $Cr(OH)^{2+}$ as the dominant species at pH < 4.5, insoluble $Cr(OH)_3$ at pH > 4.5 and soluble polyhydroxyl species such as $Cr(OH)_4^{-}$, $Cr_2(OH)_2^{4+}$, $Cr_3O_4(OH)_4^{3-}$, $Cr_4(OH)_4^{5+}$, and $Cr_2O_2(OH)_4^{2-}$ at alkaline pH values. These pH-dependent speciation features of Cr(III) determine that the modeling of Cr(III) complexation with HS becomes very difficult (Andjelković et al., 2012). Beyond their participation in Cr complexation, DOM is also involved in the photoreduction of Cr(VI) in aquatic environments (Barrera-Díaz et al., 2012). It has been suggested that solar radiation can reduce Fe(III) to Fe(II) in the presence of DOM, which gives the reduction of Cr(VI) to Cr(III) with a coupled reoxidation of Fe(II)-DOM to Fe(III)-DOM (Gaberell et al., 2003). In the absence of or with a negligible amount of Fe(III), the sunlight, mainly ultraviolet spectral fraction, can degrade oxidizable DOM that affects the reduction of Cr(VI). Under favorable conditions such as high temperature and shallow waters, the complete photodegradation of organic pollutants can be achieved (Cieśla et al., 2004). Conversely in DOM-free acidic waters, a light-induced oxidation of Cr(III) takes place when soluble Fe(III) species are present (Zhang and Bartlett, 1999). Photolysis of $FeOH^{2+}$, the major Fe(III) species that occurs at pH below 2.5–5, has been considered as the major source of hydroxyl radicals ($HO^{\cdot}$) to oxidize Cr(III) in aquatic environments with a DOM content severely depleted (Wang et al., 2009). Contrarily, it would not be expected that a DOM-free oxidation of Cr(III) by Fe(III) occurs in soils because of their universal association with organic matter. In soils, Cr(III) oxidation by Mn(IV/III) oxides has been identified as the main Cr(VI) formation pathway (Fendorf et al., 2000). Conversely, in DOM-rich soils, the oxidation of Cr(III) can be suppressed by DOM-Cr(III) complexation (Tanaka et al., 1997). Reduction of Cr(VI) to Cr(III) is more likely to occur in DOM-rich soils. Significant positive relationships between Cr(VI) reduction and amount of easily oxidizable DOM in soils have been reported, suggesting that DOM may act as an electron donor to Cr(VI) reduction with itself oxidation to CO_2 (Bolan et al., 2003).

Besides high-molecular-weight humic and fulvic acids, in naturally occurring DOM, nonhumic LMW organic acids are also present (Xiao and Wu, 2014). LMW organic acids are ubiquitous water-soluble compounds that are widely distributed in soils, waters, sediments, and atmosphere. They have been detected even in rainwater and ice cores. Soluble LMW organic acids comprise mono-, di-, and tricarboxylic acids including compounds containing unsaturated carbons and hydroxyl groups, with an arbitrary maximum molecular weight of about 300 Da. Oxalic, acetic, malonic, malic, succinic, tartaric, citric, lactic, propionic, succinic, pyruvic, and fumaric acid have been shown to be the most abundant organic acids that occur in the environment, but substituted benzoic and cinnamic acids with hydroxyl and methoxyl groups are also present (Strobel, 2001). It is commonly accepted that LMW organic acids are degradation products of various types of naturally occurring large molecules of organic matter closely related to microbial activity, but they can also originate from the Krebs cycle of microbial metabolism. Moreover, LMW organic acids are also secreted by plant roots (Ryan et al., 2001). Therefore, occurring LMW organic acids in natural environments have both abiotic and biotic origins. Although LMW organic acids are generally short-lived and are degraded by microorganisms, they are produced continually, and then their concentrations tend to be relatively stable (Xiao and Wu, 2014). LMW organic acids play a key role in the functioning of bulk soil. They increase the solubility of minerals and also enhance their dissolution rate. Monovalent organic acids such as acetic, lactic, and propionic are adsorbed weakly to the solid phase of soils, whereas divalent organic acids such as malic, fumaric, and succinic are adsorbed to a stronger degree. Greatest adsorption is observed for trivalent acids such as citric, isocitric, and aconitic (Jones et al., 2003). A major property of LMW organic acids is that they affect the size and stability of DOM because they affect both hydrophobic (van der Waals) and electronic (hydrogen bond) forces that hold humic molecules together. In this context, physicochemical characteristics of both HA and FA, such as redox potential and complexation capacity, may be affected (Piccolo et al., 1999). Among physicochemical reactions that involve DOM and are affected by LMW organic acids, both the photochemical reduction of Cr(VI) and photooxidation of Cr(III) in the presence of Fe(III) or Mn oxides are some of the most important that occur in natural environments (Hug and Laubscher, 1997; Zhang and Bartlett, 1999; Dai et al., 2011). In this way, LMW organic acids containing α-OH groups such as tartaric (with two α-OH groups) and citric (with one α-OH group) are able to reduce Cr(VI) in the presence of Fe(III) with higher reduction degree in the presence of tartaric acid, whereas n-butyric acid (without an α-OH group) practically does not reduce Cr(VI) (Tian et al., 2010). Because these three acids have one carboxylic group, it can be assumed that the number of α-OH groups instead of the number of carboxylic acids are majorly responsible for Cr(VI) reduction (Sun et al., 2009).

LMW organic acids such as formic, acetic, succinic, oxalic, tartaric, citric, and malonic are also ubiquitous components in both aqueous and gaseous tropospheric phases and in aerosol particles (Chebbi and Carlier, 1996). Lower LMW monocarboxylic acids such as formic and acetic represent the major fraction of total LMW organic acids in fog water, cloud water, and rainwater, whereas higher LMW-dicarboxylic acids such as tartaric and citric tend to be associated with atmospheric particles (Yao et al., 2002). LMW organic acids constitute an important input source of organic acids for aquatic and terrestrial environments (Löflund et al., 2002). Tropospheric LMW organic acids are mainly from anthropogenic and biogenic origins, but atmospheric transformations of acid precursors toward photogenerated $HO^{\cdot}$ also constitutes an important source (Charbouillot et al., 2012). As in surface waters, the photochemical Fe(III)-induced oxidation of Cr(III) to Cr(VI) occurs in atmospheric water in the presence of LMW organic acids of low Fe(III) photoreductivity such as acetic and succinic, or in the presence of very low levels of LMW organic acids of high Fe(III) photoreductivity such as tartaric and citric. However, Cr(III) oxidation does not occur in the presence of very high levels of low-photoreductivity LMW organic acids because of their competition with Cr(III) for available $HO^{\cdot}$, nor in the presence of mild levels of high-photoreductivity LMW organic acids because of Fe(III) catalyzed the photochemical reduction of Cr(VI) by highly reducing organic radicals generated from the direct photolysis of Fe(III)-organic acids complexes. Furthermore, the low atmospheric pH (2–5) also tends to facilitate Cr(III) oxidation in the presence of Mn(II), but to a lower extent because of the competition of Mn(II) with Cr(III) for $HO^{\cdot}$ (Zhang, 2000). Interestingly, Fe(III)-induced Cr(III) oxidation predominantly takes place in atmospheric water phase, whereas Mn(II)-induced Cr(III) oxidation mainly occurs on atmospheric particulate phase. Although the speciation of Cr in natural environments (i.e., soil, water, and atmosphere) is predominantly controlled by different physical, chemical, and biological processes such as sorption, precipitation, complexation, reduction, and oxidation (Huang et al., 2013); the behavior of Cr is also strongly influenced by external conditions, such as pH, temperature, solar radiation, aeration status, and the presence of redox inorganic and/or organic ions (Hori et al., 2015). Therefore, the fate of both Cr(III) and Cr(VI) in the environment becomes difficult to predict.

4. ABILITY OF PLANTS TO COPE WITH Cr TOXICITY

Plants growing in Cr-mineralized sites such as chromite mining areas are known to accumulate more Cr than plants growing in nonmineralized areas (Reddy et al., 2012). Furthermore, some species that grow in chromite ore rich soils have developed the ability to accumulate simultaneously Cr and other metals. For example *Dodonaea viscosa* and *Hyptis suaveolens* occurring in chromite mining areas of the India accumulate high levels of Cr, Ni, Cu, and V (Rashmi et al., 2010). Although the ability of plants to extract Cr from mineralized sites, its internal plant allocation, and cellular detoxification mechanisms are genetically determined, they can also be modified by environmental conditions (Shanker et al., 2009). Naturally occurring Cr is not a toxic element to plant tissues, but its increasing level in contaminated sites has become harmful to many plant species (Kabata-Pendias, 2011). In fact, when soils and waters contain excessive levels of Cr either as Cr(III), Cr(VI), or both, it becomes toxic to the development of plants (Shanker et al., 2005). Plant species have different responses to Cr pollution in soils and waters (Panda and Choudhury, 2005). In many plant species growing in polluted sites, linear relationships between Cr accumulation and plants tolerance are found, but in others nonlinear correlations are observed (Marbaniang and Chaturvedi, 2014; Pal and Kundu, 2014). Among plants showing nonlinear correlations, some accumulate high amounts of Cr, whereas in others Cr accumulation practically does not occur (Mei et al., 2002). According to a strategy used for plants to cope metal toxicity when grown on metal-polluted environments, they are categorized into three different groups (i.e., excluders, indicators, and accumulator/hyperaccumulators) (Dalvi and Bhalerao, 2013). Metal excluders are plants that effectively restrict the translocation of toxic metals into their higher biomass and maintain relatively low levels of metals in their shoots over a wide range of soil metal contents. However, these plants can contain high levels of metals in their roots. Excluder plants have a translocation factor (TF) < 1 and are used to stabilize soils or avoid further contaminations of waters (Mganga et al., 2011). Plants in which the uptake and translocation of metal reflect its soil/water concentration and show toxic symptoms are known as indicators. As the soil/water concentration increases, the growth of indicators plants is reduced and eventually may die. Indicator plants accumulate the metal in both roots and aerial parts, and their levels generally reflect metal level in soil or water. These plants have TF values close to 1 and can act as effective biogeochemical indicators (Kazakou et al., 2008). Accumulator plants show a tendency or ability to translocate and accumulate high metal levels in above-ground biomass from both low and high soil/water metal concentrations without toxicity symptoms. The TF in accumulator plants is greater than 1. Likewise, hyperaccumulators are plants that accumulate in excess of 100-fold more metal in their above-ground parts than excluder, indicator, and even accumulator plants. In fact, hyperaccumulator plants can simply be viewed as accumulator plants that show extreme behavior in metal uptake and translocation to the shoots. In hyperaccumulator plants, TF values are much greater than 1 (Mganga et al., 2011). TF is a useful parameter to indicate the movement of metal from soil/water to root and shoot, indicating the efficiency of plants to uptake bioavailable metals from the polluted systems (Al-Qahtan, 2012). The TF gives an idea of whether a plant is excluder, indicator, or accumulator/hyperaccumulator. Similar to other heavy metals, Cr-accumulating plants can act as excluders, indicators, or accumulator/hyperaccumulators (Table 6.1).

Even though it has been suggested that Cr accumulation and tolerance mechanisms seems to be general in accumulator and hyperaccumulator plants, interspecies and/or intervarietal differences led to the assumption that additive effects of different resistance mechanisms are operating at different organization levels (metabolic, physiological, structural, or developmental). However, distinguishing among relative contributions of each mechanism to overall phenotypic expression of Cr tolerance is very complicated because of the multiplicity of factors involved in such a process (Masarovičová and Kráľová, 2012). An overall integration of both accumulation and tolerance mechanisms does not seem to have been attempted as yet (Singh et al., 2013). Although Cr accumulation and tolerance seem to be

Table 6.1 Species That Accumulated Cr(III), Cr(VI), or Both, Either from Soils or Waters

Plant Species	E	I	A	H[a]	References
Brassica juncea				+++	Varkey et al. (2013)
Brassica napus			++		Gill et al. (2015)
Vigna radiata			++		Dheeba et al. (2014)
Leersia hexandra				+++	Liu et al. (2015)
Spartina argentinensis				+++	Redondo-Gómez (2013)
Halimione portulacoides			++		Caçador and Duarte (2015)
Dicoma niccolifera				+++	Madhavi et al. (2013)
Sutera fodina				+++	Madhavi et al. (2013)
Jamesbrittenia silenoides			++		Hossner et al. (1998)
Lolium perenne			++		Elekes (2014)
Lolium multiflorum			++		Elekes (2014)
Poa pratensis			++		Jankaitė and Vasarevičius (2007)
Azolla filiculoides				+++	Jaliseh (2011)
Azolla pinnata				+++	Shafi et al. (2015)
Azolla caroliniana				+++	Banach et al. (2012)
Bacopa monnieri			++		Lavu et al. (2013)
Wolfia globosa			++		Liu et al. (2013)
Deschampsia caespitosa			++		Hossner et al. (1998)
Eichhornia crassipes				+++	González et al. (2015)
Eragrostis bahiensis			++		Onweremadu (2014)
Eriocaulon septangulare			++		Reimer and Duthie (1993)
Helianthus annuus			++		Srivastava (2014)
Hydrocharis morsus-ranae			++		Guilizzoni et al. (1989)
Lagarosiphon major			++		Krems et al. (2013)
Hydrilla verticillata			++		Phukan et al. (2015)
Lemna minor			++		Chakraborty et al. (2014)
Lemna gibba			++		Chakraborty et al. (2014)
Festuca pratensis			++		Elekes (2014)
Nymphaea odorata			++		Choo et al. (2006)
Nymphaea spontanea			++		Choo et al. (2006)
Pellaea falcata	+				Kachenko et al. (2007)
Salsola kali			++		Gardea-Torresdey et al. (2005)
Phragmites australis			++		Sultana et al. (2015)
Pinus taeda			++		Bowers and Melhuish (1988)
Quercus falcata			++		Bowers and Melhuish (1988)
Taxodium distichum			++		Bowers and Melhuish (1988)
Pistia stratiotes			++		Chakraborty et al. (2014)
Adiantum capillus-veneris				+++	Sinam et al. (2012)
Microsorium punctatum				+++	Sinam et al. (2012)
Polygonum lapathifolium			++		Jaliseh (2011)

Table 6.1 Species That Accumulated Cr(III), Cr(VI), or Both, Either from Soils or Waters—cont'd

Plant Species	E	I	A	H[a]	References
Potamogeton crispus			++		Upadhyay et al. (2014)
Potamogeton pectinatus			++		Upadhyay et al. (2014)
Salix alba			++		Quaggiotti et al. (2007)
Salix viminalis			++		Quaggiotti et al. (2007)
Salix eleagnos			++		Quaggiotti et al. (2007)
Salix fragilis			++		Quaggiotti et al. (2007)
Salix matsudana			++		Quaggiotti et al. (2007)
Thlaspi rotundifolium			++		Giachetti and Sebastiani (2006)
Typha angustifolia			++		Chen et al. (2014)
Typha domingensis			++		Terfie and Asfaw (2015)
Vallisneria americana			++		Krems et al. (2013)
Vallisneria spiralis			++		Krems et al. (2013)
Mirabilis jalapa			++		Miao and Yan (2013)
Impatiens balsamina			++		Miao and Yan (2013)
Tagetes erecta			++		Miao and Yan (2013)
Genipa americana			++		Barbosa et al. (2007)
Fallopia japonica	+				Sołtysiak et al. (2011)
Fallopia sachalinensis	+				Sołtysiak et al. (2011)
Ipomoea aquatica			++		Ton et al. (2015)
Callitriche cophocarpa			++		Augustynowicz et al. (2014)
Spirodela polyrrhiza			++		Chakraborty et al. (2014)
Salvinia minima			++		Prado et al. (2010a)
Salvinia natans			++		Dhir and Srivastava (2011)
Salvinia herzogii			++		Suñé et al. (2007)
Salvinia molesta			++		Chakraborty et al. (2014)
Salvinia auriculata			++		Espinoza-Quiñones et al. (2009)
Fontinalis antipyretica			++		Kosior et al. (2015)
Ceratophyllum demersum			++		Liu et al. (2013)
Leptospermum scoparium		+			Madhavi et al. (2013)
Polystichum squarrosum			++		Kumari et al. (2014)
Pimelea suteri		+			Hossner et al. (1998)
Hebe odora		+			Hossner et al. (1998)
Cassinia vauvilliersii		+			Hossner et al. (1998)
Prosopis laevigata				+++	Buendía-González et al. (2010)
Nopalea cochenillifera				+++	Adki et al. (2013)
Telfairia occidentalis	+				Orhue and Uzu (2010)
Taraxacun officinale			++		Bini et al. (2012)
Elodea canadensis			++		Krems et al. (2013)
Vetiveria zizanioides			++		Pillai et al. (2013)

Continued

Table 6.1 Species That Accumulated Cr(III), Cr(VI), or Both, Either from Soils or Waters—cont'd

Plant Species	E	I	A	H[a]	References
Vetiveria nemoralis			++		Pillai et al. (2013)
Myriophyllum spicatum			++		Yabanli et al. (2014)
Tradescantia pallida			++		Sinha et al. (2014)
Gossypium hirsutum			++		Daud et al. (2014)
Scirpus lacustris			++		Singh et al. (2010)
Commelina communis	+		++		Tang et al. (2001)
Sporobolus airoides			++		Hossner et al. (1998)
Elsholtzia haichowensis	+				Tang et al. (2001)
Agave sisalana			++		Mganga et al. (2011)
Fuirena umbellata			++		Mganga et al. (2011)
Andropogon gayanus			++		Hossner et al. (1998)
Satureja abyssinica			++		Giachetti and Sebastiani (2006)
Populus × euramericana			++		Giachetti and Sebastiani (2006)
Suaeda monoica			++		Ayyappan and Ravindran (2014)
Suaeda fruticosa			++		Ayyappan and Ravindran (2014)
Cyperus esculentus			++		Krems et al. (2013)
Cynodon dactylon			++		Elekes (2014)
Zea mays			++		Dheeba et al. (2014)

A, accumulator; Cr, chromium; E, excluder; H, hyperaccumulator; I, indicator. +, ++, and +++ are not related with Cr accumulation that occurs in plants; they are only used as nominative signs.
[a]The criterion used for Cr hyperaccumulator was 1000 mg Cr kg^{-1} dry mass (Ali et al., 2013).

heritable traits related to sulfur metabolism (Quaggiotti et al., 2007; Dubey et al., 2010; Schiavon et al., 2012), differences observed among plant species can be explained by plants having different genes, alternative alleles, or the same genes but with different modifier genes (Ernst, 2006). It has been reported that both Cr(III) and Cr(VI) are found in aquatic and terrestrial systems, but the former is less soluble and available to plants than the latter (Stanin and Pirnie, 2005). Whatever the oxidation state of Cr species occurring in the environment, roots govern its uptake by plants and also the fraction that is transferred to above-ground tissues (Kim et al., 2010). It has been stated that the concentration of bioavailable and soluble Cr is considered more important to plant toxicity than total Cr concentration. In the context of Cr bioavailability, its speciation in the soil or water is important given that Cr uptake by plants is often determined by the oxidation state (Zayed and Terry, 2003). Similarly, plants have the potential to transform environmental insoluble Cr species into more soluble species for easier uptake by plants or to detoxify Cr ions through root exudation or by pH changes at the rhizosphere level (Bluskov et al., 2005).

Despite differences occurring in Cr accumulation, tolerant plants are usually efficient in avoiding either the entry of Cr ions or the occurrence of toxic concentrations of free Cr species in cytoplasmic compartments to avoid their harmful effects through the development of Cr-induced oxidative stress

(Shanker et al., 2005; Ovečka and Takáč, 2014). The main mechanism to avoid the entry of heavy metals into plants is done on the cell wall, the more important extracytoplasmic compartment (Mehes-Smith et al., 2013). Although functional groups such as −COOH, −OH, and −SH occurring in cell walls are able to bind heavy metals, cell wall polysaccharides rich in carboxyl groups (i.e., pectin) play a crucial role into the process of metal binding (Krzesłowska, 2011). Similar to pectin, lignin a highly branched polymer of phenylpropane alcohols (coumaryl, coniferyl, and sinapyl), rich in hydroxyl and phenolic-OH functional groups, is mainly responsible of heavy metal binding onto secondary cell walls (Miretzky and Fernandez Cirelli, 2010). In this way, the binding of Cr ions onto the cell wall is frequently observed in Cr-accumulating aquatic and terrestrial plants (Mohanty et al., 2006; Elangovan et al., 2008; Mangabeira et al., 2011). Because Cr is a well-known nonessential element to plants, it has been suggested that it is likely confined in roots by a barrier effect as defence strategy to cope the toxicity of this metal. High accumulations on plant cell walls of other nonessential metals such as Pb, Hg, and Cd have been also communicated (Giri and Patel, 2011; Bini et al., 2012). Protonation and deprotonation of cell wall functional groups constitutes the so-called cell wall proton-metal cation exchange activity (Meychik and Yermakov, 2001) that is the major mechanism involved in the biosorption of Cr ions onto the plant surface. At low concentrations, Cr ions can bind easily to functional groups on the cell wall; whereas at higher concentrations, free Cr ions to reach the cell wall will need to face the intraparticle diffusion and hydrolyzed ions themselves that diffuse slower (Horsfall and Spiff, 2005). However, available data of Cr accumulation onto cell surface of both aquatic and terrestrial plants are contradictory and vary according to biological and chemical properties of the rhizosphere, Cr concentration, and metabolic activity of plants (Elangovan et al., 2008; Oporto et al., 2008; Espinoza-Quiñones et al., 2009). The rhizosphere is a biodynamic zone of about 1 mm wide that surrounds plant roots. It does not have a shape and edge definable, but plays a key role in the soil/water–plant interface through the regulation of both traffic and availability of metal ions (Badri and Vivanco, 2009). Compounds exuded by roots such as amino acids, sugars, organic acids, proteins, vitamins, enzymes, inorganic ions, and an array of secondary metabolites as well as the microorganisms feeding on these compounds affect biochemical and physical properties of the rhizosphere (Walker et al., 2003). Among exuded compounds, organic acids play a major role in both the bioavailability of Cr species and the binding of Cr ions onto the cell wall through metal complexation and rhizosphere pH changes (Zeng et al., 2008). Ligand properties of exuded organic acids are responsible for the formation of mobile Cr–ligand complexes that affect both the bioavailability and transport of insoluble Cr(III) (Srivastava et al., 1999). In addition, organic acids also have a significant influence on soil chemical weathering and leaching of chromium minerals and also acts as mediator of minerals acquisition by plants in low-nutrient environments (Oburger et al., 2009). Likewise, the pH of rhizosphere strongly affects the binding of Cr ions onto cell wall. At $pH < 4$, the protonation of cell-wall functional groups becomes significant and then an increased availability of metal anion–binding sites occurs (Gadd, 2009). In this condition the biosorption of Cr anionic species such as $HCrO_4^-$, CrO_4^{2-}, and $Cr_2O_7^{2-}$ is often increased, whereas the biosorption of cationic species such as Cr^{3+} and $CrOH^{2+}$ is reduced. Conversely, under $pH > 7.6$, the deprotonation of metal binding sites occurs and the availability of metal anion binding sites decreases (Shadreck and Mugadza, 2013). Then, the biosorption of anionic Cr species onto the cell wall decreases by electrostatic repulsion. Similarly, rhizosphere microorganisms also play a major role in Cr–cell wall interaction. Interactive community of rhizotrophic microorganisms including bacteria, fungi, actinomycetes, protozoa, and algae affects physicochemical properties of the rhizosphere and participates actively in the speciation and bioavailability of Cr species in both soils and waters (Cheung and Gu, 2007; Polti et al., 2007; Singh et al., 2010). Among rhizospheric microorganisms, bacteria are by far the most common type of microorganism found in the rhizosphere of plants because of they can grow rapidly and have the ability to use a wide range of substances as either carbon or nitrogen sources (Khan, 2005). It is noteworthy that in aquatic systems the plant rhizosphere is often aerobic and oxidized, whereas in soils it is anaerobic and chemically reduced. This leads to higher Cr exposure of roots and higher accumulation of bioavailable Cr onto the cell wall in aquatic plants compared with terrestrial ones. In this context, it has been suggested that aquatic plants, through evolution, have acquired a higher "innate" tolerance to Cr compared with dryland plants (Otte et al., 2004). Cell wall–Cr absorption is not is a one-step process; it is the result of multiple processes that include rhizosphere interactions, water redox potential, organic matter, salinity, light, microbial activity, pH, and water hardness, among others. In turn, all this leads to complex patterns of Cr accumulation observed in aquatic macrophytes (Peralta-Videa et al., 2009). Even though the Cr is mainly retained in roots through binding onto the cell wall (Zayed et al., 1998; Babula et al., 2008), a fraction of root-absorbed Cr is accumulated inside the plant (Chandra and Kulshreshtha, 2004). Both Cr(VI) and Cr(III) are incorporated into roots by the symplastic pathway, with the former reduced to Cr(III) before its accumulation in the cortex (Shanker et al., 2005). Chromium(VI) moves faster than Cr(III), but two Cr species are translocated poorly to above-ground tissues (Montes-Holguin et al., 2006).

Although Cr is primarily retained inside the root, once it is translocated to aerial tissues, a pleyade of energy-dependent and non-energy-dependent physiological mechanisms are activated by the plant to counteract the Cr toxicity (Prado et al., 2013). To tolerate high levels of Cr in their tissues, plants use several metal-binding biomolecules, including LMW ligands such as amino acids, organic acids, phenolic compounds, thiol-containing compounds, and phytosiderophores, as well as small metal-binding proteins such as phytochelatins and metallothioneins, to sequester, transport, and store the accumulated Cr (Oliveira, 2012; Prado et al., 2012; Teixeira et al., 2013). Moreover, plants also enhance antioxidant enzyme activities such as superoxide dismutase, catalase, peroxidase, ascorbate peroxidase, dehydroascorbate reductase, monodehydroascorbate reductase, and glutathione reductase, among others, to protect themselves from oxidative damage caused by Cr-induced oxidative stress (Malmir, 2011; Singh et al., 2013; Kováčik et al., 2014). However, a comparison of closely related Cr-sensitive and Cr-tolerant species or ecotypes suggested that an enhanced oxidative defence is not a particular trait conferring Cr tolerance to plants (Hossain et al., 2012). Chromium-tolerant plants are usually efficient in avoiding the occurrence of toxic concentrations of free Cr ions in plasmatic compartments to avoid the development of oxidative stress (Shanker et al., 2005). Notwithstanding, the question of how plants regulate their metabolism to allow the safe uptake of Cr ions into the cytoplasm and organelles and simultaneously protect themselves from toxicity still remains (Shanker et al., 2009; Ovečka and Takáč, 2014).

5. INTERACTION OF PLANTS WITH Cr(VI) AND Cr(III): WHAT'S NEW?

Among widely distributed heavy metals, Cr has a unique feature depending on its oxidation state, that is, it can be extremely dangerous to plants or relatively innocuous. Symptoms of Cr toxicity are diverse and include, among others, leaf chlorosis and necrosis, reduction of seed germination and seedling growth, decrease of yield, impairment of photosynthesis and respiratory activities, mineral imbalance, and cytotoxic and genotoxic effects (Shanker et al., 2009). Chromium(VI) is more toxic than Cr(III) as a result of its higher oxidizing potential and higher solubility and mobility across cell membranes (Tchounwou et al., 2012). Although plants can absorb Cr ions under both oxidation states, 7 of 10 species growing in the presence of both Cr(VI) and Cr(III) show higher accumulation of the first one (Zayed et al., 1998). Highly toxic anion CrO_4^{2-} is actively incorporated by cells mainly through the nonspecific SO_4^{2-} carrier and in lesser extension through the HPO_4^{2-} carrier (Banerjee et al., 2008). In this way, a competitive uptake between CrO_4^{2-} and SO_4^{2-} anions is established (Cervantes et al., 2001). Chromate-induced competitive inhibition of SO_4^{2-} uptake can lead to cell sulfur starvation and strong decrease of sulfur-containing compounds such as methionine and cysteine, among others (Pereira et al., 2008). However, data obtained in yeast and mammalian cells suggest that the chromate-mediated sulfur depletion is not simply the result of a competitive uptake but is also the consequence of a competitive metabolism between CrO_4^{2-} and SO_4^{2-} anions, presumably at another step of the sulfur assimilation pathway (Holland and Avery, 2011). Although the molecular mechanism of deleterious effects caused by chromate-mediated sulfur depletion has yet to be fully resolved, recent evidence in yeast indicate that elevated messenger RNA mistranslation produced by methionine and cysteine depletion is the major cause of CrO_4^{2-} toxicity (Pereira et al., 2008). This messenger RNA mistranslation leads to the synthesis of aberrant proteins prone to dysfunction, oxidation, carbonylation, and aggregation (Holland et al., 2010). Results from different laboratories provide evidence toward this unified model of Cr toxicity in all eukaryotic cells. This novel mechanism does not contradict the well-known oxidative and nonoxidative mechanisms of DNA damage induced by Cr(VI) toxicity (Arakawa et al., 2012). Into the oxidative mechanism, Cr(VI), once inside the cell, is rapidly reduced to highly cytotoxic Cr(V) by certain reductants such as ascorbic acid, Cys, glutathione, and hydrogen peroxide (H_2O_2), or by other cellular metabolites such as α-lipoic acid, riboflavin, and soluble sugars and related compounds (e.g., glucose, fructose, glucuronic acid, myo-inositol, ribonucleotides), among others (Leonard et al., 2004; Sala et al., 2011). In addition, NAD(P)H-dependent flavoenzymes (glutathione reductase, lipoyl dehydrogenase, and ferredoxin-$NADP^+$ oxidoreductase) can also catalyze one electron reduction of Cr(VI) to Cr(V) (Poljsak et al., 2010). Chromium(V) is a long-lived, highly reactive intermediate and can cause DNA damage such as base oxidation and single-strand breaks and also chromosomal mutations (Liu et al., 1995). In this regard, Cr(V) reacts with H_2O_2, which is formed in the dismutation of the superoxide ion ($O_2^{\cdot-}$) to generate the deleterious $^{\cdot}OH$ and also Cr(V)-peroxo complexes via the Fenton-like and Haber–Weiss type reactions (Ding and Shi, 2002). The picture of Cr(VI) reduction and generation of $^{\cdot}OH$ can be schematized in the Eqn (6.1).

$$\begin{aligned}
&Cr(VI) + O_2^{\cdot-} \rightarrow Cr(V) + O_2 \\
&Cr(V) + H_2O_2 \rightarrow Cr(VI) + {}^{\cdot}OH + OH^{\cdot-} \\
&Cr(IV) + H_2O_2 \rightarrow Cr(V) + {}^{\cdot}OH + OH^{\cdot-} \\
&Cr(IV) + 2H_2O_2 \rightarrow Cr(VI) + 2{}^{\cdot}OH + 2OH^{\cdot-}
\end{aligned} \tag{6.1}$$

Overall, during the stepwise one-electron reduction of Cr(VI) to final stable Cr(III) a whole spectrum of reactive oxygen species is generated that can mainly target to oxidation of DNA molecule and also other biomolecules such as proteins and lipids as well as any subcellular structure or organelle (Slade et al., 2005). In addition, Cr(VI) can also induce the generation of reactive nitrogen species such as peroxynitrite anion ($ONOO^-$) from the reaction of nitric oxide with $O_2^{\cdot-}$ (Ghasemi et al., 2014).

Peroxynitrite is highly reactive and can directly react with various biological targets and components of the cell including DNA, lipids, thiols, amino acids, and LMW antioxidants, among others (Alvarez and Radi, 2003). However, under acidic pH, the peroxynitrite produces peroxynitrous acid (ONOOH), which in turn produces nitrogen dioxide ($^{\cdot}NO_2$) and $^{\cdot}OH$ radicals (Lymar et al., 2003), according to the following reaction:

$$ONOO^- + H^+ \rightarrow ONOOH \rightarrow {}^{\cdot}NO_2 + {}^{\cdot}OH \quad (6.2)$$

Additionally the peroxynitrite can react with CO_2 to form both $^{\cdot}NO_2$ and carbonate radical anion ($CO_3{}^{\cdot -}$), which are one-electron oxidants (Goldstein and Merényi, 2008). Peroxynitrite-derived radicals (i.e., $^{\cdot}OH$, $^{\cdot}NO_2$, and $CO_3{}^{\cdot -}$) are able to oxidize biomolecules such as proteins, lipids, and DNA. Thiols can also be oxidized by peroxynitrite-derived radicals and initiate radical-dependent chain reactions to produce higher oxidation states of sulfur, including sulfinic and sulfonic acid derivatives (Szabó et al., 2007). Critical plant functions such as photosynthesis, mitochondrial respiration, and carbohydrate and lipid metabolisms are also affected by reactive oxygen species and reactive nitrogen species (Prado et al., 2010a; Singh et al., 2013). In heavy metal-stressed plants, reactive oxygen species and reactive nitrogen species interact with each other (cross-talk) at two levels: (1) biosynthesis and (2) regulation of gene expression or enzyme activity (Qiao et al., 2014). In recent years, the ability of Cr(VI) to cause the oxidation of both cytosolic and mitochondrial thioredoxin isoforms (Trx1 and Trx2) in human cells has been discovered (Myers, 2012), which could contribute to its cytotoxic effects and could have important implications for cell survival and redox-sensitive cell signaling. In addition, the thioredoxin reductase (TrxR) and thioredoxin system (TrxR/Trx) has a key role in the maintenance of cellular thiol redox balance and is essential for cell survival (Cejudo et al., 2014). Because different Trxs (i.e., *f*, *m*, *h*, *x*, *z*, and *y*) are widely distributed among plants (Montrichard et al., 2009; Arsova et al., 2010), it could be expected that a similar Cr(VI)-induced oxidation of Trx also occurs in cells of Cr-exposed plants. In this regard, recently our laboratory team has cloned and obtained from submerged leaves of *Salvinia minima* exposed to 20 mg/L Cr(VI), the complete sequence of a 377-bp fragment of Trx*h* 1 type and incomplete sequences of two fragments with 266 bp and 288 bp of Trx*h* 2 type. These Trx fragments showed more than 90% sequence homology with Trx*h* 1 and Trx*h* 2 from different plant species (Pagano and Ilina, 2014, unpublished results). However, further research is needed to clarify this topic. Notwithstanding, well-characterized mechanisms of Cr(VI)-induced DNA toxicity in bacterial and mammalian cells, the Cr-dependent genotoxicity in plants is still poorly studied (Santos and Rodriguez, 2012). However, some studies carried out in recent years have addressed to this subject (Rodriguez et al., 2011; Gupta et al., 2012; Patnaik et al., 2013; Truta et al., 2014). Bioaccumulation of Cr(VI) also causes nonmediated–reactive oxygen species toxic effects on plants in terms of growth reduction, biomass accumulation, and structural alterations (Fozia et al., 2008; Mangabeira et al., 2011; Rodriguez et al., 2012). In this regard, various enzyme activities related to metabolic pathways of carbohydrates, nitrogen, and secondary metabolites as well as the concentration of soluble sugars, starch, photosynthetic pigments (chlorophyll and carotenoids), and phenolic compounds, among others, are decreased by chromium toxicity (Sankar-Ganesh et al., 2008; Prado et al., 2010b; Ozdener et al., 2011).

Chromium(III) present lower mobility than Cr(VI) and is mainly bound to organic matter in soils and waters. Moreover, it tends to form hydroxide precipitates with Fe at pH values typically occurring in aquatic environments (Mandiwana et al., 2007). Unlike the hexavalent form, Cr(III) does not use carriers to cross plant membranes (Yu et al., 2008). It mobilizes by simple diffusion and also binds easily to hydroxyl, carboxyl, phenolic-OH, amide, or sulfhydryl groups on the cell wall (Oliveira, 2012). However, once inside cells, Cr(III) either accumulated directly or from intracellular reduction of Cr(VI), exhibits a strong tendency to form stable complexes with many cellular ligands, notably with DNA and small reductive molecules such as ascorbic acid, glutathione, Cys, and histidine (Salnikow and Zhitkovich, 2008). Ternary Cr-DNA-L adducts are considered as the most relevant Cr(III)-induced nonoxidative mechanism of DNA damage in yeast and mammalian cells and probably also in plant cells (Wang et al., 2012). Although Cr(VI) has a higher redox potential than Cr(III), it may also induce oxidative stress. In this way, Cr(III) can be reduced to Cr(II) by Cys and NADH, which in turn reacts with H_2O_2 to generate $^{\cdot}OH$ radicals through the Fenton-like and Haber–Weiss reactions according to the following reactions:

$$\begin{aligned} \mathrm{Cr(III)} + O_2^{\cdot -} &\rightarrow \mathrm{Cr(II)} + O_2 \\ \mathrm{Cr(II)} + H_2O_2 &\rightarrow \mathrm{Cr(III)} + {}^{\cdot}OH + OH^{\cdot -} \end{aligned} \tag{6.3}$$

Chromium(II) can also react with lipid hydroperoxides to generate lipid hydroperoxides radicals, which also produce damaging effects on cell components (Shi et al., 1998). Hence, Cr(III)-induced oxidative stress may also have significant implications regarding the mechanism of Cr-induced toxicity in plants. Although Cr(VI) induces greater oxidative stress than that Cr(III) (Stohs and Bagchi, 1995), several studies have shown that Cr(III) appears to be as harmful as Cr(VI) in different plant and algae species (Panda and Patra, 2000; Dheeba et al., 2014; Kosior et al., 2015). In some species, Cr(III) becomes even more toxic than that of Cr(VI) (Gardea-Torresdey et al., 2005; Vignati et al., 2010). The accepted view that Cr(VI) is more toxic to plants than Cr(III) arises from studies done with the same concentration of Cr(VI) and Cr(III) in the external solution (Henriques, 2010; Joutey et al., 2013; Duman and Koca, 2014). Because Cr(VI) penetrates more easily into cells, its intracellular concentration will be greater than that of Cr(III) concentration. Thus, after cellular reduction of Cr(VI), the resulting intracellular concentration of Cr(III) becomes higher than that occurring in plants growing in the presence of Cr(III) only. In fact, this could lead to the famous Shakespeare's phrase "to be, or not to be, that is the question" in relation to Cr(VI) and Cr(III) toxicity.

6. CONCLUSIONS AND FUTURE PROSPECTS

From concepts given in this review, an intriguing question can arise on the unquestioned dogma regarding to Cr(VI) tolerance by plants that is currently accepted by entire scientific community (i.e., the conversion of Cr(VI) to "less toxic" Cr(III) as the key mechanism to cope with Cr(VI) toxicity). Future studies should address questions such as: What are the adaptive traits of plants that should be studied to gain a better understanding on the mechanisms of Cr tolerance? Omic technologies (e.g., genomic, transcriptomic, proteomic, metabolomic) will become useful tools to find vital traits making the Cr tolerance.

ACKNOWLEDGEMENTS

Dedicated to all people of the world that suffer Cr contamination. Financial support from Consejo de Investigaciones de la Universidad Nacional de Tucumán (G437) and Consejo Nacional de Investigaciones Científicas y Técnicas (PIP11/265).

REFERENCES

Adki, V.S., Jadhav, J.P., Bapat, V.A., 2013. *Nopalea cochenillifera*, a potential chromium(VI) hyperaccumulator plant. Environ. Sci. Pollut. Res. Int. 20, 1173–1180.

Aeschbacher, M., Vergari, D., Schwarzenbach, R.P., Sander, M., 2011. Electrochemical analysis of proton and electron transfer equilibria of the reducible moieties in humic acids. Environ. Sci. Technol. 45, 8385–8394.

Aeschbacher, M., Graf, C., Schwarzenbach, R.P., Sander, M., 2012. Antioxidant properties of humic substances. Environ. Sci. Technol. 46, 4916–4925.

Ali, H., Khan, E., Sajad, M.A., 2013. Phytoremediation of heavy metals—concepts and applications. Chemosphere 91, 869–881.

Al-Qahtan, K.M., 2012. Assessment of heavy metals accumulation in native plant species from soils contaminated in Riyadh city, Saudi Arabia. Life Sci. J. 9, 384–392.

Alvarez, B., Radi, R., 2003. Peroxynitrite reactivity with amino acids and proteins. Amino Acids 25, 295–311.

Alvarez-Puebla, R.A., Goulet, P.J.G., Garrido, J.J., 2005. Characterization of the porous structure of different humic fractions. Colloids Surf. A Physicochem. Eng. Aspects 256, 129–135.

Andjelković, D.H., Andjelković, T.D., Nikolić, R.S., Purenović, M.M., Blagojević, S.D., Bojić, A.L.J., Ristić, M.M., 2012. Leaching of chromium from chromium contaminated soil – a speciation study and geochemical modelling. J. Serb. Chem. Soc. 77, 119–129.

Arakawa, H., Weng, M.W., Chen, W.C., Tang, M.S., 2012. Chromium(VI) induces both bulky DNA adducts and oxidative DNA damage at adenines and guanines in the p53 gene of human lung cells. Carcinogenesis 33, 1993–2000.

Arsova, B., Hoja, U., Wimmelbacher, M., Greiner, E., Üstün, Ş., Melzer, M., Petersen, K., Lein, W., Börnke, F., 2010. Plastidial thioredoxin *z* interacts with two fructokinase-like proteins in a thiol-dependent manner: evidence for an essential role in chloroplast development in *Arabidopsis* and *Nicotiana benthamiana*. Plant Cell 22, 1498–1515.

Augustynowicz, J., Wróbel, P., Płachno, B.J., Tylko, G., Gajewski, Z., Węgrzynek, D., 2014. Chromium distribution in shoots of macrophyte *Callitriche cophocarpa* Sendtn. Planta 239, 1233–1242.

Avudainayagam, S., Megharaj, M., Owens, G., Kookana, R.S., Chittleborough, D., Naidu, R., 2003. Chemistry of chromium in soils with emphasis on tannery waste sites. Rev. Environ. Contam. Toxicol. 178, 53–91.

Ayyappan, D., Ravindran, K.C., 2014. Potentiality of *Suaeda monoica* Forsk. A salt marsh halophyte on bioaccumulation of heavy metals from tannery effluent. Int. J. Mod. Res. Rev. 2, 267–274.

Babula, P., Adam, V., Opatrilova, R., Zehnalek, J., Havel, L., Kizek, R., 2008. Uncommon heavy metals, metalloids and their plant toxicity: a review. Environ. Chem. Lett. 6, 189–213.

Badri, D.V., Vivanco, J.M., 2009. Regulation and function of root exudates. Plant Cell Environ. 32, 666–681.

Baker, H., Khalili, F., 2007. Effects of pH and temperature on the interaction of Pb(II) with Azraq humic acid studied with Schubert's ion exchange method. Ann. Environ. Sci. 1, 35–44.

Banach, A.M., Banach, K., Stępniewska, Z., 2012. Phytoremediation as a promising technology for water and soil purification: *Azolla caroliniana* Willd. As a case study. Acta Agrophys. 19, 241–252.

Banerjee, A., Nayak, D., Chakrabortty, D., Lahiri, S., 2008. Uptake studies of environmentally hazardous ^{51}Cr in mung beans. Environ. Pollut. 151, 423–427.

Barbosa, R.M.T., Almeida, A.A.F., Mielke, M.S., Loguercio, L.L., Mangabeira, P.A.O., Gomes, F.P., 2007. A physiological analysis of *Genipa americana* L.: a potential phytoremediator tree for chromium polluted watersheds. Environ. Exp. Bot. 61, 264–271.

Barrera-Díaz, C.E., Lugo-Lugo, V., Bilyeu, B., 2012. A review of chemical, electrochemical and biological methods for aqueous Cr(VI) reduction. J. Hazard. Mater. 223/224, 1–12.

Barthelmy, D., 2002. Mineralogy Database: Chromium, Web Mineral. http://webmineral.com/data/Chromium.shtml.

Bauer, I., Kappler, A., 2009. Rates and extent of reduction of Fe(III) compounds and O_2 by humic substances. Environ. Sci. Technol. 43, 4902–4908.
Bielicka, A., Bojanowska, I., Wiśniewski, A., 2005. Two faces of chromium – pollutant and bioelement. Pol. J. Environ. Stud. 14, 5–10.
Bini, C., Wahsha, M., Fontana, S., Maleci, L., 2012. Effects of heavy metals on morphological characteristics of *Taraxacum officinale* web growing on mine soils in NE Italy. J. Geochem. Explor. 123, 101–108.
Bluskov, S., Arocena, J.M., Omotoso, O.O., Young, J.P., 2005. Uptake, distribution, and speciation of chromium in *Brassica juncea*. Int. J. Phytoremediation 7, 153–165.
Bolan, N.S., Adriano, D.C., Natesan, R., Koo, B.J., 2003. Effects of organic amendments on the reduction and phytoavailability of chromate in mineral soil. J. Environ. Qual. 32, 120–128.
Borges, F., Guimarães, C., Lima, J.L.F.C., Pinto, I., Reis, S., 2005. Potentiometric studies on the complexation of copper(II) by phenolic acids as discrete ligand models of humic substances. Talanta 66, 670–673.
Bowers, L.J., Melhuish Jr., J.H., 1988. Comparison of elemental concentrations in the wood of three tree species growing adjacent to an inactive chromium smelter. Bull. Environ. Contam. Toxicol. 40, 457–461.
Brown, T.J., Shaw, R.A., Bide, T., Petavratzi, E., Raycraft, E.R., Walters, A.S., 2013. World Mineral Production 2007–11. British Geological Survey, Nottingham, England, 76 pp.
Buendía-González, L., Orozco-Villafuerte, J., Cruz-Sosa, F., 2010. *Prosopis laevigata* a potential chromium(VI) and cadmium(II) hyperaccumulator desert plant. Bioresour. Technol. 101, 5862–5867.
Caçador, I., Duarte, B., 2015. Chromium phyto-transformation in salt marshes: the role of halophytes. In: Ansari, A.A., Gill, S.S., Gill, R., Lanza, G.R., Newman, L. (Eds.), Phytoremediation: Management of Environmental Contaminants, vol. 2. Springer-Verlag, Berlin, Germany, pp. 211–217.
Campitelli, P.A., Velasco, M.I., Ceppi, S.B., 2006. Chemical and physicochemical characteristics of humic acids extracted from compost, soil and amended soil. Talanta 69, 1234–1239.
Carletti, P., Roldán, M.L., Francioso, O., Nardi, S., Sanchez-Cortes, S., 2010. Structural characterization of humic-like substances with conventional and surface-enhanced spectroscopic techniques. J. Mol. Struct. 982, 169–175.
Carrillo-González, R., Šimůnek, J., Sauvé, S., Adriano, D., 2006. Mechanisms and pathways of trace element mobility in soils. Adv. Agron. 91, 111–178.
Cejudo, F.J., Meyer, A.J., Reichheld, J.P., Rouhier, N., Traverso, J.A., 2014. Thiol-based redox homeostasis and signaling. Front. Plant Sci. 5 (266), 3. http://dx.doi.org/10.3389/fpls.2014.00266.
Cervantes, C., Campos-García, J., Devars, S., Gutiérrez-Corona, F., Loza-Tavera, H., Torres-Guzmán, J.C., Moreno-Sánchez, R., 2001. Interactions of chromium with microorganisms and plants. FEMS Microbiol. Rev. 25, 335–347.
Chakraborty, R., Muherjee, S., Kumar, S., 2014. Screening of few aquatic floating plants for chromium phytoremediation. Int. J. Environ. Technol. Manage. 17, 191–198.
Chandra, P., Kulshreshtha, K., 2004. Chromium accumulation and toxicity in aquatic vascular plants. Bot. Rev. 70, 313–327.
Chang, R.R., Mylotte, R., Hayes, M.H., McInerney, R., Tzou, Y.M., 2014. A comparison of the compositional differences between humic fractions isolated by the IHSS and exhaustive extraction procedures. Naturwissenschaften 101, 197–209.
Charbouillot, T., Gorini, S., Voyard, G., Parazols, M., Brigante, M., Deguillaume, L., Delort, A.M., Mailhot, G., 2012. Mechanism of carboxylic acid photooxidation in atmospheric aqueous phase: formation, fate and reactivity. Atmos. Environ. 56, 1–8.
Chebbi, A., Carlier, P., 1996. Carboxylic acids in the troposphere, occurrence, sources, and sinks: a review. Atmos. Environ. 30, 4233–4249.
Chen, S.Y., Huang, S.W., Chiang, P.N., Liu, J.C., Kuan, W.H., Huang, J.H., Hung, J.T., Tzou, Y.M., Chen, C.C., Wang, M.K., 2011. Influence of chemical compositions and molecular weights of humic acids on Cr(VI) photo-reduction. J. Hazard. Mater. 197, 337–344.

Chen, Y.L., Hong, X.Q., He, H., Luo, H.W., Qian, T.T., Li, R.Z., Jiang, H., Yu, H.Q., 2014. Biosorption of Cr(VI) by *Typha angustifolia*: mechanism and responses to heavy metal stress. Bioresour. Technol. 160, 89–92.

Cheung, K.H., Gu, J.D., 2007. Mechanism of hexavalent chromium detoxification by microorganisms and bioremediation application potential: a review. Int. Biodeterior. Biodegrad. 59, 8–15.

Choo, T.P., Lee, C.K., Low, K.S., Hishamuddin, O., 2006. Accumulation of chromium(VI) from aqueous solutions using water lilies (*Nymphaea spontanea*). Chemosphere 62, 961–967.

Cieśla, P., Kocot, P., Mytych, P., Stasicka, Z., 2004. Homogeneous photocatalysis by transition metal complexes in the environment. J. Mol. Catal. A Chem. 224, 17–33.

Dai, R., Yu, C., Gou, J., Lan, Y., Mao, J., 2011. Photoredox pathways of Cr(III)–tartrate complexes and their impacting factors. J. Hazard. Mater. 186, 2110–2116.

Dalvi, A.A., Bhalerao, S.A., 2013. Response of plants towards heavy metal toxicity: an overview of avoidance, tolerance and uptake mechanism. Ann. Plant Sci. 2, 362–368.

Daud, M.K., Mei, L., Variath, M.T., Ali, S., Li, C., Rafiq, M.T., Zhu, S.J., 2014. Chromium(VI) uptake and tolerance potential in cotton cultivars: effect on their root physiology, ultramorphology, and oxidative metabolism. BioMed. Res. Int. 12. http://dx.doi.org/10.1155/2014/975946. Article ID 975946.

Daulton, T.L., Little, B.J., 2006. Determination of chromium valence over the range Cr(0)–Cr(VI) by electron energy loss spectroscopy. Ultramicroscopy 106, 561–573.

Dheeba, B., Sampathkumar, P., Kannan, K., 2014. Growth, chromium accumulation potential and biochemical changes of *Vigna radiata* and *Zea mays* grown with effective microbes. Proc. Natl. Acad. Sci. India Sect. B Biol. Sci. 84, 381–387.

Dhir, B., Srivastava, S., 2011. Heavy metal removal from a multi-metal solution and wastewater by *Salvinia natans*. Ecol. Eng. 37, 893–896.

Ding, M., Shi, X., 2002. Molecular mechanisms of Cr(VI)-induced carcinogenesis. Mol. Cell. Biochem. 234/235, 293–300.

Dubey, S., Misra, P., Dwivedi, S., Chatterjee, S., Bag, S.K., Mantri, S., Asif, M.H., Rai, A., et al., 2010. Transcriptomic and metabolomic shifts in rice roots in response to Cr(VI) stress. BMC Genomics 11, 648–666.

Dudal, Y., Gérard, F., 2004. Accounting for natural organic matter in aqueous chemical equilibrium models: a review of the theories and applications. Earth-Sci. Rev. 66, 199–216.

Duman, F., Koca, F.D., 2014. Single and combined effects of exposure concentration and duration on biological responses of *Ceratophyllum demersum* L. exposed to Cr species. Int. J. Phytoremediation. 16, 1192–1208.

Elangovan, R., Philip, L., Chandraraj, K., 2008. Biosorption of chromium species by aquatic weeds: kinetics and mechanism studies. J. Hazard. Mater. 152, 100–112.

Elekes, C.C., 2014. Eco-technological solutions for the remediation of polluted soil and heavy metal recovery. In: Hernandez-Soriano, M.C. (Ed.), Environmental Risk Assessment of Soil Contamination. InTech, Rijeka, Croatia, pp. 309–335.

Emsley, J., 2011. Nature's Building Blocks: An A–Z Guide to the Elements. Oxford University Press, Oxford, 720 pp.

EPA (Environmental Protection Agency), 1998a. Toxicological Review of Hexavalent Chromium. National Center for Environmental Assessment, Office of Research and Development, Washington, USA, 70 pp.

EPA (Environmental Protection Agency), 1998b. Toxicological Review of Trivalent Chromium. National Center for Environmental Assessment, Office of Research and Development, Washington, USA, 77 pp.

Ernst, W.H.O., 2006. Evolution of metal tolerance in higher plants. For. Snow Landsc. Res. 80, 251–274.

Espinoza-Quiñones, F.R., Martin, N., Stutz, G., Tirao, G., Palácio, S.M., Rizzutto, M.A., Módenes, A.N., Silva Jr., F.G., Szymanski, N., Kroumov, A.D., 2009. Root uptake and reduction of hexavalent chromium by aquatic macrophytes as assessed by high-resolution X-ray emission. Water Res. 43, 4159–4166.

Faisal, M., Hasnain, S., 2006. Hazardous impact of chromium on environment and its appropriate remediation. J. Pharmacol. Toxicol. 1, 248–258.

Fendorf, S., Wielinga, B.W., Hansel, C.M., 2000. Chromium transformations in natural environments: the role of biological and abiological processes in chromium(VI) reduction. Int. Geol. Rev. 42, 691–701.

Fimmen, R.L., Cory, R.M., Chin, Y.P., Trouts, T.D., McKnight, D.M., 2007. Probing the oxidation–reduction properties of terrestrially and microbially derived dissolved organic matter. Geochim. Cosmochim. Acta 71, 3003–3015.

Fozia, A., Muhammad, A., Muhammad, A., Zafar, M.K., 2008. Effect of chromium on growth attributes in sunflower (*Helianthus annuus* L.). J. Environ. Sci. 20, 1475–1480.

Fukushima, M., Nakayasu, K., Tanaka, S., Nakamura, H., 1995. Chromium(III) binding abilities of humic acids. Anal. Chim. Acta 317, 195–206.

Gaberell, M., Chin, Y., Hu, S.J., Sulzberger, B., 2003. Role of dissolved organic matter composition on the photoreduction of Cr(VI) to Cr(III) in the presence of iron. Environ. Sci. Technol. 37, 4403–4409.

Gadd, G.M., 2009. Biosorption: critical review of scientific rationale, environmental importance and significance for pollution treatment. J. Chem. Technol. Biotechnol. 84, 13–28.

Gardea-Torresdey, J.L., de la Rosa, G., Peralta-Videa, J.R., Montes, M., Cruz-Jimenez, G., Cano-Aguilera, I., 2005. Differential uptake and transport of trivalent and hexavalent chromium by tumbleweed (*Salsola kali*). Arch. Environ. Contam. Toxicol. 48, 225–232.

Ghasemi, H., Rostampour, F., Ranjbar, A., 2014. The role of oxidative stress in metals toxicity/mitochondrial dysfunction as a key player. Galen Med. J. 3, 2–13.

Giachetti, G., Sebastiani, L., 2006. Development and chromium uptake in hybrid poplars cultivated on substrate polluted with industrial slags. Adv. Hort. Sci. 20, 193–198.

Gill, R.A., Zang, L., Ali, B., Farooq, M.A., Cui, P., Yang, S., Ali, S., Zhou, W., 2015. Chromium-induced physio-chemical and ultrastructural changes in four cultivars of *Brassica napus* L. Chemosphere 120, 154–164.

Giri, A.K., Patel, R.K., 2011. Toxicity and bioaccumulation potential of Cr(VI) and Hg(II) on differential concentration by *Eichhornia crassipes* in hydroponic culture. Water Sci. Technol. 63, 899–907.

Goldstein, S., Merényi, G., 2008. The chemistry of peroxynitrite: implications for biological activity. Methods Enzymol. 436, 49–61.

González, C.I., Maine, M.A., Cazenave, J., Sanchez, G.C., Benavides, M.P., 2015. Physiological and biochemical responses of *Eichhornia crassipes* exposed to Cr(III). Environ. Sci. Pollut. Res. 22, 3739–3747.

Gorshkov, A.I., Tikov, S.V., Bershov, L.V., Marfunin, A.S., 1996. The first finds of native Cr, Ni, and Fe in carbonado from the diamond deposits of Yakutia. Geochem. Int. 33, 59–63.

Guilizzoni, P., Galanti, G., Muntau, H., 1989. The aquatic macrophytes of Lake Maggiore: species composition, spatial distribution and heavy metal concentrations in tissue. Mem. Ist. Ital. Idrobiol. 46, 235–260.

Güngör, E.B.Ö., Bekbölet, M., 2010. Zinc release by humic and fulvic acid as influenced by pH, complexation and DOC sorption. Geoderma. 159, 131–138.

Gupta, K., Gaumat, S., Mishra, K., 2012. Studies on phyto-genotoxic assessment of tannery effluent and chromium on *Allium cepa*. J. Environ. Biol. 33, 557–563.

Gustafsson, J.P., 2001. Modelling the acid–base properties and metal complexation of humic substances with the Stockholm Humic Model. J. Colloid Interface Sci. 244, 102–112.

Henriques, F.S., 2010. Changes in biomass and photosynthetic parameters of tomato plants exposed to trivalent and hexavalent chromium. Biol. Plant 54, 583–586.

Holland, S.L., Avery, S.V., 2011. Chromate toxicity and the role of sulfur. Metallomics 3, 1119–1123.

Holland, S.L., Ghosh, E., Avery, S.V., 2010. Chromate-induced sulfur starvation and mRNA mistranslation in yeast are linked in a common mechanism of Cr toxicity. Toxicol. In Vitro 24, 1764–1767.

Hori, M., Shozugawa, K., Matsuo, M., 2015. Reduction process of Cr(VI) by Fe(II) and humic acid analyzed using high time resolution XAFS analysis. J. Hazard. Mater. 285, 140–147.

Horsfall Jr., M., Spiff, A.I., 2005. Effect of metal ion concentration on the biosorption of Pb^{2+} and Cd^{2+} by *Caladium bicolor* (wild cocoyam). Afr. J. Biotechnol. 4, 191–196.

Hossain, M.A., Piyatida, P., Teixeira da Silva, J.A., Fujita, M., 2012. Molecular mechanism of heavy metal toxicity and tolerance in plants: central role of glutathione in detoxification of reactive oxygen species and methylglyoxal and in heavy metal chelation. J. Bot. 37 http://dx.doi.org/10.1155/2012/872875. Article ID 872875.

Hossner, L.R., Loeppert, R.H., Newton, R.J., Szaniszlo, P.J., In collaboration with Attrep Jr., M., 1998. Literature Review: Phytoaccumulation of Chromium, Uranium, and Plutonium in Plant Systems. Amarillo National Resource Center for Plutonium, Texas, TX, 51 pp.

Huang, P.M., Hardie, A.G., 2009. Formation mechanisms of humic substances in the environment. In: Senesi, N., Xing, B., Huang, P.M. (Eds.), Biophysico-chemical Processes Involving Natural Nonliving Organic Matter in Environmental Systems. John Wiley & Sons, Inc., NJ, USA, pp. 41–109.

Huang, L., Fan, Z., Yu, C.H., Hopke, P.K., Lioy, P.J., Buckley, B.T., Lin, L., Ma, Y., 2013. Interconversion of chromium species during air sampling: effects of O_3, NO_2, SO_2, particle matrices, temperature, and humidity. Environ. Sci. Technol. 47, 4408–4415.

Hug, S.J., Laubscher, H.U., 1997. Iron(III) catalyzed photochemical reduction of chromium(VI) by oxalate and citrate in aqueous solutions. Environ. Sci. Technol. 31, 160–170.

IPCS (International Programme on Chemical Safety), World Health Organization, 2009. Inorganic Chromium(III) Compounds. Concise International Chemical Assessment Document 76. Geneva, Switzerland, 93 pp.

Jacobs, J., Testa, S.M., 2005. Overview of chromium(VI) in the environment: background and history. In: Guertin, J., Jacobs, J.A., Avakian, C.P. (Eds.), Chromium(VI) Handbook. CRC Press, FL, USA, pp. 1–21.

Jaliseh, S.A., 2011. Study of aquatic and semi-aquatic plants of aquatic ecosystems of Siahrud-e Rudbar protected area is in N. Iran and bioaccumulation of heavy metals by these plants. J. Basic Appl. Sci. Res. 1, 3466–3474.

Jankaitė, A., Vasarevičius, S., 2007. Use of *Poaceae f.* species to decontaminate soil from heavy metals. Ekologija 53, 84–89.

Jardine, P.M., Mehlhorn, T.L., Bailey, W.B., Brooks, S.C., Fendorf, S., Gentry, R.W., Phelps, T.J., Saiers, J.E., 2011. Geochemical processes governing the fate and transport of chromium(III) and chromium(VI). Vadose Zone J. 10, 1058–1070.

Jardine, P.M., Stewart, M.A., Barnett, M.O., Basta, N.T., Brooks, S.C., Fendorf, S., Mehlhorn, T.L., 2013. Influence of soil geochemical and physical properties on chromium(VI) sorption and bioaccessibility. Environ. Sci. Technol. 47, 11241–11248.

Jones, D.L., Dennis, P.G., Owen, A.G., van Hees, P.A.W., 2003. Organic acid behavior in soils – misconceptions and knowledge gaps. Plant Soil 248, 31–41.

Joutey, N.T., Bahafid, W., Sayel, H., El Ghachtouli, N., 2013. Phytotoxic effect of hexavalent chromium on germination and seedling growth of seeds of different plant species. J. Agric. Technol. 9, 361–372.

Kabata-Pendias, A., 2011. Trace Elements in Soil and Plants, fourth ed. CRC Press, FL, USA. 520 pp.

Kachenko, A.G., Singh, B., Bhatia, N.P., 2007. Heavy metal tolerance in common fern species. Aust. J. Bot. 300, 207–219.

Kappler, A., Benz, M., Schink, B., Brune, A., 2004. Electron shuttling via humic acids in microbial iron(III) reduction in a freshwater sediment. FEMS Microbiol. Ecol. 47, 85–92.

Kazakou, E., Dimitrakopoulos, P.G., Baker, A.J.M., Reeves, R.D., Troumbis, A.Y., 2008. Hypotheses, mechanisms and trade-offs of tolerance and adaptation to serpentine soils: from species to ecosystem level. Biol. Rev. 83, 495–508.

Khan, A.G., 2005. Role of soil microbes in the rhizosphere of plants growing on trace metal contaminated soils in phytoremediation. J. Trace Elem. Med. Biol. 18, 355–364.

Kim, K.R., Owens, G., Naidu, R., 2010. Effect of root-induced chemical changes on dynamics and plant uptake of heavy metals in rhizosphere soils. Pedosphere 20, 494–504.

Kim, E., Liu, Y., Leverage, W.T., Yin, J.J., White, I.M., Bentley, W.E., Payne, G.F., 2014. Context-dependent redox properties of natural phenolic materials. Biomacromolecules 15, 1653–1662.

Kimbrough, D.E., Cohen, Y., Winer, A.M., Creelman, L., Mabuni, C., 1999. A critical assessment of chromium in the environment. Crit. Rev. Environ. Sci. Technol. 29, 1–46.

Kosior, G., Samecka-Cymerman, A., Kolon, K., Brudzińska-Kosior, A., Bena, W., Kempers, A.J., 2015. Trace elements in the *Fontinalis antipyretica* from rivers receiving sewage of lignite and glass sand mining industry. Environ. Sci. Pollut. Res. 22. http://dx.doi.org/10.1007/s11356-015-4162-y.

Kotaś, J., Stasicka, Z., 2000. Chromium occurrence in the environment and methods of its speciation. Environ. Pollut. 107, 263–283.

Kováčik, J., Babula, P., Hedbavny, J., Klejdus, B., 2014. Hexavalent chromium damages chamomile plants by alteration of antioxidants and its uptake is prevented by calcium. J. Hazard. Mater. 273, 110–117.

Krems, P., Rajfur, M., Wacławek, M., Kłos, A., 2013. The use of water plants in biomonitoring and phytoremediation of waters polluted with heavy metals. Ecol. Chem. Eng. 20, 353–370.

Krzesłowska, M., 2011. The cell wall in plant cell response to trace metals: polysaccharide remodeling and its role in defense strategy. Acta Physiol. Plant 33, 35–51.

Kumari, A., Pakade, Y.B., Chand, P., Prasad, M.N.V., Lal, B., 2014. Comparative accounts of chromium accumulation in three ferns under hydroponic system. J. Sci. Ind. Res. 73, 553–558.

Lavu, R.V.S., Prasad, M.N.V., Pratti, V.L., Meißner, R., Rinklebe, J., Van De Wiele, T., Tack, F., Laing, G.D., 2013. Trace metals accumulation in *Bacopa monnieri* and their bioaccessibility. Planta Med. 79, 1081–1083.

Leita, L., Margon, A., Pastrello, A., Arčon, I., Contin, M., Mosetti, D., 2009. Soil humic acids may favour the persistence of hexavalent chromium in soil. Environ. Pollut. 157, 1862–1866.

Leonard, S.S., Harris, G.K., Shi, X., 2004. Metal-induced oxidative stress and signal transduction. Free Radical Biol. Med. 37, 1921–1942.

Liu, K.J., Jiang, J.J., Shi, X.L., Gabrys, H., Walczak, T., Swartz, H.M., 1995. Low-frequency EPR study of chromium(V) formation from chromium(VI) in living plants. Biochem. Biophys. Res. Commun. 206, 829–834.

Liu, X., Li, X., Chermaine Ong, S.M., Chu, Z., 2013. Progress of phytoremediation: focus on new plant and molecular mechanism. J. Plant Biol. Soil Health 1, 5.

Liu, J., Zhang, X.H., You, S.H., Wu, Q.X., Zhou, K.N., 2015. Function of *Leersia hexandra* Swartz in constructed wetlands for Cr(VI) decontamination: a comparative study of planted and unplanted mesocosms. Ecol. Eng. 81, 70–75.

Löflund, M., Kasper-Giebl, A., Schuster, B., Giebl, H., Hitzenberger, R., Puxbaum, H., 2002. Formic, acetic, oxalic, malonic and succinic acid concentrations and their contribution to organic carbon in cloud water. Atmos. Environ. 36, 1553–1558.

Lymar, S.V., Khairutdinov, R.F., Hurst, J.K., 2003. Hydroxyl radical formation by O–O bond homolysis in peroxynitrous acid. Inorg. Chem. 42, 5259–5266.

Madhavi, V., Vijay, A., Reddy, B., Reddy, K.G., Madhavi, G., Prasad, T.N.V.K.V., 2013. An overview on research trends in remediation of chromium. Res. J. Recent Sci. 2, 71–83.

Malmir, H.A., 2011. Comparison of antioxidant enzyme activities in leaves stems and roots of sorghum (*Sorghum bicolor* L.) exposed to chromium(VI). Afr. J. Plant Sci. 5, 436–444.

Mandiwana, K.L., Panichev, N., Kataeva, M., Siebert, S., 2007. The solubility of Cr(III) and Cr(VI) compounds in soil and their availability to plants. J. Hazard. Mater. 147, 540–545.

Mangabeira, P.A., Ferreira, A.S., de Almeida, A.A.F., Fernandes, V.F., Lucena, E., Souza, V.L., dos Santos Júnior, A.J., et al., 2011. Compartmentalization and ultrastructural alterations induced by chromium in aquatic macrophytes. BioMetals 24, 1017–1026.

Marbaniang, D., Chaturvedi, S.S., 2014. Assessment of Cr^{+6} accumulation and phytoremediation potential of three aquatic macrophytes of Meghalaya, India. Int. J. Sci. Res. 3, 36–42.

Masarovičová, E., Kráľová, K., 2012. Plant-heavy metal interaction: phytoremediation, biofortification and nanoparticles. In: Montanaro, G. (Ed.), Advances in Selected Plant Physiology Aspects. InTech, Rijeka, Croatia, pp. 75–102.

Matlins, A., 2010. Colored Gemstones, third ed. GemStone Press, VT, USA, 256 pp.
Mehes-Smith, M., Nkongolo, K., Cholewa, E., 2013. Coping mechanisms of plants to metal contaminated soil. In: Silvern, S., Young, S. (Eds.), Environmental Change and Sustainability. InTech, Rijeka, Croatia, pp. 53–90.
Mei, B., Puryear, J.D., Newton, R.J., 2002. Assessment of Cr tolerance and accumulation in selected plant species. Plant Soil 247, 223–231.
Meychik, N.R., Yermakov, I.P., 2001. Ion exchange properties of plant root cell walls. Plant Soil 234, 181–193.
Mganga, N., Manoko, M.L.K., Rulangaranga, Z.K., 2011. Classification of plants according to their heavy metal content around north Mara gold mine, Tanzania: implication for phytoremediation. Tanzania J. Sci. 37, 109–119.
Miao, Q., Yan, J., 2013. Comparison of three ornamental plants for phytoextraction potential of chromium removal from tannery sludge. J. Mater. Cycles Waste Manage. 15, 98–105.
Miretzky, P., Fernandez Cirelli, A., 2010. Cr(VI) and Cr(III) removal from aqueous solution by raw and modified lignocellulosic materials: a review. J. Hazard. Mater. 180, 1–19.
Mohanty, K., Jha, M., Meikap, B.C., Biswas, M.N., 2006. Biosorption of Cr(VI) from aqueous solutions by *Eichhornia crassipes*. Chem. Eng. J. 117, 71–77.
Montes-Holguin, M.O., Peralta-Videa, J.R., Meitzner, G., Martinez-Martinez, A., de la Rosa, G., Castillo-Michel, H.A., Gardea-Torresdey, J.L., 2006. Biochemical and spectroscopic studies of the response of *Convolvulus arvensis* L. to chromium(III) and chromium(VI) stress. Environ. Toxicol. Chem. 25, 220–226.
Montrichard, F., Alkhalfioui, F., Yano, H., Vensel, W.H., Hurkman, W.J., Buchanan, B.B., 2009. Thioredoxin targets in plants: the first 30 years. J. Proteomics 72, 452–474.
Motzer, W.E., 2005. Chemistry, geochemistry, and geology of chromium and chromium compounds. In: Guertin, J., Jacobs, J.A., Avakian, C.P. (Eds.), Chromium(VI) Handbook. CRC Press, FL, USA, pp. 23–91.
Mouvenchery, Y.K., Kučerík, J., Diehl, D., Schaumann, G.E., 2012. Cation-mediated cross-linking in natural organic matter: a review. Rev. Environ. Sci. Biotechnol. 11, 41–54.
Myers, C.R., 2012. The effects of chromium(VI) on the thioredoxin system: implications for redox regulation. Free Radical Biol. Med. 52, 2091–2107.
Nagajyoti, P.C., Lee, K.D., Sreekanth, T.V.M., 2010. Heavy metals, occurrence and toxicity for plants: a review. Environ. Chem. Lett. 8, 199–216.
Nebbioso, A., Piccolo, A., 2012. Advances in humeomics: enhanced structural identification of humic molecules after size fractionation of a soil humic acid. Anal. Chim. Acta 720, 77–90.
Nriagu, J.O., Pacyna, J.M., Milford, J.B., Davidson, C.I., 1988. Distribution and characteristic features of chromium in the atmosphere. In: Nriagu, J.O., Nieboer, E. (Eds.), Chromium in Natural and Human Environments. Wiley Interscience, NY, USA, pp. 125–173.
Oburger, E., Kirk, G.J.D., Wenzel, W.W., Puschenreiter, M., Jones, D.L., 2009. Interactive effects of organic acids in the rhizosphere. Soil Biol. Biochem. 41, 449–457.
Oliveira, H., 2012. Chromium as an environmental pollutant: insights on induced plant toxicity. J. Bot. 8 http://dx.doi.org/10.1155/2012/375843. Article ID 375843.
Onweremadu, E.U., 2014. Selected bioremediation techniques in polluted tropical soils. In: Hernandez-Soriano, M.C. (Ed.), Environmental Risk Assessment of Soil Contamination. InTech, Rijeka, Croatia, pp. 337–359.
Oporto, C., Arce, O., Van den Broeck, E., Van der Bruggen, B., Vandecasteele, C., 2008. Experimental study and modelling of Cr(VI) removal from wastewater using *Lemna minor*. Water Res. 40, 1458–1464.
Orhue, E.R., Uzu, F.O., 2010. Residual effects of chromium on early growth of fluted pumpkin (*Telfairia occidentalis* Hook F) in an ultisol. Afr. J. Gen. Agric. 6, 235–247.
Otte, M.L., Matthews, D.J., Jacob, D.L., Moran, B.M., Baker, A.J.M., 2004. Biogeochemistry of metals in the rhizosphere of wetland plants — an explanation for "innate" metal tolerance? In: Wong, M.H. (Ed.), Developments in Ecosystem, vol. 1. Elsevier B.V., Dordrecth, The Netherlands, pp. 87–94.

Ovečka, M., Takáč, T., 2014. Managing heavy metal toxicity stress in plants: biological and biotechnological tools. Biotechnol. Adv. 32, 73–86.

Ozdener, Y., Aydin, B.K., Fatma Aygün, S., Yürekli, F., 2011. Effect of hexavalent chromium on the growth and physiological and biochemical parameters on *Brassica oleracea* L. var. acephala DC. Acta Biol. Hung. 62, 463–476.

Oze, C., Bird, D.K., Fendorf, S., 2007. Genesis of hexavalent chromium from natural sources in soil and groundwater. Proc. Natl. Acad. Sci. U.S.A. 104, 6544–6549.

Pagano, E., Ilina, S., 2014. Cloning and sequencing of a complete h-type thioredoxin (Trxh1) cDNA fragment from submerged leaves of Cr-exposed Salvinia minima plants. Unpublished data.

Pal, S., Kundu, R., 2014. Study of metal resistance potential of the Cd, Cr tolerant alligator weed. J. Stress Physiol. Biochem. 10, 244–261.

Panda, S.K., Choudhury, S., 2005. Chromium stress in plants. Braz. J. Plant Physiol. 17, 95–102.

Panda, S.K., Patra, H.K., 2000. Does chromium(III) produce oxidative damage in excised wheat leaves? J. Plant Biol. 27, 105–110.

Papp, J.F., Lipin, B.R., 2010. Chromium and chromium alloys. In: Kirk-othmer Encyclopedia of Chemical Technology. John Wiley & Sons, Inc., NY, USA, pp. 1–47.

Patnaik, A.R., Achary, V.M.M., Panda, B.B., 2013. Chromium(VI)-induced hormesis and genotoxicity are mediated through oxidative stress in root cells of *Allium cepa* L. Plant Growth Regul. 71, 157–170.

Peralta-Videa, J.R., Lopez, M.L., Narayan, M., Saupe, G., Gardea-Torresdey, J., 2009. The biochemistry of environmental heavy metal uptake by plants: implications for the food chain. Int. J. Biochem. Cell Biol. 41, 1665–1677.

Pereira, Y., Lagniel, G., Godat, E., Baudouin-Cornu, P., Junot, C., Labarre, J., 2008. Chromate causes sulfur starvation in yeast. Toxicol. Sci. 106, 400–412.

Perminova, I.V., Hatfield, K., 2005. Remediation chemistry of humic substances: theory and implications for technology. In: Perminova, I.V., Hatfield, K., Hertkorn, N. (Eds.), Use of Humic Substances to Remediate Polluted Environments: From Theory to Practice. Springer, Dordrecht, The Netherlands, pp. 3–36.

Phukan, P., Phukan, R., Phukan, S.N., 2015. Heavy metal uptake capacity of *Hydrilla verticillata*: a commonly available aquatic plant. Int. Res. J. Environ. Sci. 4, 35–40.

Piccolo, A., Conte, P., Cozzolino, A., 1999. Effects of mineral and monocarboxylic acids on the molecular association of dissolved humic substances. Eur. J. Soil Sci. 50, 687–694.

Piccolo, A., 2001. The supramolecular structure of humic substances. Soil Sci. 166, 810–832.

Pillai, S.S., Girija, N., Williams, G.P., Koshy, M., 2013. Impact of organic manure on the phytoremediation potential of *Vetiveria zizanioides* in chromium-contaminated soil. Chem. Ecol. 29, 270–279.

Poljsak, B., Pócsi, I., Raspor, P., Pesti, M., 2010. Interference of chromium with biological systems in yeasts and fungi: a review. J. Basic Microbiol. 50, 21–36.

Polti, M.A., Amoroso, M.J., Abate, C.M., 2007. Chromium(VI) resistance and removal by actinomycete strains isolated from sediments. Chemosphere 67, 660–667.

Prado, C., Rodríguez-Montelongo, L., González, J.A., Pagano, E.A., Hilal, M., Prado, F.E., 2010a. Uptake of chromium by *Salvinia minima*: effect on plant growth, leaf respiration and carbohydrate metabolism. J. Hazard. Mater. 177, 546–553.

Prado, C., Rosa, M., Pagano, E., Hilal, M., Prado, F.E., 2010b. Seasonal variability of physiological and biochemical aspects of chromium accumulation in outdoor-grown *Salvinia minima*. Chemosphere 81, 584–593.

Prado, A.G.S., Pertusatti, J., Nunes, A.R., 2011. Aspects of protonation and deprotonation of humic acid surface on molecular conformation. J. Braz. Chem. Soc. 22, 1478–1483.

Prado, C., Pagano, E., Prado, F.E., Rosa, M., 2012. Detoxification of Cr(VI) in *Salvinia minima* is related to seasonal-induced changes of thiols, phenolics and antioxidative enzymes. J. Hazard. Mater. 239/240, 355–361.

Prado, C., Rosa, M., Pagano, E., Prado, F.E., 2013. Metabolic interconnectivity among alternative respiration, residual respiration, carbohydrates and phenolics in leaves of *Salvinia minima* exposed to Cr(VI). Environ. Exp. Bot. 87, 32–38.

Qiao, W., Li, C., Fan, L.M., 2014. Cross-talk between nitric oxide and hydrogen peroxide in plant responses to abiotic stresses. Environ. Exp. Bot. 100, 84–93.

Quaggiotti, S., Barcaccia, G., Schiavon, M., Nicolé, S., Galla, G., Rossignolo, V., Soattin, M., Malagoli, M., 2007. Phytoremediation of chromium using *Salix* species: cloning ESTs and candidate genes involved in the Cr response. Gene 402, 68–80.

Rashmi, B.N., Prabhakar, B.C., Gireesh, R.V., Suresh, B., Nijagunaiah, R., Ranganath, R.M., 2010. Geobotanical and biogeochemical study of chromite bearing areas of Nuggihalli schist belt, Karnataka. Curr. Sci. 99, 619–627.

Ratasuk, N., Nanny, M.A., 2007. Characterization and quantification of reversible redox sites in humic substances. Environ. Sci. Technol. 41, 7844–7850.

Reddy, L.C.S., Reddy, K.V.R., Humane, S.K., Damodaram, B., 2012. Accumulation of chromium in certain plant species growing on mine dump from Byrapur, Karnataka, India. Res. J. Chem. Sci. 2, 17–20.

Redondo-Gómez, S., 2013. Bioaccumulation of heavy metals in *Spartina*. Funct. Plant Biol. 40, 913–921.

Reimer, P., Duthie, H.C., 1993. Concentrations of zinc and chromium in aquatic macrophytes from the sudbury and muskoka regions of Ontario, Canada. Environ. Pollut. 79, 261–265.

Rimmer, D.L., 2006. Free radicals, antioxidants, and soil organic matter recalcitrance. Eur. J. Soil Sci. 57, 91–94.

Ritchie, J.D., Perdue, E.M., 2003. Proton-binding study of standard and reference fulvic acids, humic acids, and natural organic matter. Geochim. Cosmochim. Acta 67, 85–96.

Rodriguez, E., Azevedo, R., Fernandes, P., Santos, C., 2011. Cr(VI) induces DNA damage, cell cycle arrest and polyploidization: a flow cytometric and comet assay study in *Pisum sativum*. Chem. Res. Toxicol. 24, 1040–1047.

Rodriguez, E., Santos, C., Azevedo, R., Moutinho-Pereira, J., Correia, C., Dias, M.C., 2012. Chromium(VI) induces toxicity at different photosynthetic levels in pea. Plant Physiol. Biochem. 53, 94–100.

Roskill, 2014. The Economics of Chromium (Roskill Report on Metals and Minerals). Roskill Information Services Ltd, London, England, 438 pp.

Rupiasih, N.N., Vidyasagar, P.B., 2008. Humic substances: structure, function, effects and applications. Asian J. Water Environ. Pollut. 5, 39–47.

Ryan, P.R., Delhaize, E., Jones, D.L., 2001. Function and mechanism of organic anion exudation from plant roots. Annu. Rev. Plant Physiol. Plant Mol. Biol. 52, 527–560.

Sala, L.F., González, J.C., García, S.I., Frascaroli, M.I., Van Doorslaer, S., 2011. Detection and structural characterization of oxo-chromium(V)–sugar complexes by electron paramagnetic resonance. Adv. Carbohydr. Chem. Biochem. 66, 69–120.

Salnikow, K., Zhitkovich, A., 2008. Genetic and epigenetic mechanisms in metal carcinogenesis and cocarcinogenesis: nickel, arsenic and chromium. Chem. Res. Toxicol. 21, 28–44.

Sankar-Ganesh, K., Baskaran, L., Rajasekaran, S., Sumathi, K., Chidambaram, A.L.A., Sundaramoorthy, P., 2008. Chromium stress induced alterations in biochemical and enzyme metabolism in aquatic and terrestrial plants. Colloids Surf. B 63, 159–163.

Santos, C., Rodriguez, E., 2012. Review on some emerging endpoints of chromium(VI) and lead phytotoxicity. In: Mwoeia, J. (Ed.), Botany. InTech, Rijeka, Croatia, pp. 61–82.

Schaumann, G.E., Thiele-Bruhn, S., 2011. Molecular modelling of soil organic matter: squaring the circle? Geoderma 166, 1–14.

Schiavon, M., Galla, G., Wirtz, M., Pilon-Smits, E.A., Telatin, V., Quaggiotti, S., Hell, R., Barcaccia, G., Malagoli, M., 2012. Transcriptome profiling of genes differentially modulated by sulfur and chromium identifies potential targets for phytoremediation and reveals a complex S–Cr interplay on sulfate transport regulation in *B. juncea*. J. Hazard. Mater. 239/240, 192–205.

Schulten, H.R., Leinweber, P., 2000. New insights in organic mineral particles: composition, properties and models of molecular structure. Biol. Fertil. Soils 30, 399–432.

Seigneur, C., Constantinou, E., 1995. Chemical kinetic mechanism for atmospheric chromium. Environ. Sci. Technol. 29, 222–231.

Semenov, V.M., Tulina, A.S., Semenova, N.A., Ivannikova, L.A., 2013. Humification and non-humification pathways of the organic matter stabilization in soil: a review. Eurasian Soil Sci. 46, 355–368.

Shadreck, M., Mugadza, T., 2013. Chromium, an essential nutrient and pollutant: a review. Afr. J. Pure Appl. Chem. 7, 310–317.

Shafi, N., Pandit, A.K., Kamili, A.N., Mushtaq, B., 2015. Heavy metal accumulation by *Azolla pinnata* of Dal Lake ecosystem, India. J. Environ. Prot. Sustainable Dev. 1, 8–12.

Shanker, A.K., Cervantes, C., Loza-Tavera, H., Avudainayagam, S., 2005. Chromium toxicity in plants. Environ. Int. 31, 739–753.

Shanker, A.K., Djanaguiraman, M., Venkateswarlu, B., 2009. Chromium interactions in plants: current status and future strategies. Metallomics 1, 375–383.

Shi, X., Leonard, S.S., Liu, K.J., Zang, L., Gannett, P.M., Rojanasakul, Y., Castranova, V., Vallyathan, V., 1998. Cr(III)-mediated hydroxyl radical generation via Haber-Weiss cycle. J. Inorg. Biochem. 69, 263–268.

Sinam, G., Behera, S.K., Mishra, R.K., Sinha, S., Mallick, S., Khare, P.B., 2012. Comparison of two ferns (*Adiantum capillus-veneris* Linn. and *Microsorium punctatum* (Linn.) Copel) for their Cr accumulation potential and antioxidant responses. Int. J. Phytoremediation 14, 629–642.

Singh, N.K., Rai, U.N., Singh, M., Tripathi, R.D., 2010. Impact of rhizobacteria on growth and chromium accumulation in *Scirpus lacustris* L. grown under chromium supplementation. J. Environ. Biol. 31, 709–714.

Singh, H.P., Mahajan, P., Kaur, S., Batish, D.R., Kohli, R.K., 2013. Chromium toxicity and tolerance in plants. Environ. Chem. Lett. 11, 229–254.

Sinha, V., Pakshirajan, K., Chaturvedi, R., 2014. Chromium(VI) accumulation and tolerance by *Tradescantia pallida*: biochemical and antioxidant study. Appl. Biochem. Biotechnol. 173, 2297–2306.

Slade, P.G., Hailer, M.K., Martin, B.D., Sugden, K.D., 2005. Guanine specific oxidation of double-stranded DNA by Cr(VI) and ascorbic acidforms spiroiminodihydantoin and 8-oxo-2′-deoxyguanosine. Chem. Res. Toxicol. 18, 1140–1149.

Sołtysiak, J., Berchová-Bímová, K., Vach, M., Brej, T., 2011. Heavy metals content in the *Fallopia* genus in Central European Cities – study from Wroclaw and Prague. Acta Bot. Siles. 7, 209–218.

Srivastava, S., Prakash, S., Srivastava, M.M., 1999. Chromium mobilization and plant availability – the impact of organic complexing ligands. Plant Soil 212, 203–208.

Srivastava, V., 2014. Effect of chromium on growth, morphology and economic yield of *Helianthus annuus* L. varieties. Adv. Biores. 5, 42–47.

Stanin, F.T., Pirnie, M., 2005. The transport and fate of Cr(VI) in the environment. In: Guertin, J., Jacobs, J.A., Avakian, C.P. (Eds.), Chromium(VI) Handbook. CRC Press, FL, USA, pp. 161–211.

Stohs, S.J., Bagchi, D., 1995. Oxidative mechanisms in the toxicity of metal ions. Free Radical Biol. Med. 18, 321–336.

Strobel, B.W., 2001. Influence of vegetation on low-molecular-weight carboxylic acids in soil solution – a review. Geoderma 99, 169–198.

Sultana, M.Y., Chowdhurya, A.K.M.M.B., Michailides, M.K., Akratos, C.S., Tekerlekopoulou, A.G., Vayenas, D.V., 2015. Integrated Cr(VI) removal using constructed wetlands and composting. J. Hazard. Mater. 281, 106–113.

Sun, J., Mao, J.D., Gong, H., Lan, Y., 2009. Fe(III) photocatalytic reduction of Cr(VI) by low-molecular-weight organic acids with α-OH. J. Hazard. Mater. 168, 1569–1574.

Suñé, N.L., Sánchez, G.C., Caffaratti, S.E., Maine, M.A., 2007. Cadmium and chromium removal kinetics from solution by two aquatic macrophytes. Environ. Pollut. 145, 467–473.

Sutton, R., Sposito, G., 2005. Molecular structure in soil humic substances: the new view. Environ. Sci. Technol. 39, 9009–9015.

Szabó, C., Ischiropoulos, H., Radi, R., 2007. Peroxynitrite: biochemistry, pathophysiology and development of therapeutics. Nat. Rev. Drug Discov. 6, 662–680.

Tanaka, S., Nakayasu, K., Fukushima, M., 1997. Suppression effect of humic substances on oxidation of chromium(III) to chromium(VI). Toxicol. Environ. Chem. 58, 17–23.

Tang, S., Wilke, B.M., Brooks, R.R., 2001. Heavy-metal uptake by metal-tolerant *Elsholtzia haichowensis* and *Commelina communis* from China. Commun. Soil Sci. Plant Anal. 32, 895–905.

Tchounwou, P.B., Yedjou, C.G., Patlolla, A.K., Sutton, D.J., 2012. Heavy metal toxicity and the environment. Mol. Clin. Environ. Toxicol. 101, 133–164.

Teixeira, J., Ferraz, P., Almeida, A., Verde, N., Fidalgo, F., 2013. Metallothionein multigene family expression is differentially affected by chromium(III) and (VI) in *Solanum nigrum* L. plants. Food Energy Secur. 2, 130–140.

Terfie, T.A., Asfaw, S.L., 2015. Evaluation of selected wetland plants for removal of chromium from tannery wastewater in constructed wetlands, Ethiopia. Afr. J. Environ. Sci. Technol. 9, 420–427.

Theodosi, C., Markaki, Z., Tselepides, A., Mihalopoulos, N., 2010. The significance of atmospheric inputs of soluble and particulate major and trace metals to the eastern Mediterranean seawater. Mar. Chem. 120, 154–163.

Tian, X., Gao, X., Yang, F., Lan, Y., Mao, J.D., Zhou, L., 2010. Catalytic role of soils in the transformation of Cr(VI) to Cr(III) in the presence of organic acids containing α-OH groups. Geoderma 159, 270–275.

Ton, S.S., Lee, M.W., Yang, Y.H., Hoi, S.K., Cheng, W.C., Wang, K.S., Chang, H.H., Chang, S.H., 2015. Effects of reductants on phytoextraction of chromium(VI) by *Ipomoea aquatica*. Int. J. Phytoremediation 17, 429–436.

Trevisan, S., Francioso, O., Quaggiotti, S., Serenella Nardi, S., 2010. Humic substances biological activity at the plant-soil interface. From environmental aspects to molecular factors. Plant Signal. Behav. 5, 635–643.

Truta, E., Mihai, C., Gherghel, D., Vochita, G., 2014. Assessment of the cytogenetic damage induced by chromium short-term exposure in root tip meristems of barley seedlings. Water Air Soil Pollut. 225, 1–12.

Upadhyay, A.K., Singh, N.K., Rai, U.N., 2014. Comparative metal accumulation potential of *Potamogeton pectinatus* L. and *Potamogeton crispus* L.: role of enzymatic and non-enzymatic antioxidants in tolerance and detoxification of metals. Aquat. Bot. 117, 27–32.

Varkey, M., Lal, N., Khan, Z., 2013. Phytoremediation: strategies to enhance the potential for toxic metal remediation of *Brassica* oilseed species. In: Anjum, N.A., Pereira, M., Ahmad, I., Duarte, A., Umar, S., Khan, N. (Eds.), Phytotechnologies: Remediation of Environmental Contaminants. CRC Press, Taylor & Francis Group, FL, USA, pp. 293–308.

Vignati, D.A.L., Dominik, J., Beye, M.L., Pettine, M., Ferrar, B.J.D., 2010. Chromium(VI) is more toxic than chromium(III) to freshwater algae: a paradigm to revise? Ecotoxicol. Environ. Saf. 73, 743–749.

Walker, T.S., Bais, H.P., Grotewold, E., Vivanco, J.M., 2003. Root exudation and rhizosphere biology. Plant Physiol. 132, 44–51.

Wang, S.L., Chen, C.C., Tzou, Y.M., Hsu, C.L., Chen, J.H., Lin, C.F., 2009. A mechanism study of light-induced Cr(VI) reduction in an acidic solution. J. Hazard. Mater. 164, 223–228.

Wang, T.C., Song, Y.S., Wang, H., Zhang, J., Yu, S.F., Gu, Y.E., Chen, T., Wang, Y., Shen, H.Q., Jia, G., 2012. Oxidative DNA damage and global DNA hypomethylation are related to folate deficiency in chromate manufacturing workers. J. Hazard. Mater. 213, 440–446.

Xiao, M., Wu, F., 2014. A review of environmental characteristics and effects of low- molecular weight organic acids in the surface ecosystem. J. Environ. Sci. 26, 935–954.

Yabanli, M., Yozukmaz, A., Sel, F., 2014. Heavy metal accumulation in the leaves, stem and root of the invasive submerged macrophyte *Myriophyllum spicatum* L. (Haloragaceae): an example of Kadin Creek (Mugla, Turkey). Braz. Arch. Biol. Technol. 57, 434–440.

Yao, X., Fang, M., Chan, C.K., 2002. Size distributions and formation of dicarboxylic acids in atmospheric particles. Atmos. Environ. 36, 2099–2107.

Yu, X.Z., Gu, J.D., Xing, L.Q., 2008. Differences in uptake and translocation of hexavalent and trivalent chromium by two species of willows. Ecotoxicology 7, 747–755.

Zavarzina, A.G., 2011. Heterophase synthesis of humic acids in soils by immobilized phenol oxidases. In: Shukla, G., Varma, A. (Eds.), Soil Enzymology, Soil Biology, vol. 22. Springer-Verlag, Berlin, Germany, pp. 187–205.

Zayed, A.M., Terry, N., 2003. Chromium in the environment: factors affecting biological remediation. Plant Soil 249, 139–156.

Zayed, A., Lytle, C.M., Jin-Hong, Q., Terry, N., Qian, J.H., 1998. Chromium accumulation, translocation and chemical speciation in vegetable crops. Planta 206, 293–299.

Zeng, F., Chen, S., Miao, Y., Wu, F., Zhang, G., 2008. Changes of organic acid exudation and rhizosphere pH in rice plants under chromium stress. Environ. Pollut. 155, 284–289.

Zhang, H., Bartlett, R.J., 1999. Light-induced oxidation of aqueous chromium(III) in the presence of iron(III). Environ. Sci. Technol. 33, 588–594.

Zhang, H., 2000. Light and iron(III)-induced oxidation of chromium(III) in the presence of organic acids and manganese(II) in simulated atmospheric water. Atmos. Environ. 34, 1633–1640.

Zhu, M., Liu, Y., 1981. Discovery of native chromium in Xizang (Tibet). Chin. Sci. Bull. (Kexue Tongbao) 26, 1014–1017.

CHAPTER

7 SELENIUM: AN ANTIOXIDATIVE PROTECTANT IN PLANTS UNDER STRESS

Neha Handa[1], Renu Bhardwaj[1], Harpreet Kaur[1], Poonam[1], Dhriti Kapoor[1], Amandeep Rattan[2], Sukhmeen Kaur[1], Ashwani Kumar Thukral[1], Satwinderjit Kaur[1], Saroj Arora[1], Nitika Kapoor[2]

Department of Botanical and Environmental Sciences, Guru Nanak Dev University, Amritsar, Punjab, India[1]; Department of Botany, Hansraj Mahila Maha Vidyalaya, Jalandhar, Punjab, India[2]

CHAPTER OUTLINE

1. INTRODUCTION

Selenium (Se) was discovered by a Swedish Chemist J.J. Berzelius in 1817 and the element was initially named "Selene" after the Greek goddess of the moon. It is one of the most widely distributed elements of chalcogen group in Earth's crust and commonly found in rocks and soil. Se, however rarely found in its pure form, is metallic gray to black forming hexagonal crystals and is often referred to as elemental Se or Se dust. It is usually combined with other substances in rocks, such as sulfide minerals, silver, copper, lead, and nickel minerals. Earlier studies on Se were mainly focused on it toxic effects; the nutritional essentiality of this element was not discovered until 1957 when Schwartz

Plant Metal Interaction. http://dx.doi.org/10.1016/B978-0-12-803158-2.00007-2

and Foltz established the necessity of Se in animal health against liver necrosis in vitamin E–deficient rats. This report led to an upsurge in Se research related to plants and animals. This element further garnered interest when Rotruck et al. (1973) revealed that glutathione peroxidase (GPOX), an essential antioxidative enzyme, possesses Se in its active center (Thiry et al., 2012). After that, it became the subject of active research and further studies established its incorporation in various enzymes and proteins that carry essential functions in both plants and animals. Se has been explored extensively for its role in plant growth and protection from stresses, and many studies have suggested its imperative role in fortifying the defense system of plants.

2. SELENIUM IN THE ENVIRONMENT

Se is regarded as one of the most abundant and widely distributed elements present in Earth's crust with average soil concentration of 0.09 mg/kg (Afton et al., 2009). This concentration gets easily fluctuated by changes in temperature, rainfall, activity of microbes, and plant matter in the soil (Yang et al., 1983; Long et al., 1990; Brown et al., 1999; Zhang and Frankenberger, 2002; Amos et al., 2012). The variation in the concentration in the soil may also depend upon composition of ground material and leaching (Malik et al., 2012). It is usually present in four oxidation states in soil as selenate (+6), selenite (+4), selenide (−2), and elemental Se (0). The proportions of these forms in soil solution are governed by various physicochemical properties including pH, redox potential, and biological processes (Kabata-Pendias and Mukherjee, 2007). These factors determine the bioavailability of Se to living systems (Gissel-Nielsen, 1971; McNeal and Balistrieri, 1989; Mikkelsen et al., 1989). The forms that are commonly present in soil are selenate and selenite. Of these two, selenate is more soluble under oxidizing and alkaline conditions as compared with selenite, which increases its mobility and therefore increases its bioavailability to plants (Masscheleyn et al., 1990; Mayland, 1994; ATSDR, 2003). Selenite, on the other hand, is predominantly found in acidic conditions but in acidic environment it gets readily adsorbed on the surface of clay particles and iron oxide minerals which makes it less available to plants (Ylaranta, 1983, 1985; Mikkelsen et al., 1989; Environment Agency, 2007). The Se content of natural water bodies varies within a range from less than 2 ng/L up to 300 μg/L (Sager, 2006).

Se accumulation in soils and waters has usually resulted from various anthropogenic sources and geochemical processes. The rate of release of Se from geological sources has increased and its availability to life in terrestrial and aquatic ecosystems has been mainly caused by industrial and agricultural activities such as relocation of ores, industrial wastes, and mining and a variety of other industrial sources such as burning of fossil fuels (particularly coal) and refining of copper and other metals, exposure of geological strata to weather conditions, impact of plants including the release of H^+ ions from plant roots, and changeable pH. Also, drain water from agricultural sites, sewage sludge, fly ash from coal-fired power plants, oil refineries, and mining of phosphates, and metal ores all cause discharge of Se in the environment (IPCS, 1987; de Souza et al., 2002; Lemly, 2002; Shardendu et al., 2003).

3. SELENIUM UPTAKE BY PLANTS

Se enters the food chain from the environment through the plants, which have a unique and important role in delivering Se from soil to living systems. The rate of Se uptake depends on concentration and chemical form of Se in the soil solution, the rhizosphere conditions such as pH, and the presence of sulfate and phosphate, which compete with Se uptake (Blaylock and James, 1994; Dhillon and Dhillon, 2003). Se is usually taken up by plant roots in the ionic forms of selenate or selenite (Shrift and Ulrich, 1969; Asher et al., 1977; White et al., 2004). In plants, the Se metabolism is closely related to sulfur because of chemical similarities between the two elements (Anderson, 1993); therefore, absorption of Se is aided by sulfate transporters and Se competes with sulfur for ionic transport. The absorption of selenate over sulfur by a sulfate transporter depends upon the type of plant species and nutritional status (White et al., 2004). The plants that can accumulate sulfur-rich compounds also act as good accumulators for Se (Pilon-Smits and Quinn, 2010). Hence, Se accumulator species do not possess any Se-specific pathways for uptake and metabolism. However, Se hyperaccumulators, which usually belong to families *Brassicaceae*, *Fabaceae*, and *Asteraceae*, can accumulate Se up to 100 times more if grown in seleniferous soils (Pilon-Smits and Quinn, 2010). Therefore, it was suggested that these Se hyperaccumulators take up Se over sulfur and thus are able to distinguish between Se and sulfur (Neuhierl et al., 1999; Persans and Salt, 2000; Ellis et al., 2004; LeDuc et al., 2004; White et al., 2007; Pilon-Smits and Quinn, 2010).

Both selenate and selenite forms of Se are absorbed differently by the plant roots. Selenate, as with sulfate anion, is moved across the plasma membrane against its electrochemical gradient driven by cotransport of three protons for each ion (Lass and Ullrich-Eberius, 1984; Hawkesford et al., 1993). Leustek (2002) suggested that uptake is mediated by use of high- and low-affinity transporters. High-affinity transporters are mainly present in roots and are supposed to be primary transporter of selenate and sulfate from the soil (Smith et al., 1997; Shibagaki et al., 2002), whereas low-affinity transporters are located in both leaves and roots. They are involved in uptake from the soil solution into the root cells and then intracellular transfer from apoplast to symplast (Sors et al., 2005). Selenite, on the other hand, is suggested to follow passive diffusion pathways that can be easily inhibited by phosphate (Abrams et al., 1990). Li et al. (2008), however, suggested active transport by phosphate transport pathway.

After entering the plant system, Se undergoes a number of metabolic changes before being finally incorporated into biomolecules or its release in volatile form (Figures 7.1 and 7.2). The selenate form is reduced to selenite by the action of two enzymes. Adenosine triphosphate (ATP) sulfurylase combines selenate with ATP to form adenosine phosphoselenate (Wilson and Bandurski, 1958). Then adenylyl-sulfate reductase reduces adenosine phosphoselenate to selenite. These reduction reactions mainly take place in the chloroplast; however, the enzymes are also present in cytosol (Pilon-Smits and Quinn, 2010). Selenite, which is formed as a result of reduction, is further reduced to selenide. It is suggested to take place by two processes. It may be reduced enzymatically in the chloroplast in the presence of sulfite reductase (Pilon-Smits and Quinn, 2010) or nonenzymatically with the help of reduced glutathione (Anderson, 1993; Terry et al., 2000). Selenide thus formed combines with O-acetylserine to form selenocysteine in the presence of enzyme O-acetylserine thiol lyase. Selenocysteine then couples with O-phosphohomoserine in the presence of enzyme cystathionine-γ-synthase to form Se-cystathionine. Then cystathionine-β-lyase converts Se-cystathionine to Se-homocysteine. It then moves from chloroplast to cytosol and gets converted to Se-methionine by enzyme methionine synthase. This Se-methionine either moves inside chloroplast and gets incorporated into proteins or it may remain in cytosol and with the action of enzyme methionine methyltransferase, it undergoes methylation to form methyl-Se-methionine. The product thus formed gets metabolized into dimethylselenide, which is volatile in nature (Pilon-Smits and Quinn, 2010).

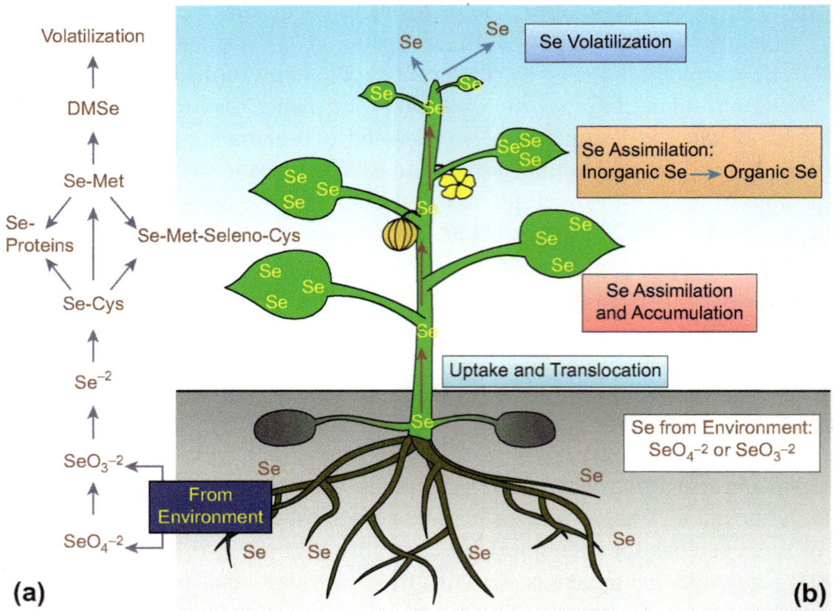

FIGURE 7.1

(a) Overview of main steps in selenium (Se) metabolism. (b) An overview of uptake, translocation, and assimilation of Se in plant system.

Part A adapted from Germ et al. (2007b).

Metabolism of Se can vary by plant species, growth stage of the plant, and the plant organ in which it is carried (Hasanuzzaman et al., 2010). There are certain plants such as various *Astragalus* species (*Fabaceae*), *Brassica oleracea*, and *Stanleya pinnata* (*Brassicaceae*) that accumulate and tolerate high concentrations of Se, whereas others do not (Feist and Parker, 2001; Lyi et al., 2005). Commonly, Se gets accumulated in tissues of shoots and leaves rather than in roots. Also, Turakainen (2007) reported that high concentrations of Se were found in young upper leaves, roots, and stolons of potato, which shows that younger parts take up and use Se efficiently.

4. ROLE OF SELENIUM IN PLANTS

Se is considered to be an essential microelement for animals, humans, plants, and microorganisms (Rotruck et al., 1973). It shows three levels of biological activity. Trace concentrations of the element are required for normal growth and development, moderate concentrations can be stored to maintain homeostatic functions, and elevated concentrations can result in toxic effects (Hamilton, 2004). Various studies on *Lolium perenne* and *Lactuca sativa* demonstrated that Se is harmful at higher concentrations and had beneficial effects at lower concentrations (Hartikainen et al., 2000; Xue et al., 2001). Various beneficial effects exerted by Se on growth, seed germination, seed production, senescence, photosynthesis, respiratory potential, and stress tolerance have been reported.

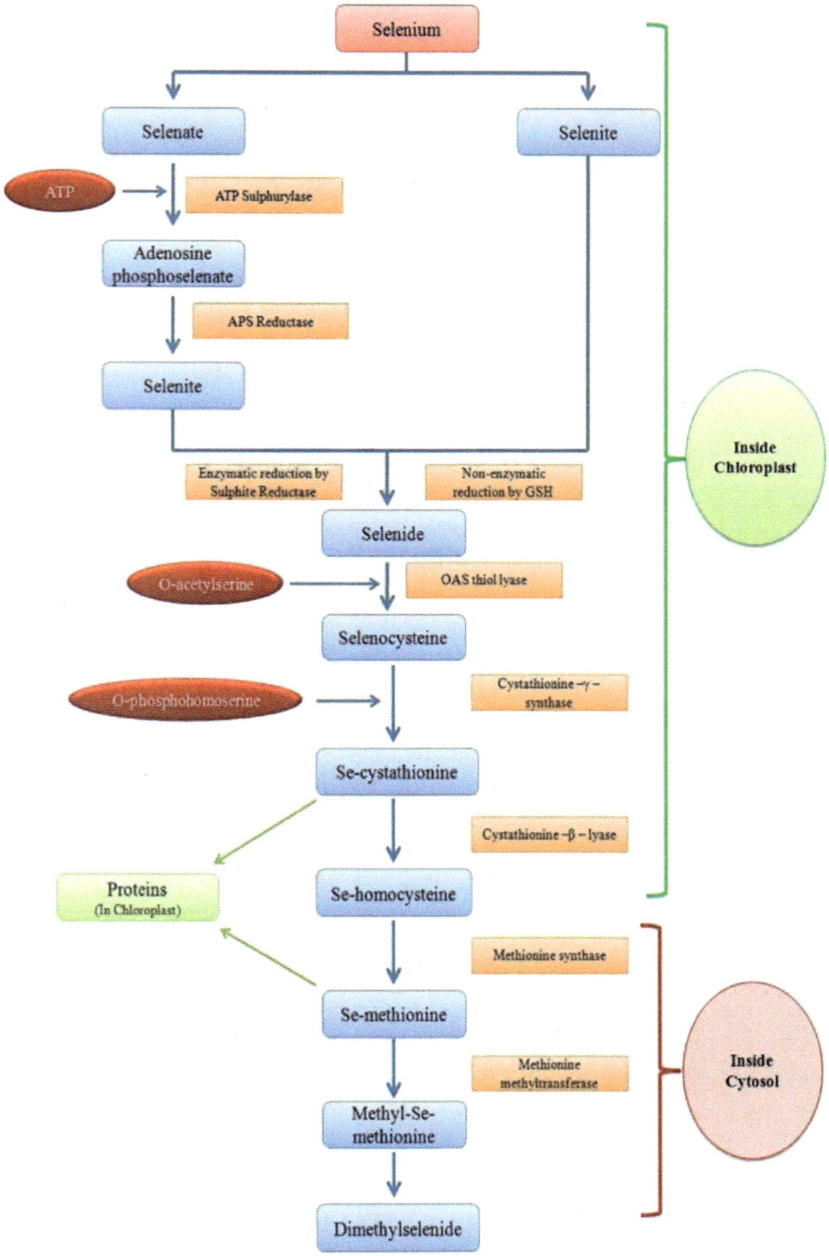

FIGURE 7.2

Schematic representation of selenium metabolism in plants.

4.1 GROWTH AND DRY MATTER PRODUCTION

Se has been shown to exert a positive effect on shoot and root length, fresh weight, and dry weight at appropriate concentrations in different plants. Also, there are reports that cite the increase in these parameters in the presence of various types of stresses, indicating the stress-protective effects of Se. Singh et al. (1980) for the first time reported that Se as selenite enhanced the growth and dry matter yield in *Brassica juncea*. A growth-promoting response was also reported in other plants such as lettuce, ryegrass, and soybean (Hartikainen and Xue, 1999; Djanaguiraman et al., 2005). It was reported by Kong et al. (2005) that exogenous application of Se promoted the growth of sorrel seedlings under salt stress. Low concentrations of Se (1–5 μM) stimulated the growth in 100 mM NaCl-stressed sorrel seedlings by enhancing the salt resistance. Ramos et al. (2010) reported increase in shoot biomass and dry matter production in lettuce plants, whereas a decrease in growth was demonstrated at higher concentrations. Similar results were shown by Zembala et al. (2010) in *Brassica napus* and wheat that showed significant increase in biomass when Se was supplemented in growth medium. *Stylosanthes humilis* seedlings, when subjected to heavy metal stress, showed increase in growth upon Se supplementation, establishing Se as ameliorative agent against heavy metal stress (Ribeiro et al., 2011). However, Se application, both in metal-treated and metal-untreated plants, helped in increasing the fresh weight. Sorghum seedlings, when subjected to heat stress, showed reduced total dry matter and leaf dry weight. Application of Se ameliorated the effect of heat stress and aided in restoring the total and leaf dry weights (Djanaguiraman et al., 2010). Increase in shoot and root lengths from Se application in ryegrass subjected to aluminum (Al) stress was reported by Cartes et al. (2010). Increase in dry weight was also observed with the application of Se in Al-treated plants, which greatly reduced plant height and dry weight. Se in the form of sodium selenite (Na_2SeO_3) showed improvement in root growth and root viability under lead (Pb) stress in *Vicia faba*. Application of Pb reduced the root growth because of its accumulation in the root tissues, which intensified the superoxide anion production; when Se treatment was given at a low concentration, improvement in root viability and growth was observed (Mroczek-Zdyrska and Wójcik, 2012). A significant enhancement of dry matter of vegetative parts, pods, and seeds dry weight was observed in *B. napus* on Se application by Hajiboland and Keivanfar (2012). Han-wen et al. (2010) observed the role of Se in promoting growth of garlic seedlings by alleviating the cadmium (Cd) stress. Similarly, additions of selenite at a concentration of 2.0 μg/kg in soil decreased the Cd content in garlic and protected the seedling growth. Debarati et al. (2011) studied the application of selenate at 2 M concentration that showed improvement in growth and metabolism in rice in comparison to 20 M concentration of selenate that showed inhibitory effects on growth. Yao et al. (2010) demonstrated the protective role of Se in wheat seedlings subjected to ultraviolet (UV)-B radiation, supplied by UVB fluorescent lamps. The application of Se (1.0 mg/kg) improved the wheat seedling growth inhibited by UVB radiation by lowering the oxidative stress generated by high UVB radiation. Thus, only optimum concentrations of Se promoted the growth of plants. In a recent study on wheat, priming of sodium selenate solutions of 25, 50, 75, and 100 μM significantly increased root length stress tolerance index, dry matter stress tolerance index, and total biomass of seedlings by inducing growth under water stress conditions (Nawaz et al., 2013). Similarly, exogenous Se application (0, 2, 4, 8, 16 μM) improved the growth in *Cucumis melo* L. subjected to 100 mM NaCl stress (KeLing et al., 2013). Application of either selenate or selenite to the nutrient solution at concentrations below 15 μM promoted the growth in lettuce plants (Hawrylak-Nowak, 2013). Foliar application of Se (30 g/ha) as sodium selenite to barley plants significantly increased the shoot dry weight under drought stress (Habibi, 2013). A recent study by Nawaz et al. (2014) established that foliar application of Se (7.06 μM) improved the drought tolerance by promoting the growth and biomass of seedlings in terms of dry matter stress tolerance index and fresh matter stress tolerance indices in wheat seedlings subjected to drought stress. Increase in growth of tomato seedlings upon Se treatment was reported to be caused by a decrease in hydrogen peroxide and malondialdehyde (MDA) and increase in antioxidative enzymes and photosynthesis efficiency (Diao et al., 2014).

4.2 SEED GERMINATION

The rate and degree of seedling establishment are extremely important factors in determining the yield. Stress reduces the germination rate. Searching for suitable ameliorants or stress alleviant is one of the challenging tasks of plant biologists. Recent researchers have identified several beneficial effects of Se including seed germination also. Chen and Sung (2001) reported that priming or soaking seeds of *Momordica charantia* in Se solution, enhanced the germination ability. The study was carried out at suboptimal temperatures at which germination of untreated seeds was reduced. But priming showed enhanced germination even at suboptimal temperatures. Pukacka et al. (2011) established the protective role of Se in recalcitrant seeds of *Acer saccharium* subjected to desiccation. Soaking of the seeds in solution of sodium selenite for 6 h increased the seed viability during drying. However, studies in which the inhibitory effects of Se with regard to germination are also reported. Carvalho et al. (2003) however, reported a significant decrease in germination of seeds of tomato, lettuce, and radish plants. These seeds were obtained from the plants grown in soil supplemented with different concentrations of Se. These seeds were tested for their germination ability and it was concluded that supplementation of more than 29 mg/kg soil will cause adverse effect on seed germination. Therefore, protective role of Se on seed germination depends upon the optimum concentration of Se. Two tomato cultivars, Shuangfeng 87-5 and Jiahe-2, were soaked under different concentrations of Se to determine the effect of Se on tomato seeds under 100 mmol/L NaCl stress. Low concentration of Se (0.004 mmo1/L) improved germination rate, germination index, and vigor in tomato cultivars by ameliorating the adverse effects of salt stress (Guangquan et al., 2010). The enhanced germination rate in *Brassica rapa* L. was observed by Lyons et al. (2009) when seeds were treated with Se. It has been reported that Se triggers the ethylene production and improved the germination rate by breaking dormancy. Pinheiro et al. (2008) reported the role of Se compounds in breaking the dormancy of scarified seeds of *S. humilis*. Se-stimulated seeds showed enhanced germination rate and produced great amounts of ethylene. Abbas (2012) observed that the low concentrations of Se (3 and 6 mg/L) improved germination (plumule length and radical length) under cold stress in sorghum plants. On the other hand, application of high Se concentration (12 mg/L) significantly reduced the germination. Pretreatment of Se reduced the negative effects of salinity stress on seed germination by promoting various seed germination characteristics such as seed germination percentage, germination rate, length and dry weight of radicle and plumule, and root/shoot dry weight ratio in chick pea subjected to salt stress (Shahbazi et al., 2013). Thus, low concentrations of Se promote the seed germination and seedling growth, whereas high concentration of Se reduced the germination rate indicate that there is positive correlation between concentration of Se and the inhibition rate of seed germination.

4.3 DELAY IN SENESCENCE

Se has the ability to delay senescence and also promote the growth of aging seedlings in lettuce and ryegrass (Hartikainen and Xue, 1999; Xue et al., 2001). The antioxidative capacity of the senescing plants is strengthened by addition of Se, which prevents the reduction of tocopherol concentration and enhances superoxide dismutase (SOD) activity (Xue et al., 2001). It was suggested by Hartikainen et al. (2000) that senescence processes are partly delayed because of enhanced antioxidation, which is associated with an increase of glutathione peroxidase activity. In ryegrass (*L. perenne*), decrease in lipid peroxidation (decreased thiobarbituric acid reactive substances) has a connection with Se-induced increase in GPOX activity. The positive relationship between the Se concentration and GPOX activity suggests the presence of Se-dependent GPOX. According to Moussa and Ahmed (2010), Se treatment ameliorated adverse effects of induced senescence by increasing activity of antioxidant enzymes. It has been reported that Se delayed senescence in 1-month-old seedlings of faba beans by increasing activity of antioxidant enzymes such as catalase and peroxidase and also chlorophyll content, total protein, free amino acids, and photosynthetic efficiency as compared with control. Pezzarossa et al. (2014) studied the effect of Se on fruit composition and ripening physiology of tomato plants grown in hydroponics. Se in the form of selenate was added, and it was observed that Se lowered the β-carotene content, which may be associated with a general delay of ripening. The authors suggested that modulation in fruit development and ripening are related to antioxidant and antisenescence properties of Se thereby improving the postharvest commercial life.

4.4 ENHANCED YIELD

Se has been widely reported to positively affect the yield of various plants necessary for human consumption. Germ et al. (2005) reported the response of *Cucurbita pepo* to foliar spraying of Se at concentration 1.5 mg/L and ambient UVB radiation. A stimulatory effect on the yield of pumpkin fruits was observed. Similar observations were made on ryegrass (Hartikainen et al., 2000), lettuce (Xue et al., 2001), and potato (Turakainen et al., 2004). The increase in yield in pumpkins was much higher in plants that were exposed to UVB-excluded solar radiation together with Se spray than those that received either treatment independently. The Se-treated potato plants have also been reported to produce high tuber yields than control plants. Also, it was observed that there is higher carbohydrate accumulation in tubers, young upper leaves, stolons, and roots. Se treatment also improved processing and storage quality in the tubers (Turakainen et al., 2004; Turakainen, 2007). Increase in yield of *B. rapa* seeds that were produced from Se-treated plants was observed by Lyons et al. (2009). An average of 43% more seeds were produced in Se-treated plants as compared with control plants. Hu et al. (2003) carried out foliar applications of selenite or selenate and determined the influence of Se on the yield and quality of green tea leaves. Significant increase in numbers of sprouts and the yield were reported because of Se application. It also promoted tea tree sprouting in advance and increased yields of green tea. Improvement of grain yield with the application of Se has been reported under stressful conditions. Djanaguiraman et al. (2010) observed the effect of Se on grain yield of sorghum plants grown under high-temperature stress in a controlled environment. Se application enhanced the activity of antioxidative defense system to overcome the adverse effects of stress. High-temperature stress increased oxidant production and membrane damage and thus reduced the grain yield by decreasing the chlorophyll content and photosynthetic rate.

Spraying of plant with sodium selenate (75 mg/L) decreased membrane damage by enhancing the antioxidant defense system and resulted in higher grain yield. A similar result was also observed by Sajedi et al. (2011) in maize (*Zea mays* L.) plants under water-deficit stress. Application of Se during vegetative growth and dough stage under water-deficit stress conditions increased the grain yield. Soleimanzadeh (2012) also observed that Se application improved the yield in sunflowers (*Helianthus annuus* L.) subjected to water stress. Plants with higher levels of Se showed higher resistance to water stress conditions; therefore, the yield increased. Zanjani et al. (2012) studied the effects of zeolite and Se application on some physiological traits and oil yield of medicinal pumpkin (*C. pepo* L.) under drought stress. Zeolite and Se application in dry lands proved helpful for oil yield improvement. Cartes et al. (2010) showed that Se in the form of selenite ranging from concentration of 5 to 60 g/ha enhanced the plant yield progressively by increasing the Se uptake in Nui, Aries, and Quartet cultivars of ryegrass. Nezami and Bybordi (2012) studied the effect of different amounts of Se (0, 3, and 6 kg/ha) on yield and yield components of two canola cultivars SLM046 and Okapi. Se application increased canola seed yield by affecting yield components, especially silique number and seed number per silique, in both the cultivars. The SLM046 cultivar showed more positive response to Se (6 kg/ha) application in comparison to Okapi. Se application on potato plants was reported to promote plant growth, tuber yield, and quality over the control treatments by enhancing protein, starch, total carbohydrates, nitrogen, phosphorous, and low-dose potassium percentages in shoots and tubers (Yassen et al., 2011). Se-treated plants enhance the distribution of photoassimilates for tuber growth, acting as a strong sink for both Se and carbohydrates. Yield of lettuce plants was also found to increase with the application of Se acting as an antioxidant (Ramos et al., 2010). This improvement in growth was a dose-dependent response and it was observed that low dose of Se enhanced plant yield. Poldma et al. (2013) reported that application of Se to leaves of *Allium cepa* helped in increasing the bulb size as well as yield. In another study on wheat, UVB exposure caused decrease in plant height, spike length, weight per spike, and grain yield; however, Se application significantly helped in overcoming the UVB inhibited yield of winter wheat plants (Yao et al., 2013). Foliar application of Se was found to be effective in drought-stressed wheat plants that showed an enhanced growth index upon exogenous Se treatment. It was established that an augmentation in physiological growth indices occurred by enhancing the grain number/spike, thereby increasing the yield (Teimouri et al., 2014). Similarly, the effects of Se on the yield components as well as grain and straw qualities of two cultivars of rice (Giza 177 and IET 1444) exposed to water-deficit condition were studied by Emam et al. (2014). It was found that Se pretreatments mitigated the adverse effects of drought and improved yield quality by increasing the levels of amylose, phenolic compounds, and flavonoid and oil content.

4.5 PHOTOSYNTHESIS

Chlorophyll aids in carrying out photosynthesis, a vital mechanism of plant physiology. Any change in environmental or nutritional status that has negative effect on the photosynthetic pigments is bound to harm the process. Se in many studies has been reported to affect chlorophyll positively. Pedrero et al. (2008) reported increase in chlorophyll content in *B. oleracea*. Concentrations of both chlorophyll a and b were shown to decrease because of cadmium stress. But when treatment of Se was given along with cadmium, chlorophyll concentrations were enhanced as compared with plants only treated with cadmium. Similar effects of Se were observed in potato plants subjected to light stress (Seppänen et al., 2003). Different light intensities for different time durations produced stress conditions in plants, thereby negatively affecting the concentrations of chlorophyll a and b and total chlorophyll content. Se supplementation to the growth medium enhanced the chlorophyll content, indicating the ameliorative effect of Se against light stress. Mane et al. (2011) studied the effect of Se on chlorophyll a, b, and total chlorophyll content in *Eichhornia crassipes*. Lower concentrations of Se enhanced the chlorophyll content, whereas higher concentrations had a negative effect. Yao et al. (2011) also studied the effect of Se on photosynthetic pigments in wheat seedlings. Wheat plants were subjected to UVB radiation, which greatly reduced the pigment contents. Se application increased the contents of chlorophyll a, b, and total chlorophyll in both control plants and UVB-treated plants.

Djanaguiraman et al. (2010) investigated the effect of Se on sorghum leaves subjected to high-temperature stress. The chlorophyll content decreased in the presence of high temperatures. Se when added to the growth medium increased the chlorophyll content, thereby increasing the photosynthetic rate. Similar results were observed by Liu et al. (2011) in barley seedlings. The content of chlorophyll was found to increase with Se concentration. The increase in pigment content was found to be dose- and time-dependent. Increase in Se concentration increased chlorophyll content; in addition, a constant increase in the content was observed up to 4 weeks after sowing. Foliar application of Se helped in overcoming drought induced stress in potato (*Solanum tuberosum* L.) by improving net photosynthesis, transpiration rate, and quantum yield of photosystem II (Germ et al., 2007a). Similar role of Se against Cd stress has also been reported in *B napus* seedlings (Filek et al., 2010). Fluorescence and electron paramagnetic resonance measurements were used to study influence of Se on photosystem activity of rape seedlings affected by Cd stress and the Cd-induced changes in the activity of both photosystems were partly diminished by Se treatment. Li et al. (2012) studied the influence of Se on photosynthetic characteristics of *Cardamine*, which became Se-enriched because of its growth in different seleniferous site conditions (hillsides, Se slag, streamsides, and streams). The plants grown in these conditions were reported to have higher photosynthetic ability because the Se concentrations in soil were suitable for the growth. Wei et al. (2011) studied the effect of Se on photosynthetic properties of *Vitis vinifera* under water stress. Water stress decreased the photosynthetic activity, which was further improved with the application of Se (0.5 mg/L). Lei and Wu (2011) observed the effect of Se on the photosynthesis and nonenzymatic antioxidants in *Cardamine enshiensis* leaves. The proper Se concentration (30–90 mg/L) improved the photosynthetic efficiency, increased the content of antioxidants, and inhibited lipid peroxidation in *C. enshiensis* leaves. Xiao et al. (2008) studied the effect of different concentrations of Se (0.1, 1, 10 mg/L) on characteristics of photosynthesis and content of soluble sugar of *Basella rubra* L. At low concentrations, net photosynthetic rate was reported to increase and then it decreased with increase of Se concentrations. Low temperature is one of the most important factors that limit photosynthetic activity.

It has been reported that chlorophyll a and b content was decreased in plants when plants were subjected to cold treatment. Abbas (2012) studied the exposure of wheat plants to low concentrations of Se (3 and 6 mg/L) that induced a significant increase in photosynthetic pigments contents at 4 °C. High concentrations of Se (12 mg/L) reduced the amount of pigments content in sorghum seedlings grown under cold stress as compared with controls. Pennanen et al. (2002) reported the increase in chlorophyll a and b contents of wheat seedlings that may be attributed to the effect of Se on protection of chloroplast enzymes and thus increasing the biosynthesis of photosynthetic pigments. Se-supplemented barley plants showed improvement in photosynthetic parameters such as stomatal conductance and CO_2 assimilation under water stress (Habibi, 2013). Similarly, foliar application of Se (50 and 150 mg/L) also increase the photosynthesis in *Olea europaea* L. subjected to drought stress

(Proietti et al., 2013). The effect of Se on photosynthetic rate and antioxidative metabolism of chloroplasts in tomato seedlings under NaCl stress were investigated by Diao et al. (2014). It was reported that Se alleviated the salt-induced oxidative stress by regulating the antioxidant defense systems in the chloroplasts of tomato seedlings, which is associated with the improvement of the photochemical efficiency of PSII and thus maintains higher photosynthetic rates. Further, Se in the form of selenate also showed the improved photosynthesis under drought conditions in two durum wheat (*Triticum durum* L.) genotypes. Se treatment increased the photosynthetic rate up to 2.3-fold in both the genotypes and thus improved the photosynthesis (Hajiboland et al., 2014).

Apart from plants, the effect of Se was also studied on photosynthetic rates of lower organism, namely *Euglena* which is photosynthetic in nature. Ekelund and Danilov (2001) studied the role of Se in "light-enhanced dark respiration" and photosynthesis in *Euglena gracilis*. It was exposed to UV radiation and then grown on growth medium containing different concentration of Se. Also, *E. gracilis* was subjected to different light intensities and period of darkness between light pulses. Both UV exposure and irradiance (higher than 600 μmol m^2/s) reduced the photosynthetic rate- and light-enhanced dark respiration. However, after UV treatment and 24 h of recovery, the presence of Se led to an increase in photosynthesis and light-enhanced dark respiration at higher irradiances. The results indicate that Se might play a role in the repair mechanisms in *E. gracilis* after UV treatments. Similar observations were also made in certain algal species. Chen et al. (2008) studied the Se induced changes in content of photosynthetic pigments in *Spirulina platensis* when exposed to various Se concentrations. Low Se concentrations ($\leq$150 mg/L) increased the content of photosynthetic pigments and activities of antioxidant enzymes whereas high Se concentrations ($\geq$175 mg/L) decreased the contents of photosynthetic pigment and induced the lipid peroxidation. Similarly, effects of Se on photosynthetic pigment contents and growth of *Chlorella vulgari* has been also studied by Chen et al. (2005). The effects of selenate on *Chlamydomonas reinhardtii* growth inhibition was studied by Geoffroy et al. (2007). It was suggested that growth inhibition at high Se concentrations may result from impaired photosynthesis. Thus, low concentration of Se was favorable for improving photosynthetic efficiency. Schiavon et al. (2012) studied the impact of Se on photosynthetic pigments in *Ulva* species, a green macroalga. The content of chlorophyll a and b in thalli cultivated for 10 days in the presence of selenate did not change significantly, although at the third day an increase of these pigments was observed at the highest selenate concentrations. On the contrary, the level of carotenoids significantly increased in the thalli exposed to selenate both at 3 and 10 days of exposure, especially when Se was supplied at high doses ranging from 50 to 100 μM.

4.6 RESPIRATORY POTENTIAL

Respiratory potential refers to the rate of carbon dioxide production by plants and is usually measured as terminal electron transport system (ETS) activity. Many studies on different plants indicated the increase in respiratory potential in the presence of Se. Smrkolj et al. (2006) observed enhanced respiratory potential in young plants of *Pisum sativum* when subjected to Se treatment. Similarly, Germ et al. (2007c) reported increased respiratory potential in chicory (*Cichorium intybus*) plants when specific concentration of Se was applied to young seedlings. *Eruca sativa* also showed similar results in the presence of Se (Germ and Osvald, 2005). Higher respiratory activity was observed in *B. rapa* when treated with Se as compared with the control plants. Leaves showed an increase of 63%, whereas flowers showed a 200% increase (Lyons et al., 2009). An increase of cellular respiration in mung bean

(*Phaseolus aureus* Roxb.) plants grown under 0.5 and 0.75 ppm Se concentration was observed by Malik et al. (2011). Possible reasons for higher ETS activity can either be its relation to increased activity of glutathione peroxidase in mitochondria or energy required by plants to repair damage caused by Se (Germ et al., 2007a). But it has been shown in plants such as ryegrass and lettuce that Se exposure leads to higher glutathione activity (Xue and Hartikainen, 2000; Hartikainen et al., 2000; Xue et al., 2001; Cartes et al., 2005). Damage from Se can be the reason that Se can mimic sulfur and replace it in certain amino acids such as methionine and cysteine, thereby changing the conformation of proteins and hence disturbing their catalytic activity (Brown and Shrift, 1982).

There are some reports also in which respiratory potential was lowered under Se presence. The general metabolic activity of individual organisms can be assessed from ETS activity in mitochondria. Under stress conditions, organisms increase their need for energy. The ability to cope with stress in vital plants is related to respiratory potential or ETS activity of certain tissues. When stress is too immense, the antioxidant metabolism is defeated. Strong stress causes reduced vitality of tissues that reflects in lower respiratory potential (Germ and Gaberščik, 2003). The effect of Se on respiratory potential in soybean (*Glycine max*) was studied by Mechora and Germ (2010) and it was observed that Se lowered the respiratory potential. Similarly, Akladious (2012) reported the lowering of respiratory potential induced by Se in wheat seedlings under low-temperature stress. Mechora et al. (2011) reported the effects of Se on electron transport system activity in red cabbage plants and established that fertilizing the soil with Se did not show any effect on the ETS activity. Germ (2005) studied the effects of ambient and filtered solar UVB radiation and of Se treatment on respiratory potential measured by ETS activity in pumpkins (*C. pepo* L.). Foliar treatment with Se, however, had no effect on the ETS activity in pumpkin plants in July, whereas it lowered ETS activity in August in solar UVB-exposed plants. Kreft et al. (2013) studied the influence of Se on mitochondrial activity of Tartary buckwheat. The plant of Tartary buckwheat were sprayed with 10 mg/L Se (VI) at the beginning of flowering. It was observed that Se-treated progeny plants showed higher ETS activity as compared with the control.

4.7 METABOLITE ACCUMULATION

Many researchers have indicated the role of Se in metabolite accumulation in food crops. Turakainen et al. (2004) studied the effect of Se on potato plants and showed that carbohydrate accumulation was enhanced when Se supplementation was provided. At the highest Se concentration (0.3 mg/kg), the maximum soluble sugar concentration was observed in upper leaves four weeks after planting, whereas in roots and stolons it was at maturity. Pennanen et al. (2002) earlier had suggested that Se-mediated plant growth in lettuce is due to increased starch accumulation in chloroplasts. Seppänen et al. (2003) had also demonstrated in their report that Se has positive effects on recovery of potato from photo-oxidative and paraquat-generated oxidative stress. The mechanism, however, is unknown but it is suggested that Se aids in protecting chloroplast during stress, which further protects the plant.

Glucosinolate concentration, on the contrary, was found to decrease upon Se fertilization to *B. oleracea*. Glucosinolates are sulfur-containing compounds found predominantly in the members of the family *Brassicaceae* and their degradation products are known to provide protection against cancer. A possible reason suggested for decreased glucosinolate concentration is that uptake of Se and sulfur is competitive and hence Se fertilization when it inhibits sulfur uptake, which will thereby aid in low production of glucosinolates (Charron et al., 2001). The plant growth is increased by Se is for the reason that of the increased starch accumulation in chloroplasts (Pennanen et al., 2002). Sharma et al.

(2008) investigated decrease in lipid content and increase in the fatty acid content in rapeseed grains grown on seleniferous soil. The decrease in lipid content might be because of breakdown of lipids into free fatty acid proportions and thus increasing the fatty acid content and decreasing lipid content.

The concentration of γ-tocopherol was found to increase, which resulted in minimizing the Cd toxicity in broccoli (*B. oleracea*) plants under Se supplementation (Pedrero et al., 2008). Sharma et al. (2010) discovered in their experimentation on rapeseed grown in greenhouse in an alkaline sandy loam soil treated with different levels of selenate and selenite ranging from 0 to 4 mg/kg that the accumulation of Se in leaves has led to the significant increase in lipid peroxidation, chlorophyll, vitamin C, and free amino acids and lowered the content of phenols, total soluble sugars, and starch. *Vicia faba* was grown on Hoagland medium containing 50 μM Pb and 1.5 and 6 μM Se for 2 weeks and it was observed that Se at low levels improved cell viability and enhanced the lipid peroxidation and cell membrane injury (Mroczek-Zdyrska and Wójcik, 2012). The seeds of *Sorghum bicolor* were given 6 h of presoaking in 4 concentrations of sodium selenate (0, 3, 6, and 12 mg/L) before sowing, and then were given cold stress (4 or 8 °C) to seedlings during the germination period for 7 days and after it allowed to restore their form at 25 °C for 3 days. In these plants, the low concentration of selenate showed the increase in the growth and levels of chlorophylls, anthocyanins, sugar, proline, ascorbic acid, and enzymatic activities but it was toxic to plant at higher concentrations. The nonenzymatic antioxidants such as ascorbic acid and carotenoids were also increased by selenate (Abbas, 2012). *Lepidium sativum* plants were given hydroponic exposure to 0, 0.5, and 2.0 mg/L of Cd and Se and it appeared in the study of Barrientos et al. (2012) that the presence of Se (IV) in the growth medium had negative effect on the grade of fatty acids unsaturation in garden grass. Bansal et al. (2012) revealed that the accumulation of Se in the grains of *Brassica* resulted in significant increase in contents of reducing sugars, starch, glucosinolate, and free and sulfur-containing amino acids and a decrease in lipid content. Three different concentrations of Se (1, 2, and 4 mg Se kg/L) were used and different parameters showed significant increase with concentrations of 2 and 4 mg/kg of soil. The 3-year-old plants of Assam tea were potted in the Se- and Al-drenched soil at concentrations of 0, 50, and 500 ppm and leaves were harvested every 30-day interval after the Se and Al treatments. The results showed that Se accumulation enhanced the proportions of different lipid classes such as glycolipids, sterols, and free fatty acids, whereas decrease in the triglyceride content, the maximum increase was expressed by the tea leaves, which received the 50 ppm concentration of Se. Aluminum and Se both promoted the ascorbic acid, chlorophylls, carotenoids, total phenolics, and proline contents of tea leaves (Sae-Lee et al., 2012). In *Lycium chinense*, treatment of Se at various concentrations (0–0.5 g/kg) significantly enhanced the content of chlorogenic acid, chlorophyll, and carotenoids (Dong et al., 2013). Another report by Owusu-Sekyere et al. (2013) established that by altering redox potential, Se upregulated the carbohydrates metabolism in *Medicago sativa*. The plants were grown on various concentrations of Se (0, 1, 5, 10, and 15 μmol/L Na_2SeO_4) and N (2 and 10 mmol/L) concentrations and it was found that application of Se increases the content of soluble sugar in leaves and stems as well as increase in the starch concentration was noted in the stems of Se-treated plants.

It was reported by Schiavon et al. (2013) that application of low doses of Se (5–10 μM) via roots in tomato plants led to the increased synthesis of phenolic compounds. Similar results were reported in potato tubers by Lei et al. (2014). Application of Se significantly increased the accumulation of various phenolic compounds such as chlorogenic acid, caffeic acid, malvidin-5-glu-3-dirhamnose-glucose, caffeic acid-acetylrhamnose ester, and caffeic acid-prenylrhamnose ester. The purple potato tuber formation was also enhanced, thus making the Se-enriched purple potato as a better functional food.

Se has also been found to improve the nutritional quality of rice grains subjected to drought stress. The rice plants were drought stressed by withholding the water and allowed to grow until full maturity and then analyzed for various parameters such as grain appearance, cooking quality, and nutritional values. It has been observed that under drought stress, total soluble carbohydrate, starch, phenols, oils, and flavonoids content decreased, whereas the coapplication of Se has mitigated the adverse effects of drought and improved the nutritional quality (Emam et al., 2014). In a recent study, Se treatment in the form of foliar spray was reported to enhance the antioxidant system as well as improvement in iron and sodium uptake in wheat plants under drought stress. The increase in the total soluble sugars and amino acids also increased with Se supplementation, which further resulted in improved yield (Nawaz et al., 2015). It has also been observed that application of Se affects the content of phenols and flavonoids in turnip plants. Further, it has been noticed that Se upregulates the genes expression involved in metabolic and biosynthetic pathways (Thiruvengadam and Chung, 2015). The role of Se and auxin was noticed under arsenic metal stress in rice plants. It has been observed that coapplication of Se and auxin improves the accumulation of various metabolites such as protein, cysteine, and proline compared with arsenic-treated rice plants alone (Pandey and Gupta, 2015).

4.8 DEFENSE AGAINST FUNGAL PATHOGENS AND HERBIVORE ATTACK

Se accumulation in plants has also played an important role in inhibiting infection by fungal pathogens. Indian mustard (*B. juncea*) is a well-studied plant in accordance to Se because of its fast-growing properties and higher transformation potential. *Brassica juncea* treated with Se was investigated for defense against fungal pathogens including the root and stem pathogen *Fusarium* sp. The plant was grown with or without Se (100 μM sodium selenate) and then treated with a suspension of *Fusarium*. The plants treated with *Fusarium* showed less weight gain than the nontreated plants. Se-treated plants showed comparatively more growth than nontreated plants upon infection. *Fusarium* also showed decreased growth when media was suspended with Se (Hanson et al., 2003). Another fungal pathogen, *Alternaria brassicicola*, showed similar results with *B. juncea*. The plants treated with Se showed significantly fewer leaf lesions than the plants not treated with Se. Also, fungi showed reduced on media supplemented with Se (Hanson et al., 2003). Both fungal pathogens showed the same response to Se in terms of reduction in growth and infection, indicating that Se in plants may function as antifungal compounds.

Apart from preventing fungal pathogens, Se has also been shown instrumental in restraining the herbivore attack on plants. Hanson et al. (2003) also studied the effect on Se application on *B. juncea* using caterpillars (*Pieris rapae*) and snails (*Mesodon ferrissi*). Se-containing plants were reported to be lethal to the caterpillars, whereas snails showed no symptoms of toxicity, although the Se levels in the tissues of both the herbivores were found to be the same. In another study by Hanson et al. (2004), it was shown that *B. juncea* on receiving Se treatments precluded the green peach aphid (*Myzus persicae*) from colonizing the treated plants. The aphids could detect Se-containing plants and avoided them in choice feeding experiments. *Brassica juncea*, which is also commonly used for phytoremediation, was revealed to be protected from fungal pathogens in leaves by the high Se levels (Hanson et al., 2003, 2004). *Astragalus bisulcatus*, which is an Se-hyperaccumulating plant species that grows naturally on seleniferous soils in the western United States and often on prairie dog colonies, is found to be protected from a variety of herbivores and is toxic to livestock (Rosenfeld and Beath, 1964; Quinn et al., 2008). It was reviewed by Barillas et al. (2011) that Se showed the ability in protecting plants from fungal pathogens and from a number of invertebrate and vertebrate herbivores because of both deterrence and toxicity. Galeas et al. (2008) has found during their field study that Se hyperaccumulator plants have been less harbored by the arthropods as compared with the non-hyperaccumulating plant species. In field conditions, study was made to test that Se can act as an arthropod deterrent. From various Se hyperaccumulator (*A. bisulcatus* and *Stanleya pinnata*) and nonhyperaccumulator plants (*Camelina microcarpa*, *Astragalus americanus*, *Descurainia pinnata*, *Medicago sativa*, and *Helianthus pumilus*) arthropod loads were measured over two growing seasons. The Se hyperaccumulating plant species were harbored significantly with fewer arthropods species in comparison to the nonhyperaccumulator species. Elemental defense hypothesis is supported by this result and it is proposed that invertebrate herbivory has contribution in the evolution of Se hyperaccumulation. The *A. bisulcatus* plants were grown in pots with two concentrations of Se (20 and 2 μM) and *S. pinnata* plants were grown in 20-μM Se concentrations and without Se as control by Quinn et al. (2010). It was investigated that Se protected the hyperaccumulators *S. pinnata* and *A. bisulcatus* from two herbivores, the two-spotted spider mite and the Western flower thrip. In a survey, Wangeline et al. (2011) collected rhizosphere fungi from different seleniferous and non-seleniferous habitats. Isolated rhizosphere fungi collected from seleniferous habitats were found significantly more Se tolerant than those from a nonseleniferous habitat. Reduction in herbivore attack by birds and aphids in Se-treated *Raphanus sativus* was also reported by Hladun et al. (2013).

4.9 PROTECTION AGAINST OXIDATIVE STRESS

Reactive oxygen species (ROS) are oxygen-containing radicals that are capable of independent existence with one or more unpaired electrons. However, the term ROS is most often expanded to include reactive oxygen-containing compounds without unpaired electrons, such as hydrogen peroxide (H_2O_2) and singlet oxygen ($^1O_2$) (Halliwell and Cross, 1994; Halliwell and Gutteridge, 1984). The consumption and utilization of oxygen in various physiological processes results in the generation of ROS. These ROS are then neutralized by the plant systems and when generation of ROS exceeds the system's ability to neutralize and eliminate them, stress conditions appear and these are defined as oxidative stress conditions (Sies, 1985, 1986; Sies and Cadenas, 1985). This unevenness in production and scavenging of ROS may occur because of lack of antioxidant capacity, which is further because of the disturbance in production, distribution, or because of excess ROS. Excess of ROS can damage cellular lipids, proteins, or DNA, thus inhibiting signal transduction pathways and normal cellular functions. Se has been demonstrated in much research (a detailed discussion appears in the following section) to promote antioxidant capacity in plants subjected to various types of stresses (Hartikainen and Xue, 1999; Djanaguiraman et al., 2005; Peng et al., 2002).

5. EFFECT OF SELENIUM ON THE ANTIOXIDATIVE DEFENSE SYSTEM

Se has been reported to affect the enzymes involved in defense system of plants. The defense system of plants gets activated in various types of stresses and Se has been reported to aid in strengthening the antioxidative enzyme activities. The antioxidative role of Se is mainly based on organoselenium because it is the bioavailable form and readily incorporated into amino acids and proteins as compared with the inorganic forms. However, the concentration at which Se has beneficial effects varies with the plant species and also with the type of stress the plant faces. Conditions such as salinity, drought, chilling temperatures, and heavy metal contaminations in soil and water are major abiotic stresses plants usually face. Many recent studies have demonstrated the protective role of Se in number of plant species subjected to such conditions.

An excess amount of salt in soil, for example, is one of the major causes of ROS production and damage to growth and development in plants. Kong et al. (2005) studied the effect of Se on defense system of sorrel seedlings. Lower concentrations of Se induced growth and increased activities of SOD and peroxidase (POD) enzymes and beneficial effects declined at higher concentrations. Similar results were observed in cucumber plants under salt-stressed conditions in which Se treatment of 5 and 10 μM enhanced growth, photosynthetic pigments, and proline contents, and reduced MDA content significantly, thereby reducing lipid peroxidation (Hawrylak-Nowak, 2009). It was observed by Feng and Wei (2012) that arsenic hyperaccumulator *Pteris vittata* L. (Chinese brake fern) is also an Se accumulator. They investigated Se accumulation mechanisms, especially the antioxidant responses of the plant to six levels of selenite (0, 1, 2, 5, 10, and 20 mg/L). Their results showed that without any significant loss in biomass and without any visible toxic symptoms, Chinese brake fern could accumulate a large amount of Se. The amount of Se taken up by roots was more than that by fronds. The production of MDA was decreased by addition of 2 mg/L of Se but was enhanced by the addition Se up to 5 mg/L or more, thereby suggesting an antioxidant role of low dosages of Se. Low dosage of Se up to 5 mg/L or less than it, increased the activity of ascorbate peroxidase (APOX), POD, and catalase (CAT) whereas a dosage more than 5 mg/L decreased their activity. Se concentration up to 5 mg/L or more stimulated enzyme glutathione reductase and increased the concentration of glutathione. SOD activity was also increased by 20 mg/L Se. Hence, their results suggest that reduced glutathione, SOD, and glutathione reductase (GR) played major roles and enzymes such as CAT, APOX, and peroxidase played limited roles in the accumulation of Se in Chinese brake fern. Hajiboland and Keivanfar (2012) observed that vegetative and reproductive growth was stimulated by Se supplementation in *B. napus* L. (canola plants). The effect of exogenously applied Se at concentrations 0, 2, 4, 8, and 16 μM was investigated on the activity of antioxidative enzymes in melon (*C. melo* L.) seedlings under stress of 100 mM NaCl (KeLing et al., 2013). The salt stress increased the activity of antioxidative enzymes such as CAT, SOD, and peroxidase. The adverse effects of salt stress in melon seedlings were ameliorated by supplementation of Se. The amelioration of NaCl stress in the seedlings varied depending on the concentration of Se. At lower concentrations (2–8 μM), it enhanced growth and acted as antioxidant by increasing the peroxidase and SOD activities and by preventing lipid peroxidation. The beneficial effects of Se on growth diminished at higher concentrations (16 μM). However, there was an increase in the activity of CAT. The results showed that the supplementation of Se induced positive effects on the growth and development of melon seedlings under salt stress.

Similar effects were observed when Se was supplemented to water-stressed plants. Water stress, particularly drought stress, is a cause of number of physiological and biochemical changes in plants. This type of stress also releases ROS as its first response (Mittler, 2002; Shigeoka et al., 2002; Apel and Hirt, 2004). It was suggested by Kuznetsov et al. (2003) that Se helps in regulating the water status of drought-stressed plants by increasing the water uptake capacity of root system, thereby indicating a physiological change by Se application. Many reports also demonstrate changes in biochemical composition for scavenging the increased production of ROS. Soleimanzadeh (2012) studied *Helianthus annus* L. (sunflower) response to Se treatment under water stress. Sunflower plants under water stress were given the treatment with four Se concentrations such as 10, 20, 30, and 40 g/ha. Plants under water stress and given Se treatment showed a significant increase in CAT, SOD, and GPOX activities as compared with control plants. Their results suggested that oxygen radicals and oxidative stress was produced in the plants because of water stress. Higher levels of Se helped in controlling the damage to membranes and in the scavenging of ROS by increasing the activity of scavenging system such as SOD, GPOX, and CAT. Yao et al. (2009) demonstrated that Se, when given in optimum concentration, boosted up the growth of wheat seedlings subjected to drought stress. Treatments of 1.0, 2.0, and 3.0 mg/kg soil significantly increased the chlorophyll content and peroxidase and CAT activities in drought-stressed wheat seedlings. Xiaoqin et al. (2009), on the other hand, found out that an extra Se supply to wheat plants under control conditions did not affect activities of CAT and peroxidase, but when Se treatment was combined with drought conditions the activities of the enzymes enhanced significantly. A detailed study by Hasanuzzaman and Fujita (2011) elaborated the defensive role of Se in *B napus* seedlings under drought stress. Se-pretreated seedlings when exposed to drought stress showed increase in ascorbate and glutathione contents. Also these seedlings showed significant increased activities of APOX, dehydroascorbate reductase, monodehydroascorbate reductase, GR, glutathione-S-transferase, GPOX, CAT, and glyoxylase I and glyoxylase II. The seedlings also showed decreased levels of hydrogen peroxide and lipid peroxidation. These results clearly suggested that Se application enhanced the tolerance of plants to oxidative damage induced by drought by enhancing antioxidant defense and methyl glyoxal detoxification systems.

Zanjani et al. (2012) studied the effect of Se and zeolite application on oil yield and physiological traits of *C. pepo* L. (medicinal pumpkin) under drought stress. Their results indicated that drought stress caused significant biochemical and physiological changes in these plants. Reduction in relative water content, stomatal conductance, total chlorophyll concentration, leaf soluble protein, and oil yield was observed. The increased activity of CAT indicated that the medicinal pumpkin possess an effective antioxidant defense mechanism to scavenge ROS. Because Se reduced oxidative conditions or free radical injuries by increasing antioxidant enzyme activities such as CAT, which ultimately have a determinate effect on plant cell and therefore, Se spraying increased relative water content and oil yield under drought conditions in *C. pepo*. Pukacka et al. (2011) also reported protective role of Se in recalcitrant *Acer saccharium* seeds exposed to desiccation conditions. Activities of GPOX and GR in embryonic axes of Se-treated seeds were found to be higher than untreated seeds under dry conditions. The effect of application of Se on physiological characteristics of drought stress–exposed barley plants (*Hordeum vulgare* L. cv. Rihane-03) was also reported by Habibi (2013). They concluded that supplementation of the drought-stressed barley plants with Se increased the activities of GPOX and CAT and lowered down lipid peroxidation. This led to the better protection of the water-deficit plants from oxidative damage. Similarly, Se pretreatment in drought-stressed *Triticum aestivum* plants decreased the content of H_2O_2 and MDA, whereas elevated the nonenzymatic antioxidants contents such as reduced glutathione and ascorbic acid and also the activities of antioxidant enzymes such as SOD and CAT. This in turn increased the membrane stability index and viability of roots (Ibrahim, 2014).

Few heavy metals are necessary components of various metabolic activities, but at higher concentrations they are instrumental in creating stress conditions leading to free radical generation. Some heavy metals are nonessential for the living systems and are toxic even at lower concentrations. They come in contact with plants via pollution from various resources in soil and water. Se has been reported in wide array of studies in increasing tolerance of plants against stress caused by heavy metals. Vorobets (2006) found an alteration in reduced glutathione content in sunflower plants upon application of Se in presence of lead. Cd toxicity has also been found to be ameliorated with Se application in broccoli. Content of MDA was reported to be significantly reduced in the presence of Se. Also level of α-tocopherol was found to increase in the plant (Pedrero et al., 2008). α-Tocopherol has been suggested as an antioxidant that favors the scavenging of singlet oxygen species (Munne-Bosch and Alegre, 2002; Munne-Bosch, 2005). In *L. sativum*, Barrientos et al. (2012) studied the Cd and Se exposure impact on fatty acids, trace elements, and oxidative stress. An edible vegetable, *L. sativum*, also called garden cress, belongs to Se-accumulating *Brassicaceae* family. To study the protective role of Se to combat Cd toxicity in plants, *L. sativum* (cv. Orgodowa) was selected because of its sensitivity to Cd and expected feasibility to incorporate Se. They used a three-level factorial design to obtain hydroponically grown young plants in exposure to Cd in the form of cadmium chloride and Se in the form of sodium selenite and their mixtures. Se (IV) and Cd (II) concentration levels 0, 0.5, and 2.0 mg/L were used for applying treatments. They evaluated growth after 7 and 14 days of exposure. They found that in the garden cress, oxidative damaged occurred from Cd exposure and that Se alleviated such adverse effects, by increasing the unsaturation ratio of fatty acids and by lowering MDA levels. When present at higher concentrations, 2.0 mg/L Se (IV), Se itself exerted similar phytotoxic effects as Cd. Kumar et al. (2012) has observed that Se efficiently ameliorated Cd toxicity by decreasing the accumulation of ROS ($\bullet O_2^-$ and H_2O_2). The Cd toxicity slowed down the growth of alga by activating the ROS but exogenously applied Se helped the alga in downregulating the ROS production.

Parameters related to oxidative stress showed improvement under simultaneous exposure to Cd (II) and Se (IV), it suggested that the protective role of Se depends on the interaction between the two elements. Filek et al. (2008) observed Se-mediated stress tolerance in rapeseed subjected to Cd stress. Concentration as low as 2 μM was found to be effective in resisting stress conditions. Similar results were shown by Srivastava et al. (2009) in *Pteris vittata* under As stress. Se significantly inhibited lipid peroxidation and increased the levels of glutathione reduced and thiols. Cd stress tolerance by Se was also studied by Sun et al. (2010) in garlic plants. Cartes et al. (2010) studied the antioxidative effect of Se on ryegrass plants under aluminum stress. A significant reduction in oxidative stress was observed through dismutation of superoxide radicals and activation of peroxidase. Se also aided in reduction of lipid peroxidation. El-Ashry and Mohamed (2012) applied different concentrations (10^{-6}, 10^{-4} M) for 2 h of some antioxidant metals such as calcium (Ca^{2+}), zinc (Zn^{2+}), and selenite (Se^{4+}) to study their chemoprotective effects on the mitotic index, chromosomal aberrations, root growth, and micro-nuclei of *Vicia faba* plants under Cd toxicity. It was found that root growth was inhibited in all the three concentrations of Cd^{2+} (10^{-5}, 10^{-3} M). The respective root growth percentage was 10.82%, 6.03%, and 4.01%. In all Cd^{2+} concentrations, treatment applied along with the selenite (10^{-6}, 10^{-4} M), there was seen improvement in the percentage of root growth although it was less than the Cd^{2+}. The selenite concentration showing highest percentage of root growth was 10^{-6} M. Also, the mitotic index showed significant reduction after treatment with all concentrations of Cd^{2+} (10^{-5}, 10^{-3} M). For all concentrations of Cd^{2+}, the different concentrations of Se (10^{-6}, 10^{-4} M) showed increase in mitotic index, especially the 10^{-6} M of selenite applied along with the 10^{-5} M Cd^{2+} treatment. This improvement was due to the protective role played by selenite. In an experiment performed by Yuan et al. (2013), application of Se (1.0 mM) to *Coleus Blumei* Benth plants under Pb stress mitigated the negative effects of Pb. Se acted as an antioxidant by increasing the glutathione content, which in turn inhibited the lipid peroxidation. Its application under Pb stress also increased the GPOX and CAT, whereas it decreased the POD and SOD activities in roots of the plants. Another study conducted on Cd-stressed sunflowers also establishes the ameliorative role of Se. Decreased biomass, chlorophyll, and carotenoids concentrations and enhanced accumulation of Cd in roots and shoots of plants were seen under exposure to 20 μM Cd. The presoaking treatment of seeds with different concentrations of Se (5, 10, and 20 μM) has reduced the phytotoxic effects of Cd on growth. CAT, APOX, and GR increased, whereas reduced SOD and GPOX activity was noticed in sunflowers under Cd stress with Se supplementation (Saidi et al., 2014). Similarly, the ameliorative effect of Se on As stress in *Oryza sativa* L. was studied by Kumar et al. (2014). Se supplementation reduced the oxidative stress induced by As in the roots and shoots of the plants. Supplementation of Se with As increased the activities of CAT, APOX, and GPOX. Cd-induced stress in *Bechmeria nivea* L. was also reported to be reduced by the exogenous application of Se. Spraying of Se at a low concentration (1.2 μmol/L) on the plants under the Cd stress, increased the activity of SOD and POD in the leaves of the plants (Wang et al., 2014).

Se treatment modulated the response of *Nicotiana tabacum* L. plants to Cd toxicity (Liu et al., 2015). Se supplementation reduced the Cd-induced accumulation of H_2O_2, $O_2^{\cdot -}$, and MDA. Activities of GPOX, APOX, GR, CAT, SOD, and POD, which were depressed by Cd alone treatment, were elevated by the Se supplementation. The interactive effects of Se (0–5 mg/L) and As (0–5 mg/L) in flue-cured tobacco (FCT) was observed by Han et al. (2015). They had observed that application of moderate concentration Se helps in the amelioration of As toxicity by improving the antioxidant capacity. Se has been found to reduce the FCT growth but effectively enhanced the activity of SOD and POD. The hydroponic experiments on *Bechmeria nivea* under cadmium stress have showed the ameliorative properties of Se. It has been observed in these experiments that application of Se reduces the content of Cd in ramie plants. Other than this, activities of various antioxidant enzymes such as SOD, POD, and APX were enhanced by the application of Se, whereas MDA content and hydrogen peroxide content was observed to decrease (Tang et al., 2015). Supplementation of growth medium of *B. campestris* L. with Se effectively declined the Cr-induced elevated amounts of $O_2^{\cdot -}$ and lipid peroxides and increase in the activities of POD and SOD was also observed (Qing et al., 2015).

Temperature is a crucial parameter that determines germination of seeds and proper growth and development of the plant. Temperatures higher or lower than the optimum temperatures can damage the growth rates. Se has also been reported to protect plants from oxidative damage caused by both heat stress and chilling stress. Chen and Sung (2001) reported enhanced germinability of cold-stressed bitter gourd (*Mimordica charantia*) seeds when primed with Se solution. It is shown that decreased temperatures reduced the seed performance, probably by peroxidative injury. But Se treatment partially restored the germinability of the seeds and this improved seed performance was linked to free radical scavenging activity of antioxidative enzymes whose activities considerably increased. GPOX activity increased up to an Se concentration of 10 mg/L, whereas all other enzymes such as SOD, CAT, peroxidase, APOX, GR, and glutathione-S-transferase showed maximum activity either at 1 or 2 mg/L concentrations of Se. MDA content was also found to decrease in stressed seeds upon Se application. Seppänen et al. (2003) grew potato plants in pots with Se concentrations 0, 0.5, and 2.0 ppm that were exposed to 600 mmol/m^2/s light intensity at low temperature (4 °C) for photooxidative stress. The results suggested by the experiments of a potato plant showed that Se acts as antioxidant and it activates protective mechanisms, which can alleviate oxidative stress in the chloroplasts. Chu et al. (2010) demonstrated similar results in wheat seedlings subjected to cold conditions and Se treatment ameliorating the stress symptoms. Se application significantly reduced MDA content and enhanced anthocyanins, flavonoids, and phenolic compounds. Activities of peroxidase and CAT were also reported to get increased in the presence of Se. Another study conducted by Hawrylak-Nowak et al. (2010) on cucumber plants also suggested a defensive role of Se against short-term chilling stress.

Se supplementation allowed no change in chlorophyll and carotenoid contents even under suboptimal temperatures. Also, increased proline content as compared with control plants was observed in the leaves. MDA content of Se-treated roots was found to be lower than the stressed plants. High-temperature stress and its Se-mediated amelioration were studied by Djanaguiraman et al. (2010) in *Sorghum bicolor.* It has been hypothesized that Se application partly alleviates high temperature–induced oxidative damage along with negative impacts of high temperatures on plant physiology, growth, and yield. A decrease in content of chlorophyll, photosynthetic rate, and anti-oxidative enzyme activities and increase in membrane damage and oxidant production was observed in heat-stressed plants, thereby resulting in lower yield. Se application decreased membrane damage by significantly enhancing the antioxidant defense of the plant system. Akladious (2012) demonstrated the protective role of Se in wheat under low-temperature stress. In wheat, low-temperature stress inhibited the growth, soluble sugars, chlorophylls, and antioxidant enzyme activities and increased membrane damage and oxidant production. Different parameters such as growth, anthocyanin, sugars, chlorophylls, enzyme activities, and proline contents were enhanced and membrane damage were decreased by soaking the seeds in Se solution for different times; all of this resulted from the enhancement of antioxidant defense system coupled with the generation of novel protein bands. In a study carried out by Abbas (2013), it was reported that Se-treated sorghum plants showed enhanced tolerance toward chilling stress. It was found that at lower concentrations of selenate (3 and 6 mg/L), the growth and the levels of anthocyanins, chlorophylls, proline, sugar, ascorbic acid, and activity of enzymes were enhanced. Carotenoids levels were decreased, whereas the levels of ascorbic acid were increased by selenate. Lipid peroxidation was reduced by lower concentrations of selenate (3 and 6 mg/L) as indicated by the low content of malondialdehyde. Different responses to selenate were shown by the activities of enzymatic antioxidants. In seedlings, selenate at lower levels (3 and 6 mg/L) enhanced the activity of guaiacol peroxidase, whereas the higher concentrations of selenate (12 mg/L) increased the activity of ascorbic acid peroxidase. That concluded that in the seedlings both of the low concentrations of selenate (3 and 6 mg/L), mainly 3 mg/L, led to cold tolerance.

Thinning of the stratospheric ozone layer is becoming a major concern these days because of increased production of ozone-depleting substances that lead to increased penetration of UVB radiation into the environment. Several studies have been conducted on the damaging effects of UVB radiation exposure to plants. Se has been documented in various reports to possess a protective role against harmful effects of UVB radiation. Studies conducted on lettuce and ryegrass showed that UVB exposure led to diminished enzyme activities. However, UVB treatment along with Se application increased CAT in both lettuce and ryegrass, SOD in lettuce, and glutathione-S-transferase in ryegrass (Xue and Hartikainen, 2000). It was reported by Yao et al. (2010) that adequate levels of Se reduce lipid peroxidation in wheat seedlings exposed to UVB radiation. Increase in total chlorophyll content, antioxidant content, and activities of antioxidative enzymes along with decrease in MDA content and superoxide radical production was observed in wheat seedlings when Se treatment was given along with UVB treatment, hence proving its defensive role in stress tolerance (Yao et al., 2011).

6. CONCLUSION AND FUTURE PROSPECTS

Several recent studies have documented both physiological and protective roles of Se against wide array of stresses and thereby proving the benefits of Se supplementation to plants. A need for intensive research pertaining to elucidation of mechanisms of antioxidative activity of Se compounds is required, which can expand our understanding about its role in oxidative stress management. Apart from all the evidence pointing toward Se being an antioxidative protectant, there are still many studies that show Se being toxic to many plant species. Hence, essentiality of Se as a micronutrient in plant systems still remains unclear. Also, suitable concentrations of exogenous Se application are a matter of a focused research that would allow the exogenous use of Se in a risk-free way. Se uptake and metabolism has always been related to behavior of sulfur in living systems. Rigorous research in the area of metabolism will provide an insight into which forms of Se are stored in food crops that will further will strengthen our knowledge regarding human nutrition perspective.

REFERENCES

Abbas, S.M., 2012. Effects of low temperature and selenium application on growth and the physiological changes in sorghum seedlings. J. Stress Physiol. Biochem. 8, 268–286.

Abbas, S.M., 2013. Low levels of selenium application attenuate low temperature stress in sorghum (*Sorghum bicolor* (L.) Moench) seedlings. Pak. J. Bot. 45, 1597–1604.

Abrams, M.M., Burau, R.G., Zasoski, R.J., 1990. Organic selenium distribution in selected California soils. Soil Sci. Soc. Am. J. 54, 979–982.

Afton, A.E., Carton, B., Caruso, J.A., 2009. Elucidating the selenium and arsenic metabolic pathways following exposure to the non-hyperaccumulating *Cholophytum comosum*, spider plant. J. Exp. Bot. 60, 1289–1297.

Akladious, S.A., 2012. Influence of different soaking times with selenium on growth, metabolic activities of wheat seedlings under low temperature stress. Afr. J. Biotechnol. 11 (82), 14792–14804.

Amos, W., Webb, S., Liu, Y., Andrews, J.C., LeDuc, D.L., 2012. Imaging translocation and transformation of bioavailable selenium by *Stanleya pinnata* with X-ray microscopy. Anal. Bioanal. Chem. 404, 1277–1285.

Anderson, J.W., 1993. Selenium interactions in sulfur metabolism. In: De Kok, L.J. (Ed.), Sulfur Nutrition and Assimilation in Higher Plants – Regulatory, Agricultural and Environmental Aspects. SPB Academic, The Netherland, pp. 49–60.

Apel, K., Hirt, H., 2004. Reactive oxygen species: metabolism, oxidative stress and signal transduction. Ann. Rev. Plant Biol. 55, 373–399.

Asher, C.J., Butler, G.W., Peterson, P.J., 1977. Selenium transport in root systems of tomato. J. Exp. Bot. 28, 279–291.

ATSDR, 2003. Toxicological Profile for Selenium. US Department of Health and Public Services, Agency for Toxic Substances and Disease Registry, Atlanta. Available from: http://www.atsdr.cdc.gov/toxprofiles/tp92-p.pdf.

Bansal, A., Sharma, S., Dhillon, S.K., Dhillon, K.S., 2012. Selenium accumulation and biochemical composition of Brassica grains grown in selenate or selenite treated alkaline sandy loam soil. Comm. Soil Sci. Plant Anal. 43, 1316–1331.

Barillas, J.V., Colin Quinn, F., Elizabeth, A.H., 2011. Selenium accumulation in plants—phytotechnological applications and ecological implications. Int. J. Phytorem. 13, 166–178.

Barrientos, E.Y., Flores, C.R., Wrobel, K., Wrobel, K., 2012. Impact of cadmium and selenium exposure on trace elements, fatty acids and oxidative stress in *Lepidium sativum*. J. Mex. Chem. Soc. 56 (1), 3–9.

Blaylock, M.J., James, B.R., 1994. Redox transformations and plant uptake of selenium resulting from root–soil interactions. Plant Soil 158, 1–12.

Brown, G.E., Foster, A.L., Ostergren, J.D., 1999. Mineral surfaces and bioavailability of heavy metals: a molecular-scale perspective. Proc. Natl. Acad. Sci. U.S.A. 96, 3388–3395.

Brown, T.A., Shrift, A., 1982. Selenium-toxicity and tolerance in higher plants. Biol. Rev. Camb. Philos. Soc. 57, 59–84.

Cartes, P., Gianfreda, L., Mora, M.L., 2005. Uptake of selenium and its antioxidant activity in ryegrass when applied as selenate and selenite forms. Plant Soil 276, 359–367.

Cartes, P., Jara, A.A., Pinilla, L., Rosas, A., Mora, M.L., 2010. Selenium improves the antioxidant ability against aluminium-induced oxidative stress in ryegrass roots. Ann. Appl. Biol. 156 (2), 297–307.

Carvalho, K.M., Gallardo-Williams, M.T., Benson, R.F., Martin, D.F., 2003. Effects of selenium supplementation on four agricultural crops. J. Agric. Food Chem. 51, 704–709.

Charron, C.S., Kopsell, D.A., Randle, W.M., Sams, C.E., 2001. Sodium selenate fertilisation increases selenium accumulation and decreases glucosinolate concentration in rapid-cycling *Brassica oleracea*. J. Sci. Food Agric. 81, 962–966.

Chen, C.C., Sung, J.M., 2001. Priming bitter gourd seeds with selenium solution enhances germinability and antioxidative responses under sub-optimal temperature. Plant Physiol. 111, 9–16.

Chen, T., Zheng, W., Luo, Y., Yang, F., Bai, Y., Tu, F., 2005. Effects of selenium stress on photosynthetic pigment contents and growth of *Chlorella vulgaris*. J. Plant Physiol. Mol. Biol. 31 (4), 369–373.

Chen, T., Zheng, W., Wong, Y., Yang, F., 2008. Selenium-induced changes in activities of antioxidant enzymes and content of photosynthetic pigments in *Spirulina platensis*. J. Integr. Plant Biol. 50 (1), 40–48.

Chu, J., Yao, X., Zhang, Z., 2010. Responses of wheat seedlings to exogenous selenium supply under cold stress. Biol. Trace Elem. Res. 136, 355–363.

Debarati, D., Saheli, P., Asok, K.B., 2011. Influence of selenium on growth and metabolism in rice (*Oryza sativa* L.) and its possible interaction with sulphate. J. Plant Physiol. 16 (1), 35–48.

Dhillon, K.S., Dhillon, S.K., 2003. Distribution and management of seleniferous soils. Adv. Agron. 79, 119–184.

Diao, M., Long, Ma, Jianwei, W., Jinxia, C., Aifei, F., Huiying, L., 2014. Selenium promotes the growth and photosynthesis of tomato seedlings under salt stress by enhancing chloroplast antioxidant defense system. J. Plant Growth Regul. 33, 671–682.

Djanaguiraman, M., Devi, D.D., Shanker, A.K., Sheeba, A., Bangarusamy, U., 2005. Selenium – an antioxidant protectant in soybean during senescence. Plant Soil 272, 77–86.

Djanaguiraman, M., Prasad, P.V.V., Seppänen, M., 2010. Selenium protects sorghum leaves from oxidative damage under high temperature stress by enhancing antioxidant defense system. Plant Physiol. Biochem. 48, 999–1007.

Dong, J.Z., Wang, Y., Wang, S.H., Yin, L.P., Xu, G.J., Zheng, C., Lei, C., Zhang, M.Z., 2013. Selenium increases chlorogenic acid, chlorophyll and carotenoids of *Lycium chinense* leaves. J. Sci. Food Agric. 93, 310–315.

Ekelund, N.G.A., Danilov, R.A., 2001. The influence of selenium on photosynthesis and "light-enhanced dark respiration" (LEDR) in the flagellate *Euglena gracilis* after exposure to ultraviolet radiation. Aqua Sci. 63, 457–465.

El Ashry, Z.M., Mohamed, F.I., 2012. Protective effects of some antioxidant metals against chromosomal damage induced by cadmium in *Vicia faba* plants. Int. J. Agric. Res. 7 (8), 376–387.

Ellis, D.R., Sors, T.G., Brunk, D.G., Albrecht, C., Orser, C., Lahner, B., Wood, K.V., Harris, H.H., Pickering, I.J., Salt, D.E., 2004. Production of Se-methylselenocysteine in transgenic plants expressing selenocysteine methyltransferase. BMC Plant Biol. 4, 1–11.

Emam, M.M., Khattab, H.E., Helal, N.M., Deraz, A.E., 2014. Effect of selenium and silicon on yield quality of rice plant grown under drought stress. Aust. J. Crop Sci. 8, 596–605.

Environment Agency, 2007. UK Soil and Herbage Pollutant Survey. Report No. 7: Environmental concentrations of heavy metals in UK soil and herbage. Environment Agency, Bristol.

Feist, L.J., Parker, D.R., 2001. Ecotypic variation in selenium accumulation among populations of *Stanleya pinnata*. New Phytol. 149, 61–69.

Feng, R.W., Wei, C.Y., 2012. Antioxidative mechanisms on selenium accumulation in *Pteris vittata* L., a potential selenium phytoremediation plant. Plant Soil Environ. 58 (3), 105–110.

Filek, M., Keskinen, R., Hartikainen, H., Szarejko, I., Janiak, A., Miszalski, Z., Golda, A., 2008. The protective role of selenium in rape seedlings subjected to cadmium stress. J. Plant Physiol. 165, 833–844.

Filek, M., Kościelniak, J., Łabanowska, M., Bednarska, E., Bidzińska, E., 2010. Selenium-induced protection of photosynthesis activity in rape (*Brassica napus*) seedlings subjected to cadmium stress. Fluorescence and EPR measurements. Photosyn. Res. 105, 27–37.

Galeas, M.L., Klamper, E.M., Bennett, L.E., Freeman, J.L., Kondratieff, B.C., Quinn, C.F., Pilon-Smits, E.A.H., 2008. Selenium hyperaccumulation reduces plant arthropod loads in the field. New Phytol. 177, 715–724.

Geoffroy, L., Gilbin, R., Simon, O., Floriani, M., Adam, C., Pradines, C., Cournac, L., Garnier-Laplace, J., 2007. Effect of selenate on growth and photosynthesis of *Chlamydomonas reinhardtii*. Aqua Toxicol. 83 (2), 149–158.

Germ, M., Gaberščik, A., 2003. Dihalni potential – kazalnik stresa pri rastlinah. Zbornik biotehniäke Fakultete Univerze v Ljubljani. Kmetijstvo 81 (2), 335–339.

Germ, M., Osvald, J., 2005. Selenium treatment affected respiratory potential in *Eruca sativa*. Acta Agric. Slov. 85, 329–335.

Germ, M., 2005. Combined impact of solar UV-B radiation and selenium treatment on respiratory potential in pumpkins (*Cucurbita pepo* L.). Acta Agric. Slov. 85 (2), 337–342.

Germ, M., Kreft, I., Osvald, J., 2005. Influence of UV-B exclusion and selenium treatment on photochemical efficiency of photosystem II, yield and respiratory potential in pumpkins (*Cucurbita pepo* L.). Plant Physiol. Biochem. 43, 445–448.

Germ, M., Kreft, I., Stibilj, V., Urbanc-Bercic, O., 2007a. Combined effects of selenium and drought on photosynthesis and mitochondrial respiration in potato. Plant Physiol. Biochem. 45 (2), 162–167.

Germ, M., Stibilj, V., Kreft, I., 2007b. Metabolic importance of selenium in plants. Eur. J. Plant Sci. Biotechnol. 1, 91–97.

Germ, M., Stibilj, V., Osvald, J., Kreft, I., 2007c. Effect of selenium foliar application on chicory (*Cichorium intybus* L.). J. Agric. Food Chem. 55, 795–798.

Gissel-Neilsen, G., 1971. Influence of pH and texture of the soil on plant uptake of added selenium. J. Agric. Food Chem. 19, 1165–1167.

Guangquan, H., Jun, L., Manman, S., Huiying, L., 2010. Effects of selenium on the germination of tomato seeds and protective system against active oxygen under salt stress. J. Shihezi Univ. Nat. Sci. (4)

Habibi, G., 2013. Effect of drought stress and selenium spraying on photosynthesis and antioxidant activity of spring barley. Acta Agric. Slov. 101, 31–39.

Hajiboland, R., Keivanfar, N., 2012. Selenium supplementation stimulates vegetative and reproductive growth in canola (*Brassica napus* L.) plants. Acta Agric. Slov. 19, 13–19.

Hajiboland, R., Sadeghzadeh, N., Sadeghzadeh, B., 2014. Effect of Se application on photosynthesis, osmolytes and water relations in two durum wheat (*Triticum durum* L.) genotypes under drought stress. Acta Agric. Slov. 103, 167–179.

Halliwell, B.H., Cross, C.E., 1994. Oxygen-derived species: their relation to human disease and environmental stress. Environ. Health Persp. 102, 5–12.

Halliwell, B.H., Gutteridge, J.M.C., 1984. Oxygen toxicity, oxygen radicals, transition metals and disease. Biochem. J. 219, 1–14.

Hamilton, S.J., 2004. Review of selenium toxicity in the aquatic food chain. Sci. Tot. Environ. 326, 1–31.

Han, D., Xiong, S., Tu, S., Liu, J., Chen, C., 2015. Interactive effects of selenium and arsenic on growth, antioxidant system, arsenic and selenium species of *Nicotiana tabacum* L. Environ. Exp. Bot. 117, 12–19.

Hanson, B., Garifullina, G.F., Lindblom, S.D., Wangeline, A., Ackley, A., Kramer, K., Norton, A.P., Lawrence, C.B., Pilon-Smits, E.A.H., 2003. Selenium accumulation protects *Brassica juncea* from invertebrate herbivory and fungal infection. New Phytol. 159, 461–469.

Hanson, B., Lindblom, S.D., Loeffler, M.L., Pilon-Smits, E.A.H., 2004. Selenium protects plants from phloem-feeding aphids due to both deterrence and toxicity. New Phytol. 162, 655–662.

Han-wen, S., Jing, H., Shu-xuan, L., Wei-jun, K., 2010. Protective role of selenium on garlic growth under cadmium stress. Soil Sci. Plant Anal. 41, 1195–1204.

Hartikainen, H., Xue, T., 1999. The promotive effect of selenium on plant growth as trigged by ultraviolet irradiation. J. Environ. Qual. 28, 1272–1275.

Hartikainen, H., Xue, T., Piironen, V., 2000. Selenium as an antioxidant and prooxidant in ryegrass. Plant Soil 225, 193–200.

Hasanuzzaman, M., Fujita, M., 2011. Selenium pretreatment upregulates the antioxidant defense and methylglyoxal detoxification system and confers enhanced tolerance to drought stress in rapeseed seedlings. Biol. Trace Elem. Res. http://dx.doi.org/10.1007/s12011-011-8998-9.

Hasanuzzaman, M., Hossain, M.A., Fujita, M., 2010. Selenium in higher plants: physiological role, antioxidant metabolism and abiotic stress tolerance. J. Plant Sci. 5, 354–375.

Hawkesford, M.J., Davidian, J.C., Grignon, C., 1993. Sulphate/proton cotransport in plasma-membrane vesicles isolated from roots of *Brassica napus* L.: increased transport in membranes isolated from sulphur-starved plants. Planta 190, 297–304.

Hawrylak-Nowak, B., 2009. Beneficial effects of exogenous selenium in cucumber seedlings subjected to salt stress. Biol. Trace Elem. Res. 132, 259–269.

Hawrylak-Nowak, B., 2013. Comparative effects of selenite and selenate on growth and selenium accumulation in lettuce plants under hydroponic conditions. Plant Growth Regul. 70, 149–157.

Hawrylak-Nowak, B., Matraszek, R., Szymanska, M., 2010. Selenium modifies the effect of short-term chilling stress on cucumber plants. Biol. Trace Elem. Res. http://dx.doi.org/10.1007/s12011-010-8613-5.

Hladun, K.R., Parker, D.R., Tran, K.D., Trumble, J.T., 2013. Effects of selenium accumulation on phytotoxicity, herbivory, and pollination ecology in radish (*Raphanus sativus* L.). Environ. Pollut. 172, 70–75.

Hu, Q., Xu, J., Pang, G., 2003. Effect of selenium on the yield and quality of green tea leaves harvested in early spring. J. Agric. Food Chem. 51, 3379–3381.

Ibrahim, H.M., 2014. Se pretreatment regulates the antioxidant defense system and reduces oxidative stress on drought-stressed wheat (*Triticum aestivum* L.) plants. Asian J. Plant Sci. 13, 120–128.

IPCS, 1987. Selenium. Environmental Health Criteria 58. World Health Organization/International Programme on Chemical Safety, Geneva. Available from: http://www.inchem.org/documents/ehc/ehc/ehc58.htm.

Kabata-Pendias, A., Mukherjee, A.B., 2007. Trace Elements from Soil to Human. Springer-Verlag, Berlin.

KeLing, H., Ling, Z., JiTao, W., Yang, Y., 2013. Influence of selenium on growth, lipid peroxidation and antioxidative enzyme activity in melon (*Cucumis melo* L.) seedlings under salt stress. Acta Soc. Bot. Pol. 82, 193–197.

Kong, L., Wang, M., Bi, D., 2005. Selenium modulates the activities of antioxidant enzymes, osmotic homeostasis and promotes the growth of sorrel seedlings under salt stress. Plant Growth Regul. 45, 155–163.

Kreft, I., Mechora, S., Germ, M., Stibilj, V., 2013. Impact of selenium on mitochondrial activity in young Tartary buckwheat plants. Plant Physiol. Biochem. 63, 196–199.

Kumar, M., Bijo, A.J., Baghel, R.S., Reddy, C.R.K., Jha, B., 2012. Selenium and spermine alleviates cadmium induced toxicity in the red seaweed *Gracilaria dura* by regulating antioxidant system and DNA methylation. Plant Physiol. Biochem. 51, 129–138.

Kumar, A., Singh, R.P., Singh, P.K., Awasthi, S., Chakrabarty, D., Trivedi, P.K., Tripathi, R.D., 2014. Selenium ameliorates arsenic induced oxidative stress through modulation of antioxidant enzymes and thiols in rice (*Oryza sativa* L.). Ecotoxicology 23, 1153–1163.

Kuznetsov, V.V., Kholodova, V.P., Kuznetsov, V.I.V., Yagodin, B.A., 2003. Selenium regulates the water status of plants exposed to drought. Dokl. Biol. Sci. 390, 266–268.

Lass, B., Ullrich-Eberius, C.I., 1984. Evidence for proton/sulfate cotransport and its kinetics in *Lemna gibba* G1. Planta 161, 53–60.

LeDuc, D.L., Tarun, A.S., Montes-Bayon, M., Meija, J., Malit, M.F., Wu, C.P., AbdelSamie, M., Chiang, C.-Y., Tagmount, A., deSouza, M.P., Neuhierl, B., Bock, A., Caruso, J.A., Terry, N., 2004. Overexpression of selenocysteine methyltransferase in *Arabidopsis* and indian mustard increases selenium tolerance and accumulation. Plant Physiol. 135, 377–383.

Lei, C., Ma, Q., Tang, Q.Y., Ai, X.R., Zhou, Z., Yao, L., Wang, Y., Wang, Q., Dong, J.Z., 2014. Sodium selenite regulates phenolics accumulation and tuber development of purple potatoes. Sci. Horticult. 165, 142–147.

Lei, H., Wu, J., 2011. Effect of selenium on the photosynthesis and nonenzymatic antioxidants in *Cardamine enshiensis* leaves. Hubei Agric. Sci. (24)

Lemly, A.D., 2002. Selenium Assessment in Aquatic Ecosystems – a Guide for Hazard Evaluation and Water Quality Criteria. Springer-Verlag, New York.

Leustek, T., 2002. Sulfate metabolism. In: Somerville, C.R., Meyerowitz, E.M. (Eds.), The Arabidopsis Book. American Society of Plant Biologists, Rockville, MD, USA, p. e0017. http://dx.doi.org/10.1199/tab.0017.

Li, D., Li, H., Wu, Y., 2012. Studies on photosynthetic characteristics of cardamine growing in different site conditions in the high selenium area in Yutangba. J. Appl. Math. Bioinf. 2, 11–21.

Li, H.-F., McGrath, S.P., Zhao, F.-J., 2008. Selenium uptake, translocation and speciation in wheat supplied with selenate or selenite. New Phytol. 178, 92–102.

Liu, K., Zhao, Y., Chen, F., Gu, Z., Bu, G., 2011. Enhanced glutathione peroxidases (GPx) activity in young barley seedlings enriched with selenium. Afr. J. Biotechnol. 10, 11483–11487.

Liu, W., Shang, S., Feng, X., Zhang, G., Wu, F., 2015. Modulation of exogenous selenium in cadmium-induced changes in antioxidative metabolism, cadmium uptake and photosynthetic performance in the 2 tobacco genotypes differing in cadmium tolerance. Environ. Toxicol. Chem. 34, 92–99.

Long, R.H.B., Benson, S.M., Tokunaga, T.K., Yee, A., 1990. Selenium immobilization in a pond sediment at Kesterson Reservoir. J. Environ. Qual. 19, 302–311.

Lyi, S.M., Heller, L.I., Rutzke, M., Welch, R.M., Kochian, L.V., Li, L., 2005. Molecular and biochemical characterization of the selenocysteine Se-methyltransferase gene and Se-methylselenocysteine synthesis in broccoli. Plant Physiol. 138, 409–420.

Lyons, G.H., Genc, Y., Soole, K., Stangoulis, J.C.R., Liu, F., Graham, R.D., 2009. Selenium increases seed production in Brassica. Plant Soil 318, 73–80.

Malik, J.A., Goel, S., Kaur, N., Sharma, S., Singh, I., Nayyar, H., 2012. Selenium antagonises the toxic effects of arsenic on mungbean (*Phaseolus aureus* Roxb.) plants by restricting its uptake and enhancing the antioxidative and detoxification mechanisms. Environ. Exp. Bot. 77, 242–248.

Malik, J.A., Kumar, S., Thakur, P., Sharma, S., Kaur, N., Kaur, R., Pathania, D., Bhandhari, K., Kaushal, N., Singh, K., Shrivasatva, A., Nayyar, H., 2011. Promotion of growth in mungbean (*Phaseolus aureus* Roxb.) by selenium is associated with stimulation of carbohydrate metabolism. Biol. Trace Elem. Res. 143 (1), 530–539.

Mane, P.C., Bhosle, A.B., Kulkarni, P.A., 2011. Biosorption and biochemical study on water hyacinth (*Eichhornia crassipes*) with reference to selenium. Arch. Appl. Sci. Res. 3, 222–229.

Masscheleyn, P.H., Delalumne, R.D., William Jr., H.P., 1990. Transformation of selenium as affected by sediment oxidation-reduction potential and pH. Environ. Sci. Technol. 24, 91–96.

Mayland, H.F., 1994. Selenium in plant and animal nutrition. In: Frankenberger Jr., W.T., Benson, S. (Eds.), Selenium in the Environment. Marcel Dekker, Inc., NY, USA, pp. 29–45.

McNeal, J.M., Balistrieri, L.S., 1989. Geochemistry and occurrence of selenium: an overview. In: Jacobs, L.W. (Ed.), Selenium in Agriculture and the Environment. American Society of Agronomy and Soil Science Society of America, Madison, WI, pp. 2–13.

Mechora, S., Germ, M., 2010. Selenium induced lower respiratory potential in *Glycine max* (L.) Merr. Acta Agric. Slov. 95 (1), 29–34.

Mechora, S., Stibilj, V., Radešček, T., Gaberščik, A., Germ, M., 2011. Impact of Se (VI) fertilization on Se concentration in different parts of red cabbage plants. J. Food Agric. Environ. 9, 357–361.

Mikkelsen, R.L., Haghnia, G.H., Page, A.L., 1989. Factors affecting selenium accumulation by crop plants. In: Jacobs, L.W. (Ed.), Selenium in Agriculture and the Environment. Am Soc Agron Soil Sci Soc Am, Madison, WI, pp. 65–93.

Mittler, R., 2002. Oxidative stress, antioxidants and stress tolerance. Trends Plant Sci. 7, 405–410.

Moussa, H.R., Ahmed, A.E.M., 2010. Protective role of selenium on development and physiological responses of Vicia faba. Int. J. Veg. Sci. 16, 174–183.

Mroczek-Zdyrska, M., Wójcik, M., 2012. The influence of selenium on root growth and oxidative stress induced by lead in *Vicia faba* L. minor plants. Biol. Trace Elem. Res. 147, 320–328.

Munne-Bosch, S., Alegre, L., 2002. The functions of tocopherols and tocotrienols in plants. Crit. Rev. Plant Sci. 21, 31–57.

Munne-Bosch, S., 2005. The role of α-tocopherol in plant stress tolerance. J. Plant Physiol. 62, 743–748.

Nawaz, F., Ashraf, M.Y., Ahmad, R., Waraich, E.A., 2013. Selenium (Se) seed priming induced growth and biochemical changes in wheat under water deficit conditions. Biol. Trace Elem. Res. 151, 284–293. http://link.springer.com/journal/12011.

Nawaz, F., Ahmad, R., Ashraf, M.Y., Waraich, E.A., Khan, S.Z., 2015. Effect of selenium foliar spray on physiological and biochemical processes and chemical constituents of wheat under drought stress. Ecotoxicol. Environ. Saf. 113, 191–200.

Nawaz, F., Ashraf, M.Y., Ahmad, R., Waraich, E.A., Shabbir, R.N., 2014. Selenium (Se) regulates seedling growth in wheat under drought stress. Adv. Chem. 2014, 1–7.

Neuhierl, B., Thanbichler, M., Lottspeich, F., Bock, A., 1999. A family of S-methylmethionine dependent thiol/selenol methyltransferases. Role in selenium tolerance and evolutionary relation. J. Biol. Chem. 274, 5407–5414.

Nezami, M.T., Bybordi, A., 2012. The effect of different amounts of selenium on yield and yield components of two canola cultivars. J. Food Agric. Environ. 10 (2), 603–606.

Owusu-Sekyere, A., Kontturi, J., Hajiboland, R., Rahmat, S., Aliasgharzad, N., Hartikainen, H., Seppänen, M., 2013. Influence of selenium (Se) on carbohydrate metabolism, nodulation and growth in alfalfa (*Medicago sativa* L.). Plant Soil 373, 541–552.

Pandey, C., Gupta, M., 2015. Selenium and auxin mitigates arsenic stress in rice (*Oryza sativa* L.) by combining the role of stress indicators, modulators and genotoxicity assay. J. Hazard. Mater. 287, 384–391.

Pedrero, Z., Madrid, Y., Hartikainen, H., Camara, C., 2008. Protective effect of selenium in broccoli (*Brassica oleracea*) plants subjected to cadmium exposure. J. Agric. Food Chem. 56, 266–271.

Peng, X.L., Liu, Y.Y., Luo, S.G., 2002. Effects of selenium on lipid peroxidation and oxidizing ability of rice roots under ferrous stress. J. Northeast Agric. Univ. 19, 9–15.

Pennanen, A., Xue, T., Hartikainen, H., 2002. Protective role of selenium in plant subjected to severe UV irradiation stress. J. Appl. Bot. 76, 66–76.

Persans, M.W., Salt, D.E., 2000. Possible molecular mechanisms involved in nickel, zinc and selenium hyperaccumulation in plants. Biotechnol. Gen. Eng. Rev. 17, 389–413.

Pezzarossaa, B., Rosellini, I., Borghesi, E., Tonutti, P., Malorgio, F., 2014. Effects of Se-enrichment on yield, fruit composition and ripening of tomato (*Solanum lycopersicum*) plants grown in hydroponics. Sci. Hortic. 165, 106–110.

Pilon-Smits, E.A.H., Quinn, C.F., 2010. Selenium metabolism in plants. In: Hell, R., Mendel, R.R. (Eds.), Cell Biology of Metals and Nutrients, Plant Cell Monographs. Springer-Verlag, Berlin, Heidelberg, pp. 225–241.

Pinheiro, F.J.A., Barros, R.S., Coelho, T.G., Souza, B.M.L., 2008. Breaking dormancy of *Stylosanthes humilis* seeds with selenium compounds. Seed Sci. Res. 18, 47–53.

Poldma, P., Moor, U., Tonutare, T., Herodes, K., Rebane, R., 2013. Selenium treatment under field conditions affects mineral nutrition, yield and antioxidant properties of bulb onion (*Allium cepa* L.). Acta Sci. Pol. 12, 167–181.

Proietti, P., Nasini, L., Buonoa, D.D., D'Amatoa, R., Tedeschinib, E., Businellia, D., 2013. Selenium protects olive (*Olea europaea* L.) from drought stress. Sci. Hortic. 164, 165–171.

Pukacka, S., Ratajczak, E., Kalemba, E., 2011. The protective role of selenium in recalcitrant *Acer saccharium* L. seeds subjected to desiccation. J. Plant Physiol. 168, 220–225.

Qing, X., Zhao, X., Hu, C., Wang, P., Zhang, Y., Zhang, X., Wang, P., Shi, H., Jia, F., Qu, C., 2015. Selenium alleviates chromium toxicity by preventing oxidative stress in cabbage (*Brassica campestris* L. ssp. *Pekinensis*) leaves. Ecotox. Environ. Safe 114, 179–189.

Quinn, C.F., Freeman, J.F., Galeas, M.L., Klamper, E.M., Pilon-Smits, E.A.H., 2008. Selenium protects plants from prairie dog herbivory – implications for the functional significance and evolution of Se hyperaccumulation. Oecologia 155, 267–275.

Quinn, C.F., Freeman, J.L., Reynolds, R.J.B., Cappa, J.J., Fakra, S.C., Marcus, M.A., Lindblom, S.D., Quinn, E.K., Bennett, L.E., Pilon-Smits, E.A.H., 2010. Selenium hyperaccumulation offers protection from cell disruptor herbivores. Plant Physiol. 153, 1630–1652.

Ramos, S.J., Faquin, V., Guilherme, L.R.G., Castro, E.M., Ávila, F.W., Carvalho, G.S., Bastos, C.E.A., Oliveira, C., 2010. Selenium biofortification and antioxidant activity in lettuce plants fed with selenate and selenite. Plant Soil Environ. 56 (12), 584–588.

Ribeiro, D.M., Mapeli, A.M., Antunes, W.C., Barros, R.S., 2011. A dual role of selenium in the growth control of seedlings of *Stylosanthes humilis*. Agric. Sci. 2, 78–85.

Rosenfeld, I., Beath, O.A., 1964. Selenium, Geobotany, Biochemistry, Toxicity, and Nutrition. Academic, New York.

Rotruck, J.T., Pope, A.L., Ganther, H.E., Hoekstra, W.G., 1973. Selenium: biochemical role as a component of glutathione peroxidase. Science 179, 588–590.

Sae-Lee, N., Kerdchoechuen, O., Laohakunjit, N., 2012. Chemical qualities and phenolic compounds of Assam tea after soil drench application of selenium and aluminium. Plant Soil 356, 381–393.

Sager, M., 2006. Selenium in food, agriculture and nutrition. Pure Appl. Chem. 78, 111–133.

Saidi, I., Chtourou, Y., Djebali, W., 2014. Selenium alleviates cadmium toxicity by preventing oxidative stress in sunflower (*Helianthus annuus*) seedlings. J. Plant Physiol. 171, 85–91.

Sajedi, N.A., Ardakani, M.R., Madani, H., Naderi, A., Miransari, M., 2011. The effects of selenium and other micronutrients on the antioxidant activities and yield of corn (*Zea mays* L.) under drought stress. Physiol. Mol. Biol. Plants 17 (3), 215–222.

Schiavon, M., Acqua, S., Mietto, A., Pilon-Smits, E.A.H., Sambo, P., Masi, A., Malagoli, M., 2013. Selenium fertilization alters the chemical composition and antioxidant constituents of tomato (*solanum lycopersicon* L.). J. Agric. Food Chem. 61, 10542–10554.

Schiavon, M., Moro, I., Pilon-Smits, E.A.H., Matozzo, V., Malagoli, M., Dalla Vecchia, F., 2012. Accumulation of selenium in *Ulva* sp. and effects on morphology, ultrastructure and antioxidant enzymes and metabolites. Aqua Toxicol. 122–123, 222–231.

Schwartz, K., Foltz, S.M., 1957. Se as an integral part of factor 3 against necrotic liver degeneration. J. Am. Chem. Soc. 79, 3292–3293.

Seppänen, M., Turakainen, M., Hartikainen, H., 2003. Selenium effects on oxidative stress in potato. Plant Sci. 165, 311–319.

Shahbazi, S., Jashni, R., Bayati, F., Jiriaie, M., 2013. Survey of selenium priming in reducing negative effects of salinity and less reserves seed on germination of chickpea (CV. Hashem). Tech. J. Eng. Appl. Sci. 3, 1371–1377.

Shardendu, S.N., Boulyga, S.F., Stengel, E., 2003. Phytoremediation of selenium by two helophyte species in subsurface flow constructed wetland. Chemosphere 50, 967–973.

Sharma, S., Bansal, A., Dhillon, S.K., Dhillon, K.S., 2010. Comparative effects of selenate and selenite on growth and biochemical composition of rapeseed (*Brassica napus* L.). Plant Soil 329, 339–348.

Sharma, S., Dhillon, K.S., Dhillon, S.K., Munshi, S.K., 2008. Changes in biochemical components of wheat and rapeseed grown on selenium-contaminated soil. Arch. Agron. Soil Sci. 54 (1), 33–40.

Shibagaki, N., Rose, A., McDermott, J.P., Fujiwara, T., Hayashi, H., Yoneyama, T., Davies, J.P., 2002. Selenate-resistant mutants of *Arabidopsis thaliana* identify Sultr1;2, a sulphate transporter required for efficient transport of sulfate into roots. Plant J. 29, 475–486.

Shigeoka, S., Ishikawa, T., Tamoi, M., Miyagawa, Y., Takeda, T., Yabuta, Y., Yoshimura, K., 2002. Regulation and functions of ascorbate peroxidase isoenzymes. J. Exp. Bot. 53, 1305–1319.

Shrift, A., Ulrich, J., 1969. Transport of selenate and selenite into Astragalus roots. Plant Physiology 44, 893–896.

Sies, H., 1985. Oxidative stress: introductory remarks. In: Sies, H. (Ed.), Oxidative Stress. Academic Press, San Diego, New York, London, pp. 1–8.

Sies, H., 1986. Biochemistry of oxidative stress. Angew. Chem. Int. Ed. Engl. 25, 1058–1071.

Sies, H., Cadenas, E., 1985. Oxidative stress: damage to intact cells and organs. Phil. Trans. R. Soc. Lond. Ser. B Biol. Sci. 311, 617–631.

Singh, M., Singh, H., Bhandari, D.K., 1980. Interaction of selenium and sulphur on the growth and chemical composition of raya. Soil Sci. 129, 238–244.

Smith, F.W., Hawkesford, M.J., Ealing, P.M., Clarkson, D.T., Vanden Berg, P.J., Belcher, A.R., Warrilow, A.G., 1997. Regulation of expression of a cDNA from barley roots encoding a high affinity sulphate transporter. Plant J. 12, 875–884.

Smrkolj, P., Germ, M., Kreft, I., Stibilj, V., 2006. Respiratory potential and Se compounds in pea (*Pisum sativum* L.) plants grown from Se-enriched seeds. J. Exp. Bot. 57, 3595–3600.

Soleimanzadeh, H., 2012. Response of sunflower (*Helianthus annus* L.) to selenium application under water stress. World Appl. Sci. J. 17 (9), 1115–1119.

Sors, T.G., Ellis, D.R., Salt, D.E., 2005. Selenium uptake, translocation, assimilation and metabolic fate in plants. Photosynth. Res. 86, 373–389.

de Souza, M.P., Pickering, I.J., Walla, M., Terry, N., 2002. Selenium assimilation and volatilization from selenocyanate-treated Indian mustard and muskgrass. Plant Physiol. 128, 625–633.

Srivastava, M., Maa, L.Q., Rathinasabapathib, B., Srivastava, P., 2009. Effects of selenium on arsenic uptake in arsenic hyperaccumulator *Pteris vittata* L. Biores Technol. 100, 1115–1121.

Sun, H.W., Ha, J., Liang, S.X., Kang, W.J., 2010. Protective role of selenium on garlic growth under cadmium stress. Comm. Soil Sci. Plant Anal. 41, 1195–1205.

Tang, H., Liu, Y., Gong, X., zeng, G., Zheng, B., Wang, D., Sun, Z., Zhou, L., Zeng, X., 2015. Effects of selenium and silicon on enhancing antioxidative capacity in ramie (*Boehmeria nivea* (L.) Gaud.) under cadmium stress. Environ. Sci. Pollut. Res. http://dx.doi.org/10.1007/s11356-015-4187-2.

Teimouri, S., Hasanpour, J., Akbar, A., 2014. Effect of selenium spraying on yield and growth indices of wheat (*Triticum aestivum* L.) under drought stress condition. Int. J Adv. Biol. Biomed. Res. 2, 2091–2103.

Terry, N., Zayed, A.M., de Souza, M.P., Tarun, A.S., 2000. Selenium in higher plants. Ann. Rev. Plant Physiol. Plant Mol. Biol. 51, 401–432.

Thiruvengadam, M., Chung, I.M., 2015. Selenium, putrescine, and cadmium influence health-promoting phytochemicals and molecular-level effects on turnip (*Brassica rapa* ssp. *rapa*). Food Chem. 173, 185–193.

Thiry, C., Ruttens, A., Temmerman, L.D., Schneider, Y.-J., Pussemier, L., 2012. Current knowledge in species-related bioavailability of selenium in food. Food Chem. 130, 767–784.

Turakainen, M., 2007. Selenium and Its Effects on Growth, Yield and Tuber Quality in Potato. University of Helsinki, Helsinki, ISBN 9521034661, pp. 50.

Turakainen, M., Hartikainen, H., Seppänen, M.M., 2004. Effects of selenium treatments on potato (*Solanum tuberosum* L.) growth and concentrations of soluble sugars and starch. J. Agric. Food Chem. 52, 5378–5382.

Vorobets, N., 2006. Glutathione peroxidase activity in sunflower shoots exposed to lead and selenium. Ann. Univ. Mariae Curie-Skłodowska 19, 151–154.

Wang, C.-L., Liu, Y.-G., Zeng, G.-M., Hu, X.-J., 2014. Mechanism of exogenous selenium alleviates cadmium induced toxicity in *Bechmeria nivea* (L.) Gaud (Ramie). Trans. Nonferrous Met. Soc. China 24, 3964–3970.

Wangeline, A.L., Valdez, J.R., Lindblom, S.D., Bowling, K.L., Reeves, F.B., Pilon-Smits, E.A.H., 2011. Selenium tolerance in rhizosphere fungi from Se hyperaccumulator and non-hyperaccumulator plants. Am. J. Bot. 98, 1139–1147.

Wei, Z., Zhu-mei, X., Jun-feng, T., 2011. Effect of selenium on photosynthetic properities of *Vitis vinifera* L.cv. Cabernet Sauvignon under water stress. North. Hortic. (22)

White, P.J., Bowen, H.C., Marshall, B., Broadley, M.R., 2007. Extraordinarily high leaf selenium to sulfur ratios define 'Se-accumulator' plants. Ann. Bot. 100, 111–118.

White, P.J., Bowen, H.C., Parmaguru, P., Fritz, M., Spracklen, W.P., Spiby, R.E., Meachan, M.C., Mead, A., Harriman, M., Trueman, L.J., Smith, B.M., Thomas, B., Broadley, M.R., 2004. Interactions between selenium and sulphur nutrition in *Arabidopsis thaliana*. J. Exp. Bot. 55, 1927–1937.

Wilson, L.G., Bandurski, R.S., 1958. Enzymatic reactions involving sulfate, sulfite, selenate and molybdate. J. Biol. Chem. 233, 975–981.

Xiao, Y.-H., He, J.-M., Wu, L.-F., Feng, Q.-Y., Zheng, J.-Y., 2008. Effect of selenium on plant growth,characteristics of photosynthesis and content of soluble sugar of *Basella rubra* L. J. Shaoguan Univ. (6)

Xiaoqin, Y., Jianzhou, C., Guangyin, W., 2009. Effects of drought stress and selenium supply on growth and physiological characteristics of wheat seedlings. Acta Physiol. Plant 31, 1031–1036.

Xue, T., Hartikainen, H., Piironen, V., 2001. Antioxidative and growth promoting effect of selenium in senescing lettuce. Plant Soil 237, 55–61.

Xue, T.L., Hartikainen, H., 2000. Association of antioxidative enzymes with the synergistic effect of selenium and UV irradiation in enhancing plant growth. Agric. Food Sci. Finland 9, 177–186.

Yang, G.Q., Wang, S.Z., Zhou, R.H., Sun, S.Z., 1983. Endemic selenium intoxication of humans in China. Am. J. Clin. Nutr. 37, 872–881.

Yao, X., Chu, J., Wang, G., 2009. Effects of selenium on wheat seedlings under drought stress. Biol. Trace Elem. Res. 130, 283–290.

Yao, X., Chu, J., He, X., Ba, C., 2011. Protective role of selenium in wheat seedlings subjected to enhanced UV-B radiation. Russ. J. Plant Physiol. 58, 283–289.

Yao, X., Jianzhou, C., Xueli, H., Binbin, L., Jingmin, L., Zhaowei, Y., 2013. Effects of selenium on agronomical characters of winter wheat exposed to enhanced ultraviolet-B. Ecotoxicol. Environ. Saf. 92, 320–326.

Yao, X.Q., Chu, J.Z., Ba, C.J., 2010. Antioxidant responses of wheat seedlings to exogenous selenium supply under enhanced ultraviolet-B. Biol. Trace Elem. Res. 136, 96–105.

Yassen, A.A., Safia, M.A., Sahar, M.Z., 2011. Impact of nitrogen fertilizer and foliar spray of selenium on growth, yield and chemical constituents of potato plants. Aus J. Basic Appl. Sci. 5 (11), 1296–1303.

Ylaranta, T., 1983. Sorption of selenite and selenate added in the soil. Ann. Agric. Fenn. 22, 29–39.

Ylaranta, T., 1985. Increasing the Selenium Content of Cereals and Grass Crops in Finland. Academic Dissertation, Agricultural Research Centre, Institute of Soil Science, Jokioinen. Yliopistopaino, Helsinki, pp. 72.

Yuan, J., Hu, M., Zhou, Z., 2013. Selenium treatment mitigates the effect of lead exposure in *Coleus blumei* benth. J. Environ. Anal. Toxicol. 3, 191. http://dx.doi.org/10.4172/2161-0525.1000191.

Zanjani, K.E., Shirani Rad, A.H., Naeemi, M., Aghdam, A.M., Taherkhani, T., 2012. Effects of zeolite and selenium application on some physiological traits and oil yield of medicinal pumpkin (*Cucurbita pepo* L.) under drought stress. Curr. Res. J. Biol. Sci. 4 (4), 462–470.

Zembala, M., Filek, M., Walas, S., Mrowiec, H., Kornaś, A., Miszalski, Z., Hartikainen, H., 2010. Effect of selenium on macro- and microelement distribution and physiological parameters of rape and wheat seedlings exposed to cadmium stress. Plant Soil 329, 457–468.

Zhang, Y., Frankenberger, W.T., 2002. Fate of dimethyldiselenide in soil. J. Environ. Qual. 31, 1124–1128.

CHAPTER

8

BRASSINOSTEROIDS ARE POTENTIAL AMELIORATORS OF HEAVY METAL STRESSES IN PLANTS

Bojjam Vidya Vardhini

Department of Botany, Telangana University, Dichpally, Nizamabad, Andhra Pradesh, India

CHAPTER OUTLINE

1. INTRODUCTION

Brassinosteroids (BRs) are a new type of polyhydroxy steroidal phytohormones with significant growth-promoting influence (Vardhini, 2012a,b; Vardhini and Anjum, 2015). Mitchell et al. (1970) discovered BRs, which were later extracted from the pollen of *Brassica napus* L. by Grove et al. (1979). BRs can be classified as C_{27}, C_{28}, or C_{29} BRs according to the number of carbons in their structure (Vardhini, 2013a,b). Haubrick and Assmann (2006) reported that 60 related compounds have

Plant Metal Interaction. http://dx.doi.org/10.1016/B978-0-12-803158-2.00008-4

been identified. However, Vardhini et al. (2006) reported that brassinolide (BL), 28-homobrassinolide (28-HomoBL) and 24-epibrassinolide (24-EpiBL) are the three bioactive BRs being widely used in most physiological and experimental studies. The commonly used bioactive BRs, such as BL, 24-EpiBL, 28-HomoBL, castasterone (CS), and 24-epicatasterone (24-EpiCS) are represented in Figure 8.1. BRs are applied to plants at different stages of the life cycle, such as vegetative stage (Vardhini and Rao, 1998), flowering stage (Vardhini, 2013a), meiosis stage (Saka et al., 2003), grain filling stage (Vardhini, 2013a), antheis stage (Liu et al., 2006) as foliar spray (Vardhini and Rao, 2000; Hayat et al., 2012), seed treatment (Kartal et al., 2009; Ekinci et al., 2012), root application (Shang et al., 2006; Song et al., 2006), and even as a shotgun approach (Hayat et al., 2010a; Yusuf et al., 2012).

BRs are considered ubiquitous in plant kingdom because they are found in almost all the phyla of the plant kingdom like alga, pteridophyte, gymnosperms, dicots, and monocots (Bajguz, 2009). The wide range of roles performed by BRs in plant parts such as cell, leaf, stem shoot callus, and so on, are shown in Table 8.1. BRs also exhibit a synergistic effect with other phytohormones in regulating plant growth and metabolism. BRs interact with auxins (Zaharah et al., 2012), cytokinins

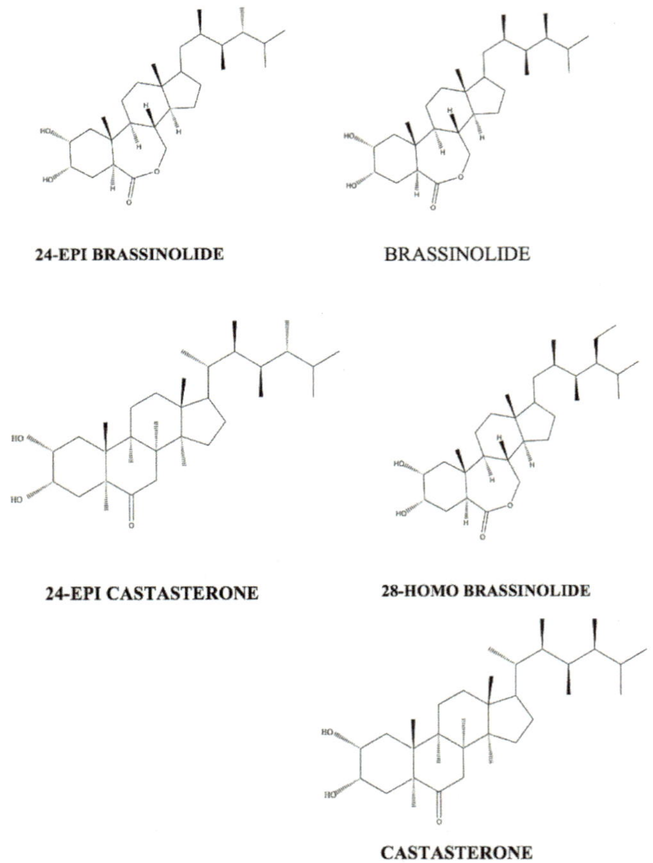

FIGURE 8.1

Bioactive brassinosteroids.

Table 8.1 Brassinosteroids and Varied Physiological/Biotechnological Responses in Plants (1995–2015)

Plant	Compound	Response	References
Arachis hypogaea	BRs	Root nodulation and nitrogenase activity	Vardhini and Rao (1999)
Arabidopsis thaliana		Induction of exaggerated growth in hydroponically grown plants	Arteca and Arteca (2001)
Zinnia elegans		Morphogenesis of tracheary elements	Yamamoto et al. (2001)
Arachis hypogeae		Changing the lycopene contents, ethylene, and carbohydrates in fruit of tomato	Vardhini and Rao (2002)
		Acceleration of cell division rate in isolated protoplasts	Ho (2003)
Phaseolus vulgaris		Growth, phytohormone content, nodulation, and nitrogenase activity	Upreti and Murthy (2004)
Brassica oleracea		Activation of the cell cycle during germination	Zadvornova et al. (2005)
Lycopersicon esculentum		Changes of lycopene contents	Weng et al. (2007)
Zea mays		Control of sex determination	Hartwig et al. (2011)
Cucurbita pepo		Control of sex expression and flower development	Manzano et al. (2011)
A. thaliana		Promotion of progression of cell cycle	González-Garcia et al. (2011)
Raphanus sativus		Qualitative changes in the storage roots	Vardhini et al. (2012)
		Cell division and expansion in the leaf	Zhiponova et al. (2013)
		Root epidermal cells	Cheng et al. (2014)
		Cytokinin-induced anthocyanin biosynthesis	Yuan et al. (2014)
Solanum lycopersicum		Stomatal closure and H_2O_2 dynamics	Xia et al. (2014)
A. thaliana		Pollen germination	Vogler et al. (2014)
Petunia hybrida	BL	Proliferation of leaf cells	Nakaya et al. (2002)
Spartina patens		Induction of callus growth and regeneration ability	Lu et al. (2003)
Vicia faba		Control of stomatal aperture and potassium currents of guard cell protoplasts along with ABA	Haubrick et al. (2006)
Gossypium hirsutum		Regeneration of plants via somatic embryogenesis	Aydin et al. (2006)
Nicotiana tabacum		Regeneration of adventitious shoots from cultured leaf discs	Kim et al. (2008)
Camellia oleifera		Leaf nutrition	Zhou et al. (2013)

Continued

Table 8.1 Brassinosteroids and Varied Physiological/Biotechnological Responses in Plants (1995–2015)—cont'd

Plant	Compound	Response	References
Pisum sativum	24-EpiBL	Regulation of Calvin cycle enzymes	Fedina et al. (2008)
Pinus wallichiana		Induction of somatic embryogenesis	Malabadi and Nataraja (2007)
Vaccinium corymbosum		In vitro micropropagation	Kudryashova et al. (2012)
Brassica napus		Lipid signaling and metabolism subjected to endophytic fungal stress	Pokotylo et al. (2014)
Liriodendron tulipifera		Reduces lignification and alters cell wall carbohydrate biosynthesis in the secondary xylem	Jin et al. (2014)
Acrocomia aculeata	Biobras-16	Germination and growth in in vitro culture of zygotic embryos and acclimatization	Borcioni and Bonato-Negrelle (2012)

24-EpiBL, 24-epibrassinolide; ABA, abscisic acid; BL, brassinolide; BR, brassinosteroid.

(Bajguz and Piotrowska-Niczyporuk, 2014), gibberellins (Domagalska et al., 2010), abscisic acid (ABA) (Liu et al., 2011), ethylene (ET) (Manazo et al., 2011), salicylic acid (SA) (El-Khallal et al., 2009; Ashraf et al., 2010), and jasmonic acid (JA) (Peng et al., 2011; Creelman and Mullet, 1997). Mutations in *Arabidopsis* BSK5 encoding a BR-signaling kinase protein resulted in varied responses to salinity and ABA (Li et al., 2012b). Treatment with 24-EpiBL increased expression of various hormone marker genes in both wild-type and mutant seedlings of *Arabidopsis* that are either deficient in or insensitive to ABA, ET, JA, and SA, thus exhibiting the ability of EpiBL to control the basic thermotolerance and salt tolerance of the mutants (Divi et al., 2010), confirming their synergistic role.

The work with BR-biosynthetic mutants in *Arabidopsis thaliana* (Tao et al., 2004) and *Pisum sativum* (Nomura et al., 1997) have provided strong evidence that BRs are essential for plant growth and development and BR-signaling plays a positive in plant growth and development (Fábregas and Caño-Delgado, 2014). A simple BR-analogue 2α, 3α-dihydroxy-17β-(3-methyl butynyloxy) 7-oxa-B-homo-5α androstan-6-1 induces bean second node splitting, which is considered as the prominent physiological feature of BRs (Strnad and Kohout, 2003). Dwarf and deetiolated phenotypes and BR-deficient species of some *Arabidopsis* mutants were rescued by application of BRs (Bishop and Yakota, 2001). Even *Pharbitis nil*, *Uzukobito* was a defective BR-biosynthetic dwarf mutant strain (Suzuki et al., 2003) that emphasized that BR-deficient and defective BR-biosynthetic species exhibited abnormal growth. Friedrichsen et al. (2002) reported that three redundant BR genes encode transcription factors that are required for normal growth, indicating the necessity of BRs for normal growth. Similarly, the inhibition of growth (Asami et al., 2000) and secondary xylem development (Nagata et al., 2001) of cress (*Lepidius sativus*) by brassinozole, a specific inhibitor of BL synthesis, was reversed by the exogenous application of BL, also indicating the necessity of BRs for normal plant growth.

Further, the study with the BR-biosynthetic mutants also revealed their ability in overcoming various abiotic (environmental) as well as biotic stresses in plants. BR biosynthetic gene *AtDWF4* opposed the inhibitory effects of ABA on seed germination and promoting cold stress tolerance in *Arabidopsis* seedlings (Divi and Krishna, 2010); the activation of stress-inducible genes in a *brassinosteroid-insensitive 1* (*bri1*) mutant resulted in higher tolerance to cold (Kim et al., 2010). Endogenous BRs are positively involved in the plant response to salt stress in *Arabidopsis* mutants such as det2-1, bin2-1, and Columbia wild-type on one hand; on the other hand, exogenously applied 24-EpiBL enhanced proline accumulation, promoted root elongation of wild-type, and partially rescued the growth of det2-1 under salt stress (Zeng et al., 2010). Goddard et al. (2014) reported that enhanced disease resistance is caused by BRI1 mutation in *Brachypodium distachyon* and barley (*Hordeum vulgare*). Thus, the ability of BRs in overcoming the negative effect of various environmental stresses is established through the BR-biosynthetic mutant studies. The present chapter is a study on the role of BRs in overcoming the hazardous effect of heavy metals on plants and gives clear insight on the role of BRs in alleviating/mitigating/ameliorating the negative impact of various heavy metals on the growth and development of plants.

2. ABIOTIC STRESSES

Plants continuously are under the influence of various environmental stresses (e.g., abiotic, biotic). Abiotic stresses are prevalent in today's plants because of various man-made environmental factors such as pollution, global warming, and greenhouse gases (Figure 8.2).

Abiotic stresses are usually caused by physical or chemical constituents of the universe which include temperature (heat, chilling), salt, water (flooding or drought), metals/metalloids, light (intense, weak), radiation (ultraviolet A, ultraviolet B), winds, allelochemicals, and organic pollutants (herbicides/pesticides/insecticides). Rapid industrialization and urbanization in the twentieth

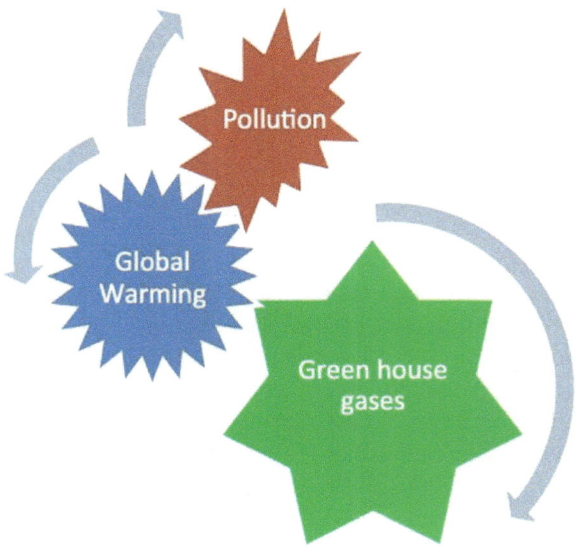

FIGURE 8.2

Reasons for heavy metal stress.

and twenty-first centuries across the globe paved the way for heavy metal and organic pollutant stresses. The ever-increasing concentrations of heavy metals in the environment directly lead to drastic and lethal changes in the plant growth and metabolism. The toxicity symptoms observed in the plants from the presence of excess amounts of heavy metals are due to various interactions at the cellular as well as molecular levels. Toxicity may be the result of binding of metals to sulfhydryl groups in proteins, thus leading to inhibition of activity or disruption of structure (Hall, 2002).

BRs are potential growth hormones that mitigate the negative effect of stresses in plants. Rao et al. (2002) stated that BRs are a new group of plant growth hormones that perform a variety of physiological roles like growth, seed germination, rhizogenesis, and senescence and also confer resistance to plants against various abiotic stresses. Du and Poovaiah (2005) reported that BRs are plant-specific steroid hormones that have an important role in coupling environmental factors, especially light, with plant growth and development. Further, Bajguz and Hayat (2009) clearly explained the various effects of BRs on the plant responses to environmental stresses. BRs have been further explored for stress-protective properties in plants against numerous biotic and abiotic stresses in plants such as high temperature, low temperature in terms of chilling as well as freezing, salt, light, water in terms of drought as well as flooding, heavy metals, herbicides, pesticides, insecticides, organic pollutants, and inorganic pollutants. The role of BRs as potential ameliorators to overcome various stresses in plants in the past 5 years (excluding heavy metal stresses) is shown in Table 8.2. There are a few odd regulations of BRs such as improving growth, photosynthesis, and yield of rapeseed (*B. napus* L.) after being frozen (Ma et al., 2009), improving yield of six wheat cultivars (*Triticum aestivum* L.) grown under newly reclaimed sandy soil (Ahmed and Shalaby, 2013). Further, application of 24-EpiBL and 28-homoBL denigrated the seasonal stress in *Brassica juncea* by increasing shoot length, dry weight, proteins, and proline and antioxidative systems (Kumar et al., 2014) and BRs even overcame the ill effects of iron deficiency in *Cucumis sativus* (Wang et al., 2012a) and *Oryza sativa* (Wang et al., 2015). Thus, although BRs were initially identified based on their growth-promoting activities, the subsequent physiological and genetic studies revealed additional functions of BRs in regulating a wide range of processes, including source/sink relationships, seed germination, photosynthesis, senescence, photomorphogenesis, flowering, and responses to different abiotic and biotic stresses (Deng et al., 2007).

3. HEAVY METAL STRESS

Heavy metal stress is considered one of the key abiotic stresses in plants that are responsible for immense damage in plant growth and metabolism. The various heavy metals represented in Figure 8.3 are not only the root causes of today's pollution, but also play damaging roles by hampering plant growth and metabolism. Heavy metals are posing a threat to the environment and are considered as essential environmental pollutants across the globe because of rapid industrialization and urbanization in the twentieth and twenty-first centuries. The ability of EpiBL to reduce the absorption of heavy metals by different plants like barley, sugar beet, tomato, and radish was reported by Khripach et al. (1996). Bajguz (2000) also reported that EpiBL blocked the heavy metals (copper (Cu), lead (Pb), cadmium (Cd), and zinc (Zn)) accumulation in *Chlorella vulgaris* cells.

The application of 24-EpiBL to *B. junceae* L. cv PBR 91 seedlings removed the negative impact of heavy metals such as Zn, manganese (Mn), cobalt (Co), and nickel (Ni) improved the growth and reduced the accumulation as well as uptake of the metals (Sharma and Bhardwaj, 2007b). Application

Table 8.2 BRs as Potential Ameliorators of Biotic and Abiotic Stresses (Excluding Heavy Metals) in Plants (2010–2015)

Stress	Plant	Compound	Effect	References
Biotic/fungi	Jujube	BRs	Antidefense enzymes under *Penicillium expansum* stress	Zhu et al. (2010)
	Barley		Resistance against *Fusarium* wilt	Ali et al. (2013)
	Tomato		Arbuscular mycorrhiza formation	Bitterlich et al. (2014)
	Atractylodes lancea	BL	Volatile oil accumulation	Ren et al. (2014)
Biotic/virus	Cotton	BRs	Virus-induced gene silencing to *Verticillium dahliae*	Gao et al. (2013a)
	Arabidopsis thaliana		Defense against virus	Zhang et al. (2015)
	Cotton	EpiBL	Carbohydrates metabolism, plasma membrane ATPases and intracellular osmolytes under *Verticillium* wilt	Bibi et al. (2014)
Biotic/nematodes	Rice	BRs	Resistance against root knot	Nahar et al. (2013)
	Tomato	28-Homo-BL	Resistance	Kaur et al. (2013a)
Abiotic/high temperature/heat	*Oryza sativa*	BRs and Ecdysone analogue	Pollen germination	Thussagunpanit et al. (2013)
		BRs	Improving photosynthesis, lipid peroxidation	Thussagunpanit et al. (2014)
	Lycopersicon esculentum		Mitigation of stress	Zhou et al. (2014)
	Hordeum vulgare	24-EpiBL	Metabolism	Janeczko et al. (2011)
	Melon		Photosynthesis	Zhang et al. (2013)
	O. sativa		Increased photosynthesis and yield	Thussagunpanit et al. (2015)
Abiotic/low temperature/chilling	*Solanum lycopersicum*	BRs	PAL activity	Aghdam et al. (2012)
	Cucumis sativus		Photosynthetic apparatus	Jiang et al. (2013)
			Alleviation of toxic effect of chlorpyrifos	Hu et al. (2013b)
	Green bell pepper	BL	Mitigation of stress	Wang et al. (2012b)
	L. esculentum		Tolerance to stress	Aghdam and Mohammadkhani (2014)

Continued

Table 8.2 BRs as Potential Ameliorators of Biotic and Abiotic Stresses (Excluding Heavy Metals) in Plants (2010–2015) —cont'd

Stress	Plant	Compound	Effect	References
	O. sativa		Yield and stress resistance	Wang et al. (2014)
	C. sativus	24-EpiBL	Photosynthesis	Hu et al. (2010)
	Brassica juncea		Antioxidative system	Kumar et al. (2010)
	Grape vines		Antioxidative defense and osmoregulation systems	Xi et al. (2013)
	C. sativus	28-Homo-BL	Growth and photosynthesis	Fariduddin et al. (2011)
Abiotic/drought	Tomato	BRs	ABA content	Yuan et al. (2010)
	Papaya		Senescence	Gomes et al. (2013)
	Raphanus sativus		Germination and seedling growth	Mahesh et al. (2013)
	Gossypium hirsutum		Fiber development	Nigam (2014)
	Zea mays	BL	Antioxidants and leaf gas exchange	Anjum et al. (2011)
	Xanthoceras sorbifolia		Overcome stress	Li and Feng (2011)
	Vigna radiata		Growth and yield	Lal et al. (2013)
	Arachis hypogeae		Enzyme activities	Savaliya et al. (2013)
			Biochemical and physiological parameters	Savaliya et al. (2014)
	Chorispora bungeana		Protective against stress	Li et al. (2012a)
	Capsicum annuum		photosynthesis	Hu et al. (2013a)
Abiotic/flooding/ hypoxia	*Z. mays*	BRs	Recovery from waterlogging	Leng et al. (2013)
	C. sativus	BL	ATPase activity and inorganic ions	Lu and Gao (2013)
Abiotic/salinity/salt	*Z. mays*	BRs	Photosystem-I	Honnerova et al. (2010)
	C. sativus		Seed germination	Wang et al. (2011)
	O. sativa		Stress tolerance	Sharma et al. (2013b)
	Trifolium alexandrinum		Growth and yield	Daur and Tatar (2013)
	Brassica napus		Enhanced tolerance	Efimova et al. (2014)
	Z. mays		Increased osmolytes	Rattan et al. (2014)
	Sorghum vulgare	BL	Improved antioxidative system	Vardhini (2011)
	V. radiata		Antioxidant activity	El-Mashad and Mohamed (2012)

	O. sativa		Physiological activities	Das et al. (2013)
	C. sativus		Alleviation of stress	Lu and Yang (2013)
	Triticum aestivum		Improved biochemical parameters	Mandavia et al. (2014)
	Wheat	24-EpiBL	Hormonal status	Avalbaev et al. (2010)
	Cajanus cajan		Overcame stress	Durigan et al. (2011)
	Fragaria x ananassa		Stomatal conductance, membrane, and ion permeability	Karlidag et al. (2011)
	Phaseolus vulgaris		Growth and yield	Rady (2011)
	Pisum sativum		Growth	Shahid et al. (2011)
	Egg plant		Amelioration of stress	Ding et al. (2012)
	Lettuce		Mitigation of stress	Ekinci et al. (2012)
	Pepper		Physiological and genetic changes	Abbas et al. (2013)
	Pigeon pea		Nitrogen metabolism	Dalio et al. (2013)
	C. sativus		Photosynthesis and proline	Fariduddin et al. (2013a,b)
	Pisum sativum		Protein tyrosine phosphorylation	Fedina (2013)
	T. aestivum		Nutrients and compatible solutes	Talaat and Shawky (2013)
	C. sativus		Physiological and biochemical responses for the salt stress mitigation	Fariduddin et al. (2014)
	V. radiata	28-Homo-BL	Photosynthesis, fluorescence	Hayat et al. (2010b)
	T. aestivum		Protection against stress	Yusuf et al. (2011b)
	B. juncea		Photosynthesis	Alyemeni et al. (2013)
	Lettuce	BR-analogue	Reversed the ill effects of salinity	Serna et al. (2015)
Abiotic/light	Tomato	BRs	Photosynthesis	Ogweno et al. (2010)
	Arabidopsis and sunflower		Shoot growth under light stress	Kurepin et al. (2012)
	L. esculentum	24-EpiBL	Photosynthesis under weak light stress	Wang et al. (2010)
Abiotic/UV radiation	*P. sativum*	24-EpiBL	Pigments under UV B stress	Dobrikova et al. (2013)
Abiotic/organic compounds	Tomato	BRs	Induce plant tolerance to phenanthrene stress	Ahammed et al. (2012a)
			Alleviates phenanthrene and pyrene phytotoxicity	Ahammed et al. (2012b)

Continued

Table 8.2 BRs as Potential Ameliorators of Biotic and Abiotic Stresses (Excluding Heavy Metals) in Plants (2010–2015)—cont'd

Stress	Plant	Compound	Effect	References
			Germination and early development of seedling	Ahammed et al. (2012c)
			Photosynthesis under phenanthrene-cadmium cocontamination stress	Ahammed et al. (2013a)
			Alleviation of polychlorinated biphenyls stress	Ahammed et al. (2013b)
			Secondary metabolism under phenanthrene stress	Ahammed et al. (2013c)
	Tomato, rice, tea, broccoli, cucumber, strawberry, etc.		Decreased pesticide residues in plants likely by promoting their metabolism through a signaling pathway involving BR-induced H_2O_2 production and cellular redox change	Zhou et al. (2015)
	Z. mays	BL	Growth under preservative stress	Hu et al. (2014)
	O. sativa	24-EpiBL	Protection against imidacloprid stress	Sharma et al. (2013a)
Abiotic/inorganic compounds	*L. esculentum*	BRs	Photosynthesis under nitric oxide stress	Hayat et al. (2010c)
	C. sativus		H_2O_2 under nitric oxide stress	Cui et al. (2011)
	Z. mays		ABA under nitric oxide stress	Zhang et al. (2011)
	Brassica chinensis	BL	Growth and quality under calcium nitrate stress	Yue et al. (2013)
	C. sativus	24-EpiBL	Photosynthesis and chloroplast ultrastructure against $Ca(NO_3)_2$ stress	Yuan et al. (2012)

24-EpiBL, 24-epibrassinolide; 28-Homo-BL, 28-homobrassinolide; ABA, abscisic acid; ATPase, adenosine triphosphatase; BL, brassinolide; BR, brassinosteroid; $Ca(NO_3)_2$, calcium nitrate; EpiBL, 24-epibrassinolide; H_2O_2, hydrogen peroxide; UV, ultraviolet.

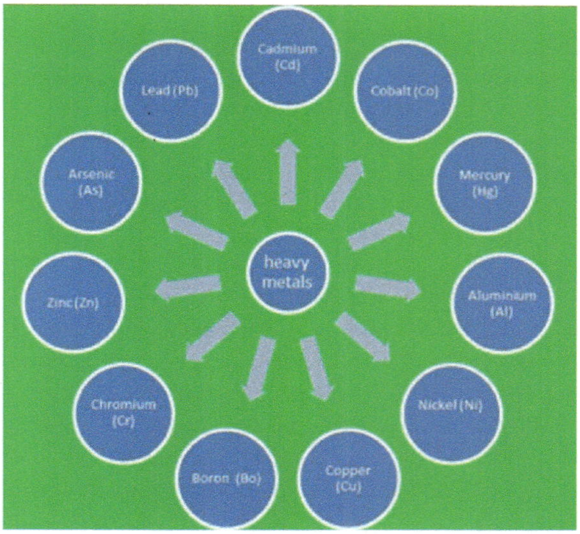

FIGURE 8.3

Heavy metals.

of three different types of BRs (24-EpiBL, 24-EpiCS, and 4154) removed the negative impact heavy metals (Cu, Cd, Pb, and Zn) in spring wheat var. Vánek cultivated in pots by increasing growth, biomass of whole plants as well grain yield (Kroutil et al., 2010). Research have shown that BL plays an important role in plant responding to heavy metals' stresses and has an antistress effect on *C. vulgaris* contaminated by heavy metals such as Cu, Pb, or chromium (Cr) by activating enzymatic (superoxide dismutase (SOD) EC 1.15.1.1, catalase (CAT) EC 1.11.1.6, ascorbate peroxidase (APX) EC 1.11.1.11, glutathione reductase (GR) EC 1.6.4.2, and nonenzymatic (reduced glutathione (GSH) total ascorbate)) systems (Bajguz, 2010). Further, application of 10^{-8} M BL to *C. vulgaris* cultures reduced the accumulation of heavy metals (Cd, Pb, and Cu) stress on growth; prevented chlorophyll, monosaccharides, and protein loss; and increased phytochelatins content (Bajguz, 2011). Recently, it was found that EpiBL (10^{-8} to 10^{-11} M) had a stimulating effect on the seed germination and seedling growth of maize affected by Zn and Ni and also reduced heavy metal-induced electrolyte leakage from maize cells Lukatkin et al. (2013).

3.1 BRs AND LEAD TOXICITY

Application of 20-hydroxyecdysone (10^{-10} to 10^{-8} M) to *C. vulgaris* cultures reduced the impact of 10^{-6} to 10^{-4} M Pb stress on growth, prevented chlorophyll, sugar and protein loss, and also increased phytochelatins synthesis (Bajguz and Godlewska-Zykiewlu, 2004). Bajguz (2002) reported that BRs (BL; 24-EpiBL; Homo-BL; CS; Homo Castasterone (homoCS); and 24-EpiCS) together with two levels of Pb (10^{-6} and 10^{-4} mol/L) showed a stronger stimulation of the synthesis of phytochelatins by *C. vulgaris* cell than the culture treated with Pb alone and the effect of BRs and Pb on *C. vulgaris* growth was very quick. Rady and Osman (2012) reported that 24-EpiBL increased the activities of the antioxidative enzymes such as SOD, CAT, APX, GR, ascorbic acid, and GSH in tomato plants grown in presence of 100 or 200 μM Pb. Supplementation of 24-EpiBL reduced Pb toxicity and enhanced the growth in radish (*Raphanus sativus* L.) seedlings by increasing the activities of antioxidant enzymes like CAT, APX, guaiacol peroxidase ((GPX) EC 1.11.1.7), SOD, and reducing peroxidase ((POD) EC 1.11.1.11) activity demonstrating the ameliorating ability of 24-EpiBL in scavenging the ROS, thereby reducing the oxidative stress induced by Pb (Anuradha and Rao, 2007a). Recently, Swamy et al. (2014) reported that application of 28-Homo-BL to *Trigonella feonu-graecum* seeds mitigated the toxicity of Pb and improved plant biomass, photosynthesis, carbohydrates, proteins, and nucleic acids.

3.2 BRs AND COPPER TOXICITY

Pregermination treatment of 24-EpiBL improved plant growth, shoot emergence, and plant biomass production in plants of *B. juncea* L. cv. PBR 91 under Cu metal stress (Sharma and Bhardwaj, 2007a). Further, Cu metal uptake and accumulation in the plants was also found reduced by 24-EpiBL (Sharma and Bhardwaj, 2007b). Fariduddin et al. (2009) reported that treatment of *B. juncea* seedlings with 10^{-10}, 10^{-8}, and 10^{-6} M homoBL improved the growth, photosynthetic parameters, and antioxidant enzymes such as CAT, POD, SOD, and proline while decreased hydrogen peroxide (H_2O_2) contents, and the elevated antioxidant enzymes and proline might have been responsible to overcome the toxic effects of Cu. Applying 24-EpiBL (10^{-11}, 10^{-9}, and 10^{-7}) ameliorated the toxic effect of Cu (0.25, 0.5, and 0.75 mM) in *B. juncea* plants by enhancing the growth, sugars, and lipid metabolism (Kaur et al., 2013b). Poonam et al. (2014) also reported that seeds of *B. juncea* plants soaked in the solution of 24-EpiBL (10^{-7}, 10^{-9}, and 10^{-11} M) and grown in Cu (0.25, 0.50, and 0.75 mM) showed increased protein content and activities of antioxidative enzymes (CAT, SOD, POD, monodehydroascorbate reductase ((MDHAR) EC 1.1.5.4), and dehydroascorbate reductase ((DHAR) EC 1.8.5.1 APOX, GR)).

Fariduddin et al. (2013a) also studied that treatment of EpiBL to the sodium chloride + Cu (100 mg/kg) stressed seeds of two cultivars (Rocket and Jumbo) of *C. sativus* plant improved growth, photosynthetic parameters, carbonic anhydrase ((CA) EC 4.2.1.1), proline and activities of various antioxidant enzymes such as CAT, POD, and SOD. Filova et al. (2013) also reported that BRs eliminated the toxic effect of Cu in six sunflower cultivars (*Helianthus annuus* L. cv. Belinda, cv. Codiwer, cv. ESPrim, cv. MAS 95, cv. MAS 97, and cv. Spirov) by decreasing the lipid peroxidation (MDA), enhancing chlorophylls, proline, and relative water content (RWC). Choudhary et al. (2010) reported that EpiBL not only induced changes in indole-3-acetic acid, ABA, and polyamine concentrations, but also enhanced the antioxidative system of radish seedlings subjected to Cu stress. Recently, Yin et al. (2014) reported that exogenous application of EpiBL mediated the plant growth, absorption, and accumulation of Cu, Fe, and Zn in tomato seedlings subjected to Cu stress. Arora et al. (2008b) reported that 28-Homo-BL ameliorated the toxic effect of Cu in *Zea mays* L. plants by enhancing seedling growth.

3.3 BRs AND CHROMIUM TOXICITY

Choudhary et al. (2011) investigated that 24-EpiBL ameliorated Cr(VI) stress in 7-day-old *R. sativus* L. cv. "Pusa chetki" seedlings by enhancing certain metabolite contents (GSH, proline, glycinebetaine, ascorbic acid, phytochelatins) and enzyme activities (GR, SOD, and APX) while reducing MDA and H_2O_2 contents. Sharma et al. (2011b) reported that seed presoaking treatment of 28-Homo-BL at $10^{(-7)}$ M was effective in ameliorating Cr(VI) stress in *R. sativus* L. (Pusa Chetaki) by increasing root/shoot length, fresh biomass, activities of antioxidant enzymes (APX, SOD, CAT, MDHAR, DHAR, GPX, and GR), lipid peroxidation, proline, and protein content while reducing MDA content. Arora et al. (2010b) also reported that 24-EpiBL regulated diminution of Cr metal toxicity in *B. juncea* L. plants by increasing protein content and activities of antioxidative enzymes (GPX, CAT, GR, APX, SOD, POD, MDHAR, and DHAR).

3.4 BRs AND BORON TOXICITY

Yusuf et al. (2011a) reported that application of 28-Homo-BL ($10^{(-8)}$ M) to boron (0.5, 1.0, and 2.0 mM) stressed *Vigna radiata* plants resulted in improved growth, water relations, photosynthesis, proline, and enhanced antioxidant enzymes (CAT, POD, and SOD). The elevated levels of antioxidant enzymes as well as proline might have played a pivotal role to confer tolerance to boron-stressed *V. radiata* plants (Yusuf et al., 2011a).

3.5 BRs AND ZINC TOXICITY

Ramakrishna and Rao (2013) reported the involvement of GSH metabolism and redox status against Zn toxicity in *R. sativus* seedlings by supplementation of 28-Homo-BL. Application of 28-Homo-BL ameliorated Zn toxicity in *Z. mays* L. (Arora et al., 2008a) and *B. junceae* (Sharma et al., 2007) by enhanced growth, reduced lipid peroxidation, and enhanced antioxidative enzyme activities, nonenzymatic antioxidants, and protein contents. Further, Arora et al. (2010a) also reported that seed treatment of 24-EpiBL mitigated the negative impact in 30-day-old leaves of *B. juncea* (var. PBR 91) under Zn metal stress in field conditions by increasing the activities of antioxidative enzymes (SOD, CAT, APOX, GPX, GR, MDHAR, and DHAR), proteins, and GSH, whereas MDH content was found reduced. Recently, Ramakrishna and Rao (2015) studied that application of 24-EpiBL and 28-Homo-BL elevated the levels of antioxidative enzymes as well as antioxidants conferring resistance to radish plants against Zn^{2+} stress resulting in improved plant growth, relative water content, and photosynthetic attributes.

3.6 BRs AND CADMIUM TOXICITY

24-EpiBL (100 nM) reduced the toxic effect of Cd (300 μM) in winter rape seedlings by increasing contents of Chl *a* and *b* carotenoids, acting on the photochemical processes by diminishing the damage caused to photochemical active reaction centers and O_2 evolving centers as well as maintaining efficient photosynthetic electron transport (Janeczko et al., 2005). Application of 0.01 μM of homoBL to 100–150 μM Cd treated *B. juncea* seedlings decreased concentration of Cd, increased plant growth, photosynthesis, CA, nitrate reductase ((NR) EC 1.6.6.1), antioxidative enzymes such as CAT, SOD, POD, and contents of proline (Hayat et al., 2007). Gao et al. (2013b) also reported that exogenous BRs enhanced growth, photosynthesis, and reduced Cd biological enrichment in *Helianthus tuberosus* seedlings subjected to Cd stress.

Rady (2011) reported that plants of *Phaseolus vulgaris* L. grown in the presence of $CdCl_2$ and sprayed with 5 μM of 24-EpiBL improved membrane stability index (MSI), RWC, antioxidative enzymes activities, proline, and minimized plant Cd^{2+} content but could not influence electrolyte leakage and lipid peroxidation. Hasan et al. (2008) reported that spraying of 28-Homo-BL protected seeds of *Cicer arietinum* (L.) cv. Uday from Cd (50, 100, or 150 μM) by increasing plant fresh mass, dry mass, number of nodules, leghemoglobin content, nitrogen, carbohydrate content, leaf chlorophyll content, proline content, NR, CA, and activities of antioxidant enzymes (CAT, POD, and SOD).

Both 24-EpiBL and 28-homoBL sprayed to seeds of two tomato cultivars (cv. K-25 and Sarvodya) supplied with 100 μM $CdCl_{(2)}$ by the shotgun approach exhibited improved growth, photosynthetic efficiency, NR, CA, proline, and antioxidative enzyme activities (Hayat et al., 2010a). Hasan et al. (2011) also reported that spraying of 10^{-8} M of 28-Homo-BL or 24-EpiBL to two tomato cultivars (K-25 and Sarvodya) subjected to Cd stress, not only improved the activity of photosynthetic machinery, antioxidant defense system in both the cultivars, but also nullified the damaging effect of the metal on these parameters. Further, application of BRs (10^{-8} M) to Cd (3, 6, 9, and 12 mg/kg)-stressed *Solanum lycopersicum* plants exhibited enhanced antioxidant system activity, photosynthetic assimilation efficiency, fruit yield, and quality of fruit (Hayat et al., 2012). Zhao et al. (2013) also reported that exogenous BRs conferred Cd tolerance to *Solatium nigrum* seedlings.

The studies conducted by Anuradha and Rao (2007b) clearly indicated the ameliorative influence of 24-EpiBL and 28-Homo-BL (3 μM) on the inhibitory effect of Cd toxicity by increased percentage of seed germination, seedling growth, length, fresh weight and dry weight, free proline, and activities of antioxidant enzymes (CAT, SOD, APX, and GPX), and reduced POD as well ascorbic acid oxidase (AAO) in *R. sativus* L. Further, seed application of 24-EpiBL ameliorated the toxic effect of Cd on plant growth, pigment content, photosynthesis, CA, and NR activities in the leaves of *R. sativus* L. under the influence of Cd stress (Anuradha and Rao, 2009). Sharma et al. (2012) reported that supplementation of 24-EpiBL (10^{-7}, 10^{-9}, 10^{-11} M) to *R. sativus* L. ameliorated the oxidative stress caused by Cd (0.5, 1.0, 1.5 mM) by enhancing the protein content and activities of glutathione S-transferase ((GST) EC 2.5.1.18) and poly phenol oxidase (PPO) enzymes. Further, Sharma et al. (2010) reported that 28-Homo-BL (10^{-11}, 10^{-9}, 10^{-7} M) ameliorated Cd (0.5, 1.0, 1.5 mM) metal stress by regulating the activities of antioxidant enzymes in *R. sativus* L. by enhancing the seedling length, biomass, protein content, and activities of antioxidant enzymes (APX, CAT, GR, POD, and SOD). A recent study revealed that application of 28-Homo-BL removed the negative impact of Cd in radish plants by improved protein content as well as activities of antioxidant enzymes, such as GST and PPO (Sharma et al., 2014). Kapoor et al. (2015) reported that 24-EpiBL mitigated the negative impact of cadmium toxicity in 7-day-old seedlings of *R. sativus* by enhancing percent germination, growth (root length, shoot length, fresh weight, and dry weight), lipid peroxidation, sodium and potassium ion concentrations, proline content, total osmolyte content, and level of antioxidants (ascorbic acid, tocopherol, and GSH).

3.7 BRs AND ALUMINUM TOXICITY

Dong et al. (2008) reported that supplementation of BRs to soybean improved the photosynthetic characteristics under aluminum (Al) toxicity. BRs ameliorated the Al toxicity in mung bean (*Phaseolus aureus*) seedlings and promoted seedling growth and chlorophyll content under Al metal stress (Bilikisu et al., 2003). Mungbean (*P. aureus* Roxb.) seedlings grown in a growth chamber in solution with combined concentrations of Al (2 and 5 mM) and BL (0.1, 10, 1000, 100,000 ng/L) showed that BL had promoted growth of mung bean seedlings under Al stress by increasing seedling height, root length, fresh weights of shoots and roots, and chlorophyll content (Abdullahi et al., 2003). Spraying of EpiBL or homoBL ameliorated the Al stress through antioxidant system in the mung bean (*V. radiata* L. Wilczek) by increasing growth (length, fresh, and dry mass of root and shoot), the activity of CA, RWC, water use efficiency, chlorophyll content, rate of photosynthesis, and activities of antioxidative enzymes (CAT, POD, and SOD) in leaves and the content of proline, both in leaves as well as roots (Ali et al., 2008b).

3.8 BRs AND NICKEL TOXICITY

The plants of *B. juncea* (L) grown in the presence of $NiCl_2$ and sprayed with 1 μM of 24-EpiBL at 15 days after sowing exhibited a significant improvement in growth, level of pigments, photosynthetic parameters, MSI, RWC, antioxidative enzymes, and the level of proline (Ali et al., 2008a). The amelioration of Ni toxicity was observed as reflected in enhanced activities of antioxidant enzymes by the application of isolated 24-EpiBL in 30-day-old *B. juncea* L. plants (Kanwar et al., 2013). BRs (24-EpiBL, CS, dolicholide, and typhasterole) mitigated *B. juncea* L. plants subjected to various concentrations of Ni metal (0.2, 0.4, and 0.6 mM) significantly by increasing the activities of stress ameliorating enzymes and lowering the metal uptake (Kanwar et al., 2012). Sharma et al. (2008) reported that the seedlings of *B. juncea* L. treated with different concentrations (0.01, 1, and 100 nM) of homoBL for 8 h and then subjected to 0, 25, 50, and 100 mg/dm^3 of Ni exhibited increased shoot and root length, protein content and activates of antioxidative enzymes (GPX, CAT, GR, APX, and SOD) indicating the alleviative effect of homoBL. Spraying with 28-Homo-BL to *B. juncea* L. cv. T-59 supplied with 50 or 100 μM Ni (Ni_{50}, Ni_{10}) partially neutralized the toxic effect of Ni on growth, net photosynthetic rate, content of chlorophyll, and the activities of NR, and CA and boosted the contents of POD and CAT in leaves and proline both in leaves and roots (Alam et al., 2007).

Archana et al. (2006) reported that supplementation of 24-EpiBL to rice seedlings ameliorated the negative impact of Ni stress by promoted growth and metabolism. Spraying 0.01 μM of 28-Homo-BL to seeds of five wheat (*T. aestivum*) cultivars (PBW-373, UP-2338, DL-LOK-01, DL-373, and HD-2338) exposed to 0, 50, and 100 μM of Ni ($NiCl_2$) ameliorated the toxic effects generated by Ni, exhibiting enhanced net photosynthetic rate, proline content, CAT, POD, and SOD, whereas proline acted as an osmolyte and reactive oxygen species scavenger (Yusuf et al., 2011b). Bhardwaj et al. (2007) studied that supplementation of 28-Homo-BL regulated the seedling growth, lipid peroxidation, and antioxidative enzyme activities of *Z. mays* L seedlings under Ni stress.

Sharma et al. (2011a) reported that *R. sativus* L. seeds presoaked in 24-EpiBL sown in Petri dishes containing heavy metal, Ni reduced the toxicity of heavy metal by enhancing the root/shoot length, fresh biomass, activities of antioxidant enzymes (APX, SOD, CAT, GPX, MDHAR, DHAR, and GR), proline, and protein content while reducing lipid peroxidation measured in terms of MDA content. Application of 10^{-6} M 24-EpiBL as a shotgun approach (presowing seed soaking) to the Ni-stressed T-44 (Ni-tolerant) and PDM-139 (Ni-sensitive) varieties of *V. radiata* plants improved their biological yield, number of nodules, leghemoglobin content, activity of various antioxidant enzymes (e.g., CAT, POD, SOD) and also the content of proline and the upregulation of antioxidant enzymes as well as proline (osmolyte) triggered by 24-EpiBL could have conferred tolerance to the Ni-stressed plants resulting in improved growth, nodulation, and yield attributes (Yusuf et al., 2012). Recently, Yusuf et al. (2014) also reported that BRs enhanced the antioxidant system and nitrogen metabolism in two contrasting cultivars of *V. radiata* under different levels of Ni.

3.9 BRs AND ARSENIC TOXICITY

Recently, Raghu et al. (2014) reported that application of BRs to *R. sativus* seedlings inhibited the negative effect of arsenic by increasing proteins, proline, SOD, CAT, and reducing the lipid peroxidation. Kanwar et al. (2015) studied that *B. juncea* L. plants exposed to different concentrations (0.1, 0.2, and 0.3 mM) of arsenic(V) and harvested after 30 and 60 days induced the synthesis of 4 BRs such as CS, teasterone, 24-epibrassinolide, and typhasterol.

3.10 BRs AND MANGANESE TOXICITY

Application of 28-Homo-BL (Bhardwaj et al., 2008) and BRs (Bhardwaj et al., 2011) regulated Mn uptake and growth of *B. juncea* L. Wang et al. (2009) reported that application of EpiBL enhanced the activities of SOD, POD, CAT, APX, DHAR, and GR; reduced ascorbate and GSH, and MDA, lipoxygenase (1.13.11.12), superoxide radical, and H_2O_2 were greatly decreased in *Z. mays* L. subjected to Mn stress, concluding the ameliorative effects of EpiBL on Mn toxicity might have been because of the upregulation of antioxidative capacity in maize under Mn stress.

3.11 BRs AND COBALT TOXICITY

Foliar spray treatment with 24-EpiBL (10^{-10}, 10^{-8}, and 10^{-6} M) alleviated the stress generated by Co (5×10^{-4}, 1.5×10^{-3}, and 2×10^{-3} M) ion in *B. juncea* L. and significantly improved shoot length, number of leaves, activities of antioxidative enzymes (SOD, CAT, POD, GR, APOX, MDHAR, and DHAR) and protein content (Arora et al., 2012).

3.12 BRs AND MERCURY TOXICITY

Sharma et al. (2012) reported that supplementation of 10^{-7}, 10^{-9}, 10^{-11} M 24-EpiBL to *R. sativus* L. ameliorated the oxidative stress caused by mercury (Hg; 0.5, 1.0, 1.5 mM) by enhancing the protein content and activities of GST and PPO enzymes. Using 28-Homo-BL removed the negative impact of Hg in radish plants and improved protein content as well as activities of antioxidant enzymes such as GST and PPO (Sharma et al., 2014). Recently, Kapoor et al. (2014) also studied that treatment of 10^{-7}, 10^{-9}, and 10^{-11} M 24-EpiBL ameliorated the toxic effect of Hg in radish plants by increasing the activities of antioxidative enzymes (SOD, CAT, APOX, GPX, GR, MDHAR, and DHAR), proteins, and GSH while reducing MDH content. Application of 24-EpiBL proved effective in overcoming mercury stress in 7-day-old seedlings of *R. sativus* by increasing growth of seedlings and altering the level of ions, osmoprotectants, and antioxidants of plant (Kapoor et al., 2015).

4. COAPPLICATION OF BRs AND OTHER COMPOUNDS UNDER HEAVY METAL STRESS

Choudhary et al. (2012) reported that coapplications of EpiBL and spermidine (polyamine) remarkably enhanced the titers of antioxidants (GSH, ascorbic acid, proline, glycine betaine, and total phenols) and activities of antioxidant enzymes (GPX, SOD, and GR) in Cr-stressed *R. sativus* L. plants. Application of BRs to tomato plants ameliorated phenanthrene + Cd cocontamination-induced stress by improving photosynthetic activity, pigment contents, enzyme-related transcript levels of the antioxidant system, secondary metabolism, and the xenobiotic detoxification system of the plants (Ahammed et al., 2013a).

5. CONCLUSION AND FUTURE PROSPECTS

Schnabl et al. (2001) reported about BR-induced stress tolerances of plants and possible mechanisms involved in it. Various data consistently provided evidence that exogenous BR treatment was effective in stressful rather than optimal conditions (Fujita et al., 2006). Gruszka (2013) studied that the BR-signaling pathway are the new key players and provides interconnections with other signaling networks that are very crucial for plant development and stress tolerance. Hao et al. (2013) reported that genetic, genomic, and proteomic studies in *Arabidopsis* have identified major BR signaling components and elucidated the signal transduction pathway from the cell-surface receptor kinase BRI1 to the BES1/BZR1 family of transcription factors wherein BRs interact with other plant hormones in coordinating gene expression, plant growth, development, and stress tolerance.

BRs were reported to provide a significant modulation of various types of stresses by upregulation of antioxidative enzyme activities and osmoprotectants in various plant species (Hayat et al., 2012; Sharma et al., 2012). Plants have a range of potential mechanisms at the cellular level that might be involved in the detoxification of heavy metals, thus increasing plant tolerance to heavy metal stress (Vázquez et al., 2013). One of the mechanisms for heavy metal detoxification in plants is the chelation of the metal ion by ligands, such as organic acids, amino acids, peptides, and polypeptides. Peptide ligands include the metallothioneins or small gene-encoded cysteine-rich polypeptides (Bajguz, 2002). Villiers et al. (2012) reported that the BR-signaling pathway plays a role in the plant response to the heavy metal Cd and gene expression where BR-regulated genes in roots of Cd-treated wild-type plants of *A. thaliana*. Divi et al. (2015) provided clear links between the molecular changes impacted by BR and its ability to confer broad-range stress tolerance, emphasizing on the importance of post-translational modification and protein turnover as BR regulatory mechanisms and also demonstrated the ability of BR transcriptome as a repertoire of new stress-related regulatory and structural genes. Bartwal et al. (2013) stated that BRs are another important category of plant metabolites that exhibit stress regulatory, growth-promoting activities and can be catapulted as highly promising and environmentally friendly natural substances, suitable for wider application in plant protection, crop yield promotion, and defense under various environmental stresses. Thus the role of BRs in upregulating various processes in plants in overcoming environmental stresses, especially different type of heavy metal stresses, is the need of the day.

REFERENCES

Abbas, S., Latif, H.H., Elsherbiny, E.A., 2013. Effect of 24-epibrassinolide on the physiological and genetic changes on two varieties of pepper under salt stress conditions. Pak. J. Bot. 45, 1273–1284.

Abdullahi, B.A., Gu, X.G., Gan, Q.L., Yang, Y.H., 2003. Brassinolide amelioration of aluminum toxicity in mung bean seedling growth. J. Plant Nutr. 26, 1725–1734.

Aghdam, M.S., Asghari, M., Farmani, B., Mohayeji, M., Moradbeygi, H., 2012. Impact of postharvest brassinosteroids treatment on PAL activity in tomato fruit in response to chilling stress. Sci. Hortic. 144, 116–120.

Aghdam, M.S., Mohammadkhani, N., 2014. Enhancement of chilling stress tolerance of tomato fruit by postharvest brassinolide treatment. Food Bioprocess Technol. 7, 909–914.

Ahammed, G.J., Gao, C.J., Ogweno, J.O., Zhou, Y.H., Xia, X.J., Mao, W.H., Shi, K., Yu, J.Q., 2012a. Brassinosteroids induce plant tolerance against phenanthrene by enhancing degradation and detoxification in *Solanum lycopersicum* L. Ecotoxicol. Environ. Saf. 80, 28–36.

Ahammed, G.J., Yuan, H.L., Ogweno, J.O., Zhou, Y.H., Xia, X.J., Mao, W.H., Shi, K., Yu, J.Q., 2012b. Brassinosteroid alleviates phenanthrene and pyrene phytotoxicity by increasing detoxification activity and photosynthesis in tomato. Chemosphere 86, 546–555.

Ahammed, G.J., Zhang, S., Shi, K., Zhou, Y.H., Yu, J.Q., 2012c. Brassinosteroid improves seed germination and early development of tomato seedling under phenanthrene stress. Plant Growth Regul. 68, 87–96.

Ahammed, G.J., Choudhary, S.P., Chen, S., Xia, X., Shi, K., Zhou, Y., Yu, J., 2013a. Role of brassinosteroids in alleviation of phenanthrene-cadmium co-contamination-induced photosynthetic inhibition and oxidative stress in tomato. J. Exp. Bot. 64, 199–213.

Ahammed, G.J., Ruan, Y.P., Zhou, J., Xia, X.J., Shi, K., Zhou, Y.H., Yu, J.Q., 2013b. Brassinosteroid alleviates polychlorinated biphenyls-induced oxidative stress by enhancing antioxidant enzymes activity in tomato. Chemosphere 90, 2645–2653.

Ahammed, G.J., Zhou, Y.H., Xia, X.J., Mao, W.H., Shi, K., Yu, J.Q., 2013c. Brassinosteroid regulates secondary metabolism in tomato towards enhanced tolerance to phenanthrene. Biol. Plant. 57, 154–158.

Ahmed, M.A., Shalaby, M.A.F., 2013. Physiological role of brassinolide in improving yield of six wheat cultivars (*Triticum aestivum* L.) grown under newly reclaimed sandy soil. J. Appl. Sci. Res. (Ma'an, Jordan) 9, 6387–6393.

Alam, M.M., Hayat, S., Ali, B., Ahmad, A., 2007. Effect of 28-homobrassinolide treatment on nickel toxicity in *Brassica juncea*. Photosynthetica 45, 139–142.

Ali, B., Hayat, S., Fariduddin, Q., Ahmad, A., 2008a. 24-Epibrassinolide protects against the stress generated by salinity and nickel in *Brassica juncea*. Chemosphere 72, 1387–1392.

Ali, B., Hasan, S.A., Hayat, S., Hayat, Q., Yadav, S., Fariduddin, Q., Ahmad, A., 2008b. A role for brassinosteroids in the amelioration of aluminium stress through antioxidant system in mung bean (*Vigna radiata* L. Wilczek). Environ. Exp. Bot. 62, 153–159.

Ali, S., Kumar, S., Doohan, F., Khan, M., 2013. Brassinosteroid enhances resistance to *Fusarium* diseases of barley. Phytopathology 103, 1260–1267.

Alyemeni, M.N., Hayat, S., Wijaya, L., Anaji, A., 2013. Foliar application of 28-homobrassinolide mitigates salinity stress by increasing the efficiency of photosynthesis in *Brassica juncea*. Acta Bot. Bras. 27, 502–505.

Anjum, S.A., Wang, L.C., Farooq, M., Hussain, M., Xue, L.L., Zou, C.M., 2011. Brassinolide application improves the drought tolerance in maize through modulation of enzymatic antioxidants and leaf gas exchange. J. Agron. Crop Sci. 197, 177–185.

Anuradha, S., Rao, S.S.R., 2007a. Effect of 24-epibrassinolide on the growth and antioxidant enzyme activities in radish seedlings under lead toxicity. Indian J. Plant Physiol. 12, 396–400.

Anuradha, S., Rao, S.S.R., 2007b. The effect of brassinosteroids on radish (*Raphanus sativus* L.) seedlings growing under cadmium stress. Plant Soil Environ. 53, 465–472.

Anuradha, S., Rao, S.S.R., 2009. Effect of 24-epibrassinolide on the photosynthetic activity of radish plants under cadmium stress. Photosynthetica 47, 317–320.

Archana, G., Anuradha, S., Rao, S.S.R., 2006. Ameliorative influence of 24-epibrassinolide on nickel toxicity-induced growth inhibition in rice (*Oryza sativa* L.) seedlings. J. Plant 33, 221–226.

Arora, N., Bhardwaj, R., Sharma, P., Arora, H.K., Arora, P., 2008a. Amelioration of zinc toxicity by 28-homobrassinolide in *Zea mays* L. Can. J. Pure Appl. Sci. 2, 503–509.

Arora, N., Bhardwaj, R., Sharma, P., Arora, H.K., 2008b. Mitigation of Cu toxicity by 28-homobrassinolide in *Zea mays* L. Plant Stress 2, 127–130.

Arora, P., Bhardwaj, R., Kanwar, M.K., 2010a. 24-Epibrassinolide induced antioxidative defense system of *Brassica juncea* L. under Zn metal stress. Physiol. Mol. Biol. Plants 16, 285–293.

Arora, P., Bhardwaj, R., Kanwar, M.K., 2010b. 24-Epibrassinoliide regulated diminution of Cr metal toxicity in *Brassica juncea* L. plants. Braz. J. Plant Physiol. 22, 159–165.

Arora, P., Bhardwaj, R., Kanwar, M.K., 2012. Effect of 24-epibrassinolide on growth, protein content and anti-oxidative defense system of *Brassica juncea* L. subjected to cobalt ion toxicity. Acta Physiol. Plant 34, 2007–2017.

Arteca, J.M., Arteca, R.N., 2001. Brassinosteroid-induced exaggerated growth in hydroponically grown *Arabidopsis* plants. Physiol. Plant 112, 104–112.

Asami, T., Mink, Y.K., Nagata, N., Yamagishi, K., Takatsuto, S., Fujioka, S., Murofushi, N., Yamaguchi, T., Yoshida, S., 2000. Characterization of brassinozole, a triazole-type brassinosteroid biosynthesis inhibitor. Plant Physiol. 123, 93–100.

Ashraf, M., Akram, N.A., Arteca, R.N., Foolad, M.R., 2010. The physiological, biochemical and molecular roles of brassinosteroids and salicylic acid in plant processes and salt tolerance. Crit. Rev. Plant Sci. 29, 162–190.

Avalbaev, A.M., Yuldashev, R.A., Fatkhutdinova, R.A., Urusov, F.A., Safutdinova, Y.V., Shakirova, F.M., 2010. The influence of 24-epibrassinolide on the hormonal status of wheat plants under sodium chloride. Appl. Biochem. Microbiol. 46, 99–102.

Aydin, Y., Talas-Ogras, T., Ipekci-Altas, Z., Gozukirmizi, N., 2006. Effects of brassinosteroid on cotton regeneration via somatic embryogenesis. Biol. Bratislava 61, 289–293.

Bibi, N., Fan, K., Dawood, M., Nawaz, G., Yuan, S., Wang, X., 2014. Exogenous application of epibrassinolide attenuated *Verticillium* wilt in upland cotton by modulating the carbohydrates metabolism, plasma membrane ATPases and intracellular osmolytes. Plant Growth Regul. 73, 155–164.

Bajguz, A., 2000. Blockade of heavy metals accumulation in *Chlorella vulgaris* cells by 24-epibrassinolide. Plant Physiol. Biochem. 38, 797–801.

Bajguz, A., 2002. Brassinosteroids and lead as stimulators of phytochelatins synthesis in *Chlorella vulgaris*. J. Plant Physiol. 159, 321–324.

Bajguz, A., 2009. Brassinosteroid enhanced the level of abscisic acid in *Chlorella vulgaris* subjected to short-term heat stress. J. Plant Physiol. 166, 882–886.

Bajguz, A., 2010. An enhancing effect of exogenous brassinolide on the growth and antioxidant activity in *Chlorella vulgaris* cultures under heavy metals stress. Environ. Exp. Bot. 68, 175–179.

Bajguz, A., 2011. Suppression of *Chlorella vulgaris* growth by cadmium, lead, and copper stress and its restoration by endogenous brassinolide. Arch. Environ. Contam. Toxicol. 60, 406–416.

Bajguz, A., Godlewska-Zykiewlu, B., 2004. Protective role of 20-hydroxyecdysone against lead stress in *Chlorella vulgaris* cultures. Phytochemistry 65, 711–720.

Bajguz, A., Hayat, S., 2009. Effects of brassinosteroids on the plant responses to environmental stresses. Plant Physiol. Biochem. 47, 1–8.

Bajguz, A., Piotrowska-Niczyporuk, A., 2014. Interactive effect of brassinosteroids and cytokinins on growth, chlorophyll, monosaccharide and protein content in the green alga *Chlorella vulgaris* (*Trebouxiophyceae*). Plant Physiol. Biochem. 80C, 176–183.

Bartwal, A., Mall, R., Lohani, P., Guru, S.K., Arora, S., 2013. Role of secondary metabolites and brassinosteroids in plant defense against environmental stresses. J. Plant Growth Regul. 32, 216–232.

Bhardwaj, R., Arora, N., Sharma, P., Arora, H.K., 2007. The effects of 28-homobrassinolide on seedling growth, lipid peroxidation and antioxidative enzyme activities under nickel stress in seedlings of *Zea mays* L. Asian J. Plant Sci. 6 (5), 765–772.

Bhardwaj, R., Sharma, P., Arora, H.K., Arora, N., 2008. 28-Homobrassinolide regulated Mn-uptake and growth of *Brassica juncea* L. Can. J. Pure Appl. Sci. 2 (1), 149–154.

Bhardwaj, R., Sharma, P., Arora, N., Arora, H.K., 2011. Amelioration of Mn toxicity by brassinosteroids in *Brassica juncea* L. Int. J. Environ. Waste Manage. 8, 114–122.

Bilikisu, A.A., Xia-gang, G., Quing-lei, G., Yong-hua, Y., 2003. Brassinolide amelioration of aluminium toxicity in mung bean. J. Plant Nutr. 26 (9), 1725–1735.

Bishop, G.J., Yakota, T., 2001. Plant steroid hormones, brassinosteroids: current highlights of molecular aspects on their synthesis, metabolism, transport, perception and response. Plant Cell Physiol. 42, 114–120.

Bitterlich, M., Krugel, U., Boldt-Burisch, K., Franken, P., Kuhn, C., 2014. The sucrose transporter SlSUT2 from tomato interacts with brassinosteroid functioning and affects arbuscular mycorrhiza formation. Plant J. 78, 877–889.

Borcioni, E., Bonato-Negrelle, R.R., 2012. Application of analogue brassinosteroide (Biobras 16 (R)) germination and growth *in vitro* culture of zygotic embryos and acclimatization of bocaiuva. Ciencia Rural 42, 270–275.

Cheng, Y., Zhu, W., Chen, Y., Ito, S., Asami, T., Wang, X., 2014. Brassinosteroids control root epidermal cell fate via direct regulation of a MYB-bHLH-WD40 complex by GSK3-like kinases. eLife 3, e02525. http://dx.doi.org/10.7554/eLife.02525.

Choudhary, S.P., Bhardwaj, R., Gupta, B.D., Dutt, P., Gupta, R.K., Biondi, S., Kanwar, M., 2010. Epibrassinolide induces changes in indole-3-acetic acid, abscisic acid and polyamine concentrations and enhances antioxidant potential of radish seedlings under copper stress. Plant Physiol. 140, 280–296.

Choudhary, S.P., Kanwar, M., Bhardwaj, R., Gupta, B.D., Gupta, R.K., 2011. Epibrassinolide ameliorates Cr (VI) stress via influencing the levels of indole-3-acetic acid, abscisic acid, polyamines and antioxidant system of radish seedlings. Chemosphere 84, 592–600.

Choudhary, S., Kanwar, M., Bhardwaj, R., Yu, J., Tran, L., 2012. Chromium stress mitigation by polyamine-brassinosteroid application involves phytohormonal and physiological strategies in *Raphanus sativus* L. PLoS One 7, e33210.

Creelman, R.A., Mullet, J.E., 1997. Oligosaccharins, brassinolides, and jasmonates: nontraditional regulators of plant growth, development, and gene expression. Plant Cell 9, 1211–1223.

Cui, J.X., Zhou, Y.H., Ding, J.G., Xia, X.J., Shi, K., Chen, S.C., Asami, T., Chen, Z., Yu, J.K., 2011. Role of nitric oxide in hydrogen peroxide-dependent induction of abiotic stress tolerance by brassinosteroids in cucumber. Plant Cell Environ. 34, 347–358.

Dalio, R.J.D., Pinheiro, H.P., Sodek, L., Haddad, C.R.B., 2013. 24-Epibrassinolide restores nitrogen metabolism of pigeon pea under saline stress. Bot. Stud. 54, 9.

Das, T., Shukla, Y.M., Poonia, T.C., Meena, M., Meena, M.D., 2013. Effects of brassinolide on physiological characteristics of rice (*Oryza sativa* L.) with different salinity levels. Ann. Biol. (Hisar, India) 29, 228–231.

Daur, I., Tatar, O., 2013. Effects of gypsum and brassinolide on soil properties, and berseem (*Trifolium alexandrinum* L.) growth, yield and chemical composition grown on saline soil. Legume Res. 36, 306–311.

Deng, Z.P., Zhang, X., Tang, W.Q., Oses-Prieto, J.A., Suzuki, N., Gendron, J.M., Chen, H.J., Guan, S.H., Chalkley, R.J., Peterman, T.K., Burlingame, A.L., Wang, Z.Y., 2007. A proteomics study of brassinosteroid response in *Arabidopsis*. Mol. Cell. Proteomics 6, 2058–2071.

Ding, H.D., Zhu, X.H., Zhu, Z.W., Yang, S.J., Zha, D.S., Wu, X.X., 2012. Amelioration of salt-induced oxidative stress in eggplant by application of 24-epibrassinolide. Biol. Plant. 56, 767–770.

Divi, U.K., Krishna, P., 2010. Overexpression of the brassinosteroid biosynthetic gene AtDWF4 in *Arabidopsis* seeds overcomes abscisic acid-induced inhibition of germination and increases cold tolerance in transgenic seedlings. J. Plant Growth Regul. 29, 385–393.

Divi, U.K., Rahman, T., Krishna, P., 2010. Brassinosteroid-mediated stress tolerance in *Arabidopsis* shows interactions with abscisic acid, ethylene and salicylic acid pathways. BMC Plant Biol. 10, 151.

Divi, U.K., Rahman, T., Krishna, P., 2015. Gene expression and functional analyses in brassinosteroid-mediated stress tolerance. Plant Biotechnol. J. http://dx.doi.org/10.1111/pbi.12396.

Dobrikova, A., Vladkova, R., Stanoeva, D., Popova, A., Velitchkova, M., 2013. Effects of 24-epibrassinolide pretreatment on UV-B-induced changes in the pigment content of pea leaves. C. R. Acad. Bulg. Sci. 66, 543–550.

Domagalska, M.A., Sarnowska, E., Nagy, F., Davis, S.J., 2010. Genetic analyses of interactions among gibberellin, abscisic acid, and brassinosteroids in the control of flowering time in *Arabidopsis thaliana*. PLOS One 5, e14012.

Dong, D., Li, Y., Jiang, L., 2008. Effects of brassinosteroids on photosynthetic characteristics in soybean under aluminum stress. Acta Agron. Sin. 34, 1673–1678.

Du, L., Poovaiah, B.W., 2005. Ca^{2+}/calmodulin is critical for brassinosteroid biosynthesis and plant growth. Nature 437, 741–745.

Durigan, D.R.J., Pinheiro, H.P., Sodek, L., Haddad, C.R.B., 2011. The effect of 24-epibrassinolide and clotrimazole on the adaptation of *Cajanus cajan* (L.) Millsp to salinity. Acta Physiol. Plant. 33, 1887–1896.

Efimova, M.V., Savchuk, A.L., Hasan, J.A.K., Litvinovskaya, R.P., Khripach, V.A., et al., 2014. Physiological mechanisms of enhancing salt tolerance of oilseed rape plants with brassinosteroids. Russ. J. Plant Physiol. 61, 733–743.

Ekinci, M., Yildirim, E., Dursun, A., Turan, M., 2012. Mitigation of salt stress in lettuce (*Lactuca sativa* L. var. Crispa) by seed and foliar 24-epibrassinolide treatments. Hort Sci. 47, 631–636.

El-Khallal, S.M., Hathout, T.A., Ashour, A.E.R.A., Kerrit, A.A.A., 2009. Brassinolide and salicylic acid induced antioxidant enzymes, hormonal balance and protein profile of maize plants grown under salt stress. Res. J. Agric. Biol. Sci. 5, 391–402.

El-Mashad, A., Mohamed, H., 2012. Brassinolide alleviates salt stress and increases antioxidant activity of cowpea plants (*Vigna sinensis*). Protoplasma 249, 625–635.

Fábregas, N., Caño-Delgado, A.I., 2014. Turning on the microscope turret: a new view for the study of brassinosteroid signaling in plant development. Physiol. Plant 151, 172–183.

Fariduddin, Q., Yusuf, M., Hayat, S., Ahmad, A., 2009. Effect of 28-homobrassinolide on antioxidant capacity and photosynthesis in *Brassica juncea* plants exposed to different levels of copper. Environ. Exp. Bot. 66, 418–424.

Fariduddin, Q., Yusuf, M., Chalkoo, S., Hayat, S., Ahmad, A., 2011. 28-Homobrassinolide improves growth and photosynthesis in *Cucumis sativus* L. through an enhanced antioxidant system in the presence of chilling stress. Photosynthetica 49, 55–64.

Fariduddin, Q., Khalil, R.R.A.E., Mir, B.A., Yusuf, M., Ahmad, A., 2013a. 24-Epibrassinolide regulates photosynthesis, antioxidant enzyme activities and proline content of *Cucumis sativus* under salt and/or copper stress. Environ. Monit. Assess. 185, 7845–7856.

Fariduddin, Q., Mir, B.A., Yusuf, M., Ahmad, A., 2013b. Comparative roles of brassinosteroids and polyamines in salt stress tolerance. Acta Physiol. Plant. 35, 2037–2053.

Fariduddin, Q., Mir, B.A., Yusuf, M., Ahmad, A., 2014. 24-Epibrassinolide and/or putrescine trigger physiological and biochemical responses for the salt stress mitigation in *Cucumis sativus* L. Photosynthetica 52, 464–474.

Fedina, E.O., 2013. Effect of 24-epibrassinolide on pea protein tyrosine phosphorylation after salinity action. Russ. J. Plant Physiol. 60, 351–358.

Fedina, E.O., Karimova, F.G., Tarchevsky, I.A., Toropygin, I.Y., Khripach, V.A., 2008. Effect of epibrassinolide on tyrosine phosphorylation of the Calvin cycle enzymes. Russ. J. Plant Physiol. 55, 193–200.

Filova, A., Sytar, O., Krivosudska, E., 2013. Effects of brassinosteroid on the induction of physiological changes in *Helianthus annuus* L. under copper stress. Acta Univ. Agric. Silvic. Mendel. Brun. 61, 623–629.

Friedrichsen, D.M., Nemhauser, J., Muramitsu, T., Maloof, N.N., Alonso, J., Ecker, J.R., Furuya, M., Chory, J., 2002. Three redundant brassinosteroids early response genes encode putative bHLH transcription factors required for normal growth. Genetics 162, 1445–1456.

Fujita, M., Fujita, Y., Noutoshi, F., Narusaka, Y., Yamaguchi-Shinozaki, K., Shinozaki, K., 2006. Cross talk between abiotic and biotic stress responses: a current view from the points of convergence in the stress signaling networks. Curr. Opin. Plant Biol. 9, 436–442.

Gao, W., Long, L., Zhu, L., Xu, L., Gao, W., et al., 2013a. Proteomic and virus-induced gene silencing (VIGS) analyses reveal that gossypol, brassinosteroids and jasmonic acid contribute to the resistance of cotton to *Verticillium dahliae*. Mol. Cell. Proteomics 12, 3690–3703.

Gao, H., Liu, J., Zheng, Q., Hong, L., Wang, C., et al., 2013b. Regulation of exogenous brassinosteroid on growth and photosynthesis of *Helianthus tuberosus* seedlings and cadmium biological enrichment under cadmium stress. Shengtai Xuebao 33, 1935–1943.

Goddard, R., Peraldi, A., Ridout, C.J., Nicholson, P., 2014. Enhanced disease resistance caused by BRI1 mutation is conserved between *Brachypodium distachyon* and barley (*Hordeum vulgare*). Mol. Plant-Microbe Ineract. 27, 1095–1106.

Gomes, M.M.A., Torres Netto, A., Campostrini, E., Bressan-Smith, R., Zullo, M.A.T., et al., 2013. Brassinosteroid analogue affects the senescence in two papaya genotypes submitted to drought stress. Theor. Exp. Plant Physiol. 25, 186–195.

González-Garcia, M.P., Vilarrasa-Blasi, J., Zhiponova, M., Divol, F., Mora-Garcia, S., Russinova, E., Caño-delgado, A.I., 2011. Brassinosteroids control meristem size by promoting cell cycle progression in *Arabidopsis* roots. Development 138, 849–859.

Grove, M.D., Spencer, G.F., Rohwededer, W.K., Mandava, N.B., Worlet, J.F., Warthen Jr., J.C., Steffens, G.L., Flippen-Andersen, J.L., Cook Jr., J.C., 1979. Brassinolide, a plant promoting steroid isolated from *Brassica napus* pollen. Nature 281, 121–124.

Gruszka, D., 2013. The brassinosteroid signaling pathway-new key players and interconnections with other signaling networks crucial for plant development and stress tolerance. Int. J. Mol. Sci. 14, 8740–8774.

Hall, J.L., 2002. Cellular mechanisms for heavy metal detoxification and tolerance. J. Exp. Bot. 53, 1–11.

Hao, J., Yin, Y., Fei, S.Z., 2013. Brassinosteroid signaling network: implications on yield and stress tolerance. Plant Cell Rep. 32, 1017–1030.

Hartwig, T., Chuck, G.S., Fujioka, S., Klempien, A., Weizbauer, R., Potluri, D.P.V., Choe, S., Johal, G.S., Schulz, B., 2011. Brassinosteroid control of sex determination in maize. Proc. Natl. Acad. Sci. U.S.A. 108, 19814–19819.

Hasan, S.A., Hayat, S., Ali, B., Ahmad, A., 2008. 28-Homobrassinolide protects chickpea (*Cicer arietinum*) from cadmium toxicity by stimulating antioxidants. Environ. Pollut. 151, 60–66.

Hasan, S.A., Hayat, S., Ahmad, A., 2011. Brassinosteroids protect photosynthetic machinery against the cadmium induced oxidative stress in two tomato cultivars. Chemosphere 84, 1446–1451.

Haubrick, L.L., Assmann, S.M., 2006. Brassinosteroids and plant function: some clues, more puzzles. Plant Cell Environ. 29, 446–457.

Haubrick, L.L., Torsethaugen, G., Assmann, S.M., 2006. Effect of bassinolide, alone and in concert with abscisic acid, on control of stomatal aperture and potassium currents of *Vicia faba* guard cell protoplasts. Physiol. Plant 128, 134–143.

Hayat, S., Ali, B., Hasan, S.A., Ahmad, A., 2007. Brassinosteroid enhanced the level of antioxidants under cadmium stress in *Brassica juncea*. Environ. Exp. Bot. 60, 33–41.

Hayat, S., Hasan, S.A., Hayat, Q., Ahmad, A., 2010a. Brassinosteroids protect *Lycopersicon esculentum* from cadmium toxicity applied as shotgun approach. Protoplasma 239, 3–14.

Hayat, S., Hasan, S.A., Yusuf, M., Hayat, Q., Ahmad, A., 2010b. Effect of 28-homobrassinolide on photosynthesis, fluorescence and antioxidant system in the presence or absence of salinity and temperature in *Vigna radiata*. Environ. Exp. Bot. 69, 105–112.

Hayat, S., Yadav, S., Ali, B., Ahmad, A., 2010c. Interactive effect of nitric oxide and brassinosteroids on photosynthesis and the antioxidant system of *Lycopersicon esculentum*. Russ. J. Plant Physiol. 57, 212–221.

Hayat, S., Alyemeni, M., Hasan, S., 2012. Foliar spray of brassinosteroid enhances yield and quality of *Solanum lycopersicum* under cadmium stress. Saudi J. Biol. Sci. 19, 325–335.

Ho, M.O., 2003. Brassinosteroids accelerate the rate of cell division in isolated protoplasts of *Petunia hybrida*. J. Plant Biotechnol. 5, 63–67.

Honnerova, J., Rothova, O., Hola, D., Kocova, M., Kohout, L., Kvasnica, M., 2010. The exogenous application of brassinosteroids to *Zea mays* (L.) stressed by long-term chilling does not affect the activities of photosystem 1 or 2. J. Plant Growth Regul. 29, 500–505.

Hu, W.H., Wu, Y., Zeng, J.Z., He, L., Zeng, Q.M., 2010. Chill-induced inhibition of photosynthesis was alleviated by 24-epibrassinolide pretreatment in cucumber during chilling and subsequent recovery. Photosynthetica 48, 537–544.

Hu, W.H., Yan, X.H., Xiao, Y.A., Zeng, J.J., Qi, H.J., Ogweno, J.O., 2013a. 24-Epibrassinosteroid alleviate drought-induced inhibition of photosynthesis in *Capsicum annuum*. Sci. Hortic. 150, 232–237.

Hu, X., Xin, L., Sun, L., Fu, Z., 2013b. Toxic effects of chlorpyrifos on cucumber and brassinosteroid-mediated responses under the conditions of chilling temperature and low light. Asian J. Ecotoxicol. 8, 513–520.

Hu, C.H., Guo, J., Chen, L., Li, J.P., Qiao, L., He, L., 2014. Influence of exogenous brassinolide on growth and resistance of maize seedling with preservative stress. Hunan Nongye Daxue Xuebao. J. Hunan Agric. Univ. 40, 113–116.

Janeczko, A., Koscielniak, J., Pilipowicz, M., Szarek-Lukaszewska, G., Skoczowski, A., 2005. Protection of winter rape photosystem 2 by 24-epibrassinolide under cadmium stress. Photosynthetica 43, 293–298.

Janeczko, A., Oklestkova, J., Pociecha, E., Koscielniak, J., Mirek, M., 2011. Physiological effects and transport of 24-epibrassinolide in heat-stressed barley. Acta Physiol. Plant 33, 1249–1259.

Jiang, Y.P., Huang, L.F., Cheng, F., Zhou, Y.H., Xia, X.J., et al., 2013. Brassinosteroids accelerate recovery of photosynthetic apparatus from cold stress by balancing the electron partitioning, carboxylation and redox homeostasis in cucumber. Physiol. Plant 148, 133–145.

Jin, H., Do, J., Shin, S.J., Choi, J.W., Choi, Y.I., et al., 2014. Exogenously applied 24-epibrassinolide reduces lignification and alters cell wall carbohydrate biosynthesis in the secondary xylem of *Liriodendron tulipifera*. Phytochemistry 101, 40–51.

Kanwar, M.K., Bhardwaj, R., Arora, P., Chowdhary, S.P., Sharma, P., Kumar, S., 2012. Plant steroid hormones produced under Ni stress are involved in the regulation of metal uptake and oxidative stress in *Brassica juncea* L. Chemosphere 86, 41–49.

Kanwar, M.K., Bhardwaj, R., Chowdhary, S.P., Arora, P., Sharma, P., Kumar, S., 2013. Isolation and characterization of 24-epibrassinolide from *Brassica juncea* L. and its effects on growth, Ni ion uptake, antioxidant defense of *Brassica* plants and *in vitro* cytotoxicity. Acta Physiol. Plant 35, 1351–1362.

Kanwar, M.K., Poonam, Bhardwaj, R., 2015. Arsenic induced modulation of antioxidative defense system and brassinosteroids in *Brassica juncea* L. Ecotoxicol. Environ. Saf. 115, 119–125.

Kapoor, D., Rattan, A., Gautam, V., Kapoor, N., Bharadwaj, R., 2014. 24-Epibrassinolide mediated photosynthetic pigments and antioxidative defense systems of radish seedling under cadmium and mercury stress. J. Stress Physiol. Biochem. 10, 110–121.

Kapoor, D., Rattan, A., Gautam, V., Bhardwaj, R., 2015. Alleviation of cadmium and mercury stress by supplementation of steroid hormone to *Raphanus sativus* seedlings. Proc. Natl. Acad. Sci. India Sect. B: Biol. Sci. http://dx.doi.org/10.1007/s40011-015-0501-5.

Karlidag, H., Yildirim, E., Turan, M., 2011. Role of 24-epibrassinolide in mitigating the adverse effects of salt stress on stomatal conductance, membrane permeability, and leaf water content, ionic composition in salt stressed strawberry (*Fragaria x ananassa*). Sci. Hortic. 130, 133–140.

Kartal, G., Temel, A., Arican, E., Gozukirmizi, N., 2009. Effects of brassinosteroids on barley root growth, antioxidant system and cell division. Plant Growth Regul. 58, 261–267.

Kaur, R., Ohri, P., Bhardwaj, R., 2013a. Effect of 28-homobrassinolide on susceptible and resistant cultivars of tomato after nematode inoculation. Plant Growth Regul. 71, 199–205.

Kaur, H., Kapoor, P.D., Kohli, S.K., Thukral, A.K., Bharadwaj, R., 2013b. Brassinosteroids treatment to *Brassica juncea* L. plants under copper metal stress effects sugar and lipid metabolism. Biospectra 8, 173–180.

Kim, S.L., Lee, Y., Lee, S.H., Kim, S.H., Han, T.J., Kim, S.K., 2008. Brassinolide influences the regeneration of adventitious shoots from cultured leaf discs of tobacco. J. Plant Biol. 51, 221–226.

Kim, S.Y., Kim, B.H., Lim, C.J., Lim, C.O., Nam, K.H., 2010. Constitutive activation of stress-inducible genes in a *brassinosteroid-insensitive 1 (bri1)* mutant results in higher tolerance to cold. Physiol. Plant 138, 191–204.

Khripach, V.A., Voronina, L.V., Malevannaya, N.N., 1996. A Preparation for Decreasing of Heavy Metals Absorbtion by Agricultural Plants. Patent.

Kroutil, M., Hejtmankova, A., Lachman, J., 2010. Effect of spring wheat (*Triticum aestivum* L.) treatment with brassinosteroids on *the content of cadmium and lead in plant aerial biomass and grain*. Plant Soil Environ. 56, 43–50.

Kudryashova, O.A., Volotovich, A.A., Vasilevskaya, T.I., Varavina, N.P., Rupasova, Z.A., Khripach, V.A., 2012. Effects of 24-epibrassinolide on in vitro micropropagation of highbush blueberry. Russ. J. Plant Physiol. 59, 586–593.

Kumar, M., Sirhindi, G., Bhardwaj, R., Kumar, S., Jain, G., 2010. Effect of exogenous H_2O_2 on antioxidant enzymes of *Brassica juncea* L. seedlings in relation to 24-epibrassinolide under chilling stress. Indian J. Biochem. Biophys. 47, 378–382.

Kumar, S., Sirhindi, G., Bhardwaj, R., Kumar, M., 2014. Brassinosteroids denigrate the seasonal stress through antioxidant defense system in *Brassica junceae* L. J. Stress Physiol. Biochem. 10, 74–83.

Kurepin, L.V., Joo, S.H., Kim, S.K., Pharis, R.P., Back, T.G., 2012. Interaction of brassinosteroids with light quality and plant hormones in regulating shoot growth of young sunflower and *Arabidopsis* seedlings. J. Plant Growth Regul. 31, 156–164.

Lal, S., Bagdi, D.L., Kakralya, B.L., Jat, M.L., Sharma, P.C., 2013. Role of brassinolide in alleviating the adverse effect of drought stress on physiology, growth and yield of green gram (*Vigna radiata* L.) genotypes. Legume Res. 36, 359–363.

Leng, Y.F., Zhang, B., Yang, J.P., He, Q.L., Kang, J.W., et al., 2013. Effects of different plant growth regulators on the recovery form waterlogging of *Zea mays* L. Anhui Acad. Agric. Sci. 41, 12015–12020.

Li, K.R., Feng, C.H., 2011. Effects of brassinolide on drought resistance of *Xanthoceras sorbifolia* seedlings under water stress. Acta Physiol. Plant 33, 1293–1300.

Li, Y.H., Liu, Y.J., Xu, X.L., Jin, M., An, L.Z., Zhang, H., 2012a. Effect of 24-epibrassinolide on drought stress-induced changes in *Chorispora bungeana*. Biol. Plant 56, 192–196.

Li, Z.Y., Xu, Z.-S., He, G.Y., Yang, G.X., Chen, M., Li, L.C., Ma, Y.Z., 2012b. A mutation in *Arabidopsis* BSK5 encoding a brassinosteroid-signaling kinase protein affects responses to salinity and abscisic acid. Biochem. Biophys. Res. Commun. 426, 522–527.

Liu, H., Guo, T., Zhu, Y., Wang, C., Kang, G., 2006. Effects of epi-brassinolide (epi-BR) application at anthesis on starch accumulation and activities of key enzymes in wheat grains. Acta Agron. Sin. 32, 924–930.

Liu, Y.J., Jiang, H.F., Zhao, Z.G., An, L.Z., 2011. Abscisic acid is involved in brassinosteroids-induced chilling tolerance in the suspension cultured cells from *Chorispora bungeana*. J. Plant Physiol. 168, 853–862.

Lu, X.M., Guo, S.R., 2013. Effects of brassinolide on the polyamines, ATPase activity and inorganic ion content in roots of cucumber seedlings under hypoxia stress. Shengtaixue Zazhi. Chin. J. Ecol. 32, 611–614.

Lu, X.M., Yang, W., 2013. Alleviation effects of brassinolide on cucumber seedlings under NaCl stress. Ying Yong Sheng Tai Xue Bao. Chin. J. Appl. Ecol. 24, 1409–1414.

Lu, Z., Huang, M., Ge, D.P., Yang, Y.H., Cao, X.N., Qin, P., She, J.M., 2003. Effect of brassinolide on callus growth and regeneration in *Spartina patens* (Poaceae). Plant Cell Tissue Organ Cult. 73, 87–89.

Lukatkin, A.S., Kashtanova, N.N., Duchovskis, P., 2013. Changes in maize seedlings growth and membrane permeability under the effect of epibrassinolide and heavy metals. Russ. Agric. Sci. 39, 307–310.

Ma, N., Liu, D., Zhang, C., Li, J., Li, G., 2009. Regulation effect of exogenous hormones on growth and photosynthesis and yield of rapeseed (*Brassica napus* L.) after frozen. Acta Agron. Sin. 35, 1336–1343.

Mahesh, B., Parshavaneni, B., Ramakrishna, B., Rao, S.S.R., 2013. Effect of brassinosteroids on germination and seedling growth of radish (*Raphanus sativus* L.) under PEG-6000 induced water stress. Am. J. Plant Sci. 4, 2305–2313.

Malabadi, R.B., Nataraja, K., 2007. 24-Epibrassinolide induces somatic embryogenesis in *Pinus wallichiana* A. B. Jacks. J. Plant Sci. 2, 171–178.

Mandavia, C., Cholke, P., Mandavia, M.K., Raval, L., Gursude, A., 2014. Influence of brassinolide and salicylic acid on biochemical parameters of wheat (*Triticum aestivum* L.) under salinity stress. Indian J. Agric. Biochem. 27, 73–76.

Manzano, S., Martinez, C., Megias, Z., Gomez, P., Garrido, D., Jamilena, M., 2011. The role of ethylene and brassinosteroids in the control of sex expression and flower development in *Cucurbita pepo*. Plant Growth Regul. 65, 213–221.

Mitchell, J.W., Mandava, N.B., Worley, J.E., Plimmer, J.R., Smith, M.V., 1970. Brassins: a family of plant hormones from rape pollen. Nature 225, 1065–1066.

Nagata, N., Asami, T., Yoshida, S., 2001. Brassinozole, an inhibitor of brassinosteroids biosynthesis, inhibits development of secondary xylem in cress plants (*Lepidium sativum*). Plant Cell Biol. 42, 1006–1011.

Nahar, K., Kyndt, T., Hause, B., Hofte, M., Gheysen, G., 2013. Brassinosteroids suppress rice defense against root-knot nematodes through antagonism with the jasmonate pathway. Mol. Plant-Microbe Interact. 26, 106–115.

Nakaya, M., Tsukaya, H., Murakami, N., Kato, M., 2002. Brassinosteroids control the proliferation of leaf cells of *Arabidopsis thaliana*. Plant Cell Physiol. 43, 239–244.

Nigam, D., 2014. Integration of brassinosteroid signal transduction with the transcription network for fiber development and drought stress in *Gossypium hirsutum* L. J. Comp. Sci. Syst. Biol. 7, 108–114.

Nomura, T., Nakayama, N., Reid, J.B., Takeuchi, Y., Yokota, T., 1997. Blockage of brassinosteroid biosynthesis and sensitivity cause dwarfism in *Pisum sativum*. Plant Physiol. 113, 31–37.

Ogweno, J.O., Hu, W.H., Song, X.S., Shi, K., Mao, W.H., Zhou, Y.H., Yu, J.Q., 2010. Photoinhibition-induced reduction in photosynthesis is alleviated by abscisic acid, cytokinin and brassinosteroid in detached tomato leaves. Plant Growth Regul. 60, 175–182.

Peng, Z., Han, C., Yuan, L., Zhang, K., Huang, H., Ren, C., 2011. Brassinosteroid enhances jasmonate-induced anthocyanin accumulation in *Arabidopsis* seedlings. J. Integr. Plant Biol. 53, 632–640.

Pokotylo, I.V., Kretynin, S.V., Khripach, V.A., Ruelland, E., Blume, Y.B., Kravets, V.S., 2014. Influence of 24-epibrassinolide on lipid signalling and metabolism in *Brassica napus*. Plant Growth Regul. 73, 9–17.

Poonam, Kaur, R., Bali, S., Singh, R., Pati, P.K., Bharadwaj, R., 2014. Treatment of 24-epiBL to *Brassica junceae* plants under Cu-metal stress lowers oxidative burst by activity of antioxidative enzymes. J. Stress Physiol. Biochem. 10, 313–327.

Rady, M.M., 2011. Effect of 24-epibrassinolide on growth, yield, antioxidant system and cadmium content of bean (*Phaseolus vulgaris* L.) plants under salinity and cadmium stress. Sci. Hortic. 129, 232–237.

Rady, M.M., Osman, A.S., 2012. Response of growth and antioxidative system of heavy metal contaminated tomato plants 24-epibrassinolide. Afr. J. Agric. Res. 7, 3249–3254.

Raghu, K., Mahesh, K., Divya Sri, N., Rao, S.S.R., 2014. Effect of brassinosteroids on the seed germination and seedling growth of radish (*Raphanus sativus* L.) under arsenic toxicity stress. Int. J. Dev. Res. 9, 1929–1933.

Ramakrishna, B., Rao, S.S.R., 2013. Preliminary studies on the involvement of glutathione metabolism and redox status against zinc toxicity in radish seedlings by 28-homobrassinolide. Environ. Exp. Bot. 96, 52–58.

Ramakrishna, B., Rao, S.S.R., 2015. Foliar application of brassinosteroids alleviates adverse effects of zinc toxicity in radish (*Raphanus sativus* L.) plants. Protoplasma 252 (2), 665–667.

Rao, S.S.R., Vardhini, B.V., Sujatha, E., Anuradha, S., 2002. Brassinosteroids – new class of phytohormones. Curr. Sci. 82, 1239–1245.

Rattan, A., Kapoor, N., Bhardwaj, B., 2014. Role of brassinosteroids in osmolytes accumulation under salinity stress in *Zea mays* plants. Int. J. Sci. Res. 3, 1822–1827.

Ren, C.G., Chen, Y., Dai, C.C., 2014. Cross-talk between calcium-calmodulin and brassinolide for fungal endophyte-induced volatile oil accumulation of *Atractylodes lancea* plantlets. J. Plant Growth Regul. 33, 285–294.

Saka, H., Fuji, S., Imakawa, A.M., Kato, N., Watanabe, S., Nishizawa, T., Yonekawa, S., 2003. Effect of brassinolide applied at the meiosis and flowering stages on the levels of endogenous plant hormones during grain filling in rice plant (*Oryza sativa* L.). Plant Prod. Sci. 6, 36–42.

Savaliya, D.D., Mandavia, C.K., Mandavia, M.K., 2013. Role of brassinolide on enzyme activities in groundnut under water deficit stress. Indian J. Agric. Biochem. 26, 92–96.

Savaliya, D.D., Mandavia, M.K., Mandavia, C., Domadiya, R.K., 2014. Effect of soaking seeds in brassinolide on biochemical and physiological parameters of some groundnut (*Arachis hypogaea* L.) cultivars under PEG induced water stress. Indian J. Agric. Biochem. 27, 30–34.

Schnabl, H., Roth, U., Friebe, A., 2001. Brassinosteroid-induced stress tolerances of plants. Recent Res. Dev. Phytochem. 5, 169–183.

Serna, M., Coll, Y., Zapata, P.J., Botella, M.A., Pretel, M.T., Amorós, A., 2015. A brassinosteroid analogue prevents the effect of salt stress on ethylene synthesis and polyamines in lettuce plants. Sci. Hortic. 185, 105–112.

Shahid, M.A., Pervez, M.A., Balal, R.M., Mattson, N.S., Rashid, A., Ahmad, R., Ayyub, C.M., Abbas, T., 2011. Brassinosteroid (24-epibrassinolide) enhances growth and alleviates the deleterious effects induced by salt stress in pea (*Pisum sativum* L.). Aust. J. Crop Sci. 5, 500–510.

Shang, Q., Song, S., Zhang, Z., Guo, S., 2006. Exogenous brassinosteroid induced salt resistance of cucumber (*Cucumis sativus* L.) seedlings. Sci. Agric. Sin. 39, 1872–1877.

Sharma, P., Bhardwaj, R., 2007a. Effects of 24-epibrassinolide on growth and metal uptake in *Brassica juncea* L. under copper metal stress. Acta Physiol. Plant. 29, 259–263.

Sharma, P., Bhardwaj, R., 2007b. Effect of 24-epibrassinolide on seed germination, seedling growth and heavy metal uptake in *Brassica junceae* L. Gen. Appl. Plant Physiol. 33, 59–73.

Sharma, P., Bhardwaj, R., Arora, N., Arora, H.K., 2007c. Effect of 28-homobrassinolide on growth, zinc metal uptake and antioxidative enzyme activities in *Brassica juncea* L. seedlings. Braz. J. Plant Physiol. 19, 203–210.

Sharma, P., Bhardwaj, R., Arora, N., Arora, H.K., Kumar, A., 2008. Effects of 28-homobrassinolide on nickel uptake, protein content and antioxidative defence system in *Brassica juncea*. Biol. Plant 52, 767–770.

Sharma, I., Pati, P.K., Bhardwaj, R., 2010. Regulation of growth and antioxidant enzyme activities by 28-homobrassinolide in seedlings of *Raphanus sativus* L. under cadmium stress. Indian J. Biochem. Biophys. 47, 172–177.

Sharma, I., Pati, P.K., Bhardwaj, R., 2011a. Effect of 24-epibrassinolide on oxidative stress markers induced by nickel-ion in *Raphanus sativus* L. Acta Physiol. Plant 33, 1723–1735.

Sharma, I., Pati, P.K., Bhardwaj, R., 2011b. Effect of 28-homobrassinolide on antioxidant defence system in *Raphanus sativus* L. under chromium toxicity. Ecotoxicology 20, 862–874.

Sharma, N., Hundal, G.S., Sharma, I., Bharadwaj, R., 2012. Effect of 24-epibrssinolide on protein content and activities of glutathione – S- transferase and poly phenol oxidase in *Raphanus sativus* L. plants under cadmium and mercury metal stress. Terr. Aquat. Environ. Toxicol. 6, 1–7.

Sharma, I., Bhardwaj, R., Pati, P.K., 2013a. Stress modulation response of 24-epibrassinolide against imidacloprid in an elite indica rice variety Pusa Basmati-1. Pestic. Biochem. Physiol. 105, 144–153.

Sharma, I., Ching, E., Saini, S., Bhardwaj, R., Pati, P.K., 2013b. Exogenous application of brassinosteroid offers tolerance to salinity by altering stress responses in rice variety Pusa Basmati-1. Plant Physiol. Biochem. 69, 17–26.

Sharma, N., Hundal, G.S., Sharma, I., Bharadwaj, R., 2014. 28-homobrassinolide alters protein content and activities of glutathione-s-transferase and polyphenol oxidase in *Raphanus sativus* L. Plants under heavy metal stress. Toxicol. Int. 21 (1), 44–50.

Song, S., Liu, W., Guo, S., Shang, Q., Zhang, Z., 2006. Salt resistance and its mechanism of cucumber under effects of exogenous chemical activator. Ying Yong Sheng Tai Xue Bao. Chin. J. Appl. Ecol. 17, 1871–1876.

Strnad, M., Kohout, L., 2003. A simple brassinolide analogue 2a, 3a-dihydroxy-17b-(3-methyl butynyloxyl) 7-oxa-B-homo-5a androstan-6-one which induces bean second node splitting. Plant Growth Regul. 40, 39–47.

Suzuki, Y., Saso, K., Fujioka, S., Yoshida, S., Nitasaka, E., Nagata, S., Nagasawa, H., Takatsuto, S., Yamaguchi, I., 2003. A dwarf mutant strain of *Pharbitis nil*, Uzukobito (kobito), has defective brassinosteroids biosynthesis. Plant J. 36, 401–410.

Swamy, K.N., Vardhini, B.V., Ramakrishna, B., Anuradha, S., Siddulu, N., Rao, S.S.R., 2014. Role of 28-homobrassinolide on growth biochemical parameters of *Trigonella foenu-graecum* L. plants subjected to lead toxicity. Int. J. Multidiscip. Curr. Res. 2, 317–321.

Talaat, N.B., Shawky, B.T., 2013. 24-Epibrassinolide alleviates salt-induced inhibition of productivity by increasing nutrients and compatible solutes accumulation and enhancing antioxidant system in wheat (*Triticum aestivum* L.). Acta Physiol. Plant. 35, 729–740.

Tao, Y.Z., Zheng, J., Xu, Z.M., Zhang, X.H., Zhang, K., Wang, G.Y., 2004. Functional analysis of ZmDWF1, a maize homolog of *Arabidopsis* brassinosteroids biosynthetic DWF1/DIM gene. Plant Sci. 167, 743–751.

Thussagunpanit, J., Jutamanee, K., Kaveeta, L., Chai-Arree, W., Pankean, P., Suksamrarn, A., 2013. Effects of a brassinosteroid and an ecdysone analogue on pollen germination of rice under heat stress. J. Pestic. Sci. (Tokyo, Jpn.) 38, 105–111.

Thussagunpanit, J., Jutamanee, K., Kaveeta, L., Chai-arree, W., Pankean, P., et al., 2014. Comparative effects of brassinosteroid and brassinosteroid mimic on improving photosynthesis, lipid peroxidation, and rice seed set under heat stress. J. Plant Growth Regul. http://dx.doi.org/10.1007/s00344-014-9467-4.

Thussagunpanit, J., Jutamanee, K., Sonjaroon, W., Kaveeta, L., Chai-Arree, W., Pankean, P., Suksamrarn, A., 2015. Effects of brassinosteroid and brassinosteroid mimic on photosynthetic efficiency and rice yield under heat stress. Phytosynthetica 53 (2), 312–320.

Upreti, K.K., Murti, G.S.R., 2004. Effects of brassinosteroids on growth, nodulation, phytohormone content and nitrogenase activity in French bean under water stress. Biol. Plant 48, 407–411.

Vardhini, B.V., 2011. Studies on the effect of brassinolide on the antioxidative system of two varieties of sorghum grown in saline soils of Karaikal. Asian Aust. J. Plant Sci. Biotechnol. 5, 31–34.

Vardhini, B.V., 2012a. Application of brassinolide mitigates Saline stress of certain metabolites of sorghum grown in Karaikal. J. Phytol. 4, 1–3.

Vardhini, B.V., 2012b. Effect of brassinolide on certain enzymes of sorghum plant grown in saline soils of Karaikal. J. Phytol. 4, 30–33.

Vardhini, B.V., 2013a. Comparative study of *Sorghum vulgare* Pers. Grown in two experimental sites by brassinolide application at vegetative, flowering and grain filling stage. Proc. A.P. Akad. Sci. 15, 75–79.

Vardhini, B.V., 2013b. Brassinosteroids, Role for amino acids, peptides and amines modulation in stressed plants – a review. In: Anjum, N.A., Gill, S.S., Gill, R. (Eds.), Plant Adaptation to Environmental Change: Significance of Amino Acids and Their Derivatives. CAB International of Nosworthy Way, Wallingford OX10 8DE, UK, pp. 300–316.

Vardhini, B.V., Rao, S.S.R., 1998. Effect of brassinosteroids on growth, metabolite content and yield of *Arachis hypogaea*. Phytochemistry 48, 927–930.

Vardhini, B.V., Rao, S.S.R., 1999. Effect of brassinosteroids on nodulation and nitrogenase activity in groundnut (*Arachis hypogaea* L.). Plant Growth Regul. 28, 165–167.

Vardhini, B.V., Rao, S.S.R., 2000. Effect of brassinosteroids on the activities of certain oxidizing and hydrolyzing enzymes of groundnut. Indian J. Plant Physiol. 5, 89–92.

Vardhini, B.V., Rao, S.S.R., 2002. Acceleration of ripening of tomato pericarp discs by brassinosteroids. Phytochemistry 61, 843–847.

Vardhini, B.V., Anjum, N.A., 2015. Brassinosteroids make plant life easier under abiotic stresses mainly by modulating major components of antioxidant defense system. Front. Environ. Sci. 2, 67. http://dx.doi.org/10.3389/fenvs.2014.0006.

Vardhini, B.V., Anuradha, S., Rao, S.S.R., 2006. Brassinosteroids – a great potential to improve crop productivity. Indian J. Plant Physiol. 11, 1–12.

Vardhini, B.V., Sujatha, E., Rao, S.S.R., 2012. Studies on the effect of brassinosteroids on the qualitative changes in the storage roots of radish. Bulg J. Agric. Sci. 18, 63–69.

Vázquez, M., Guerrero, Y., González, L., Noval, W., 2013. Brassinosteroids and plant responses to heavy metal stress. An Overview. Open J. Met. 3, 34–41.

Villiers, F., Jourdain, A., Bastien, O., Leonhardt, N., Fujioka, S., Tichtincky, G., Parcy, F., Bourguignon, J., Hugouvieux, V., 2012. Evidence for functional interaction between brassinosteroids and cadmium response in *Arabidopsis thaliana*. J. Exp. Bot. 63, 1185–1200.

Vogler, F., Schmalzl, C., Englhart, M., Bircheneder, M., Sprunck, S., 2014. Brassinosteroids promote *Arabidopsis* pollen germination and growth. Plant Reprod. 27, 153–167.
Wang, H.H., Feng, T., Peng, X.X., Yan, M.L., Zhou, P.L., Tang, X.K., 2009. Ameliorative effects of brassinosteroid on excess manganese-induced oxidative stress in *Zea mays* L. leaves. Agric. Sci. China 8, 1063–1074.
Wang, M., Jiang, W.J., Yu, H.J., 2010. Effects of exogenous epibrassinolide on photosynthetic characteristics in tomato (*Lycopersicon esculentum* Mill) seedlings under weak light stress. J. Agric. Food Chem. 58, 3642–3645.
Wang, B., Zhang, J., Xia, X., Zhang, W.-H., 2011. Ameliorative effect of brassinosteroid and ethylene on germination of cucumber seeds in the presence of sodium chloride. Plant Growth Regul. 65, 407–413.
Wang, B., Li, Y., Zhang, W.H., 2012a. Brassinosteroids are involved in response of cucumber (*Cucumis sativus*) to iron deficiency. Ann. Bot. 110, 681–688.
Wang, Q., Ding, T., Gao, L., Pang, J., Yang, N., 2012b. Effect of brassinolide on chilling injury of green bell pepper in storage. Sci. Hortic. 144, 195–200.
Wang, X.H., Shu, C., Li, H.Y., Hu, X.Q., Wang, Y.X., 2014. Effects of 0.01% brassinolide solution application on yield of rice and its resistance to autumn low-temperature damage. Acta Agric. Jiangxi 26, 36–38.
Wang, B., Li, G., Zhang, W.-H., 2015. Brassinosteroids are involved in Fe homeostasis in rice (*Oryza sativa* L.). J. Exp. Bot. http://dx.doi.org/10.1093/jxb/erv079.
Weng, Q., Zhou, B.L., Yu, Y., Fu, Y.W., 2007. Effects of exogenous ABA, BR, and ETH on changes of lycopene contents in fruit of tomato. Shenyang Nongye Daxue Xuebao 38, 784–787.
Xi, Z., Wang, Z., Fang, Y., Hu, Z., Hu, Y., et al., 2013. Effects of 24-epibrassinolide on antioxidation defense and osmoregulation systems of young grapevines (*V. vinifera* L.) under chilling stress. Plant Growth Regul. 71, 57–65.
Xia, X.J., Gao, C.J., Song, L.X., Zhou, Y.H., Shi, K., Yu, J.Q., 2014. Role of H_2O_2 dynamics in brassinosteroid-induced stomatal closure and opening in *Solanum lycopersicum*. Plant Cell Environ. 37, 2036–2050.
Yamamoto, R., Fujioka, S., Demura, T., Takatsuto, S., Yoshida, S., Fukuda, H., 2001. Brassinosteroid levels increase drastically prior to morphogenesis of tracheary elements. Plant Physiol. 125, 556–563.
Yin, B., Liang, G., Jia, W., Cui, X., 2014. Exogenous EBR mediated the plant growth, absorption and accumulation of Cu, Fe and Zn in tomato seedlings under Cu stress. Chin. J. Eco-Agric. 22, 578–584.
Yuan, G.F., Jia, C.G., Li, Z., Sun, B., Zhang, L.P., Liu, N., Wang, Q.M., 2010. Effect of brassinosteroids on drought resistance and abscisic acid concentration in tomato under water stress. Sci. Hortic. 126, 103–108.
Yuan, L., Shu, S., Sun, J., Guo, S., Tezuka, T., 2012. Effects of 24-epibrassinolide on the photosynthetic characteristics, antioxidant system, and chloroplast ultrastructure in *Cucumis sativus* L. under $Ca(NO_3)_2$ stress. Photosynth. Res. 112, 205–214.
Yuan, L.B., Peng, Z.H., Zhi, T.T., Zho, Z., Liu, Y., et al., 2014. Brassinosteroid enhances cytokinin-induced anthocyanin biosynthesis in *Arabidopsis* seedlings. Biol. Plant (ahead of print).
Yue, D., Shu, S., Sun, J., Zhu, W.M., Guo, S.R., 2013. Effects of exogenous brassinolide on the growth and quality of *Brassica chinensis* under calcium nitrate stress. J. Shenyang Agric. Univ. 44, 696–699.
Yusuf, M., Fariduddin, Q., Ahmad, A., 2011a. 28-Homobrassinolide mitigates boron induced toxicity through enhanced antioxidant system in *Vigna radiata* plants. Chemosphere 85, 1574–1584.
Yusuf, M., Fariduddin, Q., Hayat, S., Hasan, S.A., Ahmad, A., 2011b. Protective response of 28-homobrassinolide in cultivars of *Triticum aestivum* with different levels of nickel. Arch. Environ. Contam. Toxicol. 60, 68–76.
Yusuf, M., Fariduddin, Q., Ahmad, A., 2012. 24-Epibrassinolide modulates growth, nodulation, antioxidant system, and osmolyte in tolerant and sensitive varieties of *Vigna radiata* under different levels of nickel: a shotgun approach. Plant Physiol. Biochem. 57, 143–153.
Yusuf, M., Fariduddin, Q., Ahmad, I., Ahmad, A., 2014. Brassinosteroid-mediated evaluation of antioxidant system and nitrogen metabolism in two contrasting cultivars of *Vigna radiata* under different levels of nickel. Physiol. Mol. Biol. Plants. http://dx.doi.org/10.1007/s12298-014-259-x.

Zadvornova, Y.V., Alekseichuk, G.N., Laman, N.A., Khripach, V.A., Grut, S., 2005. Effect of brassinosteroids on activation of the cell cycle during germination of *Brassica oleracea* L. seeds. Dokl. Nats. Akad. Nauk Belarusi 49, 70–73.

Zaharah, S.S., Singh, Z., Symons, G.M., Reid, J.B., 2012. Role of brassinosteroids, ethylene, abscisic acid, and indole-3-acetic acid in mango fruit ripening. J. Plant Growth Regul. 31, 363–372.

Zeng, H., Tang, Q., Hue, X., 2010. *Arabidopsis* brassinosteroid mutants *del 2-1* and bin 2-1 display altered salt tolerance. J. Plant Growth Regul. 29, 44–52.

Zhang, A.Y., Zhang, J., Zhang, J.H., Ye, N.H., Zhang, H., Mingpu, T., Mingyi, J., 2011. Nitric oxide mediates brassinosteroid-induced ABA biosynthesis involved in oxidative stress tolerance in maize leaves. Plant Cell Physiol. 52, 181–192.

Zhang, Y.P., Zhu, X.H., Ding, H.D., Yang, S.J., Chen, Y.Y., 2013. Foliar application of 24-epibrassinolide alleviates high-temperature-induced inhibition of photosynthesis in seedlings of two melon cultivars. Photosynthetica 51, 341–349.

Zhang, D.W., Deng, X.G., Fu, F.Q., Lin, H.H., 2015. Induction of plant virus defense response by brassinosteroids and brassinosteroid signaling in *Arabidopsis thaliana*. Planta 241 (4), 875–85. http://dx.doi.org/10.1007/s00425-014-2218-8.

Zhao, Y., Liang, Z.Y., Yang, Y.J., 2013. Effects of exogenous brassinosteroid on Cd tolerance in *Solatium nigrum* seedlings. Zhongguo Shengtai Nongye Xuebao 21, 872–876.

Zhiponova, M.K., Vanhoutte, I., Boudolf, V., Betti, C., Dhondt, S., Coppens, F., Mylle, E., Maes, S., Gonzales Garcia, M.P., Caño-Delado, A.I., Inze, D., Beemster, G.T.S., De Veylder, L., Russinova, E., 2013. Brassinosteroid production and signaling differentially control cell division and expansion in the leaf. New Phytol. 197, 490–502.

Zhou, Y.X., Guo, X.M., Lu, S.B., Hu, Y.L., Hu, D.N., et al., 2013. Effect of water, nutrient and brassinolides on number of blossom, leaf nutrition and seed oil content of *Camellia oleifera*. Zhiwu Yingyang Yu Feiliao Xuebao. Plant Nutr. Fert. Sci. 19, 387–395.

Zhou, Y., Xia, X., Yu, G., Wang, J., Wu, J., Wang, M., Yang, y., Shi, K., Yu, Y., Chen, Z., Gan, J., Yu, J., 2015. Brassinosteroids play a critical role in the regulation of pesticide metabolism in crop plants. Sci. Rep. 5. http://dx.doi.org/10.1038/srep09018.

Zhou, J., Wang, J., Li, X., Zhou, Y.-H., Shi, K., et al., 2014. H_2O_2 mediates the crosstalk of brassinosteroid and abscisic acid in tomato responses to heat and oxidative stresses. J. Exp. Bot. 65, 4371–4383.

Zhu, Z., Zhang, Z.Q., Qin, G.Z., Tian, S.P., 2010. Effects of brassinosteroids on postharvest disease and senescence of jujube fruit in storage. Postharvest Biol. Tech. 56, 50–55.

CHAPTER

PLANT METALLOTHIONEINS: CLASSIFICATION, DISTRIBUTION, FUNCTION, AND REGULATION

9

Rohit Joshi[1], Ashwani Pareek[2], Sneh L. Singla-Pareek[1]

Plant Molecular Biology Group, International Centre for Genetic Engineering and Biotechnology, New Delhi, India[1]; Stress Physiology and Molecular Biology Laboratory, School of Life Sciences, Jawaharlal Nehru University, New Delhi, India[2]

CHAPTER OUTLINE

1. INTRODUCTION

Essential ions support a plethora of structural and physiological functions in plant cells by taking part in electron transfers, redox reactions, enzyme catalyzed reactions, and nucleic acid metabolism (Conn and Gilliham, 2010). However, despite being essential, they can become limiting factors if present at non-physiological concentrations. Even extremely potent ions including lead (Pb), arsenic (As), cadmium (Cd), molybdenum (Mo), aluminium (Al), cobalt (Co), zinc (Zn), nickel (Ni), chromium (Cr), iron (Fe),

Plant Metal Interaction. http://dx.doi.org/10.1016/B978-0-12-803158-2.00009-6

manganese (Mn) and copper (Cu) are highly poisonous, resulting in growth retardation and even death for the organism (Hossain and Komatsu, 2013; Mirza et al., 2014). Plant growth and development requires continuous uptake of various solutes from the rhizosphere by the root and subsequently their distribution to the entire plant body for maintaining the energy flux. Because plants are sessile organisms, their roots are the primary areas affected by the metal ions in a contaminated environment. Together with water, the root cells of the rhizosphere take up heavy metal ions and incorporate them into plant tissues. Similarly, the whole plant is exposed to heavy metals in aquatic plants.

Heavy metals are classified in the group that has a specific weight greater than 5 g/cm^3 (Holleman and Wiberg, 1985). Almost 40 elements are present in this class and are mainly present in the form of complexes in rocks. A steep rise in the demand of heavy metals from the enormous Industrial Revolution resulted in a massive discharge of heavy metals into the environment. After entering the atmosphere, these heavy metals settle mainly in the soil and aquatic bodies, apart from emission in the form of vapors and dust particles. Environmental pollution involving high heavy metal concentrations presents a major challenge for plants and other organisms around the world. Plants growing in these areas are exposed to high concentrations of metal ions. Some soil types naturally have quite a high heavy metal content (e.g., serpentine soils), whereas in places such as mining areas, the metal ion contamination is one of the result of human activity. Metal-induced oxidative stress has been strongly linked to early toxicity symptoms (Sharma and Dietz, 2009). Exposure to toxic metal ions or high concentrations of nontoxic ions therefore triggers stress reactions leading to adaptation and defence mechanism at the developmental, structural, physiological, and molecular levels (Hirayama and Shinozaki, 2010). Imbalances in the relative abundance of elements in the soil, their availability to plants, or their rate of uptake can all affect cellular homeostasis and can also cause irreversible damage. Plants continuously maintain their physiological concentrations of both nonessential and essential metal ions to attain ionic homeostasis.

Plants play a vital role in food chain, thus their accumulation of heavy metal ions can potentially cause severe food chain contamination. It is therefore essential to understand how they adapt to cope with heavy metal exposure. Plant defense responses are thus required to protect cells and subcellular compartments against oxidative damage. Brief exposure to moderate concentrations of metal ions and these responses can save the plant and enable its further growth and development. The adaptive exploration and the uptake of water and soluble elements are thus key drivers of physiological activity and developmental plasticity in the plant root system. The main strategies used by plants to mitigate the detrimental effects of heavy metal ions on cellular health involve avoidance of exposure, minimizing their uptake, and intracellular sequestering. If the plant cannot prevent either the internalization of toxic metal ions or their harmful effects in cells, they can have devastating effects on basic metabolism, transport processes, membranes, and cellular structure. To survive in such harsh conditions, plants have developed specific and efficient mechanisms for heavy metals intake and transformation into a physiologically endurable form critical for metabolic function in plants. Thus, an excess amount of these poisonous heavy metals and essential elements that expose the plants but do not participate in metabolism must be metabolically inactivated. However, these strategies alone are not sufficient to mitigate the effects of long-term exposure to metal ions, which will cause metal ion toxicity in cells. In susceptible plants, this generally has devastating consequences. The development of metal tolerance exhibited by some plant species is a unique and particularly interesting aspect of plant stress biology. It is achieved through a variety of mechanisms, including exclusion strategies, extracellular chelation and attenuation of metal ions, restriction of metal ions in the apoplast, and detoxification and compartmentalization of metal ions inside the plant body. The most sophisticated strategies used by plants to cope with high levels of toxic metals are based on hypertolerance and the hyperaccumulation of metal ions in the plant body without negative effects on viability (van der Ent et al., 2013). In fact, under certain ecological circumstances, heavy metal hyperaccumulation may secondarily protect plants against herbivores and pathogens (Hörger et al., 2013).

In addition, knowledge of the mechanisms employed by plants to cope with heavy metal stress can be applied in the development of practical biotechnological methods for improving plant protection, especially in crop plants. Molecular insight of plant metal accumulation regulators has several additional biotechnological applications. For instance, vital trace metal deficiencies in humans emerge to be prevalent in developing countries (Hambidge, 2000). On the other hand, most of the toxic nonessential elements (i.e., Cd) enter the human body through plant-derived material (Wagner, 1993). Significant differences in Cd accumulation occur in between plant varieties. Thus, molecular understanding of mechanisms regulating this process may result in the identification of markers for crops with reduced content of hazardous metals. Similar to other organisms plants have also developed their own regulatory network to act in response toward the uptake and distribution of both nonessential and essential metal ions. These processes comprise of sequestration and chelation of metal ions through specific ligands (Gavanji et al., 2014). Metallothioneins (MTs) and phytochelatins (PCs) are well-defined heavy metal-binding ligands in plants. In this chapter, we describe key features of our present understanding of plant metal accumulation determinants. An overview of recent experimental findings concerning plant responses to heavy metal toxicity is also provided. PCs are included in this chapter separately because currently no evidence is available for their overlapping function in plant cells. Most of the recent work includes molecular genetic studies in crops. Further, this chapter clarifies the current advances to our knowledge of the expression and function of MTs in plants and their molecular regulation.

2. PHYTOCHELATINS

PCs in plants belong to the family of enzymatically synthesized cysteine-rich polypeptides whose synthesis is induced by heavy metals (Anjum et al., 2014, 2015). They chelate heavy metals with their thiol groups, and the resulting metal–PC complexes are sequestered into vacuoles (Ovečka and Takáč, 2014). Initially, researchers thought about the functional similarity between PCs and MTs. This artificial dichotomy has been discarded only after the discovery of PC synthase gene in plants and the expression of its functional homologs in animals. However, PCs were first discovered in the yeast *Schizosaccharomyces pombe*, which, after exposure to Cd^{2+}, produces PC_2 and PC_3 (Kondo et al., 1985) and are named cadystin A and B (both cadmium and cysteine expression). Over the years, other names (including class III MTs) have also been used, but "PCs" is the most widely adopted name (Wu et al., 2014). After exposure to high concentrations of $CdSO_4$ (200 μM), *Rauvolfia serpentina* suspension cells yielded a complex that after isolation through gel filtration showed more than 90% affinity toward of the heavy metals. This compound was further separated by high-performance liquid chromatography (HPLC) analysis under acidic conditions that revealed its molecular weight of 1040 Da. Later it was discovered that all plants detoxify these metals by chelating them to peptides with varying $(\gamma\text{-Glu-Cys})_n\text{-Gly}$; $n = 2–11$ chain length (Gekeler et al., 1989). Therefore, this class of peptides was classified as PCs (PC_n, where n is the number of γ-Glu-Cys repeats). *In vivo* cell culture studies show that cadmium ions are the strongest inducers of PC formation followed by lead and zinc ions (Wagner, 1985). Similarly, other metal ions are also assembled on the basis of their PC formation strength in the cell suspension cultures of *R. serpentina* as $Cd^{2+} > Pb^{2+} > Zn^{2+} > Sb^{3+} > Ag^{+} > Hg^{2+} > As^{5-} > Cu^{+} > Sn^{2+} > Au^{3+} > Bi^{3+}$ (Grill et al., 1987).

Therefore, PCs play a dual role: first by inactivating by binding metal ions tightly to form complexes and store them in vacuoles; second, they transport the essential metal ions such as Cu^{2+} or Zn^{2+} to apoenzymes for their catalytic activity or to nucleic acid structures, like zinc fingers (Pinter and Stillman, 2014). Steffens et al. (1986) for the first time gave the experimental evidence of PC function in higher plants. They found that tomato cells were sustainably grow on the lethal concentrations of Cd^{2+} and PC accumulated much higher in cadmium tolerant cell lines than in nontolerant lines. Further, by inhibiting the PC biosynthesis by adding L-buthionine sulfoximine, an inhibitor of γ-glutamylcysteine synthetase the resistant cell lines again become sensitive. However, Howden et al. (1995) proposed most striking results signifying the importance of PC for Cd tolerance in plants. They successfully isolated a Cd-sensitive mutant (*cad1*) from *Arabidopsis* that is sensitive for Cd ions and was deficient in their capability to form Cd-PC complexes under Cd^{2+} exposure in the growth medium. Later, it was identified that *CAD1* gene is the structural gene for PC synthase (Zenk, 1996).

2.1 BIOSYNTHESIS OF PCs

Unlike glutathione (GSH), PCs are not the major gene products to be formed on the ribosome because of the absence of γ-glutamyl linkage synthesis during translation and also because of the absence of tRNA for β-alanine moiety (Zenk, 1996). The PC showed highly similar chemical configuration to GSH. In higher plants, the synthesis of PC-heavy metal complex is a vital metabolic activity. This creates a huge loss of cytosolic GSH that is later recovered by inducing GSH-synthesizing enzymes (Rüegsegger and Brunold, 1992). The γ-glutamylcysteine dipeptidyl transpeptidase (EC 2.3.2.15), also known as PC synthase, catalyzes the transfer of γ-glutamylcysteine dipeptide from GSH to its acceptor or to the elongating chain of PC oligomers forming PC. The reaction is as follows:

$$\gamma\text{-Glu-Cys-Gly} + (\gamma\text{-Glu-Cys})_n\text{-Gly} \rightarrow (\gamma\text{-Glu-Cys})_{n+1}\text{-Gly} + \text{Gly} \qquad (9.1)$$

This reaction is primarily dependent upon heavy metal ions and the formation of a stable metal–PC complex, which prevents the discharge of toxic metal ions by accelerating this reaction in forward direction. GSH is a (γ-Glu-Cys) donor (Grill et al., 1989) and is self-regulated by PC, which chelates PC synthase-activating metal leading to the termination of enzymatic reaction (Loeffler et al., 1989). This metal-activated chain elongation is also terminated by using metal chelating agents like ethylenediaminetetraacetic acid. PC synthase (M_r 95,000) is composed of four subunits having an optimal temperature of 35 °C and isoelectric point of pH 4.8.

To prevent the vital metabolic activity enzymes, it is necessary for PC synthase to chelate and inactivate each toxic metal ion present in the cytosol. These heavy metal ions enter cytosol through the permeable cell membrane. These metal ions immediately activate the inactive PC synthase that produces the variable-length PC molecules, at the expense of cytosolic GSH. Immediately after, the metal binding peptides are synthesized and form a metal–PC complex to prevent the susceptible metabolic enzymes (Kneer and Zenk, 1992). Finally these toxic metal–PC complexes are transported and deposited in the vacuole (Salt and Rauser, 1995). In vacuoles, these metals then bound to vacuolar organic acids, leading to degradation of metal–PC complex and the resulting amino acids reenter the cytosol.

Klapheck et al. (1995) reported for the first time that a Cd^{2+}-treated garden pea can synthesize PCs and iso-PC(β-Ala). Surprisingly, pea enzymes catalyze the production of iso-PC(Ser) from γ-Glu-Cys-Ser, as both are absent in the *Fabaceae* family but reported in the *Poaceae* family (Klapheck et al., 1994). However, phytochelatin synthase gene overexpression does not always have beneficial effects on heavy metal tolerance (Ovečka and Takáč, 2014).

2.2 THE OCCURRENCE OF ISO-PHYTOCHELATINS

It was observed earlier that plants belonging to the *Phaseoleae* tribe of the order *Fabales* induced a series of compounds homologous to PC when exposed to Cd^{2+}. Instead of glycine, they contain β-alanine and are analogous to h-GSH, thus they are termed homo-PCs with a general formula such as $(\gamma\text{-Glu-Cys})_n$-β-Ala (n = 2–7) (Grill et al., 1986). During the analyses of GSH derivatives in the *Poaceae* family, Klapheck et al. (1992) found hydroxymethyl-GSH (γ-Glu-Cys-Ser). Exposure of these compounds with Cd^{2+} and Cu^{2+} produce another homologous $(\gamma\text{-Glu-Cys})_n$-Ser peptide series that were later termed hydroxymethyl-PCs (Klapheck et al., 1994). After the discovery of a fourth homolog of GSH; γ-glutamylcysteinylglutamic acid (γ-Glu-Cys-Glu) in *Zea mays*, numerous GSH homologous molecules were reported (Meuwly et al., 1993). Similarly, PC peptides represented as $(\gamma\text{-Glu-Cys})_n$-Glu (n = 2–3) were discovered in *Zea mays* seedlings on exposure with Cd^{2+} (Meuwly et al., 1995).

Later investigations showed that few homologous peptides imitate the presence of PC molecules containing terminal amino acids. Because these C-terminal–modified PC peptides are rare, a new classification was introduced to define these homologous-chelating peptides. These modified C-terminal peptides with different residues than Gly were termed iso-PCs and their terminal amino acids were denoted in parentheses as an abbreviated form. The prefix iso- is added here as the term "homo" is already in use. Besides PCs and iso-PCs, peptides lacking C-terminal amino acids were represented as $(\gamma\text{-Glu-Cys})_n$ and were termed as desGly-PC. Bernhard and Kägi (1987) discovered these peptides in maize and Mehra and Winge (1988) also reported them in *S. pombe*. Further, they were characterized by Kubota et al. (1995) and Meuwly et al. (1995). Undoubtedly, all the iso-PCs as well as the desGly-PCs have role in the heavy metal detoxification.

3. METALLOTHIONEINS

Plants have evolved several adaptive strategies to deal with increasing metal ion concentrations. Besides providing metal tolerance, these mechanisms are equally efficient in detoxification of nonessential metal ions as well as excessive essential metal ions present in different organisms (Kägi, 1991). The name "metallothionein" was initially proposed in 1957 for a horse renal cortex protein that led to the discovery of first Cd^{2+}-binding protein (Margoshes and Vallee, 1957). Since their discovery of around 25 years, plant MTs are younger in comparison to animals (Hanley-Bowdoin and Lane, 1983). Metallothioneins constitute an extremely heterogeneous family of ubiquitous and transition metal ion-binding proteins in cells with a relatively low-molecular-weight ranging from 4 to 8 kDa (Koszucka and Dąbrowska, 2006). They are also characterized by high Cys residue (almost 30% of their amino acid content); these cysteines are arranged in characteristic motifs. It was observed that at the early germination stages of *Triticum aestivum*, almost 20–25% of the total cysteine residue found as nascent proteins incorporated in a single protein and is denoted as "early cysteine-labeled," or Ec, protein, respectively (Freisinger, 2011). Depending on the number and arrangement of cysteine residues, four types of plant MTs are distinguished (Freisinger, 2008). The existence of a large number of sulfhydryl groups enables MTs to participate in coordinated binding of heavy metal ions that produce metal-thiolate complexes resulting in metal ion coordination. MTs are essential in heavy metal detoxification processes; they maintain tolerance to stress generated by increased concentrations of metals (Kumar et al., 2012; Mirza et al., 2014).

4. STRUCTURE AND PROPERTIES OF MTs

The basic criteria used for the nomenclature of low-molecular-weight (almost 5–10 kDa), cysteine and metal ion-rich proteins in absence of the aromatic amino acids, characteristic Cys pattern (e.g., Cys-Cys, cys-x-cys, cys-x-y-cys motifs, in which x and y indicate noncysteine amino acids) and a typical optical features of metal-thiolate complexes (e.g., 250 nm) indicates Cd^{II}-thiolate cluster in the shoulder of peptide backbone transitions (Anjum et al., 2015; Dundar et al., 2015). The free thiol groups are absent and sulfur atoms joined divalent metals together to form thiolate clusters with tetrahedral geometry (or trigonal in Cu^{+}) (Freisinger, 2009). MTs bind to different metal ions, mainly Zn, Cd, and Cu, but their binding affinity differs among metals, with highest stability constant (10^{19}–10^{17}) in Cu, followed by Cd (10^{17}–10^{15}) and Zn (10^{14}–10^{11}) (Coyle et al., 2002). The combination of low kinetic and high thermodynamic stability are the key features of the metal–MT complexes, which hold the metals firmly, but the metal ions can easily be exchanged with other proteins (Maret, 2004). The lower metal-binding affinity could be due to its hexapeptide nature, besides this the acidic region implicated in binding with other metal ions (Carpenè et al., 2007). MT contains two subunits: a stable α-domain (C-terminal), which incorporates four divalent metal ions, and a reactive β-domain (N-terminal), which contains only three ions (Miles et al., 2000). As many as 18 different metals may associate with MTs, but only Cu^{+}, Cd^{2+}, Pb^{2+}, Ag^{+}, Hg^{2+}, and Bi^{2+} can replace Zn (Nath et al., 1988). This replacement varies with the metal species. However, MTs are stable in the Zn form or as metal-protein complexes under *in vivo* conditions (Orlowski and Piotrowski, 1998). Cu^{+} binds in several stoichiometries with at least seven Cu^{+} per molecule (Chen et al., 1996). MTs have a dynamic tertiary structure, and in the β-domain, the rapid replacement of Zn and Cd ions was observed. However, in the α-domain, the exchange is relatively slow and also occurs with other ions connected to intracellular ligands (Kägi and Kojima, 1987).

Plant MT structure, quantity of bound metal ions, and their dissociation energies were already been well-established (Freisinger, 2009). Differences between MTs structure as well as in function suggest the wide diversity of plant MT family. The dissociative mechanism involved in modulations of cellular signaling pathways and free zinc fluctuations is controlled by the oxidation of cysteine residues (Krezel et al., 2007). Metal ions were also synchronized by His residues, but their effect on MTs function and metal-binding properties is still uncertain (Blindauer, 2008). Several authors previously reported the dimerization of MTs; however, few also reported that, under aerobic conditions, intermolecular disulfide bond are developed in between C-terminal domains, whereas a nonoxidative metal-bridged dimer is formed in the N-terminal domain (Carpenè et al., 2007). The reactive properties of this metal ion bound clusters exhibit radical differences between both forms of dimer, respectively (Zangger and Armitage, 2002).

The degradation rate of MT is calculated using the similarity of the metal ion binding with the protein and their varying degradation rates among species (Richards, 1989) and differences in distribution of metal ions in between MT isoforms (Saito and Hunziker, 1996). *In vitro* studies have shown that cathepsin B, a protease, can degrade apoMT (metal-free thionein) at acidic pH. Similarly, CuMT and ZnMT show differential response in degradation pattern. The ZnMT protein rapidly dissociates Zn to contribute in cellular functions and MT synthesis, whereas CuMT forms insoluble polymers after oxidation, because of higher affinity with thionein (Richards, 1989). The metal-thiolate cluster form stabilizes at various pH optima; in spite of this, their relation to *in vivo* response of MTs is not clear yet. Primarily, the divalent metal ions Zn^{2+} and Cd^{2+} bind in tetrahedral tetrathiolate coordination spheres with MTs. The three-dimensional structural analysis clarifies that His residues can replace one or more Cys residues without changing their coordination number or geometry (Freisinger, 2011).

The information about the three-dimensional structure of plant MTs is available only for wheat Ec-1 (Loebus et al., 2011). Thus, a typical structure and folding of plant MT1, MT2, and MT3 proteins containing 30 to 45 amino acids long Cys-devoid linker regions still remain unsolved. However, Ec proteins, the linker between the γ- and the β_E-domains are almost 11 residues in length (Domènech et al., 2006). Few studies reported that these linker regions are removed through posttranslational processing (de Miranda et al., 1990). Nuclear magnetic resonance spectroscopy of wheat Ec-1 domains predicts the low (<5%) content of β-sheets and complete absence of α-helices. Similarly, using Raman spectra and infrared spectra revealed the secondary structural elements of Zn^{2+} and Cd^{2+} in cork oak MT2 (Freisinger, 2011). Despite this, the absence of helical contributions and high amounts (55–64%) of β-sheets were also virtually confirmed. In addition to this, 30% contribution of β-sheets was evaluated in the infrared spectra of chickpea Cd_5MT1. Further, detailed analysis through CD spectrum revealed a dominant single 200-nm random coil structure in addition to 10% α-helix and 30% β-sheets. The metal–thiolate clusters partially neutralize the α-helices as well as β-sheets spectral features. It is apparent from the results that plant MT1, MT2, and MT3 proteins do not contain α-helices, rather β-sheets and a mixture of β-turns and random coils contributes to a one-third portion. However, the mixture of β-turns and random coils are found in the Cys-rich regions and the linker regions are dominated by β-sheets (Zhu et al., 2000).

In comparison to mammalian MTs, plant MTs generally have a higher protein mass because of longer amino acid sequences. Without metal ions, the mammalian MTs have a mass around 6.0 kDa, whereas the largest plant MTs were reported in the MT2 subfamily (7.9 kDa), followed by MT1 (7.6 kDa) and MT3 (6.8 kDa). These differences in mass occur because of the amino acid sequences and linker regions connecting the Cys-rich regions in the proteins. Statistical analyses showed that interdomain linker regions have higher frequency of Pro and Lys residues and mostly they contain 4–12 amino acids in an even-number pattern (Tanaka et al., 2003). A highly conserved and short LysLysSer linker is found in between the two domains of mammalian MTs. This linker provides a hinge region permitting some degree of domain movement, but is not involved in interdomain contacts (Rhee et al., 1990). Both MT domains in their fully metallated form contain negative charges (i.e., −2 for α-domain and −3 for β-domain with the corresponding four and three divalent metal ions, respectively). Thus for charge neutralization, a positive charge in the linker region is required and the physical interaction between both the domains play a critical role in the protein stabilization (Jiang et al., 2000).

In the presence of metal ions, the N-terminal β-domain and the α-domain of C-terminal fold independently; hence, the structure remains same in the separately studied domains as well as in complete protein (Braun et al., 1992). However, mammalian MTs are double-domain proteins having dumbbell-like structure. Thus, it is likely that, both a dumbbell-shaped bidomain form in mammals and a single-domain hairpin-like structure is feasible. Using reverse-phase chromatography, two distinct peaks were detected from N- and C-terminal Cys-rich regions at pH 2, whereas size-exclusion chromatography showed only a single peak in the presence of Cd^{2+} ions at neutral pH. This shows that both peptides are joined by Cd^{2+} ions even after protein degradation, which confirms the presence of only one metal ion cluster. Based on the comparative analysis of circular dichroism spectra, it was proposed that chickpea, cork oak, wheat, and banana MT2 in their fully metallated form occur as a hairpin-like structure (Wan and Freisinger, 2009). This shows the possibility of a metal ion–dependent, jackknife-like domain. Thus, the fourth metal ion may bind both the clusters together to form a single one. On the basis of mammalian MT domain nomenclature, the wheat E_c-1 N-terminal domain was denominated as the γ-domain. Both the central and C-terminal Cys-rich region forms the larger domain that surrounds a mononuclear $ZnCys_2His_2$ site and Zn_3Cys_9 cluster identical to the β-domain of mammalian MTs. This forms an interleaved arrangement of two metal ion–binding sites in this domain termed as extended β-domain or the β_E-domain. However, coordination of Zn^{2+} ions by His residues is detected in cyanobacterial MTs (Blindauer et al., 2001). Taken together, it is clear that by using theoretical methods it is difficult to predict such an array of three Cys-rich regions.

5. CLASSIFICATION OF MTs

Initial classification of MTs was based upon their sequence similarities and phylogenetic relationships (Freisinger, 2009). The MT protein superfamily comprises 15 families containing various sequences obtained from both amino acids and polynucleotides inferred from all animal phyla, certain plants, fungi, and cyanobacteria. It states that any polypeptide having similarity with mammalian MTs can be classified as MT (Carpenè et al., 2007). Several MT genes have been classified in the higher plants. Plant MTs are divided into four subfamilies: p1 (class 1), p2 (class 2), p3 (class 3), and pec (class 4). The P1 subfamily includes *Cicer arietinum* MT1 along with MT1a and MT1c of *Arabidopsis thaliana*, whereas the P2 subfamily includes MT2 of *C. arietinum* and *A. thaliana* MT2a and MT2b. However, the P3 subfamily consists of *Musa acuminate* and *A. thaliana* MT3, whereas the pec subfamily comprises *A. thaliana* MT4a (Ec-2) and MT4b (Ec-1), and Ec-1 of *T. aestivum*. Several plant species, including *Arabidopsis*, rice, and sugarcane, contain all four types of MT-encoding genes (Benatti et al., 2014). This shows the divergent evolution of MT subfamilies in both monocots and dicots, with the majority of flowering plants contain all four MT subfamilies. This wide diversity among the plant MT gene family indicates the differences among their sequence as well as in function. Still, a great deal less knowledge is available about MT genes of nonflowering plants. However, type 3 MT genes were characterized from various gymnosperms (Chatthai et al., 1997), and one MT-encoding gene is reported from a brown alga, *Fucus vesiculocus* (Morris et al., 1999).

MT proteins are classified on the basis of their Cys residue arrangements (Wu et al., 2014). Plant MTs are characterized by having two Cys-rich domains divided by a long and variable (45–87 amino acids) spacer region (Wu et al., 2014). The p1 to p3 subfamilies includes approximately 40 amino acids. The Cys content differs from 10 to 17 residues, whereas the number of aromatic amino acids varies between 0 (pec) to many (p1, p2, and p3) with a low histidine content that incorporate seven divalent metal ions or 12 monovalent ions such as Cu^{+} (Nielson and Winge, 1984) in two distinct metal clusters. The four MT subfamilies occur with completely discrete cysteine arrangements in plants than in animals (Cobbett and Goldsbrough, 2002). On the basis of the position of cysteine residues in predicted proteins, Robinson et al. (1993) classified two types (types 1 and 2) of plant MTs. After that, several plant MT genes have been characterized, and additional categories have been added. MTs expressed during ripening of fruits were classified in type 3 group, whereas the first characterized MT protein (i.e., wheat E_c protein) is represented in the type 4 class. The locations of the Cys residues in mammalian MTs are invariant, and proteins from any phyla that have similar primary structures are designated class I MTs. Class I MTs consists of six Cys-Xaa-Cys motifs (where Xaa symbolizes different amino acids) that are equally distributed among the two domains. These domains are often divided by almost 40 amino acids including aromatic residues. In contrast to this, other MTs contain <10 amino acids spacer without any aromatic amino acids, in between the cysteine-rich domains. However, several distinguishing features are reported within class I MTs, especially from *Brassicaceae*, including an additional Cys residue along with a short spacer region in between the cysteine-rich domains (Zhou and Goldsbrough, 1994). Class II MTs are low-M_r metal-binding proteins that contains two Cys-rich domains separated by a much more variable spacer region. The first pair of cysteines is represented by a Cys-Cys motif that is located at three and four positions and the N-terminal end of the Cys-rich domain contains a Cys-Gly-Gly-Cys motif. In addition to this, the N-terminal domain of class II MTs contains a highly conserved (MSCCGGNCGCS) and C-terminal domain contains three Cys-Xaa-Cys motifs. These proteins were identified in cyanobacteria, yeast, nematode (*Caenorhabditis elegans*) and wheat (Kägi, 1991).

Earlier it was shown in plants that the Cd^{2+} ligands under Cd stress are composed of poly (γ-glutamylcysteinyl) glycine (Robinson et al., 1993). These polypeptides along with other γ-glutamyl isopeptides, not having Gly or substituted with β-alanine, are classified as class III MTs (Kojima, 1991). These MTs were first recognized and characterized in the fission yeast *S. pombe* and named as cadystins (Murasugi et al., 1981). Further, similar polypeptides were also isolated from plant cell cultures and termed PCs (Grill et al., 1985). These PCs were initially termed as class III MTs during classifying MTs. Since the discovery of PCs in plants as a critical metal ligand, providing tolerance toward Cd, several MT genes were characterized in plants (Cobbett and Goldsbrough, 2002). At present, class III MTs has been isolated from fungi as well as from a range of higher plants (Grill, 1989). The N-terminal domain of class III MTs contains only four Cys residues, where initially three residues had a consensus sequence of Cys-Gly-Asn-Cys-Asp-Cys. However, the fourth cysteine residue containing a highly conserved Gln-Cys-Xaa-Lys-Lys-Gly motif does not form any pair. The C-terminal cysteine-rich domain contains six cysteine residues arranged in Cys-Xaa-Cys motifs. In contrast, class IV MTs consists of three Cys-rich domains, each having five to six conserved cysteine residues separated by 10–15 amino acids. In comparison to monocots, very few members of class IV MTs are identified, but in dicots 8–10 additional amino acids were observed before the first cysteine residue in the N-terminal domain.

Recently identified genomic DNA sequences for MT genes provided an additional advantage for the classification. The N-terminal Cys-rich domain of plant MT contains an intron close to its end, but the location of this intron varies according to the type of MT. The class III MT genes contain two introns, and the first follows immediately after the N-terminal Cys-rich domain. However, in the class I MT of dicots, the first intron is localized before the last cysteine codon of the first domain. In contrast, in the class I MT of monocots, the single intron dislocates the codon in the N-terminal Cys-rich domain, after the last cysteine codon. In spite of the differences between monocot and dicot class I MT genes, the location of the first intron justifies the amino acid sequence-based classification of plant MTs (Cobbett and Goldsbrough, 2002). Genome sequencing and mapping have also confirmed that

MT genes are distributed among various chromosomes in *Arabidopsis* (Cobbett and Goldsbrough, 2002), tomato (Giritch et al., 1998), cotton (Hudspeth et al., 1996), and rice genomes (Kumar et al., 2012).

6. TISSUE DISTRIBUTION AND CELLULAR LOCALIZATION

Previous reports on the localization of MTs in particular organs, tissues, and cellular organelles explain their role during plant development and detoxification of metals. It was observed that different classes of MT showed discrete tissue-specific expression in *A. thaliana*. Expression of class IV MTs, for example, wheat E_c MT, is found to be located specifically in developing seeds. The promoter of class IV MT genes not only showed homology with the abscisic acid response elements but also abscisic acid–regulated expression (Zenk, 1996). Like other genes expressed during the maturation of embryos, MT genes also follow the RNA regulatory mechanism during imbibitions and germination (Paparella et al., 2015). In addition, MTs also provide similar mechanism for the regulation of metal ion concentration in seeds like the nutritional composition and calorific value (Lucca et al., 2001). RNA expression of class I MT genes is reported to be higher in roots than shoots, whereas class II MTs generally showed reverse response. However, class III MTs showed higher expression during the ripening of fleshy fruits, e.g., banana (Clendennen and May, 1997), apple (Reid and Ross, 1997), and kiwi (Ledger and Gardner, 1994). Generally, MT1 showed higher expression in roots, MT2 and MT3 in leaves, and MT4 in seeds (Guo et al., 2003). Using β-glucuronidase reporter gene it was found that phloem showed higher expression of MT1a and MT2b, whereas mesophyll cells of root tips and young leaves showed higher MT2a and MT3 expression (Guo et al., 2008). In trichomes, Cu highly induces the expression of MT genes and it even increased during senescence. Their expression in nuclei is also correlated with the apoptosis regulator p53 and Bax (bcl-2–like protein 4) genes (Ruttkay-Nedecky et al., 2013). Using polyclonal antibodies in immunoblot analysis, it was found that under heavy metal treatment, MT proteins are localized in roots, shoots, and germinated seedlings (Hassinen et al., 2009). MTs were further reported to be localized in the cytosol and are not transported to the vacuoles, inside the cells (Zhigang et al., 2006). Immunohistochemical studies showed an increased MT expression both in the cytoplasm and nucleus of rapidly propagating cells (Coyle et al., 2002). However, it is not still known about the importance of the retention of MT in the nucleus, yet it is hypothesized to prevent DNA from oxidative damage or regulate Zn availability for apoenzymes and transcription factors during cell division (Coyle et al., 2002).

7. FUNCTIONS OF MTs

Despite the widespread occurrence of MTs and the relatively high level of expression of several MT genes, the functions of MTs in plants remain poorly understood (Guo et al., 2008). Initial confirmation of MTs role in eukaryotic organisms for tolerance against Cd and Cu is done by using deletion mutants of mouse and yeast that are unable to express MTs (Michalska and Choo, 1993). Zhou and Goldsbrough (1994) gave the first evidence of the role of MTs in Cu and Cd tolerance by expressing two *Arabidopsis* MT genes in MT-deficient yeast. The role of MTs has been intensively studied and reported to be involved in a myriad of physiological processes such as regulation of cell growth and proliferation, toxic metal protection and homeostasis, free radical scavenging or protection from oxidative stress, DNA damage repair, and antiapoptotic defense (Gavanji et al., 2014). Further determination of the function of different plant MTs were obtained from overexpressing and knock-down lines, metal-regulated and tissue-specific expression, and by characterizing MT–metal complexes (Hassinen et al., 2009). Several abiotic and biotic stresses were also reported to influence plant MT expression, such as light (Dunaeva and Adamska, 2001; Chen et al., 2003), drought (Yang et al., 2009), low temperature (Xue et al., 2009), pathogen attack (Reymond et al., 2000), wounding (Razem and Bernards, 2002), and oxidative stress (Kumar et al., 2012; Dundar et al., 2015).

7.1 METAL DETOXICATION AND HOMEOSTASIS

Earlier, it was thought that the primary biological role of MTs would remain unsolved, but now it is clear that MTs take part in several cellular processes. Plant MTs were reported to regulate various cellular mechanisms that are related to heavy metal homeostasis (i.e., cytosolic metal binding, vacuolar sequestration, and metal efflux through plasma membrane from the cytoplasm) (Clemens, 2006). This occurs because of the characteristic structural features of MT that gave it a potent metal binding and redox capabilities (Coyle et al., 2002). Critical function of this pleiotropic protein in primitive organisms is sequestration of hazardous metals (e.g., Cd, Hg) and chemically toxic metals in their ionized form (e.g., Cu, Zn). Plant MTs are reported to provide tolerance toward Cd, Cu, and Zn after expression in metal-sensitive yeast (Zhou et al., 2014). MT genes were also found to be expressed in phloem tissues in *Arabidopsis*, indicating their role in transport of metal ions (Schiller et al., 2014). However, MTs were never reported in the phloem exudate proteins isolated from different plants.

Metal hypertolerant plant species consists of a highly specific mechanism that is quite different from the basic metal tolerance mechanism used by nonmetallicolous species (Küpper and Kochian, 2010). Different mutant as well as expression studies reported a few copper-inducing MT genes confirming the presence of copper-binding domain (van Hoof et al., 2001). Copper hypertolerance is primarily determined by the efflux of Cu ions through the plasma membrane of roots (Benatti et al., 2014). High Cu tolerance in *Silene vulgaris*, a nonhyperaccumulator, was reported to maintain elevated transcript levels of the MT-like gene (Poschenrieder et al., 2006). Moreover, Cu is the primary inducer of MT genes (Guo et al., 2003) followed by Cd and Zn (Murphy and Taiz, 1995). It has been determined through targeted mutagenesis that all four MT genes have discrete, yet overlapping, roles in the homeostasis of heavy metal and prevention of toxicity (Carpenè et al., 2007). Besides several reports on MTs being involved in the redox regulation of metabolic enzymes, production of mitochondrial energy and inactivation of toxic metals (e.g., Cd) as well as nonmetal ions, there are several reports of MT role during various stresses, including oxy-radicals, low Zn content, infection, inflammation, and toxic metals (Hernández et al., 2015). A few reports from fungi (Tucker et al., 2004) and higher plants (Yang et al., 2009) stated that MTs can release metals under oxidative stress to elicit a Zn-mediated antioxidant response.

7.2 REACTIVE OXYGEN SPECIES SCAVENGING

Several reports are available on MT expression in plants under different environmental perturbations (Kumar et al., 2012). The induction of class I MT genes in wheat and rice, by different abiotic stresses (i.e., cadmium, aluminum, heat and nutrient deficiency) shows their constitutive expression in response to stress (Cobbett and Goldsbrough, 2002). Several *in vivo* and *in vitro* reports are now available on the antioxidant activity of MTs after the initial report by Thornalley and Vasák (1985), showing its role in superoxide ($O_2^{\bullet -}$) and free hydroxyl (•OH) radicals scavenging (Carpenè et al., 2007). Guo et al. (2003) suggested their role in metal ion remobilization from senescing leaves. Similar observations were reported in the complementary DNA (cDNA) library obtained from *Populus* leaves during senescence (Bhalerao et al., 2003). Previous studies also showed that the metal binding Cys residues in MTs can also scavenge reactive oxygen species (ROS). Thus, instead of metal remobilization, the induction of MT during senescence is correlated to ROS signaling and scavenging (Leszczyszyn et al., 2013). Similarly, MT2 is reported to be induced under oxidative stress in cork tissues (Mir et al., 2004). It is also possible that a higher expression of MT during senescence is involved in metal scavenging that is produced after breakdown of proteins and pigments (Hassinen et al., 2011).

Enhanced tolerance toward drought, increased activity of antioxidant enzymes and higher zinc finger transcription factor expression in plants overexpressing MT genes suggest their role in the antioxidant system. When expressed in *Vicia faba*, the *Arabidopsis* MTs AtMT2a and AtMT3 protected guard cell chloroplasts from degradation caused by exposure to Cd. This effect was due to changes in the cells' ROS content (Lee et al., 2004). BrMT1, a MT from *Brasicca rapa*, enhances tolerance to Cd and oxidative stress when targeted to chloroplasts in *Arabidopsis* (Kim et al., 2007). Similarly, in cotton seedlings, under drought, salinity, cold, abscisic acid, ethylene, ROS, Cu, and Zn, GhMT3 was reported to show enhanced expression (Xue et al., 2009). The overexpression of GhMT3 in tobacco plants showed enhanced tolerance to oxidative stress by scavenging ROS and Zn(II) binding. Moreover, under low temperatures, *A. thaliana* MT2a elicits balance of ROS under oxidative stress (Zhu et al., 2009). Further, in the epidermal cells of rice roots, the MT2b transcript abundance was found to be downregulated during senescence (Steffens and Sauter, 2009). In addition, the knockdown mutant plants showed a negative corelationship between senescence and MT2b expression. From these studies, it is concluded that MT act as ROS scavenger, and it is to be downregulated for the development of ethylene and H_2O_2 dependent epidermal cell death.

7.3 PLANT GROWTH REGULATOR AND BIOMARKER FOR TRACE METALS

MTs play a vital role during seed and root development (Yuan et al., 2008), suberization (Guo et al., 2003), pollen germination (Guyon et al., 2000), and embryogenesis (Chatthai et al., 1997). Previous studies on rice have discovered that cytokinins downregulate MT2b (Yuan et al., 2008). The lower levels of cytokinins were observed in OsMT2b-overexpressing rice plants, whereas its mutant plants showed an inverse response. From these reports, it was concluded that OsMT2b regulates germination of seeds along with the development of roots by manipulating internal cytokinin concentrations (Hassinen et al., 2011). This indicates that MTs play a major role during several processes in plants such as developmental, metal detoxification, and scavenging of ROS.

A biomarker is defined as a measurable and/or observable change in a biological or biochemical response, ranging from the molecular to the physiological level and including behavioral changes. Responses of biomarkers indicate either exposure or toxic effects of environmental stressors and must be ecologically relevant (Cossu-Leguille and Vasseur, 2013). MT is generally thought to provide defense against nonessential metal (i.e., Cd and Hg) detoxification. Trace metals such as cadmium, copper, and mercury are assumed to induce MT biosynthesis, which proves its utilization as a probable biomarker for the detection of ecotoxicological consequence of essential, but highly toxic copper and nonessential trace metals like cadmium and mercury (Dabrio et al., 2002). Using advance techniques like HPLC-electrospray ionization-tandem mass spectometry, HPLC-inductively coupled plasma mass spectrometry (Nischwitz et al., 2003), and reverse phase HPLC coupled with fluorescence detection (Alhama et al., 2006) increases the sensitivity and accuracy for the analysis of biological samples. An increasing content of MTs has been considered to be a specific biomarker of metal exposure (Cataldo et al., 2011). High toxicity of copper in soils has also been assessed through oxidative stress biomarkers, like MTs (Gaete et al., 2010). Gillis et al. (2004) found a relationship between exposure to cadmium and MT levels in *T. tufifex*. They reported that MT levels are more sensitive to cadmium exposure that the response of population parameters (reproduction and growth), demonstrating that biomarkers are early warning signal responses.

8. TRANSCRIPTIONAL REGULATION OF MTs

To identify MTs function during development and under various environmental stresses, researchers mainly used RNA blot hybridization to study MT genes expression. In terms of transcript abundance, several MT genes show elevated expression levels in plant tissues. Further, MT genes appear most widely in randomly prepared cDNA libraries in several plants. In addition, differential screening of cDNA libraries also identified several MT genes, which show their abundance among plant species. Synthesis of MT was found to be increased after exposure to elevated concentrations of Cu^+ and Ag^+ in fungal cells (Fürst et al., 1988), Cd^{2+}, and Zn^{2+} in cyanobacteria (Olafson et al., 1988). Accumulation of MT in response to elevated metal ion concentrations, combined with its association with these ions, indicate their role in the sequestration of excess metal. Hamer (1986) first suggested the cis-acting metal-regulatory elements of animal and fungal MT, and Palmiter et al. (1983) found their role in the control of foreign gene expression in transgenic animals. Later, trans-acting metal-responsive factors have been identified (Labbé et al., 1991), purified (Labbé et al., 1993), and cloned from animals (Radtke et al., 1993), cyanobacteria (Morby et al., 1993), and yeasts (Dameron et al., 1991).

The role of MTs in the regulation of eukaryotic gene expression in relation with Zn^{2+} coordination and plant development has been reported earlier. The level of MT expression is also influenced by phytohormones such as *Gossypium hirsutum* GhMT3 transcript is upregulated by abscisic acid and ethylene (Xue et al., 2009). Steffens and Sauter (2009) showed that the *O. sativa* MT gene (OsMT2) was downregulated by ethylene (and H_2O_2) in epidermal cells undergoing cell death. The expression of types 1 and 2 MT genes decreased in *Populus trichokarpa* × *deltoides* roots treated with auxin (Kohler et al., 2004). Plant MTs are organ-specific (Dąbrowska et al., 2012). High levels of MT3 transcripts occur in ripening fruits of *Actinidia deliciosa* (Ledger and Gardner, 1994), *Malus domestica* (Reid and Ross, 1997), *Vitis vinifera* (Davies and Robinson, 2000), *Elaeis guineensis* (Abdullah et al., 2002), *Musa acuminata* (Liu et al., 2002), pineapple (Moyle et al., 2005), and *Prunus avium* (Alkio et al., 2014). MT4 presents the most specific expression pattern: it is limited to seeds and germinating pollen (Mierek-Adamska et al., 2012). During *T. aestivum* embryogenesis, an abundant level of MT4 messenger RNA gradually decreased (Kawashima et al., 1992).

The most convincing evidence to determine MTs function in plants came from the study of null mutant genotypes using transfer DNA insertional mutagenesis (Krysan et al., 1999). Another alternative strategy is RNA interference used to identify MT function without DNA insertions (Cogoni and Macino, 2000). Knock-down lines of *Arabidopsis* MT1 family were more hypersensitive than WT plants toward Cd, with a lower accumulation of Cu in roots along with As, Cd, and Zn in the leaves, confirming MT1's role in homeostasis of Cu and Zn besides tolerance toward Cd (Guo et al., 2008). Phenotypic analysis showed that under higher Cu concentrations, phloem expressed mt1a mutants, mt1a-mt2b double mutants, and MT1a-RNAi transgenic plants showed lower Cu accumulation in their roots. Yuan et al. (2008) showed that OsMT2b is expressed in the developing root and embryo and silencing of OsMT2b by RNAi causes serious defects in plant growth. Zhu et al. (2009) reported in *Arabidopsis* that ROS inducers under cold and hydrogen peroxide stress, upregulates MT2a expression. In comparison to wild-type plants, mt2a transfer DNA insertion mutants showed increased sensitivity toward cold stress and higher H_2O_2 production during germination. Similarly, under cold stress, increased MT2a expression was observed in the catalase mutant (cat2), whereas in control conditions higher H_2O_2 levels were reported (Queval et al., 2007). The heat shock transcription factor HsfA4a was reported to provide Cd tolerance in wheat and rice by inducing MT gene expression (Shim et al., 2009).

MT overexpression generally confers increased heavy metal tolerance. MT2 and MT3 overexpressed plants showed increased Cd and Cu tolerance (Zhigang et al., 2006). Similarly, TcMT2a overexpressed *A. thaliana* transgenic plants show reduced root length under heavy metals deficiency (Hassinen et al., 2009). In continuation to this, seed-specific MT4a expression in *Arabidopsis* and tobacco increased Cu and Zn tolerance (Rodríguez-Llorente et al., 2010). *Arabidopsis* plants overexpressing pigeonpea CcMT1 showed increased tolerance toward Cu and Cd (Sekhar et al., 2011), whereas the garlic MTs AsMT2b and the *Colocasia esculenta* protein CeMT2b simultaneously confer Cd tolerance and promote Cd accumulation (Zhang et al., 2006). Several genes isolated from *Thlapsi caerulescens*, a hyperaccumulator were found to be homologous to *Arabidopsis* (Hassinen et al., 2009). Microarray analysis showed that few of these genes (e.g., TcMT2a, TcMT2b, TcMT3) showed much higher level of expression in comparison to nonhyperaccumulating and nonmetallophyte genotypes. Further, in the Cu-hypertolerant populations of *T. caerulescens*, MT2b show higher expression than in nonmetallicolous populations, which suggested its role in Cu hypertolerance (van Hoof et al., 2001). However, in Zn-, Cd, and Pb-enriched soils, Jack et al. (2007) found higher expression of MT2b in *S. vulgaris*. This confirms that MT2b either enhances heavy metal tolerance or maintains Cu homeostasis under higher availability of Zn and Cd (Hassinen et al., 2011).

The MT synthesis mechanism is still not defined exactly, but it was suggested that the metal responsive element-binding transcription factor-1 (MTF-1) plays a crucial role in MT transcription. An MTF-1 zinc finger DNA-binding domain functions as a domain for zinc-sensing. The linkers in between these zinc fingers modulate the translocation of MTF-1 inside the nucleus and binding with the promoter of MT1 gene (Carpenè et al., 2007). The differential expression of MT genes during development was also shown in class I MT gene in *Brassica napus* (Buchanan-Wollaston, 1994), *Arabidopsis* (Garcia-Hernandez et al., 1998), and rice (Hsieh et al., 1995). MTs were reported to be expressed primarily in the phloem tissues of the senescing leaves, which suggest their role as a chaperone for long-distance Cu transport (Butt et al., 1998). Other genes playing role in Cu homeostasis, AtCCH (containing a metal chaperone) and AtRAN1 (copper transporter), were also reported in leaves during senescence (Himelblau and Amasino, 2000). They are part of a complex system regulating Cu uptake and its distribution (Pena et al., 1999), thus preventing the damage by ROS produced by copper ions in the membranes, proteins, and nucleic acids (Rae et al., 1999).

In silico promoter analysis of MT genes can shed light on their developmental and environmental stress responsiveness. Various internal as well as external factors were found to regulate MTs by using *in silico* analyses (Wu et al., 2014). The promoter region of several MTs shows different metal response element–like motifs, e.g., tomato (Whitelaw et al., 1997), Douglas fir (Chatthai et al., 2004), poplar (Berta et al., 2009), and rice (Lü et al., 2007). Several copies of metal response elements were observed in the MT promoter region with variable response during their metal-induced transcription. The promoter of rice MT genes (ricMT, OsMT2b) shows various regions responsible for abscisic acid, light, wounding, heavy metal responsive, and organ-specific expression (Ren and Zhao, 2009). However, in the companion cells of the *A. thaliana* MT2b gene, an evening element-related sequence was also observed as a cis-regulatory element (Tsuwamoto and Harada, 2010).

9. CONCLUSION AND FUTURE PROSPECTS

In plants, the tolerance toward heavy metal stress defines their ability to defend and effective remediation systems (Anjum et al., 2014). Recent years have offered several advantages to understand MT and PC biosynthesis, function, expression, and regulation by using model crops. Although several studies have revealed significant role of plant MTs, we are still lagging behind in identifying many functions of MTs. The wide diversity in the regions of metal binding in plant MTs in comparison to animals indicates their binding capability with broad range of metals leading to varied functions. Both because of unavailability of MT-deficient mutants in *Arabidopsis* and possible redundancy among MT gene family members, the sound knowledge about the gene function is still lacking. Extensive studies on the "phytoremediation" or detoxification of polluted environments gives an overview of the MT potential against stress tolerance. To increase the phytoremediation capacity of plants, it is essential to manipulate the expression of MTs. Further, to understand the consequence of their overexpression in specific tissues on metal accumulation and tolerance will verify their utility. The development of transgenic plants with exceptional capability to chelate different metals and avert their toxic effects is an upcoming area of research and commercial significance. The orchestrated use of conventional breeding along with molecular biology is useful in the bioremediation of contaminated sites in future. In addition, the identification of metal tolerance–related genes using genome sequencing can pave the way towards the creation of transgenics with desired characters to be used in phytoextraction technology. This together with extensive evolutionary studies across candidate genes associated with heavy metal tolerance could provide promising results.

ACKNOWLEDGEMENTS

RJ acknowledges the Start-Up research grant (Young Scientist) from the Science and Engineering Research Board, Government of India. AP and SLS-P acknowledge the support of research funds in the laboratory from the Department of Science and Technology, Department of Biotechnology, Government of India, and internal grants of ICGEB.

REFERENCES

Abdullah, S.N.A., Cheah, S.C., Murphy, D.J., 2002. Isolation and characterisation of two divergent type 3 metallothioneins from oil palm, *Elaeis guineensis*. Plant Physiol. Biochem. 40, 255–263.

Alhama, J., Romero-Ruiz, A., López-Barea, J., 2006. Metallothionein quantification in clams by reversed-phase high-performance liquid chromatography coupled to fluorescence detection after monobromobimane derivatization. J. Chromatogr. A 1107, 52–58.

Alkio, M., Jonas, U., Declercq, M., Nocker, S.V., Knoche, M., 2014. Transcriptional dynamics of the developing sweet cherry (*Prunus avium* L.) fruit: sequencing, annotation and expression profiling of exocarp-associated genes. Hort. Res. 1, 11.

Anjum, N.A., Gill, S.S., Gill, R., Hasanuzzaman, M., Duarte, A.C., Pereira, E., Ahmad, I., Tuteja, R., Tuteja, N., 2014. Metal/metalloid stress tolerance in plants: role of ascorbate, its redox couple, and associated enzymes. Protoplasma 251, 1265–1283.

Anjum, N.A., Hasanuzzaman, M., Hossain, M.A., Thangavel, P., Roychoudhury, A., Gill, S.S., Rodrigo, M.A., Adam, V., Fujita, M., Kizek, R., Duarte, A.C., Pereira, E., Ahmad, I., 2015. Jacks of metal/metalloid chelation trade in plants-an overview. Front. Plant Sci. 6, 192.

Benatti, M., Yookongkaew, N., Meetam, M., Guo, W.J., Punyasuk, N., AbuQamar, S., Goldsbrough, P., 2014. Metallothionein deficiency impacts copper accumulation and redistribution in leaves and seeds of *Arabidopsis*. New Phytol. 202, 940–951.

Bernhard, W.R., Kägi, J.H., 1987. Purification and characterization of a typical cadmium-binding polypeptides from *Zea mays*. Experientia Suppl. 52, 309–315.

Berta, M., Giovannelli, A., Potenza, E., Traversi, M.L., Racchi, M.L., 2009. Type 3 metallothioneins respond to water deficit in leaf and in the cambial zone of white poplar (*Populus alba*). J. Plant Physiol. 166, 521–530.

Bhalerao, R., Keskitalo, J., Sterky, F., Erlandsson, R., Björkbacka, H., Birve, S.J., Karlsson, J., Gardeström, P., Gustafsson, P., Lundeberg, J., Jansson, S., 2003. Gene expression in autumn leaves. Plant Physiol. 131, 430–442.

Blindauer, C.A., Harrison, M.D., Parkinson, J.A., Robinson, A.K., Cavet, J.S., Robinson, N.J., Sadler, P.J., 2001. A metallothionein containing a zinc finger within a four-metal cluster protects a bacterium from zinc toxicity. Proc. Natl. Acad. Sci. U.S.A. 98, 9593–9598.

Blindauer, C.A., 2008. Metallothioneins with unusual residues: histidines as modulators of zinc affinity and reactivity. J. Inorg. Biochem. 102, 507–521.

Braun, W., Vasák, M., Robbins, A.H., Stout, C.D., Wagner, G., Kägi, J.H., Wüthrich, K., 1992. Comparison of the NMR solution structure and the X-ray crystal structure of rat metallothionein-2. Proc. Natl. Acad. Sci. U.S.A. 89, 10124–10128.

Buchanan-Wollaston, V., 1994. Isolation of cDNA clones for genes that are expressed during leaf senescence in *Brassica napus*. Identification of a gene encoding a senescence-specific metallothionein-like protein. Plant Physiol. 105, 839–846.

Butt, A., Mousley, C., Morris, K., Beynon, J., Can, C., Holub, E., Greenberg, J.T., Buchanan-Wollaston, V., 1998. Differential expression of a senescence-enhanced metallothionein gene in *Arabidopsis* in response to isolates of *Peronospora parasitica* and *Pseudomonas syringae*. Plant J. 16, 209–221.

Carpenè, E., Andreani, G., Isani, G., 2007. Metallothionein functions and structural characteristics. J. Trace Elem. Med. Biol. 1, 35–39.

Cataldo, J.R., Hidalgo, M.E.L., Neaman, A.H., Gaete, O., 2011. Use of molecular biomarkers in *Eisenia foetida* to assess copper toxicity in agricultural soils affected by mining activities. J. Soil Sci. Plant Nutr. 11, 57–70.

Chatthai, M., Kaukinen, K.H., Tranbarger, T.J., Gupta, P.K., Misra, S., 1997. The isolation of a novel metallothionein-related cDNA expressed in somatic and zygotic embryos of Douglas-fir: regulation by ABA, osmoticum, and metal ions. Plant Mol. Biol. 34, 243–254.

Chatthai, M., Osusky, M., Osuska, L., Yevtushenko, D., Misra, S., 2004. Functional analysis of a Douglas-fir metallothionein-like gene promoter: transient assays in zygotic and somatic embryos and stable transformation in transgenic tobacco. Planta 220, 118–128.

Chen, P., Munoz, A., Nettesheim, D., Shaw 3rd, C.F., Petering, D.H., 1996. Stoichiometry and cluster specificity of copper binding to metallothionein: homogeneous metal clusters. Biochem. J. 317, 395–402.

Chen, H.J., Hou, W.C., Yang, C.Y., Huang, D.J., Liu, J.S., Lin, Y.H., 2003. Molecular cloning of two metallothionein-like protein genes with differential expression patterns from sweet potato (*Ipomoea batatas*) leaves. J. Plant Physiol. 160, 547–555.

Clemens, S., 2006. Toxic metal accumulation, responses to exposure and mechanisms of tolerance in plants. Biochimie 88, 1707–1719.

Clendennen, S.K., May, G.D., 1997. Differential gene expression in ripening banana fruit. Plant Physiol. 115, 463–469.

Cobbett, C., Goldsbrough, P., 2002. Phytochelatins and metallothioneins: roles in heavy metal detoxification and homeostasis. Annu. Rev. Plant Biol. 53, 159–182.

Cogoni, C., Macino, G., 2000. Post-transcriptional gene silencing across kingdoms. Curr. Opin. Genet. Dev. 10, 638–643.

Conn, S., Gilliham, M., 2010. Comparative physiology of elemental distributions in plants. Ann. Bot. 105, 1081–1102.

Cossu-Leguille, C., Vasseur, P., 2013. Aquatic biomarkers. In: Férard, J.F., Blaise, C. (Eds.), Encyclopedia of Aquatic Ecotoxicology, pp. 49–66.

Coyle, P., Philcox, J.C., Carey, L.C., Rofe, A.M., 2002. Metallothionein: the multipurpose protein. Cell Mol. Life Sci. 59, 627–647.

Dabrio, M., Rodríguez, A.R., Bordin, G., Bebianno, M.J., De Ley, M., Sestáková, I., Vasák, M., Nordberg, M., 2002. Recent developments in quantification methods for metallothionein. J. Inorg. Biochem. 88, 123–134.

Dąbrowska, G., Hrynkiewicz, K., Trejgell, A., 2012. Does arbuscular mycorrhizal fungi affect the growth and metallothioneins MT2 expression in the roots of *Brassica napus* L.? Acta Biol. Cracov. Ser. Bot. 54, 7–12.

Dameron, C.T., Winge, D.R., George, G.N., Sansone, M., Hu, S., Hamer, D., 1991. A copper-thiolate polynuclear cluster in the ACE1 transcription factor. Proc. Natl. Acad. Sci. U.S.A. 88, 6127–6131.

Davies, C., Robinson, S.P., 2000. Differential screening indicates a dramatic change in mRNA profiles during grape berry ripening. Cloning and characterization of cDNAs encoding putative cell wall and stress response proteins. Plant Physiol. 122, 803–812.

Domènech, J., Mir, G., Huguet, G., Capdevila, M., Molinas, M., Atrian, S., 2006. Plant metallothionein domains: functional insight into physiological metal binding and protein folding. Biochimie 88, 583–593.

Dunaeva, M., Adamska, I., 2001. Identification of genes expressed in response to light stress in leaves of *Arabidopsis thaliana* using RNA differential display. Eur. J. Biochem. 268, 5521–5529.

Dundar, E., Sonmez, G.D., Unver, T., 2015. Isolation, molecular characterization and functional analysis of OeMT2, an olive metallothionein with a bioremediation potential. Mol. Genet. Genom. 290, 187–199.

van der Ent, A., Baker, A.J.M., Reeves, R.D., Pollard, A.J., Schat, H., 2013. Hyperaccumulators of metal and metalloid trace elements: facts and fiction. Plant Soil 362, 319–334.

Freisinger, E., 2008. Plant MTs-long neglected members of the metallothionein superfamily. Dalton Trans. 47, 6663–6675.

Freisinger, E., 2009. Metallothioneins in plants. Met. Ions Life Sci. 5, 107–153.

Freisinger, E., 2011. Structural features specific to plant metallothioneins. J. Biol. Inorg. Chem. 16, 1035–1045.

Fürst, P., Hu, S., Hackett, R., Hamer, D., 1988. Copper activates metallothionein gene transcription by altering the conformation of a specific DNA binding protein. Cell 55, 705–717.

Gaete, H., Hidalgo, M., Neaman, A., Ávila, G., 2010. Evaluación de la toxicidad de cobre en suelos a través de biomarcadores de estrés oxidativo en *Eisenia foetida*. Quim. Nova 33, 566–570.

Garcia-Hernandez, M., Murphy, A., Taiz, L., 1998. Metallothioneins 1 and 2 have distinct but overlapping expression patterns in *Arabidopsis*. Plant Physiol. 118, 387–397.

Gavanji, S., Larki, B., Mojiri, M., 2014. Bioinformatics prediction of metal binding sites in metallothionein proteins *in silico* prediction of metal binding sites. J. Bioinf. 1, 20–25.

Gekeler, W., Grill, E., Winnacker, E.L., Zenk, M.H., 1989. Survey of the plant kingdom for the ability to bind heavy metals through phytochelatins. Z. Naturforsch. 44c, 361–369.

Gillis, P., Reynoldson, T., Dixon, G., 2004. Natural variation in a metallothioneinlike protein in *Tubifex tubifex* in the absence of metal exposure. Ecotoxol. Environ. Saf. 58, 22–28.

Giritch, A., Ganal, M., Stephan, U.W., Baumlein, H., 1998. Structure, expression and chromosomal localisation of the metallothionein-like gene family of tomato. Plant Mol. Biol. 37, 701–714.

Grill, E., Winnacker, E.U., Zenk, M.H., 1985. Phytochelatins: the principal heavy-metal complexing peptides of higher plants. Science 230, 674–676.

Grill, E., Gekeler, W., Winnacker, E.L., Zenk, M.H., 1986. Homophytochelatins are heavy metal-binding peptides of homo-GSH containing Fabales. FEBS Lett. 205, 47–50.

Grill, E., Winnacker, E.L., Zenk, M.H., 1987. Phytochelatins, a class of heavy-metal-binding peptides from plants, are functionally analogous to metallothioneins. Proc. Natl. Acad. Sci. U.S.A. 84, 439–443.

Grill, E., Loftier, S., Winnacker, E.L., Zenk, M.H., 1989. Phytochelatins, the heavy-metal-binding peptides of plants, are synthesized from glutathione by a specific γ-glutamylcysteine dipeptidyl transpeptidase (phytochelatin synthase). Proc. Natl. Acad. Sci. U.S.A. 86, 6838–6842.

Grill, E., 1989. Phytochelatins in plants. In: Hamer, D.H., Winge, D.R. (Eds.), Metal Ion Homeostasis: Molecular Biology and Chemistry. Alan R Liss Inc., New York, pp. 282–300.

Guo, W.J., Bundithya, W., Goldsbrough, P.B., 2003. Characterization of the *Arabidopsis* metallothionein gene family: tissue-specific expression and induction during senescence and in response to copper. New Phytol. 159, 369–381.

Guo, W., Meetam, M., Goldsbrough, P.B., 2008. Examining the specific contributions of individual *Arabidopsis* metallothioneins to copper distribution and metal tolerance. Plant Physiol. 146, 1697–1706.

Guyon, V.N., Astwood, J.D., Garner, E.C., Dunker, A.K., Taylor, L.P., 2000. Isolation and characterization of cDNAs expressed in the early stages of flavonol-induced pollen germination in petunia. Plant Physiol. 123, 699–710.

Hambidge, M., 2000. Human zinc deficiency. J. Nutr. 130, 1344S–1349S.

Hamer, D.H., 1986. Metallothionein. Annu. Rev. Biochem. 55, 913–951.

Hanley-Bowdoin, L., Lane, B.G., 1983. A novel protein programmed by the mRNA conserved in dry wheat embryos. The principal site of cysteine incorporation during early germination. Eur. J. Biochem. 135, 9–15.

Hassinen, V.H., Tuomainen, M., Peräniemi, S., Schat, H., Kärenlampi, S.O., Tervahauta, A.I., 2009. Metallothioneins 2 and 3 contribute to the metal-adapted phenotype but are not directly linked to Zn accumulation in the metal hyperaccumulator, *Thlaspi caerulescens*. J. Exp. Bot. 60, 187–196.

Hassinen, V.H., Tervahauta, A.I., Schat, H., Kärenlampi, S.O., 2011. Plant metallothioneins-metal chelators with ROS scavenging activity? Plant Biol. 13, 225–232.

Hernández, L.E., Sobrino-Plata, J., Montero-Palmero, M.B., Carrasco-Gil, S., Flores-Cáceres, M.L., Ortega-Villasante, C., Escobar, C., 2015. Contribution of glutathione to the control of cellular redox homeostasis under toxic metal and metalloid stress. J. Exp. Bot. 66, 2901–2911.

Himelblau, E., Amasino, R.M., 2000. Delivering copper within plant cells. Curr. Opin. Plant Biol. 3, 205–210.

Hirayama, T., Shinozaki, K., 2010. Research on plant abiotic stress responses in the post-genome era: past, present and future. Plant J. 61, 1041–1052.

Holleman, A.F., Wiberg, E., 1985. Lehrbuch der anorganischen chemie. Walter de Gruyter, Berlin, p. 868.

van Hoof, N.A., Hassinen, V.H., Hakvoort, H.W., Ballintijn, K.F., Schat, H., Verkleij, J.A., Ernst, W.H., Karenlampi, S.O., Tervahauta, A.I., 2001. Enhanced copper tolerance in *Silene vulgaris* (Moench) Garcke populations from copper mines is associated with increased transcript levels of a 2b-type metallothionein gene. Plant Physiol. 126, 1519–1527.

Hörger, A.C., Fones, H.N., Preston, G.M., 2013. The current status of the elemental defense hypothesis in relation to pathogens. Front. Plant Sci. 4, 395.

Hossain, Z., Komatsu, S., 2013. Contribution of proteomic studies towards understanding plant heavy metal stress response. Front. Plant Sci. 3, 310.

Howden, R., Goldsbrough, P.B., Andersen, C.R., Cobbett, C.S., 1995. Cadmium-sensitive, cad1 mutants of *Arabidopsis thaliana* are phytochelatin deficient. Plant Physiol. 107, 1059–1066.

Hsieh, H.M., Liu, W.K., Huang, P.C., 1995. A novel stress-inducible metallothionein like gene from rice. Plant Mol. Biol. 28, 381–389.

Hudspeth, R.L., Hobbs, S.L., Anderson, D.M., Rajasekaran, K., Grula, J.W., 1996. Characterization and expression of metallothionein-like genes in cotton. Plant Mol. Biol. 31, 701–705.

Jack, E., Hakvoort, H.W.J., Reumer, A., Verkleij, J.A.C., Schat, H., Ernst, W.H.O., 2007. Real-time PCR analysis of metallothionein-2b expression in metallicolous and non-metallicolous populations of *Silene vulgaris* (Moench) Garcke. Environ. Exp. Bot. 59, 84–91.

Jiang, L.J., Vasák, M., Vallee, B.L., Maret, W., 2000. Zinc transfer potentials of the alpha- and beta-clusters of metallothionein are affected by domain interactions in the whole molecule. Proc. Natl. Acad. Sci. U.S.A. 97, 2503–2508.

Kägi, J.H.R., Kojima, Y., 1987. Chemistry and biochemistry of metallothionein. Experientia Suppl. 52, 25–61.

Kägi, J.H.R., 1991. Overview of metallothionein. Methods Enzymol. 205, 613–626.

Kawashima, I., Kennedy, T.D., Chino, M., Lane, B.G., 1992. Wheat Ec metallothionein genes. Like mammalian Zn^{2+} metallothionein genes, wheat Zn^{2+} metallothionein genes are conspicuously expressed during embryogenesis. Eur. J. Biochem. 209, 971–976.

Kim, S.H., Lee, H.S., Song, W.Y., Choi, K.S., Hur, Y., 2007. Chloroplast-targeted BrMT1 (*Brassica rapa* type-1 metallothionein) enhances resistance to cadmium and ROS in transgenic *Arabidopsis* plants. J. Plant Biol. 50, 1–7.

Klapheck, S., Chrost, B., Starke, J., Zimmermann, H., 1992. γ-Glutamylcysteinylserine: a new homologue of glutathione in plants of the family Poaceae. Bot. Acta 105, 174–179.

Klapheck, S., Fliegner, W., Zimmer, I., 1994. Hydroxymethylphytochelatins [$(\gamma$-Glu-Cys$)_n$-Ser] are metal-induced peptides of the Poaceae. Plant Physiol. 104, 1325–1332.

Klapheck, S., Schlunz, S., Bergmann, L., 1995. Synthesis of phytochelatins and homo-phytochelatins in *Pisum sativum* L. Plant Physiol. 107, 515–521.

Kneer, R., Zenk, M.H., 1992. Phytochelatins protect plant enzymes from heavy metal poisoning. Phytochemistry 31, 2663–2667.

Kohler, A., Blaudez, D., Chalot, M., Martin, F., 2004. Cloning and expression of multiple metallothionein from hybrid poplar. New Phytol. 164, 83–93.

Kojima, Y., 1991. Definitions and nomenclature of metallothioneins. Methods Enzymol. 205, 8–10.

Kondo, N., lsobe, M., Imai, K., Goto, T., 1985. Synthesis of metallothionein-like peptides cadystin A and B occurring in a fission yeast, and their isomers. Agric. Biol. Chem. 49, 71–83.

Koszucka, A.M., Dąbrowska, G., 2006. Plant metallothioneins. Adv. Cell Biol. 33, 285–302.

Krezel, A., Hao, Q., Maret, W., 2007. The zinc/thiolate redox biochemistry of metallothionein and the control of zinc ion fluctuations in cell signaling. Arch. Biochem. Biophys. 463, 188–200.

Krysan, P.J., Young, J.K., Sussman, M.R., 1999. T-DNA as an insertional mutagen in *Arabidopsis*. Plant Cell 11, 2283–2290.

Kubota, H., Sato, K., Yamada, T., Maitani, T., 1995. Phytochelatins (class III metallothioneins) and their desglycyl peptides induced by cadmium in normal root cultures of *Rubia tinctorum* L. Plant Sci. 106, 157–166.

Kumar, G., Kushwaha, H.R., Punjabi-Sabharwal, V., Kumari, S., Joshi, R., Karan, R., Mittal, S., Singla-Pareek, S.L., Pareek, A., 2012. Clustered metallothionein genes are co-regulated in rice and ectopic expression of OsMT1e-P confers multiple abiotic stress tolerance in tobacco via ROS scavenging. BMC Plant Biol. 12, 107.

Küpper, H., Kochian, L.V., 2010. Transcriptional regulation of metal transport genes and mineral nutrition during acclimatization to cadmium and zinc in the Cd/Zn hyperaccumulator, *Thlaspi caerulescens* (Ganges population). New Phytol. 185, 114–129.

Labbé, S., Prévost, J., Remondelli, P., Leone, A., Séguin, C., 1991. A nuclear factor binds to the metal regulatory elements of the mouse gene encoding metallothionein-I. Nucl. Acids Res. 19, 4225–4231.

Labbé, S., Larouche, L., Mailhot, D., Séguin, C., 1993. Purification of mouse MEP-1, a nuclear protein which binds to the metal regulatory elements of genes encoding metallothionein. Nucl. Acids Res. 21, 1549–1554.

Ledger, S.E., Gardner, R.C., 1994. Cloning and characterization of five cDNAs for genes differentially expressed during fruit development of kiwifruit (*Actinidia deliciosa* var. deliciosa). Plant Mol. Biol. 25, 877–886.

Lee, J., Shim, D., Song, W.Y., Hwang, I., Lee, Y., 2004. *Arabidopsis* metallothioneins 2a and 3 enhance resistance to cadmium when expressed in *Vicia faba* guard cells. Plant Mol. Biol. 54, 805–815.

Leszczyszyn, O.I., Imam, H.T., Blindauer, C.A., 2013. Diversity and distribution of plant metallothioneins: a review of structure, properties and functions. Metallomics 5, 146–169.

Liu, P., Goh, C.J., Loh, C.S., Pua, E.C., 2002. Differential expression and characterization of three metallothionein-like genes in Cavendish banana (*Musa acuminata*). Physiol. Plant 114, 241–250.

Loebus, J., Peroza, E.A., Blüthgen, N., Fox, T., Meyer-Klaucke, W., Zerbe, O., Freisinger, E., 2011. Protein and metal cluster structure of the wheat metallothionein domain γ-E(c)-1: the second part of the puzzle. J. Biol. Inorg. Chem. 16, 683–694.

Loeffler, S., Hochberger, A., Grill, E., Winnacker, E.L., Zenk, M.H., 1989. Termination of the phytochelatin synthase reaction through sequestration of heavy metals by the reaction product. FEBS Lett. 258, 42–46.

Lü, S., Gu, H., Yuan, X., Wang, X., Wu, A.M., Qu, L., Liu, J.Y., 2007. The GUS reporter-aided analysis of the promoter activities of a rice metallothionein gene reveals different regulatory regions responsible for tissue-specific and inducible expression in transgenic *Arabidopsis*. Transgenic Res. 16, 177–191.

Lucca, P., Hurrell, R., Potrykus, I., 2001. Genetic engineering approaches to improve the bioavailability and the level of iron in rice grains. Theor. Appl. Genet. 102, 392–397.

Maret, W., 2004. Zinc and sulfur: a critical biological partnership. Biochemistry 43, 3301–3309.

Margoshes, M., Vallee, B.L., 1957. A cadmium protein from equine kidney cortex. J. Am. Chem. Soc. 79, 4813–4814.

Mehra, R.K., Winge, D.R., 1988. Cu(I) binding to the *Schizosaccharomyces pombe* γ-glutamyl peptides varying in chain lengths. Arch. Biochem. Biophys. 265, 381–389.

Meuwly, P., Thibault, P., Rauser, W.E., 1993. γ-Glutamylcysteinylglutamic acid – a new homologue of glutathione in maize seedlings exposed to cadmium. FEBS Lett. 336, 472–476.

Meuwly, P., Thibault, P., Schwan, A.L., Rauser, W.E., 1995. Three families of thiol peptides are induced by cadmium in maize. Plant J. 7, 391–400.

Michalska, A.E., Choo, K.H.A., 1993. Targeting and germ-line transmission of a null mutation at the metallothionein I and II loci in mouse. Proc. Natl. Acad. Sci. U.S.A. 90, 8088–8092.

Mierek-Adamska, A., Dąbrowska, G., Goc, A., 2012. Characterization and expression of a cDNA encoding a seed-specific metallothionein in winter rape. Acta Biol. Cracov. Ser. Bot. 54, 68.

Miles, A.T., Hawksworth, G.M., Beattie, J.H., Rodilla, V., 2000. Induction, regulation, degradation and biological significance of mammalian metallothionein. Crit. Rev. Biochem. Mol. Biol. 35, 35–70.

Mir, G., Domènech, J., Huguet, G., Guo, W.J., Goldsbrough, P., Atrian, S., Molinas, M., 2004. A plant type 2 metallothionein (MT) from cork tissue responds to oxidative stress. J. Exp. Bot. 55, 2483–2493.

de Miranda, J.R., Thomas, M.A., Thurman, D.A., Tomsett, A.B., 1990. Metallothionein genes from the flowering plant *Mimulus guttatus*. FEBS Lett. 260, 277–280.

Mirza, N., Mahmood, Q., Shah, M.M., Pervez, A., Sultan, S., 2014. Plants as useful vectors to reduce environmental toxic arsenic content. Sci. World J. 2014, 921581.

Morby, A.P., Turner, J.S., Huckle, J.W., Robinson, N.J., 1993. SmtB is a metal-dependent repressor of the cyanobacterial metallothionein gene smtA: identification of a Zn inhibited DNA-protein complex. Nucl. Acids Res. 21, 921–925.

Morris, C.A., Nicolaus, B., Sampson, V., Harwood, J.L., Kille, P., 1999. Identification and characterization of a recombinant metallothionein protein from a marine alga, *Fucus vesiculosus*. Biochem. J. 338, 53–60.

Moyle, R., Fairbairn, D.J., Ripi, J., Crowe, M., Botella, J.R., 2005. Developing pineapple fruit has a small transcriptome dominated by metallothionein. J. Exp. Bot. 56, 101–112.

Murasugi, A., Wada, C., Hayashi, Y., 1981. Purification and unique properties in UV and CD spectra of Cd-binding peptide 1 from *Schizosaccharomyces pombe*. Biochem. Biophys. Res. Commun. 103, 1021–1028.

Murphy, A., Taiz, L., 1995. Comparison of metallothionein gene expression and nonprotein thiols in ten *Arabidopsis* ecotypes. Plant Physiol. 109, 945–954.

Nath, R., Kambadur, R., Gulati, S., Paliwal, V.K., Sharma, M., 1988. Molecular aspects, physiological function and clinical significance of metallothioneins. Crit. Rev. Food Sci. Nutr. 27, 41–85.

Nielson, K.B., Winge, D.R., 1984. Preferential binding of copper to the beta domain of metallothionein. J. Biol. Chem. 259, 4941–4946.

Nischwitz, V., Michalke, B., Kettrup, A., 2003. Identification and quantification of metallothionein isoforms and superoxide dismutase in spiked liver extracts using HPLC-ESI-MS offline coupling and HPLC-ICP-MS online coupling. Anal. Bioanal. Chem. 375, 145–156.

Olafson, R.W., McCubbin, W.D., Kay, C.M., 1988. Primary- and secondary-structural analysis of a unique prokaryotic metallothionein from a *Synechococcus* sp. cyanobacterium. Biochem. J. 251, 691–699.

Orlowski, C., Piotrowski, J.K., 1998. Metal composition of human hepatic and renal metallothionein. Biol. Trace Elem. Res. 65, 133–141.

Ovečka, M., Takáč, T., 2014. Managing heavy metal toxicity stress in plants: biological and biotechnological tools. Biotechnol. Adv. 32, 73–86.

Palmiter, R.D., Norstedt, G., Gelinas, R.E., Hammer, R.E., Brinster, R.L., 1983. Metallothionein-human GH fusion genes stimulate growth of mice. Science 222, 809–814.

Paparella, S., Araújo, S.S., Rossi, G., Wijayasinghe, M., Carbonera, D., Balestrazzi, A., 2015. Seed priming: state of the art and new perspectives. Plant Cell Rep 34, 1281–1293.

Pena, M.M.O., Lee, J., Thiele, D.J., 1999. A delicate balance: homeostatic control of copper uptake and distribution. J. Nutr. 129, 1251–1260.

Pinter, T.B., Stillman, M.J., 2014. The zinc balance: competitive zinc metalation of carbonic anhydrase and metallothionein 1A. Biochemistry 53, 6276–6285.

Poschenrieder, C., Tolrá, R., Barcelo, J., 2006. Can metals defend plants against biotic stress? Trends Plant Sci. 11, 1360–1385.

Queval, G., Issakidis-Bourguet, E., Hoeberichts, F.A., Vandorpe, M., Gakière, B., Vanacker, H., Miginiac-Maslow, M., Van Breusegem, F., Noctor, G., 2007. Conditional oxidative stress responses in the *Arabidopsis* photorespiratory mutant cat2 demonstrate that redox state is a key modulator of daylength-dependent gene expression, and define photoperiod as a crucial factor in the regulation of H_2O_2-induced cell death. Plant J. 52, 640–657.

Radtke, F., Heuchel, R., Georgiev, O., Hergersberg, M., Gariglio, M., Dembic, Z., Schaffner, W., 1993. Cloned transcription factor MTF-1 activates the mouse metallothionein I promoter. EMBO J. 12, 1355–1362.

Rae, T.D., Schmidt, P.J., Pufhal, R.A., Culotta, V.C., O'Halloran, T.V., 1999. Undetectable intracellular free copper: the requirement of a copper chaperone for superoxide dismutase. Science 284, 805–808.

Razem, F.A., Bernards, M.A., 2002. Hydrogen peroxide is required for poly (phenolic) domain formation during wound-induced suberization. J. Agric. Food Chem. 50, 1009–1015.

Reid, S.J., Ross, G.S., 1997. Up-regulation of two cDNA clones encoding metallothionein-like proteins in apple fruit during cool storage. Physiol. Plant 100, 183–189.

Ren, Y., Zhao, J., 2009. Functional analysis of the rice metallothioneins gene OsMT2b promoter in transgenic *Arabidopsis* plants and rice germinated embryos. Plant Sci. 176, 528–538.

Reymond, P., Weber, H., Damond, M., Farmer, E.E., 2000. Differential gene expression to mechanical wounding and insect feeding in *Arabidopsis*. Plant Cell 12, 707–719.

Rhee, I.K., Lee, K.S., Huang, P.C., 1990. Metallothioneins with interdomain hinges expanded by insertion mutagenesis. Protein Eng. 3, 205–213.

Richards, M.P., 1989. Recent developments in trace element metabolism and function: role of metallothionein in copper and zinc metabolism. J. Nutr. 119, 1062–1070.

Robinson, N.J., Tommey, A.M., Kuske, C., Jackson, P.J., 1993. Plant metallothioneins. Biochem. J. 295, 1–10.

Rodríguez-Llorente, I.D., Pérez-Palacios, P., Doukkali, B., Caviedes, M.A., Pajuelo, E., 2010. Expression of the seed-specific metallothionein MT4a in plant vegetative tissues increases Cu and Zn tolerance. Plant Sci. 178, 327–332.

Rüegsegger, A., Brunold, C., 1992. Effect of cadmium on gamma-glutamylcysteine synthesis in maize seedlings. Plant Physiol. 99, 428–433.

Ruttkay-Nedecky, B., Nejdl, L., Gumulec, J., Zitka, O., Masarik, M., Eckschlager, T., Stiborova, M., Adam, V., Kizek, R., 2013. The role of metallothionein in oxidative stress. Int. J. Mol. Sci. 14, 6044–6066.

Saito, S., Hunziker, P.E., 1996. Differential sensitivity of metallothionein-1 and -2 in liver of zinc-injected rat toward proteolysis. Biochim. Biophys. Acta 1289, 65–70.

Salt, D.E., Rauser, W.E., 1995. MgATP-dependent transport of phytochelatins across the tonoplast of oat roots. Plant Physiol. 107, 1293–1301.

Schiller, M., Hegelund, J.N., Pedas, P., Kichey, T., Laursen, K.H., Husted, S., Schjoerring, J.K., 2014. Barley metallothioneins differ in ontogenetic pattern and response to metals. Plant Cell Environ. 37, 353–367.

Sekhar, K., Priyanka, B., Reddy, V.D., Rao, K.V., 2011. Metallothionein 1 (CcMT1) of pigeonpea (*Cajanus cajan* L.) confers enhanced tolerance to copper and cadmium in *Escherichia coli* and *Arabidopsis thaliana*. Environ. Exp. Bot. 72, 131–139.

Sharma, S.S., Dietz, K.J., 2009. The relationship between metal toxicity and cellular redox imbalance. Trends Plant Sci. 14, 43–50.

Shim, D., Hwang, J.U., Lee, J., Lee, S., Choi, Y., An, G., Martinoia, E., Lee, Y., 2009. Orthologs of the class A4 heat shock transcription factor HsfA4a confer cadmium tolerance in wheat and rice. Plant Cell 21, 4031–4043.

Steffens, B., Sauter, M., 2009. Epidermal cell death in rice is confined to cells with a distinct molecular identity and is mediated by ethylene and H_2O_2 through an autoamplified signal pathway. Plant Cell 21, 184–196.

Steffens, J.C., Hunt, D.F., Williams, B.G., 1986. Accumulation of non-protein metal-binding polypeptides (γ-glutamyl-cysteinyl)$_n$-glycine in selected cadmium-resistant tomato cells. J. Biol. Chem. 261, 13879–13882.

Tanaka, T., Kuroda, Y., Yokoyama, S., 2003. Characteristics and prediction of domain linker sequences in multi-domain proteins. J. Struct. Funct. Genomics 4, 79–85.

Thornalley, P.J., Vasák, M., 1985. Possible role for metallothionein in protection against radiation induced oxidative stress. Kinetics and mechanism of its reaction with superoxide and hydroxyl radicals. Biochim. Biophys. Acta 827, 36–44.

Tsuwamoto, R., Harada, T., 2010. Identification of a cis-regulatory element that acts in companion cell-specific expression of AtMT2B promoter through the use of *Brassica* vasculature and gene-gun-mediated transient assay. Plant Cell Physiol. 51, 80–90.

Tucker, S.L., Thornton, C.R., Tasker, K., Jacob, C., Giles, G., Egan, M., Talbot, N.J., 2004. A fungal metallothionein is required for pathogenicity of *Magnaporthe grisea*. Plant Cell 16, 1575–1588.

Wagner, G.J., 1985. Characterization of a cadmium-binding complex of cabbage leaves. Plant Physiol. 76, 797–805.

Wagner, G.J., 1993. Accumulation of cadmium in crop plants and its consequences to human health. Adv. Agron. 51, 173–212.

Wan, X., Freisinger, E., 2009. The plant metallothionein 2 from *Cicer arietinum* forms a single metal-thiolate cluster. Metallomics 1, 489–500.

Whitelaw, C.A., Le Huquet, J.A., Thurman, D.A., Tomsett, A.B., 1997. The isolation and characterization of the type II metallothionein-like genes from tomato (*Lycopersicon esculentum* L.). Plant Mol. Biol. 33, 503–511.

Wu, C.S., Chen, D.Y., Chang, C.F., Li, M.J., Hung, K.Y., Chen, L.J., Chen, P.W., 2014. The promoter and the 5′-untranslated region of rice metallothionein OsMT2b gene are capable of directing high-level gene expression in germinated rice embryos. Plant Cell Rep. 33, 793–806.

Xue, T., Li, X., Zhu, W., Wu, C., Yang, G., Zheng, C., 2009. Cotton metallothionein GhMT3a, a reactive oxygen species scavenger, increased tolerance against abiotic stress in transgenic tobacco and yeast. J. Exp. Bot. 60, 339–349.

Yang, Z., Wu, Y., Li, Y., Ling, H.Q., Chu, C., 2009. OsMT1a a type 1 metallothionein plays the pivotal role in zinc homeostasis and drought tolerance in rice. Plant Mol. Biol. 70, 219–229.

Yuan, J., Chen, D., Ren, Y., Zhang, X., Zhao, J., 2008. Characteristic and expression analysis of a metallothioneins gene OsMT2b, down-regulated by cytokinin suggests functions in root development and seed embryo germination of rice. Plant Physiol. 146, 1637–1650.

Zangger, K., Armitage, I.M., 2002. Dynamics of interdomain and intermolecular interactions in mammalian metallothioneins. J. Inorg. Biochem. 88, 135–143.

Zenk, M.H., 1996. Heavy metal detoxification in higher plants – a review. Gene 179, 21–30.

Zhang, H., Xu, W., Dai, W., He, Z., Ma, M., 2006. Functional characterization of cadmium-responsive garlic gene AsMT2b: a new member of metallothionein family. Chin. Sci. Bull. 51, 409–416.

Zhigang, A., Cuijie, L., Yuangang, Z., Yejie, D., Wachter, A., Gromes, R., Rausch, T., 2006. Expression of BjMT2, a metallothionein 2 from *Brassica juncea*, increases copper and cadmium tolerance in *Escherichia coli* and *Arabidopsis thaliana*, but inhibits root elongation in *Arabidopsis thaliana* seedlings. J. Exp. Bot. 57, 3575–3582.

Zhou, J., Goldsbrough, P.B., 1994. Functional homologs of animal and fungal metallothionein genes from *Arabidopsis*. Plant Cell 6, 875–884.

Zhou, B., Yao, W., Wang, S., Wang, X., Jiang, T., 2014. The metallothionein gene, TaMT3, from *Tamarix androssowii* confers Cd^{2+} tolerance in tobacco. Int. J. Mol. Sci. 15, 10398–10409.

Zhu, C.M., Lü, T., Zhang, R.Q., Zhao, N.M., Liu, J.Y., 2000. Modeling of kiwifruit metallothionein kiwi503. Chin. Sci. Bull. 45, 1413–1417.

Zhu, W., Zhao, D., Miao, Q., Xue, T., Li, X., Zheng, C., 2009. *Arabidopsis thaliana* metallothionein, AtMT2a, mediates ROS balance during oxidative stress. J. Plant Biol. 52, 585–592.

CHAPTER

10

RESPONSES OF PHYTOCHELATINS AND METALLOTHIONEINS IN ALLEVIATION OF HEAVY METAL STRESS IN PLANTS: AN OVERVIEW

Resham Sharma, Renu Bhardwaj, Neha Handa, Vandana Gautam, Sukhmeen Kaur Kohli, Shagun Bali, Parminder Kaur, Ashwani Kumar Thukral, Saroj Arora, Puja Ohri, Adarsh Pal Vig

Department of Botanical and Environmental Sciences, Guru Nanak Dev University, Amritsar, Punjab, India

CHAPTER OUTLINE

1. INTRODUCTION

Most heavy metals (HMs) occur naturally in the earth's crust; however, an unprecedented increase in industrial pollution (sources: agriculture, coal power plants, foundries, mines, smelters, etc.) has led to a steep upsurge in their accumulation in soil and water and caused them to dangerously disrupt our biochemical cycles and food webs (Nagajyoti et al., 2010; Singh et al., 2011; Emamverdian et al., 2015). Because HMs are nonbiodegradable, their remediation becomes impertinent to human health and environmental safety (Eide, 2006; Ali et al., 2013). For their continuous and foolproof removal

Plant Metal Interaction. http://dx.doi.org/10.1016/B978-0-12-803158-2.00010-2

from soil and water substrates, several hyperaccumulator plants have been identified and many are still being investigated for this property (Van der Ent et al., 2013; Chibuike and Obiora, 2014). Hyperaccumulators are special because of their strong metal uptake capacity and fortified physiological and molecular systems that are adept at tolerating extreme metal toxicity and they simultaneously guard their biology against deleterious effects via the synthesis of various organic ligands or metal chelators (Milner et al., 2013). These protein/nonproteinaceous metal chelators are produced within the plant cells and aid in metal transport and sequestration of HMs and have been studied for their plausible genetic and molecular aspects as well (Ovečka and Takáč, 2014). Here, we focus on the roles essayed by these cysteine, methionine, and sulfur rich metal ligands called the phytochelatins (PCs) and metallothioneins (MTs) with respect to plant tolerance of HMs. We must understand that these plant products are very much part of its various antioxidant pathways such as oxidative signaling of glutathione (GSH)-related enzymes and thiol (protein/nonprotein) compounds, etc. (Zagorchev et al., 2013). Their constituent amino acids also play a crucial role in the structural and biochemical aspects of PCs and MTs by direct participation in electron transport and redox reactions, assimilatory pathways of carbon, nitrogen, sulfur, metabolic sensing, and signaling systems (Hell and Wirtz, 2011; Romero et al., 2013, 2014). This chapter includes the occurrence and distribution, biosynthesis, and mechanistic aspects of PCs and MTs in HM tolerance and their regulation in plants under adverse conditions. Further, many prominent PC and MT genes in the past decade that have been identified and induced in nonaccumulator plant species have also been discussed in detail.

2. PHYTOCHELATINS

PCs are low-molecular-weight, cysteine-rich small polypeptides with a general structure $(\gamma\text{-Glu-Cys})_n\text{Gly}$, where $n = 2\text{–}11$(Mirza et al., 2014). To date, PCs are not only reported in plants but have also been found in fungi and other organisms (Yadav, 2010). Phytochelatins are one of the most important classes of metal chelators that counter the harmful effects of various toxic metals. PCs are known to be synthesized in the cytosol in response to HM toxicity. PC–metal and PC–metalloid complexes are very stable in nature and are formed and sequestered in the vacuolar compartments where the toxic effect metals is of less concern (Dago et al., 2014; Ray and Williams, 2011; Shen et al., 2010). Biosynthesis of PCs is catalyzed by the key enzyme phytochelatin synthase (PCS) as seen in Figure 10.1 (Kutrowska and Szelag, 2014; Yadav, 2010). PCS is constitutively synthesized and is known for its nontranscriptional activity in response to HM stress (Kamiya and Fujiwara, 2015; Chaturvedi and Archana, 2014; Rea, 2012; Shukla et al., 2012).

2.1 STRUCTURE AND BIOSYNTHESIS

Phytochelatins belong to a family of peptides which were first discovered as cadmium (Cd)-binding complexes in *Schizosaccharomyces pombe* exposed to Cd and were named as cadystins (Inouhe, 2005). The amino acids required for the synthesis of this peptide are L-glutamate (Glu), L-cysteine (Cys), and glycine (Gly). PCs are synthesized from GSH; therefore, the biosynthetic pathway overlaps with GSH biosynthesis. The general structure of PC oligomer is $(\gamma\text{-Glu-Cys})_n\text{-Gly}$ where n usually ranges from 2–5 but has been reported as high as 11 in some species (Cobbett, 2000). Certain variants from the general structure of PCs have also been identified in some plant species in which Gly can be

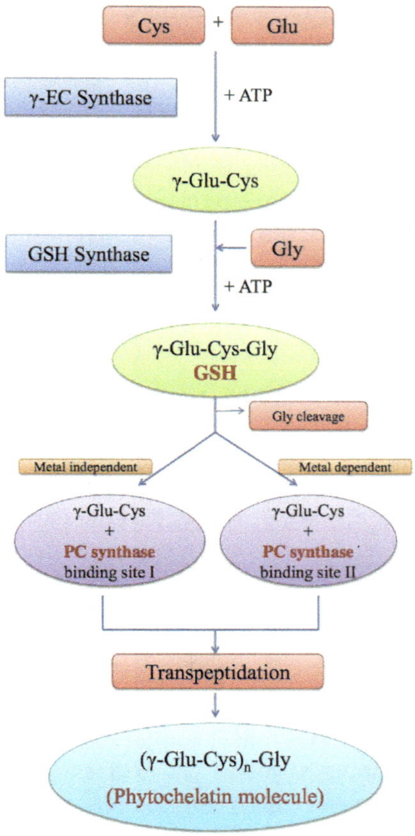

FIGURE 10.1

Schematic representation of phytochelatin biosynthetic pathway.

replaced by other amino acids like alanine (Ala), serine (Ser), and Glu, thereby making the difference in structure as $(\gamma\text{-Glu-Cys})_n$-β-Ala, $(\gamma\text{-Glu-Cys})_n$-Ser, and $(\gamma\text{-Glu-Cys})_n$-Glu, respectively (Rauser, 1995; Zenk, 1996).

During PC biosynthesis, enzyme γ-glutamylcysteine synthase catalyzes the formation of γ-glutamylcysteine from L-glutamate and L-cysteine. Further, enzyme glutathione synthase adds glycine to γ-glutamylcysteine to form GSH; in both these reactions, the presence of adenosine triphosphate (ATP) is required (Meister, 1988). After the formation of GSH, PCs are synthesized in presence of enzyme PC synthase, as seen in Figure 10.1. The enzyme was originally named γ-Glu-Cys dipeptididyl transpeptidase (EC 2.3.2.15) because it was found to carry out transpeptidation of γ-Glu-Cys moiety of GSH (Cobbett, 2000). However, before transpeptidation, glycine is cleaved from GSH and then in the next step, during transpeptidation, the resulting γ-glutamylcysteine dipeptide forms a peptide bond with either GSH to form PC2 or another PC molecule that acts as an acceptor molecule to produce $n + 1$ oligomer (Clemens, 2006).

Kinetic studies of PC synthase enzyme have indicated the presence of two substrate binding sites (sites I and II), one of which has high affinity, whereas the other has low affinity toward GSH (Vatamaniuk et al., 2000). In a medium where metal ions are absent, acylation of GSH occurs at the first site of PC synthase (Vatamaniuk et al., 2000; Rea, 2012). This site has cysteine, histidine, and aspartate, and acylation occurs at cysteine, which is a 100% conserved sequence in all PC synthases (Clemens, 2006). Hence, catalysis via site I, the acylation, and further transpeptidation to another GSH or other PC molecule is a metal-independent step (Figure 10.1). However, acylation of the second site is seen only in the presence of metal ions, specifically Cd, thereby making catalysis via site II, a metal-dependent step (Clemens, 2006; Rea, 2012).

2.2 METAL TOLERANCE STRATEGIES VIA PHYTOCHELATINS

Various exogenous as well as cytosolic strategies via PCs from the time of metal binding and uptake through the rhizosphere until its transport from roots to shoots are employed for detoxification of HMs (Polak et al., 2014). The exogenous strategies include exudation, pH of rhizosphere, specific sites of metal occurrence, and stabilization of metal ions by binding to the cell walls. Because of this stabilization, the toxicity of HM is highly reduced and transportation of metal to different cellular compartments is also affected. Cytosolic strategies include complex formation, sequestration, and conversion to less toxic and comparatively stable forms and storage (Dago et al., 2014; Shine et al., 2015). PCs are known to bind to multiple metals as well as metalloids. Several PC–metal complexes have been discovered from various plants, fungi, and microorganisms. PC–metal complexes help plants to tolerate even high levels of toxic metals by lowering the binding capacity of metal to the cell wall and detoxification by compartmentalization (Sunitha et al., 2013). In comparison to free metal ions, the PCs' metal complexes are more stable and less toxic. A large number of studies on detoxification of metal via PCS suggest the important role of PCS in HM detoxification. PCS is also involved in degradation of various GSH conjugates as well. Under natural conditions, the role of PCs and PC–metal complexes is limited to As and Cd, whereas reports for other metalloids are less evident (Franchia et al., 2015; Rigouin et al., 2013; Wünschmann et al., 2010).

Tennstedt et al. (2009) suggested involvement of PCs in Zn metal complexation and compartmentalization. It was reported that Zn detoxification mechanism was able to entirely overcome the toxic effects of metal in *Arabidopsis thaliana* plants to a great extent. It has been reported recently that PCS is induced in several plant (nonhyperaccumulators) even at a low concentration of metals and metalloids. There are several reports that suggest lack of Cad1 mutants (PCS mutant) expression resulting in an escalation of Cd sensitivity (Andresen et al., 2013). PC–metal complexes have also been reported in copper hyperaccumulating plants such as *Nicotiana caerulescenses*. Leitenmaier and Kupper (2013) reported that different ecotypes of *N. caerulescenses* were made more hypersensitive to Cd metal toxicity when the synthesis of PCS was blocked. The physiological behavior of all the ecotypes was similar to the nonsequestered plant species.

As seen in Figure 10.2, GSH (glutathione-reduced) levels in the plants under HM stress is maintained via GSH inhibitors such as buthionine sulfoxine. Afterward, PCS genes are expressed that catalyze the formation of metal–PC complexation. When two molecules of GSH ligate with an HM, they form a thiolate leading to the formation of a metal–PC complex. These metal–PC complexes are

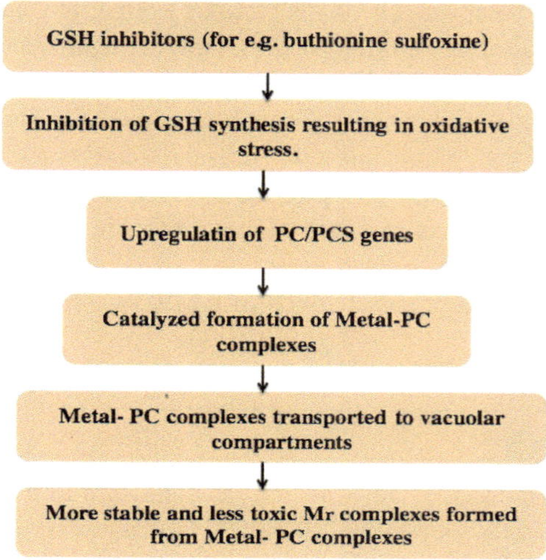

FIGURE 10.2

Metal tolerance via phytochelatins by complexation and compartmentalization strategies.

then transported to vacuolar compartments to form more stable and less toxic form of HM ions called the M_r(PC-metal) complexes (Yadav, 2010). Vacuolar compartmentalization is one of the most effective mechanisms of maintaining the accumulation and detoxification of metal ions. Vacuoles are known to lock up toxic metals in restricted compartments acting as "safe dumping sites," resulting in other part of cells being free of toxic effects of the metals (Revathi and Venugopal, 2013). Vacuolar compartmentalization is known to be most effective in Cd metal detoxification in many plants. Hypertoxicity of Cd leads to production of phytochelatins to form metal–chelate complexes (Gustin et al., 2009).

PC-mediated HM tolerance has been reported in *Ceratophyllum demersum* L. under lead toxicity. Optimal exposure to Pb metal stress leads to an elevation in the contents of glutathione and cysteine. This is also accompanied by a surge in PC levels which consequently result in reduction in GSH levels, leading to reactive oxygen species (ROS) production. Compartmentalization as well as detoxification via PCs has been reported in coontail (*C. demersum*) plants (Mishra et al., 2006, 2009; Yadav, 2010). The role of PCs in HM tolerance in *Avicennia germinans* under Cu and Cd stress at transcriptional level also seemed to suggest an increase in the expression of *AvMt2*, which might be due to its contribution in accumulation/sequestration (Yadav, 2010).

2.3 GENES FOR PHYTOCHELATINS/PCS: THEIR INDUCTION AND ROLE IN PLANTS

Tolerance to HM stress has been studied through molecular and genetic engineering tools for PC/PCS synthesis in plants. Upregulation of PCS synthase genes in various plants suggests the role of PC genes in HM tolerance (Brunetti et al., 2015; Yadav, 2010). After the cloning of the first PCS gene, a large quantity and quality of sequence data has been analyzed and deposited in a public database. Expression Sequence Tags suggest that PCS genes are ubiquitous in nature and are present in all higher plants.

PCS genes have been isolated from various plants including ferns, wheat, rice, *Arabidopsis*, *Brassica*, and so on under HM stress (Ramos et al., 2008). Homophytochelatin and phytochelatin biosynthesis has been studied in *Lotus japonicus* plants under exposure to HM stress and consequential high intracellular GSH levels (Ramos et al., 2009). Genetic manipulation-based phytoremediation studies have been described in *Arabidopsis* plants under As stress. As is transported to vacuolar compartments in the form of oxyanions, which are further converted to arsenite, where after they are sequestered in vacuole in thiol-peptide forms (Mirza et al., 2014). Increased tolerance to HM stress by overexpressing various PC and PCS genes has been tabulated extensively in Table 10.1 and it is quite clear that these are one of the most important gene pools involved in HM sequestration and detoxification (Hossain et al., 2012).

3. METALLOTHIONEINS

MTs were first reported in horse kidney cells in 1957 as metal-binding proteins (Thirumoorthy et al., 2011). These are the proteins rich in cysteine (20–30%), with a low molecular weight of ~7 kDa, are heat stable, and have a strong affinity for HMs. Structurally, they comprise two cysteine rich binding domains, α and β. They bind to metal ions through their sulfhydryl cysteine residues. The β-domain of MTs has three metal binding sites for divalent ions, which is the N-terminal part of the peptide. The α-domain, also known as the C-terminal part binds the four divalent metal ions (Nedecky et al., 2013). As shown in Figure 10.3, MTs bind to metals via thiol group of cysteine and exert a major role in detoxification and metabolism of HMs (Jia et al., 2012). These metallo-proteins defend plants against HMs through different mechanisms such as scavenging of ROS, sequestration (Hassinen et al., 2009; Huang and Wang, 2010), regulation of metalloenzymes and activation of transcription genes (Bratić et al., 2009) and metabolism of metallo-drugs under stress conditions (Gautam et al., 2012).

3.1 CLASSIFICATION AND BIOSYNTHESIS

According to various reports, the classification of MTs has been made on the basis of phylogenetic relationships and sequence similarities (Binz and Kägi, 1999; Cobbett and Goldsbrough, 2002; Zimeri et al., 2005; Freisinger, 2009). It has been classified into four subfamilies, as shown in Figure 10.4. For instance, MT1 of *Cicer arietinum* and MT1a and MT1c of *A. thaliana* are class I MTs, MT2 of *C. arietinum* and MT2a and MT2b of *A. thaliana* are class II MTs, MT3 of *A. thaliana* and *Musa acuminate* are class III MTs, and the plant PEC/pec subfamily includes MT4a (also known as Ec-2) and MT4b (also known as Ec-1) of *A. thaliana* and *Triticum aestivum* (Hassinen et al., 2011). There is a strong resemblance between the structures of GSH and class III MTs, which indicates that these metal binding polypeptides are synthesized from thiol-rich tripeptides or their precursors. Buthionine sulfoxamine, which is a powerful inhibitor of γ-glutamylcysteine synthetase, decreases cellular concentrations of γ-glutamylcysteine and GSH. Exposure of cell cultures with this compound leads to inhibition of the synthesis of class III MTs and hammers the Cd tolerance (Grill et al., 1987; Steffens, 1990). Mutants of *Schizosaccharomyces* are lacking in these enzymes that are responsible for the synthesis of GSH, which further renders them incapable of synthesizing class III MTs and are also highly sensitive to Cd (Mutoh and Hayashi, 1988). There are various alternative mechanisms that may synthesize class III MTs from GSH or γ-glutamylcysteine. In *Silene*

Table 10.1 PC and PCS Genes Expressed in Various Plants under Heavy Metal Stress

S. No.	Plant	PC/PCS Genes	Effect	Source References
1.	*Petris vittata*	*ACR2, ACR3*	Hypersensitivity to As	Indriolo et al. (2010)
2.	*Arabidopsis halleri*	*AhHMA4*	Hyperaccumulation of Zn	Barabasz et al. (2010)
3.	*Liriodendron tulipfera*	*HMA4* *PCS1*	Zinc homeostasis	Adams et al. (2011)
4.	*Noccaea caerulescens*	*HMA4*	Hypersensitivity to Cd and Zn	Ó Lochlainn et al. (2011)
5.	*Nicotiana tabaccum*	*TcPCS1*	Sequestration of Cd	Liu et al. (2011)
6.	*Oryza sativa*	*OsNRAMP1 (transporter)*	Cd accumulation	Takahashi et al. (2011)
7.	*Thlaspi caerulescens*	*TcHMA3*	Cd tolerance and hyperaccumulation	Ueno et al. (2011)
8.	*Triticum vulgaris*	*TaPCS1*	Detoxification of heavy metal	Liu et al. (2011)
9.	*Lycopersicon esculentum*	*AhHMA4*	Hyperaccumulation of Zn	Barabasz et al. (2012)
10.	*Oryza sativa*	*HMAs*	Transportation and accumulation of Zn and Cd	Takahashi et al. (2012)
11.	*Oryza sativa*	*OsIRT1*	Accumulation of Fe	Wang et al. (2013)
12.	*Arabidopsis thaliana*	*PCR2*	Accumulation of Zn	Deng et al. (2013)
13.	*Nicotiana tabaccum*	*CdPCS1*	Assimilation of Cd and As	Shukla et al. (2013a)
14.	*A. thaliana*	*AtPCS2*	Phytochelatin biosynthesis	Kuhnlenz et al. (2014)
15.	*Ceratophyllum demersum*	*CdPCS1*	Sequestration of As	Shri et al. (2014)
16.	*Nicotiana tabaccum*	*AtHMA4*	Metal homeostasis	Siemianowski et al. (2014)
17.	*Phragmites australis*	*PaPCS*	Hypersensitivity to Cd	Zhao et al. (2014)
18.	*Schizosaccharomyces pombe*	*SpHMT1*	Tolerance to heavy metal toxicity	Lee (2014)
19.	*A. thaliana*	*ACR2, ACR3*	Arsenite efflux transporters	Kamiya and Fujiwara (2015)
20.	*A. thaliana*	*XCD1/MAN3 l*	Cd accumulation and tolerance	Chen et al. (2015)
21.	*Schizosaccharomyces pombe*	*PCS1*	Tolerance to Cd metal	Shine et al. (2015)
22.	*A. thaliana*	*AtBCC3*	Cd tolerance	Brunetti et al. (2015)

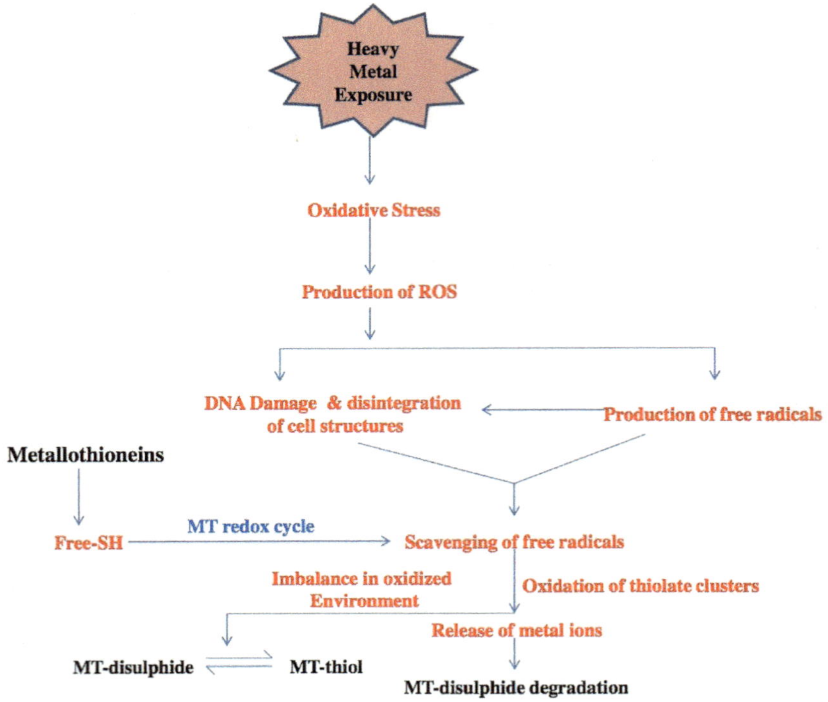

FIGURE 10.3

Heavy metal stress amelioration via metallothioneins: basic mode of action.

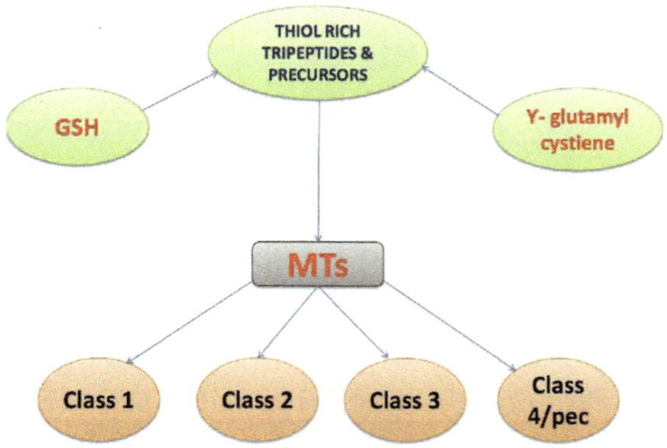

FIGURE 10.4

Heavy metal stress amelioration via metallothioneins: structure & classification.

cucubalus, class III MTs are synthesized from GSH involving γ-glutamylcysteine dipeptidyl transpeptidase (Grill et al., 1989). Biosynthesis of other MT classes involves one or two molecules of GSH that combine with previously formed class III MT. There are no additional ATP molecules required during the transfer of γ-glutamylcysteine moiety of GSH to other molecules of the previously formed class III MTs.

3.2 METAL TOLERANCE REGULATION VIA MTs

The expression of MT genes is ontogenetically regulated and changes with the type of plant tissue (Yuan et al., 2008). MT genes are activated under the influence of a wide variety of abiotic factors such as heat, cold, drought, salt, HM, and oxidative stress (Usha et al., 2007, 2009; Huang and Wang, 2009, 2010; Singh et al., 2011; Gautam et al., 2012). The stress busting properties of plant MTs are still largely unknown in comparison to those belonging to the mammals as there is difficulty in purifying plant MTs because of their instability in presence of oxygen (Guo et al., 2008; Huang et al., 2011, 2012). It has been reported by Grennan in 2011 that plant MTs play a major physiological role in metal homeostasis. MT genes are expressed in a specific tissue or organs; for instance, class MT1 genes are expressed in the root region, the class MT2 genes in the leaves, class MT3 in ripened fruit, and PEC/pec MT genes in the growing seeds (Guo et al., 2008; Huang et al., 2011). The expression of seed-specific MT4 was studied in *A. thaliana* and it was observed that the plant accumulated more Cu than Zn. When excess metal ions were applied to the host plant, the leaves and roots resulted in tolerance of both Cu and Zn through the class MT4 gene activation (Rodríguez-Llorente et al., 2010).

MT gene regulation promoter has many response elements, which help in upregulation of transcription factors. It consists of metal response elements, glucocorticoid response elements, elements activated by signal transducers and activators of transcription and the antioxidant (or electrophile) response elements. Metal-responsive transcription factor-1 activates metal response elements after HM ions bind to the MTs. Through cytokinin signaling the signal transducers and activators of transcription proteins are activated, and similarly, the antioxidant response element proteins are triggered on by associated redox imbalance. Uptake and transport of essential and nonessential metal ions are the main source of HMs, which are stably bound to MTs. When MTs are bound to HMs at all sites they serve to donate metal ions and form metalloproteins. The various metal binding and donating sites of MTs control the basic processes including gene regulation, cell proliferation and differentiation, signal transduction, and apoptosis as well as influence oxidative damage caused by oxidative stress and electrophiles (Nedecky et al., 2013). Metal tolerance via MTs probably involves mechanisms that are specific to the plant species and the metal and may be different from the mechanisms responsible for the basic levels of metal tolerance present in plants (Kupper and Kochian, 2010). Plant MTs have a major role in metal homeostasis through various cellular functions. Hence, it is evident that MTs have the following roles to play under HM stress: (1) participation in maintaining the HM homeostasis, (2) sequestration of toxic HMs by forming complexes and binding to other amino acids, and (3) protection against intracellular oxidative stress (Sunitha et al., 2013).

HM stress leads to oxidative burst in plants and is marked by release of ROS and free radicals, which instigate DNA and cellular damage. The scavenging of these harmful elements is done by the various antioxidants out of which the GSH redox complex is most crucial. The free–SH groups of MTs can be also involved in the scavenging of ROS in the MT redox cycle. Oxidation of the thiolate cluster releases the metal bound to MT and MT-disulfide as it undergoes degradation. However, when there is an imbalance in the GSH/glutathione disulfide ratio, the oxidized environment becomes prominent and the MT-disulfide bond is reduced to MT-thiol. Presence of catalysts such as selenium enhances the process of reduction. In the presence of metal ions, MTs are realigned quickly. This process constitutes the MT redox cycle, which plays a crucial role in the biologic function of MT (Nedecky et al., 2013). In *Populus tremula* × *tremuloides* the leaf MT2b transcript levels enhance Cd and Zn tolerance when grown on metal-spiked soils (Hassinen et al., 2009). The MT3 gene cloned from transgenic yeast was reported to enhance Cd and Zn absorption and tolerance, by scavenging ROS (Yang et al., 2011). However, many specifically or preferentially expressed MT genes have been identified in different plants, which affect metal tolerance; the most crucial ones have been listed in Table 10.2 and are shown in the following section.

Table 10.2 MT Genes Expressed in Various Plants under Heavy Metal Stress

S. No.	Plant	MT Genes	Effect	Source References
1.	*Populus alba* L.	*MT2*	Cd, Cu, Zn induced	Macovei et al. (2010)
2.	*Hevea brasiliensis*	*HbMT2*	Increased tolerance to both Cu and Zn ions	Zhu et al. (2010)
3.	*Avicennia marina*	*AmMT2*	Zn, Cu, Pb, and Cd	Huang and Wang (2010)
4.	*Tamarix hispida*	*ThMT3*	Enhances Cd, Zn, Cu, and NaCl tolerance in transgenic yeast	Yang et al. (2011)
5.	*Porteresia coarctata*	*PcMT3*	Tolerance against Cd-, Cu-, and Zn-induced stress	Usha et al. (2011)
6.	*Cajanus cajan,* L.	*CcMT1*	Enhanced tolerance to Cu and Cd in *Escherichia coli* and *Arabidopsis thaliana*	Sekhar et al. (2011)
7.	*Hordeum vulgare*	*MT3& MT4*	Zn binding in aleurone layer	Hegelund et al. (2012)
8.	*Bruguiera gymnorrhiza*	*BgMT2*	Zn, Cu, Pb, and Cd tolerance	Huang et al. (2011)
9.	*Kandelia candel*	*KcMT2*	Growth enhancement under Zn, Cu, Pb, or Cd exposure	Huang et al. (2012)
10.	*Oryza sativa*	*OsMT-I-IIa, OsMT-I-IIc* and *OsMT-I-IId*	Detoxification of Ar(IV)	Gautam et al. (2012)
11.	*Pisum sativum*	*PsMT2*	Enhanced tolerance and accumulation of Cu	Turchi et al. (2012)
12.	*Brassica juncea*	*BnMTs*	Characterization in seed germination	Dabrowska et al. (2013)

Table 10.2 MT Genes Expressed in Various Plants under Heavy Metal Stress—cont'd

S. No.	Plant	MT Genes	Effect	Source References
13.	*Saccharum* spp. L.	*ScMT2-1-3*	Increased the host cell's tolerance to Cd, Cu, PEG, and NaCl	Guo et al. (2013)
14.	*Brassica campestris*	*BcMT1* and *BcMT2*	Enhance tolerance to Cd and Cu and decrease production of reactive oxygen species (ROS)	Lv et al. (2013)
15.	*Sedum alfredii*	*SaMT2*	Increased Cd tolerance and accumulation	Zhang et al. (2014)
16.	*Tamarix androssowii*	*TaMT3*	Increased the ability of ROS cleaning up, indicating a stronger tolerance to Cd^{2+} stress	Zhou et al. (2014)
17.	*Oryza sativa*	*OsMT2c*	Increased ROS scavenging ability in *Arabidopsis* shows more tolerance to Cu stress	Liu et al. (2015)
18.	*Tamarix hispida*	*ThMT3*	Increases Cu tolerance and adventitious root induction in *Salix matsudana* Koidz	Yang et al. (2015)
19.	*Brassica napus* L.	*BnMP1*	Enhanced Cr(VI) tolerance by activating antioxidative defense system under S-deficient condition	Terzi and Yıldız (2015)
20.	*Iris lactea*	*IlMT2b*	Increased the Cu concentration and reduced H_2O_2 production in *A. thaliana*	Gu et al. (2015)

MT, metallothionein.

4. TRANSGENIC AND MOLECULAR PROSPECTS OF PHYTOCHELATIN- AND MT-INDUCED METAL TOLERANCE IN PLANTS

Many prominent genes involved in PC-GSH and MT cycles have been identified over the past decade: γ-glutamylcysteine synthetase (*GSH1*), glutathione synthetase (*GSH2*), ATP sulfurylase (*APS*), cystathionine synthase (*CTS*), glutathione reductase (*GR*), phytochelatin synthase (*PCS*), serine acetyltransferase (*SAT*), and glyoxalases (*glyoxalase I and II*), which are genes indoctrinating PC/PCS replicated from plants, nematodes, and fungi (Tennstedt et al., 2009; Yadav, 2010; Hassinen et al., 2011). Several MT genes (MT type I, MT type II, and MT type III) have also been identified in plants (soybean, fescue, wheat, and tomato) that can be used for transgenic purposes (Kim et al., 2013). These genes have long been isolated and characterized using different approaches including use of wild-type and mutant strains of yeast and plants such as *Arabidopsis* and *Nicotiana*, application of positional cloning strategy, and the more recent molecular markers to map the gene locations to initiate genomic sequencing (Clemens et al., 1999; Ha et al., 1999; Cobbett, 2000). But the turning point came when these isolated PC and MT genes were cloned and expressed in suitable hosts to confer them resistance against metal toxicity especially agriculturally valuable crop plants such as rice, wheat, tobacco, etc. (Dong et al., 2010; Wang et al., 2012; Lee and Hwang, 2015). Transgenic techniques have progressed well beyond all this since then. Messenger RNA accumulation of different protein genes including the *MT2* was monitored in response to Cd stress in *Vicia faba* plants. The pattern of messenger RNA transcripts for each gene was studied in relation to oxidative stress and peroxide generation (Souguir et al., 2015). The role played by MTs under Cu stress in *Iris lacteal* plants was studied through a full-length complementary DNA homologue of MT2, designated as *IlMT2b* using the rapid amplification of complementary DNA ends polymerase chain reaction method (Gu et al., 2015). *NtPCS1* (phytochelatin gene) in tobacco was used to generate transgenic tobacco lines overexpressing it in the sense or antisense direction. Thee transgenic lines were shown to have better tolerance to Cd and As (Lee and Hwang, 2015). *A. thaliana* phytochelatin synthase (*AtPCS1*) yielded mutants that have significant tolerance for Cd as compared with the nontransformed wild-type plants. The selection of mutant lines was achieved by monitoring the levels of the levels of PC precursors, GSH, and γ-glutamylcysteine upon exposure to Cd (Cahoon et al., 2015). An exhaustive collection of different reports regarding transgenic and molecular applications of PCs and MTs are presented in Table 10.3.

5. CONCLUSION AND FUTURE PROSPECTS

A lot is known about the physiological and mechanistic roles played by PCs, yet MTs remain still elusive especially in the plant system. It is factually clear that both PCs and MTs are derived from amino acid–rich precursors and are actively involved in metal binding and detoxification as observed more in hyperaccumulator plant species. This proves beyond a doubt that their role in fortifying the plant defense against metal toxicity cannot be denied. With the latest advances in biotechnology and molecular science, modulation and understanding of these amino acid–rich entities in plants at the proteomic level paves the path for future course of action. Proteomic studies provide a bird's-eye view of the plant genome and various metabolic cycles responsible for cellular detoxification and tolerance mechanism. Until now, most research has been focused at the transgenic level and protein expression under the influence of ROS scavenging, metal-chelation, and sequestration. However, deeper insights are required at the subcellular level such as identification of metal transporter proteins and vacuolar transporters involved in the cytoplasm and their role in metal efflux pathways occurring in the chloroplast and mitochondrion. Also, most reports have explained single metal/metalloid detoxification via PCs and MTs; hence, it remains to be seen how plants use these metal ligands at the time of multiple metal stress. To sum up, we need a better understanding of HM biomarkers and integrated cross- talk of various HM stress pathways inclusive of PCs and MTs via proteomics.

Table 10.3 PC/MT Genes Expressed Transgenically to Induce Metal Tolerance

S. No.	Source Plant	Gene Transferred	Transgenic Plant	Metal Tolerance Induced	Source References
1.	*Oryza sativa* L.	Rice metallothionein gene (*OsMT-I-4b*)	*Arabidopsis thaliana*	Cu, Zn, Pb, Al	Dong et al. (2010)
2.	*Schizosaccharomyces pombe, Triticum aestivum*	*S. pombe* Hmt1 and PC synthase; *T. aestivum* (*PCS1*) (*TaPCS1*) genes	*Saccharomyces cerevisiae*	Cd	Mendoza-Cózatl et al. (2010)
3.	*Saccharomyces cerevisiae*	*AtPCS1, AtABCC1*	*A. thaliana*	As	Song et al. (2010)
4.	*T. aestivum*	Phytochelatin synthase gene (*TaPCS1*)	*Populus tremula* × tremuloides cv.	Pb, Zn	Couselo et al. (2010)
5.	*A. thaliana*	Phytochelatin synthase genes: (*AtPCS1, CePCS*)	*Nicotiana tabacum*	Cd	Wojas et al. (2010)
6.	*A. thaliana*	Phytochelatin synthase (*AtPCS*)	*Oryza sativa*	Cd	Venkataramaiah et al. (2011)
7.	*Allium sativum* L.; *Saccharomyces cerevisiae*	*AsPCS1, YCF1*	*A. thaliana*	Cd, As	Guo et al. (2011)
8.	*A. thaliana*	*Arabidopsis* phytochelatin biosynthetic gene (*AtPCS1*)	*Nicotiana tabacum*	Cd	Brunetti et al. (2011)
9.	*Thlaspi caerulescens*	Phytochelatin gene (*TcPCS1*)	*Saccharomyces cerevisiae, Nicotiana tabacum*	Cd, Zn	Liu et al. (2011)
10.	*T. aestivum*	Wheat phytochelatin synthase gene (*TaPCS1*)	*Oryza sativa* L.	Cd	Wang et al. (2012)
11.	*Saccharomyces cerevisiae*	Yeast metallothionein gene (*CUP1 MT*)	*Nicotiana tabacum* L.	Cd	Krystofova et al. (2012)
12.	*Ceratophyllum demersum* L.	Phytochelatin gene cv. L. (*CdPCS1*)	*Nicotiana tobacum*	Cd, As	Shukla et al. (2012)
13.	*Saccharomyces cerevisiae*	γ-glutamyl cyclotransferase gene (*GGCT2;1*)	*A. thaliana*	As, Cd	Paulose et al. (2013)

Continued

Table 10.3 PC/MT Genes Expressed Transgenically to Induce Metal Tolerance—cont'd

S. No.	Source Plant	Gene Transferred	Transgenic Plant	Metal Tolerance Induced	Source References
14.	Synthetic phytochelatins	Synthetic phytochelatin gene (*EC*)	*A. thaliana*	As	Shukla et al. (2013b)
15.	*Ceratophyllum demersum*	Phytochelatin synthase gene (*CdPCS1*)	*Escherichia coli* and *A. thaliana*	Cd	Shukla et al. (2013a)
16.	*Oryza sativa* L.	Rice phytochelatinsynthase β-glucuronidase (*GUS*) gene	*Nicotiana tabacum*	Cd	Postrigan et al. (2013)
17.	*A. thaliana*	Serine acetyltransferase (*SAT*) and cysteine synthase (*CS*) genes	*Nicotiana tabacum*	Cd	Nakamura et al. (2014)
18.	*Tamarix androssowii*	Metallothionein gene (*TaMT3*)	*Nicotiana tobacum*	Cd	Zhou et al. (2014)
19.	*Sedum alfredii* Hance	Metallothionein gene (*SaMT2*)	*Nicotiana tobacum*	Cd, Zn	Couselo et al. (2010)
20.	*Ceratophyllum demersum*	Phytochelatin synthase gene (*CdPCS1*)	*Oryza sativa*	As	Shri et al. (2014)
21.	*Prosopis juliflora*	Metallothionein genes (*PjMT1, PjMT2*)	*Nicotiana tabacum*	Cd	Balasundaram et al. (2014)
22.	*Saccharomyces cerevisiae*	*AtABCC1* or *AtABCC2*	*Hordeum vulgare*, *A. thaliana*	Cd, As	Song et al. (2014)
23.	*A. thaliana*	*Saccharomyces cerevisiae* phytochelatin synthase (*AtPCS1*)	*Brassica juncea*	Cd	Cahoon et al. (2015)
24.	*A. thaliana*	Phytochelatin synthase1 (*AtPCS1*)	*A. thaliana*	Cd	Kühnlenz et al. (2015)
25.	*A. thaliana*	ABCC transporter (*AtABCC3*)	*A. thaliana*	Cd	Brunetti et al. (2015)
26.	*Nicotiana tabacum*	Phytochelatin synthase (*NtPCS1*)	*Nicotiana tabacum*	Cd, As	Lee and Hwang (2015)

PC, phytochelatin; MT, metallothionein.

REFERENCES

Adams, J.P., Adeli, A., Hsu, C.Y., Harkess, R.L., Page, G.P., Schultz, E.B., Yuceer, C., 2011. Poplar maintains zinc homeostasis with heavy metal genes *HMA4* and *PCS1*. J. Exp. Bot. 62, 3737–3752.

Ali, H., Khan, E., Sajad, M.A., 2013. Phytoremediation of heavy metals—concepts and applications. Chemosphere 91 (7), 869–881.

Andresen, E., Mattusch, J., Wellenreuther, G., Thomas, G., Arroyo Abad, U., Küpper, H., 2013. Different strategies of cadmium detoxification in the submerged macrophyte *Ceratophyllum demersum* L. Metallomics. http://dx.doi.org/10.1039/C3MT00088E [Epub ahead of print].

Balasundaram, U., Venkataraman, G., George, S., Parida, A., 2014. Metallothioneins from a hyperaccumulating plant *Prosopis juliflora* show difference in heavy metal accumulation in transgenic tobacco. Int. J. Agric. Environ. Biotech. 7 (2), 241–246.

Barabasz, A., Krämer, U., Hanikenne, M., Rudzka, J., Antosiewicz, D.M., 2010. Metal accumulation in tobacco expressing *Arabidopsis halleri* metal hyperaccumulation gene depends on external supply. J. Exp. Bot. 61, 3057–3067.

Barabasz, A., Wilkowska, A., Ruszczyñska, A., Bulska, E., Hanikenne, M., Czarny, M., 2012. Metal response of transgenic tomato plants expressing P_{1B}-ATPase. Physiol. Plant. 145, 315–331.

Binz, P.A., Kägi, J.H.R., 1999. Metallothionein: molecular evolution and classification. In: Klaassen, C. (Ed.), Metallothionein IV. Birkhä user Verlag, Basel, Switzerland, pp. 7–13.

Bratić, A.M., Majić, D.B., Samardžć, J.T., Maksimović, V.R., 2009. Functional analysis of the buckwheat metallothionein promoter:tissue specificity pattern and up-regulation under complex stress stimuli. J. Plant Physiol. 166, 996–1000.

Brunetti, P., Zanella, L., De Paolis, A., Di Litta, D., Cecchetti, V., Falasca, G., Barbieri, M., Altamura, M.M., Costantino, P., Cardarelli, M., 2015. Cadmium-inducible expression of the ABC-type transporter AtABCC3 increases phytochelatin-mediated cadmium tolerance in Arabidopsis. J. Exp. Bot. http://dx.doi.org/10.1093/jxb/erv185.

Brunetti, P., Zanella, L., Proia, A., De-Paolis, A., Falasca, G., Altamura, M.M., Sanità di Toppi, L., Costantino, P., Cardarelli, M., 2011. Cadmium tolerance and phytochelatin content of *Arabidopsis* seedlings over-expressing the phytochelatin synthase gene *AtPCS1*. J. Exp. Bot. 62 (15), 5509–5519.

Cahoon, R.E., Lutke, W.K., Cameron, J.C., Chen, S., Lee, S.G., Rivard, R.S., Rea, P.A., Jez, J.M., 2015. Adaptive engineering of phytochelatin-based heavy metal tolerance. J. Biol. Chem. http://dx.doi.org/10.1074/jbc.M115.652123.

Chaturvedi, R., Archana, G., 2014. Cytosolic expression of synthetic phytochelatin and bacterial metallothionein genes in *Deinococcus radiodurans* R1 for enhanced tolerance and bioaccumulation of cadmium. Biometals 27, 471–482. http://dx.doi.org/10.1007/s10534-014-9721-z.

Chen, J., Yang, L., Gu, J., Bai, X., Ren, Y., Fan, T., Han, Y., Jiang, L., Xiao, F., Liu, Y., Cao, S., 2015. *MAN3* gene regulates cadmium tolerance through the glutathione-dependent pathway in *Arabidopsis thaliana*. New Phytol. 205, 570–582.

Chibuike, G.U., Obiora, S.C., 2014. Heavy metal polluted soils: effect on plants and bioremediation methods. Appl. Environ. Soil Sci. Article ID 752708, p. 12. http://dx.doi.org/10.1155/2014/752708.

Clemens, S., 2006. Evolution and function of phtochelatin synthases. J. Plant Physiol. 163, 319–332.

Clemens, S., Kim, E.J., Neumann, D., Schroeder, J.I., 1999. Tolerance to toxic metals by a gene family of phytochelatin synthases from plants and yeast. EMBO J. 18, 3325–3333.

Cobbett, C., 2000. Phytochelatins and their roles in heavy metal detoxification. Plant Physiol. 123, 825–833.

Cobbett, C., Goldsbrough, P., 2002. Phytochelatins and metallothioneins: roles in heavy metal detoxification and homeostasis. Annu. Rev. Plant Bio. 53, 159–182.

Couselo, J.L., Navarro-Avñó, J., Ballester, A., 2010. Expression of the phytochelatin synthase TaPCS1 in transgenic aspen, insight into the problems and qualities in phytoremediation of Pb. Int. J. Phytoremediation 12 (4), 358–370.

Dabrowska, G., Adamska, A.M., Goc, A., 2013. Characterisation of *Brassica napus* L. metallothionein genes (*BnMTs*) expression in organs and during seed germination. AJCS 7, 1324–1332.

Dago, A., Gonzalez, I., Arino, C., Martinez-Coronado, A., Higueras, P., Diaz-Cruz, J.M., Esteban, M., 2014. Evaluation of mercury stress in plants from the Almadén mining district by analysis of phytochelatins and their Hg complexes. Environ. Sci. Technol. es405619y.

Deng, F., Yamaji, N., Xia, J., Ma, J.F., 2013. A member of the heavy metal P-Type ATPase OsHMA5 is involved in xylem loading of copper in rice. Plant Physiol. 163, 1353–1362.

Dong, C.J., Wang, Y., Yu, S.S., Liu, J.Y., 2010. Characterization of a novel rice metallothionein gene promoter: its tissue specificity and heavy metal responsiveness. J. Integr. Plant Biol. 52 (10), 914–924.

Eide, D.J., 2006. Zinc transporters and the cellular trafficking of zinc. Biochem. Biophys. Acta 1763, 711–722.

Emamverdian, A., Ding, Y., Mokhberdoran, F., Xie, Y., 2015. Heavy metal stress and some mechanisms of plant defense response. Sci. World J. 1–43.

Franchia, N., Piccinnia, E., Ferroa, D., Bassod, G., Spolaoree, B., Santovitoa, G., Ballarin, L., 2015. Characterization and transcription studies of a phytochelatin synthasegene from the solitary tunicate Ciona intestinalis exposed to cadmium. Aquat. Toxicol. 152, 47–56.

Freisinger, E., 2009. Metallothioneins in plants. Met. Ions Life Sci. 5, 107–153.

Gautam, N., Verma, P.K., Verma, S., Tripathi, R.D., Trivedi, P.K., Adhikari, B., Chakrabarty, D., 2012. Genome-wide identification of rice class I metallothionein gene: tissue expression patterns and induction in response to heavy metal stress. Funct. Integr. Genomics 12, 635–647.

Grennan, A.K., 2011. Metallothioneins, a diverse protein family. Plant Physiol. 155, 1750–1751.

Grill, E., Loeffler, S., Winnacker, E.L., Zenk, M.H., 1989. Phytochelatins, the heavy-metal-binding peptides of plants, are synthesized from glutathione by a specific γ-glutamylcysteine dipeptidyl transpeptidase (phytochelatin synthase). Proc. Natl. Acad. Sci. U.S.A. 86, 6838–6842.

Grill, E., Winnacker, E.L., Zenk, M.H., 1987. Phytochelatins, a class of heavy-metal-binding peptides from plants are functionally analogous to metallothioneins. Proc. Natl. Acad. Sci. U.S.A. 84, 439–443.

Gu, C.S., Liu, L.Q., Deng, Y.M., Zhu, X.D., Huang, S.Z., Lu, X.Q., 2015. The heterologous expression of the Iris lactea var. chinensis type 2 metallothionein *IlMT2b* gene enhances copper tolerance in *Arabidopsis thaliana*. Bull. Environ. Contam. Toxicol. 94, 247–253.

Guo, J., Xu, L., SU, L., Wang, H., Gao, S., Xu, J., Que, Y., 2013. ScMT2-1-3, a metallothionein gene of sugarcane, plays an important role in the regulation of heavy metal tolerance/accumulation. BioMed. Res. Int. http://dx.doi.org/10.1155/2013/904769.

Guo, J., Xua, W., Maa, M., 2011. The assembly of metals chelation by thiols and vacuolar compartmentalization conferred increased tolerance to and accumulation of cadmium and arsenic in transgenic *Arabidopsis thaliana*. J. Hazard. Mater. http://dx.doi.org/10.1016/j.jhazmat.2011.11.008.

Guo, W.J., Meetam, M., Goldsbrough, P.B., 2008. Examining the specific contributions of individual Arabidopsis metallothioneins to copper distribution and metal tolerance. Plant Physiol. 146, 1697–1706.

Gustin, J.L., Loureiro, M.E., Kim, D., Na, G., Tikhonova, M., Salt, D.E., 2009. MTP1-dependent Zn sequestration into shoot vacuoles suggests dual roles in Zn tolerance and accumulation in Zn hyperaccumulating plants. Plant J. 57, 1116–1127.

Ha, S.B., Smith, A.P., Howden, R., Dietrich, W.M., Bugg, S., O'Connell, M.J., Goldsbrough, P.B., Cobbett, C.S., 1999. Phytochelatin synthase genes from *Arabidopsis* and the yeast, *Schizosaccharomyces pombe*. Plant Cell 11, 1153–1164.

Hassinen, V.H., Tervahauta, A.I., Schat, H., Ka, S.O., 2011. Plant metallothioneins–metal chelators with ROS scavenging activity? Plant Biol. 13, 225–232.

Hassinen, V.H., Tuomainen, M., Peräniemi, S., Schat, H., Kärenlampi, S.O., Tervahauta, A.I., 2009. Metallothioneins 2 and 3 contribute to themetal-adapted phenotype but are not directly linked to Zn accumulationin the metal hyperaccumulator, *Thlaspicaerulescens*. J. Exp. Bot. 60, 187–196.

Hegelund, J.N., Schiller, M., Kichey, T., Hansen, T.H., Pedas, P., Husted, S., Schjoerring, J.K., 2012. Barley metallothioneins: MT3 and MT4 are localized in the grain aleurone layer and show differential zinc binding. Plant Physiol. 159, 1125–1137.

Hell, R., Wirtz, M., 2011. Molecular biology, biochemistry and cellular physiology of cysteine metabolism in *Arabidopsis thaliana*. Arabidopsis Book 9, e0154.

Hossain, M.A., Piyatida, P., Jaime, A., da Silva, T., Fujita, M., 2012. Molecular mechanism of heavy metal toxicity and tolerance in plants: central role of glutathione in detoxification of reactive oxygen species and methylglyoxal and in heavy metal chelation. J. Bot. http://dx.doi.org/10.1155/2012/872875.

Huang, G.Y., Wang, Y.S., 2009. Expression analysis of type 2 metallothionein gene in mangrove species (*Bruguiera gymnorrhiza*) under heavy metal stress. Chemosphere 77, 1026–1029.

Huang, G.Y., Wang, Y.S., 2010. Expression and characterization analysis of type 2 metallothionein from grey mangrove species (*Avicennia marina*) in response to metal stress. Aquat. Toxicol. 99, 86–92.

Huang, G.Y., Wang, Y.S., Ying, G.G., 2011. Cadmium-inducible BgMT2, a type 2 metallothionein gene from mangrove species (*Bruguiera gymnorrhiza*), its encoding protein shows metal-binding ability. J. Exp. Mar. Biol. Ecol. 405, 128–132.

Huang, G.Y., Wang, Y.S., Ying, G.G., Dang, A.C., 2012. Analysis of type 2 metallothionein gene from mangrove species (*Kandelia candel*). Trees 26, 1537–1544.

Indriolo, E., Na, G., Ellis, D., Salt, D.E., Banks, J.A., 2010. A vacuolar arsenite transporter necessary for arsenic tolerance in the arsenic hyperaccumulating fern *Pteris vittata* is missing in flowering plants. Plant Cell 22, 2045–2057.

Inouhe, M., 2005. Phytochelatins. Braz. J. Plant Physiol. 17, 65–78.

Jia, D.U., Jing-Li, Y., Cheng-Hao, L.I., 2012. Advances in metallotionein studies in forest trees. China POJ 5, 46–51.

Kamiya, T., Fujiwara, T., 2015. A novel allele of the *Arabidopsis* phytochelatin synthase 1 gene conferring high sensitivity to arsenic and antimony. Soil Sci. Plant Nutr. 52, 272–278.

Kim, Y.O., Jung, S., Kim, K., Bae, H.J., 2013. Role of pCeMT, a putative metallothionein from *Colocasia esculenta*, in response to metal stress. Plant Physiol. Biochem. 64, 25–32.

Krystofova, O., Zitka, O., Krizkova, S., Hynek, D., Shestivska, V., Adam, V., Hubalek, J., Mackova, M., Macek, T., Zehnalek, J., Babula, P., Havel, L., Kizek, R., 2012. Accumulation of cadmium by transgenic tobacco plants (*Nicotiana tabacum* L.) carrying yeast metallothionein gene revealed by electrochemistry. Int. J. Electrochem. Sci. 7, 886–907.

Küpper, H., Kochian, L., 2010. Transcriptional regulation of metal transport genes and mineral nutrition during acclimatization to cadmium and zinc in the Cd/Zn hyperaccumulator, *Thlaspi caerulescens* (Ganges population). New Phytol. 185, 114–129.

Kühnlenz, T., Schmidt, H., Uraguchi, S., Clemens, S., 2014. *Arabidopsis thaliana* phytochelatin synthase 2 is constitutively active in vivo and can rescue the growth defect of the *PCS1*-deficient *cad1-3* mutant on Cd-contaminated soil. J. Exp. Bot. http://dx.doi.org/10.1093/jxb/eru195.

Kühnlenz, T., Westphal, L., Schmidt, H., Scheel, D., Clemens, S., 2015. Expression of caenorhabditis elegans PCS in the AtPCS1-deficient *Arabidopsis thaliana* cad1-3 mutant separates the metal tolerance and nonhost resistance functions of phytochelatin synthases. Plant Cell Environ. http://dx.doi.org/10.1111/pce.12534.

Kutrowska, A., Szelag, M., 2014. Low-molecular weight organic acids and peptides involved in the long-distance transport of trace metals. Acta Physiol. Plant. http://dx.doi.org/10.1007/s11738-014-1576-y.

Lee, B.D., Hwang, S., 2015. Tobacco phytochelatin synthase (NtPCS1) plays important roles in cadmium and arsenic tolerance and in early plant development in tobacco. Plant Biotechnol. Rep. 9, 107–114.

Lee, S., 2014. Artificial induction of cadmium tolerance and its further enhancement via heterologous co-expression of *SpHMT1*and *AtPCS1* in the yeast cells. J. Korean Soc. Appl. Biol. Chem. 57, 307–310.

Leitenmaier, B., Kupper, H., 2013. Compartmentation and complexation of metals in hyperaccumulator plants. Front. Plant Sci. http://dx.doi.org/10.3389/fpls.2013.00374.

Liu, G.Y., Zhang, Y.X., Chai, T.Y., 2011. Phytochelatin synthase of *Thlaspi caerulescens* enhanced tolerance and accumulation of heavy metals when expressed in yeast and tobacco. Plant Cell Rep. 30, 1067–1076.

Liu, J., Shi, X., Qian, M., Zheng, L., Lian, C., Xia, Y., Shen, Z., 2015. Copper-induced hydrogen peroxide upregulation of a metallothionein gene, *OsMT2c*, from *Oryza sativa* L. confers copper tolerance in *Arabidopsis thaliana*. J. Hazard. Mater. 294, 99–108.

Lv, Y., Deng, X., Quan, L., Xia, Y., Shen, Z., 2013. Metallothioneins *BcMT1* and *BcMT2* from *Brassica campestris* enhance tolerance to cadmium and copper and decrease production of reactive oxygen species in *Arabidopsis thaliana*. Plant Soil 367, 507–519.

Macoive, A., Ventura, L., Dona, M., Fae, M., Balestrazz, A., Carbonera, D., 2010. Effect of heavy metal treatment on metallothionein expression profiles white poplar (*Populus alba* L.) cell suspension cultures. Fascicula Biol. 2, 274–279.

Meister, A., 1988. Glutathione metabolism and its selective modification. J. Biol. Chem. 263, 17205–17208.

Mendoza-Cózatl, D.G., Zhai, Z., Jobe, T.O., Akmakjian, G.Z., Yong Song, W., Limbo, O., Russell, M.R., Kozlovskyy, V.I., Martinoia, E., Vatamaniuk, O.K., Russell, P., Schroeder, J.I., 2010. Tonoplast-localized Abc2 transporter mediates phytochelatin accumulation in vacuoles and confers cadmium tolerance. J. Biol. Chem. 285, 40416–40426.

Milner, M.J., Piñeros, M., Kochian, L.V., 2013. Molecular and physiological mechanisms of plant tolerance to toxic metals. In: Jenks, M.A., Hasegawa, P.M. (Eds.), Plant Abiotic Stress, Second ed. John Wiley & Sons, Inc., Hoboken, NJ. http://dx.doi.org/10.1002/9781118764374.ch7.

Mirza, N., Mahmood, Q., Shah, M.M., Pervez, A., Sultan, S., 2014. Plants as useful vectors to reduce environmental toxic arsenic content. Sci. World J. 1–11. Article ID 921581.

Mishra, S., Srivastava, S., Tripathi, R.D., Kumar, R., Seth, C.S., Gupta, D.K., 2006. Lead detoxification by coontail (*Ceratophyllum demersum* L.) involves induction of phytochelatins and antioxidant system in response to its accumulation. Chemosphere 65 (6), 1027–1039.

Mishra, S., Tripathi, R.D., Srivastava, S., Dwivedi, S., Trevedi, P.K., Dhankher, O.P., Khare, A., 2009. Thiol metabolism paly significant role during cadmium detoxicfication by *Ceratophyllum demersum* L. Bioresour. Technol. 100, 2155–2216.

Mutoh, N., Hayashi, Y., 1988. Isolation of mutants of *Schizosacchacomyces pombe* unable to synthesise cadystins, small cadmium-binding peptides. Biochem. Biophys. Res. Commun. 151, 32–39.

Nagajyoti, P.C., Lee, K.D., Sreekanth, T.V.M., 2010. Heavy metals, occurrence and toxicity for plants: a review. Environ. Chem. Lett. 8, 199–216.

Nakamura, M., Ochiai, T., Noji, M., Ogura, Y., Suzuki, K.T., Yoshimoto, N., Yamazaki, M., Saito, K., 2014. An improved tolerance to cadmium by overexpression of two genes for cysteine synthesis in tobacco. Plant Biotechnol. 31, 141–147.

Nedecky, B.R., Nejdl, L., Gumulec, J., Zitka, O., Masarik, M., Eckschlager, T., Marie, S., Adam, V., Kizek, R., 2013. The role of metallothionein in oxidative stress. Int. J. Mol. Sci. 14, 6044–6066.

Ó Lochlainn, S.O., Bowen, H.C., Fray, R.G., Hammond, J.P., King, G.J., White, P.J., Graham, N.S., Broadley, M.R., 2011. Tandem quadruplication of HMA4 in the zinc and cadmium hyperaccumulator *Noccaea caerulescens*. PLoS One 6, e17814.

Ovečka, M., Takáč, T., 2014. Managing heavy metal toxicity stress in plants: biological and biotechnological tools. Biotechnol. Adv. 32 (1), 73–86.

Paulose, B., Chhikara, S., Coomey, J., Jung, H., Vatamaniuk, O., Dhankher, O.P., 2013. A γ-Glutamyl cyclotransferase protects *Arabidopsis* plants from heavy metal toxicity by recycling glutamate to maintain glutathione homeostasis. Plant Cell 25 (11), 4580–4595.

Polak, N., Read, D.S., Jurkschat, K., Matzke, M., Kelly, F.J., Spurgeon, D.J., Stürzenbaum, S.R., 2014. Metalloproteins and phytochelatin synthase may confer protection against zinc oxide nanoparticle induced toxicity in *Caenorhabditis elegans*. Comp. Biochem. Physiol. Part C 160, 75–85.

Postrigan, B.N., Knyazev, A.V., Kuluev, B.R., Chemeris, A.V., 2013. Effect of cadmium on promoter activity of rice phytochelatin synthase gene in transgenic tobacco plants. Russ. J. Plant Physiol. 60 (5), 701–705.

Ramos, J., Matamoros, M.A., Naya, L., James, E.K., Rouhier, N., Sato, S., 2009. The glutathione peroxidase gene family of *Lotus japonicus*: characterization of genomic clones, expression analyses and immunolocalization in legumes. New Phytol. 181, 103–114.

Ramos, J., Naya, L., Gay, M., Abian, J., Becana, M., 2008. Functional characterization of an unusual phytochelatin synthase, LjPCS3, of Lotus japonicus. Plant Physiol. 148, 536–545.

Rauser, W.E., 1995. Phytochelatins and related peptides: structure, biosynthesis, and function. Plant Physiol. 109, 1141–1149.

Ray, D., Williams, D.L., 2011. Characterization of the phytochelatin synthase of *Schistosoma mansoni*. Plos One 5 (5), e1168 doi:10.1371.

Rea, P.A., 2012. Phytochelatin synthase: of a protease a peptide polymerase made. Physiol. Plant 145, 154–164.

Revathi, S., Venugopal, S., 2013. Physiological and biochemical mechanisms of heavy metal tolerance. Int. J. Environ. Sci. 3, 1339–1354.

Rigouin, C., Nylin, E., Cogswell, A.A., Schaumlöffel, D., Dobritzsch, D., Williams, D.L., 2013. Towards an understanding of the function of the phytochelatin synthase of *Schistosoma mansoni*. PLOS Negl. Trop. Dis. 7, e2037. doi:10.1371.

Rodríguez-Llorente, I.D., Pérez-Palacios, P., Doukkali, B., Caviedes, M.A., Pajuelo, E., 2010. Expression of the seed-specific metallothionein mt4a in plant vegetative tissues increases Cu and Zn tolerance. Plant Sci. 178, 327–332.

Romero, L.C., Aroca, M.A., Laureano-Marin, A.M., Moreno, I., Garcia, I., Gotor, C., 2014. Cysteine and cysteine-related signaling pathways in *Arabidopsis thaliana*. Mol. Plant 7 (2), 264–276.

Romero, L.C., Garcia, I., Gotor, C., 2013. L-Cysteine desulfhydrase 1 modulates the generation of the signaling molecule sulfide in plant cytosol. Plant Signal. Behav. 8, e24007.

Sekhar, K., Priyanka, B., Reddy, V.D., Rao, K.V., 2011. Metallothionein 1 (*CcMT1*) of pigeon pea (*Cajanus cajan*, L.) confers enhanced tolerance to copper and cadmium in *Escherichia coli* and *Arabidopsis thaliana*. Environ. Exp. Bot. 72, 131–139.

Shen, G.-H., Zhu, C., Du, Q.-Z., 2010. Genome-wide identification of PHYTOCHELATIN and PHYTOCH_SYNTH domain-containing phytochelatin family from rice. Electron. J. Biol. 6, 73–79.

Shine, A.M., Shakya, V.P., Idnum, A., 2015. Phytochelatin synthase is required for tolerating metal toxicity in a basidiomycete yeast and is a conserved factor involved in metal homeostasis in fungi. Fungal Biol. Biotechnol. 2. Pii:3.

Shri, M., Dave, R., Diwedi, S., Shukla, D., Kesari, R., Tripathi, R.D., Trivedi, P.K., Chakrabarty, D., 2014. Heterologous expression of *Ceratophyllum demersum* phytochelatin synthase, CdPCS1, in rice leads to lower arsenic accumulation in grain. Sci. Rep. 4, 5784. http://dx.doi.org/10.1038/srep05784.

Shukla, D., Kesari, R., Mishra, S., Dwivedi, S., Tripathi, R.D., Nath, P., Trivedi, P.K., 2012. Expression of phytochelatin synthase from aquatic macrophyte *Ceratophyllum demersum* L. enhances cadmium and arsenic accumulation in tobacco. Plant Cell Rep. 31 (9), 1687–1699.

Shukla, D., Kesari, R., Tiwari, M., Dwivedi, S., Tripathi, R.D., Nath, P., Trivedi, P.K., 2013a. Expression of *Ceratophyllum demersum* phytochelatin synthase, *CdPCS1*, in *Escherichia coli* and Arabidopsis enhances heavy metal(loid)s accumulation. Protoplasma 250 (6), 1263–1272.

Shukla, D., Tiwari, M., Tripathi, R.D., Nath, P., Trivedi, P.K., 2013b. Synthetic phytochelatins complement a phytochelatin-deficient Arabidopsis mutant and enhance the accumulation of heavy metal(loid)s. Biochem. Biophys. Res. Commun. 434 (3), 664–669.

Siemianowski, O., Barabasz, A., Kendziorek, M., Ruszczyñska, A., Bulska, E., Williams, L.E., 2014. *AtHMA4*-expression in tobacco reduces Cd accumulation due to the induction of the apoplast barrier. J. Exp. Bot. 65, 1125–1139.

Singh, R.K., Anandhan, S., Singh, S., Patade, V.Y., Ahmed, Z., Pande, V., 2011. Metallothionein-like gene from *Cicer microphyllum* is regulated by multiple abiotic stresses. Protoplasma 248, 839–847.

Song, W.Y., Mendoza-Cózatl, D.G., Lee, Y., Schroeder, J.I., Ahn, S.N., Lee, H.S., Wicker, T., Martinoia, E., 2014. Phytochelatin-metal(loid) transport into vacuoles shows different substrate preferences in barley and Arabidopsis. Plant Cell Environ. 37 (5), 1192–1201.

Song, W.Y., Parka, J., Mendoza-Cózatlc, D.G., Suter-Grotemeyerd, M., Shima, D., Hörtensteinerb, S., Geislerb, M., Wederb, B., Reae, P.A., Rentschd, D., Schroederc, J.I., Leea, Y., Martinoia, E., 2010. Arsenic tolerance in Arabidopsis is mediated by two ABCC-type phytochelatin transporters. PNAS 107 (49), 21187–21192.

Souguir, D., El Ferjani, E., Ledoigt, G., Goupil, P., 2015. Transcript accumulation of stress-related genes in *Vicia faba* roots under a short exposure to cadmium. Plant Biosyst. 149 (2), 280–288.

Steffens, J.C., 1990. The heavy metal binding peptides of plants. Annu. Rev. Plant Physiol. Plant Mol. Biol. 41, 533–575.

Sunitha, M.S.L., Prashant, S., Kumar, S.A., Rao, S., Narasu, M., Kishor, P.B.K., 2013. Cellular and molecular mechanisms of heavy metal tolerance in plants: a brief overview of transgenic plants over- expressing phytochelatin synthase and metallothionein genes. Plant Cell Biotech. Mol. Biol. 14, 33–48.

Takahashi, R., Ishimaru, Y., Senoura, T., Shimo, H., Ishikawa, S., Arao, T., Nakanishi, H., Nishizawa, N.K., 2011. The OsNRAMP1 iron transporter is involved in Cd accumulation in rice. J. Exp. Bot. 62, 4843–4850.

Takahashi, R., Ishimaru, Y., Shimo, H., Ogo, Y., Senoura, T., Nishizawa, N.K., Nakanishi, H., 2012. The OsHMA2 transporter is involved in root-to-shoot translocation of Zn and Cd in rice. Plant Cell Environ. 35, 1948–1957.

Tennstedt, P., Peisker, D., Bottcher, C., Trampczynska, A., Clemens, S., 2009. Phytochelatin synthesis is essential for the detoxification of excess zinc and contributes significantly to the accumulation of zinc. Plant Physiol. 149 (2), 938–948.

Terzi, H., Yıldız, M., 2015. Interactive effects of sulfur and chromium on antioxidative defense systems and BnMP1 gene expression in canola (*Brassica napus* L.) cultivars differing in Cr(VI) tolerance. Ecotoxicology 24, 1171–1182.

Thirumoorthy, N., Sunder, A.S., Kumar, K.T.M., Kumar, M.S., Ganesh, G.N.K., Chatterjee, M., 2011. A review of metallothionein isoforms and their role in pathophysiology. World J. Surg. Oncol. 9, 54–60.

Turchi, A., Tamantini, I., Camussi, A.M., Racchi, M.L., 2012. Expression of a metallothionein A1 gene of *Pisum sativum* in white poplar enhances tolerance and accumulation of zinc and copper. Plant Sci. 183, 50–56.

Ueno, D., Koyama, E., Yamaji, N., Ma, J.F., 2011. Physiological, genetic, and molecular characterization of a high-Cd-accumulating rice cultivar, Jarjan. J. Exp. Bot. 62, 2265–2272.

Usha, B., Keeran, N.S., Harikrishnan, M., Kavitha, K., Parida, A., 2011. Characterization of a type 3 metallothionein isolated from *Porteresia coarctata*. Biol. Plant. 55, 119–124.

Usha, B., Prashanth, S.R., Parida, A., 2007. Differential expression of two metallothionein encoding genes during heavy metal stress in the mangrove species, *Avicennia marina* (Forsk.) Vierh. Curr. Sci. 93, 1215–1219.

Usha, B., Venkataraman, G., Parida, A., 2009. Heavy metal and abiotic stress inducible metallothionein isoforms from *Prosopis juliflora* (SW) D.C. show differences in binding to heavy metals in vitro. Mol. Genet. Genomics 281, 99–108.

Van der Ent, A., Baker, A.J.M., Reeves, R.D., Pollard, A.J., Schat, H., 2013. Hyperaccumulators of metal and metalloid trace elements: facts and fiction. Plant Soil 362 (1–2), 319–334.

Vatamaniuk, O.K., Mari, S., Lu, Y.P., Rea, P.A., 2000. Mechanism of heavy metal activation of phytochelatin (PC) synthase: blocked thiols are sufficient for PC synthase-catalyzed transpeptidation of glutathione and related thiol peptides. J. Biol. Chem. 275, 31451–31459.

Venkataramaiah, N., Ramakrishna, S.V., Sreevathsa, R., 2011. Overexpression of phytochelatin synthase (AtPCS) in rice for tolerance to cadmium stress. Biologia 66 (6), 1060–1073.

Wang, F., Wang, Z., Zhu, C., 2012. Heteroexpression of the wheat phytochelatin synthase gene (*TaPCS1*) in rice enhances cadmium sensitivity. Acta Biochim. Biophys. Sin. 44 (10), 886–893.

Wang, M., Gruissem, W., Bhullar, N., 2013. Nicotianamine synthase overexpression positively modulates iron homeostasis-related genes in high iron rice. Front. Plant Sci. 4, 156.

Wojas, S., Ruszczyńska, A., Bulska, E., Clemens, S., Antosiewicz, D.M., 2010. The role of subcellular distribution of cadmium and phytochelatins in the generation of distinct phenotypes of AtPCS1- and CePCS3-expressing tobacco. J. Plant Physiol. 167 (12), 981–988.

Wünschmann, J., Krajewski, M., Letzel, T., Huber, E.M., Ehrmann, A., Grill, E., 2010. Dissection of glutathione conjugate turnover in yeast. Phytochemistry 71, 54–61.

Yadav, S.K., 2010. Heavy metals toxicity in plants: an overview on the role of glutathione and phytochelatins in heavy metal stress tolerance of plants. South Afri J. Bot. 76, 167–179.

Yang, J., Chen, Z., Wu, S., Cui, Y., Zhang, L., Dong, H., Yang, C., Li, C., 2015. Overexpression of the *Tamarix hispida ThMT* 3 gene increases copper tolerance and adventitious root induction in *Salix matsudana* Koidz. Plant Cell, Tissue Organ Cult. 121, 469–479.

Yang, J.L., Wang, Y.C., Liu, G.F., Yang, C.P., Li, C.H., 2011. *Tamarix hispida* metallothionein-like *ThMT3*, a reactive oxygen species scavenger, increases tolerance against Cd^{2+}, Zn^{2+}, Cu^{2+}, and NaCl in transgenic yeast. Mol. Biol. Rep. 38, 1567–1574.

Yuan, J., Chen, D., Ren, Y., Zhang, X., Zhao, J., 2008. Characteristic and expression analysis of a metallothionein gene, *OsMT2b*, down-regulated by cytokinin suggests functions in root development and seed embryo germination of rice. Plant Physiol. 146, 1637–1650.

Zagorchev, L., Seal, C.E., Kranner, I., Odjakova, M., 2013. A Central role for thiols in plant tolerance to abiotic stress. Int. J. Mol. Sci. 14, 7405–7432.

Zenk, M.H., 1996. Heavy metal detoxification in higher plants—a review. Gene 179, 21–30.

Zhang, J., Zhang, M., Tian, S., Lu, L., Shohag, M.J.I., 2014. Metallothionein 2 (*SaMT2*) from *Sedum alfredii* Hance confers increased Cd tolerance and accumulation in yeast and tobacco. PLoS One 9 (7), e102750.

Zhao, C., Xu, J., Li, Q., Li, S., Wang, P., Xiang, F., 2014. Cloning and characterization of a *Phragmites australis* phytochelatin synthase (*PaPCS*) and achieving Cd tolerance Tall fescue. Plos One. http://dx.doi.org/10.1371/journal.pone.0103771.

Zhou, B., Yao, W., Wang, S., Wang, X., Jiang, T., 2014. The metallothionein gene, TaMT3, from *Tamarix androssowii* confers Cd^{2+} tolerance in tobacco. Int. J. Mol. Sci. 15, 10398–10409.

Zhu, J., Zhang, Q., Wu, R., Zhang, Z., 2010. HbMT2, an ethephon induced metallothionein gene from *Hevea brasiliensis* responds to H_2O_2 stress. Plant Physiol. Biochem. 48, 710–715.

Zimeri, A.M., Dhankher, O.P., McCaig, B., Meagher, R.B., 2005. The plant MT1 metallothioneins are stabilized by binding cadmium and are required for cadmium tolerance and accumulation. Plant Mol. Bio 58, 839–855.

CHAPTER

11

GLUTATHIONE AND PHYTOCHELATINS MEDIATED REDOX HOMEOSTASIS AND STRESS SIGNAL TRANSDUCTION IN PLANTS: AN INTEGRATED OVERVIEW

Shweta Singh[1], Durgesh Kumar Tripathi[1,2], Devendra Kumar Chauhan[2], Nawal Kishore Dubey[1]

D. D. Pant Interdisciplinary Research Laboratory, Department of Botany, University of Allahabad, Allahabad, India[1]; Center of Advanced Study in Botany, Banaras Hindu University, Varanasi, India[2]

CHAPTER OUTLINE

1. INTRODUCTION

Unprecedented changes from an upsurge in anthropogenic activities has strongly influenced the preexisting balance of the biosphere and has led to a disturbed imbalanced atmosphere, presenting a huge challenges to the ingenuity and expertise of science (Bhargava et al., 2012). Several industrial activities like service stations, metal fabrication shops, dust from smelters, tanneries, waste disposal sites, and paper mills are presumably responsible for polluting the environment (Wong, 2003; Freitas et al., 2004; Tripathi and Misra, 2012a). Among the major pollutants, the heavy metal contamination of water, soil, and plants is recognized as one of the most concerning/challenging issues (Singh and Chauhan, 2011; Tripathi et al., 2012a; Tripathi and Misra, 2012b, 2013; Muamar et al., 2014; Waseem et al., 2014; Tripathi et al., 2015). In this context, the past few decades have clearly witnessed a drastic environmental contamination from heavy metals.

Plant Metal Interaction. http://dx.doi.org/10.1016/B978-0-12-803158-2.00011-4

Heavy metals are precisely characterized as a member of an ill-defined subset of elements that are highly toxic with relative density higher than 5.0 g/cm^{-3} including the transition metals, some metalloids, lanthanides, and actinides (Babula et al., 2008; Bhargava et al., 2012). Metal accumulation is the most potent soil contamination that is highly devastating to living organisms and is also responsible for the major downfall in crop productivity, soil biomass, and fertility (Gratao et al., 2005; Rajkumar et al., 2009; Bhargava et al., 2012; Tripathi and Misra, 2012a; Singh et al., 2015). The increasing reliance on chemical fertilizers has negatively influenced environmental health by imposing serious long-term threats (McLaughlin et al., 1999; Wong et al., 2002; Bhargava et al., 2012). It has been observed that the high degree of agricultural soil contaminated with heavy metal not only perturbs the sustainability of crops and environmental health; it also is highly hazardous for human health because its accumulation by a plant soon gets incorporated into the food chain (Schickler and Caspi, 1999; Abhilash et al., 2014; Ovecka and Takac, 2014; Chibuike and Obiora, 2014; Nesterkova et al., 2014). It is noteworthy that a considerably high input of heavy metals disturbed the plant cellular structure, basic metabolic, and transport processes (Sharma and Dietz, 2009; Tripathi et al., 2015).

Consequently, it brings several damaging effects in plants, in particular diminishing the sustainability of agricultural productivity, which eventually will pose serious threats to food security (Hirayama and Shinozaki, 2010; Tripathi et al., 2015). Moreover, the deleterious effect of metal toxicity on protein, lipids, and thylakoid membranes could also be shown because it significantly marginalizes their cellular functions (Yadav, 2010). The preliminary response of plant toward metal toxicity is the generation of activated oxygen species (AOS), a highly toxic species, which induces oxidative injuries inside the cell (Hossain et al., 2012; Yadav, 2010). Numerous reports are available that clearly elucidate the significant role of reactive oxygen species (ROS) in altering signal transduction pathways (Wojtaszek, 1997; Mithofer et al., 2004). In addition, the plant experiences oxidative stress on prolonged metal exposure (Yadav, 2010). Hence, the maintenance of ionic homeostasis is of utmost importance in terms of attaining sustainable growth, development, and productivity.

Conversely, to combat such stresses, plants have been equipped with a unique/novel tolerance strategy that enable them to respond effectively against the adverse effect of toxic metals or metalloids; such responses involves immobilization, exclusion, chelation, and compartmentalization of the metal ions (Yadav, 2010; Ovecka and Takac, 2014; Pattanayak et al., 2014). However, debates on their tolerance scheme still prevails at various levels, with the conjecture of finding new avenues in future research.

To curtail their toxic effects, plants synthesized low-molecular-weight thiols that display higher affinity toward metal toxicity (Bricker et al., 2001). Nonenzymatic ubiquitously found low-molecular-weight tripeptide thiols include glutathione (GSH), with a general formula of γ-glutamate cysteine-glycine (May et al., 1998a,b). Recent years have seen substantial advancement in understanding the indispensable/irreplaceable role of GSH in maintaining cellular redox homeostasis and controlling stress signal signaling pathways under varied environmental insults (Noctor and Foyer, 1998). More precisely, its elementary role can be analyzed in the thiol–disulphide interactions, where the interchange of reduced GSH to oxidized form (GSSG) occurs, mediated by nicotinamide adenine dinucleotide phosphate-oxidase–dependent GSH reductase (Noctor et al., 2012). However, the specific

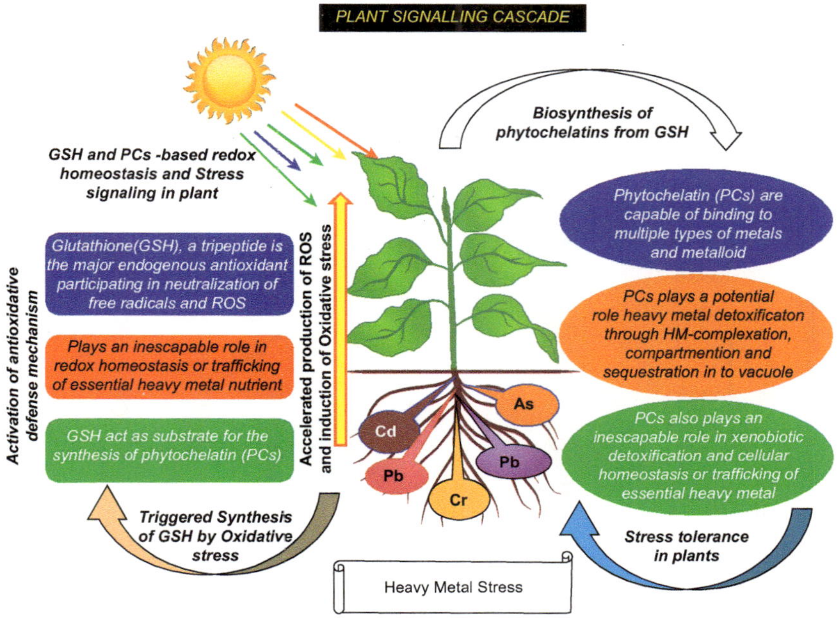

FIGURE 11.1

Figure shows glutathione (GSH)- and phytochelatin (PC)-based redox homeostasis and stress signaling in plants under heavy metal stress. ROS, reactive oxygen species.

position of GSH among oxidant and cellular reductant more precisely configured for signaling functions (Noctor et al., 2012).

Besides this endeavor of GSH, the sumptuous/eminent role of phytochelatins (PCs) in metal tolerance cannot be denied (Figure 11.1). GSH serves as the substrate for PC synthesis (Figures 11.1 and 11.2) (Freeman et al., 2004), which is a small and novel set of metal-binding peptides with the general formula of (γ-Glu-Cys)nGly (n = 2–11). A credible number of evidence is available that suggests the significance of PC peptide in xenobiotics and heavy metal sequestration (Robinson, 1989; Rauser, 1990; Steffens et al., 1986) (Figure 11.2). Numerous biochemical and genetic approaches for enhancing PC synthesis have augmented our general understanding of the significant role of PCs in metal tolerance (Figure 11.1).

The aim of the present book chapter is to assess the toxic effects of heavy metals and recent experimental findings that focus potential strategies adopted by plants against such stresses (Figure 11.1). Furthermore, it also provides strong evidence suggesting the avoidance of oxidative damage, which can be achieved more easily and efficiently by the joint effort of both GSH and PCs in plants. Taken together, these findings enhanced our understanding by prompting the intense efforts that characterizes the mechanisms underlying GSH- and PCs mediated homeostasis in plant and their unique biochemical and physiological role in metal sequestration (Figure 11.1).

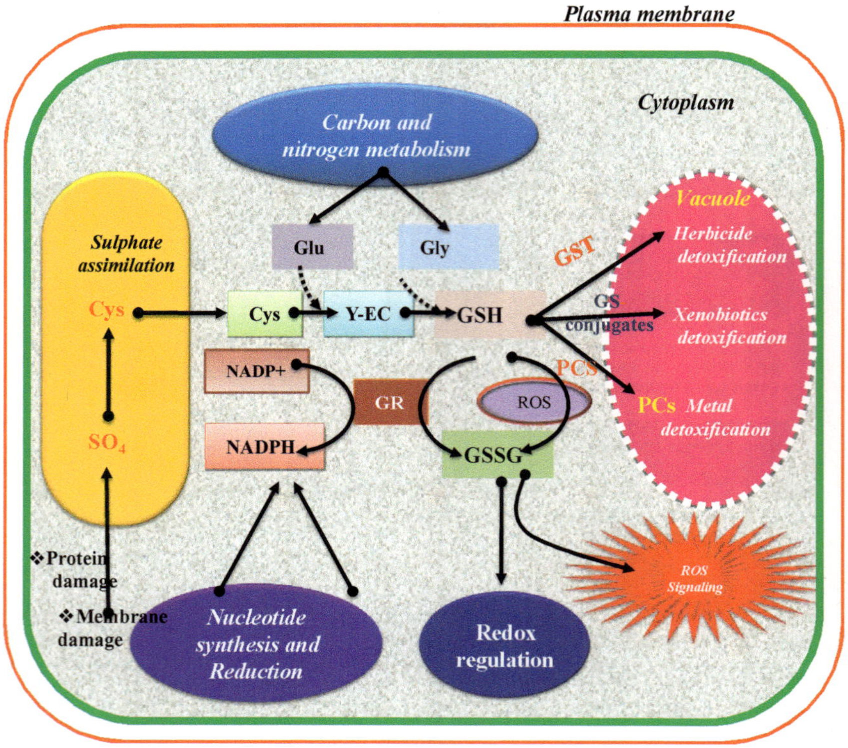

FIGURE 11.2

A general overview representing the potential role of glutathione and phytochelatins functions (synthesis, redox turnover, metabolism, signalling) in abiotic stress tolerance in plant. Cys, cysteine; g-EC, g-glutamylcysteine; GS-conjugates, glutathione S-conjugates; Glu, glutamate; Gly, glycine; ROS, reactive oxygen species.

Modified after Noctor et al., (2012) and Zagorchev et al., (2013).

2. ASSESSMENTS OF METAL-INDUCED PHYSIOLOGICAL AND BIOCHEMICAL AMENDMENTS IN HIGHER PLANTS

Over the past several decades, the accelerated/uncontrolled contamination of natural repositories with heavy metals has become a critical environmental issue; therefore, an attempt has been made to present a comprehensive introduction of the overall life-threatening processes prevails in our environment because of metal toxicity at a global level (Bowen, 1979). Explosion in anthropogenic activities, excessive accumulation of industrial waste, improper wastewater management practices, poor sanitation, serious abuse of phosphatic fertilizers, poor sanitation, and massive intensification of agriculture have led to agricultural soil being moderately contaminated with metals (Yadav, 2010; Khan et al., 2009; Shah et al., 2010; Nagajyoti et al., 2010). Because of its long residence time and irreversible nature, the contamination of agriculture soil with heavy metal now has to be resolved (Alloway and Ayres, 1997).

Despite the major efforts that been made to curb heavy metals' extension or make the environment free from metal contamination, this major problem persists, posing long-term threats to human health. Heavy metal stress is one of the most devastating abiotic stress factors that renders field unproductive (Tripathi et al., 2012a,b; Hossain et al., 2012). In a strict sense, the term "heavy metals" generally refers to any metallic element possessing a density greater than 5, such as copper (Cu), zinc (Zn), iron (Fe), cadmium (Cd), manganese, nickel, cobalt, chromium, lead (Pb), and arsenic (As) (Tripathi et al., 2012a,b; Singh et al., 2011; Hossain et al., 2012). Metal toxicity could be easily manifested upon its massive accumulation by plant cells. Further, based on redox activity, it has been segregated in two major groups: first, the redox active heavy metals comprising Fe, Cu, Cr, and cobalt; second, redox inactive heavy metal is including Cd, Zn, nickel, and aluminum (Al).

Notably, the redox active heavy metals display their direct involvement in cellular redox reactions and eventually led to the subsequent synthesis of $O_2^{\cdot -}$, H_2O_2, and $\cdot OH$ through the Haber-Weiss and Fenton reactions (Dietz et al., 1999; Schutzendubel and Polle, 2002). In contrast, the plant exposure to redox inactive heavy metals also brings oxidative stress to the plant but via indirect mechanisms including interaction with antioxidative defense machinery, disruption of electron transport system (ETS) or induction of lipid peroxidation (LPO) and the reactions is being catalyzed by metal-induced increased activity of lipoxygenase (Hossain et al., 2012).

Heavy metals are presumably the chief constituents of soil, but when their concentration exceeds that beyond the permissible/threshold limit, it becomes highly toxic and induces several visible toxicity symptoms in plants. For instance, in soil, the permissible limit for some selected heavy metals are: Cd (100 mg/kg), Zn (150–300 mg/kg), Cu (10 mg/kg), and chromium (1.5 mg/kg soil), beyond this threshold they become phytotoxic and show visible toxicity symptoms such as perturbed growth, leaf chlorosis, necrosis, receded photosynthesis, reduced transpiration (Tripathi et al., 2012a,b; Kumar et al., 2013a,b, 2014a,b). One of the most widely spread toxicity symptoms of heavy metal toxicity can be visualized in the overall degradation of the plant growth under which comes leaf chlorosis, necrosis, progressive turgor loss, arrested seed germination, crippled photosynthetic, leaf senescence, or ultimate death of plant (DalCorso et al., 2008; Dalcorso et al., 2010; Foy et al., 1978; Sandalio et al., 2001; Carrier et al., 2003). The situation is likely to be much worse by increasing anthropogenic activities that eventually intensify the area of research dealing particularly with the approaches adopted by plants, primarily to curb their toxicity.

Soil contamination with heavy metals is a concrete danger because, after its accumulation by a plant, it soon gets incorporated into the food chain. Therefore, the major objective of current research is to find new avenues for improving crops by making them resistant to heavy metal stress (Bell et al., 2001; Schwartz et al., 2001; Passariello et al., 2002; Sharma and Dubey, 2007; Sharma and Dietz, 2009; Hossain and Fujita, 2009, 2011; Yadav, 2010; Hossain et al., 2012; Tripathi et al., 2012a,b). Plant tolerance to heavy metals stress depends on the balance between the rate at which metal ions are absorbed and the efficacy with which they are being neutralized within the plant. Further, if the plant is incapable of preventing the internalization of trace elements, then in that situation they have to face inevitably the drastic effects of heavy metals on their basic metabolism, transport processes, membranes, and cellular structure. Plant exposure to heavy metal ions therefore trigger implementation of stress reactions and necessitate adaptation at the developmental, structural, physiological (Hall, 2002; Pasternak et al., 2008), and molecular (Hirayama and Shinozaki, 2010; Zhao et al., 2009) levels. Approximately 70% of yield reduction has been observed annually because of these stresses (Acquaah, 2009; Jewell et al., 2010). Inactivation and denaturation of photosynthetic enzymes, blocking of functional group of metabolically active compounds, subsequent substantial replacement or displacement of essential metal ion, and disrupted membrane integrity are among the other major physiological changes made by massive accumulation of heavy metals in soils (Ramesh and Reddy, 2008; Hossain and Fujita, 2009, 2011; Villiers et al., 2011; Hossain et al., 2012). Additionally, heavy metals are also known to disturb ionic homeostasis of plant by stimulating the formation of free radicals and AOS (Yadav et al., 2005; Anjum et al., 2012). Plants experienced oxidative stress at elevated ROS concentrations in plants, leading to LPO, DNA damage, and protein oxidation. Apart from this, plant physiologists have recently recognized another most cytotoxic metal induced precarious compound namely methylglyoxal that further intensifies the synthesis of AOS (Yadav et al., 2005; Singla-Pareek et al., 2006; Hossain et al., 2009). Hence it can be concluded that heavy metal stress hampers the overall growth of plants by altering many biochemical, physiological, and metabolic processes (Hossain et al., 2012).

3. GSH STRUCTURE, BIOSYNTHESIS, AND FUNCTION

GSH is a ubiquitously found low-molecular-weight thiol endowed with multiple functioning in plants (Noctor et al., 2012; Zechmann, 2014), such as modulation of growth and developmental processes (Anjum et al., 2010; Noctor et al., 2011, 2012) quenching of ROS (Foyer and Noctor, 2005; Mullineaux and Rausch, 2005) heavy metal and xenobiotics detoxification under varied environmental insults (Dixon et al., 1998; Cobbett and Goldsbrough, 2002; Dixon et al., 2002; Freeman et al., 2004; Son et al., 2014). Virtually, it could be found abundantly in almost all cellular components ranging from cytosol, endoplasmic reticulum, chloroplast, and vacuole to mitochondria (Yadav, 2010).

Its unique structural properties wide redox potential and relative abundance has raised considerable interest among the plant physiologists regarding their defensive function in living organisms (May et al., 1998a; Jozefczak et al., 2012; Viehweger, 2014; Xu et al., 2014). Noctor and Foyer, 1998 demonstrate the importunate role of GSH in stress-induced defense mechanisms in higher plants (Figure 11.2). The authors exploit the technology to escalate the level of GSH (at least four-fold as compared with untransformed ones) in plant, and this approach proves very significant as it has execrably enhanced the GSH-mediated conjugation and PC synthesis (Figure 11.2). Likewise, Hossain et al. (2012) elucidated the central role of GSH in stress-tolerant mechanisms.

In particular, the γ-glutamyl linkage contributes to higher stability, whereas the strong nucleophilic property of the thiol of group of GSH makes it an immensely powerful cellular reductant (Yadav, 2010; May et al., 1998a,b; Noctor and Foyer, 1998). With this nucleophilic property, GSH is capable of forming a mercaptide bond with metals and metalloids, thereby facilitating its ultimate mitigation or sequestration (Figure 11.2) (Yadav, 2010; Zagorchev et al., 2013). This chemical reactivity and relative stability make GSH a novel biochemical component, which helps in protecting the plants against varied stress conditions, including oxidative and heavy metal stress (Millar et al., 2003; Foyer and Noctor, 2005; Rausch et al., 2007; Yadav, 2010; Kumar et al., 2013b).

Being a sulfur-containing nonprotein tripeptide thiol, GSH presumably serves (Figures 11.1 and 11.2) as the major pool for the transport and storage form of reduced sulfur (Tausz et al., 2004) and therefore keeps regulating the interorgan allocation of sulfur (Herschbach and Rennenberg, 1994; Lappartient and Touraine, 1997; Noctor and Foyer, 1998). Several physiological and biological roles have been documented as belonging to GSH in plants (Rennenberg, 1995; Foyer, 1997). Because of its multifaceted role in plant, previous studies have encompassed the central role of GSH in engineering stress tolerance in plants (Figure 11.1) (Noctor et al., 2012). Moreover, its frequent stimulation/accumulation under stress conditions confirms its imperative role (Figure 11.1; Table 11.1) in the defensive

Table 11.1 Simplified Overview Representing the GSH- and PC-Linked Stress Responses and Cellular Trafficking of Some Selectively High Toxic Heavy Metals

Metals	Permissible Limit in Agricultural Soil (mg/kg Soil)	Deleterious Effect of Some of the Heavy Metals	Antioxidative Thiols (GSH and PCs)	Studied Plant Species	Probable Response of Plant Toward Antioxidative Thiols GSH and PCs	References
Cadmium (Cd)	100 mg/kg soil	Crippled photosynthesis, reduced water and nutrient uptake, chlorosis, growth inhibition, browning of root tips, leaf epinasty, alters the chloroplast ultrastructure, inactivates enzymes in CO_2 fixation, induces lipid peroxidation (LPO), inhibits pollen germination and tube growth, and disturbs the nitrogen (N) and sulfur (S) metabolism and antioxidant machinery and cause ultimate death of plants.	Glutathione (GSH)	Rice seedling *Brassica juncea* *Brassica campestris* *Lycopersicum esculentum* *Arabidopsis*	Plant exposure to Cd showed marked escalation in GSH regardless of their concentration. This elevated level of GSH improves redox homeostasis and stress tolerant mechanism.	Hossain et al. (2012), Koprivova et al. (2002), Wojcik and Tukiendorf (2004), Dong et al. (2006), Jozefczak et al. (2014), Sobrino-Plata et al. (2014a,b)
			Phytochelatins (PCs)	*Bacopa monnieri* *Thlaspi caerulescens* Maize seedling *Sedum alfredii* *Phragmites australis Trin. (cav.) ex Steudel* *Dittrichia viscosa (L.) Greuter* *Zea mays*	A positive correlation has been reported between increased level of PCs and Cd tolerance in maize, *Brassica*, *Pisum*, *Thlaspi*, *Bacopa*, *Arabidopsis*, and other plants.	Nussbaum et al. (1988), Ruegsegger and Brunold (1992), Rauser and Meuwly (1995), Haag-Kerwer et al. (1999), Howden et al. (1995a,b), Inouhe et al. (1994) and Zhang and Tyerman (1999), Pietrini et al. (2003), Zhu et al. (1999a,b), Fernández et al. (2014), Sobrino-Plata et al., 2014a,b, Szalai et al. (2013)
Zinc (Zn)	150–300 mg/kg	Limited growth of root and shoot, chlorosis in younger leaves, perturbed growth and plant metabolism, appearances of purplish-red color in leaves.	GSH	*Arabidopsis thaliana*	Plant accelerates the synthesis of GSH under Zn stress, which further protects the plant from deleterious effects of reactive oxygen species (ROS).	Tsugi et al. (2002)

Continued

Table 11.1 Simplified Overview Representing the GSH- and PC-Linked Stress Responses and Cellular Trafficking of Some Selectively High Toxic Heavy Metals—cont'd

Metals	Permissible Limit in Agricultural Soil (mg/kg Soil)	Deleterious Effect of Some of the Heavy Metals	Antioxidative Thiols (GSH and PCs)	Studied Plant Species	Probable Response of Plant Toward Antioxidative Thiols GSH and PCs	References
				Dunaliella tereiolecta	Antioxidative thiols (PCs) not only play an indispensable role in metal detoxification. but also mitigation of oxidative stress.	Hirata et al. (2001)
Cupper (Cu)	10 mg/kg soil	Retarded growth, stimulated generation of ROS and oxidative stress.	GSH	*Brassica juncea* L. *Avicennia germinans.* Rice	On stress exposure, brassica enhance the synthesis of nonprotein thiol, but after some time declination is observed in their concentration.	Wang et al. (2004) Mostofa et al. (2014)
			PCs	*Holcus lanatus* *Prunus cerasifera* (quite tolerant to Cu stress) *Silene cucubalus*	Cu and arsenic exposure led the generation of ROS species in nontolerant plants in which the excessive/rapid arsenate influx resulted in PC production and GST depletion and LPO. Accelerated PC production further enhances stress tolerance mechanism in plant and will eventually lead to metal detoxification.	Hartley-Whitaker et al. (2001), Lombardi and Sebastiani (2005), De Vos et al. (1992)
Mercury (Hg)	0.01 mg/kg soil	Highly phytotoxic to plants, induces discernible injuries and physiological disorders in plants, also accelerates	GSH	*Valusneria spiraus* L. and *Hydrilla verticillata (I.F)*	GSH and PC synthesis in these plants plays potential role in metal detoxification.	Gupta et al. (2010)

		generation of ROS on stress exposure, disruption of membrane integrity, and cellular metabolism in plants.	PCs	*Brassica napus* *Chloris barbata sw.* *Medicago sativa* *Abelmoschus esculentus*	PC synthesis on stress exposure not only plays an inescapable role in metal detoxification, but also helps in maintaining cellular redox homeostasis in combination with GST.	Iglesia-Turino et al., 2006, Patra et al., 2004 Ortega-Villasante et al., 2005, Hameed et al. (2011), Sobrino-Plata et al. (2014a,b)
Chromium (Cr)	1.5 mg/kg soil	Reduced plant growth, chlorosis in young leaves, nutrient imbalance, wilting of tops, and root injury. Inhibition of chlorophyll biosynthesis in terrestrial plants.	GSH	*Salvinia natans* *Vigna radiate* Rice	Endowed with antioxidative property that enhances chromium tolerance in salvinia and also enhance its resistant/tolerance capacity toward oxidative stress.	Dhir et al. (2011) Qiu et al. (2013)
			PCs	Maize seedling	Activation of antioxidative mechanisms from the impact of chromium-induced oxidative stress.	Maiti et al. (2012)
Lead (Pb)	1.5–251 mg/kg soil	Impediment of enzymatic activities, water imbalance, changes in membrane permeability, and muddled mineral nutrition, oxidative stress by accelerated generation of ROS in plants.	GSH	*Sedum alfredii* *Dianthus carthusianorum*	Enhanced GSH synthesis is appeared to be an intrinsic response of plant against lead stress and is crucial for determining growth under stress	Gupta et al. (2010), Wójcik and Tukiendorf (2014)
			PCs	Coontail (*Ceratophyllum demersum L.*)	An increase in GSH content has been observed at moderate exposure of Pb.	Mishra et al. (2006)
Arsenic (As)	1–40 mg/kg soil	participate for the similar uptake carriers in the root plasma lemma of plants, suppression of a high-affinity P/As uptake system	GSH	*Pteris vittata;* *Mesquite* (*Prosopis* sp) Lentil	Both GSH and PC play an imperative role in As tolerance in plants.	Cai et al. (2004) Mokgalaka-Matlala et al., 2009, Talukdar and Talukdar (2014)
			PCs	*R. serpentine;* *Pteris cretica;* *Pteris vittata* Chicken pea;	PCs perform an indispensable role in As tolerance in intact plants generally by two significant processes.	Schmöger et al. (2000), Raab et al. (2004), Cai et al. (2004) Gupta et al. (2004),

Continued

Table 11.1 Simplified Overview Representing the GSH- and PC-Linked Stress Responses and Cellular Trafficking of Some Selectively High Toxic Heavy Metals—cont'd

Metals	Permissible Limit in Agricultural Soil (mg/kg Soil)	Deleterious Effect of Some of the Heavy Metals	Antioxidative Thiols (GSH and PCs)	Studied Plant Species	Probable Response of Plant Toward Antioxidative Thiols GSH and PCs	References
				Holcus lanatus; Rubia tinctorium	Reduced uptake and accumulation of As. Intracellular detoxification. In Holcus, PCs are involved in differential arsenic tolerance.	Hartley-Whitaker et al. (2001), Maitani et al. (1996) respectively
Nickel (Ni)	1.5 mg/kg soil	Chlorosis and necrosis in different plant species, impairment of nutrient balance and resulted in disorder of cell membrane functions, decrease in water uptake is used as an indicator of the progression of Ni^{2+} toxicity in plants	GSH and PCs	*Thlaspi*	Increased GSH biosynthesis play an inescapable role in Thlaspi nickel hyperaccumulator.	Freeman et al. (2004)

mechanism (May et al., 1998a). Henceforth, the elevated level of GSH on stress exposure could be correlated with the induction of oxidative stress in plants; therefore, it could now be well-interpreted as a marker of oxidative stress (Grill, 2001; Tausz et al., 2004).

To date, several reports are available that signify the irreplaceable role of GSH in stress-tolerant mechanisms adopted by plants (Grill, 2001). Owing to its chemical reactivity of the thiol group, GSH has often been considered remarkably suitable for serving the broad range of biochemical and physiological functions (Figure 11.1; Table 11.1) of all living organisms (Grant et al., 1996; Alscher, 1989). In spite of this, it also act as a gene regulator controlling gene expression (Baier and Dietz, 1997; Wingate et al., 1988), a substrate for PCs synthesis that improves metal sequestration (Grill et al., 1987, 1989), and a precursor for the synthesis of GSH S-transferases (Figure 11.2; Table 11.1), which facilitates the conjugation of GSH with potentially dangerous xenobiotics (Marrs, 1996).

Being the major endogenous antioxidant produced peculiarly by cells under stress exposure, GSH participates directly in the neutralization of AOS and thus acts like a significant redox buffer (May et al., 1998a,b). In addition, while serving as a redox buffer, it keeps regulating the optimum level of AOS in plant, thereby protecting the cellular compartments from the hazardous/deleterious effect trace elements (Yadav, 2010). Antioxidative defense and redox homeostasis together play a remarkable role in acclimating plants to their environments; in this context, the subtle role of GSH can never be ignored (Figures 11.1 and 11.2; Table 11.1).

Tausz et al. (2004) recognized GSH as a valuable stress marker in plant eco-physiology and also established the immense significance of GSH in stress tolerance under varied environmental conditions (Figure 11.1; Table 11.1). Likewise, Grill, 2001 explored the indispensable role of GSH stimulating/modulating multiple functions such as growth, development, and varied responses of plants against biotic and abiotic stresses. In addition, May et al., 1998a suggested the potentially significant/crucial role of GSH in improving cellular processes and stress tolerant mechanism of plant (Figure 11.2; Table 11.1).

Besides this, GSH is also known to be involved in the sequential regulation of several developmental processes such as cell division and flowering that categorize it as a crucial metabolite (Vernoux et al., 2000; Ogawa et al., 2004; Yadav, 2010). In plants, GSH occurs basically in two forms: GSH and GSSG, and their reduction potential precisely depend on their intracellular GSH/GSSG ratio (Yadav, 2010; Lushchak, 2012). Recently, more convincing evidence are available that markedly signifies the prevalent role of reduced GSH in scavenging peroxide radicals (May et al., 1998a). Moreover, the relative proportion of GSH in the reduced form reflects the relative rates of reduction and oxidation which is preferably greater than 0.9 under optimal condition.

GSH synthesis is being catalyzed from its constituent amino acids (Glu, Cys, and Gly) via the incorporation of two enzymes, namely Glu–Cys ligase (g-ECS; EC 6.3.2.2; GSH1) and GSH synthetase (EC 6.3.2.3; GSH2) in the two adenosine triphosphate (ATP)-dependent steps (Grill, 2001). The first step is the rate-limiting reaction virtually encoded by the gsh1 gene, where γ-ECS catalyzes the linking of peptide bond between the γ-carboxyl group of glutamate and amino group of Cys to yield γ-glutamyl linkage, whereas the second step is been encoded by the GSH2, where GSH synthetase ligates a glycine residue with γ-EC to form GSH (May et al., 1998a,b; Tausz et al., 2004; Yadav, 2010; Noctor et al., 2012; Grill, 2001). Indeed, the compartmentation of these two stepped GSH biosynthetic pathways is the characteristic feature of the plant (Yadav, 2010).

The previously mentioned biosynthetic pathway of GSH occurs in chloroplast and nonchloroplastic compartments in both photosynthetic and nonphotosynthetic tissues (Foyer and Noctor, 2002). Of these two steps, the first one serves as the major controlling point during the chronic situation of upsurged demand for GSH (Noctor and Foyer, 1998). Exhaustive literature surveys have reported the transient increase in GSH concentration upon prolonged exposure of cadmium (Gill et al., 2012) and lead (Estrella-Gómez et al., 2012). The powerful efficacy of oxidized GSH to convert back to GSH during reductive inactivation of peroxides that establish GSH as a versatile component in an antioxidant defense system (May et al., 1998a,b; Tausz et al., 2004). Enhanced synthesis of GSH in a plant could significantly be correlated with environmental stress tolerance because it is the prime intrinsic response of plant toward stress (Noctor et al., 2012)

Keeping in view the detailed eclectic picture of GSH during the stress-tolerant mechanisms creates experimental priorities/challenges to develop an assessable framework that not only identifies the integrated response of GSH toward stress, but also explores the further aspects of GSH in metabolism and functions of plant. Further, a concerted effort should be made to enhance their synthesis in a plant so that the sessile lifestyle of the plant can function well with environmental constraints (May et al., 1998a,b).

4. PC STRUCTURE, BIOSYNTHESIS, AND FUNCTION

Among the various types of metal-binding complexes identified, the best-known metal ion chelators are metallothioneins and PCs. Metallothioneins were first discovered in 1957 by Vallee and Margoshe and are a family of Cys-rich, prevalently recognized low-molecular-weight protein with molecular weight ranging from 500 to 14,000 Da and its indispensable role in metal sequestration is always being a subject of special interest.

After the discovery of bestowed metal-binding complexes metallothioneins, research pioneers had recognized another most efficient and potentially active metal ion chelator—PC (Inouhe, 2005)—and its flawless role in metal tolerance has now been the subject of considerable interest in several studies (Figure 11.2; Table 11.1). About three decades ago, it was first discovered by Hayashi and his group as an inherent component of Cd-binding complexes in the fission yeast *Schizosaccharomyces pombe* (Kondo et al., 1984) and cell cultures of *Rauwolfia serpentine* (Grill et al., 1985) upon Cd exposure. Hence from the previous discussion, the phenomenal role of PCs in metal sequestration was well-interpreted from earlier times.

More precisely, PC was presumed to be a product of biosynthetic pathway; structurally, they were thought to be made up of three amino acids namely glutamate, cysteine, and glycine with the glutamate, and cysteine residues bonded via a g-carboxylamide bond (Cobbett, 2000). PCs were predominantly identified as a nonprotein, cysteine rich oligomer of GSH with the conventional framework of (γ-glutamyl-cysteinyl)n-glycine ($n = 2$–11) and the variants with repeated γ-glutamyl cysteinyl unit (Inouhe, 2005). A considerable debate still prevails on the major function of PCs in metal sequestration. Intriguingly, PCs have only been considered important biological players promoting metal tolerance (Figure 11.1; Table 11.1). But thereafter, a several reports have been put forward that significantly manifest the apparent involvement of PC in trace element homeostasis or its subsequent mitigation (Inouhe, 2005). Moreover, its enhanced synthesis on stress exposure gave a compelling indication of its inescapable role in cellular redox homeostasis or trafficking of essential trace elements, particularly Cu and Zn (Thumann et al., 1991). It is now touted as the one of the most efficient, preeminent, well-equipped, and nonpareil metal-binding complexes, besides metallothioneins.

Cobbett and Goldsbrough (2002) suggested the unique and apparent involvement of PCs and metallothioneins in heavy metal detoxification and redox homeostasis. Likewise, Inouhe, in 2005, well-documented the importunate/imperative role of PC in stress tolerance. The author suggested the PC-based metal detoxification in higher plants. Beyond this, it also attains much consciousness in phytoremediation programs (Pal et al., 2010; Zagorchev et al., 2012).

All the higher plants are equipped with the capacity of synthesizing PCs, and it is interesting to note that PCs show structural resemblance with GSH (°-GluCysGly) (Gekeler et al., 1989). In this context, the previous findings have already envisaged their precise synthesis from GSH (Grill et al., 1989), homo-GSH, hydroxymethyl-GSH (Klapheck et al., 1995), or g-glutamylcysteine (Hayashi et al., 1991), and elucidated the utmost significance of consecutive enzyme named PC synthase, a γ-glutamyl-cysteine transpeptidase (EC 2.3.2.15) for bringing the previously mentioned catalysis of PC from GSH provided the posttranslational activation by heavy metal particularly Cd, Ag, Pb, Cu, Hg, Zn, Sn, Au, and As, both in vivo and in vitro (Grill et al., 1987, 1989; de Knecht et al., 1995; Klapheck et al., 1995; Maitani et al., 1996; Chen et al., 1997). Currently, a plethora of progressive physiological, biochemical, and genetic studies is available that peculiarly signifies GSH as the direct substrate for PC biosynthesis (de Knecht et al., 1995; Klapheck et al., 1995; Maitani et al., 1996; Chen et al., 1997).

PC plays a magnificent/incredible role in metal sequestration because it is endowed with the capacity of binding with multiple types of metals and metalloids and it gets stimulated automatically inside the cell on metal exposure (Figure 11.2) (Grill et al., 1987; Gekeler et al., 1989). Though it is capable of binding with multiple metals such as Cd, Cu, Zn, or As via sulfhydryl and carboxyl residue, it can preferentially be controlled by the metal Cd or metalloid As, whereas its efficacy toward other metals is not so conclusive (Zagorchev et al., 2012). Rauser (1990) ascribed significantly the potential role of PCs and PC-related peptides in heavy metal detoxification. Beside this, it would also be interesting to know that PCs and PC-related peptides could be used as possible biochemical indicators for trace elements (Inouhe, 2005).

Zhao et al. (2003) investigated possible involvement of PCs in As tolerance by hyperaccumulator *Pteris vittata.* Likewise, Schat et al. (2002) demonstrated the immense role of PCs in constitutive and adaptive heavy metal tolerances peculiarly by hyperaccumulator and nonhyperaccumulator metallophytes.

For ameliorating the toxic effect of heavy metal, PC generally forms complexes with metal sequestered in the vacuole (Yadav, 2010). In spite of this, the role of PCs might not be limited to the chelation of potentially toxic metal ions (Zagorchev et al., 2013). In this context, Bhargava et al., in 2005, while working on the cyanobacterium *Anabaena doliolum,* reported the unique and non–stress-specific feature of PC. In fact, PCs are capable of providing protection against the damaging effects of ultraviolet B radiation, suggesting its specific antioxidative or protein thiol-protective role (Delalande et al., 2010). Database analysis confirmed PC as the most potent metal ion chelator, followed by GSH (Zagorchev et al., 2012). Though both GSH and PCs follow distinct physiological pathways during metal sequestration, unlike GSH, PCs never forms disulfides under oxidative stress conditions, but their dynamic role in escalating antioxidative system of plant on stress exposure is incredible. Therefore, further studies are still required for proper assessment of their putative role in stress tolerance.

5. GSH- AND PC-ASSOCIATED STRESS TOLERANCE MECHANISMS IN HIGHER PLANTS

Unfortunately, the accelerated production of AOS is the most common consequence of heavy metal exposure, perhaps resulting in oxidative stress (Noctor and Foyer, 1998). Organelles with high metabolic activities along with intense rate of electron transport predominantly serve as the major source for ROS production in plants (Apel and Hirt, 2004; Mittler et al., 2004; Mhamdi et al., 2010; Hossain et al., 2011). Generally, ROS is highly toxic to reactive species comprising unpaired electrons (Yadav, 2010). Under optimal condition, its concentration in plants is tightly regulated by complex networking of defense mechanism (Yadav, 2010; DalCorso et al., 2008; Clemens et al., 2002). So, if not properly metabolized, its upsurged concentration in plants eventually provoke oxidation of cellular components such as protein, lipid, and DNA (Foyer and Noctor, 2005; Ogawa and Iwabuchi. 2001; Ogawa et al., 2004). Besides, doing so, it likely serves as a key component for triggering onset of redox homeostasis and signal transduction pathway in plants (Figure 11.2) (Seth et al., 2012; Hossain et al., 2012).

To counterbalance the exclusively innumerable damaging effects of these species, plants have evolved an efficient and potentially significant tolerant mechanism at the cellular level to withstand excess heavy metals in metal-leached soil (Foyer and Noctor, 2005). The adaptive and consecutive tolerant strategy adopted by plant against heavy metal, however, has been regulated by a complex interrelated network of physiological, cellular, and molecular processes.

Several approaches at multiple levels are being approached for demonstrating the eclectic picture or underlying mechanism of metal accumulation and tolerance to metal stress (Meharg, 1994). The induction of an adaptive defense mechanism is a key feature, in which complexation and compartmentation with PCs, metallothioneins, GSH, and sequestration within vacuoles are utmost significance. Because of its peculiar chemical reactivity, the ubiquitously found low-molecular-weight tripeptide GSH is known to perform a wide range of biochemical functions in all living organisms during stress conditions (Yadav, 2010). Owing to its nucleophilic property, GSH is capable of forming a mercaptide bond with metals and metalloids and also for reacting with selected electrophiles. This specific chemical property, along with relatively high stability and water solubility, makes GSH a novel biochemical capable of providing protection to plants against varied stress conditions such as heavy metal stress, oxidative stress, and certain other precarious endogenous and exogenous chemicals (Millar et al., 2003; Foyer and Noctor, 2005; Rausch et al., 2007). GSH directly participates in the regulation of peroxide radicals of plant cells (Foyer and Noctor, 2005; Shao et al., 2008).

It has been well documented that participation of GSH in the regulation of intracellular redox potential predominantly depends upon its concentration and the GSH/GSSG ratio. Preliminary data indicate the significant involvement of GSH in ROS perception because it is a major component of ascorbate–GSH cycle; therefore, it prevalently acts as a direct scavenger of free radicals. Besides this, its considerably significant role has also been well-marketed in xenobiotic and heavy metal detoxification provided its prior activations and conjugation with such molecules (Figure 11.2; Table 11.1) (Marrs, 1996; Alfenito et al., 1998); this conjugation is regulated by GSH S-transferase (Edwards et al., 2000; Edwards and Dixon, 2005). Further, the subsequent transport of this conjugate into the vacuole preferably protects the plant from the damaging effects of ROS (Klein et al., 2006; Yazaki, 2006; Yadav, 2010). During this detoxification process, a transient decrease in cytosolic GSH content occurs that negatively influences the GSH/GSSG ratio, thereby creating a redox signal in stress-exposed cell (Nocito et al., 2006; Yadav, 2010). Consequently, this progressive upgrading of GSH-based detoxification indirectly influences cellular redox poise and thus, for the survival of the plant, the maintenance of the GSH/GSSG ratio becomes mandatory (Yadav, 2010). Besides, the significant role of GSH in ROS quenching, heavy metal, and xenobiotic detoxification, it also acts as a substrate for the synthesis of a metal-binding ligand PCs. Plant exposure to heavy metal such as Cd, Hg, Cu, Zn, Pb, and Ni led the synthesis of PCs from GSH and this reaction is catalyzed by constitutive enzyme PC synthase. Thereafter, several reports have put forward, which illustrates their imperative role in maintenance of redox homeostasis and heavy metal detoxification (Zenk, 1996; Hirata et al., 2005; Yadav, 2010). Biosynthesis of PCs further regulate the overall mechanism of stress tolerance that involves specifically the complexation of heavy metal with thiol compounds such as GSH and PCs in cytosol and their subsequent pumping on plasma membrane and then their sequestration in to the vacuole (Figure 11.2) (Lin and Aarts, 2012).

From the previous, it can be interpreted that GSH in particular is not at all sufficient for supporting this highly complex networking of defense mechanisms; therefore, its increased levels inside the plant should not always be linked metal tolerance (Xiang et al., 2001; Yadav, 2010).

6. CONCLUSION AND FUTURE PROSPECTS

Being cramped in distribution, plants inevitably encounter a variety of complex environmental stresses. Among these stresses, plant exposure to heavy metal stress remains the major source of environmental health risks throughout the world. The innumerable effects of heavy metal on the environment are serious matters of global concern. To survive in such difficult conditions, a plant ought to be adapted against heavy metal stress; therefore, it becomes a prerequisite to improve plants' heavy metal tolerance capacity that would eventually lessen metal incorporation in the food chain.

Though plant survival in such relentless conditions demands proper management of AOS, precariously endogenous phytochemicals and exogenous toxins, such as diverse xenobiotics, are perpetually induced in to the environment by rising anthropogenic activities. The proper neutralization or scavenging of ROS and its reaction products are very crucial because their accelerated accumulation becomes highly toxic and would lead to irreparable metabolic dysfunction and ultimately death of the plant cell.

To restrain the toxicity of heavy metals, thiol compounds such as GSH, PCs, and metallothioneins in particular serve as central or "hub" molecules in cellular metabolism, redox buffering, and substantial metal sequestration. More importantly, it can be noted here that different plants display distinct metal tolerance capacities toward heavy metal: some plants flourish well in soil enriched with an alarming level of heavy metals, whereas some are unable to even initiate propagation under such conditions.

Database analysis on some of these plants reported GSH and PCs as major biological players in terms of their relative tolerance. Being the essential component of ascorbate-GSH pathway, the tripeptide GSH is known to perform multiple functions in plants involving ROS detoxification. Besides GSH, the role of PC should also be visualized in metal sequestration.

Therefore, for enhancing metal tolerance and making plants exceptionally resistant to varied environmental insults, a concerted multidimensional effort should be made to generate transgenic plants with the help of several different genes modulating the level of thiol compounds in plants. In this context, natural heavy metal accumulators could be potential sources for genetic manipulation of other important agricultural crop plants.

To the best of our knowledge, this simplified overview provides an opportunity to assess the current understanding of the biosynthesis and function of GSH and PCs in plants, given the immense literature on this subject. Additionally, after visualizing the peculiarly indispensable role of thiols in imparting heavy metal stress tolerance in plants, there is a need for further detailed accounts of experimental validation.

ACKNOWLEDGEMENTS

Ms. Shweta Singh and Dr. D. K. Tripathi are thankful to UGC for providing the financial support in the form of fellowship.

REFERENCES

Abhilash, M.R., Srikantaswamy, S., Kumar, D.S., Kiran, B.M., 2014. Study of heavy metal uptake by the crops grown by using urban wastewater of Mysore city, India. J. Environ. Prot. Ecol. 5 (12), 1169.

Acquaah, G., 2009. Principles of Plant Genetics and Breeding. John Wiley & Sons.

Alfenito, M.R., Souer, E., Goodman, C.D., Buell, R., Mol, J., Koes, R., Walbot, V., 1998. Functional complementation of anthocyanin sequestration in the vacuole by widely divergent glutathione S-transferases. Plant Cell 10 (7), 1135–1149.

Alscher, R.G., 1989. Biosynthesis and antioxidant function of glutathione in plants. Physiol. Plantarum 77 (3), 457–464.

Alloway, B., Ayres, D.C., 1997. Chemical Principles of Environmental Pollution. CRC Press.

Anjum, N.A., Umar, S., Chan, M.T., 2010. Ascorbate-Glutathione Pathway and Stress Tolerance in Plants. Springer, Dordrecht.

Anjum, N.A., Ahmad, I., Mohmood, I., Pacheco, M., Duarte, A.C., Pereira, E., Umar, S., Ahmad, A., Khan, N.A., Iqbal, M., Prasad, M.N.V., 2012. Modulation of glutathione and its related enzymes in plants' responses to toxic metals and metalloids-a review. Environ. Exp. Bot. 75, 307–324.

Apel, K., Hirt, H., 2004. Reactive oxygen species: metabolism, oxidative stress, and signal transduction. Annu. Rev. Plant Biol. 55, 373–399.

Babula, P., Adam, V., Opatrilova, R., Zehnalek, J., Havel, L., Kizek, R., 2008. Uncommon heavy metals, metalloids and their plant toxicity: a review. Environ. Chem. Lett. 6 (4), 189–213.

Baier, M., Dietz, K.J., 1997. The plant 2-Cys peroxiredoxin BAS1 is a nuclear-encoded chloroplast protein: its expressional regulation, phylogenetic origin, and implications for its specific physiological function in plants. Plant J. 12 (1), 179–190.

Bell, F.G., Bullock, S.E.T., Hälbich, T.F.J., Lindsay, P., 2001. Environmental impacts associated with an abandoned mine in the Witbank Coalfield, South Africa. Int. J. Coal Geol. 45 (2), 195–216.

Bhargava, A., Carmona, F.F., Bhargava, M., Srivastava, S., 2012. Approaches for enhanced phytoextraction of heavy metals. J. Environ. Manage. 105, 103–120.

Bowen, H.J.M., 1979. Environmental Chemistry of the Elements. Academic Press.

Bricker, T.J., Pichtel, J., Brown, H.J., Simmons, M., 2001. Phytoextraction of Pb and Cd from superficial soil: effects of amendments and croppings. J. Environ. Sci. Health A 36, 1597–1610.

Cai, Y., Su, J., Ma, L.Q., 2004. Low molecular weight thiols in arsenic hyper accumulator *Pteris vittata* upon exposure to arsenic and other trace elements. Environ. Pollut. 129 (1), 69–78.

Carrier, P., Baryla, A., Havaux, M., 2003. Cadmium distribution and microlocalization in oilseed rape (*Brassica napus*) after long-term growth on cadmium-contaminated soil. Planta 216 (6), 939–950.

Chen, J., Zhou, J., Goldsbrough, P.B., 1997. Characterization of phytochelatin synthase from tomato. Physiol. Plantarum 101 (1), 165–172.

Chibuike, G.U., Obiora, S.C., 2014. Heavy metal polluted soils: effect on plants and bioremediation methods. Appl. Environ. Soil Sci. 12. Article ID 752708. http://dx.doi.org/10.1155/2014/752708.

Clemens, S., Palmgren, M.G., Krämer, U., 2002. A long way ahead: understanding and engineering plant metal accumulation. Trends Plant Sci. 7 (7), 309–315.

Cobbett, C., Goldsbrough, P., 2002. Phytochelatins and metallothioneins: roles in heavy metal detoxification and homeostasis. Annu. Rev. Plant Biol. 53 (1), 159–182.

Cobbett, C.S., 2000. Phytochelatin biosynthesis and function in heavy-metal detoxification. Curr. Opin. Plant Biol. 3 (3), 211–216.

DalCorso, G., Farinati, S., Maistri, S., Furini, A., 2008. How plants cope with cadmium: staking all on metabolism and gene expression. J. Integr. Plant Biol. 50 (10), 1268–1280.

DalCorso, G., Farinati, S., Furini, A., 2010. Regulatory networks of cadmium stress in plants. Plant Signa. Behav. 5 (6), 663.

Delalande, O., Desvaux, H., Godat, E., Valleix, A., Junot, C., Labarre, J., Boulard, Y., 2010. Cadmium–glutathione solution structures provide new insights into heavy metal detoxification. FEBS J. 277 (24), 5086–5096.

Dhir, B., Sharmila, P., Saradhi, P.P., Sharma, S., Kumar, R., Mehta, D., 2011. Heavy metal induced physiological alterations in *Salvinia natans*. Ecotox. Environ. Safe. 74 (6), 1678–1684.

Dietz, K.J., Baier, M., Krämer, U., 1999. Free radicals and reactive oxygen species as mediators of heavy metal toxicity in plants. In: Heavy Metal Stress in Plants. Springer, Berlin Heidelberg, pp. 73–97.

Dixon, D.P., Cummins, I., Cole, D.J., Edwards, R., 1998. Glutathione-mediated detoxification systems in plants. Curr. Opin. Plant Biol. 1 (3), 258–266.

Dixon, D.P., Davis, B.G., Edwards, R., 2002. Functional divergence in the glutathione transferase superfamily in plants identification of two classes with putative functions in redox homeostasis in *Arabidopsis thaliana*. J. Biol. Chem. 277 (34), 30859–30869.

Dong, J., Wu, F.B., Zhang, G.P., 2006. Influence of cadmium on antioxidant capacity and four microelement concentrations in tomato seedlings (*Lycopersicon esculentum*). Chemosphere 64, 1659–1666.

Edwards, R., Dixon, D.P., Walbot, V., 2000. Plant glutathione S-transferases: enzymes with multiple functions in sickness and in health. Trends Plant Sci. 5 (5), 193–198.

Edwards, R., Dixon, D.P., 2005. Plant glutathione transferases. Method Enzymol. 401, 169–186.

Estrella-Gómez, N.E., Sauri-Duch, E., Zapata-Pérez, O., Santamaría, J.M., 2012. Glutathione plays a role in protecting leaves of *Salvinia minima* from Pb^{2+} damage associated with changes in the expression of SmGS genes and increased activity of GS. Environ. Exp. Bot. 75, 188–194.

Fernández, R., Fernández-Fuego, D., Bertrand, A., González, A., 2014. Strategies for Cd accumulation in *Dittrichia viscosa* (L.) Greuter: role of the cell wall, non-protein thiols and organic acids. Plant Physiol. Biochem. 78, 63–70.

Foy, C.D., Chaney, R.T., White, M.C., 1978. The physiology of metal toxicity in plants. Ann. Rev. Plant Physiol. 29 (1), 511–566.

Foyer, C.H., 1997. Oxygen metabolism and electron transport in photosynthesis. In: Scandalios, J. (Ed.), Oxidative Stress and the Molecular Biology of Antioxidant Defenses. Cold Spring Harbor Laboratory Press, New York, pp. 587–621.

Foyer, C., Noctor, G., 2002. The molecular biology and metabolism of glutathione. In: Grill, D., Tausz, M., Kok, L. (Eds.), Significance of Glutathione to Plant Adaptation to the Environment. Springer, Netherlands, pp. 27–56. ce_2400 4542012.

Foyer, C.H., Noctor, G., 2005. Oxidant and antioxidant signalling in plants: a re-evaluation of the concept of oxidative stress in a physiological context. Plant Cell Environ. 28 (8), 1056–1071.

Freeman, J.L., Persans, M.W., Nieman, K., Albrecht, C., Peer, W., Pickering, I.J., Salt, D.E., 2004. Increased glutathione biosynthesis plays a role in nickel tolerance in Thlaspi nickel hyperaccumulators. Plant Cell 16 (8), 2176–2191.

Freitas, H., Prasad, M.N.V., Pratas, J., 2004. Plant community tolerant to trace elements growing on the degraded soils of São Domingos mine in the south east of Portugal: environmental implications. Environ. Int. 30 (1), 65–72.

Gekeler, W., Grill, E., Winnacker, E.L., Zenk, M.H., 1989. Survey of the plant kingdom for the ability to bind heavy metals through phytochelatins. Z. fuÈr Naturforschung C 44, 361–369.

Gill, S.S., Khan, N., Tuteja, N., 2012. Cadmium at high dose perturbs growth, photosynthesis and nitrogen metabolism while at low dose it up regulates sulfur assimilation and antioxidant machinery in garden cress (*Lepidium sativum* L.). Plant Sci. 182, 112–120.

Gratão, P.L., Prasad, M.N.V., Cardoso, P.F., Lea, P.J., Azevedo, R.A., 2005. Phytoremediation: green technology for the cleanup of toxic metals in the environment. Brazil J. Plant Physiol. 17 (1), 53–64.

Grant, C.M., MacIver, F.H., Dawes, I.W., 1996. Glutathione is an essential metabolite required for resistance to oxidative stress in the yeast *Saccharomyces cerevisiae*. Curr. Genet. 29 (6), 511–515.

Grill, D., 2001. Significance of Glutathione to Plant Adaptation to the Environment, vol. 2. Springer Science & Business Media.

Grill, E., Winnacker, E.L., Zenk, M.H., 1985. Phytochelatins: the principal heavy-metal complexing peptides of higher plants. Science 230 (4726), 674–676.

Grill, E., Winnacker, E.L., Zenk, M.H., 1987. Phytochelatins, a class of heavy-metal-binding peptides from plants, are functionally analogous to metallothioneins. Proc. Natl. Acad. Sci. U.S.A. 84 (2), 439–443.

Grill, E., Löffler, S., Winnacker, E.L., Zenk, M.H., 1989. Phytochelatins, the heavy-metal-binding peptides of plants, are synthesized from glutathione by a specific γ-glutamylcysteine dipeptidyl transpeptidase (phytochelatin synthase). Proc. Natl. Acad. Sci. U.S.A. 86 (18), 6838–6842.

Gupta, D.K., Tohoyama, H., Joho, M., Inouhe, M., 2004. Changes in the levels of phytochelatins and related metal-binding peptides in chickpea seedlings exposed to arsenic and different heavy metal ions. J. Plant Res. 117, 253–256.

Gupta, D.K., Huang, H.G., Yang, X.E., Razafindrabe, B.H.N., Inouhe, M., 2010. The detoxification of lead in *Sedum alfredii* H. is not related to phytochelatins but the glutathione. J. Hazard. Mater. 177 (1), 437–444.

Haag-Kerwer, A., SchaÈfer, H.J., Heiss, S., Walter, C., Rausch, T., 1999. Cadmium exposure in *Brassica juncea* causes a decline in transpiration rate and leaf expansion without effect on photosynthesis. J. Exp. Bot. 50, 1827–1835.

Hartley-Whitaker, J., Ainsworth, G., Meharg, A.A., 2001. Copper-and arsenate-induced oxidative stress in *Holcus lanatus* L. clones with differential sensitivity. Plant Cell Environ. 24 (7), 713–722.

Hall, J.L., 2002. Cellular mechanisms for heavy metal detoxification and tolerance. J. Exp. Bot. 53 (366), 1–11.

Hameed, A., Qadri, T.N., Siddiqi, T.O., Iqbal, M., 2011. Differential activation of the enzymatic antioxidant system of *Abelmoschus esculentus* L. under $CdCl_2$ and $HgCl_2$ exposure. Braz. J. Plant Physiol. 23 (1), 46–54.

Hayashi, Y., Nakagawa, C.W., Mutoh, N., Isobe, M., Goto, T., 1991. Two pathways in the biosynthesis of cadystins (γEC) n G in the cell-free system of the fission yeast. Biochem. Cell Biol. 69 (2–3), 115–121.

Herschbach, C., Rennenberg, H., 1994. Influence of glutathione (GSH) on net uptake of sulphate and sulphate transport in tobacco plants. J. Exp. Bot. 45 (8), 1069–1076.

Hirata, K., Tsujimoto, Y., Namba, T., Ohta, T., Hirayanagi, N., Miyasaka, H., Zenk, M.H., Miyamoto, K., 2001. Strong induction of phytochelatin synthesis by Zn in marine green alga, *Dunaliella tertiolecta*. J. Biosci. Bioeng. 92, 24–29.

Hirata, K., Tsuji, N., Miyamoto, K., 2005. Biosynthetic regulation of phytochelatins, heavy metal-binding peptides. J. Biosci. Bioeng. 100 (6), 593–599.

Hirayama, T., Shinozaki, K., 2010. Research on plant abiotic stress responses in the post-genome era: past, present and future. Plant J. 61 (6), 1041–1052.
Hossain, M.A., Fujita, M., 2009. Purification of glyoxalase I from onion bulbs and molecular cloning of its cDNA. Biosci. Biotech. Biochem. 73 (9), 2007–2013.
Hossain, M.A., Fujita, M., 2011. Regulatory role of components of ascorbate-glutathione (AsA-GSH) pathway in plant tolerance to oxidative stress. In: Anjum, N.A., Umar, S., Ahmed, A. (Eds.), Oxidative Stress in Plants: Causes, Consequences and Tolerance. IK International Publishing House Pvt. Ltd., New Delhi, India, 1, 1–53.
Hossain, M.A., Hossain, M.Z., Fujita, M., 2009. Stress-induced changes of methylglyoxal level and glyoxalase I activity in pumpkin seedlings and cDNA cloning of glyoxalase I gene. Aust. J. Crop Sci. 3 (2), 53–64.
Hossain, M.A., da Silva, J.A.T., Fujita, M., 2011. Glyoxalase system and reactive oxygen species detoxification system in plant abiotic stress response and tolerance: an intimate relationship. In: Agricultural and Biological Sciences Abiotic Stress in Plants - Mechanisms and Adaptations. http://dx.doi.org/10.5772/22421.
Hossain, M.A., Hossain, M.D., Rohman, M.M., Teixeira da Silva, J.A., Fujita, M., 2012. Onion Major Compounds (*Flavonoids, Organosulfurs*) and Highly Expressed Glutathione-Related Enzymes: Possible Physiological Interaction, Gene Cloning and Abiotic Stress Response. Onion Consumption and Health. Nova Science Publishers Inc., NY, USA, 49–90.
Hossain, M.A., Piyatida, P., da Silva, J.A.T., Fujita, M., 2012. Molecular mechanism of heavy metal toxicity and tolerance in plants: central role of glutathione in detoxification of reactive oxygen species and methylglyoxal and in heavy metal chelation. J. Bot. http://dx.doi.org/10.1155/2012/872875.
Howden, R., Goldsbrough, P.B., Andersen, C.R., Cobbett, C.S., 1995a. Cadmium-sensitive, cad1 mutants of *Arabidopsis thaliana* are phytochelatin deficient. Plant Physiol. 107 (4), 1059–1066.
Howden, R., Andersen, C.R., Goldsbrough, P.B., Cobbett, C.S., 1995b. A cadmium-sensitive, glutathione-deficient mutant of *Arabidopsis thaliana*. Plant Physiol. 107 (4), 1067–1073.
Iglesia-Turino, S., Febrero, A., Jauregui, O., Caldelas, C., Araus, J.L., Bort, J., 2006. Detection and quantification of unbound phytochelatin2 in plant extracts of *Brassica napus* grown with different levels of mercury. Plant Physiol. 142, 742–749.
Inouhe, M., Ninomiya, S., Tohoyama, H., Joho, M., Murayama, T., 1994. Different characteristics of roots in the cadmium-tolerance and Cd-binding complex formation between mono- and dicotyledonous plants. J. Plant Res. 107, 201–207.
Inouhe, M., 2005. Phytochelatins. Braz. J. Plant Physiol. 17 (1), 65–78.
Jewell, M.C., Campbell, B.C., Godwin, I.D., 2010. Transgenic plants for abiotic stress resistance. In: Kole, C., Michler, C.H., Abbott, A.G., Hall, T.C. (Eds.), Transgenic Crop Plants. Springer, Heidelberg, pp. 67–132.
Jozefczak, M., Remans, T., Vangronsveld, J., Cuypers, A., 2012. Glutathione is a key player in metal-induced oxidative stress defenses. Int. J. Mol. Sci. 13 (3), 3145–3175.
Jozefczak, M., Keunen, E., Schat, H., Bliek, M., Hernández, L.E., Carleer, R., Remans, T., Bohler, S., Vangronsveld, J., Cuypers, A., 2014. Differential response of Arabidopsis leaves and roots to cadmium: glutathione-related chelating capacity vs antioxidant capacity. Plant Physiol. Biochem. 83, 1–9.
Khan, M.S., Zaidi, A., Wani, P.A., Oves, M., 2009. Role of plant growth promoting rhizobacteria in the remediation of metal contaminated soils. Environ. Chem. Lett. 7 (1), 1–19.
Klapheck, S., Schlunz, S., Bergmann, L., 1995. Synthesis of phytochelatins and homo-phytochelatins in *Pisum sativum* L. Plant Physiol. 107 (2), 515–521.
Klein, M., Burla, B., Martinoia, E., 2006. The multidrug resistance-associated protein (MRP/ABCC) subfamily of ATP-binding cassette transporters in plants. FEBS Lett. 580 (4), 1112–1122.
de Knecht, J.A., van Baren, N., Ten Bookum, W.M., Sang, H.W.W.F., Koevoets, P.L., Schat, H., Verkleij, J.A., 1995. Synthesis and degradation of phytochelatins in cadmium-sensitive and cadmium-tolerant *Silene vulgaris*. Plant Sci. 106 (1), 9–18.

Kondo, N., Imai, K., Isobe, M., Goto, T., Murasugi, A., Wada-Nakagawa, C., Hayashi, Y., 1984. Cadystin a and b, major unit peptides comprising cadmium binding peptides induced in a fission yeast–separation, revision of structures and synthesis. Tetrahedron Lett. 25 (35), 3869–3872.

Koprivova, A., Kopriva, S., Jager, D., Will, B., Jouanin, L., Rennenberg, H., 2002. Evaluation of transgenic poplars overexpressing enzymes of glutathione synthesis for phytoremediation of cadmium. Plant Biol. 4 (6), 664–670.

Kumar, D., Singh, V.P., Tripathi, D.K., Prasad, S.M., Chauhan, D.K., 2013a. Effect of arsenic on growth, arsenic uptake, distribution of nutrient elements and thiols in seedlings of *Wrightia arborea* (Dennst.) Mabb. Int. J. Phytoremediat 17 (2), 128–134. http://dx.doi.org/10.1080/15226514.2013.862205.

Kumar, R.R., Goswami, S., Singh, K., Rai, G.K., Rai, R.D., 2013b. Modulation of redox signal transduction in plant system through induction of free radical/ROS scavenging redox-sensitive enzymes and metabolites. Aust. J. Crop Sci. 7 (11), 1744.

Kumar, D., Tripathi, D.K., Chauhan, D.K., 2014a. Phytoremediation potential and nutrient status of *Barringtonia acutangula* Gaerth. Tree seedlings grown under different chromium (CrVI) treatments. Biol. Trace Elem. Res. 157 (2), 164–174.

Kumar, R., Tripathi, D.K., Devanathan, A., Chauhan, D.K., Rai, A.K., 2014b. In-Situ monitoring of chromium uptake in different parts of the wheat seedling (*Triticum aestivum*) using laser-induced breakdown spectroscopy. Spectrosc. Lett. 47 (7), 554–563.

Lappartient, A.G., Touraine, B., 1997. Glutathione-mediated regulation of ATP sulfurylase activity, SO42-uptake, and oxidative stress response in intact canola roots. Plant Physiol. 114 (1), 177–183.

Lin, Y.F., Aarts, M.G., 2012. The molecular mechanism of zinc and cadmium stress response in plants. Cell Mol. Life Sci. 69 (19), 3187–3206.

Lombardi, L., Sebastiani, L., 2005. Copper toxicity in *Prunus cerasifera*: growth and antioxidant enzymes responses of in vitro grown plants. Plant Sci. 168 (3), 797–802.

Lushchak, V.I., 2012. Glutathione homeostasis and functions: potential targets for medical interventions. J. Amino Acid 1–26. http://dx.doi.org/10.1155/2012/736837.

Maiti, S., Ghosh, N., Mandal, C., Das, K., Dey, N., Adak, M.K., 2012. Responses of the maize plant to chromium stress with reference to antioxidation activity. Braz. J. Plant Physiol. 24 (3), 203–212.

Maitani, T., Kubota, H., Sato, K., Yamada, T., 1996. The composition of metals bound to class III metallothionein (phytochelatin and its desglycyl peptide) induced by various metals in root cultures of *Rubia tinctorum*. Plant Physiol. 110 (4), 1145–1150.

Marrs, K.A., 1996. The functions and regulation of glutathione S-transferases in plants. Annu. Rev. Plant Biol. 47 (1), 127–158.

May, M.J., Vernoux, T., Leaver, C., Van Montagu, M., Inzé, D., 1998a. Glutathione homeostasis in plants: implications for environmental sensing and plant development. J. Exp. Bot. 49 (321), 649–667.

May, M.J., Vernoux, T., Sánchez-Fernández, R., Van Montagu, M., Inzé, D., 1998b. Evidence for post-transcriptional activation of g-glutamylcysteine synthetase during plant stress responses. Proc. Natl. Acad. Sci. U.S.A. 95, 12049–12054.

McLaughlin, M.J., Parker, D.R., Clarke, J.M., 1999. Metals and micronutrients–food safety issues. Field Crop Res. 60 (1), 143–163.

Meharg, A.A., 1994. Integrated tolerance mechanisms: constitutive and adaptive plant responses to elevated metal concentrations in the environment. Plant Cell Environ. 17 (9), 989–993.

Mhamdi, A., Queval, G., Chaouch, S., Vanderauwera, S., Van Breusegem, F., Noctor, G., 2010. Catalase function in plants: a focus on Arabidopsis mutants as stress-mimic models. J. Exp. Bot. 61 (15), 4197–4220.

Millar, A.H., Mittova, V., Kiddle, G., Heazlewood, J.L., Bartoli, C.G., Theodoulou, F.L., Foyer, C.H., 2003. Control of ascorbate synthesis by respiration and its implications for stress responses. Plant Physiol. 133 (2), 443–447.

Mishra, S., Srivastava, S., Tripathi, R.D., Kumar, R., Seth, C.S., Gupta, D.K., 2006. Lead detoxification by coontail (*Ceratophyllum demersum* L.) involves induction of phytochelatins and antioxidant system in response to its accumulation. Chemosphere 65 (6), 1027–1039.

Mithöfer, A., Schulze, B., Boland, W., 2004. Biotic and heavy metal stress response in plants: evidence for common signals. FEBS Lett. 566 (1), 1–5.

Mittler, R., Vanderauwera, S., Gollery, M., Van Breusegem, F., 2004. Reactive oxygen gene network of plants. Trends Plant Sci. 9 (10), 490–498.

Mokgalaka-Matlala, N.S., Flores-Tavizón, E., Castillo-Michel, H., Peralta-Videa, J.R., Gardea-Torresdey, J.L., 2009. Arsenic tolerance in mesquite (Prosopis sp.): low molecular weight thiols synthesis and glutathione activity in response to arsenic. Plant Physiol. Biochem. 47, 822–826.

Mostofa, M.G., Seraj, Z.I., Fujita, M., 2014. Exogenous sodium nitroprusside and glutathione alleviate copper toxicity by reducing copper uptake and oxidative damage in rice (*Oryza sativa L.*) seedlings. Protoplasma 251 (6), 1373–1386.

Muamar, A.J., Zouahri, A., Tijane, M.H., El Housni, A., Mennane, Z., Yachou, H., Bouksaim, M., 2014. Evaluation of heavy metals pollution in groundwater, soil and some vegetables irrigated with wastewater in the Skhirat region "Morocco". J. Mater. Environ. Sci. 5 (3), 961–966.

Mullineaux, P.M., Rausch, T., 2005. Glutathione, photosynthesis and the redox regulation of stress-responsive gene expression. Photosynth. Res. 86 (3), 459–474.

Nagajyoti, P.C., Lee, K.D., Sreekanth, T.V.M., 2010. Heavy metals, occurrence and toxicity for plants: a review. Environ. Chem. Lett. 8 (3), 199–216.

Nesterkova, D.V., Vorobeichik, E.L., Reznichenko, I.S., 2014. The effect of heavy metals on the soil-earthworm-European mole food chain under the conditions of environmental pollution caused by the emissions of a copper smelting plant. Contemp. Probl. Ecol. 7 (5), 587–596.

Nocito, F.F., Lancilli, C., Crema, B., Fourcroy, P., Davidian, J.C., Sacchi, G.A., 2006. Heavy metal stress and sulfate uptake in maize roots. Plant Physiol. 141 (3), 1138–1148.

Noctor, G., Foyer, C.H., 1998. Ascorbate and glutathione: keeping active oxygen under control. Annu. Rev. Plant Biol. 49 (1), 249–279.

Noctor, G., Mhamdi, A., Chaouch, S., Han, Y.I., Neukermans, J., Marquez-Garcia, B.E.L.E.N., Queval, G., Foyer, C.H., 2012. Glutathione in plants: an integrated overview. Plant Cell Environ. 35 (2), 454–484.

Noctor, G., Queval, G., Mhamdi, A., Chaouch, S., Foyer, C.H., 2011. Glutathione. The Arabidopsis Book/ American Society of Plant Biologists, 9.

Nussbaum, S., Schmutz, D., Brunold, C., 1988. Regulation of assimilatory sulfate reduction by cadmium in *Zea mays* L. Plant Physiol. 88 (4), 1407–1410.

Ogawa, K.I., Iwabuchi, M., 2001. A mechanism for promoting the germination of Zinnia elegans seeds by hydrogen peroxide. Plant Cell Physiol. 42 (3), 286–291.

Ogawa, K.I., Hatano-Iwasaki, A., Yanagida, M., Iwabuchi, M., 2004. Level of glutathione is regulated by ATP-dependent ligation of glutamate and cysteine through photosynthesis in *Arabidopsis thaliana*: mechanism of strong interaction of light intensity with flowering. Plant Cell Physiol. 45 (1), 1–8.

Ortega-Villasante, C., Rellán-Álvarez, R., del Campo, F.F., Carpena-Ruiz, R.O., Hernández, L.E., 2005. Cellular damage induced by cadmium and mercury in *Medicago sativa*. J. Exp. Bot. 56, 2239–2251.

Ovečka, M., Takáč, T., 2014. Managing heavy metal toxicity stress in plants: biological and biotechnological tools. Biotechnol. Adv. 32 (1), 73–86.

Pal, S.H., Patra, A.K., Reza, shk., Wildi, W., Poté, J., 2010. Use of bio-resources for remediation of soil pollution. Nat. Resour. 1, 110–125. http://dx.doi.org/10.4236/nr.2010.12012.

Pasternak, M., Lim, B., Wirtz, M., Hell, R., Cobbett, C.S., Meyer, A.J., 2008. Restricting glutathione biosynthesis to the cytosol is sufficient for normal plant development. Plant J. 53 (6), 999–1012.

Passariello, B., Giuliano, V., Quaresima, S., Barbaro, M., Caroli, S., Forte, G., Carelli, G., Iavicoli, I., 2002. Evaluation of the environmental contamination at an abandoned mining site. Microchem. J. 73 (1), 245–250.

Patra, M., Bhowmik, N., Bandopadhyay, B., Sharma, A., 2004. Comparison of mercury, lead and arsenic with respect to genotoxic effects on plant systems and the development of genetic tolerance. Environ. Exp. Bot. 52 (3), 199–223.

Pattanayak, B., Padhi, S., Dhal, N.K., 2014. Genetic engineering to express metal binding proteins and peptides: implications for bioremediation. Biolife 2 (2), 442–451.

Pietrini, F., Iannelli, M.A., Pasqualini, S., Massacci, A., 2003. Interaction of cadmium with glutathione and photosynthesis in developing leaves and chloroplasts of *Phragmites australis* (Cav.) Trin. ex Steudel. Plant Physiol. 133 (2), 829–837.

Qiu, B., Zeng, F., Cai, S., Wu, X., Haider, S.I., Wu, F., Zhang, G., 2013. Alleviation of chromium toxicity in rice seedlings by applying exogenous glutathione. J. Plant Physiol. 170 (8), 772–779.

Raab, A., Feldmann, J., Meharg, A.A., 2004. The nature of arsenic-phytochelatin complexes in *Holcus lanatus* and *Pteris cretica*. Plant Physiol. 134 (3), 1113–1122.

Rajkumar, M., Vara Prasad, M.N., Freitas, H., Ae, N., 2009. Biotechnological applications of serpentine soil bacteria for phytoremediation of trace metals. CRC Cr Rev. Biotechn 29 (2), 120–130.

Ramesh, G., Reddy, M.S., 2008. Cloning and Characterization of Metallothionein Genes of *Ectomycorrhizal* Fungus *Hebeloma Cylindrosporum*. http://hdl.handle.net/10266/767.

Rausch, T., Gromes, R., Liedschulte, V., Müller, I., Bogs, J., Galovic, V., Wachter, A., 2007. Novel insight into the regulation of GSH biosynthesis in higher plants. Plant Biol. 9 (5), 565–572.

Rauser, W.E., 1990. Phytochelatins. Annu. Rev. Biochem. 59 (1), 61–86.

Rauser, W.E., Meuwly, P., 1995. Retention of cadmium in roots of maize seedlings (role of complexation by phytochelatins and related thiol peptides). Plant Physiol. 109 (1), 195–202.

Rennenberg, H., 1995. Processes involved in glutathione metabolism. In: Wallsgrove, R.M. (Ed.), Amino acids and their derivatives in higher plants. UK. Cambridge University Press, pp. 155–171.

Robinson, N.J., 1989. Metal-binding polypeptides in plants. Heavy metal tolerance higher plants. Evol. Aspen 195–214.

Ruegsegger, A., Brunold, C., 1992. Effect of cadmium on y-glutamylcysteine synthesis in maize seedlings. Plant Physiol. 99, 428–433.

Sandalio, L.M., Dalurzo, H.C., Gomez, M., Romero-Puertas, M.C., Del Rio, L.A., 2001. Cadmium induced changes in the growth and oxidative metabolism of pea plants. J. Exp. Bot. 52 (364), 2115–2126.

Schat, H., Van Nathalie, A.H., Tervahauta, A., Hakvoort, H.W., Chardonnens, A.N., Koevoets, P.L., Verkleij Jos, A.C., Ernst, W.H., 2000. Evolutionary responses to zinc and copper stress in Bladder Campion, Silene vulgaris (Moench.) Garcke. In: Plant Tolerance to Abiotic Stresses in Agriculture: Role of Genetic Engineering. Springer, Netherlands, pp. 343–360.

Schat, H., Llugany, M., VOOiJS, R., Hartley-Whitaker, J., Bleeker, P.M., 2002. The role of phytochelatins in constitutive and adaptive heavy metal tolerances in hyperaccumulator and non-hyperaccumulator metallophytes. J. Exp. Bot. 53 (379), 2381–2392.

Schickler, H., Caspi, H., 1999. Response of antioxidative enzymes to nickel and cadmium stress in hyper-accumulator plants of the genus Alyssum. Physiol. Plantarum 105, 39–44.

Schützendübel, A., Polle, A., 2002. Plant responses to abiotic stresses: heavy metal-induced oxidative stress and protection by mycorrhization. J. Exp. Bot. 53 (372), 1351–1365.

Schwartz, C., Gérard, E., Perronnet, K., Morel, J.L., 2001. Measurement of in situ phytoextraction of zinc by spontaneous metallophytes growing on a former smelter site. Sci. Total Environ. 279 (1), 215–221.

Schmöger, M., Oven, M., Grill, E., 2000. Detoxification of arsenic by phytochelatins in plants. Plant Physiol. 122, 793–801.

Seth, C.S., Remans, T., Keunen, E., Jozefczak, M., Gielen, H., Opdenakker, K., Weyens, N., Vangronsveld, J., Cuypers, A., 2012. Phytoextraction of toxic metals: a central role for glutathione. Plant Cell Environ. 35 (2), 334–346.

Shah, F.U.R., Ahmad, N., Masood, K.R., Peralta-Videa, J.R., 2010. Heavy metal toxicity in plants. In: Plant Adaptation and Phytoremediation. Springer, Netherlands, pp. 71–97.

Shao, H.B., Chu, L.Y., Lu, Z.H., Kang, C.M., 2008. Primary antioxidant free radical scavenging and redox signaling pathways in higher plant cells. Int. J. Biol. Sci. 4 (1), 8.

Sharma, S.S., Dietz, K.J., 2009. The relationship between metal toxicity and cellular redox imbalance. Trends Plant Sci. 14 (1), 43–50.

Sharma, P., Dubey, R.S., 2007. Involvement of oxidative stress and role of antioxidative defense system in growing rice seedlings exposed to toxic concentrations of aluminum. Plant Cell Rep. 26 (11), 2027–2038.

Singh, V.P., Tripathi, D.K., Kumar, D., Chauhan, D.K., 2011. Influence of exogenous silicon addition on aluminium tolerance in rice seedlings. Biol. Trace Elem. Res. 144 (1–3), 1260–1274.

Singh, D., Chauhan, S.K., 2011. Organic acids of crop plants in aluminium detoxification. Curr. Sci. 100 (10), 1509–1515.

Singh, S., Srivastava, P.K., Kumar, D., Tripathi, D.K., Chauhan, D.K., Prasad, S.M., 2015. Morpho-anatomical and biochemical adapting strategies of maize (*Zea mays* L.) seedlings against lead and chromium stresses. Biocatal. Agric. Biotech. 4, 286–295. http://dx.doi.org/10.1016/j.bcab.2015.03.004.

Singla-Pareek, S.L., Yadav, S.K., Pareek, A., Reddy, M.K., Sopory, S.K., 2006. Transgenic tobacco over expressing glyoxalase pathway enzymes grow and set viable seeds in zinc-spiked soils. Plant Physiol. 140 (2), 613–623.

Sobrino-Plata, J., Carrasco-Gil, S., Abadía, J., Escobar, C., Álvarez-Fernández, A., Hernández, L.E., 2014a. The role of glutathione in mercury tolerance resembles its function under cadmium stress in Arabidopsis. Metallomics 6 (2), 356–366.

Sobrino-Plata, J., Meyssen, D., Cuypers, A., Escobar, C., Hernández, L.E., 2014b. Glutathione is a key antioxidant metabolite to cope with mercury and cadmium stress. Plant Soil 377 (1–2), 369–381.

Son, J.A., Narayanankutty, D.P., Roh, K.S., 2014. Influence of exogenous application of glutathione on rubisco and rubisco activase in heavy metal-stressed tobacco plant grown in vitro. Saudi J. Biolog Sci. 21 (1), 89–97.

Steffens, J.C., Hunt, D.F., Williams, B.G., 1986. Accumulation of non-protein metal-binding polypeptides (gamma-glutamyl-cysteinyl) n-glycine in selected cadmium-resistant tomato cells. J. Biol. Chem. 261 (30), 13879–13882.

Szalai, G., Krantev, A., Yordanova, R., Popova, L.P., Janda, T., 2013. Influence of salicylic acid on phytochelatin synthesis in *Zea mays* during Cd stress. Turk J. Bot. 37 (4), 708–714.

Talukdar, D., Talukdar, T., 2014. Coordinated response of sulfate transport, cysteine biosynthesis, and glutathione-mediated antioxidant defense in lentil (*Lens culinaris* Medik.) genotypes exposed to arsenic. Protoplasma 251 (4), 839–855.

Tausz, M., Šircelj, H., Grill, D., 2004. The glutathione system as a stress marker in plant ecophysiology: Is a stress-response concept valid? J. Exp. Bot. 55 (404), 1955–1962.

Thumann, J., Grill, E., Winnacker, E.L., Zenk, M.H., 1991. Reactivation of metal-requiring apoenzymes by phytochelatin-metal complexes. FEBS Lett. 284 (1), 66–69.

Tripathi, A., Misra, D.R., 2012a. A study of physico-chemical properties and heavy metals in contaminated soils of municipal waste dumpsites at Allahabad, India. Int. J. Environ. Sci. 2 (4), 2024–2033.

Tripathi, A., Misra, D.R., 2012b. Contaminated sewage sludge disposal as a source of heavy metal bioavailability to native plants. Natl. Acad. Sci. Lett. 35 (2), 121–124.

Tripathi, A., Misra, D.R., 2013. Bioaccumulation of Pb, Ni and Zn in some plant species growing in and around a municipal waste dumpsite at Allahabad, India. J. Solid Waste Technol. Manage. 39 (1), 1–12.

Tripathi, D.K., Singh, V.P., Kumar, D., Chauhan, D.K., 2012a. Impact of exogenous silicon addition on chromium uptake, growth, mineral elements, oxidative stress, antioxidant capacity, and leaf and root structures in rice seedlings exposed to hexavalent chromium. Acta Physiol. Plant 34 (1), 279–289.

Tripathi, D.K., Singh, V.P., Kumar, D., Chauhan, D.K., 2012b. Rice seedlings under cadmium stress: effect of silicon on growth, cadmium uptake, oxidative stress, antioxidant capacity and root and leaf structures. Chem. Ecol. 28 (3), 281–291.

Tripathi, D.K., Singh, V.P., Chauhan, D.K., Prasad, S.M., Dubey, N.K., 2014a. Role of macronutrients in plant growth and acclimation: recent advances and future prospective. In: Improvement of Crops in the Era of Climatic Changes. Springer, New York, pp. 197–216.

Tripathi, D.K., Singh, V.P., Gangwar, S., Prasad, S.M., Maurya, J.N., Chauhan, D.K., 2014b. Role of silicon in enrichment of plant nutrients and Protection from biotic and abiotic stresses. In: Improvement of Crops in the Era of Climatic Changes. Springer, New York, pp. 39–56.

Tripathi, D.K., Singh, V.P., Prasad, S.M., Chauhan, D.K., Dubey, N.K., Rai, A.K., 2015. Silicon-mediated alleviation of Cr (VI) toxicity in wheat seedlings as evidenced by chlorophyll florescence, laser induced breakdown spectroscopy and anatomical changes. Ecotoxicol. Environ. Saf. 113, 133–144.

Tsuji, N., Hirayanagi, N., Okada, M., Miyasaka, H., Hirata, K., Zenk, M.H., Miyamoto, K., 2002. Enhancement of tolerance to heavy metals and oxidative stress in *Dunaliella tertiolecta* by Zn-induced phytochelatin synthesis. Biochem. Bioph Res. Co. 293 (1), 653–659.

Vernoux, T., Wilson, R.C., Seeley, K.A., Reichheld, J.P., Muroy, S., Brown, S., Maughand, S.C., Cobbettd, C.S., Montagua, M.V., Inzéa, D., May, M.J., Sung, Z.R., 2000. The root meristemless1/cadmium sensitive2 gene defines a glutathione-dependent pathway involved in initiation and maintenance of cell division during postembryonic root development. Plant Cell 12 (1), 97–109.

De Vos, C.R., Vonk, M.J., Vooijs, R., Schat, H., 1992. Glutathione depletion due to copper-induced phytochelatin synthesis causes oxidative stress in *Silene cucubalus*. Plant Physiol. 98 (3), 853–858.

Viehweger, K., 2014. How plants cope with heavy metals. Bot. Stud. 55 (35), 1–12.

Villiers, F., Ducruix, C., Hugouvieux, V., Jarno, N., Ezan, E., Garin, J., Junot, C., Bourguignon, J., 2011. Investigating the plant response to cadmium exposure by proteomic and metabolomic approaches. Proteomics 11 (9), 1650–1663.

Waseem, A., Arshad, J., Iqbal, F., Sajjad, A., Mehmood, Z., Murtaza, G., 2014. Pollution status of Pakistan: a retrospective review on heavy metal contamination of water, soil, and vegetables. Biomed. Res. Int. 1–29. Article ID 813206. http://dx.doi.org/10.1155/2014/813206.

Wang, S.H., Yang, Z.M., Yang, H., Lu, B., Li, S.Q., Lu, Y.P., 2004. Copper-induced stress and antioxidative responses in roots of *Brassica juncea* L. Bot. Bull. Acad. Sinica 45.

Wingate, V.P., Lawton, M.A., Lamb, C.J., 1988. Glutathione causes a massive and selective induction of plant defense genes. Plant Physiol. 87 (1), 206–210.

Wojcik, M., Tukiendorf, A., 2004. Phytochelatin synthesis and cadmium localization in wild type of *Arabidopsis thaliana*. Plant Growth Regul. 44 (1), 71–80.

Wójcik, M., Tukiendorf, A., 2014. Accumulation and tolerance of lead in two contrasting ecotypes of Dianthus carthusianorum. Phytochemistry 100, 60–65.

Wojtaszek, P., 1997. Oxidative burst: an early plant response to pathogen infection. Biochem. J. 322, 681–692.

Wong, S.C., Li, X.D., Zhang, G., Qi, S.H., Min, Y.H., 2002. Heavy metals in agricultural soils of the Pearl River Delta. South China Environ. Poll 119, 33–44.

Wong, M.H., 2003. Ecological restoration of mine degraded soils, with emphasis on metal contaminated soils. Chemosphere 50 (6), 775–780.

Xiang, C., Werner, B.L., E'Lise, M.C., Oliver, D.J., 2001. The biological functions of glutathione revisited in Arabidopsis transgenic plants with altered glutathione levels. Plant Physiol. 126 (2), 564–574.

Xu, P., Liu, L., Zeng, G., Huang, D., Lai, C., Zhao, M., Huang, C., Li, N., Wei, Z., Wu, H., Zhang, C., Lai, M., He, Y., 2014. Heavy metal-induced glutathione accumulation and its role in heavy metal detoxification in Phanerochaete chrysosporium. Appl. Microbiol. Biot. 98 (14), 6409–6418.

Yadav, S.K., Singla-Pareek, S.L., Reddy, M.K., Sopory, S.K., 2005. Transgenic tobacco plants overexpressing glyoxalase enzymes resist an increase in methylglyoxal and maintain higher reduced glutathione levels under salinity stress. FEBS Lett. 579 (27), 6265–6271.

Yadav, S.K., 2010. Heavy metals toxicity in plants: an overview on the role of glutathione and phytochelatins in heavy metal stress tolerance of plants. S. Afr. J. Bot. 76 (2), 167–179.

Yazaki, K., 2006. ABC transporters involved in the transport of plant secondary metabolites. FEBS Lett. 580, 1183–1191.

Zagorchev, L., Seal, C.E., Kranner, I., Odjakova, M., 2012. Redox state of low-molecular-weight thiols and disulphides during somatic embryogenesis of salt-treated suspension cultures of *Dactylis glomerata* L. Free Radic. Res. 46 (5), 656–664.

Zagorchev, L., Seal, C.E., Kranner, I., Odjakova, M., 2013. A central role for thiols in plant tolerance to abiotic stress. Int. J. Mol. Sci. 14 (4), 7405–7432.

Zhao, F.J., Wang, J.R., Barker, J.H.A., Schat, H., Bleeker, P.M., McGrath, S.P., 2003. The role of phytochelatins in arsenic tolerance in the hyperaccumulator *Pteris vittata*. New Phytol. 159 (2), 403–410.

Zhao, F.Y., Liu, W., Zhang, S.Y., 2009. Different responses of plant growth and antioxidant system to the combination of cadmium and heat stress in transgenic and non-transgenic rice. J. Integr. Plant Biol. 51, 942–950.

Zechmann, B., 2014. Compartment-specific importance of glutathione during abiotic and biotic stress. Front. Plant Sci. 566. http://dx.doi.org/10.3389/fpls.2014.00566.

Zenk, M.H., 1996. Heavy metal detoxification in higher plants-a review. Gene 179 (1), 21–30.

Zhang, W.H., Tyerman, S.D., 1999. Inhibition of water channels by $HgCl_2$ in intact wheat root cells. Plant Physiol. 120 (3), 849–858.

Zhu, Y.L., Pilon-Smits, E.A., Jouanin, L., Terry, N., 1999a. Over expression of glutathione synthetase in Indian mustard enhances cadmium accumulation and tolerance. Plant Physiol. 119 (1), 73–80.

Zhu, Y.L., Pilon-Smits, E.A., Tarun, A.S., Weber, S.U., Jouanin, L., Terry, N., 1999b. Cadmium tolerance and accumulation in Indian mustard is enhanced by over expressing γ-glutamylcysteine synthetase. Plant Physiol. 121 (4), 1169–1177.

FURTHER READING

Braconi, D., Bernardini, G., Santucci, A., 2011. Linking protein oxidation to environmental pollutants: redox proteomic approaches. J. Proteomics 74 (11), 2324–2337.

Campion S. Vulgaris (Moench) Garcke. In: Cherry J.H., Locy R.D., Richter A., (Eds.), Plant Tolerance to Abiotic Stresses: Role of Genetic Engineering. Dordrecht, The Netherlands.

Clemens, S., Kim, E.J., Neumann, D., Schroeder, J.I., 1999. Tolerance to toxic metals by a gene family of phytochelatin synthases from plants and yeast. EMBO J. 18 (12), 3325–3333.

Cobbett, C.S., May, M.J., Howden, R., Rolls, B., 1998. The glutathione-deficient, cadmium-sensitive mutant, cad2-1, of Arabidopsis thaliana deficient in g-glutamyl cysteine synthetase. Plant J. 16, 73–78.

Foyer, C.H., Noctor, G., 2011. Ascorbate and glutathione: the heart of the redox hub. Plant Physiol. 155 (1), 2–18.

Hajar, E.W.I., Sulaiman, A.Z.B., Sakinah, A.M., 2014. Assessment of heavy metals tolerance in leaves, stems and flowers of *Stevia Rebaudiana* plant. Plant Environ. Sci. 20, 386–393.

HongBo, S., ZongSuo, L., MingAn, S., ShiMeng, S., ZanMin, H., 2005. Investigation on dynamic changes of photosynthetic characteristics of 10 wheat (*Triticum aestivum* L.) genotypes during two vegetative-growth stages at water deficits. Colloid Surf. B 43 (3), 221–227.

Hossain, M.A., Hasanuzzaman, M., Fujita, M., 2010. Up-regulation of antioxidant and glyoxalase systems by exogenous glycinebetaine and proline in mung bean confer tolerance to cadmium stress. Physiol. Mol. Biol. Plants 16 (3), 259–272.

Margoshes, M., Vallee, B.L., 1957. A cadmium protein from equine kidney cortex. J. Am. Chem. Soc. 79 (17), 4813–4814.

Rausch, T., Wachter, A., 2005. Sulfur metabolism: a versatile platform for launching defence operations. Trends Plant Sci. 10 (10), 503–509.

Rauser, W.E., 1995. Phytochelatins and related peptides: structure, biosynthesis, and function. Plant Physiol. 109, 1141–1149.

Rauser, W.E., 1999. Structure and function of metal chelators produced by plants. Cell Biochem. Biophys. 31 (1), 19–48.

Rauser, W.E., 2003. Phytochelatin-based complexes bind various amounts of cadmium in maize seedlings depending on the time of exposure, the concentration of cadmium and the tissue. New Phytol. 158 (2), 269–278.

CHAPTER

BIOLOGICAL REMEDIATION OF MERCURY-POLLUTED ENVIRONMENTS

12

Mercedes García-Sánchez, Jiřina Száková

Department of Agro-Environmental Chemistry and Plant Nutrition, Faculty of Agrobiology, Food and Natural Resources, Kamýcká, Prague, Czech Republic

CHAPTER OUTLINE

1. INTRODUCTION

Pollution refers to the state of existence of undesirable contaminants in the environment, beyond a permissible limit, that can dangerously affect every sphere of life. Soil can be influenced by thousands of pollutants, especially heavy metals that vary in composition and concentration. Usually, soil pollution is caused by both natural and anthropogenic sources. Examples of natural sources include geothermal activities, comets, space dust, and volcanic activities. On the other hand, anthropogenic contamination is mainly due to rapid industrialization processes in combination with the intensive use of chemical substances such as hydrocarbons, pesticides, chlorinated hydrocarbons, and heavy metals (Gaur et al., 2014). Heavy metals such as lead (Pb), cadmium (Cd), arsenic (As), and mercury (Hg) are the major soil pollutants that bring about heavy metal toxicity (Vandenbosschea et al., 2014). These metals are nonbiodegradable, which means their presence in the environment is prolonged. With time, these contaminants make their way into the food chain and eventually accumulate in the human body, which can cause irreversible health problems in the population. Recent findings of bioaccumulation in

Plant Metal Interaction. http://dx.doi.org/10.1016/B978-0-12-803158-2.00012-6

potato, wheat, and rice crops have underscored the urgency of regulatory compliance worldwide. Globally, farmers and food producers must ensure the content of heavy metals in foodstuffs meets regulatory standards and compares well with products from other countries (Khan et al., 2013). For these reasons, bioremediation of polluted soils and water, especially for those used in agriculture, is in demand. In addition to the risks for humans, heavy metals are known to facilitate phytotoxicity in contaminated soil; the diversity, structure, and size of microbial communities may be considerably affected, whereas environmental and biological activities such as organic matter mineralization and leaf litter decomposition are reduced (Griffiths and Philippot, 2013). The level of soil contamination may depend on many factors, including chemical composition, toxicity, and/or the bioavailability of the heavy metals in question. Heavy metals at the soil surface are readily adsorbed. Then follows a slow desorption and subsequent distribution in the soil (Selinus and Alloway, 2013).

Conventional techniques based on physiochemical approaches, such as ion exchange, chemical precipitation, reverse osmosis, biopiles, bioslurries, incineration, soil vapor extraction, and land-filling are used for remediation of soil and water contaminated via anthropogenic activities (Dhal et al., 2013). However, these methods are neither cost-effective nor efficient because they require expensive, sophisticated infrastructure. The individual decontamination techniques that have been conventionally used suffer from recognizable drawbacks; these techniques may involve some level of risk. For example, these methods have been shown to generate toxic sludge, which itself affects the environment and may not be able to remove the metals completely (Gaur et al., 2014).

The emerging science of bioremediation and the technology used in it offer an alternative remediation method for sites contaminated with heavy metals. Bioremediation is an eco-friendly and sustainable process that can prove to be much more efficient and effective in eliminating heavy metals from the environment. Bioremediation is used as an effective means of mitigation of several pollutants such as hydrocarbons, halogenated organic solvents and organic compounds, nonchlorinated pesticides and herbicides, nitrogen compounds, and heavy metals (Dixit et al., 2015). Heavy metal remediation can be undertaken using different biological agents including plants, microorganisms, and organic amendments. Bioremediation is a natural process in which fungi, bacteria, and plants alter contaminated sites. These organisms, especially microorganisms, carry out their normal functions and metabolic activities efficiently in soils with organic amendments. In addition, soil microorganisms can use heavy metals as an energy source, which decreases the hazard presented by the metal as well as reducing their bioavailability by plants and converting them into harmless compounds in most cases (Mani and Kumar, 2014).

2. ENVIRONMENTAL IMPACT OF MERCURY CONTAMINATION IN SOIL: MERCURY DISTRIBUTION AND POTENTIAL BIOAVAILABILITY IN SOIL

Mercury is a naturally occurring metal that can exist in inorganic and organic forms (ethyl-, methyl-, alkyl-, or phenyl-mercury). The organic form of mercury is considered more hazardous than other forms. Mercury represents an element with industrial (e.g., batteries, fossil fuel emissions, paint, cosmetic products) as well as clinical applications (thermometers, sphygmomanometer, and barometers) (Patrick, 2002). In many cases, the presence of mercury in the environment is due to natural activities such as the erosion of Hg-containing ores or Hg-rich gases that dissipate from volcanic eruptions (Kocman et al., 2013). Anthropogenic activities such as mining and coal burning raise the amount of Hg in the atmosphere, soil, and water because they contribute to Hg's mobilization in the environment (Driscoll et al., 2013). The increased Hg levels in soil are mainly connected with industrial activities such as mining, smelting, chemical, and other industries and represent a small risk for environmental contamination in the impacted areas (Taylor et al., 2005; van Straaten, 2000), including for crop production (Qiu et al., 2012). The recently estimated global Hg emissions ranged from 5500 to 8900 tons, coming from natural (10%), anthropogenic (30%), and reemission and remobilization sources (60%). Anthropogenic activities contribute 1960 tons of Hg to the atmosphere from coal burning for energy (85%); mining, smelting and production (10%); cement production (9%); and artisanal and small-scale gold mining (37%) (Kurniati et al., 2014).

Contaminated soil is not uniformly polluted with Hg. Concerning the spatial distribution of Hg in contaminated soil, the highest Hg content is usually in the upper layers of the soil (Conde Bueno et al., 2009; Ettler et al., 2007). This finding was also confirmed by Hojdová et al. (2009). This study examined the impact of former Hg mining on the Hg content in soil samples taken in the vicinity of a mine shaft. Relatively high Hg concentrations, ranging from 8 to 10 mg/kg, were found in the surface horizons whereas the mineral soil horizons contained relatively low Hg concentrations ($\leq$0.6 mg/kg). In this context, soil organic matter (SOM) seemed to be the most important factor in determining the depth distribution of Hg in soil profiles (Zhang et al., 2015).

Among the different forms of Hg contaminating the environment, organic forms (mainly found as methylmercury (MeHg; CH_3Hg^{2+})) are one of the major health concerns. Methylmercury accumulates in the lipid-rich tissues of aquatic food animals, which then results in human exposure (Nam and Basu, 2011) to Hg, which is a powerful neurotoxin that primarily affects the central nervous system (Tschirhart et al., 2012). Several severe health risks may result from human exposure to MeHg, including neurological disease, risk of acute myocardial infarction, renal disorders, and changes in immune function and blood pressure regulation (National Research Council, 2000). This particular organic form is mainly formed in lakes and anaerobic sediments by the sulfate-reducing bacteria recognized as the major methylators (Yu et al., 2012). However, this methylation can also be produced by humic and fulvic acids; these are known to lead to abiotic Hg methylation (Barkay et al., 2003).

The toxicity, bioaccumulation, and environmental mobility of Hg in an area depend on the Hg forms present in soil there (Figure 12.1). For example, the organic forms of Hg such as MeHg and ethylmercury are at least an order of magnitude more mobile than inorganic forms, and thus more toxic and more readily bio-accumulated (Bloom and Porcella, 1994). The speciation of Hg in soils affected

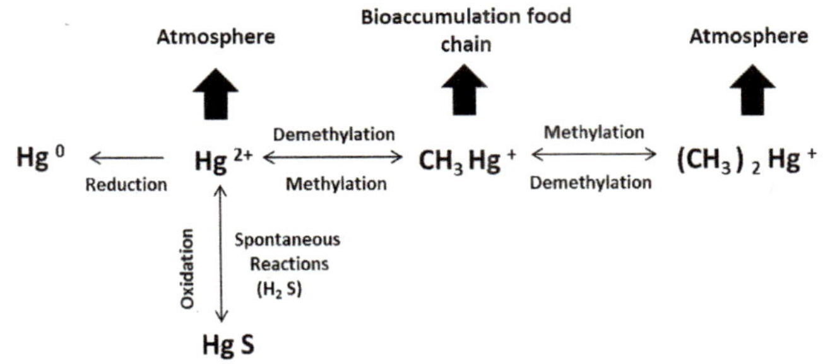

FIGURE 12.1

Fate and speciation forms of mercury in the environment.

by industrial activity connected with a chlor-alkali plant in the Thur River basin in Alsace, France, was investigated by Remy et al. (2006). In this study, concentrations of MeHg reached up to 0.027 mg/kg, and total Hg up to 29 mg/kg, showing a low but no significant MeHg pool in the Hg-contaminated soil. The study observed the effect of total Hg, SOM, and sulfur content on the soil to MeHg ratio. However, a tight correlation between MeHg and total Hg suggests equilibrium between methylation/demethylation processes in soil (Matsuyama et al., 2005). Furthermore, Fernandez-Martinez et al. (2014) determined that humic and fulvic acids and elemental Hg were the primary variables controlling Hg methylation in the soil. However, the methylation rate of Hg in soil can also be affected by soil–plant relationships. Zhou et al. (2015) documented that rice growth promotes the transformation of a residual fraction of Hg to an organic-bound fraction in soil and increases the potential risks of MeHg production. Inorganic forms of Hg are often more stable and less toxic than organic forms. The most soluble form of mercury is inorganic mercury chloride ($HgCl_2$); it is easily transported by natural processes (compared with other inorganic forms) and typically serves as substrate for the Hg methylation process (Bloom et al., 1999). Mercury species such as elemental Hg and mercury-metal amalgams are less toxic than soluble inorganic Hg species because they show less mobility during environmental processes. Other Hg species, such as Hg sulfide, are chemically stable in the soil over geologic time periods and thus are the least toxic Hg species.

To make a detailed estimate of Hg mobility and potential bioavailability, various single and sequential extraction procedures were developed. These took into account the unique chemical properties and behavior of Hg in soils. Generally, the method of sequential extraction of soil gave the most precise information about fractionation and transformation of elements in a soil, especially in relation to different soil physicochemical properties, soil pollution, and the long-term effect of soil amendments such as liming and the application of sewage sludge, coal ash, etc. (Chaudhuri et al., 2003; Tokalioğlu et al., 2003; Kaasalainen and Yli-Halla, 2003). A complete summary of the existing sequential extraction procedures, including a detailed comparison of these methods and/or extracting agents applied within individual extraction schemes, was done by Filgueiras et al. (2002) and Bacon and Davidson (2008). Bloom et al. (2003) developed a sequential extraction procedure that allowed for the special physicochemical properties of Hg releasing: (1) water-soluble Hg (F1); (2) "human stomach acid"–soluble Hg (F2); (3) organo-chelated Hg (F3); (4) elemental Hg (F4); and (5) mercuric sulfide (F5), where the applied extracting agents were deionized water for F1, 0.01 M HCl $+$ 0.1 M CH_3COOH for F2, 1 M KOH for F3, 12 M HNO_3 for F4, and aqua regia (i.e., $HNO_3 + HCl$ in ratio $1 + 3$) for F5. In the series of sediment samples, Bloom et al. (2003) found the majority of Hg presents in either the F3 or F5 fractions.

Boszke et al. (2008) applied a similar procedure for an assessment of the potential bioavailability of Hg in soil contaminated by industrial activities. This assessment found that the Hg was mainly bound to sulfides and the content of potentially bioaccessible portions was relatively low, in most cases not exceeding 5% of the total content. Similar results were observed by Šípková et al. (2014) and Rodriguez et al. (2010). In addition, Hg fractionation showed no significant differences between highly polluted soils and uncontaminated or moderately contaminated soils. The application of another sequential extraction approach to evaluate Hg distribution in contaminated soils indicated that an anthropogenic-caused Hg pool was associated with matter that was hardly oxidized or degraded, whereas the geogenic-origin Hg pool was found in the residual fraction (Feng et al., 1997). Moreover, granulometric distributions of soils can affect Hg fractionation in soils (Fernandez-Martinez et al., 2014).

To assess the risks Hg contamination can pose to humans, an assessment was performed of direct human Hg bioaccessibility from the contaminated soil from in vitro extractions by simulating gastric and intestinal solutions. These tests represented an in vitro test system for predicting the bioavailability of metals from a solid matrix via the simulation of gastric and intestinal processes. Although these methods showed linear correlation with the results from in vivo tests with a Sprague–Dawley rat model (Hettiarachchi et al., 2003; Ruby et al., 1996) and mouse model (Juhasz et al., 2010), it can reveal, for instance, the potential risk of soil ingestion in mining areas (Karadas and Kara, 2011). Similarly, Safruk et al. (2015) observed the bioaccessibility of 1.2% and 3.0% for total Hg using gastric phases and gastric intestinal phase methodologies, respectively in the in soils in the northern Canadian smelter community of Flin Flon, Manitoba.

Zagury et al. (2009) confirmed that the bioaccessible portions in soils were very low, not exceeding 3.2% of the total content, and were positively correlated with sulfate content in these soils. Zagury et al. (2006) evaluated the potential mobility and plant availability of Hg in the highly contaminated soils near chlor-alkali plants, where the total Hg content in soil reached up to 11,500 mg/kg. Although the water-extractable Hg portion was relatively low in regard to the high total content, it represented significant concentrations correlating with Hg uptake by experimental plants. As reviewed by Navarro (2008), the potential mobilization of Hg in soils is controlled by complex formation, soil pH, the primary phase mineral of Hg, and organic matter and iron oxyhydroxide content. The correlation coefficients between total Hg concentrations and the physicochemical properties of soils indicate negative correlations with pH and positive correlation with acidity, as shown in the higher Hg contamination of more acidic forest soils (Ettler et al., 2007). However, a different pattern was observed by Li et al. (2009) in soils originating from an Hg mining deposit. Extractable Hg concentrations were positively correlated with sand content and soil pH values because of the high pH level and sandy character of the Hg-containing materials in the deposit. A significant correlation between soil total Hg content and SOM and available phosphorus and total nitrogen observed by Gil et al. (2010) confirmed the role of soil physicochemical parameters. However, negative correlations between extractable Hg portions in soil and SOM confirms the significant sorption ability of the organic matter (Li et al., 2009). Also, Reis et al. (2010) observed that SOM and sulfur content contributed to Hg retention in the soil matrix and reduced the mobility of Hg, whereas the presence of Hg in the mobile phase was related to manganese and aluminum soil content. Mauclair et al. (2008) confirmed that both humic and fulvic acids were effective in eliminating Hg emissions from the soil surface. The time effect of the Hg immobilization processes was proven by Renneberg and Dudas (2001). They demonstrated that decades after the original spill, the elemental Hg had transformed and was dominantly associated with SOM (up to 85%) and also with the mineral fraction of soil, but to a lesser extent.

3. SOIL MICROORGANISMS: ENVIRONMENTAL ANALYSIS TOOLS

Soil has a heterogeneous, diverse, and porous structure made up of liquid, gaseous, and solid phases (Fredlund and Rahardjo, 1993). The liquid phase represents 25% of the total volume of soil and it carries out different kinds of relevant functions in the soil ecosystem, for example as a transporting agent, chemical solvent, available nutrient pool, and a water source for the metabolic activities of soil biota and vegetation (Porta Casanellas et al., 1998). The gaseous phase also represents 25% of the total soil volume and mainly is made up of inorganic elements (N_2, O_2, CO_2), vapors (H_2O_2 and/or NH_4), and volatile organic elements such as carbohydrates, organic acids, alcohols, oils, and pesticides (Certini and Scalenghe, 2006). In the last phase, the solid phase comprises 50% of the total soil volume. This phase is mainly represented by inorganic and organic solid components that make up 45% and 5%, respectively, of soil volume. The inorganic phase is made up of minerals classified as sand, silt, and clay particles (Schulten and Leinweber, 2003). The organic component is made up of the nonliving phase commonly known as SOM and the living phase in which soil microorganisms and soil fauna and flora reside (Nieder and Benbi, 2008). Soil is an important habitat for thousands of associated organisms.

Soil organisms can be classified as follows:

1. Microorganisms (fungi, bacteria, archaea, and viruses)
2. Fauna (protozoa, annelids, arthropods, nematodes, and mollusks)
3. Flora (plants and algae)

Soil microorganisms such as fungi, bacteria, protozoa, and viruses represent a considerable fraction of the Earth's living biomass, with surface soils estimated to contain 10^3–10^4 kg of microbial biomass per hectare (Fierer et al., 2007). There is a wide variety of microorganisms, and these can vary physiologically, in temperature requirements as well as in their use of oxygen (Pelczar et al., 1981). These microorganisms can occur in association with clay particles or organic matter, in the rhizosphere of plants, and in small colonies in the pores among the particles. From a quantitative point of view, the amount of microorganisms in soil can vary according to soil depth; microorganism diversity increases as depth decreases (Alexander, 1977). Some studies performed in arid soils have shown more abundant bacteria in the outer layers of soils compared with the inner layers. In these studies, the quantity of DNA extracted from soil microorganism was reduced compared with the increased depth (Sait et al., 2002). The diversity and abundance of microorganisms is extremely high; in fact, it is possible to estimate that 1 g of soil contains 10^{10} bacteria, whereas fungi accounts for 90% of the total living microbial biomass (Schloss and Handelsman, 2006; Bills et al., 2004). However, the knowledge of microbial soil diversity is still limited. Microbiological studies of soil are important because soil microorganisms respond rapidly to any soil disturbance, and thus are good, sensitive, reliable indicators of soil health (Griffiths and Philippot, 2013). Also, soil microorganisms are considered critical in any ecosystem because they act on the decomposition of SOM, nutrient cycling, and affect the soil chemical and physical properties. This has direct influence on soil fertility and sustainability. Microorganisms are quite susceptible to changes in the environment, and so changes in the soil community can be an early indicator of any contamination or other disturbance (Cardoso et al., 2013).

Techniques to gather information concerning soil microbial communities can be based on either culturable methods or nonculturable ones (Rincón-Florez et al., 2013). However, culturable methods are known for their selectivity. They also present limitations because more than 99% of soil microorganisms remain nonculturable and do not represent the real diversity of soil. So, because of these limitations, molecular techniques based on genomic approaches represent useful tools for identifying microorganisms and can also help us understand the interactions among microorganisms found underground and their role in different ecological process (Kavamura and Esposito, 2010).

The microbiological study of a soil sample is not only important for deepening our understanding of microbial composition but also because soil microorganisms respond rapidly to disturbances; they can provide instant data on soil health (Cardoso et al., 2013). This may be especially useful in specific systems, such as in agricultural soils in which microorganisms are threatened by agronomic operations, or in specific areas that are contaminated with pollutants such as heavy metals (Paz-Ferreiro and Fu, 2013). In the case of soils polluted with heavy metals, microorganisms play an important role because soil health and quality is shown through the functions they perform in the soil ecosystem and plant development (Xian et al., 2015). Several approaches may be used in a microbiological study to observe changes in soil microorganism functionality, abundance, and diversity. Table 12.1 summarizes

Table 12.1 Methodologies Used for Soil Microbiological Studies

Techniques	Function	Type of Measurement
Microbial biomass	Source and/or drain of C and nutrients	C and nutrient stock in cells
Soil microbial respiration	Microbial mineralization of organic carbon	Microbial quantification by C–CO_2 evolution
q CO_2	Metabolic conditions of microbial communities	Microbial quantification by C–CO_2 released per unit of microorganism in time
Microbial functional groups (soil enzymes)	Cycling nutrients (C, N, P, S)	Functional diversity of soil microorganisms
Biolog EcoPlate Systems	Metabolic diversity of microbial community	Community level physiological profile
Functional gene arrays (RNA-based)	Identification of microbial functional genes	Gene potential functionality of a soil community
Next-generation sequencing (metatranscriptomic)	Microbial gene expression	Metatranscriptome identification from different environments
Stable-isotope probing	Identification of active members of microbial communities	Functionality of soil microbial communities
Fingerprinting (DGGE, TGGE, SSSCP, T-FRLP, ARISA, RISA, LH-PCR, ARDRA, RAPD)	DNA extraction, amplification, and separation	Genetic diversity of soil microbial communities
Phospholipid fatty acid (PFLA profiling)	Quantification and determination of fatty acids	Diversity and quantification of microbial communities
Adenosine 5′-triphosphate	Determination of ATP	Quantification of the microbial biomass
Cultivation (colony-forming units)	Culture of microorganisms in selective or specific media	Diversity and recount of soil microorganisms
LIVE/DEAD BacLight kit	Fluorescence-based assay for bacterial viability	Quantification of microbial communities
Quantitative PCR	DNA extraction and amplification of target gene	Changes in microbial activity

Abbreviations: ARISA, automated ribosomal intergenic spacer analysis; ARDRA, amplified ribosomal DNA restriction analysis; ATP, adenosine triphosphate; C, carbon; CO_2, carbon dioxide; DGGE, denaturing gradient gel electrophoresis; LH-PCR, length-heterogeneity PCR; PCR, polymerase chain reaction; N, nitrogen; P, phosphorus, PFLA, phospholipids fatty acids; q CO_2, microbial metabolic quotient; RAPD, random amplified polymorphic DNA; RISA, ribosomal intergenic spacer analysis; S, sulfur; SSCP, single-strand conformation polymorphism; TGGE, temperature gradient gel electrophoresis; T-RFLP, terminal restriction fragment length polymorphism fingerprinting.

the principal techniques used to investigate soil microbiology. Nevertheless, the assessment of soil enzymes represents one of the most sensitive and reliable approaches to evaluate the soil health because enzymes respond rapidly to environmental disturbances (Rao et al., 2014). Enzymes can be used as a good bioindicators in soils affected by different xenobiotic substances, stress conditions, management practices, and those contaminated with heavy metals (Rao et al., 2014). The enzymes play an important role in soil biology as well as in nutrient cycling because they catalyze the many biochemical reactions needed for the development and growth of microorganisms and plants (Griffiths and Philippot, 2013). Also, enzymatic activities regulate the release of nutrients needed by both microorganisms and plants and are involved in the degradation of organic residue. They participate in the gas exchange between soil and the atmosphere as well as in the formation of aggregates (Cardoso et al., 2013). Enzymes can be classified into four groups: (1) oxidation-reduction oxidoreductases, (2) transferases involved in the transfer of functional groups, (3) hydrolases involved in hydrolysis, and (4) lyases: elimination of groups to form double bonds. The most important soil enzymes are oxidoreductases and hydrolases (Karaca et al., 2011, Table 12.2).

Table 12.2 Classification of Soil Enzymes

Activity	Enzyme	Cycle	Function	Microorganism
Hydrolases	B-glucosidase	C	C oxidation	Several
	Cellulase and amylase	C	Cellulose degradation with release of glucose	Mainly fungi, also bacteria
	Phosphatase (acid and alkaline)	P	Organic phosphorus cycling	Alkaline phosphatases of microorganism, and acid phosphatases with several origins
	Urease	N	Organic N mineralization to ammonia/ammonium	Several
	Protease	N	Hydrolysis of peptidic bonds	Several
	Arylsulfatase	S	Hydrolysis of aromatic ester of sulfate into phenols and sulfate	Several
Oxidoreductases	Catalase		Intracellular enzyme involved on H_2O_2 removal	Aerobic and anaerobic bacterial
	Dehydrogenase	C	Intracellular enzyme involved in the Krebs cycle	All aerobic
	Diphenol oxidase	C	Transformation of phenolic compounds into quinones (acid humic formation)	Several with potential capacity to degrade recalcitrant organic compounds
Lyases	Phenylalanine ammonia-lyase	N	Conversion of L-phenylalanine into NH_3 and cinnamic acid	Several
Transferases	Transaminase	N	Organic N mineralization	Several
	Transglycosylase	C	Organic C oxidation	Several

C, carbon; N, nitrogen; NH_3, ammonia; P, phosphorus; S, sulfur.

These enzymes are appropriate for detecting heavy metals in soils as well as for differentiating between levels of pollution (Niemeyer et al., 2012). The effect of heavy metals on soil enzyme activities and in general on soil microbial processes depends on the bioavailability of these compounds in soil. This bioavailability is related to some chemical and physical properties of soil such as pH, organic matter, clay, iron oxide content, and plant exudates as well as the nature and speciation of the heavy metal (de Santiago et al., 2013). Also, the organic matter of soils may play an important role in both the growth and activity of soil microbial communities because a heavy metal can be immobilized by humic and fulvic acids, further reducing its availability to microorganisms (Rao et al., 2014). In the case of Hg, studies performed in agronomical soil that had been heavily polluted with this metal for a long time showed no negative and/or detrimental effects of Hg on soil microbial activities. This result might be due to the insoluble form of Hg in the soil or perhaps microorganisms and their enzymatic activity had become adapted to the Hg presence (Ruggiero et al., 2011). Likewise, different types of soil polluted at different doses of Hg show different levels of impact on soil microbial biomass through time. In this study, the authors observed a reduction in microbial biomass in three types of soils with different agrochemical properties; however, over time, a positive correlation between both microbial biomass and exposure to Hg were observed in only one type of soil (Casucci, 2003). In addition to this result, we should consider that some heterotrophic bacteria are adapted to Hg, probably because of stimulation from nutrient release into the environment as a consequence of bacterial cellular lysis induced by Hg (Rasmussen and Sorensen, 2001). The extent of Hg impact on soil enzymes varies depending on the agrochemical characteristics of the soil in question, the concentration of Hg, time of exposure, and type of enzyme. Dumonet et al. (1997) proposed that the quantification of fluorescein diacetate hydrolysis might be a suitable tool for measuring the detrimental effect of xenobiotic compounds on microbial biomass. In fact, after a short exposure to Hg at the highest concentration, a decrease in this activity was observed. Likewise, phosphatase can be used as a good index of xenobiotic effects in microbial catalytic activity and arylsulfatase is an indirect indicator of fungal biomass containing ester sulfate (Doelman and Haanstra, 1989; Bandick and Dick, 1999). Casucci et al. (2003) observed low levels in phosphatase and arylsulfate enzymes in soils polluted with Hg: activity levels decreased over time and at a concentration of 10 μM. However, this effect may vary depending on the agrochemical properties of soil. This is shown by the fact that under the same conditions of Hg exposure, phosphatase can increase its activity with time of incubation. This different behavior of enzymes, phosphatase, and arylsulfatase in various types of soils can be associated with organic matter content present because it can protect enzymes from denaturalization. Other enzymes cannot modify their activity in presence of Hg—for example, *o*-diphenol-oxidase cannot because of its implication in the first step of polymerization of organic matter in soils. It is possible that the death of some microorganisms as a consequence of Hg could increase the organic matter levels in soils and stimulate the activity of *o*-diphenol-oxidase (Casucci et al., 2003).

The mechanism by which Hg provokes a decrease in some soil enzymes might be attributed to:

1. Mercury presenting an affinity for thiols groups in proteins (Robinson and Tuovinne, 1994; Velasco et al., 1999).
2. Direct inhibition of enzyme expression (i.e., hydrolases enzymes).
3. Deactivation of enzymatic reactions.
4. Mercury possibly reacting with active groups of enzymes and/or with enzyme–substrate complexes.

4. MICROORGANISMS IN MERCURY-POLLUTED SOILS

Heavy metals have important roles in some biochemical reactions because they are essential for the growth and development of microorganism flora and fauna (Higgins and Burns, 1975), but at a certain concentration they present a toxic effect. Not all heavy metals have the same biological importance. Microorganisms need essential metals for their metabolism; metals are often required in low concentrations and act as enzyme cofactors (Camargo et al., 2007). This is the case of cobalt, copper, chromium, nickel, tungsten, vanadium, and zinc. Meanwhile, other metals such as antimony, arsenic, Cd, lead, silver, uranium, and Hg have no specific physiological properties and present a high toxicity for both microorganisms and humans (Camargo et al., 2007). In the case of Hg, the cytotoxic effects of exposure to over 0.01 mM have been described (Kavamura and Espósito, 2010). Nevertheless, the level of contamination depends on factors such as chemical composition, toxicity, mobility, and varying bioavailability of the metals. Heavy metals are rapidly adsorbed into soil particles, which are followed by slow adsorption and distribution in the soil.

The presence of heavy metals, specifically Hg, in soils can affect microbiota directly though the modification of population size, diversity, and activity. Nevertheless, not all microorganisms respond in the same way to the presence of heavy metals, so fungi and bacteria are affected differently (Rajapaksha et al., 2004). Studies performed by Kandeler et al. (1996, 2000) have shown that the increased concentration of heavy metals in soils will decrease the functional diversity of soil microorganisms. The Hg toxicity for metal-tolerant *Pseudomonas aeruginosa* was higher compared with toxicity of other elements such as Cu, Zn, Cr, Cd, and Co (Hassen et al., 1998). Moreover, the activity of some soil enzymes such as dehydrogenase, acid phosphate, and β-glucosidase can be affected in the presence of these heavy metals. So, the analysis of functionality and diversity as well as the quantification of soil microorganisms using some of the approaches described previously can represent a useful strategy for understanding how the microbial population is being affected by the presence of heavy metals. Biomass and/or respiration quantification of soil microorganisms can also indicate how the activity of some microbes can be affected by high ratios of heavy metals. The presence of these compounds in soil can considerably decrease the size of the microbial community along with reducing environmental and biological activities such as organic matter mineralization and leaf litter decomposition (Kelly et al., 2009). Ruggiero et al. (2011) investigated the response of soil microorganisms already adapted to Hg stress in a long-term polluted industrial site where the total Hg content in soil reached up to 50 mg/kg. Because of the low mobile Hg levels in these soils, the different amounts of Hg present did not influence the soil microbial parameters and enzymatic activities. However, the adverse effect of MeHg on basal respiration, microbial biomass carbon, and bacterial community structures was observed in MeHg-amended soils whereas fungal communities remained unaffected (Rieder and Frey, 2013; Rieder et al., 2013). The significant release of Hg and its methylated compounds from soil to the atmosphere were documented by Tipping et al. (2011) and Hirner et al. (2000). Schlüter (2000) reviewed Hg evaporation from the soils and highlighted the role of distribution of Hg in the soil profile, the physical migration of the Hg species within the soil, and the physical and chemical sorption of Hg and its compounds. The effect of soil microbial activity should be taken into account in this case as well.

5. STRATEGIES OF MICROORGANISMS AGAINST MERCURY

To counteract the detrimental effect of heavy metals, soil microorganisms have developed strategies for tolerance and/or detoxification to protect them from contaminants (Silver and Phung, 1996). Because of their capacity to attack soil pollutants enzymatically, microorganisms represent useful tools for bioremediation purposes. The term *bioremediation* refers to the degradation and/or transformation of pollutants into their less toxic forms through the action of living organisms (Vidali, 2001). This green technology can be carried out in two different ways: in situ and ex situ. In situ treatment is done at the contamination site, whereas ex situ treatment requires the removal of soil for subsequent treatment. The *biodegradation* of xenobiotic compounds (via the production of enzymes) is allowed through the wide biochemical diversity and adaptability of microorganisms.

The mechanism of degradation can be subdivided into other techniques such as *biosorption*, *bioleaching*, *bioaccumulation*, and *biomineralization* as well as the production of a variety of products that can be used in traditional and biotechnological manipulations (Gadd, 2004). These mechanisms are also essential in biogeochemical cycles of metals as well as carbon, nitrogen, and phosphorus among others. It can be considered a suitable mechanism of bioremediation. This technology is one of the eco-friendly and sustainable processes that prove to be much more effective and efficient for eliminating heavy metals present in different spheres of the environment.

Metal speciation is due to the ability of microorganisms to mediate mobilization and immobilization processes that influence the balance between soluble and insoluble phases. The mobilization of metals can occur through the processes of *protonation*, *chelation*, and *chemical transformation*. Meanwhile, immobilization can be achieved through *precipitation* and *crystallization* of inorganic and organic forms, or through *sorption*, *uptake,* and *intracellular* sequestration (Figure 12.2).

5.1 MOBILIZATION PROCESSES

The mobilization of metals can be done through autotrophic and heterotrophic leachates and chelation by microbial metabolites and siderophores and methylation, which can result in volatilization. All these processes can lead to the dissolution and/or desorption of metal from the exchange site of clay and/or organic matter.

The bioleaching of metals, such as Hg, is highly effective using sulfur as a substrate, thus iron and sulfur-oxidizing bacteria such as *Thiobacillus ferooxidans* and *Thiobacillus thiooxidans* showed a great capacity to leach metals (e.g., Hg) present in polluted soil. Nevertheless, the metals in those leachates can be subsequently removed through the action of bacterial sulfate reduction where the soluble metals are precipitated as insoluble sulfides (White et al., 1998). The acidification of the environment by microorganism H^+-ATPasas can allow the metal release through competition with protons and the metal in a metal–anion complex or in a sorbed form resulting in the release of free metal cations.

Some bacteria are resistant to inorganic Hg salts as well as organic forms of $HgCl_2$ such as MeHg. These bacteria play an important role in the global cycling of Hg; therefore, their mechanism of resistance has been extensively studied. It is based on an operon (mer operon) that is located in the chromosome and/or plasmid (Yu et al., 2014). The mechanism of resistance includes the uptake and transport of Hg(II) by the periplasmic protein MerP and the inner membrane protein MerT. MerE is

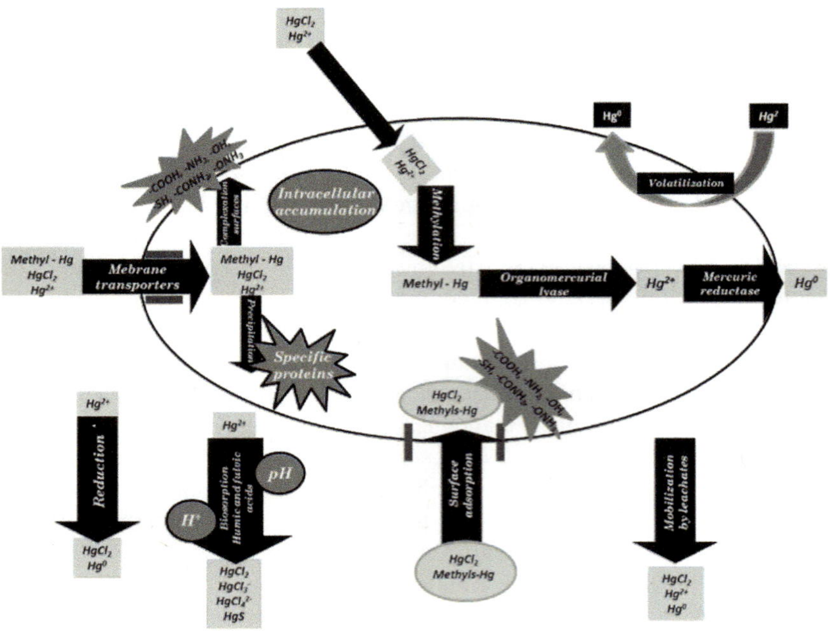

FIGURE 12.2

Main mechanism of mercury bioremediation mediated by soil microorganisms.

another membrane protein that acts as a high broad Hg transporter; it is able to transport both MeHg and Hg(II). The transformation is due to the enzymatic participation of MerA, which is a cytosolic enzyme known as mercuric reductase and is responsible for the reduction of Hg(II) to less toxic elemental Hg. The gene *merB* encodes the protein organomercurial lyase involved in cleaving C–Hg bonds producing Hg(II) and the respective organic compound. The *merG* gene product is involved in the reduction of cellular permeability to organomercurial compounds. MerD acts as a distal corepressor of transcriptional activation, whereas MerR is the repressor of *mer* genes in the absence of Hg ions or the activator of those genes in the presence of Hg (Rojas et al., 2011). Usually, the ability of bacteria to resist or to live in an environment with a high concentration of Hg is directly associated to the presence of this detoxification system based on operon *mer*. In fact, studies performed on Hg-sensitive bacteria that had been genetically modified showed that they were capable of removing Hg from polluted water (Rojas et al., 2011). Many organic and inorganic Hg-resistant bacteria have been reported as previously belonging to genera *Escherichia*, *Bacillus*, *Pseudomonas*, *Staphylococcus*, and *Brevibacterium* (De et al., 2008; Deng and Wang, 2012). Apart from these groups of bacteria, another bacterial species has been isolated from marine sponges that is able to resist $HgCl_2$ and MeHg and is also involved in the reduction of Hg^{2+} to Hg^0 by enzymatic activity encoded by operon *mer* (Santos-Galdenman et al., 2014). Areas relatively close to a river are often potentially contaminated with heavy metals such as Hg; in these sites, it is possible to discover new bacteria, such as *Staphylococcus epidermidis*, which have adapted to live in the polluted environment by degrading Hg through the presence of *mer* operon (Yu et al., 2014).

The microbial production of organic acids, commonly known as *heterotrophic leaching*, represents another important strategy for the recovery of soils polluted with heavy metals. The heterotrophic metabolism can also contribute to metal leaching as consequence of efflux of organic acids because these compounds can provide a proton source and metal-complexing organic acid anions. For example, citrate and oxalate are able to establish stable complexes with high-mobility metals and they are not readily degraded. Heterotrophic mobilization and solubilization have more effective results than autotrophic leaching because the production of certain types of organic acids such as oxalic and citric acid can be associated with the speciation, physiology, and mobility of metals in soils (Gadd and Poole, 1999).

The production of siderophores by microorganisms can also be considered a good bioremediation strategy in Hg-contaminated soils. Siderophores are low-molecular-weight molecules that are excreted to aid iron assimilation in case of deficiency (Gadd, 2000). Although primarily produced as a means of obtaining iron, siderophores are able to bind with other metals and so is another method for the treatment of metal-contaminated soils. Microorganisms, mainly bacteria, are able to solubilize these metals (or increase bioavailability) via the production of siderophores and subsequent adsorption in their biomass on metal-induced outer membrane proteins and by precipitation (Gadd, 2004).

Mercury can be removed from the environment through bio-methylation processes (mobilization) through a huge range of microorganisms such as bacteria and fungi. Methyl groups are enzymatically bound to Hg and so a given species may transform a number of different metals (e.g., Hg). This compound has a high affinity to the thiol groups of proteins, resulting in MeHg thiolate, which is then transformed into Hg thiolate and carbon compounds through the action of the organomercurial lyase enzyme. Finally, the participation of the mercuric reductase enzyme reduces MeHg thiolate to elemental Hg through volatilization processes (Bhattacharya et al., 2006). *Bacillus thurigensis* strain PW-05 is able to volatilize Hg from polluted environments and thus can be used in situ to reduce the toxic effect of Hg. It has been observed that this bacterium is able to remove up to 90% of residual Hg by a combination of volatilization and sequestration (Dash et al., 2014). *Haloferax, Halococcus*, and *Halobacterium* represent a group of archaea that live in a high-salinity environment and have potential resistance and volatilization of Hg(II) (Al-Mailen et al., 2011).

The methyl compounds formed by this process differ in volatility, solubility, and toxicity. In addition, the redox transformation of Hg by microorganisms can be considered a potential mechanism of mobilization of this metal. Thus, the reduction of Hg(II) to Hg(0) by bacteria and/or fungi results in a diffusion of elemental Hg outside cells (Hobman et al., 2000). Studies performed by Chang et al., 1999 observed the specificity and ability of mercuric reductase to reduce Hg(II) to Hg(0), showing its maximum activity of 1.2 nM Hg/mg protein/s at a higher Hg concentration of 50 μM/L.

5.2 IMMOBILIZATION PROCESSES

Biosorption is defined as the microbial uptake of organic and/or inorganic metallic forms as well as soluble and insoluble forms by physicochemical processes such as adsorption (Gupta et al., 2015). In general, biosorption can be carried out through different mechanisms, including adsorption, absorption, surface complexation, precipitation, and ion exchange. Heavy metals, especially Hg, are contaminants which can be removed by biosorption.

Modifications in pH, potential redox (E_h), organic and inorganic nutrients, and metabolites from the metabolic activity of living cells are important properties, because any changes in these factors can alter the biosorption process. In fact, pH may influence the biosorption of Hg; pH 6 stimulates sorption process of this metal. Moreover, it has been observed that at a pH of less than 6, sorption processes can be negatively affected by the complex formation of Hg with chloride. At a higher pH, Hg usually appears forming hydroxide complexes; both forms are less prone to attraction by the absorbent. However, the fraction of Hg complexes to chloride depends on the amount of Cl^- ion in the medium, thus predominant species are: $HgCl_2$, $HgCl_3^-$, and $HgCl_4^{2-}$. Also, the sorption capacity can be influenced by hydrogen concentration in the medium, thus a higher concentration of hydrogen can improve the ion-exchange potential and provide higher sorption capacity (Jafari and Cheraghi, 2014).

The sorption of Hg to cellular surfaces determines the intracellular accumulation of this metal, but the effectiveness of this mechanism depends on the affinity and specificity of the membrane transport system. Thus, the presence of peptidoglycan carboxyl groups and phosphate groups in the cellular walls of gram-positive and gram-negative bacteria provide metal binding sites. In addition, the proteinaceous S-layer and sheath represent a potential wall component that contains polysaccharides and contributes to metal binding. Pseudomurein is another cell-wall component of archaea bacteria; along with sulfonate polysaccharides and glycoproteins, it provides anionic sites (carboxyl and sulfate groups) (Kotrba et al., 2011).

The presence of chitins, glucans, mannans, and proteins in fungal cell walls represents an important component that can act as a biosorbent of Hg. Apart from these elements, it can be found polysaccharides, lipids, and pigments; these facilitate binding of many metal ions (Latgé, 2010).

The mechanism of biosorption is a highly complex process because of the complexity of the biological structures involved. Functional groups such as carboxyl, phosphate, hydroxyl, amino, and thiols are present in the structure of biomass that interacts with different metals in a variable degree which may be influenced by physicochemical factors. This includes cell metabolism as well as the location of biosorption. Cell metabolism–mediated biosorption can further be divided into metabolism-dependent and nondependent processes (Ahalya et al., 2003). The first one occurs only in viable cells and plays a vital role in the defense mechanism of microbes showing a reaction with toxic metals. It can be classified as: transport across cell membranes, which includes metabolism-independent binding where the metals bind to the cell walls; metabolism-dependent intracellular uptake that includes transport of metal ions across the cell membrane; and precipitation, which is metabolism-dependent and it is often related to a microbe's defense mechanism (Wang and Tay, 2010). In this last case, an important role is played in the synthesis of specific, low-molecular-weight metal-binding proteins, usually termed metallothioneins, which are produced in response to heavy metals. Phytochelatins represent another important group of binding proteins that contain glutamic acid and cysteine at the amino terminal position (Li et al., 2014; Chaturvedi and Archana, 2014).

A non–metabolism-dependent mechanism involves the interaction between metal and functional groups present on the microbial cell surface. Many functional groups such as carboxyl, phosphate, and sulfate are present on these surfaces because cell walls are mainly made up of polysaccharides, proteins, and lipids. This process can be classified as follows: physical adsorption (van der Waal's and electrostatic forces); ion exchange through microbial cell wall polysaccharides, which act as counterions and facilitate the exchange of bivalent metal ions; complexation of metals in the cell surface complex through interaction with active groups of polysaccharides; and precipitation independent of metabolism (Gupta et al., 2015).

The cellular localization of these mechanisms can be classified into the following types: extracellular accumulation/precipitation, cell-surface sorption/precipitation, and intracellular accumulation.

Fungi are known to tolerate and detoxify metals through various mechanisms, including transformation, intra- and extracellular precipitation, and active uptake (Gadd, 2004). The high surface-volume ratio and ability to detoxify metals are among the reasons they can be considered as potential alternatives for the bioremediation of solutions as well as soils polluted with Hg and other heavy metals. As has been described previously, fungi are able to mobilize Hg through heterotrophic leachates, but they can also perform immobilization processes, such as biosorption (Kurniati et al., 2014). Immobilization of inorganic and organic forms of Hg can be performed by fungi that are able to bind Hg in cells. An example is the case of *Aspergillus niger* strain KRP1, which showed a high affinity for Hg(II) as well as for MeHg (CH_3Hg^+) from an aqueous solution, resulting in its potential use for the removal of inorganic Hg and MeHg ions from polluted effluent (Karunasagar et al., 2003). Similar results were obtained by Das et al. (2007) using *Aspergillus versicolor.* Sorption ability was also observed in the dead biomass of *A. niger* when removing Hg(II) from wastewater. OH and NH_2^+ groups played an important role in the biosorption process on the surface of dead fungal biomass, when Hg(II) was deposited on it in a film-like structure (Khambhaty et al., 2008). Additionally, *Phanerochaete chrysosporium* showed a great affinity of Hg: Hg(II), MeHg, and ethylmercury and maximum capacities of 79, 67, and 61 mg/g of dry biomass, respectively (Saglam et al., 1999). Recent studies performed by Kurniati et al. (2014) isolated different strains of fungi that were able to grow in the presence of 25 mg/L Hg(II), but only one strain showed the highest ratio of growth diameter. The *Aspergillus flavus* strain was able to tolerate Hg(II) concentrations as high as 100 mg/L and to remove 98% of Hg(II) at 10 mg/L doses. Other studies performed in *Pleorotus ostreatus* revealed how the presence of Hg in soils was able to limit its growth at 50–100 mg/kg doses, but not inhibit its enzymatic machinery (Baldrian et al., 2000).

Other studies have been focused in the ability of some microscopic fungi, such as *Scopulariopsis brevicaulis,* to accumulate and volatilize Hg. This microscopic filamentous fungus showed a high capacity of Hg(II) sorption. Indeed, the interaction of fungi with Hg is only studied for the high affinity of fungal biomass for the sorption of Hg cations (Boriová et al., 2014). The volatilization of Hg is much more effective by archaea and bacteria (Al-Mailem et al., 2011).

In addition to the abilities of fungi for soil remediation, bacteria are also a group of microorganisms that show potential for developing biosorption processes. For example, bacteria such as *B. thurigensis* strain PW-05, which is isolated from a marine environment, is capable of accumulating Hg because of its ability to form Hg-resistant biofilm through the production of many extracellular polymeric substances like neutral polysaccharides, amyloids, extracellular enzymes, and biosurfactants (Dash et al., 2014). Studies performed with *Vibrio parahaemolyticus* PG02 highlighted its capacity for removing Hg, which suggests that the PG02 strain is a good adsorbent for Hg removal from an aqueous solution (Ali Jafari and Cheraghi, 2014).

6. SHORT SURVEY OF MERCURY IMPACT ON SOIL-BOUND ZOOCENOSES AND HIGHER PLANTS

The ability of soil-dwelling animals to accumulate and transform Hg compounds was investigated by Rieder et al. (2013). In this study, the Hg methylation ability of earthworms was observed in model conditions and the substantial role of bacteria associated with the worms' digestive tract was reported. Earthworms are suitable indicators of soil Hg contamination levels (Teršič and Gosar, 2012; Han et al.,

2012). Colacevich et al. (2011) examined the impact of Hg in extremely contaminated soils (up to 1287 mg/kg) on the biochemical processes in the earthworm organism *Lumbricus terrestris*. These extremely contaminated soils were not lethal to the animals but there was significant response of the biomarkers of oxidative stress, such as metallothioneins, and high total glutathione content. Therefore, extreme Hg soil contaminations can affect soil animal communities. Moreover, tissue concentrations of methylated Hg compounds in earthworms grown in soils treated with inorganic Hg were about six times higher than in earthworms grown in soils without Hg, whereas the methylation of Hg in soil was negligible. This suggests that Hg was mainly methylated in the earthworms (Rieder et al., 2013).

Mercury has been found to affect growth and metabolism of plants to varying degrees depending on its concentration as well as the status of Hg in the plant tissue (Patra and Sharma, 2000). Usually, Hg has a detrimental impact on plant physiology, for instance, many physiological mechanisms are negatively affected by the presence of this metal in the environment. It has described that Hg reduces photosynthetic system efficiency and uptake of nutrients and homeostasis. It inhibits plant growth and yield. Mercury is also related to genotoxicity processes because it can bind to DNA and thus cause damage to chromosomes (Patra and Sharma, 2000). Apart from these effects, Hg is able to induce oxidative stress in plants by inducing lipid peroxidation of membranes and super production of reactive oxygen species. As a consequence, plant cells respond by increasing the activity of antioxidant enzymes such as superoxide dismutase, ascorbate peroxidase, glutathione reductase, and peroxidase as well as reducing glutathione (Moreno-Jiménez et al., 2009; Lomonte et al., 2010).

Phytostabilization is a green technology that uses plant roots to immobilize heavy metal movement within the soil through the biochemical processes that occur in roots. Studies performed by Taylor et al. (2005) observed increased Hg content in soils impacted by gold mining (0.05–9.2 mg/kg), although the Hg content in vegetable and grain samples were mostly below the detection limit of 0.004 mg/kg Hg. Among the soil physicochemical properties, those connected with the Hg mobility and extractability in soil, such as SOM, pH, and manganese and aluminum oxides, are the main variables that play significant roles in plant availability of Hg (Rodrigues et al., 2012; Sierra et al., 2012). Within the plants, Hg is accumulated predominantly in the roots but has limited transport to the aboveground biomass (Zornoza et al., 2010; Kacálková et al., 2009). Sierra et al. (2011) cultivated barley (*Hordeum vulgare* L.) in soil containing 22.9 mg/kg. Concentrations of Hg in roots at plant maturity varied between 1 and 3 mg/kg; in straw and grain the concentrations ranged from 0.072 to 0.480 mg/kg and from 0.005 to 0.257 mg/kg, respectively. These concentrations did not represent serious human health intoxication risk and corresponded with the low-mobile Hg pool in soils (Šípková et al., 2014; Rodrigues et al., 2010; Rodriguez et al., 2007). Other experiments performed using indigenous herbaceous species (red fescue (*Festuca rubra*), meadow grass (*Poa pratensis*), horseradish (*Armoracia lapathifolia*), and Jerusalem sunflower (*Helianthus tuberosum*)) showed the capacity for phytostabilization of soil with moderate Hg contamination (Sas-Nowosielska et al., 2008).

The low bioavailability of Hg in soils limits some phytoremediation techniques such as phytoextraction. Wang et al. (2012) recently reviewed available Hg remediation technologies and suggested phytoextraction as an effective cleanup technology that is still under development. Similarly, Li and Tse (2015) highlighted phytoremediation as an alternative to traditional remediation techniques of Hg-contaminated soils because of the low cost and environmentally friendly approach. For effective phytoextraction, more studies to select Hg accumulator/hyperaccumulator plant species should be provided as well as more studies to understand the fundamental mechanisms involving plant uptake of Hg. There are various available environmentally friendly measures for the enhancement of Hg phytoextraction effectivity. For example, Moreno et al. (2005) described the effect of ammonium thiosulfate on the enhancement of plant Hg uptake and translocation to the above-ground biomass and Cassina et al. (2012) confirmed the effect of foliar application of phytohormone and cytokinine, in concert with soil application of ammonium thiosulfate, where cytokinine increased plant biomass and the evapotranspiration rate while ammonium thiosulfate treatment promoted soil Hg solubility and plant availability. Smolinska (2015) observed that the application of municipal green waste–based compost to the soil decreased the leaching of Hg in both acidic and neutral solutions, regardless of growing medium composition and time of analysis, and increased the efficiency of phytoextraction of Hg-contaminated soil by *Lepidium sativum* plants. The application of genetically modified plants is only one of the possible ways to enhance phytoextraction effectivity, as documented by Chang et al. (2015).

7. CONCLUSIONS AND FUTURE PROSPECTS

The accumulation of Hg in the environment increases day by day because of anthropogenic activities. Excessive use of fertilizers, herbicides, and other chemical products used in agricultural practices contributes to Hg accumulation, as do intensive industrial activities. Exposure to Hg can cause fatal consequences to human health when it builds up in the brain, kidneys, and other organs; furthermore, it is a powerful neurotoxin that affects the central nervous system.

The development of conventional strategies for its decontamination, such as stabilization/solidification, vitrification, electro-remediation, soil washing, thermal desorption, and others have several advantages but also numerous disadvantages. Conventional strategies often incur high costs and/or produce toxic sludge. These methods and technologies are still limited (compared with other heavy metals such as Cd) and ambiguous, especially in the terrestrial environment. In light of these drawbacks, bioremediation is a promising upcoming technology that offers an attractive potential alternative to these conventional methods. By using microorganisms with the capacity to enzymatically attack pollutant compounds, bioremediation transforms Hg and other heavy metals into less toxic forms.

Soil microorganisms play an important role in the Hg cycle and are possibly directly involved in Hg speciation and thus responsible for the different Hg forms in the environment. Mercury bioremediation can be carried out through processes that are mainly based on Hg mobilization and immobilization by microorganisms (bacteria and fungi). In short, Hg mobilization is produced by changes in the environment that are mostly the result of production and release of organic compounds and/or H^+ by microorganisms that contribute to leaching Hg. Bacteria are directly involved in processes of Hg mobilization; therefore, the presence of the *mer* operon confers protection to bacteria, which encodes several genes involved in the process of resistance and detoxification of Hg. Resistant bacteria may be useful for Hg remediation of polluted sites. In addition to this mechanism, the mobilization of Hg can be done through other strategies such as biomethylation, due to its high affinity to thiols groups (−SH) and volatilization, where Hg(II) is converted to Hg(0) by mercuric reductase (Mer A). Alternatively, the Hg immobilization can also be performed through biosorption processes. This mechanism can be influenced by living organisms through the release of organic acids and other compounds that can induce pH modifications affecting Hg biosorption. However, it is important to highlight that soil fungi play an important role in biosorption processes because cellular walls and membranes have a high affinity for Hg(II) and MeHg.

The unique and specific chemical properties of Hg and its behavior in the environment must be taken into account to develop new strategies for its bioremediation. Phytoremediation (phytostabilization and phytoextraction) is yet another way to decontaminate Hg-polluted areas. However, in the soil-plant system, low plant-availability of soil Hg and/or limited uptake and translocation of Hg into the aboveground biomass of plants makes the possible phytoextraction approaches extremely difficult. On the contrary, the detailed description and elucidation of the role of soil microflora in bioaccumulation and/or transformation of soil Hg can be helpful for the further development and application of the reasonable and environmentally friendly bioremediation methods.

ACKNOWLEDGEMENTS

Authors thank for financial support the GAČR project P503/12/0682, ESF, and MŠMT project.

REFERENCES

Alexander, M., 1977. Introduction to Soil Microbiology, second ed. John Wiley and Sons, New York.

Al-Mailem, D.M., Al-Awadhi, H., Sorkhoh, N.A., Eliyas, M., Radwan, S.S., 2011. Mercury resistance and volatilization by oil utilizing haloarchaea under hypersaline conditions. Extremophiles 15, 39–44.

Ahalya, N., Ramachandra, T., Kanamadi, R., 2003. Biosorption of heavy metals. Res. J. Chem. Environ. 7, 71–79.

Bacon, J.R., Davidson, C.M., 2008. Is there a future for sequential chemical extraction? Analyst 133, 25–46.

Baldrian, P., in der Wiesche, C., Gabriel, J.A., Nerud, F., Zadrazil, F., 2000. Influence of cadmium and mercury on activities of ligninolytic enzymes and degradation of polycyclic aromatic hydrocarbons by *Pleurotus ostreatus* in Soil. Appl. Environ. Microb. 66, 2471–2478.

Bandic, A.K., Dick, R.P., 1999. Field management effects on soil enzyme activities. Soil Biol. Biochem. 31, 1471–1479.

Barkay, T., Miller, S.M., Summers, A., 2003. Bacterial mercury resistance from atoms to ecosystems. FEMS Microbiol. Rev. 27, 355–384.

Bills, G.F., Christensen, M., Powell, M., Thorn, G., 2004. Saprobic soil fungi. In: Mueller, G., Bills, G.F., Foster, M.S. (Eds.), Biodiversity of Fungi, Inventory and Monitoring Methods. Elservier Academic Press, Oxford, pp. 271–302.

Bhattacharya, J., Islam, M., Cheong, Y.W., 2006. Microbial growth and action: implications for passive bioremediation of acid mine drainage. Mine Water Environ. 25, 233–240.

Boriová, K., Čerňanský, *S.*, Matus, P., Bujdos, M., Simonovicova, A., 2014. Bioaccumulation and biovolatilization of various elements using filamentous fungus *Scopulariopsis brevicaulis*. Lett. Appl. Microbiol. 59, 217–223.

Boszke, L., Kowalski, A., Astel, A., Barański, A., Gworek, B., Siepak, J., 2008. Mercury mobility and bioavailability in soil from contaminated area. Environ. Geol. 55, 1075–1087.

Bloom, S., Porcella, D.B., 1994. Less mercury. Nature 367, 694.

Bloom, N.S., Gill, G.A., Cappellino, S., Dobbs, C., Mcshea, L., Driscoll, C., Mason, R., Rudd, J., 1999. Sediment–water fluxes of mercury in Lavaca Bay. Tex. Environ. Sci. Technol. 33, 7–13.

Bloom, N.S., Preus, E., Katon, J., Hiltner, M., 2003. Selective extractions to assess the biogeochemically relevant fractionation of inorganic mercury in sediments and soils. Anal. Chim. Acta 479, 233–248.

Camargo, F.A.O., Bento, F.M., Jacques, R.J.S., 2007. Uso de microorganismos para remediacao de metais. In: Ceretta, C.A., Silvia, L.S., Reichert, J.M. (Eds.), Tópicos em Ciencia do Solo. Sociedade Brasileira de Ciencia do Solo, pp. 468–496. Minas Gerais.

Cardoso, E.J.B.N., Vasconcellos, R.L.F., Bini, D., Miyauchi, M.Y.H., Santos, C.A., Alves, P.R.L., Paula, A.M., Nakatani, A.S., Pereira, J.M., Nogueira, M.A., 2013. Soil health: looking for suitable indicators. What should be considered to assess the effects of use and management on soil health? Sci. Agric. 70, 274–289.

Cassina, L., Tassi, E., Pedron, F., Petruzzelli, G., Ambrosini, P., Barbafieri, M., 2012. Using a plant hormone and a thioligand to improve phytoremediation of Hg-contaminated soil from a petrochemical plant. J. Hazard. Mater. 231– 232, 36–42.

Casucci, C., Okeke, B., Frankenberguer, W.T., 2003. Effects of mercury on microbial biomass and enzyme activities in soil. Biol. Trace Elem. Res. 94, 179–191.

Certini, G., Scalenghe, R., 2006. Soils: basic concepts and future challenges. Cambridge University Press, Cambridge.

Chang, S., Wei, F., Yang, Y., Wang, A., Jin, Z., Li, J., He, Y., Shu, H., 2015. Engineering tobacco to remove mercury from polluted soil. Appl. Biochem. Biotechnol. 175, 3813–3827.

Chang, J.S., Hwang, Y.-P., Fong, Y.M., Lin, P.J., 1999. Detoxification of mercury by immobilized mercuric reductase. J. Chem. Technol. Biotechnol. 74, 965–973.

Chaturvedi, R., Archana, G., 2014. Cytosolic expression of synthetic phytochelatin and bacterial metallothionein genes in *Deinococcus radiodurans* R1 for enhanced tolerance and bioaccumulation of cadmium. BioMetals 27, 471–482.

Chaudhuri, D., Tripathy, S., Veeresh, H., Powell, M.A., Hart, B.R., 2003. Relationship of chemical fractions of heavy metals with microbial and enzyme activities in sludge and ash-amended acid lateritic soil from India. Environ. Geol. 44, 419–432.

Colacevich, A., Sierra, M.J., Borghinic, F., Millán, R., Sanchez-Hernandez, J.C., 2011. Oxidative stress in earthworms short- and long-term exposed to highly Hg-contaminated soils. J. Hazard. Mater. 194, 135–143.

Conde Bueno, P., Bellido, E., Martín Rubí, J.A., Jiménez Ballesta, R., 2009. Concentration and spatial variability of mercury and other heavy metals in surface soil samples of periurban waste mine tailing along a transect in the Almadén mining district (Spain). Environ. Geol. 56, 815–824.

Dash, H.R., Mangwani, N., Dash, S., 2014. Characterization and potential application in mercury bioremediation of highly mercury-resistant marine bacterium *Bacillus thuringiensis* PW-05. Environ. Sci. Pol. Res. 21, 2642–2653.

Deng, X., Wang, P., 2012. Isolation of marine bacteria highly resistant to mercury and their bioaccumulation process. Bioresour. Technol. 121, 342–347.

Dixit, R., Malaviya, D., Pandiyan, K., Singh, U.B., Sahu, A., 2015. Bioremediation of heavy metals from soil and aquatic environment: an overview of principles and criteria of fundamental processes. Sustainability 7, 2189–2212.

Dhal, B., Thatoi, H.N., Das, N.N., Pandey, B.D., 2013. Chemical and microbial remediation of hexavalent chromium from contaminated soil and mining/metallurgical solid waste: a review. J. Hazard. Mater. 251, 272–291.

Driscoll, R.P.M., Chan, H.M., Jacob, D.J., Pirrone, N., 2013. Mercury as a global pollutant: sources, pathways, and effects. Environ. Sci. Technol. 47, 4967–4983.

Doelman, P., Haanstra, L., 1989. Short- and long-term effects of heavy metals on phosphatase activity in soils: an ecological dose-response model approach. Biol. Fert. Soils 8, 235–241.

Dumontet, S., Perucci, P., Bufo, A.S., 1997. Soil organica amendments and herbicidal treatments: effect on soil microbial biomass. In: Actes du XXVII Congrès du Groupe Francais des Pesticides.

Das, S.K., Das, A.R., Guha, A.K., 2007. A study on the adsorption mechanism of mercury on Aspergillus versicolor biomass. Environ. Sci. Technol. 41, 8281–8287.

De, J., Ramaiah, N., Vardanyan, L., 2008. Detoxification of toxic heavy metals by marine bacteria highly resistant to mercury. Mar. Biotechnol. 10, 471–477.

Ettler, V., Rohovec, J., Navrátil, T., Mihaljevič, M., 2007. Mercury distribution in soil profiles polluted by lead smelting. Bull. Environ. Contam. Toxicol. 78, 13–17.

Feng, X., Chen, Y., Zhu, W., 1997. The distribution of various mercury species in soil. Chin. J. Geochem. 16, 183–188.

Fernandez-Martinez, R., Loredo, J., Ordonez, A., Rucandio, I., 2014. Mercury availability by operationally defined fractionation in granulometric distributions of soils and mine wastes from an abandoned cinnabar mine. Environ. Sci. Process Impacts 16, 1069–1075.

Fierer, N., Breitbart, M., Nulton, J., Salamon, P., Lozupone, C., Jones, R., Robeson, M., Edwards, R.A., Felts, B., Rayhawk, S., Knight, R., Rohwer, F., Jackson, R.B., 2007. Metagenomic and small-subunit rRNA analyses reveal the genetic diversity of bacteria, archaea, fungi, and viruses in soil. Appl. Environ. Microbiol. 73, 7059–7066.

Filgueiras, A.V., Lavilla, I., Bendicho, C., 2002. Chemical sequential extraction for metal partitioning in environmental solid samples. J. Environ. Monit. 4, 823–857.

Fredlund, D.G., Rahardjo, H., 1993. Soil Mechanism for Unsaturated Soils. John Wiley and Sons, New York.

Gadd, G.M., 2000. Bioremedial potential of microbial mechanisms of metal mobilization and immobilization. Curr. Opin. Biotechnol. 11, 271–279.

Gadd, G.M., 2004. Microbial influence on metal mobility and application for bioremediation. Geoderma 122, 109–119.

Gadd, G.M., Poole, R.K., 1999. Fungal Production of Citric and Oxalic Acid: Importance in Metal Speciation, Physiology and Biogeochemical Processes. In: Advances in Microbial Physiology, vol. 41. Academic Press, 47–92.

Gaur, N., Flora, G., Yadav, M., Tiwari, A., 2014. A review with recent advancements on bioremediation-based abolition of heavy metals. Environ. Sci. Process Impacts 16, 180–193.

Gil, C., Ramos-Miras, J., Roca-Pérez, L., Boluda, R., 2010. Determination and assessment of mercury content in calcareous soils. Chemosphere 78, 409–415.

Gupta, V.K., Nayak, A., Agarwal, S., 2015. Bioadsorbents for remediation of heavy metals: current status and their future prospects. Environ. Eng. Res. 20, 1–18.

Griffiths, B.S., Philippot, L., 2013. Insights into the resistance and resilience of the soil microbial community. FEMS Microbiol. Rev. http://dx.doi.org/10.1111/j.1574-6976.2012.00343.x.

Han, F.X., Su, Y., Shi, Z., Xia, Y., Tian, W., Philips, V., Monts, D.L., Gu, M., Liang, Y., 2012. Mercury distribution and speciation in floodplain soils and uptake into native earthworms (*Diplocardia* spp.). Geoderma 170, 261–268.

Hassen, A., Saidi, N., Cherif, M., Boudabous, A., 1998. Resistance of environmental bacteria to heavy metals. Bio Technol. 64, 7–15.

Hettiarachchi, G.M., Pierzynski, G.M., Oehme, F.W., Sonmez, O., Ryan, J.A., 2003. Treatment of contaminated soil with phosphorus and manganese oxide reduces lead absorption by Sprague-Dawley rats. J. Environ. Qual. 32, 1335–1345.

Higgins, I.J., Burns, R.G., 1975. The Chemistry and Microbiology of Pollution. Academic Press, London.

Hirner, A.V., Grüter, U.M., Kresimon, J., 2000. Metal(loid)organic compounds in contaminated soil. Fresenius J. Anal. Chem. 368, 263–267.

Hobman, J.L., Wilson, J.R., Brown, N.L., 2000. Microbial mercury reduction. In: Lovley, D.R. (Ed.), Environmental Microbe–metal Interactions. Am. Soc. Microbiol, Washington, pp. 177–197.

Hojdová, M., Navrátil, T., Rohovec, J., Penížek, V., Grygar, T., 2009. Mercury distribution and speciation in soils affected by historic mercury mining. Water Air Soil Pollut. 200, 89–99.

Jafari, S.A., Cheraghi, S., 2014. Mercury removal from aqueous solution by dried biomass of indigenous Vibrio parahaemolyticus PG02: Kinetic, equilibrium, and thermodynamic studies. Int. Biodeterior. Biodegrad. 92, 12–19.

Juhasz, A.L., Weber, J., Naidu, R., Gancarz, D., Rofe, A., Todor, D., Smith, E., 2010. Determination of cadmium relative bioavailability in contaminated soils and its prediction using in vitro methodologies. Environ. Sci. Technol. 44, 5240–5247.

Kaasalainen, M., Yli-Halla, M., 2003. Use of sequential extraction to assess metal partitioning in soils. Environ. Pollut. 126, 225–233.

Kacálková, L., Tlustoš, P., Száková, J., 2009. Phytoextraction of cadmium, copper, zinc and mercury by selected plants. Plant Soil Environ. 55, 295–304.

Kandeler, E., Kampichler, C., Horak, O., 1996. Influence of heavy metals on the functional diversity of soil microbial communities. Biol. Fert. Soils 23, 299–306.

Kandeler, E., Tscherko, D., Bruce, K.D., Stemmer, M., Hobbs, P.J., Bardgett, R.D., Amelung, W., 2000. Structure and function of the soil microbial community in microhabitats of a heavy metal polluted soil. Biol. Fert. Soils 32, 390–400.

Karaca, A., Cetin, S.C., Turgay, O.C., Kizilkaya, R., 2011. Soil enzymes as indication of soil quality. In: Soil Enzymology. Springer, Berlin, Heidelberg, pp. 119–148.

Karadas, C., Kara, D., 2011. In vitro gastro-intestinal method for the assessment of heavy metal bioavailability in contaminated soils. Environ. Sci. Pollut. Res. 18, 620–628.

Karunasagar, D., Arunachala, J., Rashmi, K., Naveena Lavanya Latha, J., Maruthi Mohan, P., 2003. Biosorption of inorganic and methylmercury by a biosorbent from *Aspergillus niger*. World J. Microbiol. Biotechnol. 19, 291–295.

Kavamura, V.N., Esposito, E., 2010. Biotechnological strategies applied to the decontamination of soils polluted with heavy metals. Biotechnol. Adv. 28, 61–69.

Kelly, B., Allan, C., Wilson, B.P., 2009. Soil indicators and their use by farmers in the Billabong Catchment, southern New South Wales. Aus. J. Soil Res. 47, 234–242.

Khan, K., Lu, Y., Khan, H., Ishtiaq, M., Khan, S., 2013. Heavy metals in agricultural soils and crops and their health risks in Swat District, northern Pakistan. Food Chem. Toxicol. 58, 449–458.

Khambhaty, Y., Mody, K., Basha, S., Jha, B., 2008. Hg(II) Removal from aqueous solution by dead fungal biomass of marine *Aspergillus niger*: kinetic study. Sep. Sci. Technol. 43, 1221–1238.

Kocman, D., Horvat, M., Pirrone, N., Cinnirella, S., 2013. Contribution of contaminated sites to the global mercury budget. Environ. Res. 125, 160–170.

Kotrba, P., Mackova, M., Macek, T., 2011. Microbial Biosorption of Metals–General Introduction. Microbial Biosorption of Metals. Springer, Netherlands, pp. 1–6.

Kurniati, E., Arfarita, N., Imai, T., Higuchi, T., Kanno, A., Yamamoto, K., Sekine, M., 2014. Potential bioremediation of mercury-contaminated substrate using filamentous fungi isolated from forest soil. J. Environ. Sci. 26, 1223–1231.

Latgé, J.P., 2010. Tasting the fungal cell wall. Cell Microbiol. 12, 863–872.

Li, Y., Yang, L., Ji, Y., Sun, H., Wang, W., 2009. Quantification and fractionation of mercury in soils from the Chatian mercury mining deposit, southwestern China. Environ. Geochem. Health 31, 617–628.

Li, H., Cong, Y., Ling, J., Chang, Y., 2014. Enhanced tolerance and accumulation of heavy metal ions by engineered *Escherichia coli* expressing *Pyrus calleryana* phytochelatin synthase. J. Basic Microbiol. http://dx.doi.org/10.1002/jobm.201300670.

Li, W.C., Tse, H.F., 2015. Health risk and significance of mercury in the environment. Environ. Sci. Pollut. Res. 22, 192–201.

Lomonte, C., Sgherri, C., Baker, A.J.M., Kolev, S.D., Navari-Izzo, F., 2010. Antioxidative response of *Atriplex codonocarpa* to mercury. Environ. Exp. Bot. 69, 9–16.

Mani, D., Kumar, C., 2014. Biotechnological advances in bioremediation of heavy metals contaminated ecosystems: an overview with special reference to phytoremediation. Int. J. Environ. Sci. Technol. 11, 843–872.

Matsuyama, A., Yasuda, Y., Yasutake, A., Xiaojie, L., Pin, J., Li, L., Mei, L., Yumin, A., Liya, Q., 2005. Relationship between leached total mercury and leached methylmercury from soil polluted by mercury in wastewater from an organic chemical factory in the People's Republic of China. Bull. Environ. Contam. Toxicol. 75, 1234–1240.

Mauclair, C., Layshock, J., Carpi, A., 2008. Quantifying the effect of humic matter on the emission of mercury from artificial soil surfaces. Appl. Geochem. 23, 594–601.

Moreno, F.N., Anderson, C.W.N., Stewart, R.B., Robinson, B.H., Ghomshei, M., Meech, J.A., 2005. Induced plant uptake and transport of mercury in the presence of sulphur-containing ligands and humic acid. New Phytol. 166, 445–454.

Moreno-Jimenez, E., Esteban, E., Carpena-Ruiz, R., Pealosa, J., 2009. Arsenic-and mercury-induced phytotoxicity in the Mediterranean shrubs *Pistacia lentiscus* and *Tamarix gallica* grown in hydroponic culture. Ecotoxicol. Environ. Saf. 72, 1781–1789.

Nam, D.H., Basu, N., 2011. Rapid methods to detect organic mercury and total selenium in biological samples. Chem. Cent. J. 5, 3.

National Research Council, 2000. Toxicological Effects of Methylmercury. National Academy Press, Washinghon, DC.

Navarro, A., 2008. Review of characteristics of mercury speciation and mobility from areas of mercury mining in semi-arid environments. Rev. Environ. Sci. Biotechnol. 7, 287–306.

Nieder, R., Benbi, D.K., 2008. Carbon and Nitrogen in the Terrestrial Environment. Springer, Netherlands.

Niemeyer, J.C., Bortoli Lolata, G., Martins de Carvalho, G., Mendes da Silva, E., Paulo Sousa, J., Nogueira, M.A., 2012. Microbial indicators of soil health as tools for ecological risk assessment of a metal contaminated site in Brazil. Appl. Soil Ecol. 59, 96–105.

Paz-Ferreiro, J., Fu, S., 2013. Biological índices for soil quality evaluation: perspectives and limitations. Land Degrad. Dev. http://dx.doi: 10.1002/ldr.2262.

Patra, M., Sharma, A., 2000. Mercury toxicity in plants. Bot. Rev. 66, 379–422.

Patrick, L., 2002. Eating disorders: a review of the literature with emphasis on medical complications and clinical nutrition. Altern. Med. Rev. 7, 456–471.

Porta Casanellas, J., López-Acevedo Reguerín, M., Roquero de Laburu, C., 1998. Edafología para la agricultura y el medio ambiente. Mundi-Prensa, Madrid.

Pezclar, M., Reid, R., Chan, E.C.S., 1981. MIcrobiologia, vol. 2. McGraw-Hill do Brasil, Sao Paulo.

Qiu, G., Feng, X., Meng, B., Sommar, J., Gu, C., 2012. Environmental geochemistry of an active Hg mine in Xunyang, Shaanxi Province, China. Appl. Geochem. 27, 2280–2288.

Rajapaksha, R.M.C.P., Tobor-Kaplon, M.A., Baath, E., 2004. Metal toxicity affects fungal and bacterial activities in soil differently. Appl. Environ. Microbiol. 70, 2966–2973.

Rao, M.A., Scelza, R., Acevedo, F., Diez, M.C., Gianfreda, L., 2014. Enzymes as useful tools for environmental purposes. Chemosphere 107, 145–162.

Rasmussen, L.D., Sørensen, S.J., 2001. Effects of mercury contamination on the culturable heterotrophic, functional and genetic diversity of the bacterial community in soil. FEMS Microbiol. Ecol. 36, 1–9.

Reis, A.T., Rodrigues, S.M., Davidson, C.M., Pereira, E., Duarte, A.C., 2010. Extractability and mobility of mercury from agricultural soils surrounding industrial and mining contaminated areas. Chemosphere 81, 1369–1377.

Remy, S., Prudent, P., Probst, J.L., 2006. Mercury speciation in soils of the industrialised Thur River catchment (Alsace, France). Appl. Geochem. 21, 1855–1867.

Renneberg, A.J., Dudas, M.J., 2001. Transformations of elemental mercury to inorganic and organic forms in mercury and hydrocarbon co-contaminated soils. Chemosphere 45, 1103–1109.

Rieder, S.R., Brunner, I., Daniel, O., Liu, B., Frey, B., 2013. Methylation of mercury in earthworms and the effect of mercury on the associated bacterial communities. PLoS One 8 (4), e61215.

Rieder, S.R., Frey, B., 2013. Methyl-mercury affects microbial activity and biomass, bacterial community structure but rarely the fungal community structure. Soil Biol. Biochem. 64, 164–173.

Rincon-Florez, V.A., Carvalhais, L.C., Schenk, P.M., 2013. Culture-independent molecular tools for soil and rhizosphere microbiology. Diversity 5, 581–612.

Robinson, J.B., Tuovinen, O.H., 1994. Mechanisms of microbial resistance and detoxification of mercury and organomercury compounds: physiological, biochemical, and genetic analyses. Microbiol. Rev. 48, 95–124.

Rodrigues, S.M., Henriques, B., Coimbra, J., Ferreira da Silva, E., Pereira, M.E., Duarte, A.C., 2010. Water-soluble fraction of mercury, arsenic and other potentially toxic elements in highly contaminated sediments and soils. Chemosphere 78, 1301–1312.

Rodrigues, S.M., Pereira, E., Duarte, A.C., Römkens, P.F.A.M., 2012. Derivation of soil to plant transfer functions for metals and metalloids: impact of contaminant's availability. Plant Soil 361, 329–341.

Rodriguez, L., Rincon, J., Asencio, I., Rodriguez-Castellanos, L., 2007. Capability of selected crop plants for shoot mercury accumulation from polluted soils: phytoremediation perspectives. Int. J. Phytorem. 9, 1–13.

Rojas, L.A., Yánez, C., Gonzáles, M., Lobos, S., Smalla, K., Seeger, M., 2011. Characterization of the metabolically modified heavy metal-resistant *Cupriavidus metallidurans* strain MSR33 generated for mercury bioremediation. PLoS One 6, e17555.

Ruby, M.V., Davis, A., Schoof, R., Eberle, S., Sellstone, C.M., 1996. Estimation of lead and arsenic bioavailability using a physiologically based extraction test. Environ. Sci. Technol. 30, 422–430.

Ruggiero, P., Terzano, R., Spagnuolo, M., Cavalca, L., Colombo, M., Andreoni, V., Rao, M.A., Perucci, P., Monaci, E., 2011. Hg bioavailability and impact on bacterial communities in a long-term polluted soil. J. Environ. Monit. 13, 145–156.

de Santiago-Martin, A., Cheviron, N., Quintana, J.R., Gonzalez, C., Lafuente, A.L., Mougin, C., 2013. Metal contamination disturbs biochemical and microbial properties of calcareous agricultural soils of the Mediterranean area. Arch. Environ. Contam. Toxicol. 64, 388–398.

Safruk, A.M., Berger, R.G., Jackson, B.J., Pinsent, C., Hair, A.T., Sigal, E.A., 2015. The bioaccessibility of soil-based mercury as determined by physiological based extraction tests and human biomonitoring in children. Sci. Total Environ. 518, 545–553.

Sait, M., Hugenholtz, P., Janssen, H., 2002. Cultivation of globally distributed soil bacteria from phylogenetic lineages previously only detected in cultivation-independent surveys. Environ. Microbiol. 4, 654–666.

Saglam, N., Say, R., Adil, D., Suleyman, P., Yapuc Arica, M., 1999. Biosorption of inorganic mercury and alkylmercury species on to *Phanerochaete chrysosporium* mycelium. Process Biochem. 34, 725–730.

Santos-Gandelman, J., Giambiagi-deMarval, M., Muricy, G., Barkay, T., Laport, M.S., 2014. Mercury and methylmercury detoxification potential by sponge-associated bacteria. Anton van Lee. http://dx.doi.org/10.1007/s10482-014-0224-2.

Sas-Nowosielska, A., Galimska-Stypa, R., Kucharski, R., Zielonka, U., Małkowski, E., Gray, L., 2008. Remediation aspect of microbial changes of plant rhizosphere in mercury contaminated soil. Environ. Monit. Assess. 137, 101–109.

Selinus, O., Alloway, B., 2013. Bioavailability of elements in soil. In: Selinus, O. (Ed.), Essentials of Medical Geology. Springer, Netherlands, pp. 351–373.

Sierra, M.J., Millan, R., Cardona, A.I., Schmid, T., 2011. Potential cultivation of *Hordeum vulgare* L. in soils with high mercury background concentrations. Int. J. Phytorem. 13, 765–778.

Sierra, M.J., Rodríguez-Alonso, J., Millán, R., 2012. Impact of the lavender rhizosphere on the mercury uptake in field conditions. Chemosphere 89, 1457–1466.

Silver, S., Phung, L.T., 1996. Bacterial heavy metal resistance: new surprises. Ann. Rev. Microbiol. 50, 753–789.

Šípková, A., Száková, J., Coufalík, P., Zvěřina, O., Kacálková, L., Tlustoš, P., 2014. Mercury distribution and mobility in contaminated soils from vicinity of waste incineration plant. Plant Soil Environ. 60, 87–92.

Schloss, P.D., Handelsman, J., 2006. Toward a census of bacteria in soil. PLoS Comput. Biol. 2 (0786), 0793.

Schulten, H.R., Leinweber, P., 2003. New insights into organic-mineral particles: composition, properties and models of molecular structure. Biol. Fert. Soils 30, 399–432.

Schlüter, K., 2000. Review: evaporation of mercury from soils. An integration and synthesis of current knowledge. Environ. Geol. 39, 249–271.

Smolinska, B., 2015. Green waste compost as an amendment during induced phytoextraction of mercury-contaminated soil. Environ. Sci. Pollut. Res. 22, 3528–3537.

van Straaten, P., 2000. Mercury contamination associated with small-scale gold mining in Tanzania and Zimbabwe. Sci. Total Environ. 259, 105–113.

Taylor, H., Appleton, J.D., Lister, R., Smith, B., Chitamweba, D., Mkumbo, O., Machiwa, J.F., Tesha, A.L., Beinhoff, C., 2005. Environmental assessment of mercury contamination from the Rwamagasa artisanal gold mining centre, Geita District, Tanzania. Sci. Total Environ. 343, 111–133.

Teršič, T., Gosar, M., 2012. Comparison of elemental contents in earthworm cast and soil from a mercury-contaminated site (Idrija area, Slovenia). Sci. Total Environ. 430, 28–33.

Tipping, E., Wadsworth, R.A., Norris, D.A., Hall, J.R., Ilyin, I., 2011. Long-term mercury dynamics in UK soils. Environ. Pollut. 159, 3474–3483.

Tokalioğlu, Ş., Kartal, Ş., Birol, G., 2003. Application of a three-stage sequential extraction procedure for the determination of extractable metal contents in highway soils. Turk. J. Chem. 27, 333–346.

Tschirhart, C., Handschumacher, P., Laffly, D., Bénéfice, E., 2012. Resource management, networks and spatial contrast in human mercury contamination along the Rio Beni (Bolivian Amazon). Hum. Ecol. 40, 810–818.

Vandenbosschea, M.J., Casettaa, M., Traisnela, M., 2014. Remediation of heavy metals by biomolecules—a review. Crit. Rev. Environ. Sci. Technol. http://dx.doi.org/10.1080/10643389.2014.966425.

Velasco, A., Acebo, P., Flores, N., Perera, J., 1999. The mer operon of the acidophilic bacterium Thiobacillus T3.2 diverges from its Thiobacillus ferrooxidans counterpart. Extremophiles 3, 35–43.

Vidali, M., 2001. Bioremediation. An overview. Pure Appl. Chem. 73, 1163–1172.

Wang, J., Feng, X., Anderson, C.W.N., Xing, Y., Shang, L., 2012. Remediation of mercury contaminated sites–a review. J. Hazard. Mater. 221–222, 1–18.

Wang, L.K., Tay, J.H., 2010. Environmental Bioengineering Humana.

White, C., Shaman, A.K., Gadd, G.M., 1998. An integrated microbial process for the bioremediation of soil contaminated with toxic metals. Nat. Biotechnol. 16, 572–575.

Xian, Y., Wang, M., Chen, W., 2015. Quantitative assessment on soil enzyme activities of heavy metal contaminated soils with various soil properties. Chemosphere. http://dx.doi.org/10.1016/j.chemosphere.2014.12.060.

Yu, B.S., Dong, H.L., Han, P.Y., 2012. Experimental research on microbial degradation of organic matter adsorbed in smectite internal surface area in the interlayer of the structure. Acta Petrol. Sin. 28, 949–960.

Yu, Z., Li, J., L, Y., Wang, Q., Zhai, X., Wu, G., Liu, P., Li, X., 2014. A mer operon confers mercury reduction in a *Staphylococcus epidermidis* strain isolated from Lanzhou reach of the Yellow River. Int. Biodeterior. Biodegrad. 90, 57–63.

Zagury, G.J., Bedeaux, C., Welfringer, B., 2009. Influence of mercury speciation and fractionation on bioaccessibility in soils. Arch. Environ. Contam. Toxicol. 56, 371–379.

Zagury, G.J., Neculita, C.M., Bastien, C., Deschenes, L., 2006. Mercury fractionation, bioavailability, and ecotoxicity in highly contaminated soils from chlor-alkali plants. Environ. Toxicol. Chem. 25, 1138–1147.

Zhang, H.B., Li, Y., Luo, Y.M., Christie, P., 2015. Anthropogenic mercury sequestration in different soil types on the southeast coast of China. J. Soils Sediments 15, 962–971.

Zhou, J., Liu, H.Y., Du, B.Y., Shang, L.H., Yang, J.B., Wang, Y.S., 2015. Influence of soil mercury concentration and fraction on bioaccumulation process of inorganic mercury and methylmercury in rice (*Oryza sativa* L.). Environ. Sci. Pollut. Res. 22, 6144–6154.

Zornoza, P., Millán, R., Sierra, M.J., Seco, A., Esteban, E., 2010. Efficiency of white lupin in the removal of mercury from contaminated soils: soil and hydroponic experiments. J. Environ. Sci. 22, 421–427.

CHAPTER

DETOXIFICATION AND TOLERANCE OF HEAVY METALS IN PLANTS

13

Rajesh Kumar[1], Rohit Kumar Mishra[1], Vani Mishra[2], Afifa Qidwai[1], Anand Pandey[1], Shashi Kant Shukla[1], Madhu Pandey[1], Ashutosh Pathak[1], Anupam Dikshit[1]

Biological Product Laboratory, Department of Botany, University of Allahabad, Allahabad, Uttar Pradesh, India[1]; Nanotechnology Application Centre, University of Allahabad, Allahabad, Uttar Pradesh, India[2]

CHAPTER OUTLINE

Plant Metal Interaction. http://dx.doi.org/10.1016/B978-0-12-803158-2.00013-8

1. INTRODUCTION

Metals are present naturally at various levels in the Earth's crust. Burning of fossil fuels, mining of metalliferous ores, wastes of municipal areas, and excessive use of pesticides and fertilizers lead to accelerated release of metals and metalloids into various components of ecosystems. Although many metals—such as copper (Cu), iron (Fe), manganese (Mn), nickel (Ni), zinc (Zn)—are essential for cells, they are toxic at higher concentrations (Millaleo et al., 2010; Arya and Roy, 2011; Iyaka, 2011). Metals at higher concentrations, whether of natural origin or of anthropogenic activity, may become toxic for soil microflora. The degree of heavy metal toxicity depends on the metallic element and its bioavailability in the soil. Heavy metal bioavailability is a function of abiotic factors such as metal concentration, humidity, and soil pH and biotic factors such as the presence of metal-liberating microflora. Various metals such as Zn, Cu, and Mn are essential for plants at low concentrations but become toxic at increasing concentrations, but other metals have never been shown to be essential for the development of living organisms and are toxic even at very low concentrations (e.g., Hg, lead (Pb), Cd) (Todeschini et al., 2011; Hall, 2002). Heavy metals are deposited in the environment because of use of sludge, compost of municipal, use of pesticides, fertilizers, emissions from industrial waste incinerators, vehicle exhausts, residues of metalliferous rocks or mines, and other smelting in different industries (Garbisu and Alkorta, 2003; Halim et al., 2003).

Presence of any metal and metalloid, where they are not required and in concentrations that cause toxic effects in the environment or humans, are regarded as contaminants (McIntyre, 2003). The concentration of metal in soil varies from less than 1 mg/kg to more than 100,000 mg/kg of soil (Figure 13.1), irrespective of the source of heavy metals in soil; higher concentrations of some metals degrade soil quality and reduce crop yield including poor-quality products, pose hazards to the ecosystems, and human and animal health (Blaylock and Huang, 2000; Long et al., 2002). Heavy metals create various stimuli for plants, which respond in various ways like chelation and compartmentalization of the metal ions, immobilization, exclusion, and the expression of stress response mechanisms like stress protein and ethylene (Andrades-Moreno et al., 2014). The heavy metal tolerance mechanism in plants has been

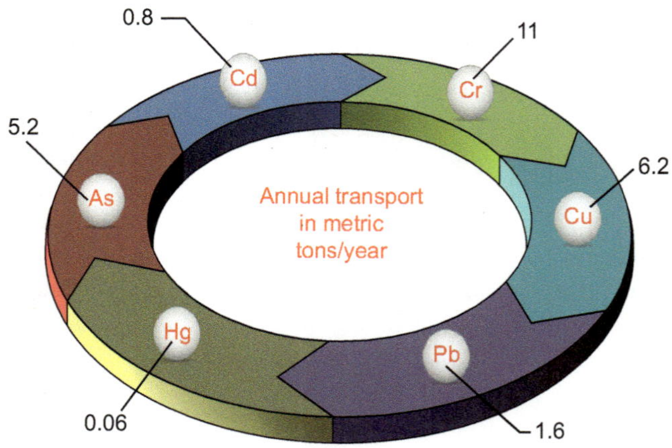

FIGURE 13.1

Annual transport of various heavy metals.

reviewed comprehensively in wide-ranging studies over many years, during which different aspects contamination sites have been studied (Salt et al., 1998).

1.1 CHELATION OF HEAVY METALS

Chelation of heavy metals by ligand is consider as a general mechanism of tolerance in plants, although in some cases the subsequent compartmentalization of the metal complex ligand. The toxicity symptoms created by metals observed occur possibly because of interaction of heavy metal at the molecular as well as the cellular level. It may be due to binding of heavy metal to sulfhydryl group of different proteins as well as generation of free radicals, leading to inhibition or destruction of cell. There are varieties of potential mechanisms of tolerance at the cellular level in plants that are reported to be involved in metals stress detoxification. It is also reported that accumulation of toxic metal within cells can be avoided by all these mechanisms, finally preventing the damaging effects by these metals. There are some examples of tolerant species that show an enhanced oxidative stress than they show homeostatic and avoidance mechanism to escape the effect of stress. Different strategies are exhibited by plants for avoiding heavy metals. As far as their occurrence is concerned, mining contributes the most, whereas urbanization contributes the least heavy metal source in the waste products of a developed countries (Figure 13.2).

FIGURE 13.2

Waste containing heavy metals produced in developed countries.

1.2 SOURCES OF HEAVY METALS

Heavy metals are introduced in plants through water uptake, which are then taken up by animals. The ingestion of these plant- and animal-based products is the main source of heavy metal accumulation in humans because they are barely metabolized. Heavy metals can pollute water sources (lakes, rivers, streams, and ground water) by leaching from consumer and industrial waste; acid rain intensifies this process by releasing heavy metals trapped in soil. Thus, the accumulation of various heavy metals from different point and non-point sources are deposited not only at the organismic level (biotic); but also in the soil and water which ultimately act as the sink for these metal contaminants (Figure 13.3).

1.3 BIOACTIVATION OF HEAVY METALS

Plants with a high capability of absorption of minerals exhibit a tremendous ability to collect and accumulate heavy metals (Yang et al., 2002; Ma et al., 2001). In most cases, heavy metals are not absorbed easily from soil because they have a low mobility in soil. However, a study showed there was

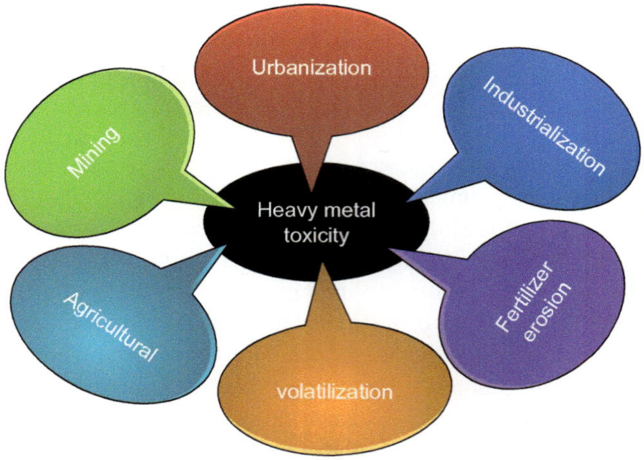

FIGURE 13.3

Source and sink of heavy metals.

no proper correlation between availability and accumulation of Zn in soil (Knight et al., 1994), whereas a significant correlation was noted between absorbed heavy metal and the availability of heavy metals in the rhizosphere. Their uptake by plant is depends on pH, organic substances, water content, metal content, and other elements in the rhizosphere. Soil and root interaction are reported to enhance availability of metals in the rhizosphere by secreting a variety of substances, for example phytochelatins (PCs), organic acids, amino acids, and enzymes (Figure 13.4). Protons accumulating by

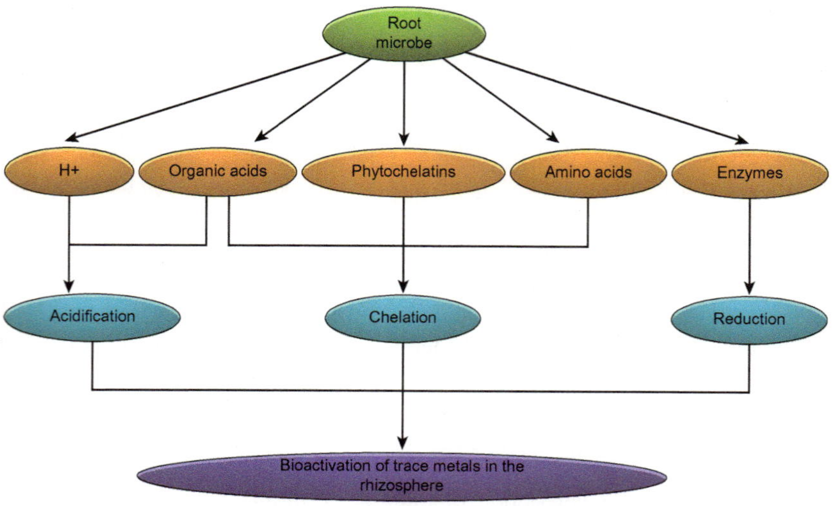

FIGURE 13.4

Steps involved in bioactivation of heavy metal.

the root can acidify rhizospheric soil, which increases the dissolution of metals (Bernal et al., 1994; Peng et al., 2005). It has been observed that the pH of rhizospheric soil of Cu-accumulating plants was lower than that of other rhizospheric soil. These protons can be removed by the plasma membrane hydrogen ion (H^+) pumps and H^+ adenosine triphosphatase (ATPase). In lupine, phosphorous promotes citrate exudation by improving plasma membrane H^+ pumps and H^+ ATPase export activity (Ligaba et al., 2004). It was reported that the protein AtHMA4 restricted at the plasma membrane and articulated in tissue surrounds roots' vascular vessels (Verret et al., 2004). Root growth can be improved by the ectopic overexpression of AtHMA4 in the presence of Zn, cadmium (Cd), and cobalt.

2. HEAVY METAL HYPERACCUMULATION IN PLANTS

The term "hyperaccumulator" was coined by Brooks (1998) for plants to differentiate from the excluder plants. These plants actively take up an essential reassessment of the cobalt and Cu hyperaccumulators (Faucon et al., 2007). The evolution of hyperaccumulation traits depends on different plants because they are distributed in a wide range of families. Why hyperaccumulating plants ascended for evolutionary reasons is still unclear and under debate. In metal-enriched soil in temperate and tropical zones, heavy accumulation does occur. New Caledonia, Latin America, South Africa, Europe, and North America all have heavy metal–dependent vegetation (Baker and Brooks, 1989). The hyperaccumulator term initially referred to plants, which are able to accumulate Ni in the shoot at more than 1 mg/g. Extremely high concentrations of heavy metal shows that Ni toxicity in vegetative organs of most plant commence at 10–15 mg/g. Hyperaccumulator plants that meet such a criterion are those that concentrate >10 mg/g Zn or Mn; more than 1 mg/g cobalt, Cu, arsenic (As), Ni, Pb, selenium, antimony (Sb), or titanium and more than 0.1 mg/g Cd in the aerial organs when they grow on native soil, without suffering phytotoxic damage (Verbruggen et al., 2009). A large number of taxa hyperaccumulate Ni (more than 75%), whereas low numbers hyperaccumulation (about five species) had been reported for Cd, which is considered a highly toxic metal. Further it has been reported that Ni reached the highest concentrations in *Sebertia acuminata* (Sapotaceae), which is endemic to serpentine soil from New Caledonia that accumulates approximately 26% Ni (dry mass) in its latex (Sagner et al., 1998). Approximately 25% of known hyperaccumulators are reported from the Brassicaceae family, particularly from genera *Thlaspi alyssum*. Brassicaceae also includes the highest number of Ni-accumulating plants (Brooks, 1998). In comparison with Ni, Zn hyperaccumulators are less frequent in Brassicaceae and Crassulaceae (Yang et al., 2004), which includes species of *Arabidopsis*, *Thlaspi*, and *Sedum alfredii*. These species also accumulate Cd. Approximately 25% of known hyperaccumulators are reported from the Brassicaceae particularly from genera *T. alyssum*. Brassicaceae also includes highest number of Ni accumulating plants (Harada et al., 2010). In comparison to Ni, Zn hyperaccumulators are less frequent in Brassicaceae and Crassulaceae (Yang et al., 2004), excluding *Arabidopsis*, *Thlaspi*, and *S. alfredii*. These species also accumulate Cd. Recently, *Solanum nigrum* (Solanaceae) had also been included as a Cd accumulator. Selenium hyperaccumulating plant species are included in different families, among which Asteraceae, Fabaceae, Rubiaceae, Scrophulariaceae, Brassicaceae, and Chenopodiaceae are common (Reeves and Baker, 2000). Despite angiosperms, such as Brassicaceae, Ma et al. (2001) and Wang et al. (2007) also reported that the genus *Pteris* and other ferns act as hyperaccumulators of As Table 13.1. Generally, the hyperaccumulators are found to be endemic to

Table 13.1 Hyperaccumulation of Various Heavy Metals and Their Effects in Some Plants

Metal	Plant	Effect	References
Arsenic	*Pteris vittata*	Inhibits photosynthesis, inhibits growth, biomass and yield, death	Dong (2005)
Cadmium	*Sesbania drummondii, Sedum alfredii*	Chlorosis, growth inhibition, reduction in photosynthesis, water and nutrient uptake, browning of root tips, death	Baker and Walker (1989)
Cromium	*Phragmites australis, Zea mays* L.	Reduction in nutrient uptake	Beyersmann and Hartwig (2008)
Lead	*Brassica napus*	Hampers metabolic pathways	Wierzbicka (1999)
Mercury	Chlor-alkali plant	Reduction in photosynthesis	Biester et al. (2002)

metalliferous soils called "strict metallophytes" that cannot survive without metalliferous soil, whereas various "facultative metallophytes" can survive on nonmetalliferous soil, even though they are ubiquitous on metal-enriched habitats (Assunc et al., 2003). Moreover, some species are reported to include both nonmetallicolous and metalicollous populations, such as in the Zn hyperaccumulators *Arabidopsis halleri* and *Thlaspi caerulescens*. Hyperaccumulation is an indigenous trait at the species level, being ubiquitous in all populations (Bert et al., 2000). Furthermore, in the Zn and in hyperaccumulators, this trait appears only in the metalicollous populations (Escarré et al., 2000; Verbruggen et al., 2009).

Classes of plants that concentrate extremely high amounts of usually toxic heavy metals through their leaves have garnered intensive interest because biologists to find the reason of this remarkable behavior, which raised a noteworthy question: Why do some plants do this? How does hyperaccumulation perform in these plants and how it is beneficial? What are the adaptive values of metal hyperaccumulation? The function of a high elemental concentration in leaves has been proposed hypothetically—drought resistance, metal tolerance, interference with neighboring plants, and defense against natural enemies.

On the basis of the tolerance hypothesis, it is assumed that the peculiar hyperaccumulation pattern is supposed to initiate plant uptake of metals from roots by cloistering them in leaf tissue, which is more tolerant (Gail et al., 2015). This clearly phases them out by detaching the high-metal aerial organ from the plant body. Another hypothesis explains that heavy metals in large amounts may enhance drought resistance of plants their water-conserving role and by behaving as osmolytes in a cell. However, these are merely hypotheses because they are not supported by any experimental verification and therefore need further investigation. The interference hypothesis, which also termed elemental allelopathy, suggests that neighboring plants may interfere with perennial hypoaccumulator plants all the way through fortification of metal on the surface of soil under the canopy. Group 1 has demonstrated that soil surface under the canopy of hyperaccumulator *S. acuminata* has higher Ni levels than that of nonhyperaccumulator species (Boyd and Jaffré, 2001). However, other groups have questioned the elemental allelopathy that *Alyssum murale*'s leaves shed with high Ni levels, but do not create a toxic zone around the Ni hyperaccumulator and are not reported to prohibit germination of competing plant seed (Zhang et al., 2007). Most Ni released from the biomass of leaves does not remain in soluble- and phyto-available forms, concluding the lack of an allelopathy effect; hence, it is rapidly bound to constituents of soil that prevent it from affecting the neighboring plants (Lacher and Smith, 2012). There is no acceptable experimental evidence that yet verifies the hypothesis of elemental allelopathy. Moreover, the hypothesis does not have satisfactory experimental verification yet (Morris et al., 2009). During the past few decades, many researchers have described the hyperaccumulation of different heavy metals in different plants. There are various factors regarding the degree of hyperaccumulation such as plant species, soil pH, temperature, organic matter, and types of heavy metals (Yang et al., 2006).

3. MAJOR STEPS INVOLVED IN HYPERACCUMULATION OF HEAVY METALS

The hyperaccumulation of heavy metals by higher plants Figure 13.5 involves three steps:

1. Transport of metals across the plasma membrane of root cells
2. Xylem loading and translocation
3. Detoxification and sequestration of metals at the whole plant and cellular levels.

Members of Brassicaceae and Fabaceae are the first hyperaccumulators of heavy metals. Most of the hyperaccumulating plants have the capability to absorb and accumulate two or more than two heavy metals (Jayakumar et al., 2013; Yang et al., 2002, 2004; He et al., 2002). However, the exact mechanism of hyperaccumulation is not understood. Usually, the accumulation ability of any heavy metals depends by its uptake capacity and intracellular transportation of plant tissues (Yang and Roemheld, 1999; Clemens et al., 2002; Yang et al., 2002).

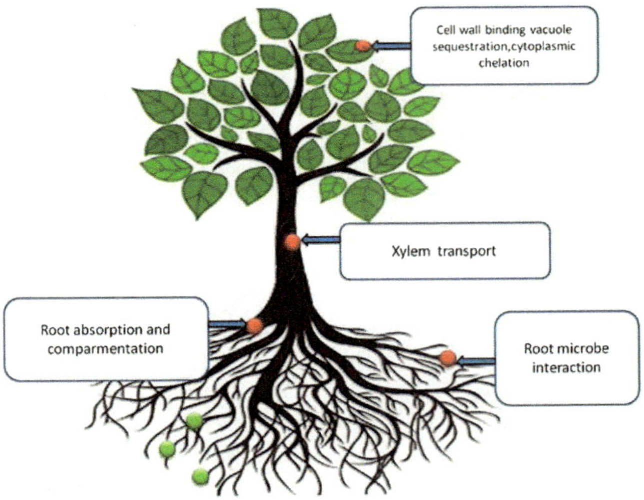

FIGURE 13.5

Major processes proposed to be involved in heavy metal hyperaccumulation by plants.

4. BIOCHEMICAL MECHANISM OF HEAVY METAL UPTAKE AND ACCUMULATION IN PLANTS

4.1 ARSENIC

This metalloid is of immense environmental apprehension because of its lethal effect and abundance, in countries such as China, Bangladesh, Hungary, India, and others. Heavy metal concentration is higher in ground water and the soil surface (Chen et al., 2006). Dissemination of As on the soil surface and in water takes place through the degeneration of rocks as well as lixiviation of different mineral elements (Luong et al., 2007). Despite the natural occurrence of heavy metals, coal mining, agricultural practices, fabrications, wood consumption, and food additives are also responsible for their accumulation. Arsenic toxicity is a global concern because it is believed to be responsible for promoting gall bladder, lung, prostate, and skin cancers. The US Environmental Protection Agency lowered the standards of As in potable water, although presently there is no definite limit of As in drinking water. Accumulation of As in higher plants is based on its chemical form bioavailability and mobility in the environment (Analia and Dundas, 2015). Arsenite—As(III) its inorganic form—is not mobile, but its toxicity is more prominent and considered to be more injurious to living organisms because it can alter the metabolic pathway (Kumaresan and Riyazuddin, 2001). Inorganic As(III) causes the detoxification of As through its own methylation, which is greatly metabolized by humans, resulting in the fabrication of organic compounds such as monomethylarsine acid (MMA), dimethylarsinic acid (DMA), and trimethylarsine that are regarded as carcinogens (Zhou et al., 2003). As per, Heikens (2006) As(V) tends to be the most recurrent and established structure of As found in aerobic soils and is therefore more available for uptake by plants. The transportation and accessibility of As in soil is depends on the texture and pH of soil (Heikens, 2006). It has been found that lower soil pH is favorable for complexation of As with Fe, whereas with a higher pH, it is calcium (Fayiga et al., 2007). It has also been reported that the presence of oxides of Fe and Mn in soil also aid in the mobility of As (Zavala and Duxbury, 2008). Various species of plants such as fern have been considered for their ability to build up As in their various forms. Ma et al. (2001), was the earliest to report that *Pteris vittata* (Chinese brake fern) is able to accumulate As. This plant takes up As(V) and reduces it to As(III) while transporting it through xylem tissues along with other mineral nutrients and finally accumulates As in As(III). Wang et al. (2007) reported that *P. vittata* can reduce and accumulate a large amount of As when treated with dydroponic media (Wang et al., 2007). There are various higher plants such as *Brassica juncea* and *Arabidopsis thaliana* that also can reduce and accumulate As, but these species show a lower accumulation capability than *P. vittata* (Singh and Ma, 2006).

Generally, plants take up and transport As through phosphate channels; nevertheless, some competition exists between As and phosphate because of their chemical similarity (Tripathi et al., 2007). Arsenic can be coordinated with a sulfur ligand forming a complex such as As(III) tris glutathione for transport (Pickering et al., 2000). The degree of accumulation and resistance of As varies in different plants because of genetic variation and or variation in the mechanism of detoxification (Wang et al., 2007). It has been reported that the As detoxification in plants entails mobilization of As from the root to the aerial part of the plant. The external As concentration controls this movement (Singh and Ma, 2006). As(V) is reduced to As(III) inside the plant tissue and hence biotransformed either to a less significant organic compound such as DMA, MMA, or as inorganic

As(III) (Pickering et al., 2000). This reduces most of the As(V) to As(III). X-ray absorption spectroscopy and liquid chromatography–mass spectrometry was used by Lopez et al. (2008) to study As speciation in honey mesquite (i.e., *Prosopis* sp.).

Researchers have treated the plant with 50 mg As(V) per liter, but the analyses showed both species of As in the roots. As(III) was found to be principally in stems, representing that mesquite reduced As(V) to As(III). However, although the plant uncovered As(III), a percentage of the As oxidized to As reduced the amount absorbed by the roots. Mobilization of As occur within plant tissue in the soluble portion and tends to chelate within cyclohexylene-dinitrotetraacetic acid, nitrilotriacetic acid or As–sulphur complexes. As(III) synchronized to sulfur and oxygen atoms are coordinated to the remaining fraction (Khang et al., 2012).

4.2 CADMIUM

Cadmium levels are extremely low in nature. It has been reported that Cd concentration varies from 0.01 to 5 mg/kg of soil in noncontaminated soil; however, the major source of Cd's spread is fertilizers, which is produced by phosphate ores (Chen et al., 2007; Ahmad et al., 2012). The emission of Cd waste has increased in populated areas throughout the world (Jarup, 2003). The chief source of Cd emission is smoking and food along with its use in various industrial applications (Lou et al., 2013). Because the alimentary canal absorbs Cd, it represents a serious environmental hazardous element and may infiltrate the placenta and destroy DNA and the cell membrane. It may cause kidney damage in humans. The recommended safe concentration of Cd by the World Health Organization is less than 10 nmol/mmol creatinine (approximately 200 mg Cd/kg kidney cortex) in the kidney. It is reported that Cd not only affects the kidneys and bones, but that it also attacks the reproductive organs of females, making it a serious threat to mammals and humans (Sun and Li, 2015). It also reported to affect the female endocrine system.

On the other hand, external factors like Fe tend to reduce Cd uptake, as reported in *Hordeum vulgare* (Sharma et al., 2004), whereas in the *T. caerulescens* Ganges ecotype, uptake of Cd is supported along with Fe uptake (Lombi et al., 2002). In *A. halleris*, Zn and Fe transport system relocates the Cd from medium to root xylem (Ueno et al., 2008). The phytosiderophore 2′-deoxymugineic acid production in maize plants is because of Cd exposure. This weak complex enables the uptake of Cd and Fe. On the other hand, in *Lactuca sativa*, the uptake of Cd is promoted by the Mn-enriched medium; 64% of the total uptake was accumulated in the cell walls and has potential to go up the food chain (Rasmos et al., 2002). In wheat, root cell vacuoles are the storage sites and radial movement of Cd is facilitated by PCs (Stolt et al., 2003; Cobbett, 2000). Diatloff et al. (2006) reported for the first time, regarding transport of Cd inside plants, that low-affinity cation transporter is responsible for calcium transport in wheat and is also accountable for Cd transport in *Pichia pastoris* (Diatloff et al., 2006). Most of the Cd hyperaccumulators tend to carry Cd by the production of low-molecular-weight thiols (LMWT). In roots and stems, LMWT production tends to hyperaccumulate Cd (Sharma et al., 2010). In leaves, LMWT contributes to half of the total Cd binding (de la Rosa et al., 2005). In rice, the Cd concentration is governed by its uptake and transfer from roots to shoots, and to a greater degree by the transfer of Cd from shoot to grain. The studies of Liu et al. (2007) on six rice cultivars demonstrates that 0.73% Cd uptake transferred to the rice grains (i.e., on an average 1.02 mg/kg) that is 100 times the concentration of Cd approved by the European Union (Liu et al., 2007; Yourtchi and Bayat, 2013).

4.3 CHROMIUM

Chromium (Cr) is one the most injurious elements of environment present naturally as chromite in different rocks as well as in complexes with other heavy metals as crocoites (Bluskov et al., 2005). Hexavalent Cr is more toxic than trivalent as it has an elevated oxidizing potential, solubility, and mobility across the living cell membrane, causing asthma, bronchitis, and pneumonitis in humans (Marqués et al., 1998). It is considered an impending carcinogen showing detrimental effects upon various organs and tissues in humans, where it stimulates the formation of reactive oxygen species that re responsible for damaging genetic material. Moreover, skin contact with Cr produces various abnormalities like allergies, dermatitis, and necrosis (Kotas and Stasicka, 2000). Particles of chromium oxide can be taken as cellular phagocytosis and solubilized inside lysosomes, resulting in liberation of Cr(III) in the cytoplasm (Beyersmann and Hartwig, 2008). After that, it is bound to either DNA molecules or hydrophobic molecules like 1,10 phenanthrolene and picolinic acid and transported across the plasma membrane, producing genomic mutations (Chigbo and Batty, 2013).

4.4 LEAD

Lead is widespread toxic element having no role in biological metabolism. The major source of Pb in the environment includes metal smelting, agriculture, industry, and urban activities (Caussy et al., 2003; Marchiol et al., 2004). Lead precipitates in the form of Pb phosphates and Pb sulfates because of their low solubility and availability (Akhtar et al., 2013). Plants do not have any special channel for lead uptake because it is not an essential element. Lead binds to the carboxylic group of uronic acids, present on root surfaces, which are the sites where Pb binds and facilitates in the uptake of Pb (Sharma and Dubey, 2005). Although some plants species like *Allium cepa*, *H. vulgare*, and *Zea mays* can tolerate Pb via complexation and inactivation, other species like *Brassica napus* and *Phaseolus vulgaris* experience toxicity because Pb hampers some metabolic pathways (Wierzbicka, 1999). In some plants, excess Pb inhibits germination of seeds, growth of plants, and synthesis of chlorophyll (Begonia et al., 2004; Xiong, 1998; Hussain et al., 2013). Roots act as a first barrier to the Pb (Figure 13.6), which results in more accumulation in roots (Blaylock and Huang, 2000). Ion-exchangeable sites present in the cell walls of the root are the binding sites for Pb, which it precipitates in the form of phosphates and carbonates (Hansda et al., 2014). Further inside movement of Pb facilitated by calcium channels accumulates near the endodermis (Kocon and Matyka, 2012; Antosiewicz, 2005).

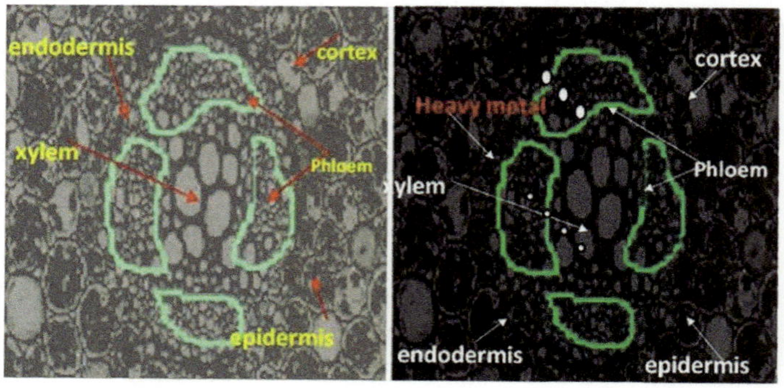

FIGURE 13.6

Cross-section of root showing comparative view of normal anatomical features along with accumulation of heavy metal (lead).

4.5 MERCURY

Emission of toxic heavy metals has increased in the past two centuries, mostly because of emanation from natural sources like volcanic eruptions. Emanation of mercury (Hg) in the environment is mainly because of chlor-alkali industries and coal-generated thermal power plants (Biester et al., 2002; Novoa-Munoz et al., 2008). Direct exposure of Hg is hazardous to human populace; soil and water connect to humans by the food chain. Uptake and accretion of Hg via crop plants represent the main access pathway for potentially health-threatening lethal metals into the food of humans and animals. Various agencies in regulatory and health sectors have ranked high on monitoring, evaluating, and reducing risk to human and other animals from exposure to Hg (Rada et al., 1990; Sample and Suter, 1999). The ranking of Hg on the Comprehensive Environmental Response, Compensation, and Liability Act list of substances was third in 1999 (ASTDR, 1999). The ranking of Hg on the Comprehensive Environmental Response, Compensation, and Liability Act list of substances was third in 1999 (ASTDR, 1999). The flow of Hg directly and through the food chain is an important concern and has several occasions resulted in remedial response activation in regions of the world (Ishtiaq and Mahmood, 2011). Mercury exposure is harmful to both terrestrial vertebrates as well as agricultural products (Doty, 2008).

Industrial and military activities have created a widespread contamination of the environment (Doty, 2008). However, different laws may prevent riparian activities in the immediate vicinity of economically poor sections of society, illegal immigrants, and migrant people on lands around the area of their existence. People used to grow different agricultural products such as grains and pulses in Hg-contaminated soil. These agricultural products are directly consumed or sold to sustain a livelihood and, finally, Hg contamination is spread across relatively broad social, economic, and geological areas. Human activity is the primary source of Hg contamination resulting in drastic health consequences. It has been noted that Zn-smelting plants resulted in a remarkable anthropogenic contributors to Hg emission with high Hg concentration in Zn ore being released through mining activities (Nriagu and Pacyna, 1988; Mukherjee et al., 2000). In some areas of China such as Hezhang, Zn smelting by local artisans has detrimentally impacted the local environment. Other sources of Hg contaminations include a thermometer manufacturing company (Krishna et al., 2003), artisanal gold mining activity (Pestana and Formoso, 2003), direct Hg mining, chlor-alkali plant emissions (Biester et al., 2002), and lead mining. Several studies have been carried out on the Hg uptake in a model of food grain sources to ascertain the impact of Hg and compounds of Hg on human health. For example, transfer of Hg, Pb, Cd, Zn, and Cu in the vegetables from soil were recorded near a Zn-smelting plant and analyzed for heavy metals concentration in the soil and vegetables (Zheng et al., 2007).

4.6 MAJOR CONSEQUENCES OF ACCUMULATION OF HEAVY METALS

Heavy metal tolerance in plants is the ability to survive virulent soil and is manifested by interaction between a genotype and the atmosphere (Macnair et al., 2000). Over time, several genetic studies on metal tolerance have been explained. The matter is whether or not there is molecular or organic chemistry that provides a metal tolerance capability; associated with this question is whether prevalence and multitolerance or cotolerance could exist wherever the plant grows in soil having many significant metals Chibuike and Obiora (2014). However, evidence for cotolerance is not much stronger than that for single tolerance (Macnair et al., 2000; Khanand and Khan, 2010). Plants are affected by different stresses generated by heavy metal accumulation like nutrient deficiency stress, oxidative stress in germinating seeds, metabolic stress, and even genetic stress (Figure 13.7).

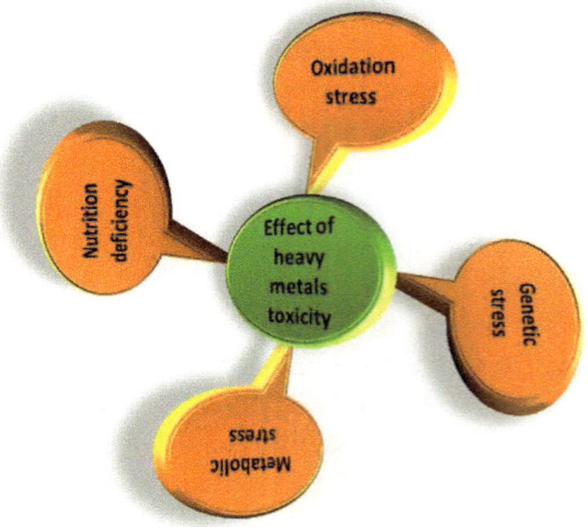

FIGURE 13.7

Major consequences of heavy metal accumulation.

5. VARIOUS CELLULAR MECHANISMS INVOLVED IN HEAVY METAL TOLERANCE

Extracellularly mycorrhizae and cell wall exudates act as protectors against heavy metals. The plasma membrane can also participate in tolerance either by reducing the uptake of heavy metals or increasing efflux metal ions. In the protoplast, different mechanisms of tolerance occur; for example, heat shock proteins (HSPs) and chelating molecules like organic acids, amino acids, or their compartmentalization away from metabolic processes by transport in the vacuoles (Figure 13.8; Table 13.2).

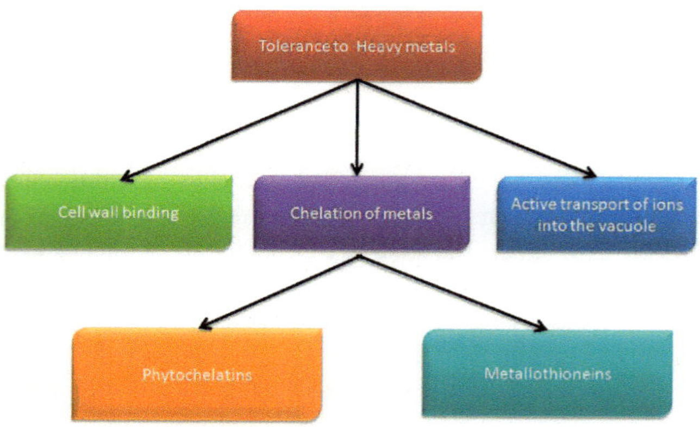

FIGURE 13.8

Biochemical mechanisms involved in heavy metal tolerance.

Table 13.2 Major Mechanisms Involved in Heavy Metal Tolerance

Metal	Mechanism Involved
Copper, zinc	Mycorrhizae
Arsenic	Plasma membrane (reduced uptake)
Cadmium	Phytochelatins
Chromium	Metallothioneins
Various other	Organic acids and amino acids
Cadmium	Heat shock proteins
Lead	Vacuolar compartmentalization

5.1 ORGANIC ACIDS

Organic acid secretion enhances root absorption by the mobilization of heavy metals. It has been reported that about 40% of the total metals of organic complexes is in the soil (Tarek and Shehata, 2015). There are many organic acids with low molecular weights that can influence the release of Cd from the soil. Solubility of Cd because of the formation of a low-molecular-weight organic acids complex—for example, acetic acid and succinate present in the rhizosphere of wheat (kyle)—is responsible for accumulating Cd genotype and nonaccumulating genotype (Arcola). Exudation of the maleate is induced by aluminum on the basis of aluminum tolerance in wheat, which is not accomplished by the changes in plasma membrane surface and in the activation of H^+ ATPase (Cieslinski et al., 1998; Ahn et al., 2004). However, in *T. caerulescens*, results were prepared with the contrast role of root exudation of metal hyperaccumulation concluding no relation between hyperaccumulation of metals and root exudation (McGrath et al., 1997; Zhao et al., 2001), whereas some researchers reported that interaction of microbes in root changes the condition of soil in the rhizosphere for increasing solubility of the Zn retained in the soil rhizosphere for hyper-accumulation in *T. caerulescens*. Characterization of the root exudates and exudation of a biochemical as well as molecular level should be further studied for increasing mobilization. In cereals a special type of structure phytosiderophores are secreted under the deficiency of iron for the increment of Fe, Zn, Cu, and Mn in the soil (Romheld, 1991). Secretion and synthesis of the phytosiderophore at the molecular level in extensively and intensively studied by distinguished workers (Yang and Roemheld, 1999; Kobayashi et al., 2003).

5.2 ROOT REDUCTASES

Reduction of Fe^{3+} or Cu^{2+} for increasing the uptake of Fe, Cu, Mn, and Mg by some dicots is done under a low supply of Fe and Cu (Welch et al., 1993). Ferric reductase genes are extracted from *Arabidopsis* and maize (Yang and Roemheld, 1999). Characteristic features of root reductase enzymes for tolerance of hyperaccumulation of metals transformed in the rhizospheric zone of soil are not exactly implicit.

5.3 PHYTOCHELATINS

PCs are enzymatically synthesized proteins involved in heavy metal detoxification in plants growing in metal-rich soil. However, there are studies showing the response of plants even at lower concentrations of heavy metals (Yadav, 2010; Clemens, 2001; Zhu et al., 2014; Goldsbrough, 2000; Pal and Rai, 2009). In plants, these metals are synthesized and regulated by a PC synthase gene encoding PC synthase enzyme. Different published reports revealed that heavy metal detoxification and accumulation take place along with advance peptide structure as well as biosynthetic regulation of PCs (Mishra and Dubey, 2006; Pal and Rai, 2009). Phytochelatins have the capability of binding to different metals including Cd, Cu, Zn, or As through sulfhydryl and carboxyl residues whose biosynthesis is controlled mainly by the metals glutathione (γ-glutamlcysteinyl-glycine), which is a precursor of PCs, involves several steps mediated by PC synthase (Olsson et al., 2015). Studies regarding the inhibition of biosynthezing of PCs via glutathione demonstrated their role in the detoxification of heavy metals in certain algae, fungi, and a group of high taxa plants. Overexpression of PC synthase genes in some bacteria and fungi increases heavy metal tolerance significantly, but in higher plants exhibiting high metal accumulating capability it does not always occur. Plants showing the hyperaccumulation of Cd, Zn, and Ni have common features of massive metal transport from root toward shoot. Hence, PC-based detoxification of heavy metals might be a defense mechanism first involved in lower plants and later established in higher vascular plants for the removal mainly of Cd (Yang and Chu, 2011). De Knecht et al. (1992) reported that deferential cadmium tolerance in *Silene vulgaris* was not because of the production of various PCs. PC production may play a small role in Cd detoxification, whereas in *S. vulgaris* PC production is higher but is not involved in Cd tolerance, even though confirmation of the role of PCs in heavy metal detoxification is sturdy in favor of Cd. Various peptides may also engage in recreation essentials in cells, like sulfur metabolism and antioxidants (Cobbett, 2000). Involvement of PCs in the detoxification of heavy metals may be a consequence of these functions. Vacuolar sequestrations of Cd in plant cell is illustrated in Figure 13.9.

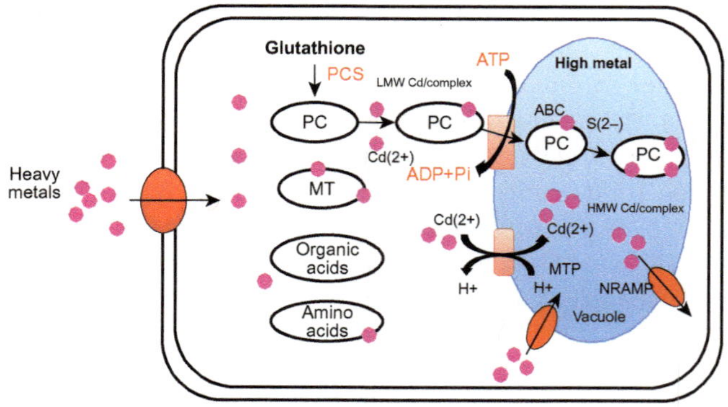

FIGURE 13.9

Uptake of heavy metal (Cd^{2+}) through zinc/iron-regulated transporters (ZIPs) enters the cytosol and stimulates the glutathione-derived synthesis of phytochelatins (PCs) by PC synthases (PCS). PCs bind cytosolic Cd^{2+} to form the low-molecular-weight (LMW) complex first, which is transported into the vacuole via a tonoplast-localized adenosine triphosphate–binding cassette (ABC) transporter. In the vacuole, a LMW Cd complex then accumulates into a high-molecular-weight (HMW) complex with more Cd^{2+}, which may enter the vacuole via a direct exchange with protons by tonoplast-localized proton exchanger transporters. Other chelators including metallothioneins (MTs), organic acids, and amino acids help buffer the cytosolic metal concentrations to a safe, low-metal state.

5.4 METALLOTHIONEINS

MTs are small, metal ion binding proteins that appear to be ubiquitous in eukaryotes and cyanobacteria and other multicellular organisms. Numerous exceedingly divergent MTs can be identified in blue green algae, algae, and higher groups of plants. Differences between plant and mammalian MTs are because of their amino acid sequence, charge, total content of cysteine, peptide spacing patterns, and length of the linker domains. Moderately higher plant groups have families with bulky complexes of MT genes, whereas small or less complex families of MTs are found in mammals. MTs of plants show significant advantageous induction properties and metal-binding activities. This may protect plants from increased level of noxious metals such as Cd and Hg. Homeostasis levels of MTs can also be affected from nutrient elements like Cu and Zn. For example, in the *Fucus vesiculosus* (a marine brown alga), a 67-amino acid MT with 16 cysteine residues was found to be Cu-inducible and bound both Cd(II) and Cu(II) in vitro.

The mammalian MT gene into *Bactris campestris*, *B. napus*, and *Nicotiana tabacum* confers Cd tolerance by increasing complexation of metal with MT and minimizing translocation of Cd to the shoots (Hall, 2002; Pal and Rai, 2009). Certain genes in plants have been found to encode MT-like proteins having sequences similar to animal MT proteins. Several MT-encoding genes were identified and isolated in *Arabidopsis.* Plant MTs have the highest complexing affinity for Cu and are induced by exposure to it (Murphy and Taiz, 1995). In *Arabidopsis*, on Cu treatment, strong induction in MT2 messenger RNA levels has been observed. When genes for MT from *Arabidopsis* were expressed in an MT-deficient yeast mutant, it complemented the mutation and provided a high level of resistance to Cu (Goldsbrough, 2000). Likewise, in Cu-sensitive mutants of *Arabidopsis*, it was shown that Cu decreases in MT could induce MT gene levels and result in increased level of free transition metal ions (Hall, 2002; Corbett, 2000; Pal and Rai, 2009).

5.5 MYCORRHIZAE

The rhizosphere is inhabited by hefty amount of microorganisms like several bacteria and fungi (mycorrhizal fungi). The bacteria colonize in roots and mycorrhizae can increase the bioavailability of diverse elements along with heavy metal ions uptake significantly (Afzal et al., 2014). First, the rapid transformation of bacteria can change the metal bioavailability in soil. For instance, a strain of *Xanthomonas maltophyla* bacterium can be exposed for catalyzing the reduction and precipitation of highly mobile Cr ions from significantly less movable environmentally less hazardous compounds (Blake et al., 1993). Induced transformation of toxic metal ions including Pb^{2+}, Hg^{2+}, Au^{3+}, Te^{4+}, and Ag^{+} can also be found in some strains (Lasat, 2002). Mobility of heavy metals can enhance the dissimilatory nonreducing activity of bacterium (Cummings et al., 1999). In addition, various soil microorganisms can show a drastically increased accumulation of Zn in the shoot of *T. caerulescens* because of the facilitation to increase the solubility of nonlabile Zn metal found in soil for increasing its biological availability for the plants (Whiting et al., 2001). Mycorrhizal symbiotic association of fungus was proposed by researchers as a mechanism of tolerance against heavy metals (Pidlisnyuk et al., 2014; Hildebrandt et al., 2007).

Evolved strategies of mycorrhizae can alleviate the accumulation of heavy metals in mixed system of culture and thus in the food chains (Hall, 2002; Kramer, 2005). These strategies can include the immobilization of metal compounds, polyphosphate granules precipitation in soil, chitin adsorption in the cell wall of fungus, and heavy metal chelation inside the fungus. Rhizospheric heavy metals can be bounded by mycorrhizae by the release of insoluble glycoproteins known as glomalin (Gonzalez-Chavez et al., 2004; Gohre and Paszkowski, 2006). It is reported that 1 g of glomalin can extract up to 4.3 mg of Cu, 0.08 mg of Cd, and 1.12 mg of Pb in contaminated soils (Gonzalez-Chavez et al., 2004). The hyphae networking of mycorrhizae present in soil is one of the important factors for the removal of heavy metals in a large surface area. For example, passive adsorption of hyphae can lead the binding of 0.5 mg of Cd in per gram of dry biomass (Joner et al., 2000). In one study, immobilization and uptake of Pb is found at greater levels in mycorrhizal roots than in nonmycorrhizal roots (Chen et al., 2005). There is various evidence for the suggestion that fungal vesicles can provide additional detoxification by absorbing the deadly substances (Gohre and Paszkowski, 2006). Heavy metals can be selectively retained in inner parenchymatous cells along with fungal cells in maize plant (Kaldorf et al., 1999). Likewise, in leguminous plants like pea and clover mycorrhizae provide buffer conditions to stress caused by Cd (Rivera-Becerril et al., 2002), which helps in the stabilization of Cd in the root systems of the colonized plants (Paradi et al., 2003).

Arbuscular mycorrhizal fungus colonizes in roots of dominant plant species flushed with highly contaminated soils (Pawlowska et al., 1996; Leung et al., 2007). These microorganisms present in soil are usually accredited as promoters of plant microbes in soil. These soil microorganisms are generally accredited as growth promoters of plants, because its extraradical mycelium pervasively participates in acquiring plant nutrients, particularly phosphorus (Smith and Read, 1997). Even though in contaminated soil, extensive inconsistency had been observed in plant responses toward vesicular arbuscular mycorrhiza (VAM) inoculation. It has been demonstrated in various studies that VAM fungi can potentially buffer heavy metal stress (Hildebrandt et al., 1999; Janousková et al., 2005; Chen et al., 2007). Therefore, its remedial properties certify that both the VAM-mediated nutritional effects as well as the effect of VAM fungi on metal allocation at the interface of soil and plant are involved in the removal of heavy metals. This alleviative authority may be recognized not only for arbuscular mycorrhiza–mediated nutritional effects, but also for the contact of arbuscular mycorrhiza fungi distributed on the interface of soil–fungus–plant as well (Kaldorf et al., 1999; Joner et al., 2000). It has been shown that in extra- and intraradical fungal tissue, the metal immobilizes, which is responsible for providing a conceivable explanation for frequently reported augmented barrier in translocation of metal in inoculated plants from the root toward the shoot (Joner and Leyval, 1997; Afzal et al., 2014; Chen et al., 2007). It has also been reported that, in some plants, VAM fungi can remarkably cope with heavy metal toxicity in contaminated plants, compared with those that are basically never exposed to elongated expression selection pressures (Malcova et al., 2003). Despite escalating understanding of VAM interaction with plants under heavy metal stress, it is still doubtful whether there is an alliance among the heavy metal tolerances between plants and fungi. That tolerant mycorrhizae may be the best addition in heavy metal tolerance on its host is merely a hypothesis, because it has a higher rate of survival as well as successful reproduction on the sites of contamination. It can also be hypothesized that the tolerance of arbuscular mycorrhiza fungus may confer additional tolerance against heavy metal on a host, leading to the higher rate of survival and successfully reproducing at the contaminated sites. On the other hand, endorsement of carbon for maintaining VAM symbiosis may cost quite a lot for the plants that are tolerant, even without being mycorrhizal, can handle well the contamination of heavy metal because it is not suitable against heavy metal–sensitive ones (Figure 13.10).

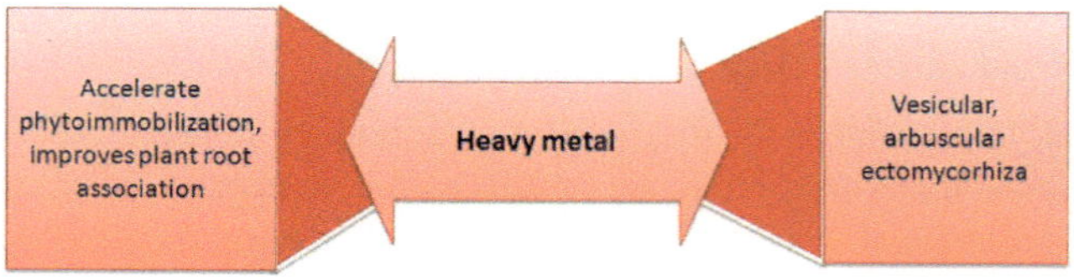

FIGURE 13.10

Heavy metal tolerance by mycorrhizae.

5.6 HEAT SHOCK PROTEINS

HSPs are a class of molecular chaperons in normal protein structure and also function to protect and repair other proteins under stress conditions of plants (Lewis et al., 2001). Activation of HSPs like HSP70, HSP80, and HSP25 by heavy metals has been reported by researchers (Neumann et al., 1994; Lin et al., 2014; Ireland et al., 2004; Zhen et al., 2007). HSPs show increased expression in response to growth of a variety of organisms at their optimal growth temperature. They are involved in preventing the denaturation of various proteins from oxidative stress resulting from heavy metal toxicity (Aritz et al., 2015). There is a great deal of evidence that an increase in heavy metal stress results in an increase of HSP expression. For example, heavy metal stress has increased the expression of messenger RNA level of HSPs in rice (Tseng et al., 1993). Neumann et al. (1995) reported that HSPs are articulated in roots of *Armeria maritime* in response to Cu stress (Neumann et al., 1995); however, a very low amount of HSPs was found in plants growing on other metalliferous soil. Moreover, HSP70 is also increased in *Enteromorpha intestinalis* after exposure to Cu stress (Lewis et al., 2001). More molecular evidence is needed to support such an important protective mechanism of plants (Figure 13.11).

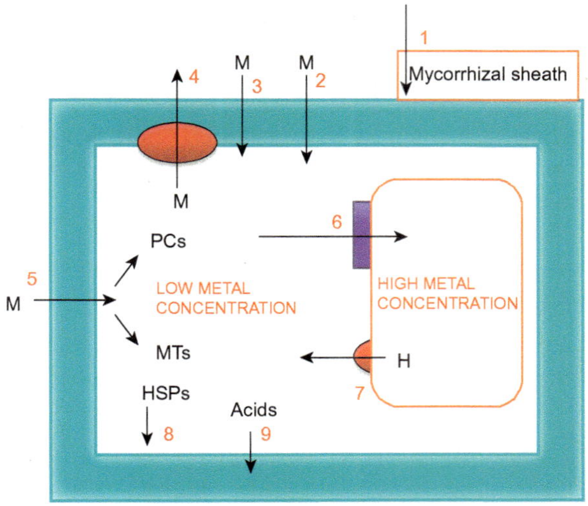

FIGURE 13.11

Cellular detoxification of heavy metals. 1, Restriction of heavy metal movement inside the root cell by mycorrhizae; 2, metal binding to cell wall; 3, reduced influx of metals; 4, active efflux from cytosol; 5, chelation of heavy metals by various ligands; 6, transport chelated complex inside a vacuole; 7, accumulation of heavy metal in a vacuole; 8, repair and protection by heat shock protein; 9, repair and protection by organic acids.

6. CONCLUSION AND FUTURE PROSPECTS

The current revisionary knowledge is focused on different evidence that categorizes impending cellular molecular mechanisms involved in resistance and tolerance of plants against prominent concentration of heavy metals. The strategies adapted by plants aim to avoid accumulation of heavy metals in cytosol and preventing toxicity symptoms. This is facilitated by using various tolerance mechanisms that present and are likely to be employed in plant homeostasis. Various mechanisms are employed for various plants. The general tolerance mechanism involves prevention of uptake of heavy metals in cytosol by directing it to an apoplastic space. It is also possible to involve more than one mechanism for reducing toxicity. The processes involved in reducing the toxicity of heavy metals are of considerable interest because an understanding of the means of manipulating tolerance could be important in developing phytoremediation purposes.

Plants developed various cellular and molecular adaptations necessary to tolerate heavy metal stress. Significant progress has been made in identifying and understanding the role of key components like metal transporters, hyperaccumulation, PCs, mycorrhizae, organic acids, amino acids, HSPs, MTs, and vascular compartmentalization that ensure heavy metal tolerance to plants. Hyperaccumulator plants are prevalent on metal soils in both tropical and temperate zones of all the continents and belong to several unrelated families. The hyperaccumulation potential has evolved more than once, although its adaptive assessment is still under deliberation. More elements and a larger number of hyperaccumulator species must be examined to authenticate the assumption of distrustful effects of heavy metals. Additionally, the investigations must to move from laboratory to field settings to make available reasonable information about elemental resistance in natural environments, where a plant can be exposed to an overabundance of herbivores with different modes of feeding as well as to pathogens and parasites. The ultimate demonstration that a particular mechanism or gene product is involved in metal tolerance will require the development of more functional tools than are available now, and many research groups are actively working on transformation systems for heavy metal tolerance in plants that should provide for the efficient overexpression or disruption of target genes. Knowledge about metal tolerance mechanisms in these plants will lead to powerful applications in bioremediation. On the whole, a multidisciplinary research effort that integrates the work of plant biologists, soil chemists, microbiologists, and environmental engineers is essential for greater success in understanding heavy metal accumulation and tolerance in plants.

ACKNOWLEDGEMENTS

Thanks are due to the head of the Department of Botany, University of Allahabad, for providing the facilities and to the University Grants Commission, New Delhi, for providing financial assistance; to Prof. A.C. Pandey, Vice-Chancellor, Bundelkhand University, Jhansi; to Dr Sanjeev Kumar Srivastava, Convener, Indian Nanoelectronics User Program (INUP), Indian Institute of Science (IISc.), Bangalore, India.

REFERENCES

Afzal, M., Shabir, G., Tahseen, R., Ejazul, I., Iqbal, S., Khan, Q.M., 2014. Endophytic *Burkholderia* sp. strain PsJN improves plant growth and phytoremediation of soil irrigated with textile effluent. Clean Soil Air Water 42, 1304–1310.

Ahmad, I., Akhtar, M.J., Zahir, Z.A., Jamil, A., 2012. Effect of cadmium on seed germination and seedling growth of four wheat (*Triticum aestivum* L.) cultivars. Pak. J. Bot. 44, 1569–1574.

Ahn, S.J., Rengel, Z., Matsumoto, H., 2004. Aluminum-induced plasma membrane surface potential and H^+-ATPase activity in near-isogenic wheat lines differing in tolerance to aluminum. New Phytol. 162, 71–79.

Akhtar, M.S., Chali, B., Azam, T., 2013. Bioremediation of arsenic and lead by plants and microbes from contaminated soil. Res. Plant Sci. 1, 68–73.

Analia, I., Dundas, G., 2015. Effect of arsenic on reflectance spectra and chlorophyll fluorescence of aquatic plants. Chemosphere 119, 697–703.

Andrades-Moreno, L., Del Castillo, I., Parra, R., Doukkali, B., Redondo-Gómez, S., Pérez-Palacios, P., 2014. Prospecting metal-resistant plant-growth promoting rhizobacteria for rhizoremediation of metal contaminated estuaries using *Spartina densiflora*. Environ. Sci. Pollut. Res. 21, 3713–3721.

Antosiewicz, D.M., 2005. Study of calcium-dependent lead-tolerance on plants differing in their level of Ca-deficiency tolerance. Environ. Pollut. 134, 23–34.

Aritz, B., Lur, E., Carlos, M., 2015. Impact of repeated single-metal and multi-metal pollution events on soil quality. Chemosphere 120, 8–15.

Arya, S.K., Roy, B.K., 2011. Manganese induced changes in growth, chlorophyll content and antioxidants activity in seedlings of broad bean (*Vicia faba* L.). J. Environ. Biol. 32, 707–711.

Assunc, A.G.L., Schat, H., Aarts, M.G.M., 2003. *Thlaspi caerulescens*, an attractive model species to study heavy metal hyperaccumulation in plants. New Phytol. 159, 351–360.

ASTDR, C., 1999. List of Priority Hazardous Substances. Agency for Toxic Substances and Disease Registry, Atlanta, GA.

Baker, A.J.M., Brooks, R.R., 1989. Terrestrial higher plants which hyperaccumulate metallic elements–a review of their distribution, ecology and phytochemistry. Biorecovery 1, 81–126.

Baker, A.J.M., Walker, P.L., 1989. Ecophysiology of metal uptake by tolerant plants. In: Shaw, A. (Ed.), Heavy metal tolerance in plants - Evolutionary aspects. CRC Press, pp. 155–177.

Begonia, M.T., Begonia, G.B., Miller, G.S., Gilliard, D., 2004. Effects of chelate application time on the phytoextraction of lead-contaminated soils. Bull. Environ. Contam. Toxicol. 73, 1033–1040.

Bernal, M.P., McGrath, S.P., Miller, A.J., Baker, A.J.M., 1994. Comparison of the chemical changes in the rhizosphere of the nickel hyperaccumulator *Alyssum murale* with the non-accumulator *Raphanus sativus*. Plant Soil 164, 251–259.

Bert, V., Macnair, M.R., De Laguerie, P., Saumitou-Laprade, P., Petit, D., 2000. Zinc tolerance and accumulation in metallicolous and nonmetallicolous populations of *Arabidopsis halleri* (Brassicaceae). New Phytol. 146, 225–233.

Beyersmann, D., Hartwig, A., 2008. Carcinogenic metal compounds: recent insight into molecular and cellular mechanisms. Arch. Toxicol. 82, 493–502.

Biester, H., Müller, G., Schöler, H.F., 2002. Binding and mobility of mercury in soils contaminated by emissions from chlor-alkali plants. Sci. Total Environ. 284, 191–193.

Blake, R.C., Choate, D.M., Bardhan, S., Revis, N., Barton, L.L., Zocco, T.G., 1993. Chemical transformation of toxic metals by a *Pseudomonas* strain from a toxic waste site. Environ. Toxicol. Chem. 12, 1365–1376.

Blaylock, M.J., Huang, J.W., 2000. Phytoremediation of Toxic Metals: Using Plants to Clean up the Environment. John Wiley, New York, vol. 9. pp. 53–71.

Bluskov, S., Arocena, J.M., Omotoso, O.O., Young, J.P., 2005. Uptake, distribution, and speciation of chromium in *Brassica Juncea*. Int. J. Phytorem. 7, 153–155.

Boyd, R.S., Jaffré, T., 2001. Phytoenrichment of soil Ni content by *Sebertia acuminate* New Caledonia and the concept of elemental allelopathy. S. Afr. J. Sci. 97, 535–538.

Brooks, R.R. (Ed.), 1998. Plants That Hyperaccumulate Heavy Metals. CAB International, Wallingford, UK p. 380.

Caussy, D., Gochfeld, M., Gurzau, E., Neagu, C., Ruedel, H., 2003. Lessons from case studies of metals: investigation exposure, bioavailability, and risk. Ecotoxicol. Environ. Saf. 56, 45–51.

Chen, B.D., Thu, Y.G., Duan, J., Xiao, X.Y., Smith, S.E., 2007. Effects of the arbuscular mycorrhizal fungus *Glomus mosseae* on growth and metal uptake by four plant species in copper mine tailings. Environ. Pollut. 147, 374–380.

Chen, Y., Graziano, J.H., Parvez, F., Hussain, I., Momotaj, H., van Geen, A., 2006. Modification of risk of arsenic-induced skin lesions by sunlight exposure, smoking, and occupational exposures in Bangladesh. Epidemiology 17, 459–467.

Chen, Z., Tang, J., Zhi, G., Hu, S., 2005. Arbuscular mycorrhizal colonization and phosphorus acquisition of plants: effects of coexisting plant species. Appl. Soil Ecol. 28, 259–269.

Chibuike, G.U., Obiora, S.C., 2014. Heavy metal polluted soils: effect on plants and bioremediation methods. Hindawi Publishing Corporation Appl. Environ. Soil Sci. 2014, 12 pp.

Chigbo, C., Batty, L., 2013. Effect of combined pollution of chromium and benzo (a) pyrene on seed growth of *Lolium perenne*. Chemosphere 90, 164–169.

Cieslinski, G.K.C., Van Rees, J., Szmigielska, A.M., Krishnamurti, G.S.R., Huang, P.M., 1998. Low-molecular-weight organic acids in rhizosphere soils of durum wheat and their effect on cadmium bioaccumulation. Plant Soil 203, 109–117.

Clemens, S., 2001. Molecular mechanism of plant metal tolerance and homeostasis. Planta 212, 475–486.

Clemens, S., Palmgren, M.G., Kramer, U., 2002. A long way ahead: understanding and engineering plant metal accumulation. Trends Plant Sci. 7, 309–315.

Cobbett, C.S., 2000. Phytochelatins and their role in heavy metal detoxification. Plant Physiol. 123, 825–883.

Cummings, D.E., Caccavo, F., Fendorf, J.S., Rosenzeig, R.F., 1999. As mobilization by the dissimilatory Fe(III)-reducing bacterium *Shewanella alga* BrY. Environ. Sci. Technol. 33, 723–729.

Diatloff, E., Forde, B.G., Robert, S.K., 2006. Expression and transport characterization of the wheat low-affinity cation transporter (LCT1) in the methylotrophic yeast *Pichia pastoris*. Bichem. Biophys. Res. Comm. 344, 807–813.

Dong, R., 2005. Molecular cloning and characterization of a phytochelatin synthase gene, *PvPCS1*, from *Pteris vittata* L. J. Ind. Microbiol. Biotechnol. 32, 527–533.

Doty, S.L., 2008. Enhancing phytoremediation through the use of transgenics and endophytes. New Phytol. 179, 318–333.

Escarré, J., Lefebvre, C., Gruber, W., 2000. Zinc and cadmium hyperaccumulation by *Thlaspi caerulescens* from metalliferous and nonmetalliferous sites in the Mediterranean area: implications for phytoremediation. New Phytol. 145, 429–437.

Faucon, M.P., Ngoy Shutcha, M., Meerts, P., 2007. Revisiting copper and cobalt concentrations in supposed hyperacumulators from SC Africa: influence of washing and metal concentration soil. Plant Soil 301, 29–36.

Fayiga, A.O., Ma, L.Q., Zhou, Q., 2007. Effects of plant arsenic uptake and heavy metals on arsenic distribution in an arsenic-contaminated soil. Environ. Pollut. 147, 737–742.

Gall, J.E., Boyd, R., Rajkaruna, N., 2015. Transfer of heavy metals through terrestrial food webs: a review. Environ. Monit. Assess. 187, 4436–4443.

Garbisu, C., Alkorta, I., 2003. Basic concepts on heavy metal soil bioremediation. Eur. J. Miner. Process. Environ. Prot. 13, 58–66.

Gohre, V., Paszkowski, U., 2006. Contribution of arbuscular mycorrhizal symbiosis to heavy metal phytoremediation. Planta 223, 1115–1122.

Goldsbrough, P., 2000. Metal tolerance in plants: the role of phytochelatins and metallothiones. In: Terry, N., Banuelos, G. (Eds.), Phytoremedation of Contaminated Soil and Water, vol. 8. CRC Press, pp. 221–233.

Gonzalez-Chavez, M.C., Carrillo-Gonzalez, R., Wright, S.F., Nichols, K.A., 2004. The role of glomalin, a protein produced by arbuscular mycorrhizal fungi, in sequestering potentially toxic elements. Environ. Pollut. 130, 317–323.

Halim, M., Conte, P., Piccolo, A., 2003. Potential availability of heavy metals to phytoextraction from contaminated soils induced by exogenous humic substances. Chemosphere 1, 265–275.

Hall, J.L., 2002. Cellular mechanism of heavy metal detoxification and tolerance. J. Exp. Bot. 53, 1–11.

Hansda, A., Anshumali, V.K., Usmani, Z., 2014. Phytoremediation of heavy metals contaminated soil using plant growth promoting rhizobacteria (PGPR): a current perspective. Recent Res. Sci. Technol. 6, 131–134.

Harada, E., Kim, J., Meyer, A.J., Hell, R., Clemens, S., Choi, Y.E., 2010. Expression profiling of tobacco leaf trichomes identifies genes for biotic and abiotic stresses. Plant Cell Physiol. 51, 1627–1637.

He, B., Yang, X.E., Wei, Y.Z., Ye, Z.Q., Ni, W.Z., 2002. A new lead resistant and accumulating ecotype—*Sedum alfredii* H. Acta Bot. Sin. 44, 1365–1370.

Heikens, A., 2006. Arsenic Contamination of Irrigation Water, Soil and Crops in Bangladesh: Risk Implications for Sustainable Agriculture and Food Safety in Asia. Food Agr Org UN, Regional Office for Asia and the Pacific. RAP Publication 2006/20. ISBN: 974-9746-88-2.

Hildebrandt, U., Kaldorf, M., Bothe, H., 1999. The zinc violer and its colonization by arbuscular mycorrhizal fungi. J. Plant Physiol. 154, 709–717.

Hildebrandt, U., Regvar, M., Bothe, H., 2007. Arbscular mycorrhiza and heavy metal tolerance. Phytochemistry 68, 139–146.

Hussain, A., Abbas, N., Arshad, F., 2013. Effects of diverse doses of lead (Pb) on different growth attributes of *Zea mays* L. Agric. Sci. 4, 262–265.

Ireland, H.E., Harding, S.J., Bonwick, G.A., Jones, M., Smith, C.J., Williams, J.H.H., 2004. Evaluation of heat shock protein 70 as a biomarker of environmental stress in *Fucus serratus* and *Lemna minor*. Biomarkers 9, 139–155.

Ishtiaq, S., Mahmood, S., 2011. Phytotoxicity of nickel and its accumulation in tissues of three *Vigna* species at their early growth stages. J. Appl. Bot. Food Qual. 84, 223–228.

Iyaka, Y.A., 2011. Nickel in soils: a review of its distribution and impacts. Sci. Res. Essays 6, 6774–6777.

Janousková, M., Pavlıkova, D., Macek, T., Vosatka, M., 2005. Arbuscular mycorrhiza decreases cadmium phytoextraction by transgenic tobacco with inserted metallothionein. Plant Soil 272, 29–40.

Jarup, L., 2003. Hazards of heavy metal contamination. Br. Med. Bull. 68, 167–182.

Jayakumar, K., Rajesh, M., Baskaran, L., Vijayarengan, P., 2013. Changes in nutritional metabolism of tomato (*Lycopersicon esculantum* Mill.) plants exposed to increasing concentration of cobalt chloride. Int. J. Food Nutr. Saf. 4, 62–69.

Joner, E.J., Brione, R., Leyvan, C., 2000. Metal binding capacity of arbuscular mycorrhizal fungi. Plant Soil 226, 227–234.

Joner, E.J., Leyval, C., 1997. Uptake of Cd by root & hyphae of *Glomus musseae/Trifolium subterraceum* mycorhiza from soil amended with high and low concentration of cadmium. New Phytol. 135, 353–360.

Kaldorf, M., Kuhn, A.J., Schroder, W.H., Hildebrandt, U., Bothe, H., 1999. Selective element deposits in maize colonized by a heavy metal tolerance conferring arbuscular mycorrhizal fungus. J. Plant Physiol. 154, 718–728.

Khanand, M.R., Khan, M.M., 2010. Effect of varying concentration of nickel and cobalt on the plant growth and yield of Chickpea. Aust. J. Basic Appl. Sci. 4, 1036–1046.

Khang, H.V., Hatayama, M., Inoue, C., 2012. Arsenic accumulation by aquatic macrophyte coontail (*Ceratophyllum demersum* L.) exposed to arsenite, and the effect of iron on the uptake of arsenite and arsenate. Environ. Exp. Bot. 83, 47–52.

de Knecht, J.A., Koevoets, P.L.M., Verkleij, J.A.C., Ernst, W.H.O., 1992. Evidence against a role for phytochelatins in naturally selected increased cadmium tolerance in *Silene vulgaris* (Moench) Garcke. New Phytol. 122, 681–688.

Knight, B., Zhao, F.J., McGrath, S.P., Shen, Z.G., 1994. Zinc and cadmium uptake by hyperaccumulator *Thlaspi caerulescens* in contaminated soils and its effects on the concentration and chemical speciation of metals in soil solution. Plant Soil 197, 71–78.

Kobayashi, T., Nakayama, Y., Itai, R.N., Nakanishi, H., Yoshihara, T., Mori, S., Nishizawa, N.K., 2003. Identification of novel cis-acting elements, IDE1 and IDE2, of the barley IDS2 gene promoter conferring iron-deficiency-inducible, root-specific expression in heterogeneous tobacco plants. Plant J. 36, 780–793.

Kocon, A., Matyka, M., 2012. Phytoextractive potential of *Miscanthus giganteus* and *Sida hermaphrodita* growing under moderate pollution of soil with Zn and Pb. J. Food Agric. Environ. 10, 1253–1256.

Kotas, J., Stasicka, Z., 2000. Chromium occurrence in the environment and methods of its speciation. Environ. Pollut. 107, 263–283.

Kramer, U., 2005. Phytoremediation: novel approaches to cleaning up polluted soils. Curr. Opin. Biotechnol. 16, 133–141.

Krishna, M.V.B., Karunasagar, D., Arunachalam, J., 2003. Study of mercury pollution near a thermometer factory using lichens and mosses. Environ. Pollut. 124, 357–360.

Kumaresan, M., Riyazuddin, P., 2001. Overview of speciation chemistry of arsenic. Curr. Sci. 80, 837–846.

Lacher, C., Smith, R.W., 2012. Sorption of Hg(II) by *Potamogeton natans* dead biomass. Miner. Eng. 92, 87–191.

Lasat, M.M., 2002. Phytoextraction of toxic metals: a review of biological mechanisms. J. Environ. Qual. 31, 109–120.

Leung, H.M., Ye, Z.H., Wong, M.H., 2007. Survival strategies of plants associated with arbuscular mycorrhizal fungi on toxic mine tailings. Chemosphere 66, 905–915.

Lewis, S., Donkin, M.E., Depledge, M.H., 2001. Hsp70 expression in *Enteromorpha intestinalis* (Chlorophyta) exposed to environmental stressors. Aquat. Toxicol. 51, 277–291.

Ligaba, A., Yamaguchi, M., Shen, H., Sasaki, T., Yamamoto, Y., Matsumoto, H., 2004. Phosphorus deficiency enhances plasma membrane H^+-ATPase activity and citrate exudation in greater purple lupin (*Lupinus pilosus*). Funct. Plant Biol. 31, 1075–1083.

Lin, Y.F., Severing, E., Te Lintel Hekkert, B., Schijlen, E., Aarts, M.G.M., 2014. A comprehensive set of transcript sequences of the heavy metal hyperaccumulator *Noccaea caerulescens*. Front. Plant Sci. 5, 261. http://dx.doi.org/10.3389/fpls.2014.00261.

Liu, J., Qian, M., Cai, G., Yang, J., Zhu, Q., 2007. Uptake and translocation of Cd in different rice cultivars and the relation with Cd accumulation in rice grain. J. Hazard. Mater. 143, 443–447.

Lombi, E., Tearall, K.L., Howarth, J.R., Zhao, F.J., Hawkesford, M.J., McGrath, S.P., 2002. Influence or iron status on cadmium and zinc uptake by different ecotypes of the hyperacumulator *Thlaspi caerulescens*. Plant Physiol. 128, 1359–1367.

Long, X.X., Yang, X.E., Ni, W.Z., 2002. Current status and perspective on phytoremediation of heavy metal polluted soils. J. Appl. Ecol. 3, 757–762.

Lopez, M.L., Peralta-Videa, J.R., Parsons, J.G., Gardea-Torresdey, J.L., 2008. Concentration and biotransformation of arsenic by *Prosopis* sp. grown in soil treated with chelating agents and phytohormones. Environ. Chem. 5, 1–12.

Lou, Y., Luo, H., Hu, T., Li, H., Fu, J., 2013. Toxic effects, uptake, and translocation of Cd and Pb in perennial ryegrass. Ecotoxicology 22, 207–214.

Luong, J.H.T., Majid, E., Male, K.B., 2007. Analytical tools for monitoring arsenic in the environment. Open Anal. Chem. J. 1, 7–14.

Ma, L.Q., Komar, K.M., Tu, C., Zhang, W.H., Cai, Y., Kennelley, E.D., 2001. A fern that hyperaccumulates arsenic—a hardy, versatile, fast-growing plant helps to remove arsenic from contaminated soils. Nature 4, 409–579.

Macnair, M.R., Tilstone, G.H., Smith, S.E., 2000. The Genetics of Metal Tolerance and Accumulation in Higher Plants, 5. CRC Press LLC, 221–233.

Malcova, R., Vosatka, M., Gryndler, M., 2003. Effects of inoculation with *Glomus intraradices* on lead uptake by *Zea mays* L. and *Agrostis capillaris* L. Appl. Soil Ecol. 23, 255–267.

Marchiol, L., Assolari, S., Sacco, P., Zerbi, G., 2004. Phytoextraction of heavy metals by canola (*Brassica napus*) and radish (*Raphanus sativus*) grownon multicontaminated soil. Environ. Pollut. 132, 21–27.

Marqués, M.J., Salvador, A., Morales-Rubio, A.E., de la Guardia, M., 1998. Analytical methodologies for chromium speciation in solid matrices: a survey of literature. Fres. J. Anal. Chem. 362, 239–248.

McGrath, S.P., Shen, Z.G., Zhao, F.J., 1997. Heavy metal uptake and chemical changes in the rhizosphere of *Thlaspi caerulescens* and *Thlaspi ochroleucum* grown in contaminated soils. Plant Soil 188, 153–159.

McIntyre, T., 2003. Phytoremediation of heavy metals from soils. Adv. Biochem. Eng. Biotechnol. 78, 97–123.

Millaleo, R., Reyes-Dıaz, M., Ivanov, A.G., Mora, M.L., Alberdi, M., 2010. Manganese as essential and toxic element for plants: transport, accumulation and resistance mechanisms. J. Soil Sci. Plant Nutr. 10, 476–494.

Mishra, S., Dubey, R.S., 2006. Heavy metal uptake and detoxification mechanisms in plants. Int. J. Agric. Res. 1, 122–141.

Morris, C., Grossi, P.R., Call, C.A., 2009. Elemental allelopathy: processes, progress and pitfalls. Plant Ecol. 202, 1–11.

Mukherjee, A.B., Melanen, U.M., Ekqvist, M., Verta, M., 2000. Assessment of atmospheric mercury emissions in Finland. Sci. Total Environ. 259, 73–83.

Murphy, A., Taiz, L., 1995. Comparison of metallothionein gene extression and non protein thiols in ten Aradiopsis ecotypes. Plant Physiol. 109, 945–954.

Neumann, D., Lichtenberger, O., Gunther, D., Tschiersch, K., Nover, L., 1994. Heat-shock proteins induce heavy-metal tolerance in higher plants. Planta 194, 360–367.

Neumann, D., ZurNieden, U., Lichtenberger, O., Leopold, I., 1995. How does *Armeria maritime* tolerate high heavy metal concentrations? J. Plant Physiol. 146, 704–717.

Novoa-Munoz, J.C., Pontevedra-Pombal, X., Martinez-Cortizas, A., Garcia-Rodeja Gayoso, E., 2008. Mercury accumulation in upland acid forest ecosystems nearby a coalfired power-plant in Southwest Europe (Galicia, NW Spain). Sci. Total Environ. 394, 303–312.

Nriagu, J.O., Pacyna, J.M., 1988. Quantitative assessment of worldwide contamination of air, water and soils by trace metals. Nature 333, 134–139.

Olsson, S., Puene-Sanchez, F., Gomez, M.J., Aquilera, A., 2015. Transcriptional response to copper excess and identification of genes involved in heavy metal tolerance in the extremophilic microalga *Chlamydomonas acidophila*. Extremophiles 19, 657–672.

Pal, R., Rai, J.P.N., 2009. Phytochelatins: peptides involved in heavy metal detoxification. Appl. Biochem. Biotech. 160, 945–963.

Paradi, I., Berecz, B., Halsz, K., Bratek, Z., 2003. Influence of arbuscular mycorrhiza and cadmium on the polyamine contents of RiT-DNA transformed *Daucus carota* L. root cultures. Acta Biol. Szegediensis 47, 31–36.

Pawlowska, T.E., Blaszkowski, J., Ruhling, A., 1996. The mycorrhizal status of plants colonizing a calamine spoil mound in southern Poland. Mycorrhiza 6, 499–505.

Peng, H.Y., Yang, X., Jiang, L., 2005. Copper phytoavailability and uptake by *Elsholtzia splendens* from contaminated soil as affected by soil amendments. J. Environ. Sci. Health 40, 839–856.

Pestana, M.H., Formoso, M.L., 2003. Mercury contamination in Lavras do Sul, south Brazil: a legacy from past and recent gold mining. Sci. Total Environ. 307, 125–140.

Pickering, I.J., Prince, R.C., George, M.J., Smith, R.D., George, G.N., Salt, D.E., 2000. Reduction and coordination of arsenic in Indian mustard. Plant Physiol. 122, 1171–1177.

Pidlisnyuk, V., Erickson, L., Kharchenko, S., Stefanovska, T., 2014. Sustainable land management: growing *Miscanthus* in soils contaminated with heavy metals. J. Environ. Prot. 5, 723–730.

Rada, R.G., Wiener, J.G., Bailey, P.A., Powell, D.E., 1990. Recent influxes of metals into Lake Pepin, a natural lake on the Upper Mississippi River. Arch. Environ. Contam. Toxicol. 19, 712–726.

Ramos, I., Esteban, E., Lucena, J.J., Garate, A., 2002. Cadmium uptake and subcellular distribution in plants of *Lactuca* sp. Cd–Mn interaction. Plant Sci. 162, 761–767.

Reeves, R.D., Baker, A.J.M., 2000. Metal-accumulating plants. In: Raskin, I., Ensley, B.D. (Eds.), Phytoremediation of Toxic Metals: Using Plants to Clean up the Environment, 16. John Wiley & Sons, pp. 193–229.

Rivera-Becerril, F., Calantzis, C., Turnau, K., Caussanel, J.P., Belimov, A.A., Gianinazzi, S., Strasser, R.J., Gianinazzi-Pearson, V., 2002. Cadmium accumulation and buffering of cadmiuminduced stress by arbuscular mycorrhiza in three *Pisum sativum* L. genotypes. J. Exp. Bot. 53, 1177–1185.

de la Rosa, G., Martinez-Martinez, A., Pelayo, H., Peralta-Videa, J.R., Sanchez-Salcido, B., Gardea-Torresdey, J.L., 2005. Production of low-molecularweight thiols as a response to cadmium uptake by tumbleweed (*Salsola kali*). Plant Physiol. Biochem. 43, 491–498.

Romheld, V., 1991. The role of phytosiderophores in acquisition of iron and other micronutrients in graminaceous species: an ecological approach. Plant Soil 130, 127–134.

Sagner, S., Kneer, R., Wanner, G., Cosson, J.P., Deus-Neumann, B., Zenk, M.H., 1998. Hyperaccumulation, complexation and distribution of nickel in *Sebertia acuminata*. Phytochemistry 47, 339–347.

Salt, D.E., Smith, R.D., Raskin, I., 1998. Phytoremediation. Annu. Rev. Plant Physiol. 49, 643–668.

Sample, B.E., Suter, G.W., 1999. Ecological risk assessment in a large river-reservoir: piscivorous wildlife. Environ. Toxicol. Chem. 18, 610–620.

Sharma, P., Dubey, R.S., 2005. Lead toxicity in plants. Braz. J. Plant Physiol. 17, 35–52.

Sharma, R.K., Devi, S., dan Dhyani, P.P., 2010. Comparative assessment of the toxic effects of copper and cypermethrinusing seeds of *Spinacia Oleracea* L. plants. Trop. Ecol. 51, 375–387.

Sharma, S.S., Kaul, S., Metwally, A., Goyal, K.C., Finkemeier, I., Dietz, K.J., 2004. Cadmium toxicity to barley (*Hordeum vulgare*) as affected by varying Fe nutritional status. Plant Sci. 166, 1287–1295.

Singh, N., Ma, L.Q., 2006. Arsenic speciation, and arsenic and phosphate distribution in arsenic hyperaccumulator *Pteris vittata* L. and non-hyperaccumulator *Pteris ensiformis* L. Environ. Pollut. 141, 238–246.

Smith, S.E., Read, D.J., 1997. Mycorrhizal Symbiosis, vol. 137 (3). Academic Press, London, pp. 563–568, http://dx.doi.org.10.1046/j.1469-8137.1997.00851-2.

Stolt, J.P., Sneller, F.E.C., Bryngelsson, T., Lundborg, T., Schat, H., 2003. Phytochelatin and cadmium accumulation in wheat. Environ. Exp. Bot. 49, 21–28.

Sun, Y., Li, Y., 2015. *In situ* stabilization remediation of cadmium (Cd) and lead (Pb) co-contaminated paddy soil using bentonite. Appl. Clay Sci. 10, 200–206.

Tarek, M., Shehata, S., 2015. Bioaccumulation and translocation of heavy metals by *Plantago major* L. grown in contaminated soils under the effect of traffic pollution. Ecol. Indicators 48, 244–251.

Todeschini, V., Lingua, G., D'Agostino, G., Carniato, F., Roc- cotiello, E., Berta, G., 2011. Effects of high zinc concentration on poplar leaves: a morphological and biochemical study. Environ. Exp. Bot. 71, 50–56.

Tripathi, R.D., Srivastava, S., Mishra, S., Singh, N., Tuli, R., Gupta, D.K., 2007. Arsenic hazards: strategies for tolerance and remediation by plants. Trends Biotechnol. 25, 158–165.

Tseng, T.S., Tzeng, S.S., Yeh, C.H., Chang, F.C., Chen, Y.M., Lin, C.Y., 1993. The heat shock response in rice seedling-isolation and expression of cDNAs that encode class-I low molecular weight heat-shock proteins. Plant Cell Physiol. 34, 165–168.

Ueno, D., Iwashita, T., Zhao, F.J., Ma, J.F., 2008. Characterization of Cd translocation and identification of Cd form in xylem sap of the Cd-hyperaccumulator *Arabidopsis halleri*. Plant Cell Physiol. 49, 540–548.

Verbruggen, N.C., Hermans, H., Schat, M., 2009. Molecular mechanisms of metal hyperaccumulation in plants. New Phytol. 181, 759–776.

Verret, F., Gravot, A., Auroy, P., Leonhardt, N., David, P., Nussaume, L., Vavasseur, A., Richaud, P., 2004. Overexpression of AtHMA4 enhances root-to-shoot translocation of zinc and cadmium and plant metal tolerance. FEBS Lett. 576, 306–312.

Wang, J., Zhao, F.J., Meharg, A.A., Raab, A., Feldmann, J., McGrath, S., 2007. Mechanisms of arsenic hyper-accumulation in *Pteris vittata*. Uptake, kinetics, interactions with phosphate, and arsenic speciation. Plant Physiol. 130, 1552–1561.

Welch, R.M., Norvell, W.A., Schaefer, S.C., Shaff, J.E., Kochian, L.V., 1993. Induction of iron(III) and copper(II) reduction in pea (*Pisumsativum* L.) roots by Fe and Cu status: does the root-cell plasmalemma Fe(III)-chelate reductase perform a general role in regulating cation uptake? Planta 190, 555–561.

Whiting, S.N., de Souza, M.P., Terry, N., 2001. Rhizosphere bacteria mobilize Zn for hyperaccumulation by *Thlaspi caerulescens*. Environ. Sci. Technol. 35, 3144–3150.

Wierzbicka, M., 1999. Comparison of lead tolerance in *Allium cepa* with other plant species. Environ. Pollut. 104, 41–52.

Xiong, Z.T., 1998. Lead uptake and effects on seed germination and plant growth in a Pb hyperaccumulator *Brassica pekinensis*. Rupr. Bull. Environ. Contam. Toxicol. 60, 285–291.

Yadav, S.K., 2010. Heavy metals toxicity in plants: an overview on the role of glutathione and phytochelatins in heavy metal stress toleranceofplants. S. Afr. J. Bot. 76, 167–179.

Yang, X.E., Long, X.X., Ni, W.Z., Fu, C.X.H., 2002. *Sedum alfredii*: a new Zn hyperaccumulating plant first found in China. China Sci. Bull. 47, 1634–1637.

Yang, X.E., Long, X.X., Ye, H.B., He, Z.L., Calvert, D.V., Stoffella, P.J., 2004. Cadmium tolerance and hyperaccumulation in a new Zn hyperaccumulating plant species (*Sedumalfredii* Hance). Plant Soil 259, 181–189.

Yang, X.E., Li, T.Q., Yang, J.C., He, Z.L., Lu, L.L., Meng, F.H., 2006. Zinc compartmentation in root, transport into xylem, and absorption into leaf cells in the hyperaccumulating species of *Sedum alfredii* Hance. Planta. 224 (1), 185–195. http://dx.doi.org/10.1007/s00425-005-0194-8.

Yang, X.E., Roemheld, V., 1999. Physiological and Genetic Aspect of Micronutrient Uptake by Higher Plants, vol. 15. Kluwer Academic Publishers, pp. 151–186.

Yang, Z., Chu, C., 2011. Towards understanding plant response to heavy metal stress. In: Shanker, Arun (Ed.), Abiotic Stress in Plants – Mechanisms and Adaptations. InTech, ISBN 978-953-307-394-1, pp. 59–78.

Yourtchi, M.S., Bayat, H.R., 2013. Effect of cadmium toxicity on growth, cadmium accumulation and macronutrient content of durum wheat (Dena CV.). Int. J. Agric. Crop Sci. 6, 1099–1103.

Zavala, Y.J., Duxbury, J.M., 2008. Arsenic in rice: estimating normal levels of total arsenic in rice grain. Environ. Sci. Technol. 42, 3856–3860.

Zhang, L., Angle, J.S., Chaney, R.L., 2007. Do high-nickel leaves shed by the nickel hyperaccumulator *Alyssum murale* inhibit seed germination of competing plants? New Phytol. 173, 509–516.

Zhao, F.J., Hamon, R.E., McLaughlin, M.J., 2001. Root exudates of the hyperaccumulator *Thlaspi caerulescens* do not enhance metal mobilization. New Phytol. 151, 613–620.

Zhen, Y., Qi, J.L., Wang, S.S., Su, J., Xu, G.H., Zhang, M.S., Yang, Y.H., 2007. Comparative proteome analysis of differentially expressed proteins induced by Al toxicity in soybean. Physiol. Plant 131, 542–554.

Zheng, N., Wang, Q., Zheng, D., 2007. Health risk of Hg, Pb, Cd, Zn, and Cu to the inhabitants around Huludao Zinc Plant in China via consumption of vegetables. Sci. Total Environ. 383, 81–89.

Zhou, X., Yoshida, K., Kuroda, K., Endo, Y., Endo, G., 2003. Effects of cyst on the cytotoxicity of arsenic compounds. Arch. Environ. Contam. Toxicol. 45, 324–330.

Zhu, L.J., Guan, D.X., Luo, J., Rathinasabapathi, B., Ma, L.Q., 2014. Characterization of arsenic-resistant endophytic bacteria from hyperaccumulators *Pteris vittata* and *Pteris ultifida*. Chemosphere 113, 9–16.

CHAPTER

PLANTS USED FOR BIOMONITORING AND PHYTOREMEDIATION OF TRACE ELEMENTS IN SOIL AND WATER

14

Oksana Sytar[1,2], Marian Brestic[1], Nataliya Taran[2], Marek Zivcak[1]

Department of Plant Physiology, Slovak University of Agriculture in Nitra, Nitra, Slovakia[1]; Plant Physiology and Ecology Department, Institute of Biology, Taras Shevchenko National University of Kyiv, Kyiv, Ukraine[2]

CHAPTER OUTLINE

1. INTRODUCTION

Phytoremediation is promising economic effective biotechnology that uses plants for the cleaning of polluted water and soil. Phytoremediation can be used for inorganic and organic contaminants present in the soil, air, and water (Salt et al., 1998; Hazrat et al., 2013). In terms of the mechanisms of remediation, several categories of phytoremediation can be distinguished, such as phytofiltration, phytoextraction, phytodegradation, phytostabilization, and phytovolatization (Ghosh and Singh, 2005; Chibuike and Obiora, 2014; Ansari et al., 2014).

Phytoextraction is a category of phytoremediation that uses plants to absorb pollutants from soil. The ions of metals stored in the roots, stems, leaves, and inflorescences of plants can be burnt and the metals can be recovered or the biomass can be removed to be safely disposed. Phytofiltration is a category of phytoremediation that use plants to remove pollutants from wastewater. Phytoremediation technology in which plants take up the contaminants from the soil and keep them in the root system is categorized as phytostabilization. Phytovolatization uses the plants to volatilize pollutants (heavy metals selenium (Se) and mercury (Hg)) by their foliage. Phytodegradation represents the use of microorganisms and plants to metabolize organic contaminants (Garbisu and Alkorta, 2001). Some plant species can function in one phytoremediation category, whereas others are able to cover two or more ways of remediation.

Plant Metal Interaction. http://dx.doi.org/10.1016/B978-0-12-803158-2.00014-X

The discovery of plants-hyperaccumulators (P-H) enhanced the progress in the phytoremediation technology. P-H are able to absorb, accumulate, and tolerate high concentrations of contaminants (especially heavy metals) that are mostly toxic to common plant species (Roosens et al., 2003; Vamerali et al., 2010; Goolsby and Mason, 2015). According to definition, the hyperaccumulators are plants that are able to accumulate more than 100 mg/kg of cadmium (Cd); 1000 mg/kg of arsenic (As), cobalt (Co), chromium (Cr), copper (Cu), nickel (Ni), lead (Pb), or Se, or; 10,000 mg/kg of zinc (Zn) and manganese (Mn) in leaves (Brooks, 1998; Ucer et al., 2013).

The accumulation of heavy metals in the crop strongly correlates with the total concentration of metals in soil, but it is significantly influenced by soil properties, such as content of macro- and micronutrients and soil acidity. The extraction potential of metal accumulator plants increases until a certain level, but after that the level of increase of metal concentration affects the bioaccumulation capacity (phytoextraction rate of metal) (Hamadouche et al., 2012). Hyperaccumulators are characterized by a translocation factor (shoot-to-root metal concentration ratio) higher than 1; it means that the heavy metals are actively transported into above-ground parts of plants (Tangahu et al., 2011; Badr et al., 2012). Nonetheless, translocation factor cannot be used as a single measure of hyperaccumulation although it can be useful for early detection of plants that can be potentially useful as P-H (van der Ent et al., 2013).

P-H have efficiently working mechanisms of metal sequestration and, sometimes, specific requirements for uptake of particular metals (Shen et al., 1997). In comparison to non-hyperaccumulating species, some hyperaccumulators are able to mobilize metals from soil fractions that are less soluble (McGrath et al., 1997).

Thanks to lower costs and lower environmental impact, the phytoextraction has been recommended as a possible alternative to the conventional practices applied for soils contaminated by heavy metals (e.g., removal and disposal of soil layers) (De Andrade et al., 2009). The key step for successful phytoremediation is examining more efficient heavy metal hyperaccumulators (Zhang et al., 2010). It was suggested that P-H have to be identified and tested in their native environments (Van der Ent et al., 2013).

More than 500 vascular plants with metal hyperaccumulating capacities have been subjected to extensive research regarding their usefulness in phytoremediation. Within this group, different strains of the crops from *Brassicaceae*, *Asteraceae*, *Fabaceae*, and *Poaceae* families have been the most studied (Fellet et al., 2007; Ciura et al., 2005; Jaffré et al., 2013). Among them, *Arabidopsis*, *Brassica*, *Thlaspi*, and *Sedum alfredii* H. have been investigated in more detail (Lone et al., 2008). These plants were found to be efficient at degradation of organic molecules in the root system thanks to extended rhizosphere with fibrous roots (Glick, 2003). It must be emphasized that many P-H would not be efficient for accumulation of multiple heavy metals, but they are selective regarding one metal only (Kamnev and van der Lelie, 2000).

The big variety of numerous classes and types of pollutants makes the removal process of many hazardous contaminants from the environment complicated. Soil and water pollution from anthropogenic activities have been considered a top issue for ecophysiology (Figure 14.1). The pollution includes numerous sources such as aerial emission from combustion of leaded petrol, industrial pollution (smelting and mining industry), vehicle emission, use of insecticides/pesticides, and sources such as natural and artificial soluble salts, excessive use of fertilizers, and wastes in agriculture (Schalscha and Ahumada, 1998; Nriagu and Pacyna, 1988; McGrath et al., 2001; Wuana and Okieimen, 2011). Individual factors, such as type of sampling strategy (which part of plants can be

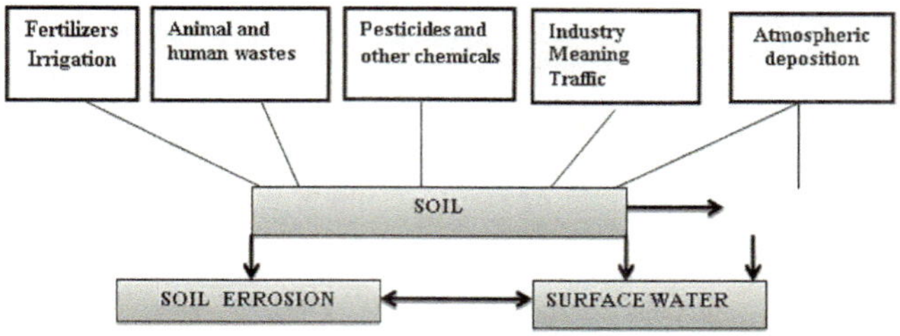

FIGURE 14.1 Soil and water contamination by heavy metals.

used, period of sample collection), representative sampling, and seasonal variability can affect the concentration of trace elements in plant tissues (Dołegowska and Migaszewski, 2015).

The most frequently heavy metals in the polluted environment have been identified as Cd, Hg, As, Pb, Cr, Ni, and Cu (Lone et al., 2008). List of heavy metals may contain also Co and Zn. It is well-known that the problems with toxicities are associated mostly with the metals Cd, Pb, Hg, As, Cu, Zn, tin, and Cr (Ghosh, 2010). The group of radioactive elements includes uranium, cesium, and strontium, often in inorganic compounds with As, sodium, nitrate, phosphate, or ammonia. Ecotoxicological risks associated with heavy metal contamination have to be evaluated by their chemical forms and soil properties as well (Galán et al., 2002). Soil and water may become contaminated with high concentrations of heavy metals by either natural reasons or as a consequence of industrial activities. In this case, to use the growth of mixed companion P-H or crop rotation can be effective tools for the cleanup of soils contaminated by heavy metals.

2. CROP PLANTS PROPOSED FOR BIOMONITORING OF TRACE ELEMENTS IN SOILS

There are common physicochemical techniques of soil remediation such as disposal as waste, incineration, soil washing, solidification, thermal desorption, and vapor extraction, but these activities may cause secondary pollution of groundwater or air. Moreover, such methods usually affect the soil properties. In addition, their extensive application is usually limited by extremely high costs, which make them useless for remediation of agricultural soils, especially in developing countries (Oh et al., 2013). Soil pollution by various trace elements and organic compounds has been a worldwide problem. Soil properties affect metal availability in diverse ways. Significant positive correlations have also been recorded between heavy metals and some soil physical properties such as moisture content and water holding capacity (Sharma and Raju, 2013). Conversely, heavy metals may modify soil properties, especially soil biological properties (Friedlová, 2010). The development phytoremediation technology for remediation of these contaminants therefore is important. Prescreening of plant species useful for phytoremediation of mixed contaminated soils showed that rye grass, sunflower, tall fescue, oat plant, and green gram were able to grow under the mixed contaminated conditions. The maximum removal of contaminants from the soil was found in sunflower (Chirakkara and Reddy, 2015). However, in many cases, the application of methods of phytoremediation has been limited because of their low remediation efficiency (Oh et al., 2014).

It is well-known that different plant species can vary regarding their uptake rate of heavy metals (Marschner, 1986; Sharma et al., 2014). In particular, plant species exist different mechanisms of metal concentration, exclusion and compartmentation (Lone et al., 2008). The nature and range of heavy metals capacities in the soil and plant system change. This variable behavior of heavy metals depends on the form and quantity of investigating elements, from the soil capacities, and type of plant (Singh and Agrawal, 2008; Nogueira et al., 2010). The heavy metals are normally immobile in the soil with neutral or alkaline pH (Hayes and Traina, 1998). Heavy metal content in plants may decrease crop yield without visible toxicity evidences on the plants (Rao and Shantaram, 1996; Thivya et al., 2014). As an example, application of enriched waste products increased heavy metal content in plants resulting to negative, but in many cases also to positive effects on plant growth (Singh and Agrawal, 2010; Santos et al., 2011). The capacities of heavy metal accumulation vary among different plant species and also between various parts of plants (Singh and Agrawal, 2010; Singh et al., 2011). For example, Zn accumulates in shoots of maize plants, but Pb and Cr can accumulate also in grains (Pierrisnard, 1996).

The phytoextraction efficiency is calculated by multiplying bioconcentration factor and factor of translocation. The heavy metals accumulating efficiency of any plant species is presented via bioconcentration factor (Ladislas et al., 2012). The willow plant with high biomass production has been identified as suitable for remediation of soil (Landberg and Greger, 1996). Some plant species such as *Thlaspi* and *Arabidopsis* have low biomass, but high hyperaccumulating characteristics. Because of the dilution factor of the plant parts, trees accumulate higher amounts of heavy metals compared with the bushes or grasses (Irshad et al., 2014). In any case, there is not a general rule, but soil-specific screening has to be applied with the aim of assessing the capability of terrestrial plants for metal extraction in certain conditions.

Efficiency to accumulate metal elements was tested in 30 plant species in hydroponics by Ebbs et al. (1997) under moderate concentrations of Cd, Cu, and Zn for 4 weeks to identify which plant species had the highest capacity for storing heavy metals in above-ground biomass. It was shown that plants of the *Brassicaceae* family, such as *Brassica juncea* L., *Brassica napus* L., and *Brassica rapa* L. were able to accumulate Zn and Cd moderately. Kumar et al. (1995) found that from monitored farmed *Brassica* species *B. juncea* and *Brassica nigra* plants have the highest ability to accumulate heavy metals in above-ground parts. *Brassica juncea* plants showed the ability to accumulate Cu (Ariyakanon and Winaipanich, 2006). Indian mustard has suitability for removing Cd from soil (Goswami and Das, 2015). At the same time, canola (*B. napus* L.) was found to be more effective in the uptake of Cu, Cd, Pb, and Zn compared with Indian mustard (*B. juncea* L.). The heavy metal uptake by the roots was higher than heavy metal uptake by shoots for both *Brassica* species. Results indicated an increase of heavy metal availability and uptake by plants in variants with ethylenediaminetetraacetic acid (EDTA) application (Turan and Esringü, 2007), which highlights the importance of organic chelates in increase of metal solubility, which can be applied to increase efficiency of phytoremediation technique. In parallel, application of exogenous citric acid could intensify Cd accumulation and reduce Cd stress in plants. This operation may be useful in stimulating the phytoextraction of Cd over P-H such as *B. napus* L. (Ehsan et al., 2014). Plants of the *Brassicaceae* family can be used also for *biofumigation*. The possibilities of joining the phytoremediation and biofumigation processes for particular species of family *Brassicaceae* have been also broadly studied (Szczygłowska et al., 2011).

During the phytoremediation process, the uptake of heavy metals by brassicas might increase the synthesis of secondary metabolites, including glucosinolates. The differences in glucosinolate levels influenced by metal accumulation are caused by indirect metal-induced changes of glucosinolate biosynthesis. Accumulation of heavy metals inducing changes in content of glucosinolates is connected with sulfur metabolism. It is still not enough information regarding the heavy metal effects on sulfur metabolism in plants (Pongrac et al., 2008; Gigolashvili and Kopriva, 2014). The relationship between the accumulation of heavy metals and glucosinolate content in plants is specific for individual plant organs, plant species, and metal; it determines the complex metabolic interactions.

The capability of sorghum (*Sorghum bicolor*), Chinese cabbage (*Brassica chinensis*), and sunflower (*Helianthus annuus*), for remediation of soil contaminated by Pb has also been studied. It was found that sunflower plants were more responsive to high concentrations of Pb in the soil and that plants reduced growth up to harvest. Chinese cabbage was found to be more capable to accumulate the Pb from the soil compared with sunflower and sorghum. The quantity of produced plant biomass can be affected by soil Pb concentration; decreasing with an increase of Pb in the soil. Accumulation of Pb in plant tissue of sunflower, Chinese cabbage, and sorghum increases with a rise of soil Pb concentration (Hamvumba et al., 2014).

Harvested sunflower (*H. annuus*) biomass can be used as adsorbent to remove Cu and Zn from aqueous solutions (Yeh, 2015). Moreover, sunflower is suggested as a plant with phytoextraction capacity to accumulate high concentrations of Cd from contaminated soils (Madejón et al., 2002; Sinegani and Khalilikhah, 2010). The claim that *H. annuus* may have a higher phytoextraction efficiency compared with *B. juncea* was not confirmed (Madejón et al., 2002). It was proved that *H. annuus* has a high intake of heavy metals into the roots, but a low relocation into the vegetative organs (Madejón et al., 2002; Kumar et al., 1995). On the other hand, De Andrade et al. (2009) found that for phytoremediation in the soils of mining and metallurgy areas should be the preferred *H. annuus* plants, and preferential accumulation of heavy metals in roots was not observed.

The potential of soybean (*Glycine max* L. Merr) to remediate (bioaccumulate) Pb in soil contaminated with different concentrations of Pb in has been studied related to environmental conditions. The results of this study revealed that pH, phosphorus, and moisture content increased, whereas nitrogen and organic carbon contents decreased in polluted soil remediated with *G. max* compared with the unpolluted soil. The results obtained suggest that the plant has phytoextraction ability and could be used in restoring soil polluted with Pb (Aransiola et al., 2013). Soybean plants typically synthesize homophytochelatins rather than phytochelatins under heavy metals exposure (Yang et al., 2005). This species had little potential to counteract the deleterious effect of Cr metal in soil at the previously mentioned treatments. Moreover, the Cr metal was extremely toxic for seeds and young seedling of *G. max* at high concentrations (Amin et al., 2014). At the same time, the exposure of soybean varieties to Pb and Co resulted in the reduction of chlorophyll, carotenoids, carbohydrate, and protein content as well as increase in proline content. Tested plants were more sensitive to Pb in comparison to Co. However, an additional supply of nitrogen not only minimized the inhibitory effect of these two heavy metals, but also decreased the proline content in plants. The effect of heavy metals with different treatments on biochemical content was significant. Contamination of soils by heavy metals evidently leads to substantial losses in seed yield of soybean canopies (Imtiyaz et al., 2014).

The application of microorganisms that stimulate the growth and yield of soybean plants to the soil can raise phytohormones and amino acids secretion and increase percentage of phosphate solubilization and nitrogen fixation by soybean (Radhakrishnan and Lee, 2013). Cotreatment with *Enterobacter asburiae* KE17 inhibited Cu and Zn uptake in soybean plants and regulated accumulation of nutrients by enhancing the quantity of nitrogen supply. It was observed that Cu and Zn activated the high accumulation of free amino acids such as alanine, glycine, threonine, aspartic acid, serine, proline, tyrosine, gamma-aminobutyric acid, leucine, and isoleucine. Thus, the application of

E. asburiae KE17 in metal-affected plants decreased the concentration of free amino acids significantly (Kang et al., 2015).

For phytoremediation technology, it is possible to also use the plants that normally do not accumulate high concentrations of heavy metals. Such plants are alfalfa (*Medicago sativa* L.), sorghum (*S. bicolor*), willow (*Salix* spp.), and corn (*Zea mays*), although some authors suggest a high heavy metal accumulating capability in these crop species (Vijayarengan, 2005). Anyway, thanks to a high biomass produced by these plants, a higher quantity of metals can be clean up from the soil even if the relative amounts within these plants may be smaller than in the common P-H. This capacity can be used for the passive monitoring of the soil contamination and as a marker of the level of contamination (Wong, 1996). In turn, the higher metals concentrations attainable in some P-H are compensated for by their lower biomass (McGrath, 1998).

Maize (*Z. mays* L.) is a widely grown cereal crop with promising attributes of a heavy metal accumulator and the tool for remediation of soil contaminated with crude oil (Liao et al., 2015). The advantage of this crop is that the maize, as a rapidly growing annual crop with extensive fibrous root system and high biomass production, is able to grow even in adverse conditions and can be easily cultivated in warm environments or warm seasons (Jadia and Fulekar, 2008; Zhang et al., 2009). The effect of six heavy metals (Cd, Cr, Hg, Co, Mn, and Pb) on the growth of maize has been studied. The results indicated that the presence of these metals in soil reduced the growth and protein content of maize. The toxicity of these metals occurred in the following order: Cd > Co > Hg > Mn > Pb > Cr. Corn was found to be an effective accumulator plant for phytoremediation of Cd and Pb polluted soils (Mojiri, 2011). It was also observed that the combined effect of two or more heavy metals was only as harmful as the effect of the most toxic heavy metal (Ghani, 2010). Though not listed as a hyperaccumulator, the previously mentioned attributes of maize can be sufficient arguments to consider its use in phytoextraction (Ebbs et al., 1997).

The removal of metal can be assessed by concentration of metal change in plant mass multiplied by the decrease of heavy metal concentration in soil (Ghosh and Singh, 2005). The result of such an assessment can be influenced by numerous cooccurring factors, such as the phytoremediation time **tp** (year) and the amount of extracted metal **Q** (mg/plant). The bioaccumulation factor **f** (soil-plant transfer coefficient) can be applied to assess the plant's phytoextraction efficiency (Zhuang et al., 2005). Phytoextraction efficiency can be calculated by different equations (see Zhang et al., 2010) taking into account the contaminated active rooting zone (soil layer), soil bulk density, and number of crop cycles per year.

The grass representative of family *Poaceae* is a sorghum (*S. bicolor*), the world's fifth major cereal crop. The seeds of sorghum are frequently used as cattle feed and for flour production, but in Central America, southern Asia, and Africa, sorghum is also an important food crop. The optimal agronomic practices supporting growth of high biomass crops represent the new way to phytoremediation of soils contaminated by heavy metals. For that reason, the studies of the phytoextraction ability of sweet sorghum (*S. bicolor* L.) under heavy metal influence were realized. It was shown that in field conditions the highest content Cd, Zn, and Cu was accumulated in the stems of sorghum, whereas the highest phytoextraction efficiency of Pb was expressed by leaves (Zhuang et al., 2009). The results of Revathi et al. (2011) revealed that with increased dosage of Cr biomass of the *S. bicolor* plant shows significant decreasing. In terms of metal accumulation, *S. bicolor* and *H. annuus* plants have positive potential of phytoremediation compared with other crop plants. For example, the removal of Zn reached about 2 kg/ha and 1 kg/ha for *S. bicolor* and *H. annuus*, respectively (Marchiol et al., 2007).

Medicago sativa (alfalfa) (family *Fabaceae*) is a flowering plant with a tetraploid genome typical of extensive root systems, sometimes stretching to 4.5 m, which makes it very strong and resistant, especially to drought (Singh et al., 2009). The ability to grow and germinate has been demonstrated for alfalfa plants under different variants of contamination: with 80 mg/kg of Ni(II), Cu(II), Cd(II), and 160 mg/kg of Zn(II) (Peralta-Videa et al., 2002). Alfalfa plants were recognized as useful to be transplanted and phytoremediate soils with high Zn, Cd, or Cu content. The research studies identified the correlation between susceptibility of living alfalfa plants to Cu(II), Zn(II), and Cd(II), and the age of the plants (Peralta-Videa et al. 2002). For phytoremediation of Cr alfalfa and sorghum plants were found effective. Alfalfa plants have higher potential for phytoremediation of Cr polluted soils than sorghum (Karimi, 2013).

The biological availability of contaminants in the biomass of harvested parts and in the soil and are the main limiting factors of phytoextraction. Contaminant bioavailability is based on the study of capacity to transfer metals into the plants and element mobility in plant tissues (Podlešáková et al., 1999). Pb is a heavy metal with a potential mobility (Podlešáková and Němeček, 2000) that can be bound on soil organic matter or as occurs in soil as part of insoluble compounds (carbonates, hydroxides, and phosphates) (Lasat, 2002; Száková et al., 2003). Elements such as Cd and Zn are good examples of contaminant bioavailability because of their pH dependent mobility and high transfer into plants.

An evaluation of phytoremediation efficiencies under the effects of heavy metals (Pb, Hg, and Cd) in combination with EDTA or ammonium addition were conducted in field experiments with three kind of plants. The tested plants included *Amaranthus retroflexus*, *S. bicolor*, and *Lolium perenne*. The maximum concentration of Pb, Cd, and Hg were detected in the shoots of *A. retroflexus, S. bicolor*, and *L. perenne* at high concentrations in $pH = 6.2$. To enhance the phytoavailability of Pb, Cd, and Hg, EDTA can be applied as a chelating agent to soil (Nejatzadeh-Barandozi and Gholami-Borujeni, 2014).

In the context of all previously mentioned research studies, for ecological soil management of phytoremediation process, we screened of results (concentration of heavy metal in the experimental substrate and in the plants from field experiment) that can be used in the future to build approximate common mathematical models for future prognosis of how much heavy metal can be removed from soil.

At the same time, after analyzing results of reported data obtained in field experiments, the following plant species capable of hyperaccumulating heavy metals were selected (Table 14.1): *Brassica carinata*, *B. juncea*, *H. annuus* L., *G. max* L. Merr, *S. bicolor*, *Z. mays* L., and *M. sativa* L.

3. CROP ROTATION FOR BIOMONITORING OF TRACE ELEMENTS IN SOIL

A great low-cost possibility to recover contaminated soils implements application of phytoremediation technologies that also can improve biological characteristics of soil. Normally, annual plants are used for absorption of metal from the soil just over a few months, usually in spring and summer, but this is not efficient enough for competent phytoextraction technologies (Sarma, 2011). As a result, the fields are not actively cleaned for most periods of the year, but they are subjected to erosion by water and wind and uncovered under phytoremediation technology for all year.

It is well-known that significantly decreasing the biomass yield depends also on reduction of nutrients, presents of pests, and appearance of diseases, which together cause a negative impact on soil

Table 14.1 Plants That Are Possible Hyperaccumulators of Heavy Metals

	Brassica carinata	*Hellianthus annus* L.	*Glycine max.* L. Merr	*Sorghum bicolor*	*Zea mays* L.	*Medicago sativa* L.
Biomass produced, kg/m^2 (DW)	81.1	94	40	102.3–107	93	57–61.6
	Pari et al. (2010)	Zürrer and Bachofen (1985)	Demirta et al. (2010)	Olufayo et al. (1997) and Theuretzbacher et al. (2012)	Theuretzbacher et al. (2012)	Slepetys et al. (2012) and Jia et al. (2009)
Metal Concentration mg/kg in the Experimental Substrate						
As	33.4	718	718	718	718	
Cd	19	7.53	7.53	7.53	7.53	0.58
Co		50.9		50.9	50.9	
Cr	85					
Cu	140	1943	1943	1943	1943	12.98
Ni	24					2.32
Pb	441	244	244	244	244	8.21
Zn	3336	1322	1322	1322	1322	43
Metal Concentration mg/kg in Herbaceous Crops from Field/Site Experiment						
As	2.81	20	230	240	30	85
Cd	12	0.64	2.4	3.7	20	0.275
Co		2.18		7.16		
Cr	9.8					
Cu	37	70	440	540	1.220	0.003
Ni	7.6					0.0013
Pb	50	5	72	100	257	0.003
Zn	1.650	150	430	580	1.200	0.01
	Marchiol et al. (2004), Sorian and Fereres (2003), and Vamerali et al. (2010)	Fellet et al. (2007), Vamerali et al. (2010), Marchiol et al. (2007), and Jatav et al. (2012)	Fellet et al. (2007) and Vamerali et al. (2010)	Fellet et al. (2007), Vamerali et al. (2010), and Marchiol et al. (2007)	Fellet et al. (2007), Vamerali et al. (2010), and Marchiol et al. (2007)	Ciura et al. (2005)

As, arsenic; Cd, cadmium; Co, cobalt; Cu, copper; Ni, nickel; Pb, lead; Zn, zinc.

fertility because of monocultural use (Lasat, 2000). In this case, plant phytoextraction efficiency as a result of biomass yield decline is significantly reduced. The use of monoculture for a period of 2 or 3 years is acceptable for soil phytoremediation. But sufficient cleanup of the soil by one plant species only for a longer duration cannot be expected. Gentle soil remediation options (GRO), which particularly include plant-based phytoremediation benefits and contaminant inactivation (stabilization) in situ are a big interest these days. The crop rotation is one of agronomic practices (selection of plant, planting methods together practices of soil management, row cropping, coppice of short rotation, plant cultivation parameters, management of fertilization process and harvesting, irrigation management, and pest and weed control) with potential positive effects on GRO (Kidd et al., 2015).

Dhillon and Dhillon (2009) recommend a rotation of plant species with the phytoremediation capacity for industrial crops. The crop rotation is economically efficient and it can improve optimal parameters of growth and yield of P-H (Dhillon and Dhillon, 2009). However, we cannot exclude the rising importance of crop rotation factor because of plants' fragility in monocultures and during long-time experiments that produce declining yields of the plant monocultures (Čechmánková et al., 2011). Based on the results of experiment with white lupin (*Lupinus albus* L.), the rotation of hyperaccumulating plant crops has been found a useful method to develop the benefits of phytoremediation of soil. Similarly, the recovery of soil quality during the phytoextraction period can be achieved with the use of crop rotation. Crop rotation is reducing some human health and ecological problems via improving the soil quality by increasing the levels of viable microorganisms and by avoiding the use of chemical pest or fertilizers (Fumagalli et al., 2014).

Energy crops need intensive cultivation characterized by high plant densities, mechanization, and short rotation (1–4 years) and plant cycles (less than 20 years) (Fiorese and Guariso, 2010). Energy crops are presented in two categories: annuals and perennials. Examples of annuals include rapeseed (*B. napus* L.) and *B. carinata*, kenaf (*Hibiscus cannabinus* L.), sweet sorghum (*S. bicolor* L. Moench), and fiber sorghum (*S. bicolor* L. Moench). Several other short rotation crops can be used in phytoremediation (Gomes, 2012).

More details on the agricultural experience in rotation among selected crop plants with hyperaccumulative activities, based on the published data, are presented in Table 14.2.

The main positive results in the field after rotation of different crops are weed regulation, pests and diseases, and conservation of soil fertility. These two positive incomes of a crop rotating process compared with monocultures may reduce a need of some inputs and improve parameters of crop yields (Smith et al., 2008).

Soil potency is often recovered in about 1 year by letting the field lie without any crop plants cultivation after 1–3 years of cultivation of oats, barley, wheat, or other cereal crops. At first, the crop rotation was perhaps proposed by the Romans in Northern and Central Europe. It was used for replacement of the fallow phase by legume forage crops or sown grass. The Romans used a 4-year rotation including root crops, forage crops, and cereal grain crops (Freyer, 2003).

Replacement of crop rotation also can be done by synthetic herbicides and mineral fertilizers, pesticides and fungicides, or farming machinery (Stoate et al., 2001). At the same time, it can be helpful for development of phytoremediation model connected with agriculture.

It was found that *Brassica* plants are companion plants for *G. max*. L. Merr., *S. bicolor*, and *M. sativa* L. (Inoue and Saeki, 2004; Nishizawa, 1978). *Medicago sativa* L. is used by farmers as crop to improve soil, fix nitrogen like legumes do, and diminish cotton pests. It also disperses some tough soils and hardpan. *Brassica* sp. plants are known because of a great biomass production, pest

Table 14.2 List of Companion Plants Suitable for Crop Rotation

	Brassica carinata	***Helianthus annuus* L.**	***Glycine max* L. Merr**	***Sorghum bicolor***	***Zea mays* L.**	***Medicago sativa* L.**
Brassica carinata						
Helianthus annuus L.	–					
Glycine max L. Merr	+	+				
Sorghum bicolor	+	+	+			
Zea mays L.	–	+	+	+		
Medicago sativa L.	+	–	–	–	+	
	El-Atrach et al. (1989), Gavito and Miller (1998), Warnke et al. (2006), and Hague et al. (2002)	Teasdale et al. (2004) and Cavigelli (2004)	Robinson (1966) and Vargas-Ayala et al. (2001)	Robinson (1966), Vargas-Ayala et al. (2001), and Calonego and Rosolem (2011)	Teasdale et al. (2004), Cavigelli (2004), and Inoue and Saeki (2004)	Latta and Blacklow (2001) and Paul et al. (1997)

management characteristics, nutrient scavenging ability, and rapid fall growth. In soil, the *Brassica* species are able to decrease concentration of chemical compounds that may be toxic to fungi, soil nematodes, and some weeds because normally *Brassica* plants have higher concentrations of these "toxic" compounds.

Brassicas are used as a rotational crop in vegetable or fruit crop production, such as potatoes and tree fruit production. Nowadays, the *Brassica* as the cover crop rotation plant is of special interest because of its possible use in biotoxic or biofumigation activity and nematode trapping, row crop production, and for nutrient capture. Some *Brassicas* have a big taproot that gives it better capacities for biotoxic or biofumigation activity than the fibrous roots of other cereal cover crops. Those *Brassicas* decompose very fast and leave a seedbed after winterkill that is easy to plant in (Calonego et al., 2011).

At the same time, *G. max*. L. Merr can be used as a cover crop for the crop rotation with the aim to decrease *H. glycines* population. A cover crop plant is an interseeded crop plant that covers soil during growth and in rotation with other crops. Results of experiments in the United States showed that cotton (*Gossypium hirsutum*), hairy indigo (*Indigofera hirsuta*), American jointvetch (*Aeschynomene americana*), sorghum (*S. bicolor*), bahiagrass (*Paspalum notatum*), wheat (*Triticum aestivum*), and velvetbean (*Mucuna pruiens*) can be used as winter/summer cover crops or as rotation crops effectively. In most cases, this increased soybean yield after crop rotation with experimental crop plants (Hague and Overstreet, 2002; Vargas-Ayala and Rodriguez-Kabana, 2001). Soybean plants are good to use for crop rotation, as are sunflowers. It was confirmed that yields after crop rotations of corn-sorghum or sunflower-soybeans exceeded those from cultivation of monoculture. Sunflower-soybean crop rotation was first-rate to continuous sunflower in sunflower disease control. Yields of the feed grains were greatly increased with sequence soybeans in a corn or sorghum. Effects of sunflowers–soybeans and sorghum–corn crop rotations on soil pH and depletion of acid-soluble phosphorus and potassium were small (Robinson, 1966).

The sporocarps of *Glomus mosseae* were used for study of crop rotation relationships of non-host plants *Brassica oleracea* and plants-host *M. sativa*. *Brassica oleracea* and *M. sativa* were cultivated in the sand-vermiculite pots inoculated with sporocarps of *G. mosseae*. In an experimental variant with cabbage acting as a crop rotation plant, the decrease in vesicular-arbuscular mycorrhizal infection of alfalfa plants was shown. The volatile compounds of cabbage roots have been affected but did not influence the process of spore germination. It was concluded that *M. sativa* and *B. oleracea* can act as companion plants (El-Atrach et al., 1989).

4. FRESHWATER VASCULAR PLANTS PROPOSED FOR BIOMONITORING OF TRACE ELEMENTS IN WATERS

The pollution of water is designed to be a dangerous ecological risk. In open oceans, the level of heavy metal pollution is low, but concerning the seas, rivers, and lakes of Earth, it rises as it approaches coastal waters (Bryan and Langston, 1992). Water pollution in lakes and rivers depends on closest point sources of pollution and on the pollution volume. The impacts of human civilizations are connected with elevated contents of Pb, Ni, Cr, Zn, and Cd, from which Cr normally has a local significance (Meybeck et al., 1989; Dixit et al., 2015).

Water plants can absorb trace elements through roots and/or shoots. In this case, direct wet or dry translocation of pollutants in water systems increases the concentrations of heavy metals

(Jackson, 1998). In case inputs of contaminants are discontinuous and show the quick dilution in water, the results of investigation of plant tissues can present integrated information about the quality of the water and process of phytoremediation (Baldantoni et al., 2005).

The accumulative effect of concentrations of metals on the trophic chains after accumulation in aquatic plants (Outridge and Noller, 1991; Tremp and Kohler, 1995; Rai, 2012; Chaudhary and Sharma, 2014). The terrestrial plants accumulate higher metal concentrations in their root when applied to an aqueous system (Caldwell et al., 2012). Aquatic macrophytes, because of their quick growth, high biomass production, and relative higher capacity of heavy metal accumulation, are more applicable for treatment of wastewater compared with the soil-grown plants. Aquatic macrophytes achieve better purification because of immediate contact with polluted water (http://www.sciencedirect.com/science/article/pii/S0045653513001914Sood et al., 2012). Macrophytes are those plants that grow near or in water and are categorized as free-floating plants, merged, and submerged plants. Water hyacinth (*Eichhornia crassipes*), water lettuce (*Pistia stratiotes*), and duckweed (*Lemna minor*) are some of the aquatic macrophytes frequently used for the heavy metal remediation in aquatic systems (Hua et al., 2012; Singh et al., 2011).

The low-cost recycling technologies for wastewater, especially contaminated by toxic trace elements and heavy metals, are of the main interest nowadays because they can be used by developing countries (Figueira and Ribeiro, 2005). Such technologies as part of conventional methods are an alternative solution for removing heavy metals from contaminated waters (Khambhaty et al., 2009; Rizwana et al., 2014). The low-cost recycling technologies can be used on the basis of their special selection for phytoremediation of different industrial events (Verma et al., 2005).

A great amount of various substances (e.g., heavy metals) is concentrated by macrophytes. In this case, macrophytes are consequently useful indicators of local pollution. Many studies concentrate on the assessment of the hyperaccumulation potential in selected freshwater vascular plants.

Pistia stratiotes is water cabbage or water lettuce and is characterized by a high growth rate. It is free-floating perennial macrophyte that can be easily handled (Hua et al., 2012; Thilakar et al., 2012). It is widespread in tropical zones, but it also exists in temperate and subtropical zones. *Pistia stratiotes*, because of its high average relative growth rate, is chosen for phytoremediation study and proposed for accumulation of As (Maine et al., 2001). The bioindicator potential of *P. stratiotes* was discovered after an investigation of effect of tannery effluent containing Cr(VI). The root system of *P. stratiotes* was found to be able to accumulate a high amount of Cr to the extent of (23.64 64 μg/g FW) with respect to the shoot system (11.22 μg/g FW) for a 2.36 ppm concentration of Cr (Gupta, 2014). Water lettuce can be a potential plant for phytoremediation for Mn-contaminated waters (Hua et al., 2012). *Pistia stratiotes* among floating aquatic macrophytes such as *Spirodela* sp. (giant duckweed), *P. stratiotes* (water lettuce), and *Salvinia* sp. was used to remove Cu (Sanches Filho et al., 2015).

Spirodela is a genus of water plant commonly called duckweed. These plants are very simple, lacking an obvious stem or leaves. Duckweed is being studied by researchers around the world as a possible source of clean energy and also for the phytoremediation process (Landesman et al., 2011). Two species of duckweed (*Spirodela polyrhiza* and *L. minor*) have capacities to phytoremediate Cd from water solution. Under 2 mg/L of Cd, *L. minor* obtained a maximum Cd accumulation (4734.56 mg/kg). In the last part of the experiment, the maximum Cd accumulation (0.711.00 mg/kg) in *S. polyrhiza* was at 3 mg/L of Cd value measured at the beginning. *Spirodela polyrhiza* and *L. minor* have been proposed as possible hyperaccumulators of Cd (Chaudhuri et al., 2014).

Lemna minor (common duckweed or lesser duckweed) is a native aquatic plant of Europe, North America, Asia, and Africa and is present everywhere in the slow-moving streams and freshwater ponds except for arctic and subarctic climates. It is naturalized in Australasia and South America and not native there. This freshwater aquatic plant has a single root that droops in the water. *Lemna minor* is a floating aquatic plant with one, two, or three leaves. The plants divide and become separate individuals when they grow more than three leaves. It has been used for the removal of Pb, Cd, and Cr from the water with the additional advantage of growing under varied climatic conditions (Goswami et al., 2009; Miretzky et al., 2004). *Lemna minor* is a potential Cd accumulator able to remediate the aquatic environment contaminated by Cd, especially at low Cd concentrations. In *L. minor* with increasing exposure time and Cd concentrations the Cd uptake in the whole plant, tolerance index (T_i), relative growth rates, and maximum efficiency of PSII (Fv/Fm) declined. But Cd accumulation by *L. minor* showed an opposite reaction (Bianconi et al., 2013).

Eichhornia crassipes (water hyacinth) is an aquatic plant that grows in the Amazon basin. Among the heavy metal hyperaccumulator plants, *E. crassipes* is widely used for tertiary treatment of waste phytoremediation water (Danilin, 2010). The other parameters are a function of the time, biosorbent dosage, and pH were studied in the experiment with absorption of Cu and Zn from water solution by hyacinth, which grows in raw water and acid-treated water (Gandhimathi et al., 2013). *Eichhornia crassipes* was found the most competent for the extraction of Cd, Cu, Zn, Fe, and Cr pursued by *Spirodela polyrrhyza* and *P. stratiotes* (Mishra and Tripathi, 2008). The broad leaves and fibrous root system of water hyacinth assist in absorbing of high concentrations of heavy metals (Mishra et al., 2008). *Typha latifolia* and *E. crassipes* are the well-known freshwater plants for cleanup of effluent that usually has a high concentration of Co, Cd, and As. Prominent cleanup of Pb from titanium sponge industry effluent was done by *E. crassipes* (Sukumaran, 2013). In acid mine drainage water, *E. crassipes* also readily reduces the level of heavy metals (Falbo and Weaks, 1990). This plant is able to remove silver from industrial wastewater in a short time (Pinto et al., 1987). Water hyacinth used on sewage was rapid in the purification process; in the vegetation period when attaining maturity, it has a low uptake of nutrients. Throughout the course of the experimental trace elements such as Cu, potassium, Mn, F, chlorine, Fe, Pb, nitrate, and sulfate were excessively reduced (Ajibade et al., 2013).

Eleocharis acicularis is known by the common names dwarf hairgrass and needle spikerush. It is a freshwater plant with worldwide distribution that grows naturally in waterways, marshes, paddy fields, and reservoirs. It is easily cultivated and controlled. The *E. acicularis* has recently been noted to be able to uptake multiple metals from drainages and mine tailings and can be a hyperaccumulator of Fe, Pb, Mn, Cr, and Zn (Sakakibara et al., 2009; Ha et al., 2009). *Eleocharis acicularis* is a great candidate for the cleanup of water polluted by heavy metals and has a hyperaccumulation capacity for Cd, As, Pb, Zn, and Cu. The highest amounts of heavy metals in green parts of *E. acicularis* were 20,239 mg Cd/kg, 1740 mg As/kg, 14,200 mg Zn/kg, 200 mg Cu/kg, and 894 mg Pb/kg (Sakakibara et al., 2011) (Table 14.3).

Myriophyllum aquaticum (parrot feather and parrot feather watermilfoil) is a flowering plant, a vascular dicot. The parrot feather is an aquatic freshwater plant with a short stem that can grow above the water surface up to 30 cm. Its reproduces by breaking fragments from the parent plant. *Myriophyllum aquaticum* is an aquatic plant with enzymes that have a role in transforming organic compounds, which make it effective for the phytoremediation process of an aquatic environment with organic chemical contamination (Susarla et al., 2002; Cano-Rodríguez et al., 2014).

Table 14.3 Freshwater Vascular Plants Proposed for Biomonitoring of Trace Elements in Waters

	Pistia stratiotes	*Spirodela* sp.	*Lemna minor*	*Eichhornia crassipes*	*Eleocharis acicularis*	*Myriophylhum aquaticum* Parrot Feather	*Ludwigia palustris* Creeping Primrose	*Mentha* Aquatic Water Mint
Biomass produced, kg/m^2 (DW)	3.670	–	1.540	14.202	3.750	2.262	3750	–
	Dwivedi and Pandey (2011)		Dwivedi and Pandey (2011)	Dwivedi and Pandey (2011)	Dwivedi and Pandey (2011)	Snow and Ghaly (2008)	Dwivedi and Pandey (2011)	
Metal Concentration mg/L in the Experimental Substrate								
Fe	3.84	3.84	3.84	5	1010	103.55	103.55	103.55
Zn	1.079	1.079	1.079/8	5	3.08	28.06	28.06	28.06
Mn	1.065	1.065	1.065		125			
Cu	1.039	1.039	1.039	5	0.952	5.56	5.56	5.56
Cr	1.015	1.015	1.015	5	1.7			
Pb	1.029	1.029	1.029		18.2			
Cd	0.05		0.099	5				
Hg						0.501	0.501	0.501
Sb					433			
As			0.091	0.15	57.1			

Accumulated Metal Concentration mg/g DW in the Experimental Freshwater Vascular Plants								
Fe	4	1.21	1.25	25.5	59.5	38.8	46.3	32.1
Zn	1	1	1/3.7	5.52	0.0295	0.549	1.498	1.243
Mn	0.5	0.5	0.7		0.338			
Cu	0.5	1.1	1	2.76	0.0061	0.304	0.314	0.848
Cr	0.45	1.1	1.05	0.27	0.265			
Pb	0.5	1.15	1.1		1.120			
Cd	0.769		20.21	0.67				
Hg						0.208	0.179	0.335
Sb					0.00618			
As			0.808	0.0018	0.00369			
	Miretzky et al. (2004), Qian et al. (1999), and Maine et al. (2001)	Miretzky et al. (2004)	Miretzky et al. (2004), Robinson et al. (2009), Jain et al. (1990), and Goswami et al. (2009)	Kumar and Tripathi (2008), Alvarado et al. (2008), and Mahmood et al. (2010)	Ha et al. (2009) and Ha et al. (2009)	Kamal et al. (2004)	Kamal et al. (2004)	Kamal et al. (2004)

As, arsenic; Cd, cadmium; Cr, chromium; Cu, copper; Fe, iron; Hg, mercury; Mn, manganese; Pb, lead; Sb, antimony; Zn, zinc.

The *Myriophyllum* species are able to accumulate such heavy metals via a rhizofiltration process (Wang et al., 1996).

The *Ludwigia palustris* (creeping primrose) is an aquatic plant that grows in shallow water with leaves oppositely organized along the stem. Leaves are approximately 0.5–2 cm long and oval-shaped to elliptic in contour. Creeping primrose stem is creeping and roots at the nodes. Kamal et al. (2004) have reported an effective removal of Cu, Fe, Hg, and Zn by water mint (*Mentha aquatica*) and creeping primrose (*L. palustris*).

Mentha aquatica (water mint) has opposite leaves and a square stem. The stems can be grown from stem cuttings or from root divisions. Stems of *M. aquatica* can have small flowers in late summer and can grow 50–80 cm high. Kamal et al. (2004) have confirmed that *L. palustris* and *M. aquatica* are hyperaccumulators of iron and Zn. These freshwater plants have capacities to accumulate Cu and Hg as well (Kamal et al., 2004). Water mint (*M. aquatica* L.) is an As-tolerant plant suitable for cultivation in the climatic conditions of Central Europe. *Mentha aquatica* seems to be able to grow in the contaminated soils without symptoms of phytotoxicity (Száková et al., 2011).

The analysis of literature data regarding biomass produced by freshwater vascular plants has been found complicated and some information is still missing. It might be due to insufficient number of experiments with freshwater vascular plants compared with crop plants. Within existing data, we decided to use the results presented by Dwivedi and Pandey (2011) as a source for the mathematical model.

5. CONCLUSIONS AND FUTURE PROSPECTS

Based on the published data, we conclude that, among crop plants selected for biomonitoring of trace elements from the soil, *B. carinata* has been shown the most efficient hyperaccumulator of As, Cu, Pb, Ni, Zn, Cd, and Cr At the same time, the positive results in phytoextraction of Co from soil were estimated for *H. annus* L. and *S. bicolor.* We suggest that cultivation of mixed companion P-H or the crop rotation can be useful for soil phytoremediation. Among freshwater vascular plants proposed for biomonitoring of trace elements in waters, the most successful results have been observed in *E. crassipes* and *L. minor.* To get a clear conclusion, a more detailed mathematical analysis to confirm the prognosis of efficient phytoextraction process in the soil and water is required.

There is a lot of literature on phytoremediation technology and P-H of heavy metals, but unfortunately, not many of them can be used for prognosis of the phytoremediation process. In this work, we collected and presented data of the concentration of metals in the experimental substrate and the concentration of accumulated metals in selected crop and freshwater vascular plants. At the same time, we provide characteristics of plant heavy metal hyperaccumulators based on the most recent research findings.

The future perspectives of phytoextraction developments are associated with additional practical issues, such as the further treatment of biomass, accumulation of pollutants in food chain, and many others. The data presented in this chapter can be used for building a simple mathematical model for water and soil purification. It may contribute to successful transfer of phytotechnologies to the agricultural or commercial sectors, and encourage exploration of ways to improve the economic viability of these technologies. Thus, presented biological data about phytoextraction incorporated into the mathematical model can serve to develop the phytotechnologies with potential economic impacts.

REFERENCES

Ajibade, F.O., Adeniran, K.A., Egbuna, C.K., 2013. Phytoremediation efficiencies of water hyacinth in removing heavy metals in domestic sewage (A case study of University of Ilorin, Nigeria). Int. J. Eng. Sci. 2 (12), 16–27.

Alvarado, S., Guédez, M., Lué-Merú, M.P., Nelson, G., Alvaro, A., Jesús, A.C., Gyula, Z., 2008. Arsenic removal from waters by bioremediation with the aquatic plants Water Hyacinth (*Eichhornia crassipes*) and Lesser Duckweed (*Lemna minor*). Bioresour. Technol. 99 (17), 8436–8440.

Amin, H., Arain, B.A., Amin, F., Surhio, M.A., 2014. Analysis of growth response and tolerance index of *Glycine max* (L.) Merr. under hexavalent chromium stress. Adv. Life Sci. 1 (4), 231–241.

Ansari, A.A., Gill, S.S., Gill, R., Lanza, G.R., Lee, N., 2014. Phytoremediation: Management of Environmental Contaminants, vol. 1. Springer, New York, USA, 792.

Aransiola, S.A., Ijah, U.J.J., Abioye, O.P., 2013. Phytoremediation of lead polluted soil by *Glycine max* L. Appl. Environ. Soil Sci. 2013, 1–7. http://dx.doi.org/10.1155/2013/631619.

Ariyakanon, N., Winaipanich, B., 2006. Phytoremediation of copper contaminated soil by *Brassica juncea* (L.) Czern and *Bidens alba* (L.) DC. var. *radiata*. J. Sci. Res. Chula Univ. 31 (1), 49–56.

De Andrade, M.G., Melo, V.D., Gabardo, J., Souza, L.C.D., Reissmann, C.B., 2009. Heavy metals in soils of a lead mining and metallurgy area. I – Phytoextraction. Rev. Bras. Cienc. Solo 33, 1879–1888.

Badr, N., Fawzy, M., Al-Qahtani, K.M., 2012. Phytoremediation: an economical solution to heavy-metal-polluted soil and evaluation of plant removal ability. World Appl. Sci. J. 16, 1292–1301.

Baldantoni, D., Maisto, G., Bartoli, G., Alfani, A., 2005. Analyses of three native aquatic plant species to assess spatial gradients of lake trace element contamination. Aquat. Bot. 83, 48–60.

Bianconi, D., Pietrini, F., Massacci, A., Iannelli, M.A., 2013. Uptake of cadmium by *Lemna minor*, a (hyper?-) accumulator plant involved in phytoremediation applications. E3S Web Conf. 1, 13002. http://dx.doi.org/10.1051/e3sconf/20130113002.

Brooks, R.R., 1998. Plants Those Hyperaccumulate Heavy Metals. CAB International, Wallingford, 380 p.

Bryan, G.W., Langston, W.J., 1992. Bioavailability, accumulation and effects of heavy metals in sediments with special reference to the United Kingdom: a review. Env. Pollut. 76, 89–131.

Caldwell, E.F., Duff, M.C., Ferguson, C.E., Coughlin, D.P., Hicks, R.A., Dixon, E., 2012. Bio-monitoring for uranium using stream-side terrestrial plants and macrophytes. J. Environ. Monit. 14 (3), 968–976.

Calonego, J.C., Rosolem, C.A., 2011. Soil water retention and s index after crop rotation and chiseling. Rev. Bras. Cienc. Solo 35, 1927–1937.

Calonego, J.C., Borghi, E., Crusciol, C.A.C., 2011. Intervalo hldrico otimo e compactagao do solo com cultivo consorciado de milho e braquiaria. Rev. Bras. Cienc. Solo 35, 2183–2190.

Cano-Rodríguez, C.T., Roa-Morales, G., Amaya-Chávez, A., Valdés-Arias, R.A., Barrera-Díaz, C.E., Balderas-Hernández, P., 2014. Tolerance of *Myriophyllum aquaticum* to exposure of industrial wastewater pretreatment with electrocoagulation and their efficiency in the removal of pollutants. J. Environ. Biol. Acad. Environ. Biol. 35 (1), 127–136.

Cavigelli, M.A., 2004. Weed seedbank dynamics in three organic farming crop rotations. Agron. J. 96, 1429–1435.

Chirakkara, R., Reddy, K., 2015. Plant species identification for phytoremediation of mixed contaminated soils. J. Hazard Toxic Radioact. Waste 1–7. http://dx.doi.org/10.1061/(ASCE)HZ.2153-5515.0000282.

Ciura, J., Poniedziałek, M., Sękara, A., Jędrszczyk, E., 2005. The possibility of using crops as metal phytoremediants. Polish J. Environ. Stud. 14 (1), 17–22.

Čechmánková, J., Vácha, R., Skála, J., Havelková, M., 2011. Heavy metals phytoextraction from heavily and moderately contaminated soil by field crops grown in monoculture and crop rotation. Soil Water Res. 6 (3), 120–130.

Chaudhary, E., Sharma, P., 2014. Duckweed plant: a better future option for phytoremediation. Int. J. Emerging Sci. Eng. 2 (7), 39–41.

Chaudhuri, D., Majumder, A., Misra, A.K., Bandyopadhyay, K., 2014. Cadmium removal by *Lemna minor* and *Spirodela polyrhiza*. Int. J. Phytorem. 16 (7–12), 1119–1132.

Chibuike, G.U., Obiora, S.C., 2014. Heavy metal polluted soils: effect on plants and bioremediation methods. Appl. Environ. Soil Sci. 2014, 1–12. http://dx.doi.org/10.1155/2014/752708.

Danilin, I., 2010. Phytoremediation of water at an urban station of water purification by water hyacinth with pre-induction of methalothioneinns. In: Proceedings of the NATO Advanced Research Workshop on Advanced Water Supply and Wastewater Treatment a Road to Safer Society and Environment, Lviv, Ukraine, pp. 287–294.

Demirta, C., Yazgan, S., Candogan, B.N., Sincik, M., Buyukcangaz, H., Goksoy, A.T., 2010. Quality and yield response of soybean (*Glycine max* L. Merrill) to osmotic stress in sub–humid environment. Afr. J. Biotech. 9 (41), 6873–6881.

Dhillon, S.K., Dhillon, K.S., 2009. Phytoremediation of selenium-contaminated soils: the efficiency of different cropping systems. Soil Use Manage. 25 (4), 441–453.

Dołęgowska, S., Migaszewski, Z.M., 2015. Plant sampling uncertainty: a critical review based on moss studies. Environ. Rev. 1–10. http://dx.doi.org/10.1139/er-2014-0052.

Dixit, R., Wasiullah, Malaviya, D., Pandiyan, K., Singh, U.B., Sahu, A., Shukla, R., Singh, B.P., Rai, J.P., Sharma, P.K., Lade, H., Paul, D., 2015. Bioremediation of heavy metals from soil and aquatic environment: an overview of principles and criteria of fundamental Processes. Sustainability 7, 2189–2212. http://dx.doi.org/10.3390/su7022189.

Dwivedi, A.K., Pandey, V.N., 2011. Morphological characterization, biomass and biofunctional yield of some aquatic weeds of North Eastern Terai Region of UP. Adv. Bio Res. 2 (2), 36–46.

Ebbs, D.S., Lasat, M.M., Brady, D.J., Cornish, J., Gordon, R., Kochin, L.V., 1997. Phytoextraction of cadmium and zinc from a contaminated site. J. Environ. Qual. 26, 1424–1430.

Ehsan, S., Ali, S., Noureen, S., Mahmood, K., Farid, M., Ishaque, W., Shakoor, M.B., Rizwan, M., 2014. Citric acid assisted phytoremediation of cadmium by *Brassica napus* L. Ecotoxicol. Environ. Saf. 106, 164–172.

El-Atrach, F., Vierheilig, H., Ocampo, J.A., 1989. Influence of non-host plants on vesicular-arbuscular mycorrhizal infection of host plants and on spore germination. Soil Biol. Biochem. 21 (1), 161–163.

Van der Ent, A., Baker, A.J.M., Reeves, R.D., Pollard, A.J., Schat, H., 2013. Hyperaccumulators of metal and metalloid trace elements: facts and fiction. Plant Soil 362, 319–334.

Falbo, M.B., Weaks, T.E., 1990. A comparison of *Eichhornia crassipes (Pontederiaceae)* and *Sphagnum quinquefarium* (*Sphagnaceae*) in treatment of acid mine water. Econ. Bot. 44, 40–49.

Fellet, G., Marchiol, L., Perosa, D., Zerbi, G., 2007. The application of phytoremediation technology in a soil contaminated by pyrite cinders. Ecol. Eng. 31, 207–214.

Figueira, R., Ribeiro, T., 2005. Transplants of aquatic mosses as biosorbent of metals released by a mine Zuent. Environ. Pollut. 136, 293–301.

Fiorese, G., Guariso, G., 2010. A GIS-based approach to evaluate biomass potential from energy crops at regional scale. Environ. Model. Software 25, 702–711.

Freyer, B., 2003. Fruchtfolgen. Konventionell. Integriert. Biologisch. Eugen Ulmer Verlag, Stuttgart, Germany.

Friedlová, M., 2010. The influence of heavy metals on soil biological and chemical properties. Soil Water Res. 5 (1), 21–27.

Fumagalli, P., Comolli, R., Ferrè, C., Ghiani, A., Gentili, R., Citterio, S., 2014. The rotation of white lupin (*Lupinus albus* L.) with metal-accumulating plant crops: a strategy to increase the benefits of soil phytoremediation. J. Environ. Manage. 145, 35–42.

Galán, E., González, I., Fernández-Caliani, J.C., 2002. Residual pollution load of soils impacted by the Aznalcóllar (Spain) mining spill after clean-up operations. Sci. Total Environ. 286, 167–179.

Gandhimathi, R., Ramesh, S.T., Arun, V.M., Nidheesh, P.V., 2013. Biosorption of Cu(II) and Zn(II) ions from aqueous solution by water hyacinth (*Eichhornia crassipes*). Int. J. Environ. Waste Manage. 11 (4), 365–386.

Garbisu, C., Alkorta, I., 2001. Phytoextraction: a cost effective plant-based technology for the removal of metals from the environment. Bioresour. Technol. 77 (3), 229–236.

Gavito, M.E., Miller, M.H., 1998. Changes in mycorrhiza development in maize induced by crop management practices. Plant Soil 198 (2), 185–192.

Ghani, A., 2010. Toxic effects of heavy metals on plant growth and metal accumulation in maize (*Zea mays* L.). Iran. J. Toxicol. 3 (3), 325–334.

Ghosh, M., Singh, V., 2005. A review on phytoremediation of heavy metals and utilization of its byproducts. Appl. Ecol. Environ. Res. 3 (1), 1–18.

Ghosh, S., 2010. Wetland macrophytes as toxic metal accumulators. Int. J. Environ. Sci. 1, 523–528.

Gigolashvili, T., Kopriva, S., 2014. Transporters in plant sulfur metabolism. Front. Plant Sci. 5, 442. http://dx.doi.org/10.3389/fpls.2014.00442.

Glick, B.R., 2003. Phytoremediation: synergistic use of plants and bacteria to clean up the environment. Biotechnol. Adv. 21, 383–393.

Goolsby, E.W., Mason, C.M., 2015. Toward a more physiologically and evolutionarily relevant definition of metal hyperaccumulation in plants. Front. Plant Sci. 6, 33. http://dx.doi.org/10.3389/fpls.2015.00033.

Gomes, H.I., 2012. Phytoremediation for bioenergy: challenges and opportunities. Environ. Technol. Rev. 1 (1), 59–66.

Goswami, R., Devi, B., Sarma, K.P., 2009. Comparative study on accumulation of cadmium in three aquatic plants. J. Environ. Res. Dev. 4 (2), 359–364.

Goswami, S., Das, S., 2015. A study on cadmium phytoremediation potential of Indian Mustard, *Brassica juncea*. Int. J. Phytorem. 17 (6), 583–588. http://dx.doi.org/10.1080/15226514.2014.935289.

Gupta, K., 2014. Bioaccumulation potential of *Pistia stratiotes* and its response to tannery effluent exposure. Int. J. Phytopharm. Res. 5 (2), 105–110.

Ha, N.T.H., Sakakibara, M., Sano, S., 2009. The hytoremediation of Sb, As, Cu, and Zn from contaminated water by the aquatic macrophyte *Eleocharis cicularis*. Clean Soil Air Water 37 (9), 720–725.

Hague, S.S., Overstreet, C., 2002. Crop rotation effects on nematode populations. In: van Santen, E. (Ed.), Making Conservation Tillage Conventional: Building a Future on 25 years of Research, Proceedings of 25th Annual Southern Conservation Tillage Conference for Sustainable Agriculture, Auburn, AL, USA, June 24–26, 2002. Alabama Agricultural Experiment Station, Auburn University, Auburn, AL, pp. 156–160.

Hamadouche, N.A., Aoumeur, H., Djediai, S., Slimani, M., Aoues, A., 2012. Phytoremediation potential of *Raphanus sativus* L. for lead contaminated soil. Acta Biologica Szegediensis 56 (1), 43–49.

Hamvumba, R., Mataa, M., Mweetwa, A.M., 2014. Evaluation of sunflower (*Helianthus annuus* L.), sorghum (*Sorghum bicolor* L.) and Chinese cabbage (*Brassica chinensis*) for phytoremediation of lead contaminated soils. Environ. Pollut. 3 (2), 65–73.

Hayes, K.F., Traina, S.J., 1998. Metal ion speciation and its significance in ecosystem health. In: Huang, P.M. (Ed.), Soil Chemistry and Ecosystems Health. SSSA, Madison, Wisconsin, USA, pp. 119–141.

Hazrat, A., Ezzat, K., Muhammad, A.S., 2013. Phytoremediation of heavy metals—concepts and applications. Chemosphere 91 (7), 869–881.

Hua, J., Zhang, C., Yin, Y., Chen, R., Wang, X., 2012. Phytoremediation potential of three aquatic macrophytes in manganese-contaminated water. Water Environ. J. 26, 335–342.

Imtiyaz, S., Agnihotri, R.K., Ganie, S.A., Sharma, R., 2014. Biochemical response of *Glycine Max* (L.) Merr. to cobalt and lead stress. J. Stress Physiol. Biochem. 10 (3), 259–272.

Inoue, H., Saeki, K., 2004. Removal of Cd from actual soils polluted with Cd and Zn and Cd-added soils by *Brassica juncea* and *Zea mays*. J. Fac. Agric. Kyushu Univ. 49 (2), 195–205.

Irshad, M., Ahmad, S., Pervez, A., 2014. Phytoaccumulation of heavy metals in natural plants thriving on wastewater effluent at hattar industrial estate, Pakistan. Int. J. Phytorem. 17 (2), 154–158.

Jackson, L.J., 1998. Paradigms of metal accumulation in rooted aquatic vascular plants. Sci. Total Environ. 219, 223–231.

Jadia, C.D., Fulekar, H., 2008. Phytotoxicity and remediation of heavy metals by fibrous root grass (Sorghum). J. Appl. Biosci. 10, 491–499.

Jaffré, T., Pillon, Y., Thomine, S., Merlot, S., 2013. The metal hyperaccumulators from New Caledonia can broaden our understanding of nickel accumulation in plants. Front. Plant Sci. 4, 279.

Jain, S.K., Vasudevan, P., Jha, N.K., 1990. *Azolla pinnata* r.br. and *Lemna minor L.* for removal of lead and zinc from polluted water. Water Res. 24 (2), 177–183.

Jatav, K.S., Singh, R.P., Mishra, J.K., Chauhan, A.K.S., 2012. Phytoremediation-an emerging green technology for removal of pollutants from soil and water resources. Indian J. L. Sci. 2 (1), 149–152.

Jia, Y., Li, F.M., Zhang, Z.H., Wang, X.L., Guo, R., Siddique, K.H.M., 2009. Productivity and water use of alfalfa and subsequent crops in the semiarid Loess Plateau with different stand ages of alfalfa and crop sequences. Field Crops Res. 114 (1), 58–65.

Kang, S.-M., Radhakrishnan, R., You, Y.-H., Khan, A.-L., Lee, K.-E., Lee, J.-D., Lee, I.-J., 2015. Enterobacter asburiae KE17 association regulates physiological changes and mitigates the toxic effects of heavy metals in soybean. Plant Biol. http://dx.doi.org/10.1111/plb.12341.

Kamal, M., Ghaly, A.E., Mahmoud, N., Cote, R., 2004. Phytoaccumulation of heavy metals by aquatic plants. Environ. Int. 29, 1029–1039.

Kamnev, A., van der Lelie, D., 2000. Chemical and biological parameters as tools to evaluate and improve heavy metal phytoremediation. Biosci. Rep. 29, 239–258.

Karimi, N., 2013. Comparative phytoremediation of chromium-contaminated soils by Alfalfa (*Medicago sativa*) and *Sorghum bicolor* (L.) Moench. Int. J. Sci. Res. Environ. Sci. (IJSRES) 1 (3), 44–49.

Khambhaty, Y., Mody, K., Basha, S., Jha, B., 2009. Biosorption of chromium (VI) onto marine *Aspergillus niger*: experimental studies and pseudosecond order kinetics. World J. Microbiol. Biotechnol. 25, 1413–1421.

Kidd, P., Mench, M., Álvarez-López, V., Bert, V., Dimitriou, I., Friesl-Hanl, W., Herzig, R., Janssen, J.O., Kolbas, A., Müller, I., Neu, S., Renella, G., Ruttens, A., Vangronsveld, J., Puschenreiter, M., 2015. Agronomic practices for improving gentle remediation of trace element-contaminated soils. Int. J. Phytorem. 01, http://dx.doi.org/10.1080/15226514.2014.1003788.

Kumar, P.B.A.N., Dushenkov, V., Motto, H., Raskin, I., 1995. Phytoextraction: the use of plants to remove heavy metals from soils. Environ. Sci. Technol. 29, 1232–1238.

Kumar, M.V., Tripathi, B.D., 2008. Concurrent removal and accumulation of heavy metals by the three aquatic macrophytes. Bioresour. Technol. 99 (15), 7091–7097.

Ladislas, S., El-Mufleh, A., Gerente, C., Chazarenc, F., Andres, Y., Bechet, B., 2012. Potential of aquatic macrophytes as bioindicators of heavy metal pollution in urban stormwater runoff. Water Air Soil Pollut. 223, 877–888.

Landberg, T., Greger, M., 1996. Differences in uptake and tolerance to heavy metals in *Salix* from unpolluted and polluted areas. Appl. Geochem. 11 (1–2), 175–180.

Landesman, L., Fedler, C., Duan, R., 2011. Plant Nutrient Phytoremediation Using Duckweed. In: Eutrophication: Causes, Consequences and Control. Springer, pp. 341–354. http://dx.doi.org/10.1007/978-90-481-9625-8_17.

Lasat, M.M., 2000. Phytoextraction of metals from contaminated soil: a review of plant/soil/metal interaction and assessment of pertinent agronomic issues. J. Hazard. Subst. Res. 2, 1–25.

Lasat, M.M., 2002. Phytoextraction of toxic metals: a review of biological mechanisms. J. Environ. Qual. 31, 109–120.

Latta, R.A., Blacklow, L.-J., 2001. Productive lucerne-crop rotations. Australian Agronomy Conference 10th AAC. Concurrent Sess. 4, 1050–1220.

Lone, M.I., He, Z.-L., Stoffella, P.J., Yang, X-e., 2008. Phytoremediation of heavy metal polluted soils and water: progresses and perspectives. J. Zhejiang Univ. Sci. B 9 (3), 210–220.

Liao, C., Xu, W., Lu, G., Liang, X., Guo, C., Yang, C., Dang, Z., 2015. Accumulation of hydrocarbons by maize (*Zea mays* L.) in remediation of soils contaminated with crude oil. Int. J. Phytorem. 17 (6), 583–588. http://dx.doi.org/10.1080/15226514.2014.935289.

Madejón, P., Murillo, J.M., Marañón, T., Cabrera, F., Soriano, M.A., 2002. Trace element and nutrient accumulation in sunflower plants two years after *Aznalcóllar* mine spill. Sci. Total Environ. 307, 239–257.

Mahmood, T., Malik, S.A., Hussain, S.T., 2010. Biosorption and recovery of heavy metals from aqueous solutions by Eichhornia crassipes (water hyacinth) ash. Bioresour. 5 (2), 1244–1256.

Maine, M.A., Duarte, M.V., Suñé, N.L., 2001. Cadmium uptake by floating macrophytes. Water Res. 35 (11), 2629–2634.

Marchiol, L., Assolari, S., Sacco, P., Zerbi, G., 2004. Phytoextraction of heavy metals by canola (*Brassica napus*) and radish (*Raphanus sativus*) grown on multicontaminated soil. Environ. Pollut. 132 (1), 21–27.

Marchiol, L., Fellet, G., Perosa, D., Zerbi, G., 2007. Removal of trace metals by *Sorghum bicolor* and *Helianthus annuus* in a site polluted by industrial wastes: a field experience. Plant Physiol. Biochem. 45, 379–387.

Marschner, H., 1986. Mineral Nutrition of Higher Plants. Academic Press, London, p. 651.

McGrath, S.P., 1998. Phytoextraction for soil remediation. In: Brooks, R.R. (Ed.), Plants that Hyperaccumulate Heavy Metals. CAB International, New York, NY, pp. 261–287.

McGrath, S.P., Shen, Z.G., Zhao, F.J., 1997. Heavy metal uptake and chemical changes in the rhizosphere of *Thlaspi caerulescens* and *Thlaspi ochroleucum* grown in contaminated soils. Plant Soil 188, 153–159.

McGrath, S.P., Zhao, F.J., Lombi, E., 2001. Plant and rhizosphere process involved in phytoremediation of metal-contaminated soils. Plant Soil 232 (1/2), 207–214.

Meybeck, M., Chapman, D., Helmer, R. (Eds.), 1989. Global Fresh Water Quality: A First Assessment. Basil Blackwell, Oxford, 307 pp.

Miretzky, P., Saralegui, A., Fernández Cirelli, A., 2004. Aquatic macrophytes potential for the simultaneous removal of heavy metals (Buenos Aires, Argentina). Chemosphere 57, 997–1005.

Mishra, V.K., Tripathi, B.D., 2008. Concurrent removal and accumulation of heavy metals by the three aquatic macrophytes. Bioresour. Technol. 99 (15), 7091–7097.

Mishra, V.K., Upadhyay, A.R., Pandey, S.K., Tripathi, B.D., 2008. Heavy metal pollution induced due to coal mining effluent on surrounding aquatic ecosystem and its management through naturally occurring aquatic macrophytes. Bioresour. Technol 99, 930–936.

Mojiri, A., 2011. The potential of corn (*Zea mays*) for phytoremediation of soil contaminated with cadmium and lead. J. Biol. Environ. Sci. 5 (13), 17–22.

Nejatzadeh-Barandozi, F., Gholami-Borujeni, F., 2014. Effectiveness of phytoremediation technologies to clean up of metalloids using three plant species in Iran. Water Environ. Res. 86 (1), 43–47.

Nishizawa, T., 1978. Annual population changes of soil nematodes in the field of continuous cropping or rotation. Kasetsart J. 12, 3–13.

Nogueira, T.A.R., Melo, W.J., Fonseca, I.M., Marques, M.O., He, Z.L., 2010. Barium uptake by maize plants as affected by sewage sludge in a long-term field study. J. Hazard. Mater. 181, 1148–1157.

Nriagu, J.O., Pacyna, J.M., 1988. Quantitative assessment of worldwide contamination of air water and soils by trace metals. Nature 333 (6169), 134–139.

Oh, K., Cao, T., Li, T., Cheng, H., 2014. Study on application of phytoremediation technology in management and remediation of contaminated soil. J. Clean Energy Technol. 2 (3), 216–220.

Oh, K., Li, T., Cheng, H.Y., Xie, Y., Yonemochi, S., 2013. Development of profitable phytoremediation of contaminated soils with biofuel crops. J. Environ. Prot. 4, 58–64.

Olufayo, A., Ruelle, P., Baldy, C., Aïdaoui, A., 1997. Biomass of grain sorghum (*Sorghum bicolor* L. Moench) under variable water regime. Biomass Bioenergy 12 (5), 383–386.

Outridge, P., Noller, B., 1991. Accumulation of toxic trace elements by freshwater vascular plants. Rev. Environ. Contam. Toxicol. 121, 2–63.

Pari, L., Assirelli, A., Suardi, A., 2010. Evaluation of *Brassica napus* and *Brassiva carinata* losses during harvesting: three years of experience. In: 18th European Biomass Conference and Exhibition, Lyon, 3–7 May 2010, France, pp. 1790–1793.

Paul, M.P., Lauer, J.G., Lueschen, W.E., Ford, H.J., Hoverstad, T.R., Oplinger, E.S., Crookston, R.K., 1997. Environment affects the corn and soybean rotation effect. Agron. J. 89 (3), 442–448.

Peralta-Videa, J.R., Gardea-Torresdey, J.L., Gomez, E., Tiemann, K.J., Parsons, J.G., Carrillo, G., 2002. Effect of mixed cadmium, copper, nickel and zinc at different pHs upon alfalfa growth and heavy metal uptake. Environ. Pollut. 119 (3), 291–301.

Pierrisnard, F., 1996. Impact de I'amendment dês boues residuaires de la ville de Marseille sur de sols a vocation agricole: comportment du Cd, Cr, Cu, Ni, Pb e Zn, des hydrocarbures et des composes polares. Dissertation. Docteur Géosciences de i' Ebvironnement – Université de Droit d'Economie et dês Sciences d'AIX-Marseille.

Pinto, C.L., Caconia, A., Souza, M., 1987. Utilization of water hyacinth for removal and recovery of silver from industrial wastewater. In: Athie, D. (Ed.), The Use of Macrophytes in Water Pollution Controls, Water Sci. Technol, vol. 19, pp. 89–102.

Podlešáková, E., Němeček, J., 2000. Trace elements with low mobility in soils. Sci. Stud. RISWC 11, 65–74.

Podlešáková, E., Němeček, J., Vácha, R., 1999. Mobility and bioavailability of trace elements in the soils of the Czech Republic. In: 5th Int. Conf. Biogeochemistry of Trace Elements. July 11–15, ISTER, Vienna, Austria, pp. 118–119.

Pongrac, P., Vogel-Mikus, K., Regvar, M., Tolrà, R., Poschenrieder, C., Barceló, J., 2008. Glucosinolate profiles change during the life cycle and mycorrhizal colonisation in a Cd/Zn hyperaccumulator *Thlaspi praecox* (*Brassicaceae*). J. Chem. Ecol. 34, 1038–1044.

Qian, J.-H., Zayed, A., Zhu, Y.-L., Yu, M., Terry, N., 1999. Phytoaccumulation of Trace Elements by Wetland Plants: III. Uptake and Accumulation of Ten Trace Elements by Twelve Plant Species. J. Environ. Qual. 28 (5), 1448–1455.

Radhakrishnan, R., Lee, I.J., 2013. Spermine promotes acclimation to osmotic-stress by modifying antioxidant, abscisic acid, and jasmonic acid signals in soybean. J. Plant Growth Regul. 32, 22–30.

Rai, P.K., 2012. An eco-sustainable green approach for heavy metals management: two case studies of developing industrial region. Environ. Monit. Assess. 184, 421–448.

Rao, K., Shantaram, M.V., 1996. Effect of urban solid wastes on dry matter yield, uptake of micronutrients and heavy metals by maize plants. J. Environ. Biol. 17, 25–32.

Revathi, K., Haribabu, T.E., Sudha, P.N., 2011. Phytoremediation of chromium contaminated soil using sorghum plant. Int. J. Environ. Sci. 2 (2), 417–428.

Rizwana, M., Darshan, M., Nilesh, D., 2014. Phytoremediation of textile waste water using potential wetland plant: eco sustainable approach. Int. J. Interdiscip. Multidiscip. Stud. (IJIMS) 1 (4), 130–138.

Robinson, R.G., 1966. Sunflower-soybean and grain sorghum-corn rotations versus monoculture. Agron. J. 58 (5), 475–477.

Robinson, A.A., Belden, J.B., Lydy, M.J., 2009. Toxicity of fluoroquinolone antibiotics to aquatic organisms. Environ. Toxicol. Chem. 24 (2), 423–430.

Roosens, N., Verbruggen, N., Meerts, P., Ximénez-Embún, P., Smith, J.A.C., 2003. Natural variation in cadmium tolerance and its relationship to metal hyperaccumulation for seven populations of Thlaspi caerulescens from western Europe. Plant Cell Environ. 26 (10), 1657–1672.

Sakakibara, M., Harada, A., Sano, S., Hori, R.S., 2009. Heavy metal tolerance and accumulation in *Eleocharis acicularis*, a heavy metal hyperaccumulating aquatic plant species. Geo-pollut. Sci. Med. Geol. Urban Geol. 5, 1–8.

Sakakibara, M., Ohmori, Y., Ha, N.T.H., Sano, S., Sera, K., 2011. Phytoremediation of heavy metal-contaminated water and sediment by *Eleocharis acicularis*. Clean Soil Air Water 39 (8), 735–741.

Salt, D.E., Smith, R.D., Raskin, L., 1998. Phytoremediation. Ann. Rev. Plant Physiol. Plant Mol. Biol. 49 (1), 643–668.

Sanches Filho, P.J., Nunes, L.V., da Rosa, N.N., Betemps, G.R., Pereira, R.S., 2015. Comparison among native floating aquatic macrophytes for bioconcentration of heavy metals. Ecotoxicol. Environ. Contam. 10 (1), 1–6. http://dx.doi.org/10.5132/eec.2015.01.01.

Santos, J.A., Nunes, L.A.P.L., Melo, W.J., Araujo, A.S.F., 2011. Tannery sludge compost amendment rates on soil microbial biomass of two different soils. Eur. J. Soil Biol. 47, 146–151.

Sarma, H., 2011. Metal hyperaccumulation in plants: a review focusing on phytoremediation technology. J. Environ. Sci. Technol. 4, 118–138.

Sharma, M.S., Raju, N.S., 2013. Correlation of heavy metal contamination with soil properties of industrial areas of Mysore, Karnataka, India by cluster analysis. Int. Res. J. Environ. Sci. 2 (10), 22–27.

Sharma, S., Singh, B., Manchanda, V.K., 2014. Phytoremediation: role of terrestrial plants and aquatic macrophytes in the remediation of radionuclides and heavy metal contaminated soil and water. Environ. Sci. Pollut. Res. 22 (2), 946–962. http://dx.doi.org/10.1007/s11356-014-3635-8.

Shen, Z.G., Zhao, F.J., McGrath, S.P., 1997. Uptake and transport of zinc in the hyperaccumulator *Thlaspi caerulescens* and the non-hyperaccumulator *Thlaspi ochroleucum*. Plant Cell Environ. 20, 898–906.

Schalscha, E., Ahumada, I., 1998. Heavy metals in rivers and soils of central Chile. Water Sci. Technol. 37 (8), 251–255.

Sinegani, A.A.S., Khalilikhah, F., 2010. Effects of EDTA, sheep manure extract and their application time on Cd uptake by *Helianthus annuus* from a calcareous mine soil. Soil Sediment Contam. 19, 378–390.

Singh, D., Gupta, R., Tiwari, A., 2011. Phytoremediation of lead from wasterwater using aquatic plants. Int. J. Biol. Res. 2 (7), 411–421.

Singh, A., Eapen, S., Fulekar, M.N., 2009. Potential of *Medicago sativa* for uptake of cadmium from contaminated environment. Rom. Biotechnol. Lett. 14 (1), 4164–4169.

Singh, R.P., Agrawal, M., 2008. Potential benefits and risks of land application of sewage sludge. Waste Manage. 28 (2), 347–358.

Singh, R.P., Agrawal, M., 2010. Biochemical and physiological responses of rice (*Oryza sativa* L.) grown on different sewage sludge amendments rates. Bull. Environ. Contam. Toxicol. 84, 606–612.

Slepetys, J., Kadziuliene, Z., Sarunaite, L., Tilvikiene, V., Kryzeviciene, A., 2012. Biomass Potential of Plants Grown for Bioenergy Production. Renewable energy and energy efficiency, Growing and processing technologies of energy crops 2012: 66–72.

Smith, R., Gross, K., Obertson, G., 2008. Effects of crop diversity on agroecosystem function: crop yield response. Ecosystems 11 (3), 355–366.

Snow, A.M., Ghaly, A.E., 2008. A comparative study of the purification of aquaculture wastewater using water hyacinth, water lettuce and parrot's feather. Am. J. Appl. Sci. 5 (4), 440–453.

Sood, A., Uniyal, P.L., Prasanna, R., Ahluwalia, A.S., 2012. Phytoremediation potential of aquatic macrophyte, *Azolla*. Ambio 41, 122–137.

Soriano, M.A., Fereres, E., 2003. Use of crops for in situ phytoremediation of polluted soils following a toxic flood from a mine spill, 256 (2), 253–264.

Stoate, C., Boatman, N.D., Borralho, R.J., Carvalho, C.R., de Snoo, G.R., Eden, P., 2001. Ecological impacts of arable intensification in Europe. J. Environ. Manag. 63 (4), 337–365.

Sukumaran, D., 2013. Phytoremediation of heavy metals from industrial effluent using constructed wetland technology. Appl. Ecol. Environ. Sci. 1.5, 92–97.

Susarla, S., Medina, V.F., McCutcheon, S.C., 2002. Phytoremediation: an ecological solution to organic chemical contamination. Ecol. Eng. 18 (5), 647–658.

Száková, J., Tlustoš, P., Goessler, W., Pokorný, T., Findenig, S., Balík, J., 2011. The effect of soil contamination level and plant origin on contents of arsenic, cadmium, zinc, and arsenic compounds in *Mentha aquatica* L. Arch. Environ. Prot. 37 (2), 109–121.

Száková, J., Tlustoš, P., Pavlíková, D., Balík, J., 2003. The element distribution into main fractions of soil amended by As, Cd, Cu, Pb, and Zn in different forms. Chem. Pap. 57, 167–168.

Szczygłowska, M., Piekarska, A., Konieczka, P., Namieśnik, J., 2011. Use of *Brassica* plants in the phytoremediation and biofumigation processes. Int. J. Mol. Sci. 12 (11), 7760–7771.

Tangahu, B.V., Abdullah, S.R.S., Basri, H., Idris, M., Anuar, N., Mukhlisin, M., 2011. A review on heavy metals (As, Pb, and Hg) uptake by plants through phytoremediation. Int. J. Chem. Eng. 2011, 1–31. http://dx.doi.org/10.1155/2011/939161.

Teasdale, J.R., Mangum, R.W., Radhakrishnan, J., Cavigelli, M.A., 2004. Weed seedbank dynamics in three organic farming crop rotations. Agron. J. 96 (5), 1429–1435.

Theuretzbacher, F., Kravanja, P., Becker, M., Bauer, A., Enguidanos, R., Amon, B., Friedl, A., Potthast, A., Amon, T., 2012. Whole plant utilization of different *Sorghum bicolor* L. Moench varieties for combined bioethanol and biogas production. In: Rovira-Más, F. (Ed.), International Conference of Agricultural Engineering, CIGR AgEng 2012. International Conference of Agricultural Engineering CIGR-AgEng 2012. Diazotec, S. L., Valencia, ISBN 84-615-9654-4, pp. 1–6.

Thilakar, R.J., Rathi, J., Pillai, P.M., 2012. Phytoaccumulation of chromium and copper by *Pistia stratiotes* L. and *Salvinia natans* (L.). J. Nat. Prod. Plant Resour. 2 (6), 725–730.

Thivya, N., Srilakshmi, K.V., Bhuvaneswari, S., Leon, T., Raj, S., 2014. Phytoaccumulation of chromium and copper by *Mentha spicata* L. J. Plant Physiol. Pathol. 2 (1), 1–3. http://dx.doi.org/10.4172/2329-955X.1000118.

Tremp, H., Kohler, A., 1995. The usefulness of macrophytemonitoring systems, exemplified on eutrophication and acidification of running waters. Acta Bot. Gall. 142, 541–550.

Turan, M., Esringü, A., 2007. Phytoremediation based on canola (*Brassica napus* L.) and Indian mustard (*Brassica juncea* L.) planted on spiked soil by aliquot amount of Cd, Cu, Pb, and Zn. Plant Soil Environ. 53 (1), 7–15.

Ucer, A., Uyanik, A., Kutbay, H.G., 2013. Removal of heavy metals using *Myriophyllum verticillatum* (Whorl-Leaf Watermilfoil) in a hydroponic system. Ekoloji 22 (87), 1–9.

Vamerali, T., Bandiera, M., Mosca, G., 2010. Field crops for phytoremediation of metal-contaminated land. A review. Environ. Chem. Lett. 8, 1–17.

Vargas-Ayala, R., Rodriguez-Kabana, R., 2001. Bioremediative management of soybean nematode population densities in crop rotations with velvetbean, cowpea, and winter crops. Nematropica 31, 37–46.

Verma, V.K., Gupta, R.K., Rai, J.P.N., 2005. Biosorption of Pb and Zn from pulp and paper industry eZuent by water hyacinth (*Eichhornia crassipes*). J. Sci. Ind. Res. 64, 778–781.

Vijayarengan, P., 2005. Nitrogen and potassium status of greengram (*Vigna radiata*) cultivars under nickel stress. Nat. Environ. Pollut. Technol. 4, 65–69.

Wang, T.C., Weissman, J.C., Ramesh, G., Varadarajan, R., Benemann, J.R., 1996. Parameters for removal of toxic heavy metals by water milfoil (*Myriophyllum spicatum*). Bull. Environ. Contam. Toxicol. 57, 779–786.

Warnke, S.A., Chen, S.Y., Wyse, D.L., Johnson, G.A., Porter, P.M., 2006. Effect of Rotation Crops on Heterodera glycines Population Density in a Greenhouse Screening Study. J. Nematol. 38 (3), 391–398.

Wong, J.W.C., 1996. Heavy metal contents in vegetables and market garden soils in Hong Kong. Environ. Technol. 17, 407–414.

Wuana, R.A., Okieimen, F.E., 2011. Heavy metals in contaminated soils: a review of sources, chemistry, risks and best available strategies for remediation. ISRN Ecol. 2011, 1–20.

Yang, X.E., Jin, X.F., Feng, Y., Islam, E., 2005. Molecular mechanisms and genetic basis of heavy metal tolerance/hyperaccumulation in plants. J. Integr. Plant Biol. Formerly Acta Bot. Sin. 47 (9), 1025–1035.

Yeh, T.Y., 2015. Biostimulator and biodegradable chelator to pytoextract not very toxic Cu and Zn. Hydrol Curr. Res. 6, 190. http://dx.doi.org/10.4172/2157-7587.1000190.

Zhang, H., Zheng, L.C., Yi, X.Y., 2009. Remediation of soil co-contaminated with pyrene and cadmium by growing maize (*Zea mays* L.). Int. J. Environ. Sci. Technol. 6 (2), 249–258.

Zhang, X., Xia, H., Li, Z., Zhang, P., Gao, B., 2010. Potential of four forage grasses in remediation of Cd and Zn contaminated soils. Bioresour. Technol. 101, 2063–2066.

Zhuang, P., Ye, Z.H., Lan, C.Y., Xie, Z.W., Hsu, W.S., 2005. Chemically assisted phytoextraction of heavy metal contaminated soils using three plant species. Plant Soil 276, 153–162.

Zürrer, H., Bachofen, R., 1985. Yields of three cultivars of sunflowers in Switzerland. Biomass 7 (4), 297–302.

CHAPTER

PHYTOEXTRACTION: THE USE OF PLANTS TO REMOVE HEAVY METALS FROM SOIL

15

Zoya Ghori[1], Hira Iftikhar[1], Muhammad F. Bhatti[1], Nasar-um-Minullah[1], Iti Sharma[2], Alvina G. Kazi[1], Parvaiz Ahmad[3]

Atta-ur-Rahman School of Applied Biosciences, National University of Sciences and Technology (NUST), Islamabad, Pakistan[1]; Department of Bioscience and Biotechnology, Banasthali University, Tonk, Rajasthan, India[2]; Department of Botany, S.P. College, Srinagar, Jammu and Kashmir, India[3]

CHAPTER OUTLINE

1. INTRODUCTION

With the industrial revolution, life has become quite easy, but every comfort comes with an expense, a darker side that in this case is the rise of global pollution. Different kinds of pollutants have been introduced into the environment by various industries, which keep on dumping their wastes into water bodies and landfills without any treatment or testing. Various pollutants induce environmental misbalancing and harmful effects on living beings. Heavy metals are among those pollutants that are extremely hazardous in excess to all life forms and can cause fatalities. They can move along with water to reach agricultural or other lands under use. Once in the soil, they can either leach down to the water table or stay in soil. This is where they enter the food chain, affecting all life forms and disturbing the ecosystems.

Heavy metals are defined as the metals that have an atomic mass greater than 20 and are transition metals, metalloids, actinides, and lanthanides. In biological processes, heavy metals are classified into two types: essential heavy metals and nonessential heavy metals. Essential heavy metals are needed by

Plant Metal Interaction. http://dx.doi.org/10.1016/B978-0-12-803158-2.00015-1

the organisms in a very small quantity and include iron (Fe), manganese (Mn), copper (Cu), zinc (Zn), nickel (Ni), cobalt (Co), and molybdenum (Mo). Contrastingly nonessential heavy metals are not needed by organisms for its living like cadmium (Cd), lead (Pb), arsenic (As), mercury (Hg), and chromium (Cr) (Ali et al., 2013). Because heavy metals are nonbiodegradable, the increase in their concentrations can be potentially toxic to the life on earth. Humans, plants, and animals are all equally affected by enhanced levels of heavy metals. Even the microorganisms in the soils are threatened by their toxicity because heavy metals hinder their biological activities by accumulating in their cells and tissues. As a result, they can increase toxicity by generating reactive oxygen species, which can lead to cell damage such as macromolecule degradation, ion uptake and usage, and DNA damages (Ahmad et al., 2008, 2010, 2011, 2015). In plants, heavy metals in particular can affect the process of photosynthesis by interfering with electron transport chain (Rascio and Navari-Izzo, 2011), water relations, and biochemical and enzymatic activities (Ahmad et al., 2008, 2010, 2011, 2015; Qadir et al., 2014).

This increase in the concentration of heavy metals is a consequence of both natural activities like volcanic eruptions and soil erosion and weathering as well as anthropogenic activities like ore mining, burning of fossil fuels, smelting, transport and industrial wastes, and use of fertilizers and other agricultural activities (Kanakaraju et al., 2008; Lias et al., 2013). Overuse of chemicals has led to the accumulation of these heavy metals, which are particularly increasing in toxicity in the soils near industrial and agricultural areas. They are affecting soil fertility and biomass yield of plants; hence, plants are the major victims of heavy metal accumulation in soils (Bhargava et al., 2012). Although developed countries are taking measures to decrease heavy metal concentrations, developing countries are still far behind. Many remediation techniques and methods have been used to address this problem and much research is being done to find alternatives that are economical and effective against heavy metal toxicity (Jadia and Fulekar, 2009; Yao et al., 2012; Dixit et al., 2015).

Heavy metal contaminated soil can be made contaminant-free by many physical, chemical, and biological methods (Jadia and Fulekar, 2009; Yao et al., 2012; Dixit et al., 2015). Based on their nature, remedial techniques can be divided into two main types: ex situ and in situ methods. The ex situ methods are those that require physical removal or immobilization of contaminated soil and its replacement by a new heavy metal free soil. This remediation technique includes physical, chemical, and thermal processes along with biological techniques like land farming, biopiling, use of bioreactors, composting, and many others. Other techniques that are included in ex situ methods include increasing the soil pH to decrease heavy metal solubility in the soil (Koning et al., 2000; Van Deuren and Center UAE, 2002). On the other hand, in situ methods involve the use of biological mechanisms at the site of contamination and do not involve the soil leaving its original site. An example of this technique is phytoremediation or green technology that uses plant's absorption mechanism to remove toxic metals and chemicals from soil (John et al., 2009; Dixit et al., 2015). The in situ methods are considered to be more promising than ex situ methods because of low cost, easy usage, and its environmental friendliness because it does not have any impact on the environment and ecosystem (Salla et al., 2011).

Ex situ remediation techniques, although expensive, have the basic advantage of removing high concentrations of heavy metals from polluted soils. These conventional remediation techniques include soil washing, landfilling and leaching, electrokinetic treatment, and soil immobilization. Another technique is washing of soil. In this case, water is added to the polluted soil which allows the heavy metals from soil to be transferred to the washing water such that the removal of water also removes the heavy metals along with it. This procedure is made efficient by the addition of chelating agents like ethylene diamine tetra acetate (EDTA), ethylene diamine disuccinate (EDDS), diethylenetriamine pentaacetic acid, and many others, which can increase the solubility of heavy metals at the site of treatment (Jiang et al., 2011; Zhao et al., 2011; Ebrahimi, 2014). Chelates function by combining with heavy metals in water and forming chelate complexes. These complexes absorb all the heavy metals from sediment and remove them with the solution. Chelates especially EDTA can remove Cu, Zn, Pb, and Cd with high efficiency, but because of its negative impact on the health of living organisms, their use is limited. Another major drawback of EDTA is that it has poor biodegradability, which can cause leaching of soil. Surfactants can also be used for the same purpose because they can also assist in better absorption of heavy metals (Ortega et al., 2008). Acid leaching is also a type of washing that uses acids like sulfuric acid to extract metals from soil.

Another type of ex situ method is electrokinetic soil remediation, which uses current to remove heavy metals from soil. The electrodes are placed in the sediment and a low direct current is allowed to flow. This results in the migration of positively charged metal ions toward cathode (negatively charged) and negatively charged metal ions toward anode (positively charged). An example is the removal of Cr(VI), which is an anion, and Cr(III), Cd, and Ni that are cations through the flow of current. These metals adhere to their respective electrodes from which they can be removed later by chemical methods (Virkutyte et al., 2002). Other methods include flotation, ultrasonic assisted extraction, and soil immobilization that can also be used to remove heavy metals form contaminated soils. Different methods are applied for differently characterized soils. Although ex situ methods are efficient for high-scale contaminant removal, they are not economical. Hence, alternate methods are now being considered for decontamination of toxic soils (Peng et al., 2009). Oilcake manure of *Calendula officinalis* is also considered as an alternative of chemical chelation of cadmium and lead contaminated alluvial soil (Mani et al., 2015).

In situ methods are more environmental friendly than ex situ methods and can be cost-effective, but the major drawback is that they are time-consuming. There are many in situ remediation technologies available like soil amendments that functions by decreasing metal solubility, sand cap and many others but the most common is phytoremediation. As already explained, phytoremediation uses plants to decontaminate heavy metal toxic soil using plants as metal accumulators. Phytoremediation is an emerging technique for soil treatment with no side effects and it can be linked to run several other processes alongside. Although it has limitations of its own, they are overweighed by its beneficial side. Several criteria have been described for careful selection of plants that can be used to treat certain heavy metals. Following these criteria, phytoremediation can be used successfully to make the land pollution free and available for use. With time, different strategies have been developed in the area to account for differences in situations for different heavy metal species and their different forms. These include phytoextraction, phytostabilization, phytofiltration, and phytodegradation for heavy metal pollution (Jadia and Fulekar, 2009; Yao et al., 2012; Dixit et al., 2015). The most common among these is phytoextraction, which allows the plants to accumulate heavy metals in their organs (Peng et al., 2009). However, even after much research in this area, there are certain issues that cannot be resolved directly, such as the use of plants native to an area in another area does not seem to be much of a practical approach. With this issue comes the need for transgenic production to combine the features of environmentally friendly species and phytoremediators to come up with efficient plant systems that can easily grow and serve the purpose. For transformation to be successful, certain criteria need to be studied in detail along with the genetic basis of phytoremediation to make selections for the best genes in the subject cases and make this technique a success. This chapter focuses on the need for phytoremediation, consideration criteria and selections, phytoextraction and how plants can accumulate metals in shoots, leaves and roots and assisted phytoextraction that uses chelants to increase the plant's accumulation capability. Furthermore the mechanisms by which uptake of heavy metals are facilitated by plants and the biotechnological approaches to develop genetically modified accumulators for phytoextraction are also discussed. Based on the genes that can be manipulated to enhance the process, some transformations made in some common model plants are also described.

2. NEED FOR REMEDIATION

Modern lifestyle and industrialization have led to many environmental problems because they generate different types of wastes and their dumping without proper treatment. These wastes contaminate the environment with the most crucial and fatal effects on the living entities marking their survival at stake. The most hazardous part of industrial effluents and other wastes is the release and accumulation of metals, especially heavy metals, in the environment and their impact on the normal metabolic and signaling activities of the organisms taking them in Ahmad et al. (2011, 2015) and Qadir et al. (2014). The release of these heavy metals in the environment results in an imbalance of the natural environment.

The wastes should be treated before they are dumped into the environment to minimize their effects on the environment by converting them into less harmful forms. However, once the water or soil is contaminated by these pollutants, proper set of processes to treat them needs to follow before they can be brought into common use. The effective treatments, however, are quite costly and are associated with production of undesirable side products, incomplete destruction of target pollutants, high energy inputs, soil structure variation and alteration, and other hazardous effects (Jadia and Fulekar, 2009; Yao et al., 2012; Dixit et al., 2015). Alternate, cost- and energy-effective ways of treating the contaminated water and soil need to be discovered, tested, and brought into use. The most efficient and economic way in this aspect has been found to be bioremediation that involves the use of living organisms to treat certain pollution causing conditions by efficient uptake of the pollutants from the desired environment (Dixit et al., 2015). Plants have been found to be quite efficient in the process, commonly termed as phytoremediation (Ma et al., 2011; Greipsson, 2011; Kamran et al., 2014).

Different techniques involving phytoremediation need to be studied in detail considering the numerous aspects to make it more efficient and feasible to use. All the physiological and biochemical events occurring in the plant that control the nutrient uptake and movement and how they get affected by different nutrient concentrations, especially heavy metals because they can have hazardous effects above permissible limits, need to be understood and studied in detail. Also, the molecular and genetic processes underlying these techniques cannot be neglected when attempting to completely understand plant behavior and efficiency in phytoremediation of various kinds of pollutants from the environment to help in keeping it a safe place to breathe and live.

3. PHYTOREMEDIATION CRITERIA AND ASPECTS: CONSIDERATIONS

Phytoremediation (phyto (Greek) for "plant," remedium (Latin) for "to clean"), as the term indicates, is the clean-up process of a contaminated area with the use of plants. The basic idea behind the use of naturally occurring plant species to restore the environment to its original unpolluted state is its simplicity, ease of processes, environmental friendliness, economical value, and side benefits (Wei et al., 2004; Iori et al., 2015).

All the beneficial aspects outweigh the challenges faced in using phytoremediation as a major treatment technology. This technology should thus be completely focused on the purpose and tested for several aspects that need to be addressed before commercial use. While considering the use of this technology at a particular polluted field or site, certain criteria should be taken under consideration to make the clean-up process more efficient and the least hazardous for the plants used. The characterization of the pollution at the target site of interest is the most important of aspect of these criteria, which points to the type of pollutant(s) or the number of different heavy metal species, the amount of each above the permissible normal range, and other characteristics associated to them such as the size of the heavy metal and its leaching capacities. The next step in the process is determining the best plant species for the target site based on its capabilities to phytoremediate the heavy metal species present in the soil and to keep them limited to the rhizosphere without letting them escape to the deeper layers and the extensive root system to spread through most of the region to completely extract the pollutants. Grasses are considered ideal because of their densely spread network of roots with extensive surface in soil for uptake (Kulakow et al., 2000). Another important aspect is the effect of the heavy metals on the plant species used in the process; this is also considered among the plant selection criteria because the tolerance level of various plant species varies with different heavy metals. Various heavy metals have varying effects on plants based on certain limits, their target location and effects inside the plant, translocation events, and accumulation in various plant parts (Barman et al., 2001). Translocation pattern of the heavy metal inside the plant should be kept strict, considering those plants that have the capability to translocate the highest concentrations of heavy metals to the aerial plant parts for accumulation to ensure their safe removal; this is important because at the harvest stage, aerial parts of the plant are completely and properly harvested, whereas sometimes the underground parts cannot be completely removed from the soil. Last but not the least is the cost estimation for the whole process of phytoremediation; it should be feasible enough according to the economy and plant selection should also be based on this criterion by determining the cultivation needs and costs. All these criteria, along with the phytoremediation technique, should be based on extensive research to make the whole procedure occur as per expected outcomes. Especially important of these research phases is the time estimates for several heavy metals to be mobilized in the soil, taken up, and properly translocated and accumulated in the plant to determine when the clean-up process will be complete (Ugya, 2015).

After the selection has been made and plants cultivated at the pollutant site, at completion of the estimated time for extraction of the heavy metal species, the plants are harvested from the field. These harvested plants then need to be ensured of a certain effect depending upon the pollutants and their form stored in them. Some are converted to less hazardous states by the plant via their internal metabolic pathways before accumulation and thus simply disposed of, others may be incinerated with the plant biomass to achieve the goal, whereas yet others may require other processes to be followed to complete the clean-up process. With incineration, energy is generated by the burning of plant biomass which can be used to undergo useful energy-requiring processes, thus linking it to other industrial processes and adding to the economy as a whole. The whole technique needs to be followed with suitable and careful selections of all steps during the process to make it more successful, resourceful, and economic.

Phytoextraction of heavy metals has garnered much attention in the past several decades since the initiation of its field trials, mainly conducted on important heavy metals, cadmium, and zinc (USEPA, 2000).

4. PHYTOREMEDIATION TECHNIQUES AND PHYTOEXTRACTION EFFICIENCY

Phytoremediation is still in its developing state, and more and more of its variations are maturing into efficient techniques for decontamination. A primary form of phytoremediation, termed phytofiltration, is based on absorption and adsorption of metal pollutants from water bodies with the help of plant roots and/or seedlings. Adsorption keeps the pollutants from escaping and moving deeper (Mukhopadhyay and Maiti, 2010).

Another form of phytoremediation involves the conversion of the pollutant into a volatile form, allowing its escape from the soil into the atmosphere; it is termed phytovolatilization. It does not involve the accumulation of the heavy metal pollutants in the plant parts (Prasad and Freitas, 2003). Phytodegradation is yet another category of phytoremediation but that does not apply to heavy metals.

The treatment of soil using plants as heavy metal uptake and storage entities via phytoremediation can be distinguished into two major techniques, phytoextraction, and phytostabilization (Salt et al., 1998; Jadia and Fulekar, 2009; Yao et al., 2012; Dixit et al., 2015).

Phytostabilization is used for highly polluted areas to provide a complete and dense vegetative cover and to avoid corrosion caused as a result of water or air. This mainly involves the immobilization and stabilization of pollutants in the soil (Singh, 2012). Because phytostabilization is based on the property of providing dense vegetation, the plants suitable for this kind of phytoremediation include those capable of producing extensive and dense root systems covering compressed soils, also restricting the pollutants to the soil zone near their roots by preventing their movement or further leaching with a direct implication of the plant being more tolerant and resistant to those pollutants (i.e., heavy metals). Plant species selection is based on their tolerance levels for certain heavy metals, initially selected as those growing in the polluted areas. The next step is the detailed study of various aspects involved in phytostabilization, including the rate and uptake pattern of selective heavy metals.

The more efficient and useful technique is phytoextraction, but is a more difficult process. This technology involves growth of tolerant plants in contaminated soils; the plant takes up large amounts of heavy metals from the soil, depleting it of the target pollutants and translocating them to the aerial parts of the plant (Rafati et al., 2011). After high level of heavy metal accumulation in the plant parts, the plants are harvested and then incinerated after drying, which generates concentrated pollutant-containing material dumped separately. This highly concentrated mass material contains even higher concentrations of pollutant than the soil, which is what makes this technology a success. The beneficial side of this technology is the energy generated as a result of incineration, which can be useful for testing different processes involving energy.

The highly concentrated accumulation of the pollutants in the plant parts is usually not a naturally favored reaction; however, some have the capability to uptake and accumulate more than necessary. This property of accumulator plants can be exploited and further modified according to need. Under normal circumstances in certain plants, because the concentration of a nutrient (in this case, a heavy metal) increases more than a certain limit, the plant defense system comes into play by performing required metabolic, physiological, and expressional changes to overcome the stressful conditions caused by the heavy metal pollutant (Chu et al., 2015). These changes can be studied in detail in the desired hyperaccumulator plants and modified according to need to enhance its clean-up efficiency (Clemens et al., 2002).

Hyperaccumulators for heavy metals found in soils have the capability of accumulating large amounts of pollutants without having any obvious physical effects or symptoms that depict toxication (Cardwell et al., 2002; Krämer, 2010; Goolsby and Mason, 2015; Emamverdian et al., 2015). The concentration in the plant parts increases as the roots uptake the heavy metals from soil and depletes it from these pollutants. The concentration of trace elements such as As, Zn, Ni, and Cd in the leaf of hyperaccumulators usually becomes up to 100 times higher than that in the contaminated soil, which clearly indicates a highly efficient cleanup of polluted soils (McGrath and Zhao, 2003). Also notable is the higher accumulation of heavy metals in shoot regions as compared with the root system (Kramer et al., 1996; Weber et al., 2004; Becher et al., 2004). Ideal phytoextraction can be carried out with plants that produce a small amount of biomass as they mature. This process can also be used to take up and concentrate economically important nutrients present in trace amounts in the lithosphere, a phenomenon sometimes referred to as phytomining (Broadhurst et al., 2004).

Phytoremediation is sometimes carried out with the aid of assisting and enhancing agents, chelators, such as EDTA. The addition of these chelating agents help to solubilize the trace metals, which are then easily taken up as the root absorbs water from the soil and translocation via the vascular system, accumulated in above-ground parts across the transpiration stream (Salt et al., 1998). This is a commercial process in some areas. Before commercialization though, field tests are carried out to weigh certain aspects and the suitability of the plant for phytoextraction of the desired metal. For example, for the phytoextraction of As, a fern cultivar has been produced and is commercially available (Salido et al., 2003).

Trees can be used for the process of phytoremediation because of their extensive root systems and minimal need of any cultivation inputs. Their woody characteristic, however, acts against the metal accumulation. Yet, poplar may be used for such processes because of its characteristics as been earlier unveiled by numerous studies on its genome and its transformation abilities (Clemens et al., 2002).

5. HYPERACCUMULATORS AND NONHYPERACCUMULATORS

As mentioned previously, phytoextraction uses plants to absorb heavy metals. The term is also confused with tolerance. Studies of evolutionary trajectories suggest that tolerance and metal accumulation are genetically different responses of plants against excess metal (Goolsby and Mason, 2015; Cappa and Pilon-Smits, 2014). The plants used for such purposes must have certain characteristics that can help them in phytoextraction of heavy metals. For example, they should have a high rate of growth and high production of above- and below-ground organs such as stems, leaves, and roots so that efficient translocation of heavy metals to all parts can be accomplished. They should also be tolerant to high concentrations of heavy metals as well as adaptable to biotic and abiotic stresses so that they can be easily cultivated and harvested (Ali et al., 2013). The plants that have all of these characteristics and are able to tolerate and accumulate high levels of heavy metals in their above-ground parts are termed hyperaccumulators (Memon and Schröder, 2009). This term was first proposed by Brooks et al. (1977) for those plants that can accumulate more than 1000 mg/kg of heavy metals. However, the most authentic definition is given by Baker and Brooks (1989), who state that the plants that can absorb and accumulate greater than 100 mg/kg Cd; 1000 mg/kg Pb, Ni, and Cu; 1000 mg/kg of Zn and Mn as their dry weight are called hyperaccumulators.

Hyperaccumulators can take up heavy metals from soils through roots and translocate them to shoots and leaves. Contrastingly, nonhyperaccumulators or accumulators are those plant species that can accumulate heavy metals in their below-ground parts and cannot translocate them to shoots and leaves except some that allow limited translocation. Figure 15.1 shows the difference between a hyperaccumulator and a nonhyperaccumulator. About 450–500 plants have been identified as hyperaccumulators. Some examples include *Thlaspi caerulescens* that can accumulate Pb, Cd, Ni, and Zn; *Alyssum bertolonii* that can uptake Ni and Co; and *Arabidopsis halleri* that can accumulate high levels of Zn and Cd but not Pb (Bert et al., 2002; Prasad, 2005; Maestri et al., 2010). There are about 45% of hyperaccumulators that belong to angiosperm families; of these, 25% of hyperaccumulators belongs to *Brassicaceae* family. Other families include members of *Asteraceae*, *Caryophyllaceae*, *Fabaceae*, *Cyperaceae*, *Poaceae*, *Cunoniaceae*, *Lamiaceae*, and many others (Padmavathiamma and Li, 2007).

To absorb such large amount of metals, it is necessary for the plant to survive in harsh conditions where the soil has abundance of heavy metals. Because plants accumulate heavy metals particularly in those organs that can allow translocation of sugars and minerals, they need a proper ratio maintained between the concentrations of heavy metals particularly in roots and shoots. This is termed a translocation factor (TF). For hyperaccumulators, it is necessary that the TF is greater than 1 (Tangahu et al., 2011). Another factor that measures the accumulation potential is bioconcentration factor (BC), which is the ratio between the amounts of heavy metals taken up by roots to that present in the soil. The BC should also be greater than 1 for hyperaccumulators (Ahmadpour et al., 2014). However, there are some disadvantages of hyperaccumulators that can reduce their heavy metal uptake efficiency.

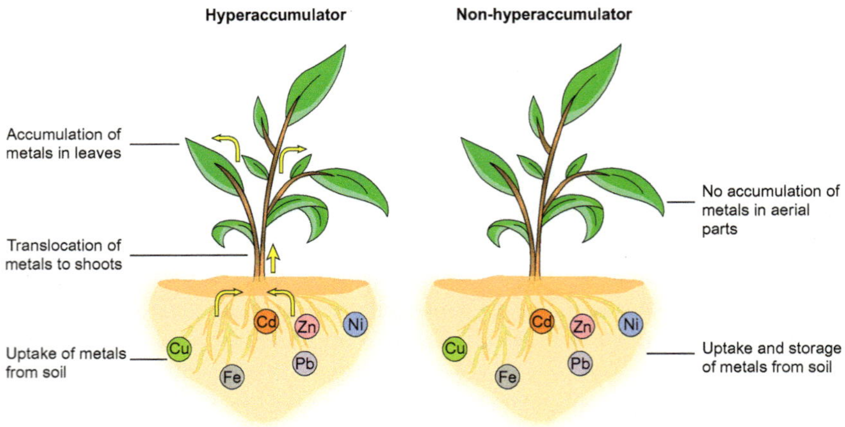

FIGURE 15.1

This figure shows how hyperaccumulators are different from nonhyperaccumulators. The hyperaccumulators allow heavy metal storage in their above-ground parts, whereas nonhyperaccumulators store heavy metals in their below-ground organs.

Hyperaccumulators have low biomass production and so heavy metals cannot be extracted easily from them. Similarly, chemically induced hyperaccumulation can cause leaching, reduction in bioavailability of heavy metals that can cause immobilization of heavy metals in soil that has toxic effects not only on soil microbes but also affects the plants and crops growing in that soil. Based on their accumulation capability, hyperaccumulators are divided into natural hyperaccumulators and transgenic hyperaccumulators. Natural hyperaccumulators are those that have a natural tendency to accumulate heavy metals in their above-ground parts but have a decreased biomass when they absorb high levels of toxic metals. Contrary transgenic hyperaccumulators are those that have been modified through biotechnological techniques to enhance the metal accumulation ability of plants and tend to have larger biomass than natural hyperaccumulators for greater metal absorption (Macek et al., 2008). Since the discovery of hyperaccumulators and nonhyperaccumulators, many studies have been done to investigate different plants and their heavy metal accumulation capacities, particularly for prevalent metals such as Cr, palladium, Ni, and Zn (Isaure et al., 2015)

Study on four crop plants—*Brassica rapa*, *Helianthus annuus*, *Zea mays*, and *Cannabis sativa*—has been reported to determine their ability as hyperaccumulators for phytoextraction. The results showed that all four plants exhibited heavy metal accumulations in shoots, particularly of Cd, Cu, Ni, Pb, and Zn. It was further observed that all metals under study had the lowest accumulation in *Z. mays*, whereas variations were observed in the highest accumulations. Zn was highly accumulated in *B. rapa*, Cu in *C. sativa*, Cd in *H. annuus*, Pb in *B. rapa*, and Ni in *C. sativa*; hence, different agronomic crops exhibit different potentials for phytoextraction against different heavy metals (Meers et al., 2005). In a similar study, five willow tree species were studied for their phytoextraction ability. These species were *Salix dasyclados* (Loden), *Salix triandra* (Noir de Villaines), *Salix schwerinii* (Christina), *Salix fragilis* (Belgisch Rood), and *Salix purpurea* × *Salix daphnoides* (Bleu). The results indicated no significant accumulation of Cr, Pb, Ni, and Cu in the shoots of all five species; however, Zn and Cd were found to accumulate in shoots and leaves in high concentrations. Among the five species under study, *S. schwerinii*, *S. dasyclados*, and *S. fragilis* showed more potential for phytoextraction of Zn and Cd (Meers et al., 2007). *Lolium multiflorum* is also used for phytoremediation of Cu, Zn, Mn, and Pb in mine tailing at La Concha (Mugica-Alvarez et al., 2015).

Spartina alterniflora (smooth cordgrass) has also been investigated for hyperaccumulation potential against soil contaminated with Cd, Pb, Cu, Fe, and Zn. *Spartina alterniflora* is a summer herbaceous grass native to southwest Louisiana. The results of the study stated that *S. alterniflora* had the ability to take heavy metals at high concentrations. Most of the accumulation took place in roots with the exception of Cu, which was found to accumulate in shoots and leaves at about 60–140 times greater than in the control. This study proved that *S. alterniflora* has the potential for phytoextraction and can serve as an efficient hyperaccumulator (Salla et al., 2011). Another perennial plant *Sedum alfredii* belonging to the family *Crassulaceae* has also been studied as a Zn and Cd hyperaccumulator and is known to uptake them in high concentrations in shoots (Yang et al., 2002, 2004). It is also a nonhyperaccumulator of Pb and stores it in the roots (He et al., 2001). Because *S. alfredii* has a high biomass and a high accumulation potential, it is considered a model plant for studying phytoextraction.

In a study done by Lotfy and Mostafa (2013), plant species *Panicum antidotale* (panicum), *Pannisetum purpureum* (Napier grass), *Curcubita pepo* (squash), *Gossypium hirsutum* (cotton), and *H. annuus* (sunflower) were found to be hyperaccumulators and nonhyperaccumulators of Co and Cr. About 30–45% of these heavy metals were removed by shoots. However, the plant's root was able to accumulate much higher concentration of these metals. Among the five species under study, the sunflower was able to accumulate the highest concentration, whereas the squash accumulated the lowest concentrations of Co and Cr in their respective organs (Lotfy and Mostafa, 2013). A similar study has been done to identify the phytoextraction potential of three varieties of sweet sorghum (*Sorghum bicolor*) namely Keller, Mary, and Rio. Sorghum is an agronomic crop that is used as a fodder. It has been observed that all the three varieties are effective Pb hyperaccumulators. Along with Pb, they are able to accumulate Cd, Zn, and Cu in stems and leaves. These heavy metals are also found to accumulate in roots and grains of sorghum, with the highest Cu concentration in grains and Cd in roots (Zhuang et al., 2009). The aquatic plant *Hydrilla verticillata* has a high uptake capacity and therefore is also considered a promising accumulator of Cd and Cr metals in water reservoirs (Phukan et al., 2015).

Xia et al. (2013) proved that *Sauropus androgynus*, a perennial plant has the ability to accumulate toxic metals from soil and is a highly effective plant for the removal of heavy metals. The plant was able to accumulate Cd, Zn, and palladium in stems and Cu in roots. The results also indicated that *S. androgynus* biomass did not significantly decrease by the accumulation of heavy metals because it has a significant phytoextraction potential and can be used commercially (Xia et al., 2013). Another study reported switchgrass (*Pannicum virgatum*) as a metal accumulator for Cd, Cr, and Zn. Switchgrass is also a perennial grass with high biomass. It also can accumulate Zn at a much higher concentration than Cd and Cr; that is, about 75 μm. As a result, growth reductions and decreased biomass were observed. Nevertheless, switchgrass can be used for phytoextraction more as an accumulator than as a hyperaccumulator (Chen et al., 2012). Apart from those mentioned previously, there are many other studies that demonstrate the potential of plants as hyperaccumulators and nonhyperaccumulators and have been reviewed in Ali et al. (2013), Bhargava et al. (2012), and Pollard et al. (2014).

6. INDUCED PHYTOEXTRACTION

As mentioned previously, hyperaccumulators and nonhyperaccumulators can absorb heavy metal contaminants in roots, stems, and leaves, but there are major drawbacks in this approach, including the significant reduction of biomass of the respective plant and the inability of the natural mechanism to absorb insoluble fractions of heavy metals in soil. To overcome this hurdle and to make sure that the plant's biomass does not decrease by high concentrations of heavy metal uptake, different chelators are used that assist the plant with metal uptake. This mechanism is termed induced phytoextraction or induced hyperaccumulation and the plants termed induced hyperaccumulators or induced nonhyperaccumulators. Figure 15.2 shows the basic process of induced phytoextraction. The basic function of chelates is to increase the heavy metal uptake capacity of the plant and also increase the potential of average hyperaccumulators like grass, yellow mustard, and lupine to uptake significant levels of heavy metals, thus facilitating phytoextraction. However, there are certain disadvantages to this approach. Chelates have the ability to increase metal solubility to such an extent that leaching of metals can occur. In addition, they can allow soil microbes to absorb these metals, thus increasing their optimum heavy metal concentration to toxic levels (Römkens et al., 2002). There are many studies that report the use of biodegradable and nonbiodegradable chelants such as EDTA and EDDS, respectively, which are discussed next.

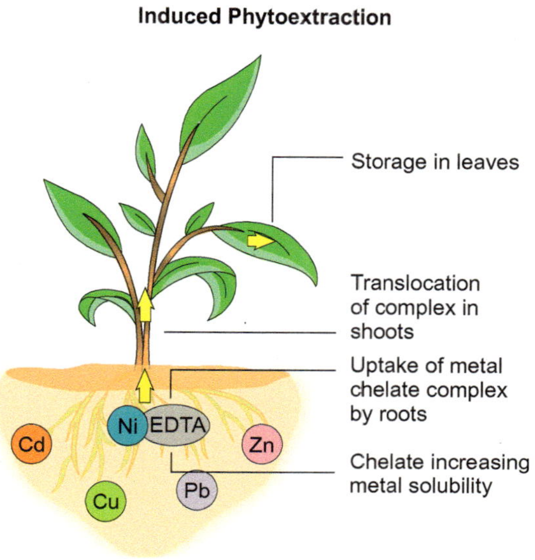

FIGURE 15.2

The process of induced phytoextraction in the presence of chelates. Chelates tend to increase the bioavailability of heavy metals in soil by forming metal chelate complexes that are then hyperaccumulated to aerial parts of the plant. EDTA, ethylene diamine tetra acetate.

Amendment effect on the heavy metal accumulation has been studied in four species: *B. rapa*, *H. annuus*, *Z. mays*, and *C. sativa*. EDTA and EDDS were used as chelates for assisting in phytoextraction. The results indicated shoot accumulation of heavy metals such as Zn, Pb, Ni, Cu, and Cd. Further observation indicated that *C. sativa* had the highest Zn, Cu, and Ni accumulation in shoots when soil was treated with EDDS and EDTA compared with control plant. EDDS was found to mobilize Cu and Ni more than EDTA. On the other hand, Cd accumulation was reported highest in *B. rapa* when EDTA was assisting as a chelate rather than EDDS. Lead accumulation was also facilitated more by EDTA, with *H. annuus* accommodating the highest concentration (Meers et al., 2005). In a similar study, willows (*Salix* sp.) were checked for induced phytoextraction potential using EDDS because of its biodegradability. Results proved that EDDS increased the willow's hyperaccumulation of heavy metals especially of Cu, Zn, Pb, and Ni in stems and leaves but did not show much potential for the uptake of Cd and Cr. This proved the EDDS ability to increase bioavailability of heavy metals without leaching the soil (Meers et al., 2007). An iron supplement also induces phytoextraction of Pb through switchgrass and corn. Similarly, green waste compost stimulates phytoextraction of Hg from Hg-rich soil (Smolinska, 2015). The As phytoextraction efficiency of *Pteris vittata* is stimulated by inoculation of arsenate-reducing bacteria *Pseudomonas* sp. P1III2, *Delftia* sp. P2III5 (A), and indoleacetic acid–producing bacteria such as *Bacillus* sp. MPV12, *Variovorax* sp. P4III4, and *Pseudoxanthomonas* sp. P4V6 (B) (Lampis et al., 2015).

In another study, *Segesbeckia orientalis* has been reported as an induced hyperaccumulator for the phytoextraction of Cd-contaminated soils. The chelants used in this study were biodegradable EDDS, nitrilotriacetic acid (NTA), and anionic polyacrylamide. The results did not indicate any toxicity to the plant and no significant decrease in biomass was observed. The concentration of Cd was observed to increase in shoots significantly by all the chelants under study when compared with the control plant. Therefore EDDS, NTA, and anionic polyacrylamide can be used as effective chelants for phytoextraction (Lan et al., 2013). Similarly, a comparison of EDDS assisted hyper-accumulators and organic fraction of municipal soil wastes assisted removal of heavy metals was compared in maize. In both cases, significant heavy metal uptake was observed but more increase in absorption was observed in organic fraction of municipal soil wastes because of the presence of dissolved organic matter, which can prove more efficient than chelates for phytoextraction (Salati et al., 2010).

A study on *Brassica napus* has been done to find the heavy metal hyperaccumulating potential of Zn, Mn, and Pb in the presence of synthetic chelates like EDTA. The results demonstrated that EDTA did not affect the biomass production of plants and significantly increased the Zn, Mn, and Pb concentrations in the shoot (Zaier et al., 2010). A similar study was conducted to determine the phytoextraction ability of *Lollium perenne* by the addition of EDTA as a phytochelatin. The results of the experiment demonstrated that the levels of Cu, Zn, and Pb extensively increased in shoots. The study was conducted in the presence of barriers made up of turf grass, sawdust, vermiculite, and zeolite and proved to be effective against EDTA-assisted leaching of soil (Zhao et al., 2011). The ability of biodegradable NTA as chelant has also been studied on *Festuca arundinacea* for the hyper-accumulation of Cu-, Pb-, and Zn-contaminated soil. The result of this study revealed an increased concentration of heavy metal accumulation in *F. arundinacea* shoots and was enhanced by NTA addition. However, slight leaching was observed that was reduced using a permeable barrier and turf grass (Zhao et al., 2013).

A similar study on *S. bicolor* (sweet sorghum) has been done to detoxify the soil contaminated with Zn, Cu, Cd, and Pb using EDTA as a chelator. The results showed effective EDTA-assisted hyper-accumulative properties of sweet sorghum, which were enhanced by the addition of ammonium sulfate and ammonium nitrate (Zhuang et al., 2009). A study by Sun et al. (2011) demonstrated that *Mirabilis jalapa* has the increased ability to phytoextract heavy metals such as Cd, Cu, Pb, and Zn in the presence of EDTA and citric acid. However, EDTA-assisted phytoextraction was able to accumulate more concentration of heavy metals in shoots than CA. Among the heavy metals, Zn was more efficiently removed by the addition of EDTA (Sun et al., 2011).

Different organic acids have different potentials for metal mobilization. This ability of chelates had been studied using maize and a comparison is drawn between naturally occurring organic acids such as oxalic acid, ascorbic acid, and acetic acid as well as synthetic chelates like EDTA. EDTA can help in the accumulation of heavy metals much more than organic acids. Among the naturally occurring organic acids, ascorbic acid has the highest potential for metal solubilization (Sabir et al., 2014). Hence, chelates—either natural or synthetic—can increase the phytoextraction ability of the plant. Although biodegradable chelates are used to prevent soil leaching, more field experiments are required before they can be proved safe for living organisms, particularly those inhabiting the soil. Because induced phytoextraction has the ability to remove heavy metal contamination on a large scale, it can be commercialized and can be used in place of ex situ methods currently used by developed countries to remove heavy metal contamination from soil.

7. HEAVY METAL UPTAKE MECHANISMS OF PLANT

Plants having the potential for phytoextraction must have the ability to store metals in their cells and organs. The process comprises multiple steps that involve the uptake of metals from the soil into the roots, from where they are translocated through different transporters to aerial parts of the plant. Once translocated to above-ground organs, they are stored in leaves (Mahmood, 2010). There are three major processes by which plants accumulate heavy metals in their organs. First is the bioavailability and uptake of heavy metals from the soil through roots, second is translocation of heavy metals from roots to shoots through xylem loading, and third is the sequestration of heavy metals in leaves particularly in vacuoles. During these processes, transport proteins are involved that allows a heavy metal to be taken up and stored in the plant. Plants are also involved in the detoxification, which occurs by combining them with natural chelates like phytochelatins (PCs) during the translocation and sequestration process (Bhargava et al., 2012). Figure 15.3 shows the mechanism by which plants undergo uptake and storage of heavy metals.

The first process required for the uptake of heavy metals is bioavailability and uptake of metals through roots. The heavy metal uptake is dependent on many characteristics of soil like pH, water content, organic substances, and the most important factor: the availability of metals in the soil. Heavy metals exist in soil in different forms. They are either present as free ions or in the form of complexes, bound to organic and inorganic matter, or as compounds like silicates or oxides and hydroxides (Salt and Rauser, 1995). Of the previously mentioned types, only those metals can accumulate naturally in a plant that exists in a soluble form that is the form of ions or complexes. Other types can also be taken up by the use of soil amendments through chelant-induced hyperaccumulation mechanism (Abollino et al., 2006).

The pH of the soil is maintained by the release of protons, which are controlled by proton pumps located in the outer membranes of the plant and can acidify the soil, thus lowering its pH. Studies on *Alyssium murale* showed the increase in Ni accumulation because of changes in pH (Bernal et al., 1994). In another study, Cu uptake by *Elsholtzia splendens* has also been affected by the decrease in pH of the rhizosphere of soil (Peng et al., 2005). Although induced phytoextraction can help in the efficient accumulation of heavy metals, plants also have an innate mechanism for the uptake of insoluble heavy metals that is through root exudates. Root exudates function as heavy metal reductases in the rhizosphere that have the inherent potential to reduce metal organic complexes formed in the soil and convert them into free ions or metal complexes, which makes it easier for the roots to absorb them (Quartacci et al., 2009).

Several studies prove root exudates as natural chelants. A study shows that root exudates can help in hyperaccumulation of Zn and Cd in *S. alfredii* (Li et al., 2005). Quartacci et al. (2009) showed the decontamination of soil from As, Cu, Zn, Cd, and Pb using *Brassica carinala*. The metallicolous plants used were found to enhance hyperaccumulation in shoots because of the release of root exudates that increased the metal solubility in soil (Quartacci et al., 2009). A similar study on *Echinochloa crus-galli* was done and showed that root exudates allow increased accumulation of Pb, Cd, and Cu. A much higher concentration of metals was absorbed by the plant in the presence of root exudates as compared with those without them. Hence they serve as potent natural chelators (Kim et al., 2010). Natural chelators have the ability to increase mobility of metals by making them uncharged. They are synthesized by the plant and are more promising than synthetic chelators that pose threats to soil. Most commonly studied natural chelators include metallothioneins (MTs) and PCs (Cobbett and

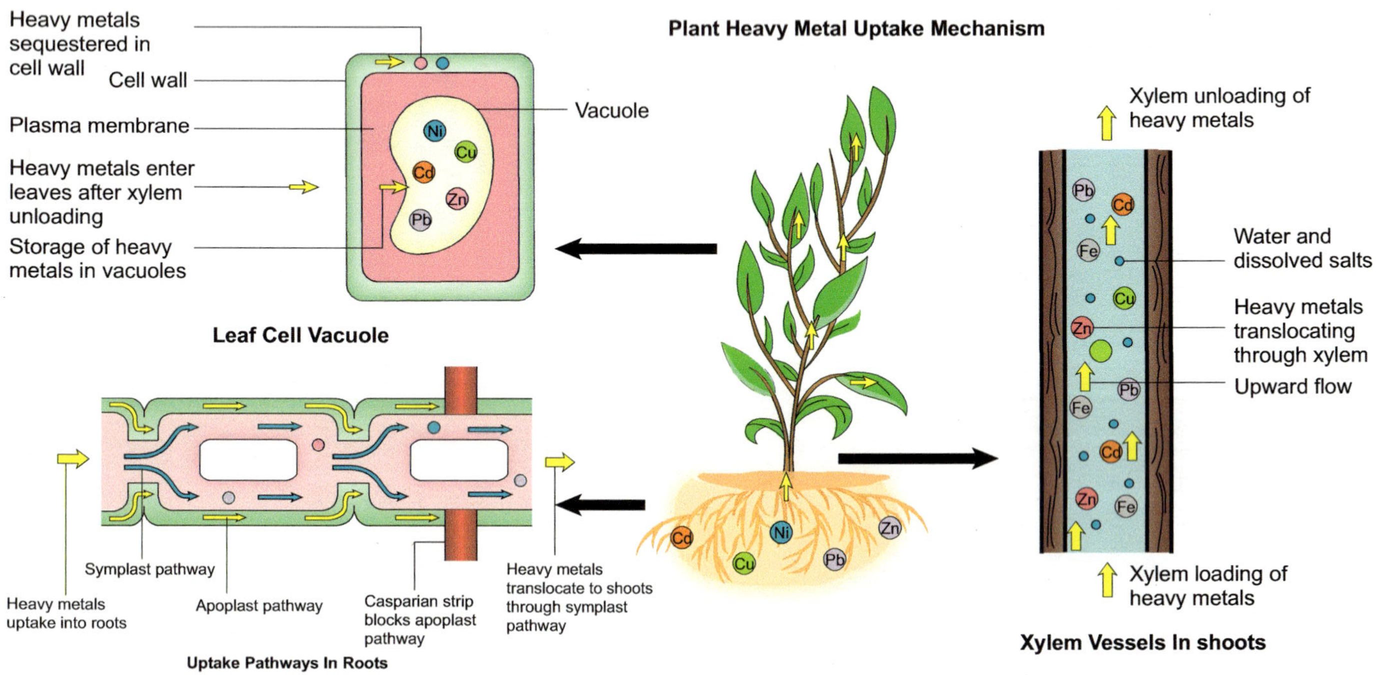

FIGURE 15.3

The three processes involved in the uptake and sequestration of heavy metals. Because of increased solubility, the heavy metals are taken up to roots from where they are translocated to shoots by means of apoplast and symplast pathway. From the shoots, the heavy metals are transported to leaves by xylem loading. In the leaves, heavy metals are stored in vacuoles.

Goldsbrough, 2002). MTs synthesis in plants is initiated when it encounters heavy metals such as Cd, Hg, Zn, Pb, and silver, but the highest levels of MTs are released in the presence of Cd toxicity. PCs are involved in providing tolerance to heavy metals and allow efficient accumulation of these metals in plants organs. They are thiol peptides with γ-Glu-Cys repeats. Their basic function is the detoxification of heavy metals before they are sequestered in leaves (Bhargava et al., 2012). The soil microbes can also increase the bioavailability of metals in the soil by secreting ions, organic acids, and other types of natural chelants. Microorganisms in rhizospheres such as bacteria and mycorrhizal fungi are involved in increasing the solubility of heavy metals in soil (Yang et al., 2005).

Organic acids like low-molecular-weight organic acids (LMWOAs) have the potential to increase heavy metal absorption as well. The addition of LMWOAs functions to increase metal mobilization by forming complexes with metals that makes the metals soluble and available for uptake. They are more efficient than synthetic chelates like EDTA, are biodegradable, and do not pose harm to the soil by allowing it to leach (Renella et al., 2004). For example, Cd solubility and uptake was increased by the addition of LMWOAs (Krishnamurti et al., 1997). Similarly, other organic acids such as acetic acid and succinate have also been proved to participate in Cd uptake in wheat (Cieśliński et al., 1998). Phytosiderophores are also organic compounds that have the ability to uptake Fe from soils (Neubauer et al., 2000). They are produced by the plant as a result of Fe deficiency. When phytosiderophores are released, they not only make Fe soluble in the soil, but other metals such as Zn, Cu, and Mn also become soluble and are taken up by the plant through membrane carriers specific for each metal (Römheld, 1991; Nair et al., 2007). A study has been conducted on a *Thlaspi* hyperaccumulator that is involved in secreting phytosiderophores. When the rhizosphere of *Thlaspi* and wheat was compared, it was found that *Thlaspi* phytosiderophores were able to remove the heavy metal contamination more than wheat phytosiderophores. According to Gulden (2014), biochar has a significant role to inhibit heavy metal uptake by plants in metal-rich soil. Another study indicates that symbiotic association of plant roots with arbuscular mycorrhizal fungi negatively affect the uptake of heavy metal in plants (Emamverdian et al., 2015). Metal uptake also depends on soil pH as reported in *Valeriana officinalis* L that high soil alkalinity favors Cu and Mn, whereas it is unfavorable to Zn uptake by plants (Adamczyk-Szabela et al., 2015).

Once heavy metals are available for uptake, there are two pathways by which they enter the roots. First is the apoplastic pathway and second is the symplastic pathway. The apoplastic pathway allows the soluble metal fraction to travel without entering the cells and through the intracellular spaces. On the other hand, the symplastic pathway allows the movement of nonessential metals like Ni, Cd, and Pb through the cytoplasm by consuming energy. Because different plant species use different mechanisms for the uptake of different heavy metals, variations in the transport pathways are observed (Lombi et al., 2002; Lu et al., 2009). Heavy metal absorption from soil is a function of Zn-regulated transporter, iron-regulated transporter proteins (ZIP). Different ZIP genes are encoded by different plants. For example, *T. caerulescens* ZIP proteins are encoded by genes ZTN1 and ZTN2 (Assunção et al., 2001). Apart from Zn accumulation, ZIP proteins are also involved in absorbing Cd and Ni from the soil (reviewed in Maestri et al., 2010).

The second process is the **translocation from roots to shoots**. Because nonhyperaccumulators do not have the ability to translocate heavy metals to above-ground parts, this process does not occur in them and the heavy metals are stored in the roots. Transport of heavy metals from roots to shoots occurs by the movement of metals into the xylem vessels that are involved in the transport of water and dissolved salts. Heavy metals enter into xylem channels through the process known as xylem loading.

This term is applied to all the materials that enter in the xylem tissues and need to be transported from roots to leaves and vice versa. It is an energy-consuming process that only allows those metals to enter that are following the symplast pathway. This is because during normal transport, the casparian strip in the endodermis blocks intracellular movement. Hence, metals that are moving through the apoplast pathway gets blocked which means that the only way to enter xylem vessels is by taking the symplast pathway (reviewed in Mahmood, 2010).

It has been demonstrated in some studies that the casparian strip barrier in the endodermis can be disrupted by chemicals. It can also be overcome by the use of natural and synthetic chelates. That is why assisted phytoextraction was observed to follow the apoplast pathway for entering the xylem vessels (Collins et al., 2002). According to Clemens et al. (2002), xylem loading is facilitated by natural chelates present within the plant that convert free metal ions into complexes, thus allowing easy and distortion-free translocation. This is because materials entering the xylem should not be reactive because they can hinder the transport. The natural chelates involved during xylem loading include PCs and organic acids as mentioned previously (Clemens et al., 2002).

Alyssum lesbiacum, a Ni hyperaccumulator and tolerant of high Ni contamination, showed the presence of histidine, a chelator ligand in the xylem vessels. This proved the functioning of chelates during root to shoot translocation (Kramer et al., 1996). In a similar study, a comparison between a hyperaccumulator and nonhyperaccumulator has been done on the presence of histidine in xylem. When *Alyssum montanum*, a nonhyperaccumulator, was allowed to express the *TP-PRT1* gene that is involved in facilitating the transport of Ni through xylem, it was observed that a large concentration of Ni and histidine complexes were present in xylem. This further proved the functioning of histidine in Ni translocation (Kerkeb and Krämer, 2003). In other hyperaccumulators such as *T. caerulescens* and *A. halleri*, nicotinamide acts as a chelator and is required for Ni and Zn translocation, respectively, by forming complexes with it (Vacchina et al., 2003; Weber et al., 2004).

As mentioned previously, the transport from root to shoots is dependent on xylem loading. This mechanism is mediated by a specific type of ATPases namely heavy metal transporting ATPases (HMAs). They function to maintain the amount of metal absorbed and help the plant survive under harsh heavy metal conditions. One of the genes controlling these protein transporters is *HMA4*, which is involved in the hyperaccumulating and translocation of Zn, Cd, and Pb. *HMA4* is also known to enhance the functioning of ZIP proteins to uptake more heavy metals from soil (Hanikenne et al., 2008). Other types of transporters involved in translocation of heavy metals include multidrug and toxin efflux encoded by *FDR3* genes, Yellow strip 1-like (YSL) proteins encoded by *TcYSL3*, *TcYSL5*, and *YSL7* genes and Nodulin 26-like intrinsic proteins that are specific for As transport (see review by Rascio and Navari-Izzo, 2011).

The third process is the detoxification and sequestration of heavy metals in leaves. When the heavy metals are transported to the leaves, they are sequestered in its cells either intracellularly or extracellularly. Before their storage, they are detoxified by chelate ligands such as MTs, PCs, and organic acids like histidine or citrate, as previously explained. Different metals are stored in different parts of leaves in different hyperaccumulators. Studies on *T. caerulescens* showed that Cd is stored in the cell wall (Cosio et al., 2005), whereas Zn accumulation was observed in vacuoles and the cell wall but particularly in the former (Küpper et al., 1999). Similarly, Ni was also found to accumulate both in the cell wall and vacuoles in *Thlaspi geosingense* (Krämer et al., 2000).

The major organelle involved in accumulating metal concentration is the vacuole. It is the storage organ that controls the level of essential and nonessential heavy metals in plant cells and do not allow metals to pose threat to it. Vacuole transport is mediated by metal ion transporters (MTPs) that specifically allow transport through vacuoles. The genes encoding MTP proteins include *TgMTP1* that is found in *T. goesingense* and can accumulate high levels of metal ions in vacuoles of stems (Persans et al., 2001). Another transporter, namely YCF1, is specific to vacuoles and helps the plant to tolerate heavy metal contamination (Li et al., 1997). Natural resistance associated macrophage proteins (NRAMP) is also a vacuolar ion transport and is mainly involved in storing of Cd and Fe. One of its genes, *NRAMP3*, has been identified in many hyperaccumulators and is involved in the accumulation of Cd, Ni, Zn, and Fe (reviewed in Maestri et al., 2010). Similarly, ABC transporters can also perform the function of carrying metal chelate complexes in vacuole. Hence, a vacuole serves to compartmentalize metals and control heavy metal distribution in plant (Yang et al., 2005).

8. DRAWBACKS: AN INSIGHT INTO THE REALITY

In lieu of all the advantageous aspects, there are certain shortcomings of this technology, including the specificity pertaining to certain conditions. Thus, use of this technology implies a massive background check carried out in the form of research to ascertain the final results as expected. The detailed behavior of plants under certain environmental conditions in response to heavy metal stress and the change in conditions associated with it need to be addressed before proceeding to the actual processes involved in the technique. Furthermore, the conditions in the locality with respect to the lithosphere are also to be studied in detail, which depends on the various soil types having different textures and sizes, and hence different characteristics or holding capacity and mobilization for the heavy metal pollutants contained in it.

The process of phytoremediation is generally quite a bit slower than other techniques involving physical and chemical procedures for treatment, making it time inefficient. It, thus cannot be used where a quick cleanup is required.

The efficiency of this technology is directly dependent on the biomass production: the greater the biomass production, the more efficient the process of phytoremediation. Thus, the efficiency is greatly reduced with the use of plants possessing lesser biomass at the end of the growth phase. It means that as the growth of the plant proceeds, its biomass increases and the chances and spots for accumulation increase accordingly and hence the limit for accumulation reaches early for plant with reduced biomass production (Tong et al., 2004).

Extensive root systems of plants are necessary to ensure the effectiveness of the procedure. Resultantly, plants producing a limited root system in the rhizosphere will not be capable of holding the heavy metals in their surrounding soil and keeping them from leaching down. This way, less dense roots easily lose the pollutants to the underground water table, affecting phytoremediation efficiency adversely (Oh et al., 2014).

Phytoremediation is highly affected by the plant's growth and response to the changing environmental conditions. Certain plants efficient in extracting the desired heavy metals from soil may not be adaptive of that particular subject site; thus, alternate plants native to that area need to be used in such places (Danh et al., 2009).

Also, presence of certain other heavy metals in addition to the target ones may cause toxicity symptoms in the plant, contributing to the inefficiency of the system. As in a real scenario, one has to deal with wide range of heavy metal species at the target site instead of just one the plant is efficient in extracting; extensive prior research is required to get the process into working with desired results otherwise it will just lead to physiological, biochemical, and morphological collapse of the plant system (Mijovilovich et al., 2009).

A major drawback in this case is the form of the heavy metal; to be taken up by the plant, the heavy metals must be present in their bio-available states so that they are in the true sense available to the plant nutrient exchange sites on the roots. In complex forms with other compounds, the roots might not be able to remove it from the soil because of unavailability. The solubility of the heavy metal is another issue in this case because accumulation might be prevented because of higher solubility.

9. CONCLUSION AND FUTURE PROSPECTS

Phytoremediation has emerged as an upcoming technology that will be introduced at commercial levels to treat lands contaminated with heavy metals coming from various sources. This technique requires cultivation cost only and generates a large amount of energy through incineration. Remediation of soil is completed via solar energy and metal uptake system of cultivated plants. However, before proceeding with the process, there should be a careful check and balance system in place to avoid any technical difficulties later. The bright side of the technology is that even a layman, if guided, initially can use it eliminating the need of trained labor. Phytoextraction can thus serve as a successful in situ method for the decontamination of heavy metal–polluted soil. It can be efficient and cost-effective and has the prospect of being a successful commercial scheme that cannot only be restricted to soils but should be used for cleaning of waters and the environment from the toxicity of heavy metals. Although a major drawback of this approach is the decreased biomass of hyper-accumulators, it can be overcome by using the induced phytoextraction technique. Among the chelants, biodegradable ones are more environmentally friendly and are effective in metal removal. Understanding plant uptake mechanisms can also facilitate a better understanding of the mechanisms of hyperaccumulation, including the molecular aspects of transportation processes involved in different organs of plant. By merging transgenic approaches with technology, many gaps have been filled and limitations overcome. As research goes on and newer and better genes are identified, they can be transferred to plant species that are better at other features to overcome any deficiencies and make the process a commercial success earlier. However, there are still a lot of aspects that warrant more research before it can be commercialized. Enhanced uptake and accumulation of the heavy metals in the aerial parts of the plant that can be easily harvested is the main goal of transgenics in this approach. Beneficial genes from fungi, bacteria, and even animals and humans have been transferred successfully into model plants for studies and have produced excellent results. This clearly indicates the success of incorporating these genes into hyperaccumulator species to come up with ideal species for the process. However, these transgenic processes also open up several other issues that cannot be neglected to keep the biosphere under natural processes from collapsing.

Although phytoextraction is a very effective technique, plants that have the ability to hyper-accumulate can also be used as bio-energy crops to fulfill agricultural needs as well as energy requirements. This technique is termed phytoattenuation. Experiments with maize as an energy-yielding crop have demonstrated that about 33,000–46,000 kWh of renewable energy is generated, which can be used as electrical or thermal energy. If this technique is commercialized, it will not only be able to decrease heavy metal contamination, but also can serve as an energy substituent (Meers et al., 2010).

Another use of hyperaccumulators is phytomining, which is the technique for recovering metals absorbed in the plants using them as mines. Although it can only be applied to plants with higher biomass content, efforts can be made to extract bio-ores that are not only environmentally safe, but also can be a potential source of energy (Rascio and Navari-Izzo, 2011). Phytoremediation has not only been studied in plants but it is also applied to microorganisms that have the ability to take up large amount of heavy metals. This is known as microremediation and can also be a potential source for removing soil pollution (Wu et al., 2010). Similarly, aquatic weeds such as *Typha latifolia* and *Eichhornia crassipes* have also been found as potential hyperaccumulators that can accumulate Pb, Cu, Cd, and As. Therefore they can be used to decontaminate effluents by using constructed wetland technology (Sukumaran, 2013).

REFERENCES

Abollino, O., Giacomino, A., Malandrino, M., Mentasti, E., Aceto, M., Barberis, R., 2006. Assessment of metal availability in a contaminated soil by sequential extraction. Water Air Soil Pollut. 173 (1–4), 315–338.

Adamczyk-Szabela, D., Justyna Markiewicz, J., Wolf, W.M., 2015. Heavy metal uptake by herbs. IV. Influence of soil pH on the content of heavy metals in *Valeriana officinalis* L. Water Air Soil Pollut. 226 (4), 106.

Ahmad, P., Jaleel, C.A., Salem, M.A., Nabi, G., Sharma, S., 2010. Roles of enzymatic and non-enzymatic antioxidants in plants during abiotic stress. Crit. Rev. Biotechnol. 30 (3), 161–175.

Ahmad, P., Nabi, G., Ashraf, M., 2011. Cadmium-induced oxidative damage in mustard [*Brassica juncea* (L.) Czern. & Coss.] plants can be alleviated by salicylic acid. South Afr. J. Bot. 77, 36–44.

Ahmad, P., Sarwat, M., Bhat, N.A., Wani, M.R., Kazi, A.G., Tran, L.S.P., 2015. Alleviation of cadmium toxicity in *Brassica juncea* L. (Czern. & Coss.) by calcium application involves various physiological and biochemical strategies. PLoS One 10 (1), e0114571.

Ahmad, P., Sarwat, M., Sharma, S., 2008. Reactive oxygen species, antioxidants and signaling in plants. J. Plant Biol. 51 (3), 167–173.

Ahmadpour, P., Ahmadpour, F., Mahmud, T., Abdu, A., Soleimani, M., Tayefeh, F.H., 2014. Phytoremediation of heavy metals: a green technology. In: Asrari, E. (Ed.), Heavy Metal Contamination of Water and Soil: Analysis, Assessment, and Remediation Strategies. CRC Press, Taylor and Francis, p. 249.

Ali, H., Khan, E., Sajad, M.A., 2013. Phytoremediation of heavy metals–concepts and applications. Chemosphere 91 (7), 869–881.

Assunção, A., Martins, P., De Folter, S., Vooijs, R., Schat, H., Aarts, M., 2001. Elevated expression of metal transporter genes in three accessions of the metal hyperaccumulator *Thlaspi caerulescens*. Plant Cell Environ. 24 (2), 217–226.

Baker, A., Brooks, R., 1989. Terrestrial higher plants which hyperaccumulate metallic elements. A review of their distribution, ecology and phytochemistry. Biorecovery 1 (2), 81–126.

Barman, S.C., Kisku, G.C., Salve, P.R., Misra, D., Sahu, R.K., Ramteke, P.W., Bhargava, S.K., 2001. Assessment of industrial effluent and its impact on soil and plants. J. Environ. 22 (4), 251–256.

Becher, M., Talke, I.N., Krall, L., Kramer, U., 2004. Cross-species microarray transcript profiling reveals high constitutive expression of metal homeostasis genes in shoots of the zinc hyperaccumulator *Arabidopsis halleri*. Plant J. 37 (2), 251–268.

Bernal, M., McGrath, S., Miller, A., Baker, A., 1994. Comparison of the chemical changes in the rhizosphere of the nickel hyperaccumulator *Alyssum murale* with the non-accumulator *Raphanus sativus*. Plant Soil 164 (2), 251–259.

Bert, V., Bonnin, I., Saumitou-Laprade, P., De Laguérie, P., Petit, D., 2002. Do *Arabidopsis halleri* from non-metallicolous populations accumulate zinc and cadmium more effectively than those from metallicolous populations? New Phytol. 155 (1), 47–57.

Bhargava, A., Carmona, F.F., Bhargava, M., Srivastava, S., 2012. Approaches for enhanced phytoextraction of heavy metals. J. Environ. Manage. 105, 103–120.

Broadhurst, C.L., Chaney, R.L., Angle, J.S., Maugel, T.K., Erbe, E.F., Murphy, C.A., 2004. Simultaneous hyperaccumulation of nickel, manganese, and calcium in *Alyssum* leaf trichomes. Environ. Sci. Technol. 38 (21), 5797–5802.

Brooks, R., Lee, J., Reeves, R.D., Jaffré, T., 1977. Detection of nickeliferous rocks by analysis of herbarium specimens of indicator plants. J. Geochem. Exp. 7, 49–57.

Cappa, J.J., Pilon-Smits, E.A.H., 2014. Evolutionary aspects of elemental hyperaccumulation. Planta 239, 267–275.

Cardwell, A.J., Hawker, D.W., Greenway, M., 2002. Metal accumulation in aquatic macrophytes from southeast Queensland, Australia. Chemosphere 48 (7), 653–663.

Chen, B.C., Lai, H.Y., Juang, K.W., 2012. Model evaluation of plant metal content and biomass yield for the phytoextraction of heavy metals by switchgrass. Ecotoxicol. Environ. Saf. 80, 393–400.

Chu, Q., Watanabe, T., Sha, Z., Osaki, M., Shinano, T., 2015. Interactions between Cs, Sr, and other nutrients and trace element accumulation in *Amaranthus* shoot in response to variety effect. J. Agric. Food Chem. 63 (8), 2355–2363.

Cieśliński, G., Van Rees, K., Szmigielska, A., Krishnamurti, G., Huang, P., 1998. Low-molecular-weight organic acids in rhizosphere soils of durum wheat and their effect on cadmium bioaccumulation. Plant Soil 203 (1), 109–117.

Clemens, S., Palmgren, M.G., Kramer, U., 2002. A long way ahead: understanding and engineering plant metal accumulation. Trend Plant Sci. 7 (7), 309–315.

Cobbett, C., Goldsbrough, P., 2002. Phytochelatins and metallothioneins: roles in heavy metal detoxification and homeostasis. Ann. Rev. Plant Biol. 53 (1), 159–182.

Collins, R.N., Merrington, G., McLaughlin, M.J., Knudsen, C., 2002. Uptake of intact zinc-ethylenediaminetetraacetic acid from soil is dependent on plant species and complex concentration. Environ. Toxicol. Chem. 21 (9), 1940–1945.

Cosio, C., DeSantis, L., Frey, B., Diallo, S., Keller, C., 2005. Distribution of cadmium in leaves of *Thlaspi caerulescens*. J. Exp. Bot. 56 (412), 765–775.

Danh, L.T., Truong, P., Mammucari, R., Tran, T., Foster, N., 2009. Vetiver grass, *Vetiveria zizanioides*: a choice plant for phytoremediation of heavy metals and organic wastes. Int. J. Phytoremediation 11 (8), 664–691.

Dixit, R., Wasiullah, M.D., Pandiyan, K., Singh, U.B., Sahu, A., Shukla, R., Singh, B.P., Rai, J.P., Sharma, P.K., Lade, H., Paul, D., 2015. Bioremediation of heavy metals from soil and aquatic environment: an overview of principles and criteria of fundamental processes. Sustainability 2015 (7), 2189–2212.

Ebrahimi, M., 2014. Effect of EDTA and DTPA on phytoremediation of pb–zn contaminated soils by *Eucalyptus camaldulensis* Dehnh and effect on treatment time. Desert 19-1, 65–73.

Emamverdian, A., Ding, Y., Mokhberdoran, F., Xie, Y., 2015. Heavy metal stress and some mechanisms of plant defense response. Sci. World J. 2015, 18. Article ID 756120.

Goolsby, E.W., Mason, C.M., 2015. Toward a more physiologically and evolutionarily relevant definition of metal hyperaccumulation in plants. Front. Plant Sci. 6, 33.

Greipsson, S., 2011. Phytoremediation. Nat. Educ. Knowl. 3, 7.

Gulden, K.T., 2014. Biochar reduces heavy metal uptake in plants. In: Article from Norwegian Institute for Agricultural and Environmental Research, vol. 11, issue. 6, p. 11.

Hanikenne, M., Talke, I.N., Haydon, M.J., Lanz, C., Nolte, A., Motte, P., Kroymann, J., Weigel, D., Krämer, U., 2008. Evolution of metal hyperaccumulation required cis-regulatory changes and triplication of HMA4. Nature 453 (7193), 391–395.

He, B., Yang, X., Ni, W., Wei, Y., Long, X., Ye, Z., 2001. Sedum alfredii: a new lead-accumulating ecotype. Acta Bot. Sin. 44 (11), 1365–1370.

Iori, V., Pietrini, F., Massacci, A., Zacchini, M., 2015. Morphophysiological responses, heavy metal accumulation and phytoremoval ability in four willow clones exposed to cadmium under hydroponics. In: Ansari, A.A., Gill, S.S., Gill, R., Lanza, G.R., Newman, L. (Eds.), Phytoremediation: Mangment of Environmental Contaminant. Springer International publishing, Switzerland, pp. 87–98.

Isaure, M.P., Huguet, S., Meyer, C.L., Castillo-Michel, H., Testemale, D., Vantelon, D., Saumitou-Laprade, P., Verbruggen, N., Sarret, G., 2015. Evidence of various mechanisms of Cd sequestration in the hyper-accumulator *Arabidopsis halleri*, the non-accumulator *Arabidopsis lyrata*, and their progenies by combined synchrotron-based techniques. J. Exp. Bot. 66 (11), 3201–3214.

Jadia, C.D., Fulekar, M.H., 2009. Phytoremediation of heavy metals: recent techniques. Afr. J. Biotechnol. 8 (6), 921–928.

Jiang, W., Tao, T., Liao, Z., 2011. Removal of heavy metal from contaminated soil with chelating agents. Open J. Soil Sci. 1, 70–76.

John, R., Ahmad, P., Gadgil, K., Sharma, S., 2009. Heavy metal toxicity: effect on plant growth, biochemical parameters and metal uptake by *Brassica juncea* L. Int. J. Plant Prod. 3 (3), 65–76.

Kamran, M.A., Amna Mufti, R., Mubariz, N., Syed, J.H., Bano, A., Javed, M.T., Munis, M.F.H., Tan, Z., Chaudhary, H.J., 2014. The potential of the flora from different regions of Pakistan in phytoremediation: a review. Environ. Sci. Pollut. Res. Int. 21, 801–812.

Kanakaraju, D., Connie, J., Long, S.M., 2008. Heavy metal concentrations in the razor clam (Solen Spp) from Muara Tebas, Sarawak. Malaysian J. Anal. Sci. 12 (1), 53–58.

Kerkeb, L., Krämer, U., 2003. The role of free histidine in xylem loading of nickel in *Alyssum lesbiacum* and *Brassica juncea*. Plant Physiol. 131 (2), 716–724.

Kim, S., Lim, H., Lee, I., 2010. Enhanced heavy metal phytoextraction by *Echinochloa crus-galli* using root exudates. J. Biosci. Bioeng. 109 (1), 47–50.

Koning, M., Hupe, K., Stegmann, R., 2000. Thermal Processes, Scrubbing/Extraction, Bioremediation and Disposal. Biotechnology Set, second ed. pp. 304–317.

Krämer, U., 2010. Metal hyperaccumulation in plants. Annu. Rev. Plant Biol. 61, 517–534.

Kramer, U., Cotter-Howells, J.D., Charnock, J.M., Baker, A.J.M., Smith, J.A.C., 1996. Free histidine as a metal chelator in plants that accumulate nickel. Nature 379, 635–638.

Krämer, U., Pickering, I.J., Prince, R.C., Raskin, I., Salt, D.E., 2000. Subcellular localization and speciation of nickel in hyperaccumulator and non-accumulator *Thlaspi* species. Plant Physiol. 122 (4), 1343–1354.

Krishnamurti, G., Cieslinski, G., Huang, P., Van Rees, K., 1997. Kinetics of cadmium release from soils as influenced by organic acids: implication in cadmium availability. J. Environ. Qual. 26 (1), 271–277.

Kulakow, P.A., Schwab, A.P., Banks, M.K., 2000. Screening plant species for growth on weathered, petroleum hydrocarbon-contaminated sediments. Int. J. Phytoremediation 2, 297–317.

Küpper, H., Zhao, F.J., McGrath, S.P., 1999. Cellular compartmentation of zinc in leaves of the hyperaccumulator *Thlaspi caerulescens*. Plant Physiol. 119 (1), 305–312.

Lampis, S., Santi, C., Ciurli, A., Andreolli, M., Vallini, G., 2015. Promotion of arsenic phytoextraction efficiency in the fern *Pteris vittata* by the inoculation of As-resistant bacteria: a soil bioremediation perspective. Front. Plant Sci. 6, 80.

Lan, J., Zhang, S., Lin, H., Li, T., Xu, X., Li, Y., Jia, Y., Gong, G., 2013. Efficiency of biodegradable EDDS, NTA and APAM on enhancing the phytoextraction of cadmium by *Siegesbeckia orientalis* L. grown in Cd-contaminated soils. Chemosphere 91 (9), 1362–1367.

Li, T.-Q., Yang, X.E., Jin, X.F., He, Z.L., Stoffella, P.J., Hu, Q.H., 2005. Root responses and metal accumulation in two contrasting ecotypes of *Sedum alfredii* Hance under lead and zinc toxic stress. J. Environ. Sci. Health A 40 (5), 1081–1096.

Li, Z.S., Lu, Y.P., Zhen, R.G., Szczypka, M., Thiele, D.J., Rea, P.A., 1997. A new pathway for vacuolar cadmium sequestration in *Saccharomyces cerevisiae*: YCF1-catalyzed transport of bis (glutathionato) cadmium. Proc. Natl. Acad. Sci. 94 (1), 42–47.

Lias, L., Jamil, T., Aliaa, S.N., 2013. A preliminary study on heavy metal concentration in the marine bivalves *Marcia marmorata* species and sediments collected from the coastal area of Kuala Perlis, North of Malaysia. IOSR J. Appl. Chem. 4 (1), 48–54.

Lombi, E., Tearall, K.L., Howarth, J.R., Zhao, F.-J., Hawkesford, M.J., McGrath, S.P., 2002. Influence of iron status on cadmium and zinc uptake by different ecotypes of the hyperaccumulator *Thlaspi caerulescens*. Plant Physiol. 128 (4), 1359–1367.

Lotfy, S., Mostafa, A., 2013. Phytoremediation of contaminated soil with cobalt and chromium. J. Geochem. Explor. 144, 367–373.

Lu, L., Tian, S.K., Yang, X.E., Li, T.Q., He, Z.I., 2009. Cadmium uptake and xylem loading are active processes in the hyperaccumulator *Sedum alfredii*. J. Plant Physiol. 166 (6), 579–587.

Ma, Y., Rajkumar, M., Luo, Y., Freitas, H., 2011. Inoculation of endophytic bacteria on host and non-host plants-effects on plant growth and Ni uptake. J. Hazard Mater. 195, 230–237.

Macek, T., Kotrba, P., Svatos, A., Novakova, M., Demnerova, K., Mackova, M., 2008. Novel roles for genetically modified plants in environmental protection. Trend Biotechnol. 26 (3), 146–152.

Maestri, E., Marmiroli, M., Visioli, G., Marmiroli, N., 2010. Metal tolerance and hyperaccumulation: costs and trade-offs between traits and environment. Environ. Exp. Bot. 68 (1), 1–13.

Mahmood, T., 2010. Phytoextraction of heavy metals-the process and scope for remediation of contaminated soils. Soil Environ. 29 (2), 91–109.

Mani, D., Kumar, C., Patel, N.K., 2015. Hyperaccumulator oilcake manure as an alternative for chelate-induced phytoremediation of heavy metals contaminated alluvial soils. Int. J. Phytoremediation 17 (3), 256–263.

McGrath, S.P., Zhao, F.J., 2003. Phytoextraction of metals and metalloids from contaminated soils. Curr. Opin. Biotechnol. 14 (3), 277–282.

Meers, E., Ruttens, A., Hopgood, M., Lesage, E., Tack, F.M., 2005. Potential of *Brassic rapa, Cannabis sativa, Helianthus annuus* and *Zea mays* for phytoextraction of heavy metals from calcareous dredged sediment derived soils. Chemosphere 61 (4), 561–572.

Meers, E., Van Slycken, S., Adriaensen, K., Ruttens, A., Vangronsveld, J., Du Laing, G., Witters, N., Thewys, T., Tack, F.M., 2010. The use of bio-energy crops (*Zea mays*) for 'phytoattenuation' of heavy metals on moderately contaminated soils: a field experiment. Chemosphere 78 (1), 35–41.

Meers, E., Vandecasteele, B., Ruttens, A., Vangronsveld, J., Tack, F.M.G., 2007. Potential of five willow species (*Salix* spp.) for phytoextraction of heavy metals. Environ. Exp. Bot. 60 (1), 57–68.

Memon, A.R., Schröder, P., 2009. Implications of metal accumulation mechanisms to phytoremediation. Environ. Sci. Pollut. Res. 16 (2), 162–175.

Mijovilovich, A., Leitenmaier, B., Meyer-Klaucke, W., Kroneck, P.M., Gotz, B., Kupper, H., 2009. Complexation and toxicity of copper in higher plants. II. Different mechanisms for copper versus cadmium detoxification in the copper-sensitive cadmium/zinc hyperaccumulator *Thlaspi caerulescens* (Ganges Ecotype). Plant Physiol. 151 (2), 715–731.

Mugica-Alvarez, V., Cortés-Jiménez, V., Vaca-Mier, M., Domínguez-Soria, V., 2015. Phytoremediation of mine tailings using *Lolium multiflorum*. Int. J. Environ. Sci. Develop. 6 (4), 246–251.

Mukhopadhyay, S., Maiti, S.K., 2010. Phytoremediation of metal enriched mine waste: a review. Glob. J. Environ. Res. 4, 135–150.

Nair, A., Juwarkar, A.A., Singh, S.K., 2007. Production and characterization of siderophores and its application in arsenic removal from contaminated soil. Water Air Soil Pollut. 180 (1–4), 199–212.

Neubauer, U., Furrer, G., Kayser, A., Schulin, R., 2000. Siderophores, NTA, and citrate: potential soil amendments to enhance heavy metal mobility in phytoremediation. Int. J. Phytoremediation 2 (4), 353–368.

Oh, K., Cao, T., Li, T., Cheng, H., 2014. Study on application of phytoremediation technology in management and remediation of contaminated soils. J. Clean Energy Technol. 2 (3), 216–220.

Ortega, L.M., Lebrun, R., Blais, J.F., Hausler, R., Drogui, P., 2008. Effectiveness of soil washing, nanofiltration and electrochemical treatment for the recovery of metal ions coming from a contaminated soil. Water Res. 42 (8), 1943–1952.

Padmavathiamma, P.K., Li, L.Y., 2007. Phytoremediation technology: hyper-accumulation metals in plants. Water Air Soil Pollut. 184 (1–4), 105–126.

Peng, H.Y., Yang, X.E., Jiang, L.Y., He, Z.L., 2005. Copper phytoavailability and uptake by *Elsholtzia Splendens* from contaminated soil as affected by soil amendments. J. Environ. Sci. Health 40 (4), 839–856.

Peng, J.F., Song, Y.H., Yuan, P., Cui, X.Y., Qiu, G.I., 2009. The remediation of heavy metals contaminated sediment. J. Hazard Mater. 161 (2), 633–640.

Persans, M.W., Nieman, K., Salt, D.E., 2001. Functional activity and role of cation-efflux family members in Ni hyperaccumulation in *Thlaspi goesingense*. Proc. Natl. Acad. Sci. 98 (17), 9995–10000.

Phukan, P., Phukan, R., Phukan, S.N., 2015. Heavy metal uptake capacity of *Hydrilla verticillata*: a commonly available aquatic plant. Int. Res. J. Environ. Sci. 4 (3), 35–40.

Pollard, A.J., Reeves, R.D., Baker, A.J., 2014. Facultative hyperaccumulation of heavy metals and metalloids. Plant Sci. 217–218, 8–17.

Prasad, M.N.V., 2005. Nickelophilous plants and their significance in phytotechnologies. Braz. J. Plant Physiol. 17 (1), 113–128.

Prasad, M.N.V., Freitas, H., 2003. Metal hyperaccumulation in plants: biodiversity prospecting for phytoremediation technology. Electron J. Biotechnol. 6, 275–321.

Qadir, S., Jamshieed, S., Rasool, S., Ashraf, M., Akram, N.A., Ahmad, P., 2014. Modulation of plant growth and metabolism in cadmium-enriched environments. Rev. Environ. Contamin Toxicol. 229, 51–88.

Quartacci, M.F., Irtelli, B., Gonnelli, C., Gabbrielli, R., Navari-Izzo, F., 2009. Naturally-assisted metal phytoextraction by *Brassica carinata*: role of root exudates. Environ. Pollut. 157 (10), 2697–2703.

Rafati, M., Khorasani, N., Moattar, F., Shirvany, A., Moraghebi, F., Hosseinzadeh, S., 2011. Phytoremediation potential of *Populus alba* and *Morus alba* for cadmium, chromuim and nickel absorption from polluted soil. Int. J. Environ. Res. 5, 961–970.

Rascio, N., Navari-Izzo, F., 2011. Heavy metal hyperaccumulating plants: how and why do they do it? And what makes them so interesting? Plant Sci. 180 (2), 169–181.

Renella, G., Landi, L., Nannipieri, P., 2004. Degradation of low molecular weight organic acids complexed with heavy metals in soil. Geoderma 122 (2), 311–315.

Römheld, V., 1991. The role of phytosiderophores in acquisition of iron and other micronutrients in graminaceous species: an ecological approach. Plant Soil 130 (1–2), 127–134.

Römkens, P., Bouwman, L., Japenga, J., Draaisma, C., 2002. Potentials and drawbacks of chelate-enhanced phytoremediation of soils. Environ. Pollut. 116 (1), 109–121.

Sabir, M., Hanafi, M.M., Zia-Ur-Rehman, M., Saifullah, A.H.R., Hakeem, K.R., Aziz, T., 2014. Comparison of low-molecular-weight organic acids and ethylenediaminetetraacetic acid to enhance phytoextraction of heavy metals by maize. Commun. Soil Sci. Plant Anal. 45 (1), 42–52.

Salati, S., Quadri, G., Tambone, F., Adani, F., 2010. Fresh organic matter of municipal solid waste enhances phytoextraction of heavy metals from contaminated soil. Environ. Pollut. 158 (5), 1899–1906.

Salido, A.L., Hasty, K.L., Lim, J.M., Butcher, D.J., 2003. Phytoremediation of arsenic and lead in contaminated soil using Chinese brake ferns (*Pteris vittata*) and Indian mustard (*Brassica juncea*). Int. J. Phytoremed. 5 (2), 89–103.

Salla, V., Hardaway, C.J., Sneddon, J., 2011. Preliminary investigation of *Spartina alterniflora* for phytoextraction of selected heavy metals in soils from Southwest Louisiana. Microchem. J. 97 (2), 207–212.

Salt, D.E., Rauser, W.E., 1995. MgATP-dependent transport of phytochelatins across the tonoplast of oat roots. Plant Physiol. 107 (4), 1293–1301.

Salt, D.E., Smith, R.D., Raskin, I., 1998. Phytoremediation. Ann. Rev. Plant Physiol. Plant Mol. Biol. 49, 643–668.

Smolinska, B., 2015. Green waste compost as an amendment during induced phytoextraction of mercury-contaminated soil. Environ. Sci. Pollut. Res. 22 (5), 3528–3537.

Singh, S., 2012. Phytoremediation: a sustainable alternative for environmental challenges. Int. J. Gr. Herb. Chem. 1, 133–139.

Sukumaran, D., 2013. Phytoremediation of heavy metals from industrial effluent using constructed wetland technology. Appl. Ecol. Environ. Sci. 1 (5), 92–97.

Sun, Y., Zhou, Q., Xu, Y., Wang, L., Liang, X., 2011. Phytoremediation for co-contaminated soils of benzo[a]pyrene (B[a]P) and heavy metals using ornamental plant *Tagetes patula*. J. Hazard Mater. 186 (2–3), 2075–2082.

Tangahu, B.V., Sheikh Abdullah, S.R., Basri, H., Idris, M., Anuar, N., Mukhlisin, M., 2011. A review on heavy metals (As, Pb, and Hg) uptake by plants through phytoremediation. Int. J. Chem. Eng. 2011, 31. Article ID 939161.

Tong, Y.P., Kneer, R., Zhu, Y.G., 2004. Vacuolar compartmentalization: a second-generation approach to engineering plants for phytoremediation. Trend Plant Sci. 9 (1), 7–9.

Ugya, A.Y., 2015. The efficiency of *Lemna minor* L. in the phytoremediation of romi stream: a case study of Kaduna refinery and petrochemical company polluted stream. J. Appl. Biol. Biotech. 3 (1), 011–014.

USEPA, 2000. Introduction to Phytoremediation. EPA, Washington, DC.

Vacchina, V., Mari, S., Czernic, P., Marquès, L., Pianelli, K., Schaumlöffel, D., Lebrun, M., Lobinski, R., 2003. Speciation of nickel in a hyperaccumulating plant by high-performance liquid chromatography-inductively coupled plasma mass spectrometry and electrospray MS/MS assisted by cloning using yeast complementation. Analy. Chem. 75 (11), 2740–2745.

Van Deuren, J., Center UAE, 2002. Remediation Technologies Screening Matrix and Reference Guide, fourth ed. Platinum International, Inc., Alexandria, VA.

Virkutyte, J., Sillanpää, M., Latostenmaa, P., 2002. Electrokinetic soil remediation—critical overview. Sci. Total Environ. 289 (1), 97–121.

Weber, M., Harada, E., Vess, C., Roepenack-Lahaye, E., Clemens, S., 2004. Comparative microarray analysis of *Arabidopsis thaliana* and *Arabidopsis halleri* roots identifies nicotianamine synthase, a ZIP transporter and other genes as potential metal hyperaccumulation factors. Plant J. 37 (2), 269–281.

Wei, S.H., Zhou, Q.X., Wang, X., Cao, W., Ren, L.P., Song, Y.F., 2004. Potential of weed species applied to remediation of soils contaminated with heavy metals. J. Environ. Sci. (China) 16 (5), 868–873.

Wu, G., Kang, H., Zhang, X., Shao, H., Chu, L., Ruan, C., 2010. A critical review on the bio-removal of hazardous heavy metals from contaminated soils: issues, progress, eco-environmental concerns and opportunities. J. Hazard Mater. 174 (1–3), 1–8.

Xia, B., Shen, S., Xue, F., 2013. Phytoextraction of heavy metals from highly contaminated soils using *Sauropus androgynus*. Soil Sed. Contamin. 22 (6), 631–640.

Yang, X., Feng, Y., He, Z., Stoffella, P.J., 2005. Molecular mechanisms of heavy metal hyperaccumulation and phytoremediation. J. Trace Elem. Med. Biol. 18 (4), 339–353.

Yang, X., Long, X., Ye, H., He, Z., Calvert, D., Stoffella, P., 2004. Cadmium tolerance and hyperaccumulation in a new Zn-hyperaccumulating plant species (*Sedum alfredii Hance*). Plant Soil 259 (1–2), 181–189.

Yang, X., Long, X., Ni, W., Fu, C., 2002. *Sedum alfredii* H: a new Zn hyperaccumulating plant first found in China. Chin. Sci. Bull. 47 (19), 1634–1637.

Yao, Z., Li, J., Xie, H., Yu, C., 2012. Review on remediation technologies of soil contaminated by heavy metals. Proc. Environ. Sci. 16, 722–729.

Zaier, H., Ghnaya, T., Ben Rejeb, K., Lakhdar, A., Rejeb, S., Jemal, F., 2010. Effects of EDTA on phytoextraction of heavy metals (Zn, Mn and Pb) from sludge-amended soil with *Brassica napus*. Bioresour. Technol. 101 (11), 3978–3983.

Zhao, H.Y., Lin, L.J., Yan, Q.L., Yang, Y.X., Zhu, X.M., Shao, J.R., 2011. Effects of EDTA and DTPA on lead and zinc accumulation of ryegrass. J. Environ. Protec. 2, 932–939.

Zhao, S., Jia, L., Duo, L., 2013. The use of a biodegradable chelator for enhanced phytoextraction of heavy metals by *Festuca arundinacea* from municipal solid waste compost and associated heavy metal leaching. Bioresour. Technol. 129, 249–255.

Zhuang, P., Shu, W., Li, Z., Liao, B., Li, J., Shao, J., 2009. Removal of metals by sorghum plants from contaminated land. J. Environ. Sci. 21 (10), 1432–1437.

CHAPTER

DUCKWEED: AN EFFICIENT HYPERACCUMULATOR OF HEAVY METALS IN WATER BODIES

16

Zeshan Ali[1], Hina Waheed[2], Alvina G. Kazi[3], Asim Hayat[4], Mushtaq Ahmad[1]

National Institute of Bioremediation, National Agricultural Research Center (NARC), Islamabad, Pakistan[1]; Department of Botany, Pir Mehr Ali Shah Arid Agriculture University, Rawalpindi, Pakistan[2]; Atta-ur-Rahman School of Applied Biosciences, National University of Sciences and Technology (NUST), Islamabad, Pakistan[3]; Soil Fertility Lab, Land Resources Research Institute, National Agricultural Research Center (NARC), Islamabad, Pakistan[4]

CHAPTER OUTLINE

Plant Metal Interaction. http://dx.doi.org/10.1016/B978-0-12-803158-2.00016-3

1. DUCKWEEDS: AN INTRODUCTION

Members of the family *Lemnaceae*, commonly called duckweeds, frequently present on the surface of still and slowly moving fresh water or wetlands (Zimmo, 2003; Zhao et al., 2015a). Regardless of their tiniest size duckweeds are the most common organisms in the freshwater ecosystems and form thick mats on the water surfaces (Godfrey and Wooten, 1979; Kutschera and Niklas, 2015). The mats are heterogeneous mixture of different genera and species. Mats can reach maximum thickness of 6 cm and create anaerobic conditions in the water body, promoting denitrification and anaerobic digestion. *Lemnaceae* plants are extremely simple and do not have an identifiable leaf or stem. The plant body of all members of this family is a thallus/frond that is a few cells thick structure containing *Aerenchyma* tissue (air pockets), which help plant float over the surface of water. Root-like structures called rootlets are present in some species. Reproduction in duckweeds is both by asexual and sexual. Asexual reproduction occurs by the meristems at the base of thallus/frond. Sexual reproduction is by three flowers (containing one pistil and two stamens). Asexual reproduction is common in duckweeds (Zhao et al., 2015a); infrequent sexual reproduction leads to the production of a fruit called the utricle, with seed produced in a sac-like structure packed with air to assist in floating over the water.

Except for deserts and permanently frozen polar regions, duckweeds are found in tropical and temperate expanses globally. Their tendency to occupy these diverse eco-zones is because of their ample genetic diversity and adaptation potential. Duckweed abundance and distribution in fresh water ecosystems is influenced by nutrient availability (Zhao et al., 2015b). Most of the species tend to be connected with the nutrient-rich waters. In the eutrophic condition, duckweeds thrive well, which is a reason for its high protein value. Duckweeds are a good protein source for animals and humans (Gupta and Prakash, 2013). Protein content can reach up to 40% of total dry weight of the plant under ideal growth conditions (Skillicorn et al., 1993). Fiber content in duckweed species is low (5%) compared with other higher plants because they do not need structural tissues to support the plant body (Gupta and Prakash, 2013). In still or slowly moving waters, duckweeds lower the light penetration to the bottom and deplete oxygen in the deep water. They are potential weeds in the fresh waters with excess nutrient supplies and are easily disseminated from their habitat by water birds and water currents to other places with no infestation. Their proliferation in stagnant waters (lentic water bodies; i.e., ponds, pools, and brooks) is fast compared with flowing waters (lotic water bodies; i.e., rivers and streams). Duckweeds, because of their extensive proliferation on fresh water surfaces, reduce evaporation and prevent algal, bacterial, and mosquito growth. They grow two times faster than other vascular plants, and each duckweed frond can reproduce at least 15–20 times before death. Their best growth is at warm temperatures (20–30 °C); at declining temperatures, their growth is hampered (Culley et al., 2009). Minimum and maximum temperature thresholds for their growth are <10 and 35–40 °C, respectively (Zirschky and Reed, 1988). Duckweeds survive harsh winters by forming starchy frond-like structures called "turion" that sink to the bottom of the water body (Appenroth and Adamec, 2015). On recommencement of normal weather, light, and temperature conditions, duckweed plants resume normal growth (Classen et al., 2000; Yin et al., 2015). A fully grown, extensive duckweed mat over a water surface can make the temperature 10 °C higher than ambient air, which is attributable to the sun's radiation effect (Reed et al., 1988).

Duckweeds are ideal experimental organisms in various disciplines of plant sciences (i.e., ecology, physiology, and toxicology) (Hillman, 1961). They are a suitable candidate for biofuel generation because of their rapid biomass production (Blinová et al., 2014; Tang et al., 2015). This characteristic has led current research to focus on their role as potential source for generating cheap, ecofriendly, clean, renewable energy. In developing countries, farmers use freshly harvested plants as fertilizers to enhance crop productivity. Also, duckweeds are used in compost manufacture to improve soil fertility. They are also used as a test organism in metal-based ecotoxicological testing in surface waters/wastewaters because of their short lifespan, quick vegetative propagation, and hyperaccumulation potential (Radić et al., 2010). Their on-time harvesting can help in nitrogen and phosphorus removal from the eutrophic waters. The way they accumulate excess minerals and metals from the water bodies therefore they are of paramount importance in phytoremediation (Gupta and Prakash, 2013). Different duckweed species are constantly reported to accumulate heavy metals from various kinds of polluted wastewaters. Their species are used in the stabilization/detention ponds for the large-scale wastewater treatment systems to improve water quality in terms of biological oxygen demand (BOD), chemical oxygen demand (COD), nutrients, pathogens, and metals (Ali et al., 2013a; Farid et al., 2014; Ugya, 2015). Heavy metal hyperaccumulation potential of duckweeds is well-established in recently reported literature.

2. SPECIFIC DIVERSITY AND TAXONOMIC CLASSIFICATION

Duckweeds have been extensively studied in the past two centuries by taxonomists i.e., Schleiden (1839), Hegelmaier (1868), Thompson (1898), Daubs (1965), and Landolt (1986, 1994). Their contributions significantly demarked generic and specific limits in *Lemnaceae*. *Lemnaceae* is a monocotyledonous family composed of five aquatic genera (i.e., *Landoltia*, *Spirodela*, *Lemna*, *Wolffia*, *Wolffiella*) and 38 species (Landolt, 1986, 1998; Les and Crawford, 1999; Kutschera and Niklas, 2015). Their structural simplicity fascinated systematists, who investigated their relationship with other monocotyledonous families.

Hegelmaier was the first systematist to provide initial complete interspecific relationships in *Lemnaceae* in a taxonomic context (Hegelmaier, 1868). Following Hegelmaier, Ivanova suggested a phylogenetic tree of species of *Lemnaceae*, which was not based on the taxonomic approaches (Ivanova, 1973). Hartog provided better information than earlier works on the intrafamilial affinities in *Lemnaceae* based on morphological and chemotaxonomic data (Hartog, 1975). In-depth interspecific and intergeneric phylogenetic relationships were comprehensively reported by Landolt (Landolt, 1986). Absence of visible morphological characters from ultrareduction in an evolutionary time frame has led to a dearth of phylogenetic information in *Lemnaceae* (Tippery et al., 2015). The morphological characteristics can otherwise serve as easy phylogenetic markers for this family. Problems in duckweed systematics led systematists to rely on chemotaxonomy (a plant's chemical constituents) to solve systematics problems. Comprehensive flavonoid surveys and comparative chemistry promoted the idea of interspecific phylogenetic relationships (Alston, 1966). Data generated from such novel chemical approaches were considered advantageous because of a lack of morphological data. Along with chemical constituents, molecular data (allozymes and chloroplast DNA) was generated to address taxonomic questions in *Lemnaceae* (Jordan et al., 1996; Crawford et al., 1997; Borisjuk et al., 2015). In the face of these advancements, basic questions related to interrelationships and delimitation of taxa still remain unclear.

Duckweeds have immense evolutionary importance because they are the Earth's tiniest angiosperms (Xu et al., 2015). Members of the genera *Wolffia* are the smallest, with a size of ~1 mm. The flower size of *Wolffia* is the smallest in angiosperms: ~0.3 mm long. Morphologically, duckweeds show little resemblance with the flowering plants (angiosperms). Size reduction in duckweeds has resulted from the simplification/alteration and loss of different anatomical structures over an evolutionary time frame. This miniature version of duckweeds has made it almost impossible to assess any homology with other flowering plants (angiosperms), even within the *Lemnaceae* family (Cusimano et al., 2011). This extreme reduction/simplicity has made duckweed thallus or frond extremely difficult for taxonomists to interpret it as a stem or a leaf. The frond size in the members of genus *Spirodela* is largest, measuring ~20 mm, whereas in *Lemna* species it is around 6–8 mm. Even among the genera, the homologies are difficult to understand. Systematists believe that *Lemnaceae* has evolved regressively, with *Spirodela* being primitive, whereas *Wolffia* is most advanced genera (Landolt, 1994). Extreme reduction, rare flowering, and small size have made duckweeds difficult for the systematic research (Tippery et al., 2015). Systematists believe that the flower-like structure in duckweeds is pseudanthium (false flower) or reduced inflorescence with three miniscule flowers that are conspicuously either male or female derived from the spadix (spike inflorescence) in *Araceae* (Cusimano et al., 2011).

3. TEST ORGANISM IN ECOTOXICOLOGICAL TESTING

Different duckweed species have been commonly used in the ecotoxicological testing of municipal and industrial wastewaters (Lakatos et al., 1993; Radić et al., 2010; Horemans et al., 2015). Improved living standards in developed countries and established industrial setups have adversely impacted environments through toxic wastewaters from urban centers. Developing countries, in comparison, are far behind in industrial development and living standards; therefore, they are not significantly contributing to global pollution via contaminated wastewaters. Generally, these wastewaters contain varying amounts of heavy metals depending on the setups in the municipal sectors and chemical operations in industry (Ali et al., 2013b, 2015). The toxicity of municipal and industrial wastewaters is influenced by different abiotic and biotic factors (Wang, 1987). Wastewaters generally have a complex composition and their toxicity can be the result of more than one substance. The substances present can act mutually to enhance toxicity level or may antagonize each other to reduce toxicity. A substance's behavior in wastewaters is a multifaceted phenomenon influenced by pH, electrical conductivity, suspended salts, total solids, and redox conditions. No matter how detailed the analytical chemical results are, their hazardous impacts cannot be directly translated as actual harm to living organisms. In this connection, toxicity testing based on a chemical-specific approach is not adequate and needs a biological system to determine an accurate response to such toxic conditions. Biomonitoring of toxic wastewaters using sensitive organisms ultimately emerges as today's need (Zhou et al., 2008). The ability of the sensitive organisms to respond to toxic substances through changes in growth patterns, fluctuating enzymes/hormonal levels, DNA damage, and pigment variations can help in categorizing wastewaters (Delmail and Labrousse, 2014). Biomonitoring has become an integral part of wastewater quality assessment programs, besides the numerical guidelines established by various regional and global environmental agencies (Böcük et al., 2013).

Biomonitoring tests commonly use either plant or animal subjects i.e., *Daphnia magna* (crustacean), *Lemna minor* (plant), *Spirodela polyrhiza* (plant), *Pimephales promelas* (fish), and *Raphidocelis subcapitata* (microalga). Plant use in the biomonitoring of wastewater is advantageous over other organisms for various reasons. Fish and daphnid usage as bioindicators of contaminated wastewaters is often not possible because of high BOD and low oxygen availability for organismal survival. Any aeration of wastewaters for bioindicator survival will change the sample integrity and is therefore not recommended. Similarly, algal biomonitoring is not feasible in turbid wastewaters, and filtration is required, which again will destroy sample integrity. High microbial contamination in wastewaters also prevents daphnid, fish, or algal usage as bioindicators. In all such inappropriate conditions (i.e., turbidity, microbial excess, and low oxygen levels), plants (*Lemna*, *Spirodela*, or other aquatic members of *Lemnaceae*) can be effectively used in biomonitoring of wastewaters (Appenroth et al., 2010). A true picture of wastewater can be attained using *Lemnaceae* plants through phytotoxicity tests without compromising sample integrity. Phytotoxicity can be documented by root elongation, seed germination, decreased pigment levels, genotoxicity, malondialdehyde formation, and higher levels of antioxidant enzymes (i.e., superoxide dismutase, peroxidase, catalase) (Khellaf and Zerdaoui, 2010a,b; Horemans et al., 2015). Their application in biomonitoring metal-containing municipal, industrial, and agricultural effluents is simple, environmentally friendly, cost-effective, and sensitive (Greenberg et al., 1992). Other benefits of using duckweeds in ecotoxicological testing include their easy storage, fast culturing, and propagation (Radić et al., 2010). For seed germination tests, their seeds are easily available, have a long shelf-life (around 5 years), and in dry state are resistant to unfavorable environmental conditions. Even wastewaters sensitive to photodegradation can be easily biomonitored by seed germination and root elongation tests using duckweed. Because both processes typically initiate and proceed in the dark, there is no photodegradation of the sample, which remains intact.

4. ROLE IN WASTEWATER RECLAMATION

Natural wastewater treatment structures integrated with aquatic macrophytes are gaining worldwide importance in wastewater treatment. Duckweeds are more effective in wastewater treatment systems because of their wide adaptability to different geographical regions and their ability to tolerate varying pH levels, temperatures, nutrients, and climatic conditions (Landolt and Kandeler, 1987; Tang et al., 2015). Therefore, *Lemnaceae* plants have been significantly used in treating wastewater of municipal, industrial, and agricultural origin. The *Lemna* species in particular have been a focus of wastewater reclamation systems/projects. These macrophytes have excellent potential of accumulating nutrients and metals from the environment (Harvey and Fox, 1973; Cheng and Stomp, 2009; Zhao et al., 2015b). Wastewater treatment by *Lemnaceae* plants is enhanced further by their mutualistic association with microbes. The pollutants reduced in the wastewater include excess nitrogen, phosphorus, COD, BOD, microbial pathogens, suspended solids, organic compounds, algal abundance, and certain metals (Ugya, 2015). It has been estimated in domestic/municipal effluents that duckweed-based treatment systems reduce 50–60% of key contaminants. Duckweed mats on the surface of open wastewater treatment facilities reduce evaporation rates by 20% (Oron et al., 1986). Conservation of precious treated water is of great significance in the arid agro-climatic regions. Impact of duckweeds on the major water quality parameters, microbial reduction, phytoplankton removal, and metals uptake is mentioned in detail later.

4.1 IMPACTS ON MAJOR WATER QUALITY CHARACTERISTICS

Nutrient (i.e., nitrogen and phosphorus) removal from wastewater is a challenging task. These nutrients are present in wastewater in the chemical form of ammonia, nitrate, nitrite, and phosphates. Duckweeds are excellent accumulators of these nutrients and the literature has many reports in this regard. In a study conducted by Körner and Vermaat (1998) on domestic shallow waters, duckweeds removed 63–99% of total phosphorus and 7–97% of total Kjeldahl nitrogen at a hydraulic retention time (HRT) of 3 days. Nutrient uptake potential of duckweeds ranges between 0.045 and 1.67 g m^2/day of nitrogen, whereas phosphorus ranges between 9% and 61% (Zirschky and Reed, 1988; Körner and Vermaat, 1998). Increased nitrogen and phosphorus uptake from wastewater has a direct impact on the development of new fronds and the rapid growth of duckweeds (Ziegler et al., 2015). Pandey (2001) reported a 50–75% reduction in nitrogen and a 17–35% reduction in phosphorus in the treated wastewater from duckweed-based treatment facility at Halisahar, India. In constructed wetlands and storm water ponds, *Lemnaceae* plants can be used to harvest nutrients (Ancell, 1998). Perniel et al. (1998) reported ammonia removal from storm waters using *L. minor* L., but with a phosphorus removal using a combination of *S. polyrhiza* and *L. minor.* Besides using duckweeds in nutrient removal, they may serve as a potential food source to the fisheries and other animals (van der Steen et al., 1999; Anderson et al., 2011). Duckweeds can be efficiently used in the treatment of wastewater generated from cattle feedlots to concentrate nutrients (Moss, 1999). Therefore, duckweeds serve a dual purpose (i.e., treating wastewater and feeding livestock with nutrient-rich duckweed). Excess nitrogen and phosphorus, if not removed from the treated wastewaters, enters fresh water bodies and can easily lead to eutrophication (Oron et al., 1988).

Thick duckweed mats have little direct impact on the BOD levels in wastewater treatment systems. Extensive mats create anoxic conditions in the water body because of low oxygen evolution by small plants and insignificant aeration. In anoxic conditions, the microbial oxidation of organic matter is negligible, leading to low BOD levels. These low oxygen levels in turn are also responsible for low COD levels (sum of BOD and natural oxidation processes). El-Kheir et al. in 2007, while working on domestic effluents reported a high reduction in BOD (90.6%) and COD (89%) concentrations on treatment with duckweeds. Another reason for low BOD/COD levels in wastewaters having extensive duckweed mats is the use of simple organic matter as food material by growing duckweed plants. Zimmo et al. (2005) reported higher BOD reduction in wastewater ponds treated with duckweeds alone compared with ponds having a mixture of duckweeds and algae. Duckweeds also reduce the total suspended salts and total dissolved solids (TDS) components of wastewater. On TDS lowering in wastewaters treated with duckweeds, electrical conductivity is also lowered because it is directly correlated with TDS (El-Kheir et al., 2007; Ali et al., 2013b). In water channels that serve for irrigation purposes, duckweed application can reduce soluble salt levels that otherwise can increase soil salinity, rendering it unfit for crop production. Higher soluble salt concentrations, if not treated biologically via duckweeds, may need demineralization by reverse osmosis, which is too costly. Sulfates and chlorides are also reduced by duckweed (*Lemna gibba* L.) in municipal effluents (El-Kheir et al., 2007).

4.2 MICROBIAL REDUCTION

Various reports have revealed duckweed role in wastewater pathogenic reduction. Falabi et al. (2002) investigated reduction of *Cryptosporidium*, *fecal coliforms*, *total coliforms*, *coliphages*, and *Giardia* in domestic wastewater by a series of duckweed-based pond system with 6 days of HRT. Influent and effluent analyses revealed reduction in *Giardia* cysts (98%), *Cryptosporidium* (89%), *coliphages*

(40%), fecal coliforms (62%), and total coliforms (61%), respectively. In a fecal coliform decay study, van der Steen et al. (2000) compared the removal efficiency of five shallow algal ponds with an integrated system containing algae and duckweed. In algal ponds, light intensity reduction by thick algal mats led to a reduction in fecal coliform decay. However, in the integrated system, duckweed attenuated algal concentration and accelerated fecal coliform decay. Results indicated better fecal coliform removal by duckweed compared with algae. El-Kheir et al. showed in 2007 that *L. gibba* L. was efficient in eliminating fecal coliforms and total coliform counts as a function of HRT. They reported a 99.8% reduction in both bacterial types at an HRT of 8 days. Similarly, Pandey (2001) described reduction in fecal coliform from 99.27% to 99.78% in treated duckweed-based pond water at an HRT of 6.4–14.2 days from Delhi, India. Duckweeds also reduce fecal coliforms in constructed wetlands. Ran et al. (2004) reported a 95% reduction in the fecal coliform in domestic effluents at an HRT of 4.26 days. Some studies have also contradicted microbial reduction by duckweed-based wastewater treatment systems. Dewedar in 1995 disclosed that the fecal coliforms levels under thick duckweed mats did not drop during the study period.

4.3 PHYTOPLANKTON REMOVAL

Excessive algal growth in water bodies has negative environmental impacts (i.e., secretion of toxic chemicals, oxygen depletion, and high BOD and COD levels) and harm to faunal diversity. Duckweeds, because of their extensive growth, outcompete algae from the water bodies. Hammouda and Abdel-Hameed (1994), in a study on a wastewater treatment facility at Beni-Suef, found a 90% reduction in algae attributable to duckweed growth. They also found that the phytoplankton abundance was reduced to only five species (i.e., *Melosira granulatus*, *Synedra ulna*, *Navicula cryptocephala*, *Nitzschia acicularis*, and *Cyclotella glomerata*). El-Kheir et al. (2007) conducted a study on phytoplankton removal from wastewater using *L. gibba* L. The wastewater treatment system contained 11 species of cyanophyceae (*Phormidium molle*, *Aphanothece clathrata*, *Spirulina major*, *Microcystis aeruginosa*, *Spirulina meneghiniana*, *Phormidium* sp., *Spirulina laxissima*, *Spirulina platensis*, *Oscillatoria amphigranulata*, *Phormidium dictyothallum*, *P. molle* var. *tenuis*), four species of chlorophyceae (*Oocystis solitaria*, *Chlamydomonas snowii*, *Chlamydomonas globosa*, *Chlorococcum texanum*), three species of bacillariophyceae (*S. ulna*, *Cyclotella meneghiniana*, *Gomphonema parvulum*), two species of euglenophyceae (*Phacus acuminatus*, *Lepocinclis salina*), and one species of Cryptophyceae (*Cryptomonas ovata*). Duckweed *L. gibba* L. reduced the population of nine phytoplankton species at an HRT of 8 days. However, some of the algal species managed to survive (i.e., *G. parvulum*, *S. major*, *C. meneghiniana*, *S. laxissima*, *M. aeruginosa*, *C. snowii*, and *P. molle*), which are tolerant to organic pollution and can be used in different pollution indices (Palmer, 1969). Suppression of phytoplankton by duckweed can be from the following reasons: (1) thick duckweed mats block light penetration to the algae, inhibiting photosynthesis; (2) duckweed rhizospheres secrete complex organic substances that inhibit algal growth; and (3) duckweeds are far more fierce competitors for food materials compared with algae, which ultimately starve and perish (Hammouda et al., 1995; Pandey, 2001).

4.4 METAL HYPERACCUMULATION

Lemnaceae plants have tremendous potential for removing heavy metals from wastewaters of different origins (i.e., municipal, agricultural, and industrial) (Zayed et al., 1998; Axtell et al., 2003). Their metals uptake and accumulation is far higher than algae and angiosperms. Treating wastewater using these aquatic macrophytes is largely effective and commonly known as phytoremediation. Researchers estimate that cost involved in decontamination of wastewater using phytoremediation technologies is cheap and uses 80% less investment than other alternative chemical/mechanical/electrical wastewater treatment technologies (Ol'shanskaya et al., 2008; Ali et al., 2013a). Phytoremediation methods for wastewater treatment using duckweeds have proven their worth against precipitation, ion exchange, adsorption, permeable reactive barriers, filtration, reverse osmosis, and other technologies. Among duckweeds, *Lemna minor*, *L. gibba*, and *S. polyrhiza* are mostly examined for metal hyperaccumulation potentials (Mkandawire and Dudel, 2005). Watanabe (1997) proposed characteristics of plants having good metal bioaccumulation potential: pest/disease resistance, ability to accumulate different metals, rapid biomass production, short lifecycle, high degree of absorbance, accumulation, and tolerance. Duckweeds mostly have these traits, which makes them suitable candidate for metal hyperaccumulation. To date, various researchers have stated duckweeds potential for metal hyperaccumulation; however, a few studies have been documented to increase hyperaccumulation potential of duckweeds. Most researchers have described only the uptake and reduction of metals from the wastewaters with less emphasis on the mechanisms involved and enhancing duckweeds metal uptake potential (Ol'shanskaya et al., 2008). Duckweed hyperaccumulation potential for different metals is described in detail in the next section.

4.4.1 Aluminum

Aluminum (Al) is a metal in the boron group and the most abundant element in the Earth's crust. In neutral pH it is not known to cause phytotoxicity; however, in acidic conditions, it is extremely phytotoxic (Stephan et al., 2008). Despite its high abundance in the Earth's crust, it is not essential for plant metabolism and has no biological function (Fodor, 2002). Its sources in wastewater include a variety of anthropogenic operations (i.e., kitchen utensils, mining actions, wastewater treatment systems, construction activities, aircraft, electrical, food, and pharmaceutical and automobile industries) (ATSDR, 1992). Aluminum behavior in wastewater is complicated and governed by pH and electrical conductivity, which determine what types of Al species are present. Information regarding Al uptake from wastewaters through aquatic macrophytes is scarce (Goulet et al., 2005). Obek and Sasmaz (2011) investigated Al uptake from secondary treated domestic effluents using *L. gibba* L. They found very high Al accumulation by the selected plant in the first 2 days (i.e., 500 and 100 μg/g, respectively). This study revealed that maximum Al bioaccumulation by *L. gibba* takes place at an HRT of 2 days. Obek and Sasmaz further suggested harvesting *L. gibba* after 2 days of HRT for promising results (Obek, 2009). Goulet et al. (2005) investigated Al accumulation potential of four aquatic plants (*L. minor*, *Potamogeton epihydrus*, *Nuphar variegatum*, and *Typha latifolia*) from Al-rich wastewaters in two consecutive summers and reported the highest Al uptake rate by *L. gibba* (0.8–17 mg/g day). Efficient Al uptake by *L. minor* and *L. gibba* from wastewaters make duckweeds ideal for Al hyperaccumulation.

4.4.2 Boron

Boron (B) is a metalloid with relatively low abundance in the Earth's crust. It is a micronutrient essentially required by plants in many metabolic processes, including plasma membrane stability, cell wall strength, nucleic acid metabolism, cell division, plant hormone synthesis, fruit and seed development, sugar synthesis, and transport (Frick, 1985). Its concentration around 0.2 mg/L is essential for seed germination, plant growth, and development. However, its elevated levels are potentially phytotoxic (i.e., teratogenic, adverse effects on growth and reproduction) (Maier and Knight, 1991). Anthropogenic sources of boron are glass industry, detergents, bleaching agents, nuclear reactors, insecticides, agrochemicals, and pharmaceutical and electronic industries, which lead to its enrichment in wastewaters (Butterwick et al., 1989). Boron-enriched wastewaters are a potential toxicological threat and need removal before discharge to environmental compartments. Reports pertaining to toxicity and accumulation of waterborne boron are limited to the *Lemna* species only. Davis et al. (2002) described for the first time waterborne boron toxicity and accumulation in *S. polyrhiza*. Frond production was markedly reduced at a B concentration of 3.55 mg/L; however, chlorosis, necrosis, and death were not observed even at $\leq$22.4 mg/L. *Lemna minor* capacity to tolerate B levels (60 mg/L) was comparatively very high, but the frond production was not affected (Wang, 1986). At a B concentration of 200 mg/L, *L. minor* showed decreased growth and failed to exist on the sixth day of exposure. *Spirodela polyrhiza* ability to uptake and accumulate boron is less than *L. minor* and the response of both species to boron is species-specific (Glandon and McNabb, 1978). *Lemna minor* uptake of B is by both active and passive transport across root cell membranes. Boron, besides its uptake and removal from wastewater by the *Lemnaceae* plants, has potentially toxic effects depending upon its concentration in the effluent media; therefore, careful selection of aquatic macrophytes should be done before introduction of these plants for phytoremediation. It is supported by the literature to use *Lemna* species for boron hyperaccumulation from wastewaters compared with *Spirodela* or other rooted aquatic macrophytes because of their sensitivity and low tolerance to high boron concentrations.

4.4.3 Cadmium

Cadmium (Cd) is a transition metal with no known biological function in living organisms. Cadmium compounds are extremely toxic even at low concentrations because of their bioavailability and solubility in wastewaters (Das et al., 1997). It is a recognized metal pollutant in wastewaters globally that originates from metal industries, pigments, paints, batteries, mining, smelting, municipal wastes and sludge, fertilizers, pesticides, and electroplating (Li and Xiong, 2004). Aquatic plant exposure to excess Cd causes growth inhibition, impaired elemental uptake, chlorosis, photosynthetic inhibition, imbalanced water relations, chloroplast disorganization, induction of heat shock proteins, necrosis, and ultimately death at levels beyond tolerance thresholds (Schützendübel and Polle, 2002; Vitória et al., 2006). These toxic effects are to some extent are related with the oxidative damage caused by reactive oxygen species (ROS; Horemans et al., 2015). Cadmium toxicity at low concentrations has made excess accumulation of this metal in aquatic plants difficult for phytoremediation purposes. *Lemnaceae* plants are well-known accumulators of Cd (Mohan and Hosetti, 1997; Prasad et al., 2001); however, little information is available regarding physiological mechanisms adopted by duckweeds for Cd tolerance. Tkalec et al. (2008) reported progressive accumulation of Cd in *L. minor* at the sixth and twelfth days of cultivation (i.e., 319.97 and 631.16 μg/g dry weights, respectively, when grown in Cd solution; 10 μM). The Cd accumulation continued to increase in the *L. minor* tissues and reached its highest levels of 631.16 μg/g (0.06% of dry weight) at which the metabolic machinery was significantly impaired. The growth rate dropped to 54% after 12 days of Cd exposure and phytotoxicity was observed by malondialdehyde (MDA) production, decreased photosynthesis, and chlorophyll pigments, lipid peroxidation, and ultrastructural deterioration. Reeves and Baker's (2000) criteria for ranking plants in hyperaccumulators category is $\geq$0.01% metal of dry weight. Hou et al. (2007) reported a *L. minor* Cd tolerance limit (<0.5 mg/L) and suggested the use of duckweed for Cd hyperaccumulation from aqueous media at concentrations within tolerance regime. On-time harvesting and precise management of *L. minor* can help in efficient removal of Cd from wastewaters.

4.4.4 Copper

Copper (Cu) is a transition metal and essential for all living organisms. Similarly, it is essential for plant growth and metabolism with an average tissue concentration of 10 μg/g of dry weight (Greger, 1999). At higher concentrations (20–30 μg/g of dry weight), it is extremely toxic to plants (Marschner, 1995). Induction of an antioxidant system because of ROS generation, chloroplast and thylakoid disorganization, decline in pigments, biomass reduction chlorosis, and lowered photosynthetic efficiency are symptoms of Cu phytotoxicity. Copper sources in wastewater include electroplating, glass manufacturing, electrical and electronic devices, metal industry, ceramics, pipe infrastructure, and plumbing activities. Duckweeds, besides using Cu as a nutrient, accumulate fair quantities in fronds and roots. Panda (2008) exposed *L. minor* to Cu concentrations of 20, 50, and 100 μM and found a higher accumulation of Cu in the *L. minor* roots compared with fronds in 48 h of exposure/treatment time. Further results indicated that the higher the Cu concentration in the aqueous media, the higher the Cu concentration in *L. minor* fronds (7.30, 9.33, and 13.27 μg/g of dry weight) and roots (13.47, 19.37, and 24.60 μg/g of dry weight) with reference to three treatments given. At 100 μM, the biochemical and physiological processes of *L. minor* were significantly affected. Khellaf and Zerdaoui (2010b) described a Cu concentration in the 0.003–0.3 mg/L range stimulated *L. gibba* growth. Concentrations beyond 0.3 mg/L have first inhibitory and then toxic effects as Cu levels further increase. *Lemna gibba* Cu removal percentage from 0.3 to 0.1 mg/L Cu solution was 60% and 80%, respectively. Cu tolerance shown by *L. gibba* at ≥0.3 mg/L can be due to Cu chelation with organic acids (i.e., citric acid and malic acid), which partially prevent toxic consequences. Babu et al. (2003) showed similar growth inhibition results for *L. gibba* at a 0.25 mg/L Cu concentration in aqueous media. Hou et al. reported in 2007 that *L. minor* impaired growth at Cu concentration of 0.5 mg/L. Razinger et al. (2007) reported Cu hyperaccumulation in tissues of *L. minor* (900 μg/g), which was 90 times higher than the average Cu concentration in plant tissues (10 μg/g). Based on these findings, the *Lemna species* can be overwhelmingly recommended for wastewater phytoremediation having low Cu levels (Greger, 1999).

4.4.5 Lead

Lead (Pb) is an important heavy metal with serious environmental implications because of its high toxicity. Lead toxicity in plants results in oxidative damage, chlorophyll biosynthesis inhibition, decreased biomass production, and stunted growth (Fargasová, 1994). In wastewaters, common sources of Pb include batteries, ammunition products, metal processing, wastewater treatment systems, pigments, paints, petroleum/coal products, automobile exhaust, electronics, and paper/pulp industries (Eick et al., 1999). Duckweeds have been reported for Pb accumulation from wastewaters (Uysal and Taner, 2011). Horvat et al. (2007) reported *L. minor* ability to accumulate heavy metals (i.e., Fe > Zn > Pb) from electroplating wastewaters. Miranda et al. (2000) described Pb uptake by *L. gibba* from solutions containing 50, 100, 200, and 300 mg/L Pb concentrations. Highest uptake at 1 day of exposure was recorded from a Pb solution containing 50 mg/L (55.2%). Pb uptake by plant decreased as the Pb concentration in solutions increased. The lowest Pb uptake (19.8%) was recorded from the solution containing 300 mg/L Pb. Similar results were obtained by Saygideger et al. (2013) while working on Pb uptake by *L. minor* from Pb-containing solutions (i.e., 1, 10, 100 mg/L Pb). *Lemna minor* fronds in response to these concentrations revealed good Pb uptake (i.e., 23.3, 153.3, and

490.5 μg/g of dry weight). Zayed et al. (1998) investigated duckweed accumulation potential for chromium, cadmium, selenium, copper, and lead in the range of 0.1–10 mg/L. Their results also confirmed Pb accumulation by duckweed, but to a lesser extent than other studied metals. In another study, Kaur et al. (2008) confirmed Pb accumulation by duckweed in the range of 1–20 mg/L and identified Pb uptake as a function of pH, treatment time, and Pb concentration in the solution media. Uysal and Taner (2009) advocated the role of temperature besides pH and Pb concentration in the solution that impacts its uptake by duckweeds. These researchers found the highest Pb(II) uptake and accumulation at temperature 30 °C and pH 4.5.

4.4.6 Arsenic

Arsenic (As) is a metalloid that is extremely toxic to plant and animal life. It is used in medicines, wood preservative, warfare agents, alloys, batteries, electronic industry, mining, pesticides, and fertilizers (Sharma and Sohn, 2009). Arsenic phytotoxicity depends on its concentration in the environment and oxidation state. In natural environments, it exists in four oxidation states: arsenic (0 oxidation state), arsine (−III oxidation state), arsenate (V oxidation state), and arsenite (III oxidation state). In toxic wastewaters, arsenate As(V), is usually the stable form; however, As(III) is the second most stable form. Both of these As species are available to aquatic plants and are potentially phytotoxic. Among the four genera of duckweeds *Wolffia*, *Spirodela*, and *Lemna* hyperaccumulate, As from wastewaters (Rahman et al., 2007; Mrak et al., 2008). Visoottiviseth et al. (2002) included aquatic plants in As hyperaccumulator category that contained 1000 μg/g As in their tissues. *Lemna gibba* potential for As(V) bioaccumulation from tailing wastewaters was studied by Mkandawire and Dudel (2005) in laboratory and field conditions. They found higher As bioaccumulation by *L. gibba* in laboratory conditions because of limited treatment time and in time harvesting. As(V) is an analog of phosphate and its increased uptake hampered phosphate uptake by *L. gibba* because both competed for same carriers sites in cell membrane. Zhang et al. (2009) investigated the uptake and accumulation of As by fronds of *Wolffia globosa* and reported 1000 mg/kg As on dry weight basis. Such level of As accumulation is 2–10 times higher than other duckweeds and earns this species a special place in the As hyperaccumulation. Duman et al. (2010) described that duckweed As accumulation potential increases with increasing As concentrations. Accumulation of different As species by the same duckweed species also varies greatly. *Lemna minor*, when exposed to 64 μM concentration of As(V) and As(III). accumulated 8674 and 17,408 μg/g dry weight, respectively, over a 6-day exposure. These results clearly indicate higher As(III) removal compared with As(V). According to Abbas and Meharg (2008), greater As(III) uptake compared with As(V) leads to rising As(III) toxicity in plants. As(III) concentration beyond 800 μg/L showed negative effects on *L. gibba* growth (Mkandawire et al., 2004). Duckweeds (i.e., *L. gibba*, *L. minor*, *Wolffia globosa*, and *S. polyrhiza*) have tremendous As phytoaccumulation potential and therefore can be used in related phytoremediation processes.

4.4.7 Nickel

Nickel (Ni) is essential element in plant metabolism; however, its excess quantities can disrupt many biological processes. In wastewaters, common Ni sources include plumbing activities, piping infrastructure for water transport, electrical, automobile, ship, mining, oil, electroplating, textile, battery, and chemical industries. Nickel removal from natural and synthetic wastewaters using *Lemnaceae* plants has been widely reported in the literature (Jain et al., 1988). Khellaf and Zerdaoui, (2010b) described tolerance range (0.003–0.5 mg/L) for *L. gibba*; Ni concentration exceeding 0.5 mg/L considerably suppressed its growth. At a still higher concentration, a ≥1 mg/L visible change in frond color was observed with the separation of colonies. At Ni concentration (0.75 mg/L), *L. gibba* growth index was reduced by 50%. At concentrations that do not inhibit *L. gibba* growth, little accumulation is witnessed (Zayed et al., 1998). Contrarily, Axtell et al. (2003) described preferential accumulation of Ni in *L. minor* fronds when treated with a solution containing Ni and Pb. The nickel removal percentage was 82% from the solution containing both metals. Kara et al. (2003) found that duckweed Ni uptake is high in the first few days of exposure and then decreases as plants reach their saturation levels. Nickel hyperaccumulation using *Lemnaceae* plants is recommended at lower Ni concentrations.

4.4.8 Silver

Silver (Ag) is nonessential element for living organisms and is highly toxic. It is also generated from many anthropogenic activities and ends up in municipal and industrial wastewaters (Xu et al., 2010). Silver phytotoxicity is due to free silver ion concentrations that induce ROS and related defense responses in plants. Average Ag levels in plant tissues are below 0.01 mg/kg (Kabata-Pendias and Pendias, 2001). Sasmaz and Obek (2012) reported preferential uptake and accumulation of silver by *L. gibba* from secondary treated municipal effluents containing silver and gold. They exposed *L. gibba* for 7 days and found the highest Ag uptake on the sixth day of treatment. Results from their experiment revealed a 2303% higher Ag accumulation in the laboratory-based studies compared with the natural environments. After reaching the saturation level on sixth day, the accumulation rate dropped on the seventh day. Substantial *L. gibba* Ag uptake and bioaccumulation make it a strong candidate for reclamation of Ag-containing wastewaters.

4.4.9 Zinc

Zinc (Zn) is a heavy metal with applications in batteries, alloys, electronics, fuels, pharmaceuticals, and fungicides. Its domestic and industrial usage has led to its elevated levels in wastewaters. At trace quantities, it is essential to plant growth and metabolism; however, at higher concentrations, it has phytotoxic effects (i.e., reduced growth, ROS production, chlorosis, decreased photosynthesis). Zn uptake and bioaccumulation in duckweeds has been described by many (Kara, 2004; Khellaf and Zerdaoui, 2009). Zinc tolerance in different duckweed species varies. Lahive et al. (2011) described Zn uptake and accumulation potential in three duckweed species (i.e., *L. minor*, *L. gibba*, and *Landoltia punctata*). *Landoltia punctata* was more efficient in Zn uptake compared with *Lemna* species and continued its uptake even at 100 mg Zn/L, whereas the other two species showed reduced Zn uptake. Reduced uptake by *Lemna* species is related with Zn toxicity. Of the two *Lemna* species, *L. gibba* is more sensitive to Zn, which ultimately leads to low Zn uptake and accumulation by this species. Zn accumulation either in roots or fronds also varied in two *Lemna* species. *Lemna minor* preferentially accumulated Zn in the roots, so that necessary metabolic functions in fronds by Zn accumulation are not disturbed (Drost et al., 2007). However, *L. gibba* accumulated Zn in the frond, and this can be a reason for low Zn uptake and accumulation by this species. *Landoltia punctata* also accumulated Zn in the fronds, but surprisingly showed higher Zn uptake and accumulation potential than both *Lemna* species. Zn uptake and accumulation is reported from other *Lemnaceae* members as well. Sharma and Gaur (1995) described Zn accumulation in *S. polyrhiza* (27 μg/mg of dry weight) when exposed to 10 mg/L Zn-containing solution over 4 days. Jafari and Akhavan (2011) reported Zn bioaccumulation potential of three species (*L. trisulca*, *L. minuta*, and *L. minor*) from Zn-contaminated wastewaters. They reported the highest Zn bioconcentration potential by *L. trisulca* (97%) followed by *L. minuta*

(89%) and *L. minor* (83%). Based on the findings of these researchers, duckweeds prove their worth in accumulating Zn from wastewaters of different origins.

5. MANAGEMENT OF METAL HYPERACCUMULATOR DUCKWEED PLANTS

In spite of the cost-effectiveness and environmentally friendly nature of phytoremediating (hyperaccumulating) plants, the fate of metal-saturated duckweed remains a burning issue. Management and safe disposal of metal-loaded plants have intrigued scientists, who have developed additional technologies to prevent any further damage by heavy metals. If phytoremediation technologies are practiced at large scales instead of laboratory-based experiments, then their safe disposal is even more critical. Data regarding safe disposal of contaminated duckweeds are scanty and currently questioned for the fruitful application of this emerging technology. However, a few methods used in this regard are anaerobic digestion for biogas production, hydrolysis, fermentation, briquetting, incineration, and carbonization.

Anaerobic digestion of contaminated duckweeds for biogas production is well-suited for the developing countries such as Pakistan that seriously suffer from a power shortage. Metal speciation may be the only concern in the anaerobic digestion processes that needs to be investigated thoroughly (Gunnarsson and Petersen, 2007). Increased biogas production can in turn reduce pressure on natural gas and oil consumption. Briquettes can be developed from the hyperaccumulator duckweeds to be used in cooking food materials. Converting contaminated duckweeds into charcoal (carbonization) and incineration (burning at high temperatures) are also good options for getting rid of hyperaccumulator plants and have limitation with respect to air pollution. Hydrolysis combined with fermentation can yield biofuels from contaminated duckweeds. All of these strategies greatly depend on the metal concentrations present in duckweed plants, and their application requires skilled operational protocols.

6. CONCLUSION AND FUTURE PROSPECTS

Duckweeds are small aquatic monocotyledonous macrophytes present in fresh water ecosystems. They have a worldwide distribution presentation of five genera and 38 species. They are commonly employed as sensitive organisms in ecotoxicological testing of wastewaters. From municipal, industrial, and agricultural wastewaters, they have successfully demonstrated the ability to remove nutrients, metals, algal, microbial, and other phytoplankton diversity. Members of the genera (*Lemna* and *Spiroldela*) are useful hyperaccumulators of heavy metals. Their sensitivity, uptake, and accumulation differ in the wastewaters because of the metal concentrations and exposure time. Based on their efficient uptake and accumulation potential, they are recommended for large-scale wastewater treatment systems. Duckweed plants containing high metal loads need proper management and safe disposal strategies (i.e., carbonization and incineration).

Because of duckweed's complete genome sequencing, considerable advances has been made in genetically modifying the *Lemnaceae* species to attain desirable traits beneficial for the agriculture, health, and environmental sectors (Yamamoto et al., 2001; Wang and Messing, 2015). The environmental sector has chiefly targeted accelerated cleanup of toxic metals and nutrients from municipal and industrial wastewaters. Genetically modified *Lemnaceae* plants focusing on specific nutrient and metal removal are the future concerns in this cutting-edge phytoremediation research. Metal-based tolerance and hyperaccumulation in duckweed species is supported by overexpression of antioxidant enzyme proteins (superoxide dismutase, peroxidase, etc.) and generation of stress compounds (proline and other organic acids). There is still a dearth of knowledge regarding molecular basis of sensitive metabolic machinery involved in physiological and biochemical responses that needs to be unraveled. Production of biofuels (biodegradable materials) using excess biomass generated by different species of this family provides a clean, sustainable, and ecofriendly solution to a global energy crisis is yet another excellent example of utility of these tiny monocot species. Besides duckweed's immense potential for energy production and phytoremediation, the *Lemnaceae* species are currently used in production of biopharmaceuticals and animal feed. Biopharmaceuticals production (i.e., proteins) will revolutionize the pharmaceutical industry and will nullify the usage of animal materials in production processes (Stomp, 2005). Continued research on *Lemnaceae* plants will unravel further applications of these tiny monocot species in health, biotechnology, environmental, and agriculture sectors.

REFERENCES

Abbas, M., Meharg, A., 2008. Arsenate arsenite and dimethyl arsinic acid (DMA) uptake and tolerance in maize (*Zea mays* L.). Plant Soil 304, 277–289.

Ali, Z., Ali, B., Mohammad, A., Ahmad, M., Ahmad, I., Napar, A.A., Kazi, A.G., Ali, A., Shah, S.S., Mujeeb-Kazi, A., 2013a. Combating water scarcity for global food security. In: Raza, A. (Ed.), Agricultural Systems in the 21st Century. Nova Science Publishers, Hauppauge, NY, pp. 1–30.

Ali, Z., Malik, R.N., Qadir, A., 2013b. Heavy metals distribution and risk assessment in soils affected by tannery effluents. Chem. Ecol. 29, 676–692.

Ali, Z., Malik, R.N., Shinwari, Z.K., Qadir, A., 2015. Enrichment risk assessment and statistical apportionment of heavy metals in tannery-affected areas. Int. J. Environ. Sci. Technol. 12, 537–550.

Alston, R.E., 1966. Chemotaxonomy or biochemical systematics? In: Swain, T. (Ed.), Comparative Phytochemistry. Academic Press, New York, pp. 33–56.

Ancell, M., 1998. Constructed Wetlands for Agricultural Wastewater Treatment. (MS) Dept of Civil Engineering, Texas Tech University.

Anderson, K., Lowman, Z., Stomp, A., Chang, J., 2011. Duckweed as a feed ingredient in laying hen diets and its effect on egg production and composition. Int. J. Poult. Sci. 10, 4–7.

Appenroth, K., Adamec, L., 2015. Specific turion yields of different clones of *Spirodela polyrhiza* depend on external phosphate thresholds. Plant Biol. 17, 125–129.

Appenroth, K., Krech, K., Keresztes, Ã., Fischer, W., Koloczek, H., 2010. Effects of nickel on the chloroplasts of the duckweeds *Spirodela polyrhiza* and *Lemna minor* and their possible use in biomonitoring and phytoremediation. Chemosphere 78, 216–223.

ATSDR, 1992. Toxicological Profile for Aluminium. US Department of Health and Human Services, Public Health Service Agency for Toxic Substances and Disease Registry, Atlanta, GA (TP-91/01).

Axtell, N., Sternberg, S., Claussen, K., 2003. Lead and nickel removal using *Microspora* and *Lemna minor*. Bioresour. Technol. 89, 41–48.

Babu, T., Akhtar, T., Lampi, M., Tripuranthakam, S., Dixon, D., Greenberg, B., 2003. Similar stress responses are elicited by copper and ultraviolet radiation in the aquatic plant *Lemna gibba*: implication of reactive oxygen species as common signals. Plant Cell Physiol. 44, 1320–1329.

Blinová, L., Bartošová, A., Sirotiak, M., 2014. Unconventional type of biomass suitable for the production of biofuels. Adv. Mater. Res. 860–863, 514–517.

Böcük, H., Yakar, A., Türker, O., 2013. Assessment of *Lemna gibba* L. (duckweed) as a potential ecological indicator for contaminated aquatic ecosystem by boron mine effluent. Ecol. Indic. 29, 538–548.

Borisjuk, N., Chu, P., Gutierrez, R., Zhang, H., Acosta, K., Friesen, N., 2015. Assessment validation and deployment strategy of a two-barcode protocol for facile genotyping of duckweed species. Plant Biol. 17, 42–49.

Butterwick, L., de Oude, N., Raymond, K., 1989. Safety assessment of boron in aquatic and terrestrial environments. Ecotoxol. Environ. Saf. 17, 339–371.

Cheng, J., Stomp, A., 2009. Growing duckweed to recover nutrients from wastewaters and for production of fuel ethanol and animal feed. Clean Soil Air Water 37, 17–26.

Classen, J.J., Cheng, J., Bergmann, B.A., Stomp, A.M., 2000. *Lemna gibba* growth and nutrient uptake in response to different nutrient levels. In: Animal Agricultural and Food Processing Waste: Proceedings of 8th International Symposium October 9–11, 2000, Des Moines, Iowa.

Crawford, D.J., Landolt, E., Les, D.H., Tepe, E., 1997. Allozyme variation and the taxonomy of *Wolffiella* (*Lemnaceae*). Aquat. Bot. 58, 43–54.

Culley, D.D., Rejmánková, E., Kvet, J., Frye, J.B., 2009. Production chemical quality and use of duckweeds (*Lemnaceae*) in aquaculture waste management and animal feeds. J. World Aquacult. Soc. 12, 27–49.

Cusimano, N., Bogner, J., Mayo, S., Boyce, P., Wong, S., Hesse, M., 2011. Relationships within the Araceae: comparison of morphological patterns with molecular phylogenies. Am. J. Bot. 98, 654–668.

Das, P., Samantaray, S., Rout, G., 1997. Studies on cadmium toxicity in plants: a review. Environ. Pollut. 98, 29–36.

Daubs, E.H., 1965. A Monograph of Lemnaceae. In: Illinois Biological Monographs, first ed., vol. 34. The University of Illinois Press, Urbana.

Davis, S., Drake, K., Maier, K., 2002. Toxicity of boron to the duckweed *Spirodella polyrrhiza*. Chemosphere 48, 615–620.

Delmail, D., Labrousse, P., 2014. Heavy-metal attack on freshwater side: physiological defense strategies of macrophytes and ecotoxicological ops. In: Ahmad, P., Wani, M.R. (Eds.), Physiological Mechanisms and Adaptation Strategies in Plants Under Changing Environment. Springer Publishing, pp. 31–54.

Dewedar, A., 1995. Fate of faecal coliform bacteria in a waste water retention reservoir containing *Lemna gibba* L. Water Res. 29, 2598–2600.

Drost, W., Matzke, M., Backhaus, T., 2007. Heavy metal toxicity to *Lemna minor*: studies on the time dependence of growth inhibition and the recovery after exposure. Chemosphere 67, 36–43.

Duman, F., Ozturk, F., Aydin, Z., 2010. Biological responses of duckweed (*Lemna minor* L.) exposed to the inorganic arsenic species as (III) and as (V): effects of concentration and duration of exposure. Ecotoxicology 19, 983–993.

Eick, M.J., Peak, J.D., Brady, P.V., Pesek, J.D., 1999. Kinetics of lead absorption/desorption on goethite: residence time effect. Soil Sci. 164, 28–39.

El-Kheir, W., Ismail, G., El-Nour, F., Tawfik, T., Hammad, D., 2007. Assessment of the efficiency of duckweed (*Lemna gibba*) in wastewater treatment. Int. J. Agric. Biol. 9, 681–687.

Falabi, J., Gerba, C., Karpiscak, M., 2002. *Giardia* and *Cryptosporidium* removal from waste-water by duckweed (*Lemna gibba* L.) covered pond. Lett. Appl. Microbiol. 34, 384–387.

Fargasová, A., 1994. Effect of Pb Cd Hg As and Cr on germination and root growth of *Sinapis alba* seeds. Bull. Environ. Contam. Toxicol. 52, 452–456.

Farid, M., Irshad, M., Fawad, M., Ali, Z., Eneji, A., Aurangzeb, N., Mohammad, A., Ali, B., 2014. Effect of cyclic phytoremediation with different wetland plants on municipal wastewater. Int. J. Phytorem. 16, 572–581.

Fodor, F., 2002. Physiological responses of vascular plants to heavy metals. In: Prasad, M.N.V., Strzatka, K. (Eds.), Physiology and Biochemistry of Metal Toxicity and Tolerance in Plants. Kluwer Academic Publisher, Netherlands, pp. 149–177.

Frick, H., 1985. Boron tolerance and accumulation in the duckweed *Lemna minor*. J. Plant Nutr. 8, 1123–1129.

Glandon, R., McNabb, C., 1978. The uptake of boron by *Lemna minor*. Aquat. Bot. 4, 53–64.

Godfrey, R., Wooten, J., 1979. Aquatic and Wetland Plants of Southeastern United States, first ed. University of Georgia Press, Athens.

Goulet, R., Lalonde, J., Munger, C., Dupuis, S., Dumont-Frenette, G., Prémont, S., Campbell, P., 2005. Phytoremediation of effluents from aluminum smelters: a study of Al retention in mesocosms containing aquatic plants. Water Res. 39, 2291–2300.

Greenberg, B., Huang, X., Dixon, D., 1992. Applications of the aquatic higher plant *Lemna gibba* for ecotoxicological assessment. Aquat. Ecosyst. Health Manage. 1, 147–155.

Greger, M., 1999. Metal availability and bioconcentration in plants. In: Prasad, M.N.V., Hagemeyer, J. (Eds.), Heavy Metal Stress in Plants-from Molecules to Ecosystems. Springer Press, Berlin, pp. 1–27.

Gunnarsson, C., Petersen, C., 2007. Water hyacinths as a resource in agriculture and energy production: a literature review. Waste Manage. 27, 117–129.

Gupta, C., Prakash, D., 2013. Duckweed: an effective tool for phyto-remediation. Toxicol. Environ. Chem. 95, 1256–1266.

Hammouda, O., Abdel-Hameed, M., 1994. Response of phytoplankton populations to aquatic treatment by *Lemna gibba*. Folia Microbiol. 39, 420–427.

Hammouda, O., Gaber, A., Abdel-Hameed, M., 1995. Assessment of the effectiveness of treatment of wastewater-contaminated aquatic systems with *Lemna gibba*. Enzym. Microb. Technol. 17, 317–323.

Hartog, C.D., 1975. Thoughts about the taxonomical relationships within the *Lemnaceae*. Aquat. Bot. 1, 407–416.

Harvey, R., Fox, J., 1973. Nutrient removal using *Lemna minor*. J. Water Pollut. Control Fed. 45, 1928–1938.

Hegelmaier, F., 1868. Die Lemnaceen Eine Monographische Untersuchung etc, first ed. Leipzig.

Hillman, William S., 1961. The *Lemnaceae* or duckweeds. A review of the descriptive and experimental literature. Bot. Rev. 27, 221–287.

Horemans, N., Van Hees, M., Van Hoeck, A., Saenen, E., De Meutter, T., Nauts, R., 2015. Uranium and cadmium provoke different oxidative stress responses in *Lemna minor* L. Plant Biol. 17, 91–100.

Horvat, T., Vidaković-Cifrek, Z., Orevsvcanin, V., Tkalec, M., Pevalek-Kozlina, B., 2007. Toxicity assessment of heavy metal mixtures by *Lemna minor* L. Sci. Total Environ. 384, 229–238.

Hou, W., Chen, X., Song, G., Wang, Q., Chi Chang, C., 2007. Effects of copper and cadmium on heavy metal polluted waterbody restoration by duckweed (*Lemna minor*). Plant Physiol. Biochem. 45, 62–69.

Ivanova, I., 1973. K sistematike semeistva *Lemnaceae* S Gray (On the taxonomy of *Lemnaceae* S Gray). Bot. Zhurn. 58, 1413–1423.

Jafari, N., Akhavan, M., 2011. Effect of pH and heavy metal concentration on phytoaccumulation of zinc by three duckweed species. Am. Eurasian J. Agric. Environ. Sci. 10, 34–41.

Jain, S., Gujral, G., Jha, N., Vasudevan, P., 1988. Heavy metal uptake by *Pleurotus* sajor-caju from metal-enriched duckweed substrate. Biol. Wastes 24, 275–282.

Jordan, W., Courtney, M., Neigel, J., 1996. Low levels of intraspecific genetic variation at a rapidly evolving chloroplast DNA locus in North American duckweeds (*Lemnaceae*). Am. J. Bot. 83, 430–439.

Kabata-Pendias, A., Pendias, H., 2001. Trace Elements in Soils and Plants. CRC Press, Washington, DC.

Kara, Y., 2004. Bioaccumulation of copper from contaminated wastewater by using *Lemna minor*. Bull. Environ. Contam. Toxicol. 72, 467–471.

Kara, Y., Basaran, D., Kara, I., Zeytunluoglu, A., Gencc, H., 2003. Bioaccumulation of nickel by aquatic macrophyta *Lemna minor* (duckweed). Int. J. Agric. Biol. 5, 281–283.

Kaur, L., Gadgil, K., Sharma, S., 2008. Phytoaccumulation of nickel from contaminated water by duckweed (*Lemna minor*). J. Environ. Res. Dev. 3, 388–394.

Khellaf, N., Zerdaoui, M., 2009. Phytoaccumulation of zinc by the aquatic plant Lemna. Bioresour. Technol. 100, 6137–6140.

Khellaf, N., Zerdaoui, M., 2010a. Growth photosynthesis and respiratory response to copper in *lemna minor*: a potential use of duckweed in biomonitoring. Iran. J. Environ. Health Sci. Eng. 7, 299–306.

Khellaf, N., Zerdaoui, M., 2010b. Growth response of the duckweed *Lemna gibba* L. to copper and nickel phytoaccumulation. Ecotoxicology 19, 1363–1368.

Körner, S., Vermaat, J., 1998. The relative importance of *Lemna gibba* L. bacteria and algae for the nitrogen and phosphorus removal in duckweed-covered domestic wastewater. Water Res. 32, 3651–3661.

Kutschera, U., Niklas, K., 2015. Darwin-Wallace Demons: survival of the fastest in populations of duckweeds and the evolutionary history of an enigmatic group of angiosperms. Plant Biol. 17, 24–32.

Lahive, E., O'Callaghan, M., Jansen, M., O'Halloran, J., 2011. Uptake and partitioning of zinc in *Lemnaceae*. Ecotoxicology 20, 1992–2002.

Lakatos, G., Mészáros, I., Bohátka, S., Szabó, S., Makádi, M., Csatlós, M., Langer, G., 1993. Application of *Lemna* species in ecotoxicological studies of heavy metals and organic biocides. Sci. Total Environ. 134, 773–778.

Landolt, E., 1986. Biosystematic Investigations in the Family of Duckweeds *Lemnaceae*, first ed. (Zürich).

Landolt, E., 1994. Taxonomy and ecology of the section *Wolffia* of the genus *Wolffia* (*Lemnaceae*). Ber. Geobot. Inst. Eidgenoss. Technol. Hochsch. Stift. Rubel 60, 137–151.

Landolt, E., 1998. *Lemna yungensis* a new duckweed species from rocks of the Andean Yungas in Bolivia. Bull. Geobot. Inst. ETH 64, 15–21.

Landolt, E., Kandeler, R., 1987. The family of *Lemnaceae*-a monographic study: phytochemistry physiology application and bibliography (vol. 2 of monograph). In: Biosystematic Investigations in the Family of Duckweeds (*Lemnaceae*), vol. 4, no 95. Publications of the Geobotanical Institute of the ETH, Stiftung Rubel, Zurich.

Les, D., Crawford, D., 1999. *Landoltia* (*Lemnaceae*) a new genus of duckweeds. Novon 9, 530–533.

Li, T., Xiong, Z., 2004. Cadmium-induced colony disintegration of duckweed (*Lemna paucicostata* Hegelm) and as biomarker of phytotoxicity. Ecotoxol. Environ. Saf. 59, 174–179.

Maier, K., Knight, A., 1991. The toxicity of waterborne boron to *Daphnia magna* and *Chironomus decorus* and the effects of water hardness and sulfate on boron toxicity. Arch. Environ. Contam. Tox. 20, 282–287.

Marschner, H., 1995. Mineral Nutrition of Higher Plants, first ed. Academic Press, London.

Miranda, G., Quiroz, A., Salazar, M., 2000. Cadmium and lead removal from water by the duckweed *Lemna gibba* L. (*Lemnaceae*). Hidrobiologica (Iztapalapa) 10, 7–12.

Mkandawire, M., Dudel, E., 2005. Accumulation of arsenic in *Lemna gibba* L. (duckweed) in tailing waters of two abandoned uranium mining sites in Saxony Germany. Sci. Total Environ. 336, 81–89.

Mkandawire, M., Lyubun, Y., Kosterin, P., Dudel, E., 2004. Toxicity of arsenic species to *Lemna gibba* L. and the influence of phosphate on arsenic bioavailability. Environ. Toxicol. 19, 26–34.

Mohan, B., Hosetti, B., 1997. Potential phytotoxicity of lead and cadmium to *Lemna minor* grown in sewage stabilization ponds. Environ. Pollut. 98, 233–238.

Moss, M., 1999. Economics and Feed Value of Integrating Duckweed Production with a Swine Operation. (MS) Texas Tech University.

Mrak, T., Slejkovec, Z., Jeran, Z., Jaćimović, R., Kastelec, D., 2008. Uptake and biotransformation of arsenate in the lichen *Hypogymnia physodes* (L.) Nyl. Environ. Pollut. 151, 300–307.

Obek, E., 2009. Bioaccumulation of heavy metals from the secondary treated municipal wastewater by *Lemna gibba*. Fresen. Environ. Bull. 18, 2159–2164.

Obek, E., Sasmaz, A., 2011. Bioaccumulation of aluminum by *Lemna gibba* L. from secondary treated municipal wastewater effluents. Bull. Environ. Contam. Toxicol. 86, 217–220.

Ol'shanskaya, L., Sobgaida, N., Tarushkina, Y., Stoyanov, A., 2008. Effect of a magnetic field on extraction of heavy metals from waste water with duckweed. Chem. Pet. Eng. 44, 475–479.

Oron, G., Devegt, A., Porath, D., 1988. Nitrogen removal and conversion by duckweed grown on waste-water. Water Res. 22, 179–184.

Oron, G., Porath, D., Wildschut, L., 1986. Wastewater treatment and renovation by different duckweed species. J. Environ. Eng. 112, 247–263.

Palmer, C., 1969. A composite rating of algae tolerating organic pollution. J. Phycol 5, 78–82.

Panda, S., 2008. Impact of copper on reactive oxygen species lipid peroxidation and antioxidants in *Lemna minor*. Biol. Plant. 52, 561–564.

Pandey, M., September–October 2001. Duckweed–based Wastewater Treatment. Central Pollution Control Board Delhi Entrepreneur Invention Intelligence.

Perniel, M., Ruan, R., Martinez, B., 1998. Nutrient removal from a stormwater detention pond using duckweed. Appl. Eng. Agric. 14, 605–609.

Prasad, M., Malec, P., Waloszek, A., Bojko, M., Strzalka, K., 2001. Physiological responses of *Lemna trisulca* L. (duckweed) to cadmium and copper bioaccumulation. Plant Sci. 161, 881–889.

Radić, S., Babić, M., Skobić, D., Roje, V., Pevalek-Kozlina, B., 2010. Ecotoxicological effects of aluminum and zinc on growth and antioxidants in *Lemna minor* L. Ecotoxicol. Environ. Saf. 73, 336–342.

Rahman, M., Hasegawa, H., Ueda, K., Maki, T., Okumura, C., Rahman, M., 2007. Arsenic accumulation in duckweed (*Spirodela polyrhiza* L.): a good option for phytoremediation. Chemosphere 69, 493–499.

Ran, N., Agami, M., Oron, G., 2004. A pilot study of constructed wetlands using duckweed (*Lemna gibba* L.) for treatment of domestic primary effluent in Israel. Water Res. 38, 2241–2248.

Razinger, J., Dermastia, M., Drinovec, L., Drobne, D., Zrimec, A., Koce, J., 2007. Antioxidative responses of duckweed (*Lemna minor* L.) to short-term copper exposure. Environ. Sci. Pollut. Res. Int. 14, 194–201.

Reed, S.C., Middlebrooks, E.J., Crites, R.W., 1988. Natural Systems for Waste Management and Treatment. McGraw-Hill, New York.

Reeves, R.D., Baker, A.J.M., 2000. Metal-accumulating plants. In: Raskin, I., Ensley, B.D. (Eds.), Phytoremediation of Toxic Metals: Using Plants to Clean-Up the Environment. Wiley, New York, pp. 193–230.

Sasmaz, A., Obek, E., 2012. The accumulation of silver and gold in *Lemna gibba* L. exposed to secondary effluents. Chem. Erde Geochem. 72, 149–152.

Saygideger, S., Keser, G., Dogan, M., 2013. Effects of lead on chlorophyll content total nitrogen and antioxidant enzyme activities in duckweed (*Lemna minor*). Int. J. Agric. Biol. 15, 145–148.

Schützendübel, A., Polle, A., 2002. Plant responses to abiotic stresses: heavy metal-induced oxidative stress and protection by mycorrhization. J. Exp. Bot. 53, 1351–1365.

Schleiden, M.J., 1839. Prodromous monographiae Lemnacearum oder Conspectus generum atque specierum. Linnaea 13, 385–392.

Sharma, S., Gaur, J., 1995. Potential of *Lemna polyrrhiza* for removal of heavy metals. Ecol. Eng. 4, 37–43.

Sharma, V., Sohn, M., 2009. Aquatic arsenic: toxicity speciation transformations and remediation. Environ. Int. 35, 743–759.

Skillicorn, P., Journey, W., Spira, W., 1993. Duckweed Aquaculture, first ed. World Bank, Washington, DC.

Stephan, C., Courchesne, F., Hendershot, W., McGrath, S., Chaudri, A., Sappin-Didier, V., Sauvé, S., 2008. Speciation of zinc in contaminated soils. Environ. Pollut. 155, 208–216.

van der Steen, P., Brenner, A., Shabtai, Y., Oron, G., 2000. Improved fecal coliform decay in integrated duckweed and algal ponds. Water Sci. Technol. 42, 363–370.

van der Steen, P., Brenner, A., Van Buuren, J., Oron, G., 1999. Post-treatment of UASB reactor effluent in an integrated duckweed and stabilization pond system. Water Res. 33, 615–620.

Stomp, A.M., 2005. The duckweeds: a valuable plant for biomanufacturing. Biotechnol. Annu. Rev. 11, 69–99.

Tang, J., Zhang, Y., Cui, Y., Ma, J., 2015. Effects of a rhizobacterium on the growth of and chromium remediation by *Lemna minor*. Environ. Sci. Pollut. Res. http://dx.doi.org/10.1007/s11356-015-4138-y.

Thompson, C., 1898. A revision of the American *Lemnaceae* occurring north of Mexico. Annu. Rep. Mo. Bot. Gard. 9, 21–42.

Tippery, N., Les, D., Crawford, D., 2015. Evaluation of phylogenetic relationships in *Lemnaceae* using nuclear ribosomal data. Plant Biol. 17, 50–58.

Tkalec, M., Prebeg, T., Roje, V., Pevalek-Kozlina, B., Ljubevsić, N., 2008. Cadmium-induced responses in duckweed *Lemna minor* L. Acta Physiol. Plant 30, 881–890.

Ugya, A., 2015. The efficiency of *Lemna minor* L. in the phytoremediation of romi stream: a case study of Kaduna refinery and petrochemical company polluted stream. J. Appl. Biol. Biotechnol. 3, 11–14.

Uysal, Y., Taner, F., 2009. Effect of pH temperature and lead concentration on the bioremoval of lead from water using *Lemna minor*. Int. J. Phytorem. 11, 591–608.

Uysal, Y., Taner, F., 2011. The evaluation of the Pb (II) removal efficiency of duckweed *Lemna minor* (L.) from aquatic mediums at different conditions. In: Gokcekus, H., et al. (Eds.), Survival and Sustainability, Environ Earth Sciences. Springer-Verlag, pp. 1107–1116.

Visoottiviseth, P., Francesconi, K., Sridokchan, W., 2002. The potential of Thai indigenous plant species for the phytoremediation of arsenic contaminated land. Environ. Pollut. 118, 453–461.

Vitória, A., Da Cunha, M., Azevedo, R., 2006. Ultrastructural changes of radish leaf exposed to cadmium. Environ. Exp. Bot. 58, 47–52.

Wang, W., 1986. Toxicity tests of aquatic pollutants by using common duckweed. Environ. Pollut. Ser. B Chem. Phys. 11, 1–14.

Wang, W., 1987. Factors affecting metal toxicity to (and accumulation by) aquatic organisms-overview. Environ. Int. 13, 437–457.

Wang, W., Messing, J., 2015. Status of duckweed genomics and transcriptomics. Plant Biol. 17, 10–15.

Watanabe, M., 1997. Phytoremediation on the brink of commercialization. Environ. Sci. Technol. 31, 182A–186A.

Xu, Q., Hu, J., Xie, K., Yang, H., Du, K., Shi, G., 2010. Accumulation and acute toxicity of silver in *Potamogeton crispus* L. J. Hazard. Mater. 173, 186–193.

Xu, Y., Ma, S., Huang, M., Peng, M., Bog, M., Sree, K., 2015. Species distribution genetic diversity and barcoding in the duckweed family (*Lemnaceae*). Hydrobiologia 743, 75–87.

Yamamoto, Y.T., Rajbhandari, N., Lin, X., Bergmann, B.A., Nishimura, Y., Stomp, A.M., 2001. Genetic transformation of duckweed *Lemna gibba* and *Lemna minor*. In Vitro Cell. Dev. Biol Plant 37, 349–353.

Yin, Y., Yu, C., Yu, L., Zhao, J., Sun, C., Ma, Y., Zhou, G., 2015. The influence of light intensity and photoperiod on duckweed biomass and starch accumulation for bioethanol production. Bioresour. Technol. 187, 84–90.

Zayed, A., Gowthaman, S., Terry, N., 1998. Phytoaccumulation of trace elements by wetland plants: I. Duckweed. J. Environ. Qual. 27, 715–721.

Zhang, X., Zhao, F., Huang, Q., Williams, P., Sun, G., Zhu, Y., 2009. Arsenic uptake and speciation in the rootless duckweed *Wolffia globosa*. New Phytol. 182, 421–428.

Zhao, Z., Shi, H., Duan, D., Li, H., Lei, T., Wang, M., 2015a. The influence of duckweed species diversity on ecophysiological tolerance to copper exposure. Aquat. Toxicol. 164, 92–98.

Zhao, Y., Fang, Y., Jin, Y., Huang, J., Bao, S., Fu, T., 2015b. Pilot-scale comparison of four duckweed strains from different genera for potential application in nutrient recovery from wastewater and valuable biomass production. Plant Biol. 17, 82–90.

Zhou, Q., Zhang, J., Fu, J., Shi, J., Jiang, G., 2008. Biomonitoring: an appealing tool for assessment of metal pollution in the aquatic ecosystem. Anal. Chim. Acta 606, 135–150.

Ziegler, P., Adelmann, K., Zimmer, S., Schmidt, C., Appenroth, K., 2015. Relative in vitro growth rates of duckweeds (*Lemnaceae*) – the most rapidly growing higher plants. Plant Biol. 17, 33–41.

Zimmo, O., 2003. Nitrogen Transformations and Removal Mechanisms in Algal and Duckweed Waste Stabilization Ponds (Ph.D. thesis). International Institute for Infrastructural Hydraulic and Environmental Engineering, Delft, Netherlands.

Zimmo, O., van der Steen, N., Gijzen, H., 2005. Effect of organic surface load on process performance of pilot-scale algae and duckweed-based waste stabilization ponds. J. Environ. Eng. 131, 587–594.

Zirschky, J., Reed, S., 1988. The use of duckweed for wastewater treatment. J. Water Pollut. Control Fed. 60, 1253–1258.

CHAPTER

GENETIC STRATEGIES FOR ADVANCING PHYTOREMEDIATION POTENTIAL IN PLANTS: A RECENT UPDATE

17

Sumira Jan[1], Bushra Rashid[2], M.M. Azooz[3,4], Mohammad Anwar Hossain[5], Parvaiz Ahmad[6]

Department of Biotechnology, ICAR – Central Institute of Temperate Horticulture, Srinagar, Jammu and Kashmir, India[1]; Center of Excellence in Molecular Biology, University of the Punjab, Lahore, Pakistan[2]; Department of Botany, Faculty of Science, South Valley University, Qena, Egypt[3]; Department of Biological Sciences, Faculty of Science, King Faisal University, Al-Hassa, Saudi Arabia[4]; Department of Genetics and Plant Breeding, Bangladesh Agricultural University, Mymensingh, Bangladesh[5]; Department of Botany, S.P. College, Srinagar, Jammu and Kashmir, India[6]

CHAPTER OUTLINE

1. INTRODUCTION

Phytoremediation represents an aesthetically agreeable, eco-friendly, and lucrative approach for chemical contamination by organics such as polycyclic aromatic hydrocarbons, herbicides, and synthetic explosives (trinitrotoluene, TNT, and Research Department explosive) or inorganic such as

Plant Metal Interaction. http://dx.doi.org/10.1016/B978-0-12-803158-2.00017-5

heavy metals mercury (Hg), lead (Pb), cadmium (Cd), selenium (Se), and manganese (Mn), among others (Abdel-Salem et al., 2015) in soils and waters. Phytoremediation technologies are emerging as cost-effective methods for remediating sites tainted with toxic metals at a fraction of the cost of conventional technologies, such as soil replacement, solidification, and washing strategies (Ali et al., 2013; Dixit et al., 2015).

Uranium concentration reduces to 350 ppm by 95% using raft-riding sunflower within 24 h (Phieler et al., 2015). This treatment proved to be economical costing at rate of $2–6$/1000 gallons of water, which is quite a bit less than microfiltration, which costs about $80/1000 gallons of water. The issues that interrogate the feasibility of phytoremediation techniques include time frame and depth limitations, the bio-amplification, atmospheric volatilization of contaminants, and doubts concerning the ultimate disposal of contaminated plants (Chen et al., 2015). Phytoremediation is more time-consuming than other technologies, but the most appropriate limitation is its applicability to the contaminants present at shallow levels within the root layer (Mani et al., 2014). Plants have an innate capability of cleaning hazardous contaminants from the environment, but the rate of bioremediation is directly proportional to plant growth rate and the biomass productivity, making the process very time-consuming (Sheron et al., 2011). Efficacy of plants in phytoremediation can be limited by processes such as the availability of toxic metal ions in the soil solution for root uptake, their rate of translocation from roots to shoots, the level of tolerance, and the rate of chemical conversion into less contaminated compounds (Prasad, 2015). Plants suitable for phytoextraction should possess exceptional qualities such as high capacities for absorption, root/shoot translocation, and detoxification of the metals or nonmetals to be extracted, and, last but not least, a high biomass and preferably a rapid growth rate. Most phytoremediative plants exhibit constrained vegetative traits like limited rooting (*Zea mays* L., *Eichhornia azurea* Kunnth, *Festuca rubra* L., *Silene vulgaris* (Moench) Garcke, *Amaranthus retroflexus* L., *Thlaspi caerulescens* L., *Helianthus annuus* L., *Cucurbita pepo*, and *Brassica juncea* L. plants), resulting in poor translation to cleanup depths and inadequate absorption capacity (Liao et al., 2015; Evangelou et al., 2015). Another potential problem associated with phytoremediation is bio-amplification of heavy metals in food chain and their atmospheric volatilization from leaf structures in the form of more volatile and toxic entity (e.g., genetically engineered strain of the yellow poplar *Liriodendron tulipifera* L.) (Weyens et al., 2015). These plants exhibit a high growth rate and produce an enzyme capable of reducing highly toxic Hg ions to insoluble Hg and highly volatile metal through bacterial gene (Mello-Farias et al., 2011).

The improvement in phytoremediation potential of plants can be attained by the amendment or variation in primary and secondary metabolism via the creation of new phenotypic and genotypic traits. For these reasons, researchers have endeavored to engineer plants with genes that can confer superior degradation abilities. A direct method for enhancing the efficacy of phytoremediation is to overexpress the genes involved in metabolism, uptake, or transport and vacuolar sequestration, and chelation of specific pollutants (Chen et al., 2013; Zhang et al., 2013). This necessitates the identification of a fast-growing (largest potential biomass and greatest nutrient responses) and more strongly metal-accumulating genotypes. Potential targets that have been or might be manipulated to improve transport, sequestration, or detoxification processes have been reviewed (Lee, 2013; Dhan et al., 2014). Successes in genetically modifying plants for phytoextraction of Cd, Pb, Se, Hg, and arsenic (As) have been developed (Zhao et al., 2012; Hunt et al., 2014; Wei et al., 2015). Thus, the present chapter will discuss the landmark laid by genetic engineering and different futuristic efforts that need to be focused for enhanced efficiency of phytoremediation. These efforts will include both boosting of existing genes involved in metal uptake, absorption processes, and introduction of novel genes from other organisms, resulting in improved phytoremediation.

2. GENETIC ENGINEERING AND PHYTOREMEDIATION

Genetic engineering is not a single step disciple; hence, phytoremediation needs to be improved under a framework of different criteria as follows.

2.1 SELECTION OF PLANT SPECIES FOR PHYTOREMEDIATION

A plant suitable for phytoremediation should possess the following characteristics: (1) ability to accumulate and tolerate metals, (2) fast growth and highly "effective" (i.e., metal accumulating) biomass, and (3) easily harvestable (Zhang et al., 2006). Though such plants with the highest tolerance and accumulation capacities are available in natural habitats such as *Liriodendron tulipifera*, *B. juncea*, *Nicotiana glaucum*, and *H. annuus*, but they usually exhibit low biomass and having constrained climatic adaptability (Jesus et al., 2015). These high biomass metallophytes with well-established genetic manipulation procedures appear to be good candidates for enhanced phytoremediation potential (Table 17.1). On the contrary, fast-growing woody plants like poplar display quite a high transformation frequency (Castro-Rodríguez et al., 2015). However, the production of transgenic lines exhibiting high biomass productivity with quite a high growth rate is difficult to be created because

Table 17.1 Hyperaccumulation of Trace Elements in Land Plants

Element	Critical Toxicity Level (μg/g)	Hyperaccumulation Concentration Criterion (μg/g)	Plants	References
Antimony	<2	230 mg/kg, 1516.5 mg/kg	*Pteris cretica*, *Pteris vittata*	Feng et al. (2011) and Muller et al. (2013)
Arsenic	<2–80	23,000	*Pteris vittata*	DiTusa et al. (2015)
Cadmium	6–10	21.46–218.9	*Noccaea caerulescens*	Simmons et al. (2015)
Cobalt	0.4–several	2325	*Lemna minor*	Sree et al. (2015)
Copper	20–30	12,300 mg/kg	*Alyssum bracteatum*	Ghasemi et al. (2015)
Lead	0.6–28	3611	*Hirschfeldia incana*	Fahr et al. (2015)
Manganese	200–3500	32,000	*Celosia argentea*	Liu et al. (2014)
Nickel	10–50	3916 mg/kg	*Psychotria grandis*	McAlister et al. (2015)
Selenium	3–100	14,900 mg/kg	*Stenyl pinnata*	Malagoli et al. (2015)
Thallium	20		*Biscutela laevigata*	Poščić et al. (2015)
Zinc	100–300	19,600 mg/kg	*Arabidopsis halleri*	Cornu et al. (2014)
			Noccaea caerulescens	Dinh et al. (2015)
Chromium	25,900 mg/kg	4825 mg/kg	*Cosmos bipinnatus*	Santiago-Cruz et al. (2014)
Mercury	0.5 ppm	327 mg/kg	*Eichhornia crassipes*	Malar et al. (2014)
Uranium	0.052 ppm	280 and 750 mg/kg	*Acacia albida* and *A. nelotica*	Fathi et al. (2014)

these characteristics are controlled by multigenes. So, it is more realistic to transfer high biomass plants to metal-hyperaccumulating genes controlled by single gene. Improvement of hyperaccumulator biomass is also visualized by the genetic manipulation of plant hormone levels (Iqbal and Ahemad, 2015).

2.2 METAL-TOLERANT MICROBIAL ASSOCIATION FOR IMPROVING TOLERANCE

Another avenue of recent research is the association of metal tolerant microorganisms like bacteria or fungi to enhance tolerance. In this alternative, the genetic engineering of plant-associated bacteria seems promising (Rajkumar et al., 2012). Microbes and nontolerant species like *Arabidopsis* have proven as reservoir of useful genes that inferred increased tolerance or better detoxification mechanism (Muehe et al., 2015). However, the insights into total mechanism underlying metal detoxification are still unclear, which slows the progress. Some functional genes in bacteria, fungi, plants, and animals have been cloned and expressed in various plants mostly via particle bombardment and *Agrobacterium tumefaciens*–mediated transformation. Goto et al. (2000) produced transgenic tobacco, lettuce, and rice plants via *A. tumefaciens*–mediated transformation to improve iron content in food crops. Bizily et al. (1999) studied the organic Hg resistance of *Arabidopsis thaliana* transferred by *merA* and *merB* and their mutant via *A. tumefaciens*–based transformation. LeDuc et al. (2004) overexpressed the gene encoding Se-Cys methyltransferase (SMT) from the *Astragalus bisulcatus* in *A. thaliana* and Indian mustard via *A. tumefaciens*–mediated transformation. Extensive work has been carried out to transfer plants with cationic transporter genes via *A. tumefaciens*–mediated transformation. van der Zaal et al. (1999) reported that a transgenic plant with *ZAT* (encoding zinc (Zn) transporter of *A. thaliana*) under control of the CaMV 35S promoter exhibited improved Zn resistance and sturdily augmented Zn content in the roots under 200 mol/L Zn exposure.

2.3 SELECTION CRITERIA OF TARGET GENES INVOLVED IN PHYTOREMEDIATION

The preference of genes to be selected for manipulation is based on its involvement in three major aspects: metal ion transportation, proteins involved in metal chelation, and metabolic enzymes involved in detoxification of metal contaminant (Hossain et al., 2012). Microbes serve as best source with known metal resistance and accumulation mechanism (Xie et al., 2015). However, the isolation of desired gene from microbes is quite complicated because metal resistance is a polygenic trait encoded by plasmid-containing operon (Doty, 2008). Table 17.2 illustrates some potential candidate genes discussed with strategies for generating hyperaccumulators using genetic engineering.

Heavy metal tolerance is a polygenic trait. Metal accumulation and tolerance are independent traits, which infer manipulation of several genes will be required to enhance phytoremediation potential (Singh et al., 2015). The quantitative trait loci (QTL) analysis in *T. caerulescens* ($2n = 14$), which is a natural Zn, Cd, and Ni hyperaccumulator, confirmed that metal accumulation, except for Cd concentration in shoot, are controlled by more than one gene. However, Cd tolerance seems to be a constitutive species-level trait in *Arabidopsis halleri* sp. *halleri* (Rascio and Navari-Izzo, 2011; Hossain et al., 2012). Characterization of native promoters of hyperaccumulators using novel tools will help in organizing the time, location, and transgenic expression level. Hyperaccumulators exhibit constitutive overexpression of genes as regular response to metalliferous environments, but similar gene expression when induced in nonaccumulator plant species lead to poor performance (Kühnlenz et al., 2015).

Table 17.2 Metallothionein (MT) Genes Identified and Their Function in Metal Accumulation and Tolerance

MT Genes	Sources	Target Plant Species	Effects	References
OsMT1a and *OsMT2b*	*Oryza sativa*	*Oryza sativa*	Plays roles in Zn homeostasis and enhance drought tolerance transgenic rice	Yang et al. (2009)
AmMT2	*Avicennia marina* L.	*Oryza sativa*	Enhances Zn, Cd, Cu, and Pb tolerances	Hunag and Wang (2010)
MT1 *MT2* *PaMT1*, *PaMT2*, *PaMT3*, *PaSPDS1*, *PaSPDS2* and *PaADC*	*Populus alba* L.	*Populus alba* L.	Cd, Cu, and Zn induced tolerance	Macovei et al. (2010)
IbMT1, type 2 (*IbMT2*), and type 3 (*IbMT3*)	*Ipomoea batatas*	*Escherichia coli*	Overexpressing IbMT1 protein exhibited enhanced tolerance to environmental stresses and heavy metals	Kim et al. (2014)
SaMT2	*Sedum alfredii*	Yeast Tobacco	Enhance Cd and Zn tolerance and accumulation in transgenic tobacco plants by chelating metals and improving antioxidant system	Zhang et al. (2014)
HbMT2a	*Hevea brasiliensis*	*Escherichia coli*	Enhances Cu and Cd tolerance	Li et al. (2015)
BnMP1	*Brassica napus*	*Brassica napus* L.	Cr tolerance under S-deficient condition	Terzi and Yıldız (2015)
OsTCTP	*Oryza sativa* L.	*Oryza sativa*	Hg stress tolerance	Xu et al. (2015)
HaMT1-2 *HaMT2-1*	*Escherichia coli*	*Helianthus annuus*	Cd detoxification	Tomas et al. (2015)

Cd, cadmium; Co, cobalt; Cu, copper; Pb, lead; Zn, zinc.

Though extensive works have been done on constitutive expression of 35S CaMV promoter in *A. thaliana*, precise and keen targeting of genes that encode transporters or enzymes involved in metal chelating would be most effective in extensive metal accumulation in harvestable parts. Transgenic plants with improved metal uptake and/or sequestration have been developed for Cd, Zn, Pb, Hg, As, and Se. Hyperaccumulators present an exceptional gene tank for novel insights into molecular mechanisms underlying their natural inclination toward extreme metallic environment. Using transcriptomic studies, hyperaccumulators provided the entrance into identification of gamut of genes, which overexpressed in extreme metallic environment, thus are involved in hyperaccumulation trait

(Craciun et al., 2012). Proteome and DNA array technologies have been used for gene probing and protein induction under metal stress, which later played significant role in metal hyperaccumulation and tolerance. Some of potential target genes that can be manipulated for enhanced phytoremediation are summarized in Table 17.3.

Table 17.3 Transporter Gene and Proteins Involved in Metal Tolerance and Their Function

Transporter Genes/ Enzymes	Sources	Target Plant Species	Effects	References
Zip transporters *ZAT1, ZAT2, ZAT3, ZAT* *ZNT1* *ZNT5* *ZnT2* *MTP 1*	*Arabidopsis thaliana, Thlaspi caerulescens*	Yeast	Fe and Cd toxicity and Zn tolerance	Kupper and Kochian (2010)
OsLCT1 *OsIRT1* *OsIRT2*	*A. thaliana*	Yeast	Cd as well as Fe accumulation	Uraguchi and Fujiwara (2012)
Mercuric ion binding protein (MerP) *merP* *Mercury transport protein* *merF*	*Gram+ve bacteria*	*A. thaliana*	High tolerance and accumulation capacity for Hg, Cd, and Pb	Hsieh et al. (2009)
P_{1B}ATPase *AtHMA4*	*A. thaliana*	Tobacco	Overexpression of AtHMA4 improved the root growth in the presence of toxic concentrations of Zn, Cd, and Co	Siemianowski et al. (2011)
Zip transporters *ZAT1, ZAT2, ZAT3, ZAT* *ZNT1* *ZNT5* *ZnT2* *MTP 1*	*A. thaliana, T. caerulescens*	Yeast	Fe and Cd toxicity and Zn tolerance	Kupper and Kochian (2010)
TgMTP1	*Thlaspi goesingense*		Zn hyperaccumulation	Gustin et al. (2009)
OsLCT1 *OsIRT1* *OsIRT2*	*A. thaliana*	Yeast	Cd as well as Fe accumulation	Uraguchi and Fujiwara (2012)

Table 17.3 Transporter Gene and Proteins Involved in Metal Tolerance and Their Function—cont'd

Transporter Genes/ Enzymes	Sources	Target Plant Species	Effects	References
CsHMA5.1 and *CsHMA5.2*	*Cucumus sativus* L.	Yeast	Increased activity of ATP-dependent Cu transport into tonoplast vesicles	Migocka et al. (2015)
CsMTP1 and *CsMTP4*	*Cucumus sativus* L.	Yeast	Elevated Zn and Cd tolerance and Zn homeostasis	Migocka et al. (2014)
OsHMA3	*Oryza sativa*	*Oryza sativa*	Cd sequestration Upregulation of Zn transporters	Sasaki et al. (2014)
SbMATE *FRD3*	*Sorghum*	*Hordeum vulgare*	Aluminum tolerance	Zhou et al. (2014)
NcNramp1	*A. thaliana*	*Noccaea caerulescens*	Cd tolerance	Milner et al. (2014)
AtABCC3 (abcc3)	Yeast	*A. thaliana*	Increased Cd tolerance	Brunetti et al. (2015)

ATP, adenosine triphosphate; Cd, cadmium; Co, cobalt; Cu, copper; Fe, iron; Hg, mercury; Pb, lead; Zn, zinc.

2.3.1 Metallothioneins, Phytochelatins, and Metal Chelators

Cloning of metallothionein (MT) genes and their introduction into suitable plant species can result in elevated metal tolerance. Tobacco and oil seeds with enhanced Cd tolerance were evolved through transfer of human MT-2 gene, whereas MT-gene transfer from pea resulted in copper (Cu) tolerance in tobacco and *A. thaliana* (Kumar et al., 2012). Expression of MT gene and elevated transcription of genes for glutathione (GSH) synthesis has been correlated with Cu and Cd tolerance in *A. thaliana*. Constitutive expression of various MT genes like MTI, human MTIA, human MTII, yeast CUPI, pea *Ps*MT A, and *TaMT3*, when transferred to *Nicotiana tabacum, Brassica* sp., and *A. thaliana* resulted in elevated Cd tolerance, as demonstrated in Table 17.2 (Zhou et al., 2014). Expression of MT gene and elevated transcription of genes of GSH synthesis has been correlated with Cu and Cd tolerance in *A. thaliana* (Yan et al., 2015). *Arabidopsis thaliana* displayed enhanced Cu and aluminum tolerance when gene-encoding GSH S-transferase was transferred from tobacco (Ezaki et al., 2000). Recent studies also showed that transgenic plants overexpressing *BcMT1* and *BcMT2* gene from *Brassica campestris* in *Arabidopsis* enhances Cd and Cu stress tolerance by increasing heavy metal concentrations and lower reactive oxygen species (ROS) accumulation (Lv et al., 2013). Likewise, Chaturvedi et al. (2014) showed that transgenic tobacco plants overexpressing SbMT-2 gene enhances metals (Zn^{2+}, Cu^{2+}, and Cd^{2+}) and abiotic stress (salt and osmotic) tolerance via elevated expression of key antioxidant enzyme encoding genes, superoxide dismutase, reducing peroxidase, and ascorbate peroxidase and by modulating ROS detoxification. Enhanced production of iron (Fe) chelator, deoxy mugineic acid in rice, was resulted via the transfer of nicotianamine transferase gene. Many researchers have reported the enhancement of MT genes and those genes encoding for the enzymes integral in sulfur metabolism and histidine biosynthesis, 1-aminocyclopropane-1-carboxylate deaminase, phytochelatin (PC) synthase, Hg^{2+} reductase, arsenate reductase, and aldolase/aldehyde reductase (Ahn et al., 2012).

2.3.2 Improvement of Metal Transporters Using Genetic Engineering

The process of metal/metalloid uptake and its transport mechanism is intricate and varied (Foucault et al., 2013). The transport of metal and its partitioning to plant organs, especially its accumulation into vacuoles, necessitate metal transporters. The effectual mode that could result in enhanced metal tolerance will be accomplished through genetic manipulation of metal transporters (Yang et al., 2015). The selective pathways that need to be focused in light of their application to phytoremediation would comprise manipulation of metal transporter at three levels: (1) metal uptake from soil to root, (2) root-to-shoot metal translocation, and (3) metal sequestration into vacuoles or cell wall. Despite an ample research on genes encoding transporters across plasma membrane of root epidermal cells, none of research in genetic engineering had focused on the enhanced absorption of toxic metals from soil solution to roots (Xie et al., 2015). Metal transport from the cytosol to the vacuole is thought to be an important component of ion tolerance and its potential for use in phytoremediation.

2.3.2.1 CPx-Type Adenosine Triphosphatases

These are type IB heavy metal transporters because they share the common feature of a conserved intramembranous cysteine-protein-cysteine, cysteine-proline-histidine, or cysteine-proline-serine motif that function in heavy metal transduction. CPx-type heavy metal adenosine triphosphatases (ATPases) use adenosine triphosphate (ATP) to pump a variety of charged substances across cell membrane and have an important function in the transport of essential as well as toxic metals, such as Cu, Zn, Cd, and Pb, across the cell membrane (Guo et al., 2015). The CPx-type ATPase was first reported in a higher plant was *Arabidopsis* P-type ATPase (PAA1). CPx-type ATPases may present in the plasma membrane and function as efflux pumps, removing potentially toxic metals from cytoplasm, or may also present at various intracellular membranes and responsible for compartmentalization of heavy metals (Hussain et al., 2004). Heavy metal transporters include CPx-type or ATPase and are involved in the overall metal ion homeostasis and tolerance found to be responsible for compartmentalization of heavy metals in plants (Hussain et al., 2004). The ectopic overexpression of AtHMA4, which is an *A. thaliana* P-1 BATPase-enhanced root growth in the presence of Zn, Cd, and cobalt (Co) (Morel et al., 2009). Amplified expression of *HMA4* in *A. halleri* or *T. caerulescens* and in tobacco showed enhanced Cd and Zn tolerance and hyperaccumulation phenotype of these species (Grispen et al., 2011).

2.3.2.2 Natural Resistance-Associated Macrophage Protein

Natural resistance-associated macrophage proteins (Nramps) are proton-driven metal ion transporters exclusively expressed in roots and shoots. These transporters transfer extensive range of metals—for instance, Mn, Zn, Cu, Fe, Cd, Ni, and Co—across membranes (Nevo and Nelson, 2006). In *Arabidopsis*, heterologous expression of *AtNRAMP1*, *AtNRAMP3*, and *AtNRAMP4* in yeast mutants signified that these proteins could transport Fe, Mn, and Cd (Ehrnstorfer et al., 2014). *AtNRAMP3* and *AtNRAMP4* are responsible for mobilization of vacuolar Fe reserves. Rice OsNRAMP1 rescues the growth of an Fe-defective yeast mutant and is related to Cd accumulation in rice (Takahashi et al.,

2012). Recent characterization of rice OsNRAMP5 revealed its involvement in Mn, Fe, and Cd uptake and transport (Ishimaru et al., 2012).

SMF—suppressor of mitochondria import function—is localized in the yeast mitochondrial membrane. SMF1, SMF2, and SMF3 are yeast homologues of the Nramp proteins with 51–54% similarity in amino acid sequence and 33–36% resemblance to DCT1 and involved in uptake of Mn and Zn into the cytoplasm (Tamayo et al., 2014).

2.3.2.3 Cation Diffusion Facilitator

The cation diffusion facilitator comprises a protein family involved in cation efflux away from cytoplasmic compartment either across cell or into cellular compartments such as vacuoles (Chen et al., 2013). The CzcD from *Ralstonia eutropha* mediates Zn^{2+} efflux and the proteins COT1 and ZRC1 from *Saccaromyces cerevisiae* when over expressed confer Co and Zn/Cd tolerance in these plants, respectively (Zhao et al., 2012). This suggests the role of these proteins in metal sequestration.

2.3.2.4 Zinc-Iron Permease

This protein family comprises metal ion transporters that transport metals such as Fe, Zn, and Cd having 15 members in the *Arabidopsis* genome. Zinc accumulation in roots of *A. thaliana* was enhanced by two-fold through transfer of Zn transporter *ZAT* gene from *Thlaspi goesingense* (van der Zaal et al., 1999). Enhanced accumulation of calcium, Cd, and Mn in tobacco was induced through the transfer of calcium vacuolar transporter *CAX-2* from *A. thaliana* (Hirschi et al., 2000). Similarly, the transfer of another transporter gene *NtCBP4* that codes for calmodulin-binding protein enhanced Ni tolerance (Baligarx, 2012). The zinc-iron permease (ZIP) transporters were characterized in *Arabidopsis* soybean and rice (Ramesh et al., 2003; Milner et al., 2013). Characterization and identification of complementary DNA in *Medicago truncatula* yielded six new metal transporters (*Mt ZIP*) displaying enhanced congruence with the ZIP family (Adams et al., 2012). *MtZIP3* and *MtZIP4* were downregulated in Mn-limited media, whereas *MtZIP5* expression was upregulated in leaves. The IRT1 transporter isolated from *Arabidopsis* exhibits broad substrate range aiding in transport of divalent ions from soil like Mn, Zn, Fe(II), and possibly Cd^{2+} (Barberon et al., 2011).

Cell walls and vacuoles play key role in detoxifying metals especially in Ni-, Zn-, or Cd-hyperaccumulating plant species (Rascio and Navari-Izzo, 2011). Metal accumulation in the vacuole may minimize exchange flank in metal and metabolic process. *Arabidopsis thaliana* possess nine high-affinity phosphate transporters (PHT) that display diverse responses toward arsenate, suggesting their interference with phosphate signaling (arsenate) (LeBlanc et al., 2013). The mutation in one of the nodulin intrinsic protein LSi1 that is a silicon transporter resulted in 60% loss in As(III) influx in rice roots (Ma and Yamaji, 2015). Another NIP family member, LSi2, exhibits the highest As(III) loading tendency into xylem, and an almost 50% decline in As accumulation of shoots has occurred as the result of transfer DNA in LSi2 locus. Mutant lines in the *Arabidopsis* population have been used extensively for isolation of genes for Pb and Mn tolerance, resulting in identification of Pb-tolerant mutants. The Fe content of tobacco plants was enhanced by 1.5-fold when FRE1 and FRE2 gene-encoding ferric reductase were transferred from yeast (Xu et al., 2014). Seeds of *B. juncea* L. were subjected to ethyl methanesulfonate and resulting plants that were self-pollinated to obtain seeds to obtain M2 seeds propagated to yield high transpiration lines of *B. juncea* capable of extracting 104% Pb. In recent years, extensive work has been reported on bioengineering of plants tolerant to alarming levels of toxic metals such as Se, Cd, As, Hg, and Pb (Table 17.2).

3. CLASSIC GENETIC STUDIES AND A MODERN APPROACH FOR IMPROVING PHYTOREMEDIATION

Genetic analysis has provided a basic understanding of the functions involved in phytoremediation, which are quite tedious to interpret at metabolic and biochemical levels. Genetic scrutiny proved an important aid for the identification of genes for production of genetically modified plants for phytoremediation using traits of specific plant varieties (Yang et al., 2015). To understand the basis of tolerance and accumulation of heavy metals and metalloids, classical genetic studies have been subjugated for the identification and interpretation of the genomic sequence in hyperaccumulators through genetic mapping and QTL. This work was initiated and plasticized in two hyperaccumulators of model plants, *A. halleri* and *T. caerulescens*, which were crossed with nonhyperaccumulator ecotypes or congeners for studies of traits segregation (Seregin et al., 2014). This study confirmed that the trait governing metal tolerance of each heavy metal is independent from accumulation of the same metal and that tolerance is a polygenic trait. Courbot et al. (2007) confirmed via the bottom-up approach that the gene HMA4 cosegregates with QTL while carrying out interspecific crosses between *A. halleri* and *Arabidopsis lyrata* where intolerance and hyperaccumulation segregate as independent traits, whereas Cd tolerance cosegregates with Zn tolerance. A research group led by Martin Broadley had shown twofold variation in cesium concentration via QTL analysis in *A. thaliana* (Lisec et al., 2008). QTL and mapping of candidate genes helped in segregation of progenies leading to mapping of putative QTLs on several chromosomes. Classical genetic studies are more or less intricate and resulting crosses always improved, but the real problem is interpreting the segregation of the traits. Three QTLs and one QTL explaining the total variance of 23.8–60.4% were mapped for Zn and Cd concentration in shoots, respectively (Deniau et al., 2006). Soric et al. (2012) analyzed variation for Cu, Fe, potassium, Mn, magnesium, and strontium concentrations in leaves of maize mapping population and determined the QTL associated with the metal concentrations. They demonstrated 11 significant QTLs for 6 metal concentrations; those QTLs for Cu, Fe, and magnesium were localized on chromosome 5 in the region of *ys1* gene supporting the involvement of *ys1* in accumulation of these metals. The isolation of the QTL regions conferring Zn tolerance (*Zntol1*, *Zntol2*, *Zntol3*) holds great promise for improving the identification of the main gene(s) responsible for the tolerance.

In recent years, identification of functional genes involved in metal tolerance and metal homeostasis has been made possible through transcriptomics. Ample information has already been produced by QTL studies during past 5 years, but transcriptomics provided precise description of many genes that play a fundamental role in metal tolerance and provide indications on the epistatic relationships that exist between those genes. Until now, transcript profiling has identified more functional genes than proteomics because of its high-resolution capacity and clear data interpretation of thousands of genes at the same time. Proteomics also leads to identification of only some proteins based on extractability and a relatively high expression level and it detects transcripts at extremely low levels (Chuh and Pratt, 2015). However, proteomics can provide specific targets whose role has been established over time. Therefore, proteins can be used in refining the genetic engineering approach by limiting the number of targets, defining precisely the targets themselves.

The major limitations with genetic engineering include the efficacy of isolated gene and its expression in its new context. However, to improve the effectiveness of expression, we must consider several factors included in so-called "expression cassette." Correct translation of coding sequence has to be executed which may require modification of codons. Moreover, the gene expression of promoter sequence has to be efficient in host species, possibly conferring tissue-specific expression and other genes have to be tailored according to host species. Genetic engineering in plants has a limited range of pyramidization and has not been used to transfer complex traits (i.e., QTLs) (Salvi and Tuberosa, 2015).

3.1 BIOTECHNOLOGICAL APPROACHES FOR CADMIUM REMEDIATION

The remediation of Cd seems to be necessary owing to its mobility in plant soil systems and its lethal effects on ecosystems (Ahmad et al., 2015). Despite various remediation procedures for Cd, phytoremediation via hyperaccumulators has proved to be effective with limited use on a commercial scale. The potential hyperaccumulators for Cd remediation include *Solanum nigrum*, *Populus* spp., *Salix calodendron, Arabis paniculata*, and *Siegesbeckia orientalis* because of their high biomass and increased extraction capacity (Zhang et al., 2013). Cadmium is detoxified in plants through the synthesis of PCs, which were first discovered as Cd-binding complexes like cadystins A and B in yeast (Hossain et al., 2012). These PCs were capable of binding to various other metals like Cu, Zn, and As via sulfhydryl and carboxyl residues. Overexpression of gene (*AtPCS1*)-encoding enzymes PC synthase and cysteine synthase in transgenic tobacco plants and *A. thaliana* resulted in accumulation of Cd, Se, and Ni (Kühnlenz et al., 2015). Gene *Escherichia coli gshll* encoding GSH synthetase in transgenic *B. juncea* showed enhanced tolerance to Cd through the accumulation of Cd metal in shoots (Zhao et al., 2014). The accumulation of Cr, Cu, and Pb increased 2.4- to 3-fold via overexpressing *E. coli gshl* gene encoding γ-glutamylcysteine synthetase.

Constitutive expression of different MTs listed in Table 17.2; mouse MT-1, human hMT-1A and hMT-II, and Chinese hamster MT-II in tobacco, cabbage, and *Arabidopsis* resulted in increased Cd accumulation compared with one lacking the MTs (Usha et al., 2011). Introduction of MT proteins AtMT2a and AtMT3 from *A. thaliana* protected chloroplast guard cells in broad beans by detoxifying through the generation of ROS. MT-1 transgenic tobacco can tolerate 300 μmol/L Cd^{2+} with normal growth, whereas wild-type could only survive in 75 μmol/L Cd^{2+} (Zhang et al., 2000a,b). Increment in MT biosynthesis was made through transformation of sunflower calli with *CUP1* gene, which led to a release of Cd-tolerant lines (Watanabe et al., 2005). Transfer of *RCS1* gene encoding cytosolic CSase from rice to tobacco plants led to a threefold increase in CSase activity resulting in increased tolerance to Cd. Simultaneous expression of spinach CSase in both cytosol and chloroplast resulted in tolerant lines with elevated concentration of Cd in the shoots (Kawashima et al., 2004). Resistance of the plants to Pb and Cd was improved through the transformation of *A. thaliana* plants with ZntA encoding Pb/Cd/Zn-transporting ATPase of *E. coli*. These transgenics showed lower accumulation of Cd and faster release of preloaded Cd than wild-type protoplasts. Root to shoot translocation of Zn, Cd, and overall metal tolerance was enhanced through overexpression of *AtHMA4* in *A. thaliana* compared with wild-type.

Amelioration of transporter genes also resulted in metal tolerance. Cadmium tolerance has been further improved through transport via vacuole sequestration. Overexpression of yeast vacuolar ATP-dependent ABC-type *YCF1* encodes protein, which accumulates Cd at greater levels than did wild-types (Wei et al., 2014). Antosiewicz and Hennig (2004) first demonstrated overexpression of *LCT1*, encoding a nonspecific transporter for Ca^{2+}, Cd^{2+}, Na^{+}, and K^{+}, resulting in amelioration of Cd toxicity by calcium. Hirschi et al. (2000) introduced *A. thaliana* Ca^{2+}/H^+ antiporter *CAX1* and *CAX2* (Ca exchanger) into tobacco, resulting in *CAX1* transformants that exhibit twofold calcium and Mn and fourfold Cd accumulation. Enhanced Cd^{2+} tolerance in *Brassica oleracea* and in sunflower was reported by inducing the overexpression of *CUP1* gene. A substantial increase in Pb and Cd tolerance was exhibited through the constitutive overexpression of *TaPCS1* in *N. glaucum*. Overexpression of NgTP1 under hydroponic conditions conferred up to 36 and 9 times more Pb^{2+} and Cd^{2+} accumulation (Martínez et al., 2006). Intracellular cysteine and GSH levels were increased in *A. thaliana* through constitutive overexpression of *Atcys-3A* encoding intrinsic OAS-TL increasing the anticipation of transgenes even at 400 μM Cd stress. The role of IRT1 in the transport of Cd and Zn has been also shown. In particular, the observation that transgenic plants for the constructs 35S:IRT1, 35S being the promoter of the CMV 35S RNA, accumulate levels of Cd and Zn higher than the wild-type plants, suggests that IRT1 is involved also in the uptake of these metals. The microbes proved as precious reservoir for MT-coding sequences as in *Mesorhizobium huakuii* subsp. *rengei* (strain B3), which has the natural ability to accumulate Cd^{2+} by 25-fold in media containing 30 μM Cd^{2+}. Cadmium uptake in roots was increased by 3 times in rice through the colonization of leguminous milk vetch *Astragalus sinicum* L. with the B3 strain (Ike et al., 2007).

3.2 BIOTECHNOLOGICAL APPROACHES FOR MERCURY TRANSFORMATION AND PHYTOREMEDIATION

Richard Meagher and his associates during early 1990s used a well-categorized mer operon from bacterial cell for the isolation of two important genes, *merA* and *merB*, to engineer new remediation system for Hg detoxification. The gene *merB* catalyzes protonolysis of carbon Hg bond and removing organic ligand and resulting in release of Hg^{2+}, which is subsequently taken up by roots and converted to Hg(0), a less toxic volatile element by mer A (Moller et al., 2014). The modification in constitutive expression of *mer*A among the variety of plants such as *A. thaliana* (Rugh et al., 1996), tobacco (Heaton et al., 2005), yellow poplar (Rugh et al., 1998), *Hibiscus tiliaceus* L. (Che et al., 2003), and rice (Heaton et al., 2003) resulted in tolerant lines that were resistant to Hg concentrations by factor of 10 than nonmodified lines. Mercury remediation in wetlands has been improvised by developing deep-rooting systems in yellow poplar and cottonwood via phytovolatilization methods. The advancement in metabolic engineering resulted in the release of transgenic plants possessing effective of the both MerB (organomercurial lyase) and MerA (mercuric reductase) enzymes for detoxifying deleterious Hg compounds (i.e., ionic mercury, methylmercury, and phenyl mercuric acetate, PMA) (Jagtap and Bapat, 2015). The primary hit points for Hg poisoning are two important autonomous cell organelles: chloroplast and endoplasmic reticulum (ER) (Das et al., 2014). Mercury detoxification was further advanced by chloroplast engineering wherein *merA* and *merB* were combined into chloroplast genome similar coexpression of *merA* and *merB* was displayed by transgenic eastern cottonwood, which was extremely tolerant to organic Hg. High levels of tolerance were reported in transgenic tobacco plants that accumulate more Hg in the shoot when grown in a medium rich in PMA or $HgCl_2$ (Pinto et al., 2014). Though the detoxification of Hg has been successfully achieved via phytovolatization, it cannot justify the escape of volatile Hg from entire globe. Therefore, the need for more concrete plans were stressed that could lead to hyper-accumulation of Hg(II) in aerial parts which could be later harvested to prevent release of Hg(0) in atmosphere. These plants were more tolerant to Hg and can accumulate high concentrations of methyl mercury than plants expressing a single transgene.

Further advancement in Hg detoxification can be carried out by working on the effectiveness of transgenic plants in a natural environment and making them native to various climatic and ecoclines where Hg contamination is risk. Alternatives should be developed for reduction of methyl Hg to Hg(II) and ultimate sequestration of Hg(II) in aerial parts. Genetic engineering could also focus on detoxifying enzymes in subcellular compartments such as ER and chloroplasts.

3.3 BIOTECHNOLOGICAL APPROACHES TO STUDY AND MANIPULATE SELENIUM METABOLISM IN PLANTS

Over expression of genes that code for enzymes, which are integral in sulfur/Se, assimilation was carried out to enhance volatilization of Se. Transgenic *B. juncea* exhibited enhanced selenate reduction and enhancement of enzymes ATP sulfurylase (APS), involved in selenate-to-selenite conversion (Pilon-Smits et al., 2014). These transgenic plants show two- to threefold higher Se volatilization rates than wild-type and 1.5-fold more sulfur. This trait was developed in Indian mustard by overexpression of a gene that coded for the first enzyme in the conversion of SeCys to SeMet, namely cystathionine gamma synthase. These transgenics accumulated 40% less Se than wild-types owing their increased volatilization. These transgenic lines were further improved by focusing on the prevention of SeCys inclusion into proteins. Expression of selenocysteine lyase was induced in *A. thaliana* and Indian mustard using a homolog called cpNifS (Pilon-Smits and Elizabeth, 2015). This enzyme splits SeCys into alanine and Se. The resulting transgenics displayed reduced Se inclusion into proteins and increased Se tolerance and twofold increments in Se accretion compared with nontransformed plants. Another strategy was developed via overexpression of enzyme SMT in *A. thaliana* or *B. juncea*, which was procured from hyperaccumulator *A. bisulcatus*. The SeCysMT transgenics demonstrate increased Se volatilization, resulting in higher tolerance and improvised Se accumulation in the form of methyl-SeCys. *Brassica juncea*, which possess both the APS and SMT transgenic trait can successfully accumulate ninefold higher Se compared with nontransgenic types. An important transgenic was developed for effective accumulation of Se was tailored through the introduction of SMT gene of *A. bisulcatus* (two-grooved poison vetch) into *A. thaliana* (Thale cress) to enhance the expression of Se-methylselenocysteine and γ-glutamylmethylselenocysteine in shoots (Pilon-Smits and LeDuc, 2009). Moreover, the new plant *Cardamine hupingshanesis* has been identified as a hyperaccumulator for Se (Yuan et al., 2013; Wu et al., 2015).

3.4 BIOTECHNOLOGICAL APPROACHES FOR ARSENIC REMEDIATION AND REDUCING ARSENIC IN FOOD CROPS

Arsenic was found to accumulate in the form of As(III) in fronds of Chinese brake fern *Pteris vittata*, which is ubiquitous in both tropical and subtropical habitats (Lessl et al., 2015). This fern was known to possess gene, *PvACR3*, encoding protein which function as arsenic effluxer restricted to vacuolar membrane signifying arsenic efflux via vacuole sequestration (Indriolo et al., 2010). *Pteris vittata* L. is an elite plant species for As accumulation, but it can conversely be used as a reservoir of genes for improvising As hyperaccumulation in other plants. Though this plant reveals substantial prospective for arsenic phytoremediation in green house experiments, but the perennial plant habit and scarcity in molecular mechanism of As detoxification paves the limitation of its potential in field remediation trails. Furthermore, the invasive nature of ferns needs to be evaluated before its large-scale use for biotechnological ventures. Arsenic tolerance was further enhanced by inducing overexpression of genes implicated in the formation of PCs or their precursor GSH. However, this attempt was found to be quite incompetent to meet high toxic levels of As accumulation. Guo et al. (2008) improvised As tolerance by inducing the coexpression of both *γ-ECS* and *PCS* in *Arabidopsis*, paving the new strategy for effective phytoremediation. Further, improvement in phytoremediation was brought with materialization of double trangenics by coexpression of two bacterial genes arsenate reductase (*arsC*) in leaves and γ-glutamylcysteine synthetase (*γ-ECS*) in both roots and shoots under constitutive expression of the *Actin 2* promoter. Arsenate reduction as the result of leaf specific expression of *arsC* and elevated biosynthesis of thiol-rich peptides from over-expression of *γ-ECS* enhanced chances for As(III) complexation and improved As tolerance. These double transgenic managed a 17-fold amplified biomass and accumulated about threefold more As in aerial parts than native types that were grown on 125 μM sodium arsenate. This work proved momentous for validating the potential of phytoremediation in the As-contaminated soil. Though *P. vittata* accumulated As in aerial parts, whereas the rest of angiosperms accumulated an ample fraction of arsenic (95%) in roots in the form of arsenite–glutathione complexes, $As(GS)_3$, as demonstrated in X-ray absorption fine structure of shoots and roots of wild-type plants (Dhankher et al., 2002). Restriction of As(III) underneath elevated the intensity of endogenous arsenate reductase activity that converts As(V) into As(III) in roots. So, a further increase in As translocation was done by blocking arsenate reductase activity. This was achieved by blocking the *AtACR2* function in roots using the various *AtACR2* RNA interference knockdown lines, which can enhance arsenate transport from roots to shoots. By means of transfer DNA insertion lines, Bleeker et al. (2006) also characterized that *AtACR2* (*Arath CDC25*) demonstrates the potential role of this enzyme. Augmented As tolerance and reduced As accumulation in shoots was induced by activating arsenate glutaredoxin from *P. vittata* (*PvGRX5*) and its heterologous overexpression in *Arabidopsis*.

All these genetic studies, though viable, have not yet been practiced in field trials, thus the future efforts should be focused on the engineering of traits that regulate arsenic uptake in roots, rapid root-to-shoot translocation, and ultimate vitalization into high biomass and fast-growing plants. Though biotechnological intervention can result in improved crops with enhanced tolerance to arsenic, their potential is yet to be evaluated in food crops like rice. Accumulation of As(III) in rice could be reduced by knocking down the expression of Lsi1 that mediates As(III) uptake in roots and Lsi2, which transports As(III) from roots to shoots. Arsenic accumulation in food crops could be decreased by amendments of soil with phosphate and Se, which are known to inhibit As uptake in plants (Kumar et al., 2014). This may be achieved by overexpression of PHT genes specialized for only phosphate transport, or transporter genes that mediate Se uptake as described in the Se section. Recently, a genome-wide transcriptome analysis identified numerous genes differentially expressed in response to As treatment in *Arabidopsis*, rice, and *Crambe abyssinica* (Paulose et al., 2010).

3.5 IMPROVING ZINC PHYTOREMEDIATION EFFICIENCY

For an effective phytoremediation, plants used should desirably possess both large biomass productivity and increased metal accumulations in shoots (Vamerali et al., 2010). However, no plant species meet such criteria. For the Zn remediation treatments, plants were supplemented with phytohormones to enhance root and shoot development for increased biomass productivity. Using biodegradable chelating agents can improve the extraction of heavy metals from soil. However, application of chelating agents for improving phytoremediation should be done with precautions to avoid their interference with microbial diversity, heavy metal leaching, and accumulation of intractable organic chelating agents in the environment. Genetic engineering is yet to explore the options for improvement of plants that could be used for Zn hyperaccumulation. Scientists are now focusing on the identification of genes involved in regulation of metal transport system across plasma membrane, enhanced production of natural chelators like MTs and PCs and modification of the Zn metabolic pathways.

Desbrosses-Fonrouge et al. (2005) made a successful attempt by characterization of one such gene in model Zn/Cd hyperaccumulator that could alter highly productive species into Zn metal tolerance. However, in-depth studies are required for better illustration of molecular and genetic mechanism in Zn hyperaccumulators meant for ecofriendly and effective phytoremediation. Most Zn-tolerant lines were developed for improving the micronutrient levels in food or staple crops like rice as biofortification (Bashir et al., 2013; Tan et al., 2015). Conventional breeding can increase Zn tolerance and its mobility across phloem of fruits, seeds, and tuber crops. Various organizations and research institutes have launched programs like Harvest Plus-Bio fortification that focus on improving staple food crops with Zn, Fe, and vitamin A using conventional breeding (Pfeiffer and McClafferty, 2007). Because natural crops possess a narrow range of genetic variation, new genotypes having elevated Zn concentration in grains were developed by inducing genetic variation like wild emmer wheat displaying high tolerance to drought and deficient Zn levels in soil (Velu et al., 2014). Besides conventional breeding, genetic engineering has focused on the identification of genes involved in control and regulation of physiological activities such as Zn uptake, translocation, distribution, and sequestration in plants, particularly in selected Zn-tolerant and Zn-hyperaccumulator plants. Studies on adaptation to Zn deficiency in *A. thaliana* revealed the potential role of transcription factors in regulation of Zn homeostasis in plant (Claus et al., 2014). The overexpression of bZIP19 and bZIP23 transcription factors could improve Zn accumulation in harvestable portions by inducing constitutive expression of a suite of Zn-deficiency responses (Assuncao et al., 2010). Elevated levels of Zn in leaf tissues of *A. thaliana* were reported when the expression of *AtHMA2* that catalyzes Zn^{2+} efflux across the membranes of root cells is reduced or overexpressing the gene encoding *AtHMA4*, which is thought to load Zn into the xylem (Takahashi et al., 2011). Overexpression of *HvNAS1* in tobacco increased Zn concentrations in leaf and seed increased from 16 to 39 mg/kg DW and from 20 to 35 mg/kg DW, respectively. Numerous crops accumulated Zn levels at higher rates like cassava with root Zn of about 40 mg/kg DW (Sayre et al., 2011) and brown rice with 56–95 mg of Zn per kg of DW (Johnson et al., 2011), have been reported. Mutilation in root absorption of Zn, reducing solubility, annual rainfall, and high pH can significantly reduce plant Zn concentrations. Zinc deficiency in soils of countries like China, India, Iran, Pakistan, and Australia hinders the accumulation of Zn in edible parts. Thus, genetic strategies should be developed for effective sequestration of Zn, resulting in development of more feasible biofortification. Accordingly, proper blending of breeding and fertilizer schemes is an excellent complementary approach to improve Zn deficiency in soil. Zinc phytoextraction can be studied further in cereal plants with high biomass and adequate concentration of Zn in edible parts, this could enhance the trace element availability in rhizosphere and its translocation toward aerial and grain tissues.

4. CONCLUSIONS AND FUTURE PROSPECTS

Genoremediation is one of the promising biotechnological approaches of phytoremediation. Recently, much research has been conducted to produce genetically modified plants for better phytoremediation of heavy metals and xenobiotics. Importantly, in most of the cases, promising results were obtained. Identification and functional characterization of novel proteins involved in vacuolar sequestration of heavy metals and cross-membrane transport had also accelerated the deliberate manipulation of plants for better phytoremediation purposes. In-depth studies of molecular mechanisms of pollutants uptake, accumulation, detoxification, and exclusion reveals that some of these mechanisms are species-specific and complete understanding of plant tolerance to pollution is still a mystery. Therefore, a better understanding of the molecular basis of the pathways and genes involved in the degradation of pollutants is crucial. Further, analysis and discovery of new genes suitable for phytoremediation is essential and accumulators need to be engineered with multiple stacked genes to produce high accumulator plants that can be used for more effective phytoremediation of contaminated sites.

Genetic engineering should focus more on vacuolar sequestration, which is an important component of metal hypertolerance and putative hyperaccumulation. Efforts should be made for improvising root uptake, xylem loading, vacuolar transport, and mobilization of heavy metals. In-depth understanding of molecular mechanism necessary for hyperaccumulation and biochemical characterization of root exudates needs to be elucidated. Numerous other plant species need to be explored for their tolerance and accumulation. Genetic engineering for improving the metal hyperaccumulators by changing oxidation states, increasing the metal sequestration proteins, and identification of transport proteins such as ZIP and ZAT could result in better and more effective phytoremediation technology. Protection and preservation of hyperaccumulators, which are often found in geographically remote areas and exposed to wreckage through mining activities, need urgent attention. Introduction of exotic and alien species often endangers their existence; hence, efforts should be made for their large-scale production and germplasm conservation. For the development of effective clean-up technology, there is exigency for the establishment of integrated network by plant biologists, soil chemists, microbiologists, and environmental engineers.

REFERENCES

Abdel-Salam, A.A., Salem, H.M., Abdel-Salam, M.A., Seleiman, M.F., 2015. Phytochemical removal of heavy metal-contaminated soils. In: Heavy Metal Contamination of Soils. Springer International Publishing, pp. 299–309.

Adams, J.P., Adeli, A., Hsu, C.Y., Harkess, R.L., Page, G.P., Pamphilis, C.W.D., Yüceer, C., 2012. Characterization of poplar ZIP family members ZIP1.2 and ZNT1. J. Plant Genet. Transgenics 3 (1), 1–13.

Ahmad, A., Hadi, F., Ali, N., 2015. Effective phytoextraction of cadmium (Cd) with increasing concentration of total phenolics and free proline in *Cannabis sativa* (L) plant under various treatments of fertilizers, plant growth regulators and sodium salt. Int. J. Phytorem. 17 (1), 56–65.

Ahn, Y.O., Kim, H.K., Lee, J., Kim, H., Lee, H.S., Kwak, S.S., 2012. Three *Brassica rapa* methallothionein genes are differentially regulated under various stress conditions. Mol. Biol. Rep. 39, 2059–2067.

Ali, H., Khan, E., Sajad, M.A., 2013. Phytoremediation of heavy metals–concepts and applications. Chemosphere 91, 869–881.

Antosiewicz, D.M., Hennig, J., 2004. Overexpression of LCT1 in tobacco enhances the protective action of calcium against cadmium toxicity. Environ. Pollut. 129, 237–245.

Assuncao, A.G.L., Herrero, E., Lin, Y.F., 2010. *Arabidopsis thaliana* transcription factors bZIP19 and bZIP23 regulate the adaptation to zinc deficiency. Proc. Natl. Acad. Sci. U.S.A. 107, 10296–10301.

Baligarx, V.C., 2012. Mechanisms of nickel uptake and hyperaccumulation by plants and implications for soil remediation. Adv. Agron. 117, 117.

Barberon, M., Zelazny, E., Robert, S., Conejero, G., Curie, C., Frim, J., Vert, G., 2011. Monoubiquitin-dependent endocytosis of the iron-regulated transporter 1 (IRT1) transporter controls iron uptake in plants. Proc. Natl. Acad. Sci. U.S.A. 108 (32), E450–E458.

Bashir, K., Takahashi, R., Hiromi, N., Nishizawa, N.K., 2013. The road to micronutrient biofortification of rice: progress and prospects. Front. Plant Sci. 4, 15.

Bizily, S.P., Rugh, C.L., Summers, A.O., Meagher, R.B., 1999. Phytoremediation of methylmercury pollution: merB expression in *Arabidopsis thaliana* confers resistance to organomercurials. Proc. Natl. Acad. Sci. U.S.A. 96, 6808–6813.

Bleeker, P.M., Hakvoort, H.W., Bliek, M., Souer, E., Schat, H., 2006. Enhanced arsenate reduction by a CDC25-like tyrosine phosphatase explains increased phytochelatin accumulation in arsenate-tolerant *Holcus lanatus*. Plant J. 45, 917–929.

Brunetti, P., Zanella, L., De Paolis, A., Di Litta, D., Cecchetti, V., Falasca, G., Cardarelli, M., 2015. Cadmium-inducible expression of the ABC-type transporter AtABCC3 increases phytochelatin-mediated cadmium tolerance in *Arabidopsis*. J. Exp. Bot. erv185.

Castro-Rodríguez, V., García-Gutiérrez, A., Canales, J., Cañas, R.A., Kirby, E.G., Avila, C., Cánovas, F.M., 2015. Poplar trees for phytoremediation of high levels of nitrate and applications in bioenergy. Plant Biotechnol. J.

Chaturvedi, A.K., Patel, M.K., Mishra, A., Tiwari, V., Jha, B., 2014. The SbMT-2 gene from a halophyte confers abiotic stress tolerance and modulates ROS scavenging in transgenic tobacco. PLoS One 9 (10), e111379. http://dx.doi.org/10.1371/journal.pone.0111379.

Che, D., Meagher, R.B., Heaton, A.C., Lima, A., Rugh, C.L., Merkle, S.A., 2003. Expression of mercuric ion reductase in Eastern cottonwood (*Populus deltoides*) confers mercuric ion reduction and resistance. Plant Biotechnol. J. 1, 311–319.

Chen, J., Chen, Y., Shi, Z.Q., Su, Y., Han, F.X., 2015. Phytoremediation to remove metals/metalloids from soils. In: Phytoremediation. Springer International Publishing, pp. 297–304.

Chen, Z., Fujii, Y., Yamaji, N., Masuda, S., Takemoto, Y., Kamiya, T., Ueno, D., 2013. Mn tolerance in rice is mediated by MTP8. 1, a member of the cation diffusion facilitator family. J. Exp. Bot. 64 (14), 4375–4387.

Chuh, K.N., Pratt, M.R., 2015. Chemical methods for the proteome-wide identification of post translationally modified proteins. Curr. Opin. Chem. Biol. 24, 27–37.

Cornu, J.Y., Deinlein, U., Höreth, S., Braun, M., Schmidt, H., Weber, M., Clemens, S., 2014. Contrasting effects of nicotianamine synthase knockdown on zinc and nickel tolerance and accumulation in the zinc/cadmium hyperaccumulator *Arabidopsis halleri*. New Phytol.

Courbot, M., Willems, G., Motte, P., Arvidsson, S., Roosens, N., 2007. A major quantitative trait locus for cadmium tolerance in *Arabidopsis halleri* colocalizes with HMA4, a gene encoding a heavy metal ATPase. Plant Physiol. 144, 1052–1065.

Craciun, A.R., Meyer, C.-L., Chen, J., Roosens, N., De Groodt, R., Hilson, P., et al., 2012. Variation in HMA4 gene copy number and expression among *Noccaea caerulescens* populations presenting different levels of Cd tolerance and accumulation. J. Exp. Bot. 63, 4179–4189.

Claus, J., Ptashnyk, M., A, Bohmann, Chavarría-Krauser, A., 2014. Global Hopf bifurcation in the ZIP regulatory system. submitted to J. Math. Biol. arXiv Preprint: 1401.5597.

Das, S., Raj, R., Mangwani, N., Dash, H.R., Chakraborty, J., 2014. Heavy metals and hydrocarbons: adverse effects and mechanism of toxicity. Microb. Biodegrad. Biorem. 23.

Deniau, A.X., Pieper, B., Ten Bookum, W.M., Lindhout, P., Aarts, M.G.M., Schat, H., 2006. QTL analysis of cadmium and zinc accumulation in the heavy metal hyperaccumulator *Thlaspi caerulescens*. Theor. Appl. Genet. 113 (5), 907–920.

Desbrosses-Fonrouge, A.G., Voigt, K., Schroder, A., 2005. *Arabidopsis thaliana* MTP l is a Zn transporter in the vacuolar membrane which mediates Zn detoxification and drives leaf Zn accumulation. Fed. Eur. Biochem. Soc. 579 (19), 4165–4174.

Dhan, L.T., Truong, P., Mammucari, R., Foster, N., 2014. A critical review of the arsenic uptake mechanisms and phytoremediation potential of *Pteris vittata*. Int. J. Phytorem. 16. http://dx.doi.org/10.1080/15226514.2013.798613.

Dhankher, O.P., Li, Y., Rosen, B.P., Shi, J., Salt, D., Senecoff, J.F., 2002. Engineered tolerance and hyperaccumulation of arsenic in plants by combining arsenate reductase and γ-glutamylcysteine synthetase expression. Nat. Biotechnol. 20, 1140–1145.

Dinh, N.T., Vu, D.T., Mulligan, D., Nguyen, A.V., 2015. Accumulation and distribution of zinc in the leaves and roots of the hyperaccumulator *Noccaea caerulescens*. Environ. Exp. Bot. 110, 85–95.

DiTusa, S.F., Fontenot, E.B., Wallace, R.W., Silvers, M.A., Steele, T.N., Elnagar, A.H., Smith, A.P., 2015. A member of the Phosphate transporter 1 (Pht1) family from the arsenic-hyperaccumulating fern *Pteris vittata* is a high-affinity arsenate transporter. New Phytol.

Dixit, R., Wasiullah, Malaviya, D., Pandiyan, K., Singh, U.B., Sahu, A., Shukla, R., Singh, B.P., Rai, J.P., Sharma, P.K., Lade, H., Paul, D., 2015. Bioremediation of heavy metals from soil and aquatic environment: an overview of principles and criteria of fundamental processes. Sustainability 7, 2189–2212.

Doty, S.L., 2008. Enhancing phytoremediation through the use of transgenics and endophytes. New Phytol. 179, 318–333.

Ehrnstorfer, I.A., Geertsma, E.R., Pardon, E., Steyaert, J., Dutzler, R., 2014. Crystal structure of a SLC11 (NRAMP) transporter reveals the basis for transition-metal ion transport. Nat. Struct. Mol. Biol. 21 (11), 990–996.

Evangelou, M.W., Papazoglou, E.G., Robinson, B.H., Schulin, R., 2015. Phytomanagement: phytoremediation and the production of biomass for economic revenue on contaminated land. In: Phytoremediation. Springer International Publishing, pp. 115–132.

Ezaki, B., Gardner, R.C., Ezaki, Y., Matsumoto, H., 2000. Expression of aluminum-induced genes in transgenic *Arabidopsis* plants can ameliorate aluminum stress and/or oxidative stress. Plant Physiol. 122, 657–665.

Fahr, M., Laplaze, L., El Mzibri, M., Doumas, P., Bendaou, N., Hocher, V., Smouni, A., 2015. Assessment of lead tolerance and accumulation in metallicolous and non-metallicolous populations of *Hirschfeldia incana*. Environ. Exp. Bot. 109, 186–192.

Fathi, R.A., Godbold, D.L., Al-Salih, H.S., Jones, D., 2014. Potential of phytoremediation to clean up uranium-contaminated soil with Acacia species. J. Environ. Earth Sci. 4 (4), 81–91.

Feng, R., Wei, C., Tu, S., Tang, S., Wu, F., 2011. Simultaneous hyperaccumulation of arsenic and antimony in Cretan brake fern: evidence of plant uptake and subcellular distributions. Microchem. J. 97 (1), 38–43.

Foucault, Y., Leveque, T., Xiong, T., Schreck, E., Austruy, A., Shahid, M., Dumat, C., 2013. Green manure plants for remediation of soils polluted by metals and metalloids: ecotoxicity and human bioavailability assessment. Chemosphere 93 (7), 1430–1435.

Ghasemi, R., Ghaderian, S.M., Ebrazeh, S., 2015. Nickel stimulates copper uptake by nickel hyperaccumulator plants in the genus Alyssum. Aust. J. Bot.

Goto, F., Yoshihara, T., Saiki, H., 2000. Iron accumulation and enhanced growth in transgenic lettuce plants expressing the iron-binding protein ferritin. Theor. Appl. Genet. 100, 658–664.

Grispen, V.M., Hakvoort, H.W., Bliek, T., Verkleij, J.A., Schat, H., 2011. Combined expression of the *Arabidopsis* metallothionein MT2b and the heavy metal transporting ATPase HMA4 enhances cadmium tolerance and the root to shoot translocation of cadmium and zinc in tobacco. Environ. Exp. Bot. 72 (1), 71–76.

Guo, J., Green, B.R., Maldonado, M.T., 2015. Sequence analysis and gene expression of potential components of copper transport and homeostasis in *Thalassiosira pseudonana*. Protist 166 (1), 58–77.

Guo, W.J., Meetam, M., Goldsbrough, P.B., 2008. Examining the specific contributions of individual *Arabidopsis* metallothioneins to copper distribution and metal tolerance. Plant Physiol. 146, 1697–1706.

Gustin, J.L., Loureiro, M.E., Kim, D., Na, G., Tikhonova, M., Salt, D.E., 2009. MTP1-dependent Zn sequestration into shoot vacuoles suggests dual roles in Zn tolerance and accumulation in Zn-hyperaccumulating plants. Plant J. 57, 1116–1127.

Heaton, A.C., Rugh, C.C., Kim, T., Meagher, R.B., 2003. Toward detoxifying mercury-polluted aquatic sediments with rice genetically engineered for mercury resistance. Environ. Toxicol. Chem. 22, 2940–2947.

Heaton, A.C.P., Rugh, C.L., Wang, N.J., Meagher, R.B., 2005. Physiological responses of Transgenic merA-tobacco (*Nicotiana tabacum*) to foliar and root mercury exposure. Water Air Soil Pollut. 161, 137–155.

Hirschi, K.D., Korenkov, V.D., Wilganowski, N.L., Wagner, G.J., 2000. Expression of *Arabidopsis CAX2* in tobacco: altered metal accumulation and increased manganese tolerance. Plant Physiol. 124, 125–133.

Hossain, M.A., Piyatida, P., da Silva, J.A.T., Fujita, M., 2012. Molecular mechanism of heavy metal toxicity and tolerance in plants: central role of glutathione in detoxification of reactive oxygen species and methylglyoxal and in heavy metal chelation. J. Bot. 37. Article ID 872875.

Hsieh, J.L., Chen, C.Y., Chiu, M.H., Chein, M.F., Chang, J.S., Endo, G., Huang, C.C., 2009. Expressing a bacterial mercuric ion binding protein in plant for phytoremediation of heavy metals. J. Hazard. Mater. 161, 920–925.

Huang, G.Y., Wang, Y.S., 2010. Expression and characterization analysis of type 2 metallothionein from grey mangrove species (*Avicennia marina*) in response to metal stress. Aquat. Toxicol. 99, 86–92.

Hunt, A.J., Anderson, C.W.N., Bruce, N., Muñoz García, A., Graedel, T.E., Hodson, M., Meech, J.A., Nassar, N.T., Parker, H.L., Rylott, E.L., Sotiriou, K., Zhang, Q., Clark, J.H., 2014. Phytoextraction as a tool for green chemistry. Green Process. Synth. 3, 3–22.

Hussain, D., Haydon, M.J., Wang, Y., Wong, E., Sherson, S.M., Young, J., Camakaris, J., Harper, J.F., Cobbett, C.S., 2004. P-Type ATPase heavy metal transporters with roles in essential zinc homeostasis in *Arabidopsis*. Plant Cell 16, 1327–1339.

Ibrahim, M., Adrees, M., Rashid, U., Raza, S.H., Abbas, F., 2015. Phytoremediation of radioactive contaminated soils. Soil Rem. Plants 599.

Ike, A., Sriprang, R., Ono, H., Murooka, Y., Yamashita, M., 2007. Bioremediation of cadmium contaminated soil using symbiosis between leguminous plant and recombinant rhizobia with the MTL4 and the PCS genes. Chemosphere 66, 1670–1676.

Indriolo, E., Na, G.N., Ellis, D., Salt, D.E., Banks, J.A., 2010. A vacuolar arsenite transporter necessary for arsenic tolerance in the arsenic hyperaccumulating fern *Pteris vittata* is missing in flowering plants. Plant Cell 22, 2045–2057.

Iqbal, J., Ahemad, M., 2015. Recent advances in bacteria-assisted phytoremediation of heavy metals from contaminated soil. Adv. Biodegrad. Biorem. Ind. Waste 401.

Ishimaru, Y., Takahashi, R., Bashir, K., Shimo, H., Senoura, T., Sugimoto, K., Nishizawa, N.K., 2012. Characterizing the role of rice NRAMP5 in manganese, iron and cadmium transport. Sci. Rep. 2.

Jagtap, U.B., Bapat, V.A., 2015. Genetic Engineering of plants for heavy metal removal from soil. In: Heavy Metal Contamination of Soils. Springer International Publishing, pp. 433–470.

Jesus, J.M., Danko, A.S., Fiuza, A., Borges, M.T., 2015. Phytoremediation of salt-affected soils: a review of processes, applicability, and the impact of climate change. Environ. Sci. Pollut. Res. 22 (9), 6511–6525.

Johnson, A.A.T., Kyriacou, B., Callahan, D.L., 2011. Constitutive over expression of the OsNAS gene family reveals single-gene strategies for effective iron- and zinc-biofortification of rice endosperm. PLoS One 6 (9), 24476.

Kawashima, C.G., Noji, M., Nakamura, M., Ogra, Y., Suzuki, K.T., Saito, K., 2004. Heavy metal tolerance of transgenic tobacco plants overexpressing cysteine synthase. Biotechnol. Lett. 26, 153–157.

Khatisashvili, G., Matchavariani, L., Gakhokidze, R., 2015. Improving phytoremediation of soil polluted with oil hydrocarbons in Georgia. Soil Rem. Plants 547.

Kim, S.H., Jeong, J.C., Ahn, Y.O., Lee, H.S., Kwak, S.S., 2014. Differential responses of three sweetpotato metallothionein genes to abiotic stress and heavy metals. Mol. Biol. Rep. 41 (10), 6957–6966.

Kühnlenz, T., Westphal, L., Schmidt, H., Scheel, D., Clemens, S., 2015. Expression of *Caenorhabditis elegans* PCS in the AtPCS1-deficient *Arabidopsis thaliana* cad1-3 mutant separates the metal tolerance and non-host resistance functions of phytochelatin synthases. Plant Cell Environ. http://dx.doi.org/10.1111/pce.12534.

Kumar, A., Singh, R.P., Singh, P.K., Awasthi, S., Chakrabarty, D., Trivedi, P.K., Tripathi, R.D., 2014. Selenium ameliorates arsenic induced oxidative stress through modulation of antioxidant enzymes and thiols in rice (*Oryza sativa* L.). Ecotoxicology 23, 1153–1163.

Kumar, G., Kushwaha, H.R., Panjabi-Sabharwal, V., Kumari, S., Joshi, R., Karan, R., Mittal, S., Singla Pareek, S.L., Pareek, A., 2012. Clustered metallothionein genes are co-regulated in rice and ectopic expression of *OsMT1e-P* confers multiple abiotic stress tolerance in tobacco via ROS scavenging. BMC Plant Biol. 12, 107.

Kupper, H., Kochian, L.V., 2010. Transcriptional regulation of metal transport genes and mineral nutrition during acclimatization to cadmium and zinc in the Cd/Zn hyperaccumulator, *Thlaspi caerulescens* (Ganges population). New Phytol. 185, 114–129.

LeBlanc, M.S., McKinney, E.C., Meagher, R.B., Smith, A.P., 2013. Hijacking membrane transporters for arsenic phytoextraction. J. Biotechnol. 163 (1), 1–9.

LeDuc, D.L., Tarun, A.S., Montes-Bayon, M., Meija, J., Malit, M.F., Wu, C.P., AbdelSamie, M., et al., 2004. Overexpression of selenocysteine methyltransferase in *Arabidopsis* and Indian mustard increases selenium tolerance and accumulation. Plant Physiol. 135 (1), 377–383.

Lee, J.H., 2013. An overview of phytoremediation as a potentially promising technology for environmental pollution control. Biotechnol. Bioprocess Eng. 18, 431–439.

Lessl, J.T., Guan, D.X., Sessa, E., Rathinasabapathi, B., Ma, L.Q., 2015. Transfer of arsenic and phosphorus from soils to the fronds and spores of arsenic hyperaccumulator *Pteris vittata* and three non-hyperaccumulators. Plant Soil. http://dx.doi.org/10.1007/s11104-014-2376-2.

Li, Y., Chen, Y.Y., Yang, S.G., Tian, W.M., 2015. Cloning and characterization of HbMT2a, a metallothionein gene from *Hevea brasiliensis* Muell. Arg differently responds to abiotic stress and heavy metals. Biochem. Biophys. Res. Commun. 461 (1), 95–101.

Liao, C., Liang, X., Lu, G., Thai, T., Xu, W., Dang, Z., 2015. Effect of surfactant amendment to PAHs-contaminated soil for phytoremediation by maize (Zea mays L.). Ecotoxicol. Environ. Saf. 112, 1–6.

Lisec, J., Meyer, R.C., Steinfath, M., Redestig, H., Becher, M., Witucka-Wall, H., Willmitzer, L., 2008. Identification of metabolic and biomass QTL in *Arabidopsis thaliana* in a parallel analysis of RIL and IL populations. Plant J. 53 (6), 960–972.

Liu, J., Shang, W., Zhang, X., Zhu, Y., Yu, K., 2014. Mn accumulation and tolerance in *Celosia argentea* Linn.: a new Mn-hyperaccumulating plant species. J. Hazard. Mater. 267, 136–141.

Lv, Y., Deng, X., Quan, L., Xia Y,Shen, Z., 2013. Metallothioneins BcMT1 and BcMT2 from *Brassica campestris* enhance tolerance to cadmium and copper and decrease production of reactive oxygen species in *Arabidopsis thaliana*. Plant Soil 367, 507–519.

Ma, J.F., Yamaji, N., 2015. A cooperated system of silicon transport in plants. Trends Plant Sci.

Macoive, A., Ventura, L., Dona, M., Fae, M., Balestrazz, A., Carbonera, D., 2010. Effect of heavy metal treatment on metallothionein expression profiles white poplar (*Populus alba* L.) cell suspension cultures. Fasc. Biol. 2, 274–279.

Malagoli, M., Schiavon, M., Pilon, M., Pilon-smits, E.A., 2015. Exploring the importance of sulphate transporters and ATP sulphurylases for selenium hyperaccumulation–A comparison of *Stanleya pinnata* and *Brassica juncea* (Brassicaceae). Front. Plant Sci. 6 (2).

Malar, S., Sahi, S.V., Favas, P.J., Venkatachalam, P., 2014. Mercury heavy-metal-induced physiochemical changes and genotoxic alterations in water hyacinths [*Eichhornia crassipes* (Mart.)]. Environ. Sci. Pollut. Res. 1–12.

Mani, D., Kumar, C., Patel, N.K., 2014. Integrated micro-biochemical approach for phytoremediation of cadmium and zinc contaminated soils. Ecotoxicol. Environ. Saf. 111, 86–95.

Martínez, M., Bernal, P., Almela, C., Vélez, D., García-Agustín, P., Serrano, R., 2006. An engineered plant that accumulates higher levels of heavy metals than *Thlaspi caerulescens*, with yields of 100 times more biomass in mine soils. Chemosphere 64, 478–485.

McAlister, R.L., Kolterman, D.A., Pollard, A.J., 2015. Nickel hyperaccumulation in populations of *Psychotria grandis* (Rubiaceae) from serpentine and non-serpentine soils of Puerto Rico. Aust. J. Bot. 63 (2), 85–91.

Mello-Farias, P.C.D., Chaves, A.L.S., Lencina, C.L., 2011. Transgenic plants for enhanced phytoremediation – physiological studies. In: Alvarez, M. (Ed.), Genetic Transformation. InTech Europe, Croatia, pp. 305–328.

Migocka, M., Kosieradzka, A., Papierniak, A., Maciaszczyk-Dziubinska, E., Posyniak, E., Garbiec, A., Filleur, S., 2014. Two metal-tolerance proteins, MTP1 and MTP4, are involved in Zn homeostasis and Cd sequestration in cucumber cells. J. Exp. Bot. eru459.

Migocka, M., Posyniak, E., Maciaszczyk-Dziubinska, E., Papierniak, A., Kosieradzka, A., 2015. Functional and biochemical characterization of cucumber genes encoding two Copper ATPases CsHMA5.1 and CsHMA5.2. J. Biol. Chem. jbc-M114.

Milner, M.J., Mitani-Ueno, N., Yamaji, N., Yokosho, K., Craft, E., Fei, Z., Kochian, L.V., 2014. Root and shoot transcriptome analysis of two ecotypes of *Noccaea caerulescens* uncovers the role of NcNramp1 in Cd hyperaccumulation. Plant J. 78 (3), 398–410.

Milner, M.J., Seamon, J., Craft, E., Kochian, L.V., 2013. Transport properties of members of the ZIP family in plants and their role in Zn and Mn homeostasis. J. Exp. Bot. 64 (1), 369–381.

Moller, A.K., Barkay, T., Hansen, M.A., Norman, A., Hansen, L.H., Sorensen, S.J., Kroer, N., 2014. Mercuric reductase genes (merA) and mercury resistance plasmids in High Arctic snow, freshwater and sea-ice brine. FEMS Microbiol. Ecol. 87 (1), 52–63.

Morel, M., Crouzet, J., Gravot, A., Auroy, P., Leonhardt, N., Vavasseur, A., Richaud, P., 2009. AtHMA3, a P1B-ATPase Allowing Cd/Zn/Co/Pb vacuolar storage in *Arabidopsis*. Plant Physiol. 149 (2), 894–904.

Muehe, E.M., Weigold, P., Adaktylou, I.J., Planer-Friedrich, B., Kraemer, U., Kappler, A., Behrens, S., 2015. Rhizosphere microbial community composition affects cadmium and zinc uptake of the metal-hyperaccumulating plant *Arabidopsis halleri*. Appl. Environ. Microbiol. AEM-03359.

Muller, K., Daus, B., Mattusch, J., Vetterlein, D., Merbach, I., Wennrich, R., 2013. Impact of arsenic on uptake and bio-accumulation of antimony by arsenic hyperaccumulator *Pteris vittata*. Environ. Pollut. 174, 128–133.

Nevo, Y., Nelson, N., 2006. The NRAMP family of metal-ion transporters. Biochim. Biophys. Acta (BBA)-Mol. Cell Res. 1763 (7), 609–620.

Paulose, B., Kandasamy, S., Dhankher, O.P., 2010. Exprerssion profiling of *Crambe abyssinica* under arsenate stress identifies genes and gene networks involved in arsenic metabolism and detoxification. BMC Plant Biol. 10, 108.

Pfeiffer, W.H., McClafferty, B., 2007. Harvest plus: breeding crops for better nutrition. Crop Sci. 47, S88–S105.

Phieler, R., Merten, D., Roth, M., Büchel, G., Kothe, E., 2015. Phytoremediation using microbially mediated metal accumulation in Sorghum bicolor. Environ. Sci. Pollut. Res. 1–9.

Pilon-Smits, E.A.H., Banuelos, G.S., Parker, D.R., 2014. Uptake, metabolism, and volatilization of selenium by terrestrial plants. In: Salinity and Drainage in San Joaquin Valley, California. Springer, Netherlands, pp. 147–164.

Pilon-Smits, E.A.H., Elizabeth, A., 2015. Selenium in plants. In: Progress in Botany. Springer International Publishing, pp. 93–107.

Pilon-Smits, E.A.H., LeDuc, D.L., 2009. Phytoremediation of selenium using transgenic plants. Curr. Opin. Biotechnol. 20, 207–212.

Pinto, A.P., de Varennes, A., Fonseca, R., Teixeira, D.M., 2014. Phytoremediation of soils contaminated with heavy metals. Phytorem. Manage. Environ. Contam. 1, 133.

Poscic, F., Fellet, G., Vischi, M., Casolo, V., Schat, H., Marchiol, L., 2015. Variation in heavy metal accumulation and genetic diversity at a regional scale among metallicolous and non-metallicolous populations of the facultative metallophyte *Biscutella laevigata* subsp. laevigata. Int. J. Phytorem. 17 (5).

Prasad, M.N.V., 2015. Engineered phyto-covers as natural caps for containment of hazardous mine and municipal solid waste dump sites–possible energy sources. In: Phytoremediation for Green Energy. Springer, Netherlands, pp. 55–68.

Qadir, M., Mateo-Sagasta, J., Jimenez, B., Siebe, C., Siemens, J., Hanjra, M.A., 2015. Environmental risks and cost-effective risk management in wastewater use systems. In: Wastewater. Springer, Netherlands, pp. 55–72.

Rajkumar, M., Sandhya, S., Prasad, M.N.V., Freitas, H., 2012. Perspectives of plant-associated microbes in heavy metal phytoremediation. Biotechnol. Adv. http://dx.doi.org/10.1016/j.biotechadv. 2012.04.011.

Ramesh, S.A., Shin, R., Eide, D.J., Schachtman, D.P., 2003. Differential metal selectivity and gene expression of two zinc transporters from Rice. Plant Physiol. 133 (1), 126–134.

Rascio, N., Navari-Izzo, F., 2011. Heavy metal hyperaccumulating plants: how and why do they do it? And what makes them so interesting? Plant Sci. 180 (2), 169–181.

Rugh, C.L., Senecoff, J.F., Meagher, R.B., Merkle, S.A., 1998. Development of transgenic yellow poplar for mercury phytoremediation. Nat. Biotechnol. 16, 925–928.

Rugh, C.L., Wilde, H.D., Stack, N.M., Thompson, D.M., Summers, A.O., Meagher, R.B., 1996. Mercuric ion reduction and resistance in transgenic *Arabidopsis thaliana* plants expressing a modified bacterial merA gene. Proc. Natl. Acad. Sci. U.S.A. 93, 3182–3187.

Salvi, S., Tuberosa, R., 2015. The crop QTLome comes of age. Curr. Opin. Biotechnol. 32, 179–185.

Santiago-Cruz, M.A., Villagrán-Vargas, E., Velázquez-Rodríguez, A.S., Vernon-Carter, E.J., Cruz-Sosa, F., Orozco-Villafuerte, J., Buendía-González, L., 2014. Exploring the Cr (VI) Phytoremediation potential of *Cosmos bipinnatus*. Water Air Soil Pollut. 225 (11), 1–8.

Sasaki, A., Yamaji, N., Ma, J.F., 2014. Overexpression of OsHMA3 enhances Cd tolerance and expression of Zn transporter genes in rice. J. Exp. Bot. 65 (20), 6013–6021.

Sayre, R., Beeching, J.R., Cahoon, E.B., 2011. The bio-cassava plus program: biofortification of cassava for Sub-Saharan Africa. Annu. Rev. Plant Biol. 62, 251–272.

Seregin, I.V., Erlikh, N.T., Kozhevnikova, A.D., 2014. Nickel and zinc accumulation capacities and tolerance to these metals in the excluder *Thlaspi arvense* and the hyperaccumulator *Noccaea caerulescens*. Russ. J. Plant Physiol. 61 (2), 204–214.

Sheron, V., Sheoran, A.S., Poonia, P., 2011. Role of hyper accumulators in phytoextraction of metals from contaminated mining sites: a review. Crit. Rev. Environ. Sci. Technol. 41, 168–214.

Shifrin, N.S., Pitts, B.S., Chow, A.C., 2015. Estimating environmental costs. Environ. Claims J. 27 (1), 9–18.

Siemianowski, O., Mills, R.F., Williams, L.E., Antosiewicz, D.M., 2011. Expression of the P_{1B}-type ATPase *AtHMA4* in tobacco modifies Zn and Cd root to shoot partitioning and metal tolerance. Plant Biotechnol. J. 9, 64–74.

Simmons, R.W., Chaney, R.L., Angle, J.S., Kruatrachue, M., Klinphoklap, S., Reeves, R.D., Bellamy, P., 2015. Towards practical cadmium phytoextraction with *Noccaea caerulescens*. Int. J. Phytorem. 17 (2), 191–199.

Singh, D., Vyas, P., Sahni, S., Sangwan, P., 2015. Phytoremediation: A biotechnological intervention. In: Applied Environmental Biotechnology: Present Scenario and Future Trends. Springer, India, pp. 59–75.

Soric, R., Ledencan, T., Zdunic, Z., Jambrovic, A., Brkic, I., Loncaric, Z., Simic, D., 2012. Quantitative trait loci for metal accumulation in maize leaf. Maydica 56 (4).

Sree, K.S., Keresztes, A., Mueller-Roeber, B., Brandt, R., Eberius, M., Fischer, W., Appenroth, K.J., 2015. Phytotoxicity of cobalt ions on the duckweed *Lemna minor*–Morphology, ion uptake, and starch accumulation. Chemosphere 131, 149–156.

Takahashi, R., Ishimaru, Y., Senoura, T., Shimo, H., Ishikawa, S., Arao T,Nakanishi, H., Nishizawa, N.K., 2011. The OsNRAMP1 iron transporter is involved in Cd accumulation in rice. J. Exp. Bot. 62, 4843–4850.

Takahashi, R., Ishimaru, Y., Shimo, H., Ogo, Y., Senoura, T., Nishizawa, N.K., Nakanishi, H., 2012. The OsHMA2 transporter is involved in root-to-shoot translocation of Zn and Cd in rice. Plant Cell Environ. 35 (11), 1948–1957.

Tamayo, E., Gomez-Gallego, T., Azcon-Aguilar, C., Ferrol, N., 2014. Genome-wide analysis of copper, iron and zinc transporters in the arbuscular mycorrhizal fungus *Rhizophagus irregularis*. Front. Plant Sci. 5.

Tan, S., Han, R., Li, P., Yang, G., Li, S., Zhang, P., Wang, W.B., Zhao, W.Z., Yin, L.P., 2015. Over-expression of the Mx*IRT1* gene increases iron and zinc content in rice seeds. Transgenic Res. 24, 109–122.

Terzi, H., Yıldız, M., 2015. Interactive effects of sulfur and chromium on antioxidative defense systems and BnMP1 gene expression in canola (*Brassica napus* L.) cultivars differing in Cr (VI) tolerance. Ecotoxicology 1–12.

Tomas, M., Pagani, M.A., Andreo, C.S., Capdevila, M., Atrian, S., Bofill, R., 2015. Sunflower metallothionein family characterisation. Study of the Zn (II)-and Cd (II)-binding abilities of the HaMT1 and HaMT2 isoforms. J. Inorg. Biochem.

Uraguchi, S., Fujiwara, T., 2012. Cadmium transport and tolerance in rice: perspectives for reducing grain cadmium accumulation. Rice 5, 5.

USEPA, 2000. Introduction to phytoremediation. United States Environment Protection Agency. Washington DC.

Usha, B., Keeran, N.S., Harikrishnan, M., Kavitha, K., Parida, A., 2011. Characterization of a type 3 metallothionein isolated from *Porteresia coarctata*. Biol. Plant 55 (1), 119–124.

Vamerali, T., Bandiera, M., Mosca, G., 2010. Field crops for phytoremediation of metal contaminated land: a review. Environ. Chem. Lett. 8, 1–17.

Velu, G., Ortiz-Monasterio, I., Cakmak, I., Hao, Y., Singh, R.P., 2014. Biofortification strategies to increase grain zinc and iron concentrations in wheat. J. Cereal Sci. 59, 365–372.

Watanabe, M., Shinmachi, F., Noguchi, A., Hasegawa, I., 2005. Introduction of yeast metallothionein gene (*CUP1*) into plant and evaluation of heavy metal tolerance of transgenic plant at the callus Stage. Soil Sci. Plant Nutr. 51, 129–133.

Wei, K., Pang, S., Yang, J., Wei, Z., 2015. Enhancement of cadmium tolerance and accumulation by introducing *Perilla frutescens* (L.) Britt var. *frutescens* genes in *Nicotiana tabacum* L. plants. Environ. Sci. Pollut. Res. http://dx.doi.org/10.1007/s11356-014-4048-4.

Wei, W., Smith, N., Wu, X., Kim, H., Seravalli, J., Khalimonchuk, O., Lee, J., 2014. YCF1-mediated cadmium resistance in yeast is dependent on copper metabolism and antioxidant enzymes. Antioxid. Redox. Signal 21 (10), 1475–1489.

Weyens, N., Beckers, B., Schellingen, K., Ceulemans, R., Van der Lelie, D., Newman, L., Vangronsveld, J., 2015. The potential of the Ni-resistant TCE-degrading *Pseudomonas putida* W619-TCE to reduce phytotoxicity and improve phytoremediation efficiency of poplar cuttings on a Ni-TCE co-contamination. Int. J. Phytorem. 17 (1), 40–48.

Wu, Z., Banuelos, G.S., Lin, Z.Q., Liu, Y., Yuan, L., Yin, X., Li, M., 2015. Biofortification and phytoremediation of selenium in China. Front. Plant Sci. 6, 136.

Xie, P., Hao, X., Herzberg, M., Luo, Y., Nies, D.H., Wei, G., 2015. Genomic analyses of metal resistance genes in three plant growth promoting bacteria of legume plants in Northwest mine tailings, China. J. Environ. Sci. 27, 179–187.

Xu, L., Wang, Y., Liu, W., Wang, J., Zhu, X., Zhang, K., Liu, L., 2015. De novo sequencing of root transcriptome reveals complex cadmium-responsive regulatory networks in radish (*Raphanus sativus* L.). Plant Sci. 236, 313–323.

Xu, N., Qian, K., Dong, Y., Chen, Y., Yu, Q., Zhang, B., Li, M., 2014. Novel role of the *Candida albicans* ferric reductase gene CFL1 in iron acquisition, oxidative stress tolerance, morphogenesis and virulence. Res. Microbiol. 165 (3), 252–261.

Yan, Z., Li, X., Chen, J., Tam, N.F.Y., 2015. Combined toxicity of cadmium and copper in *Avicennia marina* seedlings and the regulation of exogenous jasmonic acid. Ecotoxicol. Environ. Saf. 113, 124–132.

Yang, T., Chen, M.L., Wang, J.H., 2015. Genetic and chemical modification of cells for selective separation and analysis of heavy metals of biological or environmental significance. Trend Anal. Chem.

Yang, Z., Wu, Y.R., Ling, H.Q., Clu, C.C., 2009. OsMT1a, a type 1 metallothionein, plays the pivotal role in zinc homeostasis and drought tolerance in rice. Plant Mol. Biol. 70, 219–229.

Yuan, L., Zhu, Y., Lin, Z.Q., Banuelos, G., Li, W., 2013. A novel selenocystine-accumulating plant in selenium-mine drainage area in Enshi, China. PLoS One 8, 65615.

van der Zaal, B.J., NeuteboomL, W., Pinas, J.E., Chardonnens, A.N., Schat, H., Verkleij, J.A.C., Hooykaas, P.J.J., 1999. Overexpiession of a novel *Arabidopsis* gene related to putative zinctransporter genes from animals can lead to enhanced zinc resistance and accumulation. Plant Physiol. 119, 1047–1055.

Zhang, J., Zhang, M., Tian, S., Lu, L., Shohag, M.J.I., Yang, X., 2014. Metallothionein 2 (SaMT2) from *Sedum alfredii* Hance confers increased Cd tolerance and accumulation in yeast and tobacco. PloS One 9 (7), 102750.

Zhang, Q.R., Tang, C.H., Wen, S.Z., Liu, Y.G., Ke, L.L., 2006. Advances in research on genetically engineered plants for metal resistance. J. Integr. Plant Biol. 48 (11), 1257–1265.

Zhang, X.Y., Zhou, W., Ru, B.G., 2000a. Transgenic tobacco with mutant gene has higher tolerance to heavy metals. Acta Bot. Sin. 42, 416–420.

Zhang, X.Y., Zuo, X.F., Xiao, Ch.Y., Shan, L., Ru, B.G., 2000b. Expression of the mouse metallothionein mutant-c DNA improving cadmium resistance in transgenic tobacco. Chin. J. Biochem. Mol. Biol. 16, 631–636.

Zhang, Y., Liu, J., Zhu, Y., Gong, T., Wang, J., Ge, Y., 2013. Enhanced phytoremediation of mixed heavy metal (mercury)–organic pollutants (trichloroethylene) with transgenic alfalfa co-expressing glutathione S-transferase and human P450 2E1. J. Hazard. Mater. 260, 1100–1107.

Zhao, C., Xu, J., Li, Q., Li, S., Wang, P., Xiang, F., 2014. Cloning and characterization of a *Phragmites australis* phytochelatin synthase (PaPCS) and achieving Cd tolerance in tall fescue. PloS One 9 (8), 103771.

Zhao, X.Q., Wang, R.C., Lu, X.C., Lu, J.J., Li, J., Hu, H., 2012. Tolerance and biosorption of heavy metals by *Cupriavidus metallidurans* strain XXKD-1 isolated from a subsurface laneway in the Qixiashan Pb-Zn Sulfide Minery in Eastern China. Geomicrobiol. J. 29 (3), 274–286.

Zhou, G., Pereira, J.F., Delhaize, E., Zhou, M., Magalhaes, J.V., Ryan, P.R., 2014. Enhancing the aluminium tolerance of barley by expressing the citrate transporter genes SbMATE and FRD3. J. Exp. Bot. eru121.

CHAPTER

PHYTOREMEDIATION OF SALINE SOILS FOR SUSTAINABLE AGRICULTURAL PRODUCTIVITY

18

Sameen R. Imadi[1], Syed W. Shah[1], Alvina G. Kazi[1], M.M. Azooz[2,3], Parvaiz Ahmad[4]

Atta-ur-Rahman School of Applied Biosciences, National University of Sciences and Technology (NUST), Islamabad, Pakistan[1]; Department of Botany, Faculty of Science, South Valley University, Qena, Egypt[2]; Department of Biological Sciences, Faculty of Science, King Faisal University, Al-Hassa, Saudi Arabia[3]; Department of Botany, S.P. College, Srinagar, Jammu and Kashmir, India[4]

CHAPTER OUTLINE

Plant Metal Interaction. http://dx.doi.org/10.1016/B978-0-12-803158-2.00018-7

1. INTRODUCTION

Human activities have been a threat to sustainability and stability of planet Earth since the beginning of civilization. Human activities are met with potential and severe consequences toward the environment (Varallyay, 1994). Salinity is one of the major problems worldwide and affects the agricultural production severely (Ahmad et al., 2013, 2014; Hashem et al., 2015; Rasool et al., 2013). It is due to salinity that more than half billion land becomes infertile for crop production and agriculture. Current challenges of global food security can only be met if such destroyed land is made cultivable again. A permanent solution to soil salinity is a drainage system that can manage water logging and raise the underground water table efficiently. But such a system cannot be planted and cannot be used on such large scale (Qadir and Oster, 2004).

Soil conservation and restoration of degraded and saline sodic soils is one of the most serious concerns of agriculture around the globe. Phytoremediation is a biological approach to get rid of soil salinity. This approach can be employed globally. Phytoremediation refers to the term of using plant-based strategies for improving deteriorated soils (Padmavathiamma and Li, 2007). Phytoremediation is performed by planting suitable plant species and by exploiting plant root ability to improve dissolution (Padmavathiamma and Li, 2007). Phytoremediation also involves calcium nutrition to soil, which results in removal of sodium through cation exchange methods (Ashraf et al., 2010). Phytoremediation helps in ameliorating saline and sodic soils (soils that have excess sodium ions in them) through cultivation of crops that are tolerant to soil salinity and sodicity (Qadir et al., 2001).

Fertility and productivity of highly degraded sodic and saline soils can be improved and restored by successful growth and plantation of trees, which are able to decline sodium concentration in soil. It has also been observed that as the plantation age of soil increases, nitrogen (N), phosphorus (P), potassium (K), calcium (Ca) and magnesium (Mg) uptake by plants also increases, which reflects the change of nutrient status of rooting zone (Bhojvaid et al., 1996).

1.1 SALINITY: PROBLEM

Soil salinity is expected to affect the world more vigorously and extensively in coming years. Increasing land salinization is also associated with increasing salinity in water resources (MacFarlane and Williamson, 2002). Soil salinity changes the fertile land to unproductive land, which ultimately results in economical loss (Grieve et al., 1986). Irrigation of soils is directly proportional to decrease in soil salinity (Cetin and Kirda, 2003). Soil salinity is the most severe factor that affects the growth of plants. Soil salinity when combined with boron toxicity is found to be most vigorous in actions against growth of plants (Ibekwe et al., 2010). Most plants do not fully express their original genetic potential for growth, development, and yield under salt stress, which results in declining of their economic and commercial value (Abdelly et al., 2006). Salinity is a soil and water quality concern, especially in arid and semiarid areas where water demand is increasing day by day for irrigation and agriculture. Arid and semiarid areas are the regions where there are insufficient rain to leach salts and excess sodium ions out of the rhizosphere. Most of saline soils contain calcites and calcium salts of extremely low solubility (Qadir et al., 2002). Sodic and saline-sodic soils occur in almost 75 countries around the world and their extent is increasing steadily among agricultural countries. These soils are used for agriculture as there are many countries whose infrastructure is based on agriculture and irrigation (Qadir et al., 2007).

Many chemical treatments have been applied for a long time to get rid of this salinity pollution (Lu et al., 2014). These chemical treatments are harmful for the environment and are very costly. Most common chemical treatment toward saline soil is gypsum treatment. These treatments result in decrease in soil salinity and sodicity (Qadir et al., 2002).

1.2 PHYTOREMEDIATION: SOLUTION

It was a century ago when salinity and sodicity was defined as a real threat to agricultural land and their amelioration was started using calcium and calcite nutrition. After some time, chemicals were applied to soil to get it ameliorated from sodicity and salinity. These chemicals made the soil available for agriculture but the cost of chemicals increased. Phytoremediation came as a solution to all the salinity-related problems of soils (Qadir et al., 2007). Phytoremediation strategy works though plant root action and help in dissolving calcium carbonate for supplying of adequate calcium ions (Qadir et al., 2001).

Scientific research has made it possible for farmers to apply plant-based strategies to eradicate salinity (Ashraf et al., 2010; Ullah et al., 2015). By the use of phytoremediation, soils can be brought back to high fertility and high productivity. Plants, which have ability to dissolve calcite into soil through their roots, are used as phytoremediating agents (Qadir et al., 2007). Phytoextraction, a type of phytoremediation, is a technique in which heavy metal accumulator plants are grown in metal-polluted soils to clean the soils for agriculture production (Manousaki et al., 2008).

Tolerance to salt and heavy metals relies on common physiological mechanisms. For this reason, helophytes have been designated as the best and most potent plants tolerant to salt and heavy metals. Halophytes accumulate salt in their tissues or are able to dissolve soil calcite in rhizosphere to provide calcium ions, which can be exchanged with sodium ions at cation exchange sites. Phytoremediation is a low-cost reclamation method for soils (Manousaki and Kalogerakis, 2011a,b) and is gaining interest throughout the globe.

Amelioration of seashore saline soils can also be performed by planting castor bean seeds (Wu et al., 2012). Planting castor beans for 2 consecutive years results in a decrease of soil salinity by up to 0.92% and was lower than the one observed for a control experiment conducted. Amelioration of saline soil through castor bean planting is also associated with increased soil bulk density and improved soil nutrient condition. Activity and biodiversity of soil is explored with the increase in phytoremediation (Wu et al., 2012). Diversity among grown plant species and microbes also results in faster amelioration of saline and sodic saline soils. Productivity of diverse species is also associated with soil amelioration. In reclamation of soil, biodiversity plays a 42% role, whereas the remaining 58% is productivity (Singh and Garg, 2007).

Phytoremediation of saline soils result in increase of organic carbon content, total nitrogen, available phosphorus, and available potassium ions in soil. Levels of exchangeable calcium and magnesium are also observed to increase, which results in improvement of soil bulk density. Phytoremediation is also associated with increase in soil porosity, water holding capacity, field capacity, soil permeability, and increased infiltration rate. Salt-tolerant and accumulator species have a significant impact on increasing soil fertility and sustainability (Mishra et al., 2004).

Phytoremediation also provides financial benefits to farmers as the crops grown for amelioration of soil are also commercially important (Qadir et al., 2005). Phytoremediation and chemical treatment when applied on soil to ameliorate it from salinity, it was observed that both treatments result in reduction of salinity and sodicity to the same extent in same amount of time. This proves that phytoremediation can be an alternate treatment to chemical (Qadir et al., 2002).

Different types of soil pollution can be removed by phytoremediation. Saline soils polluted with cadmium can easily be phytoremediated with the planting and growth of *Salicornia europaea* (Ozawa et al., 2009).

2. SALINITY: CHALLENGE TO GLOBAL FOOD SECURITY

Exponential increase in the population and increasing demand for food is the major issue before the plant biologists worldwide (Ahmad et al., 2012). Abiotic stress including salinity are major threats to crop yield and have been involved in declining trends of food security since day one. Declining crop yield is related to soil salinity and loss of soil quality (Stocking, 2003). For reconciling biodiversity conservation and food security in environments where soils are unusable because of salinity, change in policies, institutions, and practices are necessary (Brussaard et al., 2010). World food security can only be achieved with the development of crops that are salinity tolerant (Denby and Gehring, 2005; Cominelli et al., 2013). More than 40% of cropping soils around the globe are sodic and saline and farming practices on these soils are very low. Actual yield produced from saline soils is more than 50% less than originally expected yield produced from normal soils. These soils have very low organic matter and biodiversity (Rengasamy, 2002).

2.1 SOIL SALINITY AND PLANTS: EFFECT AND RESPONSE

The major effect of soil salinity observed in plants is negative effect on osmotic potential (Ahmad and Sharma, 2008). Salinity affects dry matter production and transpiration of plants and plant yield is reduced (Ahmad, 2010). Influence of salinity on plant growth is directly dependent on the plant species and management of irrigation. Salinity significantly affects the yield, evapotranspiration, stomatal resistance, water potential, and predawn leaf water potential (Katerji et al., 2000).

In tomatoes, soluble solid content increases with increasing salinity and results in yield reduction (Navarro et al., 2005). Fruit yield reduction is directly related to reduction in water uptake (Navarro et al., 2005). Salinity affects all the morphological parameters of plants and decreases their growth performance. Increase in salinity is related to the decrease in pigment content (Rasool et al., 2013; Hashem et al., 2014, 2015).

In lettuce, soil salinity decreases the gas exchange rates and stomatal conductance (Neocleous et al., 2014). As the stress conditions are intensified in saline soils, decrease in leaf water potential and osmotic potential occurs in plants (Zhang et al., 2010; Jogaiah et al., 2014). Under saline stress, plants are observed to accumulate osmolytes and osmoprotectants such as proline, sugar, protein, and glycine betaine as a response to stress (Ahmad and Sharma, 2008; Ahmad, 2010; Ahmad et al., 2014; Rasool et al., 2013; Hashem et al., 2014, 2015; Jogaiah et al., 2014).

In *Lycium nodosum* plants, salinity has decreased the photosynthetic rate, carboxylation efficiency, and stomatal conductance by 80% (Tezara et al., 2003). As the salt concentration in soil increases, the total uptake of water decreases (Ahmad and Sharma, 2008; Ahmad, 2010; Ahmad et al., 2014; Rasool et al., 2013; Hashem et al., 2014, 2015; Jogaiah et al., 2014). Increase in soil salinity results in increased adsorbed sodium ions, decreased adsorbed calcium ions and magnesium ions, and increased precipitation of mixture of calcium and magnesium carbonates (van Hoorn et al., 1993).

Plant growth responds to salinity in two phases. The first phase is rapid, osmotic phase that is involved in inhibition of growth of young leaves of plants and a slow phase that is the ionic phase that accelerates senescence of mature leaves by making the salt ions accumulate in them. Plants can adapt to salinity by gaining osmotic tolerance, sodium ion exclusion, and tolerance to sodium ion accumulation (Munns and Tester, 2008).

Growth of *Populus alba* 1-year-old rooted cuttings was observed under saline conditions. Growth is not ceased in lower levels of salinity, but the growth starts declining as the level of salinity increases. High salt treatment results in 20% mortality. An important tolerance mechanism exists in *P. alba* as a strong sodium ion partitioning is observed in roots. Total uptake of sodium ions remains constant during high salt treatment and low salt treatment (Imada et al., 2009).

3. CHANGES IN PHYSICAL AND CHEMICAL CHARACTERISTICS OF SOIL

Saline soils have many physical changes in them as compared with normal healthy soils. These soils have low biodiversity and are known to be dead soils. They have nutrient deficiencies and toxicities because of boron, carbonate, and aluminate ions present. These soils are observed to have very low organic matter and biological activity (Rengasamy, 2002). Increase in soil salinity results in progressively smaller and more stressed microbial community, which is also seen to be least efficient toward metabolism (Rietz and Haynes, 2003).

Physical properties of saline soils are observed to be improved by growth of salt-tolerant plants. These plant varieties are shown to improve water uptake in soil, saturated hydraulic conductivity of soil, structural stability, bulk density of soil, and soil porosity. Increase in available water is associated with soil organic matter content and porosity. Hydraulic conductivity of soil is observed to increase with organic matter content, structural stability, water retention, and porosity (Akhter et al., 2004).

Salt content of soil is commonly expressed by electrical conductivity. Salinized soils are observed to have low structural stability, infiltration rate, and water-holding capacity. Soluble salts present in soil are highly mobile. For this reason, these salts are easily accumulated in plants (Tanji, 2002). Salinity and sodicity in soil, which is artificially induced by agriculture, does not affect any physical and chemical characteristics of soil (Rietz and Haynes, 2003). Salt accumulation in soil is highly dependent on quantity of irrigation water, salt concentration of irrigation water, rainfall, and amount of leaching.

Bulk density of soil is seen to be effectively increased after the saline soils are exposed to organic manure and organic matter for about 5 consecutive years. A 23% increase in bulk density has been observed as a result of this experiment (Tejada et al., 2006). Autumn and spring rainfall is expected to cause leaching of salts and accumulation of salts in deeper layers of saline soils. Percentage of exchangeable sodium ions increases with an increase in salinity (Tedeschi and Dell'Aquilla, 2005).

Saline soils are seen to possess high concentrations of potassium ions, chloride ions, and sulfate ions. Increased soil salinity is associated with increased concentrations of these ions. It is also observed that an increase in soil salinity results in decrease in concentration of magnesium ions and calcium ions in the soil (Hashem et al., 2014). Chemical characteristics of soil are adversely affected by salinity because it results in increased cadmium and heavy metal ion concentrations. Sodium ion concentration is also increased with an increase in soil salinity (Weggler-Beaton et al., 2000). In saline soils, the concentration of soluble chloride ions is highest among anion concentrations. Following the chloride ion are sulfate ions, which are followed by bicarbonate ions (Cao et al., 2012). Saline soils enhance the destroying degradation of organic matter (Wong et al., 2010). Salinity reduces soil microbial biomass and microbial activity, which in turn changes the carbon dioxide fluxes.

4. REMOVAL OF SALTS FROM SOIL

Removal of salts from the soil requires phytoremediation approaches and research knowledge. Specific plants can remove salt from soils, but it is very important to know the capacity of these plants to grow under saline conditions and the threshold of these plants. It is also important to know the critical hazardous level of saline irrigation water in which fruit trees can survive (Kilic et al., 2008).

4.1 CULTIVATING SUITABLE PLANT SPECIES

Suitable plant species can be cultivated for the purpose of phytoremediation. These species may include halophytes (the plants which grow in high salt conditions), hyperaccumulators (the plants that can accumulate salt ions in large amounts in above-ground parts), salt-tolerant plants (the plants which are highly efficient and have genes to tolerate the salt stress), and transgenic plants (which are made salt tolerant artificially by insertion of salt stress-tolerant genes). The categories of these plants are discussed in detail in the following text.

4.2 EXPLOITING THE ABILITY OF PLANT ROOTS TO IMPROVE DISSOLUTION

Increased dissolution rates of calcite through phytoremediation result in increased levels of calcium in soil solution to replace sodium for cation exchange method. This process of dissolution is driven by partial pressure of carbon dioxide in root zones and an increase of proton release from plant roots. The increase in calcite dissolution results in improved physical properties within root zones. Increased aqueous carbon dioxide levels in soil results in formation of hydrogen ions and bicarbonate ions from the calcite.

4.3 ENHANCED CALCIUM LEVELS IN SOIL SOLUTION

Calcium levels in soil can be increased by planting phytoremediating agents, which have a high ability to dissolve calcite in soil solutions and to increase the calcium concentration for exchange with sodium. Alfalfa roots are effective in dissolving calcite in soil solutions and forming calcium ions (Qadir et al., 2003a). Hydrogen ions produced as a result of calcite dissolution, further react with calcite, and release calcium ions (Qadir et al., 2003a).

4.4 REMOVE SODIUM FROM SOIL-CATION EXCHANGER

A major improvement in soil content observed by phytoremediation includes removal and exchange of sodium ions by calcium ions from soil cation exchange sites. Alfalfa plants, when used as phytoremediating agents, are a promising tool for removal of excess sodium from soil (Qadir et al., 2003b). Calcium ions released as a result of hydrogen ion reaction with calcite, work at the cation exchange site, and exchange with sodium ions (Qadir et al., 2003b).

5. IMPROVEMENT IN SOIL FERTILITY

As a result of phytoremediation, many benefits are achieved. One of the most important is improvement of soil fertility. Soil fertility is improved as a result of dissolution of calcite into calcium ions and bicarbonate ions and cationic exchange between sodium ions and calcium ions. This results in increased microbial diversity in soil, which also helps in gaining fertility (Pulford and Watson 2003; Khan, 2005). Improvement of soil fertility as a result of phytoremediation and its consequences are shown in Figure 18.1.

6. SELECTION OF PLANTS FOR PHYTOREMEDIATION

Plants best suited for phytoremediation are halophytes and hyperaccumulators. Another category of plants gaining interest for phytoremediation is transgenic plants. Plant abilities for phytoremediation are determined by ion uptake, root to shoot translocation of ion, intracellular sequestration, and chemical modification and stress resistance. These properties, if present in plants, can result in a formation of a potent phytoremediating choice. The genes, which control these mechanisms, can be manipulated and inserted into different plants for producing transgenic varieties (Pilon-Smith and Pilon, 2002).

Some plants are observed to improve the quality of saline soils and make them perfectly available for agriculture. Kallar grass is one of those plants. The growth of kallar grass in saline land for about 5 years has been highly advantageous for bringing the soil back to life (Akhter et al., 2004; reviewed by Hasanuzzaman et al., 2014). Kallar grass is effective for rehabilitation and restoration of soil fertility in saline-sodic areas on sustainable grounds (Akhter et al., 2004). Several turf grass species are found to mitigate the soil by their growth under salinity. A marker for their uptake of salts was kept as

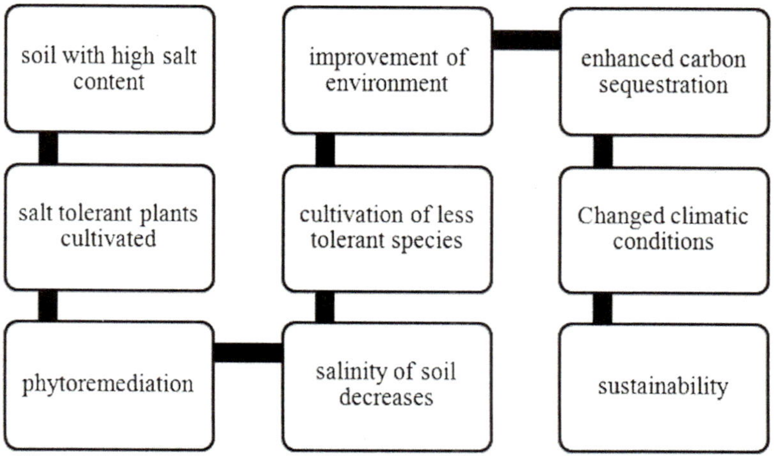

FIGURE 18.1

Improvement in soil fertility by phytoremediation and its consequences.

chloride ions, which means that if more chloride ions are present in plant leaves than normal, the plant is a potent phytoremediating agent (Beltrao et al., 2009).

Significant improvements in plants for salt tolerance have been achieved in recent years (Ahmad et al., 2013, 2014; Hashem et al., 2015; Rasool et al., 2013). The plants, which have been improved, include barley, rice, maize, sorghum, and alfalfa. Improvement in these crops has led to the increase in their economic importance (Ashraf and Wu, 1994). On classification, maize, sunflower, and potato are also placed in the list of plants that are tolerant to salt. Hence, these plants may be used for phytoremediation (Katerji et al., 2000). Soil salinity is seen to be significantly decreased when salt-tolerant species of sea orach are planted (Gharaibeh et al., 2011). Sea orach planting combined with gypsum treatment is more effective in ameliorating the saline and saline-sodic soils (Gharaibeh et al., 2011).

6.1 HYPERACCUMULATORS AND HALOPHYTES

Many plant species are able to accumulate a high quantity of salts are used as phytoremediators (Hasanuzzaman et al., 2014). Fast and luxuriantly growing halophytic herbs are observed to be high in salt accumulation. These halophytic herbs can be grown in saline soils so that they bioaccumulate the salts and helps in phytoremediation (Hasanuzzaman et al., 2014). Among the halophytic herbs *Suaeda maritama* and *Sesuvium portulacastrum* exhibit the highest accumulation of salts and can be grown easily (Ravindran et al., 2007; Hasanuzzaman et al., 2014).

Halophytes are expected to have better tolerance against environmental stresses, such as heavy metals as compared with salt-sensitive crop plants. Halophytes can prove to be potent for phytoremediation of heavy metals and decontamination of polluted soils (Hasanuzzaman et al., 2014). Halophytes are key players for phytoextraction, phytostabilization, and phytoexcretion of heavy metals in saline soils. Salt-accumulating halophytes can be used for soil desalinization in arid and semiarid environments (Manousaki and Kalogerakis, 2011a,b).

Halophytic plants have been observed to be brilliant in an ability to grasp sodium chloride in their cell vacuoles as osmoticum. Halophytes are the species that can be used to desalinize and restore saline soils and withstand highly saline water irrigation (Lokhande and Suprasanna, 2012). It was observed that rhizosphere desalinization is a result of shoot sodium accumulation. Halophytes can be used to leach salt from rhizospheres in regions where precipitation is too low (Rabhi et al., 2009). Cultivation of halophytes on salinized soil leads to marked absorption of sodium ions from roots and its accumulation in plant parts above the ground (Rabhi et al., 2010). In an experiment conducted by Rabhi et al. (2008), it was confirmed that the perennial species of halophytes when grown under saline soil and irrigated by tap water for 170 days, significant reduction in salinity is observed. The irrigation was carried out without any leaching. Soil electrical conductivity also decreases as the sodium ion concentration in soil decreases. *Sesuvium* sp. are most successful in amelioration of soil and are able to accumulate up to 30% sodium ion (Rabhi et al., 2008). Growth of alfalfa species provides evidence that salt content in saline soil profile is readily decreased with alfalfa cultivation (Cao et al., 2012).

6.2 SALT-TOLERANT PLANTS

Tamarix smyrnensis, a salt cedar widely grown in Mediterranean region, is a highly salt-tolerant plant. The plant is observed to accumulate high concentration of cadmium in saline environments in its above-ground parts, which include leaves and shoots. From the leaves, cadmium is secreted out of the plant through its salt glands. This proves that plant responds to cadmium stress as salt stress and that cadmium secretion from salt glands is a possible detoxification mechanism to prevent cadmium toxicity (Manousaki et al., 2008). *Tamarix* (salt cedar) plants, when grown for 10 weeks in saline solution, reduce salinity to a large extent. If lead and cadmium are added in this solution, the plants are also known to uptake and accumulate them resulting in their detoxification (Kadukova et al., 2008).

Common plants such as species of *Brassica* can also be used for phytoremediation of heavy metals and saline soils (Banuelos et al., 1997; Ahmad et al., 2011, 2015). *Portulaca oleracea* is salt-tolerant specie that can be grown in saline environments for phytoremediation. An experiment proved that if the water table is 1.1 m, then the growth of *Portulaca oleracea* does not require any irrigation from any means and it removes the salt to highest extent. Highest salt accumulation is observed to be 497 kg per hectare, which makes it around 16% of removed salt. This specie can be used as a potent phytoremediator (Hamidov et al., 2007a).

Chenopodium album species produce high dry biomass and have been proved that these species can be used as a rehabilitating and reclamating agent for the saline soils (Hamidov et al., 2007b). *Glycyrrhiza glabra*, commonly known as liquorice, is salt tolerant and also has been used for phytoremediating saline soils (Kushiev et al., 2005). Positive benefits to soil have been obtained on the growth of this specie. Salt content of soil on which liquorice grows has been tested to show a steep decline and fall. Abandoned soils can be brought back to life by the use of liquorice at a low cost and in less time (Kushiev et al., 2005).

7. LIMITATIONS OF PHYTOREMEDIATION

Every new technology that has many benefits for mankind is also associated with some limitations. These limitations of phytoremediation are given next.

7.1 PLANTS BECOME UNABLE TO USE AS FODDER AND FOOD

Most of the plants used for phytoremediation are fodder plants. If the soil to be ameliorated is high with salt content, then the plants may contain toxic substances and hence unable to be given to animals. Many saline soils also have heavy metal toxicity. If heavy metals are present in soil, the plants grown on it cannot be used as fodder. Phytoremediating plants also include human edible plants such as *Brassica* sp. These plants also become inedible by excess of phytoremediation (Ernst, 1998).

7.2 LONG-TERM

Phytoremediation is a long-term commitment approach because of the slow growth of plants. Bulk production of a biomass that phytoremediation of land may take years and sometimes even decades to get clarified (Joner et al., 2004). Time consumption of the technique makes it less applicable toward the soil. As compared with chemicals used, phytoremediation is an extremely slow technique (Cunningham and Ow, 1996).

7.3 LEACHING PREVENTION

With all the preventive measures taken, it is still impossible to ensure 0% leaching of heavy metals and salt ions. Because it is a plant-based system, leaching takes place in underground water, which makes this technique less efficient and less desirable because after some years those metals and salts ions may affect the fertility and productivity of soil again (Cunningham and Ow, 1996).

8. CONCLUSION AND FUTURE PROSPECTS

Phytoremediation is an emerging technology helping to clean the soil and water bodies from noxious pollutants. Phytoremediation can provide a low-cost and sustainable way to improve the economies of developing countries. Phytoremediation is one of the most potent and viable solution for poor farmers through community-based management. Phytoremediation will be a link among researchers from different institutes and from farmers. These linkages between farmers and researchers will increase the use of polluted soils for the agricultural purposes. Because this technique is a biological, the future belongs to it. The research on phytoremediation requires a proper understanding of processes and mechanisms, which foster phytoremediation, the tolerance of plant species for saline and sodic soils and markets of agricultural products, which are produced as a result of phytoremediation. Further research and knowledge is required to commercialize this technique on a large scale and will ensure the food security in a sustainable way and will make the planet Earth more beautiful place to live.

REFERENCES

Abdelly, C., Barhoumi, Z., Ghnaya, T., Debez, A., Hamed, K.B., Ksouri, R., Talbi, O., Zribi, F., Ouerghi, Z., Smaoui, A., Huchzermeyer, B., Grignon, C., 2006. Potential utilisation of halophytes for the rehabilitation and valorisation of salt-affected areas in Tunisia. In: Oztürk, M., Waisel, Y., Khan, M.A., Gork, G. (Eds.), Biosaline Agriculture and Salinity Tolerance Plants. Birkhauser Verlag, Switzerland, pp. 163–172.

Ahmad, P., 2010. Growth and antioxidant responses in mustard (*Brassica juncea* L.) plants subjected to combined effect of gibberellic acid and salinity. Arch. Agro Soil Sci. 56 (5), 575–588.

Ahmad, P., Ashraf, M., Younis, M., Hu, X., Kumar, A., Akram, N.A., Al-Qurainy, F., 2012. Role of transgenic plants in agriculture and biopharming. Biotechnol. Adv. 30, 524–540.

Ahmad, P., Azooz, M.M., Prasad, M.N.V., 2013. Salt Stress in Plants: Signalling, Omics and Adaptations. Springer Science+Business Media, LLC, 233 Spring Street, New York, NY 10013, USA.

Ahmad, P., Nabi, G., Ashraf, M., 2011. Cadmium-induced oxidative damage in mustard [*Brassica juncea* (L.) Czern. & Coss.] plants can be alleviated by salicylic acid. South Afr. J. Bot. 77, 36–44.

Ahmad, P., Ozturk, M., Sharma, S., Gucel, S., 2014. Effect of sodium carbonate-induced salinity-alkalinity on some key osmoprotectants, protein profile, antioxidant enzymes, and lipid peroxidation in two mulberry (*Morus alba* L.) cultivars. J. Plant Interact. 9 (1), 460–467.

Ahmad, P., Sarwat, M., Bhat, N.A., Wani, M.R., Kazi, A.G., Tran, L.S.P., 2015. Alleviation of cadmium toxicity in *Brassica juncea* L. (Czern. & Coss.) by calcium application involves various physiological and biochemical strategies. PloS One 10 (1), e0114571.

Ahmad, P., Sharma, S., 2008. Salt stress and phyto-biochemical responses of plants. Plant Soil Environ. 54 (3), 89–99.

Akhter, J., Murray, R., Mahmood, K., Malik, K.A., Ahmed, S., 2004. Improvement of degraded physical properties of a saline-sodic soil by reclamation with kallar grass (*Leptochloa fusca*). Plant Soil 258 (1), 207–216.

Ashraf, M., Wu, L., 1994. Breeding for salinity tolerance in plants. Crit. Rev. Plant Sci. 13 (1), 17–42.

Ashraf, M.A., Ashraf, M., Mahmood, K., Akhter, J., Hussain, F., Arshad, M., 2010. Phytoremediation of saline soils for sustainable agricultural productivity. In: Ashraf, M., Ozturk, M., Ahmad, M.S.A. (Eds.), Plant Adaptation and Phytoremediation. Springer, Berlin, Germany, pp. 335–355.

Banuelos, G.S., Ajwa, H.A., Terry, N., Zayed, A., 1997. Phytoremediation of selenium laden soils: a new technology. J. Soil Water Conserv. 52 (6), 426–430.

Beltrao, J., Neves, A., De Brito, J.C., 2009. Salt removal potential of turf grass in golf courses in the Mediterranean Basin. WSEAS Trans. Environ. Develop. 5 (5), 394–403.

Bhojvaid, P.P., Timmer, V.R., Singh, G., 1996. Reclaiming sodic soils for wheat production by *Prosopis juliflora* (Swartz) DC afforestation in India. Agroforest. Syst. 34 (2), 139–150.

Brussaard, L., Caron, B., Campbell, B., Lipper, L., Mainka, S., Rabbinge, R., Babin, D., Pulleman, M., 2010. Reconciling biodiversity conservation and food security: scientific challenges for a new agriculture. Curr. Opin. Environ. Sustain. 2 (1–2), 34–42.

Cao, J., Li, X., Kong, X., Zed, R., Dong, L., 2012. Using alfalfa (*Medicago sativa*) to ameliorate salt-affected soils in Yingda irrigation district in Northwest China. Acta Ecol. Sinica 32 (2), 68–73.

Cetin, M., Kirda, C., 2003. Spatial and temporal changes of soil salinity in a cotton field irrigated with low-quality water. J. Hydrol. 272 (1–4), 238–249.

Cominelli, E., Conti, L., Tonelli, C., Galbiati, M., 2013. Challenges and perspectives to improve crop drought and salinity tolerance. New Biotechnol. 30 (4), 355–361.

Cunningham, S.D., Ow, D.W., 1996. Promises and prospects of phytoremediation. Plant Physiol. 110 (3), 715–719.

Denby, K., Gehring, C., 2005. Engineering drought and salinity tolerance in plants: lessons from genome-wide expression profiling in Arabidopsis. Trend Biotechnol. 23 (11), 547–552.

Ernst, W.H.O., 1998. Sulfur metabolism in higher plants: potential for phytoremediation. Biodegradation 9 (3–4), 311–318.

Gharaibeh, M.A., Eltaif, N.I., Albalasmeh, A.A., 2011. Reclamation of highly calcareous saline sodic soil using atriplex halimus and by-product gypsum. Int. J. Phytoremed. 13 (9), 873–883.

Grieve, A.M., Dunford, E., Marston, D., Martin, R.E., Slavich, P., 1986. Effects of waterlogging and soil salinity on irrigated agriculture in the Murray Valley: a review. Aust. J. Exp. Agric. 26 (6), 761–777.

Hamidov, A., Khaydarova, V., Khamidov, M., Neves, A., Beltrao, J., 2007a. Remediation of saline soils using *Apocynum lancifolium* and *Chenopodium album*. In: Proc of 3rd IASME/WSEAS International Conference on Energy, Environment, Ecosystem and Sustainable Development. Agios Nikolaos, Greece, July 24–26, pp. 157–164.

Hamidov, A., Khaydarova, V., Sharipova, S., Costa, C., Beltrao, J., 2007b. Salt removal potential of *Portulaca oleracea* golden purslane. In: Proc of 3rd IASME/WSEAS International Conference on Energy, Environment, Ecosystem and Sustainable Development. Agios Nikolaos, Greece, July 24–26, pp. 151–156.

Hasanuzzaman, M., Nahar, K., Alam, M.M., Bhowmik, P.C., Hossain, M.A., Rahman, M.M., Prasad, M.N.V., Ozturk, M., Fujita, M., 2014. Potential use of halophytes to remediate saline soils. Biomed. Res. Int. 2014, 589341.

Hashem, A., Abd_Allah, E.F., Ahmad, P., 2015. Effect of AM fungi on growth, physio-biochemical attributes, lipid peroxidation, antioxidant enzymes and plant growth regulators in *Lycopersicon esculantum* Mill. subjected to different concentration of NaCl. Pak J. Bot. 47 (1), 327–340.

Hashem, A., Abd_Allah, E.F., El-Didamony, G., Alwhibi Mona, S., Egamberdieva, D., Ahmad, P., 2014. Alleviation of adverse impact of salinity on faba bean (*Vicia faba* l.) by arbuscular mycorrhizal fungi. Pak J. Bot. 46 (6), 2003–2013.

van Hoorn, J.W., Katerji, N., Hamdy, A., Mastrorilli, M., 1993. Effect of saline water on soil salinity and on water stress, growth, and yield of wheat and potatoes. Agric. Water Manage. 23 (3), 247–265.

Ibekwe, A.M., Poss, J.A., Grattan, S.R., Grieve, C.M., Suarez, D., 2010. Bacterial diversity in cucumber (*Cucumis sativus*) rhizosphere in response to salinity, soil pH, and boron. Soil Biol. Biochem. 42 (4), 567–575.

Imada, S., Yamanaka, N., Tamai, S., 2009. Effects of salinity on the growth, Na partitioning, and Na dynamics of a salt-tolerant tree, *Populus alba* L. J. Arid Environ. 73 (3), 245–251.

Jogaiah, S., Ramteke, S.D., Sharma, J., Upadhyay, A.K., 2014. Moisture and salinity stress induced changes in biochemical constituents and water relations of different grape rootstock cultivars. Int. J. Agron. 2014, 8. Article ID 789087.

Joner, E.J., Hirmann, D., Szolar, O.H.J., Todorovic, D., Leyval, C., Loibner, A.P., 2004. Priming effects on PAH degradation and ecotoxicity during a phytoremediation experiment. Environ. Pollut. 128 (3), 429–435.

Kadukova, J., Manousaki, E., Kalogerakis, N., 2008. Pb and Cd accumulation and phyto-excretion by salt cedar (*Tamarix smyrnensis* Bunge). Int. J. Phytorem. 10 (1), 31–46.

Katerji, N., van Hoorn, J.W., Hamdy, A., Mastrorilli, M., 2000. Salt tolerance classification of crops according to soil salinity and to water stress day index. Agric. Water Manage. 43 (1), 99–109.

Khan, A.G., 2005. Role of soil microbes in the rhizospheres of plants growing on trace metal contaminated soils in phytoremediation. J. Trace Elem. Med. Biol. 18 (4), 355–364.

Kilic, C.C., Kukul, Y.S., Anac, D., 2008. Performance of purslane (*Portulaca oleracea* L.) as a salt-removing crop. Agric. Water Manage. 95 (7), 854–858.

Kushiev, H., Noble, A.D., Abdullaev, I., Toshbekov, U., 2005. Remediation of abandoned saline soils using *Glycyrrhiza glabra*: a study from the hungry steppes of central Asia. Int. J. Agric. Sustain. 3 (2), 102–113.

Lokhande, V.H., Suprasanna, P., 2012. Prospects of halophytes in understanding and managing abiotic stress tolerance. In: Ahmad, P., Prasad, M.N.V. (Eds.), Environmental Adaptation and Stress Tolerance of Plants in Era of Climatic Change. Springer, Drotdretch, NY, pp. 29–56.

Lu, J., Yan, X., Ma, Y.F., Tian, C.X., Ding, J.C., 2014. Impact of salinity on treatment of saline wastewater by sequencing batch biofilm reactor process. J. Cent. South Univ. 21 (5), 1989–1994.

Manousaki, E., Kadukova, J., Papadantonakis, N., Kalogerakis, N., 2008. Phytoextraction and phytoexcretion of Cd by the leaves of *Tamarix smyrnensis* growing on contaminated non-saline and saline soils. Environ. Res. 106 (3), 326–332.

Manousaki, E., Kalogerakis, N., 2011a. Halophytes present new opportunities in phytoremediation of heavy metals and saline soils. Ind. Eng. Chem. Res. 50 (2), 656–660.

Manousaki, E., Kalogerakis, N., 2011b. Halophytes—an emerging trend in phytoremediation. Int. J. Phytorem 13 (10), 959–969.

McFarlane, D.J., Williamson, D.R., 2002. An overview of water logging and salinity in southwestern Australia as related to the 'Ucarro' experimental catchment. Agric. Water Manage. 53 (1–3), 5–29.

Mishra, A., Sharma, S.D., Pandey, R., Mishra, L., 2004. Amelioration of a highly alkaline soil by trees in northern India. Soil Use Manage. 20 (3), 325–332.

Munns, R., Tester, M., 2008. Mechanisms of salinity tolerance. Ann. Rev. Plant Biol. 59, 651–681.

Navarro, J.M., Flores, P., Carvajal, M., Martinez, V., 2005. Changes in quality and yield of tomato fruit with ammonium, bicarbonate and calcium fertilisation under saline conditions. J. Hortic. Sci. Biotechnol. 80 (3), 351–357.

Neocleous, D., Koukounaras, A., Siomos, A.S., Vasilakakis, M., 2014. Changes in photosynthesis, yield, and quality of baby lettuce under salinity stress. J. Agric. Sci. Tech. 16, 1335–1343.

Ozawa, T., Miura, M., Fukuda, M., Kakuta, S., 2009. Cadmium tolerance and accumulation in a halophyte Salicornia europaea as a new candidate for phytoremediation of saline soils. Scientific Report of the Graduate School of Life and Environmental Sciences, Osaka Prefecture Univ., vol. 60, pp. 1–8.

Padmavathiamma, P.K., Li, L.Y., 2007. Phytoremediation technology: hyper-accumulation metals in plants. Water Air Soil Pollut. 184, 105–126.

Pilon-Smits, E., Pilon, M., 2002. Phytoremediation of metals using transgenic plants. Crit. Rev. Plant Sci. 21 (5), 439–456.

Pulford, I.D., Watson, C., 2003. Phytoremediation of heavy metal-contaminated land by trees—a review. Environ. Int. 29 (4), 529–540.

Qadir, M., Ghafoor, A., Murtaza, G., 2001. Use of saline–sodic waters through phytoremediation of calcareous saline–sodic soils. Agric. Water Manage. 50 (3), 197–210.

Qadir, M., Noble, A.D., Oster, J.D., Schubert, S., Ghafoor, A., 2005. Driving forces for sodium removal during phytoremediation of calcareous sodic and saline–sodic soils: a review. Soil Use Manage. 21 (2), 173–180.

Qadir, M., Oster, J.D., Schubert, S., Noble, A.D., Sahrawat, K.L., 2007. Phytoremediation of sodic and saline-sodic soils. Adv. Agron. 96, 197–247.

Qadir, M., Oster, J.D., 2004. Crop and irrigation management strategies for saline-sodic soils and waters aimed at environmentally sustainable agriculture. Sci. Total Environ. 323 (1–3), 1–19.

Qadir, M., Qureshi, R.H., Ahmad, N., 2002. Amelioration of calcareous saline sodic soils through phytoremediation and chemical strategies. Soil Use Manage. 18 (4), 381–385.

Qadir, M., Steffens, D., Yan, F., Schubert, S., 2003a. Proton release by N_2-fixing plant roots: a possible contribution to phytoremediation of calcareous sodic soils. J. Plant Nutr. Soil Sci. 166 (1), 14–22.

Qadir, M., Steffens, D., Yan, F., Schubert, S., 2003b. Sodium removal from a calcareous saline–sodic soil through leaching and plant uptake during phytoremediation. Land Degrad. Dev. 14 (3), 301–307.

Rabhi, M., Ferchichi, S., Jouini, J., Hamrouni, M.H., Koyro, H.-W., Ranieri, A., Abdelly, C., Smaoui, A., 2010. Phytodesalination of a salt-affected soil with the halophyte *Sesuvium portulacastrum* L. to arrange in advance the requirements for the successful growth of a glycophytic crop. Biosource Technol. 101 (17), 6822–6828.

Rabhi, M., Hafsi, C., Lakhdar, A., Hajji, S., Barhoumi, Z., Hamrouni, M.H., Abdelly, C., Smaoui, A., 2009. Evaluation of the capacity of three halophytes to desalinize their rhizosphere as grown on saline soils under nonleaching conditions. Afr. J. Ecol. 47 (4), 463–468.

Rabhi, M., Talbi, O., Atia, A., Abdelly, C., Smaoui, A., 2008. Selection of a halophyte that could be used in the bioreclamation of salt-affected soils in arid and semi-arid regions. In: Abdelly, C., Oztürk, M., Ashraf, M., Grignon, C. (Eds.), Biosaline Agriculture and High Salinity Tolerance. Birkhauser Verlag, Switzerland pp. 241–246.

Rasool, S., Ahmad, A., Siddiqi, T.O., Ahmad, P., 2013. Changes in growth, lipid peroxidation and some key antioxidant enzymes in chickpea genotypes under salt stress. Acta Physiol. Plant 35 (4), 1039–1050.

Ravindran, K.C., Venkatesan, K., Balakrishnan, V., Challappan, K.P., Balasubramanian, T., 2007. Restoration of saline land by halophytes for Indian soils. Soil Biol. Biochem. 39 (10), 2661–2664.

Rengasamy, P., 2002. Transient salinity and subsoil constraints to dryland farming in Australian sodic soils: an overview. Aus J. Exp. Agric. 42 (3), 351–361.

Rietz, D.N., Haynes, R.J., 2003. Effects of irrigation-induced salinity and sodicity on soil microbial activity. Soil Biol. Biochem. 35 (6), 845–854.

Singh, B., Garg, V.K., 2007. Phytoremediation of a sodic forest ecosystem: plant community response to restoration process. Notulae Botanicae Horti Agrobotanici Cluj-Napoca 35 (1).

Stocking, M.A., 2003. Tropical soils and food security: the next 50 years. Science 302 (5649), 1356–1359.

Tanji, K.K., 2002. Salinity in the soil environment. In: Lauchli, A., Luttge, U. (Eds.), Salinity: Environment-plants- Molecules. Springer, Netherlands, pp. 21–51.

Tedeschi, A., Dell'Aquilla, R., 2005. Effects of irrigation with saline waters, at different concentrations, on soil physical and chemical characteristics. Agric. Water Manage. 77 (1–3), 308–322.

Tejada, M., Garcia, C., Gonzalez, J.L., Hernandez, M.T., 2006. Use of organic amendment as a strategy for saline soil remediation: influence on the physical, chemical and biological properties of soil. Soil Biol. Biochem. 38 (6), 1413–1421.

Tezara, W., Martinez, D., Rengifo, E., Herrera, A., 2003. Photosynthetic responses of the tropical spiny shrub *Lycium nodosum* (Solanaceae) to drought, soil salinity and saline spray. Ann. Bot. 92 (6), 757–765.

Ullah, A., Heng, S., Munis, M.F.H., Fahad, S., Yang, X., 2015. Phytoremediation of heavy metals assisted by plant growth promoting (PGP) bacteria: a review. Environ. Exp. Bot. 117, 28–40.

Varallyay, G., 1994. Climate change, soil salinity and alkalinity. In: Rounsevell, M.D.A., Loveland, P.J. (Eds.), Soil Responses to Climate Change, NATO ASI Series, vol. 23. Springer Berlin Heidelberg, pp. 39–54.

Weggler-Beaton, K., McLaughlin, M.J., Graham, R.D., 2000. Salinity increases cadmium uptake by wheat and Swiss chard from soil amended with biosolids. Aust. J. Soil Res. 38 (1), 37–46.
Wong, V.N.L., Greene, R.S.B., Dalal, R.C., Murphy, B.W., 2010. Soil carbon dynamics in saline and sodic soils: a review. Soil Use Manage. 26 (1), 2–11.
Wu, X.H., Zhang, H.S., Li, G., Liu, X.C., Qin, P., 2012. Ameliorative effect of castor bean (*Ricinus communis* L.) planting on physico-chemical and biological properties of seashore saline soil. Ecol. Engineer 38 (1), 97–100.
Zhang, H., Irving, L.J., McGill, C., Matthew, C., Zhou, D., Kemp, P., 2010. The effects of salinity and osmotic stress on barley germination rate: sodium as an osmotic regulator. Ann. Bot. 106 (6), 1027–1035.

CHAPTER

POTENTIAL OF PLANTS AND MICROBES FOR THE REMOVAL OF METALS: ECO-FRIENDLY APPROACH FOR REMEDIATION OF SOIL AND WATER

19

Naveen K. Singh, Rajeev P. Singh

Institute of Environment and Sustainable Development, Banaras Hindu University, Varanasi, Uttar Pradesh, India

CHAPTER OUTLINE

1. INTRODUCTION

An increasingly industrialized society has led to the widespread introduction of trace metals and metalloids into our environment, causing acute and diffuse contamination of soil and waters. Anthropogenic activities contributing to trace metal accumulation in soils include metalliferous mining and smelting, metallurgical industries, and use of fertilizers and soil amendments in high-production agriculture (Alloway, 1995; Adriano, 2001; Ma et al., 2011; Kumar et al., 2013; Rahimi et al., 2013; Pons-Branchu et al., 2015). Soil pollution by heavy metals is a global environmental problem and has affected about 235 million hectares of arable land worldwide (Bermudez et al., 2012). Heavy metals in soils can be taken up by plants and accumulate in their edible parts, which may eventually enter the human body through the food chain (Kaplan et al., 2011). Because of their toxicity, the enrichment of heavy metals, particularly of the ions of heavy metals, leads to decreases in

Plant Metal Interaction. http://dx.doi.org/10.1016/B978-0-12-803158-2.00019-9

biodiversity and productivity and, thereby, results in changes in structure and function of ecosystems (Mayor et al., 2013; Niemeyer et al., 2012). The conventional treatment systems used for waste reclamation has inherent limitations and cannot reduce the metals effectively (Chen et al., 2014). Because metals cannot be transferred from water through degradation by biological processes, using vegetation to remove, detoxify, or restabilize polluted sites has been a widely accepted tool in developed countries for cleaning polluted water because it regenerates the original water permanently (Ali et al., 2013). Health hazards posed by accumulation of metals in the environment and the high cost of removal and replacement of metals from polluted soil have prompted efforts to develop phytoremediation strategies. Intake of metals through the food chain by the human population has been widely reported (Pandey et al., 2015). Phytoremediation, the use of plants and their associated microbial communities to remove or inactivate pollutants from the environment, includes any of several technologies for detoxifying the environment with genetically modified or wild-type plants (Kraemer, 2005). Interactions among metals, microbes, and plants have attracted attention because of the biotechnological potential of microorganisms for metal removal directly from polluted site or the possible transfer of accumulated metals to higher plants and the diversion of heavy metals toward microbial metabolism and growth. Although many plants and bacteria have their own mechanisms for dealing with heavy metal contaminants, the interaction of plants and microorganisms may increase or decrease heavy metal accumulation in plants, depending on the nature of the plant–microbe interaction (Pattanayak and Dhal, 2014). Aquatic macrophytes, such as *Typha latifolia*, *Phragmites australis*, and *Colocasia esculenta*, possess rapid growth and high accumulated amounts of metals in their tissue at a concentration >50-fold than those found in nonhyperaccumulating plants (Rai et al., 2013; Madera-Parra et al., 2013; Rai et al., 2015). Phytoremediation of aquatic environments may be used as an alternative or in addition to conventional remediation methods, including ion exchange resins and

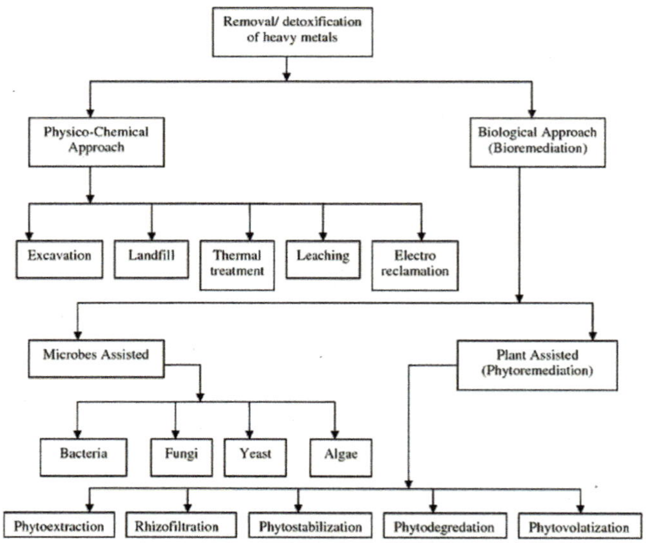

FIGURE 19.1

Approaches used for the remediation of heavy metals from contaminated soil (Khan et al., 2009).

electrodialysis, chemical precipitation, sedimentation, microfiltration, and reverse osmosis (Rai, 2009). Biological remediation techniques offer effective alternative treatments that are often less costly and are considered more environmentally friendly and publicly acceptable than conventional technologies (Figure 19.1). Various phytoremediation technologies include phytoextraction, phyto-volatilization, phytostabilization, and rhizofiltration and are summarized in several reviews (Flathman and Lanza, 1998; Prasad and Freitas, 2003; Jabeen et al., 2009). The adequate restoration of the environment requires cooperation, integration, and assimilation of biotechnological approaches along with traditional and ethical wisdom to conserve our natural resources (Mani and Kumar, 2014; Mani et al., 2015).

2. METAL TOLERANCE IN PLANTS

In plants, metal accumulation in the cells may be regulated by glutathione-phytochelatin–mediated resistance (Hossain et al., 2012; Flores-Cáceres et al., 2015). The importance of glutathione to metal tolerance is based on its dual role as an antioxidant and as a precursor of phytochelatins (PCs), which are able to chelate free metal ions (Jozefczak et al., 2012). In this system, glutathione, the cysteine containing tripeptide that also has several functions in plant cells including dealing with toxic oxygen species and amino acid transport, is used to synthesize PCs, which chelate heavy metals by formation of a thiolate. Thiolates can be transported to vacuoles for heavy metal storage (Mendoza-Cozatl and Moreno-Sanchez, 2006). PCs are activated by heavy metals and scavenge heavy metals in plant cells (Blum et al., 2007; Shahid et al., 2014). Although some plants deal with moderate levels of toxic metals by chelation, other plants have the ability to accumulate extremely high levels of heavy metals and sequester them in their tissues (Jabeen et al., 2009).

3. METAL REMOVAL BY AQUATIC AND TERRESTRIAL PLANTS

Aquatic and wetland plants including the water hyacinth *Eichhorinia crassipes* (Agunbiade et al., 2009; Mishra and Tripathi, 2009), the invasive reed *P. australis* (Ghassemzadeh et al., 2008), the duckweeds *Spirodela polyrhiza* (Rahman et al., 2007), *Lemna minor* (Hou et al., 2007; Uysal and Taner, 2009) and *Lemna gibba* (Khellaf and Zerdaoui, 2009; Megateli et al., 2009), the aquatic fern *Azolla pinnata* (Rai and Tripathi, 2009), and yellow velvetleaf *Limnocharis flava* (Abhilash et al., 2009), *T. latifolia*, *P. australis*, and *C. esculenta* (Rai et al., 2015) have recently been studied for their abilities to remove metals from aquatic systems and have shown promising results. For instance, *A. pinnata* was found to remove as much as 94% of Hg from a solution (Rai and Tripathi, 2009), whereas *Eichhornia crassipes* was found to accumulate Cr in its shoots at 223 times the concentration in the water (Agunbiade et al., 2009), and removed 84% of Cr from water and 95% of Zn from water (Mishra and Tripathi, 2009). Although metals negatively affected growth of *L. gibba*, the plants were able to remove 90% of Cd from solution after 6–8 days (Megateli et al., 2009). The number of hyperaccumulators among aquatic and wetland plants is rising (Prasad et al., 2001), and several aquatic plants have been found as hyperaccumulators, including *Salvinia minima* (Sanchez-Galvan et al., 2008), *Potamogeton natans* (Fritioff and Greger, 2006), *Ceratophyllum demersum* (Robinson et al., 2006), *Potamogeton pectinatus*, *Potamogeton crispus* (Upadhyay et al., 2014), and *S. polyrhiza* (John et al., 2008).

Plants that can accumulate high concentrations of metals are termed "hyperaccumulators." There have been more than 400 plant species identified as hyperaccumulators (Prasad and Freitas, 2003), including crop species (Vamerali et al., 2010). Hyperaccumulators may be defined based on bioconcentration factor, or the ability to accumulate metals in plant tissues. For instance, the ability to accumulate greater than 1000 times the concentration of Cd (based on concentration of metal in dry weight of plant) than that in the surrounding medium would be considered hyperaccumulation (Zayed et al., 1998). One of the most studied hyperaccumulators is the terrestrial plant *T. caerulescens*, which is a Cd/Zn hyperaccumulator (Mijovilovich et al., 2009). Bioconcentration and translocation factors are the best way to measure the efficiency of plants for trace element accumulation and translocation from contaminated environments (Madera-Parra et al., 2013; Rai et al., 2015).

4. APPLICATION OF TRANSGENIC PLANTS IN PHYTOREMEDIATION

The biotechnological potential of plant–microbe relationships for use in phytoremediation has included research on transgenic plants. Much of this research has focused on terrestrial plants, and many engineered plant systems for phytoremediation have been used to degrade organic contaminants. However, several recent studies have focused on transgenic plants and metals. Dhankher et al. (2003) found that tobacco plants expressing the bacterial arsenate reductase gene, *ars*C, were more tolerant to and accumulated more Cd. Che et al. (2003) engineered transgenic cottonwood trees (*Populus deltoides*) to express bacterial *mer*B genes, and found that plants were more resistant to organic Hg compounds than wild-type plants. Hussein et al. (2007) engineered transgenic tobacco plants through the chloroplast genome with both *mer*A and *mer*B genes, and saw fewer toxic effects of Hg and more Hg accumulation than in wild-type plants. Plant–microbe symbioses also have been exploited in transgenics. Wu et al. (2006a) manufactured a synthetic PC analog that was expressed in *Pseudomonas putida* to increase Cd binding, and this engineered bacterium was then added to sunflower roots to increase Cd accumulation and lessen Cd toxicity in plants. Although few studies so far have focused on engineering aquatic plants for decreased metal toxicity and increased metal removal (Kotrba et al., 2009; Hassan and Aarts, 2011), a study by Moontongchoon et al. focused on water spinach (*Ipomoea aquatica*). Expression of genes from sulfate assimilation pathways in these transgenes provided elevated Cd tolerance, and could be useful for remediation of metals in high sulfate environments (Moontongchoon et al., 2008). Engineered transgenic tobacco with both *mer*A and *mer*B genes showed fewer toxic effects of Hg and more Hg accumulation than in wild-type plants (Foetzki et al., 2012).

A synthetic PC analog that was expressed in *P. putida* to increase Cd binding; this engineered bacterium was then added to sunflower roots to increase Cd accumulation and lessen Cd toxicity in plants (Adams et al., 2013).

5. METAL TOLERANCE IN BACTERIA

Bacteria have demonstrated elevated tolerance to metals using many diverse mechanisms. They may maintain metal homeostasis, keeping concentrations of essential metals such as Zn from reaching toxic levels within cells (Coombs and Barkay, 2005), or they may contain resistance systems and active mechanisms for removing or sequestering metals (Gadd, 1992). Microorganism uptake metal either actively bioaccumulation and/or passively (biosorption) (Johncy et al., 2010). Many microorganisms have developed chromosomally or extrachromosomally controlled detoxification mechanisms to overcome the detrimental effects of heavy metals (Ehrlich, 1997). For heavy metals including Cd, Zn, Ni, Cr, Co and Cu, there are several types of resistance systems, including efflux pumps to remove metals from the cell and sequestration mechanisms to bind metal inside the cell. The two known types of efflux systems are adenosine triphosphatases, which pump out metals using adenosine triphosphate to drive the reaction, and proton antiports, which use the proton gradient to pump metals across the cell membrane (Nies, 2003). Another mechanism of bacterial metal resistance, best known in cyanobacteria, is sequestration by metallothioneins. Metallothioneins bind metals to sulfhydryl groups of cysteine residues (Huckle et al., 1993; Ybarra and Webb, 1999). Pages et al. (2008) recently described cadmium tolerance in *Stenotrophomonas maltophilia*, and found not only a Cd efflux pump but also accumulation of CdS particles. Some bacteria expressing metal resistance produce more extracellular polymeric substances (EPS), which bind the metal, perhaps making the microenvironment around the plant less toxic. Production of EPS has been shown to increase with increased metal resistance (van der Lelie et al., 2000). In different strains of *Rhizobium leguminosarum*, Cd-tolerant strains showed increased levels of glutathione, indicating that this tripeptide allows the bacterium to deal with heavy metals, rather than an efflux system (Figueira et al., 2005). The biomarker glutathione is an important antioxidant that may protect against metal toxicity associated with oxidative stress. Another mechanism of dealing with toxic metals may involve polyphosphates, long chains of orthophosphates, which may sequester metals (Alvarez and Jerez, 2004). Perrin et al. (2009) recently reported that Ni exposure promoted biofilm formation in *Escherichia coli* cultures, which may serve as a protective tolerance mechanism. Ni appeared to be involved in adherence by inducing transcription of genes encoding curli, the adhesive structures necessary for biofilm formation. The protective quality of EPS or these other mechanisms provided by root-associated bacteria suggests that enriching for certain bacteria may replace the technique of amending plant root zones with synthetic cross-linked polyacrylates and hydrogels to protect roots from heavy metal toxicity (Blaylock et al., 1997).

6. METAL BIOACCUMULATION AND BIOTRANSFORMATION BY BACTERIA

Biochemical processes such as bioleaching involving *Thiobacillus* spp. bacteria and *Aspergillus niger*, fungus, biosorption of low concentrations of metals in water by algal or bacterial cells, bio-oxidation or bioreduction of metal accumulation by *Bacillus subtilis* and sulfate-reducing bacteria and biomethylation of metals such as arsenic, cadmium, mercury, or lead, have shown some promises and could be used for soil sediment treatments (Mulligan et al., 2001). *Bacillus circulans* and *Bacillus megaterium* reported to bioaccumulate 34.5 and 32.0 mg/g Cr, respectively, during 24 h (Srinath et al., 2002). Similarly, soil bacterium *B. subtilis* was reported to grow and reduce chromate at concentrations ranging from 0.1 to 1 mM K_2CrO_4 (Garbisu et al., 1998). Faisal et al. (2005) reported that *Bacillus cereus* S-6 completely reduced Cr(VI), and *Oscillatoria intermedium* CrT-1 reduced Cr(VI) by 98% and 70%, respectively, after 96 h from the inoculated medium. Viti et al. (2003) reported *Corynebacterium hoagii* from a chromium-polluted soil capable of catalyzing the reduction of Cr(VI) to Cr(III), a less toxic and less water-soluble form of chromium and suggested this approach permitted selection of some bacterial strains, which could be used for bioremediation of Cr(VI)-polluted environments.

Megharaj et al. (2003) found *Arthrobacter* sp. could reduce Cr(VI) up to 50 μg/mL, whereas *Bacillus* sp. did not able to reduce Cr(VI) beyond 20 μg/mL and demonstrated *Arthrobacter* sp. has a great potential for bioremediation of Cr(VI)-containing waste. Lee et al. (2006) isolated indigenous bacteria from black, clay-like sediments from the area of the pigment manufacturing factories in Dongducheon city, Korea, contaminated by metals and found that enriched bacterial consortium reduced more than 99% of dissolved Cr(VI) in 96 h from the onset of the experiments under anaerobic condition suggested these indigenous bacteria may play a role in the treatment of Cr(VI)-contaminated sediments. Branco et al. (2004) reported *Ochrobactrum* strain 5bVl1 resists high Cr(VI) concentrations and has a high Cr(VI)-reducing ability, making it valuable to all in bioremediation. Quan et al. (2006) presented direct detoxification of chromium slag by using microorganisms and showed a bacterial consortium can efficiently accelerated Cr(VI) leaching rate and removal.

7. MICROBE-ASSISTED PHYTOREMEDIATION

Metal hyperaccumulation is an adaptive process between microbes exposed to heavy metals and plants, requiring continuous interactions among the cooccurring organisms. A recent proteomics study by Farinati et al. (2009) indicated that the presence of a rhizosphere microbial population, adapted to heavy metal–polluted sites, greatly enhanced the accumulation of metals in shoots of the hyperaccumulator *Arabidopsis halleri*. Aquatic environments include not only macrophytes, but also algae that may interact with microbes to remove contaminants from the environment. Algae can be produced in artificial systems and used to remove contaminants. Loutseti et al. (2009) used a dried mixture of microalgae and bacteria to remove Cu and Cd from wastewater. Munoz et al. successfully examined the combination of the bacterium *Ralstonia basilensis* and the microalga *Chlorella sorokiniana* on adsorption of Cd, Cu, Ni, and Zn (Munoz et al., 2006). Moreover, mycorrhizal fungi associated with plants can enhance uptake of metals when essential metal concentrations are low and vice versa—when metal quantities are too high mycorrhizae can be effective in alleviating metal toxicity decreasing plant uptake (Frey et al., 2000; Banni and Faituri, 2013). In their article describing bacterial enhancement of selenium and Hg uptake by wetland plants, De Souza et al. (1999) proposed several possible mechanisms, including bacterial stimulation of plant metal uptake compounds such as siderophores; bacterial root growth promotion increasing the root surface area; bacterial transformation of elements into more soluble forms; or bacterial stimulation of plant transporters that may transport essential elements as well as heavy metals (in the case of selenate, the sulfate transporter). van der Lelie et al. (2000) related the basis of this plant–microbe interaction to bacterial metal resistance because the bioavailability of metals could be altered by bacterial expression of resistance systems. Besides heavy metals, organic contaminants can also be removed by aquatic plants and their associated microbial communities. Unlike metals, which cannot be degraded and tend to be accumulated by plants, there is evidence that many organic contaminants are degraded in the rhizosphere by plant-associated bacteria and are transformed there rather than inside the plant. For instance, the duckweed *S. polyrhiza* and its associated bacteria may degrade the aromatic compounds phenol, aniline, and 2,4-dichlorophenol (Toyama et al., 2006), and the reed *P. australis* and its associated rhizosphere bacteria might degrade bisphenol (Toyama et al., 2009). The marine eelgrass *Zostera marina* and its associated bacteria may degrade polycyclic aromatic hydrocarbon and polychlorinated biphenyls (Huesemann et al., 2009). Inoculation with rhizobacteria had little influence on the metal concentrations in *Brassica juncea* plant tissues, but produced a much larger above-ground biomass and altered metal bioavailability in the soil and hence uptake of heavy metals from Pb-Zn mine tailings. As a consequence, higher efficiency of phytoextraction was obtained compared with control treatments (Wu et al., 2006b). The addition of either native or transformed *P. asplenii* AC to *P. australis* seeds enabled the plants (shoots and roots) to attain a greater size than noninoculated plants after growth in soil in the presence of either copper or creosote (Reed et al., 2005). Similarly, Singh et al. (2010a,b) isolated six chromium-tolerant bacterial strains from root zone of *Scirpus lacustris* reported that the inoculation of selected bacterial strain resulted to increased tolerance and accumulation of Cr in *Scirpus lacustris* and *Vigna radiata* growing under different Cr concentrations. Rai et al. (2004) conducted a revegetation trial to evaluate the feasibility of growing a legume species *Prosopis juliflora* L. on fly ash ameliorated with combination of various organic amendments, blue green algal biofertilizer, and rhizobium inoculums and reported a significant enhancements in plant biomass, photosynthetic pigments, protein content and in vivo nitrate reductase activity in the plants grown on ameliorated fly ash in comparison to the plants growing in unamended fly-ash or garden soil. Inoculation of *Lupinus luteus* grown on a nickel-enriched substrate with the engineered nickel-resistant bacterium *Burkholderia cepacia* L.S.2.4: *nccnre*, nickel concentrations in the roots increased significantly (30%) (Lodewyckx et al., 2001). Sheng et al. (2008) observed increases in biomass production and total lead uptake in *B. napus* after inoculation with *Microbacterium* sp. G16, a bacterial strain that can produce indole-3-acetic acid (IAA), 1-aminocyclopropane-1-carboxylic acid (ACC) deaminase, and siderophores.

8. MICROBIAL MECHANISMS OF PLANT GROWTH PROMOTION AND METAL UPTAKE

Bacteria promote plant growth, thus increasing surface area of the plant and allowing more metal uptake. Certain compounds produced by bacteria have been shown to promote plant growth, including siderophores. Siderophores, iron-chelating compounds, have been shown to promote plant growth even in the presence of heavy metals (Tripathi et al., 2005; Burd et al., 2000). They stimulate plant growth through mobilizing nutrients in soils, producing numerous plant growth regulators, protecting plants from phytopathogens by controlling or inhibiting them, improving soil structure and bioremediating the polluted soils by sequestering toxic heavy metal species, and degrading xenobiotic compounds (Ahemad, 2012; Ahemad and Malik, 2011). Bacterial production of siderophores may protect plants from heavy metal toxicity, increasing plant growth by providing the plants with sufficient iron and allowing them to overcome the toxic effects of heavy metals. Another plant growth promoting compound that has been studied in relation to heavy metals is ACC deaminase. ACC is an intermediate of ethylene produced by plants under stress, and bacteria that produce ACC deaminase can lower the levels of ethylene in plants, promoting plant growth (Dell'Amico et al., 2005). Belimov et al. (2001) found that bacteria containing ACC deaminase improve plant growth in metal-polluted conditions. Bacteria as well as plants can produce the auxin IAA. Rajkumar and Freitas (2008) suggested that IAA indirectly promotes metal accumulation in plants by increasing plant biomass. Grandlic et al. (2008) found that 76% of plant growth-promoting isolates from plants grown in mine tailings were able to produce IAA when supplemented with tryptophan, and these bacteria could promote growth of plants growing in mine tailings for phytostabilization.

Metal resistance has been described as a necessity for plant-associated bacteria in contaminated environments (Salt et al., 1999). van der Lelie et al. (2000) related plant metal uptake to bacterial metal resistance because the bioavailability of metals could be altered by expression of bacterial metal-resistance systems. Faisal and Hasnain (2005) inoculated sunflower plants with Cr-resistant bacteria, and found that plant growth in Cr was improved by inoculation, although inoculated plants accumulated less Cr than uninoculated plants. Kunito et al. (2001) examined rhizosphere bacteria from *Phragmites* grown in Cu. They found EPS production was greater for rhizosphere bacteria compared with nonrhizosphere bacteria. Because of Cu binding to bacterial EPS, rhizosphere soil may become less toxic to bacteria and also to plants.

Another possibility could be the lowering of pH in the rhizosphere by bacteria, which would make metals more soluble. The rhizosphere pH could be lowered by processes listed previously, or other mechanisms, such as bacterial metal resistance systems. Bravin et al. (2009) pointed out that rhizosphere pH can also be raised by biological activity; in extremely low pH environments, roots alkalized the rhizosphere, making Cu less bioavailable. Abou-Shanab et al. (2003) hypothesized that rhizosphere bacteria of *Alyssum murale* lowered rhizosphere pH, solubilizing metal for hyperaccumulation of Ni by the plant. Phytoprotection As, illustrated previously, two possibilities may exist for plant–microbe interactions in relation to metals. Bacterial mechanisms could lead to increased accumulation of metals in plants, or bacteria may keep metals from being accumulated by plants at high concentrations that are toxic to the plant. This may be the case for chelation or EPS production by bacteria (Faisal and Hasnain, 2005). Salt et al. (1999) suggested that rhizobacteria promote precipitation of Cd at the root surface, causing plants to take up less metal. Although the studies of plant–microbe interactions in the rhizosphere have been carried out in mainly terrestrial systems (although some wetland plants have been used), these principles could also apply to aquatic systems, where bacteria are still closely associated with plant roots.

9. CONCLUSION AND FUTURE PROSPECTS

Phytoremediation of metals, the use of plants to extract, contain, immobilize, or remove hazardous substances from environments is a very promising area, and several highly efficient examples have shown the applicability of this process to clean industrial waste streams, to concentrate heavy metals, and to preserve drinking water and aquatic biodiversity. Not only aquatic macrophytes, but also algae and fungi, represent a cost-effective and eco-friendly technology for environmental cleanup, a green solution often preferred in political decision-making. Rhizosphere microbes can reduce metal toxicity and enhance plant tolerance to dissolved metals and can therefore be applied to supply increased phytoprotection from harmful effects of the metals on plants. In a direct extension of this idea, the bacterial genes coding for metal resistance can be transplanted into the plant genome to confer elevated metal tolerance to plants. Further research in aquatic phytoremediation is needed to advance understanding of microbe–plant interactions. Such knowledge would increase the number of potentially widespread applications and their impact such as the treatment of heavy metals from industrial effluents in natural and constructed wetlands, or a wastewater metal stripping phase using rhizofiltration.

In the coming years, plant–microbe interactions for the removal of heavy metal contaminants may become increasingly viable options, especially in shallow wetland and estuary environments. Further, use of constructed wetlands for filtration and remediation of water is currently a popular method, and understanding the nature of plant–microbe interactions may improve this process. Developing new methods to support microbial activity to either enhance (for phytoextraction) or reduce (for phytostabilization) the bioavailability of metal contaminants in the rhizosphere could significantly improve the efficiency of these remediation techniques. Root exudates lead to selective recruitment and accumulation of a diverse range of bacterial species associated with plants. Applications have recently been extended to commercial phytomining—the recovery of precious metals such as gold, silver, platinum, and palladium in mining (Sheoran et al., 2009). The attractiveness of phytomining should increase if combined with other technologies such as biofuel production. Many possibilities exist for the large-scale application of bacterial–plant systems for removal of metals from aquatic environments.

ACKNOWLEDGEMENTS

The authors are thankful to the director of the Institute of Environment and Sustainable Development, Banaras Hindu University, Varanasi, for his continuous support and guidance. NKS thanks the Science and Engineering Research Board, New Delhi, for the Young Scientist Award and for providing a grant.

REFERENCES

Abhilash, P.C., Pandey, V.C., Srivastava, P., Rakesh, P.S., Chandran, S., Singh, N., Thomas, A.P., 2009. Phytofiltration of cadmium from water by *Limnocharis flava* (L.) Buchenau grown in free-floating culture system. J. Hazard. Mater. 170, 791–797.

Abou-Shanab, R.A., Delorme, T.A., Angle, J.S., Chaney, R.L., Ghanem, K., Moawad, H., Ghozlan, H.A., 2003. Phenotypic characterization of microbes in the rhizosphere of *Alyssum murale*. Int. J. Phytorem. 5, 367–379.

Adams, D.G., Thiel, T.D., Paula, S., 2013. Symbiosis between the cyanobacterium *Nostoc* and the liverwort *Blasia* requires a CheR-type MCP methyl transferase. Symbiosis 59, 111–120.

Adriano, D.C., 2001. Trace Elements in Terrestrial Environments: Biogeochemistry. Bioavailability and Risks of Metals. Springer-Verlag, New York.

Agunbiade, F.O., Olu-Owolabi, B.I., Adebowale, K.O., 2009. Phytoremediation potential of *Eichornia crassipes* in metal-contaminated coastal water. Bioresour. Technol. 100, 4521–4526.

Ahemad, M., 2012. Implications of bacterial resistance against heavy metals in bioremediation: a review. IIOAB J. 3, 39–46.

Ahemad, M., Malik, A., 2011. Bioaccumulation of heavy metals by zinc resistant bacteria isolated from agricultural soils irrigated with wastewater. Bacteriol. J. 2, 12–21.

Ali, H., Khan, E., Sajad, M.A., 2013. Phytoremediation of heavy metals-concepts and applications. Chemosphere 9, 869–881.

Alloway, B.J., 1995. Heavy Metals in Soils. Blackie Academic and Professional, London.

Alvarez, S., Jerez, C.A., 2004. Copper ions stimulate polyphosphate degradation and phosphate efflux in *Acidithiobacillus ferrooxidans*. Appl. Environ. Microbiol. 70, 5177–5182.

Banni, A.S., Faituri, M.Y., 2013. The role of arbuscular mycorrhizae *Glomus* spp. (mixed) and *Glomus fasciculatum* in growth and copper uptake of maize grown in soil contaminated with copper. Middle East J. Sci. Res. 17, 77–83.

Belimov, A.A., Safronova, V.I., Sergeyeva, T.A., Egorova, T.N., Matveyeva, V.A., Tsyganov, V.E., Borisov, A.Y., Tikhonovich, I.A., Kluge, C., Preisfeld, A., 2001. Characterization of plant growth promoting rhizobacteria isolated from polluted soils and containing 1-aminocyclopropane-1-carboxylate deaminase. Can. J. Microbiol. 47, 642–652.

Bermudez, G.M.A., Jasan, R., Pla, R., Pignata, M.L., 2012. Heavy metals and trace elements in atmospheric fall-out: their relationship with topsoil and wheat element composition. J. Hazard. Mater. 213, 447–456.

Blaylock, M.J., Salt, D.E., Dushenkov, S., Zakharova, O., Gussman, C., Kapulnik, Y., Ensley, B.D., Raskin, I., 1997. Enhanced accumulation of Pb in indian mustard by soil-applied chelating agents. Environ. Sci. Technol. 31, 860–865.

Blum, R., Beck, A., Korte, A., Stengel, A., Letzel, T., Lendzian, K., Grill, E., 2007. Function of phytochelatin synthase in catabolism of glutathione-conjugates. Plant J. 49, 740–749.

Branco, R., Carmen, A.M., Paula, V.M., 2004. *Ochrobacterium tritice* strain 5 bvl1-characterization of a Cr (VI) reducing strain. Can. J. Microbiol. 50, 697–703.

Bravin, M.N., Tentscher, P., Rose, J., Hinsinger, P., 2009. Rhizosphere pH gradient controls copper availability in a strongly acidic soil. Environ. Sci. Technol. 43, 5686–5691.

Burd, G.I., Dixon, D.G., Glick, B.R., 2000. Plant growth-promoting bacteria that decrease heavy metal toxicity in plants. Can. J. Microbiol. 46, 237–245.

Che, D., Meagher, R.B., Heaton, A.C.P., Lima, A., Rugh, C.L., Merkle, S.A., 2003. Expression of mercuric ion reductase in eastern cottonwood (*Populus deltoides*) confers mercuric ion resistance. Plant Biotechnol. J. 1, 311–319.

Chen, Y., Wen, Y., Zhou, Q., Vymajal, J., 2014. Effects of plant biomass on denitrifying genes in subsurface-flow constructed wetlands. Bioresour. Technol. 157, 341–345.

Coombs, J.M., Barkay, T., 2005. New findings on evolution of metal homeostasis genes: evidence from comparative genome analysis of bacteria and archaea. Appl. Environ. Microbiol. 71, 7083–7091.

De Souza, M.P., Huang, C.P.A., Chee, N., Terry, N., 1999. Rhizosphere bacteria enhance the accumulation of selenium and mercury in wetland plants. Planta 209, 259–263.

Dell'Amico, E., Cavalca, L., Andreoni, V., 2005. Analysis of rhizobacterial communities in perennial *Graminaceae* from polluted water meadow soil, and screening of metal-resistant, potentially plant growth-promoting bacteria. FEMS Microbiol. Ecol. 52, 153–162.

Dhankher, O.P., Shasti, N.A., Rosen, B.P., Fuhrmann, M., Meagher, R.B., 2003. Increased cadmium tolerance and accumulation by plants expressing bacterial arsenate reductase. New Phytol. 159, 431–441.

Ehrlich, H.L., 1997. Microbes and metals. Appl. Microbiol. Biotechnol. 48, 687–692.

Faisal, M., Hasnain, S., 2005. Chromate resistant *Bacillus cereus* augments sunflower growth by reducing toxicity of Cr(VI). J. Plant Biol. 48, 187–194.

Faisal, M., Hameed, A., Hasnain, S., 2005. Chromium-resistant bacteria and cyanobacteria: impact on Cr (VI) reduction potential and plant growth. J. Ind. Microbiol. Biotechnol. 32, 615–621.

Farinati, S., DalCorso, G., Bona, E., Corbella, M., Lampis, S., Cecconi, D., Polati, R., Berta, G., Vallini, G., Furini, A., 2009. Proteomic analysis of *Arabidopsis halleri* shoots in response to the heavy metals cadmium and zinc and rhizosphere microorganisms. Proteomics 9, 4837–4850.

Figueira, E.M., Lima, A.I., Pereira, S.I., 2005. Cadmium tolerance plasticity in *Rhizobium leguminosarum* bv. viciae: glutathione as a detoxifying agent. Can. J. Microbiol. 51, 7–14.

Flathman, P.E., Lanza, G.R., 1998. Phytoremediation: current views on an emerging green technology. J. Soil Contam. 7, 415–432.

Flores-Cáceres, M.L., Hattab, S., Hattab, S., Boussetta, H., Banni, M., Hernández, L.E., 2015. Specific mechanisms of tolerance to copper and cadmium are compromised by a limited concentration of glutathione in alfalfa plants. Plant Sci. 233, 165–173.

Foetzki, A., Quijano, C.D., Moullet, O., 2012. Surveying of pollenmediated crop-to-crop gene flow from a wheat field trial as a biosafety measure. GM Crops Food 3, 115–122.

Frey, B., Zierold, K., Brunner, I., 2000. Extracellular complexation of Cd in the Hartig net and cytosolic Zn sequestration in the fungal mantle of *Picea abies–Hebeloma crustuliniforme* ectomycorrhizas. Plant Cell Environ. 23, 1257–1265.

Fritioff, A., Greger, M., 2006. Uptake and distribution of Zn, Cu, Cd, and Pb in an aquatic plant, *Potamogeton natans*. Chemosphere 63, 220–227.

Gadd, G.M., 1992. Metals and microorganisms: a problem of definition. FEMS Microbiol. Lett. 79, 197–203.

Garbisu, C., Alkorta, I., Marıa, J.L., Serra, J.L., 1998. Aerobic chromate reduction by *Bacillus subtilis*. Biodegradation 9, 133–141.

Ghassemzadeh, F., Yousefzadeh, H., Arbab-Zavar, M.H., 2008. Removing arsenic and antimony by *Phragmites australis*: rhizofiltration technology. J. Appl. Sci. 8, 1668–1675.

Grandlic, C.J., Mendez, M.O., Chorover, J., Machado, B., Maier, R.M., 2008. Plant growth-promoting bacteria for phytostabilization of mine tailings. Environ. Sci. Technol. 42, 2079–2084.

Hassan, Z., Aarts, M.G.M., 2011. Opportunities and feasibilities for biotechnological improvement of Zn, Cd or Ni tolerance and accumulation in plants. Environ. Exp. Bot. 72, 53–63.

Hossain, M.A., Piyatida, P., Da Silva, J.A.T., Fujita, M., 2012. Molecular mechanism of heavy metal toxicity and tolerance in plants: central role of glutathione in detoxification of reactive oxygen species and methylglyoxal and in heavy metal chelation. J. Bot. http://dx.doi.org/10.1155/2012/872875.

Hou, W., Chen, X., Song, G., Wang, Q., Chi Chang, C., 2007. Effects of copper and cadmium on heavy metal polluted water body restoration by duckweed (*Lemna minor*). Plant Physiol. Biochem. 45, 62–69.

Huckle, J.W., Morby, A.P., Turner, J.S., Robinson, N.J., 1993. Isolation of a prokaryotic metallothionein locus and analysis of transcriptional control by trace metal ions. Mol. Microbiol. 7, 177–187.

Huesemann, M.H., Hausmann, T.S., Fortman, T.J., Thom, R.M., Cullinan, V., 2009. In situ phytoremediation of PAH- and PCB contaminated marine sediments with eelgrass (*Zostera marina*). Ecol. Eng. 35, 1395–1404.

Hussein, H.S., Ruiz, O.N., Terry, N., Daniell, H., 2007. Phytoremediation of mercury and organomercurials in chloroplast transgenic plants: enhanced root uptake, translocation to shoots, and volatilization. Environ. Sci. Technol. 41, 8439–8446.

Jabeen, R., Ahmad, A., Iqbal, M., 2009. Phytoremediation of heavy metals: physiological and molecular mechanisms. Bot. Rev. 75, 339–364.

John, R., Ahmad, P., Gadgil, K., Sharma, S., 2008. Effect of cadmium and lead on growth, biochemical parameters and uptake in *Lemna polyrrhiza* L. Plant Soil Environ. 54, 262–270.

Johncy, M., Hemambika, B., Hemapriya, J., Ajesh Kannan, V., 2010. Comparative assessment of heavy metal removal by immobilized and dead bacterial cells: a biosorption approach. Afr. J. Environ. Sci. Technol. 4, 77–83.

Jozefczak, M., Remans, T., Vangronsveld, J., Cuypers, A., 2012. Glutathione is a key player in metal induced oxidative stress defenses. Int. J. Mol. Sci. 13, 3145–3175.

Kaplan, O., Ince, M., Yaman, M., 2011. Sequential extraction of cadmium in different soil phases and plant parts from a former industrialized area. Environ. Chem. Lett. 9, 397–404.

Khan, M.S., Zaidi, A., Wani, P.A., Oves, M., 2009. Role of plant growth promoting rhizobacteria in the remediation of metal contaminated soils. Environ. Chem. Lett. 7, 1–9.

Khellaf, N., Zerdaoui, M., 2009. Phytoaccumulation of zinc by the aquatic plant, *Lemna gibba* L. Bioresour. Technol. 100, 6137–6140.

Kotrba, P., Najmanova, J., Macek, T., Ruml, T., Mackova, M., 2009. Genetically modified plants in phytoremediation of heavy metal and metalloid soil and sediment pollution. Biotechnol. Adv. 27, 799–810.

Kraemer, U., 2005. Phytoremediation: novel approaches to cleaning up polluted soils. Curr. Opin. Biotechnol. 16, 1–9.

Kumar, N., Mallick, S., Yadava, R.N., Singh, A.P., Sinha, S., 2013. Co-application of selenite and phosphate reduces arsenite uptake in hydroponically grown rice seedlings: toxicity and defence mechanism. Ecotoxicol. Environ. Saf. 911, 71–79.

Kunito, T., Saeki, K., Nagaoka, K., Oyaizu, H., Matsumoto, S., 2001. Characterization of copper-resistant bacterial community in rhizosphere of highly copper-contaminated soil. Eur. J. Soil Biol. 37, 95–102.

Lee, S.E., Lee, J.U., Lee, J.S., Chon, H.T., 2006. Effects of indigenous bacteria on Cr(VI) reduction in Cr-contaminated sediment with industrial wastes. J. Geochem. Explor. 88, 41–44.
van der Lelie, D., Corbisier, P., Diels, L., Gills, A., Lodewyckx, C., Mergeay, M., Taghavi, S., Spelmans, N., Vangronsveld, J., 2000. The role of bacteria in the phytoremediation of heavy metals. In: Terry, N., Banuelos, G.S. (Eds.), Phytoremediation of Contaminated Soil and Water. Lewis Publishers, pp. 265–281.
Lodewyckx, C., Mergeay, M., Vangronsveld, J., Clijsters, H., Van der Lelie, D., 2001. The effect of recombinant heavy metal resistant endophytic bacteria on heavy metal uptake by their host plant. Int. J. Phytorem. 3, 173–187.
Loutseti, S., Danielidis, D.B., Economou-Amilli, A., Katsaros, C., Santas, R., Santas, P., 2009. The application of a micro-algal/bacterial biofilter for the detoxification of copper and cadmium metal wastes. Bioresour. Technol. 100, 2099–2105.
Ma, Y., Prasad, M.N.V., Rajkumar, M., Freitas, H., 2011. Plant growth promoting rhizobacteria and endophytes accelerate phytoremediation of metalliferous soils. Biotechnol. Adv. 29, 248–258.
Madera-Parra, C.A., Peña-Salamanca, E.J., Peña, M.R., Rousseau, D.P.L., Lens, P.N.L., 2013. Phytoremediation of landfill leachate with *Colocasia esculenta*, *Gynerum sagittatum* and *Heliconia psittacorum* in constructed wetlands. Int. J. Phytorem. 17, 16–24.
Mani, D., Kumar, C., 2014. Biotechnological advances in bioremediation of heavy metals contaminated ecosystems an overview with special reference to phytoremediation. Int. J. Environ. Sci. Technol. 11, 843–887.
Mani, D., Kumar, C., Patel, N.K., 2015. Integrated micro-biochemical approach for phytoremediation of cadmium and zinc contaminated soils. Ecotoxicol. Environ. Saf. 111, 86–95.
Mayor, D.J., Gray, N.B., Elver-Evans, J., Midwood, A.J., Thornton, B., 2013. Metal-macrofauna interactions determine microbial community structure and function in copper contaminated sediments. PLoS One 8, 64–94.
Megateli, S., Semsari, S., Couderchet, M., 2009. Toxicity and removal of heavy metals (cadmium, copper, and zinc) by *Lemna gibba*. Ecotoxicol. Environ. Saf. 72, 1774–1780.
Megharaj, M., Avudainayagam, S., Naidu, R., 2003. Toxicity of hexavalent chromium and its reduction by bacteria isolated from soil contaminated with tannery waste. Curr. Microbiol. 47, 51–54.
Mendoza-Cozatl, D.G., Moreno-Sanchez, R., 2006. Control of glutathione and phytochelatin synthesis under cadmium stress. Pathway modeling for plants. J. Theor. Biol. 238, 919–936.
Mijovilovich, A., Leitenmaier, B., Meyer-Klaucke, W., Kroneck, P.M., Gotz, B., Kupper, H., 2009. Complexation and toxicity of copper in higher plants. II. Different mechanisms for copper versus cadmium detoxification in the copper-sensitive cadmium/zinc hyperaccumulator *Thlaspi caerulescens* (Ganges ecotype). Plant Physiol. 151, 715–731.
Mishra, V.K., Tripathi, B.D., 2009. Accumulation of chromium and zinc from aqueous solutions using water hyacinth (*Eichhornia crassipes*). J. Hazard. Mater. 164, 1059–1063.
Moontongchoon, P., Chadchawan, S., Leepipatpiboon, N., Akaracharanya, A., Shinmyo, A., Sano, H., 2008. Cadmium-tolerance of transgenic Ipomoea aquatica expressing serine acetyltransferase and cysteine synthase. Plant Biotechnol. 25, 201–203.
Mulligan, C.N., Yong, R.N., Gibbs, B.F., 2001. Remediation technologies for metal-contaminated soil and groundwater: an evaluation. Eng. Geol. 60, 193–207.
Munoz, R., Alvarez, M.T., Munoz, A., Terrazas, E., Guieysse, B., Mattiasson, B., 2006. Sequential removal of heavy metals ions and organic pollutants using an algal–bacterial consortium. Chemosphere 63, 903–911.
Niemeyer, J.C., Nogueira, M.A., Carvalho, G.M., Cohin-De-Pinho, S.J., Outeiro, U.S., Rodrigues, G.G., da Silva, E.M., Sousa, J.P., 2012. Functional and structural parameters to assess the ecological status of a metal contaminated area in the tropics. Ecotoxicol. Environ. Saf. 86, 188–197.
Nies, D.H., 2003. Efflux-mediated heavy metal resistance in prokaryotes. FEMS Microbiol. Rev. 781, 1–27.
Pages, D., Rose, J., Conrod, S., Cuine, S., Carrier, P., Heulin, T., Achouak, W., 2008. Heavy metal tolerance in *Stenotrophomonas maltophilia*. PLoS One 3, 1539.

Pandey, V.C., Pandey, D.N., Singh, N., 2015. Sustainable phytoremediation based on naturally colonizing and economically valuable plants. J. Cleaner Prod. 86, 37–39.

Pattanayak, B., Dhal, N.K., 2014. Biotechnological potential of plantmicrobes interaction with enhanced phytoremediation. J. Environ. Res. Dev. 8, 676–682.

Perrin, C., Briandet, R., Jubelin, G., Lejeune, P., Mandrand-Berthelot, M.A., Rodrigue, A., Dorel, C., 2009. Nickel promotes biofilm formation by *Escherichia coli* K-12 strains that produce curli. Appl. Environ. Microbiol. 75, 1723–1733.

Pons-Branchu, E., Ayrault, S., Roy-Barman, M., Bordier, L., Borst, W., Branchu, P., Douville, E., Emmanuel Dumont, E., 2015. Three centuries of heavy metal pollution in Paris (France) recorded by urban speleothems. Sci. Total Environ. 518, 86–96.

Prasad, M.N.V., Freitas, H.M.D., 2003. Metal hyperaccumulation in plants – biodiversity prospecting for phytoremediation technology. Electron. J. Biotechnol. 6, 285–321.

Prasad, M.N.V., Malec, P., Waloszek, A., Bojko, M., Strzalka, K., 2001. Physiological responses of *Lemna triscula* L. (duckweed) to cadmium and copper bioaccumulation. Plant Sci. 161, 881–889.

Quan, X., Tan, H., Zhao, Y., Hu, Y., 2006. Detoxification of chromium slag by chromate resistant bacteria. J. Hazard. Mater. 137, 836–841.

Rahimi, M., Farhadi, R., Mehdizade, R., 2013. Phytoremediation: using plants to clean up contaminated soil with heavy metals. Int. J. Agric. Res. Rev. 3, 148–152.

Rahman, M.A., Hasegawa, H., Ueda, K., Maki, T., Okumura, C., Rahman, M.M., 2007. Arsenic accumulation in duckweed (*Spirodela polyrhiza* L.): a good option for phytoremediation. Chemosphere 69, 493–499.

Rai, P.K., 2009. Heavy metal phytoremediation from aquatic ecosystems with special reference to aquatic macrophytes. Crit. Rev. Environ. Sci. Technol. 39, 697–753.

Rai, P.K., Tripathi, B.D., 2009. Comparative assessment of *Azolla pinnata* and *Vallisneria spiralis* in Hg removal from G.B. Pant Sagar of Singrauli Industrial region, India. Environ. Monit. Assess. 148, 75–84.

Rai, U.N., Pandey, K., Sinha, S., Singh, A., Saxena, R., Gupta, D.K., 2004. Revegetating fly-ash landfills with *Prosopis juliflora* L.: impact of different amendments and *Rhizobium* inoculation. J. Environ. Int. 30, 293–300.

Rai, U.N., Tripathi, R.D., Singh, N.K., Upadhyay, A.K., Dwivedi, S., Shukla, M.K., Mallick, S., Singh, S.N., Nautiyal, C.S., 2013. Constructed wetland as an ecotechnological tool for pollution treatment for conservation of Ganga river. Bioresour. Technol. 148, 535–541.

Rai, U.N., Upadhyay, A.K., Singh, N.K., Dwivedi, S., Tripathi, R.D., 2015. Seasonal applicability of horizontal sub-surface flow constructed wetland for trace elements and nutrient removal from urban wastes to conserve Ganga river water quality at Haridwar, India. Ecol. Eng. 81, 115–122.

Rajkumar, M., Freitas, H., 2008. Effects of inoculation of plant-growth promoting bacteria on Ni uptake by Indian mustard. Bioresour. Technol. 99, 3491–3498.

Reed, M.L.E., Warner, B.G., Glick, B.R., 2005. Plant growth promoting bacteria facilitate the growth of the common reed *Phragmites australis* in presence of copper or polycyclic aromatic hydrocarbon. Curr. Microbiol. 51, 425–429.

Robinson, B., Kim, N., Marchetti, M., Moni, C., Schroeter, L., van den Dijssel, C., Milne, G., Clothier, B., 2006. Arsenic hyperaccumulation by aquatic macrophytes in the Taupo Volcanic Zone, New Zealand. Environ. Exp. Bot. 58, 206–215.

Salt, D.E., Benhamou, N., Leszczyniecka, M., Raskin, I., 1999. A possible role for rhizobacteria in water treatment by plant roots. Int. J. Phytorem. 1, 67–79.

Sanchez-Galvan, G., Monroy, O., Gomez, J., Olguin, E.J., 2008. Assessment of the hyperaccumulating lead capacity of *Salvinia minima* using bioadsorption and intracellular accumulation factors. Water Air Soil Pollut. 194, 77–90.

Shahid, M., Pourrut, B., Dumat, C., Nadeem, M., Aslam, M., Pinelli, E., 2014. Heavy-metal-induced reactive oxygen species: phytotoxicity and physicochemical changes in plants. Rev. Environ. Contam. Toxicol. 232, 1–44.

Sheng, X.-F., Jiang, C.-Y., He, L.-Y., Qian, M., 2008. Characterization of heavy metal-resistant endophytic bacteria from rape (*Brassica napus*) roots and their potential in promoting growth and lead accumulation of rape. Environ. Pollut. 156, 1164–1170.

Sheoran, V., Sheoran, A.S., Poonia, P., 2009. Phytomining: a review. Miner. Eng. 22, 1007–1019.

Singh, N.K., Rai, U.N., Singh, M., Tripathi, R.D., 2010a. Impact of rhizobacteria on growth and chromium accumulation in *Scirpus lacustris* L. grown under chromium supplementation. J. Environ. Biol. 31, 709–714.

Singh, N.K., Rai, U.N., Tewari, A., Singh, M., 2010b. Metal accumulation and growth response in *Vigna radiata* L. inoculated with chromate tolerant rhizobacteria and grown on tannery sludge amended soil. Bull. Environ. Contam. Toxicol. 84, 118–124.

Srinath, T., Verma, T., Ramteke, P.W., Garg, S.K., 2002. Chromium (VI) biosorption and bioaccumulation by chrome resistance bacteria. Chemosphere 48, 427–435.

Toyama, T., Sato, Y., Inoue, D., Sei, K., Chang, Y.C., Kikuchi, S., Ike, M., 2009. Biodegradation of bisphenol A and bisphenol F in the rhizosphere sediment of *Phragmites australis*. J. Biosci. Bioeng. 108, 147–150.

Toyama, T., Yu, N., Kumada, H., Sei, K., Ike, M., Fujita, M., 2006. Accelerated aromatic compounds degradation in aquatic environment by use of interaction between *Spirodela polyrhiza* and bacteria in its rhizosphere. J. Biosci. Bioeng. 101, 346–353.

Tripathi, M., Munot, H.P., Shouche, Y., Meyer, J.M., Goel, R., 2005. Isolation and functional characterization of siderophore-producing lead- and cadmium-resistant *Pseudomonas putida* KNP9. Curr. Microbiol. 50, 233–237.

Upadhyay, A.K., Singh, N.K., Rai, U.N., 2014. Comparative metal accumulation potential of *Potamogeton pectinatus* and *Potamogeton crispus*: role of enzymatic and non-enzymatic antioxidants in tolerance and detoxification of metals. Aquat. Bot. 117, 27–32.

Uysal, Y., Taner, F., 2009. Effect of pH, temperature, and lead concentration on the bioremoval of lead from water using *Lemna minor*. Int. J. Phytorem. 11, 591–608.

Vamerali, T., Bandiera, M., Mosca, G., 2010. Field crops for phytoremediation of metal-contaminated land. A review. Environ. Chem. Lett. 8, 1–17.

Viti, C., Pace, A., Giovannetti, L., 2003. Characterization of Cr(VI)-resistant bacteria isolated from chromium-contaminated soil by tannery activity. Curr. Microbiol. 46, 1–5.

Wu, C.H., Wood, T.K., Mulchandani, A., Chen, W., 2006a. Engineering plant–microbe symbiosis for rhizoremediation of heavy metals. Appl. Environ. Microbiol. 72, 1129–1134.

Wu, S.C., Luo, Y.M., Cheung, K.C., Wong, M.H., 2006b. Influence of bacteria on Pb and Zn speciation, mobility and bioavailability in soil: a laboratory study. Environ. Pollut. 144, 765–773.

Ybarra, G.R., Webb, R., 1999. Effects of divalent metal cations and resistance mechanisms of the *Cyanobacterium synechococcus* sp. strain PCC 7942. J. Hazard. Subst. Res. 2, 1–9.

Zayed, A., Gowthaman, S., Terry, N., 1998. Phytoaccumulation of trace elements by wetland plants I. Duckweed. J. Environ. Qual. 27, 715–721.

CHAPTER

MICROBIALLY ASSISTED PHYTOREMEDIATION OF HEAVY METAL–CONTAMINATED SOILS

20

Dilfuza Egamberdieva[1], Elsayed F. Abd-Allah[2], Jaime A.T. da Silva[3]

Leibniz Centre for Agricultural Landscape Research (ZALF), Institute for Landcape Biogeochemistry, Müncheberg, Germany[1]; Department of Plant Production, Faculty of Food & Agricultural Sciences, King Saud University, Riyadh, Saudi Arabia[2]; Ikenobe, Kagawa-ken, Japan[3]

CHAPTER OUTLINE

1. INTRODUCTION

The rapidly expanding mine tailings, the disposal of metal and industrial waste, and land application of chemical fertilizers, sewage sludge, and wastewater irrigation can result in heavy metal contamination of water resources and agricultural lands (Jadia and Fulekar, 2009; Akcil et al., 2015). Soil contamination with heavy metals such as arsenic (As), nickel (Ni), cadmium (Cd), copper (Cu), zinc (Zn), lead (Pb), aluminum (Al), and mercury (Hg) is one of the world's major environmental problems, leading to significant negative consequences on ecosystems, crop yield, and human health (Kavamura and Esposito, 2010; Li et al., 2013).

Negative effects of heavy metals on soil microbial activities have long been recognized (Renella et al., 2006; Tripathi et al., 2014). For example, high Cd concentration (40 mg Cd/kg soil) inhibited microbial growth rates, soil acid phosphatase (EC 3.1.3.2), alkaline phosphatase (EC 3.1.3.1), and urease (EC 3.5.1.5) activities (Renella et al., 2006). In another study, Cd (12 mg/kg soil) and Zn (300 mg/kg soil) negatively affected the ammonifying microorganisms in loamy sand soil (Wyszkowska et al., 2008). A higher concentration of Cu (150–450 mg/kg soil) significantly inhibited the soil dehydrogenases (EC 1.1.1.), urease, and catalase (CAT) (EC 1.11.1.6) activities (Wyszkowska et al., 2009). Heavy metals cause significant decrease in plant growth and protein content at high concentrations. For example, Cd (13–26 mg/kg soil) and Hg (303–608 mg/kg soil) were the most toxic heavy metal,

Plant Metal Interaction. http://dx.doi.org/10.1016/B978-0-12-803158-2.00020-5

inhibiting dry weight of root and shoot from 13.8% to 70.5%, seed yield by 40%, nitrogen content in shoot and root by 68.6% and 79.4% (Ghani, 2010).

The remediation of heavy metal pollution from soil or water resources are not easy processes and also extremely expensive. There are several techniques for the remediation of heavy metal–contaminated sites are available, which include excavation (physical removal of the hazardous materials) (Thomasser and Rouse, 2001) and phytoremediation (heavy metal accumulation by plants) (Liu et al., 2015). Biotechnological approaches that include biostimulation of native microbial populations, bioaccumulation, biosorption, rhizoremediation (plant and microbe interaction), and phytoremediation may offer highly specific alternatives that can be used on a large scale (Sharma, 2012). Phytoremediation is a cheap, energy-efficient, and safer strategy (relative to conventional methods) for accumulating heavy metals from the soil for improved quality of the environment and human health (Sharma and Pandey, 2014; Rosa et al., 2014). For successful phytoremediation, a plant should be able to grow rapidly, reach a high biomass, and take up large concentrations of heavy metals from the contaminated soils into the roots (Liu et al., 2011; Rajkumar et al., 2012). For normal growth, plants need adequate nutrition such as water and micro- and macronutrients such as nitrogen (N), phosphorus (P), and potassium (K) and proper soil characteristics (e.g., soil texture, pH, salinity, moisture), and be tolerant to target pollutants (Ovečka and Takáč, 2014). Plants also require certain heavy metals for their growth and development, and excessive amounts of these metals can become toxic to plants (Chibuike and Obiora, 2014). Higher metal concentrations in soil likely inhibit the growth and development of plants and may thus reduce the efficiency of phytoremediation (Ali et al., 2014). For example, Agnihotri et al. (2014) observed that cobalt (Co) at 50 and 100 μM had a toxic effect on germination percentage, early seedling growth, leaf area, root development, and pigment composition (chlorophyll and carotenoids) of the mung bean (*Vigna radiata*) compared with the control. The nutrient deficiencies also negatively affect the growth and development of most hyperaccumulators (Dary et al., 2010), inhibiting their phytoextraction activities. The microbially assisted amelioration of plant stress is considered to be an important tool in metal stress management in plants (Rajkumar et al., 2013).

Plant-associated microbes stimulate root growth, nutrient uptake, and control soil-borne diseases through various mechanisms, which include production of phytohormones (Egamberdieva et al., 2014), 1-aminocyclopropane-1-carboxylate (ACC) deaminase enzyme (Shahzad et al., 2013), osmoprotectants (Berg et al., 2013), siderophores and organic acids (Ma et al., 2009), solubilization of phosphorus (Erkovan et al., 2010), and biosorption of heavy metals (Chen et al., 2014a). They also enhance resistance of plants to adverse environmental stresses such as drought, salts, heavy metals, and nutrient deficiency (Egamberdieva et al., 2015). In addition, the interaction of plants with microbes is an efficient bioremediation process for the degradation of contaminants, potential metal bioaccumulation, and/or promoting plant growth (Srivastava et al., 2013). This chapter presents an overview of the potential role of root-associated microorganisms in the bioremediation of heavy metal–polluted soils, and provides some discussion of how microbes could be exploited to enhance phytoremediation.

2. ROOT-ASSOCIATED MICROORGANISMS

The rhizosphere is colonized more intensively by microorganisms than the other regions of the soil because of its rich source of nutrients (Egamberdieva, 2005, 2012; Prashar et al., 2014). These microbes can be beneficial, neutral, or pathogenic and compete for nutrients and niches under strongly varying conditions (Berg et al., 2010; Hashem et al., 2014). Beneficial rhizosphere bacteria are of two general types: those forming a symbiotic relationship with the plant and those that are free-living in the soil and root (Lugtenberg and Kamilova, 2009; Hameed et al., 2014). Plant growth–promoting rhizobacteria (PGPR) play an important role in crop production because they: (1) stimulate root growth (Egamberdiyeva and Höflich, 2003), (2) make soil nutrients available to the plant roots (Ipek et al., 2014), (3) fix nitrogen from the air and improve soil fertility (Rejili et al., 2012), (4) suppress soil-borne pathogens and protect plants from various diseases (Egamberdieva et al., 2015), and (5) enhance plant tolerance to various environmental stresses including drought, salinity, high temperature, and heavy metals (Berg et al., 2013).

Microorganisms are able to survive extreme cold/heat and desert/salt conditions because of their potential enzymatic activities, allowing them to convert hazardous compounds into harmless forms (Ali et al., 2013; Singh et al., 2014). Some microorganisms evolved different resistance mechanisms to avoid metal toxicity and may be adapted to live in the presence of soil contaminants (Nies, 2003; Kuffner et al., 2010). Table 20.1 presents a summary of the mechanisms used by PGPR to improve phytoremediation potential of plants. Ni-resistant bacteria have been found in the rhizosphere of the Ni-hyperaccumulator *Thlaspi goesingense* (Idris et al., 2004) and *Alyssum serpyllifolium* subsp. *lusitanicum* (Becerra-Castro et al., 2009). Among bacterial species, *Arthrobacter* spp. and *Streptomyces* spp. were dominant and could tolerate as much as 10 mM Ni in plate assays. The *Phyllobacterium myrsinacearum* RC6b strain was isolated from the rhizosphere of *Sedum plumbizincicola* grown on Pb/Zn mine spoils and showed resistance to 350 mg/L Cd, 1000 mg/L Zn, and 1200 mg/L Pb (Ma et al., 2013). Wu et al. (2006) studied soil bacteria that were resistant to metals (Zn, Cu, and Pb) and found that a P-solubilizer (*Bacillus megaterium* HKP-1) and a K-solubilizer (*Bacillus mucilaginosus* HKK-1) had a high resistance to heavy metal toxicity (Zn, Cu, and Pb: 250, 100, and 300 mg/L, respectively) because of the formation of endospores, which are capable of withstanding extreme conditions. According to Naik et al. (2012), *Pseudomonas aeruginosa* strain WI-1 could tolerate up to 0.6 mM lead nitrate and also accumulated lead at 26.5 mg/g of dry cell biomass. Guo et al. (2015) isolated phosphate-solubilizing bacterium *Burkholderia metalliresistens* sp. nov., D414T from heavy metal (Pb, Cd, Cu, and Zn)-polluted paddy soils of China. The strain was able to tolerate Cd (2000 mg/L), Pb (800 mg/L), Cu (150 mg/L), and Zn (2500 mg/L).

In another study, *P. aeruginosa* was found to be highly resistant to Zn, having a minimum inhibitory concentration ranging from 1000 to 1500 mg/kg against Zn (Islam et al., 2014). The inoculation of wheat with Zn-tolerant *P. aeruginosa* improved plant biomass, the N and P uptake, increased leaf chlorophyll, and total soluble protein in plants grown in contaminated soil containing 1000 mg/kg Zn (Islam et al., 2014). The root-associated microbes could also improve the chemical and biological activities of heavy metal–contaminated soils. For example, inoculation of *Brassica juncea* with As-resistant *Staphylococcus arlettae* strain increased soil dehydrogenase, phosphatase, microbial biomass carbon, total organic carbon, and available phosphorus in arsenic-spiked (5, 10, and 15 mg/kg) soil (Srivastava et al., 2013).

The legume-rhizobia symbioses have been also reported as one of the important approaches for heavy metal remediation (Teng et al., 2015). However, their symbiotic performance was hampered by heavy metals, reducing rhizobial colonization in the root and nitrogenase activity (Chaudhary et al., 2004). According to previous studies, an inhibition of nitrogen fixation has been observed for peas (*Pisum sativum*) grown under Zn-contaminated soil (366 mg/kg Zn) (Obbard and Jones, 2001). The development of heavy metal–tolerant symbioses is an absolute necessity to enable cultivation of

Table 20.1 Summary of Reports on the Effects of PGPR on Plant in Heavy Metal–Contaminated Soils

Bacterial Strain	Metal Uptake	Mechanisms Used	References
Achromobacter. xylosoxidans Ax10	Cu (Indian mustard (*Brassica juncea*))	ACC deaminase, IAA, phosphate solubilization	Ma et al. (2009)
Acinetobacter sp. Q2BJ2, *Bacillus* sp. Q2BG1	Pb (rapeseed (*Brassica napus*))	ACC deaminase, siderophores, IAA	Zhang et al. (2011)
Acinetobacter sp.	As (chickpea (*Cicer arietinum*))	IAA production	Srivastava and Singh (2014)
Bacillus mucilaginosus HKK-1 *Bacillus megaterium* HKP-1	Zn, Pb, Cu (Indian mustard (*B. juncea*))	P, K solubilization	Wu et al. (2006)
Bacillus subtilis, *Bacillus cereus*, *B. megaterium*	Cd (Chinese violet cress (*Orychophragmus violaceus*))	IAA production	Liang et al. (2014)
Brevibacillus brevis	Cd, Ni, Pb (white clover (*Trifolium repens*))	IAA production	Vivas et al. (2006)
Burkholderia sp. J62	Pb, Cd (maize (*Zea mays*) and tomato (*Lycopersicon esculentum*))	IAA, siderophores, ACC deaminase, P solubilization	Jiang et al. (2008)
Enterobacter intermedius MH8b	Zn (white mustard [*Sinapis alba*])	ACC deaminase, IAA, hydrocyanic acid, P solubilization	Płociniczak et al. (2013)
Mesorhizobium amorphae	Cu, Zn, Cr (black locust (*Robinia pseudoacacia*))	IAA, induced stress tolerance	Hao et al. (2014)
Photobacterium spp.	Hg (reed (*Phragmites australis*))	IAA, mercury reductase activity	Mathew et al. (2015)
Phyllobacterium myrsinacearum RC6b	Cd, Zn, and Pb (stonecrop (*Sedum Plumbizincicola*))	ACC deaminase, IAA, siderophores, P solubilization	Ma et al. (2013)
Pseudomonas sp. Lk9	Cd (black nightshade (*Solanum nigrum* L.))	Siderophores, organic acids	Chen et al. (2014a)
Pseudomonas putida, *Rhodopseudomonas* sp.	Zn (*cowbane [Cicuta virosa* L.])	Metal-chelating compounds	Nagata et al. (2015)
Pseudomonas koreensis AGB-1	As, Cd, Cu, Pb and Zn (chinaschilf (*Miscanthus sinensis*))	ACC deaminase activity, IAA	Babu et al. (2015)
Pseudomonas brassicacearum Rhizobium leguminosarum	Zn (Indian mustard (*B. juncea*))	Metal-chelating molecules	Adediran et al. (2015)
P. aeruginosa	Zn (wheat (*Triticum aestivium*))	Antioxidative enzymes (catalase, peroxidase, superoxide dismutase)	Islam et al. (2014)
Staphylococcus arlettae NBRIEAG-6	As (Indian mustard (*B. juncea*))	IAA, ACC deaminase	Srivastava et al. (2013)

ACC, 1-aminocyclopropane-1-carboxylate; As, arsenic; IAA, indole acetic acid; Cd, cadmium; Cu, copper; Hg, mercury; K, potassium; Ni, nickel; P, phosphorus; Pb, lead; Zn, zinc.

leguminous plants in heavy metal–contaminated sites and stabilizing metal enriched soils (Frérot et al., 2006). Hao et al. (2014) isolated copper-resistant *Mesorhizobium amorphae* strain from root nodules of *Robinia pseudoacacia* in a zinc-lead mine tailing, which was able to induce copper, zinc, and chromium tolerance in plants. In other study, *Anthyllis vulneraria* was able to establish a symbiosis with highly metal-tolerant *Mesorhizobium metallidurans* strain and fix 80% of its total nitrogen from atmosphere (Mahieu et al., 2011).

Jebara et al. (2014) isolated *Agrobacterium tumefaciens* and *Rahnella aquatilis* strains that tolerated Pb (up to 3.35 mM Pb) from root nodules of lentil growing in heavy metal–contaminated soils. The strains induced a decline in Pb accumulation either in roots (32%) or in shoots (28%) of lentils treated with 2 mM Pb.

A mercury-reducing bacterium *Photobacterium* spp. MELD1 was isolated from reed (*Phragmites australis*) and determined their beneficial effect on plant growth and mercury uptake in Hg-contaminated soil. The strain was able to vaporize mercury (25 mg/kg, 24 h) and reduce mercury toxicity in plants (Mathew et al., 2015). Nagata et al. (2015) studied root endophytic *Pseudomonas putida* and *Rhodopseudomonas* sp. isolated from *Cicuta virosa* L. plants grown in a pond subjected to heavy metal inputs at the Hitachi mine in Japan. The strain was found to produce metal-chelating compounds that detoxify the metals in the plant tissues. de-Bashan et al. (2010) reported that *Bacillus pumilus* stimulated plant growth of a native plant Sonoran desert shrub quailbush (*Atriplex lentiformis*) in high metal content mine tailings. Sheng et al. (2008a) studied heavy metal–resistant *Bacillus edaphicus* and their plant growth-promoting and lead (Pb) uptake properties for Indian mustard (*B. juncea*) in contaminated with 400–800 mg/kg soil. They found that bacterial strain increased Pb uptake of plant (from 18% to 46%), root dry matter (from 16% to 22%), and shoot growth (from 24% to 30%) in Pb-amended soil.

3. PHYTOREMEDIATION WITH PLANT-ASSOCIATED BACTERIA

Microbial activity in the rhizosphere of plants is higher than in bulk soil. The rhizosphere, which is relatively rich in organic substrates such as free amino acids, proteins, carbohydrates, alcohols, vitamins, or hormones stimulates microbial growth and can contain up to 10^{11} bacteria per gram of root (Prashar et al., 2014). The bioavailability of heavy metals in rhizosphere soils is considered to be an important factor that determines the efficiency of the phytoextraction process (Ma et al., 2013). Rhizosphere microbes, with a high colonization ability and enzymatic activity, may improve the metal solubility and availability by reducing the soil pH or by producing chelators and siderophores (Rajkumar et al., 2012; Islam et al., 2014). Colonizing in the rhizosphere of hyperaccumulators growing in metal-polluted soils microbes improves plant growth through phytohormone production and phosphate solubilization abilities (Prapagdee et al., 2013; Sessitsch et al., 2013).

It has been previously reported by several authors that the inoculation of plants with metal-resistant, root-associated beneficial microbes could improve plant survival in metal-polluted soils because of microbial activity/action in the rhizosphere soils and enhance heavy metal extraction by plants (Ma et al., 2013; Srivastava et al., 2013; Babu et al., 2014). For example, Ma et al. (2013) isolated metal-mobilizing bacteria from the rhizosphere of *S. plumbizincicola* grown on Pb/Zn mine spoils and assessed the interactive effects of *S. plumbizincicola* and metal-mobilizing bacteria on heavy metal phytoremediation. They found that the higher metal-solubilizing bacteria *P. myrsinacearum* significantly increased the concentrations of water-soluble Cd, Zn, and Pb in soil by

16.7-, 4.6-, and 5.7-fold (control Cd 1.5, Zn 1.0, and Pb 0.5 mg/kg soil) compared with the noninoculated control treatment, respectively. In another study, Jiang et al. (2008) observed that the concentrations of water-soluble Pb and Cd increased significantly after 48 h from inoculation with a strain of *Burkholderia* sp. J62 to 5.3 and 25.8 mg/L, respectively, which was associated with a significant decrease in pH. A significant increase in lead (Pb) and cadmium (Cd) concentrations in the root of Indian mustard (*B. juncea* L.) (28–67%) and maize (*Zea mays* L.) (31–170%) were obtained in soil inoculated with *Burkholderia* sp. J62 compared with uninoculated soil (Vivas et al., 2006). Similar observations were made by Wu et al. (2006), in which inoculation of Indian mustard with metal-resistant strains of *B. megaterium* HKP-1 and *B. mucilaginosus* HKK-1 resulted in Zn, Pb, and Cu removal by plant shoots at an average of 38%, 92%, and 36%, respectively, relative to the control plant.

Plants inoculated with a Cu-resistant strain of *Achromobacter xylosoxidans* Ax10 exhibited an increase in plant dry weight of *B. juncea* in the presence of Cu (Ma et al., 2009). At 50 mg Cu/kg soil, plant dry weight increased by 97%, at 100 mg Cu/kg soil by 53%, and at 150 mg Cu/kg soil by 6%. The strain also stimulated Cu accumulation by plants, and maximum accumulation of Cu in the roots and shoots was 180 and 8 mg/kg, respectively, at the highest concentration of 150 mg Cu/kg soil, whereas noninoculated plants accumulated 42 and 4 mg/kg, respectively.

Zhang et al. (2011) isolated Pb-resistant endophytic bacteria *Acinetobacter* sp. Q2BJ2 and *Bacillus* sp. Q2BG1 from metal-tolerant *Commelina communis* plants grown on Pb and Zn mine tailing. The strains increased the dry weight of above-ground tissues (ranging from 39% to 71%) and roots (ranging from 35% to 123%) of *Brassica napus* grown in quartz sand containing 100 mg/kg of Pb, compared with the uninoculated control. The Pb content in plants increased from 58% to 62% in inoculated rape plants compared with the uninoculated control.

Similar observations were made by Srivastava et al. (2013) in which an arsenic (As)-hypertolerant bacterium *S. arlettae* strain NBRIEAG-6 significantly increased root length (93%), shoot length (58%), dry weight (100%), protein (44%), chlorophyll (95%), and carotenoids content (50%) in Indian mustard plants in the presence of 15 mg/kg arsenic (As). In addition, As concentration in shoots and roots of inoculated plants increased from 3.73% to 34.16% and 87.35–99.93%, respectively, compared with the noninoculated control.

In a follow-up study by the same group, *Acinetobacter* sp., which is able to grow at higher concentrations of As ranging up to 50,000 mg/L of As(V) (arsenate) and 1500 mg/L of As(III) (arsenite) concentration increased chlorophyll (20%), soluble protein (38%), plant biomass (106%) of chickpea (*Cicer arietinum* L.) plants in comparison to only As(V)-treated plants (10 mg/kg). The uptake of As by plant roots inoculated with bacterial strains increased significantly by 77% (Srivastava and Singh, 2014). The authors also observed that As stress was reduced in bacterially inoculated plants because of a low uptake of As in the above-soil parts of the plant. The plants treated only with As showed higher stress injury (75%) (as electrolyte leakage, EL) than the control, but the percentage of EL was less (51%) in bacterially inoculated As-treated plants in comparison to only As-treated control plants (Srivastava and Singh, 2014).

A multiple heavy metal–tolerant bacterium *Enterobacter intermedius* MH8b significantly increased the biomass of *Sinapis alba* L. plants and accumulated Zn (32%) and Cd (94%) in plant shoots (Płociniczak et al., 2013). The authors also observed that the introduced strain MH8b was able to survive in soil and had a short-term influence (24 h incubation) on the structure of indigenous microbial communities, whereas bacterial inoculation did not have a long-lasting effect (28 days) on these communities.

Liang et al. (2014) isolated Cd-tolerant bacteria associated with the roots of *Orychophragmus violaceus* L. Schultz, identified as *Bacillus subtilis*, *Bacillus cereus*, *B. megaterium*, and *P. aeruginosa*. The bacterial strains significantly enhanced the accumulation of Cd by plants. Cd concentration in the roots was significantly higher following bacterial treatments (from 52.40 to 84.79 mg/kg) than in uninoculated control plant roots (32.6 mg/kg). The addition of *B. megaterium* resulted in maximal water-extractable Cd, with a 50% increase in its concentration compared with the uninoculated control (100 mg Cd/kg soil).

Chen et al. (2014a) studied the effect of the bacterial endophyte *Pseudomonas* sp. Lk9, which can produce biosurfactants, siderophores, and organic acids, on the growth and metal uptake of Cd hyperaccumulator *Solanum nigrum* L. growing in multimetal-contaminated soil. They observed that bacterial inoculation increased soil heavy metal availability, shoot dry biomass of *S. nigrum* by 14%, and accumulation of Cd by 46.6%, Zn by 16.4%, and Cu by 16.0% compared with noninoculated control plants.

Several reports indicated a potential of microbes to reduce chromium (Cr) stress in plants and improve plant growth and development under Cr-contaminated soils (Ilias et al., 2011).

Essahale et al. (2012) reported tolerance of *Acinetobacter* sp. to Cr(VI) (400 mg/L), whereas *Staphylococcus aureus* and *Pediococcus pentosaceus* tolerated to 2000 mg/L Cr(VI) (Ilias et al., 2011), and *Mesorhizobium* sp. to 500 mg/L Cr(VI) (Wani and Khan, 2010). Though heavy metals inhibits plant growth, heavy metal–tolerant strains associated with plant help to alleviate toxic effect of stress and improve plant development. The strain *Mesorhizobium* sp., which showed Cr(VII) tolerance, was able to increase root, shoot growth, and nodulation and reduce Cr(VII) concentrations in chickpea grown under Cr-contaminated soils (Wani and Khan, 2010). Maqbool et al. (2015) studied Cr-reduced bacteria associated with okra (*Hibiscus esculentus* L.) and their effect on plant growth and development in Cr-contaminated soils. The strain *Brucella* sp. K12 showed the highest potential to reduce Cr(VI) concentrations in soils (up to 69.6%) and plant tissues (up to 56%) compared to the uninoculated control treatment. The strain K12 was able to stimulate plant growth and yield of okra in Cr-contaminated soil (Maqbool et al., 2015).

Babu et al. (2015) studied heavy metal–tolerant bacteria from roots of *Miscanthus sinensis* growing in mine-tailing soil. They found that selected strain *Pseudomonas koreensis* AGB-1 was able to tolerate to Zn (3000 mg/L), Cd (2700 mg/L), As (2700 mg/L), Pb (1.800 mg/L), and Cu (500 mg/L). The strain was able to stimulate plant height, number of leaves, dry weight, chlorophyll, and protein content by 62%, 57%, 62%, 54%, 27%, and 28%, respectively, compared with uninoculated soil. The strain was also increased heavy metal uptake in plant roots by 23% (As), 31% (Cd), 7% (Cu), 18% (Pb), and 15% (Zn) and in shoots by 31% (As), 71% (Cd), 61% (Cu), 22% (Pb), and 21% (Zn) compared with uninoculated plants (Babu et al., 2015). In a similar study, a consortium of *Rhizobium leguminosarum*, *Enterobacter clocae*, and *Pseudomonas* sp. increased *Vicia faba* growth by 40% when plants were grown in nutrient solution amended with 0.5 mM Cu (Fatnassi et al., 2015). Several authors demonstrated that heavy metal–resistant strains endowed with the heavy metal–associated genes such as ArsB, ACR3(1), aoxB, and bmtA (Fatnassi et al., 2015), i.e., *P. putida* KT2440 isolated from *Rhodopseudomonas palustris* contains *ArsM* gene encoding the As(III) S-adenosylmethionine methyltransferase that involved in arsenic removal from polluted soil (Chen et al., 2014b). Xu et al. (2015) analyzed the whole genome sequencing data of three native metal-resistant plant growth-promoting bacteria isolated from metal tailings sites of China, and showed that *M. amorphae* CCNWGS0123 contains metal transporters from P-type adenosine triphosphatase, cation diffusion facilitator, HupE/UreJ, and chromate ion transporter family involved in copper, zinc, nickel, and chromate resistance and homeostasis.

Inoculation with *P. myrsinacearum* strain RC6b improved the growth of Zn/Cd hyperaccumulator *Sedum alfredii* and alfalfa grown in mine tailing site of China. The shoot dry weight increased by 15.9–20.2% for alfalfa and 17.2–19.9% for *S. alfredii,* compared with the noninoculated plants. The inoculation also stimulated the phytoextraction efficiencies of Pb, Cd, and Zn by shoots by 16.9%, 46.3%, and 60.9%, respectively (Liu et al., 2015).

Root endophytic strains *P. putida* and *Rhodopseudomonas* sp. isolated from perennial helophyte *C. virosa* L. increased root, shoot growth, and Cu, Zn, and Pb uptake by roots compared to uninoculated plants (Nagata et al., 2015). The heavy metal-resistant strain *P. putida* CZ1 isolated from the rhizosphere of *Elsholtzia splendens* increased plant growth under copper (Cu) stress condition, whereas Cu concentration in the shoots was increased up to 211%. The strain was also able to stimulate root system, which improve nutrient utilization from soil. Similar observation reported by Kolbas et al. (2015), where endophytic bacteria from roots of Cu-tolerant *Agrostis capillaris* grown in Cu contaminated soil (13–1020 mg Cu/kg) increased shoot and root dry weight and also Cu concentration in sunflower plants by 1.3- to 2.2-fold in the 13–416 mg Cu/kg soil condition.

Hamzah et al. (2015) reported high accumulation of Pb and Hg in the root of Scirpus mucronatus by heavy metal-resistant rhizobacteria bacteria *Brevundimonas diminuta* SF-S1-5 and *Alcaligenes faecalis* SF-S1-60 compared with uninoculated control.

4. MECHANISMS BY WHICH MICROBES INFLUENCE HEAVY METAL ACCUMULATION

Root-associated beneficial microbes can use various mechanisms to stimulate plant growth and development, to protect plants from soil-borne diseases, to increase plant stress tolerance and degrade heavy metals or organic pollutants (Berg et al., 2013; Egamberdieva and Lugtenberg, 2014; Nadeem et al., 2014). These mechanisms include: (1) the production of phytohormones such as indole acetic acid (IAA), gibberellic acid, or cytokinins (Egamberdieva, 2009); (2) antifungal metabolites and/or lytic enzymes (Tu et al., 2013); (3) production of ACC deaminase (EC 4.1.99.4) to reduce the level of ethylene in the roots of developing plants (Glick, 2014); (4) solubilization of minerals such as phosphorus and potassium (Erkovan et al., 2010); (5) production of exopolysaccharides and osmoprotectants (Upadhyay and Singh, 2014); and (6) immobilization of heavy metals (Ma et al., 2013).

Environmental stresses such as drought, salinity, and heavy metals reduce the endogenous levels of phytohormones (Iqbal and Ashraf, 2010). Rhizosphere microorganisms associated with various plants synthesize and release phytohormones as secondary metabolites because of the rich supply of substrates exuded from the roots (Shahab et al., 2009; Egamberdieva et al., 2014). The microbial plant growth regulators such as gibberellins, auxins, and cytokinins can increase the root surface area, root branching, and facilitate greater absorption, thus enhancing the phytoremediation ability of hyperaccumulating plants. For example, *P. putida* TSAU produced 5.3 μg/mL IAA and *P. aureantiaca* TSAU22 produced 7.1 μg/mL when exposed to 1.5% NaCl (electrical conductivity (EC) = 25 dSm), whereas tryptophan increased IAA production by 29.5 and 31.2 μg/mL, respectively (Egamberdieva and Kucharova, 2009). Jiang et al. (2008) found that Cd-resistant *Burkholderia* sp. J62 was able to produce 3.8 μg/mL IAA and solubilize 234 μg/mL of inorganic phosphate compared with the control (32.4 μg/mL), and increased Pb and Cd contents significantly in the shoots of maize (38–192%)

compared with the uninoculated control plants. In other study, *Brevibacillus* sp. B-I was shown to produce IAA (3.95 μg/mL) and to accumulate 5.6% of Zn from the growing medium. This strain increased plant growth and Zn accumulation in *Trifolium repens* (Vivas et al., 2006). Similar observations were reported for multiple heavy metal–tolerant bacterium *E. intermedius* MH8b, which increased the accumulation of Zn (32%) and Cd (94%) in the shoots of *S. alba* plants, which could produce ACC deaminase (EC 4.1.99.4), IAA, and hydrocyanic acid and could also solubilize inorganic phosphate (Płociniczak et al., 2013). Most of these studies showed that the inoculation of heavy metal–tolerant bacteria facilitates root elongation, thereby promoting the accumulation of heavy metals by plants. For example, Belimov et al. (2005) reported that *Variovorax paradoxus*, *Rhodococcus* sp., and *Flavobacterium* sp. were able to stimulate root elongation in Indian mustard seedlings grown under Cd-contaminated soil (50 mg Cd/kg soil). In another study, inoculation of *B. napus* with a Pb-resistant strain *Pseudomonas fluorescens* G10 significantly increased the root length of seedlings by 21–35% in the presence of 5 mg/L Pb (Sheng et al., 2008a), whereas Cr-resistant *Microbacterium arborescens* HU33 stimulated the root length of ryegrass (*Prosopis juliflora* (Sw.) DC.) by 39% in the presence of 2243 mg Cr/kg soil (Khan et al., 2015).

Mercury-reducing bacterial strains isolated from the rhizosphere of reed (*P. australis*) *Photobacterium* spp. strain that increased root length, Hg uptake was found to produce from 77 to 81 μg/mL IAA and showed higher Hg reductase activity (Mathew et al., 2015). In other study, Zaidi et al. (2006) reported on the higher accumulation of Ni in stem, root, and leaf (736, 460, and 272 mg/kg, respectively) of Indian mustard plant (*B. juncea*) in the presence of bioinoculant *B. subtilis* strain SJ-101. The strain SJ-101 has also stimulated plant growth, and exhibited the capability of producing IAA (55 μg/mL), and solubilizing inorganic phosphate (90 μg/mL) in specific culture media. In the presence of higher concentrations of metals, plants synthesize the stress hormone, ethylene, which led to the inhibition of root growth (Glick, 2014). Some bacterial strains contain ACC deaminase, which can cleave the plant ethylene precursor ACC, and thereby lower the level of ethylene in stressed plants grown in metal-contaminated soil, thus preventing the inhibition of root growth by ethylene (Barnawal et al., 2012). The heavy metal–resistant and metal-mobilizing strain *P. myrsinacearum* RC6b has the potential to use ACC as the sole source of nitrogen by producing the enzyme ACC deaminase (Ma et al., 2013). A Cu-resistant PGPR *A. xylosoxidans* Ax10, which enhanced Cu uptake in Indian mustard, tolerated concentrations up to 600 mg Cu/L on a Luria Bertani agar medium and used ACC as the sole N source (Ma et al., 2009). Strain Ax10 has also exhibited an ability of producing IAA (6.4 mg/mL) and solubilizing inorganic phosphate (89.6 mg/mL) in specific culture media. Pb-resistant strains of *Acinetobacter* sp. and *Bacillus* sp., which significantly increased Pb content in the shoots (58–62%) and roots (2.1- to 3.5-fold) of *B. napus* compared with the uninoculated control, showed high ACC deaminase activity (120 and 121 α-ketobutyrate (KB)/mg h) and produced siderophores (Zhang et al., 2011). *Burkholderia* sp. J62 also showed relatively high levels of ACC deaminase activity (25.6 4 μM α KB/mg h) (Jiang et al., 2008). An As-hypertolerant bacterium *S. arlettae* strain NBRIEAG-6, which increased As concentration in the roots and shoots of inoculated *B. juncea*, produced IAA and ACC deaminase (Srivastava et al., 2013). *Pseudomonas koreensis* AGB-1, which showed improved stress tolerance in *M. sinensis* in mine-tailing soils, was able to produce ACC deaminase activity (920 ± 14 mmol a-KB per mg protein per hour) and 8.2 ± 0.4 mg IAA/L and biosurfactant (Babu et al., 2015).

Siderophores, which are iron-chelating compounds produced by microbes, are an important biological structure and can improve metal bioavailability in the rhizosphere and facilitates the increased uptake of metals from the soil by plants (Saravanan et al., 2007; Rajkumar et al., 2012). Ma et al.

(2013) observed the ability of metal-resistant RC6b to produce both catechol (654 mg/L) and hydroxamate (83.9 mg/L) type siderophores.

Metal-resistant and root-associated microbes may solubilize insoluble phosphates but this is strongly dependent on the release of various low molecular mass organic acids such as 2-ketogluconic, gluconic acids, and acetic acid produced by microbes (Muleta et al., 2013). Organic acids chelate mineral ions or decrease the pH to bring P into solution and their role in the solubilization process has been well-documented (Marra et al., 2012). Ma et al. (2013) found maximum phosphate solubilization (105 mg of P mg/L) in *P. myrsinacearum* RC6b, after 72 h of incubation along with a significant decrease in pH from 4.88 to 3.73. According to Islam et al. (2014), the amelioration of adverse effects of Zn stress on biomass production because of *P. aeruginosa* inoculation was related to enhanced antioxidative enzyme such as CAT (83%), peroxidase (EC 1.11.1.7) (180%), superoxide dismutase (SOD) (EC 1.15.1.1) (161%), and the contents of nonenzymatic components such as ascorbic acid (65%) and total phenolics (10%) compared with Zn-treated plants.

Successful phytoremediation depends not only on the interaction of plant roots with bacteria, but also on the bioavailability of heavy metals in soils. Plant roots are not able to easily absorb most heavy metals, and root-associated microbes, because of their enzymatic activities, may contribute to mobilize metal ions, increasing the bioavailable fraction (Li et al., 2009). Sheng et al. (2008b) studied the endophytic bacteria *P. fluorescens* G10 and *Microbacterium* sp. G16 isolated from the roots of rape plants growing in Pb-contaminated soil, and found that inoculation with bacterial strains significantly ($p < 0.05$) increased the water-soluble Pb in soil, compared with the control after 10 days' incubation. In other study, Jiang et al. (2008) showed that the insoluble Pb and Cd fractions were gradually solubilized during the growth of *Burkholderia* sp. strain J62. The endophytic diazotroph *Gluconacetobacter diazotrophicus* made Zn available for plant uptake through its production of 5-ketogluconic acid and dissolved various sources of Zn such as ZnO, $ZnCO_3$, or $Zn_3(PO_4)_2$ (Saravanan et al., 2007).

It has been reported that antioxidant enzymes play an important role in protection of plants from heavy metal induced oxidative injury (Adrees et al., 2015). *Pseudomonas koreensis* AGB-1 strain increased CAT and SOD activities by 42% and 33%, respectively, and reduced malondialdehyde content of *M. sinensis* by 45% grown in heavy metal–contaminated soil (Babu et al., 2015). Similar results were observed by Fatnassi et al. (2015), where inoculation of *Vicia faba* with Cu-resistant microbial consortium results in higher SOD and CAT activity in roots and leaves. The metal resistance strain *M. amorphae* CCNWGS0123 stimulated plant growth of leguminous plants grown in heavy metal–contaminated sites through nitrogen fixation and/or IAA production (Xie et al., 2015). Adediran et al. (2015) studied the mechanisms involved in plant growth stimulation of *B. juncea* grown under 400 mg Zn/kg soil and Zn phytoextraction by *Pseudomonas brassicacearum* and *R. leguminosarum* in glasshouse experiments. They suggested that the metal-chelating molecules synthesized by bacteria may play as a key mechanism in heavy metal accumulation and phytoremediation technology.

5. CONCLUSION AND FUTURE PROSPECTS

This narrow window of studies suggest that to enhance phytoextraction of metals from contaminated soils, plant roots should grow as widely as possible to absorb more heavy metals. The composition of the microbial community in soil–root interface can significantly assist plant establishment on heavy metal–contaminated soils by mediating nutrient mineralization and uptake by plants. Microbes may reduce the toxicity of metals or increase their bioavailability and thus have some potential to improve phytoextraction efficiency. Different mechanisms have been proposed to explain the beneficial effect of root-associated bacteria, which include production of phytohormones, antioxidant enzymes, biosurfactants, metal-chelating compounds, siderophores, cell wall–degrading enzymes, ACC deaminase enzyme, solubilization of minerals, and induced systemic tolerance. The metal-chelating molecules synthesized by bacteria may also play as a key mechanism in heavy metal accumulation and phytoremediation technology. The effects of heavy metals on the plant and microbe interactions are complex and may affect by various actors such as the chemical nature of metals and concentrations that should be taken into consideration. Although advances in plant–microbe systems have increased the effectiveness of heavy metal degradation and accumulation, little is known about how physical, chemical, and biological rhizosphere processes affect such interactions. Because the competition for nutrient and niches is high among indigenous microorganisms in the rhizosphere, it is important to determine potential root colonizing abilities of introduced bacteria. A synergistic interaction of bacteria in the plants may give more benefits to the plant, resulting better accumulation of heavy metals from the soil by plants. Further research is needed to investigate mechanisms involved in mobilization, degradation, various chemical aspects of metal accumulation, and transfer of metals to develop future strategies and optimize phytoextraction by microbe–plant systems. In addition, further studies are needed to assess the robustness of the microbial community for efficient bioremediation of heavy metals in natural field conditions. A more comprehensive understanding of these microbes in their natural environment is important for phytoremediation technology to reach its full potential.

ACKNOWLEDGEMENT

The research activity of Dilfuza Egamberdieva was supported by a Georg Forster Research Fellowship for experienced Researchers (HERMES), Alexander Von Humboldt Foundation.

REFERENCES

Adediran, G.A., Ngwenya, B.T., Mosselmans, J.F.W., Heal, K.V., Harvie, B.A., 2015. Mechanisms behind bacteria induced plant growth promotion and Zn accumulation in *Brassica juncea*. J. Hazard. Mater. 283, 490–499.

Adrees, M., Ali, S., Rizwan, M., Zia-ur-Rehman, M., Ibrahim, M., Abbas, F., Farid, M., Qayyum, M.F., Irshad, M.K., 2015. Mechanisms of silicon-mediated alleviation of heavy metal toxicity in plants: a review. Ecotox. Environ. Saf. 119, 186–197.

Agnihotri, R.K., Vamil, R., Singh, G., Ahmad, M., Sharma, R., 2014. Effect of molybdenum and cobalt induced heavy metal stress on seedling growth stage of *Vigna radiata*. Acta Botanica Hungarica 56 (1–2), 227–241.

Akcil, A., Erust, C., Ozdemiroglu, S., Fonti, V., Beolchini, F., 2015. A review of approaches and techniques used in aquatic contaminated sediments: metal removal and stabilization by chemical and biotechnological processes. J. Cleaner Prod. 86 (1), 24–36.

Ali, S.Z., Sandhya, V., Rao, L.V., 2013. Isolation and characterization of drought-tolerant ACC deaminase and exopolysaccharide-producing fluorescent *Pseudomonas* sp. Ann. Microbiol. 64 (2), 493–502.

Ali, H., Khan, E., Sajad, M.A., 2014. Phytoremediation of heavy metals -concepts and applications. Chemosphere 91 (7), 869–881.

Babu, A.G., Shim, J., Bang, K.S., Shea, P.J., Oh, B.T., 2014. *Trichoderma virens* PDR- 374 28: a heavy metal-tolerant and plant growth promoting fungus for remediation and bioenergy crop production on mine tailing soil. J. Environ. Manage. 132, 129–134.

Babu, A.G., Shea, P.J., Sudhakar, D., Jung, I.B., Oh, B.T., 2015. Potential use of *Pseudomonas koreensis* AGB-1 in association with *Miscanthus sinensis* to remediate heavy metal(loid)-contaminated mining site soil. J. Environ. Manage. 151, 160–166.

Barnawal, D., Bharti, N., Maji, D., Chanotiya, C.S., Kalra, A., 2012. 1-Aminocyclopropane-1-carboxylic acid (ACC) deaminase-containing rhizobacteria protect *Ocimum sanctum* plants during waterlogging stress via reduced ethylene generation. Plant Physiol. Biochem. 58, 227–235.

de-Bashana, L.E., Hernandez, J.B., Bashan, Y., Maiera, R.M., 2010. *Bacillus pumilus* ES4: candidate plant growth-promoting bacterium to enhance establishment of plants in mine tailings. Environ. Exp. Bot. 69, 343–352.

Belimov, A., Hontzeas, N., Safronova, V.I., Demchinskaya, S.V., Piluzza, G., Bullitta, S., Glick, B.R., 2005. Cadmium-tolerant plant growth promoting bacteria associated with the roots of Indian mustard (*Brassica juncea* L. Czern.). Soil Biol. Biochem. 37, 241–250.

Becerra-Castro, C., Monterroso, C., Garcia-Leston, M., Prieto-Fernandez, A., Acea, M.J., Kidd, P.S., 2009. Rhizosphere microbial densities and trace metal tolerance of the nickel hyperaccumulator *Alyssum serpyllifolium* subsp. *lusitanicum*. Intern. J. Phytorem. 11, 525–541.

Berg, G., Alavi, M., Schmidt, C.S., Zachow, C., Egamberdieva, D., Kamilova, F., Lugtenberg, B., 2013. Biocontrol and osmoprotection for plants under saline conditions. In: de Bruijn, F.J. (Ed.), Molecular Microbial Ecology of the Rhizosphere. Wiley-Blackwell, USA, pp. 587–592.

Berg, G., Egamberdieva, D., Lugtenberg, B., Hagemann, M., 2010. Symbiotic plant-microbe interactions: stress protection, plant growth promotion and biocontrol by *Stenotrophomonas*. In: Seckbach, J., Grube, M. (Eds.), Symbioses and Stress, Cellular Origin, Life in Extreme Habitats and Astrobiology, vol. 17 (4). Springer, Netherlands, pp. 445–460.

Chaudhary, P., Dudeja, S.S., Kapoor, K.K., 2004. Efectivity of host-*Rhizobium leguminosarum* symbiosis in soils receiving sewage water containing heavy metals. Microbiol. Res. 159, 121–127.

Chen, L., Luo, S.L., Li, X.J., Wan, Y., Chen, J.L., Liu, C.B., 2014a. Interaction of Cd hyperaccumulator *Solanum nigrum* L. and functional endophyte *Pseudomonas* sp. Lk9 on soil heavy metals uptake. Soil Biol. Biochem. 68, 300–308.

Chen, J., Sun, G.X., Wang, X.X., de Lorenzo, V., Rosen, B.P., Zhu, Y.G., 2014b. Volatilization of arsenic from polluted soil by *Pseudomonas putida* engineered for expression of the arsM arsenic(III) S-adenosine methyltransferase gene. Environ. Sci. Technol. 48, 10337–10344.

Chibuike, G.U., Obiora, S.C., 2014. Heavy metal polluted soils: effect on plants and bioremediation methods. Appl. Environ. Soil Sci. http://dx.doi.org/10.1155/2014/752708.

Dary, M., Chamber-Pérez, M., Palomares, A., Pajuelo, E., 2010. "In situ" phytostabilisation of heavy metal polluted soils using Lupinus luteus inoculated with metal resistant plant-growth promoting rhizobacteria. J. Hazard Mater. 177, 323–330.

Egamberdieva, D., Jabborova, D., Hashem, A., 2015. Pseudomonas induces salinity tolerance in cotton (*Gossypium hirsutum*) and resistance to Fusarium root rot through the modulation of indole-3-acetic acid. Saudi J. Biol. http://dx.doi.org/10.1016/j.sjbs.2015.04.019.

Egamberdieva, D., Botir, H., Hashem, A., Abd Allah, E.F., 2014. Characterization of salt tolerant *Enterobacter hormaechei* strain associated with tomato root grown in arid saline soil. Pure Appl. Microbiol. 8 (5), 4231–4239.

Egamberdieva, D., Lugtenberg, B., 2014. PGPR to alleviate salinity stress on plant growth. In: Miransari, M. (Ed.), Use of microbes for the alleviation of soil stresses, vol. 1. Springer, New York, pp. 73–96.

Egamberdieva, D., 2012. *Pseudomonas chlororaphis*: a salt tolerant bacterial inoculant for plant growth stimulation under saline soil conditions. Acta Physiol. Plant. 34, 751–756.

Egamberdieva, D., 2009. Alleviation of salt stress by plant growth regulators and IAA producing bacteria in wheat. Acta Physiol. Plant. 31, 861–864.

Egamberdieva, D., Kucharova, Z., 2009. Selection for root colonising bacteria stimulating wheat growth in saline soils. Biol. Fertil. Soils 45, 561–573.

Egamberdiyeva, D., Höflich, G., 2003. Influence of growth promoting bacteria on the growth of wheat at different soils and temperatures. Soil Biol. Biochem. 35, 973–978.

Egamberdiyeva, D., 2005. Characterization of *Pseudomonas* species isolated from the rhizosphere of plants grown in Serozem soil, semi-arid region of Uzbekistan. Scientific World J. 5, 501–509.

Erkovan, I., Güllap, M.K., Daşçi, M., Koç, A., 2010. Effects of phosphorus fertilizer and phosphorus solubilizing bacteria application on clover dominant meadow: yield and botanical composition. Turkish J. Field Crops 15 (1), 12–17.

Essahale, A., Malki, M., Marin, I., Moumni, M., 2012. Hexavalent chromium reduction and accumulation by *Acinetobacter* AB1 isolated from Fez Tanneries in Morocco. Indian J. Microbiol. 52, 48–53.

Fatnassi, I.C., Chiboub, M., Saadani, O., Jebara, M., Jebara, S.H., 2015. Impact of dual inoculation with *Rhizobium* and PGPR on growth and antioxidant status of *Vicia faba* L. under copper stress. C. R. Biol. 338, 241–254.

Frérot, H., Lefèbvre, C., Gruber, W., Collin, C., Dos Santos, A., Escarré, J., 2006. Specific interactions between local metallicous plants improve the phytostabilisation of mine soils. Plant Soil 282, 53–65.

Ghani, A., 2010. Toxic effects of heavy metals on plant growth and metal accumulation in maize (*Zea mays* L.). Iranian J. Toxicol. 3 (3), 325–334.

Glick, B., 2014. Bacteria with ACC deaminase can promote plant growth and help to feed the world. Microbiol. Res. 169 (1), 30–39.

Guo, J.K., Ding, Y.Z., Feng, R.W., Wang, R.G., Xu, Y.M., Chen, C., Wei, X.L., Chen, Y.M., 2015. *Burkholderia metalliresistens* sp. nov., a multiple metal resistant and phosphate-solubilising species isolated from heavy metal-polluted soil in Southeast China. Antonie Van Leeuwenhoek 107, 1591–1598.

Hashem, A., Abd Allah, E.F., Alqarawi, A.A., Al Huqail Asma, A., Egamberdieva, D., 2014. Alleviation of abiotic salt stress in *Ochradenus baccatus* (Del.) by *Trichoderma hamatum* (Bonord.) Bainier. J. Plant Interact. 9 (1), 857–868.

Hameed, A., Egamberdieva, D., Abd-Allah, E.F., Hashem, A., Kumar, A., Ahmad, P., 2014. Salinity stress and arbuscular mycorrhizal symbiosis in plants. In: Miransari, M. (Ed.), Use of Microbes for the Alleviation of Soil Stresses, vol. 1. Springer, New York, pp. 139–159.

Hamzah, A., Sarmani, S.B., Yatim, N.Y., 2015. Phytoremediation of Pb and Hg by using *Scirpus mucronatus* with addition of bacterial inoculums. J. Radioanal. Nucl. Chem. 304 (1), 151–155.

Hao, X., Mohamad, O.A., Xie, P., Rensing, C., Wei, G., 2014. Removal of zinc from aqueous solution by metal resistant symbiotic bacterium *Mesorhizobium amorphae*. Separat. Sci. Technol. 49 (3), 376–387.

Idris, R., Trifonova, R., Puschenreiter, W., Wenzel, W.W., Sessitsch, A., 2004. Bacterial communities associated with flowering plants of the Ni hyperaccumulator *Thlaspi goesingense*. Appl. Environ. Microbiol. 70 (5), 2667–2677.

Ilias, M., Rafiqullah, I.M., Debnath, B.C., Mannan, K.S.B., Hoq, M.M., 2011. Isolation and characterization of chromium(VI)-reducing bacteria from tannery effluents. Indian J. Microbiol. 51, 76–81.

Iqbal, M., Ashraf, M., 2010. Gibberellic acid mediated induction of salt tolerance in wheat plants: growth, ionic partitioning, photosynthesis, yield and hormonal homeostasis. Environ. Exp. Bot. 86, 76–85.

Ipek, M., Pirlak, L., Esitken, A., Dönmez, M.F., Turanc, M., Sahinc, F., 2014. Plant growth-promoting rhizobacteria (PGPR) increase yield, growth and nutrition of strawberry under high-calcareous soil conditions. J. Plant Nutr. 37 (7), 990–1001.

Islam, F., Yasmeen, T., Ali, Q., Ali, S., Arifa, M.S., Hussain, S., Rizvi, H., 2014. Influence of *Pseudomonas aeruginosa* as PGPR on oxidative stress tolerance in wheat under Zn stress. Ecotoxicol. Environ. Saf. 104, 285–293.

Jadia, C.D., Fulekar, M.H., 2009. Phytoremediation of heavy metals: recent techniques. Afr. J. Biotechnol. 8 (6), 921–928.

Jebara, S.H., Abdelkerim, S., Fatnassi, I.C., Chiboub, M., Saadani, O., Jebara, M., 2014. Identification of effective Pb resistant bacteria isolated from *Lens culinaris* growing in lead contaminated soils. J. Basic Microbiol. 55, 346–353.

Jiang, C.Y., Sheng, X.F., Qian, M., Wang, Q.Y., 2008. Isolation and characterization of a heavy metal-resistant *Burkholderia* sp. from heavy metal-contaminated paddy field soil and its potential in promoting plant growth and heavy metal accumulation in metal polluted soil. Chemosphere 72, 157–164.

Kavamura, V.N., Esposito, E., 2010. Biotechnological strategies applied to the decontamination of soils polluted with heavy metals. Biotechnol. Adv. 28, 61–69.

Khan, M.U., Sessitsch, A., Harris, M., Fatima, K., Imran, A., Arslan, M., Shabir, G., Khan, Q.M., Afzal, M., 2015. Cr-resistant rhizo-and endophytic bacteria associated with *Prosopis juliflora* and their potential as phytoremediation enhancing agents in metal-degraded soils. Front. Plant Sci. http://dx.doi.org/10.3389/fpls.2014.00755.

Kolbas, A., Kidd, P., Guinberteau, J., Jaunatre, R., Herzig, R., Mench, M., 2015. Endophytic bacteria take the challenge to improve Cu phytoextraction by sunflower. Environ. Sci. Pollut. Res. 22 (7), 5370–5382.

Kuffner, M., De Maria, S., Puschenreiter, M., Fallmann, K., Wieshammer, G., Gorfer, M., Strauss, J., Rivelli, A.R., Sessitsch, A., 2010. Culturable bacteria from Zn- and Cd-accumulating *Salix caprea* with differential effects on plant growth and heavy metal availability. J. Appl. Microbiol. 108, 1471–1484.

Li, W., Ye, Z., Wong, M., 2009. Metal mobilization and production of short-chain organic acids by rhizosphere bacteria associated with a Cd/Zn hyperaccumulating plant, *Sedum alfredii*. Plant Soil 326, 453–467.

Li, X., Liu, L., Wang, Y., Luo, G., Chen, X., Yang, X., Hall, M., Guo, R., Wang, H., Cui, J., He, X., 2013. Heavy metal contamination of urban soil in an old industrial city (Shenyang) in Northeast China. Geoderma 192, 50–58.

Liang, X., He, C.Q., Ni, G., Tang, G.I., Chen, X.P., Lei, Y.R., 2014. Growth and Cd accumulation of *Orychophragmus violaceus* as affected by inoculation of Cd-tolerant bacterial strains. Pedosphere 24 (3), 322–329.

Liu, L.Z., Gong, Z.Q., Zhang, Y.L., Li, P.J., 2011. Growth, cadmium accumulation and physiology of marigold (*Tagetes erecta* L.) as affected by arbuscular mycorrhizal fungi. Pedosphere 21, 319–327.

Liu, Z.F., Ge, H.G., Li, C., Zhao, Z.P., Song, F.M., Hu, S.B., 2015. Enhanced phytoextraction of heavy metals from contaminated soil by plant co-cropping associated with PGPR. Water Air Soil Pollut. 226, 29.

Lugtenberg, B., Kamilova, F., 2009. Plant-growth-promoting-rhizobacteria. Annu. Rev. Microbiol. 63, 541–556.

Ma, Y., Rajkumar, M., Luo, Y., Freitas, H., 2013. Phytoextraction of heavy metal polluted soils using *Sedum plumbizincicola* inoculated with metal mobilizing *Phyllobacterium myrsinacearum* RC6b. Chemosphere 93, 1386–1392.

Ma, Y., Rajkumar, M., Freitas, H., 2009. Inoculation of plant growth promoting bacterium *Achromobacter xylosoxidans* strain Ax10 for the improvement of copper phytoextraction by *Brassica juncea*. J. Environ. Manage. 90, 831–837.

Maqbool, Z., Asghar, Z.N., Shahzad, T., Hussain, S., Riaz, M., Ali, S., Arif, M.S., Maqsood, M., 2015. Isolating, screening and applying chromium reducing bacteria to promote growth and yield of okra (*Hibiscus esculentus* L.) in chromium contaminated soils. Ecotoxicol. Environ. Saf. 114, 343–349.

Mahieu, S., Frérot, H., Vidal, C., Galiana, A., Heulin, K., Maure, L., Brunel, B., Lefèbvre, C., Escarré, J., Cleyet-Marel, J.C., 2011. *Anthyllis vulneraria/Mesorhizobium metallidurans*, an efficient symbiotic nitrogen fixing association able to grow in mine tailings highly contaminated by Zn, Pb and Cd. Plant Soil 342, 405–417.

Marra, L.M., Soares, C.R.F.S.S., de Oliveira, S.M., Ferreira, P.A.A.A., Soares, B.L., Carvalho, R.F., Lima, J.M., Moreira, F.M., 2012. Biological nitrogen fixation and phosphate solubilization by bacteria isolated from tropical soils. Plant Soil 357, 289–307.

Mathew, D.C., Ho, Y.-N., Gicana, R.G., Mathew, G.M., Chien, M.-C., Huang, C.C., 2015. A rhizosphere associated symbiont, photobacterium spp. strain MELD1, and its targeted synergistic activity for phytoprotection against mercury. PLoS One 10 (3), e0121178. http://dx.doi.org/10.1371/journal.pone.0121178.

Muleta, D., Assefa, F., Börjesson, E., Granhall, U., 2013. Phosphate-solubilising rhizobacteria associated with *Coffea arabica* L. in natural coffee forests of southwestern Ethiopia. J. Saudi Soc. Agric. Sci. 12, 73–84.

Nies, D.H., 2003. Efflux-mediated heavy metal resistance in prokaryotes. FEMS Microbiol. Rev. 27 (2–3), 313–339.

Nadeem, S.M., Ahmad, M., Zahir, Z.A., Javaid, A., Ashraf, M., 2014. The role of mycorrhizae and plant growth promoting rhizobacteria (PGPR) in improving crop productivity under stressful environments. Biotechnol. Adv. 32 (2), 429–448.

Nagata, S., Yamaji, K., Nomura, N., Ishimoto, H., 2015. Root endophytes enhance stress-tolerance of *Cicuta virosa* L. growing in a mining pond of eastern Japan. Plant Species Biol. 30, 116–125.

Naik, M.M., Pandey, A., Dubey, S.K., 2012. *Pseudomonas aeruginosa* strain WI-1 from Mandovi estuary possesses metallothione in to alleviate lead toxicity and promotes plant growth. Ecotoxicol. Environ. Saf. 79, 129–133.

Obbard, J.P., Jones, K.C., 2001. Measurements of symbiotic nitrogen-fixation in leguminous host-plants grown in heavy metal-contaminated soils amended with sewage sludge. Environ. Pollut. 111, 311–320.

Ovečka, M., Takáč, T., 2014. Managing heavy metal toxicity stress in plants: biological and biotechnological tools. Biotechnol. Adv. 32 (1), 73–86.

Płociniczak, T., Sinkkonen, A., Romantschuk, M., Piotrowska-Seget, Z., 2013. Characterization of *Enterobacter intermedius* MH8b and its use for the enhancement of heavy metals uptake by *Sinapis alba* L. Appl. Soil Ecol. 63, 1–7.

Prapagdee, B., Chanprasert, M., Mongkolsuk, S., 2013. Bioaugmentation with cadmium-resistant plant growth-promoting rhizobacteria to assist cadmium phytoextraction by *Helianthus annuus*. Chemosphere 92 (6), 659–666.

Prashar, P., Kapoor, N., Sachdeva, S., 2014. Rhizosphere: its structure, bacterial diversity and significance. Rev. Environ. Sci. Biotechnol. 13 (1), 63–77.

Rajkumar, M., Prasad, M.N.V., Sandhya, S., Freitas, H., 2013. Climate change driven plant-metal-microbe interactions. Environ. Int. 53, 74–86.

Rajkumar, M., Sandhya, S., Prasad, M.N.V., Freitas, H., 2012. Perspectives of plant-associated microbes in heavy metal phytoremediation. Biotechnol. Adv. 30, 1562–1574.

Rejili, M., Mahdhi, M., Fterich, A., Dhaoui, A., Guefrachi, I., Abdeddayem, R., Mars, M., 2012. Symbiotic nitrogen fixation of wild legumes in Tunisia: soil fertility dynamics, field nodulation and nodules effectiveness. Agric. Ecosyst. Environ. 157, 60–69.

Renella, G., Egamberdiyeva, D., Landi, L., Mench, M., Nannipieri, P., 2006. Soil microbial activity and hydrolase activity during decomposition of model root exudates released by a model root surface in Cd-contaminated soils. Soil Biol. Biochem. 38, 702–708.

Rosa, M., Prado, F., Hilal, M., Pagano, E., Prado, C., 2014. Phytoremediation: strategies of Argentinean plants against stress by heavy metals. In: Alvarez, A., Polti, M.A. (Eds.), Bioremediation in Latin America, pp. 123–134.

Saravanan, V.S., Osborne, J., Madhaiyan, M., Mathew, I., Chung, J., Ahn, K., Sa, T.M., 2007. Zinc metal solubilization by *Gluconacetobacter diazotrophicus* and induction of pleomorphic cells. J. Microbiol. Biotechnol. 17, 1477–1482.

Sessitsch, A., Kuffner, M., Kidd, P., Vangronsveld, J., Wenzel, W.W., Fallmann, K., Puschenreiter, M., 2013. The role of plant-associated bacteria in the mobilization and phytoextraction of trace elements in contaminated soils. Soil Biol. Biochem. 60, 182–194.

Shahab, S., Ahmed, N., Khan, N.S., 2009. Indole acetic acid production and enhanced plant growth promotion by indigenous PSBs. Afr. J. Agric. Res. 4 (11), 1312–1316.

Shahzad, S.M., Arif, M.S., Riaz, M., Iqbal, Z., Ashraf, M., 2013. PGPR with varied ACC-deaminase activity induced different growth and yield response in maize (*Zea mays* L.) under fertilized conditions. Eur. J. Soil Biol. 57, 27–34.

Sharma, S., 2012. Bioremediation: features, strategies and applications. Asian J. Pharm. Life Sci. 2 (2), 202–212.

Sharma, P., Pandey, S., 2014. Status of phytoremediation in world scenario. Int. J. Environ. Biorem. Biodegrad. 2 (4), 178–191.

Sheng, X.F., He, L.Y., Wang, Q., Ye, H., Jiang, C., 2008b. Effects of inoculation of biosurfactant-producing *Bacillus* sp. J119 on plant growth and cadmium uptake in a cadmium-amended soil. J. Hazard. Mater. 155, 17–22.

Sheng, X.F., Xia, J.J., Jiang, C.Y., He, L.Y., Qian, M., 2008a. Characterization of heavy metal-resistant endophytic bacteria from rape (*Brassica napus*) roots and their potential in promoting the growth and lead accumulation of rape. Environ. Pollut. 156, 1164–1170.

Singh, A.K., Dhanjal, S., Cameotra, S.S., 2014. Surfactin restores and enhances swarming motility under heavy metal stress. Colloids Surf. B Biointerfaces 116 (1), 26–31.

Srivastava, S., Singh, N., 2014. Mitigation approach of arsenic toxicity in chickpea grown in arsenic amended soil with arsenic tolerant plant growth promoting *Acinetobacter* sp. Ecol. Eng. 70, 146–153.

Srivastava, S., Verma, P.C., Chaudhry, V., Singh, N., Abhilash, P.C., Kumar, K.V., Sharma, N., Singh, N., 2013. Influence of inoculation of arsenic-resistant *Staphylococcus arlettae* on growth and arsenic uptake in *Brassica juncea* (L.) Czern. var. R-46. J. Hazard. Mater. 262, 1039–1047.

Teng, Y., Wang, X., Li, L., Li, Z., Luo, Y., 2015. Rhizobia and their bio-partners as novel drivers for functional remediation in contaminated soils. Front. Plant Sci. http://dx.doi.org/10.3389/fpls.2015.00032.

Thomasser, R., Rouse, J.V., 2001. In-Situ Remediation of Chromium Contamination of Soil and Groundwater. Montgomery-Watson.

Tu, Q., Chen, J., Guo, J., 2013. Screening and identification of antagonistic bacteria with potential for biological control of *Penicillium italicum* of citrus fruits. Sci. Hortic. 150 (4), 125–129.

Tripathi, S., Bhattacharyya, P., Mohapatra, R., Som, A., Chowdhury, D., 2014. Influence of different fractions of heavy metals on microbial ecophysiological indicators and enzyme activities in century old municipal solid waste amended soil. Ecol. Eng. 70, 25–34.

Upadhyay, S.K., Singh, D.P., 2014. Effect of salt-tolerant plant growth-promoting rhizobacteria on wheat plants and soil health in a saline environment. Plant Biol. http://dx.doi.org/10.1111/plb.12173.

Vivas, A., Biro, B., Nemth, T., Barea, J.M., Azcon, R., 2006. Nickel-tolerant *Brevibacillus brevis* and arbuscular mycorrhizal fungus can reduce metal acquisition and nickel toxicity effects in plant growing in nickel supplemented soil. Soil Biol. Biochem. 38, 2694–2704.

Wani, P.A., Khan, M.S., 2010. *Bacillus* species enhance growth parameters of chickpea (*Cicer arietinum* L.) in chromium stressed soils. Food Chem. Toxicol. 48, 3262–3267.

Wyszkowska, J., Kucharski, M., Kucharski, J., Borowik, A., 2009. Activity of dehydrogenases, catalase and urease in copper polluted soil. J. Elementol. 14 (3), 605–617.

Wyszkowska, J., Kucharski, M., Kucharski, J., Borowik, A., 2008. Response of bacteria to soil contamination with heavy metals. J. Elementol. 13 (3), 443–453.

Wu, S.C., Cheung, K.C., Luo, Y.M., Wong, H.M., 2006. Effects of inoculation of plant growth-promoting rhizobacteria on metal uptake by *Brassica juncea*. Environ. Pollut. 140, 124–135.

Xie, P., Hao, X., Herzberg, M., Luo, Y., Nies, D.H., Wei, G., 2015. Genomic analyses of metal resistance genes in three plant growth promoting bacteria of legume plants in Northwest mine tailings, China. J. Environ. Sci. 27 (1), 179–187.

Xu, C., Chen, X.C., Duan, D.C., Peng, C., Le, T., Shi, J.Y., Jiyan, S., 2015. Effect of heavy-metal-resistant bacteria on enhanced metal uptake and translocation of the Cu-tolerant plant, *Elsholtzia splendens*. Environ. Sci. Pollut. Res. 22 (7), 5070–5081.

Zaidi, S., Usmani, S., Singh, B.R., Musarrat, J., 2006. Significance of *Bacillus subtilis* strain SJ-101 as a bioinoculant for concurrent plant growth promotion and nickel accumulation in Brassica juncea. Chemosphere 64, 991–997.

Zhang, Y., He, L., Chen, Z., Zhang, W., Wang, Q., Qian, M., Sheng, X., 2011. Characterization of lead-resistant and ACC deaminase-producing endophytic bacteria and their potential in promoting lead accumulation of rape. J. Hazard. Mater. 186, 1720–1725.

CHAPTER

LAND REFORMATION USING PLANT GROWTH–PROMOTING RHIZOBACTERIA IN THE CONTEXT OF HEAVY METAL CONTAMINATION

21

Shashi Kant Shukla[1], Rohit Kumar Mishra[1], Manisha Pandey[1], Vani Mishra[2], Ashutosh Pathak[1], Anand Pandey[1], Rajesh Kumar[1], Anupam Dikshit[1]

Biological Product Laboratory, Department of Botany, University of Allahabad, Allahabad, Uttar Pradesh, India[1]; Nanotechnology Application Centre, University of Allahabad, Allahabad, Uttar Pradesh, India[2]

CHAPTER OUTLINE

Plant Metal Interaction. http://dx.doi.org/10.1016/B978-0-12-803158-2.00021-7

1. INTRODUCTION

Agriculture is the cultivation of plants, edible fungi, and other life forms for food, fiber, biofuels, medicinal, and other products used to sustain and enhance human life; the study of agriculture is known as agricultural sciences. It is one of the most imminent sectors in the Indian economy and ranges from 18.6% to 20% of the gross domestic product of the country is the biggest industry in several states that can provide numerous job opportunities throughout the nation. Thousands of years ago in agricultural history, there was agricultural development with the help of climates, cultures, and technologies. Ample food production was not available before the green revolution because India was dependent on imports. After 1970, India made prominent increases in food assembly, but the population has also tripled since 1947, along with the increase in existing food-grain per capita with increase in the crop yield. For achieving the required yield, India reformed its agricultural policies and achieved historical production as a result of the green revolution, in which the significant contribution of scientist Prof. M.S. Swaminathan was noteworthy.

As the population increased with swift industrialization, the release of lethal waste has had an adverse effect in the form of heavy metal residues. Heavy metal pollutants are basically derived from industries in the form of waste, chemical reagents, and agricultural practices such as pesticides, chemical fertilizers, and herbicides. Metal pollutants also result from vehicle exhaust and the long durability of sewage sludge and wastewater of several sources to arable soils. These chemicals have adverse effects on human beings, plants, and soil microflora, including beneficial microbes. These contaminants accumulate in the agricultural soil where they are released and consequently reach to the food chain, creating high risks to living organisms and ultimately giving rise to food scarcity. They reach plants by the process of absorption by roots from soil and disturb the balance of the food chain in the form of particulate contaminants because of their regular presence (Tak et al., 2013).

1.1 HEAVY METALS

The environment is defined as all conditions surrounding an individual or group of individuals; in particular, the combination of external physical conditions that affect and influence the growth, development, and survival of organisms (Farlex Inc., 2005). A pollutant is any substance in the environment that causes unpleasant effects, impairing the benefit of the environment, reducing the quality of life, and ultimately causing deaths. Heavy metals are significant environmental pollutants, and their toxicity is a problem of escalating significance for ecological, evolutionary, nutritional, and environmental reasons. In general, it is a collective term that applies to groups of metals and metalloids having a density greater than 4.5 g/cm^3 or more than the water (Hawkes, 1997). The term "heavy metals" refers to any metallic element that has a relatively high density and is toxic or poisonous even at low concentrations (Lenntech Water Treatment and Air Purification, 2004). The most common heavy metal contaminants are lead (Pb), copper (Cu), magnesium (Mg), manganese (Mn), nickel (Ni), cadmium (Cd), selenium (Se), mercury (Hg), and chromium (Cr) (Allen, 2014; Street, 2012; Orisakwe, 2012). Some of these metals are micronutrients necessary for plant growth, such as zinc (Zn), iron (Fe), Mn, Ni, and cobalt (Co), whereas others have unknown biological functions, such as Cd, Pb, Hg, and As (Gaur and Adholeya, 2004). They are stable and cannot be degraded or shattered; therefore, they tend to accumulate in soils and sediments. These metals' ions can be harmful to aquatic life and water contaminated by toxic metal ions remain serious public health problem for human health (Petemele et al., 1999).

There are numerous sources of heavy metals in the environment, some of which are as follows: (1) air—including mining, smelting, sanitization of remnant fuels, production and use of metallic commercial products and vehicular exhaust; (2) water—including domestic sewage, industrial effluents, thermal power plants, and atmospheric fallout; and (3) soil—agricultural and animal wastes, municipal and industrial sewage, coal ashes, and fertilizers. These are not easily degraded and shattered and thus readily accumulate to toxic levels. Because they do not degrade, they might affect the biosphere adversely for a long time. Heavy metal toxicity in plants varies with plant species, Apart from toxic and carcinogenic effects, they also tend to accumulate in living organisms. It is well-known that heavy metals cannot be chemically degraded but need to be physically removed or be transformed into nontoxic compounds (Gaur and Adholeya, 2004).

The hazardous impact of heavy metals was reported on the human body to correlate with the level of heavy metals in the environment (Chen, 2011). Heavy metals' bioavailability in higher trophic levels of the food chain may be due to water, house dust (Zheng et al., 2013), honey (Ru et al., 2013), fish (Taweel et al., 2013), and vegetables and fruits (Liu et al., 2013; Gebrekidan et al., 2013). Heavy metal contamination causes cytotoxicity and a change in antioxidant enzyme expression (Saidi et al., 2013).

Heavy metal toxicity can be evaluated by spectrophotometric detection of heavy metals present in soil, water, plant extracts, proteomics, and biosensors (Gowd et al., 2010; Asano et al., 2010; Mansour and Gad, 2010; Yadav, 2010; Hassan et al., 2010; Soldatkin et al., 2012; Tan et al., 2011; Singh and Mittal, 2012; Luque-Garcia et al., 2011).

1.2 BIOAVAILABILITY OF HEAVY METALS

However, metals have advantages over the soil and are found in different forms.

They are available in organic fractions as well as inorganic. Unlike organic contaminants, metal(loid)s remain in the soil for longer periods and do not undergo degradation through biological or chemical processes (Bolan et al., 2014). Bioavailability of metals in soil fractions occur via both chemical extractions and mechanistic geochemical models (Kim et al., 2015). Geochemical properties of soil influence the heavy metal bioavailability (Zhang et al., 2014).

1.2.1 Types and Impact of Heavy Metals

Several heavy metals are found within nature and in a free state, such as Zn, Cu, Cd, Pb, Ni, Se, arsenic (As), and aluminum (Al). Some significant heavy metals are square measure–helpful for plant growth in a very definite quantity; however, on the other hand in an amount level but beyond a concentration level, these metals are square measuring, making them environmental disturbances. The uptake and risks of heavy metals are dependend on their bioavailability (Hare and Tessier, 1996). A lot of elaborate analysis showed that chemical properties of metals in soil and their retention within the solid section of soil is affected by pH, amount of the metal, cation-exchange capability, and geology of soil. Changes in chemical properties of soils have an effect on concentration of free metals and changes their availability for plants (Spurgeon and Hopkin, 1996).

With increasing pH, content of organic matter, and clay, the solubility of most metals decreases owing to their raised sorption. Metals of major interest in bioavailability studies, as listed by the US Environmental Protection Agency, are Al, As, beryllium, Cd, Cr, Cu, Hg, Ni, Pb, Se, and antimony (McKinney and Rogers, 1992). Different metals that are presently of lesser interest to the US Environmental Protection Agency are silver, barium, Co, Mn, molybdenum, sodium, thallium, vanadium, and Zn. These metals were selected because of their potential for human exposure and raised health risks. The bioavailability of Zn, Pb, and Cu from soil decreases with increasing pH. The minimum availableness of metals is affected by higher sorption and precipitation in alkaline and neutral environments (Moraghan and Mascani, 1991). Some highlights concerning the bioavailability of Fe, Cd, Cu, Hg, Mg, Ni, Se, and Zn in water and sediment are square measure below in Table 21.1.

Table 21.1 Important Heavy Metal Contaminants and Their Specifications

		Disease		
S.No.	**Heavy Metal**	**Acute**	**Chronic**	**Remarks**
1	Iron	Cardiac depression, metabolic acidosis	Hepatic cirrhosis	Depressed after heart attack, liver disease
2	Magnesium	Osteoporosis	Cavities	Increased risk of bone fracture
3	Nickle	Dermatitis, encephalopathy	Occupational pulmonary fibrosis, reduced sperm count	Inflammation of skin, excess fibrous connective tissue
4	Zinc	Anorexia nervosa	Neurologic degeneration	Decrease in appetite, paralysis, muscle weakness
5	Cadmium	Prostate cancer	Proteinuria, osteomalacia	Difficulty in urinating, blood in urine
6	Aluminum (high level)	Alzheimer disease		Short-term memory loss
7	Copper	Parkinson disease	Wilson disease	Shaking, rigidity, copper accumulates in tissue
8	Selenium	Pneumonitis, hypotension	Brittle hair and nails	Inflammation of lung, low blood pressure

1.2.2 Arsenic Contamination and Impact on the Environment

Arsenic's compounds were strip-mined by the first Chinese, Greek, and Egyptian civilizations. They found its poisonous properties early; however, it also has several uses such as in bronzing, wood preservation, pesticides, and fireworks and was used in a variety of semiconductor applications (ATSDR, 2000). It is a naturally occurring element that is omnipresent within the Earth's crust. Arsenic is chemically classified as nonmetallic, but it is generally considered a metal. Arsenic combined with oxygen, chlorine, and sulfur to make an inorganic arsenic compound. Arsenic and its compounds are dangerous for human beings. Arsenic cannot be destroyed within its surroundings. It undergoes modifications of its type or combines or separates from particles. The most relevant case of As toxicity by food occurred within the western areas of Japan (Kinki, Chugoku, Shikoku, and Kyushu) in 1955. These compounds were accidentally mixed into Morinaga's dried milk, made by the Tokushima plant of the Morinaga Milk Company. In the same order, of 13,389 newborns who ate the contaminated milk, 600 died, whereas 6093 suffered health difficulties, and 624 were mentally handicapped, had biological process difficulties, and had brain-damage-related paralysis (Ui, 1992). It is one of the contaminants found within the surroundings that is notoriously cyanogenic to man and other living organisms (Chutia et al., 2009). It is an extremely cyanogenic element that exists in numerous species, but the toxicity of As depends on the species.

Arsenic is nonessential and generally toxic to plants. Roots are usually the first tissue to be exposed to As where the metalloid inhibits root extension and proliferation. Upon translocation to the shoot, it can severely inhibit plant growth by slowing or arresting expansion and biomass accumulation as well as compromising plant reproductive capacity through losses in fertility, yield, and fruit production (Garg and Singla, 2011). At sufficiently high concentrations, As interferes with critical metabolic processes, which can lead to the death of plants (Figure 21.1). Biological nitrogen (N) fixation, including the contribution made by symbioses in the root nodules of legumes, supplies a large

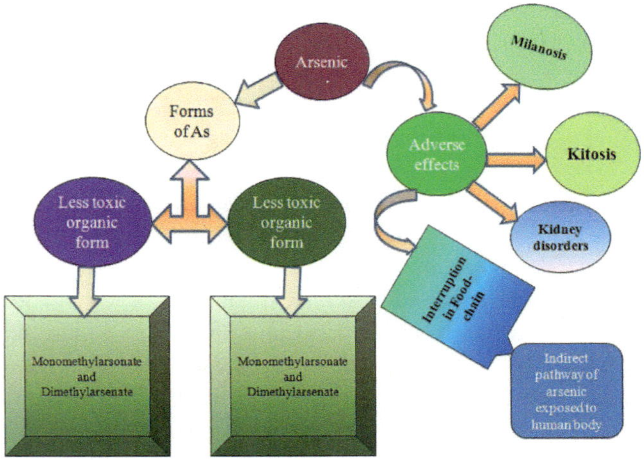

FIGURE 21.1

Forms and adverse effects of arsenic contamination on living beings.

proportion of the N in biological systems and is generally sensitive to As toxicity. Arsenic shows toxicity even at low exposure levels (Dikshit et al., 2000) and causes black foot disease (Lin et al., 1998). Arsenic contamination is also associated with coal mines (Manahan, 1997). In the Indian subcontinent, humans are severely infected with As through contaminated drinking water (Nordstrom, 2002). Arsenic becomes part of the human solid food chain when crops and fodder become contaminated through irrigation of crops that are contaminated with ground water in which As is present (Smedley and Kinniburgh, 2002). Arsenic also enters into farming systems through a variety of means that include natural geochemical processes, mining operations, and fertilization with municipal solid wastes (Meharg et al., 2009). Recently, As has been found accumulated in the new improved varieties of rice grown in Sri Lanka. About 286 μg/kg of total As was found, which was more than the recommendations of Codex Alimentarius (Jayasumana et al., 2015).

1.2.3 Mercury Contamination and Impacts on the Environment

The natural emissions of Hg occur mainly as a result of the degassing of the Earth's crust and evaporation from water bodies (Hutchinson and Meema, 1987). It exists in several forms: elemental or metallic Hg, inorganic Hg compounds, and organic Hg compounds. Human activity has significantly increased Hg levels in the environment over the past several centuries. Once emitted to the air, it falls to the Earth and builds up in our water and soils where it is transformed into methyl mercury—a highly toxic form that accumulates in the tissues of wildlife and people. It is a widespread environmental toxicant with significant potential to bio-accumulate in phytoplanktons, plants such as rice, wildlife, all kinds of fish, and in the food chain. Both manmade (70%) and natural (30%) sources account for the Hg currently found in our environment. They accumulate in the environment as a result of the normal breakdown of minerals in rocks and soil. Mercury releases from natural sources have remained relatively stable in recent history, resulting in a steady rise in environmental Hg. Mercury salts and organomercury compounds are among the most poisonous substances in our environment. The mechanism and extent of toxicity depend strongly on the type of compound and the redox state of Hg.

Terrestrial plants are generally insensible to the harmful effects of Hg compounds; however, it is known to affect photosynthesis and oxidative metabolism by interfering with electron transport in chloroplasts and mitochondria. It also inhibits the activity of aquaporins and reduces plant water uptake (Sas-Nowosielska et al., 2008). Some mushrooms, carrots, potatoes, and aquatic plants were found to be a good source of mercury accumulation. Aquatic *Posidonia oceanica* could be a biological indicator for Hg in sediments. The relationship between Hg and human beings has existed a long time (Nriagu, 1979).

Mercury and its compounds accumulate toxins and in a small quantity per area unit is dangerous to human health (Resaee et al., 2005). The highest human jeopardy in terms of Hg has been related to exposure to extremely toxicant organic alkyl radical Hg through food, primarily through the activity of aquatic organisms, principally fish. According to the United Nations, alkyl radical mercury within the human diet is nearly fully absorbed into the blood and distributed to any or all tissues, the most accumulation occurring in the brain, liver, and kidneys. Its exposure might resulted in severe medical effects which was seen in Minamata, Japan, between 1950 and 1952 (prior to the popularity of human poisonings), where birds had severe difficulties flying and exhibited alternative grossly abnormal behavior (US EPA, 1997). Symptoms of disorder included convulsions, fits, and extremely erratic movements shown among stock, particularly cats whose diets were high in food. Its poison conjointly affects the central system and therefore the areas related to the sensory, visual, auditory, and cocoordinating functions. Increasing doses end in symptom, ataxia, visual changes, dysarthria, hearing defects, loss of speech, coma, and death. The consequences of alkyl radical toxication is irreversible in most cases because of the destruction of neural cells. In humans, about one-eightieth of inhaled mercury vapor is absorbed via the respiratory tract, wherever it enters the cardiovascular system and is distributed throughout the body.

1.2.4 Cadmium Contamination and Impact on the Environment

Cadmium is a nonessential vital metal, but it attracts the most attention in soil science and plant nutrition because of its potential toxicity to humans (Figure 21.2) and its relative quality within the soil–plant system. It never occurs freely in natural environments, but chiefly is a "guest" metal with Pb (Baker et al., 1990). It is released into the surroundings by power stations, heating systems, metal operational industries, or urban traffic. Important sources of Cd to marine surroundings occur from atmospheric deposition, domestic wastewater, and industrial discharges (Benavides et al., 2005). Bioavailability of Cd to soil and crops can be from sphalerite weathering in soils (Robson et al., 2014). Wagner (1993) calculated that nonpolluted soil contaning Cd and found the varying concentrations which was ranges from 0.04 to 0.32 metric long measure and Soil solutions that have a Cd concentration varied from 0.32 to regarding one metric long measure are often considered impure to a moderate level (Benavides et al., 2005). Hence, Cd is classified as an area of intermediate toxicity, but the mechanisms of Cd toxicity do not appear to be totally understood.

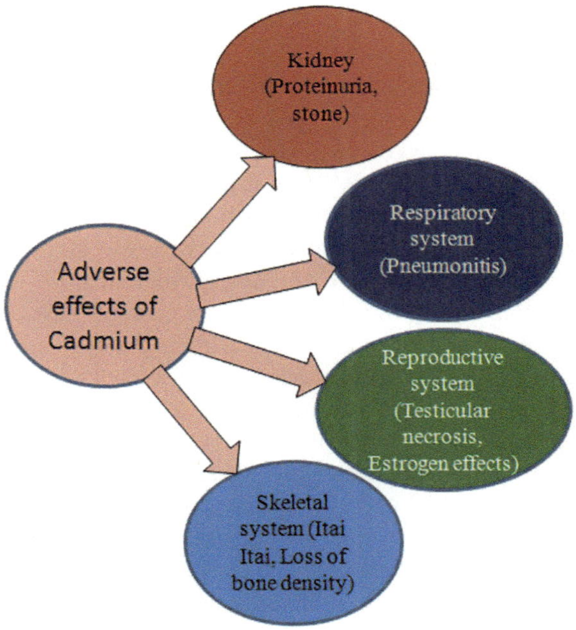

FIGURE 21.2

Adverse effects of cadmium.

Higher concentration of Cd negatively affects plant growth and development (Anita et al., 1990; Aery and Rana, 2003). Besides, it has been additionally reported that even at comparatively low concentrations, Cd alters plant metabolism (Assche and Clijsters, 1990). It is recognized in particular as a vital waste product owing to its high toxicity and solubility in water (Pinto et al., 2004). The most common symptoms of Cd toxicity in plants is chlorosis, stunning, and leaf rolls (Sandalio et al., 2001) and it inhibits plant growth, stomatal opening, transpiration, chemical process, and reduces the absorption of nitrate and its transport from root to shoot by inhibiting nitrate reductase activity in shoots (Hernandez et al., 1996). Many researchers have recommended that aerophilous stress can be a concern in Cd toxicity, by either causing chemical element radical production, or by decreasing accelerator and nonenzymatic antioxidants (Sandalio et al., 2001).

According to the US Environmental Protection Agency and the International Agency for Research on Cancer, Cd induces multiple organ damage and is a carcinogen. Long-term exposure to Cd leads to several morphopathological changes in the kidneys (Satarug and Moore, 2004). The major route of exposure to this heavy metal for the nonsmoking general population is via food; the contribution from other pathways to total uptake is small. Cadmium has a very detrimental effect on the central nervous system. Tobacco is an important source of Cd uptake in smokers because tobacco plants accumulate Cd from the soil. Data from experimental animals and humans have shown that absorption via the lungs is higher than gastrointestinal absorption (via the stomach). Up to 50% of the inhaled Cd may be absorbed and cause decreased attention and memory in humans. This is likely because Cd induces neuron cell death. Neurons are brain cells that communicate and transmit information; if they are affected, brain function is also affected (Pacini et al., 2012). Several epidemiological studies in human populations correlate the exposure to Cd with high blood pressure and cardiovascular disease (Telisman et al., 2001).

1.2.5 Lead Contamination and Impact on the Environment

Lead is a soft metal that has had many applications over the years. The main sources of this element are primarily from automobile emissions, paint chips, fertilizers, pesticides, and Pb-acid batteries or products of industry. The spreading of Pb pollution from major emission sources are mainly through air (Rosman et al., 1993; Wang et al., 2015). The history of Pb handling and health effects goes back more than 20 centuries, which are a remarkably repetitive record of Pb poisoning occurring well into this century. According to the United Nations Environment Program in 1991, Pb occurs naturally in the environment and exists in many forms such as volcanoes, windblown dust, and erosion throughout the world and is now one of the most widely and evenly distributed heavy metals. Soil and plants can be contaminated by Pb from car exhaust, dust, and gases released from various industrial sources. These sources of Pb can result from the mining and smelting of lead ores, as well as others (Figure 21.3). However, most Pb concentrations that are found in the environment are a result of human activities. It may be ingested by eating or drinking contaminated food and by swallowing large particles (diameter greater than 5 μm). Cosmetics (including lip balm) are also an important source of Pb contamination (Pirkle et al., 1998; Miller et al., 2004). Atmospheric and soil Pb can contaminate water and consequently enter the aquatic food chain (Kaste et al., 2003). After sales of unleaded gasoline started, the atmospheric Pb concentration has decreased. The concentration of Pb is higher in urban areas soil compared with suburbs and remote areas (Datko et al., 2014).

Lead in air may be transferred to plants directly through fallout or indirectly through uptake from the soil. Plants on land tend to absorb Pb from the soil and retain most of this in their roots according to

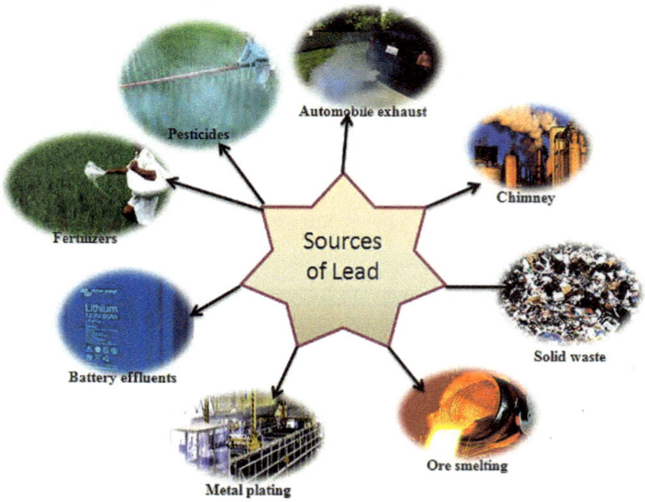

FIGURE 21.3

Sources of lead contamination in the environment.

recommendations of United Nations Environment Program, World Health Organization, and International Labor Organization. Plant leaves take in carbon dioxide, which is required for photosynthesis and liberates oxygen. Lead adheres to the outer surface of the leaf and obstructs light intensity, which reaches the inner wall. This will leads to the reduction of plant growth as well as its death by checking the velocity of photosynthesis, interruption in respiration, and at premature stage causes elongation in root cells, which affect its development. Some evidence suggests that Pb can affect population genetics. The pattern and degree of Pb accumulation are largely influenced in the developmental stage.

Lead is one out of four metals that have the most damaging effects on human health. Skin is an important exposure route for Pb contamination, but only in cases where Pb compounds can easily penetrate it (Pirkle et al., 1998; Miller et al., 2004). It can enter the human body through uptake of food (65%), such as fruit, green vegetables, meats, and crop products, edible sea life, cold drinks, and alcohol may contain significant amounts of Pb in water (20%) and air (15%). Cigarette smoke also contains small amounts of Pb, which adversely affects the lungs and causes lung cancer. It can also contaminate drinking water through corrosion of pipes and causes waterborne diseases and this is more likely to happen when the pH of water is below 7. This may be a major concern to water treatment systems, which are now required to carry out pH adjustments in water that will serve drinking purposes.

It is well-known that Pb fulfills no essential function and in the contrary is harmful to human beings and causes severe problems after uptake. It can cause several undesired problems, such as disturbance in the biosynthesis of hemoglobin and anemia, a rise in blood pressure, kidney damage, decline fertility of men through sperm damage, and can enter a fetus through the placenta of the mother and cause serious damage to the nervous systems and the brains of unborn children and causes spontaneous abortions. Some most important heavy metal and their impacts are shown in Table 21.2.

Table 21.2 Most Important Heavy Metal Contaminants and Their Impacts

Element	Toxicity Comments
Arsenic	As(III) and As(IV); organoarsenic species show decreased decrease toxicity, whereas arsenobetaine and arsenocholine are not toxic
Lead	Lead species are generally toxic; however, organic lead species are more easily absorbed by the gastrointestinal tract than the inorganic species
Mercury	Mercury is generally toxic to species, but inorganic mercury salts are less toxic than the methylated form
Cadmium	All forms of cadmium are very toxic and all need attention

2. REMEDIATION OF HEAVY METAL CONTAMINANTS

The measures of soil health became rigorous in most of the developing countries like India, as soil is one of the fundamental elements necessary for human survival and development with increased industrialization and urbanization. Heavy metal contaminants have become a severe problem in the environment that pollutes extensively and severely harmful. Soil contamination increases with urbanization in due course of time as well as the investment from countries for remediation also increases. Many efforts were undertaken by European countries (Qian and Liu, 2000) and the US Congress passed the Comprehensive Environmental Response, Compensation and Liability Act in 1980. Other laws were also passed for soil remediation, The Resource Conservation and Recovery Act and Superfund Amendments and Reauthorization Act. In the 1990s, Britain also passed the Environmental Protection Act for the same purpose, but compared with the developed countries, research and investments are not encouraged in developing countries such as India and China. Heavy metals are troublesome and persistent organic pollutants because of the three following major mechanisms of work.

1. Bio-concentration: occurs from the breathing in contaminated air continuously.
2. Bio-accumulation: occurs from the uptake of contaminated food stuffs.
3. Bio-magnification: occurs from chemical accumulations in the body that reaches from the tropic level as a result of prey and predator interaction.

These contaminants are very noxious, ultimately damaging the nation's economy. Heavy metal contaminants having the greatest share (i.e., 37%) amongst all contaminants, followed by mineral oil pollutant and polycyclic aromatic hydrocarbons. Although heavy metals cannot be destroyed completely and are very difficult to remove from soil but, their concentration will be reduced at level of nonhazardous and changed from one oxidation state to another. Several remediation technologies are used such as excavation and land fill, thermal treatment, leaching of acids, and retrieval by means of electricity, but those are not suitable due to their elevated price, brief effectiveness, and greater devastation of the rhizosphere productivity and high reliability on the specificity of metal contaminants with site and its properties. Remediation technology basically employs physical, chemical, and biological remediation.

2.1 PHYSICAL REMEDIATION

Physical remediation is the improvement of soil without addition of any other compound such as chemicals. This method is useful for a small area of arable land as the contaminated soil and done by basically two methods (i.e., surrogating contaminated soil and the thermal desorption method). The replacement of polluted soil aims to reduce the concentration of a pollutant in a particular area and improve the soil environment by remediation (Qian and Liu, 2000; Zhang et al., 2004). The soil replacement is divided into three steps.

1. Soil replacement
2. Soil spading
3. Soil importing

In soil replacement, the polluted rhizosphere is replaced by new soil. It is applicable only in a very small area. But, the new soil should be treated properly. In soil spading, the area of unhygienic soil is dug deeply and the noxious wastes are spread completely into the subterranean sites to achieve dilution and natural remediation.

In this step, a great deal of clean and pure soil is imported and mixed properly into the polluted soil to cover it completely and the concentration of pollutant is ultimately reduced. The soil substitution can efficiently segregate the soil and ecosystem and thus diminish its effect on environment. However, this method has a large functioning capacity, costs a great deal, and is suitable for soil in a small area and severe pollution only (Zhou et al., 2004). Another method by which heavy metal can be removed from the surface or remains in a very minute amount is thermal desorption, which is based on the instability of noxious wastes. It can be done by treating the tainted soil by steam, microwave, and infrared radiation for making the contaminants (e.g., As, Hg) volatile. For this method, fickle heavy metals are collected from soil using void pessimistic stress or carrier gas and accomplishes the aim of removal of heavy metals and contaminants from the soil (Li et al., 2010).

Thermal desorption can be further classified into high-temperature desorption (ranging from 320 to 560 °C) and low-temperature desorption from 90 to 320 °C based on temperature. This is an advantageous technique because of its simple procedure; tools can be transferred from one place to another and the remediated soil can be reused is most significant. The use of this device was gracefully carried out in the United States as a measure of in situ remediation. Being beneficial in so many ways, the device has its own limiting facts such as high cost and long process duration that restrict its uses for soil improvement (Aresta et al., 2008).

A field study conducted in West Bengal, India, on an electrochemical As remediation reactor provides hope for the removal of As as it reduces the concentration from ~266 to <5 μg/L in real groundwater at a low cost (Amrose et al., 2014). The new era of nanotechnology also opens opportunities in the form of magnetic nanoadsorbents for the removal of Cd from the environment (Sharma et al., 2014). Meso-iron-oxyhydroxide–impregnated beads adsorb heavy metals from wastewater and sewage (Chung et al., 2014).

2.2 CHEMICAL REMEDIATION

Chemical remediation deals with the treatment of the rhizosphere by some chemical agents and other techniques briefly discussed next.

2.2.1 Chemical Percolation

In this procedure, the impure soil is washed by hygienic water, some reagents, and some other chemical reagents or gases that filter the noxious waste from the rhizosphere (Tampouris et al., 2001). This process occurs because of ionic switch over, precipitation, chelation, and adsorption; as such, metal contaminants are transferred from soil to a liquid stage and improved by percolation. The inorganic compounds and different surfactants are used in the form of leachate for soil improvement. Tokunage and Hakuta reported in 2002 the effect of phosphoric acid, hydrogen chloride, nitric acid, and hydrogen fluoride for removal of contaminants from the rhizosphere at diverse concentrations. Among the previously mentioned chemical agents, phosphoric acid was proved to have the maximum capacity for extracting As (99%) at a 9.4% acid concentration in 6 h followed by sulfuric acid. After that, another method came into existence that was cost-effective and eco-friendly for particular contaminants' removal in the form of As (Alam et al., 2001). For example, yellow-brown forest soil was collected and then contaminated with As, then used as the standard soil for the experiment. It was observed that potassium phosphorus was the most efficient for removing As as compared with many potassium and sodium salts.

The As contamination can be also removed by a soil-washing process, which was successfully done in Korea (Lee et al., 2007). But, chemical agents used for chelation (i.e., ethylenediaminetetraacetic) were more costly with bad degradability. To reduce the problem of secondary contamination and enhance the biodegradability, biological agents can be used, such as ethylene-di-amine-di-succinic acid, which is more effective in biodegradability (Kos and Domen, 2003). The efficacy of saponin was reported by Hong et al. for remediation of heavy metals in all three types of soil (Andosol, Cambisol, and Regosol), washed in a batch experiment and found more specific for Cd and Zn (Hong et al., 2002). In another experiment, Li and coworkers in 2009 also anticipated the effect of saponin obtained from tea on contaminant elimination and observed more specificity for Pb, Cd, Zn, and Cu, respectively. The tea saponin could reduce the environmental risk by removing both soluble as well as reductive metals.

2.2.2 Chemical Obsession

Chemical obsession is the process of addition of reagents or materials into the tainted rhizosphere to make them scarcely movable and less toxic in the insoluble form, so that the immigration of noxious waste could be decreased in the soil (Zhou et al., 2004). The capacity of some finely ground crystalline apatite to stop rhizospheric contaminants and diminish metal bio-availability during the formation of metal phosphates was assessed (Hodson and Valsami-Jones, 2000). Some batch experiments also supported that bone meal with $CaCl_2$ and diethylene triamine pentaacetic acid can reduce the presence of contaminants in the soil. In the same series, another study showed the effectiveness of some chemicals such as sodium bentonite and diatoms in earth for the remediation of soil tainted with Cd (Lv et al., 2009).

Attapulgite terracotta is also used in many places for contaminated soil remediation (Fan et al., 2007) and supports the concept in the form of superb findings that state that an addition of attapulgite soil in ample amounts, 40–46% Cd contamination could be reduced with an increase in the efficacy of crop production. The effectiveness of phosphate rocks with furfural dreg and tainted soil was studied (Zhang et al., 2009) and found the agents of three habituation steps could be a very good source for soil purification against Cu, Zn, Pb, and Cd. The polluted rhizosphere can be remediated by a small amount of chemicals, though heavy metals are changed with the surroundings (Bolan et al., 2003). Finally, it is concluded that the conditioning agents can make a few changes in the rhizospheric structure at some degree, having effects on soil microbes.

2.2.3 Electrical Mediated Remediation

This is a process of soil purification with the generation of an electric field gradient by applying the current at the both sides of the rhizosphere (Luo et al., 2004). The heavy metals are passed from one pole to another pole through the passage of electricity (Acar and Alshawabkeh, 1993; Swartzbaugh et al., 1990). But, it is feasible only for soil of meager permeability with an advantage of unproblematic installation that can be operated at a very low cost (Virkutyte et al., 2002; Xu et al., 2006), not destructing the nature (Alshawabkeh and Bricka, 2000; Zhang et al., 2001). It can easily make the rhizospheric pure and protect the environment.

The nature of the rhizosphere (i.e., acidic or basic) could not be changed by instant electrokinetics, resulting in low management (cure) efficiency. This lacuna can be removed by adding some buffer solutions in anodes and cathodes to manage the pH value; it can be further modified by addition of a swap membrane to control pH with some surface active-charged particles such as ethylenediaminetetraacetic, nitrilotriacetic acid, and phosphonates for better migration (Wang et al., 2007). Moreover, there are other techniques to eliminate heavy metal pollutants from the rhizosphere such as electrokinetic oxidation and reduction-mediated remediation (Cox et al., 1996); one of the most important soil remediation methods is microbe-mediated remediation (Yu et al., 2009).

2.2.4 Vitrification

Vitrification technology is the process of heating the rhizosphere from 1400 to nearly 2000 °C to either volatized or fester the natural means organic matters. The product formed after the heat productions are collected by an off-gas management system. These pyrolyzed materials form rock-shaped glass-like structure called vitreous and surrounds the heavy metal contaminants, which are finally not allowed to move from one place to another. The vitreous is also known for its strength because it was found to be nearly 10 times stronger than concrete. The purification of polluted soil can be done by both ex situ and in situ remediation. Ex situ purification means the energy for heating the rhizosphere can be provided indirectly through blazing of fossil fuels, microwave transferring energy, and so on. In in situ methods of remediation, the heating is transferred directly by inclusion of electrodes into the rhizosphere. So, it can be concluded that by using this technique contaminants can be removed with higher efficiency from the rhizosphere. Besides these advantages, this technology has some negative points such as several complications in operating, high melting energy with low cost-benefit analysis, and limited applications (Fu, 2008).

2.3 BIOLOGICAL REMEDIATION

Until now, methods used for remediation such as excavation and filling of polluted land with pure soil, treatment by thermal means, acid leaching, and electro retrieval are not suitable for practical applications because of their high cost, brief effectiveness, large devastation of soil structure and fertility, and high reliance on the contaminants of concerned metal, rhizospheric nature, conditions of the tainted place, and so on. Thus, the developments of biological remediation strategies for heavy metal–contaminated soils are necessary. Studies conducted on various groups of plants, mushrooms, algae, and plant growth–promoting bacteria showed that the previously mentioned groups have

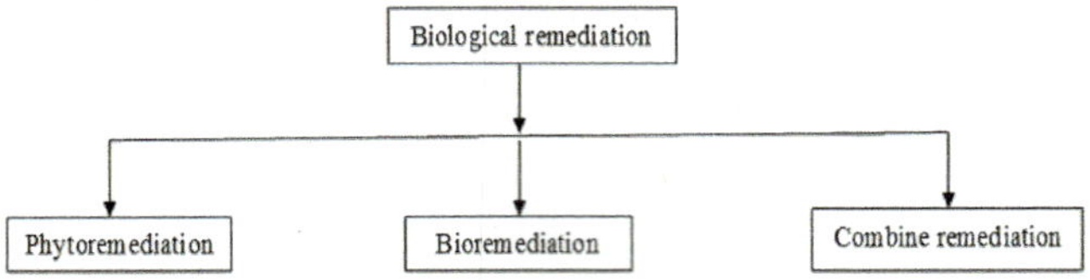

FIGURE 21.4

Systematic outline of types for biological remediation.

potential of accumulating heavy metals from the surrounding environment (Norouzi et al., 2015; Li et al., 2015; Chen et al., 2015; Liu et al., 2015; Anastopoulos and Kyzas, 2015; Ullah et al., 2015).

Biological remediation is basically of three types and are given in Figure 21.4.

2.3.1 Phytoremediation

Remediation is mainly done by the existing green plants and bacteria either to fixate or absorb and reduce their risk by cleaning the pollutant (Shen and Chen, 2000; Ullah et al., 2015). Phytoremediation (Figure 21.5) is also divided into three parts: phytostabilization, phytovolitalization, and phytoextraction (Shen and Chen, 2000).

2.3.1.1 Phytostabilization

The occurrence of heavy metal contaminants and their efficiency of voyage from the rhizosphere and ground water up to the food chain are reduced by Phytostabilization where, the pollutants are adsorbed, precipitated and finally reduced in their concentration (Wang et al., 2009).

2.3.1.2 Phytovolatilization

Phytovolatilization is the process of transforming the stages of heavy metal contaminants, either in a volatile stage or in gaseous one by root-secreted chemicals (Watanabe, 1997). Mercury is one of most concerning heavy metals. Further potentiality of a plant to neutralize the toxins of Hg was investigated by Bizily et al. (1999) in which he proposed a model using *Arabidopsis thaliana* to express a modified bacterial gene called MerBpe, which is responsible for organomercurial lyase (MerB) under controlled conditions. The expressed gene, MerB, boosts lysogenic activity for release of a less itinerant mercury (II) species as a result of breakdown of the carbon–mercury bond. MerBpe transgenic plants are growing strongly in a vast array of monomethylmercuric chloride and phenyl mercuric acetate concentrations, as opposed to the plants without MerBpe, which died or inhibited at the same concentration. The aforesaid investigation supports the hypothesis of genetically engineered plants

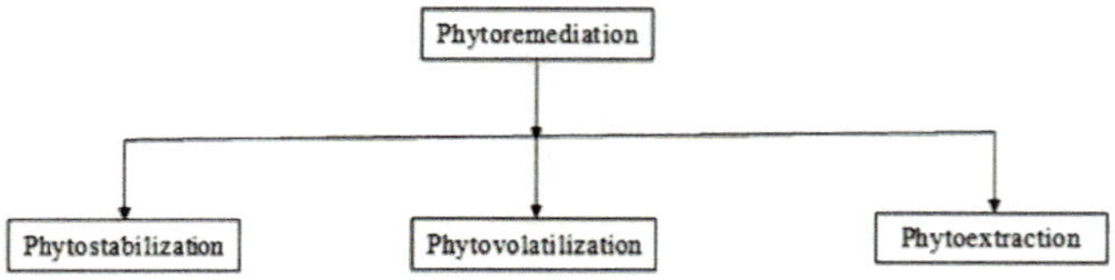

FIGURE 21.5

Outlines for types of phytoremediation.

with MerBpe gene expression to degrade the methyl mercury concentration from the tainted soil. But this methodology is limited to volatile pollutants only (Wang and Wen, 2001).

2.3.1.3 Phytoextraction

The noxious waste materials become attached to the underground part of plants having greater capacity of resistance and adsorption; then after, the pollutants are accumulated in harvested parts of used plants (Bhargava et al., 2012). This technology includes having immense uptake power as a key characteristic feature. There are some peculiar characteristics of the screened plants established by the US Department of Energy (Wang and Wen, 2001). For example, Lotus root accumulates heavy metals independently from soil (Xiong et al., 2013).

- At low pollutant concentration, the accumulating efficiency should be high.
- The accumulating amount of noxious wastes should be high.
- Accumulating diversity of heavy metal contaminants should be greater.
- Growing time along with larger biomass should be short.
- Plants should be pest- and disease-resistant (Figure 21.6).

2.3.2 Biological Remediation

Elimination of pollutants from the polluted spot and renovation of rhizospheric health is the eventual goal of any soil purification practice. To maintain the strength of rhizospheric microflora and appropriate estimation for effectiveness of soil purification procedure, steadfast indicators are required. The heavy metal pollutants could not be removed or harmed by microbes but can affect their emigration and alteration by changing their shape and size along with their chemical properties. The soil purification mechanism mainly comprises gathering a few chemical agents outside of the cell in the form of precipitation, redox reaction, and accretion inside the cell. Microbial leaching is one of easiest techniques by which the important metals are absorbed from substandard ore and mineral concentration. By the same token for industrial relevancy of raw material sharing, microbial leaching

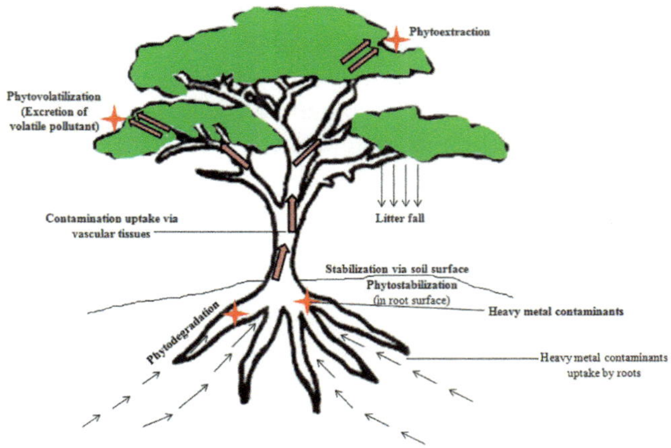

FIGURE 21.6

Different processes of heavy metal detoxification.

has shown great potential for purification of mining locations, management of noxious materials, and their removal of industry that directly falls into the sewage (Bosecker, 2001).

Few heavy metal contaminants are absorbed by mycorrhizae. Their absorbing efficacy was reported by Lamber and Weidensaul in 1991 and found that mycorrhizae can reduce the concentration of Cu and Zn contamination in sludge. Vesicular arbuscular mycorrhizae are somewhere reported as organic agents reduces the toxicity of contaminants (Abdel-Aziz et al., 1997; Miransari, 2011). Vesicular arbuscular mycorrhizae inoculants are found to be more efficient because the parameters are increased in comparison to the source with noninoculant treatment. The heavy metal concentration can be reduced by the vesicular arbuscular mycorrhizae inoculants. These findings recommend the use of mycorrhizae for reducting perilous effect of noxious waste. One of the most important heavy metal contaminants, Cd can be carried from soil to plants by spare radical hyphae of arbuscular mycorrhizal fungus, maybe because of fungal immobilization (Jones and Leyval, 1997). Resistant plant growth–promoting rhizobacteria (PGPR) assist the phytoextraction of heavy metal in *Helianthus annuus* (Prapagdee et al., 2013).

3. PLANT MECHANISMS FOR METAL DETOXIFICATION

Some metals are essential for plant growth and their nutrition such as Zn, Mn, Ni, and Cu, but they become noxious beyond a limit inside the cell. To overcome this problem, several plants show diverse type mechanisms to maintain the metabolic equilibrium of plants (Lasat, 2002; Seth et al., 2007). Ion invasion is one of the mechanisms in which the process is stimulated when ion concentration is very low in concentration and then throws the ions outside of the cell when ion concentration is very high (Pollard et al., 2002). Some higher accumulating plant species are efficient for heavy metal contaminants up to the level of thousands of parts per million and are capable of eliminating toxicity of metals. As such, Ni(II) was isolated by a reaction of organic sulfur on the peptides of cysteine residues (Meagher, 2000). Proteomics provides us a tool for examining the toxicity of heavy metals in plants, microorganisms, and animals (Luque-Garcia et al., 2011).

Several actions were studied for the deactivation of Zn contamination in plants, such as sequestration of Zn in the shoot vacuole as found in *Thlaspi caerulescens* (Lasat et al., 2000). Zinc contamination was deactivated by precipitation of Zn-phytate (Van Steyeninck et al., 1990) along with the binding to the organic acids of low molecular weight (Salt, 1999). Another severe type of toxicant, Cd, which creates several problems in plants, can be neutralized by a thiol-rich peptide (Grill et al., 1987) of plants (i.e., phytochelatins) (Vogeli-Lange and Wagner, 1990; Salt and Rauser, 1995). Along the same lines, another secondary metabolite of plants, metallothionein, a protein, has the capacity of heavy metal contamination binding properties (Tomsett and Thurman, 1988) along with numerous plants and bacteria (Kagi, 1991). The role and wide usage of phytochelatins and metallothionein and their mechanism in heavy metal detoxification was clearly reported and explained by Cobbett (2000). Plants have developed some other mechanism other than phytochelatins and metallothionein such as increased tolerance on inoculation of PGPR of plants (Dat et al., 2000; Mishra et al., 2006), aerobic microbes generated reactive oxygen species in which H_2O_2 plays a crucial role.

Apart from the previously mentioned mechanisms, plants are capable of contesting the unwanted effects of heavy metal contaminants via incidence of antioxidants. Several studies have elaborated the reduction of noxious waste materials in the presence of antioxidative enzymes (Schutzendubel and Polle, 2002). It might be due to the involvement of multiple factors in plants to tolerate the heavy metal contaminants, so that to clarify the exact mechanism of action is little bit difficult.

There are two possible mechanisms to make the polluted site clean: first by enhancing the plant growth and second by reducing metal contaminant concentration. Both actions can be taken by the involvement of rhizospheric microflora i.e., PGPR.

4. APPLICATIONS OF PLANT GROWTH–PROMOTING BACTERIA IN HEAVY METAL DETOXIFICATION

The success of phytoremediation basically depends on the degree of soil pollution along with the availability and amount of contaminants in a particular space and also the capability of plants for absorption and accumulation in the form of biomass. Normally, the growth of plants with greater accumulating efficiency is at a snail's pace at higher metal concentrations in the rhizosphere. But, this limited biomass can be increased with a safe environment by using a group of soil inhabited microflora commonly called PGPR (Dikshit et al., 2013).

This might be due to the introduction of plant growth–promoting strains in a contaminated site, enhances the efficiency of plants of a particular area. The enhanced efficacy appeared in the form of increased efficiency for reuse of nutrients, soil maintenance, and detoxification of noxious waste by alternation. Of the several actions of PGPR, plants provide root exudates in the form of amino acids, carbohydrates, alcohols, hormones, and vitamins, which are essentials for nutrition. The concentrations of root exudates are higher in the rhizosphere, attracting more PGPR and form big clumps of soil (Han et al., 2005; Babalola, 2010). These rhizospheric bacteria thus facilitate the plant growth in several manners such as detoxification of heavy metal contaminants, production of siderophores, phytohormones production, phosphates solubilization, 1-aminocyclopropane-1-carboxylate deaminase production and contains several enzymes that regulate plant growth and their production (Glick et al., 2007; Sheng and Xia, 2006; Ma et al., 2009).

The use of green plants for remediation is an old idea that can be expanded by adding an interdisciplinary approach that comprise rhizospheric flora and enhance the detoxification of heavy metal contaminants. These rhizospheric microbes are PGPR which are closely related to the plants and incorporated for the removal and immobilization of few metal pollutants from a polluted area. The working efficiency of the selected plants is increased because of PGPR strains, such as *Pseudomonas* sp., *Acinetobacter* sp., and *Zea mays* (Lippmann et al., 1995). Although numerous papers were published in the field of plant-mediated remediation of heavy metal contaminants, the knowledge about the exact mechanism of soil microflora-assisted metal extraction is still unclear (Tak et al., 2013).

4.1 LAND REFORMATION USING PGPR BY SOME OTHER MECHANISMS

Plant growth rhizobacteria are playing an important role in sustainable development by improving the crop production and quality of grains, ultimately performing land reformation. Common PGPR are *Pseudomonas fluorescens*, *Pseudomonas putida*, *Bacillus subtilis*, *Paenibacillus polymyxa*, *Paenibacillus elgii*, and *Bacillus licheniformis*, which are playing a very important role in sustainable development as well as in land reformation. There are several functions of PGPR in the improvement of soil, but some of the most important are as follows (Figure 21.8).

- Phytohormone production
- Siderophores production
- Phosphate solubilization
- Induced systemic resistance
- Nitrogen fixation (Table 21.3)

4.2 PHYTOHORMONE PRODUCTION

Plant hormones are chemical messengers that affect a plant's capability in a very minute amount. They are usually synthesized in one part of the plant and are moved from point source to other places. They are target-specific and their interaction causes physiological changes, such as growth and development, succession, and fruit ripening. Every response has the action of two or more than two phytohormones in a particular organ or tissue. Normally phytohormones have both effects (i.e., beneficial as well as harmful); several researchers reported that growth promoters are contrary to some obstructing agents to the plants. There are five recognized major groups of hormones (i.e., auxins, gibberellins, ethylene, cytokinins, and abscisic acid). A recent study shows that auxins can also be an important factor involved in the signaling pathways of bacteria and consequently can affect the microbe's physiology (Spaepen et al., 2007). There are numerous soil microbes involved in the synthesis of auxins in pure culture. The potential for auxins biosynthesis by rhizobacteria can be used as an implement for the selection of effective PGPR strain (Khalid et al., 2004). The evidence shows that PGPR influences plant growth and development by the production of phytohormone such as auxins, gibberellins, and cytokinins. The strains that produce the highest amount of auxins—indole acetic acid (IAA) and indole acetamide—in nonsterilized soil, causes maximum increase in growth

Table 21.3 Some Plant Growth–Promoting Strains and Their Specificity for Soil Improvement Observed by Workers

S.No.	PGPR Strain	Growth Regulators	References
1	*Pseudomonas chlororaphis*	Phosphate solubilization	Yu et al. (2012)
2	*Pseudomonas* spp.	IAA, siderophores, P-solubilizer	Dikshit et al., 2013
3	*Pseudomonas* and *Bacillus*	Siderophores, IAA, P-solubilization	Rajkumar et al. (2010)
4	*Azotobacter*, *Pseudomonas fluorescens*, and *Bacillus* spp.	IAA, siderophores, ammonia, HCN, P-solubilization	Ahmed et al. (2008)
5	*Bacillus subtilis*	IAA and P-solubilization	Zaidi et al. (2006)
6	*Bacillus*, *Pseudomonas*, *Azotobacter*, and *Azospirillum*	P-solubilization and IAA	Tank and Sadaf (2003)
7	*P. fluorescence*	Siderophores	Khan et al. (2002)
8	*Pseudomonas putida*	Siderophores, HCN, ammonia, exo-polysaccharides, phosphate solubilization	Ahemad and Khan (2011, 2012a,b)
9	*Bacillus* species	IAA, siderophores, HCN, ammonia	Wani and Khan (2010)

IAA, indole acetic acid; HCN, hydrogen cynide.

and consent of the wheat crop. Even the strains, which create a low amount of IAA, liberate it constantly, thus improving plant growth (Tsavkelova et al., 2007).

The best-studied example of signal exchange is the *Rhizobium*-legume symbiosis, in which the plant releases flavonoid compounds that act as signals for the bacterium to secrete Nod factors. Nod factors are perceived by plant root hairs and function in a hormone-like fashion to induce root nodules in which the *Rhizobium* bacterium can fix atmospheric N. The bacterium grows at the expense of carbohydrates from the host, but provides fixed N for amino acid biosynthesis in return (Brencic and Winans, 2005; Gray and Smith, 2005). This symbiosis is a prime example of an intimate relationship between a soil bacterium and its host plant and illustrates PGPR: in N-poor environments, the *Rhizobium* bacterium promotes legume plant growth by providing a limiting nutrient.

4.3 SIDEROPHORES PRODUCTION

Iron is a very important nutrient for almost all forms of life. Despite being available in ample amounts in the Earth's crust, the bio-availability of soluble iron is very low. Normally, iron occurs primarily in the form of Fe^{3+} is known to form insoluble hydroxides and oxyhydroxides, which makes it unreachable for both plants and animals (Rajkumar et al., 2010). For that, several PGPR secrete low molecular weight extracellular compounds (i.e., siderophores), having strong affinity to take iron from the environment. These siderophores suppress disease by competition for iron with pathogens. Siderophores are easily soluble in water and on the basis of their occurrence in the cell can be divided into two forms: extracellular and intracellular. PGPR, which are region-, session-, medium- and crop-specific, can be also grouped based on their cross-utilizing efficiencies. Several rhizospheric bacteria are not able to use them due to their complexity (Khan et al., 2009). Actually, siderophores are useful in the solubilization of iron from organic compounds under some conditions like iron deficiency (Indiragandhi et al., 2008).

Siderophores are not limited only to remediation of iron-affected soil, but also to several heavy metal contaminants by forming stable complex compound with Al, Cd, Cu, Ga, In, Pb, and Zn as well as with radionuclide (Neubauer et al., 2000; Kiss and Farkas, 1998). Chelation property of siderophores increases the solubility metal contaminants (Rajkumar et al., 2010). Consequently, siderophores helps in the alleviation of stresses, which is imposed on plants because of a high concentration of heavy metals in the rhizosphere. Plants absorb iron from PGPR siderophores by diverse mechanisms such as chelation and secretion (Schmidt, 1999). Several studies have supported the siderophores-mediated iron absorption as a result of rhizobacterial inoculation in the rhizosphere (Rajkumar et al., 2010). *Pseudomonas fluorescens*, Fe-pyoverdine complex, was absorbed by *A. thaliana* plant, which led to the accumulation of iron, ultimately improving the plant growth (Vansuyt et al., 2007). *Pseudomonas fluorescens*, an important member of PGPR, is characterized by the production of yellow-green pigments termed pyoverdines, which fluoresce under ultraviolet light and function as siderophores. The role of siderophores produced by fluorescent pseudomonades in plant growth promotion was first reported by (Kloepper et al., 1981).

4.4 PHOSPHATE SOLUBILIZATION

The enhancement of soil productiveness is one of the most common strategies to improve agricultural assembly, to which, micro- and macronutrients are required in a unique quantity. Phosphorus (P) is one of the chief macronutrients for plant growth and development. Phosphorus exists in two forms in soil: organic and inorganic phosphates. To convert insoluble phosphates (both organic and inorganic) compounds in a form accessible to the plant is an important trait for a PGPR in increasing plant yields (Igual et al., 2001; Rodríguez et al., 2006). The concentration of soluble P in soil is usually very low, normally at levels of 1 ppm or less (Goldstein, 1994). The plant takes up in several forms but most is absorbed in the forms of HPO_4^{-2} or $H_2PO_4^{-1}$ (Bhattacharyya and Jha, 2012).

The phenomenon of P fixation and precipitation in soil is generally highly dependent on pH and soil type. The biological N fixation is very important in enhancing soil fertility. In addition to biological N fixation, phosphate solubilization is similarly important. Microorganisms put forward a biological rescue system capable of solubilizing the insoluble inorganic P of soil and make it available to the plants. The capacity of some microorganisms to convert insoluble P to a reachable form, like orthophosphate, is an important trait in a PGPR for increasing plant yields (Rodríguez et al., 2006). The clear rings of phosphate solubilizer can be seen in Figure 21.7.

Among the diverse and naturally abundant microbes inhabiting the rhizosphere, the phosphate-solubilizing microorganisms, including bacteria, have provided a substitute biotechnological solution in sustainable agriculture to meet the P demands of plants. These organisms, in addition to providing P to plants, also assist plant growth by other mechanisms. Recent developments in our concern of the practical variety, rhizosphere colonizing ability, type of actions, and sensible application are likely to facilitate their use as dependable mechanism in the management of sustainable agricultural systems (Zaidi et al., 2009). Phosphate-solubilizing microorganisms include largely bacteria and fungi. The most efficient phosphate-solubilizing microorganisms belong to genera *Bacillus*, *Rhizobium*, and *Pseudomonas* among bacteria and *Aspergillus* and *Penicillium*

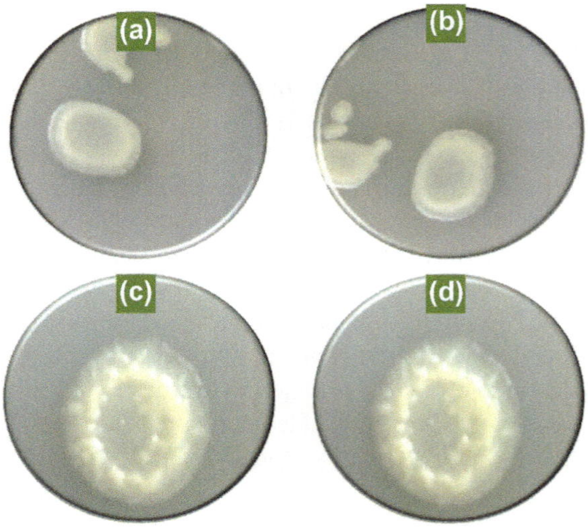

FIGURE 21.7

The phosphate solubilization activity of *Bacillus subtilis* on Pikovskaya agar plates (a–d).

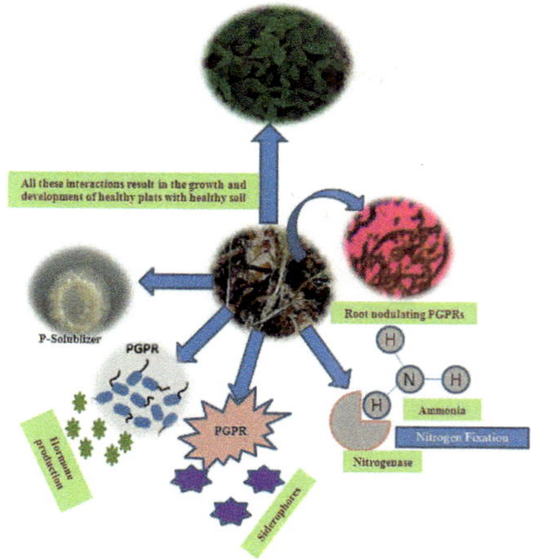

FIGURE 21.8

Diverse mechanisms of plant growth–promoting rhizobacteria resulting in good growth of plants with improved soil.

among fungi. Within rhizobia, two species of nodulating chickpea, *Mesorhizobium ciceri* and *Mesorhizobium mediterraneum*, are known as good phosphate solubilizers (Rivas et al., 2006).

However, it is known that every aspect of the process of nodule formation is limited by the availability of P. The production of organic acids, especially gluconic acid, seems to be the most frequent agent of mineral phosphate solubilization by bacteria such as *Pseudomonas* sp. *Erwinia herbicola*, *Pseudomonas cepacia*, and *Burkholderia cepacia* (Rodriguez and Fraga, 1999).

4.5 SYSTEMATICALLY INDUCED RESISTANCE IN PLANTS

Induced systemic resistance (ISR) of plants against pathogens is a common observable fact that has been intensively investigated with respect to the underlying signaling pathways as well as to its potential use in plant defense. This phenomenon was elicited by a locally contaminated plant that responded with a salicylic-dependent signaling flow that leads to the systemic appearance of a broad range and an enduring disease resistance that is proficient against fungi, bacteria and viruses. Salicylic acid has an important role in the signaling pathway leading to ISR. After infection, endogenous levels of salicylic acid increase locally and systemically, and salicylic acid levels increase in the phloem before ISR occurs. Salicylic acid is synthesized in response to infection both locally and systemically; de novo production of salicylic acid in noninfected plant parts might therefore supply to systemic expression of ISR (Heil and Bostock, 2002).

Some PGPR strains secrete a mix of unstable organic compounds that promote growth in *Arabidopsis* seedlings and induce confrontation against *Erwinia carotovora* sub sp. *carotovora* (Ryu et al., 2005). Plant growth promotion induced by the fungus, *Pythium oligandrum*, is the result of an intricate interaction that includes an indirect effect through control of pathogens in the rhizosphere and or a direct one mediated by plant-induced resistance (Floch et al., 2003).

4.6 NITROGEN FIXATION

Nitrogen (78%) is the one of major constituents of atmosphere, an essential component for plant development as well as their yield, but due to a nonutilizing form, it is unavailable to the green lungs. The unavailable N is converted into the utilizable form using PGPR by process of biological N_2 fixation by converting nitrogen in the ammonia. In the whole process, nitrogenase, an important, complex enzyme, responsible for N_2 fixation plays a very crucial role (Kim and Rees, 1994). Actually, more than two-thirds of N fixation of the whole biosphere is fixed by N fixing soil microbes; the rest are done by synthetic means (Rubio and Ludden, 2008). It requires a gentle temperature for N fixer strains to be extensively dispersed in the rhizosphere. Biological N fixation, an economical and eco-friendly process, improves the soil contamination by reducing the demand and needs of chemical fertilizers (Ladha et al., 1997) as well as pesticides, which leave toxic residues ultimately converting arable land into barren land. Many papers have been published on the types of N fixation based on the occurrence and nature of N fixing microbes, which can be categorized into two forms : symbiotic N_2 fixation and nonsymbiotic N_2 fixation.

5. CONCLUSIONS AND FUTURE PROSPECTS

An immense increase in the world population has given rise to industrialization for feeding the demands of this number of people. This has led to the discharge of several noxious waste residues that create severe problem to both plants and animals. Heavy metal contaminants are of one of these, and are basically produced in the form of effluents from industries as well as agricultural wastes, originated from overexploitation of chemical fertilizers and pesticides. These contaminants are polluting the arable land indirectly affecting nations' economies by simultaneous degradation of soil fertility, ultimately reducing the crop productivity. Because of this situation hunger is not very far from us; to avoid this and improve the health of soil, two methods can be taken into consideration.

- Increase the productivity and quality of agricultural production.
- Reduce the concentration of heavy metal contamination from the soil.

For both methodologies, soil involvement is most important step. The first one implicates a group of beneficial microflora, PGPR, which enhance plant growth and development with sustainable improvement in soil fertility. These microbes also play a crucial role in soil remediation by improving soil health. This involves several mechanisms, including:

- Phytohormone production
- Siderophores production
- Phosphate solubilization
- Induced systemic resistance
- Nitrogen fixation

The second methodology incorporates the reduction of heavy metal pollutants from the soil, ultimately improving the quality of soil by physical, chemical, and biological means of remediation. Physical remediation employs a specific area by treatment of contaminated soil such as replacement of polluted soil by pure and healthy soil. Chemical remediation deals with the inoculation of chemical agents to make contaminants less toxic, movable, and easily soluble. But both processes cannot be undertaken due to the specificities. To avoid the side effects and make the environment eco-friendly, the biological method of remediation is coming into practice. Biological remediation is also listed as phytoremediation, which is classified into three subclasses: phytoextraction, phytovolatilization, and phytostabilization. The most promising and newly growing field of biological remediation is the inoculation of PGPR along with the green plants of higher accumulation and absorption efficiency.

Despite the advancement of diverse techniques in the field of plant–microbe interactions, which reach plant growth and development along with the suppression of pathogens, their implications in the field of soil remediation are still in the nascent stage. To exploit the promising advancement of PGPR, it has become essential to raise our understanding regarding the application of PGPR–plant interaction. Recently, research has focused on the detoxification of heavy metal contaminants and their concentration in the soil by absorbing nutrients through roots. Actually, on interaction of PGPR with plants, the defense mechanisms of plants increased synergistically, improving the soil quality. These beneficial microorganisms improve the production of crops along with the betterment of soil and suppression of pathogenic organisms through diverse mechanisms, such as by producing growth hormones, bio-control agents, ISR, Systemic acquired resistance (SAR), nitrogen fixation, phosphate solubilization, and siderophores production. Some PGPR have the ability to remediate heavy metal contaminants through the green plants in the environment by phytostabilization, phytovolatilization, and phytoextraction. Some exciting results also recommended the usefulness of PGPR, specially siderophores, an important metabolite found to potentially be active in transportation and accumulation of heavy metal pollutants. No methods have reached the height of scientific hypothesis, but the concept of plant growth–promoting traits in rhizospheric/polluted soil to improve the process of phytoremediation was most suitable and eco-friendly.

ACKNOWLEDGEMENTS

Thanks are due to Head Department of Botany, University of Allahabad, for providing the facilities; to Prof. Appa Rao Podile, University of Hyderabad, for valuable suggestions; to Miss Madhu Pandey and Miss Afifa Qidwai for their helping hands; and to UGC for financial support.

REFERENCES

Abdel-Aziz, R.A., Radwan, S.M.A., Dahdoh, M.S., 1997. Reducing the heavy metals toxicity in sludge amended soil using VA mycorrhizae. Egypt J. Microbiol. 32 (2), 217–234.

Acar, Y.B., Alshawabkeh, A.N., 1993. Principles of electrokinetic remediation. Environ. Sci. Technol. 27, 2638–2647.

Aery, C., Rana, D.K., 2003. Growth and cadmium uptake in barley under the cadmium stress. J. Environ. Biol. 24, 117–123.

Agency for Toxic Substances and Disease Registry (ATSDR), 2000. Toxicological Profile for Arsenic. Atlanta, GA, USA.

Ahemad, M., Khan, M.S., 2011. Assessment of plant growth promoting activities of rhizobacterium *Pseudomonas putida* under insecticide-stress. Microbiol. J. 1, 54–64.

Ahemad, M., Khan, M.S., 2012a. Effect of fungicides on plant growth promoting activities of phosphate solubilizing *Pseudomonas putida* isolated from mustard (*Brassica compestris*) rhizosphere. Chemosphere 86, 945–950.

Ahemad, M., Khan, M.S., 2012b. Evaluation of plant growth promoting activities of rhizobacterium *Pseudomonas putida* under herbicide-stress. Ann. Microbiol. 62, 1531–1540.

Ahmad, F., Ahmad, I., Khan, M.S., 2008. Screening of free-living rhizospheric bacteria for their multiple plant growth promoting activities. Microbiol. Res. 163, 173–181.

Alam, G.M., Tokunaga, S., Maekawa, T., 2001. Extraction of arsenic in a synthetic arsenic contaminated soil using phosphate. Chemosphere 43 (8), 1035–1041.

Allen, L.H., 2014. Food Safety: Heavy Metals, third ed. Encyclopedia of Human Nutrition. 331–336.

Alshawabkeh, A.N., Bricka, R.M., 2000. Basics and Application of Electrokinetics Remediation. Marcel Dekker Inc., New York, 95–111.

Amrose, S.E., Bandaru, S.R.S., Delaire, C., Genuchten, C.M., Dutta, A., DebSarkar, A., Orr, C., Roy, J., Das, A., Gadgil, A.J., 2014. Electro-chemical arsenic remediation: field trials in West Bengal. Sci. Total Environ. 488–489, 539–546.

Anastopoulos, I., Kyzas, G.Z., 2015. Progress in batch biosorption of heavy metals onto algae. J. Mol. Liq. 209, 77–86.

Anita, R., Chugh, L.K., Sawhney, V., Sawhney, S.K., 1990. In: Arora, S.K., Singh, M., Aggarwal, R.P. (Eds.), Proceedings of Environmental Pollution. Haryana Agricultural University Press, Hisar, India, pp. 66–77.

Aresta, M., Dibenedetto, A., Fragale, C., 2008. Thermal desorption of polychlorobiphenyls from contaminated soils and their hydrodechlorination using Pd- and Rh-supported catalysts. Chemosphere 70 (6), 1052–1058.

Asano, T., Wang, P.C., Iwasaki, A., 2010. Spectrophotometric detection of labile zinc (II) released from metallothionein: a simple method to evaluate heavy metal toxicity. J. Biosci. Bioeng. 109 (6), 638–644.

Assche, V.F., Clijsters, H., 1990. Effects of metals on enzyme activity in plants. Plant Cell Environ. 13, 195–206.

Babalola, O.O., 2010. Beneficial bacteria of agricultural importance. Biotechnol. Lett. 32 (11), 1559–1570.

Baker, A.J.M., Ewart, K., Hendry, G.A.F., Thorpe, P.C., Walker, P.L., 1990. The evolutionary basis of cadmium tolerance in higher plants. In: In 4th International Conference on Environmental Contamination, pp. 23–29. Barcelona, Spain.

Benavides, M.P., Susana, M., Tomaro, M., 2005. Cadmium toxicity in plants. Braz. J. Plant. Physiol. 17 (1), 21–34.

Bhargava, A., Francisco Carmona, F., Bhargava, M., Srivastava, S., 2012. Approaches for enhanced phytoextraction of heavy metals. J. Environ. Manage. 105, 103–120.

Bhattacharyya, P.N., Jha, D.K., 2012. Plant growth-promoting rhizobacteria (PGPR): emergence in agriculture. World J. Microbiol. Biotechnol. 28, 1327–1350.

Bizily, S.P., Rugh, C.L., Summers, A.O., 1999. Phytoremediation of methylmercury pollution: merB expression in *Arabidopsis thaliana* confers resistance to organomercurial. Proc. Natl. Acad. Sci. U.S.A. 96 (12), 6808–6813.

Bolan, N., Kunhiikrishnan, A., Thangarajan, R., Kumpiene, J., Park, J., Makino, T., Kirkham, M.B., Schecker, K., 2014. Remediation of heavy metal(loid)s contaminated soils – to mobilize or to immobilize? J. Hazard. Mater. 266, 141–166.

Bolan, N.S., Adriano, D.C., Duraisamy, P., 2003. Immobilization and phyto-availability of cadmium in variable charge soils. III. Effect of bio-solid compost addition. Plant Soil 256 (1), 231–241.

Bosecker, K., 2001. Microbial leaching in environmental clean-up programmes. Hydrometallurgy 59 (2–3), 245–248.

Brencic, A., Winans, S.C., 2005. Detection of and response to signals involved in host-microbe interactions by plant associated bacteria. Microbiol. Mol. Biol. Rev. 69, 155–194.

Chen, Y.E., Cui, J.M., Yang, J.C., Zhang, Z.W., Yuan, M., Song, et al., 2015. Biomonitoring heavy metal contaminations by moss visible parameters. J. Hazard. Mater. 296, 201–209.

Chen, Y.F., 2011. Review of the research on heavy metal contamination of China's city soil and its treatment method. China Popul. Resour. Environ. 21 (3), 536–539.

Chung, S.G., Ryu, J.C., Song, M.K., An, B., Kim, S.B., Lee, S.H., Choi, J.W., 2014. Modified composites based on mesostructured iron oxyhydroxide and synthetic minerals: a potential material for the treatment of various toxic heavy metals and its toxicity. J. Hazard. Mater. 267, 161–168.

Chutia, P., Kato, S., Kojima, T., Satokawa, S., 2009. Arsenic adsorption from aqueous solution on synthetic zeolites. J. Hazard. Mater. 162, 440–447.

Cobbett, C.S., 2000. Phytochelatins and their roles in heavy metal detoxification. Plant Physiol. 123, 825–832.

Cox, C.D., Shoesmith, M.A., Ghosh, M.M., 1996. Electrokinetics remediation of mercury-contaminated soil using iodine/iodide lixiviant. Environ. Sci. Technol. 30 (6), 1933–1938.

Dat, J.F., Van, B.F., Vandenabeele, S., Vranova, E., Van, M.M., Inze, D., 2000. Dual action of active oxygen species during plant stress responses. Cell. Mol. Life Sci. 57, 779–795.

Datko, W.L., Wilkie, A., Bryant, J.R., 2014. Analysis of U.S. soil lead (Pb) studies from 1970 to 2012. Sci. Total Environ. 468–469, 854–863.

Dikshit, A.K., Pallamreddy, K., Reddy, L.V.P., Saha, J.C., 2000. Arsenic in ground water and its sorption by kimberlite tailings. J. Environ. Sci. Health 35, 65–85.

Dikshit, A., Shukla, S.K., Mishra, R.K., 2013. Plant Growth Promotion by Plant Growth Promoting Rhizobacteria (PGPR) with Nanomaterials in Current Agricultural Scenario. Lap Lambert Academic Publishing, ISBN 978-3-659-36774-8.

Fan, D.F., Huang, S.S., Liao, Q.L., 2007. Restoring experiment on cadmium polluted vegetable lands with attapulgite of varied dose. Jiangsu Geol. 31 (4), 323–328.

Farlex Inc., 2005. Environment, the Free Dictionary. Farlex Inc. Publishing, USA.

Floch, G.L., Rey, P., Benizri, E., Benhamou, N., Tirilly, Y., 2003. Impact of auxins-compounds produced by the antagonistic fungus *Pythium oligandrum* or the minor pathogen *Pythium* group F on plant growth. Plant Soil 257 (2), 459–470.

Fu, J.H., 2008. The research status of soil remediation in China. In: Annual Meeting of Chinese Society for Environmental Sciences, pp. 1056–1060.

Garg, N., Singla, P., 2011. Arsenic toxicity in crop plants: physiological effects and tolerance mechanisms. Environ. Chem. Lett. 9, 303–321.

Gaur, A., Adholeya, A., 2004. Prospects of arbuscular mycorrhizal fungi in phytoremediation of heavy metal contaminated soils. Curr. Sci. 86 (4), 528–534.

Gebrekidan, A., Weldegebriel, Y., Hadera, A., Bruggen, B.V.D., 2013. Toxicological assessment of heavy metals accumulated in vegetables and fruits grown in Ginfel river near Sheba Tannery, Tigray, Northern Ethopia. Ecotoxicol. Environ. Saf. 95, 171–178.

Glick, B.R., Cheng, Z., Czarny, J., Duan, J., 2007. Promotion of plant growth by ACC deaminase containing soil bacteria. Eur. J. Plant Pathol. 119, 329–339.

Goldstein, A.H., 1994. Involvement of the quinoprotein glucose dehydrogenises in the solubilization of exogenous phosphates by gram-negative bacteria. In: Torriani gorini, A., Yagil, E., Silver, S. (Eds.), Phosphate in Microorganisms: Cellular and Molecular Biology, pp. 197–203. Washington, DC.

Gowd, S.S., Reddy, M.R., Govil, P.K., 2010. Assessment of heavy metal contamination in soils at Jajmau (Kanpur) and Unnao industrial areas of the Ganga Plain, Uttar Pradesh, India. J. Hazard. Mater. 174 (1–3), 113–121.

Gray, E.J., Smith, D.L., 2005. Intracellular and extracellular PGPR: Commonalities and distinctions in the plant bacterium signaling process. Soil Biol. Biochem. 37, 395–412.

Grill, E., Winnacker, E.L., Zenk, M.H., 1987. Phytochelatins, a class of heavy-metal-binding peptides from plants are functionally analogous to metallothioneins. Proc. Natl. Acad. Sci. U.S.A. 84, 439–443.

Han, J., Sun, L., Dong, X., Cai, Z., Sun, X., Yang, H., Wang, Y., Song, W., 2005. Characterization of a novel plant growth-promoting bacteria strain *Delftia tsuruhatensi* HR4 both as a diazotroph and a potential bio-control agent against various plant pathogens. Syst. Appl. Microbiol. 28, 66–76.

Hare, L., Tessier, A., 1996. Predicting animal cadmium concentration in lakes. Nature 360, 430–432.

Hassan, S.H.A., Steven, W., Gonkel, V., Kim, S.M., Yoon, S.H., Joo, J.H., Shin, B., Jeon, B.H., Bae, W., Oh, S.E., 2010. Isolating and characterization of *Acidithiobacillus caldus* from a sulfur-oxidizing bacterial biosensor and its role in detection of toxic chemicals. J. Microbiol. Methods 82 (2), 151–155.

Hawkes, J.S., 1997. Heavy metals. J. Chem. Educ. 74, 1369–1374.

Heil, M., Bostock, R.M., 2002. Induced systemic resistance (ISR) against pathogens in the context of induced plant defenses. Ann. Bot. 89 (5), 503–512.

Hernandez, L.J., Gollas, G.T., Vargas, A.F., 1996. Activation of the prophenoloxidase system of the brown shrimp *Penaeus californiensis* Holmes Comp. Biochem. Physiol. 113C, 61–66.

Hodson, M.E., Valsami-Jones, É., 2000. Bonemeal additions as a remediation treatment for metal contaminated soil. Environ. Sci. Technol. 34 (16), 3501–3507.

Hong, K.J., Tokunaga, S., Kajiuchi, T., 2002. Evaluation of remediation process with plant-derived bio surfactant for recovery of heavy metals from contaminated soils. Chemosphere 49 (4), 379–387.

Hutchinson, T.C., Meema, K.M., 1987. Lead, Mercury, Cadmium, and Arsenic in the Environment. Scientific Committee on Problems of the Environment (SCOPE) 31.

Igual, J.M., Valverde, A., Cervantes, E., Velazquez, E., 2001. Phosphate-solubilizing bacteria as inoculants for agriculture: use of updated molecular techniques in their study. Agronomies 21 (6–7), 561–568.

Indiragandhi, P., Anandham, R., Madhaiyan, M., Sa, T.M., 2008. Characterization of plant growth-promoting traits of bacteria isolated from larval guts of diamond back moth *Plutella xylostella* (Lepidoptera: Plutellidae). Curr. Microbiol. 56, 327–333.

Jayasumana, C., Paranagama, P., Fonseka, S., Amarasinghe, M., Gunatilake, S., Siribaddana, Sisira, 2015. Presence of arsenic in Sri Lanka rice. Int. J. Food Contam. 2, 1.

Jones, E.J., Leyval, C., 1997. Uptake of ^{109}Cd by roots and hyphae of a *Glomus* with high and low concentration of cadmium. New Phytol. 135 (2), 353–360.

Kagi, J.H.R., 1991. Overview of metallothionein. Methods Enzymol. 205, 613–626.

Kaste, J.M., Friedland, A.J., Stürup, S., 2003. Using stable and radioactive isotopes to trace atmospherically deposited Pb in montane forest soils. Environ. Sci. Technol. 37, 3560–3567.

Khalid, A., Arshad, M., Zahir, Z.A., 2004. Screening plant growth-promoting rhizobacteria for improving growth and yield of wheat. J. Appl. Microbiol. 96 (3), 473–480.

Khan, M.S., Zaidi, A., Aamil, M., 2002. Bio-control of fungal pathogens by the use of plant growth promoting rhizobacteria and nitrogen fixing microorganisms. Indian J. Bot. Soc. 81, 255–263.

Khan, M.S., Zaidi, A., Wani, P.A., Oves, M., 2009. Role of plant growth promoting rhizobacteria in the remediation of metal contaminated soils. Environ. Chem. Lett. 7, 1–19.

Kim, J., Rees, D.C., 1994. Nitrogenase and biological nitrogen fixation. Biochemistry 33, 389–397.

Kim, R.Y., Yoon, J.Ki., Kim, T.S., Yang, J.E., Owens, G., Kim, K.R., 2015. Bioavailability of heavy metals in soils: definitions and practical implementation-a critical review. Environ. Geochem. Health http://dx.doi.org/10.1007/s10653-015-9695-y.

Kiss, T., Farkas, E., 1998. Metal-binding ability of desferrioxamine B. J. Inclusion Phenom. Mol. Recognit. Chem. 32, 385–403.

Kloepper, J.W., Scher, F.M., Laliberté, M., Tipping, B., 1981. Emergence-promoting rhizobacteria: description and implications for agriculture. In: Iron, Siderophores, and Plant Diseases, vol. 117, pp. 155–164.

Kos, B., Domen, L., 2003. Induced phytoextraction/soil washing of lead using biodegradable chelate and permeable barriers. Environ. Sci. Technol. 37, 624–629.

Ladha, J.K., Bruijn de, F.J., Malik, K.A., 1997. Introduction: assessing opportunities for nitrogen fixation in rice-a frontier project. Plant Soil 124, 1–10.

Lasat, M.M., 2002. Phytoextraction of toxic metals: a review of biological mechanisms. J. Environ. Qual. 31, 109–120.

Lasat, M.M., Pence, N.S., Garvin, D.F., Ebbs, S.D., Kochian, L.V., 2000. Molecular physiology of zinc transport in the Zn hyper accumulator *Thlaspi caerulescens*. J. Exp. Bot. 51 (342), 71–79.

Lee, M., Paik, I.S., Do, W., 2007. Soil washing of As-contaminated stream sediments in the vicinity of an abandoned mine in Korea. Environ. Geochem. Health 29 (4), 319–329.

Lenntech, 2004. Water Treatment and Air Purification. Water treatment Lenntech, Rotterdamseweg, Netherlands.

Li, J., Zhang, G.N., Li, Y., 2010. Review on the remediation technologies of POPs. Hebei Environ. Sci. 65–68.

Li, N., Kang, Y., Pan, W., Zeng, L., Zhang, Q., Luo, J., 2015. Concentration and transportation of heavy metals in vegetables and risk assessment of human exposure to bio-accessible heavy metals in soil near a waste-incinerator site, South China. Sci. Total Environ. 521–522, 144–151.

Lin, T.H., Huang, Y.L., Wang, M.Y., 1998. Arsenic species in drinking water, hair, fingernails, and urine of patients with black foot disease. J. Toxicol. Environ. Health 53, 85–93.

Lippmann, B., Leinhos, V., Bergmann, H., 1995. Influence of auxins producing rhizobacteria on root morphology and nutrient accumulation of crops. 1. Changes in root morphology and nutrient accumulation in maize (*Zea mays* L.) caused by inoculation with indole-3-acetic acid (IAA) producing *Pseudomonas* and *acinetobactor* strain or IAA applied exogenously. Angew. Bot. 69, 31–36.

Liu, B., Huang, Q., Cai, H., Guo, X., Wang, T., Gui, M., 2015. Study of heavy metal concentrations in wild edible mushrooms in Yunnan Province, China. Food Chem. 188, 294–300.

Liu, X., Song, Q., Tang, Y., Li, W., Xu, J., Wu, J., Wang, F., Brookes, P.C., 2013. Human health risk assessment of heavy metals in soil-vegetable system: a multi-medium analysis. Sci. Total Environ. 463–464, 530–540.

Luo, Q.S., Zhang, X.H., Wang, H., 2004. Influence of non-uniform electro-kinetic remediation technology on soil properties. Chin. J. Environ. Eng. 5 (4), 40–45.

Luque-Garcia, J.L., Cabezas-Sanchez, P., Camara, C., 2011. Proteomics as a tool for examining the heavy metals. Trends Anal. Chem. 30, 703–716.

Lv, L.L., Jin, M.Y., Li, B.W., 2009. Study on remediation of the soil contaminated with cadmium by applying four minerals. J. Agric. Univ. Hebei 32 (1), 1–5.

Ma, Y., Rajkumar, M., Freitas, H., 2009. Inoculation of plant growth promoting bacteria *Achromobacter xylosoxidans* strain A×10 for improvement of copper phyto-extraction by *Brassica juncea*. J. Environ. Manage. 90, 831–837.

Manahan, S.E., 1997. Environ. Sci. Technol. 11, 383–394. Boca Raton, FL.

Mansour, S.A., Gad, M.F., 2010. Risk assessment of pesticides and heavy metals contaminations in vegetables: a novel bioassay method using *Daphnia magna* Straus. Food Chem. Toxicol. 48 (1), 377–389.

McKinney, J., Rogers, R., 1992. Metal bioavailability. Environ. Sci. Technol. 26, 1298–1299.

Meagher, R.B., 2000. Phytoremediation of toxic elemental and organic pollutants. Curr. Opin. Plant Biol. 3, 153–162.

Meharg, A.A., Williams, P.N., Adomako, A., Lawgali, Y.Y., Deacon, C., Villada, A., Cambell, R.C.J., Sun, G., Zhu, Y.G., Feldmann, J., Rabb, A., Zhao, F.J., Islam, R., Hossain, S., Yanai, J., 2009. Geographical variation in total and inorganic arsenic content of polished (white) rice. Environ. Sci. Technol. 43, 1612–1617.

Miller, R.L., Tegen, I., Perlwitz, J.P., 2004. Surface radiative forcing by soil dust aerosols and the hydrologic cycle. J. Geophys. Res. http://dx.doi.org/10.1029/2003JD004085.

Miransari, M., 2011. Hyperaccumulators, arbuscular mycorrhizal fungi and stree of heavy metals. Biotechnol. Adv. 29 (6), 645–653.

Mishra, S., Srivastava, S., Tripathi, R.D., Kumar, R., Seth, C.S., Gupta, K., 2006. Lead detoxification by coontail (*Ceratophyllum demersum* L.) involves induction of phytochealatins and antioxidant system in response to its accumulation. Chemosphere 65, 1027–1039.

Moraghan, J.T., Mascani, H.J., 1991. Environmental and soil factors affecting micronutrients deficiencies and toxicities. In: Micronutrients in Agriculture Soil Sci. Soc. America, pp. 371–413.

Neubauer, U., Furrer, G., Kayser, A., Schulin, R., 2000. Siderophores, NTA, and citrate: potential soil amendments to enhance heavy metal mobility in phytoremediation. Int. J. Phytoremediation 2, 353–368.

Nordstrom, D.K., 2002. Public health-worldwide occurrences of arsenic in ground water. Science 296, 2143–2145.

Norouzi, S., Khademi, H., Cano, A.F., Acosta, J.A., 2015. Using plane tree leaves for biomonitoring of dust borne heavy metals: a case study from Isfahan, Central Iran. Ecol. Indic. 57, 64–73.

Nriagu, J.O., 1979. Production and Uses of Mercury. The Biogeochemistry of Mercury in the Environment. Topics in Environmental Health. Biomedical Press.

Orisakwe, O.E., 2012. 11-Other Heavy Metals: Antimony, Cadmium, Chromium and Mercury. Toxicity of Building Materials a Volume in Woodhead Publishing Series in Civil and Structural Engineering, 297–333.

Pacini, S., Fiore, M.G., Magherini, S., Morucci, G., Branca, J.J., Gulisano, M., Ruggiero, M., 2012. Could cadmium be responsible for some of the neurological signs and symptoms of myalgic: encephalomyelitis/chronic fatigue syndrome. Med. Hypotheses 79 (3), 403–407.

Petemele, W.S., Winkler, H.A.A., Gomej, P.E.A., 1999. Adsorption of Cd (II) Pb (II) on to functionalized formic lignin from sugar bagasse. Biores.Technol. 68, 95–100.

Pinto, A.P., Mota, A.M., Pinto, F.C., 2004. Influence of organic matter on the uptake of cadmium, zinc, copper and iron by sorghum plants. Sci. Environ. J. 326, 239–247.

Pirkle, J.L., Kaufmann, R.B., Brody, D.J., Hickman, T., Gunter, E.W., Paschal, D.C., 1998. Exposure of the U.S. population to lead, 1991–1994. Environ. Health Perspect. 106, 745–750.

Pollard, J.A., Powell, K.D., Harper, F.A., Smith, J.A.C., 2002. The genetic basis of metal hyper accumulation in plants. Crit. Rev. Plant Sci. 21, 539–566.

Prapagdee, B., Chanprasert, M., Mongkolsuk, S., 2013. Bioaugmentation with cadmium-resistant plant growth-promoting rhizobacteria to assist cadmium phytoextraction by *Helianthus annuus*. Chemosphere 92 (6), 659–666.

Qian, S.Q., Liu, Z., 2000. An overview of development in the soil-remediation technologies. Chem. Ind. Eng. Process. 4, 10–12.

Rajkumar, M., Ae, N., Prasad, M.N.V., Freitas, H., 2010. Potential of siderophores-producing bacteria for improving heavy metal phytoextraction. Trends Biotechnol. 28, 142–149.

Resaee, A., Derayat, J., Mortazavi, S.B., Yamini, Y., Jafarzadeh, M.T., 2005. Removal of Mercury from chlor-alkali industry wastewater using *Acetobacter xylinum* cellulose. Am. J. Environ. Sci. 1 (2), 102–105.

Rivas, R., Peix, A., Mateos, P.F., Trujillo, M.E., Martinez-Molina, E., Velazqueze, E., 2006. Biodiversity of populations of phosphate solubilizing rhizobia that nodulates chickpea in different Spanish soils. Plant Soil 287 (1–2), 23–33.

Robson, T.C., Braungardt, C.B., Rieuwerts, J., Worsfold, P., 2014. Cadmium contamination of agricultural soils and crops resulting from sphalerite weathering. Environ. Pollut. 184, 283–289.

Rodriguez, H., Fraga, R., 1999. Phosphate solubilizing bacteria and their role in plant growth promotion. Biotechnol. Adv. 17, 319–339.

Rodríguez, H., Fraga, R., Gonzalez, T., Bashan, T., 2006. Genetics of phosphate solubilization and its potential applications for improving plant growth-promoting bacteria. Plant Soil 287, 15–21.

Rosman, K.J.R., Chisholm, W., Boutron, C.F., Candelone, J.P., Görlach, U., 1993. Nature 362, 333–335.

Ru, Q.M., Feng, Q., He, J.Z., 2013. Risk assessment of heavy metals in honey consumed in Zhejiang province, southeastern China. Food Chem. Toxicol. 53, 256–262.

Rubio, L.M., Ludden, P.W., 2008. Biosynthesis of the iron-molybdenum cofactor of nitrogenase. Annu. Rev. Microbiol. 62, 93–111.

Ryu, C., Farag, M.A., Paré, P.W., Kloepper, J.W., 2005. Invisible signals from the underground: bacterial volatiles elicit plant growth promotion and induce systemic resistance. J. Plant Pathol. 21 (1), 7–12.

Saidi, S.A., Azaza, M.S., Windmolders, P., Pelt, J.V., Feki, A.E., 2013. Cytotoxicity evaluation and antioxidant enzyme expression related to heavy metals found in tuna by-products meal: an in vitro study in human and rat liver cell lines. Exp. Toxicol. Pathol. 65 (7–8), 1025–1033.

Salt, D.E., 1999. Zinc ligands in the metal hyperaccumulator *Thlaspi caerulescens* as determined using X-ray absorption spectroscopy. Environ. Sci. Technol. 33, 713–717.

Salt, D.E., Rauser, W.E., 1995. Mg ATP-dependent transport of phytochealatins across the tonoplast of oat roots. Plant Physiol. 107, 1293–1301.

Sandalio, L.M., Dalurzo, H.C., Gomez, M., 2001. Cadmium induced changes in the growth and oxidative metabolism of pea plants. J. Exp. Bot. 52, 2115–2126.

Sas-Nowosielska, A., Galimska-Stypa, R., Kucharski, R., Zielonka, U., Małkowski, E., Gray, L., 2008. Remediation aspect of microbial changes of plant rhizosphere in mercury contaminated soil. Environ. Monit. Assess. 137 (1–3), 101–109.

Satarug, S., Moore, M.R., 2004. Adverse health effects of chronic exposure to low-level cadmium in foodstuffs and cigarette smoke. Environ. Health Perspect. 112, 1099–1103.

Schmidt, W., 1999. Mechanisms and regulation of reduction-based iron uptake in plants. New Phytol. 141, 1–26.

Schutzendubel, A., Polle, A., 2002. Plant responses to a biotic stresses: heavy-metal induced oxidative stress and protection by mycorrhization. J. Exp. Bot. 53, 1351–1365.

Seth, C.S., Chaturvedi, P.K., Misra, V., 2007. Toxic effect of arsenate and cadmium alone and in combination on giant duckweed (*Spirodela polyrrhiza* L.) in response to its accumulation. Environ. Toxicol. 22, 539–549.

Sharma, R.K., Aditi, P., Yukti, M., Alok, A., 2014. Newly modified silica-based magnetically driven nanoadsorbent: a sustainable and versatile platform for efficient and selective recovery of cadmium from water and fly-ash ameliorated soil. Sep. Purif. Technol. 127, 121–130.

Shen, Z.G., Chen, H.M., 2000. Bioremediation of heavy metal polluted soils. Rural Eco-Environ. 16 (2), 39–44.

Sheng, X.F., Xia, J.J., 2006. Improvement of rape (*Brassica napus*) plant growth and cadmium uptake by cadmium resistant bacteria. Chemosphere 64, 1036–1042.

Singh, V., Mittal, A.K, 2012. Toxicity and treatability of leachate: Application of UASB reactor for leachate treatment from Okhla landfill, New Delhi. Water Sci. Technol. 65 (10), 1887–1894.

Smedley, P.L., Kinniburgh, D.G., 2002. A review of the source, behavior and distribution of arsenic in natural waters. Appl. Geochem. 17, 517–568.

Soldatkin, O.O., Kucherenko, I.S., Pyashkova, V.M., Kukla, A.L., Jaffrezic, R.N., El'skaya, A.V., Dzyyadevych, S.V., Soldatkin, A.P., 2012. Novel conductometric biosensor based on three-enzyme system for selective determination of heavy metal ions. Bioelectrochemistry 83, 25–30.

Spaepen, S., Vanderleyden, J., Remans, R., 2007. Indole-3-acetic acid in microbial and microorganism-plant signaling. FEMS Microbiol. Rev. 31 (4), 425–448.

Spurgeon, D.J., Hopkin, S.P., 1996. Effects of variation of the organic matter content and pH of soils on the availability and toxicity of zinc to the earthworm *Eisenia fetidia*. Pedobiology 40, 80–96.

Street, R.A., 2012. Heavy metals in medicinal plant products – an African perspective. S. Afr. J. Bot. 82, 67–74.

Swartzbaugh, J.T., Weisman, A., Gabrera-Guzman, D., 1990. The use of electro-kinetics for hazardous waste site remediation. J. Air Waste Manage. Assoc. 40 (12), 1670–1677.

Tak, H.I., Ahmad, F., Babalola, O.O., 2013. Advances in the application of plant growth-promoting rhizobacteria in phytoremediation of heavy metals. Rev. Environ. Contam. Toxicol. 223. http://dx.doi.org/10.1007/978-1-4614-5577-62.

Tampouris, S., Papassiopi, N., Paspaliaris, I., 2001. Removal of contaminant metals from fine grained soils, using agglomeration, chloride solutions and pile leaching techniques. J. Hazard. Mater. 84 (2–3), 297–319.

Tan, I., Erhan, E., Karargoz, P., Ozkan, M., 2011. Determination of mercury and nickel by amperometric biosensor prepared with thermostable lactate dehydrogenase. Trans. Nonferrous Met. Soc. China 21 (10), 2332–2338.

Tank, N., Saraf, M., 2003. Phosphate solubilization, exo-polysaccharide production and indole acetic acid secretion by rhizobacteria isolated from *Trigonella graecum*. Indian J. Microbiol. 43, 37–40.

Taweel, A., Othman, M.S., Ahmad, A.K., 2013. Assessment of heavy metals in tilapia fish (*Oreochromis niloticus*) from the Langat River and Engineering Lake in Bangi, Malaysia, and evaluation of the health risk from tilapia consumption. Ecotoxicol. Environ. Saf. 93, 45–51.

Telisman, S., Jurasovic, J., Pizent, A., Cvitkovic, P., 2001. Blood pressure in relation to biomarkers of lead, cadmium, copper, zinc, and selenium in men without occupational exposure to metals. Environ. Res. 87, 57–68.

Tomsett, A.B., Thurman, D.A., 1988. Molecular biology of metal tolerances of plants. Plant Cell. http://dx.doi.org/10.1111/j.1365-3040.1988.tb01362.

Tsavkelova, E.A., Cherdyntseva, T.A., Klimova, S.Y., Shestakov, A.I., Botina, S.G., Netrusov, A.I., 2007. Orchid-associated bacteria produce indole-3-acetic acid, promote seed germination, and increase their microbial yield in response to exogenous auxins. Arch. Microbiol. 188 (6), 655–664.

Ui, J., 1992. Industrial Pollution in Japan. The United Nations University, Tokyo.

Ullah, A., Heng, S., Munis, M.F.H., Fahad, S., Yang, X., 2015. Phytoremediation of heavy metals assisted by plant growth promoting (PGP) bacteria: a review. Environ. Exp. Bot. 117, 28–40.

US EPA, 1997. Chapter 3 of Volume III of the US EPA Mercury Study Report to Congress.

Van Steveninck, R.F.M., Van-Steveninck, M.E., Wells, A.J., Fernando, D.R., 1990. Zinc tolerance and the binding of zinc as zinc phytate in Lemna minor. X-ray microanalytical evidence. J. Plant Physiol. 137, 140–146.

Vansuyt, G., Robin, A., Briat, J.F., Curie, C., Lemanceau, P., 2007. Iron acquisition from Fe-pyoverdine by *Arabidopsis thaliana*. Mol. Plant Microbe Interact. 20, 441–447.

Virkutyte, J., Sillanpaa, M., Latostemaa, P., 2002. Electrokinetic soil remediation-critical overview. Sci. Total Environ. 289 (113), 97–121.

Vogeli-Lange, R., Wagner, G.J., 1990. Sub-cellular localization of cadmium and cadmium-binding peptides in tobacco leaves: implication of a transport function for cadmium binding peptides. Plant Physiol. 92, 1086–1093.

Wagner, G.J., 1993. Accumulation of cadmium in plants and its consequences to human health. Adv. Agron. 51, 173–212.

Wang, H., Ma, J.W., Fan, X.Y., 2007. Research progress on enhancement of *in situ* remediation of heavy metal by electro-kinetics. Ecol. Environ. Sci. 16 (1), 223–227.

Wang, H.F., Zhao, B.W., Xu, J., 2009. Technology and research progress on remediation of soils contaminated by heavy metals. Environ. Sci. Manage. 34 (11), 15–20.

Wang, J., Liu, G., Lu, L., Zhang, J., Liu, H., 2015. Geochemical normalization and assessment of heavy metals (Cu, Pb, Zn, and Ni) in sediments from the Huaihe River, Anhui, China. Catena 129, 30–38.

Wang, J.L., Wen, X.H., 2001. Environmental Biotechnology. Tsinghua University press, Beijing.

Wani, P.A., Khan, M.S., 2010. *Bacillus* species enhance growth parameters of chickpea (*Cicer arietinum* L.) in chromium stressed soils. Food Chem. Toxicol. 48, 3262–3267.

Watanabe, M.E., 1997. Phytoremediation on the brink of commercialization. Environ. Sci. Technol. 31 (4), 182–186.

Xiong, C., Zhang, Y., Xu, X., Lu, Y., Ouyang, B., Ye, Z., Li, H., 2013. Lotus roots accumulate heavy metals independently from soil in main production regions of China. Sci. Hortic. 164, 295–302.

Xu, Q., Huang, X.F., Cheng, J.J., et al., 2006. Progress on electro-kinetic remediation and its combined methods for POPs from contaminated soils. Environ. Sci. 27 (11), 2363–2368.

Yadav, S.K., 2010. Heavy metals toxicity in plants: an overview on the role of glutathione and phytochelatins in heavy metal stress tolerance of plants. S. Afr. J. Bot. 76 (2), 167–179.

Yu, X., Liu, X., Zhu, T.H., Liu, G.H., Mao, C., 2012. Co-inoculation with phosphate-solubilizing and nitrogen-fixing bacteria on solubilization of rock phosphate and their effect on growth promotion and nutrient uptake by walnut. Eur. J. Soil Biol. 50, 112–117.

Yu, Y.T., Tian, G.M., He, M.M., 2009. Comparison of two different combined bioleaching-electrokinetics remediation processes. Acta Sci. Circumstantiae 29 (1), 163–168.

Zaidi, S., Khan, M.S., Ahemad, M., Oves, M., 2009. Plant growth promotion by phosphate solublizing bacteria. Acta Microbiol. Immunol. Hung. 56 (3), 263–284.

Zaidi, S., Usmani, S., Singh, B.R., Musarrat, J., 2006. Significance of *Bacillus subtilis* strain SJ 101 as bio-inoculants for concurrent plant growth promotion and nickel accumulation in *Brassica juncea*. Chemosphere 64, 991–997.

Zhang, C., Yu, Z., Zeng, G., Jiang, M., Yang, Z., Cui, F., Zhu, M., Shen, L., Hu, L., 2014. Effects of sediment geochemical properties on heavy metal bioavailabilty. Environ. Int. 73, 270–281.

Zhang, L.J., Zhang, Y., Liu, D.H., 2009. Remediation of soils contaminated by heavy metals with different amelioration materials. Soils 41 (3), 420–424.

Zhang, X.H., Wang, H., Luo, Q.S., 2001. Electrokinetics in remediation of contaminated groundwater and soils. Adv. Water Sci. 12 (2), 249–255.

Zhang, Y.F., Sheng, J.C., Lu, Q.Y., 2004. Review on the soil remediation technologies. Gansu Agric. Sci. Technol. 10, 36–38.

Zheng, J., Chen, K.H., Yan, X., Chen, S.J., Hu, G.C., Peng, X.W., Yuan, J., Mai, B.X., Yang, Z.Y., 2013. Heavy metals in food, house dust, and water from an e-waste recycling area in South China and the potential risk to human health. Ecotoxicol. Environ. Saf. 96, 205–212.

Zhou, D.M., Hao, X.Z., Xue, Y., 2004. Advances in remediation technologies of contaminated soils. Ecol. Environ. Sci. 13 (2), 234–242.

CHAPTER

HEAVY METAL STRESS AND MOLECULAR APPROACHES IN PLANTS

22

Frederico Jacob Eutrópio[1], Alessandro Coutinho Ramos[1], Muriel da Silva Folli-Pereira[2], Natália de Aquino Portela[3], Janaina Biral dos Santos[1], Juliana Melo da Conceição[1], Amanda Azevedo Bertolazi[4], Frederico Figueira Firme[1], Sávio Bastos de Souza[1], Antônio Jesus Dorighetto Cogo[1], Nazima Rasool[5]

Centro de Biociências e Biotecnologia, Universidade Estadual do Norte Fluminense Darcy Ribeiro, Campos dos Goytacazes, Rio de Janeiro, Brazil[1]; Laboratório de Microbiologia Ambiental e Biotecnologia, Universidade Vila Velha, Vila Velha, Espírito Santo, Brazil[2]; Universidae Federal do Espírito Santo, Goiabeiras, Vitória, Espírito Santo, Brazil[3]; Centro de Ciências e Tecnologias Agrárias, Universidade Estadual do Norte Fluminense Darcy Ribeiro, Campos dos Goytacazes, Rio de Janeiro, Brazil[4]; Department of Botany, University of Kashmir, Jammu and Kashmir, India[5]

CHAPTER OUTLINE

1. INTRODUCTION

Heavy metals (HM) are classified regarding their relative density and they can be found naturally in the soil because of weathering and other pedogenic processes on rocks and the source of the material of soil. However, because of industrialization and a rapid population increase, production of anthropogenic biosolids and agrochemical waste has been enhancing the risk of soils pollution (Baker et al., 1994; Kashem et al., 2009).

Plant Metal Interaction. http://dx.doi.org/10.1016/B978-0-12-803158-2.00022-9

This is one of the main environmental problems, keeping in mind that metals reach the soil and end up depreciating the whole area (faun, flora, air, superficial and underground waters). In toxic concentrations, HMs damage plants and organisms, affecting their organs, changing their biochemical processes, organelles, cellular membranes, and causing health problems (DalCorso et al., 2010). Most of the HMs are persistent in soil because of their immobile nature, been intimately associated with soil and organic colloids. Among the metals, cadmium has a higher mobility and solubility compared with others metals in soil (Kashem et al., 2009); therefore, cadmium is important to consider in terms of contamination in the food chain (Liu et al., 2007).

The main HMs present in soil are Cadmium (Cd), copper (Cu), lead (Pb), zinc (Zn), chrome (Cr), nickel (Ni), barium (Ba), argon (Ag), cobalt (Co), mercury (Hg), and antimony (Sb), and some of these elements are essential for many physiological functions in living beings, whereas others have no known biological function (Fässler et al., 2010; Lasat, 2000). These elements in fungicides, fertilizers, animal waste, urban trash, sewage sludge in soil, and deposits of industrial dust can increase the concentration in soil, making them toxic (Fässler et al., 2010; Li et al., 2015b).

In this way soil ends up inhibiting plant growth, causing alterations in microbial and vegetal communities (Chen et al., 2003; Vangronsveld et al., 1997). Accordingly, we seek to mitigate the effect of the phytotoxicity of metals in soil by establishing the vegetation and developing an efficient process of decontamination (i.e., bioremediation), which is one technique that seeks to decontaminate the soil by living organisms, such as microorganisms and plants (Ma et al., 2011).

Phytoremediation is a well-used bioremediation technique because it keeps the biological properties and physical structure of soil; it is also inexpensive and environmentally friendly (Ali et al., 2013; Khan et al., 2000). It consists of using plants to remove inorganic contaminants, especially metals, by using the capacity of plants to accumulate high levels of pollutants (Lasat, 2000). Plants are capable of immobilizing metal in soil by forming insoluble compounds as result of the interaction of plant exudates in the rhizosphere or by adsorption in the radicular system (Kidd et al., 2009). Some species are capable of accumulating HMs in their tissues so that contamination can be removed by harvesting the plant (Maestri et al., 2010; McGrath and Zhao, 2003; Van Nevel et al., 2007). This ability to accumulate metal varies significantly between species; thus, soil contaminated with several metals requires the use of different vegetal species (Hajar et al., 2014; Hao and Jiang, 2015; Jambhulkar and Juwarkar, 2009; Quezada-Hinojosa et al., 2015).

2. HM: MICRONUTRIENTS TO SOIL CONTAMINANTS

HMs are present in soil as natural components or as a result of anthropogenic activities, such as industrial activities, application of agricultural chemical products, and inadequate removal of residues (Gil et al., 2004; Mitton et al., 2012). The conceptualization of HM encompasses many aspects, such as environmental and toxicological. HM is a term for a group of elements that have an atomic mass higher than 20 and density greater than 6 g/cm^{-3} (Rascio and Navari-Izzo, 2011). The definition of HM is also postulated in terms of its specific mass, atomic mass, atomic number, and other properties such as formation of slightly soluble sulfides and hydroxides (Duffus, 2002). Impact in environmental and human health arising from metal wastes has been associated with HM (Facchinelli et al., 2001; Wu et al., 2015). With toxicological studies, it was found that each metal shows a specific toxicological effect in a specific organism (Rascio and Navari-Izzo, 2011).

The term heavy metal is frequently used to reference a group of metals and semimetals (metalloids) that have been associated with the potential contamination, toxicity, or ecotoxicity. Other names are used to denominate elements in this group, such as trace elements, because of the concentration they are found in naturally, in parts per million or lower, beyond toxic metals and micronutrients. The term toxic metal must be used with caution, because some metals are only considered toxic at high levels. The term micronutrient is more convenient for elements that are proved to be essential to plants and animals in certain concentrations (Duffus, 2002).

For soils, the term contaminated by HM is used when an increase in concentration occurs when compared with natural conditions, whereas the term pollution is used when this increased concentration harms the environment (Rodríguez Martín et al., 2006).

In an analysis of health risk caused by HMs in soils, plants in the food chain introduced a concept of classification of these metals into four groups. Group 1 includes the elements Ag, Cr, Ti, Zr, and Y and represents a low risk because these elements are not absorbed by the extensions of plants because of their low solubility in soil and consequent negligible uptake by plants. High concentrations of these elements in food generally indicate direct contamination on soil or another form of contamination. Group 2 includes As, Hg, and Pb, which are strongly absorbed by soil colloids. Although they may be absorbed by plant root, they are not readily transferred into edible tissues and therefore represent minimal risk to human health. Group 3 consists of B, Cu, Mn, Mo, Ni, and Zn, which are readily absorbed by plants, and depending on the concentration may represent a risk to human health. Group 4, Cd, Co, and Se, is a human health risk when concentrated in plant tissues because of soil contamination (Ali et al., 2013; Alvarez et al., 2003; Cobbett, 2000a; Evangelou et al., 2007; Gonçalves et al., 2011; Maksymiec, 2007; Rascio and Navari-Izzo, 2011; Sharma and Dietz, 2009; Yang and Chu, 2008).

Of all potentially toxic elements present in soil, Cd and Se are the primary concerns in terms of contaminating the food chain. Problems from contamination by Se are predominantly the result of natural soil enrichment. Cadmium concentration in soil in many countries is increasing because of inadvertent additions in fertilizers, biosolids, and soil alteration from the atmosphere. Health problem from contamination by Cd have occurred only in cases of serious soil contamination (McLaughlin et al., 1999).

Some metals are classified as essential metals because they play an important biological function in the metabolism of many organisms. For example, Fe, Zn, Cu, Ni, and Mn act as cofactors for enzymes in plants, microorganisms, and mammals (Rascio and Navari-Izzo, 2011).

One factor that significantly affects metal toxicity is the chemical form in which the element is found. The problem associated with soil contamination by metals is the existence of bioavailable forms of these elements (Li et al., 2015a). The metals found in soluble forms and exchangeable forms are those that have greater bioavailability, thereby being the most worrying. In the soluble form, the metal is in an ionic form or organic complex and thereby is easily absorbed by plants or leachates, reaching underground water bodies. The most toxic form is not a free metal, but a cation or being connected with carbon chains. In organisms, the main mechanism of toxic action is the affinity for sulfur. Therefore, when present in its cationic form, metals react with the sulfhydryl radical present in the enzymatic protein structure, changing its properties and leading to harmful consequences to the living beings (Maestri et al., 2010; Yadav, 2010).

HM normally does not exist individually in soil, but coexists with another contaminant, thereby causing a combined effect of contaminants by HM in the environment. The adsorption, absorption, and coordination among the different HMs may affect its distribution, bioavailability, and toxicity in soil (Wang et al., 2009). Therefore, the effect of combined pollution, which depends on the metal mixture, may vary significantly (Jensen and Sverdrup, 2002).

The content of a given element in a given soil is highly dependent on the mineralogical composition of the source material (Roca et al., 2012). Knowledge of the dynamics of metals in soil allows the use of preventive practices to control the availability of these elements, distribution in the environment, and transfer from the soil to the food chain. Unlike organic contaminants, most metals do not suffer from microbial degradation or chemicals; therefore, the metal concentration in soil persists for a long period after its entry. The most important parameters that control HM adsorption and its distribution between soil and water are the type of soil, metal species and concentration, soil pH, and time of contact. In general, a better retention of the metal and lower solubility occurs more in alkaline soils; increasing the pH is one way to precipitate hydroxides that are less soluble from these metals (Al-Wabel et al., 2014).

3. PLANT STRATEGIES TO TOLERATE HM TOXICITY

Plants can present different tolerance mechanisms in response to excess of HMs, including a reduction in transport through the membrane, exclusion, phytochelatins (PCs), metallothionein (MT) formation, chelation by organic acids and amino acids, and metal compartmentalization in subcellular structures (Hall, 2002; Ovečka and Takáč, 2014; Pal and Rai, 2010).

3.1 TRANSPORT REDUCTION THROUGH THE MEMBRANE

The availability of metals to plants is strongly dependent on the chemical and physiological conditions in the rhizosphere. One of the main factors is soil pH, which has a significant effect over the solubility of HM ions. The availability of bivalent cations to the plant roots slightly increases in acidic conditions and decreases in alkaline soils (forming insoluble precipitates such as hydroxides) (Härdtle et al., 2004; Zhao et al., 2011). Furthermore, when the rhizosphere is exposed to acidic conditions ($pH < 5.0$), plant growth is significantly reduced; however, the plant can neutralize this effect by increasing the pH of the rhizosphere. The plant increases the degree of precipitation and complexation of metal ions in the vicinity of the roots and, thus, helps prevent the absorption of HM ions, thereby reducing the impact of metal toxicity. Besides pH, soil organic matter content, nitrogen forms present in the rhizosphere, and the availability of phosphorus and iron helps plants to buffer the rhizosphere (Hinsinger et al., 2011). All of these factors have significant effects on the uptake of HMs.

The first barrier against the entry of HMs, acting mainly at the root level, is the immobilization of HMs in the cell wall by extracellular carbohydrates such as mucilage and callose (Revathi and Venugopal, 2013; Shu et al., 2007; Waraich et al., 2011). This avoids the presence of free ions in the root tissue and, consequently, the translocation of ions to shoots, thus reducing phytotoxicity. In this context, pectins and histidines stand out for the immobilization of HMs in the cell wall (Baetz and Martinoia, 2014; Watt and Evans, 1999).

It is important to highlight that, among a number of ions associated with roots, only a portion is absorbed by the cells. A significant fraction is adsorbed by negatively charged groups in the cell wall of roots. Thus, it is possible to have plants that accumulate a significant concentration of metal in the roots, but express a limited concentration in shoots (Ali et al., 2013; Rascio and Navari-Izzo, 2011; Visioli and Marmiroli, 2013).

3.2 EXCLUSION

Preventing entry of metals in the cytosol through exudation of compounds by the action of plasma membrane may theoretically be the best defense strategy. Some plants, known as exclusionary plants, possess specialized mechanisms to reduce the entry of HMs in the roots. Malate, citrate, and oxalate have been identified as important chelating agents secreted by roots that are involved in plant resistance to HMs (McGrath and Zhao, 2003; Sas et al., 2001; Zorrig et al., 2010).

Malate, citrate, and oxalate are weak organic acids, have a low molecular weight, possess at least one carboxyl group, and are termed oxygen-donor metal ligands. Some of these compounds are present in all plant cells as intermediates of the tricarboxylic acid cycle, the main respiratory pathway involved in the oxidation of pyruvate (Adrees et al., 2015; Akhter et al., 2014; Rascio and Navari-Izzo, 2011; Revathi and Venugopal, 2013).

The dicarboxylic acid malate, an important plant metabolite, can be present in all cell types and involved in ion balance, P and Fe uptake, and Al and Zn tolerance. In particular, it is a cytosolic Zn chelator in zinc-tolerant plants; Zn is bound to malate in the cytoplasm and transferred as a complex across the tonoplast to the vacuole. There, Zn-malate may disassociate and finally malate may be retranslocated to the cytoplasm (Bian et al., 2013; Ibrahim, 2015; Komal et al., 2014; Oliva et al., 2012; Tsednee et al., 2014; Viehweger, 2014).

Citrate, which is synthesized in plants by the enzyme citrate synthase, has a higher capacity for metal ions than malate and oxalate, and although its principal role is to chelate Fe^{2+}, it also has a strong affinity also for Ni^{2+}, Cd^{2+}, Zn^{2+}, and other divalent and trivalent ions (Oliva et al., 2012; Palmer and Guerinot, 2011; Schneider et al., 2013). Citrate metabolism in plants is carried out by several metabolic pathways located in different cellular compartments. Present in considerable amounts, citrate belongs to the key metabolites in plant cells (Anjum et al., 2015; Ezaki et al., 2013; Komal et al., 2014).

Oxalate is a common constituent of plants. Although it can be found in relatively small amounts in plants, it can be accumulated at high levels in several plants species (such as *Caryophyllaceae*, *Chenopodiaceae*, and *Polygonaceae*) (Norman et al., 2013; Tuazon-nartea and Savage, 2013). The role of oxalic acid in Al tolerance has been well-documented (Bian et al., 2013; Brunner and Sperisen, 2013; Ezaki et al., 2013; Simões et al., 2012; Yang et al., 2013; Zhou et al., 2014).

3.3 CHELATION

Chelation of HMs is a ubiquitous detoxification strategy described in a wide variety of plants. Inside the cell, HM ions that are not immediately required metabolically may reach toxic concentrations, and plant cells have evolved various mechanisms to store excess metals to prevent their participation in unwanted toxic reactions. If the toxic metal concentration exceeds a certain threshold inside the cells, an active metabolic process contributes to the production of chelating compounds(Akhter et al., 2014; Ali et al., 2013; Kobayashi and Nishizawa, 2012; Merlot et al., 2014; Oliva et al., 2012; Rascio and Navari-Izzo, 2011; Viehweger, 2014; Yruela, 2013).

Chelation contributes to metal detoxification by reducing the concentration of free metal in the cytosol, limiting its reactivity and solubility. In plants, the major known classes of HMs chelates include PCs, MTs, organic acids, and amino acids (Cobbett and Goldsbrough, 2002; Hall, 2002; Hodson and Brookes, 2012; Soudek et al., 2014).

3.4 PHYTOCHELATINS

One of the mechanisms of tolerance to HMs in plants is related to the synthesis of thiolic peptides called PCs, which form complexes with HMs, in particular Cd, in the free sulfur (S) present in cysteine (Flores-Cáceres et al., 2015; Komal et al., 2014).

The PCs are formed by three amino acids: glutamate (Glu), cysteine (Cys), and glycine (Gly) with Glu and Cys connected via a γ-carboxylamide. The structure of the PCs is formed with an increase in the repetitions of the dipeptide γGlu-Cys followed by a Gly terminal. It has a general structure of $(\gamma\text{-Glu-Cys})_n$-Gly, where $n = 2–11$, but is usually found with the n varying from 2 to 5. The PCs have been identified in a wide variety of species of plants and in some microorganisms (Cobbett and Goldsbrough, 2002; Cobbett, 2000a, 2000b; Pal and Rai, 2010; Pérez-Clemente et al., 2013).

These peptides are enzymatically synthesized, using glutathione (GSH) as a substrate, through a reaction catalyzed by the enzyme γ-glutamylcysteine dipeptidyl transpeptidase known as PC synthase, which is activated by the presence of HMs. PC synthase is activated after a few minutes of exposure to a variety of metals and metalloids. In vitro, the activity of PC synthase was activated only in the presence of metal ions and the best activator studied was Cd followed by Ag, Bi, Pb, Zn, Cu, Hg, and Au. These metals also induced the synthesis of PCs in vivo in plant cell cultures.

The PCs are structurally related to GSH (γ-Glu-Cys-Gly), and numerous physiological, biochemical, and genetic studies have confirmed that GSH (or, in many cases, related compounds) is the substrate for the biosynthesis of PCs (Hossain et al., 2012; Weydert and Cullen, 2011; Yadav, 2010).

In addition to the PCs, some plants may have other peptides related to PCs in the presence of HMs. Legumes produce peptides with the structure $(\gamma\text{-Glu-Cys})_n$-βAla (Grill et al., 1986), which are formed by homo-GSH that can partially or completely replace GSH in these plants. Some species of the family *Poaceae* (*Gramineae*) produce peptides containing serine as an amino terminal, with the structure $(\gamma\text{-Glu-Cys})_n$-Ser, called hydroxymethyl-PCs (Pal and Rai, 2010). These peptides are formed from the presence of hydroxymethyl-GSH in addition to GSH in these plants.

Cd is the strongest inducer of PC in vivo and is the element that forms the most stable complexes with PCs because of its high affinity for sulfur (Lux et al., 2011). However, PC synthesis is not only related to that element. Grill and Zenk (1987), studying the synthesis of PC in a cell culture suspension of *Rauvolfia serpentina* exposed to metals, concluded that the metals induce the synthesis of PCs in the following decreasing order: Cd^{2+}, Pb^{2+}, Zn^{2+}, Sb^{3+}, Ag^{+}, Hg^{2+}, As^{5-}, Cu^{+}, Sn^{2+}, Au^{3+}, and Bi^{3+}. According to these same authors, Ni, Te, W, and Se ions did not induce PC synthesis.

3.5 METALLOTHIONEINS

MTs are low-molecular-weight proteins that are nonenzymatic, rich in cysteine (Cys), and efficient in complexing the metals by affinity with sulfur present in the Cys (Cobbett and Goldsbrough, 2002).

MTs are classified based on the arrangement of Cys. Class I MTs have more than 20 conserved Cys, are common in mammals and vertebrates, and known to confer tolerance to Cd^{2+}. The MTs without a specific arrangement of Cys are classified as class II MTs and include all those found in plants, fungi, and invertebrates (Carpenè et al., 2007).

Despite MTs being more common in animals, there are four types of MTs in plants, classified according to the arrangement of Cys in the protein formation. The Cys are present in plants MTs like Cys-x-Cys, Cys-x-x-Cys (where x is an amino acid different from Cys), or groups of Cys–Cys (Carpenè et al., 2007).

The diversity of MTs in plants suggests that they may not only differ in amino acid sequence, but also in function and specificity of a particular metal (Ali et al., 2013). However, there is still no information about the true function of each MT in a plant because of the difficulty in obtaining purified MT. The MT has a tendency to hydrolyze itself, particularly in the region between Cys in the protein sequence. Nevertheless, various functions have been proposed for MTs in plants, such as detoxification of metals (mainly Cu), complexation of cytosolic Zn, and metal secretion via trichome on the leaves (Coyle et al., 2002; Grill and Zenk, 1987; Komal et al., 2014).

3.6 ORGANIC ACIDS AND AMINO ACIDS

Because of the reactivity of metal ions with sulfur, nitrogen, and oxygen, carboxylic acids and amino acids represent potential ligands of HMs.

Citrate, malate, and oxalate have been implicated in several processes, including tolerance to HMs, metal transport through the xylem, and vacuolar sequestration (McGrath and Zhao, 2003; Rascio and Navari-Izzo, 2011; Zorrig et al., 2010). Citric acid is the largest Cd^{2+} ligand when it is in low concentrations; it forms complexes with Ni^{2+} in hyperaccumulator plants and contributes to the accumulation and tolerance to Zn^{2+} (Labanowski et al., 2008; Soudek et al., 2014; Zorrig et al., 2010).

Histidine, an amino acid produced by plants in response to the presence of metals, is involved in a mechanism of tolerance to Ni, and in low concentrations it is involved with Co tolerance (McGrath and Zhao, 2003; Rascio and Navari-Izzo, 2011; Visioli and Marmiroli, 2013).

3.7 COMPARTMENTALIZATION

Cells use a combination of compartmentalization, chelation, and exclusion to restrict the deleterious effects of HMs and to ensure that the correct kind of metal is delivered to its target protein. The main storage compartment of HMs in plant cells is the vacuole and there is evidence of vacuolar sequestration of metal ions in plants, which prevents the movement of HMs in the cytosol and transports them to a limited area (Ernst, 2006; Hall, 2002; Lux et al., 2011; Rascio and Navari-Izzo, 2011; Soudek et al., 2014; Yang and Chu, 2008).

Transporters potentially related to this process have been identified in *Arabidopsis thaliana*. The superfamily of P-type adenosine triphosphatase (ATPase) uses the energy of adenosine triphosphate hydrolysis to translocate cations through biological membranes and can be divided into several subfamilies, including carriers of HMs P_{1B}-ATPase (Colangelo and Guerinot, 2006).

The vacuole is a key cellular compartment where the metal may be stored when in excess, providing a source during periods of deficiency. For example, Fe is known to be transported to the vacuole by the VIT1 transporter. The remobilization of Fe from the vacuole is mediated by the actions of NRAMP3 and

NRAMP4, which are regulated under Fe deficiency and are located in the vacuolar membrane. Although NRAMP3 or NRAMP4 mutants show no phenotype, nramp3 and Nramp4 mutants have a lethality rate of 90% when grown in Fe-deficient soils, suggesting that these proteins are functionally redundant and necessary for Fe mobilization under Fe deficiency (Palmer and Guerinot, 2011).

4. EFFECT OF HMS ON PRODUCTIVITY

The toxic effects of HMs in plants depends on the plant species, the concentration and type of metal, the chemical form in which the HM is present, soil composition, and pH (Nagajyoti et al., 2010; Nagata et al., 2013). The prolonged use of phosphate fertilizers, application of sewage sludge, industrial waste, pesticides, and poor agricultural practices have been associated with this contamination by HMs in the soil (Bell et al., 2001; Kashem et al., 2009; Passariello et al., 2002; Schwartz et al., 2001).

Reduced growth is one of the most common physiological consequences of exposure to HMs in plants (Hu et al., 2013; Tamás et al., 2008). The changes induced by the metal ions in the physiological and structural integrity of the leaves reduce the rates of photosynthesis and respiration, which has deleterious effects on the power supply and the efficiency of other metabolic processes. Transpiration and transport processes between several organs are also affected (Ying et al., 2010). Furthermore, HM toxicity has adverse effects on the roots, reducing the ability of the plant to absorb water and nutrients (Sharma and Dietz, 2009). Consequently, major changes in the functioning of roots and leaves will affect developmental processes, such as formation of flowering, embryogenesis, and seed formation.

5. CLIMATIC CHANGES AND HM TOXICITY

The most common methods of remediation of contaminated soils are physical, chemical, and biological, although the methods of physical and chemical remediation are both very difficult and expensive, in addition to negatively affecting the soil ecosystem. One potential method for the remediation of soils contaminated with HM is phytoremediation, which is an economical and environmentally friendly technique that uses tolerant plants and associated microorganisms to remove, turn, stabilize, volatilize, or absorb HMs in soil using biochemical processes (Rajkumar et al., 2013).

The evidence available in the literature regarding climate change and increased CO_2 concentration will have positive effects on phytoremediation of HMs and plant growth. An increase in root exudate and microbial diversity caused by increased CO_2 concentration could facilitate the phytoremediation of HMs (depending on the plant and metal in question), an increase in cycling, availability of nutrients, mobilization and immobilization of metals, and on plant growth and in soil quality. However, the impacts caused by global warming could hamper the process of phytoremediation, although the findings of recent studies are inconsistent.

6. CONCLUSIONS AND FUTURE PROSPECTS

Contamination by HMs is a major problem affecting crop yield. The ability to tolerate HMs is different according to the plant species because they differ in their ability to tolerate HMs. Some plants grow well in soil enriched with toxic levels of HMs, whereas others do not grow at all. Recent studies indicate that GSH plays an important role in HMs tolerance. The HM stress induces production of reactive oxygen species(ROS) generated by oxidative stress and GSH detoxifies ROS by the GSH–ascorbate cycle. In addition to the metal ions accumulated, high levels of ROS negatively affects plant development. The metal ions are detoxified by PCs synthesized by GSH upon plant exposure to HMs. The PCs form complexes with metal ions and are stored in the vacuole. This tolerance mechanism to HMs in plants strongly suggests that the plants are not limited to GSH. In particular, the role of GSH, PCs, cysteine synthesis, and glyoxalase has been related to HM tolerance. Several plant species have been identified as HMs accumulators; however, plants of agricultural importance are not good accumulators and continue to suffer from the stress caused by exposure to HMs. A viable and affordable alternative to reduce stress in agricultural plants would be the use of bioremediators such as mycorrhizal fungi. Mycorrhizae use different mechanisms to tolerate HMs and act as a physical barrier preventing the entry of a metal into the plant. In addition to decreasing the plant stress caused by HMs, mycorrhizae may stimulate greater productivity.

REFERENCES

Adrees, M., Ali, S., Rizwan, M., Zia-ur-Rehman, M., Ibrahim, M., Abbas, F., Farid, M., Qayyum, M.F., Irshad, M.K., 2015. Mechanisms of silicon-mediated alleviation of heavy metal toxicity in plants: a review. Ecotoxicol. Environ. Saf. 119, 186–197.

Akhter, M.F., Omelon, C.R., Gordon, R.a., Moser, D., Macfie, S.M., 2014. Localization and chemical speciation of cadmium in the roots of barley and lettuce. Environ. Exp. Bot. 100, 10–19.

Ali, H., Khan, E., Sajad, M.A., 2013. Phytoremediation of heavy metals–concepts and applications. Chemosphere 91, 869–881.

Alvarez, E., Fernández Marcos, M.L., Vaamonde, C., Fernández-Sanjurjo, M.J., 2003. Heavy metals in the dump of an abandoned mine in Galicia (NW Spain) and in the spontaneously occurring vegetation. Sci. Total Environ. 313, 185–197.

Al-Wabel, M.I., Usman, A.R.A., El-Naggar, A.H., Aly, A.A., Ibrahim, H.M., Elmaghraby, S., Al-Omran, A., 2014. *Conocarpus* biochar as a soil amendment for reducing heavy metal availability and uptake by maize plants. Saudi J. Biol. Sci. 22, 503–511.

Anjum, N.A., Hasanuzzaman, M., Hossain, M.A., Thangavel, P., Roychoudhury, A., Gill, S.S., Rodrigo, M.A.M., Adam, V., Fujita, M., Kizek, R., Duarte, A.C., Pereira, E., Ahmad, I., 2015. Jacks of metal/metalloid chelation trade in plants–an overview. Front. Plant Sci. 6, 1–17.

Baetz, U., Martinoia, E., 2014. Root exudates: the hidden part of plant defense. Trends Plant Sci. 19, 90–98.

Baker, A.J.M., McGrath, S.P., Sidoli, C.M.D., Reeves, R.D., 1994. The possibility of in situ heavy metal decontamination of polluted soils using crops of metal-accumulating plants. Resour. Conserv. Recycl 11, 41–49.

Bell, F.G., Bullock, S.E.T., Hälbich, T.F.J., Lindsay, P., 2001. Environmental impacts associated with an abandoned mine in the Witbank Coalfield, South Africa. Int. J. Coal Geol. 45, 195–216.

Bian, M., Zhou, M., Sun, D., Li, C., 2013. Molecular approaches unravel the mechanism of acid soil tolerance in plants. Crop J. 1, 91–104.

Brunner, I., Sperisen, C., 2013. Aluminum exclusion and aluminum tolerance in woody plants. Front. Plant Sci. 4, 172.

Carpenè, E., Andreani, G., Isani, G., 2007. Metallothionein functions and structural characteristics. J. Trace Elem. Med. Biol. 21 (Suppl. 1), 35–39.

Chen, Y., He, Y., Luo, Y., Yu, Y., Lin, Q., Wong, M., 2003. Physiological mechanism of plant roots exposed to cadmium. Chemosphere 50, 789–793.

Cobbett, C., Goldsbrough, P., 2002. Phytochelatins and metallothioneins: roles in heavy metal detoxification and homeostasis. Annu. Rev. Plant Biol. 53, 159–182.

Cobbett, C.S., 2000a. Update on heavy metal stress phytochelatins and their roles in heavy metal detoxification. Plant Physiol. 123, 825–832.

Cobbett, C.S., 2000b. Heavy metal detoxification in plants : phytochelatin biosynthesis and function. Aust. Biochem. 31, 15–19.

Colangelo, E.P., Guerinot, M.L., 2006. Put the metal to the petal: metal uptake and transport throughout plants. Curr. Opin. Plant Biol. 9, 322–330.

Coyle, P., Philcox, J.C., Carey, L.C., Rofe, A.M., 2002. Cellular and molecular Life Sciences Metallothionein : the multipurpose protein. Cell. Mol. Life Sci. 59, 627–647.

DalCorso, G., Farinati, S., Furini, A., 2010. Regulatory networks of cadmium stress in plants. Plant Signal. Behav. 5, 663–667.

Duffus, J.H., 2002. "Heavy Metals"—a meaningless term? Pure Appl. Chem. 74, 793–807.

Ernst, W.H.O., 2006. Evolution of metal tolerance in higher plants. For. Snow Landsc. Res. 80, 251–274.

Evangelou, M.W.H., Ebel, M., Schaeffer, A., 2007. Chelate assisted phytoextraction of heavy metals from soil. Effect, mechanism, toxicity, and fate of chelating agents. Chemosphere 68, 989–1003.

Ezaki, B., Jayaram, K., Higashi, A., Takahashi, K., 2013. A combination of five mechanisms confers a high tolerance for aluminum to a wild species of Poaceae, *Andropogon virginicus* L. Environ. Exp. Bot. 93, 35–44.

Facchinelli, A., Sacchi, E., Mallen, L., 2001. Multivariate statistical and GIS-based approach to identify heavy metal sources in soils. Environ. Pollut. 114, 313–324.

Fässler, E., Robinson, B.H., Gupta, S.K., Schulin, R., 2010. Uptake and allocation of plant nutrients and Cd in maize, sunflower and tobacco growing on contaminated soil and the effect of soil conditioners under field conditions. Nutr. Cycl. Agroecosyst 87, 339–352.

Flores-Cáceres, M.L., Hattab, S., Hattab, S., Boussetta, H., Banni, M., Hernández, L.E., 2015. Specific mechanisms of tolerance to copper and cadmium are compromised by a limited concentration of glutathione in alfalfa plants. Plant Sci. 233, 165–173.

Gil, C., Boluda, R., Ramos, J., 2004. Determination and evaluation of cadmium, lead and nickel in greenhouse soils of Almería (Spain). Chemosphere 55, 1027–1034.

Gonçalves, A.C., Nacke, H., Schwantes, D., Nava, I.A., Strey, L., 2011. Phytoavailability of toxic heavy metals and productivity in wheat cultivated under residual effect of fertilization in soybean culture. Water Air Soil Pollut. 220, 205–211.

Grill, E., Gekeler, W., Winnacker, E.L., Zenk, M.H., 1986. Homo-phyto-chelatins are heavy metal binding peptides of homo-glutathione containing fabales. FEBS Lett 205, 47–50.

Grill, E., Zenk, M.H., 1987. Phytochelatins, a class of heavy-metal-binding peptides from plants, are functionally analogous to metallothioneins. Proc. Natl. Acad. Sci. U.S.A 84, 439–443.

Hajar, E.W.I., Sulaiman, A.Z.B., Sakinah, A.M.M., 2014. Assessment of heavy metals tolerance in leaves, stems and flowers of *Stevia rebaudiana* plant. Procedia Environ. Sci. 20, 386–393.

Hall, J.L., 2002. Cellular mechanisms for heavy metal detoxification and tolerance. J. Exp. Bot. 53, 1–11.

Hao, Q., Jiang, C., 2015. Heavy metal concentrations in soils and plants in Rongxi Manganese Mine of Chongqing, Southwest of China. Acta Ecol. Sin 35, 46–51.

Härdtle, W., von Oheimb, G., Friedel, A., Meyer, H., Westphal, C., 2004. Relationship between pH-values and nutrient availability in forest soils–the consequences for the use of ecograms in forest ecology. Flora-Morphol. Distrib. Funct. Ecol. Plants 199, 134–142.

Hinsinger, P., Brauman, A., Devau, N., Gérard, F., Jourdan, C., Laclau, J.-P., Le Cadre, E., Jaillard, B., Plassard, C., 2011. Acquisition of phosphorus and other poorly mobile nutrients by roots. Where do plant nutrition models fail? Plant Soil 348, 29–61.

Hodson, M.J., Brookes, O., 2012. Metal Toxicity and Tolerance in Plants, 28–32.

Hossain, M.A., Piyatida, P., da Silva, J.A.T., Fujita, M., 2012. Molecular mechanism of heavy metal toxicity and tolerance in plants: central role of glutathione in detoxification of reactive oxygen species and methylglyoxal and in heavy metal chelation. J. Bot. 2012, 1–37.

Hu, Y.F., Zhou, G., Na, X.F., Yang, L., Nan, W.Bin, Liu, X., Zhang, Y.Q., Li, J.L., Bi, Y.R., 2013. Cadmium interferes with maintenance of auxin homeostasis in *Arabidopsis* seedlings. J. Plant Physiol. 170, 965–975.

Ibrahim, L.A., 2015. Experimental Study for Heavy Metals Mitigation in Polluted Soil. J. Agric. Sci. Rev. 4, 119–126.

Jambhulkar, H.P., Juwarkar, A.A., 2009. Assessment of bioaccumulation of heavy metals by different plant species grown on fly ash dump. Ecotoxicol. Environ. Saf. 72, 1122–1128.

Jensen, J., Sverdrup, L.E., 2002. Joint toxicity of linear alkylbenzene sulfonates and pyrene on *Folsomia fimetaria*. Ecotoxicol. Environ. Saf. 52, 75–81.

Kashem, M.A., Kawai, S., Kikuchi, N., Takahashi, H., Sugawara, R., Singh, B.R., 2009. Effect of Lherzolite on chemical fractions of Cd and Zn and their uptake by plants in contaminated soil. Water Air Soil Pollut. 207, 241–251.

Khan, A., Kuek, C., Chaudhry, T., Khoo, C., Hayes, W., 2000. Role of plants, mycorrhizae and phytochelators in heavy metal contaminated land remediation. Chemosphere 41, 197–207.

Kidd, P., Barceló, J., Bernal, M.P., Navari-Izzo, F., Poschenrieder, C., Shilev, S., Clemente, R., Monterroso, C., 2009. Trace element behaviour at the root–soil interface: Implications in phytoremediation. Environ. Exp. Bot. 67, 243–259.

Kobayashi, T., Nishizawa, N.K., 2012. Iron uptake, translocation, and regulation in higher plants. Annu. Rev. Plant Biol. 63, 131–152.

Komal, T., Mustafa, M., Kazi, A.G., 2014. Heavy metal induced adaptation strategies and repair mechanisms plants. Endocytobiosis Cell Res. 25, 33–41.

Labanowski, J., Monna, F., Bermond, A., Cambier, P., Fernandez, C., Lamy, I., van Oort, F., 2008. Kinetic extractions to assess mobilization of Zn, Pb, Cu, and Cd in a metal-contaminated soil: EDTA vs. citrate. Environ. Pollut. 152, 693–701.

Lasat, M.M., 2000. Phytoextraction of metals from contaminated soil: a review of plant/soil/metal interaction and assessment of pertinent agronomic issues. J. Hazard. Subst. Res. 2, 1–25.

Li, H., Shi, A., Zhang, X., 2015a. Particle size distribution and characteristics of heavy metals in road-deposited sediments from Beijing Olympic Park. J. Environ. Sci. 32, 228–237.

Li, J., Jia, C., Lu, Y., Tang, S., Shim, H., 2015b. Multivariate analysis of heavy metal leaching from urban soils following simulated acid rain. Microchem. J. 122, 89–95.

Liu, G., Greenshields, D.L., Sammynaiken, R., Hirji, R.N., Selvaraj, G., Wei, Y., 2007. Targeted alterations in iron homeostasis underlie plant defense responses. J. Cell Sci. 120, 596–605.

Lux, A., Martinka, M., Vaculík, M., White, P.J., 2011. Root responses to cadmium in the rhizosphere: a review. J. Exp. Bot. 62, 21–37.

Ma, Y., Prasad, M.N.V., Rajkumar, M., Freitas, H., 2011. Plant growth promoting rhizobacteria and endophytes accelerate phytoremediation of metalliferous soils. Biotechnol. Adv. 29, 248–258.

Maestri, E., Marmiroli, M., Visioli, G., Marmiroli, N., 2010. Metal tolerance and hyperaccumulation: costs and trade-offs between traits and environment. Environ. Exp. Bot. 68, 1–13.

Maksymiec, W., 2007. Signaling responses in plants to heavy metal stress. Acta Physiol. Plant 29, 177–187.

McGrath, S.P., Zhao, F.-J., 2003. Phytoextraction of metals and metalloids from contaminated soils. Curr. Opin. Biotechnol. 14, 277–282.

McLaughlin, M., Parker, D., Clarke, J., 1999. Metals and micronutrients–food safety issues. F. Crop Res. 60, 143–163.

Merlot, S., Hannibal, L., Martins, S., Martinelli, L., Amir, H., Lebrun, M., Thomine, S., 2014. The metal transporter PgIREG1 from the hyperaccumulator *Psychotria gabriellae* is a candidate gene for nickel tolerance and accumulation. J. Exp. Bot. 65, 1551–1564.

Mitton, F.M., Gonzalez, M., Peña, A., Miglioranza, K.S.B., 2012. Effects of amendments on soil availability and phytoremediation potential of aged p,p'-DDT, p,p'-DDE and p,p'-DDD residues by willow plants (*Salix* sp.). J. Hazard. Mater. 203-204, 62–68.

Nagajyoti, P.C., Lee, K.D., Sreekanth, T.V.M., 2010. Heavy metals, occurrence and toxicity for plants: a review. Environ. Chem. Lett. 8, 199–216.

Nagata, T., Oobo, T., Aozasa, O., 2013. Efficacy of a bacterial siderophore, pyoverdine, to supply iron to *Solanum lycopersicum* plants. J. Biosci. Bioeng. 115, 1–5.

Norman, H.C., Masters, D.G., Barrett-Lennard, E.G., 2013. Halophytes as forages in saline landscapes: interactions between plant genotype and environment change their feeding value to ruminants. Environ. Exp. Bot. 92, 96–109.

Oliva, S.R., Mingorance, M.D., Leidi, E.O., 2012. Tolerance to high Zn in the metallophyte *Erica andevalensis* Cabezudo & Rivera. Ecotoxicology 21, 2012–2021.

Ovečka, M., Takáč, T., 2014. Managing heavy metal toxicity stress in plants: biological and biotechnological tools. Biotechnol. Adv. 32, 73–86.

Pal, R., Rai, J.P.N., 2010. Phytochelatins: peptides involved in heavy metal detoxification. Appl. Biochem. Biotechnol. 160, 945–963.

Palmer, C., Guerinot, M.L., 2011. A question of balance: facing the challenges of Cu, Fe and Zn homeostasis. Nat. Chem. Biol. 5, 333–340.

Passariello, B., Giuliano, V., Quaresima, S., Barbaro, M., Caroli, S., Forte, G., Carelli, G., Iavicoli, I., 2002. Evaluation of the environmental contamination at an abandoned mining site. Microchem. J. 73, 245–250.

Pérez-Clemente, R.M., Vives, V., Zandalinas, S.I., López-Climent, M.F., Muñoz, V., Gómez-Cadenas, A., 2013. Biotechnological approaches to study plant responses to stress. Biomed. Res. Int. 2013, 654120.

Quezada-Hinojosa, R., Föllmi, K.B., Gillet, F., Matera, V., 2015. Cadmium accumulation in six common plant species associated with soils containing high geogenic cadmium concentrations at Le Gurnigel, Swiss Jura Mountains. Catena 124, 85–96.

Rajkumar, M., Prasad, M.N.V., Swaminathan, S., Freitas, H., 2013. Climate change driven plant-metal-microbe interactions. Environ. Int. 53, 74–86.

Rascio, N., Navari-Izzo, F., 2011. Heavy metal hyperaccumulating plants: how and why do they do it? and what makes them so interesting? Plant Sci. 180, 169–181.

Revathi, S., Venugopal, S., 2013. Physiological and biochemical mechanisms of heavy metal tolerance. Int. J. Environ. Sci. 3, 1339–1354.

Roca, N., Pazos, M.S., Bech, J., 2012. Background levels of potentially toxic elements in soils: a case study in Catamarca (a semiarid region in Argentina). Catena 92, 55–66.

Rodríguez Martín, J.A., Arias, M.L., Grau Corbí, J.M., 2006. Heavy metals contents in agricultural topsoils in the Ebro basin (Spain). Application of the multivariate geoestatistical methods to study spatial variations. Environ. Pollut. 144, 1001–1012.

Sas, L., Rengel, Z., Tang, C., 2001. Excess cation uptake, and extrusion of protons and organic acid anions by *Lupinus albus* under phosphorus deficiency. Plant Sci. 160, 1191–1198.

Schneider, T., Persson, D.P., Husted, S., Schellenberg, M., Gehrig, P., Lee, Y., Martinoia, E., Schjoerring, J.K., Meyer, S., 2013. A proteomics approach to investigate the process of Zn hyperaccumulation in *Noccaea caerulescens* (J and C. Presl) F.K. Meyer. Plant J. 73, 131–142.

Schwartz, C., Gerard, E., Perronnet, K., Morel, J.L., 2001. Measurement of in situ phytoextraction of zinc by spontaneous metallophytes growing on a former smelter site. Sci. Total Environ. 279, 215–221.

Sharma, S.S., Dietz, K.-J., 2009. The relationship between metal toxicity and cellular redox imbalance. Trends Plant Sci. 14, 43–50.

Shu, L., Shen, J., Rengel, Z., Tang, C., Zhang, F., Cawthray, G.R., 2007. Formation of cluster roots and citrate exudation by *Lupinus albus* in response to localized application of different phosphorus sources. Plant Sci. 172, 1017–1024.

Simões, C.C., Melo, J.O., Magalhaes, J.V., Guimarães, C.T., 2012. Genetic and molecular mechanisms of aluminum tolerance in plants. Genet. Mol. Res. 11, 1949–1957.

Soudek, P., Petrová, Š., Vaňková, R., Song, J., Vaněk, T., 2014. Accumulation of heavy metals using *Sorghum* sp. Chemosphere 104, 15–24.

Tamás, L., Dudíková, J., Ďurčeková, K., Huttová, J., Mistrík, I., Zelinová, V., 2008. The impact of heavy metals on the activity of some enzymes along the barley root. Environ. Exp. Bot. 62, 86–91.

Tsednee, M., Yang, S.-C., Lee, D.-C., Yeh, K.-C., 2014. Root-secreted nicotianamine from *Arabidopsis halleri* facilitates zinc hypertolerance by regulating zinc bioavailability. Plant Physiol. 166, 839–852.

Tuazon-nartea, J., Savage, G., 2013. Investigation of oxalate levels in sorrel plant parts and sorrel-based products. FNS 4, 838–843.

Van Nevel, L., Mertens, J., Oorts, K., Verheyen, K., 2007. Phytoextraction of metals from soils: how far from practice? Environ. Pollut. 150, 34–40.

Vangronsveld, J., Colpaer, J.V., Tichelen, K.K.Van, 1997. Reclamation of a bare industrial area contaminated by non-ferrous metals: physicochemical and biological evaluation of the durability of soil treatment and revegetation. Environ. Pollut. 94, 131–140.

Viehweger, K., 2014. How plants cope with heavy metals. Bot. Stud. 55, 35.

Visioli, G., Marmiroli, N., 2013. The proteomics of heavy metal hyperaccumulation by plants. J. Proteomics 79, 133–145.

Wang, J., Zhang, C.B., Jin, Z.X., 2009. The distribution and phytoavailability of heavy metal fractions in rhizosphere soils of *Paulowniu fortunei* (seem) Hems near a Pb/Zn smelter in Guangdong, PR China. Geoderma 148, 299–306.

Waraich, E.A., Ahmad, R., Ashraf, M.Y., 2011. Role of mineral nutrition in alleviation of drought stress in plants. Aust. J. Crop Sci. 5, 764–777.

Watt, M., Evans, J.R., 1999. Update on root biology proteoid roots. Physiology and development. Plant Physiol. 121, 317–323.

Weydert, C.J., Cullen, J.J., 2011. Measurement of superoxide dismutase, catalase, and glutathione peroxidase in cultured cells and tissue. Natl. Inst. Heal 5, 51–66.

Wu, Q., Leung, J.Y.S., Geng, X., Chen, S., Huang, X., Li, H., Huang, Z., Zhu, L., Chen, J., Lu, Y., 2015. Heavy metal contamination of soil and water in the vicinity of an abandoned e-waste recycling site: Implications for dissemination of heavy metals. Sci. Total Environ. 506–507, 217–225.

Yadav, S.K., 2010. Heavy metals toxicity in plants: an overview on the role of glutathione and phytochelatins in heavy metal stress tolerance of plants. South Afr. J. Bot 76, 167–179.

Yang, L.T., Qi, Y.P., Jiang, H.X., Chen, L.S., 2013. Roles of organic acid anion secretion in aluminium tolerance of higher plants. Biomed. Res. Int. 2013, 9–15.

Yang, Z., Chu, C., 2008. Towards Understanding Plant Response to Heavy Metal Stress.

Ying, R.-R., Qiu, R.-L., Tang, Y.-T., Hu, P.-J., Qiu, H., Chen, H.-R., Shi, T.-H., Morel, J.-L., 2010. Cadmium tolerance of carbon assimilation enzymes and chloroplast in Zn/Cd hyperaccumulator *Picris divaricata*. J. Plant Physiol. 167, 81–87.

Yruela, I., 2013. Transition metals in plant photosynthesis. Metallomics 5, 1090–1109.

Zhao, J., Dong, Y., Xie, X., Li, X., Zhang, X., Shen, X., 2011. Effect of annual variation in soil pH on available soil nutrients in pear orchards. Acta Ecol. Sin 31, 212–216.

Zhou, G., Pereira, J.F., Delhaize, E., Zhou, M., Magalhaes, J.V., Ryan, P.R., 2014. Enhancing the aluminium tolerance of barley by expressing the citrate transporter genes SbMATE. J. Exp. Bot. 65, 2381–2390.

Zorrig, W., Rouached, A., Shahzad, Z., Abdelly, C., Davidian, J.-C., Berthomieu, P., 2010. Identification of three relationships linking cadmium accumulation to cadmium tolerance and zinc and citrate accumulation in lettuce. J. Plant Physiol. 167, 1239–1247.

CHAPTER

HEAVY METAL ATPase (*HMA2*, *HMA3*, AND *HMA4*) GENES IN HYPERACCUMULATION MECHANISM OF HEAVY METALS

23

Khushboo Chaudhary[1], Sumira Jan[2], Suphiya Khan[1]

Department of Bioscience and Biotechnology, Banasthali University, Tonk, Rajasthan, India[1]; Department of Biotechnology, ICAR – Central Institute of Temperate Horticulture, Srinagar, Jammu and Kashmir, India[2]

CHAPTER OUTLINE

1. INTRODUCTION

Heavy metal contaminated soils cause an increasing problem to human and animal health. The plants that accumulate specific metals at high concentration have played a crucial role in heavy metal remediation over the past 20 years (Rocca et al., 2009). Heavy metals, such as As, Cd, Hg, Pb, or Se, are not essential because they do not perform any known physiological and metabolic functions in plants. Others, such as Co, Cu, Fe, Mn, Mo, Ni, and Zn, are essential elements required for normal growth and metabolism of plants. These heavy elements can easily lead to poisoning when their concentration rises to high values (Park et al., 2014). Heavy metal phytotoxicity may result from alteration of numerous physiological processes caused at cellular/molecular levels by inactivating enzymes, blocking functional groups of metabolically important molecules, displacing or substituting for essential elements, and disrupting membrane integrity (Prasad, 2013; Lori et al., 2015).

Phytoremediation is popular because of its low cost and effectiveness for decontamination of land (Stephenson and Black, 2014; Ahmadpour et al., 2014). The processes of phytoremediation include the number of techniques (i.e., phytodegradation, phytoaccumulation, phytostabilization, phytovolatilization, and rhizofiltration) (Ali et al., 2013). The basis of the genetics of metal hyperaccumulation in woody plant species is still unknown, but the molecular mechanisms of the adaptation to metals in model plants viz *Arabidopsis* are well-understood (Babst-Kostecka et al., 2014; Poscic et al., 2015).

Plant Metal Interaction. http://dx.doi.org/10.1016/B978-0-12-803158-2.00023-0

It is hypothesized that metal specificity and localization of different metal transporters in various plants are essential for developing the genetically modified plants to accumulate specific metals, with subsequent uses in either phytoremediation to decontaminate land or to improve human nutrition by biofortification (Rascio and Navari-Izzo, 2011). Enhancement of phytoremediation efficiency could be done with the help of plant growth–promoting bacteria (Ullah et al., 2015). Plant–microbe combination has promoted the activity of the effective microbes for the plant growth. It is also suggested that inoculation of mycorrhizae to some plants may promote the uptake, translocation, and accumulation of soil metals (Khan et al., 2012; Chanda et al., 2014; Belliturk et al., 2015). Soil amendments may need to be added to enhance plant growth and biomass or contaminant uptake or degradation depending on the different stages occurring in the plant.

Detoxification of heavy metals includes different metabolic modulations including metal binding via chelation, metal sequestration into vacuoles, and efflux of toxic metals (Lee and Hwang, 2015). The strategy adopted by diverse plants in response to varied metal toxicity is intricate, and the in-depth analysis of the approach can be assessed through the molecular genetic approach (Guerra et al., 2015). Heavy metal poisoning increases the production of reactive oxygen species (ROS), which cause oxidative stress to plants (Rocca et al., 2009; Ahmad et al., 2015). This oxidative stress leads to lipid peroxidation, biological macromolecule deterioration, membrane dismantling, ion leakage, and DNA-strand cleavage (Palma et al., 2013; Dubey et al., 2014; Rout and Panigrahi, 2015). Plants possess defense mechanisms that control uptake, accumulation, and translocation of these dangerous elements and detoxify them.

Plants have developed various cellular and molecular adaptations necessary to tolerate the heavy metal stress. Different genes contribute to heavy metal hyperaccumulation and are more highly expressed in the hyperaccumulators compared with nonhyperaccumulators (Iqbal et al., 2013). Significant progress has been made in the identifying and understanding the role of key components such as hyperaccumulation transporter genes such as *HMA2*, *HMA3*, and *HMA4* that ensure heavy metal tolerance to plants (Park et al., 2014; Komal et al., 2015). Overexpression of genes responsible for heavy metal uptake, translocation, and sequestration may allow the production of plants that can be successfully exploited in phytoremediation. The cloning of different transporter genes is the foundation of genetically engineering a plant for effective phytoremediation approach (Yamaji et al., 2013). During the past few decades, phytoremediation technologies have advanced at an immense pace and numerous novel hyperaccumulator plants have been identified. The focus of the present study on the heavy metal transporters genes (*HMA2*, *HMA3*, and *HMA4*) paved the way toward the development of more lucid the phytoremediation approach.

2. GENES ASSOCIATED WITH HEAVY METAL TOLERANCE

There are diverse reports documenting the characterization of molecular mechanism conferring the differential tolerance in plants to array of metal toxicities (Migocka et al., 2014). Several molecular tools are being used for elucidating different traits from a genomic approach (Gustin et al., 2011). Transcriptome studies in various metal-tolerant lines have been used for complementary DNA arrays to elucidate the underlying mechanism of metal detoxification (Kabir et al., 2012). The possible role of identified genes in metal tolerance has been discussed as well. The assembly of cysteine-rich protein metallothionein (MT) and overexpression of BcMT1 and BcMT2 in *Brassica campestris* enhances tolerance to cadmium and declines generation of ROS in *Arabidopsis thaliana* (Lv et al., 2012). Several reports have demonstrated increased expression of GSH synthesis genes *GSH1* and *GSH2* in *A. thaliana* resulted in Cd tolerance (Semane et al., 2007; Wojcik and Tukiendorf, 2011). There are numerous reports signifying that overexpression of phytochelatin (PC) synthase genes improve tolerance to heavy metal stress (Liu et al., 2011). Overexpression of *A. thaliana* phytochelatin synthase (*AtPCS1*) improved Cd tolerance, and its accumulation in tobacco plants was directly related to GSH availability (Pomponi et al., 2006). Overexpression of bacterial-glutamylcysteine synthetase intercedes alterations in cadmium influx, allocation, and detoxification in poplar (He et al., 2015). Koren'kov et al. (2007) demonstrated overexpression of *AtCAX2* or *AtCAX4* in tobacco enhanced Cd and Zn transport into root tonoplast vesicles as well as Cd accumulation in roots of plants exposed to Cd. Plant tolerance to Cd was increased in *AtHMA3*-expressing plants, and Cd accumulation was enhanced about two to three times compared with wild-type plants (Morel et al., 2009). Overexpression of *NRAMP1* resulted in a slight increase in Cd in leaves (Takahashi et al., 2011). Lee and Hwang (2015) reported the tobacco gene *NtHb1* encoding a nonsymbiotic class 1 hemoglobin to confer cadmium (Cd) tolerance in yeast *Saccharomyces cerevisiae* by expressing complementary DNA library of *Nicotiana tabacum* in wild-type yeast Y800. *NtHb1*-expressing tobacco displayed increased tolerance to Cd, and the increase of *Hb* by *NtHb1* expression promotes Cd tolerance and reduces Cd accretion by assuaging the level of nitric oxide induced by Cd in transgenic tobacco. Merlot et al. (2014) characterized the activity of three metal transporters from the *NRAMP* and *IREG/FPN* families and identified potential role of *PgIREG1* in nickel tolerance when expressed in yeast and in transgenic plants, where it localizes in the tonoplast. Moreover, *PgIREG1* exhibited higher expression in *Psychotria gabriellae* than in the related nonaccumulator species *Psychotria semperflorens*. This study confirmed *PgIREG1* as a candidate gene for nickel tolerance and hyperaccumulation in *P. gabriellae*. These results paved a way to advancement into next-generation sequencing technologies to access the transcriptome of nonmodel nickel hyperaccumulators to identify the underlying molecular mechanisms. Migocka et al. (2014) reported the increased expression of *CsMTP8* in *S. cerevisiae* leading to Mn accumulation in yeast cells and in *A. thaliana* enhanced plant tolerance to Mn rich in nutrition media as well as the accumulation of Mn in plant tissues. Balasundaram et al. (2014) reported two MT genes from *Prosopis juliflora*, *PjMT1* and *PjMT2*, and cloned them separately in pCAMBIA 1301 and transformed them into *N. tabacum.* These transformants, when investigated for cadmium accumulation and tolerance, *PjMT1* transformants showed better performance than *PjMT2* counterparts.

3. HEAVY METAL HYPERACCUMULATING PLANTS

Improvement in the hyperaccumulator growth rate through selective breeding or by the transfer of metal hyperaccumulation genes to high biomass species has been worked out. Many researchers have reported the hyperaccumulation of heavy metals in some important plants. Hyperaccumulation depends on the plant species, soil physiochemical properties (pH, cation exchange capacity, organic matter content, electrical conductivity (E.C)), and different types of heavy metals (Van der and Reeves, 2015). In hyperaccumulators, there is a rapid and active translocation of the heavy metals to the shoot via the xylem, which could probably be upregulated by transpiration (Rees et al., 2015). Most of the heavy metal uptake occurs through root cells, where they are detoxified via complexation with amino acids, organic acids or metal-binding peptides, and/or sequestered into vacuoles (Hall, 2002). This immensely restricts translocation to the above-ground organs, thus protecting the leaf tissues and

Table 23.1 Metal Hyperaccumulating Plants

Metal	Plant	References
Arsenic	*Pteris vittata, Arabidopsis bisulcatus*	Mehdawi et al. (2011)
Copper	*Eichhornia crassipes, Euphorbia macroclada*	Mokhtar et al. (2011)
Chromium	*Phragmites australis, Zea mays* L. *Cv Ganga 5*	Calheiros et al. (2008)
Manganese	*Phytolacca americana*	Pollard et al. (2009)
Nickel	*Berkheya coddii, Alyssum* and *Thlaspi*	Bani et al. (2010)
Selenium	*Astragalus racemosus, Cardamine hupingshanesis*	Tong et al. (2014)
Titanium	*Iberis intermedia*	Leblanc et al. (1999)
Zinc	*Sedum alfredii, Euphorbia macroclada*	Nematian and Kazemeini (2013)

particularly the metabolically active photosynthetic cells from heavy metal damage. The high metal tolerance may also be due to highly efficient intracellular compartmentalization and chelation process (Sunitha et al., 2013; Sacky et al., 2014; Zhao et al., 2015). Different chelators such as EDTA, organic acid, citric acid, or nicotianamine may be involved in translocation of metal cations through the xylem (Ghnaya et al., 2013; Luo et al., 2015). Uptake of metal ions from the xylem apoplast into the shoot symplast is mediated by metal transporters in the shoot cell membrane (Socha and Guerinot, 2014). A list of hyperaccumulating plants is shown in Table 23.1.

4. HEAVY METALS TOLERANCE TRANSPORTERS GENE

Heavy metals are nonbiodegradable pollutants and they affect human health and the environment. Progress in the molecular mechanism of plant stress response to heavy metals has been made, especially in herbaceous plants such as *Arabidopsis halleri, A. Thaliana*, and *Triticum caerulescens* (Claire-Lise and Nathalie, 2012). *Arabidopsis thaliana* has always played a very important role in uncovering the molecular mechanism of plant response to pollutants because its genome information is available. The most closely connected to members of the *A. thaliana* is P1B-type. *HMA2* is also involved in the translocation of Zn and Cd in *A. thaliana*, barley, rice, and wheat (Table 23.2). In *Arabidopsis*, the cellular and subcellular patterns of *AtHMA2* expression were similar to those of *AtHMA4*. HMA2p-GUS expression was observed primarily in the vascular tissues of the leaf, root, and stem; HMA2-GFP protein was also localized in the plasma membrane. The *HMA2* gene from different plants has been reported for possible applications in phytoremediation approach (Hussain et al., 2004). The ATPase family of integral membrane transporter proteins that help to uptake transition metals are involved in mediating metal-resistant and metal-hyperaccumulating traits. Dose-dependence of metal root/shoot distribution, different in transgenic and wild-types has also been reported for plants expressing the 35S promoter *AtHMA4* as well as the metal transporters such as HvHMA2 (Barabasz et al., 2013). Although *HMA3* may involve metal detoxification by sequestering Cd into the vacuole (Figure 23.1), *HMA4* acts as a physiological master switch during the process of hyperaccumulation metal, and "*HMA2* and *HMA4* play roles in root to shoot metal translocation." It is hypothesized that description of the roles of several types of these metal transporters in plants will be essential for the development to genetically modify

Table 23.2 Heavy Metal ATPase (*HMA2*, *HMA3*, and *HMA4*) Transporters in Different Plants

Gene	Plant	Metal	References
HMA2	*Arabidopsis thaliana*	*HMA2* has N and C terminal domains that can bind to Zn ions with high affinity.	Wong and Cobbett (2009)
	Hordeum vulgare	HMA2 functions as a Zn and Cd pump and plays a role in root to shoot Zn and Cd transport.	Mills et al. (2012)
	Oryza sativa	*HMA2* is a major transporter of Zn and Cd from root to shoot at the vegetative stage, and the role of *HMA2* in the nodes is to preferentially transport Zn and Cd to the grain at the reproductive stage.	Satoh-Nagasawa et al. (2012) and Takahashi et al. (2012)
	Triticum aestivum	TaHMA2 overexpression improved root-shoot Cd and Zn translocation in rice.	Tan et al. (2013)
HMA3	*A. thaliana*	Sequestration of Cd into vacuoles to limit Cd transport to the xylem.	Morel et al. (2009)
	A. thaliana	No functional, efficient root to shoot translocation of Cd.	Park et al. (2012)
	Glycine max (AC Hime; a high Cd accumulator) *G. max (Westag 97; a low Cd accumulator)*	Single nucleotide change from G to A at nucleotide position 1823 in GmHMA3, high seed Cd levels glycine (G) at position 608 is critical for the function of GmHMA3, low seed Cd levels.	Wang et al. (2012)
	Oryza sativa (Choko-koku, Jarjan, Anjana Dhan) *O. sativa (Nipponbare, Koshihikari, Sasanishiki)*	High Cd concentrations in grains and shoots low-Cd concentrations in grains and shoots.	Ueno et al. (2010) and Miyadate et al. (2011)

Continued

Table 23.2 Heavy Metal ATPase (*HMA2*, *HMA3*, and *HMA4*) Transporters in Different Plants—cont'd

Gene	Plant	Metal	References
HMA4	*A. thaliana (Wassilewskija and Columbia)*	The *HMA4* expression level in Col-0 was induced, but those of Ws tended to decrease as compared to the control.	Park et al. (2012) and Chao et al. (2012)
	A. thaliana (Columbia)	C-terminal domain of AtHMA4 serves a dual role as $Zn2^+$ and $Cd2^+$ chelator (sensor) and as a regulator for the efficient export of $Zn2^+$ and $Cd2^+$.	Baekgaard et al. (2010)
	Thlaspi caerulescens	Key role in the extreme Cd tolerance. Cd tolerance seems to derive from lower uptake and efficient translocation to the shoots, perhaps by *HMA4*.	Lochlainn et al. (2011), Craciun et al. (2012), and Ueno et al. (2011)
	Populus trichocarpa	At early stages of development, *HMA4* relieves some negative effects of Zn.	Adams et al. (2011)
	Lycopersicon esculentum	Between transgenic and wild-type plants exposed to low and high Zn (leaves and roots).	Barabasz et al. (2010)
	Noccaea caerulescens	Zn and Cd tolerance gene in root.	Halimaa et al. (2014)

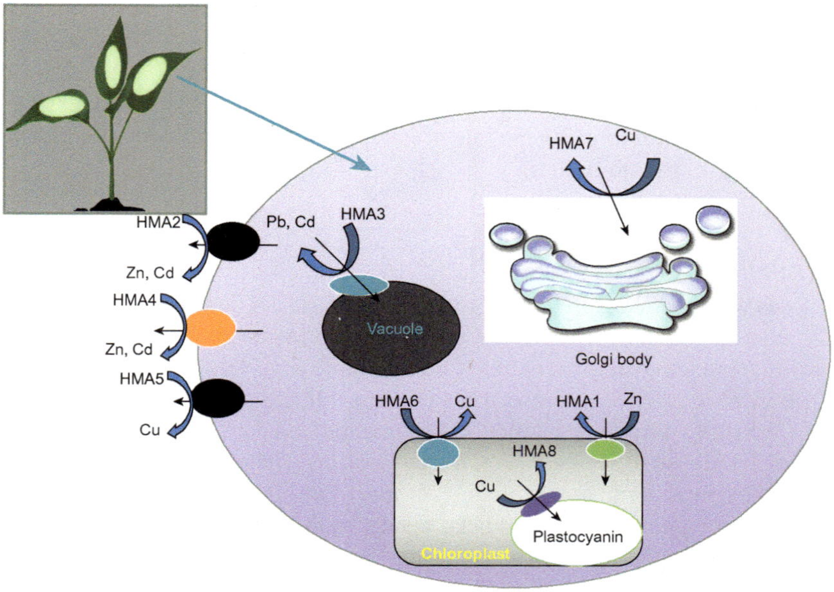

FIGURE 23.1

Heavy metal ATPase gene contributes in hyperaccumulation of heavy metals (cell organelles for translocation and sequestration).

plants that accumulate specific metals, with subsequent use in either phytoremediation or in improving human nutrition.

The efficiency increased by *HMA3* and *HMA4* is a prerequisite for hyperaccumulation and hyperresistance in hyperaccumulating plants. Through genetic engineering, these genes are able to generate nontoxic food. Park et al. (2014) reported that overexpression of CsHMA3 might increase Pb and Zn tolerance and uptake. Also, the transgenic lines showed a wider leaf shape compared with the wild-type plant because of a stimulation of genes related to leaf growth, and displayed a greater total seed production compared with the wild-type species under heavy metal contamination. These include the *HMA3* and *HMA4* encoded proteins that belong to the P1B-type ATPase family, which have been characterized in various organisms (Table 23.2). *HMA4* was by far the most highly expressed metal transporter in the three *Noccaea caerulescens* accessions (Halimaa et al., 2014). Orthologous HMA genes in metal hyperaccumulators may function similarly.

Many genes from different organisms have been identified and characterized that are involved in the acquisition, allocation, and decontamination of metals. It is well-known that *HMA3* is a tonoplast-localized transporter involved in metal detoxification by sequestering specific metals such as cadmium into the vacuole (Song et al., 2014). *HMA2* and *HMA4* are plasma-membrane proteins contribute in translocating Zn and Cd ions from the root to shoot and detoxifying these metals in above-ground tissues (Wu et al., 2015). Although the sequences suggest similarities in function, *HMA2*, *HMA3*, and *HMA4* do not function the same depending on the species (Migocka et al., 2015). Many species of plants have the ability to adapt to harsh environmental conditions. Gene functions may develop because of selection from local climates and environments. In particular, these gene alterations may increase *HMA2*, *HMA3*, and *HMA4* substrate affinity depending on the type of heavy metal contamination (Park et al., 2014). We recommend the cooverexpression of *HMA3* with *HMA4* "with an *HMA4* overexpressing plant using a root-specific promoter and an *HMA3* overexpressing plant using a shoot-specific promoter" for improving the rate of translocation, compartmentation, and detoxification of toxic metals in above-ground tissues.

5. CONCLUSION AND FUTURE PROSPECTS

The chapter closes with an aphorism that overexpression of *HMA3* and *HMA4* genes can induce heavy metal stress tolerance in plants. The heterologous overexpression of *HMA3* and *HMA4* in plants is probably a useful approach to engineer altered metal supply in tissues for phytoremediation purposes because these transporter genes increase the ability to tolerate heavy metal stress to clean up the environment successfully. Phytoremediation has been approved commercially as well for cleaning up of the environment and also to increase biomass production. There is also an urgent need to find and characterize other hyperaccumulators, to cultivate them, and better assess agronomic practices and management to enhance plant growth and metal uptake by selective breeding and gene manipulation. Even then, metal uptake might pose environmental risks, unless the biomass produced during the phytoremediation process could be rendered economical by burning it to produce bio-ore or converting it into bioenergy. Understanding the signaling mechanism of transporter genes will be an important tool to understand the genetics of hyperaccumulation. Further, it will help to improve the phytoremediation efficiency via transgenic approach. Metabolic engineering strategies are exceedingly used for environmental applications. Strategies directed toward manipulating the biosynthesis of peptides such as PCs hold promise for the remediation of soils polluted with heavy metals. Evolution of *AtPCS1* mutants that exhibit high levels of cadmium tolerance has been a milestone in the field of transgenic. Phytoremediation shows a potential way of upgrading heavy metal pollution through the use of transgenic plants as artificial hyperaccumulators. Discovery of three transgenic lines of sugar beet (*Beta vulgaris* L.) with enhanced tolerance to different concentrations of cadmium, zinc, and copper was engineered through overexpression of *glutamylcysteine* synthetase-glutathione synthetase isolated from *Streptococcus thermophilus*. Advancement in development of transgenic lines might bring the newest milestone in the evolution of metal tolerance.

REFERENCES

Adams, J.P., Adeli, A., Hsu, C.Y., Harkess, R.L., Page, G.P., dePamphilis, C.W., Schultz, E.,B., Yuceer, C., 2011. Poplar maintains zinc homeostasis with heavy metal genes *HMA4* and PCS1. J. Exp. Bot. 62, 3737–3752.

Ahmad, P., Sarwat, M., Bhat, N.A., Wani, M.R., Kazi, A.G., Tran, L.S., 2015. Alleviation of cadmium toxicity in *Brassica juncea* L. (Czern. & Coss.) by calcium application involves various physiological and biochemical strategies. PloS One 10 (1).

Ahmadpour, P., Ahmadpour, F., Mahmud, T.M.M., Abdu, A., Soleimani, M., Tayefeh, F.H., 2014. Phytoremediation of heavy metals: a green technology. In: Heavy Metal Contamination of Water and Soil: Analysis, Assessment, and Remediation Strategies, p. 249.

Ali, H., Khan, E., Sajad, M.A., 2013. Phytoremediation of heavy metals-Concepts and applications. Chemosphere 91, 869–881.

Babst-Kostecka, A.A., Parisod, C., Gode, C., Vollenweider, P., Pauwels, M., 2014. Patterns of genetic divergence among populations of the pseudometallophyte *Biscutella laevigata* from southern Poland. Plant Soil 383 (1–2), 245–256.

Baekgaard, L., Mikkelsen, M.D., Sorensen, D.M., Hegelund, J.N., Persson, D.P., Mills, R.F., Yang, Z., Husted, S., Andersen, J.P., Buch-Pedersen, M.J., Schjoerring, J.K., Williams, L.E., Palmgren, M.G., 2010. A combined zinc/cadmium sensor and zinc/cadmium export regulator in a heavy metal pump. J. Biol. Chem. 285 (4), 31243–31252.

Bani, A., Pavlova, D., Echevarria, G., Mullaj, A., Reeves, R.D., Morel, J.L., Sulçe, S., 2010. Nickel hyperaccumulation by the species of *Alyssum* and *Thlaspi* (Brassicaceae) from the ultramafic soils of the Balkans. Bot. Serbic. 34 (1), 3–14.

Barabasz, A., Krämer, U., Hanikenne, M., Rudzka, J., Antosiewicz, D.M., 2010. Metal accumulation into tobacco expressing *Arabidopsis halleri* metal hyperaccumulation gene depends on external supply. J. Exp. Bot. 61, 3057–3067.

Barabasz, A., Wilkowska, A., Tracz, K., Ruszczyñska, A., Bulska, E., Mills, R.F., 2013. Expression of HvHMA2 in tobacco modifies Zn–Fe–Cd homeostasis. J. Plant Physiol. 170, 1176–1186.

Balasundaram, U., Venkataraman, G., George, S., Parida, A., 2014. Metallothioneins from a hyperaccumulating plant prosopis juliflora show difference in heavy metal accumulation in transgenic tobacco. International J. Agricul. Environ. Biotech. 7, 241–246.

Belliturk, K., Shrestha, P., Gorres, J.H., 2015. The importance of phytoremediation of heavy metal contaminated soil using vermicompost for sustainable agriculture. J. Rice Res. 3, 114.

Calheiros, C.S.C., Rangel, A.O.S.S., Castro, P.M.L., 2008. The effects of tannery waste water on the development of different plant species and chromium accumulation in *Phragmites australis*. Arch. Environ. Cont. Tox. 55, 404–414.

Chanda, D., Sharma, G.D., Jha, D.K., Hijri, M., 2014. Associations of arbuscular mycorrhizal (AM) fungi in the phytoremediation of trace metal (TM) contaminated soils. J. Res. Bio. 4 (2).

Chao, D.Y., Silva, A., Baxter, I., Huang, Y.S., Nordborg, M., Danku, J., Lahner, B., Yakubova, E., Salt, D.E., 2012. Genome-wide association studies identify heavy metal ATPase3 as the primary determinant of natural variation in leaf cadmium in *Arabidopsis thaliana*. PLoS Genet. 8 (9), 1002923.

Craciun, A.R., Meyer, C.L., Chen, J., Roosens, N., De Groodt, R., Hilson, P., Verbruggen, N., 2012. Variation in *HMA4* gene copy number and expression among *Noccaea caerulescens* populations presenting different levels of Cd tolerance and accumulation. J. Exp. Bot. 63, 4179–4189.

Claire-Lise, M., Nathalie, V., 2012. The use of the model species *Arabidopsis halleri* towards phytoextraction of cadmium polluted soils. New biotechnology 30, 9–14.

Dubey, S., Shri, M., Misra, P., Lakhwani, D., Bag, S.K., Asif, M.H., Chakrabarty, D., 2014. Heavy metals induce oxidative stress and genome-wide modulation in transcriptome of rice root. Funct. Integr. Gen. 14 (2), 401–417.

Ghnaya, T., Zaier, H., Baioui, R., Sghaier, S., Lucchini, G., Sacchi, G.A., Abdelly, C., 2013. Implication of organic acids in the long-distance transport and the accumulation of lead in *Sesuvium portulacastrum* and *Brassica juncea*. Chemosphere 90 (4), 1449–1454.

Guerra, F.P., Reyes, L., Vergara-Jaque, A., Campos-Hernandez, C., Gutierrez, A., Perez-Diaz, J., Ruiz-Lara, S., 2015. *Populus deltoides* Kunitz trypsin inhibitor 3 confers metal tolerance and binds copper, revealing a new defensive role against heavy metal stress. Environ. Exper. Bot. 115, 28–37.

Gustin, J.L., Zanis, M.J., Salt, D.E., 2011. Structure and evolution of the plant cation diffusion facilitator family of ion transporters. BMC Evol. Bio. 11, 76.

Halimaa, P., Lin, Y.F., Ahonen, V.H., Blande, D., Clemens, S., Gyenesei, A., Haikio, E., Karenlampi, S.O., Laiho, A., Mark Aarts, G.M., Pursiheimo, J.P., Schat, H., Schmidt, H., Tuomainen, M.H., Tervahauta, A.I., 2014. Gene expression differences between *Noccaea caerulescens* Ecotypes help to identify candidate genes for metal phytoremediation. Environ. Sci. Technol. 48, 3344–3353.

Hall, J.L., 2002. Cellular mechanisms for heavy metal detoxification and tolerance. J. Exp. Bot. 53, 1–11.
He, J., Li, H., Ma, C., Zhang, Y., Polle, A., Rennenberg, H., Cheng, X., Luo, Z.B., 2015. Overexpression of bacterial g-glutamylcysteine synthetase mediates changes in cadmium influx allocation and detoxification in poplar. New Phytol. 205, 240–254.
Hussain, D., Haydon, M.J., Wang, Y., Wong, E., Sherson, S.M., Young, J., Camakaris, J., Harper, J.F., Cobbett, C.S., 2004. P-Type ATPase heavy metal transporters with roles in essential zinc homeostasis in *Arabidopsis*. Plant Cell 16 (5), 1327–1339.
Iqbal, M., Nawaz, I., Hassan, Z., Hakvoort, H.W.J., Bliek, M., Aarts, M.G.A., Schat, H., 2013. Expression of *HMA4* cDNAs of the zinc hyperaccumulator *Noccaea caerulescens* from endogenous NcHMA4 promoters does not complement the zinc deficiency phenotype of the *Arabidopsis thaliana* hma2 hma4 double mutant. Plant Sci. 4 (404), 1–12.
Kabir, A.H., Paltridge, N.G., Able, A.J., Paull, J.G., Stangoulis, J.C.R., 2012. Natural variation for Fe-efficiency is associated with upregulation of Strategy I mechanisms and enhanced citrate and ethylene synthesis in *Pisum sativum* L. Planta 235, 1409–1419.
Khan, A.G., Kuek, T.M., Chaudhary, C.S., Khoo, Hayes, W.J., 2012. Role of plants mycorrhizae and phytochelators in heavy metal contaminated land remediation. Chemosphere 41, 197–207.
Komal, T., Mustafa, M., Ali, Z., Kazi, A.G., 2015. Heavy metal uptake and transport in plants. In: Heavy Metal Contamination of Soils. Springer International Publishing, pp. 181–194.
Korenkov, V., Hirschi, K., Crutchfield, J.D., Wagner, G.J., 2007. Enhancing tonoplast Cd/H antiporter activity increases Cd, Zn and Mn tolerance: and impacts root/shoot Cd partitioning in *Nicotiana tabacum* L. Planta 226, 1379–1387.
Leblanc, M., Petit, D., Deram, A., Robinson, B., Brooks, R.R., 1999. The phytomining and environmental significance of hyperaccumulation of thallium by *Iberisintermedia* from southern France. Econ. Geol. 94, 109–113.
Lee, B.D., Hwang, S., 2015. Overexpression of NtUBQ2 encoding Ub-extension protein enhances cadmium tolerance by activating 20S and 26S proteasome in tobacco (*Nicotiana tabacum*). Acta Physio. Plant. 37 (2), 1–8.
Liu, G.Y., Zhang, Y.X., Chai, T.Y., 2011. Phytochelatin synthase of *Thlaspi caerulescens* enhanced tolerance and accumulation of heavy metals when expressed in yeast and tobacco. Plant Cell Rep. 30, 1067–1076.
Lochlainn, S.O., Bowen, H.C., Fray, R.G., Hammond, J.P., King, G.J., White, P.J., Graham, N.S., Broadley, M.R., 2011. Tandem quadruplication of HMA4 in the zinc (Zn) and cadmium (Cd) hyper- accumulator *Noccaea caerulescens*. PLoS One 6 (3), 1–9.
Lori, V., Pietrini, F., Massacci, A., Zacchini, M., 2015. Morphophysiological responses, heavy metal accumulation and phytoremoval ability in four Willow clones exposed to cadmium under hydroponics. In: Phytoremediation. Springer International Publishing, pp. 87–98.
Luo, Q., Sun, L.N., Wang, H., Hu, X.M., 2015. Metabolic profiling analysis of root exudates from the Cd hyperaccumulator *Sedum alfredii* under different Cd exposure concentrations and times. Anal. Methods 7 (9), 3793–3800.
Lv, Y., Deng, X., Quan, L., Xia, Y., Shen, Z., 2012. Metallothioneins BcMT1 and BcMT2 from *Brassica campestris* enhance tolerance to cadmium and copper and decrease production of reactive oxygen species in *Arabidopsis thaliana*. Plant Soil 367, 507–519.
Mehdawi El, F.A., Quinn, C.F., Pilon-Smits, E.A.H., 2011. Selenium hyperaccumulators facilitate selenium-tolerant neighbors via phytoenrichment and reduced herbivory curt. Curr Biol 21, 1440–1449.
Merlot, S., Hannibal, L., Martins, S., Martinelli, L., Amir, H., Lebrun, M., Thomine, S., 2014. The metal transporter PgIREG1 from the hyperaccumulator *Psychotria gabriellae* is a candidate gene for nickel tolerance and accumulation. J. Exp. Bot. 025.
Migocka, M., Papierniak, A., Maciaszczyk-Dziubinska, E., Pozdzik, P., Posyniak, E., Garbiec, A., Filleur, S., 2014. Cucumber metal transport protein MTP8 confers increased tolerance to manganese when expressed in yeast and *Arabidopsis thaliana*. J. Exper. Bot. 295.

Migocka, M., Papierniak, A., Maciaszczyk-dziubinska, E.W.A., Posyniak, E., Kosieradzka, A., 2015. Molecular and biochemical properties of two P1B2-ATPases, CsHMA3 and CsHMA4, from cucumber. Plant Cell Environ. 38 (6), 1127–1141.

Mills, R.F., Peaston, K.A., Runions, J., Williams, L.E., 2012. HvHMA2, a P1B-ATPase from barley, is highly conserved among cereals and functions in Zn and Cd transport. PLoS One 7 (8), 1–14.

Miyadate, H., Adachi, S., Hiraizumi, A., Tezuka, K., Nakazawa, N., Kawamoto, T., Katou, K., Kodama, I., Sakurai, K., Takahashi, H., Satoh-Nagasawa, N., Watanabe, A., Fujimura, T., Akagi, H., 2011. OsHMA3, a P1B-type of ATPase affects root-to-shoot cadmium translocation in rice by mediating efflux into vacuoles. New Phytol. 189 (1), 190–199.

Mokhtar, H., Morad, N., Fizri, F.F.A., 2011. Phytoaccumulation of copper from aqueous solutions using *Eichhornia Crassipes* and *Centella asiatica*. Inter. J. Environ. Sci. Develop. 3 (1), 89–95.

Morel, M., Crouzet, J., Gravot, A., Auroy, P., Leonhardt, N., Vavasseur, A., Richaud, P., 2009. AtHMA3: a P1B-ATPase allowing Cd/Zn/Co/Pb vacuolar storage in *Arabidopsis*. Plant Physiol. 149, 894–904.

Nematian, M.A., Kazemeini, F., 2013. Accumulation of Pb, Zn, Cu and Fe in plants and hyperaccumulator choice in Galali iron mine area. Iran. Inter. J. Agricult. Crop Sci. 5–4, 426–432.

Palma, J.M., Gupta, D.K., Corpas, F.J., 2013. Metalloenzymes involved in the metabolism of reactive oxygen species and heavy metal stress. In: Heavy Metal Stress in Plants. Springer, Berlin Heidelberg, pp. 1–17.

Park, W., Han, K.H., Ahn, S.J., 2012. Differences in root-to-shoot Cd and Zn translocation and by HMA3 and 4 could influence chlorophyll and anthocyanin content in *Arabidopsis* Ws and Col-0 ecotypes under excess metals. Soil Sci. Plant Nutr. 58, 334–348.

Park, W., Feng, Y., Ahn, S.-J.U., 2014. Alteration of leaf shape, improved metal tolerance, and productivity of seed by overexpression of CsHMA3 in *Camelina sativa*. Biotech. Biofuels 7, 1–17.

Pollard, A.J., Stewart, H.L., Roberson, C.B., 2009. Manganese hyper accumulation in *Phytolacca americana* L. from the southeastern United States. Northeast. Nat. 16, 155–162.

Pomponi, M., Censi, V., Di girolamo, V., De Paolis, A., Sanitadi Toppi, L., Aromolo, R., Costantino, P., Cardarelli, M., 2006. Overexpression of *Arabidopsis* phytochelatin synthase in tobacco plants enhances Cd^{2+} tolerance and accumulation but not translocation to the shoot. Planta 223, 180–190.

Poscic, F., Fellet, G., Vischi, M., Casolo, V., Schat, H., Marchiol, L., 2015. Variation in heavy metal accumulation and genetic diversity at a regional scale among metallicolous and non-metallicolous populations of the facultative metallophyte *Biscutella laevigata* subsp. *laevigata*. Intern. J. Phytoremed. 17 (5), 464–475.

Prasad, M.N.V. (Ed.), 2013. Heavy Metal Stress in Plants: From Biomolecules to Ecosystems. Springer Science & Business Media.

Rascio, N., Navari-Izzo, F., 2011. Heavy metal hyperaccumulating plants: how and why do they do it? And what makes them so interesting? Plant Sci. 180, 169–181.

Rees, F., Sterckeman, T., Morel, J.L., 2015. Root development of non-accumulating and hyperaccumulating plants in metal-contaminated soils amended with biochar. Chemosphere.

Rocca La, N., Andreoli, C., Giacometti, G.M., Rascio, N., Moro, I., 2009. Responses of the Antarctic microalga *Koliella antartica* (Trebouxiophyceae, Chlorophyta) to cadmium contamination. Photosynthetica 47, 471–479.

Rout, G.R., Panigrahi, J., 2015. Analysis of signaling pathways during heavy metal toxicity: a functional genomics Perspective. In: Elucidation of Abiotic Stress Signaling in Plants. Springer, New York, pp. 295–322.

Sacky, J., Leonhardt, T., Borovicka, J., Gryndler, M., Briksí, A., Kotrba, P., 2014. Intracellular sequestration of zinc, cadmium and silver in *Hebeloma mesophaeum* and characterization of its metallothionein genes. Fun. Gen. Bio. 67, 3–14.

Satoh-Nagasawa, N., Mori, M., Nakazawa, N., Kawamoto, T., Nagato, Y., Sakurai, K., Takahashi, H., Watanabe, A., Akagi, H., 2012. Mutations in rice (*Oryza sativa*) heavy metal ATPase 2(OsHMA2) restrict the translocation of zinc and cadmium. Plant Cell Physiol. 53 (1), 213–224.

Semane, B., Cuypers, A., Smeets, K., Van Belleghem, F., Horemans, N., Schat, H., Vangronsveld, J., 2007. Cadmium responses in *Arabidopsis thaliana*: glutathione metabolism and antioxidative defence system. Physiol. Plant. 129, 519–528.

Socha, A.L., Guerinot, M.L., 2014. Mn-euvering manganese: the role of transporter gene family members in manganese uptake and mobilization in plants. Fron. Plant Sci. 5.

Song, W.Y., Park, J., Eisenach, C., Maeshima, M., Lee, Y., Martinoia, E., 2014. ABC transporters and heavy metals. In: Plant ABC Transporters. Springer International Publishing, pp. 1–17.

Stephenson, C., Black, C.R., 2014. One step forward, two steps back: the evolution of phytoremediation into commercial technologies. Biosci. Horiz. 7, 009.

Sunitha, M.S., Prashant, S., Kumar, A., Rao, S., Narasu, M.L., Ravi Kishor, P.B., 2013. Cellular and molecular mechanisms of heavy metal tolerance in plants: a brief overview of transgenic plants overexpressing phytochelatin synthase and metallothionein genes. Plant Cell Biotechnol. Mol. Biol. 14 (1–2), 33–48.

Takahashi, R., Ishimaru, Y., Senoura, T., Shimo, H., Ishikawa, S., Arao, T., Nakanishi, H., Nishizawa, N.K., 2011. The OsNRAMP1 iron transporter is involved in Cd accumulation in rice. J. Exp. Bot. 62, 4843–4850.

Takahashi, R., Ishimaru, Y., Shimo, H., Ogo, Y., Senoura, T., Nishizawa, N.K., Nakanishi, H., 2012. The OsHMA2 transporter is involved in root-to-shoot translocation of Zn and Cd in rice. Plant Cell Environ. 35, 1948–1957.

Tan, J., Wang, J., Chai, T., Zhang, Y., Feng, S., Li, Y., Zhao, H., Liu, H., Chai, X., 2013. Functional analyses of TaHMA2, a P1B-typeATPase in wheat. Plant Biotechnol. J. 11, 420–431.

Tong, X., Yuan, L., Luo, L., Yin, X., 2014. Characterization of a selenium-tolerant rhizosphere strain from a novel Se-hyperaccumulating plant *Cardamine hupingshanesis*. Scientific World J. 1-8, 108562.

Ueno, D., Milner, M.J., Yamaji, N., Yokosho, K., Koyama, E., Clemencia Zambrano, M., Kaskie, M., Ebbs, S., Kochian, L.V., Ma, J.F., 2011. Elevated expression of TcHMA3 plays a key role in the extreme Cd tolerance in a Cd-hyperaccumulating ecotype of *Thlaspi caerulescens*. Plant J. 66 (5), 852–862.

Ueno, D., Yamaji, N., Kono, I., Huang, C.F., Ando, T., Yano, M., Ma, J.F., 2010. Gene limiting cadmium accumulation in rice. Proc. Natl. Acad. Sci. U.S.A. 107, 16500–16505.

Ullah, A., Heng, S., Munis, M.F.H., Fahad, S., Yang, X., 2015. Phytoremediation of heavy metals assisted by plant growth promoting (PGP) bacteria: a review. Environ. Exper. Bot. 117, 28–40.

Van der Ent, A., Reeves, R.D., 2015. Foliar metal accumulation in plants from copper-rich ultramafic outcrops: case studies from Malaysia and Brazil. Plant Soil 389 (1–2), 401–418.

Wang, Y., Yu, K.F., Poysa, V., Shi, C., Zhou, Y.H., 2012. A single point mutation in GmHMA3 affects cadimum (Cd) translocation and accumulation in soybean seeds. Mol. Plant 5, 1154–1156.

Wojcik, M., Tukiendorf, A., 2011. Glutathione in adaptation of *Arabidopsis thaliana* to cadmium stress. Biol. Plant 55, 125–132.

Wong, C.K.E., Cobbett, C.S., 2009. HMA P-type ATPases are the major mechanism for root-to-shoot Cd translocation in *Arabidopsis thaliana*. New Phytol. 181, 71–78.

Wu, Z., Zhao, X., Sun, X., Tan, Q., Tang, Y., Nie, Z., Hu, C., 2015. Xylem transport and gene expression play decisive roles in cadmium accumulation in shoots of two oilseed rape cultivars (*Brassica napus*). Chemosphere 119, 1217–1223.

Yamaji, N., Xia, J., Mitani-Ueno, N., Yokosho, K., Ma, J.F., 2013. Preferential delivery of zinc to developing tissues in rice is mediated by P-type heavy metal ATPase OsHMA2. Plant Physiol. 162 (2), 927–939.

Zhao, D., Li, T., Wang, J., Zhao, Z., 2015. Diverse strategies conferring extreme cadmium (Cd) tolerance in the dark septate endophyte (DSE), Exophiala pisciphila: evidence from RNA-seq data. Microbio. Res. 170, 27–35.

CHAPTER

HEAVY METAL STRESS: PLANT RESPONSES AND SIGNALING

24

Asiya Hameed[1], Saiema Rasool[2], M.M. Azooz[3,4], Mohammad Anwar Hossain[5], Mohammad Abass Ahanger[6], Parvaiz Ahmad[7]

Department of Botany, Faculty of Science, Jamia Hamdard, New Delhi, India[1]; Department of Forest Management, Faculty of Forestry, University Putra Malaysia, Serdang, Selangor, Malaysia[2]; Department of Botany, Faculty of Science, South Valley University, Qena, Egypt[3]; Department of Biological Sciences, Faculty of Science, King Faisal University, Al-Hassa, Saudi Arabia[4]; Department of Genetics and Plant Breeding, Bangladesh Agricultural University, Mymensingh, Bangladesh[5]; Stress Physiology Laboratory, School of Studies in Botany, Jiwaji University Gwalior, MP, India[6]; Department of Botany, S.P. College, Srinagar, Jammu and Kashmir, India[7]

CHAPTER OUTLINE

Plant Metal Interaction. http://dx.doi.org/10.1016/B978-0-12-803158-2.00024-2

1. INTRODUCTION

Heavy metals (HMs) are metals with a density higher than 5 g/cm^3. Of 90 naturally occurring elements, there are 53 HMs, but all of them are not biologically important. Under physiological conditions based on their solubility, 17 HMs may be available for living cells and are important for organisms and ecosystem (Schützendubel and Polle, 2002). Among these metals, Fe, Mo, and Mn are important as micronutrients, whereas Zn, Ni, Cu, Co, V, W, and Cr are trace elements but at higher concentrations are toxic and the other metals like As, Hg, Ag, Sb, Cd, Pd, and U have no function and seems to be toxic to microorganisms and plants (Nazar et al., 2012).

HM pollution is a not only a big problem in most parts of the world, but it is also responsible for the loss of agricultural productivity. Many industries and agricultural activities contribute to HM contamination in urban areas. HMs can directly influence growth, senescence, and energy synthesis processes because of their high reactivity. They adversely affect the absorption and transport of essential elements, thereby disturbing the metabolism and having an impact on growth and reproduction (Xu et al., 2008). More than 30 μg of lead (Pb) leads to a reduction in growth and decline in chlorophyll synthesis in leaf (Ruley et al., 2004). Sharma and Dubey (2005) have reported that Pb causes many problems like reduction in growth as well as crop production, yellowing of young leaves, reduction in absorption of essential elements such as Fe, and decline in rate of photosynthesis. When plants are subjected to HM stress production, and quenching of reactive oxygen species (ROS) were found imbalanced. ROS, such as superoxide radical (O_2^-) and hydrogen peroxide (H_2O_2) and the hydroxyl radical (OH) can have detrimental effects on normal metabolism because of oxidative damage to lipids, proteins, and nucleic acids. The toxicity and tolerance to HMs vary from plant to plant, and several efforts to their tolerance mechanism at the molecular level have not been fully understood (Thapa et al., 2012). So to cope with the stress, plants need a coordination of complex physiological and biochemical processes. Gene expression, protein modifications, and changes in metabolite composition will lead to proper stress signal perception and tolerance (Urano et al., 2010).

2. PLANT GROWTH RESPONSE TO HM STRESS

Growth is a complex process; therefore, HMs can affect it at several points. The toxic effect of HMs is obvious at different levels of cell structures and function of the plants. HMs use wide range of variations in the physiological and metabolic processes (Ahmad et al., 2011; Asgher et al., 2014; Khan et al., 2015).

Inhibition of germination and root extension can be the result of interference with cell division or with cell elongation. It has been observed that the main influence of HMs (Hg, Cd, Co, Cu, Pb, and Zn) is the inhibition of germination, root elongation, and shoot and leaf growth (Munzuroglu and Geckil, 2002; Suzuki, 2005; Rascio et al., 2008). The elasticity of cell walls is so much reduced by it that they may break under mechanical stress (Gall et al., 2015). The inhibition of root elongation in many instances is the most sensitive parameter of HM toxicity (Schützendübel et al., 2001). Because of the binding of HM cations to plant constituents, the order of toxicity usually conforms to the stability of the metal–organic complexes. In crop production, Al toxicity is one of the major growth-limiting factors in acidic mineral soils (Panda et al., 2009). The root system becomes stubby as a result of inhibition of elongation of the main axis and lateral roots because of restricted cell division. The root becomes stunted and brittle and apices become swollen and damaged (Panda et al., 2009). Al causes extensive root injury, especially in root cap region, and hampers the mineral and water uptake (Panda et al., 2009). The severity of inhibition of root growth is a suitable indicator of genotypical differences in Al toxicity (reviewed by Rout et al., 2001). HMs induce the toxicity by binding with the organic compounds called complexes. These in turn lose their capacity to function and eventually lead to death of the affected cells.

HMs interfere with the cell division and thereby reduce the growth of both root and shoot meristem. Inhibition of mobilization of nitrogen and phosphorous during seedling growth have been observed in spinach on exposure to mercury (Hg) (Gothberg et al., 2004). Similarly, Ashger et al. (2014) have observed reduced sulfur assimilation because of cadmium stress in *Brassica* spp. Toxicity of Hg is more pronounced at lower concentrations than others. On exposure to HMs, root and shoot growth declines and could be positively correlated with low nitrogen associated with reduced nitrate reductase (NR) activity or low pigment content (reviewed in Nazar et al., 2012).

Seedlings contaminated with HMs are likely to have stunted and poorly developed root system that inhibits the potential to absorb nutrients and moisture from soil and thereby hamper further growth. Significant reduction in number and leaf area indicate an interference of HMs on plants and can be attributed to the reduced photosynthetic activity and low nitrogen content (Faizan et al., 2011). The declining leaf area indicates not only retarded growth, but also a morphological adaptation to stressed conditions. This could also be attributed to the effect of Cd treatment on the leaf area that modifies the relative transpiration surface as found in *Phaseolus coccineus* exposed to Cu stress.

Cd, Al, and Hg are highly toxic and inhibit the uptake of water and nutrients in peas and are closely coincided with the extensive accumulation especially in roots. This blocks the entry or binding of essential ions like Ca, Mg, and K to ion-carriers (Wang and Greger, 2004). John et al. (2009) also reported the decline in shoot and root length in *Brassica juncea* subjected to Cd and Pd stress. Decline in growth in mustard plants because of higher levels of Cd and Pb is also reported by Ahmad et al. (2012), and this decline is attributed to less nutrient uptake and lowering of water potential. Arsenic (As) also affects root growth more severely than shoot growth possibly because of the retention of As in the shoots in higher amount than in the stem (Stoeva et al., 2005).

The inhibiting effects of HMs such as Cd and Hg on morphological parameters have been reported in *Abelmoschus esculentus*. The phytotoxic symptoms in response to HMs may be due to a range of interactions at the cellular level. Excess HMs may stimulate the formation of free radicals that damage the cellular functioning of the plant (Ahmad et al., 2011, 2012).

HMs resulted in the decreased dry matter and seed yield, lowered protein content in seeds, indicating an adverse effects exerted by HMs. Cd has been found to be the most toxic metal for crop yield and caused the most severe reduction in the dry weight of shoot, root, and seed yield (Ahmad et al., 2011, 2012; Ashger et al., 2014). Several studies have also demonstrated a relatively higher phytotoxicity of Cd and Co than that of Mn. The reduction in the fresh and dry weight of shoot and root because of HM stress has been reported by many workers (Ahmad et al., 2011; Qui et al., 2013, 2014).

Amendment of soil with HMs at concentrations higher than the normal levels resulted in a conspicuous decrease of root and shoot biomass expressed in terms of dry weight (Ghani, 2010). It has earlier been reported that increasing Co supply resulted in decreased root biomass indicating an alteration of physiology and metabolism of plants. Biomass loss (fresh weight) under metal treatment has also been reported by many workers (Ahmad et al., 2011, 2012; Mostofa and Fujita, 2013; Mostofa et al., 2014).

The majority of the responses in relation to environmental stress in plants are linked with growth, differentiation, and physiological aspects such as photosynthesis, ion uptake, and transport. Under Cd stress, plants show a number of symptoms like chlorosis, leaf rolling, and stunting. In certain species, this resulted in structural disorders and thereby growth restrictions (Djebali et al., 2010; Jin et al., 2008). Cadmium stress leads to decrease in shoot dry weight that is concomitant with a significant reduction in both length and diameter of the internodes as well as in leaf area and thickness, but it is dependent on tissue age. Cd that reached the leaves resulted in further physiological and structural damages. Leaf growth was inhibited and blade thickness was diminished because of the reduced enlargement of mesophyll cells resulting in increasing tissue dry weight/fresh weight ratio, especially in young leaves. Under Cd stress, stomatal conductance declined sharply, suggesting that stomatal functionality may be compromised (Perfus-Barbeoch et al., 2002). Decreased stomatal conductance and photosynthetic rate have been reported by Ahmad et al. (2011) and Asgher et al. (2014). Changes in plant water relations with a decline in transpiration rate have been observed in other Cd-exposed species (Ahmad et al., 2011) and were ascribed to decreased leaf blade expansion. Exposure of aquatic plant *Eladea canadensis* to HM was found to inhibit cell division, but induced a significant enlargement of only one of the two layers constituting the leaf blade. This suggests that the changes in leaf cell enlargement caused by Cd may actually be due to specific morphogenetic effects rather than to impaired water balance (Rascio et al., 2008).

Cu or Cd applied alone or in combination caused significant reduction in root diameter, width, and thickness of the leaf midrib (Kasim, 2006). The diameter of xylem vessel, parenchyma cell area, pith, cortex of root, dimensions of stem, vascular bundles, number of xylem arms in root, frequency of stomata on abaxial leaf surface, and reduction in grain yield have also been observed because of HM stress. The increase in the number of xylem tissues from an increased rate of transpiration and thickening of the cell wall with reduction in number of cells in the cortical region of the roots because of accumulation of HMs was also reported by Rauser and Ackerley (1987).

Extensive inhibitions in the number and diameter of xylem vessels have been reported in many plant species. This could be due to inhibition in the water flow transportation to shoots in plants thus contributing the imbalance in the water transport. Cd causes damages in the leaves and thus renders loss in transpiration, leading to decreases in stomatal conductance, which is concomitant with a decreased leaf blade expansion. Arsenic (As) stress resulted in loss of root hairs and damage to epidermal cells and the cortex with those cells losing their shape and showing signs of shriveling and disintegration. Incomplete differentiation of stele and formation of pith is also reported in As-treated plants. As causes permanent damage to the meristem but root border cells accumulates high levels of As and limiting its movement in to the root.

The major responses elicited by stem to Cd and Hg are disintegration of cells, inhibition, or alteration of stem area (Parmar et al., 2013). ROS attacks various cellular components, including cell wall and membranes resulting into different alterations, which ultimately lead to their disorganization as has also been reported (Janicka-Russak et al., 2008; Ahmad et al., 2011, 2012, 2015; Tran and Popova, 2013). The alteration in stem structure could be correlated with structural changes vis-a-vis dysfunctioning of root system following HM treatments. An insufficient supply of essential nutrients and hormones from the root adversely influences the differentiation of tissues in stem (Aloni, 2010). The stem diameter decrease may be the result of interference with cell division or cell elongation.

3. BIOCHEMICAL RESPONSE

The biochemical basis of HM toxicity is not always clear. Cd, Cu, and Hg affect sulfhydryl groups in proteins and inactivate them. For a radioactive metal such as copper, an excess supply may result in uncontrolled redox reactions, giving rise to the formation of toxic free radicals. This may lead to lipid peroxidation and membrane leakage (Qadir et al., 2004; Ahmad et al., 2011, 2012). Other HMs may inactivate major enzymes by replacing the activating cation. For example, Zn may replace the Mg in ribulose-1,5-bisphosphate carboxylase/oxygenase (Rubisco), reducing the activity of this enzyme and hence the photosynthetic capacity. Like Zn, Cd also affects photosynthesis. Fluorescence measurements indicate that the Calvin cycle is the major process affected and that this subsequently leads to a "downregulation" of photosystem II (PSII) (Nagajyoti et al., 2010). Cd affects the mineral composition even in Cd-resistant species. It reduces the concentration of Mn, Cu, and chlorophyll in the leaves, even at a concentration in solution that has no effect on biomass production (Li et al., 2008).

Various environmental stresses decrease the chlorophyll pigments and have been reported in many plants (Ahmad et al., 2011, 2012; Tomar et al., 2015; Ahanger et al., 2015). Loss in chlorophyll pigment can lead to damage in the photosynthetic apparatus. Net photosynthesis is also sensitive to Cd, which directly affects chlorophyll biosynthesis (Li et al., 2008; Ahmad et al., 2011, 2012; Asgher et al., 2014) and proper development of chloroplasts (Djebali et al., 2010; Jin et al., 2008). *Pisum sativum* and *Zea mays* showed an impaired photosynthesis because of lower PSII activity under metal stress conditions (Hattab et al., 2009; Wang et al., 2009). HM stress shows negative influence on the photosynthetic attributes and the enzymes like Rubisco (Asgher et al., 2014).

The photosynthetic apparatus reveals a sensitive parameter against HM stress. The changes in this mechanism have found to be dose-dependent and vary from one plant to other. Cd can be accumulated at excessive levels in plants, which in turn adversely affects plant growth and different biochemical and physiological processes (John et al., 2009; Irfan et al., 2014; Ahmad et al., 2011, 2012, 2015). Cu plays a central part in physiological processes in plants, including photosynthesis related plastocyanin and membrane structure, respiration, and necessary cofactors for oxidative enzymes. Not all elements are essential for plant growth but some play a key role at low levels because the HMs stimulate the metabolism and hence growth. The chlorophyll analysis proved that there is a significant decrease in pigment contents because of the higher Cu exposure (Chu et al., 2006; Tanyolac et al., 2006; Singh et al., 2007).

A remarkable decrease in the chlorophyll has been found because of elevated HM concentrations that lead to many consequences such as inhibition of enzyme activity, replacement of central metal Mg with HMs, decrease in source of essential metals involved in chlorophyll synthesis such as Fe and Zn (Mostofa and Fujita, 2013; Mostofa et al., 2014), and retardation of metabolic pathways. Certain metals are very important for the normal functioning of the cell; for example, Zn serves an activator of various enzymes and encodes DNA. Likewise, Mn is cofactor and also part of the oxygen-evolving complex in chloroplast. Molybdenum is an axial component of regulation of nitrogen. Nickel is an essential component of some enzymes and has been reported in some beans and tea plants. When these metals are in excess, they inhibit a large number of plant enzymes such as Calvin cycle and chlorophyll biosynthesis and consequently decrease the photosynthetic activity. Despite this, Ni alters the plant water relations and increases the defense mechanism by activating the antioxidative enzymes in plants. Displacement of essential metal ions inhibits the enzyme activity during chlorophyll biosynthesis (Shakya et al., 2008).

HMs have been found to decrease chlorophyll a, chlorophyll b, and total chlorophylls in various plant species (Ahmad et al., 2011, 2012). The inhibit various enzymes that affect the metabolic processes responsible for chlorophyll biosynthesis (Pourraut et al., 2011). The decrease in the chlorophyll content of the cells may be either because of reduced synthesis or of accelerated degradation of light-harvesting pigments. HMs affect both light and dark reactions of photosynthesis, inhibiting the photosynthetic electron transport chain and PSII being the most sensitive target (Gajewska and Skłodowska, 2008). Mercury ions complexed with organic compounds also cause deterioration of 33 KDa polypeptide of PII submembrane. Decreased photosynthetic pigments are also because of the prolonged duration of HM exposure. There could also be a consequence of cadmium-induced change in Mg and Fe content in leaf (Benavides et al., 2005). Significant reduction in chlorophyll content may also be due to strong oxidation of the photochemical apparatus (Mishra and Dubey, 2006) and a reduction in chloroplast size and density. Photosynthesis of woody plants shows a decreasing trend under HMs before the symptoms become visible (Huang et al., 2010). These effects are directly or indirectly related to biochemical and molecular changes in response to HMs (Semane et al., 2010).

Lipids and proteins provide an important indicator of oxidative stress as they rapidly get damaged under environmental stress. This decrease may be due to increased proteolytic activity (reference), various membrane modifications by protein denaturation, or interaction with thiol group proteins (Anjum et al., 2015). HMs are also responsible for inhibiting uptake of Mg^{2+} and K^{+}, decreasing synthesis, and increasing protein degradation and prevention of Rubisco activity (Jiang et al., 2009). Decrease in protein content because of HM–induced oxidation of proteins mediated by H_2O_2 to increase proteolytic activity and have been proposed as an index of oxidative stress (Romer-Puertas et al., 2002). It could be due the interference of metals with sulfur containing amino acids and protein or it is caused either by reduced de novo synthesis or by increased decomposition of protein to amino acids (Anjum et al., 2015). Reports pertaining to reduced protein content because of metal stress are well-documented (Zhang et al., 2005; John et al., 2009; Sethy and Ghosh, 2013; Tiwari et al., 2013).

Presence of HMs caused a decrease in the nitrate content indicating the least accumulation of nitrate in the leaves and might be due to reduced nitrate transport from root to shoot (Chaffei et al., 2004). Cd toxicity seems to reduce the mobilization of endogenous nitrate because it interferes with the metal Mo, which acts as the key enzyme to regulate the nitrate (Khudsar et al., 2001).

Decrease in NR activity might be related to the reduction of nitrate concentration in plant tissue as a result of HM stress. It plays a central part in the maintenance of the whole plant system and any interference with it decreases enzyme activity and hence changes its conformation and stability by binding with −S−S− and −SH groups of protein and −COOH and $-NH_2$ group of enzymes. Cd-induced supraoptimal generation of ROS could interfere with the active state of NR, rendering it inactive. The NR activity exhibited a progressive decline in response to increasing dose of Cd (Irfan et al., 2014). The inhibition in NR activity is suggested to be from a reduced supply of nicotinamide adenine dinucleotide (NADH), disorganization of chloroplast, decreased nitrate supply to the site of enzyme synthesis, water stress created by metals, and direct effects on protein synthesis because it has a strong affinity with the thiol group of enzymes. Gill et al. (2012) reported a differential response of leaf N and NR activity in response to Cd stress garden cress (*Lepidium sativum* L.). NR activity and N content in the leaves were reduced under higher Cd treatments, which resulted in disruption of the coordination between carbon, S, and N metabolism and thus poor plant growth and susceptibility to Cd stress. Unaltered activity of NR complemented the carbon and S metabolism and antioxidant machinery for possible tolerance to Cd toxicity at a lower Cd level.

Nitrate reduction in vivo depends upon the supply of NADH as source of electrons and the inhibition of NR activity may be due to decreased rate of photosynthetic supply, respiration, and NADH availability. Decline in NR activity because of stress might also be due to the reduced uptake of nitrogen or adverse effect of NR enzyme activity itself (Iqbal et al., 2015). Nitrate and NR in response to As and Al have been observed in rice because of reduced assimilation of nitrogen (Sharma and Dubey, 2005).

Sugar acts as a regulatory molecule to control seed and embryo development. Vinod et al. (2012) reported that HM modified the carbohydrate accumulation in wheat leaves and its level was significantly reduced in leaf tissues in response to Cu and Zn stress. Similar decrease in sugar content was also reported by Singh et al. (2007). Manivasagaperumal et al. (2011) reported that sugar and starch content showed a decreasing trend with progressive increase in Zn content in cluster Bean. The observed decline in total sugar with respect to the high level of Zn may be due to its role on the enzymatic reactions related to the cycles of carbohydrate catabolism. Recently, the importance of glucose as a direct and central signal molecule in modulating plant development has been recognized; for example, glucose acts as a key regulator in many vital processes such as germination; root, stem, and shoot growth; photosynthesis; carbon and nitrogen metabolism; flowering; and senescence (Rolland et al., 2002). Additionally, sugars such as glucose have been identified to be involved in these physiological processes through the regulation of transcription factors (Li et al., 2009). On the other hand, as an important component of cell wall polysaccharides, glucose also plays important roles in cell wall formation (Moore et al., 2003). The cell wall is a pivotal site for Cd storage in plants (Lozano-Rodríguez et al., 1997; Carrier et al., 2003), and up to 70–90% of the total Cd is accumulated in the cell walls in *Athyrium yokoscense* (Nishizono et al., 1987). Recently, Shi et al. (2015) showed that exogenous glucose alleviated Cd toxicity is mediated through increasing Cd fixation in the root cell wall and sequestration into the vacuoles.

Proline (Pro) acts as an indicator under HM stress and plays an important role in the osmoprotection, reconstruction of chlorophyll, protection of enzyme from denaturation, tolerance to stress by osmoregulation, and stabilization of protein synthesis and also acts as a source of energy to restore the growth. Proline enhancement is also because of de novo synthesis or decreased degradation of Pro (Liang et al., 2013). Proline accumulation is quite pronounced under HM stress (Ahmad et al., 2011, 2012) and may contribute to osmotic adjustment, enzyme protection, and stabilizing the structure of organelles and macromolecules. Ni inactivates the proteins by suppressing the activity of ribonuclease and protease (Maheshwari and Dubey, 2011). In response to Cd stress, Pro content was found to increase in *B. juncea*, *Triticum aestivum*, and *Vigna radiata* (Dhir et al., 2004), but Pro accumulation decreased on exposure to Cd^{2+} in hydrophytes (*Ceratophyllum*, *Wolffia*, and *Hydrilla*). Similarly, increasing Pro content in response to Cd^{2+} in sunflower was also reported (Zengin and Munzuroglu, 2006). Increased level of Pro was observed in *Brassica oleracea* in the presence of Cd and Hg as reported by Theriappan et al. (2011). In many plant species, elevated levels of Pro were correlated with metal tolerance and metal-induced Pro accumulation seems to be very common in a wide variety of plants. In this case, Pro acts as an osmoprotectant and prevented metal-induced osmotic stress. Although Pro and betaine do not take part directly in metal chelation, they stabilize native state of proteins by regulating their water and hence these osmoprotectants help in maintaining the conformational characteristics and integrity of proteins. Recently studies in plants have demonstrated the capacity of Pro to detoxify ROS. In many plant studies, exogenous application of Pro and glycine betaine (GB) showed enhanced tolerance against metal-induced oxidative stress that was due to the upregulation of antioxidant defense system and glyoxalase systems (Hossain et al., 2010; Kumchai et al., 2013; Irfan et al., 2014). In response to Cd stress in mung beans, inhibition or insufficient activation of antioxidant enzymes and glyoxalase pathway enzymes were observed. Importantly Pro or betaine pretreated seedlings modulated the activities of antioxidative and glyoxalase pathway enzymes and rendered the plant more tolerant to Cd-induced oxidative stress (Hossain et al., 2010). Recently, Irfan et al. (2014) found a good correlation between Pro accumulation and Cd stress tolerance in *B. juncea*. In another study, Ahmad and Gupta (2013) reported that stimulation of Pro accumulation was found to be associated with enhanced antioxidant defense system and regulation in expression of phytochelatin (PC) synthase (PCS), metallothionine-2, glutathione reductase (GR), and glutathione synthetase genes under As stresses that rendered the plants more tolerant to As stress. In vitro grown cabbage (*B. oleracea* var. *capitata*) showed enhanced tolerance to molybdenum (Mo) toxicity when the seedlings were supplemented with exogenous Pro, and it was observed that supplementation of 80 mM Pro to the Mo-treated medium could help 50% seedlings to overcome the toxicity (Kumchai et al., 2013). Addition of Pro in Mo-stressed seedlings upregulated the antioxidant enzyme (superoxide dismutase (SOD), ascorbate peroxidase (APX), glutathione *S*-transferase (GST), and catalase (CAT)) activities that provided metal chelation and enhanced tolerance. One of the important roles of Pro is the enhancement of endogenous glutathione (GSH) level from which PCs are synthesized, where the metal binds to the constitutively expressed enzyme PCS, thereby activating it to catalyze the conversion of GSH to PCs (Zenk, 1996).

4. GENERATION OF ROS IN CELLS

Oxygen is an important constituent for the sustainability of all the organisms. The reduction of O_2 to H_2O provides the energy to the organism, but when it is incompletely reduced, ROS is generated, which is highly reactive and oxidizes biological molecules. Chloroplasts, mitochondria, and peroxisomes are the main sites for ROS production. ROS can be produced from both endogenous and exogenous substances. Potential endogenous sources include mitochondria, cytochrome P450 metabolism, peroxisomes, and inflammatory cell activation.

4.1 ROS GENERATION IN MITOCHONDRIA

Mitochondria are also potential sites for ROS production primarily generating O_2^- and H_2O_2 because of the overreduction of electron transport chain that harbors electrons with sufficient free energy to directly reduce O_2. Direct reduction of O_2 to O_2^- takes place in the flavoprotein region of NADH dehydrogenase segment of the respiratory chain. During mitochondrial electron transport, the oxygen radical is markedly enhanced in the presence of antimycin A, which blocks electron flow after ubiquinone. This results in the accumulation of reduced ubiquinone that may undergo autooxidation, resulting in the production of O_2^- (Mittler, 2002; Gill and Tuteja, 2010). Several observations reveal ubiquinone as a major H_2O_2-generating location of the mitochondrial electron transport chain in vitro and it would appear that O_2^- is a major precursor of H_2O_2.

4.2 ROS GENERATION IN CHLOROPLASTS

Photosynthesizing plants are especially at the risk of oxidative damage, because of their oxygenic conditions and the abundance of the photosensitizers and polyunsaturated fatty acids (PUFAs) in the chloroplast envelope. In photosynthetic electron transport, the photoreduction of O_2 to O_2^- takes place and is called the Mehler reaction. Although photoreduction of O_2 is an important alternative sink for the consumption of excess energy because it is always associated with the generation of toxic ROS. The photoproduction of ROS rate is enhanced under the conditions in which photon intensity is in excess of that required for the CO_2 assimilation. Under the conditions of photon excess the relaxation systems suppresses the photoproduction of ROS in chloroplast, such as photorespiration, the cyclic electron flows through either PSI or PSII, and the downregulation of PSII quantum yield as regulated by the xanthophylls cycle and the photon gradient across thylakoid membrane (Asada, 1999).

4.3 ROS GENERATION IN MICROSOMES AND PEROXISOMES

In plants, O_2^- is known to be produced during nicotinamide adenine dinucleotide phosphate-oxidase (NADPH)-dependent microsomal electron transport. In the presence of NADPH or related reduced compounds, a chain reaction starts that provides the basis for the H_2O_2 producing NADH-oxidase activity of peroxidases (Mittler, 2002). H_2O_2 production is distinguished from that by the phagocyte-type NADPH-oxidase by (1) different K_m values for oxygen, (2) different requirements for NADH and NADPH, and (3) different sensitivities of the two enzymes to inhibitors such as cyanide, azide, and diphenyleneiodonium. Based on these differences, rapid H_2O_2 production in some plant species triggered by pathogen attack and elicitor treatment has been attributed to the NAD(P) H-oxidase activity of apoplastic peroxidases. In addition to its NAD(P)H-oxidase activity, horseradish peroxidase can reduce H_2O_2 to OH^- (Chen et al., 2014). During the infection of barley by *Erysiphe graminis*, a germin-like oxalate oxidase was induced and is another alternative source of ROS (Dumas et al., 1995). Allan and Fluhr (1997) reported that addition of fungal elicitor gives rise to two distinct ROS-producing mechanisms in epidermal cells of tobacco. The two sources of ROS production are (1) NADPH oxidase and/or a xanthine oxidase and (2) peroxidase and/or amine oxidase.

4.4 OTHER SOURCES OF ROS GENERATION IN PLANTS

Other important sources of ROS in plants that have received little attention are detoxification reactions catalyzed by cytochrome P450 in cytoplasm and endoplasmic reticulum. ROS are also generated at plasma membrane level or extracellularly in apoplast in plants. pH-dependent cell wall peroxidases, germin-like oxalate oxidases, and amine oxidases have been proposed as a source of H_2O_2 in apoplast of plant cells. pH-dependent cell wall peroxidases are activated by alkaline pH that, in the presence of a reductant, produces H_2O_2. Alkalization of apoplast upon elicitor recognition precedes the oxidative burst and production of H_2O_2 by a pH-dependent cell wall peroxidase has been proposed as an alternative way of ROS production during biotic stress (Mittler, 2002; Ahmad et al., 2010).

5. OXIDATIVE DAMAGE TO BIOMOLECULES

5.1 OXIDATIVE DAMAGE TO DNA

Plant genome is very stable, but its exposure to biotic and abiotic stress might damage DNA. The higher concentration of ROS can be important mediators of damage to cell structures, nucleic acids, lipids, and proteins (Tuteja et al., 2009). The OH^- is known to react with all components of the DNA molecule, damaging both the purine and pyrimidine bases and also the deoxyribose backbone.

ROS-induced DNA damage involves single or double-stranded DNA breaks, purine, pyrimidine, or deoxyribose modifications, and DNA cross-links. DNA damage can result either in arrest or induction of transcription, induction of signal transduction pathways, replication errors, cell membrane destruction and damage to photosynthetic proteins and genomic instability (Tuteja et al., 2009). Mitochondrial DNA (mtDNA) and cytoplasmic DNA (ctDNA) are close to major sites of ROS production and high rates of modification and equally high repair rates have been reported in mammalian mitochondria. No comparable information is available for plant cells. However, the presence of multiple copies of mtDNA and ctDNA enables the cell to select against negative mutations. 8-hydroxyguanine (8-OH-G) is the most commonly observed modification in DNA and was first reported in human urine. This oxidized DNA product is important because it is both relatively easily formed and is mutagenic. It is a good biomarker of oxidative stress of an organism. It has been noted that 8-OH-G undergoes keto-enol tautomerism and therefore, 8-OH-G is often called 8-oxoguanine or 8-oxo-G; however, 8-oxo-G and 8-OH-G are equivalent.

The major type of DNA damage caused by exposure to ultraviolet B is the formation of dimers between adjacent pyrimidines (i.e., ultraviolet photoproducts consist primarily of cyclobutane pyrimidine dimmers and 6-4PPs dimers). Several mechanisms are available for repairing DNA damage both in the nucleus and in the mitochondria. These include direct reversal of the damage, replacement of the base, and replacement of the whole nucleotide (Tuteja et al., 2009). Specific and general repair mechanisms that can repair DNA base modifications have also been observed. The efficiency of repair mechanisms may be enhanced following exposure to ROS because expression of many DNA repair enzymes is up regulated following oxidative stress.

5.2 LIPID PEROXIDATION

Metal-induced generation of ROS results in the attack of not only DNA in the cell nucleus, but also other cellular components involving PUFA residues of phospholipids, which are extremely sensitive to oxidation (Abd-Allah et al., 2015). PUFAs linoleic acid (18:2) and linolenic acid (18:3) are particularly susceptible to attack to $^1O_2$ and $HO^•$, giving rise to complex mixtures of lipid hydroperoxides (Ahmad et al., 2010). The peroxidation of lipids is considered the most damaging process known to occur in every living organism. Membrane damage is sometimes taken as a single parameter to determine the level of lipid destruction (i.e., lipid peroxidation). Now, it has been recognized that during lipid peroxidation, products are formed from polyunsaturated precursors that include small hydrocarbon fragments such as ketones, malondialdehyde (MDA), etc. and compounds related to them. Some of these compounds react with thiobarbituric acid to form colored products called thiobarbituric acid reactive substances that can be measured by monitoring their absorption at around 530 nm. Extensive PUFA peroxidation decreases the fluidity of the membrane, increases leakiness, and causes secondary damage to membrane proteins. Several aldehydes (e.g., 4-hydroxy-2-nonenal (HNE) and MDA), as well as hydroxyl and keto fatty acids, are formed as a result of PUFA peroxidation. The aldehyde breakdown products can form conjugates with DNA and proteins. Aldehydes formed in the mitochondria may be involved in causing cytoplasmic male sterility in maize because a restorer gene in this species encodes a mitochondrial aldehyde dehydrogenase (Miller et al., 2010).

In plant cells, some of the PUFA oxidation products function as secondary messengers either directly or after enzymatic modification (Mittler, 2002; Miller et al., 2010). The overall process of lipid peroxidation consists of three stages: initiation, propagation, and termination (Gill and Tuteja, 2010).

Once formed, peroxyl radicals can be rearranged via a cyclization reaction to endoperoxides (precursors of MDA), with the final product of the peroxidation process being MDA. Peroxidation of lipids is an autocatalytic process that is terminated for example by the recombination of radicals ($R^{\bullet} + R^{\bullet} \rightarrow$ nonradical product) or depletion of the substrate.

MDA can react with DNA bases guanine (G), adenine (A), and cytosine (C) to form adducts M1G, M1A, and M1C, respectively. M1G adducts were found to range in tissue at levels ranging from below the limit of detection to as high as 1.2 adducts per 106 nucleosides (which corresponds to approximately 6000 adducts per cell). Site-specific experiments confirmed that M1G is mutagenic in *Escherichia coli*, inducing transversions to thiamine (T) and transitions to A (Fink et al., 1997). The mutation frequencies are comparable with those reported for 8-oxo-G in similar systems. M1G is repaired by both bacterial and mammalian nucleotide excision repair pathways and is also repaired in *E. coli* by mismatch repair.

Plants exposed to HM stress exhibited an increase in lipid peroxidation because of the generation of free radicals. This was related to blockage of electron flow in PSII by metal ions that lead to the formation of excited chlorophyll that in turn causes the production of free radicals. Treatment with Cd significantly increased the accumulation of lipid peroxides in *P. sativum* (Metwally et al., 2005), *Arabidopsis* (Guo et al., 2008), *B. juncea* (Ahmad et al., 2011; Asgher et al., 2014), and *Bechmeria nivea* (Liu et al., 2007).

5.3 OXIDATIVE DAMAGE TO PROTEINS

Protein oxidation is defined here as a covalent modification of a protein induced by ROS or byproducts of oxidative stress. Most types of protein oxidations are essentially irreversible, whereas a few involving sulfur-containing amino acids are reversible. Protein oxidation is widespread and often used as a diagnostic marker for oxidative stress.

The amino acids cysteine (Cys) and methionine (Met) are very reactive, especially with $^1O_2$ and $HO^{\bullet}$. ROS causes the oxidation of thiol to disulfide (R1–S–S–R2, cysteine) and is a very important metabolic redox regulation mechanism. Cysteine sulfhydryl groups can be generated from the disulfides by endogenous cellular reducing agents, such as GSH and protein disulfide isomerase, which reverse the formation of non-native disulfide bridges. Chloroplasts, mitochondria, and other cellular compartments of the cell contain large number of potential Trx-regulated proteins that reduces cysteine. The further oxidation of cysteine via cysteine sulfenic acid (R–SOH) to cysteine sulfinic acid ($R{-}SO_2H$) is also enzymatically reversible and probably involved in signaling pathways. Another reversible modification is oxidation of methionine to methionine sulfoxide. In chloroplasts, the small heat shock protein is inactivated by methionine sulfoxidation, but revitalized by reduction catalyzed by the enzyme peptide methionine sulfoxide reductase using Trx as the reductant. Met has been suggested to act as an endogenous antioxidant.

Carbonylation is the most commonly occurring oxidative protein modification and is irreversible. Protein carbonylation may occur because of direct oxidation of amino acid side chains (e.g., Pro and arginine to γ-glutamyl semialdehyde, lysine to amino adipic semialdehyde, threonine to amino ketobutyrate). It can also be induced through the interaction of proteins with oxidative byproducts of other molecules, such as lipid peroxidation derivatives such as HNE and MDA. Protein carbonylation can also be observed in glycation reactions, in which lysine can be nonenzymatically derivatized with oxidized sugars.

In stressed tissues, a significant proportion of all proteins are carbonylated. Plant protein analysis by Nguyen and Donaldson (2005) gave an estimation of 4 nmol carbonyl groups per milligram of protein. Cadmium (Cd)-treated pea plants raised the carbonylation level from 4 to 5.6 nmol/mg protein. The carbonylation level increases from 6.9 to 16.3 nmol/mg of peroxisomal protein as a result of Cd treatment of the intact plant; this can be the result of the higher local ROS concentration in the peroxisomes (Romero-Puertas et al., 2002). It has been reported that protein carbonylation increased during vegetative phase and decreased during reproductive phase in *Arabidopsis* to limit the transfer of oxidatively damaged components to the offspring. Protein carbonylation is higher in mitochondria than in chloroplasts and peroxisomes. This indicates that the mitochondria are more susceptible to oxidative damage and/or the removal of modified proteins is less efficient in the mitochondria. Several carbonylated proteins have been identified in the mitochondrial matrix. In dry *Arabidopsis* seeds, most of the carbonylation was found in the storage proteins, but carbonylation of a number of other proteins increased strongly during seed germination. The oxidized proteins derived from several cellular compartments including the cytosol, chloroplasts, and mitochondria (Ahmad et al., 2010; Gill and Tuteja, 2010).

Oxidative modification of proteins can have multiple adverse effects on the cellular physiology. For example, oxidative modification of enzymes has been shown to inhibit their activities. Enzymes that have metals on or near the active sites are particularly sensitive to metal catalyzed oxidation. Depending on the extent of oxidative damage, such as percentage of molecules that are modified and how long the modification has remained in the system, the affronts can have mild to severe effects on cellular functions.

6. ANTIOXIDANT AND HM STRESS

HM generates the ROS including H_2O_2 leading to oxidative stress in plants (Hossain et al., 2012; Mostofa et al., 2014; Shahid et al., 2014; Asgher et al., 2014). To cope up with free radicals, plants have developed a strong defense mechanism called antioxidant defense systems composed of both enzymatic and nonenzymatic components. Enzymes such as CAT, APX, monodehydroascorbate reductase (MDHAR), dehydroascorbate reductase (DHAR), glutathione peroxidase (GPX), GST, GR, peroxiredoxin, and nonenzymatic compounds such as ascorbate (AsA), GSH, α-tocopherol, and flavonoids are constantly alert to regulating the concentration of ROS, including H_2O_2 (Asada, 1999; Miller et al., 2010; Kapoor et al., 2015). The delicate balance between ROS production and scavenging that allows this duality in function to exist in plants is thought to be organized by a large network of genes that tightly regulates ROS production and scavenging (Petrov and Van Breusegem, 2012). HMs have been found to decrease CAT activities and increase APX in *B. juncea* (Ahmad et al., 2011). Varying responses to Cd-induced oxidative stress are probably related both to the levels of Cd supplied and to the concentration of thiolic groups already present or induced by Cd treatment. Thiols possess strong antioxidative properties and they are consequently able to counteract oxidative stress (Mostofa and Fujita, 2013; Mostofa et al., 2014; Shahid et al., 2014).

The AsA-GSH cycle is essential in HM–induced plants to combat oxidative damage (Mostofa and Fujita, 2013; Mostofa et al., 2014). The increase in APX and GR activities in metal stresses plants maintains AsA and GSH turnover and activation of the H_2O_2 scavenging AsA-GSH cycle. CAT often decreases following exposure to HM concentration (Hossain et al., 2010; Ahmad et al., 2011). Differential responses of SOD showed increases in both leaves and roots. Cd-induced increase in SOD have been found in *B. juncea* (Ahmad et al., 2011) and *Morus alba* (Ahmad et al., 2012). Increase in SOD activity may be attributed to increased production of ROS or increased expression of gene encoding SOD. Reduction of SOD confers to inactivation of enzyme by H_2O_2 produced in different cellular compartments. SOD is a key enzyme in scavenging ROS by converting oxygen radicals to hydrogen peroxide in chloroplast, mitochondria, and cytosol. Increase in SOD activity shows high ROS accumulation that in turn is removed by antioxidative enzymes by inhibiting oxygen radical accumulation. Efficient functioning of SOD blocks O_2 driven cell damage by converting it to H_2O_2, which is then reduced to water and molecular oxygen by action of enzymes (i.e., APX and CAT) working at different locations in the cell (Mittler, 2002; Foyer and Noctor, 2005). Increased APX under Hg stress confers the role in detoxification of H_2O_2 into water using AsA as an electron donor, resulting in the formation of dehydroascorbate. GR is also associated in this pathway. Enhancement in the activities of SOD, APX, and GR under HM stress (Cd, Zn, Cu, Pb, and Fe) may be considered as circumstantial evidence for tolerance mechanism developed by this plant.

Increased GR activity could be explained by transcriptional or translational modifications to keep surplus GR level to protect against HM stress (Romero-Puertas et al., 2002). The susceptibility to oxidative stress is a function of the overbalance between the factors that increase oxidant generation and those substances that exhibit antioxidant capability (Ahmad et al., 2010, 2011). GR catalyses the reduction of glutathione disulfide (GSSG) to GSH at the expanse of NADPH and plays an important role in the maintenance of GSH pool.

Increase in GPX occurs as a result of de novo protein synthesis or the activation enzymes that persists in plant cells to combat the ROS effects. CAT plays an effective role in protecting the mechanisms occurring in mitochondria and peroxisomes by scavenging the free radicals, especially hydrogen peroxide, during photorespiration as well as under stress conditions. Pb has been shown to inhibit the CAT activities. This decrease in CAT led to an increase in lipid peroxidation as a result of decrease in H_2O_2 detoxifications. Decrease in CAT is also from modification and inactivation of CAT enzymes and also because of binding of nonessential elements.

SOD, APX, and CAT are adapted to environmental stress and are called a plant-protective enzymatic system. These enzyme activities, except for CAT, increase under HM stress and have been reported in *A. esculentus*. Despite SOD, APX, and CAT, HMs also affect the activities of other enzymes such as amylase, alcohol dehydrogenase, and NR. Most of the effects are inhibitory.

Enzymatic mechanisms involved in metabolic pathways represent the major defense strategy to combat against ROS production. SOD catalyses the dismutation of superoxide molecules into H_2O_2 and O_2. H_2O_2 is subsequently converted to water by peroxidases and CAT and in the way the enzyme activity is balanced. Extensive studies have been carried out on ROS activities of enzymes regarding the defense system on plants against HM stress.

Among the water-soluble antioxidants found in plants, AsA is one of the most abundant and serves as a major redox buffer, cofactor for enzymes, regulator of cell division and growth, and signaling molecule (Gallie, 2013). Different type of ROS ($^1O_2$, O_2^-, and OH) can be neutralized by AsA or it can act as a cofactor of APX. Under HM stress conditions, plant cells must maintain higher levels of AsA to counter the adverse effects of oxidative stress. Differential regulation of AsA levels was observed in response to Cd stress. Decrease in AsA content and AsA redox state observed in response to Cd stress was observed in rapeseed, rice, and *Arabidopsis* (Hsu and Kao, 2007; Smeets et al., 2009; Chao et al., 2010; Hasanuzzaman et al., 2012; Keunen et al., 2013). Liu et al. (2007) showed that AsA contents increased in *B. nivea* plants under low-level Cd stress, whereas the contents decreased under high-level Cd stress. These findings also demonstrated that higher endogenous AsA is essential to detoxify ROS directly by proper functioning of APX. Importantly, exogenous pretreatment with AsA or its precursor can modulate the gene expression related with HM stress tolerance (Zhao et al., 2005; Paradiso et al., 2008; Chao et al., 2010).

GSH is one of the most powerful antioxidant involved in both signal transduction and defense against ROS and methylglyoxal (MG) (Singla-Pareek et al., 2006; Jin et al., 2008; Hossain et al., 2012). ROS-induced oxidation of protein is inhibited by GHS. GSH also indirectly played roles in protecting membranes by maintaining α-tocopherol and zeaxanthin in reduced states. The broad range of functions of GSH in modulating HM stress tolerance is well-documented (Hossain et al., 2012) and as a first-line defense response against Cd stress, plants synthesize GSH and PCs to form Cd-SH complexes and to diminish free Cd ions in the cytoplasm (Guo et al., 2008; Mishra and Dubey, 2006; Khan et al., 2015). Two important enzymes such GST and GPX involved in ROS and endogenous toxic compounds are dependent on endogenous GSH level (Foyer and Noctor, 2005; Anjum et al., 2014). GSH also an played important role in the detoxification of MG, a cytotoxic compound accumulated in large amounts under abiotic stress conditions, including HM stress (Hossain and Fujita, 2009; Hossain et al., 2010, 2012). Along with its oxidized form (GSSG), GSH acts as a redox couple vital for upholding cellular homeostasis and plays key roles in diverse signaling systems in plants. An increased level of GSH was found to detoxify ROS under Cd stress conditions in various crop plants (Qadir et al., 2004; Anjum et al., 2008; Metwally et al., 2003; Ahammed et al., 2013). Sharp increases in GSH level were found to increase in leaves, stems, and root tissues exposed to multiple HM (Cd^{2+}, Pb^{2+}, and Hg^{2+}) stress (Huang et al., 2010). *Brassica juncea* seedlings treated with Cd stress showed higher GSH and PC synthesis (Mohamed et al., 2012). The expression of genes associated with GSH metabolism responds to HM stress (Xiang and Oliver, 1998). However, depletion of the GSH pool has also been attributed to increased utilization of GSH (Pietrini et al., 2003). Decreased GSH levels ultimately resulted in deteriorated ROS metabolism; defenses against Cd have been found in sensitive genotypes of pea (Metwally et al., 2005). In-depth studies on Cd hyperaccumulating plants have validated the significance of GSH and its metabolizing enzymes for Cd tolerance (reviewed in Hossain et al., 2012).

Carotenoids play a significant role as scavengers to protect the chloroplast against photo-oxidative damage. Carotenoids are simple unsaturated hydrocarbons based on lycopene and their oxygenated derivatives like xanthophylls. ROS detoxification is an essential mechanism of carotenoid that contributes toward the protection of enzymes under HM stress.

7. TRANSLOCATION OF HMS

Micronutrients at optimum concentrations act as the sole energy regulators for the plant body. However, the processes associated with the movement of metals from root to shoot should be in a regulated manner. It involves the movement of toxic metals into the root epidermal cells from the soil solution where from ions move radially either through the root symplast and/or apoplast to the root stele where xylem loading occurs. These metal ions, including the solutes moving along with it, must cross the endodermis, whereby the Casparian strip effectively blocks the further radial apoplastic movement of ions and other solutes. At this point, the metal ions cross the plasma membrane and enter the root symplasm to reach the stele, allowing the plant to regulate the movement of ions into the xylem and restrict transport of ions to the rest of the plant. One of the important steps is loading of the ions into the xylem and subsequent translocation to the shoots via transpirational pull of water and solutes in the xylem. The process is supposed to be mediated by the regulated efflux of metal ions from xylem parenchyma cells into the lumen of nonliving xylem vessels. The last step is movement of metal ions from the xylem sap into the leaf with subsequent transport across the plasma membrane of leaf cells.

Plant transporters are classified into three main groups—channels, carriers, and pumps—with distinctive characteristics in terms of substrate specificity and energy source used to mediate the transport of different types of organic and inorganic solutes across biological membranes. Channels are described by their extremely high rate of transport across the membrane (Sanders and Bethke, 2000). Carriers, on the other hand, undergo conformational changes in protein structure that facilitate transport across the membrane. They bind to the substrate on one side of the membrane and after the conformational change; the substrate is released on the other side of the membrane.

To protect the plants against HMs, roots secrete exudates into the soil matrix. The main role of root exudates is to chelate metals and thereby stops the uptake of HMs to prevent their uptake inside the cells (Marschner, 1995). For example, Ni-chelating histidine and citrate in root exudates reduce the uptake of Ni from soil. The binding of metal ions such as Cu and Zn in the apoplast also helps to control the metal content of root cells. Cation binding sites present on the root cell wall allows the exchange of metal thus influencing the availability of ions for uptake and diffusion into the apoplast. The cell wall acts as a central hub for immobilization of toxic HM ions by providing pectic sites and histidyl groups as well as extracellular carbohydrates such as callose and mucilage. Thus, different tobacco genotypes with chemically distinct root cell wall surfaces have different sensitivities to Mn toxicity. The data suggest that the chemical properties of the cell wall may alter and modify the plant metal uptake and hence metal tolerance. However, the role of the cell wall in metal tolerance is not completely understood. The cell wall is in direct contact with metal ions in the soil, but only a limited number of absorption sites are available, suggesting the cell wall has only a minor impact on metal tolerance. However, *Silene vulgaris* ssp. humilis is a HM–tolerant plant that accumulates different HMs by binding them to proteins or silicates in the epidermal cell walls (Bringezu et al., 1997).

Metal ions absorbed by the roots are loaded into the xylem and translocated to the shoots as complexes with various chelators. Organic acids, especially citrate, are the major chelators for Fe and Ni in the xylem. Ni may also be chelated by histidine and translocated and the methionine derivative nicotianamine (NA) is involved in the transport of Cu. Several types of transporter proteins are involved in the root-to-shoot transport of metals. Whereas metal ions are translocated from source to sink tissue via phloem, NA is involved in the long-distance transport of metals inside the xylem and phloem, but other chelators are required for loading. High-molecular-weight compounds that chelate Ni, Co, and Fe are found in the phloem. Furthermore, transporters help to bind the complexes with the organic compounds and are transported as needed for a wide range of processes. Also, turgor pressure is accomplished by balancing the internal and external concentrations of inorganic and organic ions (Sanders and Bethke, 2000).

HM transporter proteins are classified as zinc-regulated transporter, iron-regulated transporter–like protein ZIP family, ATP-binding cassette (ABC) transporters, the P-type metal ATPases, the natural resistance-associated macrophage protein family, ABC transporters of the mitochondria, cation diffusion facilitator family of proteins, copper transporter family proteins, and many others (Kramer et al., 2007). Plant cell walls are continuous and uninterrupted and act as a cation exchanger, holding large numbers of metals, and providing for some exclusion. Proper homeostasis in barley (*Hordeum vulgare* L.) leaves with the increase in cellular Zn with increasing exposure to external Zn was fully accounted by an increase in vacuolar Zn with the cytoplasm. To locate the position of Cd and the potential Cd-binding peptides, protoplasts and vacuoles were isolated from leaves of Cd-exposed tobacco (*Nicotiana rustica* var Pavonii) seedlings and purified vacuoles contained almost all of the Cd-binding peptides and Cd found in protoplasts. Computer simulation studies with Zn and Cd showed that the vacuole is the probable site of HM sequestration and detoxification in *Nicotiana tabacum*. However, many plant HM transporters remain to be identified at the molecular level; the transport function, specificity, and molecular location of most of these proteins in plants remain unknown.

Mechanism of uptake and transportation helps to alleviate the HM toxicity in plants. Plants accumulate HM ions either by selective uptake of ions through roots or by diffusion of elements in the soil (Peralta-Videa et al., 2009). Concentration in the cortex is then moved apoplastically by concentration gradient and gets accumulated in the cell wall. Two types of uptake systems (i.e., active and passive transport) have been reported for Cd in roots of spruce and soybean.

Transport process provides a central role for the mechanism of detoxification and tolerance. With the use of latest genetic and molecular techniques, a wide range of gene families have been found in transition metal transport. They are ATPase, natural resistance-associated macrophage protein family, cation diffusion facilitator family, ZIP family, and cation antiporters. These metal transporters maintain the internal metal concentration within the physiological limits. Resistance to Zn associated with ion transport was found to enhance expressions of the Zn transporter in *Arabidopsis*.

ABC transporter AtPDR8 have shown to mediate the extrusion of Cd out of selective permeable membrane of root epidermal cells. Detoxification is also accomplished through members of the ZIP family. They are plasma proteins induced in roots and shoots of *Arabidopsis* under Zn-limiting uptake being potentially implicated in Cd root to shoot transport (Kramer et al., 2006).

8. SIGNALING IN PLANTS UNDER HM STRESS

As with all other organisms, plants have also evolved different mechanisms to maintain physiological concentrations of essential metal ions and to reduce contact to nonessential HMs. Some mechanisms are required for homeostasis and are ubiquitous, whereas other mechanisms target individual metal ion for the exclusion of particular metals from the intracellular environment. When these two mechanisms are exhausted, then the plants activate oxidative stress defense mechanisms and synthesis of stress-related proteins and signaling molecules such as heat shock proteins, hormones, and ROS and finally the transcriptional activation of specific metal-responsive genes to work against the stress (Maksymiec, 2007). However, different signaling pathways may be used to respond different HMs. The signal pathways comprise the calcium calmodulin system, hormones, ROS signaling, and the mitogen-activated protein kinase (MAPK) phosphorylation cascade which congregate by activating the previously mentioned stress-related genes. Shao et al. (2008, 2009) and Thapa et al. (2012) have reported that the HM stress causes calcium level changes, MAPK cascade, and transcriptional modulation of the stress-responsive genes.

8.1 THE CALCIUM-CALMODULIN SYSTEM

Calcium is an essential macronutrient taken by the plants through roots and is delivered to the shoots via xylem (Tuteja et al., 2009; Ahmad et al., 2012). Calcium ions regulate a range of activities within the cell such as cell division and elongation, cytoplasmic streaming, photomorphogenesis, and plant defense against environmental stresses (Tuteja et al., 2009; Kader and Lindberg, 2010). They function as the central node in overall signaling web and have a promising role in stress tolerance (Tuteja and Sopory, 2008). Surplus HM alters the stability of Ca channels by increasing the calcium flux into the cell. Intracellular Ca is a secondary messenger that interacts with calmodulin to transmit the signal and regulate the downstream flux genes involved in HM transport, metabolism, and tolerance (Yang and Poovaiah, 2003; Choong et al., 2014). In plants exposed to Cd, higher intracellular Ca levels have been observed, inducing adaptive mechanisms that alleviate the toxic effects of the HM (Yang and Poovaiah, 2003). Calmodulin, an important Ca^{2+}-binding protein, is a small acidic protein and is responsible for the regulation of intracellular Ca^{2+} levels. Increased Ca^{2+} concentration activates calmodulin that then induces specific kinases. Calmodulin is a very important calcium-binding protein in Ca^{2+} signaling and has been found to be involved in biotic and abiotic stresses (Tuteja and Sopory, 2008). In the response of other HM toxicity such as Ni and Pb, the Ca-calmodulin system is also involved. Transgenic tobacco plant expressing NtCBP4 (*Nicotiana tobacum* calmodulin-binding protein) can tolerate higher levels of Ni^{2+} but are sensitive to Pb^{2+}, showing the omission of Ni^{2+} and the accumulation of more Pb^{2+} than wild-type plants.

8.2 CONNECTION WITH HORMONE SIGNAL PATHWAYS

In many physiological and developmental processes, plant hormones play a vital role in the adaptation to abiotic stress, as shown by the regulation of hormone synthesis in the presence of HMs (Ahmad et al., 2011). For instance, an increase in ethylene biosynthesis in response to Cd stress has been observed in mustard (Masood et al., 2012), pea (Rodriguez-Serrano et al., 2009), and in soybean (Chmielowska-Bak et al., 2013). Arteca and Arteca (2007) also reported that, except for Zn, both Cu and Cd were found to elicit the greatest amount of ethylene produced by *Arabidopsis* plants. Maksymiec and Krupa (2006) have reported that ethylene may be involved in the inhibitory action of copper on roots and leaves of dicotyledonous plants. Some researchers have reported that the plants exposed to Cd, Cu, Fe, and Zn stresses produce the higher levels of ethylene, but the copper may not have the same effect. In *Phaseolus coccineus*, copper and cadmium induce the accumulation of jasmonic acid (Maksymiec et al., 2005; Maksymiec, 2011), and the copper has shown this effect on rice (Rakwal et al., 1996) and *Arabidopsis thaliana* (Maksymiec et al., 2005) as well. However, decrease in jasmonic acid content in Cd-stressed roots of soybean leads to an increase in abscisic acid (ABA) and metabolites at different times of Cd stress, suggesting their involvement in Cd response (Perez Chaca et al., 2014) and showing that in some experimental conditions jasmonate accumulation after HM stress may not occur. Under Cd stress, pea seedlings showed increase in salicylic acid (SA) and its exogenous ability helps to protect leaves from lipid peroxidation caused by Cd toxicity (Popova et al., 2009). They suggested that high endogenous levels of SA after treatment with Cd may act directly as an antioxidant to scavenge the ROS and/or indirectly modulate redox balance through activation of antioxidant responses. Zhang and Chen (2011) also showed that exogenously applied SA prevented the Cd-induced photochemical efficiency decrease and mitigated Cd toxicity in *Arabidopsis*. Additionally, SA pretreatment could alleviate Cd-induced ROS overproduction. Very recently, Mostofa and Fujita (2013) showed Cu stress leads to severe oxidative stress in rice seedlings, as indicated by chlorosis, necrosis, and rolling of leaves and higher levels of oxidative parameters such as level of hydrogen peroxide, lipid peroxidation, and lipoxygenase activity. Importantly, SA pretreatment restrained the ROS and MG detoxification systems and rendered the plants more tolerant to Cd-induced oxidative stress tolerance. Yan et al. (2013) studied the effects of methyl jasmonate (MeJA) on *Capsicum frutescens* seedlings exposed to Cd stress. Cd stress caused an impaired growth and increased lipid peroxidation and decreased chlorophyll b, whereas the negative effects of Cd stress were significantly mitigated by MeJA. Exogenous application of MeJA increases the level of endogenous MeJA and the activities of APX and CAT at low MeJA concentrations. These results suggested that low exogenous MeJA concentrations exhibited protective effects on the growth and physiology of *Capsicum frutescens* seedlings under Cd stress. Taken together, all this evidence could explain to some extent the protective role of SA on photochemical activity of chloroplast membranes and photosynthetic carboxylation reactions in Cd-stressed pea plants. Recently, Chen et al. (2014) studied the roles of exogenous MeJA in Cd stress tolerance in the leave of *Kandelia obovata* seedlings. Cd stress lead to a decrease in chlorophyll content and increase in lipid peroxidation. Exogenous MeJA alleviates oxidative stress by increasing the activities of CAT and APX and also modulated the expression of type-2 metallothionein gene (*KoMT2*) expression in the leaves of *K. obovata* seedlings exposed to Cd stress. Methyl jasmonate also inhibited the uptake of Cd to the aboveground part (leaves) of the seedlings, which helped to reduce the direct damages caused by Cd to the photosynthetic organ of the plant. The reduced uptake of Cd might be a result of stomatal closure and decreased transpiration by exogenous MeJA.

8.3 ROLE OF ROS

One of the major consequences of the HM accumulation is the production of ROS as stated previously, leading to oxidative stress and the effect of HMs on oxidative processes is the base for other connections with signaling response (Tuteja and Sopory, 2008). HMs such as Cd can produce ROS via Fenton and Haber–Weiss reactions and indirectly inhibit the antioxidant enzymes (Romero-Puertas et al., 2007). Hydrogen peroxide (H_2O_2) acts as a signaling molecule in response to HMs and other stresses (Dat et al., 2000). In *A. thaliana*, Cu and Cd treatment increases the level of H_2O_2 (Maksymiec and Krupa, 2006). Lin et al. (2005) have reported that Cu can act through changes in H_2O_2-dependent peroxidase activity by stiffening the cell wall from the formation of cross-linking among the cell wall polymers. As per reports, Cd also enhances H_2O_2 accumulation (Maksymiec and Krupa, 2006). Thus, H_2O_2 accumulation changes the redox status of the cell (Foyer and Noctor, 2005) and induces the production of antioxidants and the activation of antioxidant mechanism. H_2O_2 priming was also found to increase the plant tolerance to HM stress. Hu et al. (2009) showed H_2O_2 pretreatment induced Cd tolerance in rice seedlings. Plant growth (length and biomass of shoot and root) was significantly repressed by Cd exposure. However, pretreatment with 100 μM H_2O_2 for 1 day mitigated Cd stress by inducing the antioxidant enzymes (SOD, CAT, GPX, APX, and GST) and elevated the contents of GSH and AsA. As a result, the H_2O_2 and MDA contents decreased in plants and seedling growth was less inhibited. In contrast, H_2O_2 pretreatment decreased Cd concentration in shoots, thus lowering the ratio of Cd concentration in shoots and roots, indicating that H_2O_2 may affect Cd distribution in rice seedlings. The improved Cd tolerance is partly from an enhanced antioxidant system that efficiently prevents the accumulation of H_2O_2 during Cd stress. Increased Cd sequestration in rice roots may contribute to the decline of Cd translocation. Xu et al. (2010) showed H_2O_2-induced AsA and GSH metabolism in inducing Al-induced oxidative stress tolerance in wheat seedlings. Al stress increased the O_2^- and H_2O_2 levels leading to more predominant lipid peroxidation, programmed cell death, and inhibited root elongation in both Al-tolerant and sensitive genotypes. Al-stress increased the activities of SOD, POD, CAT, MDHAR, DHAR, GR, GPX, AsA, and GSH content and their redox state. However, Al-stressed seedlings pretreated with H_2O_2 showed higher SOD, POD, CAT, MDHAR, DHAR, GR, and GPX activities and AsA and GSH content and their redox state than nontreated Al-stressed seedlings. Importantly, antioxidant capacity was more enhanced in the Al-sensitive genotype than in the tolerant one. Therefore, H_2O_2 pretreatment makes the plant more tolerant to Al-induced oxidative stress by inducing AsA and GSH levels and their metabolizing enzymes. To support these findings, Bai et al. (2011) studied the effect of H_2O_2 pretreatment on Cd tolerance and translocation by using two rice cultivars (N07-6 and N07-63) differing in Cd tolerance. Imposition of Cd stress significantly inhibited rice growth, enhanced the production of GSH, non-protein thiols (NPT), PCs, and MDA, and increased the activity of GST. The H_2O_2 pretreatment alleviated Cd toxicity by further increasing the GSH, NPT, and PCs contents as well as the GST activity in root tissues. These findings indicated that endogenous H_2O_2 act as a stress signal and regulated the HM stress tolerance by modulating multiple stress signaling pathways and genes.

8.4 THE MAPK CASCADE

When plants come across with excessive HMs, the gene expression pattern changes. HM responsive changes to Cd and Cu have shown to code for signal transduction components in rice, maize, and *Arabidopsis* mitogen-activated protein kinase kinase kinase (MAPKKK), transcription factors, stress-induced proteins, proteins participating in protein folding, and sulfur and GSH metabolism (Yeh et al., 2007; Liu et al., 2010; Wang et al., 2010). The MAPK class plays a remarkable role in plant signaling of a variety of abiotic and biotic stresses and is an essential step in the establishment of resistance to pathogens (Pitzschke et al., 2009). This pathway is also used in hormone signal transduction and in response to developmental stimuli (Jonak et al., 2002). The MAPK cascade consists of three kinases consecutively activated by phosphorylation: MAPKKK, MAPKK, and MAPK. Last, in this cascade of phosphorylation, the MAPKs phosphorylate different substrates in different cellular compartments including transcription factors in the nucleus. The MAPKKK is a Ser/Thr protein kinase that phosphorylates and thereby activates the MAPKKs. These MAPKKs are dual-specific kinases that phosphorylate MAPKs on a Thr and Tyr residue. And the dual phosphorylation of MAPKs renders the enzyme active and MAPKs are predirected Ser/Thr kinases phosphorylating numerous substrates in various cellular compartments. In this way, this pathway allows the transduction of the information to downstream targets. In alfalfa (*Medicago sativa*) seedlings, four isoforms of MAPK were activated when exposed to Cu or Cd stress (Jonak et al., 2004) and the MAPK gene is also activated by Cd treatments in rice (Yeh et al., 2007). Hence, recognition of these MAPKs and their expression kinetics led researchers to propose the distinct metal-specific signaling pathways exist in plants (Jonak et al., 2004). Finally, all of these pathways congregate in the regulation of transcription factors that activate genes required for stress adaptation, mainly in the perspective of HMs meaning genes for the activation of metal transporters and the biosynthesis of chelating compounds.

9. CONCLUSIONS AND FUTURE PERSPECTIVES

Global detrimental effect on health because of the HM exposure is a matter of serious concern. Nowadays, researchers are much more concerned in developing new technologies at low-cost and ecofriendly land retrieval techniques. Understanding the mechanisms that help plants to cope with

HM stress would help in making new tools that are used in phytoremediation. Hence, in response to HM stress, molecular and cellular adaptation of plant cells seems obligatory to improve HM plant tolerance, which will reduce the chances of HM entrance into the food chain. Being a multigenic trait, recent findings have revealed besides other mechanisms plants protect themselves from HM toxicity through an elevated level of enzymatic and nonenzymatic components of antioxidant system. In our chapter, it is clear that the HMs irrespective of their redox-related mode of action are also efficient in disturbing the equilibrium of antioxidants in cells and slanting the balance that induces the ROS reacting with the functioning of cellular macromolecules and organelles. Consequently, functional diversity and molecular adaptability of PCs and MTs are becoming fascinating regarding the HM detoxification and sustaining cellular ion balance. They act as cellular homeostatic agents and interact directly or indirectly with antioxidant defense system and get tangled in translocating and distributing excessive metal ion between root and shoot in a time- or tissue-specific manner. GSH also plays a vital role in HM uptake, transport, sequestration, and formation of specific metal-binding ligands such as PCs. MAPK cascades transduce environmental and developmental signals into adaptive and programmed responses. Physiological and developmental processes including stress and hormonal responses and innate immunity are regulated by MAPK cascades. More research is needed to identify substrates of MAPKs and cross-talk with other signaling molecules.

REFERENCES

Abd-Allah, E.F., Abeer, H., Alqarawi, A.A., Alwathnani, H.A., 2015. Alleviation of adverse impact of cadmium stress in sunflower (*Helianthus annuus* L.) by arbuscular mycorrhizal fungi. Pak. J. Bot. 47 (2), 785–795.

Ahammed, G.J., Choudhary, S.P., Chen, S., Xia, X., Shi, K., Zhu, Y., Yu, J., 2013. Role of brassinosteriods in alleviation of phenanthrene-cadmium co-contamination-induced photosynthetic inhibition and oxidative stress in tomato. J. Exp. Bot. 64, 199–213.

Ahanger, M.A., Agarwal, R.M., Tomar, N.S., Shrivastava, M., 2015. Potassium and water stress induced changes in oat (*Avena sativa* L. cultivar Kent) with particular reference to antioxidant components. J. Plant Interact. http://dx.doi.org/10.1080/17429145.2015.1056260.

Ahmad, M.A., Gupta, M., 2013. Exposure of *Brassica juncea* (L.) to arsenic species in hydroponic medium: comparative analysis in accumulation and iochemical and transcriptional alterations. Environ. Sci. Pollut. Res. Int. 20 (11), 8141–8150.

Ahmad, P., Jaleel, C.A., Salem, M.A., Nabi, G., Sharma, S., 2010. Roles of enzymatic and nonenzymatic antioxidants in plants during abiotic stress. Crit. Rev. Biotechnol. 30 (3), 161–175.

Ahmad, P., Ozturk, M., Gucel, S., 2012. Oxidative damage and antioxidants induced by heavy metal stress in two cultivars of mustard (*Brassica juncea* L.) plants. Fresenius Environ. Bull. 21 (10), 2953–2961.

Ahmad, P., Sarwat, M., Bhat, N.A., Wani, M.R., Kazi, A.G., Tran, L.S., January 28, 2015. Alleviation of cadmium toxicity in *Brassica juncea* L. (Czern. & Coss.) by calcium application involves various physiological and biochemical strategies. PLoS One 10 (1). http://dx.doi.org/10.1371/journal.pone.0114571.

Ahmad, P., Nabi, G., Ashraf, M., 2011. Cadmium-induced oxidative damage in mustard [*Brassica juncea* (L.) Czern. & Coss.] plants can be alleviated by salicylic acid. S. Afr. J. Bot. 77, 36–44.

Allan, A.C., Fluhr, R., 1997. Two distinct sources of elicited reactive oxygen species in tobacco epidermal cells. Plant Cell 9, 1559–1572.

Aloni, R., 2010. The induction of vascular tissues by auxin. In: Davies, P.J. (Ed.), Plant Hormones: Biosynthesis, Signal Transduction, Action! Springer, pp. 485–518.

Anjum, N.A., Sofo, A., Scopa, A., Roychoudhury, A., Gill, S.S., Iqbal, M., Lukatkin, A.S., Pereira, E., Duarte, A.C., Ahmad, I., 2015. Environ. Sci. Pollut. Res. 22, 4099–4121.

Anjum, N.A., Umar, S., Ahmad, A., Iqbal, M., Khan, N.A., 2008. Ontogenic variation in response of *Brassica campestris* L. to cadmium toxicity. J. Plant Interact. 3, 189–198.

Anjum, N.A., Umar, S., Iqbal, M., 2014. Assessment of cadmium accumulation, toxicity, and tolerance in Brassicaceae and Fabaceae plants-implications for phytoremediation. Environ. Sci. Pollut. Res. 21, 10286–10293.

Arteca, R.N., Arteca, J.M., 2007. Heavy-metal-induced ethylene production in *Arabidopsis thaliana*. J. Plant Physiol. 164, 1480–1488.

Asada, K., 1999. The water-water cycle in chloroplasts: scavenging of active oxygens and dissipation of excess photons. Ann. Rev. Plant Physiol. Plant Mol. Biol. 50, 601–639.

Asgher, M., Khan, N.A., Khan, M.I.R., Fatma, M., Masood, A., 2014. Ethylene production is associated with alleviation of cadmium-induced oxidative stress by sulfur in mustard types differing in ethylene sensitivity. Ecotoxicol. Environ. Saf. 106, 54–61.

Bai, X.J., Liu, L.J., Zhang, C.H., Ge, Y., Cheng, W., 2011. Effect of H_2O_2 pretreatment on Cd tolerance of different rice cultivars. Rice Sci. 18, 29–35.

Benavides, M.P., Gallego, S.M., Tomaro, M.L., 2005. Cadmium toxicity in plants. Braz. J. Plant Physiol. 17, 21–34.

Bringezu, S., Fischer-Kowalski, M., Klein, R., Palm, V., 1997. Regional and National Material Flow Accounting: From Paradigm to Practice of Sustainability. Leiden.

Carrier, P., Baryla, A., Havaux, M., 2003. Cadmium distribution and microlocalization in oilseed rape (*Brassica napus*) after long-term growth on cadmium-contaminated soil. Planta 216, 939–950.

Chaffei, C., Pageau, K., Suzuki, A., Gouia, H., Ghorbel, M.H., Masclaux-Daubresse, C., 2004. Cadmium toxicity induced changes in nitrogen management in *Lycopersicum esculentum* leading to a metabolic safeguard through an amino acid storage strategy. Plant Cell Physiol. 45, 1681–1693.

Chao, Y.Y., Hong, C.Y., Kao, C.H., 2010. The decline in ascorbic acid is associated with cadmium toxicity of rice seedlings. Plant Physiol. Biochem. 48, 374–381.

Chen, J., Yan, Z., Li, X., 2014. Effect of methyl jasmonate on cadmium uptake and antioxidative capacity in *Kandelia obovata* seedlings under cadmium stress. Ecotoxicol. Environ. Saf. 104, 349–356.

Chmielowska-Bak, J., Lefevre, I., Lutts, S., Deckert, J., 2013. Short term signaling responses in roots of young soybean seedlings exposed to cadmium stress. J. Plant Physiol. 170, 1585–1594.

Choong, G., Liu, Y., Templeton, A.M., 2014. Interplay of calcium and cadmium in mediating cadmium toxicity. Chem. Biol. Interact. 211, 54–65.

Chu, L., Liu, D.Y., Wang, Y.B., Ding, J.H., Wang, L.L., 2006. Separate and combined effects of Cu and Cd on seedling growth and active oxygen metabolism system of *Trifolium repens* L. Front. Biosci. 11, 2861–2867.

Dat, J., Vandenabeele, S., Vranova, E., Van Montagu, M., Inzé, D., Van Breusegem, F., 2000. Dual action of the active oxygen species during plant stress responses. Cell Mol. Life Sci. 57, 779–795.

Dhir, B., Sharmila, P., Saradhi, P.P., 2004. Hydrophytes lack potential to exhibit cadmium stress induced enhancement in lipid peroxidation and accumulation of proline. Aquat. Toxicol. 66, 141–147.

Djebali, W., Hédiji, H., Abbes, Z., Barhoumi, Z., Yaakoubi, H., Zoghlami, L.B., Bi, W.C., 2010. Aspects on growth and anatomy of internodes and leaves of cadmium-treated *Solanum lycopersicum* L. plants. J. Biol. Res. Thessaloniki 13, 75–84.

Dumas, B., Freyssinet, G., Pallett, K.E., 1995. Tissue-specific expression of germin-like oxalate oxidase during development and fungal infection of barley seedlings. Plant Physiol. 107, 1091–1096.

Faizan, S., Kauser, S., Parveen, R., 2011. Varietal differences for cadmium-induced seedling mortality, foliar toxicity symptoms, plant growth, proline and nitrate reductase activity in chickpea (*Cicer arietinum* L.). Biol. Med. 3 (2), 196–206.

Fink, S.P., Reddy, G.R., Marnett, L.J., 1997. Mutagenicity in Escherichia coli of the major DNA adduct derived from the endogenous mutagen malondialdehyde. Proc. Natl. Acad. Sci. USA. 5 94 (16), 8652–8657.

Foyer, C.H., Noctor, G., 2005. Redox homeostasis and antioxidant signaling: a metabolic interface between stress perception and physiological responses. Plant Cell 17, 1866–1875.

Gajewska, E., Skłodowska, M., 2008. Differential biochemical responses of wheat shoots and roots to nickel stress: antioxidative reactions and proline accumulation. Plant Growth Regul. 54, 179–188.

Gall, H.L., Philippe, F., Domon, J.M., Gillet, F., Pelloux, J., Rayon, C., 2015. Plants cell wall metabolism in response to abiotic stress. Plants 4, 112–166.

Gallie, D.R., 2013. The role of L-ascorbic acid recycling in responding to environmental stress and in promoting plant growth. J. Exp. Bot. 64, 433–443.

Ghani, A., 2010. Toxic effects of heavy metals on plant growth and metal accumulation in maize (Zea mays L.). Ind. J. Toxicol. 3, 325–334.

Gill, S.S., Khan, N., Tuteja, N., 2012. Cadmium at high dose perturbs growth, photosynthesis and nitrogen metabolism while at low dose it up regulates sulfur assimilation and antioxidant machinery in garden cress (*Lepidium sativum* L.). Plant Sci. 182, 112–120.

Gill, S.S., Tuteja, N., 2010. Reactive oxygen species and antioxidant machinery in abiotic stress tolerance in crop plants. Plant Physiol. Biochem. 48, 909–930.

Gothberg, A., Greger, M., Holm, K., Bengstsson, B.E., 2004. Influence of nutrient levels on uptake and effects of mercury, cadmium, and lead in water spinach. J. Environ. Qual. 33, 1247–1255.

Guo, J.B., Dai, X., Xu, W.Z., Ma, M., 2008. Over-expressing *GSH1* and *AsPCS1* simultaneously increases the tolerance and accumulation of cadmium and arsenic in *Arabidopsis thaliana*. Chemosphere 72, 1020–1026.

Hasanuzzaman, M., Hossain, M.A., Fujita, M., 2012. Exogenous selenium pretreatment protects rapeseed seedlings from cadmium-induced oxidative stress by upregulating antioxidant defense and methylglyoxal detoxification systems. Biol. Trace Elem. Res. 149, 248–261.

Hattab, S., Dridi, B., Chouba, L., Ben, K.M., Bousetta, H., 2009. Photosynthesis and growth responses of pea *Pisum sativum* L. under heavy metals stress. J. Environ. Sci. 221, 1552–1556.

Hossain, M.A., Fujita, M., 2009. Purification of glyoxalase I from onion bulbs and molecular cloning of its cDNA. Biosci. Biotechnol. Biochem. 73 (9), 2007–2013.

Hossain, M.A., Piyatida, P., Teixeira da Silva, J.A., Fujita, M., 2012. Molecular mechanism of heavy metal toxicity and tolerance in plants: central role of glutathione in detoxification of reactive oxygen species and methylglyoxal and in heavy metal chelation. J. Bot. 2012, 37. Article ID 872875.

Hossain, M.A., Hasanuzzaman, M., Fujita, M., 2010. Up-regulation of antioxidant and glyoxalase systems by exogenous glycinebetaine and proline in mung bean confer tolerance to cadmium stress. Physiol. Mol. Biol. Plants 16, 259–272.

Hsu, Y.T., Kao, C.H., 2007. Cadmium induced oxidative damage in rice leaves reduced by polyamines. Plant Soil 291, 27–37.

Hu, Y., Ge, Y., Zhang, C., Ju, T., Cheng, W., 2009. Cadmium toxicity and translocation in rice seedlings are reduced by hydrogen peroxide pretreatment. Plant Growth Regul. 59, 51–61.

Huang, G.Y., Wang, Y.S., Sun, C.C., Dong, J.D., Sun, Z.X., 2010. The effect of multiple heavy metals on ascorbate, glutathione and related enzymes in two mangrove plant seedlings (*Kandelia candel* and *Bruguiera gymnorrhiza*). Oceanol. Hydrobiol. Stud. 39, 11–25.

Iqbal, N., Umar, S., Khan, N.A., 2015. Nitrogen availability regulates proline and ethylene production and alleviates salinity stress in mustard (*Brassica juncea*). J. Plant Physiol. 178, 84–91.

Irfan, M., Ahmad, A., Hayat, S., 2014. Effect of cadmium on the growth and antioxidant enzymes in two varieties of *Brassica juncea*. Saudi J. Biol. Sci. 21 (2), 125–131.

Janicka-Russak, M., Katarzyna, K., Marek, B., Grazyna, K., 2008. Response of plasma membrane H^+-ATPase to heavy metal stress in *Cucumis sativus* roots. J. Exp. Bot. 59, 3721–3728. http://dx.doi.org/10.1093/jxb/ern219.

Jiang, Y., Cukic, B., Adjeroh, D.A., Skinner, H.D., Lin, J., Shen, Q.J., Jiang, B.H., 2009. An algorithm for identifying novel targets of transcription factor families: application to hypoxia-inducible factor 1 targets. Cancer Inform. 7, 75–89.

Jin, X., Yang, X., Mahmood, Q., Islam, E., Liu, D., Li, H., 2008. Response of antioxidant enzymes, ascorbate and glutathione metabolism towards Cd in hyperaccumulator and nonhyperaccumulator ecotypes of *Sedum Alfredii* H. Environ. Toxicol. 23, 517–529.

John, R., Ahmad, P., Gadgil, K., Sharma, S., 2009. Cadmium and lead-induced changes in lipid peroxidation, antioxidative enzymes and metal accumulation in *Brassica juncea* L. at three different growth stages. Arch. Agron. Soil Sci. 55 (4), 395–405.

Jonak, C., Okresz, L., Bogre, L., Hirt, H., 2002. Complexity, cross talk and integration of plant MAP kinase signalling. Curr. Opin. Plant Biol. 5, 415–424.

Jonak, C., Nakagami, H., Hirt, H., 2004. Heavy metal stress. Activation of distinct mitogen-activated protein kinase pathways by copper and cadmium. Plant Physiol. 136 (2), 3276–3283.

Kader, M.A., Lindberg, S., 2010. Cytosolic calcium and pH signaling in plants under salinity stress. Plant Signal Behav. 5 (3), 233–238.

Kapoor, D., Sharma, R., Handa, N., Kaur, H., Rattan, A., Yadav, P., Gautam, V., Kaur, R., Bhardwaj, R., 2015. Redox homeostasis in plants under abiotic stress: role of electron carriers, energy metabolism mediators and proteinaceous thiols. Front. Environ. Sci. 3, 13.

Kasim, W.A., 2006. Changes induced by copper and cadmium stress in the anatomy and grain yield of Sorghum bicolor (L.) Moench. Int. J. Agri. Biol. 1, 123–128.

Keunen, E., Remans, T., Opdenakker, K., Jozefczak, M., Gielen, H., Guisez, Y., Vangronsveld, J., Cuypers, A., 2013. A mutant of the *Arabidopsis thaliana* LIPOXYGENASE1 gene shows altered signalling and oxidative stress related responses after cadmium exposure. Plant Physiol. Biochem. 63, 272–280.

Khan, M.I.R., Nazir, F., Asgher, M., Per, T.S., Khan, N.A., 2015. Selenium and sulfur influence ethylene formation and alleviate cadmium-induced oxidative stress by improving proline and glutathione production in wheat. J. Plant Physiol. 173, 9–18.

Khudsar, T., Mahmooduzzafar, M., Iqbal, M., 2001. Cadmium induced changes in leaf epidermes, photosynthetic rate and pigment concentrations in *Cajanus cajan*. Biol. Plant. 44, 59–64.

Kramer, E.R., Knott, L., Su, F., Dessaud, E., Krull, C.E., et al., 2006. Cooperation between GDNF/Ret and ephrinA/EphA4 signals for motor-axon pathway selection in the limb. Neuron 50, 35–47.

Kramer, U., Talke, I.N., Hanikenne, M., 2007. Transition metal transport. FEBS Lett. 581, 2263–2272.

Kumchai, J., Huang, J.Z., Lee, C.Y., Chen, F.C., Chin, S.W., 2013. Proline partially overcomes excess molybdenum toxicity in cabbage seedlings grown in vitro. Genet. Mol. Res. 12, 5589–5601.

Li, A., Zhang, Z.J., Wang, X.C., Huang, R.F., 2009. Ethylene response factor terf1 enhances glucose sensitivity in tobacco through activating the expression of sugar-related genes. J. Integr. Plant Biol. 51, 184–193.

Li, M., Zhang, L.J., Tao, L., Li, W., 2008. Ecophysiological responses of *Jussiaea repens* to cadmium exposure. Aquat. Bot. 88, 347–352.

Liang, X., Zhang, L., Natarajan, S.K., Becker, D.F., 2013. Proline mechanisms of stress survival. Antioxid. Redox Signal. 19, 998–1011.

Lin, C.H., Chen, L.M., Liu, Z.H., 2005. Rapid effect of copper on lignin biosynthesis in soybean roots. Plant Sci. 168, 855–861.

Liu, Y., Wang, X., Zeng, G., Qu, D., Gu, J., Zhou, M., Chai, L., 2007. Cadmium-induced oxidative stress and response of the ascorbate–glutathione cycle in *Bechmeria nivea* (L.) Gaud. Chemosphere 69, 99–107.

Liu, X.M., Kim, K.E., Kim, K.C., Nguyen, X.C., Han, H.J., Jung, M.S., 2010. Cadmium activates *Arabidopsis* MPK3 and MPK6 via accumulation of reactive oxygen species. Phytochemistry 71, 614–618.

Lozano-Rodríguez, E., Hernandez, L.E., Bonay, P., Carpena, R.O., 1997. Distribution of Cd in shoot and root tissues of maize and pea plants: physiological disturbances. J. Exp. Bot. 48, 123–128.

Maheshwari, R., Dubey, R.S., 2011. Effect of nickel toxicity on the alteration of phosphate pool and the suppressing activity of phosphorolytic enzymes in germinating seeds and growing seedlings of rice. Int. J. Plant Physiol. Biochem. 3 (3), 50–59.

Maksymiec, W., Wianowska, D., Dawidowicz, A.L., Radkiewicz, S., Mardarowicz, M., Krupa, Z., 2005. The level of jasmonic acid in *Arabidopsis thaliana* and *Phaseolus coccineus* plants under heavy metal stress. J. Plant Physiol. 162, 1338–1346.

Maksymiec, 2007. Signaling responses in plants to heavy metal stress. Acta. Physiol. Plant 29, 177–187.

Maksymiec, W., Krupa, Z., 2006. The effects of short-term exposition to Cd, excess Cu ions and jasmonate on oxidative stress appearing in *Arabidopsis thaliana*. Environ. Exp. Bot. 57, 187–194.

Maksymiec, W., 2011. Effects of jasmonate and some other signalling factors on bean and onion growth during the initial phase of cadmium action. Biol. Plant. 55, 112–118.

Manivasagaperumal, R., Balamurugan, S., Thiyagarajan, G., Sekar, J., 2011. Effect of zinc on germination, seedling growth and biochemical content of cluster bean (*Cyamopsis tetragonoloba* (L.) Taub). Curr. Bot. 2, 11–15.

Marschner, H., 1995. Mineral Nutrition of Higher Plants.

Masood, A., Iqbal, N., Khan, N.A., 2012. Role of ethylene in alleviation of cadmium-induced photosynthetic capacity inhibition by sulphur in mustard. Plant Cell Environ. 35, 524–533.

Metwally, A., Safronova, V.I., Belimov, A.A., Dietz, K.J., 2005. Genotypic variation of the response to cadmium toxicity in *Pisum sativum* L. J. Exp. Bot. 56, 167–178.

Metwally, A., Finkemeier, I., Georgi, M., Dietz, K.J., 2003. Salicylic acid alleviates the cadmium toxicity in barley seedling. Plant Physiol. 132, 272–281.

Miller, G., Suzuki, N., Ciftci-Yilmaz, S., Mittler, R., 2010. Reactive oxygen species homeostasis and signalling during drought and salinity stresses. Plant Cell Environ. 33, 453–467.

Mishra, S., Dubey, R.S., 2006. Inhibition of ribonuclease and protease activities in arsenic-exposed rice seedlings: role of proline as enzyme protectant. J. Plant Physiol. 163, 927–936.

Mittler, R., 2002. Oxidative stress, antioxidants and stress tolerance. Trends Plant Sci. 7 (9), 405–410.

Mohamed, A.A., Castagna, A., Ranieri, A., di Toppi, L.S., 2012. Cadmium tolerance in *Brassica juncea* roots and shoots is affected by antioxidant status and phytochelatin biosynthesis. Plant Physiol. Biochem. 57, 15–22.

Moore, B., Zhou, L., Rolland, F., Hall, Q., Cheng, W.H., Liu, Y.X., Hwang, I., Jones, T., Sheen, J., 2003. Role of the *Arabidopsis* glucose sensor HXK1 in nutrient, light, and hormonal signaling. Science 300, 332–336.

Mostofa, M.G., Fujita, M., 2013. Salicylic acid alleviates copper toxicity in rice (*Oryza sativa* L.) seedlings by up-regulating antioxidative and glyoxalase systems. Ecotoxicology 22, 959–973.

Mostofa, M.G., Seraj, Z.I., Fujita, M., 2014. Exogenous sodium nitroprusside and glutathione alleviate copper toxicity by reducing copper uptake and oxidative damage in rice (*Oryza sativa* L.) seedlings. Protoplasma. http://dx.doi.org/10.1007/s00709-014-0639-7.

Munzuroglu, O., Geckil, H., 2002. Effects of metals on seed germination, root elongation, and coleoptile and hypocotyl growth in *Triticum aestivum* and *Cucumis sativus*. Arch. Environ. Contam. Toxicol. 43, 203–213.

Nagajyoti, P.C., Lee, K.D., Sreekanth, T.V.M., 2010. Heavy metals, occurrence and toxicity for plants: a review. Environ. Chem. Lett. 8, 199–216.

Nazar, R., Iqbal, N., Masood, A., Khan, M.I.R., Syeed, S., Khan, N.A., 2012. Cadmium toxicity in plants and role of mineral nutrients in its alleviation. Am. J. Plant Sci. 3, 1476–1489.

Nguyen, A.T., Donaldson, R.P., 2005. Metal-catalyzed oxidation induces carbonylation of peroxisomal proteins and loss of enzymatic activities. Arch. Biochem. Biophys. 439 (1), 25–31.

Nishizono, H., Ichikawa, H., Suziki, S., Ishii, F., 1987. The role of the root cell wall in the heavy metal tolerance of Athyrium yokoscense. Plant Soil. 101 (1), 15–20.

Panda, S.K., Baluska, F., Matsumoto, H., 2009. Aluminum stress signaling in plants. Plant Signal Behav 4 (7), 592–597.

Paradiso, A., Berardino, R., de Pinto, M.C., Sanita di Toppi, L., Storelli, M.M., De Gara, L., 2008. Increase in ascorbate-glutathione metabolism as local and precocious systemic responses induced by cadmium in durum wheat plants. Plant Cell Physiol. 49, 362–374.

Parmar, P., Kumari, N., Sharma, V., 2013. Structural and functional alterations in photosynthetic apparatus of plants under cadmium stress. Bot. Stud. 54, 45.

Peralta-Videa, J.R., Zhao, L., Lopez-Moreno, M.L., de la Rosa, G., Hong, J., Gardea-Torresdey, J.L., 2009. Nanomaterials and the environment: a review for the biennium. J. Hazard. Mater. 186, 1–15.

Perez Chaca, M.V., Viligocco, A., Reinoso, H., Molina, A., Abdala, G., Zirulnik, F., Pedranzani, H., 2014. Effects of cadmium stress on growth, anatomy and hormone contents in *Glycine max* (L.) Merr. Acta Physiol. Plant. 36, 2815–2826.

Perfus-Barbeoch, L., Leonhardt, N., Vavasseur, A., Forestier, C., 2002. Heavy metal toxicity: cadmium permeates through calcium channels and disturbs the plant water status. Plant J. 32, 539–548.

Petrov, V.D., Van Breusegem, F., 2012. Hydrogen peroxide-a central hub for information flow in plant cells. AoB Plants. pls014. http://dx.doi.org/10.1093/aobpla/pls014.

Pietrini, F., Innelli, M.A., Pasqualini, S., Massacci, A., 2003. interaction of cadmium with glutathione and photosynthesis in developing leaves & chloroplast of *Phragmites australis* (Cav.)Trin. ex. Steudel. Plant Physiol. 133, 829–837.

Pitzschke, A., Schikora, A., Hirt, H., 2009. MAPK cascade signalling networks in plant defence. Curr. Opin. Plant Biol. 12, 1–6.

Popova, L.P., Maslenkova, L.T., Yordanova, R.Y., Ivanova, A.P., Krantev, A.P., Szalai, G., Janda, T., 2009. Exogenous treatment with salicylic acid attenuates cadmium toxicity in pea seedlings. Plant Physiol. Biochem. 47 (3), 224–231.

Pourraut, B., Shahid, M., Dumat, C., Winterton, P., Pinelli, E., 2011. Lead uptake, toxicity and detoxification in plants. Rev. Environ. Contam. Toxicol. 213, 113–136.

Qadir, S., Qureshi, M.I., Javed, S., Abdin, M.Z., 2004. Genotypic variation in phytoremediation potential of *Brassica juncea* cultivars exposed to Cd stress. Plant Sci. 167, 1171–1181.

Qui, Z., Guo, J., Zhu, A., Zhang, L., Zhang, M., 2014. Exogenous jasmonic acid can enhance tolerance of wheat seedlings to salt stress. Ecotoxicol. Environ. Saf. http://dx.doi.org/10.1016/j.ecoenv.2014.03.014.

Qui, Z., Li, J., Zhang, M., Li, Z., 2013. He-Ne laser pretreatment protects wheat seedlings against cadmium-induced oxidative stress. Ecotoxicol. Environ. Saf. 88, 135–141.

Rakwal, R., Tamogami, S., Kodama, O., 1996. Role of jasmonic acid as a signaling molecule in copper chloride-elicited rice phytoalexin production. Biosci. Biotech. Bioch. 60, 1046–1048.

Rascio, N., Vecchia, F.D., La Rocca, N., Barbato, R., Pagliano, C., Raviolo, M., Gonnelli, C., Gabbrielli, R., 2008. Metal accumulation and damage in rice (cv. *Vialone nano*) seedlings exposed to cadmium. Environ. Exp. Bot. 62, 267–278.

Rauser, W.E., Ackerley, C.A., 1987. Localization of cadmium in granules within differentiating and mature root cells; Can. J. Bot. 65, 643–646.

Rodriguez-Serrano, M., Romero-Puertas, M.C., Pazmiño, D.M., Testillano, P.S., Risueño, M.C., del Río, L.A., Sandalio, L.M., 2009. Cellular response of pea plants to cadmium toxicity: cross talk between reactive oxygen species, nitric oxide, and calcium. Plant Physiol. 150 (1), 229–243.

Rolland, F., Moore, B., Sheen, J., 2002. Sugar sensing and signaling in plants. Plant Cell 14, 185–205.

Romero-Puertas, M.C., Corpas, F.J., Rodriguez-Serrano, M., Gomez, M., del Río, L.A., Sandalio, L.M., 2007. Differential expression and regulation of antioxidative enzymes by Cd in pea plants. J. Plant Physiol. 164, 1346–1357.

Romero-Puertas, M.C., Palma, J.M., Gomez, M., del Río, L.A., Sandalio, L.M., 2002. Cadmium causes the oxidative modification of proteins in pea plants. Plant Cell Environ. 25, 677–686.

Rout, G.R., Samantaray, S., Das, P., 2001. Aluminium toxicity in plants: a review. Agronomie 21, 3–21.

Ruley, A.T., Sharma, N.C., Sahi, S.V., 2004. Antioxidant defense in a lead accumulating plant, *Sesbania drummondii*. Plant Physiol. Biochem. 42, 899–906.

Sanders, D., Bethke, P., 2000. Membrane transport. In: Buchanan, B.B., Gruissem, W., Jones, R.L. (Eds.), Biochemistry and Molecular Biology of Plants. American Society of Plant Biologists, Rockville, MD, pp. 110–158.

Schützendübel, A., Schwanz, P., Terchmann, T., Grossk Langeenfeld-Heyger, R., Godbold, D.L., Polle, A., 2001. Cadmium-induced changes in antioxidative systems, hydrogen peroxide content, and differentiation in Scots pine roots. Plant Physiol. 75, 887–898.

Schützendubel, A., Polle, A., 2002. Plant responses to abiotic stresses: heavy metal-induced oxidative stress and protection by mycorrhization. J. Exp. Bot. 53 (372), 1351–1365.

Semane, B., Dupae, J., Cuypers, A., Noben, J.P., Tuomainen, M., Tervahauta, A., Kärenlampi, S., Belleghem, F.V., Smeets, K., Vangronsveld, J., 2010. Leaf proteome responses of *Arabidopsis thaliana* exposed to mild cadmium stress. Plant Physiol. 167, 247–254.

Sethy, S.K., Ghosh, S., 2013. Effect of heavy metals on germination of seeds. J. Nat. Sci. Biol. Med. 4 (2), 272–275.

Shahid, M., Pourrut, B., Dumat, C., Nadeem, M., Aslam, M., Pinelli, E., 2014. Heavy-metal-induced reactive oxygen species: phytotoxicity and physicochemical changes in plants. Rev. Environ. Contam. Toxicol. 232, 1–44.

Shakya, K., Chettri, M.K., Sawidis, T., 2008. Impact of heavy metals (copper, zinc and lead) on the chlorophyll content of some mosses. Arch. Environ. Contam. Toxicol. 54, 412–421.

Shao, H.B., Chu, L.Y., Jaleel, C.A., Manivannan, P., Panneerselvam, R., Shao, M.A., 2009. Understanding water deficit stress-induced changes in the basic metabolism of higher plants-biotechnologically and sustainably improving agriculture and the ecoenvironment in arid regions of the globe. Crit. Rev. Biotechnol. 29 (2), 131–151.

Shao, H.B., Chu, L.Y., Shao, M.A., 2008. Calcium as a versatile plant signal transducer under soil water stress. BioEssays 30 (7), 634–641.

Sharma, P., Dubey, R.S., 2005. Lead toxicity in plants. Braz. J. Plant Physiol. 17 (1), 35–52.

Shi, Y.Z., Zhu, X.F., Wan, J.X., Li, G.X., Zheng, S.J., 2015. Glucose alleviates cadmium toxicity by increasing cadmium fixation in root cell wall and sequestration into vacuole in *Arabidopsis*. J. Integr. Plant Biol. http://dx.doi.org/10.1111/jipb.12312.

Singh, D., Nath, K., Sharma, Y.K., 2007. Response of wheat seed germination and seedling growth under copper stress. J. Environ. Biol. 28, 409–414.

Singla-Pareek, S.L., Yadav, S.K., Pareek, A., Reddy, M.K., Sopory, S.K., 2006. Transgenic tobacco overexpressing glyoxalase pathway enzymes grow and set viable seeds in zinc-spiked soils. Plant Physiol. 140, 613–623.

Smeets, K., Opdenakker, K., Remans, T., Van Sanden, S., Van Belleghem, F., Semane, B., Horemans, N., Guisez, Y., Vangronsveld, J., Cuypers, A., 2009. Oxidative stress-related responses at transcriptional and enzymatic levels after exposure to Cd or Cu in a multipollution context. J. Plant Physiol. 166, 1982–1992.

Stoeva, N., Berova, M., Zlatev, Z., 2005. Effect of arsenic on some physiological parameters in bean plants. Biol. Plant. 49 (2), 293–296.

Suzuki, N., 2005. Alleviation by calcium of cadmium-induced root growth inhibition in *Arabidopsis* seedlings. Plant Biotechnol. 22 (1), 19–25.

Tanyolac, D., Ekmekc, Y., Unalan, S., 2006. Changes in photochemical and antioxidant enzyme activities in maize (*Zea mays* L.) leaves exposed to excess copper. Chemosphere 67, 89–98.

Thapa, S.S., Paudyal, I., Khanal, S., Paudel, N., Mansberger, S.L., van Rens, G.H., 2012. Central corneal thickness and intraocular pressure in a Nepalese population: the Bhaktapur Glaucoma Study. J. Glaucoma 21 (7), 481–485.

Theriappan, P., Gupta, A.K., Dhasarathan, P., 2011. Accumulation of proline under salinity and heavy metal stress in cauliflower seedlings. J. Appl. Sci. Environ. Manage. 15, 251–255.

Tiwari, U., Agnihotri, R.K., Swati, S., Sharma, R., 2013. Effect of Lead nitrate induced heavy metal toxicity on some biochemical constituents of wheat (*Triticum aestivum* L.). Res. J. Agric. Sci. 4 (2), 283–285.

Tomar, N.S., Sharma, M., Agarwal, R.M., 2015. Phytochemical analysis of *Jatropha curcas* L. during different seasons and developmental stages and seedling growth of wheat *Triticum aestivum* L.) as affected by extracts/leachates of *Jatropha curcas* L. Physiol. Mol. Biol. Plants 21 (1), 83–92.

Tran, T.A., Popova, L.P., 2013. Functions and toxicity of cadmium in plants: recent advances and future prospects. Turkish J. Bot. 37, 01–13.

Tuteja, N., Ahmad, P., Panda, B.B., Tuteja, R., 2009. Genotoxic stress in plants: shedding light on DNA damage, repair and DNA repair helicases. Mutat. Res. 681, 134–149.

Tuteja, N., Sopory, S.K., 2008. Chemical signaling under abiotic stress environment in plants. Plant Signal Behav. 3 (8), 525–536.

Urano, K., Kurihara, Y., Seki, M., Shinozaki, K., 2010. 'Omics' analyses of regulatory networks in plant abiotic stress responses. Curr. Opin. Plant Biol. 13, 132–138.

Vinod, K., Awasthi, G., Chauhan, P.K., 2012. Cu and Zn tolerance and responses of the biochemical and physiochemical system of wheat. J. Stress Physiol. Biochem. 8 (3), 203–213.

Wang, H., Zhao, S.C., Liu, R.L., Zhou, W., Jin, J.Y., 2009. Changes of photosynthetic activities of maize (*Zea mays* L.) seedlings in response to cadmium stress. Photosynthetica 47 (2), 277–328.

Wang, J., Ding, H., Zhang, A., Ma, F., Cao, J., Jiang, M., 2010. A novel mitogen-activated protein kinase gene in maize (*Zea mays*), ZmMPK3, is involved in response to diverse environmental cues. J. Integr. Plant Biol. 52, 442–452.

Wang, Y., Greger, M., 2004. Clonal differences in mercury tolerance, accumulation, and distribution in willow. J. Environ. Qual. 33 (5), 1779–1785.

Xiang, C., Oliver, D.J., 1998. Glutathione metabolic genes coordinately respond to heavy metals and jasmonic acid in Arabidopsis. Plant Cell. 10, 1539–1550.

Xu, Y., Feng, L., Jeffrey, P.D., Shi, Y.G., Morel, F.M.M., 2008. Structure and metal exchange in the cadmium carbonic anhydrase of marine diatoms. Nature 452, 56–61.

Xu, F.J., Jin, C.W., Liu, W.J., Zhang, Y.S., Lin, X.Y., 2010. Pretreatment with H_2O_2 alleviates aluminum-induced oxidative stress in wheat seedlings. J. Integr. Plant Biol. 54, 44–53.

Yan, Z., Chen, J., Li, X., 2013. Methyl jasmonate as modulator of Cd toxicity in *Capsicum frutescens* var. fasciculatum seedlings. Ecotoxicol. Environ. Saf. 98 (1), 203–209.

Yang, T., Poovaiah, B.W., 2003. Calcium/calmodulin-mediated signal network in plants. Trends Plant Sci. 8 (10), 505–512.

Yeh, C.M., Chien, P.S., Huang, H.J., 2007. Distinct signalling pathways for induction of MAP kinase activities by cadmium and copper in rice roots. J. Exp. Bot. 58 (3), 659–667.

Zengin, F.K., Munzuroglu, O., 2006. Toxic effects of cadmium (Cd^{++}) on metabolism of sunflower (*Helianthus annuus* L.) seedlings. Acta Agric. Scand. Sect. B 56, 224–229.

Zenk, M.H., 1996. Heavy metal detoxification in higher plants – a review. Gene 179, 21–30.

Zhang, H., Jiang, Y., He, Z., Ma, M., 2005. Cadmium accumulation and oxidative burst in garlic (*Allium sativum*). J. Plant Physiol. 162, 977–984.

Zhang, W., Chen, W., 2011. Role of salicylic acid in alleviating photochemical damage and autophagic cell death induction of cadmium stress in *Arabidopsis thaliana*. Photochem. Photobiol. Sci. 10 (6), 947–955.

Zhao, Z., Cai, Y., Zhu, Y., Kneer, R., 2005. Cadmium-induced oxidative stress and protection by L-galactono-1, 4-lactone in winter wheat (*Triticum aestivum* L.). J. Plant Nutr. Soil Sci. 168, 759–763.

CHAPTER

HEAVY METAL STRESS SIGNALING IN PLANTS

25

Smita Kumar[1], **Prabodh K. Trivedi**[2]

Department of Biochemistry, University of Lucknow, Lucknow, Uttar Pradesh, India[1]; *CSIR-National Botanical Research Institute (CSIR-NBRI), Lucknow, Uttar Pradesh, India*[2]

CHAPTER OUTLINE

1. INTRODUCTION

Heavy metal toxicity is a longstanding and increasingly serious threat to the environment. Geogenic sources like leaching and weathering of rocks and anthropogenic activities such as intensive metal mining, smelting, combustion of fossil fuels, use of pesticides, and sewage sludge have burdened the soil with number of heavy metals (Kumar et al., 2015a). These metals not only negatively affect growth physiology of plants but also cause chronic and epidemic human health hazards when they get entry into the food chain (Peralta-Videa et al., 2009; Chakraborti et al., 2015). Heavy metals stress cause biochemical, physiological, and morphological changes in plants. Few heavy metals act as ligands for the catalytic activity of enzymes and actively bind to the sulfhydryl groups of proteins that disrupt the biochemical functioning of the cell (Tripathi et al., 2007; Kumar et al., 2015a). The phytotoxic effect of heavy metals generates reactive oxygen species (ROS) and causes oxidative stress peroxidating membrane lipids, protein oxidation, and damage to the cellular and subcellular organelles (Hayat et al., 2012; Ali et al., 2013; Qadir et al., 2014). Heavy metals are taken up by the plants and accumulate in the edible parts, reducing yield and productivity and thus contaminating the food chain (Zhao et al., 2010). Consequently, there is an urgent need to minimize the accumulation of heavy metals in food crops for food security. For decades, studies have been carried out in this respect to understand the molecular mechanisms underlying the metal uptake, accumulation, and detoxification in plants. Studies suggest that plants have evolved sophisticated signaling mechanisms for adaptation

Plant Metal Interaction. http://dx.doi.org/10.1016/B978-0-12-803158-2.00025-4

to adverse environmental conditions that include stress signals perception, transduction, and transmission (Golldack et al., 2014). Plants sense stress stimuli, which trigger signaling cascades leading to the plant tolerance and adaptation.

Signaling pathways document occurrence of receptors/ion channels, signaling molecules, and transcription factors that trigger cellular responses in plants. Research findings have shown that the inducible effect of the heavy metal stress generates varied cellular responses. Though understanding of signaling pathways and the existing cross-talk among the networks is extremely complex to reveal, several studies have elucidated the stress-responsive networks (Figure 25.1) (Lin and Aarts, 2012; Islam et al., 2015). This chapter sheds light on the intricate nexus of signaling pathways operating in plants and conferring heavy metal stress tolerance and adaptation. Presumably, the component of signaling network during heavy metal stress may include receptors for the perception of the signal and nonprotein messengers such as cyclic nucleotides, calcium, hydrogen ions, and oxylipids for signal transduction. Several enzymes including kinases and phosphatases relay the signals and gene expression of various transcription factors and the synthesis of metal-detoxifying peptides (Rao et al., 2011). Heavy metal stress activates distinct signaling pathways in plants such as calcium-dependent signaling, ROS signaling, hormone signaling, and a constellation of signaling molecules that mediate signal transduction and enhance the expression of stress-responsive genes or transcription factors (Dubey et al., 2014). Here, we have described these signaling pathways/molecules and their cross-talk under heavy metal stress.

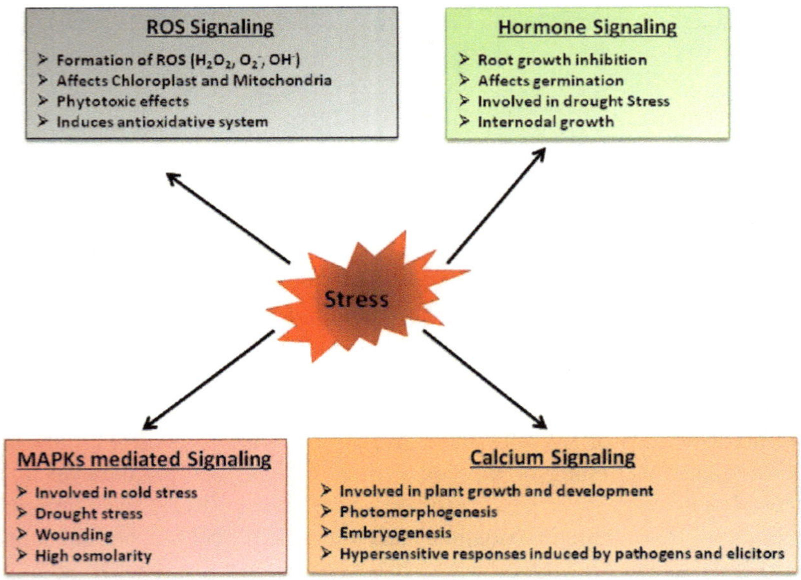

FIGURE 25.1

Signaling pathways involved in response to different stress conditions in plants.

2. CALCIUM-DEPENDENT SIGNALING PATHWAY

Various studies demonstrated that calcium ion (Ca^{2+}) acts as a messenger in the normal physiology of plants as well as in plant responses to different environmental stresses (Tuteja and Mahajan 2007a; Himschoot et al., 2015). Ca^{2+} channels are present on the plasma membrane, tonoplast, and on different organelles of the plant cell (Tuteja, 2009; Edel and Kudla, 2014). It has been observed that the stress stimuli evoke modulation in the concentration of the cytosolic free Ca^{2+} for the complex interaction and signal transduction (Rudd and Franklin-Tong, 2001). Ca^{2+} transmits various internal and external signals and is important for plants' response process. Studies suggest that under stress conditions, the cytosolic Ca^{2+} concentration changes because of the influx of Ca^{2+} from outside the cell, or via release of the Ca^{2+} from the intracellular stores of the cell (Steinhorst and Kudla, 2014).

Biological sensors that sense the transient increase in the Ca^{2+} concentration are required for the biological conversion of a signal mediated by a chemical messenger (Figure 25.2). The Ca^{2+}-sensing mechanisms are evolutionary conserved across the plant species. In plants, an array of diverse and extensive set of Ca^{2+} sensors such as calmodulins (CaMs), calmodulin-like proteins, calcineurin B-like proteins (CBLs), and Ca^{2+}-dependent protein kinases (CDPKs) exist that sense, decode calcium signatures, and convey the changes in the cytosolic Ca^{2+} concentration for the response (Sanders et al., 2002; Luan et al., 2002; Dodd et al., 2010; Conde et al., 2011; Steinhorst and Kudla, 2013). Most of the Ca^{2+} sensors bind Ca^{2+} using a helix-loop-helix motif designated as the "EF hand" (Tuteja and Mahajan 2007b). However, there are a few proteins that bind Ca^{2+} but do not contain EF hand motifs. Some of these proteins include phospholipase D, annexins, calreticulin, and calnexin (Tuteja and Mahajan 2007b). The CDPK regulates plant development and constitutive expression of CDPK gene in rice has been observed to be involved in grain filling (Manimaran et al., 2015). Though studies have demonstrated the regulatory role of Ca^{2+} and Ca^{2+}-binding proteins in different stress conditions including anomalous temperature and oxidative stress (Knight and Knight, 2001), the understanding of the mechanisms by which Ca^{2+} regulates varied cellular responses during heavy metal stress is still limited.

Against the stimuli, the cytoplasmic Ca^{2+} concentration changes and the CDPKs function as a potential sensor and relay the downstream signaling cascade (Figure 25.2) (Asano et al., 2012). CDPKs are multigene family plant sensory proteins that bind Ca^{2+} directly before phosphorylating substrates that are involved in signaling response against stress (Schulz et al., 2013; Hamel et al., 2014). The kinetics and magnitude of Ca^{2+} concentration varies with the signals and determines the specificity of the cellular response. Several studies have been conducted to understand the role of calcium binding–like proteins in plants against abiotic stresses. The functional characterization of CBLs in different plant species such as chickpea (Tripathi et al., 2009), *Glycine max* (Li et al., 2012), *Vitis vinifera* (Tripathi et al., 2009), and tomato (de la Torre F et al., 2013) has shown enhanced tolerance toward abiotic stresses. The manipulation of the Ca^{2+} sensors has been shown to confer tolerance against multiple stresses (Goswami et al., 2015). The major information about the Ca^{2+} fluctuations, Ca^{2+} sensors, relay of downstream signaling, and their definitive targets has been revealed in terms of abiotic stresses such as drought and salinity. For instance, the overexpression of *AtCBL1* enhances tolerance against drought, salt, nutrient deficiency, and temperature (Cheong et al., 2003).

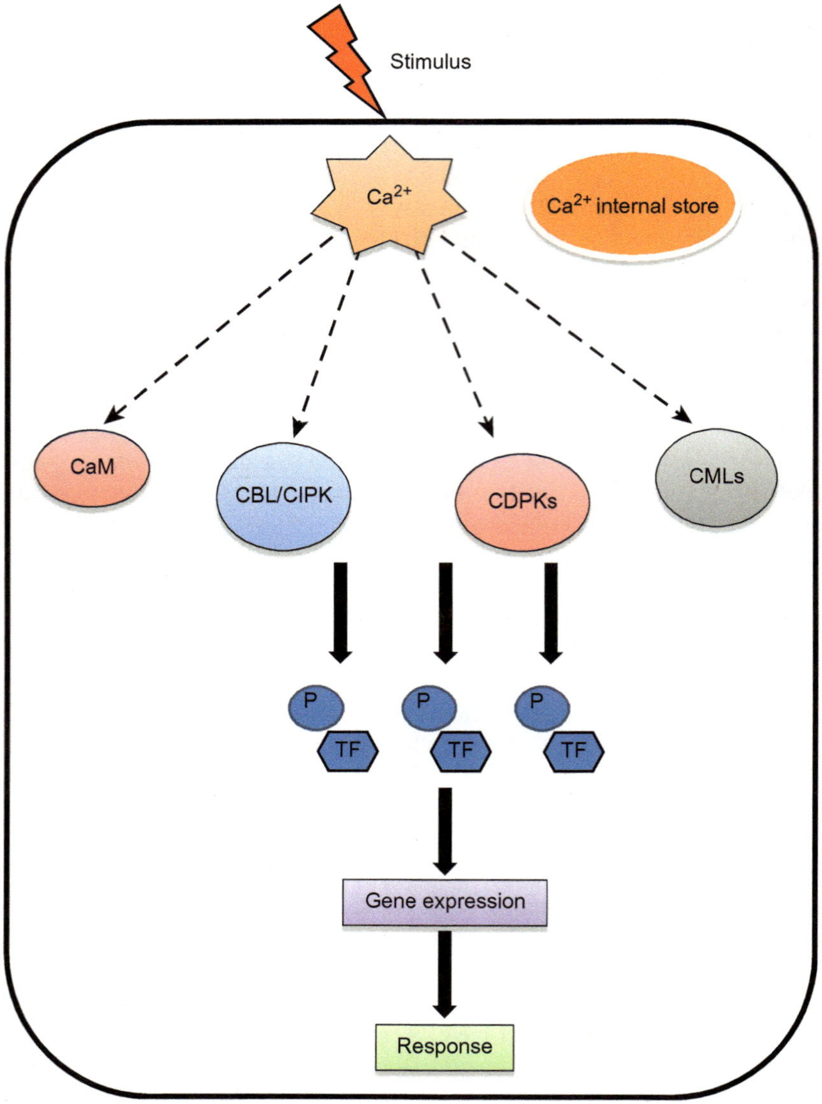

FIGURE 25.2

Diagrammatical representation of calcium signaling in response to stimulus in plants via biological sensors.

Various studies have proposed the role of Ca^{2+} and Ca^{2+}-dependent signaling pathways in response to heavy metal stress. Studies have suggested that the application of exogenous Ca^{2+} helps in the alleviation of heavy metal stress by affecting the physiological and biochemical processes. Moreover, the exogenous Ca^{2+} enhances the activity of antioxidant enzymes such as ascorbate peroxidase, glutathione reductase, and superoxide dismutase (Ahmad et al., 2015). In foxtail millet, Cr stress activates exogenous H_2S synthesis and Ca^{2+} signaling. It has been observed that Ca^{2+} enhances the activation of antioxidant enzymes and provides tolerance against Cr stress (Fang et al., 2014). Transcriptional profiling of the rice roots exposed to long- and short-term Cr stress suggested the involvement of CDPKs. The CDPK-like protein activity increased with the increasing chromium (Cr)(VI) concentration, which reflects the role of Ca^{2+}-dependent signaling pathway in response to heavy metal stress (Huang et al., 2014). Studies on the marine alga, *Ulva compressa*, also suggest that Cu induces the release of intracellular Ca^{2+} that activates the Ca^{2+} channels and generates H_2O_2 in organelles. Furthermore, it has been demonstrated that the cross-talk among the various intracellular signals and the level of Ca^{2+}, NO, and H_2O_2 regulates the gene expression of antioxidant system (Gonzalez et al., 2012). In response to As stress, differential expression of CAMs suggest the possible involvement of Ca^{2+}-dependent signaling in plants (Chakrabarty et al., 2009). It has been reported that Pb^{2+} stress induces the activation of Ca^{2+}-dependent signaling pathway via ROS production and Ca^{2+} accumulation. Study suggested that Ca^{2+} participates in the Pb^{2+}-induced cell death and mitogen-activated protein kinase (MAPK) activities through CDPK pathway by increasing the activity of CDPK-like kinase (Huang and Huang, 2008). All these studies suggest an important role for the regulatory circuit of Ca^{2+}, which is critical for maintaining cellular redox homeostasis and ion balance in response to heavy metal stress. Therefore, the information about the diverse expression pattern of the Ca^{2+}-binding proteins in response to heavy metal stress in plants will be useful for the further investigations of the function and regulation of Ca^{2+}-dependent signaling in plants.

3. MAPK CASCADE

The canonical MAPK are important signaling module involved in many cellular and physiological processes in plants (Hamel et al., 2006; Rao et al., 2010). MAPKs are serine/threonine kinases, and this three-tier phospho-relay signaling module is evolutionary conserved in the plant kingdom (Jonak et al., 2002). These cascades phosphorylate numerous transcription factors such as DREB, bZIP, NAC, and WRKY and thus influence cellular functions involving differentiation, growth, and developmental processes and stress response (Lin and Aarts, 2012). MAPKs have been implicated in signaling the stress stimuli and subsequently enabling the activation of resistance mechanism in plants (Figure 25.3). This signaling module helps in initiating the adaptive downstream signaling and activates plant response to cues of environmental stresses. MAPK cascades comprise MAP kinase kinase kinases, MAP kinase kinases, and MAP kinases (MAPKs/MPKs) (Nakagami et al., 2005). During evolution in the plant kingdom, this cascade has been conserved (Janitza et al., 2012) and shows an unprecedented complexity. The intertwined MAPK module transduces diverse environmental cues and the activated MAPKs lead to a myriad of cellular and physiological responses via activation of transcription factors and cytosolic proteins (Figure 25.3) (Samajova et al., 2013). Transcriptional activation of these protein kinases in rice demonstrates the regulatory role of MAPKs in plant development, defense, and stress signaling (Kim et al., 2003). Genome analysis of various plants suggests that members of this signaling cascade are encoded by multigene families. The *Arabidopsis* genome contains at least 20 MAPKs, 10 MAP kinase kinases, and 60–80 MAP kinase kinase kinases (Chen et al., 2012b); a similar number has been reported in other plants, including rice and maize (Rao et al., 2010).

The role of MAPKs has been studied in various abiotic stresses including heavy metal stress. Differential regulation of MAPKs under cold, heat, drought, and salinity have been observed in

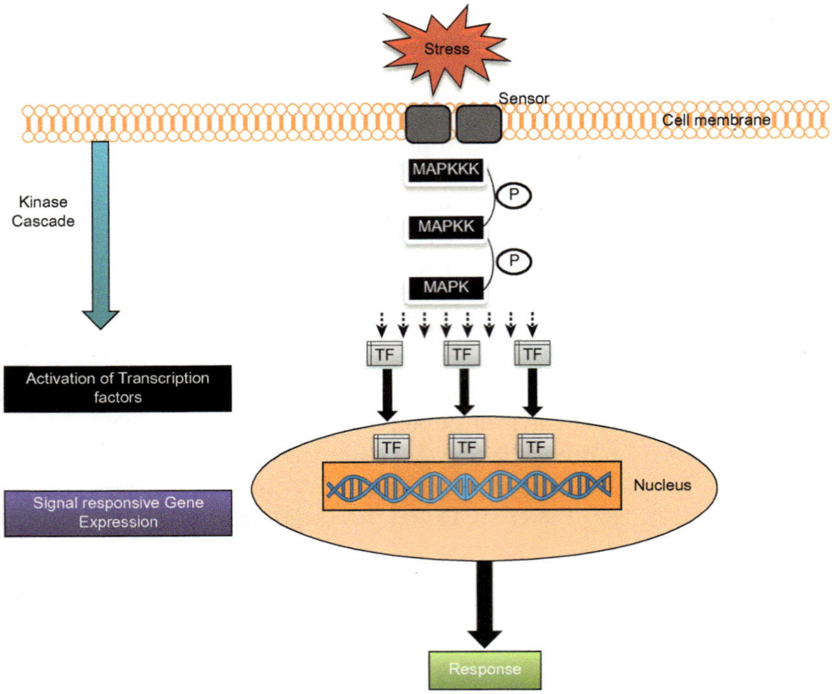

FIGURE 25.3

Relay of mitogen-activated protein kinase (MAPK) signaling cascades under stress conditions.

rice, suggesting the involvement of gene family in stress signaling as well as cross-talk (Kumar et al., 2008; Chen et al., 2012a; Zang et al., 2012). It has been reported that under elevated temperature, mitogen-activated proteins phosphorylate heat shock factors regulating heat stress response (Link et al., 2002). Excess heavy metals (Cu and Cd) activate distinct MAPKs in *Medicago sativa* (Jonak et al., 2004). Similarly, Cd activates MAPKs (OsMAPK2) and the MBP kinase gene in rice (Yeh et al., 2004). Under As(III) stress, the activation of MAPK cascades transduce the metal stress-mediated signals in *Brassica juncea*, which leads to enhanced tolerance against As stress (Gupta et al., 2009). Additionally, studies suggest that heavy metals activate MAPKs via ROS generation and accumulation in different plant species such as *Arabidopsis* and rice (Lin et al., 2005; Liu et al., 2010). Also, in response to Cu stress, MAPKs are activated, which further interacts with heat shock proteins and enhances tolerance against metal stress (Chen et al., 2008). Cd and Cu induce distinct signaling pathways by activating MAPKs in rice roots (Yeh et al., 2007). It has been observed that Cd accumulation perturbates MAPK signaling in rice (Zhao et al., 2013). Under Cd stress, MAPKs modulate expression of most of the key genes involved in auxin signaling. Recently, genomic profiling of rice also suggested possible involvement of MAPK signaling cascade under Cr stress (Huang et al., 2014; Trinh et al., 2014). Recently, the functional analyses of *Arabidopsis* mutants have broadened the view of the MAPK components and their role in providing tolerance against mechanistically distinct abiotic stresses (Rodriquez et al., 2010). For example, reduced tolerance toward salinity stress is observed in the *mkk2* mutant, whereas the overexpression of MKK2 enhances the resistance against salt and cold stress (Teige et al., 2004).

Apart from the information available about these protein kinases in heavy metal stress, there is still a need to generate more information, develop a better understanding, and elucidate the role of uncharacterized MAPKs involved in plant defense. Hence, the functional analysis of these modules will help to elucidate the signaling mechanisms operating in plants against multiple stresses and in achieving stress including heavy metal–tolerant crop plants. Collectively, it is important to envisage the role of MAPK cascades in downstream signaling in response to different heavy metals.

4. ROS SIGNALING

The perception of stress stimuli and the signal transduction mechanism switches on the plant adaptive responses that determine the survival of plants under severely adverse conditions. One such signal is the production and accumulation of ROS (mostly superoxide anion, hydrogen peroxide, hydroxyl radical, singlet oxygen) in different organelles including chloroplast, mitochondria, and peroxisomes (Choudhury et al., 2013; Wrzaczek et al., 2013). In response to heavy metal stress, enhanced ROS production termed "oxidative burst" occurs, which acts as an alarm signal and triggers the gene expression of transcription factors and many defense-related genes (Figure 25.4) (Sharma et al., 2012). The generation of ROS cripples normal functioning of the cell and causes oxidative damage to biomolecules such as DNA and proteins (Kapoor et al., 2015; Karuppanapandian et al., 2011). In vivo, ROS possesses a dual role that depends on duration, site, and concentration. The lower concentration of ROS acts as a signal and evokes a plant stress response, whereas the higher concentration of ROS poses a threat to the plant cell and causes damage and programmed cell death (PCD). ROS receptors include the histidine kinases, redox-sensitive transcription factors, ROS-sensitive phosphatases, and redox-regulated ion channels (Steffens, 2014). Several studies have shown that heavy metal stress induces ROS generation, which is seen by the accumulation of malondialdehyde causing lipid peroxidation in the plants (Shri et al., 2009; Sharma et al., 2012; Dubey et al., 2014; Rai et al., 2015). Studies also suggest that different genotypes of rice differ in generation of ROS, which leads to differential response under arsenic exposure (Rai et al., 2015).

Heavy metals such as Cd have been reported to cause lipid peroxidation and PCD in many plants species because of perturbation in the chloroplast and mitochondria metabolism (Cho and Seo, 2005; Bi et al., 2009; Metwally et al., 2005). Al toxicity also disrupts cellular redox homeostasis and causes oxidative burst, mitochondrial dysfunction, and PCD in plants (Li and Xing, 2010). It has been reported that Hg exposure induces Ca^{2+} accumulation, reduces ROS production, and activates MAPKs, which contribute to the plant defense against Hg stress (Chen et al., 2014). Arsenic severely affects plant growth and development, and various studies have reported that As exposure induces oxidative stress in plants (Shri et al., 2009; Dubey et al., 2014; Islam et al., 2015). Differential modulation of antioxidant system has been observed in many plant species under As stress (Dwivedi et al., 2010; Kanwar and Poonam, 2015). As a line of defense, two types of detoxification mechanisms operate in the plant system; the primary detoxification mechanism includes the enhanced activity of antioxidant isoenzymes which maintain ROS homeostasis by scavenging ROS and in the secondary detoxification mechanism, the heavy metals are chelated by peptides and are sequestered into the vacuoles (Kumar et al., 2015a).

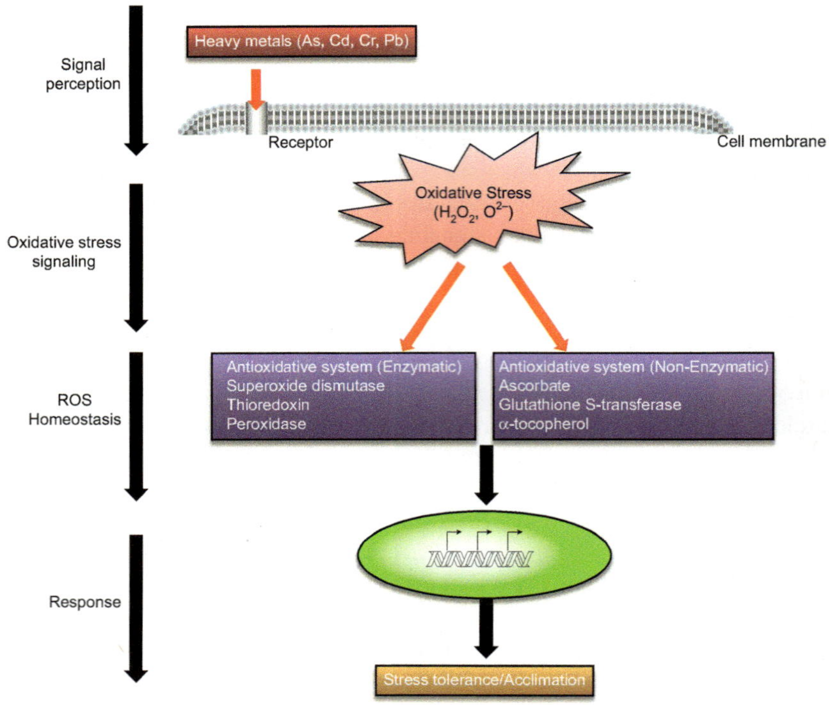

FIGURE 25.4

Downstream reactive oxygen system (ROS) signaling in response to heavy metal stress. As, arsenic; Cd, cadmium; Cr, chromium; Pb, lead.

The heavy metal detoxification via thiol metabolism and chelation mechanism has been well-studied and described in several plant species such as *Arabidopsis* (Howden et al., 1995; Shukla et al., 2013), *Brassica* (Xue Mie et al., 2007), and *Zea mays* (Nocito et al., 2006). The role of glutathione and the sulfur assimilatory pathway has been determined in providing tolerance against metal stress (Gratao et al., 2005; Kumar et al., 2011, 2015b). In response to ROS generation, the antioxidant machinery, which includes the enzymatic as well as nonenzymatic system, is activated to alleviate the inhibitory effect of oxidative species in plants (Rai et al., 2011). The antioxidant enzyme system comprises superoxide dismutase, ascorbate, peroxidase, glutathione reductase, and catalase. The components of the nonenzymatic system consist of compounds such as ascorbate, glutathiones, glutathione S-transferase, and α-tocopherol. The increased activity of the isozymes including ascorbate peroxidase, glutathione reductase peroxidase, and superoxide dismutase has been observed in the rice seedlings exposed to As stress (Shri et al., 2009). Cr and Cd are considered to be redox-active heavy metals and induce oxidative stress in plants (Dubey et al., 2010; Zeng et al., 2012). Studies suggest that Cr(VI)-induced oxidative stress from ROS generation is a major problem for seedling growth in rice (Zeng et al., 2012). Genome-wide expression analysis after application of Cr(VI) leads to modulation of several signaling cascades and ROS level modulation (Dubey et al., 2010, 2014;

Huang et al., 2014; Trinh et al., 2014). Also, ROS-induced signaling cascade modulates expression of several genes that play important roles in auxin signaling (Zhao et al., 2012).

Identification and overexpression of the genes involved in the detoxification mechanism of plants helps in conferring tolerance against metal stress (Venkataramaiah et al., 2011; Kumar et al., 2013a,b; Tiwari et al., 2014). For instance, the overexpression of genes encoding antioxidant enzymes; dehydro ascorbate reductase, glutathione S-transferase, and glutathione reductase provide tolerance against metal-induced oxidative stress (Le Martret et al., 2011). The expression of rice glutathione S-transferase in *Arabidopsis (OsGSTL2)* provides tolerance against heavy metal stress and other abiotic stresses (Kumar et al., 2013b). The expression of *OsMATE1* and *OsMATE2* influenced plant development and combat heavy metal stress and pathogen susceptibility in *Arabidopsis* (Tiwari et al., 2014). Thus, enhancing antioxidant potential of the plant is essential for improving plants tolerance against heavy metal stress. Summing, the perception and transmission of stress signals modulate the production and accumulation of ROS, which further trigger the signaling cascades and target the repertoire of stress-responsive genes or the transcription factors to combat different stresses.

Intensive research on the ROS generation under heavy metal stress has been conducted and the distinct signaling involved in stress response has been explored; however, there is still a lacuna about the relay of specific signals, cross-talk, and targets. Therefore, to study the plausible ROS signaling mechanism leading to metal tolerance, development of transgenic plants with the aid of manipulation techniques might be considerably useful.

5. HORMONE SIGNALING

Phytohormones are ubiquitously present in plants and play prominent role in growth and development (Divi and Krishna, 2009). In general, auxin has a profound effect on the growth, abscisic acid (ABA) influence bud dormancy, and stomatal closure; cytokinin delays senescence; gibberellin is involved in seed germination; and brassinosteroids regulate plant growth and differentiation (Jaillais and Chory, 2010; Sun et al., 2005; Vert et al., 2005). In addition, the synergistic and antagonistic effect of plant hormones such as ethylene (ET), jasmonic acid (JA), salicylic acid (SA), and ABA regulate plant's response against environmental stresses (Fujita et al., 2006; Mauch-Mani and Mauch, 2005). These are low-molecular-weight, endogenous molecules that are involved in cross-talk between signaling networks and govern plant response to stress conditions (Bostock, 2005). Importantly, ET, JA, and SA are primarily associated with the plant defense against abiotic stresses (Lorenzo and Solano, 2005). These phytohormones regulate the C-repeat binding factor pathway (Kazan, 2015). Major signaling components and pathways are activated by ET and JA, which include EIN2, EIN3, and JAZ proteins and the AP2/ERF transcription factor gene family (Kazan, 2015). A recent study by Singh and Shah (2014) demonstrated that JA exposure enhances antioxidant response and accumulation of antioxidants, leading to Cd stress tolerance in rice. This clearly suggests that modulation in levels of phytohormones changes the response of plants to metal stress.

Emerging evidence has identified the role of hormone signaling in response to heavy metal stress in plants (Figure 25.5). It has been observed that, in response to Cd, ET levels are elevated in different plants species such as *Arabidopsis* and pea (Vassilev et al., 2004). Under Cr stress, expression of the aminocyclopropane-1-carboxylic acid synthase (ACS) gene family in rice indicates the role of ET in

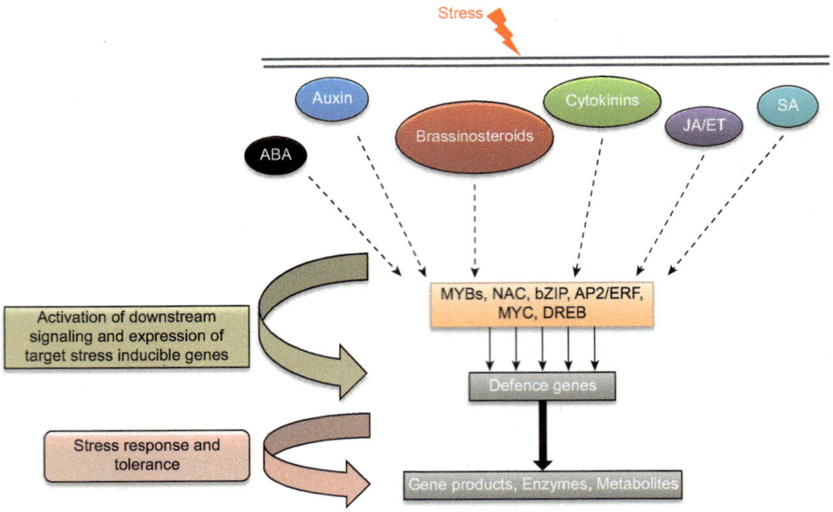

FIGURE 25.5

Hormone signal transduction under stress conditions. ABA, abscisic acid; ET, ethylene; JA, jasmonic acid; SA, salicylic acid.

heavy metal stress (Rzewuski and Sauter, 2008; Trinh et al., 2014). During Cr stress, several-fold upregulation of AP2/ERF genes in rice indicates that ET mediates growth response in plants under heavy metal stress conditions (Steffens, 2014; Trinh et al., 2014). Comparative transcriptome analysis of As(V)- and As(III)-treated rice seedlings identified that As modulates the pathways related to signal transduction and plant defense and indicated that As(III) mainly affects hormonal signaling processes such as ABA metabolism (Chakraborty et al., 2009). Genome-wide expression analysis in *Arabidopsis* in response to gold exposure suggested a role of ABA in providing a stress response. Analysis demonstrated modulation in pathways responsible for glutathione-mediated detoxification and plant hormone biosynthesis. Interestingly, *cis*-acting element search in genes with significant change in expression identified a highly significant enriched motif, ACGT, which is an ABA-responsive core element (Shukla et al., 2014). In another study, through genome-wide expression analysis, the role of ET signaling has been shown to play important role during Hg stress. A study carried out on *Medicago truncatula* exposed to different concentrations of Hg followed by microarray analysis identified differentially expressed genes associated with hormone metabolism, especially ET metabolism and signaling (Montero-Palmero et al., 2014). Comparative transcriptome analysis carried out on rice exposed to As stress demonstrated the important role of JA signaling and lipid metabolism and their regulation by microRNA (Yu et al., 2012).

The expression of inducible genes have been studied that confer tolerance against different abiotic stress conditions such as salinity and drought in different plant species *Arabidopsis*, *Brassica*, soybean, tomato, wheat, and rice (Chen et al., 2012a; Pan et al., 2012; Schmidt et al., 2013; Zhai et al., 2013). It has been demonstrated that the activation of endogenous MPK3 and MPK6 kinases induce the expression of MAPK 9, which further induces ET and camalexin biosynthesis and confer tolerance against salt stress in *Arabidopsis* (Xu et al., 2008). In-depth investigations are required to explore the hormone-induced signaling in plants in response to heavy metal stress. Moreover, the identification and characterization of hormone mediated signaling pathways can contribute to the development of plants with enhanced heavy metal stress tolerance. Further, the challenges and unanswered questions in response to metal stress need to be studied in detail to develop strategies for combating heavy metal stress and tolerance in plants.

6. SIGNALING CASCADES CROSS-TALK DURING HEAVY METAL STRESS

In recent years, various studies have explored the complex interaction and new paradigms for the signaling networks and cross-talk during plant growth, development, and in response to stress. Studies indicate that in response to stress, cross-talk between the signaling pathways occur that can mitigate the effect of heavy metal stress and restores the redox/ion homeostasis (Figure 25.6). The interaction

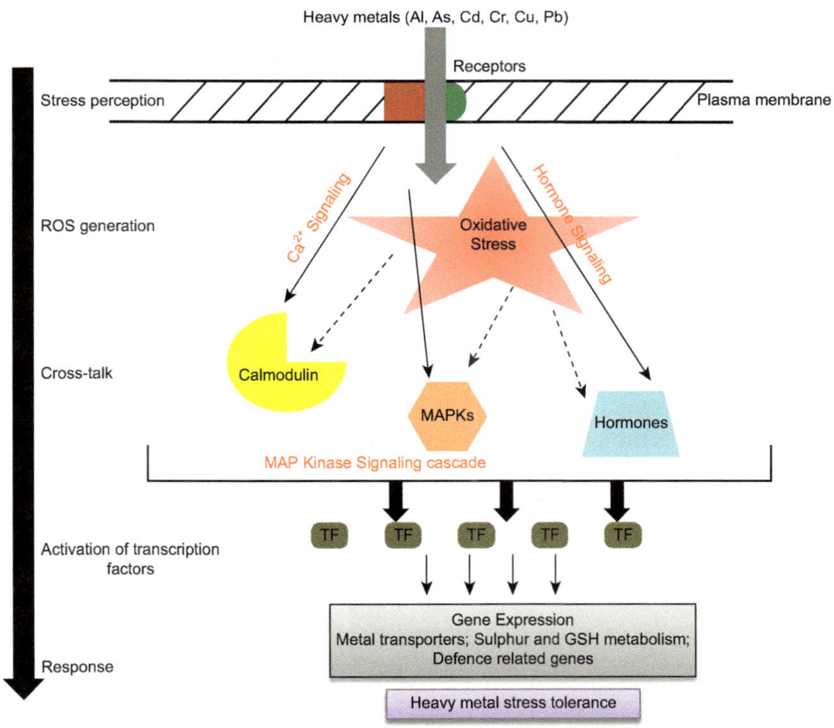

FIGURE 25.6

Schematic representation of heavy metal induced signal transduction in plants. Al, aluminum; As, arsenic; Cd, cadmium; Cu, copper; GSH, glutathione; MAPK, mitogen-activated protein kinase; Pb, lead; ROS, reactive oxygen species.

between the complex multicomponent signaling pathways orchestrates growth and development of plant under different stress conditions. The integration of phytohormones JA and ET and their cross-talk with other hormones such as auxin, ABA, and gibberellic acid have been reported under abiotic stresses (Santino et al., 2013) and might exist during heavy metal exposure. In response to abiotic stress, the effect of phytohormones depends on the secondary signals such as Ca^{2+} and ROS (Kazan, 2015). Studies have reported the collaborative effect of the ET and JA pathways under different abiotic stress conditions such as ozone-induced cell death in *Arabidopsis*, photo-oxidative stress, and essential nutrient deficiency (Troufflard et al., 2010; Iqbal et al., 2013). Moreover, the secondary messengers function in a concert to transduce the signal leading to the gene expression of defense-related genes. It has been observed that Ca^{2+} increases ROS accumulation, and Ca^{2+}-dependent signaling pathways and ROS signaling pathways are interconnected to transmit the signals (Steinhorst and Kudla, 2013). Also, hydrogen sulfide collaborates with the Ca^{2+}-signaling pathway to combat Cr stress in *Setaria italica*, suggesting cross-talk between the Ca^{2+} signaling and ROS pathways (Fang et al., 2014). More interestingly, it has been demonstrated that various hormones are located downstream of the ROS in signal cascade and ROS themselves act as secondary messenger in many hormone signaling pathways (Orozco-Cardenas et al., 2001).

Additionally, MAPK signaling cascades facilitate cross-talk in hormonal signaling and activate plant response against different stress stimuli. MAPK module has an important role in ET signaling. ET-modulated MAPKs have been studied in *Arabidopsis*, which depicts that the modulation in the protein kinase activity influences the ET signal transduction (Novikova et al., 2000). MKK9 activates the MPK3/MPK6 pathway, which regulates EIN3 and activates ACS genes (Yoo et al., 2008; Zao and Guo, 2011). The role of ET and ROS has been observed in salinity, flooding, and heavy metal stress (Steffens, 2014). The MAPK-mediated phosphorylation affects ET and ROS homeostasis during salt stress signaling by modulating gene expression (Teige et al., 2004; Dong et al., 2011; Steffens, 2014). In rice, under salt stress, *OsMPK3* and *OsMPK6* activate an ET-signaling pathway, which further leads to ROS accumulation (Li et al., 2014). Studies have also demonstrated that ET responsive transcription factors, AP2/ERF, and regulate the ROS generation and accumulation (Steffens, 2014). The transcriptome profiling of the roots of rice seedlings exposed to Cr showed modulation in the ET biosynthesis and ROS signaling indicating the cross-talk in the two pathways under heavy metal stress (Huang et al., 2014; Trinh et al., 2014).

ROS disrupt the auxin homeostasis and signalling as well as affects the morphogenetic responses in plants demonstrating the intricate cross-talk between the pathways (Blomster et al., 2011). Under metal stress, H_2O_2 acts as a signaling molecule and enhances the expression of several transcription factors—WRKY, NAM, and Zat11—stimulating PCD in the plants (Gechev and Hille, 2005). In soybeans, Cd stress induces the expression of genes involved in ET signaling and MAPK pathway suggesting the cross-talk between the two pathways (Chmielowska-Bak et al., 2013). During Cd stress, modulation in the expression of auxin signaling pathway and MAPK multicomponent suggest tight regulation of these two pathways in plants in response to heavy metal stress (Zhao et al., 2013). Moreover, it has been reported that ROS and CDPKs trigger the MAPK signaling pathway in rice roots under Pb stress (Huang and Huang, 2008). Based on limited studies in the area of heavy metal stress, the role of various signaling cascades has been suggested; however, detailed studies are required to establish these cross-talks using mutants or developing transgenic lines with modulating specific signaling cascades.

7. CONCLUSIONS AND FUTURE PROSPECTS

Environmental stresses, including heavy metal stress, affect various physiological processes and cause alterations in plant growth and development. To combat the stress, plants have developed sophisticated mechanisms, including signal transduction, which transmit the signals and activate the expression of defense-related genes and provide tolerance against the stress stimuli. The basics of signal decoding, amplification, and relay of the information encoded within the stress signatures are critical for the plant response to diverse stress stimuli including heavy metal stress. It is important to understand the mechanisms involved in maintaining the specificity of the signal transmission and generation of plant response under heavy metal stress. Studies on plant "signalome," which include the identification, characterization, and regulation of all the signaling components and cross-talk among these components, have helped in understanding the inner workings of the plant cell responses to varied stress signals. However, limited information is available about signaling cascades operating under heavy metal stress. With the advent of the elucidation of signal transduction pathways and the genetic manipulation, the capacity of heavy metal tolerance has increased in plants to some extent. Moreover, to enhance the intrinsic resilience of plants to the metal stress and for mitigation purposes, the functional analysis of various components involved in complex signaling network interactions is prerequisite. Therefore, a high-throughput analysis using various tools is required to decipher the missing links in the heavy metal stress signaling in plants. Furthermore, to generate more information, development of better understanding and elucidation of the role of uncharacterized genes involved in plant defense, the functional analysis of different modules will be useful in determining the signaling mechanisms operating in plants against multiple stresses. Collectively, it is important to envisage the role of different signal components in downstream signaling in response to different heavy metals. Therefore, integrated studies on the coordinated interplay and relay of the signals may provide compelling evidence for the metal acclimation/response in the near future.

REFERENCES

Ahmad, P., Sarwat, M., Bhat, N.A., Wani, M.R., Kazi, A.G., Tran, L.S.P., 2015. Alleviation of cadmium toxicity in *Brassica juncea* L. (Czern. & Coss.) by calcium application involves various physiological and biochemical strategies. PLoS One 10, e0114571.

Ali, H., Khan, E., Sajad, M.A., 2013. Phytoremediation of heavy metals—Concepts and applications. Chemosphere 91, 869–881.

Asano, T., Hayashi, N., Kobayashi, M., Aoki, N., et al., 2012. A rice calcium-dependent protein kinase OsCPK12 oppositely modulates salt-stress tolerance and blast disease resistance. Plant J. 69, 26–36.

Bi, Y.H., Chen, W.L., Zhang, W.N., Zhou, Q., Yun, L., Xing, D., 2009. Production of reactive oxygen species, impairment of photosynthetic function and dynamic changes in mitochondria are early events in cadmium-induced cell death in *Arabidopsis thaliana*. Biol. Cell 101, 629–643.

Blomster, T., Salojärvi, J., Sipari, N., Brosche, M., Ahlfors, R., Keinänen, M., et al., 2011. Apoplastic reactive oxygen species transiently decrease auxin signaling and cause stress-induced morphogenic response in *Arabidopsis*. Plant Physiol. 157, 1866–1883.

Bostock, R.M., 2005. Signal crosstalk and induced resistance: straddling the line between cost and benefit. Annu. Rev. Phytopathol. 43, 545–580.

Chakrabarty, D., Trivedi, P.K., Misra, P., Tiwari, M., et al., 2009. Comparative transcriptome analysis of arsenate and arsenite stresses in rice seedlings. Chemosphere 74, 688–702.

Chakraborty, D., Rahman, M.M., Mukherjee, A., Alauddin, M., et al., 2015. Groundwater arsenic contamination in Bangladesh- 21years of research. J. Trace Elem. Med. Biol. 31, 237–248.

Chen, J.H., Jiang, H.W., Hsieh, E.J., Chen, H.Y., et al., 2012a. Drought and salt stress tolerance of an *Arabidopsis* glutathione *S*-transferase U17 knockout mutant are attributed to the combined effect of glutathione and abscisic acid. Plant Physiol. 158, 340–351.

Chen, L., Hu, W., Tan, S., Wang, M., et al., 2012b. Genome-wide identification and analysis of MAPK and MAPKK gene families in *Brachypodium distachyon*. Plos One 7, e46774.

Chen, P.Y., Lee, K.T., Chi, W.C., Hirt, H., Chang, C.C., Huang, H.J., 2008. Possible involvement of MAP kinase pathways in acquired metal-tolerance induced by heat in plants. Planta 228, 499–509.

Chen, Y.A., Chi, W.C., Trinh, N.N., Huang, L.Y., et al., 2014. Transcriptome profiling and physiological studies reveal a major role for aromatic amino acids in mercury stress tolerance in rice seedlings. PloS One 9, e95163.

Cheong, Y.H., Kim, K.N., Pandey, G.K., Gupta, R., Grant, J.J., Luan, S., 2003. CBL1, a calcium sensor that differentially regulates salt, drought and cold responses in *Arabidopsis*. Plant Cell 15, 1833–1845.

Chmielowska-Bak, J., Lefevre, I., Lutts, S., Deckert, J., 2013. Short term signaling responses in roots of young soybean seedlings exposed to cadmium stress. J. Plant Physiol. 170, 1585–1594.

Cho, U.H., Seo, N.H., 2005. Oxidative stress in *Arabidopsis thaliana* exposed to cadmium is due to hydrogen peroxide accumulation. Plant Sci. 168, 113–120.

Choudhury, S., Panda, P., Sahoo, L., Panda, S.K., 2013. Reactive oxygen species signaling in plants under abiotic stress. Plant Signal Behav. 8, e23681.

Conde, A., Chaves, M.M., Geros, H., 2011. Membrane transport, sensing and signaling in plant adaptation to environmental stress. Plant Cell Physiol. 52, 1583–1602.

Divi, U.K., Krishna, P., 2009. Brassinosteroid: a biotechnological target for enhancing crop yield and stress tolerance. New Biotechnol. 26, 131–136.

Dodd, A.N., Kudla, J., Sanders, D., 2010. The language of calcium signaling. Ann. Rev. Plant Biol. 61, 593–620.

Dong, H., Zhen, Z., Peng, J., Chang, L., Gong, Q., Wang, N.N., 2011. Loss of ACS7 confers abiotic stress tolerance by modulating ABA sensitivity and accumulation in *Arabidopsis*. J. Exp. Bot. 62, 4875–4887.

Dubey, S., Misra, P., Dwivedi, S., Chatterjee, S., et al., 2010. Transcriptomic and metabolomic shifts in rice roots in response to Cr (VI) stress. BMC Genomics 11, 648.

Dubey, S., Shri, M., Misra, P., Lakhwani, D., et al., 2014. Heavy metals induce oxidative stress and genome-wide modulation in transcriptome of rice root. Funct. Integr. Genomics 14, 401–417.

Dwivedi, S., Tripathi, R.D., Tripathi, P., Kumar, A., 2010. Arsenate exposure affects amino acids, mineral nutrient status and antioxidants in rice (*Oryza sativa* L.) genotypes. Environ. Sci. Technol. 44, 9542–9549.

Edel, K.H., Kudla, J., 2014. Increasing complexity and versatility: How the calcium signaling toolkit was shaped during plant land colonization. Cell Calcium 57, 231–246.

Fang, H., Jing, T., Liu, Z., Zhang, L., Jin, Z., Pei, Y., 2014. Hydrogen sulfide interacts with calcium signaling to enhance the chromium tolerance in *Setaria italic*. Cell Calcium 56, 472–481.

Fujita, M., Fujita, Y., Noutoshi, Y., Takahashi, F., et al., 2006. Crosstalk between abiotic and biotic stress responses: a current view from the points of convergence in the stress signaling networks. Curr. Opin. Plant Biol. 9, 436–442.

Gechev, T.S., Hille, J., 2005. Hydrogen peroxide as a signal controlling plant programmed cell death. J. Cell Biol. 168, 17–20.

Golldack, D., Li, C., Mohan, H., Probst, N., 2014. Tolerance to drought and salt stress in plants: unraveling the signaling networks. Front. Plant Sci. 5, 151.

Gonzalez, A., Cabrera Mde, L., Henriquez, M.J., Contreras, R.A., Morales, B., Moenne, A., 2012. Cross talk among calcium, hydrogen peroxide, and nitric oxide and activation of gene expression involving calmodulins and calcium-dependent protein kinases in *Ulva compressa* exposed to copper excess. Plant Physiol. 15, 1451–1462.

Goswami, S., Kumar, R.R., Sharma, S.K., Kaa, Y.K., et al., 2015. Calcium triggers protein kinases-induced signal transduction for augmenting the thermotolerance of developing wheat (*Triticum aestivum*) grain under the heat stress. J. Plant Biochem. Biotechnol. http://dx.doi.org/10.1007/s13562-014-0295-1.

Gratao, P.L., Polle, A., Lea, P.J., Azevedo, R.A., 2005. Making the life of heavy metal-stressed plants a little easier. Funct. Plant Biol. 32, 481–494.

Gupta, M., Sharma, P., Sarin, N.B., Sinha, A.K., 2009. Differential response of arsenic stress in two varieties of *Brassica juncea* L. Chemosphere 74, 1201–1208.

Hamel, L.P., Nicole, M.C., Sritubtim, S., Morency, M.J., et al., 2006. Ancient signals: comparative genomics of plant MAPK and MAPKK gene families. Trends Plant Sci. 11, 192–198.

Hamel, L.P., Sheen, J., Seguin, A., 2014. Ancient signals: comparative genomics of green plant CDPKs. Trends Plant Sci. 19, 79–89.

Hayat, S., Khalique, G., Irfan, M., Wani, A.S., et al., 2012. Physiological changes induced by chromium stress in plants: an overview. Protoplasma 249, 599–611.

Himschoot, E., Beeckman, T., Friml, J., Vanneste, S., 2015. Calcium is an organizer of cell polarity in plants. Biochim. Biophys. Acta. http://dx.doi.org/10.1016/j.bbamcr.2015.02.017.

Howden, R., Goldsbrough, P.B., Andersen, C.R., Cobbett, C.S., 1995. Cadmium-sensitive, cad1 mutants of *Arabidopsis thaliana* are phytochelatin deficient. Plant Physiol. 107, 1059–1066.

Huang, T.L., Huang, H.J., 2008. ROS and CDPK-like-kinase-mediated activation of MAP kinase in rice roots exposed to lead. Chemosphere 71, 1377–1385.

Huang, T.L., Huang, L.Y., Fu, S.F., Trinh, N.N., Huang, H.J., 2014. Genomic profiling of rice roots with short- and long- term chromium stress. Plant Mol. Biol. 86, 157–170.

Iqbal, N., Masood, A., Khan, M.I.R., Asgher, M., Fatma, M., Khan, N.A., 2013. Cross-talk between sulfur assimilation and ethylene signaling in plants. Plant Signal Behav. 8, e22478.

Islam, E., Khan, M.T., Irem, S., 2015. Biochemical mechanisms of signaling: perspectives in plant under arsenic stress. Ecotoxicol Environ. Saf. 114C, 126–133.

Jaillais, Y., Chory, J., 2010. Unraveling the paradoxes of plant hormone signaling integration. Nat. Struct. Mol. Biol. 17, 642–645.

Janitza, P., Ullrich, K.K., Quint, M., 2012. Toward a comprehensive phylogenetic reconstruction of the evolutionary history of mitogen-activated protein kinases in the plant kingdom. Front. Plant Sci. 3, 271.

Jonak, C., Nakagami, H., Hirt, H., 2004. Heavy metal stress. Activation of distinct mitogen-activated protein kinase pathways by copper and cadmium. Plant Physiol. 136, 3276–3283.

Jonak, C., Okresz, L., Bogre, L., Hirt, H., 2002. Complexity, cross talk and integration of plant MAP kinase signalling. Curr. Opin. Plant Biol. 5, 415–424.

Kanwar, M.K., Poonam, B.R., 2015. Arsenic induced modulation of antioxidative defense system and brassinosteroids in *Brassica juncea* L. Ecotoxicol. Environ. Saf. 115, 119–125.

Kapoor, D., Sharma, R., Handa, N., Kaur, H., et al., 2015. Redox homeostasis in plants under abiotic stress: role of electron carriers, energy metabolism mediators and proteinaceous thiols. Front. Environ. Sci. 3, 1–12.

Karuppanapanadian, T., Moon, J.C., Kim, C., Manoharan, K., Kim, W., 2011. Reactive oxygen species in plants: their generation, signal transduction, and scavenging mechanisms. Aust. J. Crop Sci. 5, 709–725.

Kazan, K., 2015. Diverse roles of jasmonates and ethylene in abiotic stress tolerance. Trends Plant Sci. 20, 219–229.

Kim, C.Y., Liu, Y., Thorne, E.T., Yang, H., Fukushige, H., Gassmann, W., Hildebrand, D., Sharp, R.E., Zhang, S., 2003. Activation of a stress-responsive mitogen-activated protein kinase cascade induces the biosynthesis of ethylene in plants. Plant Cell 15, 2707–2718.

Knight, H., Knight, M.R., 2001. Abiotic stress signalling pathways: specificity and cross-talk. Trends Plant Sci. 6, 262–267.

Kumar, K., Rao, K.P., Sharma, P., Sinha, A.K., 2008. Differential regulation of rice mitogen activated protein kinase kinase (MKK) by abiotic stress. Plant Physiol. Biochem. 46, 891–897.

Kumar, S., Asif, M.H., Chakrabarty, D., Tripathi, R.D., Dubey, R.S., Trivedi, P.K., 2015b. Comprehensive analysis of regulatory elements of the promoters of rice sulphate transporter gene family and functional characterisation of *OsSul1;1* promoter under different metal stress. Plant Signal Behav. http://dx.doi.org/10.4161/15592324.2014.990843.

Kumar, S., Asif, M.H., Chakrabarty, D., Tripathi, R.D., Dubey, R.S., Trivedi, P.K., 2013a. Differential expression of rice lambda class GST gene family members during plant growth, development, and in response to stress conditions. Plant Mol. Biol. Rep. 31, 569–580.

Kumar, S., Asif, M.H., Chakrabarty, D., Tripathi, R.D., Dubey, R.S., Trivedi, P.K., 2013b. Expression of a rice Lambda class of glutathione S-transferase, *OsGSTL2*, in *Arabidopsis* provides tolerance to heavy metal and other abiotic stresses. J. Hazard Mat. 248, 228–237.

Kumar, S., Asif, M.H., Chakrabarty, D., Tripathi, R.D., Trivedi, P.K., 2011. Differential expression and alternative splicing of rice sulphate transporter family members regulate sulphur status during plant growth, development and stress conditions. Funct. Integr. Genomics 11, 259–273.

Kumar, S., Dubey, R.S., Tripathi, R.D., Chakrabarty, D., Trivedi, P.K., 2015a. Omics and biotechnology of arsenic stress and detoxification in plants: current updates and prospective. Environ. Int. 74, 221–230.

Le Martret, B., Poage, M., Shiel, K., Nugent, G.D., Dix, P.J., 2011. Tobacco chloroplast transformants expressing genes encoding dehydroascorbate reductase, glutathione reductase, and glutathione-S-transferase, exhibit altered anti-oxidant metabolism and improved abiotic stress tolerance. Plant Biotech. J. 9, 661–673.

Li, C.H., Wang, G., Zhao, J.L., Zhang, L.Q., Ai, L.F., Han, Y.F., et al., 2014. The receptor-like kinase SIT1 mediates salt sensitivity by activating MAPK3/6 and regulating ethylene homeostasis in rice. Plant Cell 26, 2538–2553.

Li, Z., Xing, D., 2010. Mitochondrial pathway leading to programmed cell death induced by aluminum phytotoxicty in *Arabidopsis*. Plant Signal Behav. 5, 1660–1662.

Li, Z.Y., Xu, Z.S., He, G.Y., Yang, G.X., Chen, M., Li, L.C., Ma, Y.Z., 2012. Overexpression of soybean GmCBL1 enhances abiotic stress tolerance and promotes hypocotyl elongation in *Arabidopsis*. Biochem. Biophys. Res. Commun. 427, 731–736.

Lin, C.W., Chang, H.B., Huang, H.J., 2005. Zinc induces mitogen-activated protein kinase activation mediated by reactive oxygen species in rice roots. Plant Physiol. Biochem. 43, 963–968.

Lin, Y.F., Aarts, M.G.M., 2012. The molecular mechanism of zinc and cadmium stress response in plants. Cell Mol. Life Sci. 69, 3187–3206.

Link, V., Sinha, A.K., Vashista, P., Hoffman, M.G., et al., 2002. A heat-activated MAP kinase in tomato: a possible regulator of the heat stress response. FEBS Lett. 531, 179–183.

Liu, X.M., Kim, K.E., Kim, K.C., et al., 2010. Cadmium activates *Arabidopsis* MPK3 and MPK6 via accumulation of reactive oxygen species. Phytochemistry 71, 614–618.

Lorenzo, O., Solano, R., 2005. Molecular players regulating the jasmonate signalling network. Curr. Opin. Plant Biol. 8, 532–540.

Luan, S., Kudla, J., Rodriguez-Concepcion, M., Yalovsky, S., Gruissem, W., 2002. Calmodulins and calcineurin B-like proteins calcium sensors for specific signal response coupling in plants. Plant Cell 14, 389–400.

Manimaran, P., Mangrauthia, S.K., Sundaram, R.M., Balachandran, S.M., 2015. Constitutive expression and silencing of a novel seed specific calcium dependent protein kinase gene in rice reveals its role in grain filling. J. Plant Physiol. 174, 41–48.

Mauch-Mani, B., Mauch, F., 2005. The role of abscisic acid in plant–pathogen interactions. Curr. Opin. Plant Biol. 8, 409–414.

Metwally, A., Safronova, V.I., Belimov, A.A., Dietz, K.J., 2005. Genotypic variation of the response to cadmium toxicity in *Pisum sativum* L. J. Exp. Bot. 56, 167–178.

Montero-Palmero, M.B., Martin-Barranco, A., Escobar, C., Hernandez, L.E., 2014. Early transcriptional responses to mercury: a role for ethylene in mercury-induced stress. New Phytol. 201, 116–130.

Nakagami, H., Pitzschke, A., Hirt, H., 2005. Emerging MAP kinase pathways in plant stress signalling. Trends Plant Sci. 10, 339–346.

Nocito, F.F., Lancilli, C., Crema, B., Fourcroy, P., Davidian, J.C., Sacchi, G.A., 2006. Heavy metal stress and sulfate uptake in maize roots. Plant Physiol. 141, 1138–1148.

Novikova, G.V., Moshkov, I.E., Smith, A.R., Hall, M.A., 2000. The effect of ethylene on MAPKinase-like activity in *Arabidopsis thaliana*. FEBS Lett. 474, 29–32.

Orozco-Cardenas, M.L., Narvez-Vasquez, J., Ryan, C.A., 2001. Hydrogen peroxide acts as a second messenger for the induction of defense genes in tomato plants in response to wounding, systemin, and methyl jasmonate. Plant Cell 13, 179–191.

Pan, Y., Seymour, G.B., Lu, C., Hu, Z., Chen, X., Chen, G., 2012. An ethylene response factor (ERF5) promoting adaptation to drought and salt tolerance in tomato. Plant Cell Rep. 31, 349–360.

Peralta-Videa, J.R., Lopez, M.L., Narayan, M., Saupe, G., Gardea-Torresdey, J., 2009. The biochemistry of environmental heavy metal uptake by plants: Implications for the food chain. Int. J. Biochem. Cell Biol. 41, 1665–1677.

Qadir, S., Jamshieed, S., Rasool, S., Ashraf, S., et al., 2014. Modulation of plant growth and metabolism in cadmium-enriched environments. Rev. Environ. Contam. Toxicol. 229, 51–88.

Rai, A., Bhardwaj, A., Misra, P., Bag, S.K., et al., 2015. Comparative transcriptional profiling of contrasting rice genotypes shows expression differences during arsenic stress. Plant Genome 8, 1–14.

Rai, A., Tripathi, P., Dwivedi, S., Dubey, S., Shri, M., Kumar, S., et al., 2011. Arsenic tolerances in rice (*Oryza sativa*) have a predominant role in transcriptional regulation of a set of genes including sulphur assimilation pathway and antioxidant system. Chemosphere 82, 986–995.

Rao, K.P., Richa, T., Kumar, K., Raghuram, B., Sinha, A.K., 2010. *In silico* analysis reveals 75 members of protein kinsase kinase kinase gene family in rice. DNA Res. 17, 139–153.

Rao, K.P., Vani, G., Kumar, K., Wankhede, D.P., Misra, M., Gupta, M., Sinha, A.K., 2011. Arsenic stress activates MAP kinase in rice roots and leaves. Arch. Biochem. Biophys. 506, 73–82.

Rodriquez, M.C., Petersen, M., Mundy, J., 2010. Mitogen-activated protein kinase signaling in plants. Annu. Rev. Plant Biol. 61, 621–649.

Rudd, J.J., Franklin-Tong, V.E., 2001. Unravelling response-specificity in Ca^{2+} signalling pathways in plant cells. New Phytol. 151, 7–33.

Rzewuski, G., Sauter, M., 2008. Ethylene biosynthesis and signaling in rice. Plant Sci. 175, 32–42.

Samajova, O., Plihal, O., Al-Yousif, M., Hirt, H., Samaj, J., 2013. Improvement of stress tolerance in plants by genetic manipulation of mitogen-activated protein kinases. Biotech. Adv. 31, 118–128.

Sanders, D., Pelloux, J., Brownlee, C., Harper, J.F., 2002. Calcium at the crossroads of signaling. Plant Cell 14, 401–417.

Santino, A., Taurino, M., De Domenico, S., Bonsegna, S., Poltronieri, P., Pastor, V., Flors, V., 2013. Jasmonate signaling in plant development and defense response to multiple (a)biotic stresses. Plant Cell Rep. 32, 1085–1098.

Schmidt, R., Mieulet, D., Hubberten, H.M., Obata, T., et al., 2013. SALT-RESPONSIVE ERF1 regulates reactive oxygen species–dependent signaling during the initial response to salt stress in rice. Plant Cell 25, 2115–2131.

Schulz, P., Herde, M., Romeis, T., 2013. Calcium-dependent protein kinases: hubs in plant stress signaling and development. Plant Physiol. 163, 523–530.

Sharma, P., Jha, A.B., Dubey, R.S., Pessarakli, M., 2012. Reactive oxygen species, oxidative damage, and antioxidative defense mechanism in plants under stressful conditions. J. Bot. 2012, 1–26.

Shri, M., Kumar, S., Chakrabarty, D., Trivedi, P.K., et al., 2009. Effect of arsenic on growth, oxidative stress, and antioxidant system in rice seedlings. Ecotoxicol Environ. Saf. 72, 1102–1110.

Shukla, D., Krishnamurthy, S., Sahi, S.V., 2014. Genome wide transcriptome analysis reveals ABA mediated response in *Arabidopsis* during gold ($AuCl_4^-$) treatment. Front. Plant Sci. 5, 652.

Shukla, D., Tiwari, M., Tripathi, R.D., Nath, P., Trivedi, P.K., 2013. Synthetic phytochelatins complement a phytochelatin-deficient *Arabidopsis* mutant and enhance the accumulation of heavy metal(loid)s. Biochem. Biophys. Res. Commun. 434, 664–669.

Singh, I., Shah, K., 2014. Exogenous application of methyl jasmonate lowers the effect of cadmium-induced oxidative injury in rice seedlings. Phytochemistry 108, 57–66.

Steffens, B., 2014. The role of ethylene and ROS in salinity, heavy metal, and flooding responses in rice. Front. Plant Sci. 5, 685.

Steinhorst, L., Kudla, J., 2013. Calcium and reactive oxygen species rule the waves of signaling. Plant Physiol. 163, 471–485.

Steinhorst, L., Kudla, J., 2014. Signaling in cells and organisms- calcium holds the line. Curr. Opin. Plant Biol. 22, 14–21.

Sun, Y., Veerabomma, S., Abdel-Mageed, H.A., Foker, M., et al., 2005. Brassinosteroid regulates fiber development on cultured cotton ovules. Plant Cell Physiol. 46, 1384–1391.

Teige, M., Schkiel, E., Eulgem, T., et al., 2004. The MKK2 pathway mediates cold and salt stress signaling in *Arabidopsis*. Mol. Cell 15, 141–152.

Tiwari, M., Sharma, D., Singh, M., Tripathi, R.D., Trivedi, P.K., 2014. Expression of OsMATE1 and OsMATE2 alters development, stress responses and pathogen susceptibility in *Arabidopsis*. Sci. Rep. 4, 1–12.

de la Torre, F., Gutierrez-Beltran, E., Pareja-Jaime, Y., Chakravarthy, S., Martin, G.B., del Pozo, O., 2013. The tomato calcium sensor Cbl10 and its interacting protein kinase Cipk6 define a signaling pathway in plant immunity. Plant Cell 25, 2748–2764.

Trinh, N.N., Huang, T.L., Chi, W.C., Fu, S.F., Chen, C.C., Huang, H.J., 2014. Chromium stress response effect on signal transduction and expression of signaling genes in rice. Physiol. Plant 150, 205–224.

Tripathi, R.D., Srivastava, S., Mishra, S., Singh, N., et al., 2007. Arsenic hazards: strategies for tolerance and remediation by plants. Trends Biotechnol. 25, 158–165.

Tripathi, V., Parasuraman, B., Laxmi, A., Chattopadhyay, 2009. CIPK6, a CBL-interacting protein kinase is required for development and salt tolerance in plants. Plant J. 58, 778–790.

Troufflard, S., Mullen, W., Larson, T.R., Graham, I.A., Crozier, A., Amtmann, A., Armengaud, P., 2010. Potassium deficiency induces the biosynthesis of oxylipins and glucosinolates in *Arabidopsis thaliana*. BMC Plant Biol. 10, 172.

Tuteja, N., 2009. Integrated calcium signaling in plants. Signal Commun. Plants 2009, 29–49.

Tuteja, N., Mahajan, S., 2007a. Calcium signaling network in plants: an overview. Plant Signal Behav. 2, 79–85.

Tuteja, N., Mahajan, S., 2007b. Further characterization of calcineurin-B like protein and its interacting partner CBL-interacting protein kinase from *Pisum sativum*. Plant Signal Behav. 2, 358–361.

Vassilev, A., Lidon, F., Scotti, P., Graca, M.D., Yordanov, I., 2004. Cadmium-induced changes in chloroplast lipids and photosystem activities in barley plants. Biol. Plantarum 48, 153–156.

Venkataramaiah, N., Ramakrishna, S.V., Sreevathsa, R., 2011. Overexpression of phytochelatin synthase (*AtPCS*) in rice for tolerance to cadmium stress. Biologia 66, 1060–1073.

Vert, G., Nemhauser, J.L., Geldner, N., Hong, F., Chory, J., 2005. Molecular mechanisms of steroid hormone signaling in plants. Annu. Rev. Cell Dev. Biol. 21, 177–201.

Wrzaczek, M., Brosche, M., Kangasjarvi, J., 2013. ROS signaling loops—production, perception, regulation. Curr. Opin. Plant Biol. 16, 575–582.

Xu, J., Li, Y., Wang, Y., Liu, H., Lei, L., Yang, H., Liu, G., Ren, D., 2008. Activation of MAPK kinase 9 induces ethylene and camalexin biosynthesis and enhances sensitivity to salt stress in *Arabidopsis*. J. Biol. Chem. 283, 26996–27006.

Xue Mei, S., Bo, L., Si Qi, H., SK, M., Lai Lang, X.U., 2007. Coordinated expression of sulfate transporters and its relation with sulfur metabolites in *Brassica napus* exposed to cadmium. Bot. Stud. 48, 43–54.

Yeh, C.M., Chien, P.S., Huang, H.J., 2007. Distinct signalling pathways for induction of MAP kinase activities by cadmium and copper in rice roots. J. Exp. Bot. 58, 659–671.

Yeh, C.M., Hsiao, L.-J., Huang, H.-J., 2004. Cadmium activates a mitogen-activated protein kinase gene and MBP kinases in rice. Plant Cell Physiol. 45, 1306–1312.

Yoo, S.D., Cho, Y.H., Tena, G., Xiong, Y., Sheen, J., 2008. Dual control of nuclear EIN3 by bifurcate MAPK cascades in C_2H_4 signalling. Nature 451, 789–795.

Yu, L.J., Luo, Y.F., Liao, B., Xie, L.J., Chen, L., et al., 2012. Comparative transcriptome analysis of transporters, phytohormone and lipid metabolism pathways in response to arsenic stress in rice (*Oryza sativa*). New Phytol. 195, 97–112.

Zeng, F., Qiu, B., Wu, X., Niu, S., Wu, F., Zhang, G., 2012. Glutathione-mediated alleviation of chromium toxicity in rice plants. Biol. Trace Elem. Res. 148, 255–263.

Zhai, Y., Wang, Y., Li, Y., Lei, T., et al., 2013. Isolation and molecular characterization of *GmERF7*, a soybean ethylene-response factor that increases salt stress tolerance in tobacco. Gene 513, 174–183.

Zhang, M., Pan, J., Kong, X., Zhou, Y., Liu, Y., et al., 2012. *ZmMKK3*, a novel maize group B mitogen-activated protein kinase kinase gene, mediates osmotic stress and ABA signal responses. J. Plant Physiol. 169, 1501–1510.

Zhao, F.J., McGrath, S.P., Meharg, A.A., 2010. Arsenic as a food chain contaminant: mechanisms of plant uptake and metabolism and mitigation strategies. Annu. Rev. Plant Biol. 61, 535–559.

Zhao, F.Y., Han, M.M., Zhang, S.Y., Wang, K., Zhang, C.R., Liu, T., Liu, W., 2012. Hydrogen peroxide-mediated growth of the root system occurs via auxin signaling modification and variations in the expression of cell-cycle genes in rice seedlings exposed to cadmium stress. J. Integr. Plant Biol. 54, 911–1006.

Zhao, F.Y., Hu, F., Zhang, S.Y., et al., 2013. MAPKs regulate root growth by influencing auxin signaling and cell cycle-related gene expression in cadmium-stressed rice. Environ. Sci. Pollu Res. 20, 5449–5460.

Zhao, Q., Guo, H.W., 2011. Paradigms and paradox in the ethylene signaling pathway and interaction network. Mol. Plant 4, 626–634.

Index

Note: Page numbers followed by "f" and "t" indicates figures and tables, respectively.

A

C

D

E

F

G

I

N

T

U

V

W

X

Y

Z